Unit Conversions (Equivalents)

Length

1 in. = 2.54 cm (defined)
1 cm = 0.3937 in.
1 ft = 30.48 cm
1 m = 39.37 in. = 3.281 ft
1 mi = 5280 ft = 1.609 km
1 km = 0.6214 mi
1 nautical mile (U.S.) = 1.151 mi = 6076 ft = 1.852 km
1 fermi = 1 femtometer (fm) = 10^{-15} m
1 angstrom (Å) = 10^{-10} m = 0.1 nm
1 light-year (ly) = 9.461×10^{15} m
1 parsec = 3.26 ly = 3.09×10^{16} m

Volume

1 liter (L) = 1000 mL = 1000 cm^3 = 1.0×10^{-3} m^3 =
    1.057 qt (U.S.) = 61.02 in.^{3}
1 gal (U.S.) = 4 qt (U.S.) = 231 in.^{3} = 3.785 L =
    0.8327 gal (British)
1 quart (U.S.) = 2 pints (U.S.) = 946 mL
1 pint (British) = 1.20 pints (U.S.) = 568 mL
1 m^3 = 35.31 ft^3

Speed

1 mi/h = 1.4667 ft/s = 1.6093 km/h = 0.4470 m/s
1 km/h = 0.2778 m/s = 0.6214 mi/h
1 ft/s = 0.3048 m/s (exact) = 0.6818 mi/h = 1.0973 km/h
1 m/s = 3.281 ft/s = 3.600 km/h = 2.237 mi/h
1 knot = 1.151 mi/h = 0.5144 m/s

Angle

1 radian (rad) = 57.30° = 57°18′
1° = 0.01745 rad
1 rev/min (rpm) = 0.1047 rad/s

Time

1 day = 8.640×10^4 s
1 year = 3.156×10^7 s

Mass

1 atomic mass unit (u) = 1.6605×10^{-27} kg
1 kg = 0.06852 slug
[1 kg has a weight of 2.20 lb where $g = 9.80$ m/s^2.]

Force

1 lb = 4.448 N
1 N = 10^5 dyne = 0.2248 lb

Energy and Work

1 J = 10^7 ergs = 0.7376 ft·lb
1 ft·lb = 1.356 J = 1.29×10^{-3} Btu = 3.24×10^{-4} kcal
1 kcal = 4.19×10^3 J = 3.97 Btu
1 eV = 1.602×10^{-19} J
1 kWh = 3.600×10^6 J = 860 kcal
1 Btu = 1.056×10^3 J

Power

1 W = 1 J/s = 0.7376 ft·lb/s = 3.41 Btu/h
1 hp = 550 ft·lb/s = 746 W

Pressure

1 atm = 1.01325 bar = 1.01325×10^5 N/m^2
    = 14.7 lb/in.^{2} = 760 torr
1 lb/in.^{2} = 6.895×10^3 N/m^2
1 Pa = 1 N/m^2 = 1.450×10^{-4} lb/in.^{2}

SI Derived Units and Their Abbreviations

Quantity	Unit	Abbreviation	In Terms of Base Units†
Force	newton	N	kg·m/s^2
Energy and work	joule	J	kg·m^2/s^2
Power	watt	W	kg·m^2/s^3
Pressure	pascal	Pa	kg/(m·s^2)
Frequency	hertz	Hz	s^{-1}
Electric charge	coulomb	C	A·s
Electric potential	volt	V	kg·m^2/(A·s^3)
Electric resistance	ohm	Ω	kg·m^2/(A^2·s^3)
Capacitance	farad	F	A^2·s^4/(kg·m^2)
Magnetic field	tesla	T	kg/(A·s^2)
Magnetic flux	weber	Wb	kg·m^2/(A·s^2)
Inductance	henry	H	kg·m^2/(s^2·A^2)

† kg = kilogram (mass), m = meter (length), s = second (time), A = ampere (electric current).

Metric (SI) Multipliers

Prefix	Abbreviation	Value
yotta	Y	10^{24}
zeta	Z	10^{21}
exa	E	10^{18}
peta	P	10^{15}
tera	T	10^{12}
giga	G	10^{9}
mega	M	10^{6}
kilo	k	10^{3}
hecto	h	10^{2}
deka	da	10^{1}
deci	d	10^{-1}
centi	c	10^{-2}
milli	m	10^{-3}
micro	μ	10^{-6}
nano	n	10^{-9}
pico	p	10^{-12}
femto	f	10^{-15}
atto	a	10^{-18}
zepto	z	10^{-21}
yocto	y	10^{-24}

Force   Displacement

Volume I

FOURTH EDITION

PHYSICS
for
SCIENTISTS & ENGINEERS
with Modern Physics

DOUGLAS C. GIANCOLI

PEARSON
Prentice Hall

Upper Saddle River, New Jersey 07458

Library of Congress Cataloging-in-Publication Data

Giancoli, Douglas C.
  Physics for scientists and engineers with modern physics / Douglas C.
Giancoli.—4th ed.
    p. cm.
  Includes bibliographical references and index.
  ISBN 0-13-227358-6
  1. Physics—Textbooks.   I. Title.
QC21.3.G539 2008
530—dc22
                                              2006039431

President, ESM: Paul Corey
Sponsoring Editor: Christian Botting
Production Editor: Frank Weihenig, Prepare Inc.
Executive Managing Editor: Kathleen Schiaparelli
Art Director and Interior & Cover Designer: John Christiana
Manager, Art Production: Sean Hogan
Senior Development Editor: Karen Karlin
Copy Editor: Jocelyn Phillips
Proofreader: Marne Evans
Buyer: Alan Fischer
Art Production Editor: Connie Long
Illustrators: Audrey Simonetti and Mark Landis
Photo Researchers: Mary Teresa Giancoli and Truitt & Marshall
Senior Administrative Coordinator: Trisha Tarricone
Composition: Emilcomp/Prepare Inc.
Photo credits appear on page A-44 which constitutes
  a continuation of the copyright page.

© 2008, 2000, 1989, 1984 by Douglas C. Giancoli
Published by Pearson Education, Inc.
Pearson Prentice Hall
Pearson Education, Inc.
Upper Saddle River, NJ 07458

Pearson Prentice Hall™ is a trademark of Pearson Education, Inc.

Printed in the United States of America
10 9 8 7 6 5 4 3 2 1

ISBN   0-13-227358-6

Pearson Education LTD., *London*
Pearson Education Australia PTY, Limited, *Sydney*
Pearson Education Singapore, Pte. Ltd
Pearson Education North Asia Ltd, *Hong Kong*
Pearson Education Canada, Ltd., *Toronto*
Pearson Educación de Mexico, S.A. de C.V.
Pearson Education—Japan, *Tokyo*
Pearson Education Malaysia, Pte. Ltd

Contents

iii

FOURTH EDITION

Volume I

PHYSICS

for

SCIENTISTS & ENGINEERS

with Modern Physics

DOUGLAS C. GIANCOLI

PEARSON

Prentice
Hall

Upper Saddle River, New Jersey 07458

Library of Congress Cataloging-in-Publication Data

Giancoli, Douglas C.
   Physics for scientists and engineers with modern physics / Douglas C.
Giancoli.—4th ed.
      p. cm.
   Includes bibliographical references and index.
   ISBN 0-13-227358-6
   1. Physics—Textbooks.   I. Title.
   QC21.3.G539 2008
   530—dc22

                                          2006039431

President, ESM: Paul Corey
Sponsoring Editor: Christian Botting
Production Editor: Frank Weihenig, Prepare Inc.
Executive Managing Editor: Kathleen Schiaparelli
Art Director and Interior & Cover Designer: John Christiana
Manager, Art Production: Sean Hogan
Senior Development Editor: Karen Karlin
Copy Editor: Jocelyn Phillips
Proofreader: Marne Evans
Buyer: Alan Fischer
Art Production Editor: Connie Long
Illustrators: Audrey Simonetti and Mark Landis
Photo Researchers: Mary Teresa Giancoli and Truitt & Marshall
Senior Administrative Coordinator: Trisha Tarricone
Composition: Emilcomp/Prepare Inc.
Photo credits appear on page A-44 which constitutes
   a continuation of the copyright page.

© 2008, 2000, 1989, 1984 by Douglas C. Giancoli
Published by Pearson Education, Inc.
Pearson Prentice Hall
Pearson Education, Inc.
Upper Saddle River, NJ 07458

Printed in the United States of America
10 9 8 7 6 5 4 3 2 1

ISBN   0-13-227358-6

Pearson Education LTD., *London*
Pearson Education Australia PTY, Limited, *Sydney*
Pearson Education Singapore, Pte. Ltd
Pearson Education North Asia Ltd, *Hong Kong*
Pearson Education Canada, Ltd., *Toronto*
Pearson Educación de Mexico, S.A. de C.V.
Pearson Education—Japan, *Tokyo*
Pearson Education Malaysia, Pte. Ltd

Contents

Volume 1

Force   Displacement

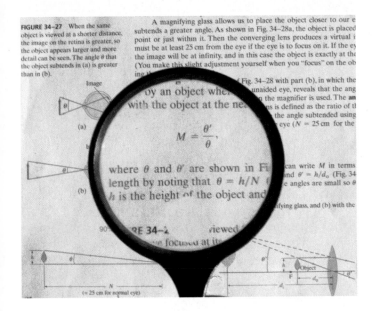

Contents of Volume 3

APPLICATIONS (SELECTED)

Preface

I was motivated from the beginning to write a textbook different from others that present physics as a sequence of facts, like a Sears catalog: "here are the facts and you better learn them." Instead of that approach in which topics are begun formally and dogmatically, I have sought to begin each topic with concrete observations and experiences students can relate to: start with specifics and only then go to the great generalizations and the more formal aspects of a topic, showing *why* we believe what we believe. This approach reflects how science is actually practiced.

Why a Fourth Edition?

Two recent trends in physics textbooks are disturbing: (1) their revision cycles have become short—they are being revised every 3 or 4 years; (2) the books are getting larger, some over 1500 pages. I don't see how either trend can be of benefit to students. My response: (1) It has been 8 years since the previous edition of this book. (2) This book makes use of physics education research, although it avoids the detail a Professor may need to say in class but in a book shuts down the reader. And this book still remains among the shortest.

This new edition introduces some important new pedagogic tools. It contains new physics (such as in cosmology) and many new appealing applications (list on previous page). Pages and page breaks have been carefully formatted to make the physics easier to follow: no turning a page in the middle of a derivation or Example. Great efforts were made to make the book attractive so students will want to *read* it.

Some of the new features are listed below.

What's New

Chapter-Opening Questions: Each Chapter begins with a multiple-choice question, whose responses include common misconceptions. Students are asked to answer before starting the Chapter, to get them involved in the material and to get any preconceived notions out on the table. The issues reappear later in the Chapter, usually as Exercises, where the material has been covered. The Chapter-Opening Questions also show students the power and usefulness of Physics.

APPROACH paragraph in worked-out numerical Examples: A short introductory paragraph before the Solution, outlining an approach and the steps we can take. Brief NOTES after the Solution may remark on the Solution, may give an alternate approach, or mention an application.

Step-by-Step Examples: After many Problem Solving Strategies, the next Example is done step-by-step following precisely the steps just seen.

Exercises within the text, after an Example or derivation, give students a chance to see if they have understood enough to answer a simple question or do a simple calculation. Many are multiple choice.

Greater clarity: No topic, no paragraph in this book was overlooked in the search to improve the clarity and conciseness of the presentation. Phrases and sentences that may slow down the principal argument have been eliminated: keep to the essentials at first, give the elaborations later.

$\vec{F}, \vec{v}, \vec{B}$ *Vector notation, arrows*: The symbols for vector quantities in the text and Figures now have a tiny arrow over them, so they are similar to what we write by hand.

Cosmological Revolution: With generous help from top experts in the field, readers have the latest results.

Page layout: more than in the previous edition, serious attention has been paid to how each page is formatted. Examples and all important derivations and arguments are on facing pages. Students then don't have to turn back and forth. Throughout, readers see, on two facing pages, an important slice of physics.

New Applications: LCDs, digital cameras and CCDs, electric hazards, GFCIs, photocopiers, inkjet and laser printers, metal detectors, underwater vision, curve balls, airplane wings, DNA, how we actually *see* images. (Turn back a page to see a longer list.)

Examples modified: more math steps are spelled out, and many new Examples added. About 10% are Estimation Examples.

This Book is Shorter than other complete full-service books at this level. Shorter explanations are easier to understand and more likely to be read.

Content and Organizational Changes

- **Rotational Motion**: Chapters 10 and 11 have been reorganized. All of angular momentum is now in Chapter 11.
- **First law of thermodynamics**, in Chapter 19, has been rewritten and extended. The full form is given: $\Delta K + \Delta U + \Delta E_{int} = Q - W$, where internal energy is E_{int}, and U is potential energy; the form $Q - W$ is kept so that $dW = P\,dV$.
- Kinematics and Dynamics of Circular Motion are now treated together in Chapter 5.
- Work and Energy, Chapters 7 and 8, have been carefully revised.
- Work done by friction is discussed now with energy conservation (energy terms due to friction).
- Chapters on Inductance and AC Circuits have been combined into one: Chapter 30.
- Graphical Analysis and Numerical Integration is a new optional Section 2–9. Problems requiring a computer or graphing calculator are found at the end of most Chapters.
- Length of an object is a script ℓ rather than normal l, which looks like 1 or I (moment of inertia, current), as in $F = I\ell B$. Capital L is for angular momentum, latent heat, inductance, dimensions of length $[L]$.
- Newton's law of gravitation remains in Chapter 6. Why? Because the $1/r^2$ law is too important to relegate to a late chapter that might not be covered at all late in the semester; furthermore, it is one of the basic forces in nature. In Chapter 8 we can treat real gravitational potential energy and have a fine instance of using $U = -\int \vec{F} \cdot d\vec{\ell}$.
- New Appendices include the differential form of Maxwell's equations and more on dimensional analysis.
- Problem Solving Strategies are found on pages 30, 58, 64, 96, 102, 125, 166, 198, 229, 261, 314, 504, and 551, of this first volume.

Organization

Some instructors may find that this book contains more material than can be covered in their courses. The text offers great flexibility. Sections marked with a star * are considered optional. These contain slightly more advanced physics material, or material not usually covered in typical courses and/or interesting applications; they contain no material needed in later Chapters (except perhaps in later optional Sections). For a brief course, all optional material could be dropped as well as major parts of Chapters 1, 13, 16, 26, 30, and 35, and selected parts of Chapters 9, 12, 19, 20, 33, and the modern physics Chapters. Topics not covered in class can be a valuable resource for later study by students. Indeed, this text can serve as a useful reference for years because of its wide range of coverage.

Versions of this Book

Complete version: 44 Chapters including 9 Chapters of modern physics.

Classic version: 37 Chapters including one each on relativity and quantum theory.

3 Volume version: Available separately or packaged together (Vols. 1 & 2 or all 3 Volumes):

Volume 1: Chapters 1–20 on mechanics, including fluids, oscillations, waves, plus heat and thermodynamics.

Volume 2: Chapters 21–35 on electricity and magnetism, plus light and optics.

Volume 3: Chapters 36–44 on modern physics: relativity, quantum theory, atomic physics, condensed matter, nuclear physics, elementary particles, cosmology and astrophysics.

Thanks

Many physics professors provided input or direct feedback on every aspect of this textbook. They are listed below, and I owe each a debt of gratitude.

Mario Affatigato, Coe College
Lorraine Allen, United States Coast Guard Academy
Zaven Altounian, McGill University
Bruce Barnett, Johns Hopkins University
Michael Barnett, Lawrence Berkeley Lab
Anand Batra, Howard University
Cornelius Bennhold, George Washington University
Bruce Birkett, University of California Berkeley
Dr. Robert Boivin, Auburn University
Subir Bose, University of Central Florida
David Branning, Trinity College
Meade Brooks, Collin County Community College
Bruce Bunker, University of Notre Dame
Grant Bunker, Illinois Institute of Technology
Wayne Carr, Stevens Institute of Technology
Charles Chiu, University of Texas Austin
Robert Coakley, University of Southern Maine
David Curott, University of North Alabama
Biman Das, SUNY Potsdam
Bob Davis, Taylor University
Kaushik De, University of Texas Arlington
Michael Dennin, University of California Irvine
Kathy Dimiduk, University of New Mexico
John DiNardo, Drexel University
Scott Dudley, United States Air Force Academy
John Essick, Reed College
Cassandra Fesen, Dartmouth College
Alex Filippenko, University of California Berkeley
Richard Firestone, Lawrence Berkeley Lab
Mike Fortner, Northern Illinois University
Tom Furtak, Colorado School of Mines
Edward Gibson, California State University Sacramento
John Hardy, Texas A&M
J. Erik Hendrickson, University of Wisconsin Eau Claire
Laurent Hodges, Iowa State University
David Hogg, New York University
Mark Hollabaugh, Normandale Community College
Andy Hollerman, University of Louisiana at Lafayette
Bob Jacobsen, University of California Berkeley
Teruki Kamon, Texas A&M
Daryao Khatri, University of the District of Columbia
Jay Kunze, Idaho State University

Jim LaBelle, Dartmouth College
M.A.K. Lodhi, Texas Tech
Bruce Mason, University of Oklahoma
Dan Mazilu, Virginia Tech
Linda McDonald, North Park College
Bill McNairy, Duke University
Raj Mohanty, Boston University
Giuseppe Molesini, Istituto Nazionale di Ottica Florence
Lisa K. Morris, Washington State University
Blaine Norum, University of Virginia
Alexandria Oakes, Eastern Michigan University
Michael Ottinger, Missouri Western State University
Lyman Page, Princeton and WMAP
Bruce Partridge, Haverford College
R. Daryl Pedigo, University of Washington
Robert Pelcovitz, Brown University
Vahe Peroomian, UCLA
James Rabchuk, Western Illinois University
Michele Rallis, Ohio State University
Paul Richards, University of California Berkeley
Peter Riley, University of Texas Austin
Larry Rowan, University of North Carolina Chapel Hill
Cindy Schwarz, Vassar College
Peter Sheldon, Randolph-Macon Woman's College
Natalia A. Sidorovskaia, University of Louisiana at Lafayette
George Smoot, University of California Berkeley
Mark Sprague, East Carolina University
Michael Strauss, University of Oklahoma
Laszlo Takac, University of Maryland Baltimore Co.
Franklin D. Trumpy, Des Moines Area Community College
Ray Turner, Clemson University
Som Tyagi, Drexel University
John Vasut, Baylor University
Robert Webb, Texas A&M
Robert Weidman, Michigan Technological University
Edward A. Whittaker, Stevens Institute of Technology
John Wolbeck, Orange County Community College
Stanley George Wojcicki, Stanford University
Edward Wright, UCLA
Todd Young, Wayne State College
William Younger, College of the Albemarle
Hsiao-Ling Zhou, Georgia State University

I owe special thanks to Prof. Bob Davis for much valuable input, and especially for working out all the Problems and producing the Solutions Manual for all Problems, as well as for providing the answers to odd-numbered Problems at the end of this book. Many thanks also to J. Erik Hendrickson who collaborated with Bob Davis on the solutions, and to the team they managed (Profs. Anand Batra, Meade Brooks, David Currott, Blaine Norum, Michael Ottinger, Larry Rowan, Ray Turner, John Vasut, William Younger). Many thanks to Katherine Whatley and Judith Beck who provided answers to the conceptual Questions at the end of each Chapter. I am grateful to Profs. John Essick, Bruce Barnett, Robert Coakley, Biman Das, Michael Dennin, Kathy Dimiduk, John DiNardo, Scott Dudley, David Hogg, Cindy Schwarz, Ray Turner, and Som Tyagi, who inspired many of the Examples, Questions, Problems, and significant clarifications.

Crucial for rooting out errors, as well as providing excellent suggestions, were Profs. Kathy Dimiduk, Ray Turner, and Lorraine Allen. A huge thank you to them and to Prof. Giuseppe Molesini for his suggestions and his exceptional photographs for optics.

For Chapter 44 on Cosmology and Astrophysics, I was fortunate to receive generous input from some of the top experts in the field, to whom I owe a debt of gratitude: George Smoot, Paul Richards, and Alex Filippenko (UC Berkeley), Lyman Page (Princeton and WMAP), Edward Wright (UCLA and WMAP), and Michael Strauss (University of Oklahoma).

I especially wish to thank Profs. Howard Shugart, Chair Marjorie Shapiro, and many others at the University of California, Berkeley, Physics Department for helpful discussions, and for hospitality. Thanks also to Prof. Tito Arecchi and others at the Istituto Nazionale di Ottica, Florence, Italy.

Finally, I am grateful to the many people at Prentice Hall with whom I worked on this project, especially Paul Corey, Christian Botting, Sean Hogan, Frank Weihenig, John Christiana, and Karen Karlin.

The final responsibility for all errors lies with me. I welcome comments, corrections, and suggestions as soon as possible to benefit students for the next reprint.

D.C.G.

email: Paul_Corey@Prenhall.com

Post: Paul Corey
      One Lake Street
      Upper Saddle River, NJ 07458

About the Author

Douglas C. Giancoli obtained his BA in physics (summa cum laude) from the University of California, Berkeley, his MS in physics at the Massachusetts Institute of Technology, and his PhD in elementary particle physics at the University of California, Berkeley. He then spent 2 years as a post-doctoral fellow at UC Berkeley's Virus lab developing skills in molecular biology and biophysics. His mentors include Nobel winners Emilio Segré and Donald Glaser.

He has taught a wide range of traditional undergraduate courses, as well as innovative ones, and continues to update his texbooks meticulously, seeking ways to better provide an understanding of physics for students.

Doug's favorite spare-time activity is the outdoors, especially climbing peaks (here on a summit in the Dolomites, winter ascent 2007). He says climbing peaks is like learning physics: it takes effort and the rewards are great.

Online Supplements (partial list)

MasteringPhysics™ (www.masteringphysics.com)
is a sophisticated online tutoring and homework system developed specially for courses using calculus-based physics. Originally developed by David Pritchard and collaborators at MIT, MasteringPhysics provides **students** with individualized online tutoring by responding to their wrong answers and providing hints for solving multi-step problems when they get stuck. It gives them immediate and up-to-date assessment of their progress, and shows where they need to practice more. MasteringPhysics provides **instructors** with a fast and effective way to assign tried-and-tested online homework assignments that comprise a range of problem types. The powerful post-assignment diagnostics allow instructors to assess the progress of their class as a whole as well as individual students, and quickly identify areas of difficulty.

WebAssign (www.webassign.com)
CAPA and LON-CAPA (www.lon-capa.org/)

Student Supplements (partial list)

Student Study Guide & Selected Solutions Manual (Volume I: 0-13-2273241, Volumes II & III: 0-13-227325X) by Frank Wolfs

Student Pocket Companion (0-13-2273268) by Biman Das

Tutorials in Introductory Physics (0-13-097069-7)
by Lillian C. McDermott, Peter S. Schaffer, and the Physics Education Group at the University of Washington

Physlet® Physics (0-13-101969-4)
by Wolfgang Christian and Mario Belloni

Ranking Task Exercises in Physics, Student Edition (0-13-144851-X)
by Thomas L. O'Kuma, David P. Maloney, and Curtis J. Hieggelke

E&M TIPERs: Electricity & Magnetism Tasks Inspired by Physics Education Research (0-13-185499-2) by Curtis J. Hieggelke, David P. Maloney, Stephen E. Kanim, and Thomas L. O'Kuma

Mathematics for Physics with Calculus (0-13-1913360)
by Biman Das

To Students

HOW TO STUDY

1. Read the Chapter. Learn new vocabulary and notation. Try to respond to questions and exercises as they occur.

2. Attend all class meetings. Listen. Take notes, especially about aspects you do not remember seeing in the book. Ask questions (everyone else wants to, but maybe you will have the courage). You will get more out of class if you read the Chapter first.

3. Read the Chapter again, paying attention to details. Follow derivations and worked-out Examples. Absorb their logic. Answer Exercises and as many of the end of Chapter Questions as you can.

4. Solve 10 to 20 end of Chapter Problems (or more), especially those assigned. In doing Problems you find out what you learned and what you didn't. Discuss them with other students. Problem solving is one of the great learning tools. Don't just look for a formula—it won't cut it.

NOTES ON THE FORMAT AND PROBLEM SOLVING

1. Sections marked with a star (*) are considered **optional**. They can be omitted without interrupting the main flow of topics. No later material depends on them except possibly later starred Sections. They may be fun to read, though.

2. The customary **conventions** are used: symbols for quantities (such as m for mass) are italicized, whereas units (such as m for meter) are not italicized. Symbols for vectors are shown in boldface with a small arrow above: $\vec{F}$.

3. Few equations are valid in all situations. Where practical, the **limitations** of important equations are stated in square brackets next to the equation. The equations that represent the great laws of physics are displayed with a tan background, as are a few other indispensable equations.

4. At the end of each Chapter is a set of **Problems** which are ranked as Level I, II, or III, according to estimated difficulty. Level I Problems are easiest, Level II are standard Problems, and Level III are "challenge problems." These ranked Problems are arranged by Section, but Problems for a given Section may depend on earlier material too. There follows a group of General Problems, which are not arranged by Section nor ranked as to difficulty. Problems that relate to optional Sections are starred (*). Most Chapters have 1 or 2 Computer/Numerical Problems at the end, requiring a computer or graphing calculator. Answers to odd-numbered Problems are given at the end of the book.

5. Being able to solve **Problems** is a crucial part of learning physics, and provides a powerful means for understanding the concepts and principles. This book contains many aids to problem solving: (a) worked-out **Examples** and their solutions in the text, which should be studied as an integral part of the text; (b) some of the worked-out Examples are **Estimation Examples**, which show how rough or approximate results can be obtained even if the given data are sparse (see Section 1–6); (c) special **Problem Solving Strategies** placed throughout the text to suggest a step-by-step approach to problem solving for a particular topic—but remember that the basics remain the same; most of these "Strategies" are followed by an Example that is solved by explicitly following the suggested steps; (d) special problem-solving Sections; (e) "Problem Solving" marginal notes which refer to hints within the text for solving Problems; (f) **Exercises** within the text that you should work out immediately, and then check your response against the answer given at the bottom of the last page of that Chapter; (g) the Problems themselves at the end of each Chapter (point 4 above).

6. **Conceptual Examples** pose a question which hopefully starts you to think and come up with a response. Give yourself a little time to come up with your own response before reading the Response given.

7. **Math** review, plus some additional topics, are found in Appendices. Useful data, conversion factors, and math formulas are found inside the front and back covers.

USE OF COLOR

Vectors

A general vector

    resultant vector (sum) is slightly thicker

    components of any vector are dashed

Displacement ($\vec{\mathbf{D}}, \vec{\mathbf{r}}$)

Velocity ($\vec{\mathbf{v}}$)

Acceleration ($\vec{\mathbf{a}}$)

Force ($\vec{\mathbf{F}}$)

    Force on second or

    third object in same figure

Momentum ($\vec{\mathbf{p}}$ or $m\vec{\mathbf{v}}$)

Angular momentum ($\vec{\mathbf{L}}$)

Angular velocity ($\vec{\boldsymbol{\omega}}$)

Torque ($\vec{\boldsymbol{\tau}}$)

Electric field ($\vec{\mathbf{E}}$)

Magnetic field ($\vec{\mathbf{B}}$)

Electricity and magnetism

Electric field lines

Equipotential lines

Magnetic field lines

Electric charge (+)   + or • +

Electric charge (−)   − or • −

Electric circuit symbols

Wire, with switch S

Resistor

Capacitor

Inductor

Battery

Ground

Optics

Light rays

Object

Real image (dashed)

Virtual image (dashed and paler)

Other

Energy level (atom, etc.)

Measurement lines |←1.0 m→|

Path of a moving object

Direction of motion or current

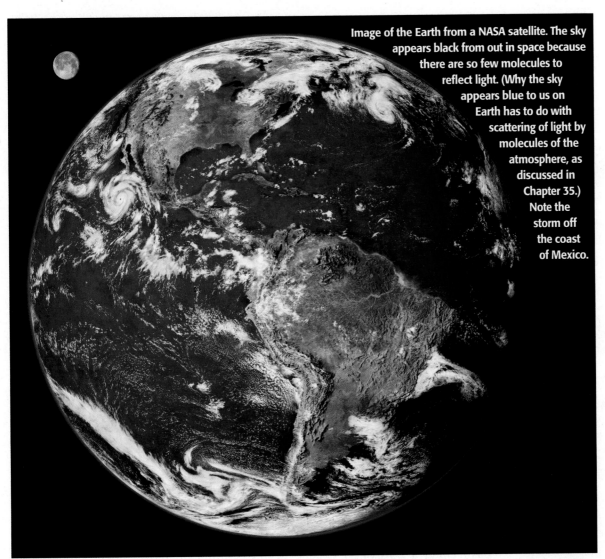

Image of the Earth from a NASA satellite. The sky appears black from out in space because there are so few molecules to reflect light. (Why the sky appears blue to us on Earth has to do with scattering of light by molecules of the atmosphere, as discussed in Chapter 35.) Note the storm off the coast of Mexico.

Introduction, Measurement, Estimating

CHAPTER-OPENING QUESTION—Guess now!

Suppose you wanted to actually measure the radius of the Earth, at least roughly, rather than taking other people's word for what it is. Which response below describes the best approach?

(a) Give up; it is impossible using ordinary means.
(b) Use an extremely long measuring tape.
(c) It is only possible by flying high enough to see the actual curvature of the Earth.
(d) Use a standard measuring tape, a step ladder, and a large smooth lake.
(e) Use a laser and a mirror on the Moon or on a satellite.

[*We start each Chapter with a Question, like the one above. Try to answer it right away. Don't worry about getting the right answer now—the idea is to get your preconceived notions out on the table. If they are misconceptions, we expect them to be cleared up as you read the Chapter. You will usually get another chance at the Question later in the Chapter when the appropriate material has been covered. These Chapter-Opening Questions will also help you to see the power and usefulness of physics.*]

CONTENTS

1

P hysics is the most basic of the sciences. It deals with the behavior and structure of matter. The field of physics is usually divided into *classical physics* which includes motion, fluids, heat, sound, light, electricity and magnetism; and *modern physics* which includes the topics of relativity, atomic structure, condensed matter, nuclear physics, elementary particles, and cosmology and astrophysics. We will cover all these topics in this book, beginning with motion (or mechanics, as it is often called) and ending with the most recent results in our study of the cosmos.

An understanding of physics is crucial for anyone making a career in science or technology. Engineers, for example, must know how to calculate the forces within a structure to design it so that it remains standing (Fig. 1–1a). Indeed, in Chapter 12 we will see a worked-out Example of how a simple physics calculation—or even intuition based on understanding the physics of forces—would have saved hundreds of lives (Fig. 1–1b). We will see many examples in this book of how physics is useful in many fields, and in everyday life.

1–1 The Nature of Science

The principal aim of all sciences, including physics, is generally considered to be the search for order in our observations of the world around us. Many people think that science is a mechanical process of collecting facts and devising theories. But it is not so simple. Science is a creative activity that in many respects resembles other creative activities of the human mind.

One important aspect of science is **observation** of events, which includes the design and carrying out of experiments. But observation and experiment require imagination, for scientists can never include everything in a description of what they observe. Hence, scientists must make judgments about what is relevant in their observations and experiments.

Consider, for example, how two great minds, Aristotle (384–322 B.C.) and Galileo (1564–1642), interpreted motion along a horizontal surface. Aristotle noted that objects given an initial push along the ground (or on a tabletop) always slow down and stop. Consequently, Aristotle argued that the natural state of an object is to be at rest. Galileo, in his reexamination of horizontal motion in the 1600s, imagined that if friction could be eliminated, an object given an initial push along a horizontal surface would continue to move indefinitely without stopping. He concluded that for an object to be in motion was just as natural as for it to be at rest. By inventing a new approach, Galileo founded our modern view of motion (Chapters 2, 3, and 4), and he did so with a leap of the imagination. Galileo made this leap conceptually, without actually eliminating friction.

Observation, with careful experimentation and measurement, is one side of the scientific process. The other side is the invention or creation of theories to explain and order the observations. Theories are never derived directly from observations. Observations may help inspire a theory, and theories are accepted or rejected based on the results of observation and experiment.

The great theories of science may be compared, as creative achievements, with great works of art or literature. But how does science differ from these other creative activities? One important difference is that science requires **testing** of its ideas or theories to see if their predictions are borne out by experiment.

Although the testing of theories distinguishes science from other creative fields, it should not be assumed that a theory is "proved" by testing. First of all, no measuring instrument is perfect, so exact confirmation is not possible. Furthermore, it is not possible to test a theory in every single possible circumstance. Hence a theory cannot be absolutely verified. Indeed, the history of science tells us that long-held theories can be replaced by new ones.

1–2 Models, Theories, and Laws

When scientists are trying to understand a particular set of phenomena, they often make use of a **model**. A model, in the scientist's sense, is a kind of analogy or mental image of the phenomena in terms of something we are familiar with. One

(a)

(b)

FIGURE 1–1 (a) This Roman aqueduct was built 2000 years ago and still stands. (b) The Hartford Civic Center collapsed in 1978, just two years after it was built.

example is the wave model of light. We cannot see waves of light as we can water waves. But it is valuable to think of light as made up of waves because experiments indicate that light behaves in many respects as water waves do.

The purpose of a model is to give us an approximate mental or visual picture—something to hold on to—when we cannot see what actually is happening. Models often give us a deeper understanding: the analogy to a known system (for instance, water waves in the above example) can suggest new experiments to perform and can provide ideas about what other related phenomena might occur.

You may wonder what the difference is between a theory and a model. Usually a model is relatively simple and provides a structural similarity to the phenomena being studied. A **theory** is broader, more detailed, and can give quantitatively testable predictions, often with great precision.

It is important, however, not to confuse a model or a theory with the real system or the phenomena themselves.

Scientists give the title **law** to certain concise but general statements about how nature behaves (that energy is conserved, for example). Sometimes the statement takes the form of a relationship or equation between quantities (such as Newton's second law, $F = ma$).

To be called a law, a statement must be found experimentally valid over a wide range of observed phenomena. For less general statements, the term **principle** is often used (such as Archimedes' principle).

Scientific laws are different from political laws in that the latter are *prescriptive*: they tell us how we ought to behave. Scientific laws are *descriptive*: they do not say how nature *should* behave, but rather are meant to describe how nature *does* behave. As with theories, laws cannot be tested in the infinite variety of cases possible. So we cannot be sure that any law is absolutely true. We use the term "law" when its validity has been tested over a wide range of cases, and when any limitations and the range of validity are clearly understood.

Scientists normally do their research as if the accepted laws and theories were true. But they are obliged to keep an open mind in case new information should alter the validity of any given law or theory.

1–3 Measurement and Uncertainty; Significant Figures

In the quest to understand the world around us, scientists seek to find relationships among physical quantities that can be measured.

Uncertainty

Reliable measurements are an important part of physics. But no measurement is absolutely precise. There is an uncertainty associated with every measurement. Among the most important sources of uncertainty, other than blunders, are the limited accuracy of every measuring instrument and the inability to read an instrument beyond some fraction of the smallest division shown. For example, if you were to use a centimeter ruler to measure the width of a board (Fig. 1–2), the result could be claimed to be precise to about 0.1 cm (1 mm), the smallest division on the ruler, although half of this value might be a valid claim as well. The reason is that it is difficult for the observer to estimate (or interpolate) between the smallest divisions. Furthermore, the ruler itself may not have been manufactured to an accuracy very much better than this.

When giving the result of a measurement, it is important to state the **estimated uncertainty** in the measurement. For example, the width of a board might be written as 8.8 ± 0.1 cm. The ± 0.1 cm ("plus or minus 0.1 cm") represents the estimated uncertainty in the measurement, so that the actual width most likely lies between 8.7 and 8.9 cm. The **percent uncertainty** is the ratio of the uncertainty to the measured value, multiplied by 100. For example, if the measurement is 8.8 and the uncertainty about 0.1 cm, the percent uncertainty is

$$\frac{0.1}{8.8} \times 100\% \approx 1\%,$$

where ≈ means "is approximately equal to."

FIGURE 1–2 Measuring the width of a board with a centimeter ruler. The uncertainty is about ± 1 mm.

Often the uncertainty in a measured value is not specified explicitly. In such cases, the uncertainty is generally assumed to be one or a few units in the last digit specified. For example, if a length is given as 8.8 cm, the uncertainty is assumed to be about 0.1 cm or 0.2 cm. It is important in this case that you do not write 8.80 cm, for this implies an uncertainty on the order of 0.01 cm; it assumes that the length is probably between 8.79 cm and 8.81 cm, when actually you believe it is between 8.7 and 8.9 cm.

Significant Figures

The number of reliably known digits in a number is called the number of **significant figures**. Thus there are four significant figures in the number 23.21 cm and two in the number 0.062 cm (the zeros in the latter are merely place holders that show where the decimal point goes). The number of significant figures may not always be clear. Take, for example, the number 80. Are there one or two significant figures? We need words here: If we say it is *roughly* 80 km between two cities, there is only one significant figure (the 8) since the zero is merely a place holder. If there is no suggestion that the 80 is a rough approximation, then we can often assume (as we will in this book) that it is 80 km within an accuracy of about 1 or 2 km, and then the 80 has two significant figures. If it is precisely 80 km, to within ± 0.1 km, then we write 80.0 km (three significant figures).

When making measurements, or when doing calculations, you should avoid the temptation to keep more digits in the final answer than is justified. For example, to calculate the area of a rectangle 11.3 cm by 6.8 cm, the result of multiplication would be 76.84 cm². But this answer is clearly not accurate to 0.01 cm², since (using the outer limits of the assumed uncertainty for each measurement) the result could be between 11.2 cm × 6.7 cm = 75.04 cm² and 11.4 cm × 6.9 cm = 78.66 cm². At best, we can quote the answer as 77 cm², which implies an uncertainty of about 1 or 2 cm². The other two digits (in the number 76.84 cm²) must be dropped because they are not significant. As a rough general rule (i.e., in the absence of a detailed consideration of uncertainties), we can say that *the final result of a multiplication or division should have only as many digits as the number with the least number of significant figures used in the calculation.* In our example, 6.8 cm has the least number of significant figures, namely two. Thus the result 76.84 cm² needs to be rounded off to 77 cm².

EXERCISE A The area of a rectangle 4.5 cm by 3.25 cm is correctly given by (*a*) 14.625 cm²; (*b*) 14.63 cm²; (*c*) 14.6 cm²; (*d*) 15 cm².

When adding or subtracting numbers, the final result is no more precise than the least precise number used. For example, the result of subtracting 0.57 from 3.6 is 3.0 (and not 3.03).

Keep in mind when you use a calculator that all the digits it produces may not be significant. When you divide 2.0 by 3.0, the proper answer is 0.67, and not some such thing as 0.666666666. Digits should not be quoted in a result, unless they are truly significant figures. However, to obtain the most accurate result, you should normally *keep one or more extra significant figures throughout a calculation, and round off only in the final result.* (With a calculator, you can keep all its digits in intermediate results.) Note also that calculators sometimes give too few significant figures. For example, when you multiply 2.5 × 3.2, a calculator may give the answer as simply 8. But the answer is accurate to two significant figures, so the proper answer is 8.0. See Fig. 1–3.

| CONCEPTUAL EXAMPLE 1–1 | **Significant figures.** Using a protractor (Fig. 1–4), you measure an angle to be 30°. (*a*) How many significant figures should you quote in this measurement? (*b*) Use a calculator to find the cosine of the angle you measured.

RESPONSE (*a*) If you look at a protractor, you will see that the precision with which you can measure an angle is about one degree (certainly not 0.1°). So you can quote two significant figures, namely, 30° (not 30.0°). (*b*) If you enter cos 30° in your calculator, you will get a number like 0.866025403. However, the angle you entered is known only to two significant figures, so its cosine is correctly given by 0.87; you must round your answer to two significant figures.

NOTE Cosine and other trigonometric functions are reviewed in Appendix A.

FIGURE 1–3 These two calculators show the wrong number of significant figures. In (a), 2.0 was divided by 3.0. The correct final result would be 0.67. In (b), 2.5 was multiplied by 3.2. The correct result is 8.0.

PROBLEM SOLVING

Significant figure rule: Number of significant figures in final result should be same as the least significant input value

⚠ **CAUTION**

Calculators err with significant figures

 PROBLEM SOLVING

Report only the proper number of significant figures in the final result. Keep extra digits during the calculation

FIGURE 1–4 Example 1–1. A protractor used to measure an angle.

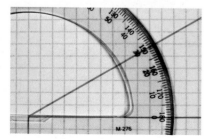

EXERCISE B Do 0.00324 and 0.00056 have the same number of significant figures?

Be careful not to confuse significant figures with the number of decimal places.

EXERCISE C For each of the following numbers, state the number of significant figures and the number of decimal places: (*a*) 1.23; (*b*) 0.123; (*c*) 0.0123.

Scientific Notation

We commonly write numbers in "powers of ten," or "scientific" notation—for instance 36,900 as 3.69×10^4, or 0.0021 as 2.1×10^{-3}. One advantage of scientific notation is that it allows the number of significant figures to be clearly expressed. For example, it is not clear whether 36,900 has three, four, or five significant figures. With powers of ten notation the ambiguity can be avoided: if the number is known to three significant figures, we write 3.69×10^4, but if it is known to four, we write 3.690×10^4.

EXERCISE D Write each of the following in scientific notation and state the number of significant figures for each: (*a*) 0.0258, (*b*) 42,300, (*c*) 344.50.

Percent Uncertainty versus Significant Figures

The significant figures rule is only approximate, and in some cases may underestimate the accuracy (or uncertainty) of the answer. Suppose for example we divide 97 by 92:

$$\frac{97}{92} = 1.05 \approx 1.1.$$

Both 97 and 92 have two significant figures, so the rule says to give the answer as 1.1. Yet the numbers 97 and 92 both imply an uncertainty of ± 1 if no other uncertainty is stated. Now 92 ± 1 and 97 ± 1 both imply an uncertainty of about 1% ($1/92 \approx 0.01 = 1\%$). But the final result to two significant figures is 1.1, with an implied uncertainty of ± 0.1, which is an uncertainty of $0.1/1.1 \approx 0.1 \approx 10\%$. In this case it is better to give the answer as 1.05 (which is three significant figures). Why? Because 1.05 implies an uncertainty of ± 0.01 which is $0.01/1.05 \approx 0.01 \approx 1\%$, just like the uncertainty in the original numbers 92 and 97.

SUGGESTION: Use the significant figures rule, but consider the % uncertainty too, and add an extra digit if it gives a more realistic estimate of uncertainty.

Approximations

Much of physics involves approximations, often because we do not have the means to solve a problem precisely. For example, we may choose to ignore air resistance or friction in doing a Problem even though they are present in the real world, and then our calculation is only an approximation. In doing Problems, we should be aware of what approximations we are making, and be aware that the precision of our answer may not be nearly as good as the number of significant figures given in the result.

Accuracy versus Precision

There is a technical difference between "precision" and "accuracy." **Precision** in a strict sense refers to the repeatability of the measurement using a given instrument. For example, if you measure the width of a board many times, getting results like 8.81 cm, 8.85 cm, 8.78 cm, 8.82 cm (interpolating between the 0.1 cm marks as best as possible each time), you could say the measurements give a *precision* a bit better than 0.1 cm. **Accuracy** refers to how close a measurement is to the true value. For example, if the ruler shown in Fig. 1–2 was manufactured with a 2% error, the accuracy of its measurement of the board's width (about 8.8 cm) would be about 2% of 8.8 cm or about ± 0.2 cm. Estimated uncertainty is meant to take both accuracy and precision into account.

TABLE 1–1 Some Typical Lengths or Distances (order of magnitude)

Length (or Distance)	Meters (approximate)
Neutron or proton (diameter)	10^{-15} m
Atom (diameter)	10^{-10} m
Virus [see Fig. 1–5a]	10^{-7} m
Sheet of paper (thickness)	10^{-4} m
Finger width	10^{-2} m
Football field length	10^{2} m
Height of Mt. Everest [see Fig. 1–5b]	10^{4} m
Earth diameter	10^{7} m
Earth to Sun	10^{11} m
Earth to nearest star	10^{16} m
Earth to nearest galaxy	10^{22} m
Earth to farthest galaxy visible	10^{26} m

FIGURE 1–5 Some lengths: (a) viruses (about 10^{-7} m long) attacking a cell; (b) Mt. Everest's height is on the order of 10^{4} m (8850 m, to be precise).

(a)

(b)

1–4 Units, Standards, and the SI System

The measurement of any quantity is made relative to a particular standard or **unit**, and this unit must be specified along with the numerical value of the quantity. For example, we can measure length in British units such as inches, feet, or miles, or in the metric system in centimeters, meters, or kilometers. To specify that the length of a particular object is 18.6 is meaningless. The unit *must* be given; for clearly, 18.6 meters is very different from 18.6 inches or 18.6 millimeters.

For any unit we use, such as the meter for distance or the second for time, we need to define a **standard** which defines exactly how long one meter or one second is. It is important that standards be chosen that are readily reproducible so that anyone needing to make a very accurate measurement can refer to the standard in the laboratory.

Length

The first truly international standard was the **meter** (abbreviated m) established as the standard of **length** by the French Academy of Sciences in the 1790s. The standard meter was originally chosen to be one ten-millionth of the distance from the Earth's equator to either pole,[†] and a platinum rod to represent this length was made. (One meter is, very roughly, the distance from the tip of your nose to the tip of your finger, with arm and hand stretched out to the side.) In 1889, the meter was defined more precisely as the distance between two finely engraved marks on a particular bar of platinum–iridium alloy. In 1960, to provide greater precision and reproducibility, the meter was redefined as 1,650,763.73 wavelengths of a particular orange light emitted by the gas krypton-86. In 1983 the meter was again redefined, this time in terms of the speed of light (whose best measured value in terms of the older definition of the meter was 299,792,458 m/s, with an uncertainty of 1 m/s). The new definition reads: "The meter is the length of path traveled by light in vacuum during a time interval of 1/299,792,458 of a second."[‡]

British units of length (inch, foot, mile) are now defined in terms of the meter. The inch (in.) is defined as precisely 2.54 centimeters (cm; 1 cm = 0.01 m). Other conversion factors are given in the Table on the inside of the front cover of this book. Table 1–1 presents some typical lengths, from very small to very large, rounded off to the nearest power of ten. See also Fig. 1–5. [Note that the abbreviation for inches (in.) is the only one with a period, to distinguish it from the word "in".]

Time

The standard unit of **time** is the **second** (s). For many years, the second was defined as 1/86,400 of a mean solar day (24 h/day × 60 min/h × 60 s/min = 86,400 s/day). The standard second is now defined more precisely in terms of the frequency of radiation emitted by cesium atoms when they pass between two particular states. [Specifically, one second is defined as the time required for 9,192,631,770 periods of this radiation.] There are, by definition, 60 s in one minute (min) and 60 minutes in one hour (h). Table 1–2 presents a range of measured time intervals, rounded off to the nearest power of ten.

Mass

The standard unit of **mass** is the **kilogram** (kg). The standard mass is a particular platinum–iridium cylinder, kept at the International Bureau of Weights and Measures near Paris, France, whose mass is defined as exactly 1 kg. A range of masses is presented in Table 1–3. [For practical purposes, 1 kg weighs about 2.2 pounds on Earth.]

[†]Modern measurements of the Earth's circumference reveal that the intended length is off by about one-fiftieth of 1%. Not bad!

[‡]The new definition of the meter has the effect of giving the speed of light the exact value of 299,792,458 m/s.

TABLE 1–2 Some Typical Time Intervals

Time Interval	Seconds (approximate)
Lifetime of very unstable subatomic particle	10^{-23} s
Lifetime of radioactive elements	10^{-22} s to 10^{28} s
Lifetime of muon	10^{-6} s
Time between human heartbeats	10^{0} s (= 1 s)
One day	10^{5} s
One year	3×10^{7} s
Human life span	2×10^{9} s
Length of recorded history	10^{11} s
Humans on Earth	10^{14} s
Life on Earth	10^{17} s
Age of Universe	10^{18} s

TABLE 1–3 Some Masses

Object	Kilograms (approximate)
Electron	10^{-30} kg
Proton, neutron	10^{-27} kg
DNA molecule	10^{-17} kg
Bacterium	10^{-15} kg
Mosquito	10^{-5} kg
Plum	10^{-1} kg
Human	10^{2} kg
Ship	10^{8} kg
Earth	6×10^{24} kg
Sun	2×10^{30} kg
Galaxy	10^{41} kg

When dealing with atoms and molecules, we usually use the **unified atomic mass unit** (u). In terms of the kilogram,

$$1 \text{ u} = 1.6605 \times 10^{-27} \text{ kg}.$$

The definitions of other standard units for other quantities will be given as we encounter them in later Chapters. (Precise values of this and other numbers are given inside the front cover.)

Unit Prefixes

In the metric system, the larger and smaller units are defined in multiples of 10 from the standard unit, and this makes calculation particularly easy. Thus 1 kilometer (km) is 1000 m, 1 centimeter is $\frac{1}{100}$ m, 1 millimeter (mm) is $\frac{1}{1000}$ m or $\frac{1}{10}$ cm, and so on. The prefixes "centi-," "kilo-," and others are listed in Table 1–4 and can be applied not only to units of length but to units of volume, mass, or any other metric unit. For example, a centiliter (cL) is $\frac{1}{100}$ liter (L), and a kilogram (kg) is 1000 grams (g).

Systems of Units

When dealing with the laws and equations of physics it is very important to use a consistent set of units. Several systems of units have been in use over the years. Today the most important is the **Système International** (French for International System), which is abbreviated SI. In SI units, the standard of length is the meter, the standard for time is the second, and the standard for mass is the kilogram. This system used to be called the MKS (meter-kilogram-second) system.

A second metric system is the **cgs system**, in which the centimeter, gram, and second are the standard units of length, mass, and time, as abbreviated in the title. The **British engineering system** has as its standards the foot for length, the pound for force, and the second for time.

We use SI units almost exclusively in this book.

Base versus Derived Quantities

Physical quantities can be divided into two categories: *base quantities* and *derived quantities*. The corresponding units for these quantities are called *base units* and *derived units*. A **base quantity** must be defined in terms of a standard. Scientists, in the interest of simplicity, want the smallest number of base quantities possible consistent with a full description of the physical world. This number turns out to be seven, and those used in the SI are given in Table 1–5. All other quantities can be defined in terms of these seven base quantities,[†] and hence are referred to as **derived quantities**. An example of a derived quantity is speed, which is defined as distance divided by the time it takes to travel that distance. A Table inside the front cover lists many derived quantities and their units in terms of base units. To define any quantity, whether base or derived, we can specify a rule or procedure, and this is called an **operational definition**.

[†] The only exceptions are for angle (radians—see Chapter 8) and solid angle (steradian). No general agreement has been reached as to whether these are base or derived quantities.

TABLE 1–4 Metric (SI) Prefixes

Prefix	Abbreviation	Value
yotta	Y	10^{24}
zetta	Z	10^{21}
exa	E	10^{18}
peta	P	10^{15}
tera	T	10^{12}
giga	G	10^{9}
mega	M	10^{6}
kilo	k	10^{3}
hecto	h	10^{2}
deka	da	10^{1}
deci	d	10^{-1}
centi	c	10^{-2}
milli	m	10^{-3}
micro[†]	μ	10^{-6}
nano	n	10^{-9}
pico	p	10^{-12}
femto	f	10^{-15}
atto	a	10^{-18}
zepto	z	10^{-21}
yocto	y	10^{-24}

[†] μ is the Greek letter "mu."

TABLE 1–5 SI Base Quantities and Units

Quantity	Unit	Unit Abbreviation
Length	meter	m
Time	second	s
Mass	kilogram	kg
Electric current	ampere	A
Temperature	kelvin	K
Amount of substance	mole	mol
Luminous intensity	candela	cd

1–5 Converting Units

Any quantity we measure, such as a length, a speed, or an electric current, consists of a number *and* a unit. Often we are given a quantity in one set of units, but we want it expressed in another set of units. For example, suppose we measure that a table is 21.5 inches wide, and we want to express this in centimeters. We must use a **conversion factor**, which in this case is (by definition) exactly

$$1 \text{ in.} = 2.54 \text{ cm}$$

or, written another way,

$$1 = 2.54 \text{ cm/in.}$$

Since multiplying by one does not change anything, the width of our table, in cm, is

$$21.5 \text{ inches} = (21.5 \text{ in.}) \times \left(2.54 \frac{\text{cm}}{\text{in.}} \right) = 54.6 \text{ cm.}$$

Note how the units (inches in this case) cancelled out. A Table containing many unit conversions is found inside the front cover of this book. Let's consider some Examples.

FIGURE 1–6 The world's second highest peak, K2, whose summit is considered the most difficult of the "8000-ers." K2 is seen here from the north (China).

TABLE 1–6
The 8000-m Peaks

Peak	Height (m)
Mt. Everest	8850
K2	8611
Kangchenjunga	8586
Lhotse	8516
Makalu	8462
Cho Oyu	8201
Dhaulagiri	8167
Manaslu	8156
Nanga Parbat	8125
Annapurna	8091
Gasherbrum I	8068
Broad Peak	8047
Gasherbrum II	8035
Shisha Pangma	8013

EXAMPLE 1–2 **The 8000-m peaks.** The fourteen tallest peaks in the world (Fig. 1–6 and Table 1–6) are referred to as "eight-thousanders," meaning their summits are over 8000 m above sea level. What is the elevation, in feet, of an elevation of 8000 m?

APPROACH We need simply to convert meters to feet, and we can start with the conversion factor 1 in. = 2.54 cm, which is exact. That is, 1 in. = 2.5400 cm to any number of significant figures, because it is *defined* to be.

SOLUTION One foot is 12 in., so we can write

$$1 \text{ ft} = (12 \text{ in.})\left(2.54 \frac{\text{cm}}{\text{in.}} \right) = 30.48 \text{ cm} = 0.3048 \text{ m,}$$

which is exact. Note how the units cancel (colored slashes). We can rewrite this equation to find the number of feet in 1 meter:

$$1 \text{ m} = \frac{1 \text{ ft}}{0.3048} = 3.28084 \text{ ft.}$$

We multiply this equation by 8000.0 (to have five significant figures):

$$8000.0 \text{ m} = (8000.0 \text{ m})\left(3.28084 \frac{\text{ft}}{\text{m}} \right) = 26{,}247 \text{ ft.}$$

An elevation of 8000 m is 26,247 ft above sea level.

NOTE We could have done the conversion all in one line:

$$8000.0 \text{ m} = (8000.0 \text{ m})\left(\frac{100 \text{ cm}}{1 \text{ m}} \right)\left(\frac{1 \text{ in.}}{2.54 \text{ cm}} \right)\left(\frac{1 \text{ ft}}{12 \text{ in.}} \right) = 26{,}247 \text{ ft.}$$

The key is to multiply conversion factors, each equal to one (= 1.0000), and to make sure the units cancel.

EXERCISE E There are only 14 eight-thousand-meter peaks in the world (see Example 1–2), and their names and elevations are given in Table 1–6. They are all in the Himalaya mountain range in India, Pakistan, Tibet, and China. Determine the elevation of the world's three highest peaks in feet.

EXAMPLE 1–3 **Apartment area.** You have seen a nice apartment whose floor area is 880 square feet (ft^2). What is its area in square meters?

APPROACH We use the same conversion factor, 1 in. = 2.54 cm, but this time we have to use it twice.

SOLUTION Because 1 in. = 2.54 cm = 0.0254 m, then $1\,\text{ft}^2 = (12\,\text{in.})^2(0.0254\,\text{m/in.})^2 = 0.0929\,\text{m}^2$. So $880\,\text{ft}^2 = (880\,\text{ft}^2)(0.0929\,\text{m}^2/\text{ft}^2) \approx 82\,\text{m}^2$.

NOTE As a rule of thumb, an area given in ft^2 is roughly 10 times the number of square meters (more precisely, about $10.8\times$).

EXAMPLE 1–4 **Speeds.** Where the posted speed limit is 55 miles per hour (mi/h or mph), what is this speed (a) in meters per second (m/s) and (b) in kilometers per hour (km/h)?

APPROACH We again use the conversion factor 1 in. = 2.54 cm, and we recall that there are 5280 ft in a mile and 12 inches in a foot; also, one hour contains $(60\,\text{min/h}) \times (60\,\text{s/min}) = 3600\,\text{s/h}$.

SOLUTION (a) We can write 1 mile as

$$1\,\text{mi} = (5280\,\text{ft})\left(12\,\frac{\text{in.}}{\text{ft}}\right)\left(2.54\,\frac{\text{cm}}{\text{in.}}\right)\left(\frac{1\,\text{m}}{100\,\text{cm}}\right) = 1609\,\text{m}.$$

We also know that 1 hour contains 3600 s, so

$$55\,\frac{\text{mi}}{\text{h}} = \left(55\,\frac{\text{mi}}{\text{h}}\right)\left(1609\,\frac{\text{m}}{\text{mi}}\right)\left(\frac{1\,\text{h}}{3600\,\text{s}}\right) = 25\,\frac{\text{m}}{\text{s}},$$

where we rounded off to two significant figures.
(b) Now we use 1 mi = 1609 m = 1.609 km; then

$$55\,\frac{\text{mi}}{\text{h}} = \left(55\,\frac{\text{mi}}{\text{h}}\right)\left(1.609\,\frac{\text{km}}{\text{mi}}\right) = 88\,\frac{\text{km}}{\text{h}}.$$

NOTE Each conversion factor is equal to one. You can look up most conversion factors in the Table inside the front cover.

PROBLEM SOLVING
Conversion factors = 1

EXERCISE F Would a driver traveling at 15 m/s in a 35 mi/h zone be exceeding the speed limit?

When changing units, you can avoid making an error in the use of conversion factors by checking that units cancel out properly. For example, in our conversion of 1 mi to 1609 m in Example 1–4(a), if we had incorrectly used the factor $\left(\frac{100\,\text{cm}}{1\,\text{m}}\right)$ instead of $\left(\frac{1\,\text{m}}{100\,\text{cm}}\right)$, the centimeter units would not have cancelled out; we would not have ended up with meters.

PROBLEM SOLVING
Unit conversion is wrong if units do not cancel

1–6 Order of Magnitude: Rapid Estimating

We are sometimes interested only in an approximate value for a quantity. This might be because an accurate calculation would take more time than it is worth or would require additional data that are not available. In other cases, we may want to make a rough estimate in order to check an accurate calculation made on a calculator, to make sure that no blunders were made when the numbers were entered.

A rough estimate is made by rounding off all numbers to one significant figure and its power of 10, and after the calculation is made, again only one significant figure is kept. Such an estimate is called an **order-of-magnitude estimate** and can be accurate within a factor of 10, and often better. In fact, the phrase "order of magnitude" is sometimes used to refer simply to the power of 10.

PROBLEM SOLVING
How to make a rough estimate

(a)

(b)

FIGURE 1–7 Example 1–5. (a) How much water is in this lake? (Photo is of one of the Rae Lakes in the Sierra Nevada of California.) (b) Model of the lake as a cylinder. [We could go one step further and estimate the mass or weight of this lake. We will see later that water has a density of $1000 \, \text{kg/m}^3$, so this lake has a mass of about $(10^3 \, \text{kg/m}^3)(10^7 \, \text{m}^3) \approx 10^{10} \, \text{kg}$, which is about 10 billion kg or 10 million metric tons. (A metric ton is 1000 kg, about 2200 lbs, slightly larger than a British ton, 2000 lbs.)]

PHYSICS APPLIED
Estimating the volume (or mass) of a lake; see also Fig. 1–7

EXAMPLE 1–5 | ESTIMATE | **Volume of a lake.** Estimate how much water there is in a particular lake, Fig. 1–7a, which is roughly circular, about 1 km across, and you guess it has an average depth of about 10 m.

APPROACH No lake is a perfect circle, nor can lakes be expected to have a perfectly flat bottom. We are only estimating here. To estimate the volume, we can use a simple model of the lake as a cylinder: we multiply the average depth of the lake times its roughly circular surface area, as if the lake were a cylinder (Fig. 1–7b).

SOLUTION The volume V of a cylinder is the product of its height h times the area of its base: $V = h\pi r^2$, where r is the radius of the circular base.† The radius r is $\frac{1}{2}$ km $= 500$ m, so the volume is approximately

$$V = h\pi r^2 \approx (10 \, \text{m}) \times (3) \times (5 \times 10^2 \, \text{m})^2 \approx 8 \times 10^6 \, \text{m}^3 \approx 10^7 \, \text{m}^3,$$

where π was rounded off to 3. So the volume is on the order of $10^7 \, \text{m}^3$, ten million cubic meters. Because of all the estimates that went into this calculation, the order-of-magnitude estimate $(10^7 \, \text{m}^3)$ is probably better to quote than the $8 \times 10^6 \, \text{m}^3$ figure.

NOTE To express our result in U.S. gallons, we see in the Table on the inside front cover that $1 \, \text{liter} = 10^{-3} \, \text{m}^3 \approx \frac{1}{4}$ gallon. Hence, the lake contains $(8 \times 10^6 \, \text{m}^3)(1 \, \text{gallon}/4 \times 10^{-3} \, \text{m}^3) \approx 2 \times 10^9$ gallons of water.

EXAMPLE 1–6 | ESTIMATE | **Thickness of a page.** Estimate the thickness of a page of this book.

APPROACH At first you might think that a special measuring device, a micrometer (Fig. 1–8), is needed to measure the thickness of one page since an ordinary ruler clearly won't do. But we can use a trick or, to put it in physics terms, make use of a *symmetry*: we can make the reasonable assumption that all the pages of this book are equal in thickness.

PROBLEM SOLVING
Use symmetry when possible

SOLUTION We can use a ruler to measure hundreds of pages at once. If you measure the thickness of the first 500 pages of this book (page 1 to page 500), you might get something like 1.5 cm. Note that 500 numbered pages,

†Formulas like this for volume, area, etc., are found inside the back cover of this book.

counted front and back, is 250 separate sheets of paper. So one page must have a thickness of about

$$\frac{1.5\,\text{cm}}{250\,\text{pages}} \approx 6 \times 10^{-3}\,\text{cm} = 6 \times 10^{-2}\,\text{mm},$$

or less than a tenth of a millimeter (0.1 mm).

EXAMPLE 1–7 **ESTIMATE** **Height by triangulation.** Estimate the height of the building shown in Fig. 1–9, by "triangulation," with the help of a bus-stop pole and a friend.

APPROACH By standing your friend next to the pole, you estimate the height of the pole to be 3 m. You next step away from the pole until the top of the pole is in line with the top of the building, Fig. 1–9a. You are 5 ft 6 in. tall, so your eyes are about 1.5 m above the ground. Your friend is taller, and when she stretches out her arms, one hand touches you, and the other touches the pole, so you estimate that distance as 2 m (Fig. 1–9a). You then pace off the distance from the pole to the base of the building with big, 1-m-long steps, and you get a total of 16 steps or 16 m.

SOLUTION Now you draw, to scale, the diagram shown in Fig. 1–9b using these measurements. You can measure, right on the diagram, the last side of the triangle to be about $x = 13$ m. Alternatively, you can use similar triangles to obtain the height x:

$$\frac{1.5\,\text{m}}{2\,\text{m}} = \frac{x}{18\,\text{m}}, \quad \text{so} \quad x \approx 13\tfrac{1}{2}\,\text{m}.$$

Finally you add in your eye height of 1.5 m above the ground to get your final result: the building is about 15 m tall.

EXAMPLE 1–8 **ESTIMATE** **Estimating the radius of Earth.** Believe it or not, you can estimate the radius of the Earth without having to go into space (see the photograph on page 1). If you have ever been on the shore of a large lake, you may have noticed that you cannot see the beaches, piers, or rocks at water level across the lake on the opposite shore. The lake seems to bulge out between you and the opposite shore—a good clue that the Earth is round. Suppose you climb a stepladder and discover that when your eyes are 10 ft (3.0 m) above the water, you can just see the rocks at water level on the opposite shore. From a map, you estimate the distance to the opposite shore as $d \approx 6.1$ km. Use Fig. 1–10 with $h = 3.0$ m to estimate the radius R of the Earth.

APPROACH We use simple geometry, including the theorem of Pythagoras, $c^2 = a^2 + b^2$, where c is the length of the hypotenuse of any right triangle, and a and b are the lengths of the other two sides.

SOLUTION For the right triangle of Fig. 1–10, the two sides are the radius of the Earth R and the distance $d = 6.1$ km $= 6100$ m. The hypotenuse is approximately the length $R + h$, where $h = 3.0$ m. By the Pythagorean theorem,

$$R^2 + d^2 \approx (R + h)^2$$
$$\approx R^2 + 2hR + h^2.$$

We solve algebraically for R, after cancelling R^2 on both sides:

$$R \approx \frac{d^2 - h^2}{2h} = \frac{(6100\,\text{m})^2 - (3.0\,\text{m})^2}{6.0\,\text{m}} = 6.2 \times 10^6\,\text{m} = 6200\,\text{km}.$$

NOTE Precise measurements give 6380 km. But look at your achievement! With a few simple rough measurements and simple geometry, you made a good estimate of the Earth's radius. You did not need to go out in space, nor did you need a very long measuring tape. Now you know the answer to the Chapter-Opening Question on p. 1.

FIGURE 1–8 Example 1–6. Micrometer used for measuring small thicknesses.

FIGURE 1–9 Example 1–7. Diagrams are really useful!

(a)

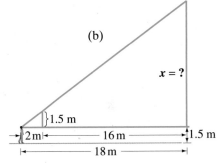

(b)

$x = ?$

FIGURE 1–10 Example 1–8, but not to scale. You can see small rocks at water level on the opposite shore of a lake 6.1 km wide if you stand on a stepladder.

EXAMPLE 1–9 **ESTIMATE** **Total number of heartbeats.** Estimate the total number of beats a typical human heart makes in a lifetime.

APPROACH A typical resting heart rate is 70 beats/min. But during exercise it can be a lot higher. A reasonable average might be 80 beats/min.

SOLUTION One year in terms of seconds is $(24\,\text{h})(3600\,\text{s/h})(365\,\text{d}) \approx 3 \times 10^7\,\text{s}$. If an average person lives 70 years $= (70\,\text{yr})(3 \times 10^7\,\text{s/yr}) \approx 2 \times 10^9\,\text{s}$, then the total number of heartbeats would be about

$$\left(80\,\frac{\text{beats}}{\text{min}}\right)\left(\frac{1\,\text{min}}{60\,\text{s}}\right)(2 \times 10^9\,\text{s}) \approx 3 \times 10^9,$$

or 3 trillion.

Another technique for estimating, this one made famous by Enrico Fermi to his physics students, is to estimate the number of piano tuners in a city, say, Chicago or San Francisco. To get a rough order-of-magnitude estimate of the number of piano tuners today in San Francisco, a city of about 700,000 inhabitants, we can proceed by estimating the number of functioning pianos, how often each piano is tuned, and how many pianos each tuner can tune. To estimate the number of pianos in San Francisco, we note that certainly not everyone has a piano. A guess of 1 family in 3 having a piano would correspond to 1 piano per 12 persons, assuming an average family of 4 persons. As an order of magnitude, let's say 1 piano per 10 people. This is certainly more reasonable than 1 per 100 people, or 1 per every person, so let's proceed with the estimate that 1 person in 10 has a piano, or about 70,000 pianos in San Francisco. Now a piano tuner needs an hour or two to tune a piano. So let's estimate that a tuner can tune 4 or 5 pianos a day. A piano ought to be tuned every 6 months or a year—let's say once each year. A piano tuner tuning 4 pianos a day, 5 days a week, 50 weeks a year can tune about 1000 pianos a year. So San Francisco, with its (very) roughly 70,000 pianos, needs about 70 piano tuners. This is, of course, only a rough estimate.[†] It tells us that there must be many more than 10 piano tuners, and surely not as many as 1000.

PROBLEM SOLVING
Estimating how many piano tuners there are in a city

*1–7 Dimensions and Dimensional Analysis

When we speak of the **dimensions** of a quantity, we are referring to the type of base units or base quantities that make it up. The dimensions of area, for example, are always length squared, abbreviated $[L^2]$, using square brackets; the units can be square meters, square feet, cm^2, and so on. Velocity, on the other hand, can be measured in units of km/h, m/s, or mi/h, but the dimensions are always a length $[L]$ divided by a time $[T]$: that is, $[L/T]$.

The formula for a quantity may be different in different cases, but the dimensions remain the same. For example, the area of a triangle of base b and height h is $A = \frac{1}{2}bh$, whereas the area of a circle of radius r is $A = \pi r^2$. The formulas are different in the two cases, but the dimensions of area are always $[L^2]$.

Dimensions can be used as a help in working out relationships, a procedure referred to as **dimensional analysis**. One useful technique is the use of dimensions to check if a relationship is *incorrect*. Note that we add or subtract quantities only if they have the same dimensions (we don't add centimeters and hours); and the quantities on each side of an equals sign must have the same dimensions. (In numerical calculations, the units must also be the same on both sides of an equation.)

For example, suppose you derived the equation $v = v_0 + \frac{1}{2}at^2$, where v is the speed of an object after a time t, v_0 is the object's initial speed, and the object undergoes an acceleration a. Let's do a dimensional check to see if this equation

---

[†]A check of the San Francisco Yellow Pages (done after this calculation) reveals about 50 listings. Each of these listings may employ more than one tuner, but on the other hand, each may also do repairs as well as tuning. In any case, our estimate is reasonable.

*Some Sections of this book, such as this one, may be considered *optional* at the discretion of the instructor, and they are marked with an asterisk (*). See the Preface for more details.

could be correct or is surely incorrect. Note that numerical factors, like the $\frac{1}{2}$ here, do not affect dimensional checks. We write a dimensional equation as follows, remembering that the dimensions of speed are $[L/T]$ and (as we shall see in Chapter 2) the dimensions of acceleration are $[L/T^2]$:

$$\left[\frac{L}{T}\right] \stackrel{?}{=} \left[\frac{L}{T}\right] + \left[\frac{L}{T^2}\right][T^2] = \left[\frac{L}{T}\right] + [L].$$

The dimensions are incorrect: on the right side, we have the sum of quantities whose dimensions are not the same. Thus we conclude that an error was made in the derivation of the original equation.

A dimensional check can only tell you when a relationship is wrong. It can't tell you if it is completely right. For example, a dimensionless numerical factor (such as $\frac{1}{2}$ or 2π) could be missing.

Dimensional analysis can also be used as a quick check on an equation you are not sure about. For example, suppose that you can't remember whether the equation for the period of a simple pendulum T (the time to make one back-and-forth swing) of length ℓ is $T = 2\pi\sqrt{\ell/g}$ or $T = 2\pi\sqrt{g/\ell}$, where g is the acceleration due to gravity and, like all accelerations, has dimensions $[L/T^2]$. (Do not worry about these formulas—the correct one will be derived in Chapter 14; what we are concerned about here is a person's recalling whether it contains ℓ/g or g/ℓ.) A dimensional check shows that the former (ℓ/g) is correct:

$$[T] = \sqrt{\frac{[L]}{[L/T^2]}} = \sqrt{[T^2]} = [T],$$

whereas the latter (g/ℓ) is not:

$$[T] \neq \sqrt{\frac{[L/T^2]}{[L]}} = \sqrt{\frac{1}{[T^2]}} = \frac{1}{[T]}.$$

Note that the constant 2π has no dimensions and so can't be checked using dimensions.

Further uses of dimensional analysis are found in Appendix C.

EXAMPLE 1-10 **Planck length.** The smallest meaningful measure of length is called the "Planck length," and is defined in terms of three fundamental constants in nature, the speed of light $c = 3.00 \times 10^8$ m/s, the gravitational constant $G = 6.67 \times 10^{-11}$ m^3/kg·s^2, and Planck's constant $h = 6.63 \times 10^{-34}$ kg·m^2/s. The Planck length λ_P (λ is the Greek letter "lambda") is given by the following combination of these three constants:

$$\lambda_P = \sqrt{\frac{Gh}{c^3}}.$$

Show that the dimensions of λ_P are length $[L]$, and find the order of magnitude of λ_P.

APPROACH We rewrite the above equation in terms of dimensions. The dimensions of c are $[L/T]$, of G are $[L^3/MT^2]$, and of h are $[ML^2/T]$.

SOLUTION The dimensions of λ_P are

$$\sqrt{\frac{[L^3/MT^2][ML^2/T]}{[L^3/T^3]}} = \sqrt{[L^2]} = [L]$$

which is a length. The value of the Planck length is

$$\lambda_P = \sqrt{\frac{Gh}{c^3}} = \sqrt{\frac{(6.67 \times 10^{-11}\,\text{m}^3/\text{kg}\cdot\text{s}^2)(6.63 \times 10^{-34}\,\text{kg}\cdot\text{m}^2/\text{s})}{(3.0 \times 10^8\,\text{m/s})^3}} \approx 4 \times 10^{-35}\,\text{m},$$

which is on the order of 10^{-34} or 10^{-35} m.

NOTE Some recent theories (Chapters 43 and 44) suggest that the smallest particles (quarks, leptons) have sizes on the order of the Planck length, 10^{-35} m. These theories also suggest that the "Big Bang," with which the Universe is believed to have begun, started from an initial size on the order of the Planck length.

Summary

[The Summary that appears at the end of each Chapter in this book gives a brief overview of the main ideas of the Chapter. The Summary *cannot* serve to give an understanding of the material, which can be accomplished only by a detailed reading of the Chapter.]

Physics, like other sciences, is a creative endeavor. It is not simply a collection of facts. Important **theories** are created with the idea of explaining **observations**. To be accepted, theories are **tested** by comparing their predictions with the results of actual experiments. Note that, in general, a theory cannot be "proved" in an absolute sense.

Scientists often devise models of physical phenomena. A **model** is a kind of picture or analogy that helps to describe the phenomena in terms of something we already know. A **theory**, often developed from a model, is usually deeper and more complex than a simple model.

A scientific **law** is a concise statement, often expressed in the form of an equation, which quantitatively describes a wide range of phenomena.

Measurements play a crucial role in physics, but can never be perfectly precise. It is important to specify the **uncertainty** of a measurement either by stating it directly using the ± notation, and/or by keeping only the correct number of **significant figures**.

Physical quantities are always specified relative to a particular standard or **unit**, and the unit used should always be stated. The commonly accepted set of units today is the **Système International** (SI), in which the standard units of length, mass, and time are the **meter**, **kilogram**, and **second**.

When converting units, check all **conversion factors** for correct cancellation of units.

Making rough, **order-of-magnitude estimates** is a very useful technique in science as well as in everyday life.

[*The **dimensions** of a quantity refer to the combination of base quantities that comprise it. Velocity, for example, has dimensions of [length/time] or [L/T]. **Dimensional analysis** can be used to check a relationship for correct form.]

Questions

1. What are the merits and drawbacks of using a person's foot as a standard? Consider both (*a*) a particular person's foot, and (*b*) any person's foot. Keep in mind that it is advantagous that fundamental standards be accessible (easy to compare to), invariable (do not change), indestructible, and reproducible.

2. Why is it incorrect to think that the more digits you represent in your answer, the more accurate it is?

3. When traveling a highway in the mountains, you may see elevation signs that read "914 m (3000 ft)." Critics of the metric system claim that such numbers show the metric system is more complicated. How would you alter such signs to be more consistent with a switch to the metric system?

4. What is wrong with this road sign:
   Memphis 7 mi (11.263 km)?

5. For an answer to be complete, the units need to be specified. Why?

6. Discuss how the notion of symmetry could be used to estimate the number of marbles in a 1-liter jar.

7. You measure the radius of a wheel to be 4.16 cm. If you multiply by 2 to get the diameter, should you write the result as 8 cm or as 8.32 cm? Justify your answer.

8. Express the sine of 30.0° with the correct number of significant figures.

9. A recipe for a soufflé specifies that the measured ingredients must be exact, or the soufflé will not rise. The recipe calls for 6 large eggs. The size of "large" eggs can vary by 10%, according to the USDA specifications. What does this tell you about how exactly you need to measure the other ingredients?

10. List assumptions useful to estimate the number of car mechanics in (*a*) San Francisco, (*b*) your hometown, and then make the estimates.

11. Suggest a way to measure the distance from Earth to the Sun.

*12. Can you set up a complete set of base quantities, as in Table 1–5, that does not include length as one of them?

Problems

[The Problems at the end of each Chapter are ranked I, II, or III according to estimated difficulty, with (I) Problems being easiest. Level (III) Problems are meant mainly as a challenge for the best students, for "extra credit." The Problems are arranged by Sections, meaning that the reader should have read up to and including that Section, but not only that Section—Problems often depend on earlier material. Each Chapter also has a group of General Problems that are not arranged by Section and not ranked.]

1–3 Measurement, Uncertainty, Significant Figures

(*Note:* In Problems, assume a number like 6.4 is accurate to ± 0.1; and 950 is ± 10 unless 950 is said to be "precisely" or "very nearly" 950, in which case assume 950 ± 1.)

1. (I) The age of the universe is thought to be about 14 billion years. Assuming two significant figures, write this in powers of ten in (*a*) years, (*b*) seconds.

2. (I) How many significant figures do each of the following numbers have: (*a*) 214, (*b*) 81.60, (*c*) 7.03, (*d*) 0.03, (*e*) 0.0086, (*f*) 3236, and (*g*) 8700?

3. (I) Write the following numbers in powers of ten notation: (*a*) 1.156, (*b*) 21.8, (*c*) 0.0068, (*d*) 328.65, (*e*) 0.219, and (*f*) 444.

4. (I) Write out the following numbers in full with the correct number of zeros: (*a*) 8.69×10^4, (*b*) 9.1×10^3, (*c*) 8.8×10^{-1}, (*d*) 4.76×10^2, and (*e*) 3.62×10^{-5}.

5. (II) What is the percent uncertainty in the measurement 5.48 ± 0.25 m?

6. (II) Time intervals measured with a stopwatch typically have an uncertainty of about 0.2 s, due to human reaction time at the start and stop moments. What is the percent uncertainty of a hand-timed measurement of (*a*) 5 s, (*b*) 50 s, (*c*) 5 min?

7. (II) Add $(9.2 \times 10^3 \text{ s}) + (8.3 \times 10^4 \text{ s}) + (0.008 \times 10^6 \text{ s})$.

8. (II) Multiply 2.079×10^2 m by 0.082×10^{-1}, taking into account significant figures.

9. (III) For small angles θ, the numerical value of $\sin \theta$ is approximately the same as the numerical value of $\tan \theta$. Find the largest angle for which sine and tangent agree to within two significant figures.

10. (III) What, roughly, is the percent uncertainty in the volume of a spherical beach ball whose radius is $r = 0.84 \pm 0.04$ m?

1–4 and 1–5 Units, Standards, SI, Converting Units

11. (I) Write the following as full (decimal) numbers with standard units: (a) 286.6 mm, (b) 85 μV, (c) 760 mg, (d) 60.0 ps, (e) 22.5 fm, (f) 2.50 gigavolts.

12. (I) Express the following using the prefixes of Table 1–4: (a) 1×10^6 volts, (b) 2×10^{-6} meters, (c) 6×10^3 days, (d) 18×10^2 bucks, and (e) 8×10^{-8} seconds.

13. (I) Determine your own height in meters, and your mass in kg.

14. (I) The Sun, on average, is 93 million miles from Earth. How many meters is this? Express (a) using powers of ten, and (b) using a metric prefix.

15. (II) What is the conversion factor between (a) ft^2 and yd^2, (b) m^2 and ft^2?

16. (II) An airplane travels at 950 km/h. How long does it take to travel 1.00 km?

17. (II) A typical atom has a diameter of about 1.0×10^{-10} m. (a) What is this in inches? (b) Approximately how many atoms are there along a 1.0-cm line?

18. (II) Express the following sum with the correct number of significant figures: 1.80 m + 142.5 cm + 5.34×10^5 μm.

19. (II) Determine the conversion factor between (a) km/h and mi/h, (b) m/s and ft/s, and (c) km/h and m/s.

20. (II) How much longer (percentage) is a one-mile race than a 1500-m race ("the metric mile")?

21. (II) A *light-year* is the distance light travels in one year (at speed = 2.998×10^8 m/s). (a) How many meters are there in 1.00 light-year? (b) An astronomical unit (AU) is the average distance from the Sun to Earth, 1.50×10^8 km. How many AU are there in 1.00 light-year? (c) What is the speed of light in AU/h?

22. (II) If you used only a keyboard to enter data, how many years would it take to fill up the hard drive in your computer that can store 82 gigabytes (82×10^9 bytes) of data? Assume "normal" eight-hour working days, and that one byte is required to store one keyboard character, and that you can type 180 characters per minute.

23. (III) The diameter of the Moon is 3480 km. (a) What is the surface area of the Moon? (b) How many times larger is the surface area of the Earth?

1–6 Order-of-Magnitude Estimating

(*Note*: Remember that for rough estimates, only round numbers are needed both as input to calculations and as final results.)

24. (I) Estimate the order of magnitude (power of ten) of: (a) 2800, (b) 86.30×10^2, (c) 0.0076, and (d) 15.0×10^8.

25. (II) Estimate how many books can be shelved in a college library with 3500 m^2 of floor space. Assume 8 shelves high, having books on both sides, with corridors 1.5 m wide. Assume books are about the size of this one, on average.

26. (II) Estimate how many hours it would take a runner to run (at 10 km/h) across the United States from New York to California.

27. (II) Estimate the number of liters of water a human drinks in a lifetime.

28. (II) Estimate how long it would take one person to mow a football field using an ordinary home lawn mower (Fig. 1–11). Assume the mower moves with a 1-km/h speed, and has a 0.5-m width.

FIGURE 1–11
Problem 28.

29. (II) Estimate the number of dentists (a) in San Francisco and (b) in your town or city.

30. (III) The rubber worn from tires mostly enters the atmosphere as *particulate pollution*. Estimate how much rubber (in kg) is put into the air in the United States every year. To get started, a good estimate for a tire tread's depth is 1 cm when new, and rubber has a mass of about 1200 kg per m^3 of volume.

31. (III) You are in a hot air balloon, 200 m above the flat Texas plains. You look out toward the horizon. How far out can you see—that is, how far is your horizon? The Earth's radius is about 6400 km.

32. (III) I agree to hire you for 30 days and you can decide between two possible methods of payment: either (1) $1000 a day, or (2) one penny on the first day, two pennies on the second day and continue to double your daily pay each day up to day 30. Use quick estimation to make your decision, and justify it.

33. (III) Many sailboats are moored at a marina 4.4 km away on the opposite side of a lake. You stare at one of the sailboats because, when you are lying flat at the water's edge, you can just see its deck but none of the side of the sailboat. You then go to that sailboat on the other side of the lake and measure that the deck is 1.5 m above the level of the water. Using Fig. 1–12, where $h = 1.5$ m, estimate the radius R of the Earth.

FIGURE 1–12 Problem 33.
You see a sailboat across a lake (not to scale). R is the radius of the Earth. You are a distance $d = 4.4$ km from the sailboat when you can see only its deck and not its side. Because of the curvature of the Earth, the water "bulges out" between you and the boat.

34. (III) Another experiment you can do also uses the radius of the Earth. The Sun sets, fully disappearing over the horizon as you lie on the beach, your eyes 20 cm above the sand. You immediately jump up, your eyes now 150 cm above the sand, and you can again see the top of the Sun. If you count the number of seconds ($= t$) until the Sun fully disappears again, you can estimate the radius of the Earth. But for this Problem, use the known radius of the Earth and calculate the time t.

*35. (I) What are the dimensions of density, which is mass per volume?

*36. (II) The speed v of an object is given by the equation $v = At^3 - Bt$, where t refers to time. (a) What are the dimensions of A and B? (b) What are the SI units for the constants A and B?

*37. (II) Three students derive the following equations in which x refers to distance traveled, v the speed, a the acceleration (m/s^2), t the time, and the subscript zero $(_0)$ means a quantity at time $t = 0$: (a) $x = vt^2 + 2at$, (b) $x = v_0 t + \frac{1}{2}at^2$, and (c) $x = v_0 t + 2at^2$. Which of these could possibly be correct according to a dimensional check?

*38. (II) Show that the following combination of the three fundamental constants of nature that we used in Example 1–10 (that is G, c, and h) forms a quantity with the dimensions of time:

$$t_P = \sqrt{\frac{Gh}{c^5}}.$$

This quantity, t_P, is called the *Planck time* and is thought to be the earliest time, after the creation of the Universe, at which the currently known laws of physics can be applied.

General Problems

39. *Global positioning satellites* (GPS) can be used to determine positions with great accuracy. If one of the satellites is at a distance of 20,000 km from you, what percent uncertainty in the distance does a 2-m uncertainty represent? How many significant figures are needed in the distance?

40. *Computer chips* (Fig. 1–13) etched on circular silicon wafers of thickness 0.300 mm are sliced from a solid cylindrical silicon crystal of length 25 cm. If each wafer can hold 100 chips, what is the maximum number of chips that can be produced from one entire cylinder?

FIGURE 1–13   Problem 40.
The wafer held by the hand (above) is shown below, enlarged and illuminated by colored light. Visible are rows of integrated circuits (chips).

41. (a) How many seconds are there in 1.00 year? (b) How many nanoseconds are there in 1.00 year? (c) How many years are there in 1.00 second?

42. American football uses a field that is 100 yd long, whereas a regulation soccer field is 100 m long. Which field is longer, and by how much (give yards, meters, and percent)?

43. A typical adult human lung contains about 300 million tiny cavities called alveoli. Estimate the average diameter of a single alveolus.

44. One hectare is defined as $1.000 \times 10^4 \, m^2$. One acre is $4.356 \times 10^4 \, ft^2$. How many acres are in one hectare?

45. Estimate the number of gallons of gasoline consumed by the total of all automobile drivers in the United States, per year.

46. Use Table 1–3 to estimate the total number of protons or neutrons in (a) a bacterium, (b) a DNA molecule, (c) the human body, (d) our Galaxy.

47. An average family of four uses roughly 1200 L (about 300 gallons) of water per day $(1 \, L = 1000 \, cm^3)$. How much depth would a lake lose per year if it uniformly covered an area of 50 km² and supplied a local town with a population of 40,000 people? Consider only population uses, and neglect evaporation and so on.

48. Estimate the number of gumballs in the machine of Fig. 1–14.

FIGURE 1–14   Problem 48. Estimate the number of gumballs in the machine.

49. Estimate how many kilograms of laundry soap are used in the U.S. in one year (and therefore pumped out of washing machines with the dirty water). Assume each load of laundry takes 0.1 kg of soap.

50. How big is a ton? That is, what is the volume of something that weighs a ton? To be specific, estimate the diameter of a 1-ton rock, but first make a wild guess: will it be 1 ft across, 3 ft, or the size of a car? [*Hint*: Rock has mass per volume about 3 times that of water, which is 1 kg per liter $(10^3 \, cm^3)$ or 62 lb per cubic foot.]

51. A certain audio compact disc (CD) contains 783.216 megabytes of digital information. Each byte consists of exactly 8 bits. When played, a CD player reads the CD's digital information at a constant rate of 1.4 megabits per second. How many minutes does it take the player to read the entire CD?

52. Hold a pencil in front of your eye at a position where its blunt end just blocks out the Moon (Fig. 1–15). Make appropriate measurements to estimate the diameter of the Moon, given that the Earth–Moon distance is 3.8×10^5 km.

FIGURE 1–15   Problem 52. How big is the Moon?

53. A heavy rainstorm dumps 1.0 cm of rain on a city 5 km wide and 8 km long in a 2-h period. How many metric tons $(1 \text{ metric ton} = 10^3 \text{ kg})$ of water fell on the city? $(1 \, cm^3$ of water has a mass of $1 \, g = 10^{-3} \, kg$.) How many gallons of water was this?

54. Noah's ark was ordered to be 300 cubits long, 50 cubits wide, and 30 cubits high. The cubit was a unit of measure equal to the length of a human forearm, elbow to the tip of the longest finger. Express the dimensions of Noah's ark in meters, and estimate its volume (m^3).

55. Estimate how many days it would take to walk around the world, assuming 10 h walking per day at 4 km/h.

56. One liter $(1000 \, cm^3)$ of oil is spilled onto a smooth lake. If the oil spreads out uniformly until it makes an oil slick just one molecule thick, with adjacent molecules just touching, estimate the diameter of the oil slick. Assume the oil molecules have a diameter of $2 \times 10^{-10} \, m$.

57. Jean camps beside a wide river and wonders how wide it is. She spots a large rock on the bank directly across from her. She then walks upstream until she judges that the angle between her and the rock, which she can still see clearly, is now at an angle of 30° downstream (Fig. 1–16). Jean measures her stride to be about 1 yard long. The distance back to her camp is 120 strides. About how far across, both in yards and in meters, is the river?

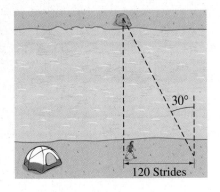

FIGURE 1–16
Problem 57.

58. A watch manufacturer claims that its watches gain or lose no more than 8 seconds in a year. How accurate is this watch, expressed as a percentage?

59. An angstrom (symbol Å) is a unit of length, defined as $10^{-10} \, m$, which is on the order of the diameter of an atom. (a) How many nanometers are in 1.0 angstrom? (b) How many femtometers or fermis (the common unit of length in nuclear physics) are in 1.0 angstrom? (c) How many angstroms are in 1.0 m? (d) How many angstroms are in 1.0 light-year (see Problem 21)?

60. The diameter of the Moon is 3480 km. What is the volume of the Moon? How many Moons would be needed to create a volume equal to that of Earth?

61. Determine the percent uncertainty in θ, and in $\sin \theta$, when (a) $\theta = 15.0° \pm 0.5°$, (b) $\theta = 75.0° \pm 0.5°$.

62. If you began walking along one of Earth's lines of longitude and walked north until you had changed latitude by 1 minute of arc (there are 60 minutes per degree), how far would you have walked (in miles)? This distance is called a "nautical mile."

63. Make a rough estimate of the volume of your body (in m^3).

64. Estimate the number of bus drivers (a) in Washington, D.C., and (b) in your town.

65. The American Lung Association gives the following formula for an average person's expected lung capacity V (in liters, where $1 \, L = 10^3 \, cm^3$):

$$V = 4.1H - 0.018A - 2.69,$$

where H and A are the person's height (in meters), and age (in years), respectively. In this formula, what are the units of the numbers 4.1, 0.018, and 2.69?

66. The density of an object is defined as its mass divided by its volume. Suppose the mass and volume of a rock are measured to be 8 g and 2.8325 cm^3. To the correct number of significant figures, determine the rock's density.

67. To the correct number of significant figures, use the information inside the front cover of this book to determine the ratio of (a) the surface area of Earth compared to the surface area of the Moon; (b) the volume of Earth compared to the volume of the Moon.

68. One mole of atoms consists of 6.02×10^{23} individual atoms. If a mole of atoms were spread uniformly over the surface of the Earth, how many atoms would there be per square meter?

69. Recent findings in astrophysics suggest that the observable Universe can be modeled as a sphere of radius $R = 13.7 \times 10^9$ light-years with an average mass density of about $1 \times 10^{-26} \, kg/m^3$, where only about 4% of the Universe's total mass is due to "ordinary" matter (such as protons, neutrons, and electrons). Use this information to estimate the total mass of ordinary matter in the observable Universe. (1 light-year = $9.46 \times 10^{15} \, m$.)

Answers to Exercises

A: (d).

B: No: they have 3 and 2, respectively.

C: All three have three significant figures, although the number of decimal places is (a) 2, (b) 3, (c) 4.

D: (a) 2.58×10^{-2}, 3; (b) 4.23×10^4, 3 (probably); (c) 3.4450×10^2, 5.

E: Mt. Everest, 29,035 ft; K2, 28,251 ft; Kangchenjunga, 28,169 ft.

F: No: 15 m/s ≈ 34 mi/h.

A high-speed car has released a parachute to reduce its speed quickly. The directions of the car's velocity and acceleration are shown by the green ($\vec{v}$) and gold ($\vec{a}$) arrows.

Motion is described using the concepts of velocity and acceleration. In the case shown here, the acceleration $\vec{a}$ is in the opposite direction from the velocity $\vec{v}$, which means the object is slowing down. We examine in detail motion with constant acceleration, including the vertical motion of objects falling under gravity.

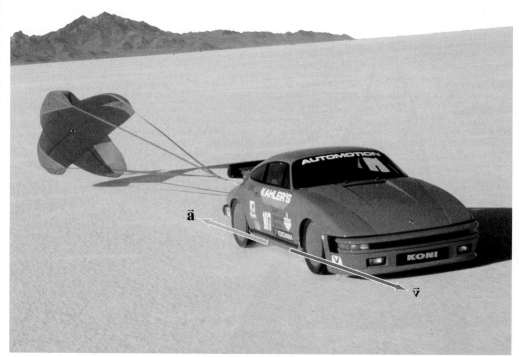

C H A P T E R

2

Describing Motion: Kinematics in One Dimension

CHAPTER-OPENING QUESTION—Guess now!

[*Don't worry about getting the right answer now—you will get another chance later in the Chapter. See also p. 1 of Chapter 1 for more explanation.*]

Two small heavy balls have the same diameter but one weighs twice as much as the other. The balls are dropped from a second-story balcony at the exact same time. The time to reach the ground below will be:

(a) twice as long for the lighter ball as for the heavier one.

(b) longer for the lighter ball, but not twice as long.

(c) twice as long for the heavier ball as for the lighter one.

(d) longer for the heavier ball, but not twice as long.

(e) nearly the same for both balls.

T he motion of objects—baseballs, automobiles, joggers, and even the Sun and Moon—is an obvious part of everyday life. It was not until the sixteenth and seventeenth centuries that our modern understanding of motion was established. Many individuals contributed to this understanding, particularly Galileo Galilei (1564–1642) and Isaac Newton (1642–1727).

The study of the motion of objects, and the related concepts of force and energy, form the field called **mechanics**. Mechanics is customarily divided into two parts: **kinematics**, which is the description of how objects move, and **dynamics**, which deals with force and why objects move as they do. This Chapter and the next deal with kinematics.

For now we only discuss objects that move without rotating (Fig. 2–1a). Such motion is called **translational motion**. In this Chapter we will be concerned with describing an object that moves along a straight-line path, which is one-dimensional translational motion. In Chapter 3 we will describe translational motion in two (or three) dimensions along paths that are not straight.

We will often use the concept, or *model*, of an idealized **particle** which is considered to be a mathematical **point** with no spatial extent (no size). A point particle can undergo only translational motion. The particle model is useful in many real situations where we are interested only in translational motion and the object's size is not significant. For example, we might consider a billiard ball, or even a spacecraft traveling toward the Moon, as a particle for many purposes.

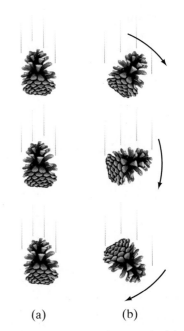

(a)   (b)

FIGURE 2–1   The pinecone in (a) undergoes pure translation as it falls, whereas in (b) it is rotating as well as translating.

2–1 Reference Frames and Displacement

Any measurement of position, distance, or speed must be made with respect to a **reference frame**, or **frame of reference**. For example, while you are on a train traveling at 80 km/h, suppose a person walks past you toward the front of the train at a speed of, say, 5 km/h (Fig. 2–2). This 5 km/h is the person's speed with respect to the train as frame of reference. With respect to the ground, that person is moving at a speed of 80 km/h + 5 km/h = 85 km/h. It is always important to specify the frame of reference when stating a speed. In everyday life, we usually mean "with respect to the Earth" without even thinking about it, but the reference frame must be specified whenever there might be confusion.

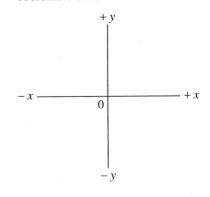

FIGURE 2–2   A person walks toward the front of a train at 5 km/h. The train is moving 80 km/h with respect to the ground, so the walking person's speed, relative to the ground, is 85 km/h.

When specifying the motion of an object, it is important to specify not only the speed but also the direction of motion. Often we can specify a direction by using the cardinal points, north, east, south, and west, and by "up" and "down." In physics, we often draw a set of **coordinate axes**, as shown in Fig. 2–3, to represent a frame of reference. We can always place the origin 0, and the directions of the x and y axes, as we like for convenience. The x and y axes are always perpendicular to each other. Objects positioned to the right of the origin of coordinates (0) on the x axis have an x coordinate which we usually choose to be positive; then points to the left of 0 have a negative x coordinate. The position along the y axis is usually considered positive when above 0, and negative when below 0, although the reverse convention can be used if convenient. Any point on the plane can be specified by giving its x and y coordinates. In three dimensions, a z axis perpendicular to the x and y axes is added.

For one-dimensional motion, we often choose the x axis as the line along which the motion takes place. Then the **position** of an object at any moment is given by its x coordinate. If the motion is vertical, as for a dropped object, we usually use the y axis.

FIGURE 2–3   Standard set of xy coordinate axes.

The displacement may not equal the total distance traveled

FIGURE 2–4 A person walks 70 m east, then 30 m west. The total distance traveled is 100 m (path is shown dashed in black); but the displacement, shown as a solid blue arrow, is 40 m to the east.

FIGURE 2–5 The arrow represents the displacement $x_2 - x_1$. Distances are in meters.

FIGURE 2–6 For the displacement $\Delta x = x_2 - x_1 = 10.0\,\text{m} - 30.0\,\text{m}$, the displacement vector points to the left.

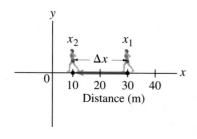

We need to make a distinction between the *distance* an object has traveled and its **displacement**, which is defined as the *change in position* of the object. That is, *displacement is how far the object is from its starting point.* To see the distinction between total distance and displacement, imagine a person walking 70 m to the east and then turning around and walking back (west) a distance of 30 m (see Fig. 2–4). The total *distance* traveled is 100 m, but the *displacement* is only 40 m since the person is now only 40 m from the starting point.

Displacement is a quantity that has both magnitude and direction. Such quantities are called **vectors**, and are represented by arrows in diagrams. For example, in Fig. 2–4, the blue arrow represents the displacement whose magnitude is 40 m and whose direction is to the right (east).

We will deal with vectors more fully in Chapter 3. For now, we deal only with motion in one dimension, along a line. In this case, vectors which point in one direction will have a positive sign, whereas vectors that point in the opposite direction will have a negative sign, along with their magnitude.

Consider the motion of an object over a particular time interval. Suppose that at some initial time, call it t_1, the object is on the x axis at the position x_1 in the coordinate system shown in Fig. 2–5. At some later time, t_2, suppose the object has moved to position x_2. The displacement of our object is $x_2 - x_1$, and is represented by the arrow pointing to the right in Fig. 2–5. It is convenient to write

$$\Delta x = x_2 - x_1,$$

where the symbol Δ (Greek letter delta) means "change in." Then Δx means "the change in x," or "change in position," which is the displacement. Note that the "change in" any quantity means the final value of that quantity, minus the initial value.

Suppose $x_1 = 10.0\,\text{m}$ and $x_2 = 30.0\,\text{m}$. Then

$$\Delta x = x_2 - x_1 = 30.0\,\text{m} - 10.0\,\text{m} = 20.0\,\text{m},$$

so the displacement is 20.0 m in the positive direction, Fig. 2–5.

Now consider an object moving to the left as shown in Fig. 2–6. Here the object, say, a person, starts at $x_1 = 30.0\,\text{m}$ and walks to the left to the point $x_2 = 10.0\,\text{m}$. In this case her displacement is

$$\Delta x = x_2 - x_1 = 10.0\,\text{m} - 30.0\,\text{m} = -20.0\,\text{m},$$

and the blue arrow representing the vector displacement points to the left. For one-dimensional motion along the x axis, a vector pointing to the right has a positive sign, whereas a vector pointing to the left has a negative sign.

EXERCISE A An ant starts at $x = 20\,\text{cm}$ on a piece of graph paper and walks along the x axis to $x = -20\,\text{cm}$. It then turns around and walks back to $x = -10\,\text{cm}$. What is the ant's displacement and total distance traveled?

2–2 Average Velocity

The most obvious aspect of the motion of a moving object is how fast it is moving—its speed or velocity.

The term "speed" refers to how far an object travels in a given time interval, regardless of direction. If a car travels 240 kilometers (km) in 3 hours (h), we say its average speed was 80 km/h. In general, the **average speed** of an object is defined as *the total distance traveled along its path divided by the time it takes to travel this distance*:

$$\text{average speed} = \frac{\text{distance traveled}}{\text{time elapsed}}. \tag{2–1}$$

The terms "velocity" and "speed" are often used interchangeably in ordinary language. But in physics we make a distinction between the two. Speed is simply a positive number, with units. **Velocity**, on the other hand, is used to signify both the *magnitude* (numerical value) of how fast an object is moving and also the *direction* in which it is moving. (Velocity is therefore a vector.) There is a second difference between speed and velocity: namely, the **average velocity** is defined in terms of *displacement*, rather than total distance traveled:

$$\text{average velocity} = \frac{\text{displacement}}{\text{time elapsed}} = \frac{\text{final position} - \text{initial position}}{\text{time elapsed}}.$$

Average speed and average velocity have the same magnitude when the motion is all in one direction. In other cases, they may differ: recall the walk we described earlier, in Fig. 2–4, where a person walked 70 m east and then 30 m west. The total distance traveled was 70 m + 30 m = 100 m, but the displacement was 40 m. Suppose this walk took 70 s to complete. Then the average speed was:

$$\frac{\text{distance}}{\text{time elapsed}} = \frac{100 \text{ m}}{70 \text{ s}} = 1.4 \text{ m/s}.$$

The magnitude of the average velocity, on the other hand, was:

$$\frac{\text{displacement}}{\text{time elapsed}} = \frac{40 \text{ m}}{70 \text{ s}} = 0.57 \text{ m/s}.$$

This difference between the speed and the magnitude of the velocity can occur when we calculate *average* values.

To discuss one-dimensional motion of an object in general, suppose that at some moment in time, call it t_1, the object is on the x axis at position x_1 in a coordinate system, and at some later time, t_2, suppose it is at position x_2. The **elapsed time** is $\Delta t = t_2 - t_1$; during this time interval the displacement of our object is $\Delta x = x_2 - x_1$. Then the average velocity, defined as *the displacement divided by the elapsed time*, can be written

$$\bar{v} = \frac{x_2 - x_1}{t_2 - t_1} = \frac{\Delta x}{\Delta t}, \qquad \textbf{(2–2)}$$

where v stands for velocity and the bar ($^-$) over the v is a standard symbol meaning "average."

For the usual case of the $+x$ axis to the right, note that if x_2 is less than x_1, the object is moving to the left, and then $\Delta x = x_2 - x_1$ is less than zero. The sign of the displacement, and thus of the average velocity, indicates the direction: the average velocity is positive for an object moving to the right along the $+x$ axis and negative when the object moves to the left. The direction of the average velocity is always the same as the direction of the displacement.

Note that it is always important to choose (and state) the *elapsed time*, or *time interval*, $t_2 - t_1$, the time that passes during our chosen period of observation.

⚠ CAUTION

Average speed is not necessarily equal to the magnitude of the average velocity

PROBLEM SOLVING

+ or − sign can signify the direction for linear motion

EXAMPLE 2–1 **Runner's average velocity.** The position of a runner as a function of time is plotted as moving along the x axis of a coordinate system. During a 3.00-s time interval, the runner's position changes from $x_1 = 50.0$ m to $x_2 = 30.5$ m, as shown in Fig. 2–7. What was the runner's average velocity?

APPROACH We want to find the average velocity, which is the displacement divided by the elapsed time.

SOLUTION The displacement is $\Delta x = x_2 - x_1 = 30.5 \text{ m} - 50.0 \text{ m} = -19.5 \text{ m}$. The elapsed time, or time interval, is $\Delta t = 3.00$ s. The average velocity is

$$\bar{v} = \frac{\Delta x}{\Delta t} = \frac{-19.5 \text{ m}}{3.00 \text{ s}} = -6.50 \text{ m/s}.$$

The displacement and average velocity are negative, which tells us that the runner is moving to the left along the x axis, as indicated by the arrow in Fig. 2–7. Thus we can say that the runner's average velocity is 6.50 m/s to the left.

FIGURE 2–7 Example 2–1. A person runs from $x_1 = 50.0$ m to $x_2 = 30.5$ m. The displacement is −19.5 m.

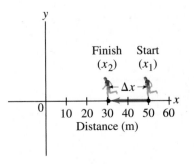

EXAMPLE 2–2 **Distance a cyclist travels.** How far can a cyclist travel in 2.5 h along a straight road if her average velocity is 18 km/h?

APPROACH We want to find the distance traveled, so we solve Eq. 2–2 for Δx.

SOLUTION We rewrite Eq. 2–2 as $\Delta x = \bar{v}\,\Delta t$, and find

$$\Delta x = \bar{v}\,\Delta t = (18\,\text{km/h})(2.5\,\text{h}) = 45\,\text{km}.$$

EXERCISE B A car travels at a constant 50 km/h for 100 km. It then speeds up to 100 km/h and is driven another 100 km. What is the car's average speed for the 200 km trip? (*a*) 67 km/h; (*b*) 75 km/h; (*c*) 81 km/h; (*d*) 50 km/h.

2–3 Instantaneous Velocity

If you drive a car along a straight road for 150 km in 2.0 h, the magnitude of your average velocity is 75 km/h. It is unlikely, though, that you were moving at precisely 75 km/h at every instant. To describe this situation we need the concept of *instantaneous velocity*, which is the velocity at any instant of time. (Its magnitude is the number, with units, indicated by a speedometer, Fig. 2–8.) More precisely, the **instantaneous velocity** at any moment is defined as *the average velocity over an infinitesimally short time interval*. That is, Eq. 2–2 is to be evaluated in the limit of Δt becoming extremely small, approaching zero. We can write the definition of instantaneous velocity, v, for one-dimensional motion as

$$v = \lim_{\Delta t \to 0} \frac{\Delta x}{\Delta t}. \tag{2–3}$$

The notation $\lim_{\Delta t \to 0}$ means the ratio $\Delta x / \Delta t$ is to be evaluated in the limit of Δt approaching zero. But we do not simply set $\Delta t = 0$ in this definition, for then Δx would also be zero, and we would have an undefined number. Rather, we are considering the *ratio* $\Delta x / \Delta t$, as a whole. As we let Δt approach zero, Δx approaches zero as well. But the ratio $\Delta x / \Delta t$ approaches some definite value, which is the instantaneous velocity at a given instant.

In Eq. 2–3, the limit as $\Delta t \to 0$ is written in calculus notation as dx/dt and is called the *derivative* of x with respect to t:

$$v = \lim_{\Delta t \to 0} \frac{\Delta x}{\Delta t} = \frac{dx}{dt}. \tag{2–4}$$

This equation is the definition of instantaneous velocity for one-dimensional motion.

For instantaneous velocity we use the symbol v, whereas for average velocity we use $\bar{v}$, with a bar above. In the rest of this book, when we use the term "velocity" it will refer to instantaneous velocity. When we want to speak of the average velocity, we will make this clear by including the word "average."

Note that the *instantaneous* speed always equals the magnitude of the instantaneous velocity. Why? Because distance traveled and the magnitude of the displacement become the same when they become infinitesimally small.

If an object moves at a uniform (that is, constant) velocity during a particular time interval, then its instantaneous velocity at any instant is the same as its average velocity (see Fig. 2–9a). But in many situations this is not the case. For example, a car may start from rest, speed up to 50 km/h, remain at that velocity for a time, then slow down to 20 km/h in a traffic jam, and finally stop at its destination after traveling a total of 15 km in 30 min. This trip is plotted on the graph of Fig. 2–9b. Also shown on the graph is the average velocity (dashed line), which is $\bar{v} = \Delta x / \Delta t = 15\,\text{km}/0.50\,\text{h} = 30\,\text{km/h}$.

FIGURE 2–8 Car speedometer showing mi/h in white, and km/h in orange.

FIGURE 2–9 Velocity of a car as a function of time: (a) at constant velocity; (b) with varying velocity.

To better understand instantaneous velocity, let us consider a graph of the position of a particular particle versus time (x vs. t), as shown in Fig. 2–10. (Note that this is different from showing the "path" of a particle on an x vs. y plot.) The particle is at position x_1 at a time t_1, and at position x_2 at time t_2. P_1 and P_2 represent these two points on the graph. A straight line drawn from point $P_1(x_1, t_1)$ to point $P_2(x_2, t_2)$ forms the hypotenuse of a right triangle whose sides are Δx and Δt. The ratio $\Delta x / \Delta t$ is the **slope** of the straight line $P_1 P_2$. But $\Delta x / \Delta t$ is also the average velocity of the particle during the time interval $\Delta t = t_2 - t_1$. Therefore, we conclude that the average velocity of a particle during any time interval $\Delta t = t_2 - t_1$ is equal to the slope of the straight line (or *chord*) connecting the two points (x_1, t_1) and (x_2, t_2) on an x vs. t graph.

Consider now a time t_i, intermediate between t_1 and t_2, at which time the particle is at x_i (Fig. 2–11). The slope of the straight line $P_1 P_i$ is less than the slope of $P_1 P_2$ in this case. Thus the average velocity during the time interval $t_i - t_1$ is less than during the time interval $t_2 - t_1$.

Now let us imagine that we take the point P_i in Fig. 2–11 to be closer and closer to point P_1. That is, we let the interval $t_i - t_1$, which we now call Δt, to become smaller and smaller. The slope of the line connecting the two points becomes closer and closer to the slope of a line tangent to the curve at point P_1. The average velocity (equal to the slope of the chord) thus approaches the slope of the tangent at point P_1. The definition of the instantaneous velocity (Eq. 2–3) is the limiting value of the average velocity as Δt approaches zero. Thus the *instantaneous velocity equals the slope of the tangent to the curve* at that point (which we can simply call "the slope of the curve" at that point).

Because the velocity at any instant equals the slope of the tangent to the x vs. t graph at that instant, we can obtain the velocity at any instant from such a graph. For example, in Fig. 2–12 (which shows the same curve as in Figs. 2–10 and 2–11), as our object moves from x_1 to x_2, the slope continually increases, so the velocity is increasing. For times after t_2, however, the slope begins to decrease and in fact reaches zero (so $v = 0$) where x has its maximum value, at point P_3 in Fig. 2–12. Beyond this point, the slope is negative, as for point P_4. The velocity is therefore negative, which makes sense since x is now decreasing—the particle is moving toward decreasing values of x, to the left on a standard xy plot.

If an object moves with constant velocity over a particular time interval, its instantaneous velocity is equal to its average velocity. The graph of x vs. t in this case will be a straight line whose slope equals the velocity. The curve of Fig. 2–10 has no straight sections, so there are no time intervals when the velocity is constant.

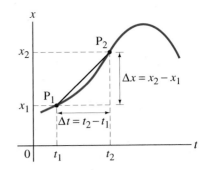

FIGURE 2–10 Graph of a particle's position x vs. time t. The slope of the straight line $P_1 P_2$ represents the average velocity of the particle during the time interval $\Delta t = t_2 - t_1$.

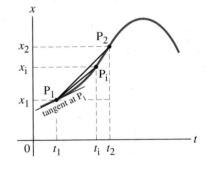

FIGURE 2–11 Same position vs. time curve as in Fig. 2–10, but note that the average velocity over the time interval $t_i - t_1$ (which is the slope of $P_1 P_i$) is less than the average velocity over the time interval $t_2 - t_1$. The slope of the thin line tangent to the curve at point P_1 equals the instantaneous velocity at time t_1.

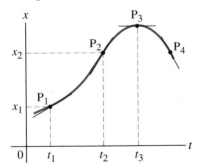

FIGURE 2–12 Same x vs. t curve as in Figs. 2–10 and 2–11, but here showing the slope at four different points: At P_3, the slope is zero, so $v = 0$. At P_4 the slope is negative, so $v < 0$.

EXERCISE C What is your speed at the instant you turn around to move in the opposite direction? (*a*) Depends on how quickly you turn around; (*b*) always zero; (*c*) always negative; (*d*) none of the above.

The derivatives of various functions are studied in calculus courses, and this book gives a summary in Appendix B. The derivatives of polynomial functions (which we use a lot) are:

$$\frac{d}{dt}(Ct^n) = nCt^{n-1} \quad \text{and} \quad \frac{dC}{dt} = 0,$$

where C is any constant.

(a)

Tangent at P_2 whose slope is $v_2 = 21.0$ m/s

Slope of straight line
$\bar{v} = 16.8$ m/s

P_2

$\Delta x = 33.6$ m

P_1

$\Delta t = 2.00$ s

(b)

FIGURE 2–13 Example 2–3.
(a) Engine traveling on a straight track.
(b) Graph of x vs. t: $x = At^2 + B$.

EXAMPLE 2–3 **Given x as a function of t.** A jet engine moves along an experimental track (which we call the x axis) as shown in Fig. 2–13a. We will treat the engine as if it were a particle. Its position as a function of time is given by the equation $x = At^2 + B$, where $A = 2.10$ m/s^2 and $B = 2.80$ m, and this equation is plotted in Fig. 2–13b. (a) Determine the displacement of the engine during the time interval from $t_1 = 3.00$ s to $t_2 = 5.00$ s. (b) Determine the average velocity during this time interval. (c) Determine the magnitude of the instantaneous velocity at $t = 5.00$ s.

APPROACH We substitute values for t_1 and t_2 in the given equation for x to obtain x_1 and x_2. The average velocity can be found from Eq. 2–2. We take the derivative of the given x equation with respect to t to find the instantaneous velocity, using the formulas just given.

SOLUTION (a) At $t_1 = 3.00$ s, the position (point P_1 in Fig. 2–13b) is

$$x_1 = At_1^2 + B = (2.10 \text{ m/s}^2)(3.00 \text{ s})^2 + 2.80 \text{ m} = 21.7 \text{ m}.$$

At $t_2 = 5.00$ s, the position (P_2 in Fig. 2–13b) is

$$x_2 = (2.10 \text{ m/s}^2)(5.00 \text{ s})^2 + 2.80 \text{ m} = 55.3 \text{ m}.$$

The displacement is thus

$$x_2 - x_1 = 55.3 \text{ m} - 21.7 \text{ m} = 33.6 \text{ m}.$$

(b) The magnitude of the average velocity can then be calculated as

$$\bar{v} = \frac{\Delta x}{\Delta t} = \frac{x_2 - x_1}{t_2 - t_1} = \frac{33.6 \text{ m}}{2.00 \text{ s}} = 16.8 \text{ m/s}.$$

This equals the slope of the straight line joining points P_1 and P_2 shown in Fig. 2–13b.

(c) The instantaneous velocity at $t = t_2 = 5.00$ s equals the slope of the tangent to the curve at point P_2 shown in Fig. 2–13b. We could measure this slope off the graph to obtain v_2. But we can calculate v more precisely for any time t, using the given formula

$$x = At^2 + B,$$

which is the engine's position x as a function of time t. We take the derivative of x with respect to time (see formulas at bottom of previous page):

$$v = \frac{dx}{dt} = \frac{d}{dt}(At^2 + B) = 2At.$$

We are given $A = 2.10$ m/s^2, so for $t = t_2 = 5.00$ s,

$$v_2 = 2At = 2(2.10 \text{ m/s}^2)(5.00 \text{ s}) = 21.0 \text{ m/s}.$$

2–4 Acceleration

An object whose velocity is changing is said to be accelerating. For instance, a car whose velocity increases in magnitude from zero to 80 km/h is accelerating. Acceleration specifies how rapidly the velocity of an object is changing.

Average Acceleration

Average acceleration is defined as the change in velocity divided by the time taken to make this change:

$$\text{average acceleration} = \frac{\text{change of velocity}}{\text{time elapsed}}.$$

In symbols, the average acceleration over a time interval $\Delta t = t_2 - t_1$ during which the velocity changes by $\Delta v = v_2 - v_1$, is defined as

$$\bar{a} = \frac{v_2 - v_1}{t_2 - t_1} = \frac{\Delta v}{\Delta t}. \tag{2–5}$$

Because velocity is a vector, acceleration is a vector too. But for one-dimensional motion, we need only use a plus or minus sign to indicate acceleration direction relative to a chosen coordinate axis.

EXAMPLE 2–4 **Average acceleration.** A car accelerates along a straight road from rest to 90 km/h in 5.0 s, Fig. 2–14. What is the magnitude of its average acceleration?

APPROACH Average acceleration is the change in velocity divided by the elapsed time, 5.0 s. The car starts from rest, so $v_1 = 0$. The final velocity is $v_2 = 90$ km/h $= 90 \times 10^3$ m/3600 s $= 25$ m/s.

SOLUTION From Eq. 2–5, the average acceleration is

$$\bar{a} = \frac{v_2 - v_1}{t_2 - t_1} = \frac{25 \text{ m/s} - 0 \text{ m/s}}{5.0 \text{ s}} = 5.0 \frac{\text{m/s}}{\text{s}}.$$

This is read as "five meters per second per second" and means that, on average, the velocity changed by 5.0 m/s during each second. That is, assuming the acceleration was constant, during the first second the car's velocity increased from zero to 5.0 m/s. During the next second its velocity increased by another 5.0 m/s, reaching a velocity of 10.0 m/s at $t = 2.0$ s, and so on. See Fig. 2–14.

$t_1 = 0$
$v_1 = 0$

Acceleration
[$a = 5.0$ m/s²]

at $t = 1.0$ s
$v = 5.0$ m/s

at $t = 2.0$ s
$v = 10.0$ m/s

at $t = t_2 = 5.0$ s
$v = v_2 = 25$ m/s

FIGURE 2–14 Example 2–4. The car is shown at the start with $v_1 = 0$ at $t_1 = 0$. The car is shown three more times, at $t = 1.0$ s, $t = 2.0$ s, and at the end of our time interval, $t_2 = 5.0$ s. We assume the acceleration is constant and equals 5.0 m/s². The green arrows represent the velocity vectors; the length of each arrow represents the magnitude of the velocity at that moment. The acceleration vector is the orange arrow. Distances are not to scale.

We almost always write the units for acceleration as m/s² (meters per second squared) instead of m/s/s. This is possible because:

$$\frac{\text{m/s}}{\text{s}} = \frac{\text{m}}{\text{s} \cdot \text{s}} = \frac{\text{m}}{\text{s}^2}.$$

According to the calculation in Example 2–4, the velocity changed on average by 5.0 m/s during each second, for a total change of 25 m/s over the 5.0 s; the average acceleration was 5.0 m/s².

Note that *acceleration tells us how quickly the velocity changes*, whereas *velocity tells us how quickly the position changes*.

Velocity and acceleration. (*a*) If the velocity of an object is zero, does it mean that the acceleration is zero? (*b*) If the acceleration is zero, does it mean that the velocity is zero? Think of some examples.

RESPONSE A zero velocity does not necessarily mean that the acceleration is zero, nor does a zero acceleration mean that the velocity is zero. (*a*) For example, when you put your foot on the gas pedal of your car which is at rest, the velocity starts from zero but the acceleration is not zero since the velocity of the car changes. (How else could your car start forward if its velocity weren't changing—that is, accelerating?) (*b*) As you cruise along a straight highway at a constant velocity of 100 km/h, your acceleration is zero: $a = 0, v \neq 0$.

EXERCISE D A powerful car is advertised to go from zero to 60 mi/h in 6.0 s. What does this say about the car: (*a*) it is fast (high speed); or (*b*) it accelerates well?

EXAMPLE 2–6 **Car slowing down.** An automobile is moving to the right along a straight highway, which we choose to be the positive x axis (Fig. 2–15). Then the driver puts on the brakes. If the initial velocity (when the driver hits the brakes) is $v_1 = 15.0$ m/s, and it takes 5.0 s to slow down to $v_2 = 5.0$ m/s, what was the car's average acceleration?

APPROACH We put the given initial and final velocities, and the elapsed time, into Eq. 2–5 for $\bar{a}$.

SOLUTION In Eq. 2–5, we call the initial time $t_1 = 0$, and set $t_2 = 5.0$ s. (Note that our choice of $t_1 = 0$ doesn't affect the calculation of $\bar{a}$ because only $\Delta t = t_2 - t_1$ appears in Eq. 2–5.) Then

$$\bar{a} = \frac{5.0 \text{ m/s} - 15.0 \text{ m/s}}{5.0 \text{ s}} = -2.0 \text{ m/s}^2.$$

The negative sign appears because the final velocity is less than the initial velocity. In this case the direction of the acceleration is to the left (in the negative x direction)—even though the velocity is always pointing to the right. We say that the acceleration is 2.0 m/s² to the left, and it is shown in Fig. 2–15 as an orange arrow.

at $t_1 = 0$
$v_1 = 15.0$ m/s

Acceleration
$a = -2.0$ m/s²

at $t_2 = 5.0$ s
$v_2 = 5.0$ m/s

FIGURE 2–15 Example 2–6, showing the position of the car at times t_1 and t_2, as well as the car's velocity represented by the green arrows. The acceleration vector (orange) points to the left as the car slows down while moving to the right.

CAUTION

Deceleration means the magnitude of the velocity is decreasing; a is not necessarily negative

Deceleration

When an object is slowing down, we can say it is **decelerating**. But be careful: deceleration does *not* mean that the acceleration is necessarily negative. The velocity of an object moving to the right along the positive x axis is positive; if the object is slowing down (as in Fig. 2–15), the acceleration *is* negative. But the same car moving to the left (decreasing x), and slowing down, has positive acceleration that points to the right, as shown in Fig. 2–16. We have a deceleration whenever the magnitude of the velocity is decreasing, and then the velocity and acceleration point in opposite directions.

FIGURE 2–16 The car of Example 2–6, now moving to the *left* and decelerating. The acceleration is

$$
\begin{aligned}
a &= \frac{v_2 - v_1}{\Delta t} \\
&= \frac{(-5.0 \text{ m/s}) - (-15.0 \text{ m/s})}{5.0 \text{ s}} \\
&= \frac{-5.0 \text{ m/s} + 15.0 \text{ m/s}}{5.0 \text{ s}} = +2.0 \text{ m/s}.
\end{aligned}
$$

$v_2 = -5.0$ m/s $v_1 = -15.0$ m/s

a

EXERCISE E A car moves along the x axis. What is the sign of the car's acceleration if it is moving in the positive x direction with (*a*) increasing speed or (*b*) decreasing speed? What is the sign of the acceleration if the car moves in the negative direction with (*c*) increasing speed or (*d*) decreasing speed?

Instantaneous Acceleration

The **instantaneous acceleration**, a, is defined as the *limiting value of the average acceleration as we let Δt approach zero:*

$$a = \lim_{\Delta t \to 0} \frac{\Delta v}{\Delta t} = \frac{dv}{dt}. \qquad (2\text{--}6)$$

This limit, dv/dt, is the derivative of v with respect to t. We will use the term "acceleration" to refer to the instantaneous value. If we want to discuss the average acceleration, we will always include the word "average."

If we draw a graph of the velocity, v, vs. time, t, as shown in Fig. 2–17, then the average acceleration over a time interval $\Delta t = t_2 - t_1$ is represented by the slope of the straight line connecting the two points P_1 and P_2 as shown. [Compare this to the position vs. time graph of Fig. 2–10 for which the slope of the straight line represents the average velocity.] The instantaneous acceleration at any time, say t_1, is the slope of the tangent to the v vs. t curve at that time, which is also shown in Fig. 2–17. Let us use this fact for the situation graphed in Fig. 2–17; as we go from time t_1 to time t_2 the velocity continually increases, but the acceleration (the rate at which the velocity changes) is decreasing since the slope of the curve is decreasing.

FIGURE 2–17 A graph of velocity v vs. time t. The average acceleration over a time interval $\Delta t = t_2 - t_1$ is the slope of the straight line $P_1 P_2$: $\bar{a} = \Delta v/\Delta t$. The instantaneous acceleration at time t_1 is the slope of the v vs. t curve at that instant.

EXAMPLE 2–7 Acceleration given $x(t)$. A particle is moving in a straight line so that its position is given by the relation $x = (2.10\ \text{m/s}^2)t^2 + (2.80\ \text{m})$, as in Example 2–3. Calculate (a) its average acceleration during the time interval from $t_1 = 3.00\ \text{s}$ to $t_2 = 5.00\ \text{s}$, and (b) its instantaneous acceleration as a function of time.

APPROACH To determine acceleration, we first must find the velocity at t_1 and t_2 by differentiating x: $v = dx/dt$. Then we use Eq. 2–5 to find the average acceleration, and Eq. 2–6 to find the instantaneous acceleration.

SOLUTION (a) The velocity at any time t is

$$v = \frac{dx}{dt} = \frac{d}{dt}\left[(2.10\ \text{m/s}^2)t^2 + 2.80\ \text{m}\right] = (4.20\ \text{m/s}^2)t,$$

as we saw in Example 2–3c. Therefore, at $t_1 = 3.00\ \text{s}$, $v_1 = (4.20\ \text{m/s}^2)(3.00\ \text{s}) = 12.6\ \text{m/s}$ and at $t_2 = 5.00\ \text{s}$, $v_2 = 21.0\ \text{m/s}$. Therefore,

$$\bar{a} = \frac{\Delta v}{\Delta t} = \frac{21.0\ \text{m/s} - 12.6\ \text{m/s}}{5.00\ \text{s} - 3.00\ \text{s}} = 4.20\ \text{m/s}^2.$$

(b) With $v = (4.20\ \text{m/s}^2)t$, the instantaneous acceleration at any time is

$$a = \frac{dv}{dt} = \frac{d}{dt}\left[(4.20\ \text{m/s}^2)t\right] = 4.20\ \text{m/s}^2.$$

The acceleration in this case is constant; it does not depend on time. Figure 2–18 shows graphs of (a) x vs. t (the same as Fig. 2–13b), (b) v vs. t, which is linearly increasing as calculated above, and (c) a vs. t, which is a horizontal straight line because $a = $ constant.

FIGURE 2–18 Example 2–7. Graphs of (a) x vs. t, (b) v vs. t, and (c) a vs. t for the motion $x = At^2 + B$. Note that v increases linearly with t and that the acceleration a is constant. Also, v is the slope of the x vs. t curve, whereas a is the slope of the v vs. t curve.

(a)

(b)

(c)

Like velocity, acceleration is a rate. The velocity of an object is the rate at which its displacement changes with time; its acceleration, on the other hand, is the rate at which its velocity changes with time. In a sense, acceleration is a "rate of a rate." This can be expressed in equation form as follows: since $a = dv/dt$ and $v = dx/dt$, then

$$a = \frac{dv}{dt} = \frac{d}{dt}\left(\frac{dx}{dt}\right) = \frac{d^2x}{dt^2}.$$

Here d^2x/dt^2 is the *second derivative* of x with respect to time: we first take the derivative of x with respect to time (dx/dt), and then we again take the derivative with respect to time, $(d/dt)(dx/dt)$, to get the acceleration.

EXERCISE F The position of a particle is given by the following equation:
$$x = (2.00\ \text{m/s}^3)t^3 + (2.50\ \text{m/s})t.$$
What is the acceleration of the particle at $t = 2.00\ \text{s}$? (a) $13.0\ \text{m/s}^2$; (b) $22.5\ \text{m/s}^2$; (c) $24.0\ \text{m/s}^2$; (d) $2.00\ \text{m/s}^2$.

FIGURE 2–19 Example 2–8.

CONCEPTUAL EXAMPLE 2–8 Analyzing with graphs. Figure 2–19 shows the velocity as a function of time for two cars accelerating from 0 to 100 km/h in a time of 10.0 s. Compare (a) the average acceleration; (b) instantaneous acceleration; and (c) total distance traveled for the two cars.

RESPONSE (a) Average acceleration is $\Delta v/\Delta t$. Both cars have the same Δv (100 km/h) and the same Δt (10.0 s), so the average acceleration is the same for both cars. (b) Instantaneous acceleration is the slope of the tangent to the v vs. t curve. For about the first 4 s, the top curve is steeper than the bottom curve, so car A has a greater acceleration during this interval. The bottom curve is steeper during the last 6 s, so car B has the larger acceleration for this period. (c) Except at $t = 0$ and $t = 10.0$ s, car A is always going faster than car B. Since it is going faster, it will go farther in the same time.

2–5 Motion at Constant Acceleration

We now examine the situation when the magnitude of the acceleration is constant and the motion is in a straight line. In this case, the instantaneous and average accelerations are equal. We use the definitions of average velocity and acceleration to derive a set of valuable equations that relate x, v, a, and t when a is constant, allowing us to determine any one of these variables if we know the others.

To simplify our notation, let us take the initial time in any discussion to be zero, and we call it t_0: $t_1 = t_0 = 0$. (This is effectively starting a stopwatch at t_0.) We can then let $t_2 = t$ be the elapsed time. The initial position (x_1) and the initial velocity (v_1) of an object will now be represented by x_0 and v_0, since they represent x and v at $t = 0$. At time t the position and velocity will be called x and v (rather than x_2 and v_2). The average velocity during the time interval $t - t_0$ will be (Eq. 2–2)

$$\bar{v} = \frac{\Delta x}{\Delta t} = \frac{x - x_0}{t - t_0} = \frac{x - x_0}{t}$$

since we chose $t_0 = 0$. The acceleration, assumed constant in time, is (Eq. 2–5)

$$a = \frac{v - v_0}{t}.$$

A common problem is to determine the velocity of an object after any elapsed time t, when we are given the object's constant acceleration. We can solve such problems by solving for v in the last equation to obtain:

$$v = v_0 + at. \qquad \text{[constant acceleration]} \quad \textbf{(2–7)}$$

If an object starts from rest $(v_0 = 0)$ and accelerates at 4.0 m/s², after an elapsed time $t = 6.0$ s its velocity will be $v = at = (4.0 \text{ m/s}^2)(6.0 \text{ s}) = 24$ m/s.

Next, let us see how to calculate the position x of an object after a time t when it undergoes constant acceleration. The definition of average velocity (Eq. 2–2) is $\bar{v} = (x - x_0)/t$, which we can rewrite as

$$x = x_0 + \bar{v}t. \qquad\qquad \textbf{(2–8)}$$

Because the velocity increases at a uniform rate, the average velocity, $\bar{v}$, will be midway between the initial and final velocities:

$$\bar{v} = \frac{v_0 + v}{2}. \qquad \text{[constant acceleration]} \quad \textbf{(2–9)}$$

(Careful: Equation 2–9 is not necessarily valid if the acceleration is not constant.) We combine the last two Equations with Eq. 2–7 and find

$$\begin{aligned}
x &= x_0 + \bar{v}t \\
&= x_0 + \left(\frac{v_0 + v}{2}\right)t \\
&= x_0 + \left(\frac{v_0 + v_0 + at}{2}\right)t
\end{aligned}$$

or

$$x = x_0 + v_0 t + \tfrac{1}{2}at^2. \qquad \text{[constant acceleration]} \quad \textbf{(2–10)}$$

⚠ **CAUTION**

Average velocity, but only if $a = $ constant

Equations 2–7, 2–9, and 2–10 are three of the four most useful equations for motion at constant acceleration. We now derive the fourth equation, which is useful in situations where the time t is not known. We substitute Eq. 2–9 into Eq. 2–8:

$$x = x_0 + \bar{v}t = x_0 + \left(\frac{v + v_0}{2}\right)t.$$

Next we solve Eq. 2–7 for t, obtaining

$$t = \frac{v - v_0}{a},$$

and substituting this into the previous equation we have

$$x = x_0 + \left(\frac{v + v_0}{2}\right)\left(\frac{v - v_0}{a}\right) = x_0 + \frac{v^2 - v_0^2}{2a}.$$

We solve this for v^2 and obtain

$$v^2 = v_0^2 + 2a(x - x_0), \qquad \text{[constant acceleration]} \quad \textbf{(2–11)}$$

which is the useful equation we sought.

We now have four equations relating position, velocity, acceleration, and time, when the acceleration a is constant. We collect these kinematic equations here in one place for future reference (the tan background screen emphasizes their usefulness):

$v = v_0 + at$	$[a = \text{constant}]$ **(2–12a)**
$x = x_0 + v_0 t + \frac{1}{2}at^2$	$[a = \text{constant}]$ **(2–12b)**
$v^2 = v_0^2 + 2a(x - x_0)$	$[a = \text{constant}]$ **(2–12c)**
$\bar{v} = \dfrac{v + v_0}{2}.$	$[a = \text{constant}]$ **(2–12d)**

Kinematic equations for constant acceleration (we'll use them a lot)

These useful equations are not valid unless a is a constant. In many cases we can set $x_0 = 0$, and this simplifies the above equations a bit. Note that x represents position, not distance, that $x - x_0$ is the displacement, and that t is the elapsed time.

EXAMPLE 2–9 **Runway design.** You are designing an airport for small planes. One kind of airplane that might use this airfield must reach a speed before takeoff of at least 27.8 m/s (100 km/h), and can accelerate at 2.00 m/s². (*a*) If the runway is 150 m long, can this airplane reach the required speed for takeoff? (*b*) If not, what minimum length must the runway have?

PHYSICS APPLIED
Airport design

APPROACH The plane's acceleration is constant, so we can use the kinematic equations for constant acceleration. In (*a*), we want to find v, and we are given:

Known	Wanted
$x_0 = 0$	v
$v_0 = 0$	
$x = 150 \text{ m}$	
$a = 2.00 \text{ m/s}^2$	

SOLUTION (*a*) Of the above four equations, Eq. 2–12c will give us v when we know v_0, a, x, and x_0:

$$v^2 = v_0^2 + 2a(x - x_0)$$
$$= 0 + 2(2.00 \text{ m/s}^2)(150 \text{ m}) = 600 \text{ m}^2/\text{s}^2$$
$$v = \sqrt{600 \text{ m}^2/\text{s}^2} = 24.5 \text{ m/s}.$$

PROBLEM SOLVING
Equations 2–12 are valid only when the acceleration is constant, which we assume in this Example

This runway length is *not* sufficient.

(*b*) Now we want to find the minimum length of runway, $x - x_0$, given $v = 27.8 \text{ m/s}$ and $a = 2.00 \text{ m/s}^2$. So we again use Eq. 2–12c, but rewritten as

$$(x - x_0) = \frac{v^2 - v_0^2}{2a} = \frac{(27.8 \text{ m/s})^2 - 0}{2(2.00 \text{ m/s}^2)} = 193 \text{ m}.$$

A 200-m runway is more appropriate for this plane.

NOTE We did this Example as if the plane were a particle, so we round off our answer to 200 m.

EXERCISE G A car starts from rest and accelerates at a constant 10 m/s² during a $\frac{1}{4}$ mile (402 m) race. How fast is the car going at the finish line? (a) 8090 m/s; (b) 90 m/s; (c) 81 m/s; (d) 809 m/s.

2–6 Solving Problems

Before doing more worked-out Examples, let us look at how to approach problem solving. First, it is important to note that physics is *not* a collection of equations to be memorized. Simply searching for an equation that might work can lead you to a wrong result and will surely not help you understand physics. A better approach is to use the following (rough) procedure, which we put in a special "Problem Solving Strategy." (Other such Problem Solving Strategies, as an aid, will be found throughout the book.)

PROBLEM SOLVING

1. Read and **reread** the whole problem carefully before trying to solve it.

2. Decide what **object** (or objects) you are going to study, and for what **time interval**. You can often choose the initial time to be $t = 0$.

3. **Draw** a **diagram** or picture of the situation, with coordinate axes wherever applicable. [You can place the origin of coordinates and the axes wherever you like to make your calculations easier.]

4. Write down what quantities are "**known**" or "given," and then what you *want* to know. Consider quantities both at the beginning and at the end of the chosen time interval.

5. Think about which **principles of physics** apply in this problem. Use common sense and your own experiences. Then plan an approach.

6. Consider which **equations** (and/or definitions) relate the quantities involved. Before using them, be sure their **range of validity** includes your problem (for example, Eqs. 2–12 are valid only when the acceleration is constant). If you find an applicable equation that involves only known quantities and one desired unknown, **solve** the equation algebraically for the unknown. Sometimes several sequential calculations, or a combination of equations, may be needed. It is often preferable to solve algebraically for the desired unknown before putting in numerical values.

7. Carry out the **calculation** if it is a numerical problem. Keep one or two extra digits during the calculations, but round off the final answer(s) to the correct number of significant figures (Section 1–3).

8. Think carefully about the result you obtain: Is it **reasonable**? Does it make sense according to your own intuition and experience? A good check is to do a rough **estimate** using only powers of ten, as discussed in Section 1–6. Often it is preferable to do a rough estimate at the *start* of a numerical problem because it can help you focus your attention on finding a path toward a solution.

9. A very important aspect of doing problems is keeping track of **units**. An equals sign implies the units on each side must be the same, just as the numbers must. If the units do not balance, a mistake has no doubt been made. This can serve as a **check** on your solution (but it only tells you if you're wrong, not if you're right). Always use a consistent set of units.

EXAMPLE 2–10 **Acceleration of a car.** How long does it take a car to cross a 30.0-m-wide intersection after the light turns green, if the car accelerates from rest at a constant 2.00 m/s^2?

APPROACH We follow the Problem Solving Strategy above, step by step.

SOLUTION

1. **Reread** the problem. Be sure you understand what it asks for (here, a time interval).

2. The **object** under study is the car. We choose the **time interval**: $t = 0$, the initial time, is the moment the car starts to accelerate from rest $(v_0 = 0)$; the time t is the instant the car has traveled the full 30.0-m width of the intersection.

3. **Draw** a **diagram**: the situation is shown in Fig. 2–20, where the car is shown moving along the positive x axis. We choose $x_0 = 0$ at the front bumper of the car before it starts to move.

FIGURE 2–20 Example 2–10.

$a = 2.00 \text{ m/s}^2$ $a = 2.00 \text{ m/s}^2$

$x_0 = 0$ $x =$
$v_0 = 0$ 30.0 m

4. The "**knowns**" and the "wanted" are shown in the Table in the margin, and we choose $x_0 = 0$. Note that "starting from rest" means $v = 0$ at $t = 0$; that is, $v_0 = 0$.

5. The **physics**: the motion takes place at constant acceleration, so we can use the kinematic equations, Eqs. 2–12.

6. **Equations**: we want to find the time, given the distance and acceleration; Eq. 2–12b is perfect since the only unknown quantity is t. Setting $v_0 = 0$ and $x_0 = 0$ in Eq. 2–12b $\left(x = x_0 + v_0 t + \frac{1}{2}at^2\right)$, we can solve for t:

$$x = \tfrac{1}{2}at^2,$$

$$t^2 = \frac{2x}{a},$$

so

$$t = \sqrt{\frac{2x}{a}}.$$

7. The **calculation**:

$$t = \sqrt{\frac{2x}{a}} = \sqrt{\frac{2(30.0\,\text{m})}{2.00\,\text{m/s}^2}} = 5.48\,\text{s}.$$

This is our answer. Note that the units come out correctly.

8. We can check the **reasonableness** of the answer by calculating the final velocity $v = at = (2.00\,\text{m/s}^2)(5.48\,\text{s}) = 10.96\,\text{m/s}$, and then finding $x = x_0 + \bar{v}t = 0 + \frac{1}{2}(10.96\,\text{m/s} + 0)(5.48\,\text{s}) = 30.0\,\text{m}$, which is our given distance.

9. We checked the **units**, and they came out perfectly (seconds).

NOTE In steps 6 and 7, when we took the square root, we should have written $t = \pm\sqrt{2x/a} = \pm 5.48\,\text{s}$. Mathematically there are two solutions. But the second solution, $t = -5.48\,\text{s}$, is a time *before* our chosen time interval and makes no sense physically. We say it is "unphysical" and ignore it.

We explicitly followed the steps of the Problem Solving Strategy for Example 2–10. In upcoming Examples, we will use our usual "Approach" and "Solution" to avoid being wordy.

EXAMPLE 2–11 **ESTIMATE** **Air bags.** Suppose you want to design an air-bag system that can protect the driver at a speed of 100 km/h (60 mph) if the car hits a brick wall. Estimate how fast the air bag must inflate (Fig. 2–21) to effectively protect the driver. How does the use of a seat belt help the driver?

APPROACH We assume the acceleration is roughly constant, so we can use Eqs. 2–12. Both Eqs. 2–12a and 2–12b contain t, our desired unknown. They both contain a, so we must first find a, which we can do using Eq. 2–12c if we know the distance x over which the car crumples. A rough estimate might be about 1 meter. We choose the time interval to start at the instant of impact with the car moving at $v_0 = 100$ km/h, and to end when the car comes to rest $(v = 0)$ after traveling 1 m.

SOLUTION We convert the given initial speed to SI units: $100\,\text{km/h} = 100 \times 10^3\,\text{m}/3600\,\text{s} = 28\,\text{m/s}$. We then find the acceleration from Eq. 2–12c:

$$a = -\frac{v_0^2}{2x} = -\frac{(28\,\text{m/s})^2}{2.0\,\text{m}} = -390\,\text{m/s}^2.$$

This enormous acceleration takes place in a time given by (Eq. 2–12a):

$$t = \frac{v - v_0}{a} = \frac{0 - 28\,\text{m/s}}{-390\,\text{m/s}^2} = 0.07\,\text{s}.$$

To be effective, the air bag would need to inflate faster than this.

What does the air bag do? It spreads the force over a large area of the chest (to avoid puncture of the chest by the steering wheel). The seat belt keeps the person in a stable position against the expanding air bag.

Known	Wanted
$x_0 = 0$	t
$x = 30.0$ m	
$a = 2.00$ m/s^2	
$v_0 = 0$	

PHYSICS APPLIED
Car safety—air bags

FIGURE 2–21 Example 2–11. An air bag deploying on impact.

FIGURE 2–22 Example 2–12: stopping distance for a braking car.

Travel during reaction time

$v = $ constant $ = 14$ m/s
$t = 0.50$ s
$a = 0$

Travel during braking

v decreases from 14 m/s to zero
$a = -6.0$ m/s^2

x

⊕ PHYSICS APPLIED

Braking distances

Part 1: Reaction time

Known	Wanted
$t = 0.50$ s	x
$v_0 = 14$ m/s	
$v = 14$ m/s	
$a = 0$	
$x_0 = 0$	

Part 2: Braking

Known	Wanted
$x_0 = 7.0$ m	x
$v_0 = 14$ m/s	
$v = 0$	
$a = -6.0$ m/s^2	

FIGURE 2–23 Example 2–12. Graphs of (a) v vs. t and (b) x vs. t.

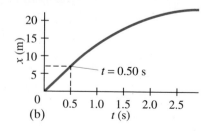

EXAMPLE 2–12 ESTIMATE Braking distances. Estimate the minimum stopping distance for a car, which is important for traffic safety and traffic design. The problem is best dealt with in two parts, two separate time intervals. (1) The first time interval begins when the driver decides to hit the brakes, and ends when the foot touches the brake pedal. This is the "reaction time" during which the speed is constant, so $a = 0$. (2) The second time interval is the actual braking period when the vehicle slows down ($a \neq 0$) and comes to a stop. The stopping distance depends on the reaction time of the driver, the initial speed of the car (the final speed is zero), and the acceleration of the car. For a dry road and good tires, good brakes can decelerate a car at a rate of about 5 m/s^2 to 8 m/s^2. Calculate the total stopping distance for an initial velocity of 50 km/h ($= 14$ m/s ≈ 31 mi/h) and assume the acceleration of the car is -6.0 m/s^2 (the minus sign appears because the velocity is taken to be in the positive x direction and its magnitude is decreasing). Reaction time for normal drivers varies from perhaps 0.3 s to about 1.0 s; take it to be 0.50 s.

APPROACH During the "reaction time," part (1), the car moves at constant speed of 14 m/s, so $a = 0$. Once the brakes are applied, part (2), the acceleration is $a = -6.0$ m/s^2 and is constant over this time interval. For both parts a is constant, so we can use Eqs. 2–12.

SOLUTION Part (1). We take $x_0 = 0$ for the first time interval, when the driver is reacting (0.50 s): the car travels at a constant speed of 14 m/s so $a = 0$. See Fig. 2–22 and the Table in the margin. To find x, the position of the car at $t = 0.50$ s (when the brakes are applied), we cannot use Eq. 2–12c because x is multiplied by a, which is zero. But Eq. 2–12b works:

$$x = v_0 t + 0 = (14 \text{ m/s})(0.50 \text{ s}) = 7.0 \text{ m}.$$

Thus the car travels 7.0 m during the driver's reaction time, until the instant the brakes are applied. We will use this result as input to part (2).

Part (2). During the second time interval, the brakes are applied and the car is brought to rest. The initial position is $x_0 = 7.0$ m (result of part (1)), and other variables are shown in the second Table in the margin. Equation 2–12a doesn't contain x; Eq. 2–12b contains x but also the unknown t. Equation 2–12c, $v^2 - v_0^2 = 2a(x - x_0)$, is what we want; after setting $x_0 = 7.0$ m, we solve for x, the final position of the car (when it stops):

$$x = x_0 + \frac{v^2 - v_0^2}{2a}$$

$$= 7.0 \text{ m} + \frac{0 - (14 \text{ m/s})^2}{2(-6.0 \text{ m/s}^2)} = 7.0 \text{ m} + \frac{-196 \text{ m}^2/\text{s}^2}{-12 \text{ m/s}^2}$$

$$= 7.0 \text{ m} + 16 \text{ m} = 23 \text{ m}.$$

The car traveled 7.0 m while the driver was reacting and another 16 m during the braking period before coming to a stop, for a total distance traveled of 23 m. Figure 2–23 shows graphs of (a) v vs. t and (b) x vs. t.

NOTE From the equation above for x, we see that the stopping distance after the driver hit the brakes ($= x - x_0$) increases with the *square* of the initial speed, not just linearly with speed. If you are traveling twice as fast, it takes four times the distance to stop.

EXAMPLE 2–13 **ESTIMATE** **Two Moving Objects: Police and Speeder.**
A car speeding at 150 km/h passes a still police car which immediately takes off in hot pursuit. Using simple assumptions, such as that the speeder continues at constant speed, estimate how long it takes the police car to overtake the speeder. Then estimate the police car's speed at that moment and decide if the assumptions were reasonable.

APPROACH When the police car takes off, it accelerates, and the simplest assumption is that its acceleration is constant. This may not be reasonable, but let's see what happens. We can estimate the acceleration if we have noticed automobile ads, which claim cars can accelerate from rest to 100 km/h in 5.0 s. So the average acceleration of the police car could be approximately

$$a_P = \frac{100 \text{ km/h}}{5.0 \text{ s}} = 20 \frac{\text{km/h}}{\text{s}} \left(\frac{1000 \text{ m}}{1 \text{ km}}\right)\left(\frac{1 \text{ h}}{3600 \text{ s}}\right) = 5.6 \text{ m/s}^2.$$

SOLUTION We need to set up the kinematic equations to determine the unknown quantities, and since there are two moving objects, we need two separate sets of equations. We denote the speeding car's position by x_S and the police car's position by x_P. Because we are interested in solving for the time when the two vehicles arrive at the same position on the road, we use Eq. 2–12b for each car:

$$x_S = v_{0S}t + \tfrac{1}{2}a_S t^2 = (150 \text{ km/h})t = (42 \text{ m/s})t$$
$$x_P = v_{0P}t + \tfrac{1}{2}a_P t^2 = \tfrac{1}{2}(5.6 \text{ m/s}^2)t^2,$$

where we have set $v_{0P} = 0$ and $a_S = 0$ (speeder assumed to move at constant speed). We want the time when the cars meet, so we set $x_S = x_P$ and solve for t:

$$(42 \text{ m/s})t = (2.8 \text{ m/s}^2)t^2.$$

The solutions are

$$t = 0 \quad \text{and} \quad t = \frac{42 \text{ m/s}}{2.8 \text{ m/s}^2} = 15 \text{ s}.$$

The first solution corresponds to the instant the speeder passed the police car. The second solution tells us when the police car catches up to the speeder, 15 s later. This is our answer, but is it reasonable? The police car's speed at $t = 15$ s is

$$v_P = v_{0P} + a_P t = 0 + (5.6 \text{ m/s}^2)(15 \text{ s}) = 84 \text{ m/s}$$

or 300 km/h ($\approx$ 190 mi/h). Not reasonable, and highly dangerous.

NOTE More reasonable is to give up the assumption of constant acceleration. The police car surely cannot maintain constant acceleration at those speeds. Also, the speeder, if a reasonable person, would slow down upon hearing the police siren. Figure 2–24 shows (a) x vs. t and (b) v vs. t graphs, based on the original assumption of a_P = constant, whereas (c) shows v vs. t for more reasonable assumptions.

⚠ **CAUTION**
Initial assumptions need to be checked out for reasonableness

  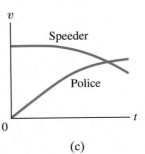

FIGURE 2–24 Example 2–13.

2–7 Freely Falling Objects

One of the most common examples of uniformly accelerated motion is that of an object allowed to fall freely near the Earth's surface. That a falling object is accelerating may not be obvious at first. And beware of thinking, as was widely believed before the time of Galileo (Fig. 2–25), that heavier objects fall faster than lighter objects and that the speed of fall is proportional to how heavy the object is.

Galileo made use of his new technique of imagining what would happen in idealized (simplified) cases. For free fall, he postulated that all objects would fall with the *same constant acceleration* in the absence of air or other resistance. He showed that this postulate predicts that for an object falling from rest, the distance traveled will be proportional to the square of the time (Fig. 2–26); that is, $d \propto t^2$. We can see this from Eq. 2–12b; but Galileo was the first to derive this mathematical relation.

To support his claim that falling objects increase in speed as they fall, Galileo made use of a clever argument: a heavy stone dropped from a height of 2 m will drive a stake into the ground much further than will the same stone dropped from a height of only 0.2 m. Clearly, the stone must be moving faster in the former case.

Galileo claimed that *all* objects, light or heavy, fall with the *same* acceleration, at least in the absence of air. If you hold a piece of paper horizontally in one hand and a heavier object—say, a baseball—in the other, and release them at the same time as in Fig. 2–27a, the heavier object will reach the ground first. But if you repeat the experiment, this time crumpling the paper into a small wad (see Fig. 2–27b), you will find that the two objects reach the floor at nearly the same time.

Galileo was sure that air acts as a resistance to very light objects that have a large surface area. But in many ordinary circumstances this air resistance is negligible. In a chamber from which the air has been removed, even light objects like a feather or a horizontally held piece of paper will fall with the same acceleration as any other object (see Fig. 2–28). Such a demonstration in vacuum was not possible in Galileo's time, which makes Galileo's achievement all the greater. Galileo is often called the "father of modern science," not only for the content of his science (astronomical discoveries, inertia, free fall) but also for his approach to science (idealization and simplification, mathematization of theory, theories that have testable consequences, experiments to test theoretical predictions).

Galileo's specific contribution to our understanding of the motion of falling objects can be summarized as follows:

> **at a given location on the Earth and in the absence of air resistance, all objects fall with the same constant acceleration.**

We call this acceleration the **acceleration due to gravity** on the surface of the Earth, and we give it the symbol g. Its magnitude is approximately

$$g = 9.80 \text{ m/s}^2. \qquad \text{[at surface of Earth]}$$

In British units g is about 32 ft/s². Actually, g varies slightly according to latitude and elevation, but these variations are so small that we will ignore them for most

FIGURE 2–25 Galileo Galilei (1564–1642).

⚠ **CAUTION**

A freely falling object increases in speed, but not *in proportion to its mass or weight*

FIGURE 2–26 Multiflash photograph of a falling apple, at equal time intervals. The apple falls farther during each successive interval, which means it is accelerating.

Acceleration due to gravity

(a)   (b)

FIGURE 2–27 (a) A ball and a light piece of paper are dropped at the same time. (b) Repeated, with the paper wadded up.

FIGURE 2–28 A rock and a feather are dropped simultaneously (a) in air, (b) in a vacuum.

Air-filled tube   Evacuated tube

(a)   (b)

purposes. The effects of air resistance are often small, and we will neglect them for the most part. However, air resistance will be noticeable even on a reasonably heavy object if the velocity becomes large.[†] Acceleration due to gravity is a vector as is any acceleration, and its direction is downward, toward the center of the Earth.

When dealing with freely falling objects we can make use of Eqs. 2–12, where for a we use the value of g given above. Also, since the motion is vertical we will substitute y in place of x, and y_0 in place of x_0. We take $y_0 = 0$ unless otherwise specified. *It is arbitrary whether we choose y to be positive in the upward direction or in the downward direction; but we must be consistent about it throughout a problem's solution.*

PROBLEM SOLVING

You can choose y to be positive either up or down

EXERCISE H Return to the Chapter-Opening Question, page 18, and answer it again now. Try to explain why you may have answered differently the first time.

EXAMPLE 2–14 **Falling from a tower.** Suppose that a ball is dropped $(v_0 = 0)$ from a tower 70.0 m high. How far will it have fallen after a time $t_1 = 1.00$ s, $t_2 = 2.00$ s, and $t_3 = 3.00$ s? Ignore air resistance.

APPROACH Let us take y as positive downward, so the acceleration is $a = g = +9.80 \text{ m/s}^2$. We set $v_0 = 0$ and $y_0 = 0$. We want to find the position y of the ball after three different time intervals. Equation 2–12b, with x replaced by y, relates the given quantities (t, a, and v_0) to the unknown y.

SOLUTION We set $t = t_1 = 1.00$ s in Eq. 2–12b:
$$y_1 = v_0 t_1 + \tfrac{1}{2} a t_1^2 = 0 + \tfrac{1}{2} a t_1^2 = \tfrac{1}{2}(9.80 \text{ m/s}^2)(1.00 \text{ s})^2 = 4.90 \text{ m}.$$
The ball has fallen a distance of 4.90 m during the time interval $t = 0$ to $t_1 = 1.00$ s. Similarly, after 2.00 s ($= t_2$), the ball's position is
$$y_2 = \tfrac{1}{2} a t_2^2 = \tfrac{1}{2}(9.80 \text{ m/s}^2)(2.00 \text{ s})^2 = 19.6 \text{ m}.$$
Finally, after 3.00 s ($= t_3$), the ball's position is (see Fig. 2–29)
$$y_3 = \tfrac{1}{2} a t_3^2 = \tfrac{1}{2}(9.80 \text{ m/s}^2)(3.00 \text{ s})^2 = 44.1 \text{ m}.$$

EXAMPLE 2–15 **Thrown down from a tower.** Suppose the ball in Example 2–14 is *thrown* downward with an initial velocity of 3.00 m/s, instead of being dropped. (*a*) What then would be its position after 1.00 s and 2.00 s? (*b*) What would its speed be after 1.00 s and 2.00 s? Compare with the speeds of a dropped ball.

APPROACH Again we use Eq. 2–12b, but now v_0 is not zero, it is $v_0 = 3.00$ m/s.

SOLUTION (*a*) At $t = 1.00$ s, the position of the ball as given by Eq. 2–12b is
$$y = v_0 t + \tfrac{1}{2} a t^2 = (3.00 \text{ m/s})(1.00 \text{ s}) + \tfrac{1}{2}(9.80 \text{ m/s}^2)(1.00 \text{ s})^2 = 7.90 \text{ m}.$$
At $t = 2.00$ s, (time interval $t = 0$ to $t = 2.00$ s), the position is
$$y = v_0 t + \tfrac{1}{2} a t^2 = (3.00 \text{ m/s})(2.00 \text{ s}) + \tfrac{1}{2}(9.80 \text{ m/s}^2)(2.00 \text{ s})^2 = 25.6 \text{ m}.$$
As expected, the ball falls farther each second than if it were dropped with $v_0 = 0$.
(*b*) The velocity is obtained from Eq. 2–12a:
$$\begin{aligned} v = v_0 + at \\ = 3.00 \text{ m/s} + (9.80 \text{ m/s}^2)(1.00 \text{ s}) = 12.8 \text{ m/s} \quad [\text{at } t_1 = 1.00 \text{ s}] \\ = 3.00 \text{ m/s} + (9.80 \text{ m/s}^2)(2.00 \text{ s}) = 22.6 \text{ m/s}. \quad [\text{at } t_2 = 2.00 \text{ s}] \end{aligned}$$
In Example 2–14, when the ball was dropped ($v_0 = 0$), the first term (v_0) in these equations was zero, so
$$\begin{aligned} v = 0 + at \\ = (9.80 \text{ m/s}^2)(1.00 \text{ s}) = 9.80 \text{ m/s} \quad [\text{at } t_1 = 1.00 \text{ s}] \\ = (9.80 \text{ m/s}^2)(2.00 \text{ s}) = 19.6 \text{ m/s}. \quad [\text{at } t_2 = 2.00 \text{ s}] \end{aligned}$$

NOTE For both Examples 2–14 and 2–15, the speed increases linearly in time by 9.80 m/s during each second. But the speed of the downwardly thrown ball at any instant is always 3.00 m/s (its initial speed) higher than that of a dropped ball.

FIGURE 2–29 Example 2–14. (a) An object dropped from a tower falls with progressively greater speed and covers greater distance with each successive second. (See also Fig. 2–26.) (b) Graph of y vs. t.

Acceleration due to gravity

$y = 0$
$y_1 = 4.90$ m (After 1.00 s)
$y_2 = 19.6$ m (After 2.00 s)
$+y$
$y_3 = 44.1$ m (After 3.00 s)
$+y$

(a)

(b)

[†]The speed of an object falling in air (or other fluid) does not increase indefinitely. If the object falls far enough, it will reach a maximum velocity called the **terminal velocity** due to air resistance.

FIGURE 2–30 An object thrown into the air leaves the thrower's hand at A, reaches its maximum height at B, and returns to the original position at C. Examples 2–16, 2–17, 2–18, and 2–19.

EXAMPLE 2–16 **Ball thrown upward, I.** A person throws a ball *upward* into the air with an initial velocity of 15.0 m/s. Calculate (a) how high it goes, and (b) how long the ball is in the air before it comes back to the hand. Ignore air resistance.

APPROACH We are not concerned here with the throwing action, but only with the motion of the ball *after* it leaves the thrower's hand (Fig. 2–30) and until it comes back to the hand again. Let us choose y to be positive in the upward direction and negative in the downward direction. (This is a different convention from that used in Examples 2–14 and 2–15, and so illustrates our options.) The acceleration due to gravity is downward and so will have a negative sign, $a = -g = -9.80 \text{ m/s}^2$. As the ball rises, its speed decreases until it reaches the highest point (B in Fig. 2–30), where its speed is zero for an instant; then it descends, with increasing speed.

SOLUTION (a) We consider the time interval from when the ball leaves the thrower's hand until the ball reaches the highest point. To determine the maximum height, we calculate the position of the ball when its velocity equals zero ($v = 0$ at the highest point). At $t = 0$ (point A in Fig. 2–30) we have $y_0 = 0$, $v_0 = 15.0 \text{ m/s}$, and $a = -9.80 \text{ m/s}^2$. At time t (maximum height), $v = 0$, $a = -9.80 \text{ m/s}^2$, and we wish to find y. We use Eq. 2–12c, replacing x with y: $v^2 = v_0^2 + 2ay$. We solve this equation for y:

$$y = \frac{v^2 - v_0^2}{2a} = \frac{0 - (15.0 \text{ m/s})^2}{2(-9.80 \text{ m/s}^2)} = 11.5 \text{ m}.$$

The ball reaches a height of 11.5 m above the hand.

(b) Now we need to choose a different time interval to calculate how long the ball is in the air before it returns to the hand. We could do this calculation in two parts by first determining the time required for the ball to reach its highest point, and then determining the time it takes to fall back down. However, it is simpler to consider the time interval for the entire motion from A to B to C (Fig. 2–30) in one step and use Eq. 2–12b. We can do this because y represents position or displacement, and not the total distance traveled. Thus, at both points A and C, $y = 0$. We use Eq. 2–12b with $a = -9.80 \text{ m/s}^2$ and find

$$y = y_0 + v_0 t + \tfrac{1}{2} a t^2$$
$$0 = 0 + (15.0 \text{ m/s}) t + \tfrac{1}{2}(-9.80 \text{ m/s}^2) t^2.$$

This equation is readily factored (we factor out one t):

$$(15.0 \text{ m/s} - 4.90 \text{ m/s}^2 \, t) t = 0.$$

There are two solutions:

$$t = 0 \quad \text{and} \quad t = \frac{15.0 \text{ m/s}}{4.90 \text{ m/s}^2} = 3.06 \text{ s}.$$

The first solution ($t = 0$) corresponds to the initial point (A) in Fig. 2–30, when the ball was first thrown from $y = 0$. The second solution, $t = 3.06 \text{ s}$, corresponds to point C, when the ball has returned to $y = 0$. Thus the ball is in the air 3.06 s.

NOTE We have ignored air resistance, which could be significant, so our result is only an approximation to a real, practical situation.

We did not consider the throwing action in this Example. Why? Because during the throw, the thrower's hand is touching the ball and accelerating the ball at a rate unknown to us—the acceleration is *not* g. We consider only the time when the ball is in the air and the acceleration is equal to g.

Every quadratic equation (where the variable is squared) mathematically produces two solutions. In physics, sometimes only one solution corresponds to the real situation, as in Example 2–10, in which case we ignore the "unphysical" solution. But in Example 2–16, both solutions to our equation in t^2 are physically meaningful: $t = 0$ and $t = 3.06 \text{ s}$.

⚠ **CAUTION**

Quadratic equations have two solutions. Sometimes only one corresponds to reality, sometimes both

CONCEPTUAL EXAMPLE 2–17 **Two possible misconceptions.** Give examples to show the error in these two common misconceptions: (1) that acceleration and velocity are always in the same direction, and (2) that an object thrown upward has zero acceleration at the highest point (B in Fig. 2–30).

RESPONSE Both are wrong. (1) Velocity and acceleration are *not* necessarily in the same direction. When the ball in Example 2–16 is moving upward, its velocity is positive (upward), whereas the acceleration is negative (downward). (2) At the highest point (B in Fig. 2–30), the ball has zero velocity for an instant. Is the acceleration also zero at this point? No. The velocity near the top of the arc points upward, then becomes zero (for zero time) at the highest point, and then points downward. Gravity does not stop acting, so $a = -g = -9.80 \text{ m/s}^2$ even there. Thinking that $a = 0$ at point B would lead to the conclusion that upon reaching point B, the ball would stay there: if the acceleration (= rate of change of velocity) were zero, the velocity would stay zero at the highest point, and the ball would stay up there without falling. In sum, the acceleration of gravity always points down toward the Earth, even when the object is moving up.

⚠ **CAUTION**

(1) Velocity and acceleration are not always in the same direction; the acceleration (of gravity) always points down
(2) $a \neq 0$ even at the highest point of a trajectory

EXAMPLE 2–18 **Ball thrown upward, II.** Let us consider again the ball thrown upward of Example 2–16, and make more calculations. Calculate (*a*) how much time it takes for the ball to reach the maximum height (point B in Fig. 2–30), and (*b*) the velocity of the ball when it returns to the thrower's hand (point C).

APPROACH Again we assume the acceleration is constant, so we can use Eqs. 2–12. We have the height of 11.5 m from Example 2–16. Again we take y as positive upward.

SOLUTION (*a*) We consider the time interval between the throw $(t = 0, v_0 = 15.0 \text{ m/s})$ and the top of the path $(y = +11.5 \text{ m}, v = 0)$, and we want to find t. The acceleration is constant at $a = -g = -9.80 \text{ m/s}^2$. Both Eqs. 2–12a and 2–12b contain the time t with other quantities known. Let us use Eq. 2–12a with $a = -9.80 \text{ m/s}^2$, $v_0 = 15.0 \text{ m/s}$, and $v = 0$:

$$v = v_0 + at;$$

setting $v = 0$ and solving for t gives

$$t = -\frac{v_0}{a} = -\frac{15.0 \text{ m/s}}{-9.80 \text{ m/s}^2} = 1.53 \text{ s}.$$

This is just half the time it takes the ball to go up and fall back to its original position [3.06 s, calculated in part (*b*) of Example 2–16]. Thus it takes the same time to reach the maximum height as to fall back to the starting point.

(*b*) Now we consider the time interval from the throw $(t = 0, v_0 = 15.0 \text{ m/s})$ until the ball's return to the hand, which occurs at $t = 3.06 \text{ s}$ (as calculated in Example 2–16), and we want to find v when $t = 3.06 \text{ s}$:

$$v = v_0 + at = 15.0 \text{ m/s} - (9.80 \text{ m/s}^2)(3.06 \text{ s}) = -15.0 \text{ m/s}.$$

NOTE The ball has the same speed (magnitude of velocity) when it returns to the starting point as it did initially, but in the opposite direction (this is the meaning of the negative sign). And, as we saw in part (*a*), the time is the same up as down. Thus the motion is symmetrical about the maximum height.

The acceleration of objects such as rockets and fast airplanes is often given as a multiple of $g = 9.80 \text{ m/s}^2$. For example, a plane pulling out of a dive and undergoing 3.00 g's would have an acceleration of $(3.00)(9.80 \text{ m/s}^2) = 29.4 \text{ m/s}^2$.

| **EXERCISE I** If a car is said to accelerate at 0.50 g, what is its acceleration in m/s²?

FIGURE 2–30
(Repeated for Example 2–19)

EXAMPLE 2–19 **Ball thrown upward, III; the quadratic formula.** For the ball in Example 2–18, calculate at what time t the ball passes a point 8.00 m above the person's hand. (See repeated Fig. 2–30 here).

APPROACH We choose the time interval from the throw $(t = 0, v_0 = 15.0 \text{ m/s})$ until the time t (to be determined) when the ball is at position $y = 8.00$ m, using Eq. 2–12b.

SOLUTION We want to find t, given $y = 8.00$ m, $y_0 = 0$, $v_0 = 15.0$ m/s, and $a = -9.80 \text{ m/s}^2$. We use Eq. 2–12b:

$$y = y_0 + v_0 t + \tfrac{1}{2} a t^2$$
$$8.00 \text{ m} = 0 + (15.0 \text{ m/s}) t + \tfrac{1}{2}(-9.80 \text{ m/s}^2) t^2.$$

To solve any quadratic equation of the form $at^2 + bt + c = 0$, where a, b, and c are constants (a is *not* acceleration here), we use the **quadratic formula**:

$$t = \frac{-b \pm \sqrt{b^2 - 4ac}}{2a}.$$

We rewrite our y equation just above in standard form, $at^2 + bt + c = 0$:

$$(4.90 \text{ m/s}^2) t^2 - (15.0 \text{ m/s}) t + (8.00 \text{ m}) = 0.$$

So the coefficient a is 4.90 m/s^2, b is -15.0 m/s, and c is 8.00 m. Putting these into the quadratic formula, we obtain

$$t = \frac{15.0 \text{ m/s} \pm \sqrt{(15.0 \text{ m/s})^2 - 4(4.90 \text{ m/s}^2)(8.00 \text{ m})}}{2(4.90 \text{ m/s}^2)},$$

which gives us $t = 0.69$ s and $t = 2.37$ s. Are both solutions valid? Yes, because the ball passes $y = 8.00$ m when it goes up ($t = 0.69$ s) and again when it comes down ($t = 2.37$ s).

NOTE Figure 2–31 shows graphs of (a) y vs. t and (b) v vs. t for the ball thrown upward in Fig. 2–30, incorporating the results of Examples 2–16, 2–18, and 2–19.

FIGURE 2–31 Graphs of (a) y vs. t, (b) v vs. t for a ball thrown upward, Examples 2–16, 2–18, and 2–19.

EXAMPLE 2–20 **Ball thrown upward at edge of cliff.** Suppose that the person of Examples 2–16, 2–18, and 2–19 is standing on the edge of a cliff, so that the ball can fall to the base of the cliff 50.0 m below as in Fig. 2–32. (a) How long does it take the ball to reach the base of the cliff? (b) What is the total distance traveled by the ball? Ignore air resistance (likely to be significant, so our result is an approximation).

APPROACH We again use Eq. 2–12b, but this time we set $y = -50.0$ m, the bottom of the cliff, which is 50.0 m below the initial position $(y_0 = 0)$.

SOLUTION (a) We use Eq. 2–12b with $a = -9.80\,\text{m/s}^2$, $v_0 = 15.0\,\text{m/s}$, $y_0 = 0$, and $y = -50.0\,\text{m}$:

$$y = y_0 + v_0 t + \tfrac{1}{2}at^2$$
$$-50.0\,\text{m} = 0 + (15.0\,\text{m/s})t - \tfrac{1}{2}(9.80\,\text{m/s}^2)t^2.$$

Rewriting in the standard form we have

$$(4.90\,\text{m/s}^2)t^2 - (15.0\,\text{m/s})t - (50.0\,\text{m}) = 0.$$

Using the quadratic formula, we find as solutions $t = 5.07\,\text{s}$ and $t = -2.01\,\text{s}$. The first solution, $t = 5.07\,\text{s}$, is the answer we are seeking: the time it takes the ball to rise to its highest point and then fall to the base of the cliff. To rise and fall back to the top of the cliff took 3.06 s (Example 2–16); so it took an additional 2.01 s to fall to the base. But what is the meaning of the other solution, $t = -2.01\,\text{s}$? This is a time before the throw, when our calculation begins, so it isn't relevant here.[†]

(b) From Example 2–16, the ball moves up 11.5 m, falls 11.5 m back down to the top of the cliff, and then down another 50.0 m to the base of the cliff, for a total distance traveled of 73.0 m. Note that the *displacement*, however, was −50.0 m. Figure 2–33 shows the y vs. t graph for this situation.

EXERCISE J Two balls are thrown from a cliff. One is thrown directly up, the other directly down, each with the same initial speed, and both hit the ground below the cliff. Which ball hits the ground at the greater speed: (a) the ball thrown upward, (b) the ball thrown downward, or (c) both the same? Ignore air resistance.

*2–8 Variable Acceleration; Integral Calculus

In this brief optional Section we use integral calculus to derive the kinematic equations for constant acceleration, Eqs. 2–12a and b. We also show how calculus can be used when the acceleration is not constant. If you have not yet studied simple integration in your calculus course, you may want to postpone reading this Section until you have. We discuss integration in more detail in Section 7–3, where we begin to use it in the physics.

First we derive Eq. 2–12a, assuming as we did in Section 2–5 that an object has velocity v_0 at $t = 0$ and a constant acceleration a. We start with the definition of instantaneous acceleration, $a = dv/dt$, which we rewrite as

$$dv = a\,dt.$$

We take the definite integral of both sides of this equation, using the same notation we did in Section 2–5:

$$\int_{v=v_0}^{v} dv = \int_{t=0}^{t} a\,dt$$

which gives, since $a = \text{constant}$,

$$v - v_0 = at.$$

This is Eq. 2–12a, $v = v_0 + at$.

Next we derive Eq. 2–12b starting with the definition of instantaneous velocity, Eq. 2–4, $v = dx/dt$. We rewrite this as

$$dx = v\,dt$$

or

$$dx = (v_0 + at)\,dt$$

where we substituted in Eq. 2–12a.

[†]The solution $t = -2.01\,\text{s}$ could be meaningful in a different physical situation. Suppose that a person standing on top of a 50.0-m-high cliff sees a rock pass by him at $t = 0$ moving upward at 15.0 m/s; at what time did the rock leave the base of the cliff, and when did it arrive back at the base of the cliff? The equations will be precisely the same as for our original Example, and the answers $t = -2.01\,\text{s}$ and $t = 5.07\,\text{s}$ will be the correct answers. Note that we cannot put all the information for a problem into the mathematics, so we have to use common sense in interpreting results.

FIGURE 2–32 Example 2–20. The person in Fig. 2–30 stands on the edge of a cliff. The ball falls to the base of the cliff, 50.0 m below.

FIGURE 2–33 Example 2–20, the y vs. t graph.

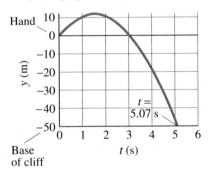

Now we integrate:

$$\int_{x=x_0}^{x} dx = \int_{t=0}^{t} (v_0 + at)\, dt$$

$$x - x_0 = \int_{t=0}^{t} v_0\, dt + \int_{t=0}^{t} at\, dt$$

$$x - x_0 = v_0 t + \tfrac{1}{2} at^2$$

since v_0 and a are constants. This result is just Eq. 2–12b, $x = x_0 + v_0 t + \tfrac{1}{2} at^2$.

Finally let us use calculus to find velocity and displacement, given an acceleration that is not constant but varies in time.

EXAMPLE 2–21 **Integrating a time-varying acceleration.** An experimental vehicle starts from rest $(v_0 = 0)$ at $t = 0$ and accelerates at a rate given by $a = (7.00\ \text{m/s}^3)t$. What is (a) its velocity and (b) its displacement 2.00 s later?

APPROACH We cannot use Eqs. 2–12 because a is not constant. We integrate the acceleration $a = dv/dt$ over time to find v as a function of time; and then integrate $v = dx/dt$ to get the displacement.

SOLUTION From the definition of acceleration, $a = dv/dt$, we have

$$dv = a\, dt.$$

We take the integral of both sides from $v = 0$ at $t = 0$ to velocity v at an arbitrary time t:

$$\int_{0}^{v} dv = \int_{0}^{t} a\, dt$$

$$v = \int_{0}^{t} (7.00\ \text{m/s}^3)t\, dt$$

$$= (7.00\ \text{m/s}^3)\left(\frac{t^2}{2}\right)\Big|_{0}^{t} = (7.00\ \text{m/s}^3)\left(\frac{t^2}{2} - 0\right) = (3.50\ \text{m/s}^3)t^2.$$

At $t = 2.00\ \text{s}$, $v = (3.50\ \text{m/s}^3)(2.00\ \text{s})^2 = 14.0\ \text{m/s}$.

(b) To get the displacement, we assume $x_0 = 0$ and start with $v = dx/dt$ which we rewrite as $dx = v\, dt$. Then we integrate from $x = 0$ at $t = 0$ to position x at time t:

$$\int_{0}^{x} dx = \int_{0}^{t} v\, dt$$

$$x = \int_{0}^{2.00\ \text{s}} (3.50\ \text{m/s}^3)t^2\, dt = (3.50\ \text{m/s}^3)\frac{t^3}{3}\Big|_{0}^{2.00\ \text{s}} = 9.33\ \text{m}.$$

In sum, at $t = 2.00\ \text{s}$, $v = 14.0\ \text{m/s}$ and $x = 9.33\ \text{m}$.

*2–9 Graphical Analysis and Numerical Integration

This Section is optional. It discusses how to solve certain Problems numerically, often needing a computer to do the sums. Some of this material is also covered in Chapter 7, Section 7–3.

If we are given the velocity v of an object as a function of time t, we can obtain the displacement, x. Suppose the velocity as a function of time, $v(t)$, is given as a graph (rather than as an equation that could be integrated as discussed in Section 2–8), as shown in Fig 2–34a. If we are interested in the time interval from t_1 to t_2, as shown, we divide the time axis into many small subintervals, Δt_1, Δt_2, Δt_3, ... , which are indicated by the dashed vertical lines. For each subinterval, a horizontal dashed line is drawn to indicate the average velocity during that time interval. The displacement during any subinterval is given by Δx_i, where the subscript i represents the particular subinterval

$(i = 1, 2, 3, \dots)$. From the definition of average velocity (Eq. 2–2) we have

$$\Delta x_i = \overline{v}_i \, \Delta t_i.$$

Thus the displacement during each subinterval equals the product of $\overline{v}_i$ and Δt_i, and equals the area of the dark rectangle in Fig. 2–34a for that subinterval. The total displacement between times t_1 and t_2 is the sum of the displacements over all the subintervals:

$$x_2 - x_1 = \sum_{t_1}^{t_2} \overline{v}_i \, \Delta t_i, \qquad \textbf{(2–13a)}$$

where x_1 is the position at t_1 and x_2 is the position at t_2. This sum equals the area of all the rectangles shown.

It is often difficult to estimate $\overline{v}_i$ with precision for each subinterval from the graph. We can get greater accuracy in our calculation of $x_2 - x_1$ by breaking the interval $t_2 - t_1$ into more, but narrower, subintervals. Ideally, we can let each Δt_i approach zero, so we approach (in principle) an infinite number of subintervals. In this limit the area of all these infinitesimally thin rectangles becomes exactly equal to the area under the curve (Fig. 2–34b). Thus *the total displacement between any two times is equal to the area between the velocity curve and the t axis between the two times t_1 and t_2.* This limit can be written

$$x_2 - x_1 = \lim_{\Delta t \to 0} \sum_{t_1}^{t_2} \overline{v}_i \, \Delta t_i$$

or, using standard calculus notation,

$$x_2 - x_1 = \int_{t_1}^{t_2} v(t) \, dt. \qquad \textbf{(2–13b)}$$

We have let $\Delta t \to 0$ and renamed it dt to indicate that it is now infinitesimally small. The average velocity, $\overline{v}$, over an infinitesimal time dt is the instantaneous velocity at that instant, which we have written $v(t)$ to remind us that v is a function of t. The symbol $\int$ is an elongated S and indicates a sum over an infinite number of infinitesimal subintervals. We say that we are taking the *integral* of $v(t)$ over dt from time t_1 to time t_2, and this is equal to the area between the $v(t)$ curve and the t axis between the times t_1 and t_2 (Fig. 2–34b). The integral in Eq. 2–13b is a *definite integral*, since the limits t_1 and t_2 are specified.

Similarly, if we know the acceleration as a function of time, we can obtain the velocity by the same process. We use the definition of average acceleration (Eq. 2–5) and solve for Δv:

$$\Delta v = \overline{a} \, \Delta t.$$

If a is known as a function of t over some time interval t_1 to t_2, we can subdivide this time interval into many subintervals, Δt_i, just as we did in Fig. 2–34a. The change in velocity during each subinterval is $\Delta v_i = \overline{a}_i \, \Delta t_i$. The total change in velocity from time t_1 until time t_2 is

$$v_2 - v_1 = \sum_{t_1}^{t_2} \overline{a}_i \, \Delta t_i, \qquad \textbf{(2–14a)}$$

where v_2 represents the velocity at t_2 and v_1 the velocity at t_1. This relation can be written as an integral by letting $\Delta t \to 0$ (the number of intervals then approaches infinity)

$$v_2 - v_1 = \lim_{\Delta t \to 0} \sum_{t_1}^{t_2} \overline{a}_i \, \Delta t_i$$

or

$$v_2 - v_1 = \int_{t_1}^{t_2} a(t) \, dt. \qquad \textbf{(2–14b)}$$

Equations 2–14 will allow us to determine the velocity v_2 at some time t_2 if the velocity is known at t_1 and a is known as a function of time.

If the acceleration or velocity is known at discrete intervals of time, we can use the summation forms of the above equations, Eqs. 2–13a and 2–14a, to estimate velocity or displacement. This technique is known as **numerical integration**. We now take an Example that can also be evaluated analytically, so we can compare the results.

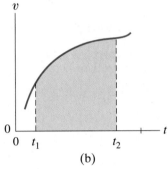

FIGURE 2–34 Graph of v vs. t for the motion of a particle. In (a), the time axis is broken into subintervals of width Δt_i, the average velocity during each Δt_i is $\overline{v}_i$, and the area of all the rectangles, $\Sigma \overline{v}_i \Delta t_i$, is numerically equal to the total displacement $(x_2 - x_1)$ during the total time $(t_2 - t_1)$. In (b), $\Delta t_i \to 0$ and the area under the curve is equal to $(x_2 - x_1)$.

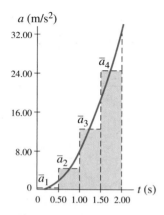

FIGURE 2–35   Example 2–22.

EXAMPLE 2–22 **Numerical integration.** An object starts from rest at $t = 0$ and accelerates at a rate $a(t) = (8.00 \text{ m/s}^4)t^2$. Determine its velocity after 2.00 s using numerical methods.

APPROACH Let us first divide up the interval $t = 0.00$ s to $t = 2.00$ s into four subintervals each of duration $\Delta t_i = 0.50$ s (Fig. 2–35). We use Eq. 2–14a with $v_2 = v$, $v_1 = 0$, $t_2 = 2.00$ s, and $t_1 = 0$. For each of the subintervals we need to estimate $\bar{a}_i$. There are various ways to do this and we use the simple method of choosing $\bar{a}_i$ to be the acceleration $a(t)$ at the midpoint of each interval (an even simpler but usually less accurate procedure would be to use the value of a at the start of the subinterval). That is, we evaluate $a(t) = (8.00 \text{ m/s}^4)t^2$ at $t = 0.25$ s (which is midway between 0.00 s and 0.50 s), 0.75 s, 1.25 s, and 1.75 s.

SOLUTION The results are as follows:

i	1	2	3	4
$\bar{a}_i\,(\text{m/s}^2)$	0.50	4.50	12.50	24.50

Now we use Eq. 2–14a, and note that all Δt_i equal 0.50 s (so they can be factored out):

$$v(t = 2.00 \text{ s}) = \sum_{t=0}^{t=2.00\text{ s}} \bar{a}_i \,\Delta t_i$$
$$= (0.50 \text{ m/s}^2 + 4.50 \text{ m/s}^2 + 12.50 \text{ m/s}^2 + 24.50 \text{ m/s}^2)(0.50 \text{ s})$$
$$= 21.0 \text{ m/s}.$$

We can compare this result to the analytic solution given by Eq. 2–14b since the functional form for a is integrable analytically:

$$v = \int_0^{2.00\text{ s}} (8.00 \text{ m/s}^4)t^2 \, dt = \left. \frac{8.00 \text{ m/s}^4}{3} t^3 \right|_0^{2.00\text{ s}}$$
$$= \frac{8.00 \text{ m/s}^4}{3} \left[(2.00 \text{ s})^3 - (0)^3 \right] = 21.33 \text{ m/s}$$

or 21.3 m/s to the proper number of significant figures. This analytic solution is precise, and we see that our numerical estimate is not far off even though we only used four Δt intervals. It may not be close enough for purposes requiring high accuracy. If we use more and smaller subintervals, we will get a more accurate result. If we use 10 subintervals, each with $\Delta t = 2.00 \text{ s}/10 = 0.20$ s, we have to evaluate $a(t)$ at $t = 0.10$ s, 0.30 s, ..., 1.90 s to get the $\bar{a}_i$, and these are as follows:

i	1	2	3	4	5	6	7	8	9	10
$\bar{a}_i\,(\text{m/s}^2)$	0.08	0.72	2.00	3.92	6.48	9.68	13.52	18.00	23.12	28.88

Then, from Eq. 2–14a we obtain

$$v(t = 2.00 \text{ s}) = \sum \bar{a}_i \,\Delta t_i = \left(\sum \bar{a}_i \right)(0.200 \text{ s})$$
$$= (106.4 \text{ m/s}^2)(0.200 \text{ s}) = 21.28 \text{ m/s},$$

where we have kept an extra significant figure to show that this result is much closer to the (precise) analytic one but still is not quite identical to it. The percentage difference has dropped from 1.4% $(0.3 \text{ m/s}^2/21.3 \text{ m/s}^2)$ for the four-subinterval computation to only 0.2% $(0.05/21.3)$ for the 10-subinterval one.

In the Example above we were given an analytic function that was integrable, so we could compare the accuracy of the numerical calculation to the known precise one. But what do we do if the function is not integrable, so we can't compare our numerical result to an analytic one? That is, how do we know if we've taken enough subintervals so that we can trust our calculated estimate to be accurate to within some desired uncertainty, say 1 percent? What we can do is compare two successive numerical calculations: the first done with n subintervals and the second with, say, twice as many subintervals $(2n)$. If the two results are within the desired uncertainty (say 1 percent), we can usually assume that the calculation with more subintervals is within the desired uncertainty of the true value. If the two calculations are not that close, then a third calculation, with more subintervals (maybe double, maybe 10 times as many, depending on how good the previous approximation was) must be done, and compared to the previous one. The procedure is easy to automate using a computer spreadsheet application.

If we wanted to also obtain the displacement x at some time, we would have to do a second numerical integration over v, which means we would first need to calculate v for many different times. Programmable calculators and computers are very helpful for doing the long sums.

Problems that use these numerical techniques are found at the end of many Chapters of this book; they are labeled Numerical/Computer and are given an asterisk to indicate that they are optional.

Summary

[The Summary that appears at the end of each Chapter in this book gives a brief overview of the main ideas of the Chapter. The Summary *cannot* serve to give an understanding of the material, which can be accomplished only by a detailed reading of the Chapter.]

Kinematics deals with the description of how objects move. The description of the motion of any object must always be given relative to some particular **reference frame**.

The **displacement** of an object is the change in position of the object.

Average speed is the distance traveled divided by the elapsed time or time interval, Δt, the time period over which we choose to make our observations. An object's **average velocity** over a particular time interval Δt is its displacement Δx during that time interval, divided by Δt:

$$\bar{v} = \frac{\Delta x}{\Delta t}. \qquad (2\text{--}2)$$

The **instantaneous velocity**, whose magnitude is the same as the *instantaneous speed*, is defined as the average velocity taken over an infinitesimally short time interval ($\Delta t \to 0$):

$$v = \lim_{\Delta t \to 0} \frac{\Delta x}{\Delta t} = \frac{dx}{dt}, \qquad (2\text{--}4)$$

where dx/dt is the derivative of x with respect to t.

On a graph of position vs. time, the *slope* is equal to the instantaneous velocity.

Acceleration is the change of velocity per unit time. An object's **average acceleration** over a time interval Δt is

$$\bar{a} = \frac{\Delta v}{\Delta t}, \qquad (2\text{--}5)$$

where Δv is the change of velocity during the time interval Δt.

Instantaneous acceleration is the average acceleration taken over an infinitesimally short time interval:

$$a = \lim_{\Delta t \to 0} \frac{\Delta v}{\Delta t} = \frac{dv}{dt}. \qquad (2\text{--}6)$$

If an object moves in a straight line with *constant acceleration*, the velocity v and position x are related to the acceleration a, the elapsed time t, the initial position x_0, and the initial velocity v_0 by Eqs. 2–12:

$$v = v_0 + at, \qquad x = x_0 + v_0 t + \tfrac{1}{2}at^2,$$
$$v^2 = v_0^2 + 2a(x - x_0), \qquad \bar{v} = \frac{v + v_0}{2}. \qquad (2\text{--}12)$$

Objects that move vertically near the surface of the Earth, either falling or having been projected vertically up or down, move with the constant downward **acceleration due to gravity**, whose magnitude is $g = 9.80 \text{ m/s}^2$ if air resistance can be ignored.

[*The kinematic Equations 2–12 can be derived using integral calculus.]

Questions

1. Does a car speedometer measure speed, velocity, or both?
2. Can an object have a varying speed if its velocity is constant? Can it have varying velocity if its speed is constant? If yes, give examples in each case.
3. When an object moves with constant velocity, does its average velocity during any time interval differ from its instantaneous velocity at any instant?
4. If one object has a greater speed than a second object, does the first necessarily have a greater acceleration? Explain, using examples.
5. Compare the acceleration of a motorcycle that accelerates from 80 km/h to 90 km/h with the acceleration of a bicycle that accelerates from rest to 10 km/h in the same time.
6. Can an object have a northward velocity and a southward acceleration? Explain.
7. Can the velocity of an object be negative when its acceleration is positive? What about vice versa?
8. Give an example where both the velocity and acceleration are negative.
9. Two cars emerge side by side from a tunnel. Car A is traveling with a speed of 60 km/h and has an acceleration of 40 km/h/min. Car B has a speed of 40 km/h and has an acceleration of 60 km/h/min. Which car is passing the other as they come out of the tunnel? Explain your reasoning.
10. Can an object be increasing in speed as its acceleration decreases? If so, give an example. If not, explain.
11. A baseball player hits a ball straight up into the air. It leaves the bat with a speed of 120 km/h. In the absence of air resistance, how fast would the ball be traveling when the catcher catches it?
12. As a freely falling object speeds up, what is happening to its acceleration—does it increase, decrease, or stay the same? (a) Ignore air resistance. (b) Consider air resistance.
13. You travel from point A to point B in a car moving at a constant speed of 70 km/h. Then you travel the same distance from point B to another point C, moving at a constant speed of 90 km/h. Is your average speed for the entire trip from A to C 80 km/h? Explain why or why not.
14. Can an object have zero velocity and nonzero acceleration at the same time? Give examples.
15. Can an object have zero acceleration and nonzero velocity at the same time? Give examples.
16. Which of these motions is *not* at constant acceleration: a rock falling from a cliff, an elevator moving from the second floor to the fifth floor making stops along the way, a dish resting on a table?
17. In a lecture demonstration, a 3.0-m-long vertical string with ten bolts tied to it at equal intervals is dropped from the ceiling of the lecture hall. The string falls on a tin plate, and the class hears the clink of each bolt as it hits the plate. The sounds will not occur at equal time intervals. Why? Will the time between clinks increase or decrease near the end of the fall? How could the bolts be tied so that the clinks occur at equal intervals?

18. Describe in words the motion plotted in Fig. 2–36 in terms of v, a, etc. [*Hint:* First try to duplicate the motion plotted by walking or moving your hand.]

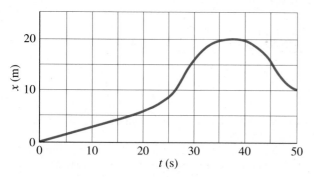

FIGURE 2–36 Question 18, Problems 9 and 86.

19. Describe in words the motion of the object graphed in Fig. 2–37.

FIGURE 2–37 Question 19, Problem 23.

Problems

[The Problems at the end of each Chapter are ranked I, II, or III according to estimated difficulty, with (I) Problems being easiest. Level III are meant as challenges for the best students. The Problems are arranged by Section, meaning that the reader should have read up to and including that Section, but not only that Section—Problems often depend on earlier material. Finally, there is a set of unranked "General Problems" not arranged by Section number.]

2–1 to 2–3 Speed and Velocity

1. (I) If you are driving 110 km/h along a straight road and you look to the side for 2.0 s, how far do you travel during this inattentive period?

2. (I) What must your car's average speed be in order to travel 235 km in 3.25 h?

3. (I) A particle at $t_1 = -2.0$ s is at $x_1 = 4.3$ cm and at $t_2 = 4.5$ s is at $x_2 = 8.5$ cm. What is its average velocity? Can you calculate its average speed from these data?

4. (I) A rolling ball moves from $x_1 = 3.4$ cm to $x_2 = -4.2$ cm during the time from $t_1 = 3.0$ s to $t_2 = 5.1$ s. What is its average velocity?

5. (II) According to a rule-of-thumb, every five seconds between a lightning flash and the following thunder gives the distance to the flash in miles. Assuming that the flash of light arrives in essentially no time at all, estimate the speed of sound in m/s from this rule. What would be the rule for kilometers?

6. (II) You are driving home from school steadily at 95 km/h for 130 km. It then begins to rain and you slow to 65 km/h. You arrive home after driving 3 hours and 20 minutes. (a) How far is your hometown from school? (b) What was your average speed?

7. (II) A horse canters away from its trainer in a straight line, moving 116 m away in 14.0 s. It then turns abruptly and gallops halfway back in 4.8 s. Calculate (a) its average speed and (b) its average velocity for the entire trip, using "away from the trainer" as the positive direction.

8. (II) The position of a small object is given by $x = 34 + 10t - 2t^3$, where t is in seconds and x in meters. (a) Plot x as a function of t from $t = 0$ to $t = 3.0$ s. (b) Find the average velocity of the object between 0 and 3.0 s. (c) At what time between 0 and 3.0 s is the instantaneous velocity zero?

9. (II) The position of a rabbit along a straight tunnel as a function of time is plotted in Fig. 2–36. What is its instantaneous velocity (a) at $t = 10.0$ s and (b) at $t = 30.0$ s? What is its average velocity (c) between $t = 0$ and $t = 5.0$ s, (d) between $t = 25.0$ s and $t = 30.0$ s, and (e) between $t = 40.0$ s and $t = 50.0$ s?

10. (II) On an audio compact disc (CD), digital bits of information are encoded sequentially along a spiral path. Each bit occupies about 0.28 μm. A CD player's readout laser scans along the spiral's sequence of bits at a constant speed of about 1.2 m/s as the CD spins. (a) Determine the number N of digital bits that a CD player reads every second. (b) The audio information is sent to each of the two loudspeakers 44,100 times per second. Each of these samplings requires 16 bits and so one would (at first glance) think the required bit rate for a CD player is

$$N_0 = 2\left(44{,}100 \; \frac{\text{samplings}}{\text{second}}\right)\left(16 \; \frac{\text{bits}}{\text{sampling}}\right) = 1.4 \times 10^6 \; \frac{\text{bits}}{\text{second}},$$

where the 2 is for the 2 loudspeakers (the 2 stereo channels). Note that N_0 is less than the number N of bits actually read per second by a CD player. The excess number of bits $(= N - N_0)$ is needed for encoding and error-correction. What percentage of the bits on a CD are dedicated to encoding and error-correction?

11. (II) A car traveling 95 km/h is 110 m behind a truck traveling 75 km/h. How long will it take the car to reach the truck?

12. (II) Two locomotives approach each other on parallel tracks. Each has a speed of 95 km/h with respect to the ground. If they are initially 8.5 km apart, how long will it be before they reach each other? (See Fig. 2–38).

FIGURE 2–38 Problem 12.

13. (II) Digital bits on a 12.0-cm diameter audio CD are encoded along an outward spiraling path that starts at radius $R_1 = 2.5$ cm and finishes at radius $R_2 = 5.8$ cm. The distance between the centers of neighboring spiral-windings is $1.6 \mu m \, (= 1.6 \times 10^{-6}$ m$)$. (a) Determine the total length of the spiraling path. [Hint: Imagine "unwinding" the spiral into a straight path of width $1.6 \mu m$, and note that the original spiral and the straight path both occupy the same area.] (b) To read information, a CD player adjusts the rotation of the CD so that the player's readout laser moves along the spiral path at a constant speed of 1.25 m/s. Estimate the maximum playing time of such a CD.

14. (II) An airplane travels 3100 km at a speed of 720 km/h, and then encounters a tailwind that boosts its speed to 990 km/h for the next 2800 km. What was the total time for the trip? What was the average speed of the plane for this trip? [Hint: Does Eq. 2–12d apply, or not?]

15. (II) Calculate the average speed and average velocity of a complete round trip in which the outgoing 250 km is covered at 95 km/h, followed by a 1.0-h lunch break, and the return 250 km is covered at 55 km/h.

16. (II) The position of a ball rolling in a straight line is given by $x = 2.0 - 3.6t + 1.1t^2$, where x is in meters and t in seconds. (a) Determine the position of the ball at $t = 1.0$ s, 2.0 s, and 3.0 s. (b) What is the average velocity over the interval $t = 1.0$ s to $t = 3.0$ s? (c) What is its instantaneous velocity at $t = 2.0$ s and at $t = 3.0$ s?

17. (II) A dog runs 120 m away from its master in a straight line in 8.4 s, and then runs halfway back in one-third the time. Calculate (a) its average speed and (b) its average velocity.

18. (III) An automobile traveling 95 km/h overtakes a 1.10-km-long train traveling in the same direction on a track parallel to the road. If the train's speed is 75 km/h, how long does it take the car to pass it, and how far will the car have traveled in this time? See Fig. 2–39. What are the results if the car and train are traveling in opposite directions?

` **FIGURE 2–39** Problem 18.

19. (III) A bowling ball traveling with constant speed hits the pins at the end of a bowling lane 16.5 m long. The bowler hears the sound of the ball hitting the pins 2.50 s after the ball is released from his hands. What is the speed of the ball, assuming the speed of sound is 340 m/s?

2–4 Acceleration

20. (I) A sports car accelerates from rest to 95 km/h in 4.5 s. What is its average acceleration in m/s²?

21. (I) At highway speeds, a particular automobile is capable of an acceleration of about 1.8 m/s². At this rate, how long does it take to accelerate from 80 km/h to 110 km/h?

22. (I) A sprinter accelerates from rest to 9.00 m/s in 1.28 s. What is her acceleration in (a) m/s²; (b) km/h²?

23. (I) Figure 2–37 shows the velocity of a train as a function of time. (a) At what time was its velocity greatest? (b) During what periods, if any, was the velocity constant? (c) During what periods, if any, was the acceleration constant? (d) When was the magnitude of the acceleration greatest?

24. (II) A sports car moving at constant speed travels 110 m in 5.0 s. If it then brakes and comes to a stop in 4.0 s, what is the magnitude of its acceleration in m/s², and in g's $(g = 9.80 \, \text{m/s}^2)$?

25. (II) A car moving in a straight line starts at $x = 0$ at $t = 0$. It passes the point $x = 25.0$ m with a speed of 11.0 m/s at $t = 3.00$ s. It passes the point $x = 385$ m with a speed of 45.0 m/s at $t = 20.0$ s. Find (a) the average velocity and (b) the average acceleration between $t = 3.00$ s and $t = 20.0$ s.

26. (II) A particular automobile can accelerate approximately as shown in the velocity vs. time graph of Fig. 2–40. (The short flat spots in the curve represent shifting of the gears.) Estimate the average acceleration of the car in (a) second gear; and (b) fourth gear. (c) What is its average acceleration through the first four gears?

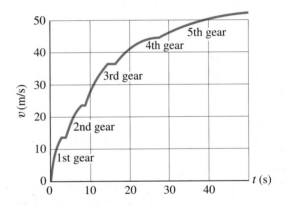

FIGURE 2–40 Problem 26. The velocity of a high-performance automobile as a function of time, starting from a dead stop. The flat spots in the curve represent gear shifts.

27. (II) A particle moves along the x axis. Its position as a function of time is given by $x = 6.8t + 8.5t^2$, where t is in seconds and x is in meters. What is the acceleration as a function of time?

28. (II) The position of a racing car, which starts from rest at $t = 0$ and moves in a straight line, is given as a function of time in the following Table. Estimate (a) its velocity and (b) its acceleration as a function of time. Display each in a Table and on a graph.

t(s)	0	0.25	0.50	0.75	1.00	1.50	2.00	2.50
x(m)	0	0.11	0.46	1.06	1.94	4.62	8.55	13.79

t(s)	3.00	3.50	4.00	4.50	5.00	5.50	6.00
x(m)	20.36	28.31	37.65	48.37	60.30	73.26	87.16

29. (II) The position of an object is given by $x = At + Bt^2$, where x is in meters and t is in seconds. (a) What are the units of A and B? (b) What is the acceleration as a function of time? (c) What are the velocity and acceleration at $t = 5.0$ s? (d) What is the velocity as a function of time if $x = At + Bt^{-3}$?

2–5 and 2–6 Motion at Constant Acceleration

30. (I) A car slows down from 25 m/s to rest in a distance of 85 m. What was its acceleration, assumed constant?

31. (I) A car accelerates from 12 m/s to 21 m/s in 6.0 s. What was its acceleration? How far did it travel in this time? Assume constant acceleration.

32. (I) A light plane must reach a speed of 32 m/s for takeoff. How long a runway is needed if the (constant) acceleration is $3.0 \, \text{m/s}^2$?

33. (II) A baseball pitcher throws a baseball with a speed of 41 m/s. Estimate the average acceleration of the ball during the throwing motion. In throwing the baseball, the pitcher accelerates the ball through a displacement of about 3.5 m, from behind the body to the point where it is released (Fig. 2–41).

FIGURE 2–41
Problem 33.

34. (II) Show that $\bar{v} = (v + v_0)/2$ (see Eq. 2–12d) is not valid when the acceleration $a = A + Bt$, where A and B are constants.

35. (II) A world-class sprinter can reach a top speed (of about 11.5 m/s) in the first 15.0 m of a race. What is the average acceleration of this sprinter and how long does it take her to reach that speed?

36. (II) An inattentive driver is traveling 18.0 m/s when he notices a red light ahead. His car is capable of decelerating at a rate of $3.65 \, \text{m/s}^2$. If it takes him 0.200 s to get the brakes on and he is 20.0 m from the intersection when he sees the light, will he be able to stop in time?

37. (II) A car slows down uniformly from a speed of 18.0 m/s to rest in 5.00 s. How far did it travel in that time?

38. (II) In coming to a stop, a car leaves skid marks 85 m long on the highway. Assuming a deceleration of $4.00 \, \text{m/s}^2$, estimate the speed of the car just before braking.

39. (II) A car traveling 85 km/h slows down at a constant $0.50 \, \text{m/s}^2$ just by "letting up on the gas." Calculate (a) the distance the car coasts before it stops, (b) the time it takes to stop, and (c) the distance it travels during the first and fifth seconds.

40. (II) A car traveling at 105 km/h strikes a tree. The front end of the car compresses and the driver comes to rest after traveling 0.80 m. What was the magnitude of the average acceleration of the driver during the collision? Express the answer in terms of "g's," where $1.00 \, g = 9.80 \, \text{m/s}^2$.

41. (II) Determine the stopping distances for an automobile with an initial speed of 95 km/h and human reaction time of 1.0 s: (a) for an acceleration $a = -5.0 \, \text{m/s}^2$; (b) for $a = -7.0 \, \text{m/s}^2$.

42. (II) A space vehicle accelerates uniformly from 65 m/s at $t = 0$ to 162 m/s at $t = 10.0$ s. How far did it move between $t = 2.0$ s and $t = 6.0$ s?

43. (II) A 75-m-long train begins uniform acceleration from rest. The front of the train has a speed of 23 m/s when it passes a railway worker who is standing 180 m from where the front of the train started. What will be the speed of the last car as it passes the worker? (See Fig. 2–42.)

FIGURE 2–42 Problem 43.

44. (II) An unmarked police car traveling a constant 95 km/h is passed by a speeder traveling 135 km/h. Precisely 1.00 s after the speeder passes, the police officer steps on the accelerator; if the police car's acceleration is $2.00 \, \text{m/s}^2$, how much time passes before the police car overtakes the speeder (assumed moving at constant speed)?

45. (III) Assume in Problem 44 that the speeder's speed is not known. If the police car accelerates uniformly as given above and overtakes the speeder after accelerating for 7.00 s, what was the speeder's speed?

46. (III) A runner hopes to complete the 10,000-m run in less than 30.0 min. After running at constant speed for exactly 27.0 min, there are still 1100 m to go. The runner must then accelerate at $0.20 \, \text{m/s}^2$ for how many seconds in order to achieve the desired time?

47. (III) Mary and Sally are in a foot race (Fig. 2–43). When Mary is 22 m from the finish line, she has a speed of 4.0 m/s and is 5.0 m behind Sally, who has a speed of 5.0 m/s. Sally thinks she has an easy win and so, during the remaining portion of the race, decelerates at a constant rate of $0.50 \, \text{m/s}^2$ to the finish line. What constant acceleration does Mary now need during the remaining portion of the race, if she wishes to cross the finish line side-by-side with Sally?

FIGURE 2–43 Problem 47.

2–7 Freely Falling Objects

[Neglect air resistance.]

48. (I) A stone is dropped from the top of a cliff. It is seen to hit the ground below after 3.75 s. How high is the cliff?

49. (I) If a car rolls gently $(v_0 = 0)$ off a vertical cliff, how long does it take it to reach 55 km/h?

50. (I) Estimate (a) how long it took King Kong to fall straight down from the top of the Empire State Building (380 m high), and (b) his velocity just before "landing."

51. (II) A baseball is hit almost straight up into the air with a speed of about 20 m/s. (a) How high does it go? (b) How long is it in the air?

52. (II) A ball player catches a ball 3.2 s after throwing it vertically upward. With what speed did he throw it, and what height did it reach?

53. (II) A kangaroo jumps to a vertical height of 1.65 m. How long was it in the air before returning to Earth?

54. (II) The best rebounders in basketball have a vertical leap (that is, the vertical movement of a fixed point on their body) of about 120 cm. (a) What is their initial "launch" speed off the ground? (b) How long are they in the air?

55. (II) A helicopter is ascending vertically with a speed of 5.10 m/s. At a height of 105 m above the Earth, a package is dropped from a window. How much time does it take for the package to reach the ground? [Hint: v_0 for the package equals the speed of the helicopter.]

56. (II) For an object falling freely from rest, show that the distance traveled during each successive second increases in the ratio of successive odd integers (1, 3, 5, etc.). (This was first shown by Galileo.) See Figs. 2–26 and 2–29.

57. (II) A baseball is seen to pass upward by a window 23 m above the street with a vertical speed of 14 m/s. If the ball was thrown from the street, (a) what was its initial speed, (b) what altitude does it reach, (c) when was it thrown, and (d) when does it reach the street again?

58. (II) A rocket rises vertically, from rest, with an acceleration of 3.2 m/s² until it runs out of fuel at an altitude of 950 m. After this point, its acceleration is that of gravity, downward. (a) What is the velocity of the rocket when it runs out of fuel? (b) How long does it take to reach this point? (c) What maximum altitude does the rocket reach? (d) How much time (total) does it take to reach maximum altitude? (e) With what velocity does it strike the Earth? (f) How long (total) is it in the air?

59. (II) Roger sees water balloons fall past his window. He notices that each balloon strikes the sidewalk 0.83 s after passing his window. Roger's room is on the third floor, 15 m above the sidewalk. (a) How fast are the balloons traveling when they pass Roger's window? (b) Assuming the balloons are being released from rest, from what floor are they being released? Each floor of the dorm is 5.0 m high.

60. (II) A stone is thrown vertically upward with a speed of 24.0 m/s. (a) How fast is it moving when it reaches a height of 13.0 m? (b) How much time is required to reach this height? (c) Why are there two answers to (b)?

61. (II) A falling stone takes 0.33 s to travel past a window 2.2 m tall (Fig. 2–44). From what height above the top of the window did the stone fall?

FIGURE 2–44 Problem 61.

2.2 m

To travel this distance took 0.33 s

62. (II) Suppose you adjust your garden hose nozzle for a hard stream of water. You point the nozzle vertically upward at a height of 1.5 m above the ground (Fig. 2–45). When you quickly turn off the nozzle, you hear the water striking the ground next to you for another 2.0 s. What is the water speed as it leaves the nozzle?

1.5 m

FIGURE 2–45
Problem 62.

63. (III) A toy rocket moving vertically upward passes by a 2.0-m-high window whose sill is 8.0 m above the ground. The rocket takes 0.15 s to travel the 2.0 m height of the window. What was the launch speed of the rocket, and how high will it go? Assume the propellant is burned very quickly at blastoff.

64. (III) A ball is dropped from the top of a 50.0-m-high cliff. At the same time, a carefully aimed stone is thrown straight up from the bottom of the cliff with a speed of 24.0 m/s. The stone and ball collide part way up. How far above the base of the cliff does this happen?

65. (III) A rock is dropped from a sea cliff and the sound of it striking the ocean is heard 3.4 s later. If the speed of sound is 340 m/s, how high is the cliff?

66. (III) A rock is thrown vertically upward with a speed of 12.0 m/s. Exactly 1.00 s later, a ball is thrown up vertically along the same path with a speed of 18.0 m/s. (a) At what time will they strike each other? (b) At what height will the collision occur? (c) Answer (a) and (b) assuming that the order is reversed: the ball is thrown 1.00 s before the rock.

2–8 Variable Acceleration; Calculus

*67.** (II) Given $v(t) = 25 + 18t$, where v is in m/s and t is in s, use calculus to determine the total displacement from $t_1 = 1.5$ s to $t_2 = 3.1$ s.

*68.** (III) The acceleration of a particle is given by $a = A\sqrt{t}$ where $A = 2.0$ m/s$^{5/2}$. At $t = 0, v = 7.5$ m/s and $x = 0$. (a) What is the speed as a function of time? (b) What is the displacement as a function of time? (c) What are the acceleration, speed and displacement at $t = 5.0$ s?

*69.** (III) Air resistance acting on a falling body can be taken into account by the approximate relation for the acceleration:

$$a = \frac{dv}{dt} = g - kv,$$

where k is a constant. (a) Derive a formula for the velocity of the body as a function of time assuming it starts from rest ($v = 0$ at $t = 0$). [Hint: Change variables by setting $u = g - kv$.] (b) Determine an expression for the terminal velocity, which is the maximum value the velocity reaches.

2–9 Graphical Analysis and Numerical Integration

[See Problems 95–97 at the end of this Chapter.]

Problems **47**

General Problems

70. A fugitive tries to hop on a freight train traveling at a constant speed of 5.0 m/s. Just as an empty box car passes him, the fugitive starts from rest and accelerates at $a = 1.2 \text{ m/s}^2$ to his maximum speed of 6.0 m/s. (a) How long does it take him to catch up to the empty box car? (b) What is the distance traveled to reach the box car?

71. The acceleration due to gravity on the Moon is about one-sixth what it is on Earth. If an object is thrown vertically upward on the Moon, how many times higher will it go than it would on Earth, assuming the same initial velocity?

72. A person jumps from a fourth-story window 15.0 m above a firefighter's safety net. The survivor stretches the net 1.0 m before coming to rest, Fig. 2–46. (a) What was the average deceleration experienced by the survivor when she was slowed to rest by the net? (b) What would you do to make it "safer" (that is, to generate a smaller deceleration): would you stiffen or loosen the net? Explain.

15.0 m

1.0 m

FIGURE 2–46 Problem 72.

73. A person who is properly restrained by an over-the-shoulder seat belt has a good chance of surviving a car collision if the deceleration does not exceed 30 "g's" $(1.00\, g = 9.80 \text{ m/s}^2)$. Assuming uniform deceleration of this value, calculate the distance over which the front end of the car must be designed to collapse if a crash brings the car to rest from 100 km/h.

74. Pelicans tuck their wings and free-fall straight down when diving for fish. Suppose a pelican starts its dive from a height of 16.0 m and cannot change its path once committed. If it takes a fish 0.20 s to perform evasive action, at what minimum height must it spot the pelican to escape? Assume the fish is at the surface of the water.

75. Suppose a car manufacturer tested its cars for front-end collisions by hauling them up on a crane and dropping them from a certain height. (a) Show that the speed just before a car hits the ground, after falling from rest a vertical distance H, is given by $\sqrt{2gH}$. What height corresponds to a collision at (b) 50 km/h? (c) 100 km/h?

76. A stone is dropped from the roof of a high building. A second stone is dropped 1.50 s later. How far apart are the stones when the second one has reached a speed of 12.0 m/s?

77. A bicyclist in the Tour de France crests a mountain pass as he moves at 15 km/h. At the bottom, 4.0 km farther, his speed is 75 km/h. What was his average acceleration (in m/s²) while riding down the mountain?

78. Consider the street pattern shown in Fig. 2–47. Each intersection has a traffic signal, and the speed limit is 50 km/h. Suppose you are driving from the west at the speed limit. When you are 10.0 m from the first intersection, all the lights turn green. The lights are green for 13.0 s each. (a) Calculate the time needed to reach the third stoplight. Can you make it through all three lights without stopping? (b) Another car was stopped at the first light when all the lights turned green. It can accelerate at the rate of 2.00 m/s² to the speed limit. Can the second car make it through all three lights without stopping? By how many seconds would it make it or not?

FIGURE 2–47 Problem 78.

79. In putting, the force with which a golfer strikes a ball is planned so that the ball will stop within some small distance of the cup, say 1.0 m long or short, in case the putt is missed. Accomplishing this from an uphill lie (that is, putting the ball downhill, see Fig. 2–48) is more difficult than from a downhill lie. To see why, assume that on a particular green the ball decelerates constantly at 1.8 m/s² going downhill, and constantly at 2.8 m/s² going uphill. Suppose we have an uphill lie 7.0 m from the cup. Calculate the allowable range of initial velocities we may impart to the ball so that it stops in the range 1.0 m short to 1.0 m long of the cup. Do the same for a downhill lie 7.0 m from the cup. What in your results suggests that the downhill putt is more difficult?

FIGURE 2–48 Problem 79.

80. A robot used in a pharmacy picks up a medicine bottle at $t = 0$. It accelerates at 0.20 m/s² for 5.0 s, then travels without acceleration for 68 s and finally decelerates at -0.40 m/s^2 for 2.5 s to reach the counter where the pharmacist will take the medicine from the robot. From how far away did the robot fetch the medicine?

81. A stone is thrown vertically upward with a speed of 12.5 m/s from the edge of a cliff 75.0 m high (Fig. 2–49). (a) How much later does it reach the bottom of the cliff? (b) What is its speed just before hitting? (c) What total distance did it travel?

FIGURE 2–49
Problem 81.

82. Figure 2–50 is a position versus time graph for the motion of an object along the x axis. Consider the time interval from A to B. (a) Is the object moving in the positive or negative direction? (b) Is the object speeding up or slowing down? (c) Is the acceleration of the object positive or negative? Next, consider the time interval from D to E. (d) Is the object moving in the positive or negative direction? (e) Is the object speeding up or slowing down? (f) Is the acceleration of the object positive or negative? (g) Finally, answer these same three questions for the time interval from C to D.

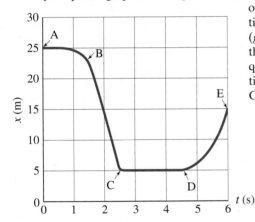

FIGURE 2–50
Problem 82.

83. In the design of a *rapid transit system*, it is necessary to balance the average speed of a train against the distance between stops. The more stops there are, the slower the train's average speed. To get an idea of this problem, calculate the time it takes a train to make a 9.0-km trip in two situations: (a) the stations at which the trains must stop are 1.8 km apart (a total of 6 stations, including those at the ends); and (b) the stations are 3.0 km apart (4 stations total). Assume that at each station the train accelerates at a rate of 1.1 m/s² until it reaches 95 km/h, then stays at this speed until its brakes are applied for arrival at the next station, at which time it decelerates at -2.0 m/s². Assume it stops at each intermediate station for 22 s.

84. A person jumps off a diving board 4.0 m above the water's surface into a deep pool. The person's downward motion stops 2.0 m below the surface of the water. Estimate the average deceleration of the person while under the water.

85. Bill can throw a ball vertically at a speed 1.5 times faster than Joe can. How many times higher will Bill's ball go than Joe's?

86. Sketch the v vs. t graph for the object whose displacement as a function of time is given by Fig. 2–36.

87. A person driving her car at 45 km/h approaches an intersection just as the traffic light turns yellow. She knows that the yellow light lasts only 2.0 s before turning to red, and she is 28 m away from the near side of the intersection (Fig. 2–51). Should she try to stop, or should she speed up to cross the intersection before the light turns red? The intersection is 15 m wide. Her car's maximum deceleration is -5.8 m/s², whereas it can accelerate from 45 km/h to 65 km/h in 6.0 s. Ignore the length of her car and her reaction time.

FIGURE 2–51   Problem 87.

88. A car is behind a truck going 25 m/s on the highway. The driver looks for an opportunity to pass, guessing that his car can accelerate at 1.0 m/s², and he gauges that he has to cover the 20-m length of the truck, plus 10-m clear room at the rear of the truck and 10 m more at the front of it. In the oncoming lane, he sees a car approaching, probably also traveling at 25 m/s. He estimates that the car is about 400 m away. Should he attempt the pass? Give details.

89. Agent Bond is standing on a bridge, 13 m above the road below, and his pursuers are getting too close for comfort. He spots a flatbed truck approaching at 25 m/s, which he measures by knowing that the telephone poles the truck is passing are 25 m apart in this country. The bed of the truck is 1.5 m above the road, and Bond quickly calculates how many poles away the truck should be when he jumps down from the bridge onto the truck, making his getaway. How many poles is it?

90. A police car at rest, passed by a speeder traveling at a constant 130 km/h, takes off in hot pursuit. The police officer catches up to the speeder in 750 m, maintaining a constant acceleration. (a) Qualitatively plot the position vs. time graph for both cars from the police car's start to the catch-up point. Calculate (b) how long it took the police officer to overtake the speeder, (c) the required police car acceleration, and (d) the speed of the police car at the overtaking point.

91. A fast-food restaurant uses a conveyor belt to send the burgers through a grilling machine. If the grilling machine is 1.1 m long and the burgers require 2.5 min to cook, how fast must the conveyor belt travel? If the burgers are spaced 15 cm apart, what is the rate of burger production (in burgers/min)?

92. Two students are asked to find the height of a particular building using a barometer. Instead of using the barometer as an altitude-measuring device, they take it to the roof of the building and drop it off, timing its fall. One student reports a fall time of 2.0 s, and the other, 2.3 s. What % difference does the 0.3 s make for the estimates of the building's height?

93. Figure 2–52 shows the position vs. time graph for two bicycles, A and B. (a) Is there any instant at which the two bicycles have the same velocity? (b) Which bicycle has the larger acceleration? (c) At which instant(s) are the bicycles passing each other? Which bicycle is passing the other? (d) Which bicycle has the highest instantaneous velocity? (e) Which bicycle has the higher average velocity?

FIGURE 2–52 Problem 93.

94. You are traveling at a constant speed v_M, and there is a car in front of you traveling with a speed v_A. You notice that $v_M > v_A$, so you start slowing down with a constant acceleration a when the distance between you and the other car is x. What relationship between a and x determines whether or not you run into the car in front of you?

*Numerical/Computer

*95. (II) The Table below gives the speed of a particular drag racer as a function of time. (a) Calculate the average acceleration (m/s^2) during each time interval. (b) Using numerical integration (see Section 2–9) estimate the total distance traveled (m) as a function of time. [*Hint*: for $\bar{v}$ in each interval sum the velocities at the beginning and end of the interval and divide by 2; for example, in the second interval use $\bar{v} = (6.0 + 13.2)/2 = 9.6$] (c) Graph each of these.

$t(s)$	0	0.50	1.00	1.50	2.00	2.50	3.00	3.50	4.00	4.50	5.00
$v(km/h)$	0.0	6.0	13.2	22.3	32.2	43.0	53.5	62.6	70.6	78.4	85.1

*96. (III) The acceleration of an object (in m/s^2) is measured at 1.00-s intervals starting at $t = 0$ to be as follows: 1.25, 1.58, 1.96, 2.40, 2.66, 2.70, 2.74, 2.72, 2.60, 2.30, 2.04, 1.76, 1.41, 1.09, 0.86, 0.51, 0.28, 0.10. Use numerical integration (see Section 2–9) to estimate (a) the velocity (assume that $v = 0$ at $t = 0$) and (b) the displacement at $t = 17.00$ s.

*97. (III) A lifeguard standing at the side of a swimming pool spots a child in distress, Fig. 2–53. The lifeguard runs with average speed v_R along the pool's edge for a distance x, then jumps into the pool and swims with average speed v_S on a straight path to the child. (a) Show that the total time t it takes the lifeguard to get to the child is given by

$$t = \frac{x}{v_R} + \frac{\sqrt{D^2 + (d - x)^2}}{v_S}.$$

(b) Assume $v_R = 4.0$ m/s and $v_S = 1.5$ m/s. Use a graphing calculator or computer to plot t vs. x in part (a), and from this plot determine the optimal distance x the lifeguard should run before jumping into the pool (that is, find the value of x that minimizes the time t to get to the child).

$d = 10.0$ m

$D = 8.0$ m

x

FIGURE 2–53 Problem 97.

Answers to Exercises

A: -30 cm; 50 cm.

B: (a).

C: (b).

D: (b).

E: (a) +; (b) −; (c) −; (d) +.

F: (c).

G: (b).

H: (e).

I: 4.9 m/s^2.

J: (c).

This snowboarder flying through the air shows an example of motion in two dimensions. In the absence of air resistance, the path would be a perfect parabola. The gold arrow represents the downward acceleration of gravity, $\vec{g}$. Galileo analyzed the motion of objects in 2 dimensions under the action of gravity near the Earth's surface (now called "projectile motion") into its horizontal and vertical components.

We will discuss how to manipulate vectors and how to add them. Besides analyzing projectile motion, we will also see how to work with relative velocity.

3

Kinematics in Two or Three Dimensions; Vectors

CHAPTER-OPENING QUESTION—Guess now!

[*Don't worry about getting the right answer now—you will get another chance later in the Chapter. See also p. 1 of Chapter 1 for more explanation.*]

A small heavy box of emergency supplies is dropped from a moving helicopter at point A as it flies along in a horizontal direction. Which path in the drawing below best describes the path of the box (neglecting air resistance) as seen by a person standing on the ground?

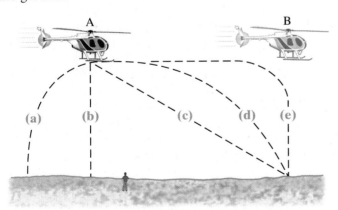

In Chapter 2 we dealt with motion along a straight line. We now consider the description of the motion of objects that move in paths in two (or three) dimensions. To do so, we first need to discuss vectors and how they are added. We will examine the description of motion in general, followed by an interesting special case, the motion of projectiles near the Earth's surface. We also discuss how to determine the relative velocity of an object as measured in different reference frames.

3–1 Vectors and Scalars

We mentioned in Chapter 2 that the term *velocity* refers not only to how fast an object is moving but also to its direction. A quantity such as velocity, which has *direction* as well as *magnitude*, is a **vector** quantity. Other quantities that are also vectors are displacement, force, and momentum. However, many quantities have no direction associated with them, such as mass, time, and temperature. They are specified completely by a number and units. Such quantities are called **scalar** quantities.

Drawing a diagram of a particular physical situation is always helpful in physics, and this is especially true when dealing with vectors. On a diagram, each vector is represented by an arrow. The arrow is always drawn so that it points in the direction of the vector quantity it represents. The length of the arrow is drawn proportional to the magnitude of the vector quantity. For example, in Fig. 3–1, green arrows have been drawn representing the velocity of a car at various places as it rounds a curve. The magnitude of the velocity at each point can be read off Fig. 3–1 by measuring the length of the corresponding arrow and using the scale shown (1 cm = 90 km/h).

When we write the symbol for a vector, we will always use boldface type, with a tiny arrow over the symbol. Thus for velocity we write $\vec{v}$. If we are concerned only with the magnitude of the vector, we will write simply v, in italics, as we do for other symbols.

FIGURE 3–1 Car traveling on a road, slowing down to round the curve. The green arrows represent the velocity vector at each position.

Scale for velocity:
1 cm = 90 km/h

3–2 Addition of Vectors—Graphical Methods

Because vectors are quantities that have direction as well as magnitude, they must be added in a special way. In this Chapter, we will deal mainly with displacement vectors, for which we now use the symbol $\vec{D}$, and velocity vectors, $\vec{v}$. But the results will apply for other vectors we encounter later.

We use simple arithmetic for adding scalars. Simple arithmetic can also be used for adding vectors if they are in the same direction. For example, if a person walks 8 km east one day, and 6 km east the next day, the person will be 8 km + 6 km = 14 km east of the point of origin. We say that the *net* or *resultant* displacement is 14 km to the east (Fig. 3–2a). If, on the other hand, the person walks 8 km east on the first day, and 6 km west (in the reverse direction) on the second day, then the person will end up 2 km from the origin (Fig. 3–2b), so the resultant displacement is 2 km to the east. In this case, the resultant displacement is obtained by subtraction: 8 km − 6 km = 2 km.

But simple arithmetic cannot be used if the two vectors are not along the same line. For example, suppose a person walks 10.0 km east and then walks 5.0 km north. These displacements can be represented on a graph in which the positive y axis points north and the positive x axis points east, Fig. 3–3. On this graph, we draw an arrow, labeled $\vec{D}_1$, to represent the 10.0-km displacement to the east. Then we draw a second arrow, $\vec{D}_2$, to represent the 5.0-km displacement to the north. Both vectors are drawn to scale, as in Fig. 3–3.

FIGURE 3–2 Combining vectors in one dimension.

Resultant = 14 km (east)

8 km   6 km   East

(a)

Resultant = 2 km (east)

6 km

8 km   East

(b)

FIGURE 3–3 A person walks 10.0 km east and then 5.0 km north. These two displacements are represented by the vectors $\vec{D}_1$ and $\vec{D}_2$, which are shown as arrows. The resultant displacement vector, $\vec{D}_R$, which is the vector sum of $\vec{D}_1$ and $\vec{D}_2$, is also shown. Measurement on the graph with ruler and protractor shows that $\vec{D}_R$ has a magnitude of 11.2 km and points at an angle $\theta = 27°$ north of east.

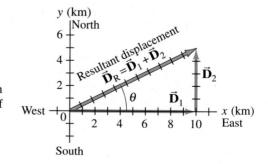

After taking this walk, the person is now 10.0 km east and 5.0 km north of the point of origin. The **resultant displacement** is represented by the arrow labeled $\vec{\mathbf{D}}_R$ in Fig. 3–3. Using a ruler and a protractor, you can measure on this diagram that the person is 11.2 km from the origin at an angle $\theta = 27°$ north of east. In other words, the resultant displacement vector has a magnitude of 11.2 km and makes an angle $\theta = 27°$ with the positive x axis. The magnitude (length) of $\vec{\mathbf{D}}_R$ can also be obtained using the theorem of Pythagoras in this case, since D_1, D_2, and D_R form a right triangle with D_R as the hypotenuse. Thus

$$D_R = \sqrt{D_1^2 + D_2^2} = \sqrt{(10.0\,\text{km})^2 + (5.0\,\text{km})^2}$$
$$= \sqrt{125\,\text{km}^2} = 11.2\,\text{km}.$$

You can use the Pythagorean theorem, of course, only when the vectors are *perpendicular* to each other.

The resultant displacement vector, $\vec{\mathbf{D}}_R$, is the sum of the vectors $\vec{\mathbf{D}}_1$ and $\vec{\mathbf{D}}_2$. That is,

$$\vec{\mathbf{D}}_R = \vec{\mathbf{D}}_1 + \vec{\mathbf{D}}_2.$$

This is a *vector* equation. An important feature of adding two vectors that are not along the same line is that the magnitude of the resultant vector is not equal to the sum of the magnitudes of the two separate vectors, but is smaller than their sum. That is,

$$D_R \le D_1 + D_2,$$

where the equals sign applies only if the two vectors point in the same direction. In our example (Fig. 3–3), $D_R = 11.2\,\text{km}$, whereas $D_1 + D_2$ equals 15 km, which is the total distance traveled. Note also that we cannot set $\vec{\mathbf{D}}_R$ equal to 11.2 km, because we have a vector equation and 11.2 km is only a part of the resultant vector, its magnitude. We could write something like this, though: $\vec{\mathbf{D}}_R = \vec{\mathbf{D}}_1 + \vec{\mathbf{D}}_2 = (11.2\,\text{km}, 27°\,\text{N of E})$.

EXERCISE A Under what conditions can the magnitude of the resultant vector above be $D_R = D_1 + D_2$?

Figure 3–3 illustrates the general rules for graphically adding two vectors together, no matter what angles they make, to get their sum. The rules are as follows:

1. On a diagram, draw one of the vectors—call it $\vec{\mathbf{D}}_1$—to scale.
2. Next draw the second vector, $\vec{\mathbf{D}}_2$, to scale, placing its tail at the tip of the first vector and being sure its direction is correct.
3. The arrow drawn from the tail of the first vector to the tip of the second vector represents the *sum*, or **resultant**, of the two vectors.

The length of the resultant vector represents its magnitude. Note that vectors can be translated parallel to themselves (maintaining the same length and angle) to accomplish these manipulations. The length of the resultant can be measured with a ruler and compared to the scale. Angles can be measured with a protractor. This method is known as the **tail-to-tip method of adding vectors**.

The resultant is not affected by the order in which the vectors are added. For example, a displacement of 5.0 km north, to which is added a displacement of 10.0 km east, yields a resultant of 11.2 km and angle $\theta = 27°$ (see Fig. 3–4), the same as when they were added in reverse order (Fig. 3–3). That is, now using $\vec{\mathbf{V}}$ to represent any type of vector,

$$\vec{\mathbf{V}}_1 + \vec{\mathbf{V}}_2 = \vec{\mathbf{V}}_2 + \vec{\mathbf{V}}_1, \qquad \text{[commutative property]} \quad \textbf{(3–1a)}$$

which is known as the *commutative* property of vector addition.

FIGURE 3–4 If the vectors are added in reverse order, the resultant is the same. (Compare to Fig. 3–3.)

FIGURE 3–5 The resultant of three vectors: $\vec{V}_R = \vec{V}_1 + \vec{V}_2 + \vec{V}_3$.

The tail-to-tip method of adding vectors can be extended to three or more vectors. The resultant is drawn from the tail of the first vector to the tip of the last one added. An example is shown in Fig. 3–5; the three vectors could represent displacements (northeast, south, west) or perhaps three forces. Check for yourself that you get the same resultant no matter in which order you add the three vectors; that is,

$$(\vec{V}_1 + \vec{V}_2) + \vec{V}_3 = \vec{V}_1 + (\vec{V}_2 + \vec{V}_3), \quad \text{[associative property]} \quad \textbf{(3–1b)}$$

which is known as the *associative* property of vector addition.

A second way to add two vectors is the **parallelogram method**. It is fully equivalent to the tail-to-tip method. In this method, the two vectors are drawn starting from a common origin, and a parallelogram is constructed using these two vectors as adjacent sides as shown in Fig. 3–6b. The resultant is the diagonal drawn from the common origin. In Fig. 3–6a, the tail-to-tip method is shown, and it is clear that both methods yield the same result.

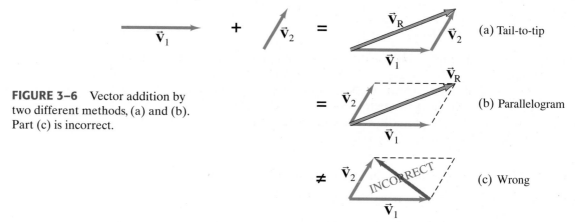

FIGURE 3–6 Vector addition by two different methods, (a) and (b). Part (c) is incorrect.

(a) Tail-to-tip

(b) Parallelogram

(c) Wrong

⚠ **C A U T I O N**

Be sure to use the correct diagonal on parallelogram to get the resultant

It is a common error to draw the sum vector as the diagonal running between the tips of the two vectors, as in Fig. 3–6c. *This is incorrect*: it does not represent the sum of the two vectors. (In fact, it represents their difference, $\vec{V}_2 - \vec{V}_1$, as we will see in the next Section.)

CONCEPTUAL EXAMPLE 3–1 **Range of vector lengths.** Suppose two vectors each have length 3.0 units. What is the range of possible lengths for the vector representing the sum of the two?

RESPONSE The sum can take on any value from 6.0 (= 3.0 + 3.0) where the vectors point in the same direction, to 0 (= 3.0 − 3.0) when the vectors are antiparallel.

EXERCISE B If the two vectors of Example 3–1 are perpendicular to each other, what is the resultant vector length?

3–3 Subtraction of Vectors, and Multiplication of a Vector by a Scalar

FIGURE 3–7 The negative of a vector is a vector having the same length but opposite direction.

Given a vector $\vec{V}$, we define the *negative* of this vector $(-\vec{V})$ to be a vector with the same magnitude as $\vec{V}$ but opposite in direction, Fig. 3–7. Note, however, that no vector is ever negative in the sense of its magnitude: the magnitude of every vector is positive. Rather, a minus sign tells us about its direction.

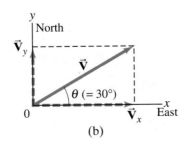

FIGURE 3–8 Subtracting two vectors: $\vec{V}_2 - \vec{V}_1$.

We can now define the subtraction of one vector from another: the difference between two vectors $\vec{V}_2 - \vec{V}_1$ is defined as

$$\vec{V}_2 - \vec{V}_1 = \vec{V}_2 + (-\vec{V}_1).$$

That is, the difference between two vectors is equal to the sum of the first plus the negative of the second. Thus our rules for addition of vectors can be applied as shown in Fig. 3–8 using the tail-to-tip method.

A vector $\vec{V}$ can be multiplied by a scalar c. We define their product so that $c\vec{V}$ has the same direction as $\vec{V}$ and has magnitude cV. That is, multiplication of a vector by a positive scalar c changes the magnitude of the vector by a factor c but doesn't alter the direction. If c is a negative scalar, the magnitude of the product $c\vec{V}$ is still $|c|V$ (where $|c|$ means the magnitude of c), but the direction is precisely opposite to that of $\vec{V}$. See Fig. 3–9.

FIGURE 3–9 Multiplying a vector $\vec{V}$ by a scalar c gives a vector whose magnitude is c times greater and in the same direction as $\vec{V}$ (or opposite direction if c is negative).

EXERCISE C What does the "incorrect" vector in Fig. 3–6c represent? (a) $\vec{V}_2 - \vec{V}_1$, (b) $\vec{V}_1 - \vec{V}_2$, (c) something else (specify).

3–4 Adding Vectors by Components

Adding vectors graphically using a ruler and protractor is often not sufficiently accurate and is not useful for vectors in three dimensions. We discuss now a more powerful and precise method for adding vectors. But do not forget graphical methods—they are useful for visualizing, for checking your math, and thus for getting the correct result.

Consider first a vector $\vec{V}$ that lies in a particular plane. It can be expressed as the sum of two other vectors, called the **components** of the original vector. The components are usually chosen to be along two perpendicular directions, such as the x and y axes. The process of finding the components is known as **resolving the vector into its components**. An example is shown in Fig. 3–10; the vector $\vec{V}$ could be a displacement vector that points at an angle $\theta = 30°$ north of east, where we have chosen the positive x axis to be to the east and the positive y axis north. This vector $\vec{V}$ is resolved into its x and y components by drawing dashed lines out from the tip (A) of the vector (lines AB and AC) making them perpendicular to the x and y axes. Then the lines OB and OC represent the x and y components of $\vec{V}$, respectively, as shown in Fig. 3–10b. These *vector components* are written $\vec{V}_x$ and $\vec{V}_y$. We generally show vector components as arrows, like vectors, but dashed. The *scalar components*, V_x and V_y, are the magnitudes of the vector components, with units, accompanied by a positive or negative sign depending on whether they point along the positive or negative x or y axis. As can be seen in Fig. 3–10, $\vec{V}_x + \vec{V}_y = \vec{V}$ by the parallelogram method of adding vectors.

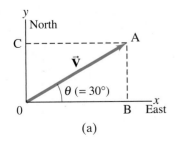

(a)

(b)

FIGURE 3–10 Resolving a vector $\vec{V}$ into its components along an arbitrarily chosen set of x and y axes. The components, once found, themselves represent the vector. That is, the components contain as much information as the vector itself.

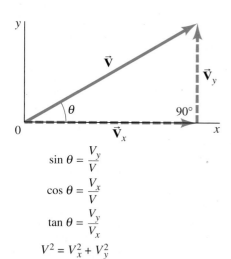

$$\sin \theta = \frac{V_y}{V}$$

$$\cos \theta = \frac{V_x}{V}$$

$$\tan \theta = \frac{V_y}{V_x}$$

$$V^2 = V_x^2 + V_y^2$$

FIGURE 3–11 Finding the components of a vector using trigonometric functions.

Space is made up of three dimensions, and sometimes it is necessary to resolve a vector into components along three mutually perpendicular directions. In rectangular coordinates the components are $\vec{V}_x$, $\vec{V}_y$, and $\vec{V}_z$. Resolution of a vector in three dimensions is merely an extension of the above technique.

The use of trigonometric functions for finding the components of a vector is illustrated in Fig. 3–11, where a vector and its two components are thought of as making up a right triangle. (See also Appendix A for other details on trigonometric functions and identities.) We then see that the sine, cosine, and tangent are as given in Fig. 3–11. If we multiply the definition of $\sin \theta = V_y/V$ by V on both sides, we get

$$V_y = V \sin \theta. \tag{3–2a}$$

Similarly, from the definition of $\cos \theta$, we obtain

$$V_x = V \cos \theta. \tag{3–2b}$$

Note that θ is chosen (by convention) to be the angle that the vector makes with the positive x axis, measured positive counterclockwise.

The components of a given vector will be different for different choices of coordinate axes. It is therefore crucial to specify the choice of coordinate system when giving the components.

There are two ways to specify a vector in a given coordinate system:

1. We can give its components, V_x and V_y.
2. We can give its magnitude V and the angle θ it makes with the positive x axis.

We can shift from one description to the other using Eqs. 3–2, and, for the reverse, by using the theorem of Pythagoras[†] and the definition of tangent:

$$V = \sqrt{V_x^2 + V_y^2} \tag{3–3a}$$

$$\tan \theta = \frac{V_y}{V_x} \tag{3–3b}$$

as can be seen in Fig. 3–11.

We can now discuss how to add vectors using components. The first step is to resolve each vector into its components. Next we can see, using Fig. 3–12, that the addition of any two vectors $\vec{V}_1$ and $\vec{V}_2$ to give a resultant, $\vec{V} = \vec{V}_1 + \vec{V}_2$, implies that

$$V_x = V_{1x} + V_{2x}$$
$$V_y = V_{1y} + V_{2y}. \tag{3–4}$$

That is, the sum of the x components equals the x component of the resultant, and the sum of the y components equals the y component of the resultant, as can be verified by a careful examination of Fig. 3–12. Note that we do *not* add x components to y components.

[†]In three dimensions, the theorem of Pythagoras becomes $V = \sqrt{V_x^2 + V_y^2 + V_z^2}$, where V_z is the component along the third, or z, axis.

FIGURE 3–12 The components of $\vec{V} = \vec{V}_1 + \vec{V}_2$ are

$$V_x = V_{1x} + V_{2x}$$
$$V_y = V_{1y} + V_{2y}.$$

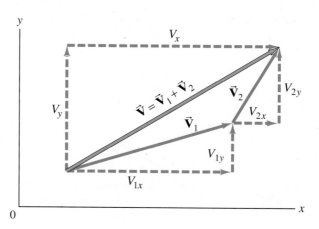

If the magnitude and direction of the resultant vector are desired, they can be obtained using Eqs. 3–3.

The components of a given vector depend on the choice of coordinate axes. You can often reduce the work involved in adding vectors by a good choice of axes—for example, by choosing one of the axes to be in the same direction as one of the vectors. Then that vector will have only one nonzero component.

EXAMPLE 3–2 **Mail carrier's displacement.** A rural mail carrier leaves the post office and drives 22.0 km in a northerly direction. She then drives in a direction 60.0° south of east for 47.0 km (Fig. 3–13a). What is her displacement from the post office?

APPROACH We choose the positive x axis to be east and the positive y axis to be north, since those are the compass directions used on most maps. The origin of the xy coordinate system is at the post office. We resolve each vector into its x and y components. We add the x components together, and then the y components together, giving us the x and y components of the resultant.

SOLUTION Resolve each displacement vector into its components, as shown in Fig. 3–13b. Since $\vec{D}_1$ has magnitude 22.0 km and points north, it has only a y component:

$$D_{1x} = 0, \qquad D_{1y} = 22.0 \text{ km}.$$

$\vec{D}_2$ has both x and y components:

$$D_{2x} = +(47.0 \text{ km})(\cos 60°) = +(47.0 \text{ km})(0.500) = +23.5 \text{ km}$$
$$D_{2y} = -(47.0 \text{ km})(\sin 60°) = -(47.0 \text{ km})(0.866) = -40.7 \text{ km}.$$

Notice that D_{2y} is negative because this vector component points along the negative y axis. The resultant vector, $\vec{D}$, has components:

$$D_x = D_{1x} + D_{2x} = 0 \text{ km} + 23.5 \text{ km} = +23.5 \text{ km}$$
$$D_y = D_{1y} + D_{2y} = 22.0 \text{ km} + (-40.7 \text{ km}) = -18.7 \text{ km}.$$

This specifies the resultant vector completely:

$$D_x = 23.5 \text{ km}, \qquad D_y = -18.7 \text{ km}.$$

We can also specify the resultant vector by giving its magnitude and angle using Eqs. 3–3:

$$D = \sqrt{D_x^2 + D_y^2} = \sqrt{(23.5 \text{ km})^2 + (-18.7 \text{ km})^2} = 30.0 \text{ km}$$

$$\tan \theta = \frac{D_y}{D_x} = \frac{-18.7 \text{ km}}{23.5 \text{ km}} = -0.796.$$

A calculator with an INV TAN, an ARC TAN, or a TAN^{-1} key gives $\theta = \tan^{-1}(-0.796) = -38.5°$. The negative sign means $\theta = 38.5°$ below the x axis, Fig. 3–13c. So, the resultant displacement is 30.0 km directed at 38.5° in a southeasterly direction.

NOTE Always be attentive about the quadrant in which the resultant vector lies. An electronic calculator does not fully give this information, but a good diagram does.

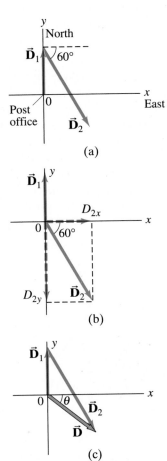

FIGURE 3–13 Example 3–2. (a) The two displacement vectors, $\vec{D}_1$ and $\vec{D}_2$. (b) $\vec{D}_2$ is resolved into its components. (c) $\vec{D}_1$ and $\vec{D}_2$ are added graphically to obtain the resultant $\vec{D}$. The component method of adding the vectors is explained in the Example.

The signs of trigonometric functions depend on which "quadrant" the angle falls in: for example, the tangent is positive in the first and third quadrants (from 0° to 90°, and 180° to 270°), but negative in the second and fourth quadrants; see Appendix A. The best way to keep track of angles, and to check any vector result, is always to draw a vector diagram. A vector diagram gives you something tangible to look at when analyzing a problem, and provides a check on the results.

The following Problem Solving Strategy should not be considered a prescription. Rather it is a summary of things to do to get you thinking and involved in the problem at hand.

PROBLEM SOLVING

Identify the correct quadrant by drawing a careful diagram

Adding Vectors

Here is a brief summary of how to add two or more vectors using components:

1. **Draw a diagram**, adding the vectors graphically by either the parallelogram or tail-to-tip method.
2. **Choose *x* and *y* axes.** Choose them in a way, if possible, that will make your work easier. (For example, choose one axis along the direction of one of the vectors so that vector will have only one component.)
3. **Resolve** each vector into its *x* and *y* **components**, showing each component along its appropriate (*x* or *y*) axis as a (dashed) arrow.
4. **Calculate each component** (when not given) using sines and cosines. If θ_1 is the angle that vector $\vec{V}_1$ makes with the positive *x* axis, then:

$$V_{1x} = V_1 \cos\theta_1, \qquad V_{1y} = V_1 \sin\theta_1.$$

Pay careful attention to **signs**: any component that points along the negative *x* or *y* axis gets a minus sign.

5. **Add** the *x* **components** together to get the *x* component of the resultant. Ditto for *y*:

$$V_x = V_{1x} + V_{2x} + \text{any others}$$
$$V_y = V_{1y} + V_{2y} + \text{any others}.$$

This is the answer: the components of the resultant vector. Check signs to see if they fit the quadrant shown in your diagram (point 1 above).

6. If you want to know the **magnitude and direction** of the resultant vector, use Eqs. 3–3:

$$V = \sqrt{V_x^2 + V_y^2}, \qquad \tan\theta = \frac{V_y}{V_x}.$$

The vector diagram you already drew helps to obtain the correct position (quadrant) of the angle θ.

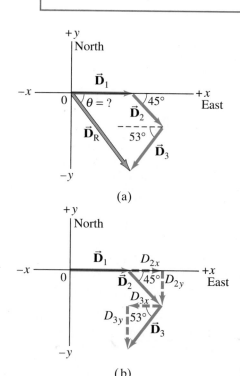

FIGURE 3–14 Example 3–3.

Vector	Components	
	x (km)	*y* (km)
$\vec{D}_1$	620	0
$\vec{D}_2$	311	−311
$\vec{D}_3$	−331	−439
$\vec{D}_R$	600	−750

EXAMPLE 3–3 **Three short trips.** An airplane trip involves three legs, with two stopovers, as shown in Fig. 3–14a. The first leg is due east for 620 km; the second leg is southeast (45°) for 440 km; and the third leg is at 53° south of west, for 550 km, as shown. What is the plane's total displacement?

APPROACH We follow the steps in the Problem Solving Strategy above.
SOLUTION
1. **Draw a diagram** such as Fig. 3–14a, where $\vec{D}_1$, $\vec{D}_2$, and $\vec{D}_3$ represent the three legs of the trip, and $\vec{D}_R$ is the plane's total displacement.
2. **Choose axes**: Axes are also shown in Fig. 3–14a: *x* is east, *y* north.
3. **Resolve components**: It is imperative to draw a good diagram. The components are drawn in Fig. 3–14b. Instead of drawing all the vectors starting from a common origin, as we did in Fig. 3–13b, here we draw them "tail-to-tip" style, which is just as valid and may make it easier to see.
4. **Calculate the components**:

$$\vec{D}_1: D_{1x} = +D_1\cos 0° = D_1 = 620\,\text{km}$$
$$D_{1y} = +D_1\sin 0° = 0\,\text{km}$$
$$\vec{D}_2: D_{2x} = +D_2\cos 45° = +(440\,\text{km})(0.707) = +311\,\text{km}$$
$$D_{2y} = -D_2\sin 45° = -(440\,\text{km})(0.707) = -311\,\text{km}$$
$$\vec{D}_3: D_{3x} = -D_3\cos 53° = -(550\,\text{km})(0.602) = -331\,\text{km}$$
$$D_{3y} = -D_3\sin 53° = -(550\,\text{km})(0.799) = -439\,\text{km}.$$

We have given a minus sign to each component that in Fig. 3–14b points in the −*x* or −*y* direction. The components are shown in the Table in the margin.
5. **Add the components**: We add the *x* components together, and we add the *y* components together to obtain the *x* and *y* components of the resultant:

$$D_x = D_{1x} + D_{2x} + D_{3x} = 620\,\text{km} + 311\,\text{km} - 331\,\text{km} = 600\,\text{km}$$
$$D_y = D_{1y} + D_{2y} + D_{3y} = 0\,\text{km} - 311\,\text{km} - 439\,\text{km} = -750\,\text{km}.$$

The *x* and *y* components are 600 km and −750 km, and point respectively to the east and south. This is one way to give the answer.
6. **Magnitude and direction**: We can also give the answer as

$$D_R = \sqrt{D_x^2 + D_y^2} = \sqrt{(600)^2 + (-750)^2}\,\text{km} = 960\,\text{km}$$
$$\tan\theta = \frac{D_y}{D_x} = \frac{-750\,\text{km}}{600\,\text{km}} = -1.25, \qquad \text{so } \theta = -51°.$$

Thus, the total displacement has magnitude 960 km and points 51° below the *x* axis (south of east), as was shown in our original sketch, Fig. 3–14a.

3-5 Unit Vectors

Vectors can be conveniently written in terms of *unit vectors*. A **unit vector** is defined to have a magnitude exactly equal to one (1). It is useful to define unit vectors that point along coordinate axes, and in an x, y, z rectangular coordinate system these unit vectors are called $\hat{\mathbf{i}}, \hat{\mathbf{j}}$, and $\hat{\mathbf{k}}$. They point, respectively, along the positive x, y, and z axes as shown in Fig. 3–15. Like other vectors, $\hat{\mathbf{i}}, \hat{\mathbf{j}}$, and $\hat{\mathbf{k}}$ do not have to be placed at the origin, but can be placed elsewhere as long as the direction and unit length remain unchanged. It is common to write unit vectors with a "hat": $\hat{\mathbf{i}}, \hat{\mathbf{j}}, \hat{\mathbf{k}}$ (and we will do so in this book) as a reminder that each is a unit vector.

Because of the definition of multiplication of a vector by a scalar (Section 3–3), the components of a vector $\vec{\mathbf{V}}$ can be written $\vec{\mathbf{V}}_x = V_x\hat{\mathbf{i}},\quad \vec{\mathbf{V}}_y = V_y\hat{\mathbf{j}}$, and $\vec{\mathbf{V}}_z = V_z\hat{\mathbf{k}}$. Hence any vector $\vec{\mathbf{V}}$ can be written in terms of its components as

$$\vec{\mathbf{V}} = V_x\hat{\mathbf{i}} + V_y\hat{\mathbf{j}} + V_z\hat{\mathbf{k}}. \tag{3-5}$$

Unit vectors are helpful when adding vectors analytically by components. For example, Eq. 3–4 can be seen to be true by using unit vector notation for each vector (which we write for the two-dimensional case, with the extension to three dimensions being straightforward):

$$\begin{aligned}
\vec{\mathbf{V}} = (V_x)\hat{\mathbf{i}} + (V_y)\hat{\mathbf{j}} &= \vec{\mathbf{V}}_1 + \vec{\mathbf{V}}_2 \\
&= (V_{1x}\hat{\mathbf{i}} + V_{1y}\hat{\mathbf{j}}) + (V_{2x}\hat{\mathbf{i}} + V_{2y}\hat{\mathbf{j}}) \\
&= (V_{1x} + V_{2x})\hat{\mathbf{i}} + (V_{1y} + V_{2y})\hat{\mathbf{j}}.
\end{aligned}$$

Comparing the first line to the third line, we get Eq. 3–4.

EXAMPLE 3–4 **Using unit vectors.** Write the vectors of Example 3–2 in unit vector notation, and perform the addition.

APPROACH We use the components we found in Example 3–2,
$$D_{1x} = 0,\quad D_{1y} = 22.0\,\text{km}, \quad\text{and}\quad D_{2x} = 23.5\,\text{km},\quad D_{2y} = -40.7\,\text{km},$$
and we now write them in the form of Eq. 3–5.

SOLUTION We have
$$\vec{\mathbf{D}}_1 = 0\hat{\mathbf{i}} + 22.0\,\text{km}\,\hat{\mathbf{j}}$$
$$\vec{\mathbf{D}}_2 = 23.5\,\text{km}\,\hat{\mathbf{i}} - 40.7\,\text{km}\,\hat{\mathbf{j}}.$$
Then
$$\begin{aligned}
\vec{\mathbf{D}} = \vec{\mathbf{D}}_1 + \vec{\mathbf{D}}_2 &= (0 + 23.5)\,\text{km}\,\hat{\mathbf{i}} + (22.0 - 40.7)\,\text{km}\,\hat{\mathbf{j}} \\
&= 23.5\,\text{km}\,\hat{\mathbf{i}} - 18.7\,\text{km}\,\hat{\mathbf{j}}.
\end{aligned}$$

The components of the resultant displacement, $\vec{\mathbf{D}}$, are $D_x = 23.5\,\text{km}$ and $D_y = -18.7\,\text{km}$. The magnitude of $\vec{\mathbf{D}}$ is $D = \sqrt{(23.5\,\text{km})^2 + (18.7\,\text{km})^2} = 30.0\,\text{km}$, just as in Example 3–2.

3-6 Vector Kinematics

We can now extend our definitions of velocity and acceleration in a formal way to two- and three-dimensional motion. Suppose a particle follows a path in the xy plane as shown in Fig. 3–16. At time t_1, the particle is at point P_1, and at time t_2, it is at point P_2. The vector $\vec{\mathbf{r}}_1$ is the position vector of the particle at time t_1 (it represents the displacement of the particle from the origin of the coordinate system). And $\vec{\mathbf{r}}_2$ is the position vector at time t_2.

In one dimension, we defined displacement as the *change in position* of the particle. In the more general case of two or three dimensions, the **displacement vector** is defined as the vector representing change in position. We call it $\Delta\vec{\mathbf{r}}$,[†] where

$$\Delta\vec{\mathbf{r}} = \vec{\mathbf{r}}_2 - \vec{\mathbf{r}}_1.$$

This represents the displacement during the time interval $\Delta t = t_2 - t_1$.

[†]We used $\vec{\mathbf{D}}$ for the displacement vector earlier in the Chapter for illustrating vector addition. The new notation here, $\Delta\vec{\mathbf{r}}$, emphasizes that it is the difference between two position vectors.

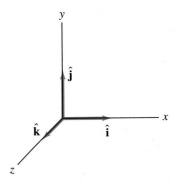

FIGURE 3–15 Unit vectors $\hat{\mathbf{i}}, \hat{\mathbf{j}}$, and $\hat{\mathbf{k}}$ along the x, y, and z axes.

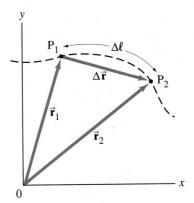

FIGURE 3–16 Path of a particle in the xy plane. At time t_1 the particle is at point P_1 given by the position vector $\vec{\mathbf{r}}_1$; at t_2 the particle is at point P_2 given by the position vector $\vec{\mathbf{r}}_2$. The displacement vector for the time interval $t_2 - t_1$ is $\Delta\vec{\mathbf{r}} = \vec{\mathbf{r}}_2 - \vec{\mathbf{r}}_1$.

(a)

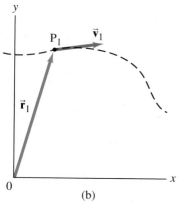

(b)

FIGURE 3–17 (a) As we take Δt and $\Delta\vec{r}$ smaller and smaller [compare to Fig. 3–16] we see that the direction of $\Delta\vec{r}$ and of the instantaneous velocity ($\Delta\vec{r}/\Delta t$, where $\Delta t \to 0$) is (b) tangent to the curve at P_1.

FIGURE 3–18 (a) Velocity vectors $\vec{v}_1$ and $\vec{v}_2$ at instants t_1 and t_2 for a particle at points P_1 and P_2, as in Fig. 3–16. (b) The direction of the average acceleration is in the direction of $\Delta\vec{v} = \vec{v}_2 - \vec{v}_1$.

(a)

(b)

In unit vector notation, we can write

$$\vec{r}_1 = x_1\hat{i} + y_1\hat{j} + z_1\hat{k}, \tag{3–6a}$$

where x_1, y_1, and z_1 are the coordinates of point P_1. Similarly,

$$\vec{r}_2 = x_2\hat{i} + y_2\hat{j} + z_2\hat{k}.$$

Hence

$$\Delta\vec{r} = (x_2 - x_1)\hat{i} + (y_2 - y_1)\hat{j} + (z_2 - z_1)\hat{k}. \tag{3–6b}$$

If the motion is along the x axis only, then $y_2 - y_1 = 0$, $z_2 - z_1 = 0$, and the magnitude of the displacement is $\Delta r = x_2 - x_1$, which is consistent with our earlier one-dimensional equation (Section 2–1). Even in one dimension, displacement is a vector, as are velocity and acceleration.

The **average velocity vector** over the time interval $\Delta t = t_2 - t_1$ is defined as

$$\text{average velocity} = \frac{\Delta\vec{r}}{\Delta t}. \tag{3–7}$$

Now let us consider shorter and shorter time intervals—that is, we let Δt approach zero so that the distance between points P_2 and P_1 also approaches zero, Fig. 3–17. We define the **instantaneous velocity vector** as the limit of the average velocity as Δt approaches zero:

$$\vec{v} = \lim_{\Delta t \to 0} \frac{\Delta\vec{r}}{\Delta t} = \frac{d\vec{r}}{dt}. \tag{3–8}$$

The direction of $\vec{v}$ at any moment is along the line tangent to the path at that moment (Fig. 3–17).

Note that the magnitude of the average velocity in Fig. 3–16 is not equal to the average speed, which is the actual distance traveled along the path, $\Delta\ell$, divided by Δt. In some special cases, the average speed and average velocity are equal (such as motion along a straight line in one direction), but in general they are not. However, in the limit $\Delta t \to 0$, Δr always approaches $\Delta\ell$, so the instantaneous speed *always* equals the magnitude of the instantaneous velocity at any time.

The instantaneous velocity (Eq. 3–8) is equal to the derivative of the position vector with respect to time. Equation 3–8 can be written in terms of components starting with Eq. 3–6a as:

$$\vec{v} = \frac{d\vec{r}}{dt} = \frac{dx}{dt}\hat{i} + \frac{dy}{dt}\hat{j} + \frac{dz}{dt}\hat{k} = v_x\hat{i} + v_y\hat{j} + v_z\hat{k}, \tag{3–9}$$

where $v_x = dx/dt$, $v_y = dy/dt$, $v_z = dz/dt$ are the x, y, and z components of the velocity. Note that $d\hat{i}/dt = d\hat{j}/dt = d\hat{k}/dt = 0$ since these unit vectors are constant in both magnitude and direction.

Acceleration in two or three dimensions is treated in a similar way. The **average acceleration vector**, over a time interval $\Delta t = t_2 - t_1$ is defined as

$$\text{average acceleration} = \frac{\Delta\vec{v}}{\Delta t} = \frac{\vec{v}_2 - \vec{v}_1}{t_2 - t_1}, \tag{3–10}$$

where $\Delta\vec{v}$ is the change in the instantaneous velocity vector during that time interval: $\Delta\vec{v} = \vec{v}_2 - \vec{v}_1$. Note that $\vec{v}_2$ in many cases, such as in Fig. 3–18a, may not be in the same direction as $\vec{v}_1$. Hence the average acceleration vector may be in a different direction from either $\vec{v}_1$ or $\vec{v}_2$ (Fig. 3–18b). Furthermore, $\vec{v}_2$ and $\vec{v}_1$ may have the same magnitude but different directions, and the difference of two such vectors will not be zero. Hence acceleration can result from either a change in the magnitude of the velocity, or from a change in direction of the velocity, or from a change in both.

The **instantaneous acceleration vector** is defined as the limit of the average acceleration vector as the time interval Δt is allowed to approach zero:

$$\vec{a} = \lim_{\Delta \to 0} \frac{\Delta\vec{v}}{\Delta t} = \frac{d\vec{v}}{dt}, \tag{3–11}$$

and is thus the derivative of $\vec{v}$ with respect to t.

We can write $\vec{\mathbf{a}}$ using components:

$$\vec{\mathbf{a}} = \frac{d\vec{\mathbf{v}}}{dt} = \frac{dv_x}{dt}\hat{\mathbf{i}} + \frac{dv_y}{dt}\hat{\mathbf{j}} + \frac{dv_z}{dt}\hat{\mathbf{k}}$$

$$= a_x\hat{\mathbf{i}} + a_y\hat{\mathbf{j}} + a_z\hat{\mathbf{k}}, \tag{3-12}$$

where $a_x = dv_x/dt$, etc. Because $v_x = dx/dt$, then $a_x = dv_x/dt = d^2x/dt^2$, as we saw in Section 2–4. Thus we can also write the acceleration as

$$\vec{\mathbf{a}} = \frac{d^2x}{dt^2}\hat{\mathbf{i}} + \frac{d^2y}{dt^2}\hat{\mathbf{j}} + \frac{d^2z}{dt^2}\hat{\mathbf{k}}. \tag{3-12c}$$

The instantaneous acceleration will be nonzero not only when the magnitude of the velocity changes but also if its direction changes. For example, a person riding in a car traveling at constant speed around a curve, or a child riding on a merry-go-round, will both experience an acceleration because of a change in the direction of the velocity, even though the speed may be constant. (More on this in Chapter 5.)

In general, we will use the terms "velocity" and "acceleration" to mean the instantaneous values. If we want to discuss average values, we will use the word "average."

EXAMPLE 3–5 **Position given as a function of time.** The position of a particle as a function of time is given by

$$\vec{\mathbf{r}} = \left[(5.0\,\text{m/s})t + (6.0\,\text{m/s}^2)t^2\right]\hat{\mathbf{i}} + \left[(7.0\,\text{m}) - (3.0\,\text{m/s}^3)t^3\right]\hat{\mathbf{j}},$$

where r is in meters and t is in seconds. (a) What is the particle's displacement between $t_1 = 2.0\,\text{s}$ and $t_2 = 3.0\,\text{s}$? (b) Determine the particle's instantaneous velocity and acceleration as a function of time. (c) Evaluate $\vec{\mathbf{v}}$ and $\vec{\mathbf{a}}$ at $t = 3.0\,\text{s}$.

APPROACH For (a), we find $\Delta\vec{\mathbf{r}} = \vec{\mathbf{r}}_2 - \vec{\mathbf{r}}_1$, inserting $t_1 = 2.0\,\text{s}$ for finding $\vec{\mathbf{r}}_1$, and $t_2 = 3.0\,\text{s}$ for $\vec{\mathbf{r}}_2$. For (b), we take derivatives (Eqs. 3–9 and 3–11), and for (c) we substitute $t = 3.0\,\text{s}$ into our results in (b).

SOLUTION (a) At $t_1 = 2.0\,\text{s}$,

$$\vec{\mathbf{r}}_1 = \left[(5.0\,\text{m/s})(2.0\,\text{s}) + (6.0\,\text{m/s}^2)(2.0\,\text{s})^2\right]\hat{\mathbf{i}} + \left[(7.0\,\text{m}) - (3.0\,\text{m/s}^3)(2.0\,\text{s})^3\right]\hat{\mathbf{j}}$$

$$= (34\,\text{m})\hat{\mathbf{i}} - (17\,\text{m})\hat{\mathbf{j}}.$$

Similarly, at $t_2 = 3.0\,\text{s}$,

$$\vec{\mathbf{r}}_2 = (15\,\text{m} + 54\,\text{m})\hat{\mathbf{i}} + (7.0\,\text{m} - 81\,\text{m})\hat{\mathbf{j}} = (69\,\text{m})\hat{\mathbf{i}} - (74\,\text{m})\hat{\mathbf{j}}.$$

Thus

$$\Delta\vec{\mathbf{r}} = \vec{\mathbf{r}}_2 - \vec{\mathbf{r}}_1 = (69\,\text{m} - 34\,\text{m})\hat{\mathbf{i}} + (-74\,\text{m} + 17\,\text{m})\hat{\mathbf{j}} = (35\,\text{m})\hat{\mathbf{i}} - (57\,\text{m})\hat{\mathbf{j}}.$$

That is, $\Delta x = 35\,\text{m}$, and $\Delta y = -57\,\text{m}$.

(b) To find velocity, we take the derivative of the given $\vec{\mathbf{r}}$ with respect to time, noting (Appendix B–2) that $d(t^2)/dt = 2t$, and $d(t^3)/dt = 3t^2$:

$$\vec{\mathbf{v}} = \frac{d\vec{\mathbf{r}}}{dt} = \left[5.0\,\text{m/s} + (12\,\text{m/s}^2)t\right]\hat{\mathbf{i}} + \left[0 - (9.0\,\text{m/s}^3)t^2\right]\hat{\mathbf{j}}.$$

The acceleration is (keeping only two significant figures):

$$\vec{\mathbf{a}} = \frac{d\vec{\mathbf{v}}}{dt} = (12\,\text{m/s}^2)\hat{\mathbf{i}} - (18\,\text{m/s}^3)t\,\hat{\mathbf{j}}.$$

Thus $a_x = 12\,\text{m/s}^2$ is constant; but $a_y = -(18\,\text{m/s}^3)t$ depends linearly on time, increasing in magnitude with time in the negative y direction.

(c) We substitute $t = 3.0\,\text{s}$ into the equations we just derived for $\vec{\mathbf{v}}$ and $\vec{\mathbf{a}}$:

$$\vec{\mathbf{v}} = (5.0\,\text{m/s} + 36\,\text{m/s})\hat{\mathbf{i}} - (81\,\text{m/s})\hat{\mathbf{j}} = (41\,\text{m/s})\hat{\mathbf{i}} - (81\,\text{m/s})\hat{\mathbf{j}}$$

$$\vec{\mathbf{a}} = (12\,\text{m/s}^2)\hat{\mathbf{i}} - (54\,\text{m/s}^2)\hat{\mathbf{j}}.$$

Their magnitudes at $t = 3.0\,\text{s}$ are $v = \sqrt{(41\,\text{m/s})^2 + (81\,\text{m/s})^2} = 91\,\text{m/s}$, and $a = \sqrt{(12\,\text{m/s}^2)^2 + (54\,\text{m/s}^2)^2} = 55\,\text{m/s}^2$.

Constant Acceleration

In Chapter 2 we studied the important case of one-dimensional motion for which the acceleration is constant. In two or three dimensions, if the acceleration vector, $\vec{a}$, is constant in magnitude and direction, then $a_x = $ constant, $a_y = $ constant, $a_z = $ constant. The average acceleration in this case is equal to the instantaneous acceleration at any moment. The equations we derived in Chapter 2 for one dimension, Eqs. 2–12a, b, and c, apply separately to each perpendicular component of two- or three-dimensional motion. In two dimensions we let $\vec{v}_0 = v_{x0}\hat{\mathbf{i}} + v_{y0}\hat{\mathbf{j}}$ be the initial velocity, and we apply Eqs. 3–6a, 3–9, and 3–12b for the position vector, $\vec{r}$, velocity, $\vec{v}$, and acceleration, $\vec{a}$. We can then write Eqs. 2–12a, b, and c, for two dimensions as shown in Table 3–1.

TABLE 3–1 Kinematic Equations for Constant Acceleration in 2 Dimensions

x Component (horizontal)		y Component (vertical)
$v_x = v_{x0} + a_x t$	(Eq. 2–12a)	$v_y = v_{y0} + a_y t$
$x = x_0 + v_{x0}t + \frac{1}{2}a_x t^2$	(Eq. 2–12b)	$y = y_0 + v_{y0}t + \frac{1}{2}a_y t^2$
$v_x^2 = v_{x0}^2 + 2a_x(x - x_0)$	(Eq. 2–12c)	$v_y^2 = v_{y0}^2 + 2a_y(y - y_0)$

The first two of the equations in Table 3–1 can be written more formally in vector notation.

$$\vec{v} = \vec{v}_0 + \vec{a}t \qquad [\vec{a} = \text{constant}] \quad \textbf{(3–13a)}$$
$$\vec{r} = \vec{r}_0 + \vec{v}_0 t + \frac{1}{2}\vec{a}t^2. \qquad [\vec{a} = \text{constant}] \quad \textbf{(3–13b)}$$

Here, $\vec{r}$ is the position vector at any time, and $\vec{r}_0$ is the position vector at $t = 0$. These equations are the vector equivalent of Eqs. 2–12a and b. In practical situations, we usually use the component form given in Table 3–1.

3–7 Projectile Motion

In Chapter 2, we studied one-dimensional motion of an object in terms of displacement, velocity, and acceleration, including purely vertical motion of a falling object undergoing acceleration due to gravity. Now we examine the more general translational motion of objects moving through the air in two dimensions near the Earth's surface, such as a golf ball, a thrown or batted baseball, kicked footballs, and speeding bullets. These are all examples of **projectile motion** (see Fig. 3–19), which we can describe as taking place in two dimensions.

Although air resistance is often important, in many cases its effect can be ignored, and we will ignore it in the following analysis. We will not be concerned now with the process by which the object is thrown or projected. We consider only its motion *after* it has been projected, and *before* it lands or is caught—that is, we analyze our projected object only when it is moving freely through the air under the action of gravity alone. Then the acceleration of the object is that due to gravity, which acts downward with magnitude $g = 9.80 \text{ m/s}^2$, and we assume it is constant.[†]

Galileo was the first to describe projectile motion accurately. He showed that it could be understood by analyzing the horizontal and vertical components of the motion separately. For convenience, we assume that the motion begins at time $t = 0$ at the origin of an xy coordinate system (so $x_0 = y_0 = 0$).

Let us look at a (tiny) ball rolling off the end of a horizontal table with an initial velocity in the horizontal (x) direction, v_{x0}. See Fig. 3–20, where an object falling vertically is also shown for comparison. The velocity vector $\vec{v}$ at each instant points in the direction of the ball's motion at that instant and is always tangent to the path. Following Galileo's ideas, we treat the horizontal and vertical components of the velocity, v_x and v_y, separately, and we can apply the kinematic equations (Eqs. 2–12a through 2–12c) to the x and y components of the motion.

First we examine the vertical (y) component of the motion. At the instant the ball leaves the table's top ($t = 0$), it has only an x component of velocity. Once the

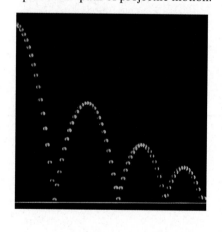

FIGURE 3–19 This strobe photograph of a ball making a series of bounces shows the characteristic "parabolic" path of projectile motion.

[†]This restricts us to objects whose distance traveled and maximum height above the Earth are small compared to the Earth's radius (6400 km).

FIGURE 3–20 Projectile motion of a small ball projected horizontally. The dashed black line represents the path of the object. The velocity vector $\vec{v}$ at each point is in the direction of motion and thus is tangent to the path. The velocity vectors are green arrows, and velocity components are dashed. (A vertically falling object starting at the same point is shown at the left for comparison; v_y is the same for the falling object and the projectile.)

ball leaves the table (at $t = 0$), it experiences a vertically downward acceleration g, the acceleration due to gravity. Thus v_y is initially zero $(v_{y0} = 0)$ but increases continually in the downward direction (until the ball hits the ground). Let us take y to be positive upward. Then $a_y = -g$, and from Eq. 2–12a we can write $v_y = -gt$ since we set $v_{y0} = 0$. The vertical displacement is given by $y = -\frac{1}{2}gt^2$.

In the horizontal direction, on the other hand, the acceleration is zero (we are ignoring air resistance). With $a_x = 0$, the horizontal component of velocity, v_x, remains constant, equal to its initial value, v_{x0}, and thus has the same magnitude at each point on the path. The horizontal displacement is then given by $x = v_{x0}t$. The two vector components, $\vec{v}_x$ and $\vec{v}_y$, can be added vectorially at any instant to obtain the velocity $\vec{v}$ at that time (that is, for each point on the path), as shown in Fig. 3–20.

One result of this analysis, which Galileo himself predicted, is that *an object projected horizontally will reach the ground in the same time as an object dropped vertically*. This is because the vertical motions are the same in both cases, as shown in Fig. 3–20. Figure 3–21 is a multiple-exposure photograph of an experiment that confirms this.

FIGURE 3–21 Multiple-exposure photograph showing positions of two balls at equal time intervals. One ball was dropped from rest at the same time the other was projected horizontally outward. The vertical position of each ball is seen to be the same at each instant.

EXERCISE D Return to the Chapter-Opening Question, page 51, and answer it again now. Try to explain why you may have answered differently the first time.

If an object is projected at an upward angle, as in Fig. 3–22, the analysis is similar, except that now there is an initial vertical component of velocity, v_{y0}. Because of the downward acceleration of gravity, the upward component of velocity v_y gradually decreases with time until the object reaches the highest point on its path, at which point $v_y = 0$. Subsequently the object moves downward (Fig. 3–22) and v_y increases in the downward direction, as shown (that is, becoming more negative). As before, v_x remains constant.

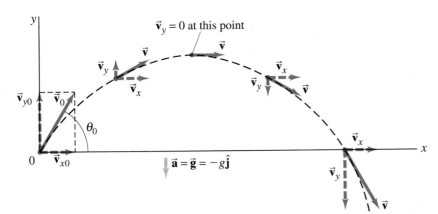

FIGURE 3–22 Path of a projectile fired with initial velocity $\vec{v}_0$ at angle θ_0 to the horizontal. Path is shown dashed in black, the velocity vectors are green arrows, and velocity components are dashed. The acceleration $\vec{a} = d\vec{v}/dt$ is downward. That is, $\vec{a} = \vec{g} = -g\hat{j}$ where $\hat{j}$ is the unit vector in the positive y direction.

3–8 Solving Problems Involving Projectile Motion

We now work through several Examples of projectile motion quantitatively.

We can simplify Eqs. 2–12 (Table 3–1) for the case of projectile motion because we can set $a_x = 0$. See Table 3–2, which assumes y is positive upward, so $a_y = -g = -9.80 \text{ m/s}^2$. Note that if θ is chosen relative to the $+x$ axis, as in Fig. 3–22, then

$$v_{x0} = v_0 \cos \theta_0,$$

$$v_{y0} = v_0 \sin \theta_0.$$

In doing problems involving projectile motion, we must consider a time interval for which our chosen object is in the air, influenced only by gravity. We do not consider the throwing (or projecting) process, nor the time after the object lands or is caught, because then other influences act on the object, and we can no longer set $\vec{\mathbf{a}} = \vec{\mathbf{g}}$.

TABLE 3–2 Kinematic Equations for Projectile Motion
(y positive upward; $a_x = 0$, $a_y = -g = -9.80 \text{ m/s}^2$)

Horizontal Motion ($a_x = 0, v_x = \text{constant}$)		Vertical Motion† ($a_y = -g = \text{constant}$)
$v_x = v_{x0}$	(Eq. 2–12a)	$v_y = v_{y0} - gt$
$x = x_0 + v_{x0}t$	(Eq. 2–12b)	$y = y_0 + v_{y0}t - \frac{1}{2}gt^2$
	(Eq. 2–12c)	$v_y^2 = v_{y0}^2 - 2g(y - y_0)$

† If y is taken positive downward, the minus $(-)$ signs in front of g become plus $(+)$ signs.

PROBLEM SOLVING

Projectile Motion

Our approach to solving problems in Section 2–6 also applies here. Solving problems involving projectile motion can require creativity, and cannot be done just by following some rules. Certainly you must avoid just plugging numbers into equations that seem to "work."

1. As always, **read** carefully; **choose** the **object** (or objects) you are going to analyze.
2. **Draw** a careful **diagram** showing what is happening to the object.
3. **Choose** an origin and an xy **coordinate system**.
4. Decide on the **time interval**, which for projectile motion can only include motion under the effect of gravity alone, not throwing or landing. The time interval must be the same for the x and y analyses.

The x and y motions are connected by the common time.

5. **Examine** the horizontal (x) and vertical (y) **motions** separately. If you are given the initial velocity, you may want to resolve it into its x and y components.
6. List the **known** and **unknown** quantities, choosing $a_x = 0$ and $a_y = -g$ or $+g$, where $g = 9.80 \text{ m/s}^2$, and using the $+$ or $-$ sign, depending on whether you choose y positive down or up. Remember that v_x never changes throughout the trajectory, and that $v_y = 0$ at the highest point of any trajectory that returns downward. The velocity just before landing is generally not zero.
7. Think for a minute before jumping into the equations. A little planning goes a long way. **Apply** the **relevant equations** (Table 3–2), combining equations if necessary. You may need to combine components of a vector to get magnitude and direction (Eqs. 3–3).

EXAMPLE 3–6 **Driving off a cliff.** A movie stunt driver on a motorcycle speeds horizontally off a 50.0-m-high cliff. How fast must the motorcycle leave the cliff top to land on level ground below, 90.0 m from the base of the cliff where the cameras are? Ignore air resistance.

APPROACH We explicitly follow the steps of the Problem Solving Strategy above.

SOLUTION

1. and 2. **Read, choose the object, and draw a diagram**. Our object is the motorcycle and driver, taken as a single unit. The diagram is shown in Fig. 3–23.

3. **Choose a coordinate system**. We choose the y direction to be positive upward, with the top of the cliff as $y_0 = 0$. The x direction is horizontal with $x_0 = 0$ at the point where the motorcycle leaves the cliff.

4. **Choose a time interval**. We choose our time interval to begin $(t = 0)$ just as the motorcycle leaves the cliff top at position $x_0 = 0$, $y_0 = 0$; our time interval ends just before the motorcycle hits the ground below.

5. **Examine x and y motions**. In the horizontal (x) direction, the acceleration $a_x = 0$, so the velocity is constant. The value of x when the motorcycle reaches the ground is $x = +90.0$ m. In the vertical direction, the acceleration is the acceleration due to gravity, $a_y = -g = -9.80$ m/s^2. The value of y when the motorcycle reaches the ground is $y = -50.0$ m. The initial velocity is horizontal and is our unknown, v_{x0}; the initial vertical velocity is zero, $v_{y0} = 0$.

6. **List knowns and unknowns**. See the Table in the margin. Note that in addition to not knowing the initial horizontal velocity v_{x0} (which stays constant until landing), we also do not know the time t when the motorcycle reaches the ground.

7. **Apply relevant equations**. The motorcycle maintains constant v_x as long as it is in the air. The time it stays in the air is determined by the y motion—when it hits the ground. So we first find the time using the y motion, and then use this time value in the x equations. To find out how long it takes the motorcycle to reach the ground below, we use Eq. 2–12b (Table 3–2) for the vertical (y) direction with $y_0 = 0$ and $v_{y0} = 0$:

$$y = y_0 + v_{y0}t + \tfrac{1}{2}a_y t^2$$
$$= 0 + 0 + \tfrac{1}{2}(-g)t^2$$

or

$$y = -\tfrac{1}{2}gt^2.$$

We solve for t and set $y = -50.0$ m:

$$t = \sqrt{\frac{2y}{-g}} = \sqrt{\frac{2(-50.0\ \text{m})}{-9.80\ \text{m/s}^2}} = 3.19\ \text{s}.$$

To calculate the initial velocity, v_{x0}, we again use Eq. 2–12b, but this time for the horizontal (x) direction, with $a_x = 0$ and $x_0 = 0$:

$$x = x_0 + v_{x0}t + \tfrac{1}{2}a_x t^2$$
$$= 0 + v_{x0}t + 0$$

or

$$x = v_{x0}t.$$

Then

$$v_{x0} = \frac{x}{t} = \frac{90.0\ \text{m}}{3.19\ \text{s}} = 28.2\ \text{m/s},$$

which is about 100 km/h (roughly 60 mi/h).

NOTE In the time interval of the projectile motion, the only acceleration is g in the negative y direction. The acceleration in the x direction is zero.

FIGURE 3–23 Example 3–6.

Known	Unknown
$x_0 = y_0 = 0$	v_{x0}
$x = 90.0$ m	t
$y = -50.0$ m	
$a_x = 0$	
$a_y = -g = -9.80$ m/s^2	
$v_{y0} = 0$	

FIGURE 3–24 Example 3–7.

$\vec{\mathbf{v}}_y = 0$ at this point

$37.0°$

$\vec{\mathbf{a}} = \vec{\mathbf{g}} = -g\hat{\mathbf{j}}$

PHYSICS APPLIED

Sports

EXAMPLE 3–7 **A kicked football.** A football is kicked at an angle $\theta_0 = 37.0°$ with a velocity of 20.0 m/s, as shown in Fig. 3–24. Calculate (a) the maximum height, (b) the time of travel before the football hits the ground, (c) how far away it hits the ground, (d) the velocity vector at the maximum height, and (e) the acceleration vector at maximum height. Assume the ball leaves the foot at ground level, and ignore air resistance and rotation of the ball.

APPROACH This may seem difficult at first because there are so many questions. But we can deal with them one at a time. We take the y direction as positive upward, and treat the x and y motions separately. The total time in the air is again determined by the y motion. The x motion occurs at constant velocity. The y component of velocity varies, being positive (upward) initially, decreasing to zero at the highest point, and then becoming negative as the football falls.

SOLUTION We resolve the initial velocity into its components (Fig. 3–24):

$$v_{x0} = v_0 \cos 37.0° = (20.0 \text{ m/s})(0.799) = 16.0 \text{ m/s}$$
$$v_{y0} = v_0 \sin 37.0° = (20.0 \text{ m/s})(0.602) = 12.0 \text{ m/s}.$$

(a) We consider a time interval that begins just after the football loses contact with the foot until it reaches its maximum height. During this time interval, the acceleration is g downward. At the maximum height, the velocity is horizontal (Fig. 3–24), so $v_y = 0$; and this occurs at a time given by $v_y = v_{y0} - gt$ with $v_y = 0$ (see Eq. 2–12a in Table 3–2). Thus

$$t = \frac{v_{y0}}{g} = \frac{(12.0 \text{ m/s})}{(9.80 \text{ m/s}^2)} = 1.224 \text{ s} \approx 1.22 \text{ s}.$$

From Eq. 2–12b, with $y_0 = 0$, we have

$$y = v_{y0}t - \tfrac{1}{2}gt^2$$
$$= (12.0 \text{ m/s})(1.224 \text{ s}) - \tfrac{1}{2}(9.80 \text{ m/s}^2)(1.224 \text{ s})^2 = 7.35 \text{ m}.$$

Alternatively, we could have used Eq. 2–12c, solved for y, and found

$$y = \frac{v_{y0}^2 - v_y^2}{2g} = \frac{(12.0 \text{ m/s})^2 - (0 \text{ m/s})^2}{2(9.80 \text{ m/s}^2)} = 7.35 \text{ m}.$$

The maximum height is 7.35 m.

(b) To find the time it takes for the ball to return to the ground, we consider a different time interval, starting at the moment the ball leaves the foot ($t = 0$, $y_0 = 0$) and ending just before the ball touches the ground ($y = 0$ again). We can use Eq. 2–12b with $y_0 = 0$ and also set $y = 0$ (ground level):

$$y = y_0 + v_{y0}t - \tfrac{1}{2}gt^2$$
$$0 = 0 + v_{y0}t - \tfrac{1}{2}gt^2.$$

This equation can be easily factored:

$$t\left(\tfrac{1}{2}gt - v_{y0}\right) = 0.$$

There are two solutions, $t = 0$ (which corresponds to the initial point, y_0), and

$$t = \frac{2v_{y0}}{g} = \frac{2(12.0 \text{ m/s})}{(9.80 \text{ m/s}^2)} = 2.45 \text{ s},$$

which is the total travel time of the football.

NOTE The time needed for the whole trip, $t = 2v_{y0}/g = 2.45\,\text{s}$, is double the time to reach the highest point, calculated in (a). That is, the time to go up equals the time to come back down to the same level (ignoring air resistance).

(c) The total distance traveled in the x direction is found by applying Eq. 2–12b with $x_0 = 0$, $a_x = 0$, $v_{x0} = 16.0\,\text{m/s}$:

$$x = v_{x0}t = (16.0\,\text{m/s})(2.45\,\text{s}) = 39.2\,\text{m}.$$

(d) At the highest point, there is no vertical component to the velocity. There is only the horizontal component (which remains constant throughout the flight), so $v = v_{x0} = v_0 \cos 37.0° = 16.0\,\text{m/s}$.

(e) The acceleration vector is the same at the highest point as it is throughout the flight, which is $9.80\,\text{m/s}^2$ downward.

NOTE We treated the football as if it were a particle, ignoring its rotation. We also ignored air resistance. Because air resistance is significant on a football, our results are only estimates.

EXERCISE E Two balls are thrown in the air at different angles, but each reaches the same height. Which ball remains in the air longer: the one thrown at the steeper angle or the one thrown at a shallower angle?

CONCEPTUAL EXAMPLE 3–8 **Where does the apple land?** A child sits upright in a wagon which is moving to the right at constant speed as shown in Fig. 3–25. The child extends her hand and throws an apple straight upward (from her own point of view, Fig. 3–25a), while the wagon continues to travel forward at constant speed. If air resistance is neglected, will the apple land (a) behind the wagon, (b) in the wagon, or (c) in front of the wagon?

RESPONSE The child throws the apple straight up from her own reference frame with initial velocity $\vec{v}_{y0}$ (Fig. 3–25a). But when viewed by someone on the ground, the apple also has an initial horizontal component of velocity equal to the speed of the wagon, $\vec{v}_{x0}$. Thus, to a person on the ground, the apple will follow the path of a projectile as shown in Fig. 3–25b. The apple experiences no horizontal acceleration, so $\vec{v}_{x0}$ will stay constant and equal to the speed of the wagon. As the apple follows its arc, the wagon will be directly under the apple at all times because they have the same horizontal velocity. When the apple comes down, it will drop right into the outstretched hand of the child. The answer is (b).

(a) Wagon reference frame

(b) Ground reference frame

FIGURE 3–25 Example 3–8.

CONCEPTUAL EXAMPLE 3–9 **The wrong strategy.** A boy on a small hill aims his water-balloon slingshot horizontally, straight at a second boy hanging from a tree branch a distance d away, Fig. 3–26. At the instant the water balloon is released, the second boy lets go and falls from the tree, hoping to avoid being hit. Show that he made the wrong move. (He hadn't studied physics yet.) Ignore air resistance.

RESPONSE Both the water balloon and the boy in the tree start falling at the same instant, and in a time t they each fall the same vertical distance $y = \frac{1}{2}gt^2$, much like Fig. 3–21. In the time it takes the water balloon to travel the horizontal distance d, the balloon will have the same y position as the falling boy. Splat. If the boy had stayed in the tree, he would have avoided the humiliation.

FIGURE 3–26 Example 3–9.

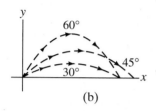

FIGURE 3–27 Example 3–10.
(a) The range R of a projectile;
(b) there are generally two angles θ_0 that will give the same range. Can you show that if one angle is θ_{01}, the other is $\theta_{02} = 90° - \theta_{01}$?

EXAMPLE 3–10 **Level horizontal range.** (a) Derive a formula for the horizontal range R of a projectile in terms of its initial speed v_0 and angle θ_0. The horizontal *range* is defined as the horizontal distance the projectile travels before returning to its original height (which is typically the ground); that is, y (final) $= y_0$. See Fig. 3–27a. (b) Suppose one of Napoleon's cannons had a muzzle speed, v_0, of 60.0 m/s. At what angle should it have been aimed (ignore air resistance) to strike a target 320 m away?

APPROACH The situation is the same as in Example 3–7, except we are now not given numbers in (a). We will algebraically manipulate equations to obtain our result.

SOLUTION (a) We set $x_0 = 0$ and $y_0 = 0$ at $t = 0$. After the projectile travels a horizontal distance R, it returns to the same level, $y = 0$, the final point. We choose our time interval to start $(t = 0)$ just after the projectile is fired and to end when it returns to the same vertical height. To find a general expression for R, we set both $y = 0$ and $y_0 = 0$ in Eq. 2–12b for the vertical motion, and obtain

$$y = y_0 + v_{y0}t + \tfrac{1}{2}a_y t^2$$

so

$$0 = 0 + v_{y0}t - \tfrac{1}{2}gt^2.$$

We solve for t, which gives two solutions: $t = 0$ and $t = 2v_{y0}/g$. The first solution corresponds to the initial instant of projection and the second is the time when the projectile returns to $y = 0$. Then the range, R, will be equal to x at the moment t has this value, which we put into Eq. 2–12b for the *horizontal* motion $(x = v_{x0}t$, with $x_0 = 0)$. Thus we have:

$$R = v_{x0}t = v_{x0}\left(\frac{2v_{y0}}{g}\right) = \frac{2v_{x0}v_{y0}}{g} = \frac{2v_0^2 \sin\theta_0 \cos\theta_0}{g}, \qquad [y = y_0]$$

where we have written $v_{x0} = v_0 \cos\theta_0$ and $v_{y0} = v_0 \sin\theta_0$. This is the result we sought. It can be rewritten, using the trigonometric identity $2\sin\theta\cos\theta = \sin 2\theta$ (Appendix A or inside the rear cover):

$$R = \frac{v_0^2 \sin 2\theta_0}{g}. \qquad [\text{only if } y \text{ (final)} = y_0]$$

We see that the maximum range, for a given initial velocity v_0, is obtained when $\sin 2\theta$ takes on its maximum value of 1.0, which occurs for $2\theta_0 = 90°$; so

$$\theta_0 = 45° \text{ for maximum range, and } R_{\max} = v_0^2/g.$$

[When air resistance is important, the range is less for a given v_0, and the maximum range is obtained at an angle smaller than 45°.]

NOTE The maximum range increases by the square of v_0, so doubling the muzzle velocity of a cannon increases its maximum range by a factor of 4.

(b) We put $R = 320$ m into the equation we just derived, and (assuming, unrealistically, no air resistance) we solve it to find

$$\sin 2\theta_0 = \frac{Rg}{v_0^2} = \frac{(320 \text{ m})(9.80 \text{ m/s}^2)}{(60.0 \text{ m/s})^2} = 0.871.$$

We want to solve for an angle θ_0 that is between 0° and 90°, which means $2\theta_0$ in this equation can be as large as 180°. Thus, $2\theta_0 = 60.6°$ is a solution, but $2\theta_0 = 180° - 60.6° = 119.4°$ is also a solution (see Appendix A–9). In general we will have two solutions (see Fig. 3–27b), which in the present case are given by

$$\theta_0 = 30.3° \text{ or } 59.7°.$$

Either angle gives the same range. Only when $\sin 2\theta_0 = 1$ (so $\theta_0 = 45°$) is there a single solution (that is, both solutions are the same).

EXERCISE F The maximum range of a projectile is found to be 100 m. If the projectile strikes the ground a distance of 82 m away, what was the angle of launch? (a) 35° or 55°; (b) 30° or 60°; (c) 27.5° or 72.5°; (d) 13.75° or 76.25°.

The level range formula derived in Example 3–10 applies only if takeoff and landing are at the same height $(y = y_0)$. Example 3–11 below considers a case where they are not equal heights $(y \neq y_0)$.

EXAMPLE 3–11   A punt. Suppose the football in Example 3–7 was punted and left the punter's foot at a height of 1.00 m above the ground. How far did the football travel before hitting the ground? Set $x_0 = 0$, $y_0 = 0$.

 PHYSICS APPLIED
Sports

PROBLEM SOLVING
Do not use any formula unless you are sure its range of validity fits the problem; the range formula does not apply here because $y \neq y_0$

APPROACH The x and y motions are again treated separately. But we cannot use the range formula from Example 3–10 because it is valid only if y (final) $= y_0$, which is not the case here. Now we have $y_0 = 0$, and the football hits the ground where $y = -1.00$ m (see Fig. 3–28). We choose our time interval to start when the ball leaves his foot ($t = 0$, $y_0 = 0$, $x_0 = 0$) and end just before the ball hits the ground ($y = -1.00$ m). We can get x from Eq. 2–12b, $x = v_{x0}t$, since we know that $v_{x0} = 16.0$ m/s from Example 3–7. But first we must find t, the time at which the ball hits the ground, which we obtain from the y motion.

FIGURE 3–28 Example 3–11: the football leaves the punter's foot at $y = 0$, and reaches the ground where $y = -1.00$ m.

SOLUTION With $y = -1.00$ m and $v_{y0} = 12.0$ m/s (see Example 3–7), we use the equation

$$y = y_0 + v_{y0}t - \tfrac{1}{2}gt^2,$$

and obtain

$$-1.00 \text{ m} = 0 + (12.0 \text{ m/s})t - (4.90 \text{ m/s}^2)t^2.$$

We rearrange this equation into standard form $(ax^2 + bx + c = 0)$ so we can use the quadratic formula:

$$(4.90 \text{ m/s}^2)t^2 - (12.0 \text{ m/s})t - (1.00 \text{ m}) = 0.$$

The quadratic formula (Appendix A–1) gives

$$t = \frac{12.0 \text{ m/s} \pm \sqrt{(-12.0 \text{ m/s})^2 - 4(4.90 \text{ m/s}^2)(-1.00 \text{ m})}}{2(4.90 \text{ m/s}^2)}$$

$$= 2.53 \text{ s} \quad \text{or} \quad -0.081 \text{ s}.$$

The second solution would correspond to a time prior to our chosen time interval that begins at the kick, so it doesn't apply. With $t = 2.53$ s for the time at which the ball touches the ground, the horizontal distance the ball traveled is (using $v_{x0} = 16.0$ m/s from Example 3–7):

$$x = v_{x0}t = (16.0 \text{ m/s})(2.53 \text{ s}) = 40.5 \text{ m}.$$

Our assumption in Example 3–7 that the ball leaves the foot at ground level would result in an underestimate of about 1.3 m in the distance our punt traveled.

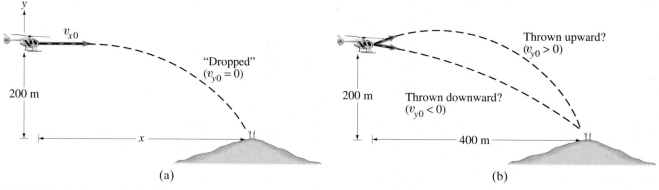

FIGURE 3–29 Example 3–12.

PHYSICS APPLIED

Reaching a target from a moving helicopter

EXAMPLE 3–12 **Rescue helicopter drops supplies.** A rescue helicopter wants to drop a package of supplies to isolated mountain climbers on a rocky ridge 200 m below. If the helicopter is traveling horizontally with a speed of 70 m/s (250 km/h), (a) how far in advance of the recipients (horizontal distance) must the package be dropped (Fig. 3–29a)? (b) Suppose, instead, that the helicopter releases the package a horizontal distance of 400 m in advance of the mountain climbers. What vertical velocity should the package be given (up or down) so that it arrives precisely at the climbers' position (Fig. 3–29b)? (c) With what speed does the package land in the latter case?

APPROACH We choose the origin of our xy coordinate system at the initial position of the helicopter, taking $+y$ upward, and use the kinematic equations (Table 3–2).

SOLUTION (a) We can find the time to reach the climbers using the vertical distance of 200 m. The package is "dropped" so initially it has the velocity of the helicopter, $v_{x0} = 70$ m/s, $v_{y0} = 0$. Then, since $y = -\frac{1}{2}gt^2$, we have

$$t = \sqrt{\frac{-2y}{g}} = \sqrt{\frac{-2(-200\text{ m})}{9.80\text{ m/s}^2}} = 6.39\text{ s}.$$

The horizontal motion of the falling package is at constant speed of 70 m/s. So

$$x = v_{x0}t = (70\text{ m/s})(6.39\text{ s}) = 447\text{ m} \approx 450\text{ m},$$

assuming the given numbers were good to two significant figures.

(b) We are given $x = 400$ m, $v_{x0} = 70$ m/s, $y = -200$ m, and we want to find v_{y0} (see Fig. 3–29b). Like most problems, this one can be approached in various ways. Instead of searching for a formula or two, let's try to reason it out in a simple way, based on what we did in part (a). If we know t, perhaps we can get v_{y0}. Since the horizontal motion of the package is at constant speed (once it is released we don't care what the helicopter does), we have $x = v_{x0}t$, so

$$t = \frac{x}{v_{x0}} = \frac{400\text{ m}}{70\text{ m/s}} = 5.71\text{ s}.$$

Now let's try to use the vertical motion to get v_{y0}: $y = y_0 + v_{y0}t - \frac{1}{2}gt^2$. Since $y_0 = 0$ and $y = -200$ m, we can solve for v_{y0}:

$$v_{y0} = \frac{y + \frac{1}{2}gt^2}{t} = \frac{-200\text{ m} + \frac{1}{2}(9.80\text{ m/s}^2)(5.71\text{ s})^2}{5.71\text{ s}} = -7.0\text{ m/s}.$$

Thus, in order to arrive at precisely the mountain climbers' position, the package must be thrown *downward* from the helicopter with a speed of 7.0 m/s.

(c) We want to know v of the package at $t = 5.71$ s. The components are:

$$v_x = v_{x0} = 70\text{ m/s}$$
$$v_y = v_{y0} - gt = -7.0\text{ m/s} - (9.80\text{ m/s}^2)(5.71\text{ s}) = -63\text{ m/s}.$$

So $v = \sqrt{(70\text{ m/s})^2 + (-63\text{ m/s})^2} = 94$ m/s. (Better not to release the package from such an altitude, or use a parachute.)

Projectile Motion Is Parabolic

We now show that the path followed by any projectile is a *parabola*, if we can ignore air resistance and can assume that $\vec{\mathbf{g}}$ is constant. To do so, we need to find y as a function of x by eliminating t between the two equations for horizontal and vertical motion (Eq. 2–12b in Table 3–2), and for simplicity we set $x_0 = y_0 = 0$:

$$x = v_{x0}t$$

$$y = v_{y0}t - \tfrac{1}{2}gt^2$$

From the first equation, we have $t = x/v_{x0}$, and we substitute this into the second one to obtain

$$y = \left(\frac{v_{y0}}{v_{x0}}\right)x - \left(\frac{g}{2v_{x0}^2}\right)x^2. \qquad \textbf{(3–14)}$$

We see that y as a function of x has the form

$$y = Ax - Bx^2,$$

where A and B are constants for any specific projectile motion. This is the well-known equation for a parabola. See Figs. 3–19 and 3–30.

The idea that projectile motion is parabolic was, in Galileo's day, at the forefront of physics research. Today we discuss it in Chapter 3 of introductory physics!

FIGURE 3–30 Examples of projectile motion—sparks (small hot glowing pieces of metal), water, and fireworks. The parabolic path characteristic of projectile motion is affected by air resistance.

3–9 Relative Velocity

We now consider how observations made in different frames of reference are related to each other. For example, consider two trains approaching one another, each with a speed of 80 km/h with respect to the Earth. Observers on the Earth beside the train tracks will measure 80 km/hr for the speed of each of the trains. Observers on either one of the trains (a different frame of reference) will measure a speed of 160 km/h for the train approaching them.

Similarly, when one car traveling 90 km/h passes a second car traveling in the same direction at 75 km/h, the first car has a speed relative to the second car of 90 km/h − 75 km/h = 15 km/h.

When the velocities are along the same line, simple addition or subtraction is sufficient to obtain the relative velocity. But if they are not along the same line, we must make use of vector addition. We emphasize, as mentioned in Section 2–1, that when specifying a velocity, it is important to specify what the reference frame is.

FIGURE 3–31 To move directly across the river, the boat must head upstream at an angle θ. Velocity vectors are shown as green arrows:

$\vec{\mathbf{v}}_{BS}$ = velocity of **B**oat with respect to the **S**hore,

$\vec{\mathbf{v}}_{BW}$ = velocity of **B**oat with respect to the **W**ater,

$\vec{\mathbf{v}}_{WS}$ = velocity of the **W**ater with respect to the **S**hore (river current).

When determining relative velocity, it is easy to make a mistake by adding or subtracting the wrong velocities. It is important, therefore, to draw a diagram and use a careful labeling process. Each velocity is labeled by *two subscripts: the first refers to the object, the second to the reference frame in which it has this velocity.* For example, suppose a boat is to cross a river to the opposite side, as shown in Fig. 3–31. We let $\vec{\mathbf{v}}_{BW}$ be the velocity of the **B**oat with respect to the **W**ater. (This is also what the boat's velocity would be relative to the shore if the water were still.) Similarly, $\vec{\mathbf{v}}_{BS}$ is the velocity of the **B**oat with respect to the **S**hore, and $\vec{\mathbf{v}}_{WS}$ is the velocity of the **W**ater with respect to the **S**hore (this is the river current). Note that $\vec{\mathbf{v}}_{BW}$ is what the boat's motor produces (against the water), whereas $\vec{\mathbf{v}}_{BS}$ is equal to $\vec{\mathbf{v}}_{BW}$ plus the effect of the current, $\vec{\mathbf{v}}_{WS}$. Therefore, the velocity of the boat relative to the shore is (see vector diagram, Fig. 3–31)

$$\vec{\mathbf{v}}_{BS} = \vec{\mathbf{v}}_{BW} + \vec{\mathbf{v}}_{WS}. \tag{3–15}$$

By writing the subscripts using this convention, we see that the inner subscripts (the two W's) on the right-hand side of Eq. 3–15 are the same, whereas the outer subscripts on the right of Eq. 3–15 (the B and the S) are the same as the two subscripts for the sum vector on the left, $\vec{\mathbf{v}}_{BS}$. By following this convention (first subscript for the object, second for the reference frame), you can write down the correct equation relating velocities in different reference frames.† Figure 3–32 gives a derivation of Eq. 3–15.

Equation 3–15 is valid in general and can be extended to three or more velocities. For example, if a fisherman on the boat walks with a velocity $\vec{\mathbf{v}}_{FB}$ relative to the boat, his velocity relative to the shore is $\vec{\mathbf{v}}_{FS} = \vec{\mathbf{v}}_{FB} + \vec{\mathbf{v}}_{BW} + \vec{\mathbf{v}}_{WS}$. The equations involving relative velocity will be correct when adjacent inner subscripts are identical and when the outermost ones correspond exactly to the two on the velocity on the left of the equation. But this works only with plus signs (on the right), not minus signs.

It is often useful to remember that for any two objects or reference frames, A and B, the velocity of A relative to B has the same magnitude, but opposite direction, as the velocity of B relative to A:

$$\vec{\mathbf{v}}_{BA} = -\vec{\mathbf{v}}_{AB}. \tag{3–16}$$

For example, if a train is traveling 100 km/h relative to the Earth in a certain direction, objects on the Earth (such as trees) appear to an observer on the train to be traveling 100 km/h in the opposite direction.

†We thus would know by inspection that (for example) the equation $\vec{\mathbf{V}}_{BW} = \vec{\mathbf{V}}_{BS} + \vec{\mathbf{V}}_{WS}$ is wrong.

FIGURE 3–32 Derivation of relative velocity equation (Eq. 3–15), in this case for a person walking along the corridor in a train. We are looking down on the train and two reference frames are shown: xy on the Earth and $x'y'$ fixed on the train. We have:

$\vec{\mathbf{r}}_{PT}$ = position vector of person (P) relative to train (T),

$\vec{\mathbf{r}}_{PE}$ = position vector of person (P) relative to Earth (E),

$\vec{\mathbf{r}}_{TE}$ = position vector of train's coordinate system (T) relative to Earth (E).

From the diagram we see that

$$\vec{\mathbf{r}}_{PE} = \vec{\mathbf{r}}_{PT} + \vec{\mathbf{r}}_{TE}.$$

We take the derivative with respect to time to obtain

$$\frac{d}{dt}(\vec{\mathbf{r}}_{PE}) = \frac{d}{dt}(\vec{\mathbf{r}}_{PT}) + \frac{d}{dt}(\vec{\mathbf{r}}_{TE}).$$

or, since $d\vec{\mathbf{r}}/dt = \vec{\mathbf{v}}$,

$$\vec{\mathbf{v}}_{PE} = \vec{\mathbf{v}}_{PT} + \vec{\mathbf{v}}_{TE}.$$

This is the equivalent of Eq. 3–15 for the present situation (check the subscripts!).

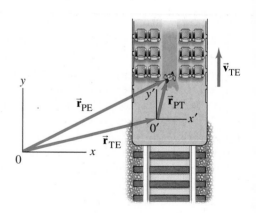

CONCEPTUAL EXAMPLE 3–13 **Crossing a river.** A woman in a small motor boat is trying to cross a river that flows due west with a strong current. The woman starts on the south bank and is trying to reach the north bank directly north from her starting point. Should she (a) head due north, (b) head due west, (c) head in a north-westerly direction, (d) head in a northeasterly direction?

RESPONSE If the woman heads straight across the river, the current will drag the boat downstream (westward). To overcome the river's westward current, the boat must acquire an eastward component of velocity as well as a northward compo-nent. Thus the boat must (d) head in a northeasterly direction (see Fig. 3–33). The actual angle depends on the strength of the current and how fast the boat moves relative to the water. If the current is weak and the motor is strong, then the boat can head almost, but not quite, due north.

EXAMPLE 3–14 **Heading upstream.** A boat's speed in still water is $v_{BW} = 1.85$ m/s. If the boat is to travel directly across a river whose current has speed $v_{WS} = 1.20$ m/s, at what upstream angle must the boat head? (See Fig. 3–33.)

APPROACH We reason as in Example 3–13, and use subscripts as in Eq. 3–15. Figure 3–33 has been drawn with $\vec{v}_{BS}$, the velocity of the **B**oat relative to the **S**hore, pointing directly across the river because this is how the boat is supposed to move. (Note that $\vec{v}_{BS} = \vec{v}_{BW} + \vec{v}_{WS}$.) To accomplish this, the boat needs to head upstream to offset the current pulling it downstream.

SOLUTION Vector $\vec{v}_{BW}$ points upstream at an angle θ as shown. From the diagram,

$$\sin\theta = \frac{v_{WS}}{v_{BW}} = \frac{1.20 \text{ m/s}}{1.85 \text{ m/s}} = 0.6486.$$

Thus $\theta = 40.4°$, so the boat must head upstream at a 40.4° angle.

FIGURE 3–33 Examples 3–13 and 3–14.

EXAMPLE 3–15 **Heading across the river.** The same boat $(v_{BW} = 1.85$ m/s$)$ now heads directly across the river whose current is still 1.20 m/s. (a) What is the velocity (magnitude and direction) of the boat relative to the shore? (b) If the river is 110 m wide, how long will it take to cross and how far downstream will the boat be then?

APPROACH The boat now heads directly across the river and is pulled down-stream by the current, as shown in Fig. 3–34. The boat's velocity with respect to the shore, $\vec{v}_{BS}$, is the sum of its velocity with respect to the water, $\vec{v}_{BW}$, plus the velocity of the water with respect to the shore, $\vec{v}_{WS}$:

$$\vec{v}_{BS} = \vec{v}_{BW} + \vec{v}_{WS},$$

just as before.

SOLUTION (a) Since $\vec{v}_{BW}$ is perpendicular to $\vec{v}_{WS}$, we can get v_{BS} using the theorem of Pythagoras:

$$v_{BS} = \sqrt{v_{BW}^2 + v_{WS}^2} = \sqrt{(1.85 \text{ m/s})^2 + (1.20 \text{ m/s})^2} = 2.21 \text{ m/s}.$$

We can obtain the angle (note how θ is defined in the diagram) from:

$$\tan\theta = v_{WS}/v_{BW} = (1.20 \text{ m/s})/(1.85 \text{ m/s}) = 0.6486.$$

Thus $\theta = \tan^{1}(0.6486) = 33.0°$. Note that this angle is not equal to the angle calculated in Example 3–14.

(b) The travel time for the boat is determined by the time it takes to cross the river. Given the river's width $D = 110$ m, we can use the velocity component in the direction of D, $v_{BW} = D/t$. Solving for t, we get $t = 110 \text{ m}/1.85 \text{ m/s} = 59.5$ s. The boat will have been carried downstream, in this time, a distance

$$d = v_{WS}t = (1.20 \text{ m/s})(59.5 \text{ s}) = 71.4 \text{ m} \approx 71 \text{ m}.$$

NOTE There is no acceleration in this Example, so the motion involves only constant velocities (of the boat or of the river).

FIGURE 3–34 Example 3–15. A boat heading directly across a river whose current moves at 1.20 m/s.

FIGURE 3–35 Example 3–16.

(a)   (b)   (c)

EXAMPLE 3–16 **Car velocities at 90°.** Two automobiles approach a street corner at right angles to each other with the same speed of 40.0 km/h ($= 11.11$ m/s), as shown in Fig. 3–35a. What is the relative velocity of one car with respect to the other? That is, determine the velocity of car 1 as seen by car 2.

APPROACH Figure 3–35a shows the situation in a reference frame fixed to the Earth. But we want to view the situation from a reference frame in which car 2 is at rest, and this is shown in Fig. 3–35b. In this reference frame (the world as seen by the driver of car 2), the Earth moves toward car 2 with velocity $\vec{v}_{E2}$ (speed of 40.0 km/h), which is of course equal and opposite to $\vec{v}_{2E}$, the velocity of car 2 with respect to the Earth (Eq. 3–16):

$$\vec{v}_{2E} = -\vec{v}_{E2}.$$

Then the velocity of car 1 as seen by car 2 is (see Eq. 3–15)

$$\vec{v}_{12} = \vec{v}_{1E} + \vec{v}_{E2}$$

SOLUTION Because $\vec{v}_{E2} = -\vec{v}_{2E}$, then

$$\vec{v}_{12} = \vec{v}_{1E} - \vec{v}_{2E}.$$

That is, the velocity of car 1 as seen by car 2 is the difference of their velocities, $\vec{v}_{1E} - \vec{v}_{2E}$, both measured relative to the Earth (see Fig. 3–35c). Since the magnitudes of $\vec{v}_{1E}$, $\vec{v}_{2E}$, and $\vec{v}_{E2}$ are equal (40.0 km/h = 11.11 m/s), we see (Fig. 3–35b) that $\vec{v}_{12}$ points at a 45° angle toward car 2; the speed is

$$v_{12} = \sqrt{(11.11 \text{ m/s})^2 + (11.11 \text{ m/s})^2} = 15.7 \text{ m/s} \ (= 56.6 \text{ km/h}).$$

Summary

A quantity that has both a magnitude and a direction is called a **vector**. A quantity that has only a magnitude is called a **scalar**.

Addition of vectors can be done graphically by placing the tail of each successive arrow (representing each vector) at the tip of the previous one. The sum, or **resultant vector**, is the arrow drawn from the tail of the first to the tip of the last. Two vectors can also be added using the parallelogram method.

Vectors can be added more accurately using the analytical method of adding their **components** along chosen axes with the aid of trigonometric functions. A vector of magnitude V making an angle θ with the x axis has components

$$V_x = V \cos\theta \qquad V_y = V \sin\theta. \qquad \textbf{(3–2)}$$

Given the components, we can find the magnitude and direction from

$$V = \sqrt{V_x^2 + V_y^2}, \qquad \tan\theta = \frac{V_y}{V_x}. \qquad \textbf{(3–3)}$$

It is often helpful to express a vector in terms of its components along chosen axes using **unit vectors**, which are vectors of unit length along the chosen coordinate axes; for Cartesian coordinates the unit vectors along the x, y, and z axes are called $\hat{\mathbf{i}}$, $\hat{\mathbf{j}}$, and $\hat{\mathbf{k}}$.

The general definitions for the **instantaneous velocity**, $\vec{v}$, and **acceleration**, $\vec{a}$, of a particle (in one, two, or three dimensions) are

$$\vec{v} = \frac{d\vec{r}}{dt} \qquad \textbf{(3–8)}$$

$$\vec{a} = \frac{d\vec{v}}{dt}, \qquad \textbf{(3–11)}$$

where $\vec{r}$ is the position vector of the particle. The kinematic equations for motion with constant acceleration can be written for each of the x, y, and z components of the motion and have the same form as for one-dimensional motion (Eqs. 2–12). Or they can be written in the more general vector form:

$$\vec{v} = \vec{v}_0 + \vec{a}t$$
$$\vec{r} = \vec{r}_0 + \vec{v}_0 t + \tfrac{1}{2}\vec{a}t^2 \qquad \textbf{(3–13)}$$

Projectile motion of an object moving in the air near the Earth's surface can be analyzed as two separate motions if air resistance can be ignored. The horizontal component of the motion is at constant velocity, whereas the vertical component is at constant acceleration, g, just as for an object falling vertically under the action of gravity.

The velocity of an object relative to one frame of reference can be found by vector addition if its velocity relative to a second frame of reference, and the **relative velocity** of the two reference frames, are known.

Questions

1. One car travels due east at 40 km/h, and a second car travels north at 40 km/h. Are their velocities equal? Explain.
2. Can you conclude that a car is not accelerating if its speedometer indicates a steady 60 km/h?
3. Can you give several examples of an object's motion in which a great distance is traveled but the displacement is zero?
4. Can the displacement vector for a particle moving in two dimensions ever be longer than the length of path traveled by the particle over the same time interval? Can it ever be less? Discuss.
5. During baseball practice, a batter hits a very high fly ball and then runs in a straight line and catches it. Which had the greater displacement, the player or the ball?
6. If $\vec{V} = \vec{V}_1 + \vec{V}_2$, is V necessarily greater than V_1 and/or V_2? Discuss.
7. Two vectors have length $V_1 = 3.5$ km and $V_2 = 4.0$ km. What are the maximum and minimum magnitudes of their vector sum?
8. Can two vectors, of unequal magnitude, add up to give the zero vector? Can *three* unequal vectors? Under what conditions?
9. Can the magnitude of a vector ever (*a*) equal, or (*b*) be less than, one of its components?
10. Can a particle with constant speed be accelerating? What if it has constant velocity?
11. Does the odometer of a car measure a scalar or a vector quantity? What about the speedometer?
12. A child wishes to determine the speed a slingshot imparts to a rock. How can this be done using only a meter stick, a rock, and the slingshot?
13. In archery, should the arrow be aimed directly at the target? How should your angle of aim depend on the distance to the target?

14. A projectile is launched at an upward angle of 30° to the horizontal with a speed of 30 m/s. How does the horizontal component of its velocity 1.0 s after launch compare with its horizontal component of velocity 2.0 s after launch, ignoring air resistance?
15. A projectile has the least speed at what point in its path?
16. It was reported in World War I that a pilot flying at an altitude of 2 km caught in his bare hands a bullet fired at the plane! Using the fact that a bullet slows down considerably due to air resistance, explain how this incident occurred.
17. Two cannonballs, A and B, are fired from the ground with identical initial speeds, but with θ_A larger than θ_B. (*a*) Which cannonball reaches a higher elevation? (*b*) Which stays longer in the air? (*c*) Which travels farther?
18. A person sitting in an enclosed train car, moving at constant velocity, throws a ball straight up into the air in her reference frame. (*a*) Where does the ball land? What is your answer if the car (*b*) accelerates, (*c*) decelerates, (*d*) rounds a curve, (*e*) moves with constant velocity but is open to the air?
19. If you are riding on a train that speeds past another train moving in the same direction on an adjacent track, it appears that the other train is moving backward. Why?
20. Two rowers, who can row at the same speed in still water, set off across a river at the same time. One heads straight across and is pulled downstream somewhat by the current. The other one heads upstream at an angle so as to arrive at a point opposite the starting point. Which rower reaches the opposite side first?
21. If you stand motionless under an umbrella in a rainstorm where the drops fall vertically you remain relatively dry. However, if you start running, the rain begins to hit your legs even if they remain under the umbrella. Why?

Problems

3–2 to 3–5 Vector Addition; Unit Vectors

1. (I) A car is driven 225 km west and then 78 km southwest (45°). What is the displacement of the car from the point of origin (magnitude and direction)? Draw a diagram.
2. (I) A delivery truck travels 28 blocks north, 16 blocks east, and 26 blocks south. What is its final displacement from the origin? Assume the blocks are equal length.
3. (I) If $V_x = 7.80$ units and $V_y = -6.40$ units, determine the magnitude and direction of $\vec{V}$.
4. (II) Graphically determine the resultant of the following three vector displacements: (1) 24 m, 36° north of east; (2) 18 m, 37° east of north; and (3) 26 m, 33° west of south.
5. (II) $\vec{V}$ is a vector 24.8 units in magnitude and points at an angle of 23.4° above the negative x axis. (*a*) Sketch this vector. (*b*) Calculate V_x and V_y. (*c*) Use V_x and V_y to obtain (again) the magnitude and direction of $\vec{V}$. [*Note*: Part (*c*) is a good way to check if you've resolved your vector correctly.]

6. (II) Figure 3–36 shows two vectors, $\vec{A}$ and $\vec{B}$, whose magnitudes are $A = 6.8$ units and $B = 5.5$ units. Determine $\vec{C}$ if (*a*) $\vec{C} = \vec{A} + \vec{B}$, (*b*) $\vec{C} = \vec{A} - \vec{B}$, (*c*) $\vec{C} = \vec{B} - \vec{A}$. Give the magnitude and direction for each.

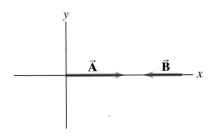

FIGURE 3–36 Problem 6.

7. (II) An airplane is traveling 835 km/h in a direction 41.5° west of north (Fig. 3–37). (a) Find the components of the velocity vector in the northerly and westerly directions. (b) How far north and how far west has the plane traveled after 2.50 h?

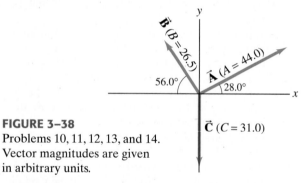

FIGURE 3–37
Problem 7.

8. (II) Let $\vec{V}_1 = -6.0\hat{i} + 8.0\hat{j}$ and $\vec{V}_2 = 4.5\hat{i} - 5.0\hat{j}$. Determine the magnitude and direction of (a) $\vec{V}_1$, (b) $\vec{V}_2$, (c) $\vec{V}_1 + \vec{V}_2$ and (d) $\vec{V}_2 - \vec{V}_1$.

9. (II) (a) Determine the magnitude and direction of the sum of the three vectors $\vec{V}_1 = 4.0\hat{i} - 8.0\hat{j}$, $\vec{V}_2 = \hat{i} + \hat{j}$, and $\vec{V}_3 = -2.0\hat{i} + 4.0\hat{j}$. (b) Determine $\vec{V}_1 - \vec{V}_2 + \vec{V}_3$.

10. (II) Three vectors are shown in Fig. 3–38. Their magnitudes are given in arbitrary units. Determine the sum of the three vectors. Give the resultant in terms of (a) components, (b) magnitude and angle with x axis.

FIGURE 3–38
Problems 10, 11, 12, 13, and 14.
Vector magnitudes are given in arbitrary units.

11. (II) (a) Given the vectors $\vec{A}$ and $\vec{B}$ shown in Fig. 3–38, determine $\vec{B} - \vec{A}$. (b) Determine $\vec{A} - \vec{B}$ without using your answer in (a). Then compare your results and see if they are opposite.

12. (II) Determine the vector $\vec{A} - \vec{C}$, given the vectors $\vec{A}$ and $\vec{C}$ in Fig. 3–38.

13. (II) For the vectors shown in Fig. 3–38, determine (a) $\vec{B} - 2\vec{A}$, (b) $2\vec{A} - 3\vec{B} + 2\vec{C}$.

14. (II) For the vectors given in Fig. 3–38, determine (a) $\vec{A} - \vec{B} + \vec{C}$, (b) $\vec{A} + \vec{B} - \vec{C}$, and (c) $\vec{C} - \vec{A} - \vec{B}$.

15. (II) The summit of a mountain, 2450 m above base camp, is measured on a map to be 4580 m horizontally from the camp in a direction 32.4° west of north. What are the components of the displacement vector from camp to summit? What is its magnitude? Choose the x axis east, y axis north, and z axis up.

16. (III) You are given a vector in the xy plane that has a magnitude of 90.0 units and a y component of −55.0 units. (a) What are the two possibilities for its x component? (b) Assuming the x component is known to be positive, specify the vector which, if you add it to the original one, would give a resultant vector that is 80.0 units long and points entirely in the −x direction.

3–6 Vector Kinematics

17. (I) The position of a particular particle as a function of time is given by $\vec{r} = (9.60t\,\hat{i} + 8.85\hat{j} - 1.00t^2\,\hat{k})$ m. Determine the particle's velocity and acceleration as a function of time.

18. (I) What was the average velocity of the particle in Problem 17 between $t = 1.00$ s and $t = 3.00$ s? What is the magnitude of the instantaneous velocity at $t = 2.00$ s?

19. (II) What is the shape of the path of the particle of Problem 17?

20. (II) A car is moving with speed 18.0 m/s due south at one moment and 27.5 m/s due east 8.00 s later. Over this time interval, determine the magnitude and direction of (a) its average velocity, (b) its average acceleration. (c) What is its average speed. [Hint: Can you determine all these from the information given?]

21. (II) At $t = 0$, a particle starts from rest at $x = 0$, $y = 0$, and moves in the xy plane with an acceleration $\vec{a} = (4.0\hat{i} + 3.0\hat{j})$ m/s². Determine (a) the x and y components of velocity, (b) the speed of the particle, and (c) the position of the particle, all as a function of time. (d) Evaluate all the above at $t = 2.0$ s.

22. (II) (a) A skier is accelerating down a 30.0° hill at 1.80 m/s² (Fig. 3–39). What is the vertical component of her acceleration? (b) How long will it take her to reach the bottom of the hill, assuming she starts from rest and accelerates uniformly, if the elevation change is 325 m?

FIGURE 3–39 Problem 22.

23. (II) An ant walks on a piece of graph paper straight along the x axis a distance of 10.0 cm in 2.00 s. It then turns left 30.0° and walks in a straight line another 10.0 cm in 1.80 s. Finally, it turns another 70.0° to the left and walks another 10.0 cm in 1.55 s. Determine (a) the x and y components of the ant's average velocity, and (b) its magnitude and direction.

24. (II) A particle starts from the origin at $t = 0$ with an initial velocity of 5.0 m/s along the positive x axis. If the acceleration is $(-3.0\hat{i} + 4.5\hat{j})$ m/s², determine the velocity and position of the particle at the moment it reaches its maximum x coordinate.

25. (II) Suppose the position of an object is given by $\vec{r} = (3.0t^2\hat{i} - 6.0t^3\hat{j})$ m. (a) Determine its velocity $\vec{v}$ and acceleration $\vec{a}$, as a function of time. (b) Determine $\vec{r}$ and $\vec{v}$ at time $t = 2.5$ s.

26. (II) An object, which is at the origin at time $t = 0$, has initial velocity $\vec{v}_0 = (-14.0\hat{i} - 7.0\hat{j})$ m/s and constant acceleration $\vec{a} = (6.0\hat{i} + 3.0\hat{j})$ m/s². Find the position $\vec{r}$ where the object comes to rest (momentarily).

27. (II) A particle's position as a function of time t is given by $\vec{r} = (5.0t + 6.0t^2)\,\text{m}\,\hat{i} + (7.0 - 3.0t^3)\,\text{m}\,\hat{j}$. At $t = 5.0\,\text{s}$, find the magnitude and direction of the particle's displacement vector $\Delta\vec{r}$ relative to the point $\vec{r}_0 = (0.0\hat{i} + 7.0\hat{j})\,\text{m}$.

3–7 and 3–8 Projectile Motion (neglect air resistance)

28. (I) A tiger leaps horizontally from a 7.5-m-high rock with a speed of 3.2 m/s. How far from the base of the rock will she land?

29. (I) A diver running 2.3 m/s dives out horizontally from the edge of a vertical cliff and 3.0 s later reaches the water below. How high was the cliff and how far from its base did the diver hit the water?

30. (II) Estimate how much farther a person can jump on the Moon as compared to the Earth if the takeoff speed and angle are the same. The acceleration due to gravity on the Moon is one-sixth what it is on Earth.

31. (II) A fire hose held near the ground shoots water at a speed of 6.5 m/s. At what angle(s) should the nozzle point in order that the water land 2.5 m away (Fig. 3–40)? Why are there two different angles? Sketch the two trajectories.

FIGURE 3–40
Problem 31.

|← 2.5 m →|

32. (II) A ball is thrown horizontally from the roof of a building 9.0 m tall and lands 9.5 m from the base. What was the ball's initial speed?

33. (II) A football is kicked at ground level with a speed of 18.0 m/s at an angle of 38.0° to the horizontal. How much later does it hit the ground?

34. (II) A ball thrown horizontally at 23.7 m/s from the roof of a building lands 31.0 m from the base of the building. How high is the building?

35. (II) A shot-putter throws the shot (mass = 7.3 kg) with an initial speed of 14.4 m/s at a 34.0° angle to the horizontal. Calculate the horizontal distance traveled by the shot if it leaves the athlete's hand at a height of 2.10 m above the ground.

36. (II) Show that the time required for a projectile to reach its highest point is equal to the time for it to return to its original height if air resistance is neglible.

37. (II) You buy a plastic dart gun, and being a clever physics student you decide to do a quick calculation to find its maximum horizontal range. You shoot the gun straight up, and it takes 4.0 s for the dart to land back at the barrel. What is the maximum horizontal range of your gun?

38. (II) A baseball is hit with a speed of 27.0 m/s at an angle of 45.0°. It lands on the flat roof of a 13.0-m-tall nearby building. If the ball was hit when it was 1.0 m above the ground, what horizontal distance does it travel before it lands on the building?

39. (II) In Example 3–11 we chose the x axis to the right and y axis up. Redo this problem by defining the x axis to the left and y axis down, and show that the conclusion remains the same—the football lands on the ground 40.5 m to the right of where it departed the punter's foot.

40. (II) A grasshopper hops down a level road. On each hop, the grasshopper launches itself at angle $\theta_0 = 45°$ and achieves a range $R = 1.0$ m. What is the average horizontal speed of the grasshopper as it progresses down the road? Assume that the time spent on the ground between hops is negligible.

41. (II) Extreme-sports enthusiasts have been known to jump off the top of El Capitan, a sheer granite cliff of height 910 m in Yosemite National Park. Assume a jumper runs horizontally off the top of El Capitan with speed 5.0 m/s and enjoys a freefall until she is 150 m above the valley floor, at which time she opens her parachute (Fig. 3–41). (*a*) How long is the jumper in freefall? Ignore air resistance. (*b*) It is important to be as far away from the cliff as possible before opening the parachute. How far from the cliff is this jumper when she opens her chute?

5.0 m/s

910 m

150 m

FIGURE 3–41
Problem 41.

42. (II) Here is something to try at a sporting event. Show that the maximum height h attained by an object projected into the air, such as a baseball, football, or soccer ball, is approximately given by

$$h \approx 1.2\,t^2\,\text{m},$$

where t is the total time of flight for the object in seconds. Assume that the object returns to the same level as that from which it was launched, as in Fig. 3–42. For example, if you count to find that a baseball was in the air for $t = 5.0\,\text{s}$, the maximum height attained was $h = 1.2 \times (5.0)^2 = 30\,\text{m}$. The beauty of this relation is that h can be determined without knowledge of the launch speed v_0 or launch angle θ_0.

v_0

θ_0

h

FIGURE 3–42   Problem 42.

43. (II) The pilot of an airplane traveling 170 km/h wants to drop supplies to flood victims isolated on a patch of land 150 m below. The supplies should be dropped how many seconds before the plane is directly overhead?

44. (II) (a) A long jumper leaves the ground at 45° above the horizontal and lands 8.0 m away. What is her "takeoff" speed v_0? (b) Now she is out on a hike and comes to the left bank of a river. There is no bridge and the right bank is 10.0 m away horizontally and 2.5 m, vertically below. If she long jumps from the edge of the left bank at 45° with the speed calculated in (a), how long, or short, of the opposite bank will she land (Fig. 3–43)?

FIGURE 3–43 Problem 44.

45. (II) A high diver leaves the end of a 5.0-m-high diving board and strikes the water 1.3 s later, 3.0 m beyond the end of the board. Considering the diver as a particle, determine (a) her initial velocity, $\vec{v}_0$; (b) the maximum height reached; and (c) the velocity $\vec{v}_f$ with which she enters the water.

46. (II) A projectile is shot from the edge of a cliff 115 m above ground level with an initial speed of 65.0 m/s at an angle of 35.0° with the horizontal, as shown in Fig. 3–44. (a) Determine the time taken by the projectile to hit point P at ground level. (b) Determine the distance X of point P from the base of the vertical cliff. At the instant just before the projectile hits point P, find (c) the horizontal and the vertical components of its velocity, (d) the magnitude of the velocity, and (e) the angle made by the velocity vector with the horizontal. (f) Find the maximum height above the cliff top reached by the projectile.

FIGURE 3–44 Problem 46.

47. (II) Suppose the kick in Example 3–7 is attempted 36.0 m from the goalposts, whose crossbar is 3.00 m above the ground. If the football is directed perfectly between the goalposts, will it pass over the bar and be a field goal? Show why or why not. If not, from what horizontal distance must this kick be made if it is to score?

48. (II) Exactly 3.0 s after a projectile is fired into the air from the ground, it is observed to have a velocity $\vec{v} = (8.6\hat{i} + 4.8\hat{j})$ m/s, where the x axis is horizontal and the y axis is positive upward. Determine (a) the horizontal range of the projectile, (b) its maximum height above the ground, and (c) its speed and angle of motion just before it strikes the ground.

49. (II) Revisit Example 3–9, and assume that the boy with the slingshot is *below* the boy in the tree (Fig. 3–45) and so aims *upward*, directly at the boy in the tree. Show that again the boy in the tree makes the wrong move by letting go at the moment the water balloon is shot.

FIGURE 3–45 Problem 49.

50. (II) A stunt driver wants to make his car jump over 8 cars parked side by side below a horizontal ramp (Fig. 3–46). (a) With what minimum speed must he drive off the horizontal ramp? The vertical height of the ramp is 1.5 m above the cars and the horizontal distance he must clear is 22 m. (b) If the ramp is now tilted upward, so that "takeoff angle" is 7.0° above the horizontal, what is the new minimum speed?

FIGURE 3–46 Problem 50.

51. (II) A ball is thrown horizontally from the top of a cliff with initial speed v_0 (at $t = 0$). At any moment, its direction of motion makes an angle θ to the horizontal (Fig. 3–47). Derive a formula for θ as a function of time, t, as the ball follows a projectile's path.

FIGURE 3–47 Problem 51.

52. (II) At what projection angle will the range of a projectile equal its maximum height?

53. (II) A projectile is fired with an initial speed of 46.6 m/s at an angle of 42.2° above the horizontal on a long flat firing range. Determine (*a*) the maximum height reached by the projectile, (*b*) the total time in the air, (*c*) the total horizontal distance covered (that is, the range), and (*d*) the velocity of the projectile 1.50 s after firing.

54. (II) An athlete executing a long jump leaves the ground at a 27.0° angle and lands 7.80 m away. (*a*) What was the takeoff speed? (*b*) If this speed were increased by just 5.0%, how much longer would the jump be?

55. (III) A person stands at the base of a hill that is a straight incline making an angle ϕ with the horizontal (Fig. 3–48). For a given initial speed v_0, at what angle θ (to the horizontal) should objects be thrown so that the distance d they land up the hill is as large as possible?

FIGURE 3–48 Problem 55. Given ϕ and v_0, determine θ to make d maximum.

56. (III) Derive a formula for the horizontal range R, of a projectile when it lands at a height h above its initial point. (For $h < 0$, it lands a distance $-h$ below the starting point.) Assume it is projected at an angle θ_0 with initial speed v_0.

3–9 Relative Velocity

57. (I) A person going for a morning jog on the deck of a cruise ship is running toward the bow (front) of the ship at 2.0 m/s while the ship is moving ahead at 8.5 m/s. What is the velocity of the jogger relative to the water? Later, the jogger is moving toward the stern (rear) of the ship. What is the jogger's velocity relative to the water now?

58. (I) Huck Finn walks at a speed of 0.70 m/s across his raft (that is, he walks perpendicular to the raft's motion relative to the shore). The raft is traveling down the Mississippi River at a speed of 1.50 m/s relative to the river bank (Fig. 3–49). What is Huck's velocity (speed and direction) relative to the river bank?

FIGURE 3–49 Problem 58.

59. (II) Determine the speed of the boat with respect to the shore in Example 3–14.

60. (II) Two planes approach each other head-on. Each has a speed of 780 km/h, and they spot each other when they are initially 12.0 km apart. How much time do the pilots have to take evasive action?

61. (II) A child, who is 45 m from the bank of a river, is being carried helplessly downstream by the river's swift current of 1.0 m/s. As the child passes a lifeguard on the river's bank, the lifeguard starts swimming in a straight line until she reaches the child at a point downstream (Fig. 3–50). If the lifeguard can swim at a speed of 2.0 m/s relative to the water, how long does it take her to reach the child? How far downstream does the lifeguard intercept the child?

FIGURE 3–50 Problem 61.

62. (II) A passenger on a boat moving at 1.70 m/s on a still lake walks up a flight of stairs at a speed of 0.60 m/s, Fig. 3–51. The stairs are angled at 45° pointing in the direction of motion as shown. Write the vector velocity of the passenger relative to the water.

FIGURE 3–51 Problem 62.

63. (II) A person in the passenger basket of a hot-air balloon throws a ball horizontally outward from the basket with speed 10.0 m/s (Fig. 3–52). What initial velocity (magnitude and direction) does the ball have relative to a person standing on the ground (*a*) if the hot-air balloon is rising at 5.0 m/s relative to the ground during this throw, (*b*) if the hot-air balloon is descending at 5.0 m/s relative to the ground.

FIGURE 3–52 Problem 63.

64. (II) An airplane is heading due south at a speed of 580 km/h. If a wind begins blowing from the southwest at a speed of 90.0 km/h (average), calculate (*a*) the velocity (magnitude and direction) of the plane, relative to the ground, and (*b*) how far from its intended position it will be after 11.0 min if the pilot takes no corrective action. [*Hint:* First draw a diagram.]

65. (II) In what direction should the pilot aim the plane in Problem 64 so that it will fly due south?

66. (II) Two cars approach a street corner at right angles to each other (see Fig. 3–35). Car 1 travels at 35 km/h and car 2 at 45 km/h. What is the relative velocity of car 1 as seen by car 2? What is the velocity of car 2 relative to car 1?

67. (II) A swimmer is capable of swimming 0.60 m/s in still water. (*a*) If she aims her body directly across a 55-m-wide river whose current is 0.50 m/s, how far downstream (from a point opposite her starting point) will she land? (*b*) How long will it take her to reach the other side?

68. (II) (*a*) At what upstream angle must the swimmer in Problem 67 aim, if she is to arrive at a point directly across the stream? (*b*) How long will it take her?

69. (II) A motorboat whose speed in still water is 3.40 m/s must aim upstream at an angle of 19.5° (with respect to a line perpendicular to the shore) in order to travel directly across the stream. (*a*) What is the speed of the current? (*b*) What is the resultant speed of the boat with respect to the shore? (See Fig. 3–31.)

70. (II) A boat, whose speed in still water is 2.70 m/s, must cross a 280-m-wide river and arrive at a point 120 m upstream from where it starts (Fig. 3–53). To do so, the pilot must head the boat at a 45.0° upstream angle. What is the speed of the river's current?

FIGURE 3–53
Problem 70.

71. (III) An airplane, whose air speed is 580 km/h, is supposed to fly in a straight path 38.0° N of E. But a steady 72 km/h wind is blowing from the north. In what direction should the plane head?

General Problems

72. Two vectors, $\vec{V}_1$ and $\vec{V}_2$, add to a resultant $\vec{V} = \vec{V}_1 + \vec{V}_2$. Describe $\vec{V}_1$ and $\vec{V}_2$ if (*a*) $V = V_1 + V_2$, (*b*) $V^2 = V_1^2 + V_2^2$, (*c*) $V_1 + V_2 = V_1 - V_2$.

73. A plumber steps out of his truck, walks 66 m east and 35 m south, and then takes an elevator 12 m into the subbasement of a building where a bad leak is occurring. What is the displacement of the plumber relative to his truck? Give your answer in components; also give the magnitude and angles, with respect to the *x* axis, in the vertical and horizontal plane. Assume *x* is east, *y* is north, and *z* is up.

74. On mountainous downhill roads, escape routes are sometimes placed to the side of the road for trucks whose brakes might fail. Assuming a constant upward slope of 26°, calculate the horizontal and vertical components of the acceleration of a truck that slowed from 110 km/h to rest in 7.0 s. See Fig. 3–54.

FIGURE 3–54
Problem 74.

75. A light plane is headed due south with a speed relative to still air of 185 km/h. After 1.00 h, the pilot notices that they have covered only 135 km and their direction is not south but southeast (45.0°). What is the wind velocity?

76. An Olympic long jumper is capable of jumping 8.0 m. Assuming his horizontal speed is 9.1 m/s as he leaves the ground, how long is he in the air and how high does he go? Assume that he lands standing upright—that is, the same way he left the ground.

77. Romeo is chucking pebbles gently up to Juliet's window, and he wants the pebbles to hit the window with only a horizontal component of velocity. He is standing at the edge of a rose garden 8.0 m below her window and 9.0 m from the base of the wall (Fig. 3–55). How fast are the pebbles going when they hit her window?

FIGURE 3–55
Problem 77.

78. Raindrops make an angle θ with the vertical when viewed through a moving train window (Fig. 3–56). If the speed of the train is v_T, what is the speed of the raindrops in the reference frame of the Earth in which they are assumed to fall vertically?

FIGURE 3–56
Problem 78.

79. Apollo astronauts took a "nine iron" to the Moon and hit a golf ball about 180 m. Assuming that the swing, launch angle, and so on, were the same as on Earth where the same astronaut could hit it only 32 m, estimate the acceleration due to gravity on the surface of the Moon. (We neglect air resistance in both cases, but on the Moon there is none.)

80. A hunter aims directly at a target (on the same level) 68.0 m away. (*a*) If the bullet leaves the gun at a speed of 175 m/s, by how much will it miss the target? (*b*) At what angle should the gun be aimed so the target will be hit?

81. The cliff divers of Acapulco push off horizontally from rock platforms about 35 m above the water, but they must clear rocky outcrops at water level that extend out into the water 5.0 m from the base of the cliff directly under their launch point. See Fig. 3–57. What minimum pushoff speed is necessary to clear the rocks? How long are they in the air?

FIGURE 3–57
Problem 81.

82. When Babe Ruth hit a homer over the 8.0-m-high right-field fence 98 m from home plate, roughly what was the minimum speed of the ball when it left the bat? Assume the ball was hit 1.0 m above the ground and its path initially made a 36° angle with the ground.

83. The speed of a boat in still water is v. The boat is to make a round trip in a river whose current travels at speed u. Derive a formula for the time needed to make a round trip of total distance D if the boat makes the round trip by moving (*a*) upstream and back downstream, and (*b*) directly across the river and back. We must assume $u < v$; why?

84. At serve, a tennis player aims to hit the ball horizontally. What minimum speed is required for the ball to clear the 0.90-m-high net about 15.0 m from the server if the ball is "launched" from a height of 2.50 m? Where will the ball land if it just clears the net (and will it be "good" in the sense that it lands within 7.0 m of the net)? How long will it be in the air? See Fig. 3–58.

FIGURE 3–58   Problem 84.

85. Spymaster Chris, flying a constant 208 km/h horizontally in a low-flying helicopter, wants to drop secret documents into her contact's open car which is traveling 156 km/h on a level highway 78.0 m below. At what angle (with the horizontal) should the car be in her sights when the packet is released (Fig. 3–59)?

FIGURE 3–59
Problem 85.

86. A basketball leaves a player's hands at a height of 2.10 m above the floor. The basket is 3.05 m above the floor. The player likes to shoot the ball at a 38.0° angle. If the shot is made from a horizontal distance of 11.00 m and must be accurate to ± 0.22 m (horizontally), what is the range of initial speeds allowed to make the basket?

87. A particle has a velocity of $\vec{v} = \left(-2.0\hat{i} + 3.5t\hat{j}\right)$ m/s. The particle starts at $\vec{r} = \left(1.5\hat{i} - 3.1\hat{j}\right)$ m at $t = 0$. Give the position and acceleration as a function of time. What is the shape of the resulting path?

88. A projectile is launched from ground level to the top of a cliff which is 195 m away and 135 m high (see Fig. 3–60). If the projectile lands on top of the cliff 6.6 s after it is fired, find the initial velocity of the projectile (magnitude and direction). Neglect air resistance.

FIGURE 3–60
Problem 88.

89. In hot pursuit, Agent Logan of the FBI must get directly across a 1200-m-wide river in minimum time. The river's current is 0.80 m/s, he can row a boat at 1.60 m/s, and he can run 3.00 m/s. Describe the path he should take (rowing plus running along the shore) for the minimum crossing time, and determine the minimum time.

90. A boat can travel 2.20 m/s in still water. (*a*) If the boat points its prow directly across a stream whose current is 1.30 m/s, what is the velocity (magnitude and direction) of the boat relative to the shore? (*b*) What will be the position of the boat, relative to its point of origin, after 3.00 s?

91. A boat is traveling where there is a current of 0.20 m/s east (Fig. 3–61). To avoid some offshore rocks, the boat must clear a buoy that is NNE (22.5°) and 3.0 km away. The boat's speed through still water is 2.1 m/s. If the boat wants to pass the buoy 0.15 km on its right, at what angle should the boat head?

FIGURE 3–61
Problem 91.

92. A child runs down a 12° hill and then suddenly jumps upward at a 15° angle above horizontal and lands 1.4 m down the hill as measured along the hill. What was the child's initial speed?

93. A basketball is shot from an initial height of 2.4 m (Fig. 3–62) with an initial speed $v_0 = 12$ m/s directed at an angle $\theta_0 = 35°$ above the horizontal. (*a*) How far from the basket was the player if he made a basket? (*b*) At what angle to the horizontal did the ball enter the basket?

FIGURE 3–62
Problem 93.

94. You are driving south on a highway at 25 m/s (approximately 55 mi/h) in a snowstorm. When you last stopped, you noticed that the snow was coming down vertically, but it is passing the windows of the moving car at an angle of 37° to the horizontal. Estimate the speed of the snowflakes relative to the car and relative to the ground.

95. A rock is kicked horizontally at 15 m/s from a hill with a 45° slope (Fig. 3–63). How long does it take for the rock to hit the ground?

FIGURE 3–63
Problem 95.

96. A batter hits a fly ball which leaves the bat 0.90 m above the ground at an angle of 61° with an initial speed of 28 m/s heading toward centerfield. Ignore air resistance. (*a*) How far from home plate would the ball land if not caught? (*b*) The ball is caught by the centerfielder who, starting at a distance of 105 m from home plate, runs straight toward home plate at a constant speed and makes the catch at ground level. Find his speed.

97. A ball is shot from the top of a building with an initial velocity of 18 m/s at an angle $\theta = 42°$ above the horizontal. (*a*) What are the horizontal and vertical components of the initial velocity? (*b*) If a nearby building is the same height and 55 m away, how far below the top of the building will the ball strike the nearby building?

98. At $t = 0$ a batter hits a baseball with an initial speed of 28 m/s at a 55° angle to the horizontal. An outfielder is 85 m from the batter at $t = 0$ and, as seen from home plate, the line of sight to the outfielder makes a horizontal angle of 22° with the plane in which the ball moves (see Fig. 3–64). What speed and direction must the fielder take to catch the ball at the same height from which it was struck? Give the angle with respect to the outfielder's line of sight to home plate.

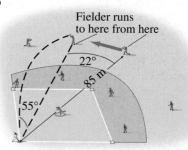

FIGURE 3–64
Problem 98.

*Numerical/Computer

***99.** (II) Students shoot a plastic ball horizontally from a projectile launcher. They measure the distance x the ball travels horizontally, the distance y the ball falls vertically, and the total time t the ball is in the air for six different heights of the projectile launcher. Here is their data.

Time, t (s)	Horizontal distance, x (m)	Vertical distance, y (m)
0.217	0.642	0.260
0.376	1.115	0.685
0.398	1.140	0.800
0.431	1.300	0.915
0.478	1.420	1.150
0.491	1.480	1.200

(*a*) Determine the best-fit straight line that represents x as a function of t. What is the initial speed of the ball obtained from the best-fit straight line? (*b*) Determine the best-fit quadratic equation that represents y as a function of t. What is the acceleration of the ball in the vertical direction?

***100.** (III) A shot-putter throws from a height $h = 2.1$ m above the ground as shown in Fig. 3–65, with an initial speed of $v_0 = 13.5$ m/s. (*a*) Derive a relation that describes how the distance traveled d depends on the release angle θ_0. (*b*) Using the given values for v_0 and h, use a graphing calculator or computer to plot d vs. θ_0. According to your plot, what value for θ_0 maximizes d?

FIGURE 3–65 Problem 100.

Answers to Exercises

A: When the two vectors D_1 and D_2 point in the same direction.

B: $3\sqrt{2} = 4.24$.

C: (*a*).

D: (*d*).

E: Both balls reach the same height, so are in the air for the same length of time.

F: (*c*).

The space shuttle Discovery is carried out into space by powerful rockets. They are accelerating, increasing in speed rapidly. To do so, a force must be exerted on them according to Newton's second law, $\Sigma \vec{\mathbf{F}} = m\vec{\mathbf{a}}$. What exerts this force? The rocket engines exert a force on the gases they push out (expel) from the rear of the rockets (labeled $\vec{\mathbf{F}}_{GR}$). According to Newton's third law, these ejected gases exert an equal and opposite force on the rockets in the forward direction. It is this "reaction" force exerted on the rockets by the gases, labeled $\vec{\mathbf{F}}_{RG}$, that accelerates the rockets forward.

Dynamics: Newton's Laws of Motion

CHAPTER-OPENING QUESTIONS—Guess now!

A 150-kg football player collides head-on with a 75-kg running back. During the collision, the heavier player exerts a force of magnitude F_A on the smaller player. If the smaller player exerts a force F_B back on the heavier player, which response is most accurate?

(a) $F_B = F_A$.
(b) $F_B < F_A$.
(c) $F_B > F_A$.
(d) $F_B = 0$.
(e) We need more information.

Second Question:

A line by the poet T. S. Eliot (from *Murder in the Cathedral*) has the women of Canterbury say "the earth presses up against our feet." What force is this?

(a) Gravity.
(b) The normal force.
(c) A friction force.
(d) Centrifugal force.
(e) No force—they are being poetic.

CONTENTS

83

We have discussed how motion is described in terms of velocity and acceleration. Now we deal with the question of *why* objects move as they do: What makes an object at rest begin to move? What causes an object to accelerate or decelerate? What is involved when an object moves in a curved path? We can answer in each case that a force is required. In this Chapter[†], we will investigate the connection between force and motion, which is the subject called **dynamics**.

4–1 Force

Intuitively, we experience **force** as any kind of a push or a pull on an object. When you push a stalled car or a grocery cart (Fig. 4–1), you are exerting a force on it. When a motor lifts an elevator, or a hammer hits a nail, or the wind blows the leaves of a tree, a force is being exerted. We often call these *contact forces* because the force is exerted when one object comes in contact with another object. On the other hand, we say that an object falls because of the *force of gravity*.

If an object is at rest, to start it moving requires force—that is, a force is needed to accelerate an object from zero velocity to a nonzero velocity. For an object already moving, if you want to change its velocity—either in direction or in magnitude—a force is required. In other words, to accelerate an object, a force is always required. In Section 4–4 we discuss the precise relation between acceleration and net force, which is Newton's second law.

One way to measure the magnitude (or strength) of a force is to use a spring scale (Fig. 4–2). Normally, such a spring scale is used to find the weight of an object; by weight we mean the force of gravity acting on the object (Section 4–6). The spring scale, once calibrated, can be used to measure other kinds of forces as well, such as the pulling force shown in Fig. 4–2.

A force exerted in a different direction has a different effect. Force has direction as well as magnitude, and is indeed a vector that follows the rules of vector addition discussed in Chapter 3. We can represent any force on a diagram by an arrow, just as we did with velocity. The direction of the arrow is the direction of the push or pull, and its length is drawn proportional to the magnitude of the force.

FIGURE 4–1 A force exerted on a grocery cart—in this case exerted by a person.

FIGURE 4–2 A spring scale used to measure a force.

4–2 Newton's First Law of Motion

What is the relationship between force and motion? Aristotle (384–322 B.C.) believed that a force was required to keep an object moving along a horizontal plane. To Aristotle, the natural state of an object was at rest, and a force was believed necessary to keep an object in motion. Furthermore, Aristotle argued, the greater the force on the object, the greater its speed.

Some 2000 years later, Galileo disagreed: he maintained that it is just as natural for an object to be in motion with a constant velocity as it is for it to be at rest.

To understand Galileo's idea, consider the following observations involving motion along a horizontal plane. To push an object with a rough surface along a

[†]We treat everyday objects in motion here; the treatment of the submicroscopic world of atoms and molecules, and when velocities are extremely high, close to the speed of light $(3.0 \times 10^8 \, \text{m/s})$, are treated using quantum theory (Chapter 37 ff), and the theory of relativity (Chapter 36).

tabletop at constant speed requires a certain amount of force. To push an equally heavy object with a very smooth surface across the table at the same speed will require less force. If a layer of oil or other lubricant is placed between the surface of the object and the table, then almost no force is required to keep the object moving. Notice that in each successive step, less force is required. As the next step, we imagine that the object does not rub against the table at all—or there is a perfect lubricant between the object and the table—and theorize that once started, the object would move across the table at constant speed with *no* force applied. A steel ball bearing rolling on a hard horizontal surface approaches this situation. So does a puck on an air table, in which a thin layer of air reduces friction almost to zero.

It was Galileo's genius to imagine such an idealized world—in this case, one where there is no friction—and to see that it could lead to a more accurate and richer understanding of the real world. This idealization led him to his remarkable conclusion that if no force is applied to a moving object, it will continue to move with constant speed in a straight line. An object slows down only if a force is exerted on it. Galileo thus interpreted friction as a force akin to ordinary pushes and pulls.

To push an object across a table at constant speed requires a force from your hand that can balance out the force of friction (Fig. 4–3). When the object moves at constant speed, your pushing force is equal in magnitude to the friction force, but these two forces are in opposite directions, so the *net* force on the object (the vector sum of the two forces) is zero. This is consistent with Galileo's viewpoint, for the object moves with constant speed when no net force is exerted on it.

Upon this foundation laid by Galileo, Isaac Newton (Fig. 4–4) built his great theory of motion. Newton's analysis of motion is summarized in his famous "three laws of motion." In his great work, the *Principia* (published in 1687), Newton readily acknowledged his debt to Galileo. In fact, **Newton's first law of motion** is close to Galileo's conclusions. It states that

> **Every object continues in its state of rest, or of uniform velocity in a straight line, as long as no net force acts on it.**

The tendency of an object to maintain its state of rest or of uniform velocity in a straight line is called **inertia**. As a result, Newton's first law is often called the **law of inertia**.

CONCEPTUAL EXAMPLE 4–1 | **Newton's first law.** A school bus comes to a sudden stop, and all of the backpacks on the floor start to slide forward. What force causes them to do that?

RESPONSE It isn't "force" that does it. By Newton's first law, the backpacks continue their state of motion, maintaining their velocity. The backpacks slow down if a force is applied, such as friction with the floor.

Inertial Reference Frames

Newton's first law does not hold in every reference frame. For example, if your reference frame is fixed in an accelerating car, an object such as a cup resting on the dashboard may begin to move toward you (it stayed at rest as long as the car's velocity remained constant). The cup accelerated toward you, but neither you nor anything else exerted a force on it in that direction. Similarly, in the reference frame of the decelerating bus in Example 4–1, there was no force pushing the backpacks forward. In accelerating reference frames, Newton's first law does not hold. Reference frames in which Newton's first law does hold are called **inertial reference frames** (the law of inertia is valid in them). For most purposes, we usually make the approximation that a reference frame fixed on the Earth is an inertial frame. This is not precisely true, due to the Earth's rotation, but usually it is close enough.

Any reference frame that moves with constant velocity (say, a car or an airplane) relative to an inertial frame is also an inertial reference frame. Reference frames where the law of inertia does *not* hold, such as the accelerating reference frames discussed above, are called **noninertial** reference frames. How can we be sure a reference frame is inertial or not? By checking to see if Newton's first law holds. Thus Newton's first law serves as the definition of inertial reference frames.

FIGURE 4–3 $\vec{F}$ represents the force applied by the person and $\vec{F}_{fr}$ represents the force of friction.

NEWTON'S FIRST LAW OF MOTION

FIGURE 4–4
Isaac Newton (1642–1727).

4-3 Mass

Newton's second law, which we come to in the next Section, makes use of the concept of mass. Newton used the term *mass* as a synonym for *quantity of matter*. This intuitive notion of the mass of an object is not very precise because the concept "quantity of matter" is not very well defined. More precisely, we can say that **mass** is a *measure of the inertia* of an object. The more mass an object has, the greater the force needed to give it a particular acceleration. It is harder to start it moving from rest, or to stop it when it is moving, or to change its velocity sideways out of a straight-line path. A truck has much more inertia than a baseball moving at the same speed, and a much greater force is needed to change the truck's velocity at the same rate as the ball's. The truck therefore has much more mass.

To quantify the concept of mass, we must define a standard. In SI units, the unit of mass is the **kilogram** (kg) as we discussed in Chapter 1, Section 1–4.

The terms *mass* and *weight* are often confused with one another, but it is important to distinguish between them. Mass is a property of an object itself (a measure of an object's inertia, or its "quantity of matter"). Weight, on the other hand, is a force, the pull of gravity acting on an object. To see the difference, suppose we take an object to the Moon. The object will weigh only about one-sixth as much as it did on Earth, since the force of gravity is weaker. But its mass will be the same. It will have the same amount of matter as on Earth, and will have just as much inertia—for in the absence of friction, it will be just as hard to start it moving on the Moon as on Earth, or to stop it once it is moving. (More on weight in Section 4–6.)

⚠ CAUTION
Distinguish mass from weight

4-4 Newton's Second Law of Motion

Newton's first law states that if no net force is acting on an object at rest, the object remains at rest; or if the object is moving, it continues moving with constant speed in a straight line. But what happens if a net force *is* exerted on an object? Newton perceived that the object's velocity will change (Fig. 4–5). A net force exerted on an object may make its velocity increase. Or, if the net force is in a direction opposite to the motion, the force will reduce the object's velocity. If the net force acts sideways on a moving object, the *direction* of the object's velocity changes (and the magnitude may as well). Since a change in velocity is an acceleration (Section 2–4), we can say that *a net force causes acceleration*.

What precisely is the relationship between acceleration and force? Everyday experience can suggest an answer. Consider the force required to push a cart when friction is small enough to ignore. (If there is friction, consider the *net* force, which is the force you exert minus the force of friction.) If you push the cart with a gentle but constant force for a certain period of time, you will make the cart accelerate from rest up to some speed, say 3 km/h. If you push with twice the force, the cart will reach 3 km/h in half the time. The acceleration will be twice as great. If you triple the force, the acceleration is tripled, and so on. Thus, the acceleration of an object is directly proportional to the net applied force. But the acceleration depends on the mass of the object as well. If you push an empty grocery cart with the same force as you push one that is filled with groceries, you will find that the full cart accelerates more slowly. The greater the mass, the less the acceleration for the same net force. The mathematical relation, as Newton argued, is that the acceleration of an object is inversely proportional to its mass. These relationships are found to hold in general and can be summarized as follows:

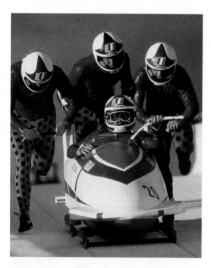
FIGURE 4–5 The bobsled accelerates because the team exerts a force.

NEWTON'S SECOND LAW OF MOTION

> **The acceleration of an object is directly proportional to the net force acting on it, and is inversely proportional to the object's mass. The direction of the acceleration is in the direction of the net force acting on the object.**

This is **Newton's second law of motion**.

Newton's second law can be written as an equation:

$$\vec{\mathbf{a}} = \frac{\Sigma\vec{\mathbf{F}}}{m},$$

where $\vec{\mathbf{a}}$ stands for acceleration, m for the mass, and $\Sigma\vec{\mathbf{F}}$ for the *net force* on the object. The symbol Σ (Greek "sigma") stands for "sum of"; $\vec{\mathbf{F}}$ stands for force, so $\Sigma\vec{\mathbf{F}}$ means the *vector sum of all forces* acting on the object, which we define as the **net force**.

We rearrange this equation to obtain the familiar statement of Newton's second law:

$$\Sigma\vec{\mathbf{F}} = m\vec{\mathbf{a}}. \tag{4–1a}$$

NEWTON'S SECOND LAW OF MOTION

Newton's second law relates the description of motion (acceleration) to the cause of motion (force). It is one of the most fundamental relationships in physics. From Newton's second law we can make a more precise definition of **force** as *an action capable of accelerating an object.*

Every force $\vec{\mathbf{F}}$ is a vector, with magnitude and direction. Equation 4–1a is a vector equation valid in any inertial reference frame. It can be written in component form in rectangular coordinates as

$$\Sigma F_x = ma_x, \qquad \Sigma F_y = ma_y, \qquad \Sigma F_z = ma_z, \tag{4–1b}$$

where

$$\vec{\mathbf{F}} = F_x\hat{\mathbf{i}} + F_y\hat{\mathbf{j}} + F_z\hat{\mathbf{k}}.$$

The component of acceleration in each direction is affected only by the component of the net force in that direction.

In SI units, with the mass in kilograms, the unit of force is called the **newton** (N). One newton, then, is the force required to impart an acceleration of 1 m/s^2 to a mass of 1 kg. Thus $1 \text{ N} = 1 \text{ kg} \cdot \text{m/s}^2$.

In cgs units, the unit of mass is the gram (g) as mentioned earlier.[†] The unit of force is the *dyne*, which is defined as the net force needed to impart an acceleration of 1 cm/s^2 to a mass of 1 g. Thus $1 \text{ dyne} = 1 \text{ g} \cdot \text{cm/s}^2$. It is easy to show that $1 \text{ dyne} = 10^{-5} \text{ N}$.

In the British system, the unit of force is the *pound* (abbreviated lb), where $1 \text{ lb} = 4.448222 \text{ N} \approx 4.45 \text{ N}$. The unit of mass is the *slug*, which is defined as that mass which will undergo an acceleration of 1 ft/s^2 when a force of 1 lb is applied to it. Thus $1 \text{ lb} = 1 \text{ slug} \cdot \text{ft/s}^2$. Table 4–1 summarizes the units in the different systems.

It is very important that only one set of units be used in a given calculation or problem, with the SI being preferred. If the force is given in, say, newtons, and the mass in grams, then before attempting to solve for the acceleration in SI units, we must change the mass to kilograms. For example, if the force is given as 2.0 N along the x axis and the mass is 500 g, we change the latter to 0.50 kg, and the acceleration will then automatically come out in m/s^2 when Newton's second law is used:

$$a_x = \frac{\Sigma F_x}{m} = \frac{2.0 \text{ N}}{0.50 \text{ kg}} = \frac{2.0 \text{ kg} \cdot \text{m/s}^2}{0.50 \text{ kg}} = 4.0 \text{ m/s}^2.$$

TABLE 4–1
Units for Mass and Force

System	Mass	Force
SI	kilogram (kg)	newton (N) ($= \text{kg} \cdot \text{m/s}^2$)
cgs	gram (g)	dyne ($= \text{g} \cdot \text{cm/s}^2$)
British	slug	pound (lb)

Conversion factors: $1 \text{ dyne} = 10^{-5} \text{ N}$;
$1 \text{ lb} \approx 4.45 \text{ N}$.

 PROBLEM SOLVING
Use a consistent set of units

EXAMPLE 4–2 **ESTIMATE** **Force to accelerate a fast car.** Estimate the net force needed to accelerate (a) a 1000-kg car at $\frac{1}{2}g$; (b) a 200-g apple at the same rate.

APPROACH We use Newton's second law to find the net force needed for each object. This is an estimate (the $\frac{1}{2}$ is not said to be precise) so we round off to one significant figure.

SOLUTION (a) The car's acceleration is $a = \frac{1}{2}g = \frac{1}{2}(9.8 \text{ m/s}^2) \approx 5 \text{ m/s}^2$. We use Newton's second law to get the net force needed to achieve this acceleration:

$$\Sigma F = ma \approx (1000 \text{ kg})(5 \text{ m/s}^2) = 5000 \text{ N}.$$

(If you are used to British units, to get an idea of what a 5000-N force is, you can divide by 4.45 N/lb and get a force of about 1000 lb.)

(b) For the apple, $m = 200 \text{ g} = 0.2 \text{ kg}$, so

$$\Sigma F = ma \approx (0.2 \text{ kg})(5 \text{ m/s}^2) = 1 \text{ N}.$$

[†]Be careful not to confuse g for gram with g for the acceleration due to gravity. The latter is always italicized (or boldface when a vector).

EXAMPLE 4–3 **Force to stop a car.** What average net force is required to bring a 1500-kg car to rest from a speed of 100 km/h within a distance of 55 m?

APPROACH We use Newton's second law, $\Sigma F = ma$, to determine the force, but first we need to calculate the acceleration a. We assume the acceleration is constant, so we can use the kinematic equations, Eqs. 2–12, to calculate it.

FIGURE 4–6
Example 4–3.

$v_0 = 100$ km/h $v = 0$

$x = 0$ $x = 55$ m x (m)

SOLUTION We assume the motion is along the $+x$ axis (Fig. 4–6). We are given the initial velocity $v_0 = 100$ km/h $= 27.8$ m/s (Section 1–5), the final velocity $v = 0$, and the distance traveled $x - x_0 = 55$ m. From Eq. 2–12c, we have

$$v^2 = v_0^2 + 2a(x - x_0),$$

so

$$a = \frac{v^2 - v_0^2}{2(x - x_0)} = \frac{0 - (27.8 \text{ m/s})^2}{2(55 \text{ m})} = -7.0 \text{ m/s}^2.$$

The net force required is then

$$\Sigma F = ma = (1500 \text{ kg})(-7.1 \text{ m/s}^2) = -1.1 \times 10^4 \text{ N}.$$

The force must be exerted in the direction *opposite* to the initial velocity, which is what the negative sign means.

NOTE If the acceleration is not precisely constant, then we are determining an "average" acceleration and we obtain an "average" net force.

Newton's second law, like the first law, is valid only in inertial reference frames (Section 4–2). In the noninertial reference frame of an accelerating car, for example, a cup on the dashboard starts sliding—it accelerates—even though the net force on it is zero; thus $\Sigma \vec{F} = m\vec{a}$ doesn't work in such an accelerating reference frame ($\Sigma \vec{F} = 0$, but $\vec{a} \neq 0$ in this noninertial frame).

EXERCISE A Suppose you watch a cup slide on the (smooth) dashboard of an accelerating car as we just discussed, but this time from an inertial reference frame outside the car, on the street. From your inertial frame, Newton's laws are valid. What force pushes the cup off the dashboard?

Precise Definition of Mass

As mentioned in Section 4–3, we can quantify the concept of mass using its definition as a measure of inertia. How to do this is evident from Eq. 4–1a, where we see that the acceleration of an object is inversely proportional to its mass. If the same net force ΣF acts to accelerate each of two masses, m_1 and m_2, then the ratio of their masses can be defined as the inverse ratio of their accelerations:

$$\frac{m_2}{m_1} = \frac{a_1}{a_2}.$$

If one of the masses is known (it could be the standard kilogram) and the two accelerations are precisely measured, then the unknown mass is obtained from this definition. For example, if $m_1 = 1.00$ kg, and for a particular force $a_1 = 3.00$ m/s² and $a_2 = 2.00$ m/s², then $m_2 = 1.50$ kg.

4-5 Newton's Third Law of Motion

Newton's second law of motion describes quantitatively how forces affect motion. But where, we may ask, do forces come from? Observations suggest that a force exerted on any object is always exerted *by another object.* A horse pulls a wagon, a person pushes a grocery cart, a hammer pushes on a nail, a magnet attracts a paper clip. In each of these examples, a force is exerted *on* one object, and that force is exerted *by* another object. For example, the force exerted *on* the nail is exerted *by* the hammer.

But Newton realized that things are not so one-sided. True, the hammer exerts a force on the nail (Fig. 4–7). But the nail evidently exerts a force back on the hammer as well, for the hammer's speed is rapidly reduced to zero upon contact. Only a strong force could cause such a rapid deceleration of the hammer. Thus, said Newton, the two objects must be treated on an equal basis. The hammer exerts a force on the nail, and the nail exerts a force back on the hammer. This is the essence of **Newton's third law of motion**:

> **Whenever one object exerts a force on a second object, the second exerts an equal force in the opposite direction on the first.**

This law is sometimes paraphrased as "to every action there is an equal and opposite reaction." This is perfectly valid. But to avoid confusion, it is very important to remember that the "action" force and the "reaction" force are acting on *different* objects.

As evidence for the validity of Newton's third law, look at your hand when you push against the edge of a desk, Fig. 4–8. Your hand's shape is distorted, clear evidence that a force is being exerted on it. You can *see* the edge of the desk pressing into your hand. You can even *feel* the desk exerting a force on your hand; it hurts! The harder you push against the desk, the harder the desk pushes back on your hand. (You only feel forces exerted *on* you; when you exert a force on another object, what you feel is that object pushing back on you.)

FIGURE 4–7 A hammer striking a nail. The hammer exerts a force on the nail and the nail exerts a force back on the hammer. The latter force decelerates the hammer and brings it to rest.

NEWTON'S THIRD LAW OF MOTION

⚠ **CAUTION**

Action and reaction forces act on different objects

Force exerted on hand by desk

Force exerted on desk by hand

FIGURE 4–8 If your hand pushes against the edge of a desk (the force vector is shown in red), the desk pushes back against your hand (this force vector is shown in a different color, violet, to remind us that this force acts on a different object).

The force the desk exerts on your hand has the same magnitude as the force your hand exerts on the desk. This is true not only if the desk is at rest but is true even if the desk is accelerating due to the force your hand exerts.

As another demonstration of Newton's third law, consider the ice skater in Fig. 4–9. There is very little friction between her skates and the ice, so she will move freely if a force is exerted on her. She pushes against the wall; and then *she* starts moving backward. The force she exerts on the wall cannot make *her* start moving, for that force acts on the wall. Something had to exert a force *on her* to start her moving, and that force could only have been exerted by the wall. The force with which the wall pushes on her is, by Newton's third law, equal and opposite to the force she exerts on the wall.

When a person throws a package out of a small boat (initially at rest), the boat starts moving in the opposite direction. The person exerts a force on the package. The package exerts an equal and opposite force back on the person, and this force propels the person (and the boat) backward slightly.

FIGURE 4–9 An example of Newton's third law: when an ice skater pushes against the wall, the wall pushes back and this force causes her to accelerate away.

Force on skater   Force on wall

FIGURE 4–10 Another example of Newton's third law: the launch of a rocket. The rocket engine pushes the gases downward, and the gases exert an equal and opposite force upward on the rocket, accelerating it upward. (A rocket does *not* accelerate as a result of its propelling gases pushing against the ground.)

FIGURE 4–11 We can walk forward because, when one foot pushes backward against the ground, the ground pushes forward on that foot (Newton's third law). The two forces shown *act on different objects*.

Horizontal force exerted on the ground by person's foot
$\vec{\mathbf{F}}_{GP}$

Horizontal force exerted on the person's foot by the ground
$\vec{\mathbf{F}}_{PG}$

NEWTON'S THIRD LAW OF MOTION

Rocket propulsion also is explained using Newton's third law (Fig. 4–10). A common misconception is that rockets accelerate because the gases rushing out the back of the engine push against the ground or the atmosphere. Not true. What happens, instead, is that a rocket exerts a strong force on the gases, expelling them; and the gases exert an equal and opposite force *on the rocket*. It is this latter force that propels the rocket forward—the force exerted *on* the rocket *by* the gases (see Chapter-Opening photo, page 83). Thus, a space vehicle is maneuvered in empty space by firing its rockets in the direction opposite to that in which it needs to accelerate. When the rocket pushes on the gases in one direction, the gases push back on the rocket in the opposite direction. Jet aircraft too accelerate because the gases they thrust out backwards exert a forward force on the engines (Newton's third law).

Consider how we walk. A person begins walking by pushing with the foot backward against the ground. The ground then exerts an equal and opposite force forward on the person (Fig. 4–11), and it is this force, *on the person*, that moves the person forward. (If you doubt this, try walking normally where there is no friction, such as on very smooth slippery ice.) In a similar way, a bird flies forward by exerting a backward force on the air, but it is the air pushing forward (Newton's third law) on the bird's wings that propels the bird forward.

CONCEPTUAL EXAMPLE 4–4 | **What exerts the force to move a car?** What makes a car go forward?

RESPONSE A common answer is that the engine makes the car move forward. But it is not so simple. The engine makes the wheels go around. But if the tires are on slick ice or deep mud, they just spin. Friction is needed. On firm ground, the tires push backward against the ground because of friction. By Newton's third law, the ground pushes on the tires in the opposite direction, accelerating the car forward.

We tend to associate forces with active objects such as humans, animals, engines, or a moving object like a hammer. It is often difficult to see how an inanimate object at rest, such as a wall or a desk, or the wall of an ice rink (Fig. 4–9), can exert a force. The explanation is that every material, no matter how hard, is elastic (springy) at least to some degree. A stretched rubber band can exert a force on a wad of paper and accelerate it to fly across the room. Other materials may not stretch as readily as rubber, but they do stretch or compress when a force is applied to them. And just as a stretched rubber band exerts a force, so does a stretched (or compressed) wall, desk, or car fender.

From the examples discussed above, we can see how important it is to remember *on* what object a given force is exerted and *by* what object that force is exerted. A force influences the motion of an object only when it is applied *on* that object. A force exerted *by* an object does not influence that same object; it only influences the other object *on* which it is exerted. Thus, to avoid confusion, the two prepositions *on* and *by* must always be used—and used with care.

One way to keep clear which force acts on which object is to use double subscripts. For example, the force exerted on the **P**erson by the **G**round as the person walks in Fig. 4–11 can be labeled $\vec{\mathbf{F}}_{PG}$. And the force exerted on the ground by the person is $\vec{\mathbf{F}}_{GP}$. By Newton's third law

$$\vec{\mathbf{F}}_{GP} = -\vec{\mathbf{F}}_{PG}. \tag{4–2}$$

$\vec{\mathbf{F}}_{GP}$ and $\vec{\mathbf{F}}_{PG}$ have the same magnitude (Newton's third law), and the minus sign reminds us that these two forces are in opposite directions.

Note carefully that the two forces shown in Fig. 4–11 act on different objects—hence we used slightly different colors for the vector arrows representing these forces. These two forces would never appear together in a sum of forces in Newton's second law, $\Sigma \vec{\mathbf{F}} = m\vec{\mathbf{a}}$. Why not? Because they act on different objects: $\vec{\mathbf{a}}$ is the acceleration of one particular object, and $\Sigma \vec{\mathbf{F}}$ must include *only* the forces on that *one* object.

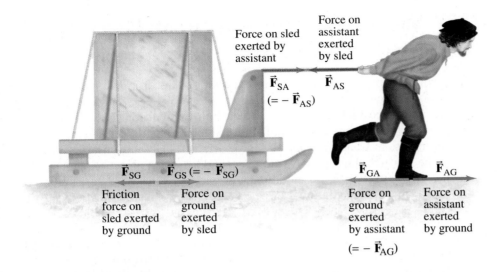

Force on sled
exerted by
assistant

Force on
assistant
exerted
by sled

$\vec{\mathbf{F}}_{SA}$
$(= -\vec{\mathbf{F}}_{AS})$

$\vec{\mathbf{F}}_{AS}$

$\vec{\mathbf{F}}_{SG}$ $\vec{\mathbf{F}}_{GS} (= -\vec{\mathbf{F}}_{SG})$

Friction
force on
sled exerted
by ground

Force on
ground
exerted
by sled

$\vec{\mathbf{F}}_{GA}$

Force on
ground
exerted
by assistant
$(= -\vec{\mathbf{F}}_{AG})$

$\vec{\mathbf{F}}_{AG}$

Force on
assistant
exerted
by ground

FIGURE 4–12 Example 4–5, showing only horizontal forces. Michelangelo has selected a fine block of marble for his next sculpture. Shown here is his assistant pulling it on a sled away from the quarry. Forces on the assistant are shown as red (magenta) arrows. Forces on the sled are purple arrows. Forces acting on the ground are orange arrows. Action–reaction forces that are equal and opposite are labeled by the same subscripts but reversed (such as $\vec{\mathbf{F}}_{GA}$ and $\vec{\mathbf{F}}_{AG}$) and are of different colors because they act on different objects.

| CONCEPTUAL EXAMPLE 4–5 | **Third law clarification.** Michelangelo's assistant has been assigned the task of moving a block of marble using a sled (Fig. 4–12). He says to his boss, "When I exert a forward force on the sled, the sled exerts an equal and opposite force backward. So how can I ever start it moving? No matter how hard I pull, the backward reaction force always equals my forward force, so the net force must be zero. I'll never be able to move this load." Is he correct?

RESPONSE No. Although it is true that the action and reaction forces are equal in magnitude, the assistant has forgotten that they are exerted on different objects. The forward ("action") force is exerted by the assistant on the sled (Fig. 4–12), whereas the backward "reaction" force is exerted by the sled on the assistant. To determine if the *assistant* moves or not, we must consider only the forces *on the assistant* and then apply $\Sigma\vec{\mathbf{F}} = m\vec{\mathbf{a}}$, where $\Sigma\vec{\mathbf{F}}$ is the net force *on the assistant*, $\vec{\mathbf{a}}$ is the acceleration of the assistant, and m is the assistant's mass. There are two forces on the assistant that affect his forward motion; they are shown as bright red (magenta) arrows in Figs. 4–12 and 4–13: they are (1) the horizontal force $\vec{\mathbf{F}}_{AG}$ exerted on the assistant by the ground (the harder he pushes backward against the ground, the harder the ground pushes forward on him—Newton's third law), and (2) the force $\vec{\mathbf{F}}_{AS}$ exerted on the assistant by the sled, pulling backward on him; see Fig. 4–13. If he pushes hard enough on the ground, the force on him exerted by the ground, $\vec{\mathbf{F}}_{AG}$, will be larger than the sled pulling back, $\vec{\mathbf{F}}_{AS}$, and the assistant accelerates forward (Newton's second law). The sled, on the other hand, accelerates forward when the force on *it* exerted by the assistant is greater than the frictional force exerted backward on it by the ground (that is, when $\vec{\mathbf{F}}_{SA}$ has greater magnitude than $\vec{\mathbf{F}}_{SG}$ in Fig. 4–12).

PROBLEM SOLVING

A study of Newton's second and third laws

Force on
assistant
exerted
by sled

$\vec{\mathbf{F}}_{AS}$

$\vec{\mathbf{F}}_{AG}$

Force on
assistant
exerted
by ground

FIGURE 4–13 Example 4–5. The horizontal forces on the assistant.

Using double subscripts to clarify Newton's third law can become cumbersome, and we won't usually use them in this way. We will usually use a single subscript referring to what exerts the force on the object being discussed. Nevertheless, if there is any confusion in your mind about a given force, go ahead and use two subscripts to identify *on* what object and *by* what object the force is exerted.

EXERCISE B Return to the first Chapter-Opening Question, page 83, and answer it again now. Try to explain why you may have answered differently the first time.

EXERCISE C A massive truck collides head-on with a small sports car. (*a*) Which vehicle experiences the greater force of impact? (*b*) Which experiences the greater acceleration during the impact? (*c*) Which of Newton's laws are useful to obtain the correct answers?

EXERCISE D If you push on a heavy desk, does it always push back on you? (*a*) Not unless someone else also pushes on it. (*b*) Yes, if it is out in space. (*c*) A desk never pushes to start with. (*d*) No. (*e*) Yes.

4–6 Weight—the Force of Gravity; and the Normal Force

As we saw in Chapter 2, Galileo claimed that all objects dropped near the surface of the Earth would fall with the same acceleration, $\vec{g}$, if air resistance was negligible. The force that causes this acceleration is called the *force of gravity* or *gravitational force*. What exerts the gravitational force on an object? It is the Earth, as we will discuss in Chapter 6, and the force acts vertically[†] downward, toward the center of the Earth. Let us apply Newton's second law to an object of mass m falling freely due to gravity. For the acceleration, $\vec{a}$, we use the downward acceleration due to gravity, $\vec{g}$. Thus, the **gravitational force** on an object, $\vec{F}_G$, can be written as

$$\vec{F}_G = m\vec{g}. \tag{4–3}$$

The direction of this force is down toward the center of the Earth. The magnitude of the force of gravity on an object, mg, is commonly called the object's **weight**.

In SI units, $g = 9.80 \text{ m/s}^2 = 9.80 \text{ N/kg},$[‡] so the weight of a 1.00-kg mass on Earth is $1.00 \text{ kg} \times 9.80 \text{ m/s}^2 = 9.80 \text{ N}$. We will mainly be concerned with the weight of objects on Earth, but we note that on the Moon, on other planets, or in space, the weight of a given mass will be different than it is on Earth. For example, on the Moon the acceleration due to gravity is about one-sixth what it is on Earth, and a 1.0-kg mass weighs only 1.6 N. Although we will not use British units, we note that for practical purposes on the Earth, a mass of 1 kg weighs about 2.2 lb. (On the Moon, 1 kg weighs only about 0.4 lb.)

The force of gravity acts on an object when it is falling. When an object is at rest on the Earth, the gravitational force on it does not disappear, as we know if we weigh it on a spring scale. The same force, given by Eq. 4–3, continues to act. Why, then, doesn't the object move? From Newton's second law, the net force on an object that remains at rest is zero. There must be another force on the object to balance the gravitational force. For an object resting on a table, the table exerts this upward force; see Fig. 4–14a. The table is compressed slightly beneath the object, and due to its elasticity, it pushes up on the object as shown. The force exerted by the table is often called a **contact force**, since it occurs when two objects are in contact. (The force of your hand pushing on a cart is also a contact force.) When a contact force acts *perpendicular* to the common surface of contact, it is referred to as the **normal force** ("normal" means perpendicular); hence it is labeled $\vec{F}_N$ in Fig. 4–14a.

The two forces shown in Fig. 4–14a are both acting on the statue, which remains at rest, so the vector sum of these two forces must be zero (Newton's second law). Hence $\vec{F}_G$ and $\vec{F}_N$ must be of equal magnitude and in opposite directions. But they are *not* the equal and opposite forces spoken of in Newton's third law. The action and reaction forces of Newton's third law act on *different objects*, whereas the two forces shown in Fig. 4–14a act on the *same* object. For each of the forces shown in Fig. 4–14a, we can ask, "What is the reaction force?" The upward force, $\vec{F}_N$, on the statue is exerted by the table. The reaction to this force is a force exerted by the statue downward on the table. It is shown in Fig. 4–14b, where it is labeled $\vec{F}'_N$. This force, $\vec{F}'_N$, exerted on the table by the statue, is the reaction force to $\vec{F}_N$ in accord with Newton's third law. What about the other force on the statue, the force of gravity $\vec{F}_G$ exerted by the Earth? Can you guess what the reaction is to this force? We will see in Chapter 6 that the reaction force is also a gravitational force, exerted on the Earth by the statue.

EXERCISE E Return to the second Chapter-Opening Question, page 83, and answer it again now. Try to explain why you may have answered differently the first time.

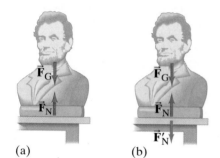

(a)   (b)

FIGURE 4–14 (a) The net force on an object at rest is zero according to Newton's second law. Therefore the downward force of gravity ($\vec{F}_G$) on an object at rest must be balanced by an upward force (the normal force $\vec{F}_N$) exerted by the table in this case. (b) $\vec{F}'_N$ is the force exerted on the table by the statue and is the reaction force to $\vec{F}_N$ by Newton's third law. ($\vec{F}'_N$ is shown in a different color to remind us it acts on a different object.) The reaction force to $\vec{F}_G$ is not shown.

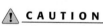

⚠ **CAUTION**

*Weight and normal force are **not** action–reaction pairs*

[†]The concept of "vertical" is tied to gravity. The best definition of *vertical* is that it is the direction in which objects fall. A surface that is "horizontal," on the other hand, is a surface on which a round object won't start rolling: gravity has no effect. Horizontal is perpendicular to vertical.

[‡]Since $1 \text{ N} = 1 \text{ kg} \cdot \text{m/s}^2$ (Section 4–4), then $1 \text{ m/s}^2 = 1 \text{ N/kg}$.

EXAMPLE 4–6 **Weight, normal force, and a box.** A friend has given you a special gift, a box of mass 10.0 kg with a mystery surprise inside. The box is resting on the smooth (frictionless) horizontal surface of a table (Fig. 4–15a). (*a*) Determine the weight of the box and the normal force exerted on it by the table. (*b*) Now your friend pushes down on the box with a force of 40.0 N, as in Fig. 4–15b. Again determine the normal force exerted on the box by the table. (*c*) If your friend pulls upward on the box with a force of 40.0 N (Fig. 4–15c), what now is the normal force exerted on the box by the table?

APPROACH The box is at rest on the table, so the net force on the box in each case is zero (Newton's second law). The weight of the box has magnitude mg in all three cases.

SOLUTION (*a*) The weight of the box is $mg = (10.0\,\text{kg})(9.80\,\text{m/s}^2) = 98.0\,\text{N}$, and this force acts downward. The only other force on the box is the normal force exerted upward on it by the table, as shown in Fig. 4–15a. We chose the upward direction as the positive y direction; then the net force ΣF_y on the box is $\Sigma F_y = F_N - mg$; the minus sign means mg acts in the negative y direction (m and g are magnitudes). The box is at rest, so the net force on it must be zero (Newton's second law, $\Sigma F_y = ma_y$, and $a_y = 0$). Thus

$$\Sigma F_y = ma_y$$
$$F_N - mg = 0,$$

so we have

$$F_N = mg.$$

The normal force on the box, exerted by the table, is 98.0 N upward, and has magnitude equal to the box's weight.

(*b*) Your friend is pushing down on the box with a force of 40.0 N. So instead of only two forces acting on the box, now there are three forces acting on the box, as shown in Fig. 4–15b. The weight of the box is still $mg = 98.0\,\text{N}$. The net force is $\Sigma F_y = F_N - mg - 40.0\,\text{N}$, and is equal to zero because the box remains at rest ($a = 0$). Newton's second law gives

$$\Sigma F_y = F_N - mg - 40.0\,\text{N} = 0.$$

We solve this equation for the normal force:

$$F_N = mg + 40.0\,\text{N} = 98.0\,\text{N} + 40.0\,\text{N} = 138.0\,\text{N},$$

which is greater than in (*a*). The table pushes back with more force when a person pushes down on the box. The normal force is not always equal to the weight!

(*c*) The box's weight is still 98.0 N and acts downward. The force exerted by your friend and the normal force both act upward (positive direction), as shown in Fig. 4–15c. The box doesn't move since your friend's upward force is less than the weight. The net force, again set to zero in Newton's second law because $a = 0$, is

$$\Sigma F_y = F_N - mg + 40.0\,\text{N} = 0,$$

so

$$F_N = mg - 40.0\,\text{N} = 98.0\,\text{N} - 40.0\,\text{N} = 58.0\,\text{N}.$$

The table does not push against the full weight of the box because of the upward pull exerted by your friend.

NOTE The weight of the box ($= mg$) does not change as a result of your friend's push or pull. Only the normal force is affected.

Recall that the normal force is elastic in origin (the table in Fig. 4–15 sags slightly under the weight of the box). The normal force in Example 4–6 is vertical, perpendicular to the horizontal table. The normal force is not always vertical, however. When you push against a wall, for example, the normal force with which the wall pushes back on you is horizontal (Fig. 4–9). For an object on a plane inclined at an angle to the horizontal, such as a skier or car on a hill, the normal force acts perpendicular to the plane and so is not vertical.

(a) $\Sigma F_y = F_N - mg = 0$

(b) $\Sigma F_y = F_N - mg - 40.0\,\text{N} = 0$

(c) $\Sigma F_y = F_N - mg + 40.0\,\text{N} = 0$

FIGURE 4–15 Example 4–6.
(a) A 10-kg gift box is at rest on a table. (b) A person pushes down on the box with a force of 40.0 N. (c) A person pulls upward on the box with a force of 40.0 N. The forces are all assumed to act along a line; they are shown slightly displaced in order to be distinguishable. Only forces acting on the box are shown.

⚠ **CAUTION**
The normal force is not always equal to the weight

⚠ **CAUTION**
The normal force, $\vec{F}_N$, is not necessarily vertical

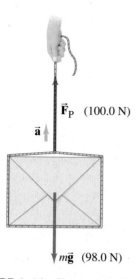

$\vec{\mathbf{F}}_P$ (100.0 N)

$\vec{\mathbf{a}}$

$m\vec{\mathbf{g}}$ (98.0 N)

FIGURE 4–16 Example 4–7. The box accelerates upward because $F_P > mg$.

EXAMPLE 4–7 **Accelerating the box.** What happens when a person pulls upward on the box in Example 4–6c with a force equal to, or greater than, the box's weight? For example, let $F_P = 100.0 \, \text{N}$ (Fig. 4–16) rather than the 40.0 N shown in Fig. 4–15c.

APPROACH We can start just as in Example 4–6, but be ready for a surprise.

SOLUTION The net force on the box is

$$\Sigma F_y = F_N - mg + F_P$$
$$= F_N - 98.0 \, \text{N} + 100.0 \, \text{N},$$

and if we set this equal to zero (thinking the acceleration might be zero), we would get $F_N = -2.0 \, \text{N}$. This is nonsense, since the negative sign implies F_N points downward, and the table surely cannot *pull* down on the box (unless there's glue on the table). The least F_N can be is zero, which it will be in this case. What really happens here is that the box accelerates upward because the net force is not zero. The net force (setting the normal force $F_N = 0$) is

$$\Sigma F_y = F_P - mg = 100.0 \, \text{N} - 98.0 \, \text{N}$$
$$= 2.0 \, \text{N}$$

upward. See Fig. 4–16. We apply Newton's second law and see that the box moves upward with an acceleration

$$a_y = \frac{\Sigma F_y}{m} = \frac{2.0 \, \text{N}}{10.0 \, \text{kg}}$$
$$= 0.20 \, \text{m/s}^2.$$

FIGURE 4–17 Example 4–8. The acceleration vector is shown in gold to distinguish it from the red force vectors.

$\vec{\mathbf{a}}$

$m\vec{\mathbf{g}}$

$\vec{\mathbf{F}}_N$

EXAMPLE 4–8 **Apparent weight loss.** A 65-kg woman descends in an elevator that briefly accelerates at 0.20g downward. She stands on a scale that reads in kg. (*a*) During this acceleration, what is her weight and what does the scale read? (*b*) What does the scale read when the elevator descends at a constant speed of 2.0 m/s?

APPROACH Figure 4–17 shows all the forces that act on the woman (and *only* those that act on her). The direction of the acceleration is downward, so we choose the positive direction as down (this is the opposite choice from Examples 4–6 and 4–7).

SOLUTION (*a*) From Newton's second law,

$$\Sigma F = ma$$
$$mg - F_N = m(0.20g).$$

We solve for F_N:

$$F_N = mg - 0.20mg = 0.80mg,$$

and it acts upward. The normal force $\vec{\mathbf{F}}_N$ is the force the scale exerts on the person, and is equal and opposite to the force she exerts on the scale: $F_N' = 0.80mg$ downward. Her weight (force of gravity on her) is still $mg = (65 \, \text{kg})(9.8 \, \text{m/s}^2) = 640 \, \text{N}$. But the scale, needing to exert a force of only $0.80mg$, will give a reading of $0.80m = 52 \, \text{kg}$.

(*b*) Now there is no acceleration, $a = 0$, so by Newton's second law, $mg - F_N = 0$ and $F_N = mg$. The scale reads her true mass of 65 kg.

NOTE The scale in (*a*) may give a reading of 52 kg (as an "apparent mass"), but her mass doesn't change as a result of the acceleration: it stays at 65 kg.

4–7 Solving Problems with Newton's Laws: Free-Body Diagrams

Newton's second law tells us that the acceleration of an object is proportional to the *net force* acting on the object. The **net force**, as mentioned earlier, is the *vector sum* of all forces acting on the object. Indeed, extensive experiments have shown that forces do add together as vectors precisely according to the rules we developed in Chapter 3. For example, in Fig. 4–18, two forces of equal magnitude (100 N each) are shown acting on an object at right angles to each other. Intuitively, we can see that the object will start moving at a 45° angle and thus the net force acts at a 45° angle. This is just what the rules of vector addition give. From the theorem of Pythagoras, the magnitude of the resultant force is $F_R = \sqrt{(100\,\text{N})^2 + (100\,\text{N})^2} = 141\,\text{N}$.

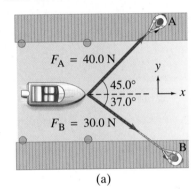

(a)   (b)

FIGURE 4–18 (a) Two forces, $\vec{\mathbf{F}}_A$ and $\vec{\mathbf{F}}_B$, exerted by workers A and B, act on a crate. (b) The sum, or resultant, of $\vec{\mathbf{F}}_A$ and $\vec{\mathbf{F}}_B$ is $\vec{\mathbf{F}}_R$.

EXAMPLE 4–9 **Adding force vectors.** Calculate the sum of the two forces exerted on the boat by workers A and B in Fig. 4–19a.

APPROACH We add force vectors like any other vectors as described in Chapter 3. The first step is to choose an *xy* coordinate system (see Fig. 4–19a), and then resolve vectors into their components.

SOLUTION The two force vectors are shown resolved into components in Fig. 4–19b. We add the forces using the method of components. The components of $\vec{\mathbf{F}}_A$ are

$$F_{Ax} = F_A \cos 45.0° = (40.0\,\text{N})(0.707) = 28.3\,\text{N},$$
$$F_{Ay} = F_A \sin 45.0° = (40.0\,\text{N})(0.707) = 28.3\,\text{N}.$$

The components of $\vec{\mathbf{F}}_B$ are

$$F_{Bx} = +F_B \cos 37.0° = +(30.0\,\text{N})(0.799) = +24.0\,\text{N},$$
$$F_{By} = -F_B \sin 37.0° = -(30.0\,\text{N})(0.602) = -18.1\,\text{N}.$$

F_{By} is negative because it points along the negative *y* axis. The components of the resultant force are (see Fig. 4–19c)

$$F_{Rx} = F_{Ax} + F_{Bx} = 28.3\,\text{N} + 24.0\,\text{N} = 52.3\,\text{N},$$
$$F_{Ry} = F_{Ay} + F_{By} = 28.3\,\text{N} - 18.1\,\text{N} = 10.2\,\text{N}.$$

To find the magnitude of the resultant force, we use the Pythagorean theorem

$$F_R = \sqrt{F_{Rx}^2 + F_{Ry}^2} = \sqrt{(52.3)^2 + (10.2)^2}\,\text{N} = 53.3\,\text{N}.$$

The only remaining question is the angle θ that the net force $\vec{\mathbf{F}}_R$ makes with the *x* axis. We use:

$$\tan \theta = \frac{F_{Ry}}{F_{Rx}} = \frac{10.2\,\text{N}}{52.3\,\text{N}} = 0.195,$$

and $\tan^{-1}(0.195) = 11.0°$. The net force on the boat has magnitude 53.3 N and acts at an 11.0° angle to the *x* axis.

FIGURE 4–19 Example 4–9: Two force vectors act on a boat.

(a)

(b)

(c)

When solving problems involving Newton's laws and force, it is very important to draw a diagram showing all the forces acting *on* each object involved. Such a diagram is called a **free-body diagram**, or **force diagram**: choose one object, and draw an arrow to represent each force acting on it. Include *every* force acting on that object. Do not show forces that the chosen object exerts on *other* objects. To help you identify each and every force that is exerted on your chosen object, ask yourself what other objects could exert a force on it. If your problem involves more than one object, a separate free-body diagram is needed for each object. For now, the likely forces that could be acting are *gravity* and *contact forces* (one object pushing or pulling another, normal force, friction). Later we will consider air resistance, drag, buoyancy, pressure, as well as electric and magnetic forces.

PROBLEM SOLVING

Free-body diagram

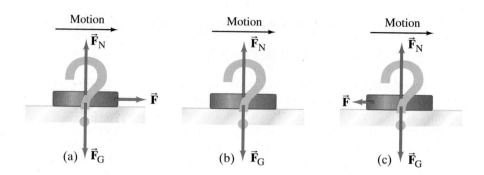

FIGURE 4–20 Example 4–10. Which is the correct free-body diagram for a hockey puck sliding across frictionless ice?

CONCEPTUAL EXAMPLE 4–10 **The hockey puck.** A hockey puck is sliding at constant velocity across a flat horizontal ice surface that is assumed to be frictionless. Which of the sketches in Fig. 4–20 is the correct free-body diagram for this puck? What would your answer be if the puck slowed down?

RESPONSE Did you choose (*a*)? If so, can you answer the question: what exerts the horizontal force labeled $\vec{\mathbf{F}}$ on the puck? If you say that it is the force needed to maintain the motion, ask yourself: what exerts this force? Remember that another object must exert any force—and there simply isn't any possibility here. Therefore, (*a*) is wrong. Besides, the force $\vec{\mathbf{F}}$ in Fig. 4–20a would give rise to an acceleration by Newton's second law. It is (*b*) that is correct. No net force acts on the puck, and the puck slides at constant velocity across the ice.

In the real world, where even smooth ice exerts at least a tiny friction force, then (*c*) is the correct answer. The tiny friction force is in the direction opposite to the motion, and the puck's velocity decreases, even if very slowly.

Here now is a brief summary of how to approach solving problems involving Newton's laws.

Newton's Laws; Free-Body Diagrams

1. **Draw a sketch** of the situation.
2. Consider only one object (at a time), and draw a **free-body diagram** for that object, showing *all* the forces acting *on* that object. Include any unknown forces that you have to solve for. Do not show any forces that the chosen object exerts on other objects.

   Draw the arrow for each force vector reasonably accurately for direction and magnitude. Label each force acting on the object, including forces you must solve for, as to its source (gravity, person, friction, and so on).

   If several objects are involved, draw a free-body diagram for each object *separately*, showing all the forces acting *on that object* (and *only* forces acting on that

object). For each (and every) force, you must be clear about: *on* what object that force acts, and *by* what object that force is exerted. Only forces acting *on* a given object can be included in $\Sigma\vec{\mathbf{F}} = m\vec{\mathbf{a}}$ for that object.

3. Newton's second law involves vectors, and it is usually important to **resolve vectors** into components. **Choose** *x* and *y* **axes** in a way that simplifies the calculation. For example, it often saves work if you choose one coordinate axis to be in the direction of the acceleration.

4. For each object, **apply Newton's second law** to the *x* and *y* components separately. That is, the *x* component of the net force on that object is related to the *x* component of that object's acceleration: $\Sigma F_x = ma_x$, and similarly for the *y* direction.

5. **Solve** the equation or equations for the unknown(s).

This Problem Solving Strategy should not be considered a prescription. Rather it is a summary of things to do that will start you thinking and getting involved in the problem at hand.

When we are concerned only about translational motion, all the forces on a given object can be drawn as acting at the center of the object, thus treating the object as a *point particle*. However, for problems involving rotation or statics, the place *where* each force acts is also important, as we shall see in Chapters 10, 11, and 12.

In the Examples that follow, we assume that all surfaces are very smooth so that friction can be ignored. (Friction, and Examples using it, are discussed in Chapter 5).

⚠ **CAUTION**
Treating an object as a particle

EXAMPLE 4–11 **Pulling the mystery box.** Suppose a friend asks to examine the 10.0-kg box you were given (Example 4–6, Fig. 4–15), hoping to guess what is inside; and you respond, "Sure, pull the box over to you." She then pulls the box by the attached cord, as shown in Fig. 4–21a, along the smooth surface of the table. The magnitude of the force exerted by the person is $F_P = 40.0$ N, and it is exerted at a 30.0° angle as shown. Calculate (a) the acceleration of the box, and (b) the magnitude of the upward force F_N exerted by the table on the box. Assume that friction can be neglected.

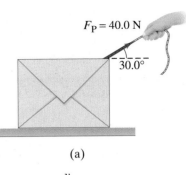

APPROACH We follow the Problem Solving Strategy on the previous page.

SOLUTION

1. **Draw a sketch**: The situation is shown in Fig. 4–21a; it shows the box and the force applied by the person, F_P.

2. **Free-body diagram**: Figure 4–21b shows the free-body diagram of the box. To draw it correctly, we show *all* the forces acting on the box and *only* the forces acting on the box. They are: the force of gravity $m\vec{g}$; the normal force exerted by the table $\vec{F}_N$; and the force exerted by the person $\vec{F}_P$. We are interested only in translational motion, so we can show the three forces acting at a point, Fig. 4–21c.

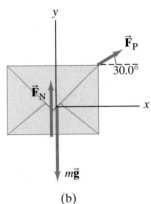

3. **Choose axes and resolve vectors**: We expect the motion to be horizontal, so we choose the x axis horizontal and the y axis vertical. The pull of 40.0 N has components

$$F_{Px} = (40.0 \text{ N})(\cos 30.0°) = (40.0 \text{ N})(0.866) = 34.6 \text{ N},$$
$$F_{Py} = (40.0 \text{ N})(\sin 30.0°) = (40.0 \text{ N})(0.500) = 20.0 \text{ N}.$$

In the horizontal (x) direction, $\vec{F}_N$ and $m\vec{g}$ have zero components. Thus the horizontal component of the net force is F_{Px}.

4. (a) **Apply Newton's second law** to determine the x component of the acceleration:

$$F_{Px} = ma_x.$$

5. (a) **Solve**:

$$a_x = \frac{F_{Px}}{m} = \frac{(34.6 \text{ N})}{(10.0 \text{ kg})} = 3.46 \text{ m/s}^2.$$

The acceleration of the box is 3.46 m/s^2 to the right.

(b) Next we want to find F_N.

4′. (b) **Apply Newton's second law** to the vertical (y) direction, with upward as positive:

$$\Sigma F_y = ma_y$$
$$F_N - mg + F_{Py} = ma_y.$$

5′. (b) **Solve**: We have $mg = (10.0 \text{ kg})(9.80 \text{ m/s}^2) = 98.0 \text{ N}$ and, from point 3 above, $F_{Py} = 20.0$ N. Furthermore, since $F_{Py} < mg$, the box does not move vertically, so $a_y = 0$. Thus

$$F_N - 98.0 \text{ N} + 20.0 \text{ N} = 0,$$

so

$$F_N = 78.0 \text{ N}.$$

NOTE F_N is less than mg: the table does not push against the full weight of the box because part of the pull exerted by the person is in the upward direction.

FIGURE 4–21 (a) Pulling the box, Example 4–11; (b) is the free-body diagram for the box, and (c) is the free-body diagram considering all the forces to act at a point (translational motion only, which is what we have here).

EXERCISE F A 10.0-kg box is dragged on a horizontal frictionless surface by a horizontal force of 10.0 N. If the applied force is doubled, the normal force on the box will (a) increase; (b) remain the same; (c) decrease.

Tension in a Flexible Cord

When a flexible cord pulls on an object, the cord is said to be under **tension**, and the force it exerts on the object is the tension F_T. If the cord has negligible mass, the force exerted at one end is transmitted undiminished to each adjacent piece of cord along the entire length to the other end. Why? Because $\Sigma\vec{F} = m\vec{a} = 0$ for the cord if the cord's mass m is zero (or negligible) no matter what $\vec{a}$ is. Hence the forces pulling on the cord at its two ends must add up to zero (F_T and $-F_T$). Note that flexible cords and strings can only pull. They can't push because they bend.

PROBLEM SOLVING
Cords can pull but can't push; tension exists throughout a cord

FIGURE 4–22 Example 4–12. (a) Two boxes, A and B, are connected by a cord. A person pulls horizontally on box A with force $F_P = 40.0$ N. (b) Free-body diagram for box A. (c) Free-body diagram for box B.

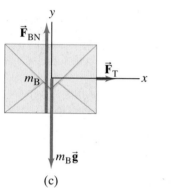

Our next Example involves two boxes connected by a cord. We can refer to this group of objects as a system. A *system* is any group of one or more objects we choose to consider and study.

EXAMPLE 4–12 **Two boxes connected by a cord.** Two boxes, A and B, are connected by a lightweight cord and are resting on a smooth (frictionless) table. The boxes have masses of 12.0 kg and 10.0 kg. A horizontal force F_P of 40.0 N is applied to the 10.0-kg box, as shown in Fig. 4–22a. Find (a) the acceleration of each box, and (b) the tension in the cord connecting the boxes.

APPROACH We streamline our approach by not listing each step. We have two boxes so we draw a free-body diagram for each. To draw them correctly, we must consider the forces on *each* box by itself, so that Newton's second law can be applied to each. The person exerts a force F_P on box A. Box A exerts a force F_T on the connecting cord, and the cord exerts an opposite but equal magnitude force F_T back on box A (Newton's third law). These two horizontal forces on box A are shown in Fig. 4–22b, along with the force of gravity $m_A \vec{\mathbf{g}}$ downward and the normal force $\vec{\mathbf{F}}_{AN}$ exerted upward by the table. The cord is light, so we neglect its mass. The tension at each end of the cord is thus the same. Hence the cord exerts a force F_T on the second box. Figure 4–22c shows the forces on box B, which are $\vec{\mathbf{F}}_T$, $m_B \vec{\mathbf{g}}$, and the normal force $\vec{\mathbf{F}}_{BN}$. There will be only horizontal motion. We take the positive x axis to the right.

SOLUTION (a) We apply $\Sigma F_x = m a_x$ to box A:

$$\Sigma F_x = F_P - F_T = m_A a_A. \qquad \text{[box A]}$$

For box B, the only horizontal force is F_T, so

$$\Sigma F_x = F_T = m_B a_B. \qquad \text{[box B]}$$

The boxes are connected, and if the cord remains taut and doesn't stretch, then the two boxes will have the same acceleration a. Thus $a_A = a_B = a$. We are given $m_A = 10.0$ kg and $m_B = 12.0$ kg. We can add the two equations above to eliminate an unknown (F_T) and obtain

$$(m_A + m_B)a = F_P - F_T + F_T = F_P$$

or

$$a = \frac{F_P}{m_A + m_B} = \frac{40.0 \text{ N}}{22.0 \text{ kg}} = 1.82 \text{ m/s}^2.$$

This is what we sought.

Alternate Solution We would have obtained the same result had we considered a single system, of mass $m_A + m_B$, acted on by a net horizontal force equal to F_P. (The tension forces F_T would then be considered internal to the system as a whole, and summed together would make zero contribution to the net force on the *whole* system.)

(b) From the equation above for box B $(F_T = m_B a_B)$, the tension in the cord is

$$F_T = m_B a = (12.0 \text{ kg})(1.82 \text{ m/s}^2) = 21.8 \text{ N}.$$

Thus, F_T is less than $F_P (= 40.0$ N), as we expect, since F_T acts to accelerate only m_B.

NOTE It might be tempting to say that the force the person exerts, F_P, acts not only on box A but also on box B. It doesn't. F_P acts only on box A. It affects box B via the tension in the cord, F_T, which acts on box B and accelerates it.

⚠ **CAUTION**

For any object, use only the forces on that object in calculating $\Sigma F = ma$

EXAMPLE 4–13 **Elevator and counterweight (Atwood's machine).** A system of two objects suspended over a pulley by a flexible cable, as shown in Fig. 4–23a, is sometimes referred to as an *Atwood's machine*. Consider the real-life application of an elevator (m_E) and its counterweight (m_C). To minimize the work done by the motor to raise and lower the elevator safely, m_E and m_C are made similar in mass. We leave the motor out of the system for this calculation, and assume that the cable's mass is negligible and that the mass of the pulley, as well as any friction, is small and ignorable. These assumptions ensure that the tension F_T in the cable has the same magnitude on both sides of the pulley. Let the mass of the counterweight be $m_C = 1000$ kg. Assume the mass of the empty elevator is 850 kg, and its mass when carrying four passengers is $m_E = 1150$ kg. For the latter case ($m_E = 1150$ kg), calculate (*a*) the acceleration of the elevator and (*b*) the tension in the cable.

APPROACH Again we have two objects, and we will need to apply Newton's second law to each of them separately. Each mass has two forces acting on it: gravity downward and the cable tension pulling upward, $\vec{F}_T$. Figures 4–23b and c show the free-body diagrams for the elevator (m_E) and for the counterweight (m_C). The elevator, being the heavier, will accelerate downward, whereas the counterweight will accelerate upward. The magnitudes of their accelerations will be equal (we assume the cable doesn't stretch). For the counterweight, $m_C g = (1000 \text{ kg})(9.80 \text{ m/s}^2) = 9800 \text{ N}$, so F_T must be greater than 9800 N (in order that m_C will accelerate upward). For the elevator, $m_E g = (1150 \text{ kg})(9.80 \text{ m/s}^2) = 11,300 \text{ N}$, which must have greater magnitude than F_T so that m_E accelerates downward. Thus our calculation must give F_T between 9800 N and 11,300 N.

SOLUTION (*a*) To find F_T as well as the acceleration a, we apply Newton's second law, $\Sigma F = ma$, to each object. We take upward as the positive y direction for both objects. With this choice of axes, $a_C = a$ because m_C accelerates upward, and $a_E = -a$ because m_E accelerates downward. Thus

$$F_T - m_E g = m_E a_E = -m_E a$$
$$F_T - m_C g = m_C a_C = +m_C a.$$

We can subtract the first equation from the second to get

$$(m_E - m_C)g = (m_E + m_C)a,$$

where a is now the only unknown. We solve this for a:

$$a = \frac{m_E - m_C}{m_E + m_C} g = \frac{1150 \text{ kg} - 1000 \text{ kg}}{1150 \text{ kg} + 1000 \text{ kg}} g = 0.070g = 0.68 \text{ m/s}^2.$$

The elevator (m_E) accelerates downward (and the counterweight m_C upward) at $a = 0.070g = 0.68 \text{ m/s}^2$.

(*b*) The tension in the cable F_T can be obtained from either of the two $\Sigma F = ma$ equations, setting $a = 0.070g = 0.68 \text{ m/s}^2$:

$$F_T = m_E g - m_E a = m_E(g - a)$$
$$= 1150 \text{ kg} (9.80 \text{ m/s}^2 - 0.68 \text{ m/s}^2) = 10,500 \text{ N},$$

or

$$F_T = m_C g + m_C a = m_C(g + a)$$
$$= 1000 \text{ kg} (9.80 \text{ m/s}^2 + 0.68 \text{ m/s}^2) = 10,500 \text{ N},$$

which are consistent. As predicted, our result lies between 9800 N and 11,300 N.

NOTE We can check our equation for the acceleration a in this Example by noting that if the masses were equal ($m_E = m_C$), then our equation above for a would give $a = 0$, as we should expect. Also, if one of the masses is zero (say, $m_C = 0$), then the other mass ($m_E \neq 0$) would be predicted by our equation to accelerate at $a = g$, again as expected.

(a)

(b)   (c)

FIGURE 4–23 Example 4–13. (a) Atwood's machine in the form of an elevator–counterweight system. (b) and (c) Free-body diagrams for the two objects.

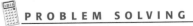
PROBLEM SOLVING
Check your result by seeing if it works in situations where the answer is easily guessed

FIGURE 4–24 Example 4–14.

CONCEPTUAL EXAMPLE 4–14 **The advantage of a pulley.** A mover is trying to lift a piano (slowly) up to a second-story apartment (Fig. 4–24). He is using a rope looped over two pulleys as shown. What force must he exert on the rope to slowly lift the piano's 2000-N weight?

RESPONSE The magnitude of the tension force F_T within the rope is the same at any point along the rope if we assume we can ignore its mass. First notice the forces acting on the lower pulley at the piano. The weight of the piano pulls down on the pulley via a short cable. The tension in the rope, looped through this pulley, pulls up *twice*, once on each side of the pulley. Let us apply Newton's second law to the pulley–piano combination (of mass m), choosing the upward direction as positive:

$$2F_T - mg = ma.$$

To move the piano with constant speed (set $a = 0$ in this equation) thus requires a tension in the rope, and hence a pull on the rope, of $F_T = mg/2$. The mover can exert a force equal to half the piano's weight. We say the pulley has given a **mechanical advantage** of 2, since without the pulley the mover would have to exert twice the force.

FIGURE 4–25 Example 4–15.

(a)

(b)

EXAMPLE 4–15 **Accelerometer.** A small mass m hangs from a thin string and can swing like a pendulum. You attach it above the window of your car as shown in Fig. 4–25a. When the car is at rest, the string hangs vertically. What angle θ does the string make (a) when the car accelerates at a constant $a = 1.20 \text{ m/s}^2$, and (b) when the car moves at constant velocity, $v = 90 \text{ km/h}$?

APPROACH The free-body diagram of Fig. 4–25b shows the pendulum at some angle θ and the forces on it: $m\vec{g}$ downward, and the tension $\vec{F}_T$ in the cord. These forces do not add up to zero if $\theta \neq 0$, and since we have an acceleration a, we therefore expect $\theta \neq 0$. Note that θ is the angle relative to the vertical.

SOLUTION (a) The acceleration $a = 1.20 \text{ m/s}^2$ is horizontal, so from Newton's second law,

$$ma = F_T \sin\theta$$

for the horizontal component, whereas the vertical component gives

$$0 = F_T \cos\theta - mg.$$

Dividing these two equations, we obtain

$$\tan\theta = \frac{F_T \sin\theta}{F_T \cos\theta} = \frac{ma}{mg} = \frac{a}{g}$$

or

$$\tan\theta = \frac{1.20 \text{ m/s}^2}{9.80 \text{ m/s}^2}$$

$$= 0.122,$$

so

$$\theta = 7.0°.$$

(b) The velocity is constant, so $a = 0$ and $\tan\theta = 0$. Hence the pendulum hangs vertically $(\theta = 0°)$.

NOTE This simple device is an **accelerometer**—it can be used to measure acceleration.

Inclines

Now we consider what happens when an object slides down an incline, such as a hill or ramp. Such problems are interesting because gravity is the accelerating force, yet the acceleration is not vertical. Solving such problems is usually easier if we choose the xy coordinate system so that one axis points in the direction of the acceleration. Thus we often take the x axis to point along the incline and the y axis perpendicular to the incline, as shown in Fig. 4–26a. Note also that the normal force is not vertical, but is perpendicular to the plane, Fig. 4–26b.

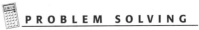
EXAMPLE 4–16 **Box slides down an incline.** A box of mass m is placed on a smooth (frictionless) incline that makes an angle θ with the horizontal, as shown in Fig. 4–26a. (a) Determine the normal force on the box. (b) Determine the box's acceleration. (c) Evaluate for a mass $m = 10\,\text{kg}$ and an incline of $\theta = 30°$.

APPROACH We expect the motion to be along the incline, so we choose the x axis along the slope, positive down the slope (the direction of motion). The y axis is perpendicular to the incline, upward. The free-body diagram is shown in Fig. 4–26b. The forces on the box are its weight mg vertically downward, which is shown resolved into its components parallel and perpendicular to the incline, and the normal force F_N. The incline acts as a constraint, allowing motion along its surface. The "constraining" force is the normal force.

SOLUTION (a) There is no motion in the y direction, so $a_y = 0$. Applying Newton's second law we have

$$F_y = ma_y$$

$$F_N - mg\cos\theta = 0,$$

where F_N and the y component of gravity ($mg\cos\theta$) are all the forces acting on the box in the y direction. Thus the normal force is given by

$$F_N = mg\cos\theta.$$

Note carefully that unless $\theta = 0°$, F_N has magnitude less than the weight mg.
(b) In the x direction the only force acting is the x component of $m\vec{g}$, which we see from the diagram is $mg\sin\theta$. The acceleration a is in the x direction so

$$F_x = ma_x$$

$$mg\sin\theta = ma,$$

and we see that the acceleration down the plane is

$$a = g\sin\theta.$$

Thus the acceleration along an incline is always less than g, except at $\theta = 90°$, for which $\sin\theta = 1$ and $a = g$. This makes sense since $\theta = 90°$ is pure vertical fall. For $\theta = 0°$, $a = 0$, which makes sense because $\theta = 0°$ means the plane is horizontal so gravity causes no acceleration. Note too that the acceleration does not depend on the mass m.
(c) For $\theta = 30°$, $\cos\theta = 0.866$ and $\sin\theta = 0.500$, so

$$F_N = 0.866mg = 85\,\text{N},$$

and

$$a = 0.500g = 4.9\,\text{m/s}^2.$$

We will discuss more Examples of motion on an incline in the next Chapter, where friction will be included.

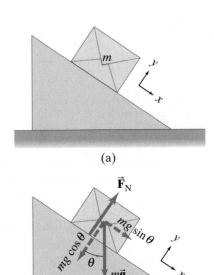

(a)

(b)

FIGURE 4–26 Example 4–16. (a) Box sliding on inclined plane. (b) Free-body diagram of box.

4–8 Problem Solving—A General Approach

A basic part of a physics course is solving problems effectively. The approach discussed here, though emphasizing Newton's laws, can be applied generally for other topics discussed throughout this book.

PROBLEM SOLVING

In General

1. **Read** and reread written problems carefully. A common error is to skip a word or two when reading, which can completely change the meaning of a problem.

2. **Draw** an accurate picture or diagram of the situation. (This is probably the most overlooked, yet most crucial, part of solving a problem.) Use arrows to represent vectors such as velocity or force, and label the vectors with appropriate symbols. When dealing with forces and applying Newton's laws, make sure to include all forces on a given object, including unknown ones, and make clear what forces act on what object (otherwise you may make an error in determining the *net force* on a particular object).

3. A separate **free-body diagram** needs to be drawn for each object involved, and it must show *all* the forces acting on a given object (and only on that object). Do not show forces that act on other objects.

4. Choose a convenient *xy* **coordinate system** (one that makes your calculations easier, such as one axis in the direction of the acceleration). Vectors are to be resolved into components along the coordinate axes. When using Newton's second law, apply $\Sigma \vec{F} = m\vec{a}$ separately to *x* and *y* components, remembering that *x* direction forces are related to a_x, and similarly for *y*. If more than one object is involved, you can choose different (convenient) coordinate systems for each.

5. List the knowns and the unknowns (what you are trying to determine), and decide what you need in order to find the unknowns. For problems in the present Chapter, we use Newton's laws. More generally, it may help to see if one or more **relationships** (or **equations**) relate the unknowns to the knowns.

But be sure each relationship is applicable in the given case. It is very important to know the limitations of each formula or relationship—when it is valid and when not. In this book, the more general equations have been given numbers, but even these can have a limited range of validity (often stated in brackets to the right of the equation).

6. Try to solve the problem approximately, to see if it is doable (to check if enough information has been given) and reasonable. Use your intuition, and make **rough calculations**—see "Order of Magnitude Estimating" in Section 1–6. A rough calculation, or a reasonable guess about what the range of final answers might be, is very useful. And a rough calculation can be checked against the final answer to catch errors in calculation, such as in a decimal point or the powers of 10.

7. **Solve** the problem, which may include algebraic manipulation of equations and/or numerical calculations. Recall the mathematical rule that you need as many independent equations as you have unknowns; if you have three unknowns, for example, then you need three independent equations. It is usually best to work out the algebra symbolically before putting in the numbers. Why? Because (*a*) you can then solve a whole class of similar problems with different numerical values; (*b*) you can check your result for cases already understood (say, $\theta = 0°$ or 90°); (*c*) there may be cancellations or other simplifications; (*d*) there is usually less chance for numerical error; and (*e*) you may gain better insight into the problem.

8. Be sure to keep track of **units**, for they can serve as a check (they must balance on both sides of any equation).

9. Again consider if your answer is **reasonable**. The use of dimensional analysis, described in Section 1–7, can also serve as a check for many problems.

Summary

Newton's three laws of motion are the basic classical laws describing motion.

Newton's first law (the **law of inertia**) states that if the net force on an object is zero, an object originally at rest remains at rest, and an object in motion remains in motion in a straight line with constant velocity.

Newton's second law states that the acceleration of an object is directly proportional to the net force acting on it, and inversely proportional to its mass:

$$\Sigma \vec{F} = m\vec{a}. \tag{4–1a}$$

Newton's second law is one of the most important and fundamental laws in classical physics.

Newton's third law states that whenever one object exerts a force on a second object, the second object always exerts a force on the first object which is equal in magnitude but opposite in direction:

$$\vec{F}_{AB} = -\vec{F}_{BA}, \qquad (4\text{--}2)$$

where $\vec{F}_{BA}$ is the force on object B exerted by object A. This is true even if objects are moving and accelerating, and/or have different masses.

The tendency of an object to resist a change in its motion is called **inertia**. **Mass** is a measure of the inertia of an object.

Weight refers to the **gravitational force** on an object, and is equal to the product of the object's mass m and the acceleration of gravity $\vec{g}$:

$$\vec{F}_G = m\vec{g}. \qquad (4\text{--}3)$$

Force, which is a vector, can be considered as a push or pull; or, from Newton's second law, force can be defined as an action capable of giving rise to acceleration. The **net force** on an object is the vector sum of all forces acting on that object.

For solving problems involving the forces on one or more objects, it is essential to draw a **free-body diagram** for each object, showing all the forces acting on only that object. Newton's second law can be applied to the vector components for each object.

Questions

1. Why does a child in a wagon seem to fall backward when you give the wagon a sharp pull forward?

2. A box rests on the (frictionless) bed of a truck. The truck driver starts the truck and accelerates forward. The box immediately starts to slide toward the rear of the truck bed. Discuss the motion of the box, in terms of Newton's laws, as seen (a) by Andrea standing on the ground beside the truck, and (b) by Jim who is riding on the truck (Fig. 4–27).

FIGURE 4–27 Question 2.

3. If the acceleration of an object is zero, are no forces acting on it? Explain.

4. If an object is moving, is it possible for the net force acting on it to be zero?

5. Only one force acts on an object. Can the object have zero acceleration? Can it have zero velocity? Explain.

6. When a golf ball is dropped to the pavement, it bounces back up. (a) Is a force needed to make it bounce back up? (b) If so, what exerts the force?

7. If you walk along a log floating on a lake, why does the log move in the opposite direction?

8. Why might your foot hurt if you kick a heavy desk or a wall?

9. When you are running and want to stop quickly, you must decelerate quickly. (a) What is the origin of the force that causes you to stop? (b) Estimate (using your own experience) the maximum rate of deceleration of a person running at top speed to come to rest.

10. (a) Why do you push down harder on the pedals of a bicycle when first starting out than when moving at constant speed? (b) Why do you need to pedal at all when cycling at constant speed?

11. A father and his young daughter are ice skating. They face each other at rest and push each other, moving in opposite directions. Which one has the greater final speed?

12. Suppose that you are standing on a cardboard carton that just barely supports you. What would happen to it if you jumped up into the air? It would (a) collapse; (b) be unaffected; (c) spring upward a bit; (d) move sideways.

13. A stone hangs by a fine thread from the ceiling, and a section of the same thread dangles from the bottom of the stone (Fig. 4–28). If a person gives a sharp pull on the dangling thread, where is the thread likely to break: below the stone or above it? What if the person gives a slow and steady pull? Explain your answers.

FIGURE 4–28
Question 13.

14. The force of gravity on a 2-kg rock is twice as great as that on a 1-kg rock. Why then doesn't the heavier rock fall faster?

15. Would a spring scale carried to the Moon give accurate results if the scale had been calibrated on Earth, (a) in pounds, or (b) in kilograms?

16. You pull a box with a constant force across a frictionless table using an attached rope held horizontally. If you now pull the rope with the same force at an angle to the horizontal (with the box remaining flat on the table), does the acceleration of the box (a) remain the same, (b) increase, or (c) decrease? Explain.

17. When an object falls freely under the influence of gravity there is a net force mg exerted on it by the Earth. Yet by Newton's third law the object exerts an equal and opposite force on the Earth. Does the Earth move?

18. Compare the effort (or force) needed to lift a 10-kg object when you are on the Moon with the force needed to lift it on Earth. Compare the force needed to throw a 2-kg object horizontally with a given speed on the Moon and on Earth.

19. Which of the following objects weighs about 1 N: (*a*) an apple, (*b*) a mosquito, (*c*) this book, (*d*) you?

20. According to Newton's third law, each team in a tug of war (Fig. 4–29) pulls with equal force on the other team. What, then, determines which team will win?

FIGURE 4–29 Question 20. A tug of war. Describe the forces on each of the teams and on the rope.

21. When you stand still on the ground, how large a force does the ground exert on you? Why doesn't this force make you rise up into the air?

22. Whiplash sometimes results from an automobile accident when the victim's car is struck violently from the rear. Explain why the head of the victim seems to be thrown backward in this situation. Is it really?

23. Mary exerts an upward force of 40 N to hold a bag of groceries. Describe the "reaction" force (Newton's third law) by stating (*a*) its magnitude, (*b*) its direction, (*c*) *on* what object it is exerted, and (*d*) *by* what object it is exerted.

24. A bear sling, Fig. 4–30, is used in some national parks for placing backpackers' food out of the reach of bears. Explain why the force needed to pull the backpack up increases as the backpack gets higher and higher. Is it possible to pull the rope hard enough so that it doesn't sag at all?

FIGURE 4–30 Question 24.

Problems

4–4 to 4–6 Newton's Laws, Gravitational Force, Normal Force

1. (I) What force is needed to accelerate a child on a sled (total mass = 55 kg) at 1.4 m/s^2?

2. (I) A net force of 265 N accelerates a bike and rider at 2.30 m/s^2. What is the mass of the bike and rider together?

3. (I) What is the weight of a 68-kg astronaut (*a*) on Earth, (*b*) on the Moon ($g = 1.7 \text{ m/s}^2$), (*c*) on Mars ($g = 3.7 \text{ m/s}^2$), (*d*) in outer space traveling with constant velocity?

4. (I) How much tension must a rope withstand if it is used to accelerate a 1210-kg car horizontally along a frictionless surface at 1.20 m/s^2?

5. (II) Superman must stop a 120-km/h train in 150 m to keep it from hitting a stalled car on the tracks. If the train's mass is 3.6×10^5 kg, how much force must he exert? Compare to the weight of the train (give as %). How much force does the train exert on Superman?

6. (II) What average force is required to stop a 950-kg car in 8.0 s if the car is traveling at 95 km/h?

7. (II) Estimate the average force exerted by a shot-putter on a 7.0-kg shot if the shot is moved through a distance of 2.8 m and is released with a speed of 13 m/s.

8. (II) A 0.140-kg baseball traveling 35.0 m/s strikes the catcher's mitt, which, in bringing the ball to rest, recoils backward 11.0 cm. What was the average force applied by the ball on the glove?

9. (II) A fisherman yanks a fish vertically out of the water with an acceleration of 2.5 m/s^2 using very light fishing line that has a breaking strength of 18 N (≈ 4 lb). The fisherman unfortunately loses the fish as the line snaps. What can you say about the mass of the fish?

10. (II) A 20.0-kg box rests on a table. (*a*) What is the weight of the box and the normal force acting on it? (*b*) A 10.0-kg box is placed on top of the 20.0-kg box, as shown in Fig. 4–31. Determine the normal force that the table exerts on the 20.0-kg box and the normal force that the 20.0-kg box exerts on the 10.0-kg box.

FIGURE 4–31 Problem 10.

11. (II) What average force is needed to accelerate a 9.20-gram pellet from rest to 125 m/s over a distance of 0.800 m along the barrel of a rifle?

12. (II) How much tension must a cable withstand if it is used to accelerate a 1200-kg car vertically upward at 0.70 m/s^2?

13. (II) A 14.0-kg bucket is lowered vertically by a rope in which there is 163 N of tension at a given instant. What is the acceleration of the bucket? Is it up or down?

14. (II) A particular race car can cover a quarter-mile track (402 m) in 6.40 s starting from a standstill. Assuming the acceleration is constant, how many "g's" does the driver experience? If the combined mass of the driver and race car is 535 kg, what horizontal force must the road exert on the tires?

15. (II) A 75-kg petty thief wants to escape from a third-story jail window. Unfortunately, a makeshift rope made of sheets tied together can support a mass of only 58 kg. How might the thief use this "rope" to escape? Give a quantitative answer.

16. (II) An elevator (mass 4850 kg) is to be designed so that the maximum acceleration is 0.0680g. What are the maximum and minimum forces the motor should exert on the supporting cable?

17. (II) Can cars "stop on a dime"? Calculate the acceleration of a 1400-kg car if it can stop from 35 km/h on a dime (diameter = 1.7 cm.) How many g's is this? What is the force felt by the 68-kg occupant of the car?

18. (II) A person stands on a bathroom scale in a motionless elevator. When the elevator begins to move, the scale briefly reads only 0.75 of the person's regular weight. Calculate the acceleration of the elevator, and find the direction of acceleration.

19. (II) High-speed elevators function under two limitations: (1) the maximum magnitude of vertical acceleration that a typical human body can experience without discomfort is about $1.2 \, \text{m/s}^2$, and (2) the typical maximum speed attainable is about 9.0 m/s. You board an elevator on a skyscraper's ground floor and are transported 180 m above the ground level in three steps: acceleration of magnitude $1.2 \, \text{m/s}^2$ from rest to 9.0 m/s, followed by constant upward velocity of 9.0 m/s, then deceleration of magnitude $1.2 \, \text{m/s}^2$ from 9.0 m/s to rest. (a) Determine the elapsed time for each of these 3 stages. (b) Determine the change in the magnitude of the normal force, expressed as a % of your normal weight during each stage. (c) What fraction of the total transport time does the normal force not equal the person's weight?

20. (II) Using focused laser light, *optical tweezers* can apply a force of about 10 pN to a 1.0-μm diameter polystyrene bead, which has a density about equal to that of water: a volume of $1.0 \, \text{cm}^3$ has a mass of about 1.0 g. Estimate the bead's acceleration in g's.

21. (II) A rocket with a mass of 2.75×10^6 kg exerts a vertical force of 3.55×10^7 N on the gases it expels. Determine (a) the acceleration of the rocket, (b) its velocity after 8.0 s, and (c) how long it takes to reach an altitude of 9500 m. Assume g remains constant, and ignore the mass of gas expelled (not realistic).

22. (II) (a) What is the acceleration of two falling sky divers (mass = 132 kg including parachute) when the upward force of air resistance is equal to one-fourth of their weight? (b) After popping open the parachute, the divers descend leisurely to the ground at constant speed. What now is the force of air resistance on the sky divers and their parachute? See Fig. 4–32.

FIGURE 4–32 Problem 22.

23. (II) An exceptional standing jump would raise a person 0.80 m off the ground. To do this, what force must a 68-kg person exert against the ground? Assume the person crouches a distance of 0.20 m prior to jumping, and thus the upward force has this distance to act over before he leaves the ground.

24. (II) The cable supporting a 2125-kg elevator has a maximum strength of 21,750 N. What maximum upward acceleration can it give the elevator without breaking?

25. (III) The 100-m dash can be run by the best sprinters in 10.0 s. A 66-kg sprinter accelerates uniformly for the first 45 m to reach top speed, which he maintains for the remaining 55 m. (a) What is the average horizontal component of force exerted on his feet by the ground during acceleration? (b) What is the speed of the sprinter over the last 55 m of the race (i.e., his top speed)?

26. (III) A person jumps from the roof of a house 3.9-m high. When he strikes the ground below, he bends his knees so that his torso decelerates over an approximate distance of 0.70 m. If the mass of his torso (excluding legs) is 42 kg, find (a) his velocity just before his feet strike the ground, and (b) the average force exerted on his torso by his legs during deceleration.

4–7 Using Newton's Laws

27. (I) A box weighing 77.0 N rests on a table. A rope tied to the box runs vertically upward over a pulley and a weight is hung from the other end (Fig. 4–33). Determine the force that the table exerts on the box if the weight hanging on the other side of the pulley weighs (a) 30.0 N, (b) 60.0 N, and (c) 90.0 N.

FIGURE 4–33
Problem 27.

28. (I) Draw the free-body diagram for a basketball player (a) just before leaving the ground on a jump, and (b) while in the air. See Fig. 4–34.

FIGURE 4–34
Problem 28.

29. (I) Sketch the free-body diagram of a baseball (a) at the moment it is hit by the bat, and again (b) after it has left the bat and is flying toward the outfield.

30. (I) A 650-N force acts in a northwesterly direction. A second 650-N force must be exerted in what direction so that the resultant of the two forces points westward? Illustrate your answer with a vector diagram.

31. (II) Christian is making a Tyrolean traverse as shown in Fig. 4–35. That is, he traverses a chasm by stringing a rope between a tree on one side of the chasm and a tree on the opposite side, 25 m away. The rope must sag sufficiently so it won't break. Assume the rope can provide a tension force of up to 29 kN before breaking, and use a "safety factor" of 10 (that is, the rope should only be required to undergo a tension force of 2.9 kN) at the center of the Tyrolean traverse. (a) Determine the distance x that the rope must sag if it is to be within its recommended safety range and Christian's mass is 72.0 kg. (b) If the Tyrolean traverse is incorrectly set up so that the rope sags by only one-fourth the distance found in (a), determine the tension force in the rope. Will the rope break?

FIGURE 4–35   Problem 31.

32. (II) A window washer pulls herself upward using the bucket–pulley apparatus shown in Fig. 4–36. (a) How hard must she pull downward to raise herself slowly at constant speed? (b) If she increases this force by 15%, what will her acceleration be? The mass of the person plus the bucket is 72 kg.

FIGURE 4–36
Problem 32.

33. (II) One 3.2-kg paint bucket is hanging by a massless cord from another 3.2-kg paint bucket, also hanging by a massless cord, as shown in Fig. 4–37. (a) If the buckets are at rest, what is the tension in each cord? (b) If the two buckets are pulled upward with an acceleration of 1.25 m/s² by the upper cord, calculate the tension in each cord.

FIGURE 4–37
Problems 33 and 34.

34. (II) The cords accelerating the buckets in Problem 33b, Fig. 4–37, each has a weight of 2.0 N. Determine the tension in each cord at the three points of attachment.

35. (II) Two snowcats in Antarctica are towing a housing unit to a new location, as shown in Fig. 4–38. The sum of the forces $\vec{F}_A$ and $\vec{F}_B$ exerted on the unit by the horizontal cables is parallel to the line L, and $F_A = 4500$ N. Determine F_B and the magnitude of $\vec{F}_A + \vec{F}_B$.

FIGURE 4–38
Problem 35.

36. (II) A train locomotive is pulling two cars of the same mass behind it, Fig. 4–39. Determine the ratio of the tension in the coupling (think of it as a cord) between the locomotive and the first car (F_{T1}), to that between the first car and the second car (F_{T2}), for any nonzero acceleration of the train.

FIGURE 4–39   Problem 36.

37. (II) The two forces $\vec{F}_1$ and $\vec{F}_2$ shown in Fig. 4–40a and b (looking down) act on a 18.5-kg object on a frictionless tabletop. If $F_1 = 10.2$ N and $F_2 = 16.0$ N, find the net force on the object and its acceleration for (a) and (b).

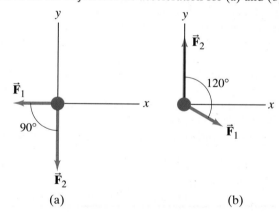

FIGURE 4–40   Problem 37.

38. (II) At the instant a race began, a 65-kg sprinter exerted a force of 720 N on the starting block at a 22° angle with respect to the ground. (a) What was the horizontal acceleration of the sprinter? (b) If the force was exerted for 0.32 s, with what speed did the sprinter leave the starting block?

39. (II) A mass m is at rest on a horizontal frictionless surface at $t = 0$. Then a constant force F_0 acts on it for a time t_0. Suddenly the force doubles to $2F_0$ and remains constant until $t = 2t_0$. Determine the total distance traveled from $t = 0$ to $t = 2t_0$.

40. (II) A 3.0-kg object has the following two forces acting on it:
$$\vec{F}_1 = \left(16\hat{i} + 12\hat{j}\right) \text{N}$$
$$\vec{F}_2 = \left(-10\hat{i} + 22\hat{j}\right) \text{N}$$

If the object is initially at rest, determine its velocity $\vec{v}$ at $t = 3.0$ s.

41. (II) Uphill escape ramps are sometimes provided to the side of steep downhill highways for trucks with overheated brakes. For a simple 11° upward ramp, what length would be needed for a runaway truck traveling 140 km/h? Note the large size of your calculated length. (If sand is used for the bed of the ramp, its length can be reduced by a factor of about 2.)

42. (II) A child on a sled reaches the bottom of a hill with a velocity of 10.0 m/s and travels 25.0 m along a horizontal straightaway to a stop. If the child and sled together have a mass of 60.0 kg, what is the average retarding force on the sled on the horizontal straightaway?

43. (II) A skateboarder, with an initial speed of 2.0 m/s, rolls virtually friction free down a straight incline of length 18 m in 3.3 s. At what angle θ is the incline oriented above the horizontal?

44. (II) As shown in Fig. 4–41, five balls (masses 2.00, 2.05, 2.10, 2.15, 2.20 kg) hang from a crossbar. Each mass is supported by "5-lb test" fishing line which will break when its tension force exceeds 22.2 N (= 5 lb). When this device is placed in an elevator, which accelerates upward, only the lines attached to the 2.05 and 2.00 kg masses do not break. Within what range is the elevator's acceleration?

FIGURE 4–41
Problem 44.

45. (II) A 27-kg chandelier hangs from a ceiling on a vertical 4.0-m-long wire. (a) What horizontal force would be necessary to displace its position 0.15 m to one side? (b) What will be the tension in the wire?

46. (II) Three blocks on a frictionless horizontal surface are in contact with each other as shown in Fig. 4–42. A force $\vec{F}$ is applied to block A (mass m_A). (a) Draw a free-body diagram for each block. Determine (b) the acceleration of the system (in terms of m_A, m_B, and m_C), (c) the net force on each block, and (d) the force of contact that each block exerts on its neighbor. (e) If $m_A = m_B = m_C = 10.0$ kg and $F = 96.0$ N, give numerical answers to (b), (c), and (d). Explain how your answers make sense intuitively.

FIGURE 4–42   Problem 46.

47. (II) Redo Example 4–13 but (a) set up the equations so that the direction of the acceleration $\vec{a}$ of each object is in the direction of motion of that object. (In Example 4–13, we took $\vec{a}$ as positive upward for both masses.) (b) Solve the equations to obtain the same answers as in Example 4–13.

48. (II) The block shown in Fig. 4–43 has mass $m = 7.0$ kg and lies on a fixed smooth frictionless plane tilted at an angle $\theta = 22.0°$ to the horizontal. (a) Determine the acceleration of the block as it slides down the plane. (b) If the block starts from rest 12.0 m up the plane from its base, what will be the block's speed when it reaches the bottom of the incline?

FIGURE 4–43
Block on inclined
plane. Problems 48
and 49.

49. (II) A block is given an initial speed of 4.5 m/s up the 22° plane shown in Fig. 4–43. (a) How far up the plane will it go? (b) How much time elapses before it returns to its starting point? Ignore friction.

50. (II) An object is hanging by a string from your rearview mirror. While you are accelerating at a constant rate from rest to 28 m/s in 6.0 s, what angle θ does the string make with the vertical? See Fig. 4–44.

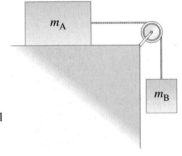

FIGURE 4–44
Problem 50.

51. (II) Figure 4–45 shows a block (mass m_A) on a smooth horizontal surface, connected by a thin cord that passes over a pulley to a second block (m_B), which hangs vertically. (a) Draw a free-body diagram for each block, showing the force of gravity on each, the force (tension) exerted by the cord, and any normal force. (b) Apply Newton's second law to find formulas for the acceleration of the system and for the tension in the cord. Ignore friction and the masses of the pulley and cord.

FIGURE 4–45
Problems 51, 52,
and 53. Mass m_A rests
on a smooth horizontal
surface, m_B hangs
vertically.

52. (II) (a) If $m_A = 13.0$ kg and $m_B = 5.0$ kg in Fig. 4–45, determine the acceleration of each block. (b) If initially m_A is at rest 1.250 m from the edge of the table, how long does it take to reach the edge of the table if the system is allowed to move freely? (c) If $m_B = 1.0$ kg, how large must m_A be if the acceleration of the system is to be kept at $\frac{1}{100}g$?

53. (III) Determine a formula for the acceleration of the system shown in Fig. 4–45 (see Problem 51) if the cord has a non-negligible mass m_C. Specify in terms of ℓ_A and ℓ_B, the lengths of cord from the respective masses to the pulley. (The total cord length is $\ell = \ell_A + \ell_B$.)

54. (III) Suppose the pulley in Fig. 4–46 is suspended by a cord C. Determine the tension in this cord after the masses are released and before one hits the ground. Ignore the mass of the pulley and cords.

FIGURE 4–46
Problem 54.

55. (III) A small block of mass m rests on the sloping side of a triangular block of mass M which itself rests on a horizontal table as shown in Fig. 4–47. Assuming all surfaces are frictionless, determine the magnitude of the force $\vec{\mathbf{F}}$ that must be applied to M so that m remains in a fixed position relative to M (that is, m doesn't move on the incline). [*Hint:* Take x and y axes horizontal and vertical.]

FIGURE 4–47
Problem 55.

56. (III) The double Atwood machine shown in Fig. 4–48 has frictionless, massless pulleys and cords. Determine (*a*) the acceleration of masses m_A, m_B, and m_C, and (*b*) the tensions F_{TA} and F_{TC} in the cords.

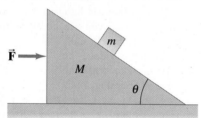

FIGURE 4–48
Problem 56.

57. (III) Suppose two boxes on a frictionless table are connected by a heavy cord of mass 1.0 kg. Calculate the acceleration of each box and the tension at each end of the cord, using the free-body diagrams shown in Fig. 4–49. Assume $F_P = 35.0$ N, and ignore sagging of the cord. Compare your results to Example 4–12 and Fig. 4–22.

58. (III) The two masses shown in Fig. 4–50 are each initially 1.8 m above the ground, and the massless frictionless pulley is 4.8 m above the ground. What maximum height does the lighter object reach after the system is released? [*Hint:* First determine the acceleration of the lighter mass and then its velocity at the moment the heavier one hits the ground. This is its "launch" speed. Assume the mass doesn't hit the pulley. Ignore the mass of the cord.]

FIGURE 4–50
Problem 58.

59. (III) Determine a formula for the magnitude of the force $\vec{\mathbf{F}}$ exerted on the large block (m_C) in Fig. 4–51 so that the mass m_A does not move relative to m_C. Ignore all friction. Assume m_B does not make contact with m_C.

FIGURE 4–51   Problem 59.

60. (III) A particle of mass m, initially at rest at $x = 0$, is accelerated by a force that increases in time as $F = Ct^2$. Determine its velocity v and position x as a function of time.

61. (III) A heavy steel cable of length ℓ and mass M passes over a small massless, frictionless pulley. (*a*) If a length y hangs on one side of the pulley (so $\ell - y$ hangs on the other side), calculate the acceleration of the cable as a function of y. (*b*) Assuming the cable starts from rest with length y_0 on one side of the pulley, determine the velocity v_f at the moment the whole cable has fallen from the pulley. (*c*) Evaluate v_f for $y_0 = \frac{2}{3}\ell$. [*Hint:* Use the chain rule, $dv/dt = (dv/dy)(dy/dt)$, and integrate.]

FIGURE 4–49   Problem 57. Free-body diagrams for each of the objects of the system shown in Fig. 4–22a. Vertical forces, $\vec{\mathbf{F}}_N$ and $\vec{\mathbf{F}}_G$, are not shown.

General Problems

62. A person has a reasonable chance of surviving an automobile crash if the deceleration is no more than 30 *g*'s. Calculate the force on a 65-kg person accelerating at this rate. What distance is traveled if brought to rest at this rate from 95 km/h?

63. A 2.0-kg purse is dropped 58 m from the top of the Leaning Tower of Pisa and falls 55 m before reaching the ground with a speed of 27 m/s. What was the average force of air resistance?

64. Tom's hang glider supports his weight using the six ropes shown in Fig. 4–52. Each rope is designed to support an equal fraction of Tom's weight. Tom's mass is 74.0 kg. What is the tension in each of the support ropes?

FIGURE 4–52 Problem 64.

65. A wet bar of soap (*m* = 150 g) slides freely down a ramp 3.0 m long inclined at 8.5°. How long does it take to reach the bottom? How would this change if the soap's mass were 300 g?

66. A crane's trolley at point P in Fig. 4–53 moves for a few seconds to the right with constant acceleration, and the 870-kg load hangs at a 5.0° angle to the vertical as shown. What is the acceleration of the trolley and load?

FIGURE 4–53 Problem 66.

67. A block (mass m_A) lying on a fixed frictionless inclined plane is connected to a mass m_B by a cord passing over a pulley, as shown in Fig. 4–54. (*a*) Determine a formula for the acceleration of the system in terms of m_A, m_B, θ, and *g*. (*b*) What conditions apply to masses m_A and m_B for the acceleration to be in one direction (say, m_A down the plane), or in the opposite direction? Ignore the mass of the cord and pulley.

FIGURE 4–54
Problems 67 and 68.

68. (*a*) In Fig. 4–54, if $m_A = m_B = 1.00$ kg and $\theta = 33.0°$, what will be the acceleration of the system? (*b*) If $m_A = 1.00$ kg and the system remains at rest, what must the mass m_B be? (*c*) Calculate the tension in the cord for (*a*) and (*b*).

69. The masses m_A and m_B slide on the smooth (frictionless) inclines fixed as shown in Fig. 4–55. (*a*) Determine a formula for the acceleration of the system in terms of m_A, m_B, θ_A, θ_B, and *g*. (*b*) If $\theta_A = 32°$, $\theta_B = 23°$, and $m_A = 5.0$ kg, what value of m_B would keep the system at rest? What would be the tension in the cord (negligible mass) in this case? (*c*) What ratio, m_A/m_B, would allow the masses to move at constant speed along their ramps in either direction?

FIGURE 4–55
Problem 69.

70. A 75.0-kg person stands on a scale in an elevator. What does the scale read (in N and in kg) when (*a*) the elevator is at rest, (*b*) the elevator is climbing at a constant speed of 3.0 m/s, (*c*) the elevator is descending at 3.0 m/s, (*d*) the elevator is accelerating upward at 3.0 m/s², (*e*) the elevator is accelerating downward at 3.0 m/s²?

71. A city planner is working on the redesign of a hilly portion of a city. An important consideration is how steep the roads can be so that even low-powered cars can get up the hills without slowing down. A particular small car, with a mass of 920 kg, can accelerate on a level road from rest to 21 m/s (75 km/h) in 12.5 s. Using these data, calculate the maximum steepness of a hill.

72. If a bicyclist of mass 65 kg (including the bicycle) can coast down a 6.5° hill at a steady speed of 6.0 km/h because of air resistance, how much force must be applied to climb the hill at the same speed (and the same air resistance)?

73. A bicyclist can coast down a 5.0° hill at a constant speed of 6.0 km/h. If the force of air resistance is proportional to the speed *v* so that $F_{air} = cv$, calculate (*a*) the value of the constant *c*, and (*b*) the average force that must be applied in order to descend the hill at 18.0 km/h. The mass of the cyclist plus bicycle is 80.0 kg.

74. Francesca dangles her watch from a thin piece of string while the jetliner she is in accelerates for takeoff, which takes about 16 s. Estimate the takeoff speed of the aircraft if the string makes an angle of 25° with respect to the vertical, Fig. 4–56.

FIGURE 4–56
Problem 74.

75. (*a*) What minimum force *F* is needed to lift the piano (mass *M*) using the pulley apparatus shown in Fig. 4–57? (*b*) Determine the tension in each section of rope: F_{T1}, F_{T2}, F_{T3}, and F_{T4}.

F_{T3}

F_{T2} F_{T1}

F_{T4}

F

FIGURE 4–57
Problem 75.

76. In the design of a supermarket, there are to be several ramps connecting different parts of the store. Customers will have to push grocery carts up the ramps and it is obviously desirable that this not be too difficult. The engineer has done a survey and found that almost no one complains if the force required is no more than 18 N. Ignoring friction, at what maximum angle θ should the ramps be built, assuming a full 25-kg grocery cart?

77. A jet aircraft is accelerating at 3.8 m/s^2 as it climbs at an angle of 18° above the horizontal (Fig. 4–58). What is the total force that the cockpit seat exerts on the 75-kg pilot?

FIGURE 4–58
Problem 77.

18°

78. A 7650-kg helicopter accelerates upward at 0.80 m/s^2 while lifting a 1250-kg frame at a construction site, Fig. 4–59. (*a*) What is the lift force exerted by the air on the helicopter rotors? (*b*) What is the tension in the cable (ignore its mass) that connects the frame to the helicopter? (*c*) What force does the cable exert on the helicopter?

$\vec{\mathbf{a}}$ $\vec{\mathbf{F}}_T$

$m\vec{\mathbf{g}}$

FIGURE 4–59
Problem 78.

79. A super high-speed 14-car Italian train has a mass of 640 metric tons (640,000 kg). It can exert a maximum force of 400 kN horizontally against the tracks, whereas at maximum constant velocity (300 km/h), it exerts a force of about 150 kN. Calculate (*a*) its maximum acceleration, and (*b*) estimate the force of friction and air resistance at top speed.

80. A fisherman in a boat is using a "10-lb test" fishing line. This means that the line can exert a force of 45 N without breaking (1 lb = 4.45 N). (*a*) How heavy a fish can the fisherman land if he pulls the fish up vertically at constant speed? (*b*) If he accelerates the fish upward at 2.0 m/s^2, what maximum weight fish can he land? (*c*) Is it possible to land a 15-lb trout on 10-lb test line? Why or why not?

81. An elevator in a tall building is allowed to reach a maximum speed of 3.5 m/s going down. What must the tension be in the cable to stop this elevator over a distance of 2.6 m if the elevator has a mass of 1450 kg including occupants?

82. Two rock climbers, Bill and Karen, use safety ropes of similar length. Karen's rope is more elastic, called a *dynamic rope* by climbers. Bill has a *static rope*, not recommended for safety purposes in pro climbing. (*a*) Karen falls freely about 2.0 m and then the rope stops her over a distance of 1.0 m (Fig. 4–60). Estimate how large a force (assume constant) she will feel from the rope. (Express the result in multiples of her weight.) (*b*) In a similar fall, Bill's rope stretches by only 30 cm. How many times his weight will the rope pull on him? Which climber is more likely to be hurt?

FIGURE 4–60
Problem 82.

83. Three mountain climbers who are roped together in a line are ascending an icefield inclined at 31.0° to the horizontal (Fig. 4–61). The last climber slips, pulling the second climber off his feet. The first climber is able to hold them both. If each climber has a mass of 75 kg, calculate the tension in each of the two sections of rope between the three climbers. Ignore friction between the ice and the fallen climbers.

31.0°

FIGURE 4–61   Problem 83.

84. A "doomsday" asteroid with a mass of $1.0 \times 10^{10} \text{ kg}$ is hurtling through space. Unless the asteroid's speed is changed by about 0.20 cm/s, it will collide with Earth and cause tremendous damage. Researchers suggest that a small "space tug" sent to the asteroid's surface could exert a gentle constant force of 2.5 N. For how long must this force act?

85. A 450-kg piano is being unloaded from a truck by rolling it down a ramp inclined at 22°. There is negligible friction and the ramp is 11.5 m long. Two workers slow the rate at which the piano moves by pushing with a combined force of 1420 N parallel to the ramp. If the piano starts from rest, how fast is it moving at the bottom?

86. Consider the system shown in Fig. 4–62 with $m_A = 9.5$ kg and $m_B = 11.5$ kg. The angles $\theta_A = 59°$ and $\theta_B = 32°$. (a) In the absence of friction, what force $\vec{F}$ would be required to pull the masses at a constant velocity up the fixed inclines? (b) The force $\vec{F}$ is now removed. What is the magnitude and direction of the acceleration of the two blocks? (c) In the absence of $\vec{F}$, what is the tension in the string?

FIGURE 4–62
Problem 86.

87. A 1.5-kg block rests on top of a 7.5-kg block (Fig. 4–63). The cord and pulley have negligible mass, and there is no significant friction anywhere. (a) What force F must be applied to the bottom block so the top block accelerates to the right at 2.5 m/s²? (b) What is the tension in the connecting cord?

FIGURE 4–63
Problem 87.

88. You are driving home in your 750-kg car at 15 m/s. At a point 45 m from the beginning of an intersection, you see a green traffic light change to yellow, which you expect will last 4.0 s, and the distance to the far side of the intersection is 65 m (Fig. 4–64). (a) If you choose to accelerate, your car's engine will furnish a forward force of 1200 N. Will you make it completely through the intersection before the light turns red? (b) If you decide to panic stop, your brakes will provide a force of 1800 N. Will you stop before entering the intersection?

FIGURE 4–64 Problem 88.

*Numerical/Computer

*89. (II) A large crate of mass 1500 kg starts sliding from rest along a frictionless ramp, whose length is ℓ and whose inclination with the horizontal is θ. (a) Determine as a function of θ: (i) the acceleration a of the crate as it goes downhill, (ii) the time t to reach the bottom of the incline, (iii) the final velocity v of the crate when it reaches the bottom of the ramp, and (iv) the normal force F_N on the crate. (b) Now assume $\ell = 100$ m. Use a spreadsheet to calculate and graph a, t, v, and F_N as functions of θ from $\theta = 0°$ to 90° in 1° steps. Are your results consistent with the known result for the limiting cases $\theta = 0°$ and $\theta = 90°$?

Answers to Exercises

A: No force is needed. The car accelerates out from under the cup. Think of Newton's first law (see Example 4–1).

B: (a).

C: (a) The same; (b) the sports car; (c) third law for part (a), second law for part (b).

D: (e).

E: (b).

F: (b).

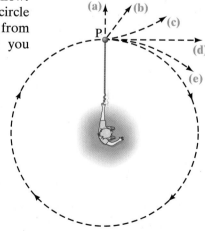

Newton's laws are fundamental in physics. These photos show two situations of using Newton's laws which involve some new elements in addition to those discussed in the previous Chapter. The downhill skier illustrates *friction* on an incline, although at this moment she is not touching the snow, and so is retarded only by air resistance which is a velocity-dependent force (an optional topic in this Chapter). The people on the rotating amusement park ride below illustrate the dynamics of circular motion.

5

Using Newton's Laws: Friction, Circular Motion, Drag Forces

CHAPTER-OPENING QUESTION—Guess now!
You revolve a ball around you in a horizontal circle at constant speed on a string, as shown here from above. Which path will the ball follow if you let go of the string at point P?

CONTENTS

This chapter continues our study of Newton's laws and emphasizes their fundamental importance in physics. We cover some important applications of Newton's laws, including friction and circular motion. Although some material in this Chapter may seem to repeat topics covered in Chapter 4, in fact, new elements are involved.

5–1 Applications of Newton's Laws Involving Friction

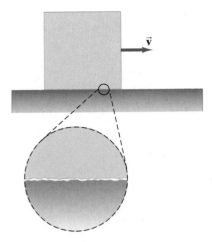

FIGURE 5–1 An object moving to the right on a table or floor. The two surfaces in contact are rough, at least on a microscopic scale.

Until now we have ignored friction, but it must be taken into account in most practical situations. Friction exists between two solid surfaces because even the smoothest looking surface is quite rough on a microscopic scale, Fig. 5–1. When we try to slide an object across another surface, these microscopic bumps impede the motion. Exactly what is happening at the microscopic level is not yet fully understood. It is thought that the atoms on a bump of one surface may come so close to the atoms of the other surface that attractive electric forces between the atoms could "bond" as a tiny weld between the two surfaces. Sliding an object across a surface is often jerky, perhaps due to the making and breaking of these bonds. Even when a round object rolls across a surface, there is still some friction, called *rolling friction*, although it is generally much less than when objects slide across a surface. We focus our attention now on sliding friction, which is usually called **kinetic friction** (*kinetic* is from the Greek for "moving").

When an object slides along a rough surface, the force of kinetic friction acts opposite to the direction of the object's velocity. The magnitude of the force of kinetic friction depends on the nature of the two sliding surfaces. For given surfaces, experiment shows that the friction force is approximately proportional to the *normal force* between the two surfaces, which is the force that either object exerts on the other and is perpendicular to their common surface of contact (see Fig. 5–2). The force of friction between hard surfaces in many cases depends very little on the total surface area of contact; that is, the friction force on this book is roughly the same whether it is being slid on its wide face or on its spine, assuming the surfaces have the same smoothness. We consider a simple model of friction in which we make this assumption that the friction force is independent of area. Then we write the proportionality between the magnitudes of the friction force F_{fr} and the normal force F_N as an equation by inserting a constant of proportionality, μ_k:

$$F_{fr} = \mu_k F_N . \qquad \text{[kinetic friction]}$$

This relation is not a fundamental law; it is an experimental relation between the magnitude of the friction force F_{fr}, which acts parallel to the two surfaces, and the magnitude of the normal force F_N, which acts perpendicular to the surfaces. It is *not* a vector equation since the two forces have directions perpendicular to one another. The term μ_k is called the *coefficient of kinetic friction*, and its value depends on the nature of the two surfaces. Measured values for a variety of surfaces are given in Table 5–1. These are only approximate, however, since μ depends on whether the surfaces are wet or dry, on how much they have been sanded or rubbed, if any burrs remain, and other such factors. But μ_k is roughly independent of the sliding speed, as well as the area in contact.

FIGURE 5–2 When an object is pulled along a surface by an applied force ($\vec{F}_A$), the force of friction $\vec{F}_{fr}$ opposes the motion. The magnitude of $\vec{F}_{fr}$ is proportional to the magnitude of the normal force (F_N).

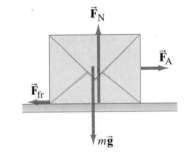

TABLE 5–1 Coefficients of Friction[†]

Surfaces	Coefficient of Static Friction, μ_s	Coefficient of Kinetic Friction, μ_k
Wood on wood	0.4	0.2
Ice on ice	0.1	0.03
Metal on metal (lubricated)	0.15	0.07
Steel on steel (unlubricated)	0.7	0.6
Rubber on dry concrete	1.0	0.8
Rubber on wet concrete	0.7	0.5
Rubber on other solid surfaces	1–4	1
Teflon® on Teflon in air	0.04	0.04
Teflon on steel in air	0.04	0.04
Lubricated ball bearings	<0.01	<0.01
Synovial joints (in human limbs)	0.01	0.01

[†]Values are approximate and intended only as a guide.

What we have been discussing up to now is *kinetic friction*, when one object slides over another. There is also **static friction**, which refers to a force parallel to the two surfaces that can arise even when they are not sliding. Suppose an object such as a desk is resting on a horizontal floor. If no horizontal force is exerted on the desk, there also is no friction force. But now suppose you try to push the desk, and it doesn't move. You are exerting a horizontal force, but the desk isn't moving, so there must be another force on the desk keeping it from moving (the net force is zero on an object at rest). This is the force of *static friction* exerted by the floor on the desk. If you push with a greater force without moving the desk, the force of static friction also has increased. If you push hard enough, the desk will eventually start to move, and kinetic friction takes over. At this point, you have exceeded the maximum force of static friction, which is given by $(F_{fr})_{max} = \mu_s F_N$, where μ_s is the *coefficient of static friction* (Table 5–1). Because the force of static friction can vary from zero to this maximum value, we write

$$F_{fr} \le \mu_s F_N. \qquad\qquad \text{[static friction]}$$

You may have noticed that it is often easier to keep a heavy object sliding than it is to start it sliding in the first place. This is consistent with μ_s generally being greater than μ_k (see Table 5–1).

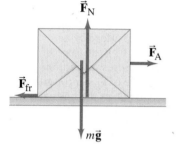

FIGURE 5–2 Repeated for Example 5–1.

FIGURE 5–3 Example 5–1. Magnitude of the force of friction as a function of the external force applied to an object initially at rest. As the applied force is increased in magnitude, the force of static friction increases linearly to just match it, until the applied force equals $\mu_s F_N$. If the applied force increases further, the object will begin to move, and the friction force drops to a roughly constant value characteristic of kinetic friction.

EXAMPLE 5–1 **Friction: static and kinetic.** Our 10.0-kg mystery box rests on a horizontal floor. The coefficient of static friction is $\mu_s = 0.40$ and the coefficient of kinetic friction is $\mu_k = 0.30$. Determine the force of friction, F_{fr}, acting on the box if a horizontal external applied force F_A is exerted on it of magnitude: (*a*) 0, (*b*) 10 N, (*c*) 20 N, (*d*) 38 N, and (*e*) 40 N.

APPROACH We don't know, right off, if we are dealing with static friction or kinetic friction, nor if the box remains at rest or accelerates. We need to draw a free-body diagram, and then determine in each case whether or not the box will move: the box starts moving if F_A is greater than the maximum static friction force (Newton's second law). The forces on the box are gravity $m\vec{g}$, the normal force exerted by the floor $\vec{F}_N$, the horizontal applied force $\vec{F}_A$, and the friction force $\vec{F}_{fr}$, as shown in Fig. 5–2.

SOLUTION The free-body diagram of the box is shown in Fig. 5–2. In the vertical direction there is no motion, so Newton's second law in the vertical direction gives $\Sigma F_y = ma_y = 0$, which tells us $F_N - mg = 0$. Hence the normal force is

$$F_N = mg = (10.0\,\text{kg})(9.80\,\text{m/s}^2) = 98.0\,\text{N}.$$

(*a*) Because $F_A = 0$ in this first case, the box doesn't move, and $F_{fr} = 0$.
(*b*) The force of static friction will oppose any applied force up to a maximum of

$$\mu_s F_N = (0.40)(98.0\,\text{N}) = 39\,\text{N}.$$

When the applied force is $F_A = 10\,\text{N}$, the box will not move. Newton's second law gives $\Sigma F_x = F_A - F_{fr} = 0$, so $F_{fr} = 10\,\text{N}$.
(*c*) An applied force of 20 N is also not sufficient to move the box. Thus $F_{fr} = 20\,\text{N}$ to balance the applied force.
(*d*) The applied force of 38 N is still not quite large enough to move the box; so the friction force has now increased to 38 N to keep the box at rest.
(*e*) A force of 40 N will start the box moving since it exceeds the maximum force of static friction, $\mu_s F_N = (0.40)(98\,\text{N}) = 39\,\text{N}$. Instead of static friction, we now have kinetic friction, and its magnitude is

$$F_{fr} = \mu_k F_N = (0.30)(98.0\,\text{N}) = 29\,\text{N}.$$

There is now a net (horizontal) force on the box of magnitude $F = 40\,\text{N} - 29\,\text{N} = 11\,\text{N}$, so the box will accelerate at a rate

$$a_x = \frac{\Sigma F}{m} = \frac{11\,\text{N}}{10.0\,\text{kg}} = 1.1\,\text{m/s}^2$$

as long as the applied force is 40 N. Figure 5–3 shows a graph that summarizes this Example.

Friction can be a hindrance. It slows down moving objects and causes heating and binding of moving parts in machinery. Friction can be reduced by using lubricants such as oil. More effective in reducing friction between two surfaces is to maintain a layer of air or other gas between them. Devices using this concept, which is not practical for most situations, include air tracks and air tables in which the layer of air is maintained by forcing air through many tiny holes. Another technique to maintain the air layer is to suspend objects in air using magnetic fields ("magnetic levitation"). On the other hand, friction can be helpful. Our ability to walk depends on friction between the soles of our shoes (or feet) and the ground. (Walking involves static friction, not kinetic friction. Why?) The movement of a car, and also its stability, depend on friction. When friction is low, such as on ice, safe walking or driving becomes difficult.

| CONCEPTUAL EXAMPLE 5–2 | **A box against a wall.** You can hold a box against a rough wall (Fig. 5–4) and prevent it from slipping down by pressing hard horizontally. How does the application of a horizontal force keep an object from moving vertically?

RESPONSE This won't work well if the wall is slippery. You need friction. Even then, if you don't press hard enough, the box will slip. The horizontal force you apply produces a normal force on the box exerted by the wall (net force horizontally is zero since box doesn't move horizontally.) The force of gravity mg, acting downward on the box, can now be balanced by an upward static friction force whose maximum magnitude is proportional to the normal force. The harder you push, the greater F_N is and the greater F_{fr} can be. If you don't press hard enough, then $mg > \mu_s F_N$ and the box begins to slide down.

FIGURE 5–4 Example 5–2.

EXERCISE A If $\mu_s = 0.40$ and $mg = 20\,\text{N}$, what minimum force F will keep the box from falling: (a) 100 N; (b) 80 N; (c) 50 N; (d) 20 N; (e) 8 N?

| EXAMPLE 5–3 | **Pulling against friction.** A 10.0-kg box is pulled along a horizontal surface by a force F_P of 40.0 N applied at a 30.0° angle above horizontal. This is like Example 4–11 except now there is friction, and we assume a coefficient of kinetic friction of 0.30. Calculate the acceleration.

APPROACH The free-body diagram is shown in Fig. 5–5. It is much like that in Fig. 4–21, but with one more force, that of friction.

FIGURE 5–5 Example 5–3.

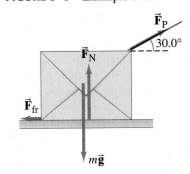

SOLUTION The calculation for the vertical (y) direction is just the same as in Example 4–11, $mg = (10.0\,\text{kg})(9.80\,\text{m/s}^2) = 98.0\,\text{N}$ and $F_{Py} = (40.0\,\text{N})(\sin 30.0°) = 20.0\,\text{N}$. With y positive upward and $a_y = 0$, we have

$$F_N - mg + F_{Py} = ma_y$$
$$F_N - 98.0\,\text{N} + 20.0\,\text{N} = 0,$$

so the normal force is $F_N = 78.0\,\text{N}$. Now we apply Newton's second law for the horizontal (x) direction (positive to the right), and include the friction force:

$$F_{Px} - F_{fr} = ma_x.$$

The friction force is kinetic as long as $F_{fr} = \mu_k F_N$ is less than $F_{Px} = (40.0\,\text{N})\cos 30.0° = 34.6\,\text{N}$, which it is:

$$F_{fr} = \mu_k F_N = (0.30)(78.0\,\text{N}) = 23.4\,\text{N}.$$

Hence the box does accelerate:

$$a_x = \frac{F_{Px} - F_{fr}}{m} = \frac{34.6\,\text{N} - 23.4\,\text{N}}{10.0\,\text{kg}} = 1.1\,\text{m/s}^2.$$

In the absence of friction, as we saw in Example 4–11, the acceleration would be much greater than this.

NOTE Our final answer has only two significant figures because our least significant input value $(\mu_k = 0.30)$ has two.

EXERCISE B If $\mu_k F_N$ were greater than F_{Px}, what would you conclude?

(a)

(b)

(c) $m\vec{g}$ (d) $m\vec{g}$

FIGURE 5–6   Example 5–4.

FIGURE 5–7   Example 5–5.

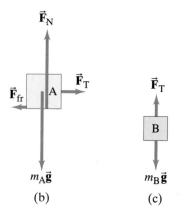

(a)

(b)   (c)

CONCEPTUAL EXAMPLE 5–4

CONCEPTUAL EXAMPLE 5–4   **To push or to pull a sled?** Your little sister wants a ride on her sled. If you are on flat ground, will you exert less force if you push her or pull her? See Figs. 5–6a and b. Assume the same angle θ in each case.

RESPONSE Let us draw free-body diagrams for the sled–sister combination, as shown in Figs. 5–6c and d. They show, for the two cases, the forces exerted by you, $\vec{F}$ (an unknown), by the snow, $\vec{F}_N$ and $\vec{F}_{fr}$, and gravity $m\vec{g}$. (a) If you push her, and $\theta > 0$, there is a vertically downward component to your force. Hence the normal force upward exerted by the ground (Fig. 5–6c) will be larger than mg (where m is the mass of sister plus sled). (b) If you pull her, your force has a vertically upward component, so the normal force F_N will be less than mg, Fig. 5–6d. Because the friction force is proportional to the normal force, F_{fr} will be less if you pull her. So you exert less force if you pull her.

EXAMPLE 5–5   **Two boxes and a pulley.** In Fig. 5–7a, two boxes are connected by a cord running over a pulley. The coefficient of kinetic friction between box A and the table is 0.20. We ignore the mass of the cord and pulley and any friction in the pulley, which means we can assume that a force applied to one end of the cord will have the same magnitude at the other end. We wish to find the acceleration, a, of the system, which will have the same magnitude for both boxes assuming the cord doesn't stretch. As box B moves down, box A moves to the right.

APPROACH The free-body diagrams for each box are shown in Figs. 5–7b and c. The forces on box A are the pulling force of the cord F_T, gravity $m_A g$, the normal force exerted by the table F_N, and a friction force exerted by the table F_{fr}; the forces on box B are gravity $m_B g$, and the cord pulling up, F_T.

SOLUTION Box A does not move vertically, so Newton's second law tells us the normal force just balances the weight,
$$F_N = m_A g = (5.0\,\text{kg})(9.8\,\text{m/s}^2) = 49\,\text{N}.$$
In the horizontal direction, there are two forces on box A (Fig. 5–7b): F_T, the tension in the cord (whose value we don't know), and the force of friction
$$F_{fr} = \mu_k F_N = (0.20)(49\,\text{N}) = 9.8\,\text{N}.$$
The horizontal acceleration is what we wish to find; we use Newton's second law in the x direction, $\Sigma F_{Ax} = m_A a_x$, which becomes (taking the positive direction to the right and setting $a_{Ax} = a$):
$$\Sigma F_{Ax} = F_T - F_{fr} = m_A a.\qquad\text{[box A]}$$
Next consider box B. The force of gravity $m_B g = (2.0\,\text{kg})(9.8\,\text{m/s}^2) = 19.6\,\text{N}$ pulls downward; and the cord pulls upward with a force F_T. So we can write Newton's second law for box B (taking the downward direction as positive):
$$\Sigma F_{By} = m_B g - F_T = m_B a.\qquad\text{[box B]}$$
[Notice that if $a \neq 0$, then F_T is not equal to $m_B g$.]

We have two unknowns, a and F_T, and we also have two equations. We solve the box A equation for F_T:
$$F_T = F_{fr} + m_A a,$$
and substitute this into the box B equation:
$$m_B g - F_{fr} - m_A a = m_B a.$$
Now we solve for a and put in numerical values:
$$a = \frac{m_B g - F_{fr}}{m_A + m_B} = \frac{19.6\,\text{N} - 9.8\,\text{N}}{5.0\,\text{kg} + 2.0\,\text{kg}} = 1.4\,\text{m/s}^2,$$
which is the acceleration of box A to the right, and of box B down.

If we wish, we can calculate F_T using the third equation up from here:
$$F_T = F_{fr} + m_A a = 9.8\,\text{N} + (5.0\,\text{kg})(1.4\,\text{m/s}^2) = 17\,\text{N}.$$

NOTE Box B is not in free fall. It does not fall at $a = g$ because an additional force, F_T, is acting upward on it.

116   CHAPTER 5   Using Newton's Laws: Friction, Circular Motion, Drag Forces

In Chapter 4 we examined motion on ramps and inclines, and saw that it is usually an advantage to choose the x axis along the plane, in the direction of acceleration. There we ignored friction, but now we take it into account.

EXAMPLE 5–6 **The skier.** The skier in Fig. 5–8a is descending a 30° slope, at constant speed. What can you say about the coefficient of kinetic friction μ_k?

APPROACH We choose the x axis along the slope, positive pointing downslope in the direction of the skier's motion. The y axis is perpendicular to the surface as shown in Fig. 5–8b, which is the free-body diagram for our system which we choose as the skier and her skis (total mass m). The forces acting are gravity, $\vec{\mathbf{F}}_G = m\vec{\mathbf{g}}$, which points vertically downward (not perpendicular to the slope), and the two forces exerted on her skis by the snow—the normal force perpendicular to the snowy slope (not vertical), and the friction force parallel to the surface. These three forces are shown acting at one point in Fig. 5–8b, for convenience.

SOLUTION We have to resolve only one vector into components, the weight $\vec{\mathbf{F}}_G$, and its components are shown as dashed lines in Fig. 5–8c:

$$F_{Gx} = mg \sin\theta,$$
$$F_{Gy} = -mg \cos\theta,$$

where we have stayed general by using θ rather than 30° for now. There is no acceleration, so Newton's second law applied to the x and y components gives

$$\Sigma F_y = F_N - mg \cos\theta = ma_y = 0$$
$$\Sigma F_x = mg \sin\theta - \mu_k F_N = ma_x = 0.$$

From the first equation, we have $F_N = mg \cos\theta$. We substitute this into the second equation:

$$mg \sin\theta - \mu_k (mg \cos\theta) = 0.$$

Now we solve for μ_k:

$$\mu_k = \frac{mg \sin\theta}{mg \cos\theta} = \frac{\sin\theta}{\cos\theta} = \tan\theta$$

which for $\theta = 30°$ is

$$\mu_k = \tan\theta = \tan 30° = 0.58.$$

Notice that we could use the equation

$$\mu_k = \tan\theta$$

to determine μ_k under a variety of conditions. All we need to do is observe at what slope angle the skier descends at constant speed. Here is another reason why it is often useful to plug in numbers only at the end: we obtained a general result useful for other situations as well.

In problems involving a slope or "inclined plane," avoid making errors in the directions of the normal force and gravity. The normal force is *not* vertical: it is perpendicular to the slope or plane. And gravity is *not* perpendicular to the slope—gravity acts vertically downward toward the center of the Earth.

PHYSICS APPLIED
Skiing

FIGURE 5–8 Example 5–6. A skier descending a slope; $\vec{\mathbf{F}}_G = m\vec{\mathbf{g}}$ is the force of gravity (weight) on the skier.

(a)

(b)

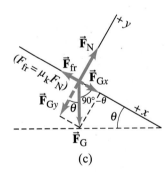

(c)

⚠ **CAUTION**
Directions of gravity and the normal force

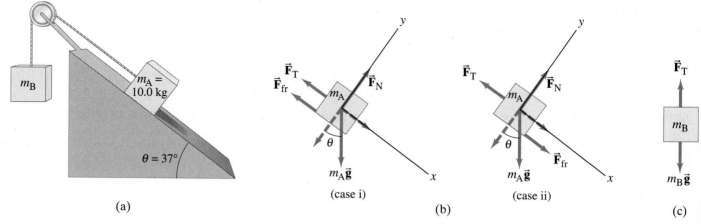

FIGURE 5-9 Example 5-7. Note choice of x and y axes.

EXAMPLE 5-7 **A ramp, a pulley, and two boxes.** A box of mass $m_A = 10.0$ kg rests on a surface inclined at $\theta = 37°$ to the horizontal. It is connected by a light-weight cord, which passes over a massless and frictionless pulley, to a second box of mass m_B, which hangs freely as shown in Fig. 5-9a. (a) If the coefficient of static friction is $\mu_s = 0.40$, determine what range of values for mass m_B will keep the system at rest. (b) If the coefficient of kinetic friction is $\mu_k = 0.30$, and $m_B = 10.0$ kg, determine the acceleration of the system.

APPROACH Figure 5-9b shows two free-body diagrams for box m_A because the force of friction can be either up or down the slope, depending on which direction the box slides: (i) if $m_B = 0$ or is sufficiently small, m_A would tend to slide down the incline, so $\vec{F}_{fr}$ would be directed up the incline; (ii) if m_B is large enough, m_A will tend to be pulled up the plane, so $\vec{F}_{fr}$ would point down the plane. The tension force exerted by the cord is labeled $\vec{F}_T$.

SOLUTION (a) For both cases (i) and (ii), Newton's second law for the y direction (perpendicular to the plane) is the same:

$$F_N - m_A g \cos\theta = m_A a_y = 0$$

since there is no y motion. So

$$F_N = m_A g \cos\theta.$$

Now for the x motion. We consider case (i) first for which $\Sigma F = ma$ gives

$$m_A g \sin\theta - F_T - F_{fr} = m_A a_x.$$

We want $a_x = 0$ and we solve for F_T since F_T is related to m_B (whose value we are seeking) by $F_T = m_B g$ (see Fig. 5-9c). Thus

$$m_A g \sin\theta - F_{fr} = F_T = m_B g.$$

We solve this for m_B and set F_{fr} at its maximum value $\mu_s F_N = \mu_s m_A g \cos\theta$ to find the minimum value that m_B can have to prevent motion $(a_x = 0)$:

$$m_B = m_A \sin\theta - \mu_s m_A \cos\theta$$

$$= (10.0 \text{ kg})(\sin 37° - 0.40 \cos 37°) = 2.8 \text{ kg}.$$

Thus if $m_B < 2.8$ kg, then box A will slide down the incline.

Now for case (ii) in Fig. 5-9b, box A being pulled *up* the incline. Newton's second law is

$$m_A g \sin\theta + F_{fr} - F_T = m_A a_x = 0.$$

Then the maximum value m_B can have without causing acceleration is given by

$$F_T = m_B g = m_A g \sin \theta + \mu_s m_A g \cos \theta$$

or

$$m_B = m_A \sin \theta + \mu_s m_A \cos \theta$$

$$= (10.0 \text{ kg})(\sin 37° + 0.40 \cos 37°) = 9.2 \text{ kg}.$$

Thus, to prevent motion, we have the condition

$$2.8 \text{ kg} < m_B < 9.2 \text{ kg}.$$

(b) If $m_B = 10.0 \text{ kg}$ and $\mu_k = 0.30$, then m_B will fall and m_A will rise up the plane (case ii). To find their acceleration a, we use $\Sigma F = ma$ for box A:

$$m_A a = F_T - m_A g \sin \theta - \mu_k F_N.$$

Since m_B accelerates downward, Newton's second law for box B (Fig. 5–9c) tells us $m_B a = m_B g - F_T$, or $F_T = m_B g - m_B a$, and we substitute this into the equation above:

$$m_A a = m_B g - m_B a - m_A g \sin \theta - \mu_k F_N.$$

We solve for the acceleration a and substitute $F_N = m_A g \cos \theta$, and then $m_A = m_B = 10.0 \text{ kg}$, to find

$$a = \frac{m_B g - m_A g \sin \theta - \mu_k m_A g \cos \theta}{m_A + m_B}$$

$$= \frac{(10.0 \text{ kg})(9.80 \text{ m/s}^2)(1 - \sin 37° - 0.30 \cos 37°)}{20.0 \text{ kg}}$$

$$= 0.079 g = 0.78 \text{ m/s}^2.$$

NOTE It is worth comparing this equation for acceleration a with that obtained in Example 5–5: if here we let $\theta = 0$, the plane is horizontal as in Example 5–5, and we obtain $a = (m_B g - \mu_k m_A g)/(m_A + m_B)$ just as in Example 5–5.

5–2 Uniform Circular Motion—Kinematics

An object moves in a straight line if the net force on it acts in the direction of motion, or the net force is zero. If the net force acts at an angle to the direction of motion at any moment, then the object moves in a curved path. An example of the latter is projectile motion, which we discussed in Chapter 3. Another important case is that of an object moving in a circle, such as a ball at the end of a string revolving around one's head, or the nearly circular motion of the Moon about the Earth.

An object that moves in a circle at constant speed v is said to experience **uniform circular motion**. The *magnitude* of the velocity remains constant in this case, but the *direction* of the velocity continuously changes as the object moves around the circle (Fig. 5–10). Because acceleration is defined as the rate of change of velocity, a change in direction of velocity constitutes an acceleration, just as a change in magnitude of velocity does. Thus, an object revolving in a circle is continuously accelerating, even when the speed remains constant $(v_1 = v_2 = v)$. We now investigate this acceleration quantitatively.

FIGURE 5–10 A small object moving in a circle, showing how the velocity changes. At each point, the instantaneous velocity is in a direction tangent to the circular path.

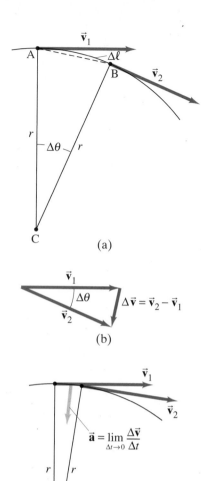

Acceleration is defined as

$$\vec{a} = \lim_{\Delta t \to 0} \frac{\Delta \vec{v}}{\Delta t} = \frac{d\vec{v}}{dt},$$

where $\Delta \vec{v}$ is the change in velocity during the short time interval Δt. We will eventually consider the situation in which Δt approaches zero and thus obtain the instantaneous acceleration. But for purposes of making a clear drawing (Fig. 5–11), we consider a nonzero time interval. During the time interval Δt, the particle in Fig. 5–11a moves from point A to point B, covering a distance $\Delta \ell$ *along the arc* which subtends an angle $\Delta \theta$. The change in the velocity vector is $\vec{v}_2 - \vec{v}_1 = \Delta \vec{v}$, and is shown in Fig. 5–11b.

Now we let Δt be very small, approaching zero. Then $\Delta \ell$ and $\Delta \theta$ are also very small, and $\vec{v}_2$ will be almost parallel to $\vec{v}_1$ (Fig. 5–11c); $\Delta \vec{v}$ will be essentially perpendicular to them. Thus $\Delta \vec{v}$ points toward the center of the circle. Since $\vec{a}$, by definition, is in the same direction as $\Delta \vec{v}$, it too must point toward the center of the circle. Therefore, this acceleration is called **centripetal acceleration** ("center-pointing" acceleration) or **radial acceleration** (since it is directed along the radius, toward the center of the circle), and we denote it by $\vec{a}_R$.

We next determine the magnitude of the radial (centripetal) acceleration, a_R. Because CA in Fig. 5–11a is perpendicular to $\vec{v}_1$, and CB is perpendicular to $\vec{v}_2$, it follows that the angle $\Delta \theta$, defined as the angle between CA and CB, is also the angle between $\vec{v}_1$ and $\vec{v}_2$. Hence the vectors $\vec{v}_1, \vec{v}_2$, and $\Delta \vec{v}$ in Fig. 5–11b form a triangle that is geometrically similar[†] to triangle CAB in Fig. 5–11a. If we take $\Delta \theta$ to be very small (letting Δt be very small) and setting $v = v_1 = v_2$ because the magnitude of the velocity is assumed not to change, we can write

$$\frac{\Delta v}{v} \approx \frac{\Delta \ell}{r},$$

or

$$\Delta v \approx \frac{v}{r} \Delta \ell.$$

This is an exact equality when Δt approaches zero, for then the arc length $\Delta \ell$ equals the chord length AB. We want to find the instantaneous acceleration, a_R, so we use the expression above to write

$$a_R = \lim_{\Delta t \to 0} \frac{\Delta v}{\Delta t} = \lim_{\Delta t \to 0} \frac{v}{r} \frac{\Delta \ell}{\Delta t}.$$

Then, because

$$\lim_{\Delta t \to 0} \frac{\Delta \ell}{\Delta t}$$

is just the linear speed, v, of the object, we have for the centripetal (radial) acceleration

$$a_R = \frac{v^2}{r}. \qquad \text{[centripetal (radial) acceleration]} \quad \textbf{(5–1)}$$

Equation 5–1 is valid even when v is not constant.

To summarize, *an object moving in a circle of radius r at constant speed v has an acceleration whose direction is toward the center of the circle and whose magnitude is* $a_R = v^2/r$. It is not surprising that this acceleration depends on v and r. The greater the speed v, the faster the velocity changes direction; and the larger the radius, the less rapidly the velocity changes direction.

The acceleration vector points toward the center of the circle. But the velocity vector always points in the direction of motion, which is tangential to the circle. Thus the velocity and acceleration vectors are perpendicular to each other at every point in the path for uniform circular motion (Fig. 5–12). This is another example that illustrates the error in thinking that acceleration and velocity are always in the same direction. For an object falling vertically, $\vec{a}$ and $\vec{v}$ are indeed parallel. But in circular motion, $\vec{a}$ and $\vec{v}$ are perpendicular, not parallel (nor were they parallel in projectile motion, Section 3–7).

EXERCISE C Can Equations 2–12, the kinematic equations for constant acceleration, be used for uniform circular motion? For example, could Eq. 2–12b be used to calculate the time for the revolving ball in Fig. 5–12 to make one revolution?

FIGURE 5–11 Determining the change in velocity, $\Delta \vec{v}$, for a particle moving in a circle. The length $\Delta \ell$ is the distance along the arc, from A to B.

⚠ **C A U T I O N**

In uniform circular motion, the speed is constant, but the acceleration is not zero

FIGURE 5–12 For uniform circular motion, $\vec{a}$ is always perpendicular to $\vec{v}$.

[†]Appendix A contains a review of geometry.

Circular motion is often described in terms of the **frequency** f, the number of revolutions per second. The **period** T of an object revolving in a circle is the time required for one complete revolution. Period and frequency are related by

$$T = \frac{1}{f}. \qquad\qquad (5\text{--}2)$$

For example, if an object revolves at a frequency of 3 rev/s, then each revolution takes $\frac{1}{3}$ s. For an object revolving in a circle (of circumference $2\pi r$) at constant speed v, we can write

$$v = \frac{2\pi r}{T},$$

since in one revolution the object travels one circumference.

EXAMPLE 5–8 **Acceleration of a revolving ball.** A 150-g ball at the end of a string is revolving uniformly in a horizontal circle of radius 0.600 m, as in Fig. 5–10 or 5–12. The ball makes 2.00 revolutions in a second. What is its centripetal acceleration?

APPROACH The centripetal acceleration is $a_R = v^2/r$. We are given r, and we can find the speed of the ball, v, from the given radius and frequency.

SOLUTION If the ball makes two complete revolutions per second, then the ball travels in a complete circle in a time interval equal to 0.500 s, which is its period T. The distance traveled in this time is the circumference of the circle, $2\pi r$, where r is the radius of the circle. Therefore, the ball has speed

$$v = \frac{2\pi r}{T} = \frac{2\pi(0.600\ \text{m})}{(0.500\ \text{s})} = 7.54\ \text{m/s}.$$

The centripetal acceleration[†] is

$$a_R = \frac{v^2}{r} = \frac{(7.54\ \text{m/s})^2}{(0.600\ \text{m})} = 94.7\ \text{m/s}^2.$$

EXERCISE D If the radius is doubled to 1.20 m but the period stays the same, by what factor will the centripetal acceleration change? (a) 2, (b) 4, (c) $\frac{1}{2}$, (d) $\frac{1}{4}$, (e) none of these.

EXAMPLE 5–9 **Moon's centripetal acceleration.** The Moon's nearly circular orbit about the Earth has a radius of about 384,000 km and a period T of 27.3 days. Determine the acceleration of the Moon toward the Earth.

APPROACH Again we need to find the velocity v in order to find a_R. We will need to convert to SI units to get v in m/s.

SOLUTION In one orbit around the Earth, the Moon travels a distance $2\pi r$, where $r = 3.84 \times 10^8$ m is the radius of its circular path. The time required for one complete orbit is the Moon's period of 27.3 d. The speed of the Moon in its orbit about the Earth is $v = 2\pi r/T$. The period T in seconds is $T = (27.3\ \text{d})(24.0\ \text{h/d})(3600\ \text{s/h}) = 2.36 \times 10^6$ s. Therefore,

$$a_R = \frac{v^2}{r} = \frac{(2\pi r)^2}{T^2 r} = \frac{4\pi^2 r}{T^2} = \frac{4\pi^2(3.84 \times 10^8\ \text{m})}{(2.36 \times 10^6\ \text{s})^2}$$

$$= 0.00272\ \text{m/s}^2 = 2.72 \times 10^{-3}\ \text{m/s}^2.$$

We can write this acceleration in terms of $g = 9.80\ \text{m/s}^2$ (the acceleration of gravity at the Earth's surface) as

$$a = 2.72 \times 10^{-3}\ \text{m/s}^2 \left(\frac{g}{9.80\ \text{m/s}^2} \right) = 2.78 \times 10^{-4} g.$$

NOTE The centripetal acceleration of the Moon, $a = 2.78 \times 10^{-4}\ g$, is *not* the acceleration of gravity for objects at the Moon's surface due to the Moon's gravity. Rather, it is the acceleration due to the *Earth's* gravity for any object (such as the Moon) that is 384,000 km from the Earth. Notice how small this acceleration is compared to the acceleration of objects near the Earth's surface.

⚠ CAUTION

Distinguish the Moon's gravity on objects at its surface, from the Earth's gravity acting on the Moon (this Example)

[†]Differences in the final digit can depend on whether you keep all digits in your calculator for v (which gives $a_R = 94.7\ \text{m/s}^2$), or if you use $v = 7.54\ \text{m/s}$ in which case you get $a_R = 94.8\ \text{m/s}^2$. Both results are valid since our assumed accuracy is about $\pm 0.1\ \text{m/s}$ (see Section 1–3).

*Centrifugation

Centrifuges and very high speed ultracentrifuges, are used to sediment materials quickly or to separate materials. Test tubes held in the centrifuge rotor are accelerated to very high rotational speeds: see Fig. 5–13, where one test tube is shown in two positions as the rotor turns. The small green dot represents a small particle, perhaps a macromolecule, in a fluid-filled test tube. At position A the particle has a tendency to move in a straight line, but the fluid resists the motion of the particles, exerting a centripetal force that keeps the particles moving nearly in a circle. The resistive force exerted by the fluid (liquid, gas, or gel, depending on the application) usually does not quite equal mv^2/r, and the particles move slowly toward the bottom of the tube. A centrifuge provides an "effective gravity" much larger than normal gravity because of the high rotational speeds, thus causing more rapid sedimentation.

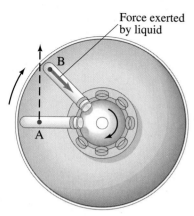

FIGURE 5–13 Two positions of a rotating test tube in a centrifuge (top view). At A, the green dot represents a macromolecule or other particle being sedimented. It would tend to follow the dashed line, heading toward the bottom of the tube, but the fluid resists this motion by exerting a force on the particle as shown at point B.

Force exerted by liquid

EXAMPLE 5–10 **Ultracentrifuge.** The rotor of an ultracentrifuge rotates at 50,000 rpm (revolutions per minute). A particle at the top of a test tube (Fig. 5–13) is 6.00 cm from the rotation axis. Calculate its centripetal acceleration, in "g's."

APPROACH We calculate the centripetal acceleration from $a_R = v^2/r$.

SOLUTION The test tube makes 5.00×10^4 revolutions each minute, or, dividing by 60 s/min, 833 rev/s. The time to make one revolution, the period T, is

$$T = \frac{1}{(833 \text{ rev/s})} = 1.20 \times 10^{-3} \text{ s/rev.}$$

At the top of the tube, a particle revolves in a circle of circumference $2\pi r = (2\pi)(0.0600 \text{ m}) = 0.377$ m per revolution. The speed of the particle is then

$$v = \frac{2\pi r}{T} = \left(\frac{0.377 \text{ m/rev}}{1.20 \times 10^{-3} \text{ s/rev}} \right) = 3.14 \times 10^2 \text{ m/s.}$$

The centripetal acceleration is

$$a_R = \frac{v^2}{r} = \frac{(3.14 \times 10^2 \text{ m/s})^2}{0.0600 \text{ m}} = 1.64 \times 10^6 \text{ m/s}^2,$$

which, dividing by $g = 9.80 \text{ m/s}^2$, is 1.67×10^5 g's = 167,000 g's.

5–3 Dynamics of Uniform Circular Motion

FIGURE 5–14 A force is required to keep an object moving in a circle. If the speed is constant, the force is directed toward the circle's center.

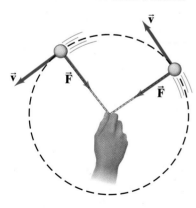

! **CAUTION**

Centripetal force is not a new kind of force (Every force must be exerted by an object)

According to Newton's second law $(\Sigma\vec{F} = m\vec{a})$, an object that is accelerating must have a net force acting on it. An object moving in a circle, such as a ball on the end of a string, must therefore have a force applied to it to keep it moving in that circle. That is, a net force is necessary to give it centripetal acceleration. The magnitude of the required force can be calculated using Newton's second law for the radial component, $\Sigma F_R = ma_R$, where a_R is the centripetal acceleration, $a_R = v^2/r$, and ΣF_R is the total (or net) force in the radial direction:

$$\Sigma F_R = ma_R = m\frac{v^2}{r}. \qquad \text{[circular motion]} \quad \textbf{(5–3)}$$

For uniform circular motion $(v = \text{constant})$, the acceleration is a_R, which is directed toward the center of the circle at any moment. Thus the *net force too must be directed toward the center of the circle*, Fig. 5–14. A net force is necessary because if no net force were exerted on the object, it would not move in a circle but in a straight line, as Newton's first law tells us. The direction of the net force is continually changing so that it is always directed toward the center of the circle. This force is sometimes called a centripetal ("pointing toward the center") force. But be aware that "centripetal force" does not indicate some new kind of force. The term merely describes the *direction* of the net force needed to provide a circular path: the net force is directed toward the circle's center. The force *must be applied by other objects*. For example, to swing a ball in a circle on the end of a string, you pull on the string and the string exerts the force on the ball. (Try it.)

There is a common misconception that an object moving in a circle has an outward force acting on it, a so-called centrifugal ("center-fleeing") force. This is incorrect: *there is no outward force* on the revolving object. Consider, for example, a person swinging a ball on the end of a string around her head (Fig. 5–15). If you have ever done this yourself, you know that you feel a force pulling outward on your hand. The misconception arises when this pull is interpreted as an outward "centrifugal" force pulling on the ball that is transmitted along the string to your hand. This is not what is happening at all. To keep the ball moving in a circle, you pull *inwardly* on the string, and the string exerts this force on the ball. The ball exerts an equal and opposite force on the string (Newton's third law), and *this* is the outward force your hand feels (see Fig. 5–15).

The force *on the ball* is the one exerted *inwardly* on it by you, via the string. To see even more convincing evidence that a "centrifugal force" does not act on the ball, consider what happens when you let go of the string. If a centrifugal force were acting, the ball would fly outward, as shown in Fig. 5–16a. But it doesn't; the ball flies off tangentially (Fig. 5–16b), in the direction of the velocity it had at the moment it was released, because the inward force no longer acts. Try it and see!

EXERCISE E Return to the Chapter-Opening Question, page 112, and answer it again now. Try to explain why you may have answered differently the first time.

FIGURE 5–15 Swinging a ball on the end of a string.

FIGURE 5–16 If centrifugal force existed, the revolving ball would fly outward as in (a) when released. In fact, it flies off tangentially as in (b). For example, in (c) sparks fly in straight lines tangentially from the edge of a rotating grinding wheel.

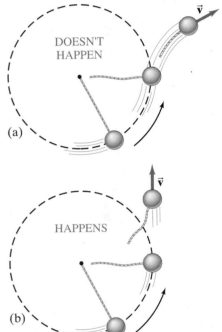

(a)

(b)

EXAMPLE 5–11 **ESTIMATE** **Force on revolving ball (horizontal).** Estimate the force a person must exert on a string attached to a 0.150-kg ball to make the ball revolve in a horizontal circle of radius 0.600 m. The ball makes 2.00 revolutions per second $(T = 0.500\,\text{s})$, as in Example 5–8. Ignore the string's mass.

APPROACH First we need to draw the free-body diagram for the ball. The forces acting on the ball are the force of gravity, $m\vec{g}$ downward, and the tension force $\vec{F}_T$ that the string exerts toward the hand at the center (which occurs because the person exerts that same force on the string). The free-body diagram for the ball is as shown in Fig. 5–17. The ball's weight complicates matters and makes it impossible to revolve a ball with the cord perfectly horizontal. We assume the weight is small, and put $\phi \approx 0$ in Fig. 5–17. Thus $\vec{F}_T$ will act nearly horizontally and, in any case, provides the force necessary to give the ball its centripetal acceleration.

SOLUTION We apply Newton's second law to the radial direction, which we assume is horizontal:

$$(\Sigma F)_R = ma_R,$$

where $a_R = v^2/r$ and $v = 2\pi r/T = 2\pi(0.600\,\text{m})/(0.500\,\text{s}) = 7.54\,\text{m/s}$. Thus

$$F_T = m\frac{v^2}{r} = (0.150\,\text{kg})\frac{(7.54\,\text{m/s})^2}{(0.600\,\text{m})} \approx 14\,\text{N}.$$

NOTE We keep only two significant figures in the answer because we ignored the ball's weight; it is $mg = (0.150\,\text{kg})(9.80\,\text{m/s}^2) = 1.5\,\text{N}$, about $\frac{1}{10}$ of our result, which is small but not so small as to justify stating a more precise answer for F_T.

NOTE To include the effect of $m\vec{g}$, resolve $\vec{F}_T$ in Fig. 5–17 into components, and set the horizontal component of $\vec{F}_T$ equal to mv^2/r and its vertical component equal to mg.

(c)

FIGURE 5–17 Example 5–11.

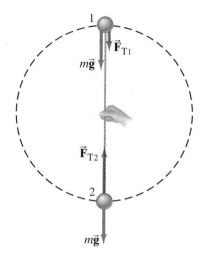

FIGURE 5–18 Example 5–12. Free-body diagrams for positions 1 and 2.

⚠ CAUTION

Circular motion only if cord is under tension

EXAMPLE 5–12 **Revolving ball (vertical circle).** A 0.150-kg ball on the end of a 1.10-m-long cord (negligible mass) is swung in a *vertical* circle. (a) Determine the minimum speed the ball must have at the top of its arc so that the ball continues moving in a circle. (b) Calculate the tension in the cord at the bottom of the arc, assuming the ball is moving at twice the speed of part (a).

APPROACH The ball moves in a vertical circle and is *not* undergoing uniform circular motion. The radius is assumed constant, but the speed v changes because of gravity. Nonetheless, Eq. 5–1 is valid at each point along the circle, and we use it at the top and bottom points. The free-body diagram is shown in Fig. 5–18 for both positions.

SOLUTION (a) At the top (point 1), two forces act on the ball: $m\vec{\mathbf{g}}$, the force of gravity, and $\vec{\mathbf{F}}_{T1}$, the tension force the cord exerts at point 1. Both act downward, and their vector sum acts to give the ball its centripetal acceleration a_R. We apply Newton's second law, for the vertical direction, choosing downward as positive since the acceleration is downward (toward the center):

$$(\Sigma F)_R = ma_R$$
$$F_{T1} + mg = m\frac{v_1^2}{r}. \qquad \text{[at top]}$$

From this equation we can see that the tension force F_{T1} at point 1 will get larger if v_1 (ball's speed at top of circle) is made larger, as expected. But we are asked for the *minimum* speed to keep the ball moving in a circle. The cord will remain taut as long as there is tension in it. But if the tension disappears (because v_1 is too small) the cord can go limp, and the ball will fall out of its circular path. Thus, the minimum speed will occur if $F_{T1} = 0$, for which we have

$$mg = m\frac{v_1^2}{r}. \qquad \text{[minimum speed at top]}$$

We solve for v_1, keeping an extra digit for use in (b):

$$v_1 = \sqrt{gr} = \sqrt{(9.80\,\text{m/s}^2)(1.10\,\text{m})} = 3.283\,\text{m/s}.$$

This is the minimum speed at the top of the circle if the ball is to continue moving in a circular path.

(b) When the ball is at the bottom of the circle (point 2 in Fig. 5–18), the cord exerts its tension force F_{T2} upward, whereas the force of gravity, $m\vec{\mathbf{g}}$, still acts downward. Choosing *upward* as positive, Newton's second law gives:

$$(\Sigma F)_R = ma_R$$
$$F_{T2} - mg = m\frac{v_2^2}{r}. \qquad \text{[at bottom]}$$

The speed v_2 is given as twice that in (a), namely 6.566 m/s. We solve for F_{T2}:

$$F_{T2} = m\frac{v_2^2}{r} + mg$$
$$= (0.150\,\text{kg})\frac{(6.566\,\text{m/s})^2}{(1.10\,\text{m})} + (0.150\,\text{kg})(9.80\,\text{m/s}^2) = 7.35\,\text{N}.$$

EXERCISE F A rider on a Ferris wheel moves in a vertical circle of radius r at constant speed v (Fig. 5–19). Is the normal force that the seat exerts on the rider at the top of the wheel (a) less than, (b) more than, or (c) the same as, the force the seat exerts at the bottom of the wheel?

FIGURE 5–19 Exercise F.

EXAMPLE 5-13 **Conical pendulum.** A small ball of mass m, suspended by a cord of length ℓ, revolves in a circle of radius $r = \ell \sin \theta$, where θ is the angle the string makes with the vertical (Fig. 5–20). (a) In what direction is the acceleration of the ball, and what causes the acceleration? (b) Calculate the speed and period (time required for one revolution) of the ball in terms of ℓ, θ, g, and m.

APPROACH We can answer (a) by looking at Fig. 5–20, which shows the forces on the revolving ball at one instant: the acceleration points horizontally toward the center of the ball's circular path (not along the cord). The force responsible for the acceleration is the *net* force which here is the vector sum of the forces acting on the mass m: its weight $\vec{F}_G$ (of magnitude $F_G = mg$) and the force exerted by the tension in the cord, $\vec{F}_T$. The latter has horizontal and vertical components of magnitude $F_T \sin \theta$ and $F_T \cos \theta$, respectively.

SOLUTION (b) We apply Newton's second law to the horizontal and vertical directions. In the vertical direction, there is no motion, so the acceleration is zero and the net force in the vertical direction is zero:

$$F_T \cos \theta - mg = 0.$$

In the horizontal direction there is only one force, of magnitude $F_T \sin \theta$, that acts toward the center of the circle and gives rise to the acceleration v^2/r. Newton's second law tells us:

$$F_T \sin \theta = m \frac{v^2}{r}.$$

We solve the second equation for v, and substitute for F_T from the first equation (and use $r = \ell \sin \theta$):

$$v = \sqrt{\frac{rF_T \sin \theta}{m}} = \sqrt{\frac{r}{m}\left(\frac{mg}{\cos \theta}\right) \sin \theta}$$

$$= \sqrt{\frac{\ell g \sin^2 \theta}{\cos \theta}}.$$

The period T is the time required to make one revolution, a distance of $2\pi r = 2\pi \ell \sin \theta$. The speed v can thus be written $v = 2\pi \ell \sin \theta / T$; then

$$T = \frac{2\pi \ell \sin \theta}{v} = \frac{2\pi \ell \sin \theta}{\sqrt{\dfrac{\ell g \sin^2 \theta}{\cos \theta}}}$$

$$= 2\pi \sqrt{\frac{\ell \cos \theta}{g}}.$$

NOTE The speed and period do not depend on the mass m of the ball. They do depend on ℓ and θ.

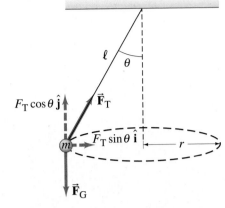

FIGURE 5–20 Example 5–13. Conical pendulum.

PROBLEM SOLVING

Uniform Circular Motion

1. **Draw a free-body diagram**, showing all the forces acting on each object under consideration. Be sure you can identify the source of each force (tension in a cord, Earth's gravity, friction, normal force, and so on). Don't put in something that doesn't belong (like a centrifugal force).

2. **Determine** which of the forces, or which of their components, act to provide the centripetal acceleration—that is, all the **forces or components that act radially**, toward or away from the center of the circular path. The sum of these forces (or components) provides the centripetal acceleration, $a_R = v^2/r$.

3. **Choose a convenient coordinate system**, preferably with one axis along the acceleration direction.

4. **Apply Newton's second law** to the radial component:

$$(\Sigma F)_R = ma_R = m\frac{v^2}{r}. \qquad \text{[radial direction]}$$

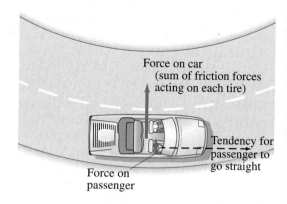

Force on car
(sum of friction forces
acting on each tire)

Tendency for
passenger to
go straight

Force on
passenger

FIGURE 5–22 Race car heading into a curve. From the tire marks we see that most cars experienced a sufficient friction force to give them the needed centripetal acceleration for rounding the curve safely. But, we also see tire tracks of cars on which there was not sufficient force—and which unfortunately followed more nearly straight-line paths.

FIGURE 5–23 Example 5–14. Forces on a car rounding a curve on a flat road. (a) Front view, (b) top view.

(a)

(b)

5–4 Highway Curves: Banked and Unbanked

An example of circular dynamics occurs when an automobile rounds a curve, say to the left. In such a situation, you may feel that you are thrust outward toward the right side door. But there is no mysterious centrifugal force pulling on you. What is happening is that you tend to move in a straight line, whereas the car has begun to follow a curved path. To make you go in the curved path, the seat (friction) or the door of the car (direct contact) exerts a force on you (Fig. 5–21). The car also must have a force exerted on it toward the center of the curve if it is to move in that curve. On a flat road, this force is supplied by friction between the tires and the pavement.

If the wheels and tires of the car are rolling normally without slipping or sliding, the bottom of the tire is at rest against the road at each instant; so the friction force the road exerts on the tires is static friction. But if the static friction force is not great enough, as under icy conditions or high speed, sufficient friction force cannot be applied and the car will skid out of a circular path into a more nearly straight path. See Fig. 5–22. Once a car skids or slides, the friction force becomes kinetic friction, which is less than static friction.

EXAMPLE 5–14 Skidding on a curve. A 1000-kg car rounds a curve on a flat road of radius 50 m at a speed of 15 m/s (54 km/h). Will the car follow the curve, or will it skid? Assume: (a) the pavement is dry and the coefficient of static friction is $\mu_s = 0.60$; (b) the pavement is icy and $\mu_s = 0.25$.

APPROACH The forces on the car are gravity mg downward, the normal force F_N exerted upward by the road, and a horizontal friction force due to the road. They are shown in Fig. 5–23, which is the free-body diagram for the car. The car will follow the curve if the maximum static friction force is greater than the mass times the centripetal acceleration.

SOLUTION In the vertical direction there is no acceleration. Newton's second law tells us that the normal force F_N on the car is equal to the weight mg:

$$F_N = mg = (1000\,\text{kg})(9.80\,\text{m/s}^2) = 9800\,\text{N}.$$

In the horizontal direction the only force is friction, and we must compare it to the force needed to produce the centripetal acceleration to see if it is sufficient. The net horizontal force required to keep the car moving in a circle around the curve is

$$(\Sigma F)_R = ma_R = m\frac{v^2}{r} = (1000\,\text{kg})\frac{(15\,\text{m/s})^2}{(50\,\text{m})} = 4500\,\text{N}.$$

Now we compute the maximum total static friction force (the sum of the friction forces acting on each of the four tires) to see if it can be large enough to provide a safe centripetal acceleration. For (a), $\mu_s = 0.60$, and the maximum friction force attainable (recall from Section 5–1 that $F_{fr} \leq \mu_s F_N$) is

$$(F_{fr})_{max} = \mu_s F_N = (0.60)(9800\,\text{N}) = 5880\,\text{N}.$$

Since a force of only 4500 N is needed, and that is, in fact, how much will be exerted by the road as a static friction force, the car can follow the curve. But in

(b) the maximum static friction force possible is

$$(F_{fr})_{max} = \mu_s F_N = (0.25)(9800 \, N) = 2450 \, N.$$

The car will skid because the ground cannot exert sufficient force (4500 N is needed) to keep it moving in a curve of radius 50 m at a speed of 54 km/h.

The banking of curves can reduce the chance of skidding. The normal force exerted by a banked road, acting perpendicular to the road, will have a component toward the center of the circle (Fig. 5–24), thus reducing the reliance on friction. For a given banking angle θ, there will be one speed for which no friction at all is required. This will be the case when the horizontal component of the normal force toward the center of the curve, $F_N \sin\theta$ (see Fig. 5–24), is just equal to the force required to give a vehicle its centripetal acceleration—that is, when

$$F_N \sin\theta = m\frac{v^2}{r}. \qquad \text{[no friction required]}$$

The banking angle of a road, θ, is chosen so that this condition holds for a particular speed, called the "design speed."

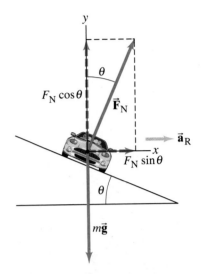

FIGURE 5–24 Normal force on a car rounding a banked curve, resolved into its horizontal and vertical components. The centripetal acceleration is horizontal (*not* parallel to the sloping road). The friction force on the tires, not shown, could point up or down along the slope, depending on the car's speed. The friction force will be zero for one particular speed.

> ⚠ **CAUTION**
> F_N *is not always equal to mg*

EXAMPLE 5–15 Banking angle. (a) For a car traveling with speed v around a curve of radius r, determine a formula for the angle at which a road should be banked so that no friction is required. (b) What is this angle for an expressway off-ramp curve of radius 50 m at a design speed of 50 km/h?

APPROACH Even though the road is banked, the car is still moving along a horizontal circle, so the centripetal acceleration needs to be horizontal. We choose our x and y axes as horizontal and vertical so that a_R, which is horizontal, is along the x axis. The forces on the car are the Earth's gravity mg downward, and the normal force F_N exerted by the road perpendicular to its surface. See Fig. 5–24, where the components of F_N are also shown. We don't need to consider the friction of the road because we are designing a road to be banked so as to eliminate dependence on friction.

SOLUTION (a) Since there is no vertical motion, $\Sigma F_y = ma_y$ gives us

$$F_N \cos\theta - mg = 0.$$

Thus,

$$F_N = \frac{mg}{\cos\theta}.$$

[Note in this case that $F_N \geq mg$ since $\cos\theta \leq 1$.]
We substitute this relation for F_N into the equation for the horizontal motion,

$$F_N \sin\theta = m\frac{v^2}{r},$$

and obtain

$$\frac{mg}{\cos\theta}\sin\theta = m\frac{v^2}{r}$$

or

$$\tan\theta = \frac{v^2}{rg}.$$

This is the formula for the banking angle θ: no friction needed at speed v.
(b) For $r = 50$ m and $v = 50$ km/h (or 14 m/s),

$$\tan\theta = \frac{(14 \, m/s)^2}{(50 \, m)(9.8 \, m/s^2)} = 0.40,$$

so $\theta = 22°$.

EXERCISE G The banking angle of a curve for a design speed v is θ_1. What banking angle θ_2 is needed for a design speed of $2v$? (a) $\theta_2 = 4\theta_1$; (b) $\theta_2 = 2\theta_1$; (c) $\tan\theta_2 = 4\tan\theta_1$; (d) $\tan\theta_2 = 2\tan\theta_1$.

EXERCISE H Can a heavy truck and a small car travel safely at the same speed around an icy banked-curve road?

(a)

(b)

FIGURE 5–25 The speed of an object moving in a circle changes if the force on it has a tangential component, F_{tan}. Part (a) shows the force $\vec{F}$ and its vector components; part (b) shows the acceleration vector and its vector components.

*5–5 Nonuniform Circular Motion

Circular motion at constant speed occurs when the net force on an object is exerted toward the center of the circle. If the net force is not directed toward the center but is at an angle, as shown in Fig. 5–25a, the force has two components. The component directed toward the center of the circle, $\vec{F}_R$, gives rise to the centripetal acceleration, $\vec{a}_R$, and keeps the object moving in a circle. The component tangent to the circle, $\vec{F}_{tan}$, acts to increase (or decrease) the speed, and thus gives rise to a component of the acceleration tangent to the circle, $\vec{a}_{tan}$. When the speed of the object is changing, a tangential component of force is acting.

When you first start revolving a ball on the end of a string around your head, you must give it tangential acceleration. You do this by pulling on the string with your hand displaced from the center of the circle. In athletics, a hammer thrower accelerates the hammer tangentially in a similar way so that it reaches a high speed before release.

The tangential component of the acceleration, a_{tan}, has magnitude equal to the rate of change of the *magnitude* of the object's velocity:

$$a_{tan} = \frac{dv}{dt}. \tag{5–4}$$

The radial (centripetal) acceleration arises from the change in *direction* of the velocity and, as we have seen, has magnitude

$$a_R = \frac{v^2}{r}.$$

The tangential acceleration always points in a direction tangent to the circle, and is in the direction of motion (parallel to $\vec{v}$, which is always tangent to the circle) if the speed is increasing, as shown in Fig. 5–25b. If the speed is decreasing, $\vec{a}_{tan}$ points antiparallel to $\vec{v}$. In either case, $\vec{a}_{tan}$ and $\vec{a}_R$ are always perpendicular to each other; and *their directions change* continually as the object moves along its circular path. The total vector acceleration $\vec{a}$ is the sum of the two components:

$$\vec{a} = \vec{a}_{tan} + \vec{a}_R. \tag{5–5}$$

Since $\vec{a}_R$ and $\vec{a}_{tan}$ are always perpendicular to each other, the magnitude of $\vec{a}$ at any moment is

$$a = \sqrt{a_{tan}^2 + a_R^2}.$$

EXAMPLE 5–16 **Two components of acceleration.** A race car starts from rest in the pit area and accelerates at a uniform rate to a speed of 35 m/s in 11 s, moving on a circular track of radius 500 m. Assuming constant tangential acceleration, find (a) the tangential acceleration, and (b) the radial acceleration, at the instant when the speed is $v = 15$ m/s.

APPROACH The tangential acceleration relates to the change in speed of the car, and can be calculated as $a_{tan} = \Delta v / \Delta t$. The centripetal acceleration relates to the change in the *direction* of the velocity vector and is calculated using $a_R = v^2/r$.

SOLUTION (a) During the 11-s time interval, we assume the tangential acceleration a_{tan} is constant. Its magnitude is

$$a_{tan} = \frac{\Delta v}{\Delta t} = \frac{(35 \text{ m/s} - 0 \text{ m/s})}{11 \text{ s}} = 3.2 \text{ m/s}^2.$$

(b) When $v = 15$ m/s, the centripetal acceleration is

$$a_R = \frac{v^2}{r} = \frac{(15 \text{ m/s})^2}{(500 \text{ m})} = 0.45 \text{ m/s}^2.$$

NOTE The radial acceleration increases continually, whereas the tangential acceleration stays constant.

EXERCISE I When the speed of the race car in Example 5–16 is 30 m/s, how are (a) a_{tan} and (b) a_R changed?

These concepts can be used for an object moving along any curved path, such as that shown in Fig. 5–26. We can treat any portion of the curve as an arc of a circle with a "radius of curvature" r. The velocity at any point is always tangent to the path. The acceleration can be written, in general, as a vector sum of two components: the tangential component $a_{tan} = dv/dt$, and the radial (centripetal) component $a_R = v^2/r$.

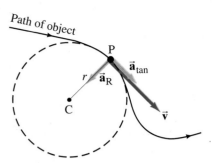

*5–6 Velocity-Dependent Forces: Drag and Terminal Velocity

FIGURE 5–26 Object following a curved path (solid line). At point P the path has a radius of curvature r. The object has velocity $\vec{v}$, tangential acceleration $\vec{a}_{tan}$ (the object is here increasing in speed), and radial (centripetal) acceleration $\vec{a}_R$ (magnitude $a_R = v^2/r$) which points toward the center of curvature C.

When an object slides along a surface, the force of friction acting on the object is nearly independent of how fast the object is moving. But other types of resistive forces do depend on the object's velocity. The most important example is for an object moving through a liquid or gas, such as air. The fluid offers resistance to the motion of the object, and this resistive force, or **drag force**, depends on the velocity of the object.[†]

The way the drag force varies with velocity is complicated in general. But for small objects at very low speeds, a good approximation can often be made by assuming that the drag force, F_D, is directly proportional to the magnitude of the velocity, v:

$$F_D = -bv. \tag{5–6}$$

The minus sign is necessary because the drag force opposes the motion. Here b is a constant (approximately) that depends on the viscosity of the fluid and on the size and shape of the object. Equation 5–6 works well for small objects moving at low speed in a viscous liquid. It also works for very small objects moving in air at very low speeds, such as dust particles. For objects moving at high speeds, such as an airplane, a sky diver, a baseball, or an automobile, the force of air resistance can be better approximated as being proportional to v^2:

$$F_D \propto v^2.$$

For accurate calculations, however, more complicated forms and numerical integration generally need to be used. For objects moving through liquids, Eq. 5–6 works well for everyday objects at normal speeds (e.g., a boat in water).

Let us consider an object that falls from rest, through air or other fluid, under the action of gravity and a resistive force proportional to v. The forces acting on the object are the force of gravity, mg, acting downward, and the drag force, $-bv$, acting upward (Fig. 5–27a). Since the velocity $\vec{v}$ points downward, let us take the positive direction as downward. Then the net force on the object can be written

$$\Sigma F = mg - bv.$$

From Newton's second law $\Sigma F = ma$, we have

$$mg - bv = m\frac{dv}{dt}, \tag{5–7}$$

where we have written the acceleration according to its definition as rate of change of velocity, $a = dv/dt$. At $t = 0$, we set $v = 0$ and the acceleration $dv/dt = g$. As the object falls and increases in speed, the resistive force increases, and this reduces the acceleration, dv/dt (see Fig. 5–27b). The velocity continues to increase, but at a slower rate. Eventually, the velocity becomes so large that the magnitude of the resistive force, bv, approaches that of the gravitational force, mg; when the two are equal, we have

$$mg - bv = 0. \tag{5–8}$$

At this point $dv/dt = 0$ and the object no longer increases in speed. It has reached its **terminal velocity** and continues to fall at this constant velocity until it hits the ground. This sequence of events is shown in the graph of Fig. 5–27b. The value of the terminal velocity v_T can be obtained from Eq. 5–8.

$$v_T = \frac{mg}{b}. \tag{5–9}$$

If the resistive force is assumed proportional to v^2, or an even higher power of v, the sequence of events is similar and a terminal velocity reached, although it will not be given by Eq. 5–9.

FIGURE 5–27 (a) Forces acting on an object falling downward. (b) Graph of the velocity of an object falling due to gravity when the air resistance drag force is $F_D = -bv$. Initially, $v = 0$ and $dv/dt = g$, but as time goes on dv/dt (= slope of curve) decreases because of F_D. Eventually, v approaches a maximum value, v_T, the terminal velocity, which occurs when F_D has magnitude equal to mg.

$$\vec{F}_D = -b\vec{v}$$

$$m\vec{g}$$

(a)

(b)

[†]Any buoyant force (Chapter 13) is ignored in this Section.

Force proportional to velocity. Determine the velocity as a function of time for an object falling vertically from rest when there is a resistive force linearly proportional to v.

APPROACH This is a derivation and we start with Eq. 5–7, which we rewrite as

$$\frac{dv}{dt} = g - \frac{b}{m}v.$$

SOLUTION In this equation there are two variables, v and t. We collect variables of the same type on one or the other side of the equation:

$$\frac{dv}{g - \dfrac{b}{m}v} = dt \quad \text{or} \quad \frac{dv}{v - \dfrac{mg}{b}} = -\frac{b}{m}\,dt.$$

Now we can integrate, remembering $v = 0$ at $t = 0$:

$$\int_0^v \frac{dv}{v - \dfrac{mg}{b}} = -\frac{b}{m}\int_0^t dt$$

which gives

$$\ln\left(v - \frac{mg}{b}\right) - \ln\left(-\frac{mg}{b}\right) = -\frac{b}{m}t$$

or

$$\ln\frac{v - mg/b}{-mg/b} = -\frac{b}{m}t.$$

We raise each side to the exponential [note that the natural log and the exponential are inverse operations of each other: $e^{\ln x} = x$, or $\ln(e^x) = x$] and obtain

$$v - \frac{mg}{b} = -\frac{mg}{b}e^{-\frac{b}{m}t}$$

so

$$v = \frac{mg}{b}\left(1 - e^{-\frac{b}{m}t}\right).$$

This relation gives the velocity v as a function of time and corresponds to the graph of Fig. 5–27b. As a check, note that at $t = 0$, and $v = 0$

$$a(t = 0) = \frac{dv}{dt} = \frac{mg}{b}\frac{d}{dt}\left(1 - e^{-\frac{b}{m}t}\right) = \frac{mg}{b}\left(\frac{b}{m}\right) = g,$$

as expected (see also Eq. 5–7). At large t, $e^{-\frac{b}{m}t}$ approaches zero, so v approaches mg/b, which is the terminal velocity, v_T, as we saw earlier. If we set $\tau = m/b$, then $v = v_T(1 - e^{-t/\tau})$. So $\tau = m/b$ is the time required for the velocity to reach 63% of the terminal velocity (since $e^{-1} = 0.37$). Figure 5–27b shows a plot of speed v vs. time t, where the terminal velocity $v_T = mg/b$.

Summary

When two objects slide over one another, the force of **friction** that each exerts on the other can be written approximately as $F_{fr} = \mu_k F_N$, where F_N is the **normal force** (the force each object exerts on the other perpendicular to their contact surfaces), and μ_k is the coefficient of **kinetic friction**. If the objects are at rest relative to each other, even though forces act, then F_{fr} is just large enough to hold them at rest and satisfies the inequality $F_{fr} \leq \mu_s F_N$, where μ_s is the coefficient of **static friction**.

An object moving in a circle of radius r with constant speed v is said to be in **uniform circular motion**. It has a **radial acceleration** a_R that is directed radially toward the center of the circle (also called **centripetal acceleration**), and has magnitude

$$a_R = \frac{v^2}{r}. \tag{5–1}$$

The direction of the velocity vector and that of the accelera-

tion $\vec{a}_R$ are continually changing in direction, but are perpendicular to each other at each moment.

A force is needed to keep an object revolving uniformly in a circle, and the direction of this force is toward the center of the circle. This force may be gravity (as for the Moon), or tension in a cord, or a component of the normal force, or another type of force or a combination of forces.

[*When the speed of circular motion is not constant, the acceleration has two components, tangential as well as radial. The force too has tangential and radial components.]

[*A **drag force** acts on an object moving through a fluid, such as air or water. The drag force F_D can often be approximated by $F_D = -bv$ or $F_D \propto v^2$, where v is the speed of the object relative to the fluid.]

Questions

1. A heavy crate rests on the bed of a flatbed truck. When the truck accelerates, the crate remains where it is on the truck, so it, too, accelerates. What force causes the crate to accelerate?

2. A block is given a push so that it slides up a ramp. After the block reaches its highest point, it slides back down, but the magnitude of its acceleration is less on the descent than on the ascent. Why?

3. Why is the stopping distance of a truck much shorter than for a train going the same speed?

4. Can a coefficient of friction exceed 1.0?

5. Cross-country skiers prefer their skis to have a large coefficient of static friction but a small coefficient of kinetic friction. Explain why. [*Hint*: Think of uphill and downhill.]

6. When you must brake your car very quickly, why is it safer if the wheels don't lock? When driving on slick roads, why is it advisable to apply the brakes slowly?

7. When attempting to stop a car quickly on dry pavement, which of the following methods will stop the car in the least time? (*a*) Slam on the brakes as hard as possible, locking the wheels and *skidding* to a stop. (*b*) Press the brakes as hard as possible without locking the wheels and *rolling* to a stop. Explain.

8. You are trying to push your stalled car. Although you apply a horizontal force of 400 N to the car, it doesn't budge, and neither do you. Which force(s) must also have a magnitude of 400 N: (*a*) the force exerted by the car on you; (*b*) the friction force exerted by the car on the road; (*c*) the normal force exerted by the road on you; (*d*) the friction force exerted by the road on you?

9. It is not easy to walk on an icy sidewalk without slipping. Even your gait looks different than on dry pavement. Describe what you need to do differently on the icy surface and why.

10. A car rounds a curve at a steady 50 km/h. If it rounds the same curve at a steady 70 km/h, will its acceleration be any different? Explain.

11. Will the acceleration of a car be the same when a car travels around a sharp curve at a constant 60 km/h as when it travels around a gentle curve at the same speed? Explain.

12. Describe all the forces acting on a child riding a horse on a merry-go-round. Which of these forces provides the centripetal acceleration of the child?

13. A child on a sled comes flying over the crest of a small hill, as shown in Fig. 5–28. His sled does not leave the ground, but he feels the normal force between his chest and the sled decrease as he goes over the hill. Explain this decrease using Newton's second law.

FIGURE 5–28
Question 13.

14. Sometimes it is said that water is removed from clothes in a spin dryer by centrifugal force throwing the water outward. Is this correct? Discuss.

15. Technical reports often specify only the rpm for centrifuge experiments. Why is this inadequate?

16. A girl is whirling a ball on a string around her head in a horizontal plane. She wants to let go at precisely the right time so that the ball will hit a target on the other side of the yard. When should she let go of the string?

17. The game of tetherball is played with a ball tied to a pole with a string. When the ball is struck, it whirls around the pole as shown in Fig. 5–29. In what direction is the acceleration of the ball, and what causes the acceleration?

FIGURE 5–29
Problem 17.

18. Astronauts who spend long periods in outer space could be adversely affected by weightlessness. One way to simulate gravity is to shape the spaceship like a cylindrical shell that rotates, with the astronauts walking on the inside surface (Fig. 5–30). Explain how this simulates gravity. Consider (*a*) how objects fall, (*b*) the force we feel on our feet, and (*c*) any other aspects of gravity you can think of.

FIGURE 5–30
Question 18.

19. A bucket of water can be whirled in a vertical circle without the water spilling out, even at the top of the circle when the bucket is upside down. Explain.

20. A car maintains a constant speed v as it traverses the hill and valley shown in Fig. 5–31. Both the hill and valley have a radius of curvature R. At which point, A, B, or C, is the normal force acting on the car (*a*) the largest, (*b*) the smallest? Explain. (*c*) Where would the driver feel heaviest and (*d*) lightest? Explain. (*e*) How fast can the car go without losing contact with the road at A?

FIGURE 5–31 Question 20.

21. Why do bicycle riders lean in when rounding a curve at high speed?

22. Why do airplanes bank when they turn? How would you compute the banking angle given the airspeed and radius of the turn? [*Hint*: Assume an aerodynamic "lift" force acts perpendicular to the wings.]

*23. For a drag force of the form $F = -bv$, what are the units of b?

*24. Suppose two forces act on an object, one force proportional to v and the other proportional to v^2. Which force dominates at high speed?

Problems

5–1 Friction and Newton's Laws

1. (I) If the coefficient of kinetic friction between a 22-kg crate and the floor is 0.30, what horizontal force is required to move the crate at a steady speed across the floor? What horizontal force is required if μ_k is zero?

2. (I) A force of 35.0 N is required to start a 6.0-kg box moving across a horizontal concrete floor. (a) What is the coefficient of static friction between the box and the floor? (b) If the 35.0-N force continues, the box accelerates at 0.60 m/s². What is the coefficient of kinetic friction?

3. (I) Suppose you are standing on a train accelerating at 0.20 g. What minimum coefficient of static friction must exist between your feet and the floor if you are not to slide?

4. (I) The coefficient of static friction between hard rubber and normal street pavement is about 0.90. On how steep a hill (maximum angle) can you leave a car parked?

5. (I) What is the maximum acceleration a car can undergo if the coefficient of static friction between the tires and the ground is 0.90?

6. (II) (a) A box sits at rest on a rough 33° inclined plane. Draw the free-body diagram, showing all the forces acting on the box. (b) How would the diagram change if the box were sliding down the plane. (c) How would it change if the box were sliding up the plane after an initial shove?

7. (II) A 25.0-kg box is released on a 27° incline and accelerates down the incline at 0.30 m/s². Find the friction force impeding its motion. What is the coefficient of kinetic friction?

8. (II) A car can decelerate at −3.80 m/s² without skidding when coming to rest on a level road. What would its deceleration be if the road is inclined at 9.3° and the car moves uphill? Assume the same static friction coefficient.

9. (II) A skier moves down a 27° slope at constant speed. What can you say about the coefficient of friction, μ_k? Assume the speed is low enough that air resistance can be ignored.

10. (II) A wet bar of soap slides freely down a ramp 9.0 m long inclined at 8.0°. How long does it take to reach the bottom? Assume $\mu_k = 0.060$.

11. (II) A box is given a push so that it slides across the floor. How far will it go, given that the coefficient of kinetic friction is 0.15 and the push imparts an initial speed of 3.5 m/s?

12. (II) (a) Show that the minimum stopping distance for an automobile traveling at speed v is equal to $v^2/2\,\mu_s\,g$, where μ_s is the coefficient of static friction between the tires and the road, and g is the acceleration of gravity. (b) What is this distance for a 1200-kg car traveling 95 km/h if $\mu_s = 0.65$? (c) What would it be if the car were on the Moon (the acceleration of gravity on the Moon is about $g/6$) but all else stayed the same?

13. (II) A 1280-kg car pulls a 350-kg trailer. The car exerts a horizontal force of 3.6×10^3 N against the ground in order to accelerate. What force does the car exert on the trailer? Assume an effective friction coefficient of 0.15 for the trailer.

14. (II) Police investigators, examining the scene of an accident involving two cars, measure 72-m-long skid marks of one of the cars, which nearly came to a stop before colliding. The coefficient of kinetic friction between rubber and the pavement is about 0.80. Estimate the initial speed of that car assuming a level road.

15. (II) Piles of snow on slippery roofs can become dangerous projectiles as they melt. Consider a chunk of snow at the ridge of a roof with a slope of 34°. (a) What is the minimum value of the coefficient of static friction that will keep the snow from sliding down? (b) As the snow begins to melt the coefficient of static friction decreases and the snow finally slips. Assuming that the distance from the chunk to the edge of the roof is 6.0 m and the coefficient of kinetic friction is 0.20, calculate the speed of the snow chunk when it slides off the roof. (c) If the edge of the roof is 10.0 m above ground, estimate the speed of the snow when it hits the ground.

16. (II) A small box is held in place against a rough vertical wall by someone pushing on it with a force directed upward at 28° above the horizontal. The coefficients of static and kinetic friction between the box and wall are 0.40 and 0.30, respectively. The box slides down unless the applied force has magnitude 23 N. What is the mass of the box?

17. (II) Two crates, of mass 65 kg and 125 kg, are in contact and at rest on a horizontal surface (Fig. 5–32). A 650-N force is exerted on the 65-kg crate. If the coefficient of kinetic friction is 0.18, calculate (a) the acceleration of the system, and (b) the force that each crate exerts on the other. (c) Repeat with the crates reversed.

FIGURE 5–32
Problem 17.

18. (II) The crate shown in Fig. 5–33 lies on a plane tilted at an angle $\theta = 25.0°$ to the horizontal, with $\mu_k = 0.19$. (a) Determine the acceleration of the crate as it slides down the plane. (b) If the crate starts from rest 8.15 m up the plane from its base, what will be the crate's speed when it reaches the bottom of the incline?

FIGURE 5–33
Crate on inclined plane.
Problems 18 and 19.

19. (II) A crate is given an initial speed of 3.0 m/s up the 25.0° plane shown in Fig. 5–33. (a) How far up the plane will it go? (b) How much time elapses before it returns to its starting point? Assume $\mu_k = 0.17$.

20. (II) Two blocks made of different materials connected together by a thin cord, slide down a plane ramp inclined at an angle θ to the horizontal as shown in Fig. 5–34 (block B is above block A). The masses of the blocks are m_A and m_B, and the coefficients of friction are μ_A and μ_B. If $m_A = m_B = 5.0$ kg, and $\mu_A = 0.20$ and $\mu_B = 0.30$, determine (a) the acceleration of the blocks and (b) the tension in the cord, for an angle $\theta = 32°$.

FIGURE 5–34
Problems 20 and 21.

21. (II) For two blocks, connected by a cord and sliding down the incline shown in Fig. 5–34 (see Problem 20), describe the motion (a) if $\mu_A < \mu_B$, and (b) if $\mu_A > \mu_B$. (c) Determine a formula for the acceleration of each block and the tension F_T in the cord in terms of m_A, m_B, and θ; interpret your results in light of your answers to (a) and (b).

22. (II) A flatbed truck is carrying a heavy crate. The coefficient of static friction between the crate and the bed of the truck is 0.75. What is the maximum rate at which the driver can decelerate and still avoid having the crate slide against the cab of the truck?

23. (II) In Fig. 5–35 the coefficient of static friction between mass m_A and the table is 0.40, whereas the coefficient of kinetic friction is 0.30 (a) What minimum value of m_A will keep the system from starting to move? (b) What value(s) of m_A will keep the system moving at constant speed?

FIGURE 5–35 Problems 23 and 24.

24. (II) Determine a formula for the acceleration of the system shown in Fig. 5–35 in terms of m_A, m_B, and the mass of the cord, m_C. Define any other variables needed.

25. (II) A small block of mass m is given an initial speed v_0 up a ramp inclined at angle θ to the horizontal. It travels a distance d up the ramp and comes to rest. (a) Determine a formula for the coefficient of kinetic friction between block and ramp. (b) What can you say about the value of the coefficient of static friction?

26. (II) A 75-kg snowboarder has an initial velocity of 5.0 m/s at the top of a 28° incline (Fig. 5–36). After sliding down the 110-m long incline (on which the coefficient of kinetic friction is $\mu_k = 0.18$), the snowboarder has attained a velocity v. The snowboarder then slides along a flat surface (on which $\mu_k = 0.15$) and comes to rest after a distance x. Use Newton's second law to find the snowboarder's acceleration while on the incline and while on the flat surface. Then use these accelerations to determine x.

FIGURE 5–36 Problem 26.

27. (II) A package of mass m is dropped vertically onto a horizontal conveyor belt whose speed is $v = 1.5$ m/s, and the coefficient of kinetic friction between the package and the belt is $\mu_k = 0.70$. (a) For how much time does the package slide on the belt (until it is at rest relative to the belt)? (b) How far does the package move during this time?

28. (II) Two masses $m_A = 2.0$ kg and $m_B = 5.0$ kg are on inclines and are connected together by a string as shown in Fig. 5–37. The coefficient of kinetic friction between each mass and its incline is $\mu_k = 0.30$. If m_A moves up, and m_B moves down, determine their acceleration.

FIGURE 5–37 Problem 28.

29. (II) A child slides down a slide with a 34° incline, and at the bottom her speed is precisely half what it would have been if the slide had been frictionless. Calculate the coefficient of kinetic friction between the slide and the child.

30. (II) (a) Suppose the coefficient of kinetic friction between m_A and the plane in Fig. 5–38 is $\mu_k = 0.15$, and that $m_A = m_B = 2.7$ kg. As m_B moves down, determine the magnitude of the acceleration of m_A and m_B, given $\theta = 34°$. (b) What smallest value of μ_k will keep the system from accelerating?

FIGURE 5–38 Problem 30.

31. (III) A 3.0-kg block sits on top of a 5.0-kg block which is on a horizontal surface. The 5.0-kg block is pulled to the right with a force $\vec{F}$ as shown in Fig. 5–39. The coefficient of static friction between all surfaces is 0.60 and the kinetic coefficient is 0.40. (a) What is the minimum value of F needed to move the two blocks? (b) If the force is 10% greater than your answer for (a), what is the acceleration of each block?

FIGURE 5–39 Problem 31.

32. (III) A 4.0-kg block is stacked on top of a 12.0-kg block, which is accelerating along a horizontal table at $a = 5.2 \text{ m/s}^2$ (Fig. 5–40). Let $\mu_k = \mu_s = \mu$. (a) What minimum coefficient of friction μ between the two blocks will prevent the 4.0-kg block from sliding off? (b) If μ is only half this minimum value, what is the acceleration of the 4.0-kg block with respect to the table, and (c) with respect to the 12.0-kg block? (d) What is the force that must be applied to the 12.0-kg block in (a) and in (b), assuming that the table is frictionless?

FIGURE 5–40
Problem 32.

33. (III) A small block of mass m rests on the rough, sloping side of a triangular block of mass M which itself rests on a horizontal frictionless table as shown in Fig. 5–41. If the coefficient of static friction is μ, determine the minimum horizontal force F applied to M that will cause the small block m to start moving up the incline.

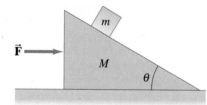

FIGURE 5–41
Problem 33.

5–2 to 5–4 Uniform Circular Motion

34. (I) What is the maximum speed with which a 1200-kg car can round a turn of radius 80.0 m on a flat road if the coefficient of friction between tires and road is 0.65? Is this result independent of the mass of the car?

35. (I) A child sitting 1.20 m from the center of a merry-go-around moves with a speed of 1.30 m/s. Calculate (a) the centripetal acceleration of the child and (b) the net horizontal force exerted on the child (mass = 22.5 kg).

36. (I) A jet plane traveling 1890 km/h (525 m/s) pulls out of a dive by moving in an arc of radius 4.80 km. What is the plane's acceleration in g's?

37. (II) Is it possible to whirl a bucket of water fast enough in a vertical circle so that the water won't fall out? If so, what is the minimum speed? Define all quantities needed.

38. (II) How fast (in rpm) must a centrifuge rotate if a particle 8.00 cm from the axis of rotation is to experience an acceleration of 125,000 g's?

39. (II) Highway curves are marked with a suggested speed. If this speed is based on what would be safe in wet weather, estimate the radius of curvature for a curve marked 50 km/h. Use Table 5–1.

40. (II) At what minimum speed must a roller coaster be traveling when upside down at the top of a circle (Fig. 5–42) so that the passengers do not fall out? Assume a radius of curvature of 7.6 m.

FIGURE 5–42
Problem 40.

41. (II) A sports car crosses the bottom of a valley with a radius of curvature equal to 95 m. At the very bottom, the normal force on the driver is twice his weight. At what speed was the car traveling?

42. (II) How large must the coefficient of static friction be between the tires and the road if a car is to round a level curve of radius 85 m at a speed of 95 km/h?

43. (II) Suppose the space shuttle is in orbit 400 km from the Earth's surface, and circles the Earth about once every 90 min. Find the centripetal acceleration of the space shuttle in its orbit. Express your answer in terms of g, the gravitational acceleration at the Earth's surface.

44. (II) A bucket of mass 2.00 kg is whirled in a vertical circle of radius 1.10 m. At the lowest point of its motion the tension in the rope supporting the bucket is 25.0 N. (a) Find the speed of the bucket. (b) How fast must the bucket move at the top of the circle so that the rope does not go slack?

45. (II) How many revolutions per minute would a 22-m-diameter Ferris wheel need to make for the passengers to feel "weightless" at the topmost point?

46. (II) Use dimensional analysis (Section 1–7) to obtain the form for the centripetal acceleration, $a_R = v^2/r$.

47. (II) A jet pilot takes his aircraft in a vertical loop (Fig. 5–43). (a) If the jet is moving at a speed of 1200 km/h at the lowest point of the loop, determine the minimum radius of the circle so that the centripetal acceleration at the lowest point does not exceed 6.0 g's. (b) Calculate the 78-kg pilot's effective weight (the force with which the seat pushes up on him) at the bottom of the circle, and (c) at the top of the circle (assume the same speed).

FIGURE 5–43
Problem 47.

48. (II) A proposed space station consists of a circular tube that will rotate about its center (like a tubular bicycle tire), Fig. 5–44. The circle formed by the tube has a diameter of about 1.1 km. What must be the rotation speed (revolutions per day) if an effect equal to gravity at the surface of the Earth (1.0 g) is to be felt?

FIGURE 5–44
Problem 48.

49. (II) On an ice rink two skaters of equal mass grab hands and spin in a mutual circle once every 2.5 s. If we assume their arms are each 0.80 m long and their individual masses are 60.0 kg, how hard are they pulling on one another?

50. (II) Redo Example 5–11, precisely this time, by not ignoring the weight of the ball which revolves on a string 0.600 m long. In particular, find the magnitude of $\vec{F}_T$, and the angle it makes with the horizontal. [*Hint*: Set the horizontal component of $\vec{F}_T$ equal to ma_R; also, since there is no vertical motion, what can you say about the vertical component of $\vec{F}_T$?]

51. (II) A coin is placed 12.0 cm from the axis of a rotating turntable of variable speed. When the speed of the turntable is slowly increased, the coin remains fixed on the turntable until a rate of 35.0 rpm (revolutions per minute) is reached, at which point the coin slides off. What is the coefficient of static friction between the coin and the turntable?

52. (II) The design of a new road includes a straight stretch that is horizontal and flat but that suddenly dips down a steep hill at 22°. The transition should be rounded with what minimum radius so that cars traveling 95 km/h will not leave the road (Fig. 5–45)?

FIGURE 5–45
Problem 52.

53. (II) A 975-kg sports car (including driver) crosses the rounded top of a hill (radius = 88.0 m) at 12.0 m/s. Determine (a) the normal force exerted by the road on the car, (b) the normal force exerted by the car on the 72.0-kg driver, and (c) the car speed at which the normal force on the driver equals zero.

54. (II) Two blocks, with masses m_A and m_B, are connected to each other and to a central post by cords as shown in Fig. 5–46. They rotate about the post at frequency f (revolutions per second) on a frictionless horizontal surface at distances r_A and r_B from the post. Derive an algebraic expression for the tension in each segment of the cord (assumed massless).

FIGURE 5–46   Problem 54.

55. (II) Tarzan plans to cross a gorge by swinging in an arc from a hanging vine (Fig. 5–47). If his arms are capable of exerting a force of 1350 N on the rope, what is the maximum speed he can tolerate at the lowest point of his swing? His mass is 78 kg and the vine is 5.2 m long.

FIGURE 5–47
Problem 55.

56. (II) A pilot performs an evasive maneuver by diving vertically at 310 m/s. If he can withstand an acceleration of 9.0 *g*'s without blacking out, at what altitude must he begin to pull out of the dive to avoid crashing into the sea?

57. (III) The position of a particle moving in the *xy* plane is given by $\vec{r} = 2.0 \cos(3.0 \text{ rad/s } t)\hat{i} + 2.0 \sin(3.0 \text{ rad/s } t)\hat{j}$, where *r* is in meters and *t* is in seconds. (a) Show that this represents circular motion of radius 2.0 m centered at the origin. (b) Determine the velocity and acceleration vectors as functions of time. (c) Determine the speed and magnitude of the acceleration. (d) Show that $a = v^2/r$. (e) Show that the acceleration vector always points toward the center of the circle.

58. (III) If a curve with a radius of 85 m is properly banked for a car traveling 65 km/h, what must be the coefficient of static friction for a car not to skid when traveling at 95 km/h?

59. (III) A curve of radius 68 m is banked for a design speed of 85 km/h. If the coefficient of static friction is 0.30 (wet pavement), at what range of speeds can a car safely make the curve? [*Hint*: Consider the direction of the friction force when the car goes too slow or too fast.]

5–5 Nonuniform Circular Motion

*60. (II) A particle starting from rest revolves with uniformly increasing speed in a clockwise circle in the *xy* plane. The center of the circle is at the origin of an *xy* coordinate system. At $t = 0$, the particle is at $x = 0.0$, $y = 2.0$ m. At $t = 2.0$ s, it has made one-quarter of a revolution and is at $x = 2.0$ m, $y = 0.0$. Determine (a) its speed at $t = 2.0$ s, (b) the average velocity vector, and (c) the average acceleration vector during this interval.

*61. (II) In Problem 60 assume the tangential acceleration is constant and determine the components of the instantaneous acceleration at (a) $t = 0.0$, (b) $t = 1.0$ s, and (c) $t = 2.0$ s.

*62. (II) An object moves in a circle of radius 22 m with its speed given by $v = 3.6 + 1.5t^2$, with *v* in meters per second and *t* in seconds. At $t = 3.0$ s, find (a) the tangential acceleration and (b) the radial acceleration.

*63. (III) A particle rotates in a circle of radius 3.80 m. At a particular instant its acceleration is 1.15 m/s² in a direction that makes an angle of 38.0° to its direction of motion. Determine its speed (a) at this moment and (b) 2.00 s later, assuming constant tangential acceleration.

*64. (III) An object of mass *m* is constrained to move in a circle of radius *r*. Its tangential acceleration as a function of time is given by $a_{tan} = b + ct^2$, where *b* and *c* are constants. If $v = v_0$ at $t = 0$, determine the tangential and radial components of the force, F_{tan} and F_R, acting on the object at any time $t > 0$.

5–6 Velocity-Dependent Forces

*65. (I) Use dimensional analysis (Section 1–7) in Example 5–17 to determine if the time constant τ is $\tau = m/b$ or $\tau = b/m$.

*66. (II) The terminal velocity of a 3×10^{-5} kg raindrop is about 9 m/s. Assuming a drag force $F_D = -bv$, determine (a) the value of the constant *b* and (b) the time required for such a drop, starting from rest, to reach 63% of terminal velocity.

*67. (II) An object moving vertically has $\vec{v} = \vec{v}_0$ at $t = 0$. Determine a formula for its velocity as a function of time assuming a resistive force $F = -bv$ as well as gravity for two cases: (a) $\vec{v}_0$ is downward and (b) $\vec{v}_0$ is upward.

*68. (III) The drag force on large objects such as cars, planes, and sky divers moving through air is more nearly $F_D = -bv^2$. (a) For this quadratic dependence on v, determine a formula for the terminal velocity v_T, of a vertically falling object. (b) A 75-kg sky diver has a terminal velocity of about 60 m/s; determine the value of the constant b. (c) Sketch a curve like that of Fig. 5–27b for this case of $F_D \propto v^2$. For the same terminal velocity, would this curve lie above or below that in Fig. 5–27? Explain why.

*69. (III) A bicyclist can coast down a 7.0° hill at a steady 9.5 km/h. If the drag force is proportional to the square of the speed v, so that $F_D = -cv^2$, calculate (a) the value of the constant c and (b) the average force that must be applied in order to descend the hill at 25 km/h. The mass of the cyclist plus bicycle is 80.0 kg. Ignore other types of friction.

*70. (III) Two drag forces act on a bicycle and rider: F_{D1} due to rolling resistance, which is essentially velocity independent; and F_{D2} due to air resistance, which is proportional to v^2. For a specific bike plus rider of total mass 78 kg, $F_{D1} \approx 4.0$ N; and for a speed of 2.2 m/s, $F_{D2} \approx 1.0$ N. (a) Show that the total drag force is

$$F_D = 4.0 + 0.21v^2,$$

where v is in m/s, and F_D is in N and opposes the motion. (b) Determine at what slope angle θ the bike and rider can coast downhill at a constant speed of 8.0 m/s.

*71. (III) Determine a formula for the position and acceleration of a falling object as a function of time if the object starts from rest at $t = 0$ and undergoes a resistive force $F = -bv$, as in Example 5–17.

*72. (III) A block of mass m slides along a horizontal surface lubricated with a thick oil which provides a drag force proportional to the square root of velocity:

$$F_D = -bv^{\frac{1}{2}}.$$

If $v = v_0$ at $t = 0$, determine v and x as functions of time.

*73. (III) Show that the maximum distance the block in Problem 72 can travel is $2m\,v_0^{3/2}/3b$.

*74. (III) You dive straight down into a pool of water. You hit the water with a speed of 5.0 m/s, and your mass is 75 kg. Assuming a drag force of the form $F_D = -(1.00 \times 10^4 \text{ kg/s})\,v$, how long does it take you to reach 2% of your original speed? (Ignore any effects of buoyancy.)

*75. (III) A motorboat traveling at a speed of 2.4 m/s shuts off its engines at $t = 0$. How far does it travel before coming to rest if it is noted that after 3.0 s its speed has dropped to half its original value? Assume that the drag force of the water is proportional to v.

General Problems

76. A coffee cup on the horizontal dashboard of a car slides forward when the driver decelerates from 45 km/h to rest in 3.5 s or less, but not if she decelerates in a longer time. What is the coefficient of static friction between the cup and the dash? Assume the road and the dashboard are level (horizontal).

77. A 2.0-kg silverware drawer does not slide readily. The owner gradually pulls with more and more force, and when the applied force reaches 9.0 N, the drawer suddenly opens, throwing all the utensils to the floor. What is the coefficient of static friction between the drawer and the cabinet?

78. A roller coaster reaches the top of the steepest hill with a speed of 6.0 km/h. It then descends the hill, which is at an average angle of 45° and is 45.0 m long. What will its speed be when it reaches the bottom? Assume $\mu_k = 0.12$.

79. An 18.0-kg box is released on a 37.0° incline and accelerates down the incline at 0.220 m/s². Find the friction force impeding its motion. How large is the coefficient of friction?

80. A flat puck (mass M) is revolved in a circle on a frictionless air hockey table top, and is held in this orbit by a light cord which is connected to a dangling mass (mass m) through a central hole as shown in Fig. 5–48. Show that the speed of the puck is given by $v = \sqrt{mgR/M}$.

FIGURE 5–48 Problem 80.

81. A motorcyclist is coasting with the engine off at a steady speed of 20.0 m/s but enters a sandy stretch where the coefficient of kinetic friction is 0.70. Will the cyclist emerge from the sandy stretch without having to start the engine if the sand lasts for 15 m? If so, what will be the speed upon emerging?

82. In a "Rotor-ride" at a carnival, people rotate in a vertical cylindrically walled "room." (See Fig. 5–49). If the room radius was 5.5 m, and the rotation frequency 0.50 revolutions per second when the floor drops out, what minimum coefficient of static friction keeps the people from slipping down? People on this ride said they were "pressed against the wall." Is there really an outward force pressing them against the wall? If so, what is its source? If not, what is the proper description of their situation (besides nausea)? [Hint: Draw a free-body diagram for a person.]

FIGURE 5–49 Problem 82.

83. A device for training astronauts and jet fighter pilots is designed to rotate the trainee in a horizontal circle of radius 11.0 m. If the force felt by the trainee is 7.45 times her own weight, how fast is she rotating? Express your answer in both m/s and rev/s.

84. A 1250-kg car rounds a curve of radius 72 m banked at an angle of 14°. If the car is traveling at 85 km/h, will a friction force be required? If so, how much and in what direction?

85. Determine the tangential and centripetal components of the net force exerted on a car (by the ground) when its speed is 27 m/s, and it has accelerated to this speed from rest in 9.0 s on a curve of radius 450 m. The car's mass is 1150 kg.

86. The 70.0-kg climber in Fig. 5–50 is supported in the "chimney" by the friction forces exerted on his shoes and back. The static coefficients of friction between his shoes and the wall, and between his back and the wall, are 0.80 and 0.60, respectively. What is the minimum normal force he must exert? Assume the walls are vertical and that the static friction forces are both at their maximum. Ignore his grip on the rope.

FIGURE 5–50
Problem 86.

87. A small mass m is set on the surface of a sphere, Fig. 5–51. If the coefficient of static friction is $\mu_s = 0.70$, at what angle ϕ would the mass start sliding?

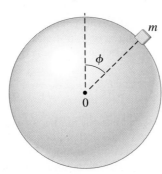

FIGURE 5–51
Problem 87.

88. A 28.0-kg block is connected to an empty 2.00-kg bucket by a cord running over a frictionless pulley (Fig. 5–52). The coefficient of static friction between the table and the block is 0.45 and the coefficient of kinetic friction between the table and the block is 0.32. Sand is gradually added to the bucket until the system just begins to move. (a) Calculate the mass of sand added to the bucket. (b) Calculate the acceleration of the system.

28.0 kg

FIGURE 5–52
Problem 88.

89. A car is heading down a slippery road at a speed of 95 km/h. The minimum distance within which it can stop without skidding is 66 m. What is the sharpest curve the car can negotiate on the icy surface at the same speed without skidding?

90. What is the acceleration experienced by the tip of the 1.5-cm-long sweep second hand on your wrist watch?

91. An airplane traveling at 480 km/h needs to reverse its course. The pilot decides to accomplish this by banking the wings at an angle of 38°. (a) Find the time needed to reverse course. (b) Describe any additional force the passengers experience during the turn. [Hint: Assume an aerodynamic "lift" force that acts perpendicularly to the flat wings; see Fig. 5–53.]

Lift force

FIGURE 5–53
Problem 91.

38°

92. A banked curve of radius R in a new highway is designed so that a car traveling at speed v_0 can negotiate the turn safely on glare ice (zero friction). If a car travels too slowly, then it will slip toward the center of the circle. If it travels too fast, it will slip away from the center of the circle. If the coefficient of static friction increases, it becomes possible for a car to stay on the road while traveling at a speed within a range from $v_{\min}$ to $v_{\max}$. Derive formulas for $v_{\min}$ and $v_{\max}$ as functions of μ_s, v_0, and R.

93. A small bead of mass m is constrained to slide without friction inside a circular vertical hoop of radius r which rotates about a vertical axis (Fig. 5–54) at a frequency f. (a) Determine the angle θ where the bead will be in equilibrium—that is, where it will have no tendency to move up or down along the hoop. (b) If $f = 2.00$ rev/s and $r = 22.0$ cm, what is θ? (c) Can the bead ride as high as the center of the circle ($\theta = 90°$)? Explain.

FIGURE 5–54
Problem 93.

94. *Earth is not quite an inertial frame.* We often make measurements in a reference frame fixed on the Earth, assuming Earth is an inertial reference frame. But the Earth rotates, so this assumption is not quite valid. Show that this assumption is off by 3 parts in 1000 by calculating the acceleration of an object at Earth's equator due to Earth's daily rotation, and compare to $g = 9.80$ m/s^2, the acceleration due to gravity.

95. While fishing, you get bored and start to swing a sinker weight around in a circle below you on a 0.45-m piece of fishing line. The weight makes a complete circle every 0.50 s. What is the angle that the fishing line makes with the vertical? [Hint: See Fig. 5–20.]

96. Consider a train that rounds a curve with a radius of 570 m at a speed of 160 km/h (approximately 100 mi/h). (a) Calculate the friction force needed on a train passenger of mass 75 kg if the track is not banked and the train does not tilt. (b) Calculate the friction force on the passenger if the train tilts at an angle of 8.0° toward the center of the curve.

97. A car starts rolling down a 1-in-4 hill (1-in-4 means that for each 4 m traveled along the road, the elevation change is 1 m). How fast is it going when it reaches the bottom after traveling 55 m? (a) Ignore friction. (b) Assume an effective coefficient of friction equal to 0.10.

98. The sides of a cone make an angle ϕ with the vertical. A small mass m is placed on the inside of the cone and the cone, with its point down, is revolved at a frequency f (revolutions per second) about its symmetry axis. If the coefficient of static friction is μ_s, at what positions on the cone can the mass be placed without sliding on the cone? (Give the maximum and minimum distances, r, from the axis).

99. A 72-kg water skier is being accelerated by a ski boat on a flat ("glassy") lake. The coefficient of kinetic friction between the skier's skis and the water surface is $\mu_k = 0.25$ (Fig. 5–55). (a) What is the skier's acceleration if the rope pulling the skier behind the boat applies a horizontal tension force of magnitude $F_T = 240\,\text{N}$ to the skier ($\theta = 0°$)? (b) What is the skier's horizontal acceleration if the rope pulling the skier exerts a force of $F_T = 240\,\text{N}$ on the skier at an upward angle $\theta = 12°$? (c) Explain why the skier's acceleration in part (b) is greater than that in part (a).

$F_T = 240\,\text{N}$

$\mu_k = 0.25$

FIGURE 5–55 Problem 99.

100. A ball of mass $m = 1.0\,\text{kg}$ at the end of a thin cord of length $r = 0.80\,\text{m}$ revolves in a vertical circle about point O, as shown in Fig. 5–56. During the time we observe it, the only forces acting on the ball are gravity and the tension in the cord. The motion is circular but not uniform because of the force of gravity. The ball increases in speed as it descends and decelerates as it rises on the other side of the circle. At the moment the cord makes an angle $\theta = 30°$ below the horizontal, the ball's speed is 6.0 m/s. At this point, determine the tangential acceleration, the radial acceleration, and the tension in the cord, F_T. Take θ increasing downward as shown.

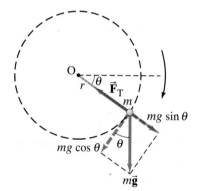

FIGURE 5–56 Problem 100.

101. A car drives at a constant speed around a banked circular track with a diameter of 127 m. The motion of the car can be described in a coordinate system with its origin at the center of the circle. At a particular instant the car's acceleration in the horizontal plane is given by

$$\vec{a} = (-15.7\hat{i} - 23.2\hat{j})\,\text{m/s}^2.$$

(a) What is the car's speed? (b) Where (x and y) is the car at this instant?

*Numerical/Computer

***102.** (III) The force of air resistance (drag force) on a rapidly falling body such as a skydiver has the form $F_D = -kv^2$, so that Newton's second law applied to such an object is

$$m\frac{dv}{dt} = mg - kv^2,$$

where the downward direction is taken to be positive. (a) Use numerical integration [Section 2–9] to estimate (within 2%) the position, speed, and acceleraton, from $t = 0$ up to $t = 15.0\,\text{s}$, for a 75-kg skydiver who starts from rest, assuming $k = 0.22\,\text{kg/m}$. (b) Show that the diver eventually reaches a steady speed, the *terminal speed*, and explain why this happens. (c) How long does it take for the skydiver to reach 99.5% of the terminal speed?

***103.** (III) The coefficient of kinetic friction μ_k between two surfaces is not strictly independent of the velocity of the object. A possible expression for μ_k for wood on wood is

$$\mu_k = \frac{0.20}{(1 + 0.0020v^2)^2},$$

where v is in m/s. A wooden block of mass 8.0 kg is at rest on a wooden floor, and a constant horizontal force of 41 N acts on the block. Use numerical integration [Section 2–9] to determine and graph (a) the speed of the block, and (b) its position, as a function of time from 0 to 5.0 s. (c) Determine the percent difference for the speed and position at 5.0 s if μ_k is constant and equal to 0.20.

***104.** (III) Assume a net force $F = -mg - kv^2$ acts during the upward vertical motion of a 250-kg rocket, starting at the moment ($t = 0$) when the fuel has burned out and the rocket has an upward speed of 120 m/s. Let $k = 0.65\,\text{kg/m}$. Estimate v and y at 1.0-s intervals for the upward motion only, and estimate the maximum height reached. Compare to free-flight conditions without air resistance ($k = 0$).

Answers to Exercises

A: (c).

B: F_{Px} is insufficient to keep the box moving for long.

C: No—the acceleration is not constant (in direction).

D: (a), it doubles.

E: (d).

F: (a).

G: (c).

H: Yes.

I: (a) No change; (b) 4 times larger.

The astronauts in the upper left of this photo are working on the Space Shuttle. As they orbit the Earth—at a rather high speed—they experience apparent weightlessness. The Moon, in the background, also is orbiting the Earth at high speed. What keeps the Moon and the space shuttle (and its astronauts) from moving off in a straight line away from Earth? It is the force of gravity. Newton's law of universal gravitation states that all objects attract all other objects with a force proportional to their masses and inversely proportional to the square of the distance between them.

Gravitation and Newton's Synthesis

C H A P T E R

6

CHAPTER-OPENING QUESTION—Guess now!

A space station revolves around the Earth as a satellite, 100 km above Earth's surface. What is the net force on an astronaut at rest inside the space station?

(a) Equal to her weight on Earth.
(b) A little less than her weight on Earth.
(c) Less than half her weight on Earth.
(d) Zero (she is weightless).
(e) Somewhat larger than her weight on Earth.

S ir Isaac Newton not only put forth the three great laws of motion that serve as the foundation for the study of dynamics. He also conceived of another great law to describe one of the basic forces in nature, gravitation, and he applied it to understand the motion of the planets. This new law, published in 1687 in his book *Philosophiae Naturalis Principia Mathematica* (the *Principia*, for short), is called Newton's law of universal gravitation. It was the capstone of Newton's analysis of the physical world. Indeed, Newtonian mechanics, with its three laws of motion and the law of universal gravitation, was accepted for centuries as a mechanical basis for the way the universe works.

6-1 Newton's Law of Universal Gravitation

Among his many great accomplishments, Sir Isaac Newton examined the motion of the heavenly bodies—the planets and the Moon. In particular, he wondered about the nature of the force that must act to keep the Moon in its nearly circular orbit around the Earth.

Newton was also thinking about the problem of gravity. Since falling objects accelerate, Newton had concluded that they must have a force exerted on them, a force we call the force of gravity. Whenever an object has a force exerted *on* it, that force is exerted *by* some other object. But what exerts the force of gravity? Every object on the surface of the Earth feels the force of gravity, and no matter where the object is, the force is directed toward the center of the Earth (Fig. 6–1). Newton concluded that it must be the Earth itself that exerts the gravitational force on objects at its surface.

According to legend, Newton noticed an apple drop from a tree. He is said to have been struck with a sudden inspiration: If gravity acts at the tops of trees, and even at the tops of mountains, then perhaps it acts all the way to the Moon! With this idea that it is Earth's gravity that holds the Moon in its orbit, Newton developed his great theory of gravitation. But there was controversy at the time. Many thinkers had trouble accepting the idea of a force "acting at a distance." Typical forces act through contact—your hand pushes a cart and pulls a wagon, a bat hits a ball, and so on. But gravity acts without contact, said Newton: the Earth exerts a force on a falling apple and on the Moon, even though there is no contact, and the two objects may even be very far apart.

Newton set about determining the magnitude of the gravitational force that the Earth exerts on the Moon as compared to the gravitational force on objects at the Earth's surface. At the surface of the Earth, the force of gravity accelerates objects at 9.80 m/s^2. The centripetal acceleration of the Moon is calculated from $a_R = v^2/r$ (see Example 5–9) and gives $a_R = 0.00272 \text{ m/s}^2$. In terms of the acceleration of gravity at the Earth's surface, g, this is equivalent to

$$a_R = \frac{0.00272 \text{ m/s}^2}{9.80 \text{ m/s}^2} g \approx \frac{1}{3600} g.$$

That is, the acceleration of the Moon toward the Earth is about $\frac{1}{3600}$ as great as the acceleration of objects at the Earth's surface. The Moon is 384,000 km from the Earth, which is about 60 times the Earth's radius of 6380 km. That is, the Moon is 60 times farther from the Earth's center than are objects at the Earth's surface. But $60 \times 60 = 60^2 = 3600$. Again that number 3600! Newton concluded that the gravitational force F exerted by the Earth on any object decreases with the square of its distance, r, from the Earth's center:

$$F \propto \frac{1}{r^2}.$$

The Moon is 60 Earth radii away, so it feels a gravitational force only $\frac{1}{60^2} = \frac{1}{3600}$ times as strong as it would if it were a point at the Earth's surface.

Newton realized that the force of gravity on an object depends not only on distance but also on the object's mass. In fact, it is directly proportional to its mass, as we have seen. According to Newton's third law, when the Earth exerts its gravitational force on any object, such as the Moon, that object exerts an equal and opposite force on the Earth (Fig. 6–2). Because of this symmetry, Newton reasoned, the magnitude of the force of gravity must be proportional to *both* the masses. Thus

$$F \propto \frac{m_E m_B}{r^2},$$

where m_E is the mass of the Earth, m_B the mass of the other object, and r the distance from the Earth's center to the center of the other object.

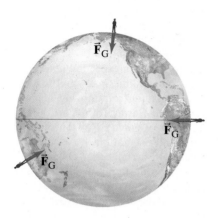

FIGURE 6–1 Anywhere on Earth, whether in Alaska, Australia, or Peru, the force of gravity acts downward toward the center of the Earth.

FIGURE 6–2 The gravitational force one object exerts on a second object is directed toward the first object, and is equal and opposite to the force exerted by the second object on the first.

Moon

Gravitational force exerted on Moon by Earth

Earth

Gravitational force exerted on Earth by the Moon

Newton went a step further in his analysis of gravity. In his examination of the orbits of the planets, he concluded that the force required to hold the different planets in their orbits around the Sun seems to diminish as the inverse square of their distance from the Sun. This led him to believe that it is also the gravitational force that acts between the Sun and each of the planets to keep them in their orbits. And if gravity acts between these objects, why not between all objects? Thus he proposed his **law of universal gravitation**, which we can state as follows:

> **Every particle in the universe attracts every other particle with a force that is proportional to the product of their masses and inversely proportional to the square of the distance between them. This force acts along the line joining the two particles.**

The magnitude of the gravitational force can be written as

$$F = G\frac{m_1 m_2}{r^2}, \tag{6-1}$$

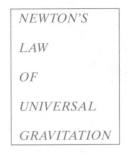

where m_1 and m_2 are the masses of the two particles, r is the distance between them, and G is a universal constant which must be measured experimentally and has the same numerical value for all objects.

The value of G must be very small, since we are not aware of any force of attraction between ordinary-sized objects, such as between two baseballs. The force between two ordinary objects was first measured by Henry Cavendish in 1798, over 100 years after Newton published his law. To detect and measure the incredibly small force between ordinary objects, he used an apparatus like that shown in Fig. 6–3. Cavendish confirmed Newton's hypothesis that two objects attract one another and that Eq. 6–1 accurately describes this force. In addition, because Cavendish could measure F, m_1, m_2, and r accurately, he was able to determine the value of the constant G as well. The accepted value today is

$$G = 6.67 \times 10^{-11}\,\text{N}\cdot\text{m}^2/\text{kg}^2.$$

(See Table inside front cover for values of all constants to highest known precision.)

Strictly speaking, Eq. 6–1 gives the magnitude of the gravitational force that one particle exerts on a second particle that is a distance r away. For an extended object (that is, not a point), we must consider how to measure the distance r. You might think that r would be the distance between the centers of the objects. This is true for two spheres, and is often a good approximation for other objects. A correct calculation treats each extended body as a collection of particles, and the total force is the sum of the forces due to all the particles. The sum over all these particles is often best done using integral calculus, which Newton himself invented. When extended bodies are small compared to the distance between them (as for the Earth–Sun system), little inaccuracy results from considering them as point particles.

Newton was able to show (see derivation in Appendix D) that the *gravitational force exerted on a particle outside a sphere, with a spherically symmetric mass distribution, is the same as if the entire mass of the sphere was concentrated at its center.* Thus Eq. 6–1 gives the correct force between two uniform spheres where r is the distance between their centers.

EXAMPLE 6–1 ESTIMATE Can you attract another person gravitationally?
A 50-kg person and a 70-kg person are sitting on a bench close to each other. Estimate the magnitude of the gravitational force each exerts on the other.

APPROACH This is an estimate: we let the distance between the centers of the two people be $\frac{1}{2}$ m (about as close as you can get).

SOLUTION We use Eq. 6–1, which gives

$$F = \frac{(6.67 \times 10^{-11}\,\text{N}\cdot\text{m}^2/\text{kg}^2)(50\,\text{kg})(70\,\text{kg})}{(0.5\,\text{m})^2} \approx 10^{-6}\,\text{N},$$

rounded off to an order of magnitude. Such a force is unnoticeably small unless extremely sensitive instruments are used.

NOTE As a fraction of their weight, this force is $(10^{-6}\,\text{N})/(70\,\text{kg})(9.8\,\text{m/s}^2) \approx 10^{-9}$.

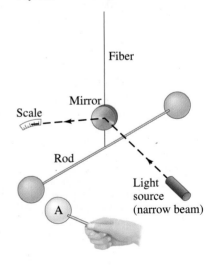

FIGURE 6–3 Schematic diagram of Cavendish's apparatus. Two spheres are attached to a light horizontal rod, which is suspended at its center by a thin fiber. When a third sphere (labeled A) is brought close to one of the suspended spheres, the gravitational force causes the latter to move, and this twists the fiber slightly. The tiny movement is magnified by the use of a narrow light beam directed at a mirror mounted on the fiber. The beam reflects onto a scale. Previous determination of how large a force will twist the fiber a given amount then allows the experimenter to determine the magnitude of the gravitational force between two objects.

Motion

$2r_E$

r_E

FIGURE 6–4 Example 6–2.

FIGURE 6–5 Example 6–3. Orientation of Sun (S), Earth (E), and Moon (M) at right angles to each other (not to scale).

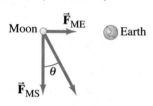

Moon $\vec{F}_{ME}$ Earth

θ

$\vec{F}_{MS}$

Sun

⚠ CAUTION

Distinguish between Newton's second law and the law of universal gravitation

EXAMPLE 6–2 **Spacecraft at $2r_E$.** What is the force of gravity acting on a 2000-kg spacecraft when it orbits two Earth radii from the Earth's center (that is, a distance $r_E = 6380\,\text{km}$ above the Earth's surface, Fig. 6–4)? The mass of the Earth is $m_E = 5.98 \times 10^{24}\,\text{kg}$.

APPROACH We could plug all the numbers into Eq. 6–1, but there is a simpler approach. The spacecraft is twice as far from the Earth's center as when it is at the surface of the Earth. Therefore, since the force of gravity decreases as the square of the distance (and $\frac{1}{2^2} = \frac{1}{4}$), the force of gravity on the satellite will be only one-fourth its weight at the Earth's surface.

SOLUTION At the surface of the Earth, $F_G = mg$. At a distance from the Earth's center of $2r_E$, F_G is $\frac{1}{4}$ as great:

$$F_G = \tfrac{1}{4}mg = \tfrac{1}{4}(2000\,\text{kg})(9.80\,\text{m/s}^2) = 4900\,\text{N}.$$

EXAMPLE 6–3 **Force on the Moon.** Find the net force on the Moon $(m_M = 7.35 \times 10^{22}\,\text{kg})$ due to the gravitational attraction of both the Earth $(m_E = 5.98 \times 10^{24}\,\text{kg})$ and the Sun $(m_S = 1.99 \times 10^{30}\,\text{kg})$, assuming they are at right angles to each other as in Fig. 6–5.

APPROACH The forces on our object, the Moon, are the gravitational force exerted on the Moon by the Earth F_{ME} and the force exerted by the Sun F_{MS}, as shown in the free-body diagram of Fig. 6–5. We use the law of universal gravitation to find the magnitude of each force, and then add the two forces as vectors.

SOLUTION The Earth is $3.84 \times 10^5\,\text{km} = 3.84 \times 10^8\,\text{m}$ from the Moon, so F_{ME} (the gravitational force on the Moon due to the Earth) is

$$F_{ME} = \frac{(6.67 \times 10^{-11}\,\text{N}\cdot\text{m}^2/\text{kg}^2)(7.35 \times 10^{22}\,\text{kg})(5.98 \times 10^{24}\,\text{kg})}{(3.84 \times 10^8\,\text{m})^2} = 1.99 \times 10^{20}\,\text{N}.$$

The Sun is $1.50 \times 10^8\,\text{km}$ from the Earth and the Moon, so F_{MS} (the gravitational force on the Moon due to the Sun) is

$$F_{MS} = \frac{(6.67 \times 10^{-11}\,\text{N}\cdot\text{m}^2/\text{kg}^2)(7.35 \times 10^{22}\,\text{kg})(1.99 \times 10^{30}\,\text{kg})}{(1.50 \times 10^{11}\,\text{m})^2} = 4.34 \times 10^{20}\,\text{N}.$$

The two forces act at right angles in the case we are considering (Fig. 6–5), so we can apply the Pythagorean theorem to find the magnitude of the total force:

$$F = \sqrt{(1.99 \times 10^{20}\,\text{N})^2 + (4.34 \times 10^{20}\,\text{N})^2} = 4.77 \times 10^{20}\,\text{N}.$$

The force acts at an angle θ (Fig. 6–5) given by $\theta = \tan^{-1}(1.99/4.34) = 24.6°$.

NOTE The two forces, F_{ME} and F_{MS}, have the same order of magnitude $(10^{20}\,\text{N})$. This may be surprising. Is it reasonable? The Sun is much farther from Earth than the Moon (a factor of $10^{11}\,\text{m}/10^8\,\text{m} \approx 10^3$), but the Sun is also much more massive (a factor of $10^{30}\,\text{kg}/10^{23}\,\text{kg} \approx 10^7$). Mass divided by distance squared $(10^7/10^6)$ comes out within an order of magnitude, and we have ignored factors of 3 or more. Yes, it is reasonable.

Note carefully that the law of universal gravitation describes a *particular* force (gravity), whereas Newton's second law of motion $(F = ma)$ tells how an object accelerates due to *any* type of force.

*Spherical Shells

Newton was able to show, using the calculus he invented for the purpose, that a thin uniform spherical shell exerts a force on a particle *outside* it as if all the shell's mass were at its center; and that such a thin uniform shell exerts *zero* force on a particle *inside* the shell. (The derivation is given in Appendix D.) The Earth can be modelled as a series of concentric shells starting at its center, each shell uniform but perhaps having a different density to take into account Earth's varying density in various layers. As a simple example, suppose the Earth were uniform throughout; what is the gravitational force on a particle placed exactly halfway from Earth's center to its surface? Only the mass *inside* this radius $r = \frac{1}{2} r_E$ would exert a net force on this particle. The mass of a sphere is proportional to its volume $V = \frac{4}{3} \pi r^3$, so the mass m inside $r = \frac{1}{2} r_E$ is $\left(\frac{1}{2}\right)^3 = \frac{1}{8}$ the mass of the entire Earth. The gravitational force on the particle at $r = \frac{1}{2} r_E$, which is proportional to m/r^2 (Eq. 6–1), is reduced to $\left(\frac{1}{8}\right)/\left(\frac{1}{2}\right)^2 = \frac{1}{2}$ the gravitational force it would experience at Earth's surface.

6–2 Vector Form of Newton's Law of Universal Gravitation

We can write Newton's law of universal gravitation in vector form as

$$\vec{\mathbf{F}}_{12} = -G \frac{m_1 m_2}{r_{21}^2} \hat{\mathbf{r}}_{21}, \qquad (6\text{–}2)$$

where $\vec{\mathbf{F}}_{12}$ is the vector force on particle 1 (of mass m_1) exerted by particle 2 (of mass m_2), which is a distance r_{21} away; $\hat{\mathbf{r}}_{21}$ is a unit vector that points from particle 2 toward particle 1 along the line joining them so that $\hat{\mathbf{r}}_{21} = \vec{\mathbf{r}}_{21}/r_{21}$, where $\vec{\mathbf{r}}_{21}$ is the displacement vector as shown in Fig. 6–6. The minus sign in Eq. 6–2 is necessary because the force on particle 1 due to particle 2 points toward m_2, in the direction opposite to $\hat{\mathbf{r}}_{21}$. The displacement vector $\vec{\mathbf{r}}_{12}$ is a vector of the same magnitude as $\vec{\mathbf{r}}_{21}$, but it points in the opposite direction so that

$$\vec{\mathbf{r}}_{12} = -\vec{\mathbf{r}}_{21}.$$

By Newton's third law, the force $\vec{\mathbf{F}}_{21}$ acting on m_2 exerted by m_1 must have the same magnitude as $\vec{\mathbf{F}}_{12}$ but acts in the opposite direction (Fig. 6–7), so that

$$\vec{\mathbf{F}}_{21} = -\vec{\mathbf{F}}_{12} = G \frac{m_1 m_2}{r_{21}^2} \hat{\mathbf{r}}_{21}$$
$$= -G \frac{m_2 m_1}{r_{12}^2} \hat{\mathbf{r}}_{12}.$$

The force of gravity exerted on one particle by a second particle is always directed toward the second particle, as in Fig. 6–6. When many particles interact, the total gravitational force on a given particle is the vector sum of the forces exerted by each of the others. For example, the total force on particle number 1 is

$$\vec{\mathbf{F}}_1 = \vec{\mathbf{F}}_{12} + \vec{\mathbf{F}}_{13} + \vec{\mathbf{F}}_{14} + \cdots + \vec{\mathbf{F}}_{1n} = \sum_{i=2}^{n} \vec{\mathbf{F}}_{1i} \qquad (6\text{–}3)$$

where $\vec{\mathbf{F}}_{1i}$ means the force on particle 1 exerted by particle i, and n is the total number of particles.

This vector notation can be very helpful, especially when sums over many particles are needed. However, in many cases we do not need to be so formal and we can deal with directions by making careful diagrams.

6–3 Gravity Near the Earth's Surface; Geophysical Applications

When Eq. 6–1 is applied to the gravitational force between the Earth and an object at its surface, m_1 becomes the mass of the Earth m_E, m_2 becomes the mass of the object m, and r becomes the distance of the object from the Earth's center, which is the radius of the Earth r_E. This force of gravity due to the Earth is the weight of the object, which we have been writing as mg. Thus,

$$mg = G \frac{m m_E}{r_E^2}.$$

We can solve this for g, the acceleration of gravity at the Earth's surface:

$$g = G \frac{m_E}{r_E^2}. \qquad (6\text{–}4)$$

Thus, the acceleration of gravity at the surface of the Earth, g, is determined by m_E and r_E. (Don't confuse G with g; they are very different quantities, but are related by Eq. 6–4.)

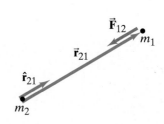

FIGURE 6–6 The displacement vector $\vec{\mathbf{r}}_{21}$ points from particle of mass m_2 to particle of mass m_1. The unit vector shown, $\hat{\mathbf{r}}_{21}$ is in the same direction as $\vec{\mathbf{r}}_{21}$, but is defined as having length one.

FIGURE 6–7 By Newton's third law, the gravitational force on particle 1 exerted by particle 2, $\vec{\mathbf{F}}_{12}$, is equal and opposite to that on particle 2 exerted by particle 1, $\vec{\mathbf{F}}_{21}$; that is $\vec{\mathbf{F}}_{21} = -\vec{\mathbf{F}}_{12}$.

⚠ C A U T I O N

Distinguish G from g

Until G was measured, the mass of the Earth was not known. But once G was measured, Eq. 6–4 could be used to calculate the Earth's mass, and Cavendish was the first to do so. Since $g = 9.80\,\text{m/s}^2$ and the radius of the Earth is $r_E = 6.38 \times 10^6\,\text{m}$, then, from Eq. 6–4, we obtain

$$m_E = \frac{g r_E^2}{G} = \frac{(9.80\,\text{m/s}^2)(6.38 \times 10^6\,\text{m})^2}{6.67 \times 10^{-11}\,\text{N} \cdot \text{m}^2/\text{kg}^2}$$

$$= 5.98 \times 10^{24}\,\text{kg}$$

for the mass of the Earth.

Equation 6–4 can be applied to other planets, where g, m, and r would refer to that planet.

FIGURE 6–8 Example 6–4. Mount Everest, 8850 m (29,035 ft) above sea level; in the foreground, the author with sherpas at 5500 m (18,000 ft).

EXAMPLE 6–4 **ESTIMATE** **Gravity on Everest.** Estimate the effective value of g on the top of Mt. Everest, 8850 m (29,035 ft) above sea level (Fig. 6–8). That is, what is the acceleration due to gravity of objects allowed to fall freely at this altitude?

APPROACH The force of gravity (and the acceleration due to gravity g) depends on the distance from the center of the Earth, so there will be an effective value g' on top of Mt. Everest which will be smaller than g at sea level. We assume the Earth is a uniform sphere (a reasonable "estimate").

SOLUTION We use Eq. 6–4, with r_E replaced by $r = 6380\,\text{km} + 8.9\,\text{km} = 6389\,\text{km} = 6.389 \times 10^6\,\text{m}$:

$$g = G \frac{m_E}{r^2} = \frac{(6.67 \times 10^{-11}\,\text{N} \cdot \text{m}^2/\text{kg}^2)(5.98 \times 10^{24}\,\text{kg})}{(6.389 \times 10^6\,\text{m})^2}$$

$$= 9.77\,\text{m/s}^2,$$

which is a reduction of about 3 parts in a thousand (0.3%).

NOTE This is an estimate because, among other things, we ignored the mass accumulated under the mountaintop.

TABLE 6–1
Acceleration Due to Gravity at Various Locations on Earth

Location	Elevation (m)	g (m/s²)
New York	0	9.803
San Francisco	0	9.800
Denver	1650	9.796
Pikes Peak	4300	9.789
Sydney, Australia	0	9.798
Equator	0	9.780
North Pole (calculated)	0	9.832

PHYSICS APPLIED
Geology—mineral and oil exploration

Note that Eq. 6–4 does not give precise values for g at different locations because the Earth is not a perfect sphere. The Earth not only has mountains and valleys, and bulges at the equator, but also its mass is not distributed precisely uniformly (see Table 6–1). The Earth's rotation also affects the value of g (see Example 6–5). However, for most practical purposes when an object is near the Earth's surface, we will simply use $g = 9.80\,\text{m/s}^2$ and write the weight of an object as mg.

EXERCISE A Suppose you could double the mass of a planet but kept its volume the same. How would the acceleration of gravity, g, at the surface change?

The value of g can vary locally on the Earth's surface because of the presence of irregularities and rocks of different densities. Such variations in g, known as "gravity anomalies," are very small—on the order of 1 part per 10^6 or 10^7 in the value of g. But they can be measured ("gravimeters" today can detect variations in g to 1 part in 10^9). Geophysicists use such measurements as part of their investigations into the structure of the Earth's crust, and in mineral and oil exploration. Mineral deposits, for example, often have a greater density than surrounding material; because of the greater mass in a given volume, g can have a slightly greater value on top of such a deposit than its flanks. "Salt domes," under which petroleum is often found, have a lower than average density and searches for a slight reduction in the value of g in certain locales have led to the discovery of oil.

EXAMPLE 6-5 **Effect of Earth's rotation on g.** Assuming the Earth is a perfect sphere, determine how the Earth's rotation affects the value of g at the equator compared to its value at the poles.

APPROACH Figure 6–9 shows a person of mass m standing on a doctor's scale at two places on the Earth. At the North Pole there are two forces acting on the mass m: the force of gravity, $\vec{F}_G = m\vec{g}$, and the force with which the scale pushes up on the mass, $\vec{w}$. We call this latter force w because it is what the scale reads as the weight of the object, and by Newton's third law it equals the force with which the mass pushes down on the scale. Since the mass is not accelerating, Newton's second law tells us

$$mg - w = 0,$$

so $w = mg$. Thus the weight w that the spring registers equals mg, which is no surprise. Next, at the equator, there *is* an acceleration because the Earth is rotating. The same magnitude of the force of gravity $F_G = mg$ acts downward (we are letting g represent the acceleration of gravity in the absence of rotation and we ignore the slight bulging of the equator). The scale pushes upward with a force w'; w' is also the force with which the person pushes down on the scale (Newton's third law) and hence is the weight registered on the scale. From Newton's second law we now have (see Fig. 6–9)

$$mg - w' = m\frac{v^2}{r_E},$$

because the person of mass m now has a centripetal acceleration due to Earth's rotation; $r_E = 6.38 \times 10^6$ m is the Earth's radius and v is the speed of m due to the Earth's daily rotation.

SOLUTION First we determine the speed v of an object at rest on the Earth's equator, rembering that Earth makes one rotation (distance = circumference of Earth = $2\pi r_E$) in 1 day = $(24\,\text{h})(60\,\text{min/h})(60\,\text{s/min}) = 8.64 \times 10^4$ s:

$$v = \frac{2\pi r_E}{1\,\text{day}} = \frac{(6.283)(6.38 \times 10^6\,\text{m})}{(8.64 \times 10^4\,\text{s})}$$
$$= 4.640 \times 10^2\,\text{m/s}.$$

The effective weight is $w' = mg'$ where g' is the effective value of g, and so $g' = w'/m$. Solving the equation above for w', we have

$$w' = m\left(g - \frac{v^2}{r_E}\right),$$

so

$$g' = \frac{w'}{m} = g - \frac{v^2}{r_E}.$$

Hence

$$\Delta g = g - g' = \frac{v^2}{r_E} = \frac{(4.640 \times 10^2\,\text{m/s})^2}{(6.38 \times 10^6\,\text{m})}$$
$$= 0.0337\,\text{m/s}^2,$$

which is about $\Delta g \approx 0.003g$, a difference of 0.3%.

NOTE In Table 6–1 we see that the difference in g at the pole and equator is actually greater than this: $(9.832 - 9.780)\,\text{m/s}^2 = 0.052\,\text{m/s}^2$. This discrepancy is due mainly to the Earth being slightly fatter at the equator (by 21 km) than at the poles.

NOTE The calculation of the effective value of g at latitudes other than at the poles or equator is a two-dimensional problem because $\vec{F}_G$ acts radially toward the Earth's center whereas the centripetal acceleration is directed perpendicular to the axis of rotation, parallel to the equator and that means that a plumb line (the effective direction of g) is not precisely vertical except at the equator and the poles.

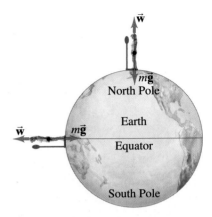

FIGURE 6–9 Example 6–5.

Earth as Inertial Reference Frame

We often make the assumption that reference frames fixed on the Earth are inertial reference frames. Our calculation in Example 6–5 above shows that this assumption can result in errors no larger than about 0.3% in the use of Newton's second law, for example. We discuss the effects of Earth's rotation and reference frames in more detail in Chapter 11, including the Coriolis effect.

6–4 Satellites and "Weightlessness"

Satellite Motion

Artificial satellites circling the Earth are now commonplace (Fig. 6–10). A satellite is put into orbit by accelerating it to a sufficiently high tangential speed with the use of rockets, as shown in Fig. 6–11. If the speed is too high, the spacecraft will not be confined by the Earth's gravity and will escape, never to return. If the speed is too low, it will return to Earth. Satellites are usually put into circular (or nearly circular) orbits, because such orbits require the least takeoff speed.

FIGURE 6–10 A satellite, the International Space Station, circling the Earth.

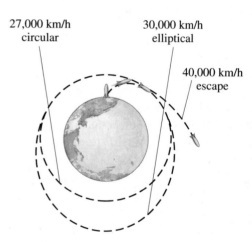

27,000 km/h circular

30,000 km/h elliptical

40,000 km/h escape

FIGURE 6–11 Artificial satellites launched at different speeds.

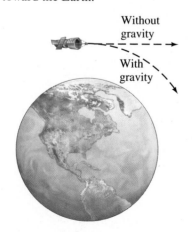

FIGURE 6–12 A moving satellite "falls" out of a straight-line path toward the Earth.

Without gravity

With gravity

It is sometimes asked: "What keeps a satellite up?" The answer is: its high speed. If a satellite stopped moving, it would fall directly to Earth. But at the very high speed a satellite has, it would quickly fly out into space (Fig. 6–12) if it weren't for the gravitational force of the Earth pulling it into orbit. In fact, a satellite *is* falling (accelerating toward Earth), but its high tangential speed keeps it from hitting Earth.

For satellites that move in a circle (at least approximately), the needed acceleration is centripetal and equals v^2/r. The force that gives a satellite this acceleration is the force of gravity exerted by the Earth, and since a satellite may be at a considerable distance from the Earth, we must use Newton's law of universal gravitation (Eq. 6–1) for the force acting on it. When we apply Newton's second law, $\Sigma F_R = ma_R$ in the radial direction, we find

$$G\frac{mm_E}{r^2} = m\frac{v^2}{r}, \qquad (6\text{–}5)$$

where m is the mass of the satellite. This equation relates the distance of the satellite from the Earth's center, r, to its speed, v, in a circular orbit. Note that only one force—gravity—is acting on the satellite, and that r is the sum of the Earth's radius r_E plus the satellite's height h above the Earth: $r = r_E + h$.

EXAMPLE 6–6 **Geosynchronous satellite.** A *geosynchronous* satellite is one that stays above the same point on the Earth, which is possible only if it is above a point on the equator. Such satellites are used for TV and radio transmission, for weather forecasting, and as communication relays. Determine (*a*) the height above the Earth's surface such a satellite must orbit, and (*b*) such a satellite's speed. (*c*) Compare to the speed of a satellite orbiting 200 km above Earth's surface.

APPROACH To remain above the same point on Earth as the Earth rotates, the satellite must have a period of 24 hours. We can apply Newton's second law, $F = ma$, where $a = v^2/r$ if we assume the orbit is circular.

SOLUTION (*a*) The only force on the satellite is the force of universal gravitation due to the Earth. (We can ignore the gravitational force exerted by the Sun. Why?) We apply Eq. 6–5, assuming the satellite moves in a circle:

$$G\frac{m_{Sat}\, m_E}{r^2} = m_{Sat}\frac{v^2}{r}.$$

This equation has two unknowns, r and v. But the satellite revolves around the Earth with the same period that the Earth rotates on its axis, namely once in 24 hours. Thus the speed of the satellite must be

$$v = \frac{2\pi r}{T},$$

where $T = 1\,\text{day} = (24\,\text{h})(3600\,\text{s/h}) = 86{,}400\,\text{s}$. We substitute this into the "satellite equation" above and obtain (after canceling m_{Sat} on both sides)

$$G\frac{m_E}{r^2} = \frac{(2\pi r)^2}{rT^2}.$$

After cancelling an r, we can solve for r^3:

$$r^3 = \frac{Gm_E T^2}{4\pi^2} = \frac{(6.67 \times 10^{-11}\,\text{N}\cdot\text{m}^2/\text{kg}^2)(5.98 \times 10^{24}\,\text{kg})(86{,}400\,\text{s})^2}{4\pi^2}$$
$$= 7.54 \times 10^{22}\,\text{m}^3.$$

We take the cube root and find

$$r = 4.23 \times 10^7\,\text{m},$$

or 42,300 km from the Earth's center. We subtract the Earth's radius of 6380 km to find that a geosynchronous satellite must orbit about 36,000 km (about $6\,r_E$) above the Earth's surface.

(*b*) We solve for v in the satellite equation, Eq. 6–5:

$$v = \sqrt{\frac{Gm_E}{r}} = \sqrt{\frac{(6.67 \times 10^{-11}\,\text{N}\cdot\text{m}^2/\text{kg}^2)(5.98 \times 10^{24}\,\text{kg})}{(4.23 \times 10^7\,\text{m})}} = 3070\,\text{m/s}.$$

We get the same result if we use $v = 2\pi r/T$.

(*c*) The equation in part (*b*) for v shows $v \propto \sqrt{1/r}$. So for $r = r_E + h = 6380\,\text{km} + 200\,\text{km} = 6580\,\text{km}$, we get

$$v' = v\sqrt{\frac{r}{r'}} = (3070\,\text{m/s})\sqrt{\frac{(42{,}300\,\text{km})}{(6580\,\text{km})}} = 7780\,\text{m/s}.$$

NOTE The center of a satellite orbit is always at the center of the Earth; so it is not possible to have a satellite orbiting above a fixed point on the Earth at any latitude other than 0°.

CONCEPTUAL EXAMPLE 6–7 **Catching a satellite.** You are an astronaut in the space shuttle pursuing a satellite in need of repair. You find yourself in a circular orbit of the same radius as the satellite, but 30 km behind it. How will you catch up with it?

RESPONSE We saw in Example 6–6 (or see Eq. 6–5) that the velocity is proportional to $1/\sqrt{r}$. Thus you need to aim for a smaller orbit in order to increase your speed. Note that you cannot just increase your speed without changing your orbit. After passing the satellite, you will need to slow down and rise upward again.

(a) $\vec{\mathbf{a}} = 0$

(b) $\uparrow \vec{\mathbf{a}} = \frac{1}{2}\vec{\mathbf{g}}$ (up)

(c) $\downarrow \vec{\mathbf{a}} = \vec{\mathbf{g}}$ (down)

FIGURE 6–13 (a) An object in an elevator at rest exerts a force on a spring scale equal to its weight. (b) In an elevator accelerating upward at $\frac{1}{2}g$, the object's apparent weight is $1\frac{1}{2}$ times larger than its true weight. (c) In a freely falling elevator, the object experiences "weightlessness": the scale reads zero.

EXERCISE B Two satellites orbit the Earth in circular orbits of the same radius. One satellite is twice as massive as the other. Which of the following statements is true about the speeds of these satellites? (*a*) The heavier satellite moves twice as fast as the lighter one. (*b*) The two satellites have the same speed. (*c*) The lighter satellite moves twice as fast as the heavier one. (*d*) The heavier satellite moves four times as fast as the lighter one.

Weightlessness

People and other objects in a satellite circling the Earth are said to experience *apparent weightlessness.* Let us first look at a simpler case, that of a falling elevator. In Fig. 6–13a, an elevator is at rest with a bag hanging from a spring scale. The scale reading indicates the downward force exerted on it by the bag. This force, exerted *on* the scale, is equal and opposite to the force exerted *by* the scale upward on the bag, and we call its magnitude w. Two forces act on the bag: the downward gravitational force and the upward force exerted by the scale equal to w. Because the bag is not accelerating ($a = 0$) when we apply $\Sigma F = ma$ to the bag in Fig. 6–13a we obtain

$$w - mg = 0,$$

where mg is the weight of the bag. Thus, $w = mg$, and since the scale indicates the force w exerted on it by the bag, it registers a force equal to the weight of the bag, as we expect.

Now let the elevator have an acceleration, a. Applying Newton's second law, $\Sigma F = ma$, to the bag as seen from an inertial reference frame (the elevator itself is not an inertial frame) we have

$$w - mg = ma.$$

Solving for w, we have

$$w = mg + ma. \qquad \qquad [a \text{ is } + \text{ upward}]$$

We have chosen the positive direction up. Thus, if the acceleration a is up, a is positive; and the scale, which measures w, will read more than mg. We call w the *apparent weight* of the bag, which in this case would be greater than its actual weight (mg). If the elevator accelerates downward, a will be negative and w, the apparent weight, will be less than mg. The direction of the velocity $\vec{\mathbf{v}}$ doesn't matter. Only the direction of the acceleration $\vec{\mathbf{a}}$ (and its magnitude) influences the scale reading.

Suppose, for example, the elevator's acceleration is $\frac{1}{2}g$ upward; then we find

$$w = mg + m(\tfrac{1}{2}g) = \tfrac{3}{2}mg.$$

That is, the scale reads $1\frac{1}{2}$ times the actual weight of the bag (Fig. 6–13b). The apparent weight of the bag is $1\frac{1}{2}$ times its real weight. The same is true of the person: her apparent weight (equal to the normal force exerted on her by the elevator floor) is $1\frac{1}{2}$ times her real weight. We can say that she is experiencing $1\frac{1}{2}$ g's, just as astronauts experience so many g's at a rocket's launch.

If, instead, the elevator's acceleration is $a = -\frac{1}{2}g$ (downward), then $w = mg - \frac{1}{2}mg = \frac{1}{2}mg$. That is, the scale reads half the actual weight. If the elevator is in *free fall* (for example, if the cables break), then $a = -g$ and $w = mg - mg = 0$. The scale reads zero. See Fig. 6–13c. The bag appears weightless. If the person in the elevator accelerating at $-g$ let go of a pencil, say, it would not fall to the floor. True, the pencil would be falling with acceleration g. But so would the floor of the elevator and the person. The pencil would hover right in front of the person. This phenomenon is called *apparent weightlessness* because in the reference frame of the person, objects don't fall or seem to have weight—yet gravity does not disappear. Gravity is still acting on each object, whose weight is still mg. The person and other objects seem weightless only because the elevator is accelerating in free fall, and there is no contact force on the person to make her feel the weight.

The "weightlessness" experienced by people in a satellite orbit close to the Earth (Fig. 6–14) is the same apparent weightlessness experienced in a freely falling elevator. It may seem strange, at first, to think of a satellite as freely falling. But a satellite is indeed falling toward the Earth, as was shown in Fig. 6–12. The force of gravity causes it to "fall" out of its natural straight-line path. The acceleration of the satellite must be the acceleration due to gravity at that point, since the only force acting on it is gravity. (We used this to obtain Eq. 6–5.) Thus, although the force of gravity acts on objects within the satellite, the objects experience an apparent weightlessness because they, and the satellite, are accelerating together as in free fall.

EXERCISE C Return to the Chapter-Opening Question, page 139, and answer it again now. Try to explain why you may have answered differently the first time.

Figure 6–15 shows some examples of "free fall," or apparent weightlessness, experienced by people on Earth for brief moments.

A completely different situation occurs if a spacecraft is out in space far from the Earth, the Moon, and other attracting bodies. The force of gravity due to the Earth and other heavenly bodies will then be quite small because of the distances involved, and persons in such a spacecraft would experience real weightlessness.

EXERCISE D Could astronauts in a spacecraft far out in space easily play catch with a bowling ball ($m = 7\,\text{kg}$)?

FIGURE 6–14 This astronaut is moving outside the International Space Station. He must feel very free because he is experiencing apparent weightlessness.

FIGURE 6–15 Experiencing "weightlessness" on Earth.

(a)

(b)

(c)

6–5 Kepler's Laws and Newton's Synthesis

More than a half century before Newton proposed his three laws of motion and his law of universal gravitation, the German astronomer Johannes Kepler (1571–1630) had worked out a detailed description of the motion of the planets about the Sun. Kepler's work resulted in part from the many years he spent examining data collected by Tycho Brahe (1546–1601) on the positions of the planets in their motion through the heavens.

FIGURE 6–16 *Kepler's first law.* An ellipse is a closed curve such that the sum of the distances from any point P on the curve to two fixed points (called the foci, F_1 and F_2) remains constant. That is, the sum of the distances, $F_1P + F_2P$, is the same for all points on the curve. A circle is a special case of an ellipse in which the two foci coincide, at the center of the circle. The semimajor axis is *s* (that is, the long axis is 2*s*) and the semiminor axis is *b*, as shown. The *eccentricity, e,* is defined as the ratio of the distance from either focus to the center divided by the semimajor axis *a*. Thus *es* is the distance from the center to either focus, as shown. For a circle, $e = 0$. The Earth and most of the other planets have nearly circular orbits. For Earth $e = 0.017$.

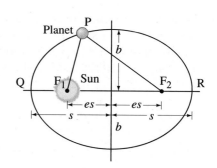

Among Kepler's writings were three empirical findings that we now refer to as **Kepler's laws of planetary motion**. These are summarized as follows, with additional explanation in Figs. 6–16 and 6–17.

Kepler's first law: The path of each planet about the Sun is an ellipse with the Sun at one focus (Fig. 6–16).

Kepler's second law: Each planet moves so that an imaginary line drawn from the Sun to the planet sweeps out equal areas in equal periods of time (Fig. 6–17).

Kepler's third law: The ratio of the squares of the periods of any two planets revolving about the Sun is equal to the ratio of the cubes of their semimajor axes. [The semimajor axis is half the long (major) axis of the orbit, as shown in Fig. 6–16, and represents the planet's mean distance from the Sun.[†]] That is, if T_1 and T_2 represent the periods (the time needed for one revolution about the Sun) for any two planets, and s_1 and s_2 represent their semimajor axes, then

$$\left(\frac{T_1}{T_2}\right)^2 = \left(\frac{s_1}{s_2}\right)^3.$$

We can rewrite this as

$$\frac{s_1^3}{T_1^2} = \frac{s_2^3}{T_2^2},$$

meaning that s^3/T^2 should be the same for each planet. Present-day data are given in Table 6–2; see the last column.

Kepler arrived at his laws through careful analysis of experimental data. Fifty years later, Newton was able to show that Kepler's laws could be derived mathematically from the law of universal gravitation and the laws of motion. He also showed that for any reasonable form for the gravitational force law, only one that depends on the inverse square of the distance is fully consistent with Kepler's laws. He thus used Kepler's laws as evidence in favor of his law of universal gravitation, Eq. 6–1.

We will derive Kepler's second law later, in Chapter 11. Here we derive Kepler's third law, and we do it for the special case of a circular orbit, in which case the semimajor axis is the radius *r* of the circle. (Most planetary orbits are close to a circle.) First, we write Newton's second law of motion, $\Sigma F = ma$. For *F* we use the law of universal gravitation (Eq. 6–1) for the force between the Sun and a planet of mass m_1, and for *a* the centripetal acceleration, v^2/r. We assume the mass of the Sun M_S is much greater than the mass of its planets, so we ignore the effects of the planets on each other. Then

$$\Sigma F = ma$$
$$G\frac{m_1 M_S}{r_1^2} = m_1\frac{v_1^2}{r_1}.$$

FIGURE 6–17 *Kepler's second law.* The two shaded regions have equal areas. The planet moves from point 1 to point 2 in the same time as it takes to move from point 3 to point 4. Planets move fastest in that part of their orbit where they are closest to the Sun. Exaggerated scale.

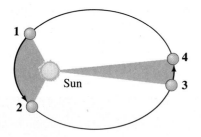

TABLE 6–2 Planetary Data Applied to Kepler's Third Law

Planet	Mean Distance from Sun, *s* (10^6 km)	Period, *T* (Earth yr)	$\frac{s^3/T^2}{\left(10^{24}\frac{\text{km}^3}{\text{yr}^2}\right)}$
Mercury	57.9	0.241	3.34
Venus	108.2	0.615	3.35
Earth	149.6	1.0	3.35
Mars	227.9	1.88	3.35
Jupiter	778.3	11.86	3.35
Saturn	1427	29.5	3.34
Uranus	2870	84.0	3.35
Neptune	4497	165	3.34
Pluto	5900	248	3.34

[†]The semimajor axis is equal to the planet's mean distance from the Sun in the sense that it equals half the sum of the planet's nearest and farthest distances from the Sun (points Q and R in Fig. 6–16). Most planetary orbits are close to circles, and for a circle the semimajor axis is the radius of the circle.

Here m_1 is the mass of a particular planet, r_1 its distance from the Sun, and v_1 its average speed in orbit; M_S is the mass of the Sun, since it is the gravitational attraction of the Sun that keeps each planet in its orbit. The period T_1 of the planet is the time required for one complete orbit, which is a distance equal to $2\pi r_1$, the circumference of a circle. Thus

$$v_1 = \frac{2\pi r_1}{T_1}.$$

We substitute this formula for v_1 into the equation above:

$$G\frac{m_1 M_S}{r_1^2} = m_1 \frac{4\pi^2 r_1}{T_1^2}.$$

We rearrange this to get

$$\frac{T_1^2}{r_1^3} = \frac{4\pi^2}{GM_S}. \qquad\qquad (6\text{-}6)$$

We derived this for planet 1 (say, Mars). The same derivation would apply for a second planet (say, Saturn) orbiting the Sun,

$$\frac{T_2^2}{r_2^3} = \frac{4\pi^2}{GM_S},$$

where T_2 and r_2 are the period and orbit radius, respectively, for the second planet. Since the right sides of the two previous equations are equal, we have $T_1^2/r_1^3 = T_2^2/r_2^3$ or, rearranging,

$$\left(\frac{T_1}{T_2}\right)^2 = \left(\frac{r_1}{r_2}\right)^3, \qquad\qquad (6\text{-}7)$$

which is Kepler's third law. Equations 6–6 and 6–7 are valid also for elliptical orbits if we replace r with the semimajor axis s.

The derivations of Eqs. 6–6 and 6–7 (Kepler's third law) compared two planets revolving around the Sun. But they are general enough to be applied to other systems. For example, we could apply Eq. 6–6 to our Moon revolving around Earth (then M_S would be replaced by M_E, the mass of the Earth). Or we could apply Eq. 6–7 to compare two moons revolving around Jupiter. But Kepler's third law, Eq. 6–7, applies only to objects orbiting the same attracting center. Do not use Eq. 6–7 to compare, say, the Moon's orbit around the Earth to the orbit of Mars around the Sun because they depend on different attracting centers.

In the following Examples, we assume the orbits are circles, although it is not quite true in general.

⚠ CAUTION

Compare orbits of objects only around the same center

EXAMPLE 6–8 **Where is Mars?** Mars' period (its "year") was first noted by Kepler to be about 687 days (Earth-days), which is $(687\text{ d}/365\text{ d}) = 1.88$ yr (Earth years). Determine the mean distance of Mars from the Sun using the Earth as a reference.

APPROACH We are given the ratio of the periods of Mars and Earth. We can find the distance from Mars to the Sun using Kepler's third law, given the Earth–Sun distance as 1.50×10^{11} m (Table 6–2; also Table inside front cover).

SOLUTION Let the distance of Mars from the Sun be r_{MS}, and the Earth–Sun distance be $r_{ES} = 1.50 \times 10^{11}$ m. From Kepler's third law (Eq. 6–7):

$$\frac{r_{MS}}{r_{ES}} = \left(\frac{T_M}{T_E}\right)^{\frac{2}{3}} = \left(\frac{1.88\text{ yr}}{1\text{ yr}}\right)^{\frac{2}{3}} = 1.52.$$

So Mars is 1.52 times the Earth's distance from the Sun, or 2.28×10^{11} m.

EXAMPLE 6–9 **The Sun's mass determined.** Determine the mass of the Sun given the Earth's distance from the Sun as $r_{ES} = 1.5 \times 10^{11}$ m.

APPROACH Equation 6–6 relates the mass of the Sun M_S to the period and distance of any planet. We use the Earth.

SOLUTION The Earth's period is $T_E = 1 \text{ yr} = (365\frac{1}{4}\text{ d})(24\text{ h/d})(3600\text{ s/h}) = 3.16 \times 10^7$ s. We solve Eq. 6–6 for M_S:

$$M_S = \frac{4\pi^2 r_{ES}^3}{G T_E^2} = \frac{4\pi^2 (1.5 \times 10^{11}\text{ m})^3}{(6.67 \times 10^{-11}\text{ N}\cdot\text{m}^2/\text{kg}^2)(3.16 \times 10^7\text{ s})^2} = 2.0 \times 10^{30} \text{ kg.}$$

EXERCISE E Suppose there were a planet in circular orbit exactly halfway between the orbits of Mars and Jupiter. What would its period be in Earth-years? Use Table 6–2.

Accurate measurements on the orbits of the planets indicated that they did not precisely follow Kepler's laws. For example, slight deviations from perfectly elliptical orbits were observed. Newton was aware that this was to be expected because any planet would be attracted gravitationally not only by the Sun but also (to a much lesser extent) by the other planets. Such deviations, or **perturbations**, in the orbit of Saturn were a hint that helped Newton formulate the law of universal gravitation, that all objects attract gravitationally. Observation of other perturbations later led to the discovery of Neptune and Pluto. Deviations in the orbit of Uranus, for example, could not all be accounted for by perturbations due to the other known planets. Careful calculation in the nineteenth century indicated that these deviations could be accounted for if another planet existed farther out in the solar system. The position of this planet was predicted from the deviations in the orbit of Uranus, and telescopes focused on that region of the sky quickly found it; the new planet was called Neptune. Similar but much smaller perturbations of Neptune's orbit led to the discovery of Pluto in 1930.

Starting in the mid 1990s, planets revolving about distant stars (Fig. 6–18) were inferred from the regular "wobble" of each star due to the gravitational attraction of the revolving planet(s). Many such "extrasolar" planets are now known.

The development by Newton of the law of universal gravitation and the three laws of motion was a major intellectual achievement: with these laws, he was able to describe the motion of objects on Earth and in the heavens. The motions of heavenly bodies and objects on Earth were seen to follow the same laws (not recognized previously). For this reason, and also because Newton integrated the results of earlier scientists into his system, we sometimes speak of **Newton's synthesis**.

The laws formulated by Newton are referred to as **causal laws**. By **causality** we mean the idea that one occurrence can cause another. When a rock strikes a window, we infer that the rock *causes* the window to break. This idea of "cause and effect" relates to Newton's laws: the acceleration of an object was seen to be *caused* by the net force acting on it.

As a result of Newton's theories the universe came to be viewed by many scientists and philosophers as a big machine whose parts move in a *deterministic* way. This deterministic view of the universe, however, had to be modified by scientists in the twentieth century (Chapter 38).

FIGURE 6–18 Our solar system (a) is compared to recently discovered planets orbiting (b) the star 47 Ursae Majoris and (c) the star Upsilon Andromedae with at least three planets. m_J is the mass of Jupiter. (Sizes not to scale.)

EXAMPLE 6–10 **Lagrange Point.** The mathematician Joseph-Louis Lagrange discovered five special points in the vicinity of the Earth's orbit about the Sun where a small satellite (mass m) can orbit the Sun with the same period T as Earth's (= 1 year). One of these "Lagrange Points," called L1, lies between the Earth (mass M_E) and Sun (mass M_S), on the line connecting them (Fig. 6–19). That is, the Earth and the satellite are always separated by a distance d. If the Earth's orbital radius is R_{ES}, then the satellite's orbital radius is $(R_{ES} - d)$. Determine d.

APPROACH We use Newton's law of universal gravitation and set it equal to the mass times the centripetal acceleration. But how could an object with a smaller orbit than Earth's have the same period as Earth? Kepler's third law clearly tells us a smaller orbit around the Sun results in a smaller period. But that law depends on only the Sun's gravitational attraction. Our mass m is pulled by both the Sun and the Earth.

SOLUTION Because the satellite is assumed to have negligible mass in comparison to the masses of the Earth and Sun, to an excellent approximation the Earth's orbit will be determined solely by the Sun. Applying Newton's second law to the Earth gives

$$\frac{GM_E M_S}{R_{ES}^2} = M_E \frac{v^2}{R_{ES}} = \frac{M_E}{R_{ES}} \frac{(2\pi R_{ES})^2}{T^2}$$

or

$$\frac{GM_S}{R_{ES}^2} = \frac{4\pi^2 R_{ES}}{T^2}. \tag{i}$$

Next we apply Newton's second law to the satellite m (which has the same period T as Earth), including the pull of both Sun and Earth (see simplified form, Eq. (i))

$$\frac{GM_S}{(R_{ES} - d)^2} - \frac{GM_E}{d^2} = \frac{4\pi^2(R_{ES} - d)}{T^2},$$

which we rewrite as

$$\frac{GM_S}{R_{ES}^2}\left(1 - \frac{d}{R_{ES}}\right)^{-2} - \frac{GM_E}{d^2} = \frac{4\pi^2 R_{ES}}{T^2}\left(1 - \frac{d}{R_{ES}}\right).$$

We now use the binomial expansion $(1 + x)^n \approx 1 + nx$, if $x \ll 1$. Setting $x = d/R_{ES}$ and assuming $d \ll R_{ES}$, we have

$$\frac{GM_S}{R_{ES}^2}\left(1 + 2\frac{d}{R_{ES}}\right) - \frac{GM_E}{d^2} = \frac{4\pi^2 R_{ES}}{T^2}\left(1 - \frac{d}{R_{ES}}\right). \tag{ii}$$

Substituting GM_S/R_{ES}^2 from Eq. (i) into Eq. (ii) we find

$$\frac{GM_S}{R_{ES}^2}\left(1 + 2\frac{d}{R_{ES}}\right) - \frac{GM_E}{d^2} = \frac{GM_S}{R_{ES}^2}\left(1 - \frac{d}{R_{ES}}\right).$$

Simplifying, we have

$$\frac{GM_S}{R_{ES}^2}\left(3\frac{d}{R_{ES}}\right) = \frac{GM_E}{d^2}.$$

We solve for d to find

$$d = \left(\frac{M_E}{3 M_S}\right)^{\frac{1}{3}} R_{ES}.$$

Substituting in values we find

$$d = 1.0 \times 10^{-2} R_{ES} = 1.5 \times 10^6 \text{ km}.$$

NOTE Since $d/R_{ES} = 10^{-2}$, we were justified in using the binomial expansion.

NOTE Placing a satellite at L1 has two advantages: the satellite's view of the Sun is never eclipsed by the Earth, and it is always close enough to Earth to transmit data easily. The L1 point of the Earth–Sun system is currently home to the Solar and Heliospheric Observatory (SOHO) satellite, Fig. 6–20.

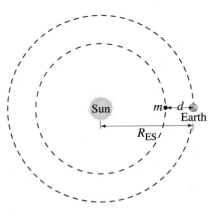

FIGURE 6–19 Finding the position of the Lagrange Point L1 for a satellite that can remain along the revolving line between the Sun and Earth, at distance d from the Earth. Thus a mass m at L1 has the same period around the Sun as the Earth has. (Not to scale.)

FIGURE 6–20 Artist's rendition of the Solar and Heliospheric Observatory (SOHO) satellite in orbit.

*6–6 Gravitational Field

Most of the forces we meet in everyday life are contact forces: you push or pull on a lawn mower, a tennis racket exerts a force on a tennis ball when they make contact, or a ball exerts a force on a window when they make contact. But the gravitational force acts over a distance: there is a force even when the two objects are not in contact. The Earth, for example, exerts a force on a falling apple. It also exerts a force on the Moon, 384,000 km away. And the Sun exerts a gravitational force on the Earth. The idea of a force *acting at a distance* was a difficult one for early thinkers. Newton himself felt uneasy with this concept when he published his law of universal gravitation.

Another point of view that helps with these conceptual difficulties is the concept of the **field**, developed in the nineteenth century by Michael Faraday (1791–1867) to aid understanding of electric and magnetic forces which also act over a distance. Only later was it applied to gravity. According to the field concept, a **gravitational field** surrounds every object that has mass, and this field permeates all of space. A second object at a particular location near the first object experiences a force because of the gravitational field that exists there. Because the gravitational field at the location of the second mass is considered to act directly on this mass, we are a little closer to the idea of a contact force.

To be quantitative, we can define the **gravitational field** as the gravitational force per unit mass at any point in space. If we want to measure the gravitational field at any point, we place a small "test" mass m at that point and measure the force $\vec{F}$ exerted on it (making sure only gravitational forces are acting). Then the gravitational field, $\vec{g}$, at that point is defined as

$$\vec{g} = \frac{\vec{F}}{m}.$$

[gravitational field] **(6–8)**

The units of $\vec{g}$ are N/kg.

From Eq. 6–8 we see that the gravitational field an object experiences has magnitude equal to the acceleration due to gravity at that point. (When we speak of acceleration, however, we use units m/s², which is equivalent to N/kg, since $1\,\text{N} = 1\,\text{kg}\cdot\text{m/s}^2$.)

If the gravitational field is due to a single spherically symmetric (or small) object of mass M, such as when m is near the Earth's surface, then the gravitational field at a distance r from M has magnitude

$$g = \frac{1}{m} G \frac{mM}{r^2} = G \frac{M}{r^2}.$$

In vector notation we write

$$\vec{g} = -\frac{GM}{r^2}\hat{\mathbf{r}},$$

$$\begin{bmatrix} \text{due to a} \\ \text{single mass } M \end{bmatrix}$$

where $\hat{\mathbf{r}}$ is a unit vector pointing radially outward from mass M, and the minus sign reminds us that the field points toward mass M (see Eqs. 6–1, 6–2, and 6–4). If several different bodies contribute significantly to the gravitational field, then we write the gravitational field $\vec{g}$ as the vector sum of all these contributions. In interplanetary space, for example, $\vec{g}$ at any point in space is the vector sum of terms due to the Earth, Sun, Moon, and other bodies that contribute. The gravitational field $\vec{g}$ at any point in space does not depend on the value of our test mass, m, placed at that point; $\vec{g}$ depends only on the masses (and locations) of the bodies that create the field there.

6–7 Types of Forces in Nature

We have already discussed that Newton's law of universal gravitation, Eq. 6–1, describes how a particular type of force—gravity—depends on the distance between, and masses of, the objects involved. Newton's second law, $\Sigma\vec{F} = m\vec{a}$, on the other hand, tells how an object will accelerate due to *any* type of force. But what are the types of forces that occur in nature besides gravity?

In the twentieth century, physicists came to recognize four different fundamental forces in nature: (1) the gravitational force; (2) the electromagnetic force (we shall see later that electric and magnetic forces are intimately related); (3) the strong nuclear force; and (4) the weak nuclear force. In this Chapter, we discussed the gravitational force in detail. The nature of the electromagnetic force will be discussed in detail in Chapters 21 to 31. The strong and weak nuclear forces operate at the level of the atomic nucleus; although they manifest themselves in such phenomena as radioactivity and nuclear energy (Chapters 41 to 43), they are much less obvious in our daily lives.

Physicists have been working on theories that would unify these four forces—that is, to consider some or all of these forces as different manifestations of the same basic force. So far, the electromagnetic and weak nuclear forces have been theoretically united to form *electroweak* theory, in which the electromagnetic and weak forces are seen as two different manifestations of a single *electroweak force*. Attempts to further unify the forces, such as in *grand unified theories* (GUT), are hot research topics today.

But where do everyday forces fit into this scheme? Ordinary forces, other than gravity, such as pushes, pulls, and other contact forces like the normal force and friction, are today considered to be due to the electromagnetic force acting at the atomic level. For example, the force your fingers exert on a pencil is the result of electrical repulsion between the outer electrons of the atoms of your finger and those of the pencil.

*6–8 Principle of Equivalence; Curvature of Space; Black Holes

We have dealt with two aspects of mass. In Chapter 4, we defined mass as a measure of the inertia of a body. Newton's second law relates the force acting on a body to its acceleration and its **inertial mass**, as we call it. We might say that inertial mass represents a resistance to any force. In this Chapter we have dealt with mass as a property related to the gravitational force—that is, mass as a quantity that determines the strength of the gravitational force between two bodies. This we call the **gravitational mass**.

It is not obvious that the inertial mass of a body should be equal to its gravitational mass. The force of gravity might have depended on a different property of a body, just as the electrical force depends on a property called electric charge. Newton's and Cavendish's experiments indicated that the two types of mass are equal for a body, and modern experiments confirm it to a precision of about 1 part in 10^{12}.

Albert Einstein (1879–1955) called this equivalence between gravitational and inertial masses the **principle of equivalence**, and he used it as a foundation for his *general theory of relativity* (c. 1916). The principle of equivalence can be stated in another way: there is no experiment observers can perform to distinguish if an acceleration arises because of a gravitational force or because their reference frame is accelerating. If you were far out in space and an apple fell to the floor of your spacecraft, you might assume a gravitational force was acting on the apple. But it would also be possible that the apple fell because your spacecraft accelerated upward (relative to an inertial system). The effects would be indistinguishable, according to the principle of equivalence, because the apple's inertial and gravitational masses—that determine how a body "reacts" to outside influences—are indistinguishable.

FIGURE 6–21 (a) Light beam goes straight across an elevator that is not accelerating. (b) The light beam bends (exaggerated) in an elevator accelerating in an upward direction.

(a)　(b)

FIGURE 6–22 (a) Three stars in the sky. (b) If the light from one of these stars passes very near the Sun, whose gravity bends the light beam, the star will appear higher than it actually is.

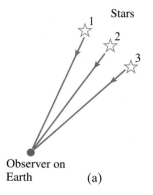

Stars

Observer on Earth　(a)

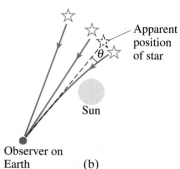

Apparent position of star

Sun

Observer on Earth　(b)

The principle of equivalence can be used to show that light ought to be deflected due to the gravitational force of a massive object. Let us consider a thought experiment in an elevator in free space where virtually no gravity acts. If a light beam enters a hole in the side of the elevator, the beam travels straight across the elevator and makes a spot on the opposite side if the elevator is at rest (Fig. 6–21a). If the elevator is accelerating upward as in Fig. 6–21b, the light beam still travels straight as observed in the original reference frame at rest. In the upwardly accelerating elevator, however, the beam is observed to curve downward. Why? Because during the time the light travels from one side of the elevator to the other, the elevator is moving upward at ever-increasing speed.

According to the equivalence principle, an upwardly accelerating reference frame is equivalent to a downward gravitational field. Hence, we can picture the curved light path in Fig. 6–21b as being the effect of a gravitational field. Thus we expect gravity to exert a force on a beam of light and to bend it out of a straight-line path!

Einstein's general theory of relativity predicts that light should be affected by gravity. It was calculated that light from a distant star would be deflected by 1.75″ of arc (tiny but detectable) as it passed near the Sun, as shown in Fig. 6–22. Such a deflection was measured and confirmed in 1919 during an eclipse of the Sun. (The eclipse reduced the brightness of the Sun so that the stars in line with its edge at that moment would be visible.)

That a light beam can follow a curved path suggests that *space itself is curved* and that it is gravitational mass that causes the curvature. The curvature is greatest near very massive objects. To visualize this curvature of space, we might think of space as being like a thin rubber sheet; if a heavy weight is hung from it, it curves as shown in Fig. 6–23. The weight corresponds to a huge mass that causes space (space itself!) to curve.

The extreme curvature of space-time shown in Fig. 6–23 could be produced by a **black hole**, a star that becomes so dense and massive that gravity would be so strong that even light could not escape it. Light would be pulled back in by the force of gravity. Since no light could escape from such a massive star, we could not see it—it would be black. An object might pass by it and be deflected by its gravitational field, but if the object came too close it would be swallowed up, never to escape. Hence the name black holes. Experimentally there is good evidence for their existence. One likely possibility is a giant black hole at the center of our Galaxy and probably at the center of other galaxies.

FIGURE 6–23 Rubber-sheet analogy for space (technically space-time) curved by matter.

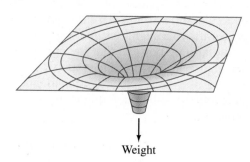

Weight

Summary

Newton's **law of universal gravitation** states that every particle in the universe attracts every other particle with a force proportional to the product of their masses and inversely proportional to the square of the distance between them:

$$F = G\frac{m_1 m_2}{r^2}. \qquad (6-1)$$

The direction of this force is along the line joining the two particles, and is always attractive. It is this gravitational force that keeps the Moon revolving around the Earth and the planets revolving around the Sun.

The total gravitational force on any object is the vector sum of the forces exerted by all other objects; frequently the effects of all but one or two objects can be ignored.

Satellites revolving around the Earth are acted on by gravity, but "stay up" because of their high tangential speed.

Newton's three laws of motion, plus his law of universal gravitation, constituted a wide-ranging theory of the universe. With them, motion of objects on Earth and in the heavens could be accurately described. And they provided a theoretical base for **Kepler's laws** of planetary motion.

[*According to the **field** concept, a **gravitational field** surrounds every object that has mass, and it permeates all of space. The gravitational field at any point in space is the vector sum of the fields due to all massive objects and can be defined as

$$\vec{g} = \frac{\vec{F}}{m} \qquad (6-8)$$

where $\vec{F}$ is the force acting on a small "test" mass m placed at that point.]

The four fundamental forces in nature are (1) the gravitational force, (2) electromagnetic force, (3) strong nuclear force, and (4) weak nuclear force. The first two fundamental forces are responsible for nearly all "everyday" forces.

Questions

1. Does an apple exert a gravitational force on the Earth? If so, how large a force? Consider an apple (*a*) attached to a tree and (*b*) falling.

2. The Sun's gravitational pull on the Earth is much larger than the Moon's. Yet the Moon's is mainly responsible for the tides. Explain. [*Hint*: Consider the difference in gravitational pull from one side of the Earth to the other.]

3. Will an object weigh more at the equator or at the poles? What two effects are at work? Do they oppose each other?

4. Why is more fuel required for a spacecraft to travel from the Earth to the Moon than it does to return from the Moon to the Earth?

5. The gravitational force on the Moon due to the Earth is only about half the force on the Moon due to the Sun (see Example 6–3). Why isn't the Moon pulled away from the Earth?

6. How did the scientists of Newton's era determine the distance from the Earth to the Moon, despite not knowing about spaceflight or the speed of light? [*Hint*: Think about why two eyes are useful for depth perception.]

7. If it were possible to drill a hole all the way through the Earth along a diameter, then it would be possible to drop a ball through the hole. When the ball was right at the center of the Earth, what would be the total gravitational force exerted on it by the Earth?

8. Why is it not possible to put a satellite in geosynchronous orbit above the North Pole?

9. Which pulls harder gravitationally, the Earth on the Moon, or the Moon on the Earth? Which accelerates more?

10. Would it require less speed to launch a satellite (*a*) toward the east or (*b*) toward the west? Consider the Earth's rotation direction.

11. An antenna loosens and becomes detached from a satellite in a circular orbit around the Earth. Describe the antenna's motion subsequently. If it will land on the Earth, describe where; if not, describe how it could be made to land on the Earth.

12. Describe how careful measurements of the variation in g in the vicinity of an ore deposit might be used to estimate the amount of ore present.

13. The Sun is below us at midnight, nearly in line with the Earth's center. Are we then heavier at midnight, due to the Sun's gravitational force on us, than we are at noon? Explain.

14. When will your apparent weight be the greatest, as measured by a scale in a moving elevator: when the elevator (*a*) accelerates downward, (*b*) accelerates upward, (*c*) is in free fall, or (*d*) moves upward at constant speed? In which case would your apparent weight be the least? When would it be the same as when you are on the ground?

15. If the Earth's mass were double what it actually is, in what ways would the Moon's orbit be different?

16. The source of the Mississippi River is closer to the center of the Earth than is its outlet in Louisiana (since the Earth is fatter at the equator than at the poles). Explain how the Mississippi can flow "uphill."

17. People sometimes ask, "What keeps a satellite up in its orbit around the Earth?" How would you respond?

18. Explain how a runner experiences "free fall" or "apparent weightlessness" between steps.

19. If you were in a satellite orbiting the Earth, how might you cope with walking, drinking, or putting a pair of scissors on a table?

20. Is the centripetal acceleration of Mars in its orbit around the Sun larger or smaller than the centripetal acceleration of the Earth?

21. The mass of the planet Pluto was not known until it was discovered to have a moon. Explain how this enabled an estimate of Pluto's mass.

22. The Earth moves faster in its orbit around the Sun in January than in July. Is the Earth closer to the Sun in January, or in July? Explain. [*Note*: This is not much of a factor in producing the seasons—the main factor is the tilt of the Earth's axis relative to the plane of its orbit.]

23. Kepler's laws tell us that a planet moves faster when it is closer to the Sun than when it is farther from the Sun. What causes this change in speed of the planet?

*24. Does your body directly sense a gravitational field? (Compare to what you would feel in free fall.)

*25. Discuss the conceptual differences between $\vec{g}$ as acceleration due to gravity and $\vec{g}$ as gravitational field.

Problems

6–1 to 6–3 Law of Universal Gravitation

1. (I) Calculate the force of Earth's gravity on a spacecraft 2.00 Earth radii above the Earth's surface if its mass is 1480 kg.

2. (I) Calculate the acceleration due to gravity on the Moon. The Moon's radius is 1.74×10^6 m and its mass is 7.35×10^{22} kg.

3. (I) A hypothetical planet has a radius 2.3 times that of Earth, but has the same mass. What is the acceleration due to gravity near its surface?

4. (I) A hypothetical planet has a mass 1.80 times that of Earth, but the same radius. What is g near its surface?

5. (I) If you doubled the mass and tripled the radius of a planet, by what factor would g at its surface change?

6. (II) Calculate the effective value of g, the acceleration of gravity, at (a) 6400 m, and (b) 6400 km, above the Earth's surface.

7. (II) You are explaining to friends why astronauts feel weightless orbiting in the space shuttle, and they respond that they thought gravity was just a lot weaker up there. Convince them and yourself that it isn't so by calculating how much weaker gravity is 300 km above the Earth's surface.

8. (II) Every few hundred years most of the planets line up on the same side of the Sun. Calculate the total force on the Earth due to Venus, Jupiter, and Saturn, assuming all four planets are in a line, Fig. 6–24. The masses are $M_V = 0.815\, M_E$, $M_J = 318\, M_E$, $M_{Sat} = 95.1\, M_E$, and the mean distances of the four planets from the Sun are 108, 150, 778, and 1430 million km. What fraction of the Sun's force on the Earth is this?

FIGURE 6–24 Problem 8 (not to scale).

9. (II) Four 8.5-kg spheres are located at the corners of a square of side 0.80 m. Calculate the magnitude and direction of the gravitational force exerted on one sphere by the other three.

10. (II) Two objects attract each other gravitationally with a force of 2.5×10^{-10} N when they are 0.25 m apart. Their total mass is 4.00 kg. Find their individual masses.

11. (II) Four masses are arranged as shown in Fig. 6–25. Determine the x and y components of the gravitational force on the mass at the origin (m). Write the force in vector notation ($\hat{\mathbf{i}}, \hat{\mathbf{j}}$).

FIGURE 6–25
Problem 11.

12. (II) Estimate the acceleration due to gravity at the surface of Europa (one of the moons of Jupiter) given that its mass is 4.9×10^{22} kg and making the assumption that its density is the same as Earth's.

13. (II) Suppose the mass of the Earth were doubled, but it kept the same density and spherical shape. How would the weight of objects at the Earth's surface change?

14. (II) Given that the acceleration of gravity at the surface of Mars is 0.38 of what it is on Earth, and that Mars' radius is 3400 km, determine the mass of Mars.

15. (II) At what distance from the Earth will a spacecraft traveling directly from the Earth to the Moon experience zero net force because the Earth and Moon pull with equal and opposite forces?

16. (II) Determine the mass of the Sun using the known value for the period of the Earth and its distance from the Sun. [Hint: The force on the Earth due to the Sun is related to the centripetal acceleration of the Earth.] Compare your answer to that obtained using Kepler's laws, Example 6–9.

17. (II) Two identical point masses, each of mass M, always remain separated by a distance of $2R$. A third mass m is then placed a distance x along the perpendicular bisector of the original two masses, as shown in Fig. 6–26. Show that the gravitational force on the third mass is directed inward along the perpendicular bisector and has a magnitude of

$$F = \frac{2GMmx}{\left(x^2 + R^2\right)^{\frac{3}{2}}}.$$

FIGURE 6–26
Problem 17.

18. (II) A mass M is ring shaped with radius r. A small mass m is placed at a distance x along the ring's axis as shown in Fig. 6–27. Show that the gravitational force on the mass m due to the ring is directed inward along the axis and has magnitude

$$F = \frac{GMmx}{\left(x^2 + r^2\right)^{\frac{3}{2}}}.$$

[Hint: Think of the ring as made up of many small point masses dM; sum over the forces due to each dM, and use symmetry.]

FIGURE 6–27
Problem 18.

19. (III) (a) Use the binomial expansion

$$(1 \pm x)^n = 1 \pm nx + \frac{n(n-1)}{2} x^2 \pm \cdots$$

to show that the value of g is altered by approximately

$$\Delta g \approx -2g \frac{\Delta r}{r_E}$$

at a height Δr above the Earth's surface, where r_E is the radius of the Earth, as long as $\Delta r \ll r_E$. (b) What is the meaning of the minus sign in this relation? (c) Use this result to compute the effective value of g at 125 km above the Earth's surface. Compare to a direct use of Eq. 6–1.

20. (III) The center of a 1.00 km diameter spherical pocket of oil is 1.00 km beneath the Earth's surface. Estimate by what percentage g directly above the pocket of oil would differ from the expected value of g for a uniform Earth? Assume the density of oil is 8.0×10^2 kg/m³.

21. (III) Determine the magnitude and direction of the effective value of $\vec{g}$ at a latitude of 45° on the Earth. Assume the Earth is a rotating sphere.

*22. (III) It can be shown (Appendix D) that for a uniform sphere the force of gravity at a point inside the sphere depends only on the mass closer to the center than that point. The net force of gravity due to points outside the radius of the point cancels. How far would you have to drill into the Earth, to reach a point where your weight is reduced by 5.0%? Approximate the Earth as a uniform sphere.

6–4 Satellites and Weightlessness

23. (I) The space shuttle releases a satellite into a circular orbit 680 km above the Earth. How fast must the shuttle be moving (relative to Earth's center) when the release occurs?

24. (I) Calculate the speed of a satellite moving in a stable circular orbit about the Earth at a height of 5800 km.

25. (II) You know your mass is 65 kg, but when you stand on a bathroom scale in an elevator, it says your mass is 76 kg. What is the acceleration of the elevator, and in which direction?

26. (II) A 13.0-kg monkey hangs from a cord suspended from the ceiling of an elevator. The cord can withstand a tension of 185 N and breaks as the elevator accelerates. What was the elevator's minimum acceleration (magnitude and direction)?

27. (II) Calculate the period of a satellite orbiting the Moon, 120 km above the Moon's surface. Ignore effects of the Earth. The radius of the Moon is 1740 km.

28. (II) Two satellites orbit Earth at altitudes of 5000 km and 15,000 km. Which satellite is faster, and by what factor?

29. (II) What will a spring scale read for the weight of a 53-kg woman in an elevator that moves (a) upward with constant speed 5.0 m/s, (b) downward with constant speed 5.0 m/s, (c) upward with acceleration 0.33 g, (d) downward with acceleration 0.33 g, and (e) in free fall?

30. (II) Determine the time it takes for a satellite to orbit the Earth in a circular "near-Earth" orbit. A "near-Earth" orbit is at a height above the surface of the Earth that is very small compared to the radius of the Earth. [Hint: You may take the acceleration due to gravity as essentially the same as that on the surface.] Does your result depend on the mass of the satellite?

31. (II) What is the apparent weight of a 75-kg astronaut 2500 km from the center of the Earth's Moon in a space vehicle (a) moving at constant velocity and (b) accelerating toward the Moon at 2.3 m/s²? State "direction" in each case.

32. (II) A Ferris wheel 22.0 m in diameter rotates once every 12.5 s (see Fig. 5–19). What is the ratio of a person's apparent weight to her real weight (a) at the top, and (b) at the bottom?

33. (II) Two equal-mass stars maintain a constant distance apart of 8.0×10^{11} m and rotate about a point midway between them at a rate of one revolution every 12.6 yr. (a) Why don't the two stars crash into one another due to the gravitational force between them? (b) What must be the mass of each star?

34. (III) (a) Show that if a satellite orbits very near the surface of a planet with period T, the density (= mass per unit volume) of the planet is $\rho = m/V = 3\pi/GT^2$. (b) Estimate the density of the Earth, given that a satellite near the surface orbits with a period of 85 min. Approximate the Earth as a uniform sphere.

35. (III) Three bodies of identical mass M form the vertices of an equilateral triangle of side ℓ and rotate in circular orbits about the center of the triangle. They are held in place by their mutual gravitation. What is the speed of each?

36. (III) An inclined plane, fixed to the inside of an elevator, makes a 32° angle with the floor. A mass m slides on the plane without friction. What is its acceleration relative to the plane if the elevator (a) accelerates upward at $0.50\,g$, (b) accelerates downward at $0.50\,g$, (c) falls freely, and (d) moves upward at constant speed?

6–5 Kepler's Laws

37. (I) Use Kepler's laws and the period of the Moon (27.4 d) to determine the period of an artificial satellite orbiting very near the Earth's surface.

38. (I) Determine the mass of the Earth from the known period and distance of the Moon.

39. (I) Neptune is an average distance of 4.5×10^9 km from the Sun. Estimate the length of the Neptunian year using the fact that the Earth is 1.50×10^8 km from the Sun on the average.

40. (II) Planet A and planet B are in circular orbits around a distant star. Planet A is 9.0 times farther from the star than is planet B. What is the ratio of their speeds v_A/v_B?

41. (II) Our Sun rotates about the center of our Galaxy $(m_G \approx 4 \times 10^{41}$ kg) at a distance of about 3×10^4 light-years $[1\,\text{ly} = (3.00 \times 10^8\,\text{m/s}) \cdot (3.16 \times 10^7\,\text{s/yr}) \cdot (1.00\,\text{yr})]$. What is the period of the Sun's orbital motion about the center of the Galaxy?

42. (II) Table 6–3 gives the mean distance, period, and mass for the four largest moons of Jupiter (those discovered by Galileo in 1609). (a) Determine the mass of Jupiter using the data for Io. (b) Determine the mass of Jupiter using data for each of the other three moons. Are the results consistent?

TABLE 6–3 Principal Moons of Jupiter
(Problems 42, 43, and 47)

Moon	Mass (kg)	Period (Earth days)	Mean distance from Jupiter (km)
Io	8.9×10^{22}	1.77	422×10^3
Europa	4.9×10^{22}	3.55	671×10^3
Ganymede	15×10^{22}	7.16	1070×10^3
Callisto	11×10^{22}	16.7	1883×10^3

43. (II) Determine the mean distance from Jupiter for each of Jupiter's moons, using Kepler's third law. Use the distance of Io and the periods given in Table 6–3. Compare your results to the values in the Table.

44. (II) The asteroid belt between Mars and Jupiter consists of many fragments (which some space scientists think came from a planet that once orbited the Sun but was destroyed). (a) If the mean orbital radius of the asteroid belt (where the planet would have been) is about three times farther from the Sun than the Earth is, how long would it have taken this hypothetical planet to orbit the Sun? (b) Can we use these data to deduce the mass of this planet?

45. (III) The comet Hale–Bopp has a period of 2400 years. (a) What is its mean distance from the Sun? (b) At its closest approach, the comet is about 1.0 AU from the Sun (1 AU = distance from Earth to the Sun). What is the farthest distance? (c) What is the ratio of the speed at the closest point to the speed at the farthest point?

46. (III) (a) Use Kepler's second law to show that the ratio of the speeds of a planet at its nearest and farthest points from the Sun is equal to the inverse ratio of the near and far distances: $v_N/v_F = d_F/d_N$. (b) Given that the Earth's distance from the Sun varies from 1.47 to 1.52×10^{11} m, determine the minimum and maximum velocities of the Earth in its orbit around the Sun.

47. (III) The orbital periods T and mean orbital distances r for Jupiter's four largest moons are given in Table 6–3, on the previous page. (a) Starting with Kepler's third law in the form

$$T^2 = \left(\frac{4\pi^2}{Gm_J}\right)r^3,$$

where m_J is the mass of Jupiter, show that this relation implies that a plot of $\log(T)$ vs. $\log(r)$ will yield a straight line. Explain what Kepler's third law predicts about the slope and y-intercept of this straight-line plot. (b) Using the data for Jupiter's four moons, plot $\log(T)$ vs. $\log(r)$ and show that you get a straight line. Determine the slope of this plot and compare it to the value you expect if the data are consistent with Kepler's third law. Determine the y-intercept of the plot and use it to compute the mass of Jupiter.

*6–6 Gravitational Field

* 48. (II) What is the magnitude and direction of the gravitational field midway between the Earth and Moon? Ignore effects of the Sun.

* 49. (II) (a) What is the gravitational field at the surface of the Earth due to the Sun? (b) Will this affect your weight significantly?

* 50. (III) Two identical particles, each of mass m, are located on the x axis at $x = +x_0$ and $x = -x_0$. (a) Determine a formula for the gravitational field due to these two particles for points on the y axis; that is, write $\vec{\mathbf{g}}$ as a function of y, m, x_0, and so on. (b) At what point (or points) on the y axis is the magnitude of $\vec{\mathbf{g}}$ a maximum value, and what is its value there? [Hint: Take the derivative $d\vec{\mathbf{g}}/dy$.]

General Problems

51. How far above the Earth's surface will the acceleration of gravity be half what it is at the surface?

52. At the surface of a certain planet, the gravitational acceleration g has a magnitude of $12.0 \, \text{m/s}^2$. A 13.0-kg brass ball is transported to this planet. What is (a) the mass of the brass ball on the Earth and on the planet, and (b) the weight of the brass ball on the Earth and on the planet?

53. A certain white dwarf star was once an average star like our Sun. But now it is in the last stage of its evolution and is the size of our Moon but has the mass of our Sun. (a) Estimate gravity on the surface on this star. (b) How much would a 65-kg person weigh on this star? (c) What would be the speed of a baseball dropped from a height of 1.0 m when it hit the surface?

54. What is the distance from the Earth's center to a point outside the Earth where the gravitational acceleration due to the Earth is $\frac{1}{10}$ of its value at the Earth's surface?

55. The rings of Saturn are composed of chunks of ice that orbit the planet. The inner radius of the rings is 73,000 km, while the outer radius is 170,000 km. Find the period of an orbiting chunk of ice at the inner radius and the period of a chunk at the outer radius. Compare your numbers with Saturn's mean rotation period of 10 hours and 39 minutes. The mass of Saturn is 5.7×10^{26} kg.

56. During an *Apollo* lunar landing mission, the command module continued to orbit the Moon at an altitude of about 100 km. How long did it take to go around the Moon once?

57. Halley's comet orbits the Sun roughly once every 76 years. It comes very close to the surface of the Sun on its closest approach (Fig. 6–28). Estimate the greatest distance of the comet from the Sun. Is it still "in" the solar system? What planet's orbit is nearest when it is out there?

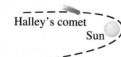

FIGURE 6–28
Problem 57.

58. The Navstar Global Positioning System (GPS) utilizes a group of 24 satellites orbiting the Earth. Using "triangulation" and signals transmitted by these satellites, the position of a receiver on the Earth can be determined to within an accuracy of a few centimeters. The satellite orbits are distributed evenly around the Earth, with four satellites in each of six orbits, allowing continuous navigational "fixes." The satellites orbit at an altitude of approximately 11,000 nautical miles [1 nautical mile = 1.852 km = 6076 ft]. (a) Determine the speed of each satellite. (b) Determine the period of each satellite.

59. Jupiter is about 320 times as massive as the Earth. Thus, it has been claimed that a person would be crushed by the force of gravity on a planet the size of Jupiter since people can't survive more than a few g's. Calculate the number of g's a person would experience at the equator of such a planet. Use the following data for Jupiter: mass $= 1.9 \times 10^{27}$ kg, equatorial radius $= 7.1 \times 10^4$ km, rotation period = 9 hr 55 min. Take the centripetal acceleration into account.

60. The Sun rotates about the center of the Milky Way Galaxy (Fig. 6–29) at a distance of about 30,000 light-years from the center $(1 \, \text{ly} = 9.5 \times 10^{15} \, \text{m})$. If it takes about 200 million years to make one rotation, estimate the mass of our Galaxy. Assume that the mass distribution of our Galaxy is concentrated mostly in a central uniform sphere. If all the stars had about the mass of our Sun $(2 \times 10^{30} \, \text{kg})$, how many stars would there be in our Galaxy?

FIGURE 6–29 Edge-on view of our galaxy. Problem 60.

61. Astronomers have observed an otherwise normal star, called S2, closely orbiting an extremely massive but small object at the center of the Milky Way Galaxy called SgrA. S2 moves in an elliptical orbit around SgrA with a period of 15.2 yr and an eccentricity $e = 0.87$ (Fig. 6–16). In 2002, S2 reached its closest approach to SgrA, a distance of only 123 AU $(1 \, \text{AU} = 1.50 \times 10^{11} \, \text{m}$ is the mean Earth–Sun distance). Determine the mass M of SgrA, the massive compact object (believed to be a supermassive black hole) at the center of our Galaxy. State M in kg and in terms of the mass of our Sun.

62. A satellite of mass 5500 kg orbits the Earth and has a period of 6200 s. Determine (a) the radius of its circular orbit, (b) the magnitude of the Earth's gravitational force on the satellite, and (c) the altitude of the satellite.

63. Show that the rate of change of your weight is

$$-2G\frac{m_E m}{r^3}v$$

if you are traveling directly away from Earth at constant speed v. Your mass is m, and r is your distance from the center of the Earth at any moment.

64. Astronomers using the Hubble Space Telescope deduced the presence of an extremely massive core in the distant galaxy M87, so dense that it could be a black hole (from which no light escapes). They did this by measuring the speed of gas clouds orbiting the core to be 780 km/s at a distance of 60 light-years $(5.7 \times 10^{17} \text{ m})$ from the core. Deduce the mass of the core, and compare it to the mass of our Sun.

65. Suppose all the mass of the Earth were compacted into a small spherical ball. What radius must the sphere have so that the acceleration due to gravity at the Earth's new surface was equal to the acceleration due to gravity at the surface of the Sun?

66. A plumb bob (a mass m hanging on a string) is deflected from the vertical by an angle θ due to a massive mountain nearby (Fig. 6–30). (a) Find an approximate formula for θ in terms of the mass of the mountain, m_M, the distance to its center, D_M, and the radius and mass of the Earth. (b) Make a rough estimate of the mass of Mt. Everest, assuming it has the shape of a cone 4000 m high and base of diameter 4000 m. Assume its mass per unit volume is 3000 kg per m³. (c) Estimate the angle θ of the plumb bob if it is 5 km from the center of Mt. Everest.

FIGURE 6–30 Problem 66.

67. A geologist searching for oil finds that the gravity at a certain location is 2 parts in 10^7 smaller than average. Assume that a deposit of oil is located 2000 m directly below. Estimate the size of the deposit, assumed spherical. Take the density (mass per unit volume) of rock to be 3000 kg/m³ and that of oil to be 800 kg/m³.

68. You are an astronaut in the space shuttle pursuing a satellite in need of repair. You are in a circular orbit of the same radius as the satellite (400 km above the Earth), but 25 km behind it. (a) How long will it take to overtake the satellite if you reduce your orbital radius by 1.0 km? (b) By how much must you reduce your orbital radius to catch up in 7.0 h?

69. A science-fiction tale describes an artificial "planet" in the form of a band completely encircling a sun (Fig. 6–31). The inhabitants live on the inside surface (where it is always noon). Imagine that this sun is exactly like our own, that the distance to the band is the same as the Earth–Sun distance (to make the climate temperate), and that the ring rotates quickly enough to produce an apparent gravity of g as on Earth. What will be the period of revolution, this planet's year, in Earth days?

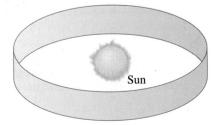

FIGURE 6–31
Problem 69.

70. How long would a day be if the Earth were rotating so fast that objects at the equator were apparently weightless?

71. An asteroid of mass m is in a circular orbit of radius r around the Sun with a speed v. It has an impact with another asteroid of mass M and is kicked into a new circular orbit with a speed of $1.5\,v$. What is the radius of the new orbit in terms of r?

72. Newton had the data listed in Table 6–4, plus the relative sizes of these objects: in terms of the Sun's radius R, the radii of Jupiter and Earth were $0.0997R$ and $0.0109R$. Newton used this information to determine that the average density $\rho\,(= \text{mass/volume})$ of Jupiter is slightly less than that of the Sun, while the average density of the Earth is four times that of the Sun. Thus, without leaving his home planet, Newton was able to predict that the composition of the Sun and Jupiter is markedly different than that of Earth. Reproduce Newton's calculation and find his values for the ratios ρ_J/ρ_{Sun} and ρ_E/ρ_{Sun} (the modern values for these ratios are 0.93 and 3.91, respectively).

TABLE 6–4 Problem 72

	Orbital Radius, R (in AU = 1.50×10^{11} m)	Orbital Period, T (Earth days)
Venus about Sun	0.724	224.70
Callisto about Jupiter	0.01253	16.69
Moon about Earth	0.003069	27.32

73. A satellite circles a spherical planet of unknown mass in a circular orbit of radius 2.0×10^7 m. The magnitude of the gravitational force exerted on the satellite by the planet is 120 N. (a) What would be the magnitude of the gravitational force exerted on the satellite by the planet if the radius of the orbit were increased to 3.0×10^7 m? (b) If the satellite circles the planet once every 2.0 h in the larger orbit, what is the mass of the planet?

74. A uniform sphere has mass M and radius r. A spherical cavity (no mass) of radius $r/2$ is then carved within this sphere as shown in Fig. 6–32 (the cavity's surface passes through the sphere's center and just touches the sphere's outer surface). The centers of the original sphere and the cavity lie on a straight line, which defines the x axis. With what gravitational force will the hollowed-out sphere attract a point mass m which lies on the x axis a distance d from the sphere's center? [*Hint*: Subtract the effect of the "small" sphere (the cavity) from that of the larger entire sphere.]

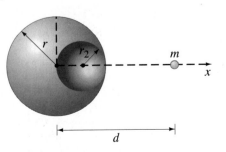

FIGURE 6–32 Problem 74.

75. The gravitational force at different places on Earth due to the Sun and the Moon depends on each point's distance from the Sun or Moon, and this variation is what causes the **tides**. Use the values inside the front cover of this book for the Earth–Moon distance R_{EM}, the Earth–Sun distance R_{ES}, the Moon's mass M_M, the Sun's mass, M_S, and the Earth's radius R_E. (a) First consider two small pieces of the Earth, each of mass m, one on the side of the Earth nearest the Moon, the other on the side farthest from the Moon. Show that the ratio of the Moon's gravitational forces on these two masses is

$$\left(\frac{F_{near}}{F_{far}}\right)_M = 1.0687.$$

(b) Next consider two small pieces of the Earth, each of mass m, one on the nearest point of Earth to the Sun, the other at the farthest point from the Sun. Show that the ratio of the Sun's gravitational forces on these two masses is

$$\left(\frac{F_{near}}{F_{far}}\right)_S = 1.000171.$$

(c) Show that the ratio of the Sun's average gravitational force on the Earth compared to that of the Moon's is

$$\left(\frac{F_S}{F_M}\right)_{avg} = 178.$$

Note that the Moon's smaller force varies much more across the Earth's diameter than the Sun's larger force. (d) Estimate the resulting "force difference" (the cause of the tides)

$$\Delta F = F_{near} - F_{far} = F_{far}\left(\frac{F_{near}}{F_{far}} - 1\right) \approx F_{avg}\left(\frac{F_{near}}{F_{far}} - 1\right)$$

for the Moon and for the Sun. Show that the ratio of the tide-causing force differences due to the Moon compared to the Sun is

$$\frac{\Delta F_M}{\Delta F_S} \approx 2.3.$$

Thus the Moon's influence on tide production is over two times as great as the Sun's.

*76. A particle is released at a height r_E (radius of Earth) above the Earth's surface. Determine its velocity when it hits the Earth. Ignore air resistance. [*Hint*: Use Newton's second law, the law of universal gravitation, the chain rule, and integrate.]

77. Estimate the value of the gravitational constant G in Newton's law of universal gravitation using the following data: the acceleration due to gravity at the Earth's surface is about $10\ m/s^2$; the Earth has a circumference of about $40 \times 10^6\ m$; rocks found on the Earth's surface typically have densities of about $3000\ kg/m^3$ and assume this density is constant throughout (even though you suspect it is not true).

78. Between the orbits of Mars and Jupiter, several thousand small objects called asteroids move in nearly circular orbits around the Sun. Consider an asteroid that is spherically shaped with radius r and density $2700\ kg/m^3$. (a) You find yourself on the surface of this asteroid and throw a baseball at a speed of $22\ m/s$ (about $50\ mi/h$). If the baseball is to travel around the asteroid in a circular orbit, what is the largest radius asteroid on which you are capable of accomplishing this feat? (b) After you throw the baseball, you turn around and face the opposite direction and catch the baseball. How much time T elapses between your throw and your catch?

*Numerical/Computer

*79. (II) The accompanying table shows the data for the mean distances of planets (except Pluto) from the Sun in our solar system, and their periods of revolution about the Sun.

Planet	Mean Distance (AU)	Period (Years)
Mercury	0.387	0.241
Venus	0.723	0.615
Earth	1.000	1.000
Mars	1.524	1.881
Jupiter	5.203	11.88
Saturn	9.539	29.46
Uranus	19.18	84.01
Neptune	30.06	164.8

(a) Graph the square of the periods as a function of the cube of the average distances, and find the best-fit straight line. (b) If the period of Pluto is 247.7 years, estimate the mean distance of Pluto from the Sun from the best-fit line.

Answers to Exercises

A: g would double.

B: (b).

C: (b).

D: No; even though they are experiencing weightlessness, the massive ball would require a large force to throw and to decelerate when caught (inertial mass, Newton's second law).

E: 6.17 yr.

Force   Displacement

This baseball pitcher is about to accelerate the baseball to a high velocity by exerting a force on it. He will be doing work on the ball as he exerts the force over a displacement of several meters, from behind his head until he releases the ball with arm outstretched in front of him. The total work done on the ball will be equal to the kinetic energy $\left(\frac{1}{2}mv^2\right)$ acquired by the ball, a result known as the work-energy principle.

Work and Energy

CHAPTER-OPENING QUESTION—Guess now!

You push very hard on a heavy desk, trying to move it. You do work on the desk:

(a) Whether or not it moves, as long as you are exerting a force.
(b) Only if it starts moving.
(c) Only if it doesn't move.
(d) Never—it does work on you.
(e) None of the above.

Until now we have been studying the translational motion of an object in terms of Newton's three laws of motion. In that analysis, *force* has played a central role as the quantity determining the motion. In this Chapter and the two that follow, we discuss an alternative analysis of the translational motion of objects in terms of the quantities *energy* and *momentum*. The significance of energy and momentum is that they are *conserved*. In quite general circumstances they remain constant. That conserved quantities exist gives us not only a deeper insight into the nature of the world but also gives us another way to approach solving practical problems.

The conservation laws of energy and momentum are especially valuable in dealing with systems of many objects, in which a detailed consideration of the forces involved would be difficult or impossible. These laws are applicable to a wide range of phenomena, including the atomic and subatomic worlds, where Newton's laws cannot be applied.

This Chapter is devoted to the very important concept of *energy* and the closely related concept of *work*. These two quantities are scalars and so have no direction associated with them, which often makes them easier to work with than vector quantities such as acceleration and force.

CONTENTS

FIGURE 7–1 A person pulling a crate along the floor. The work done by the force $\vec{\mathbf{F}}$ is $W = Fd \cos \theta$, where $\vec{\mathbf{d}}$ is the displacement.

$\vec{\mathbf{F}}$

θ

$F \cos \theta = F_{\parallel}$

$\vec{\mathbf{d}}$

7–1 Work Done by a Constant Force

The word *work* has a variety of meanings in everyday language. But in physics, work is given a very specific meaning to describe what is accomplished when a force acts on an object, and the object moves through a distance. We consider only translational motion for now and, unless otherwise explained, objects are assumed to be rigid with no complicating internal motion, and can be treated like particles. Then the **work** done on an object by a constant force (constant in both magnitude and direction) is defined to be *the product of the magnitude of the displacement times the component of the force parallel to the displacement*. In equation form, we can write

$$W = F_{\parallel} d,$$

where $F_{\parallel}$ is the component of the constant force $\vec{\mathbf{F}}$ parallel to the displacement $\vec{\mathbf{d}}$. We can also write

$$W = Fd \cos \theta, \tag{7–1}$$

where F is the magnitude of the constant force, d is the magnitude of the displacement of the object, and θ is the angle between the directions of the force and the displacement (Fig. 7–1). The $\cos \theta$ factor appears in Eq. 7–1 because $F \cos \theta \, (= F_{\parallel})$ is the component of $\vec{\mathbf{F}}$ that is parallel to $\vec{\mathbf{d}}$. Work is a scalar quantity—it has only magnitude, which can be positive or negative.

Let us consider the case in which the motion and the force are in the same direction, so $\theta = 0$ and $\cos \theta = 1$; in this case, $W = Fd$. For example, if you push a loaded grocery cart a distance of 50 m by exerting a horizontal force of 30 N on the cart, you do $30 \, \text{N} \times 50 \, \text{m} = 1500 \, \text{N} \cdot \text{m}$ of work on the cart.

As this example shows, in SI units work is measured in newton-meters (N·m). A special name is given to this unit, the **joule** (J): $1 \, \text{J} = 1 \, \text{N} \cdot \text{m}$.

[In the cgs system, the unit of work is called the *erg* and is defined as $1 \, \text{erg} = 1 \, \text{dyne} \cdot \text{cm}$. In British units, work is measured in foot-pounds. It is easy to show that $1 \, \text{J} = 10^7 \, \text{erg} = 0.7376 \, \text{ft} \cdot \text{lb}$.]

A force can be exerted on an object and yet do no work. If you hold a heavy bag of groceries in your hands at rest, you do no work on it. You do exert a force on the bag, but the displacement of the bag is zero, so the work done by you on the bag is $W = 0$. You need both a force and a displacement to do work. You also do no work on the bag of groceries if you carry it as you walk horizontally across the floor at constant velocity, as shown in Fig. 7–2. No horizontal force is required to move the bag at a constant velocity. The person shown in Fig. 7–2 does exert an upward force $\vec{\mathbf{F}}_P$ on the bag equal to its weight. But this upward force is perpendicular to the horizontal displacement of the bag and thus is doing no work. This conclusion comes from our definition of work, Eq. 7–1: $W = 0$, because $\theta = 90°$ and $\cos 90° = 0$.

FIGURE 7–2 The person does no work on the bag of groceries since $\vec{\mathbf{F}}_P$ is perpendicular to the displacement $\vec{\mathbf{d}}$.

$\vec{\mathbf{F}}_P$

$\vec{\mathbf{d}}$

$m\vec{\mathbf{g}}$

Thus, when a particular force is perpendicular to the displacement, no work is done by that force. When you start or stop walking, there is a horizontal acceleration and you do briefly exert a horizontal force, and thus do work on the bag.

When we deal with work, as with force, it is necessary to specify whether you are talking about work done *by* a specific object or done *on* a specific object. It is also important to specify whether the work done is due to one particular force (and which one), or the total (net) work done by the *net force* on the object.

EXAMPLE 7–1 **Work done on a crate.** A person pulls a 50-kg crate 40 m along a horizontal floor by a constant force $F_P = 100$ N, which acts at a 37° angle as shown in Fig. 7–3. The floor is smooth and exerts no friction force. Determine (a) the work done by each force acting on the crate, and (b) the net work done on the crate.

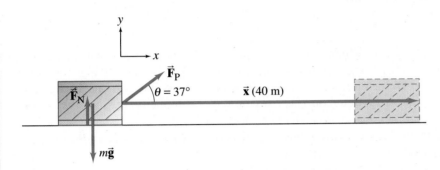

FIGURE 7–3 Example 7–1. A 50-kg crate is pulled along a smooth floor.

APPROACH We choose our coordinate system so that $\vec{x}$ can be the vector that represents the 40-m displacement (that is, along the x axis). Three forces act on the crate, as shown in Fig. 7–3: the force exerted by the person $\vec{F}_P$; the gravitational force exerted by the Earth, $m\vec{g}$; and the normal force $\vec{F}_N$ exerted upward by the floor. The net force on the crate is the vector sum of these three forces.

SOLUTION (a) The work done by the gravitational and normal forces is zero, since they are perpendicular to the displacement $\vec{x}$ ($\theta = 90°$ in Eq. 7–1):

$$W_G = mgx \cos 90° = 0$$
$$W_N = F_N x \cos 90° = 0.$$

The work done by $\vec{F}_P$ is

$$W_P = F_P x \cos \theta = (100 \text{ N})(40 \text{ m}) \cos 37° = 3200 \text{ J}.$$

(b) The net work can be calculated in two equivalent ways:
(1) The net work done on an object is the algebraic sum of the work done by each force, since work is a scalar:

$$W_{net} = W_G + W_N + W_P$$
$$= 0 + 0 + 3200 \text{ J} = 3200 \text{ J}.$$

(2) The net work can also be calculated by first determining the net force on the object and then taking its component along the displacement: $(F_{net})_x = F_P \cos \theta$. Then the net work is

$$W_{net} = (F_{net})_x x = (F_P \cos \theta) x$$
$$= (100 \text{ N})(\cos 37°)(40 \text{ m}) = 3200 \text{ J}.$$

In the vertical (y) direction, there is no displacement and no work done.

EXERCISE A A box is dragged a distance d across a floor by a force $\vec{F}_P$ which makes an angle θ with the horizontal as in Fig. 7–1 or 7–3. If the magnitude of $\vec{F}_P$ is held constant but the angle θ is increased, the work done by $\vec{F}_P$ (a) remains the same; (b) increases; (c) decreases; (d) first increases, then decreases.

EXERCISE B Return to the Chapter-Opening Question, page 163, and answer it again now. Try to explain why you may have answered differently the first time.

Work

1. **Draw a free-body diagram** showing all the forces acting on the object you choose to study.

2. **Choose** an *xy* **coordinate system**. If the object is in motion, it may be convenient to choose one of the coordinate directions as the direction of one of the forces, or as the direction of motion. [Thus, for an object on an incline, you might choose one coordinate axis to be parallel to the incline.]

3. **Apply Newton's laws** to determine any unknown forces.

4. Find the **work done by a specific force** *on* the object by using $W = Fd\cos\theta$ for a constant force. Note that the work done is negative when a force tends to oppose the displacement.

5. To find the **net work** done on the object, either (*a*) find the work done by each force and add the results algebraically; or (*b*) find the net force on the object, F_{net}, and then use it to find the net work done, which for constant net force is:
$$W_{net} = F_{net}\, d\cos\theta.$$

FIGURE 7–4 Example 7–2.

(a)

(b)

(c)

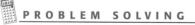

P R O B L E M   S O L V I N G

Work done by gravity depends on the height of the hill and not on the angle of incline

EXAMPLE 7–2 **Work on a backpack.** (*a*) Determine the work a hiker must do on a 15.0-kg backpack to carry it up a hill of height $h = 10.0$ m, as shown in Fig. 7–4a. Determine also (*b*) the work done by gravity on the backpack, and (*c*) the net work done on the backpack. For simplicity, assume the motion is smooth and at constant velocity (i.e., acceleration is zero).

APPROACH We explicitly follow the steps of the Problem Solving Strategy above.
SOLUTION

1. **Draw a free-body diagram.** The forces on the backpack are shown in Fig. 7–4b: the force of gravity, $m\vec{\mathbf{g}}$, acting downward; and $\vec{\mathbf{F}}_H$, the force the hiker must exert upward to support the backpack. The acceleration is zero, so horizontal forces on the backpack are negligible.

2. **Choose a coordinate system.** We are interested in the vertical motion of the backpack, so we choose the y coordinate as positive vertically upward.

3. **Apply Newton's laws.** Newton's second law applied in the vertical direction to the backpack gives
$$\Sigma F_y = ma_y$$
$$F_H - mg = 0$$
since $a_y = 0$. Hence,
$$F_H = mg = (15.0\,\text{kg})(9.80\,\text{m/s}^2) = 147\,\text{N}.$$

4. **Work done by a specific force.** (*a*) To calculate the work done by the hiker on the backpack, we write Eq. 7–1 as
$$W_H = F_H(d\cos\theta),$$
and we note from Fig. 7–4a that $d\cos\theta = h$. So the work done by the hiker is
$$W_H = F_H(d\cos\theta) = F_H h = mgh$$
$$= (147\,\text{N})(10.0\,\text{m}) = 1470\,\text{J}.$$
Note that the work done depends only on the change in elevation and not on the angle of the hill, θ. The hiker would do the same work to lift the pack vertically the same height h.

(*b*) The work done by gravity on the backpack is (from Eq. 7–1 and Fig. 7–4c)
$$W_G = F_G d\cos(180° - \theta).$$
Since $\cos(180° - \theta) = -\cos\theta$, we have
$$W_G = F_G d(-\cos\theta) = mg(-d\cos\theta)$$
$$= -mgh$$
$$= -(15.0\,\text{kg})(9.80\,\text{m/s}^2)(10.0\,\text{m}) = -1470\,\text{J}.$$

NOTE The work done by gravity (which is negative here) doesn't depend on the angle of the incline, only on the vertical height h of the hill. This is because gravity acts vertically, so only the vertical component of displacement contributes to work done.

5. **Net work done.** (*c*) The *net* work done on the backpack is $W_{net} = 0$, since the net force on the backpack is zero (it is assumed not to accelerate significantly). We can also determine the net work done by adding the work done by each force:
$$W_{net} = W_G + W_H = -1470\,\text{J} + 1470\,\text{J} = 0.$$

NOTE Even though the *net* work done by all the forces on the backpack is zero, the hiker *does do* work on the backpack equal to 1470 J.

CONCEPTUAL EXAMPLE 7–3 **Does the Earth do work on the Moon?** The Moon revolves around the Earth in a nearly circular orbit, with approximately constant tangential speed, kept there by the gravitational force exerted by the Earth. Does gravity do (a) positive work, (b) negative work, or (c) no work at all on the Moon?

RESPONSE The gravitational force $\vec{F}_G$ on the Moon (Fig. 7–5) acts toward the Earth and provides its centripetal force, inward along the radius of the Moon's orbit. The Moon's displacement at any moment is tangent to the circle, in the direction of its velocity, perpendicular to the radius and perpendicular to the force of gravity. Hence the angle θ between the force $\vec{F}_G$ and the instantaneous displacement of the Moon is 90°, and the work done by gravity is therefore zero ($\cos 90° = 0$). This is why the Moon, as well as artificial satellites, can stay in orbit without expenditure of fuel: no work needs to be done against the force of gravity.

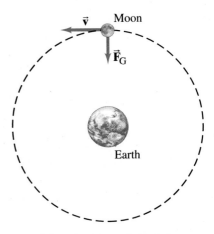

FIGURE 7–5 Example 7–3.

7–2 Scalar Product of Two Vectors

Although work is a scalar, it involves the product of two quantities, force and displacement, both of which are vectors. Therefore, we now investigate the multiplication of vectors, which will be useful throughout the book, and apply it to work.

Because vectors have direction as well as magnitude, they cannot be multiplied in the same way that scalars are. Instead we must *define* what the operation of vector multiplication means. Among the possible ways to define how to multiply vectors, there are three ways that we find useful in physics: (1) multiplication of a vector by a scalar, which was discussed in Section 3–3; (2) multiplication of one vector by a second vector to produce a scalar; (3) multiplication of one vector by a second vector to produce another vector. The third type, called the *vector product*, will be discussed later, in Section 11–2.

We now discuss the second type, called the *scalar product*, or *dot product* (because a dot is used to indicate the multiplication). If we have two vectors, $\vec{A}$ and $\vec{B}$, then their **scalar (or dot) product** is defined to be

$$\vec{A} \cdot \vec{B} = AB \cos \theta, \tag{7–2}$$

where A and B are the magnitudes of the vectors and θ is the angle ($< 180°$) between them when their tails touch, Fig. 7–6. Since A, B, and $\cos \theta$ are scalars, then so is the scalar product $\vec{A} \cdot \vec{B}$ (read "A dot B").

This definition, Eq. 7–2, fits perfectly with our definition of the work done by a constant force, Eq. 7–1. That is, we can write the work done by a constant force as the scalar product of force and displacement:

$$W = \vec{F} \cdot \vec{d} = Fd \cos \theta. \tag{7–3}$$

Indeed, the definition of scalar product, Eq. 7–2, is so chosen because many physically important quantities, such as work (and others we will meet later), can be described as the scalar product of two vectors.

An equivalent definition of the scalar product is that it is the product of the magnitude of one vector (say B) and the component (or projection) of the other vector along the direction of the first ($A \cos \theta$). See Fig. 7–6.

Since A, B, and $\cos \theta$ are scalars, it doesn't matter in what order they are multiplied. Hence the scalar product is **commutative**:

$$\vec{A} \cdot \vec{B} = \vec{B} \cdot \vec{A}. \qquad \text{[commutative property]}$$

It is also easy to show that it is **distributive** (see Problem 33 for the proof):

$$\vec{A} \cdot (\vec{B} + \vec{C}) = \vec{A} \cdot \vec{B} + \vec{A} \cdot \vec{C}. \qquad \text{[distributive property]}$$

FIGURE 7–6 The scalar product, or dot product, of two vectors $\vec{A}$ and $\vec{B}$ is $\vec{A} \cdot \vec{B} = AB \cos \theta$. The scalar product can be interpreted as the magnitude of one vector (B in this case) times the projection of the other vector, $A \cos \theta$, onto $\vec{B}$.

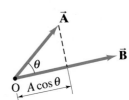

Let us write our vectors $\vec{A}$ and $\vec{B}$ in terms of their rectangular components using unit vectors (Section 3–5, Eq. 3–5) as

$$\vec{A} = A_x\hat{i} + A_y\hat{j} + A_z\hat{k}$$
$$\vec{B} = B_x\hat{i} + B_y\hat{j} + B_z\hat{k}.$$

We will take the scalar product, $\vec{A} \cdot \vec{B}$, of these two vectors, remembering that the unit vectors, $\hat{i}, \hat{j},$ and $\hat{k}$, are perpendicular to each other

$$\hat{i} \cdot \hat{i} = \hat{j} \cdot \hat{j} = \hat{k} \cdot \hat{k} = 1$$
$$\hat{i} \cdot \hat{j} = \hat{i} \cdot \hat{k} = \hat{j} \cdot \hat{k} = 0.$$

Thus the scalar product equals

$$\vec{A} \cdot \vec{B} = \left(A_x\hat{i} + A_y\hat{j} + A_z\hat{k}\right) \cdot \left(B_x\hat{i} + B_y\hat{j} + B_z\hat{k}\right)$$
$$= A_xB_x + A_yB_y + A_zB_z. \qquad \text{(7–4)}$$

Equation 7–4 is very useful.

If $\vec{A}$ is perpendicular to $\vec{B}$, then Eq. 7–2 tells us $\vec{A} \cdot \vec{B} = AB\cos 90° = 0$. But the converse, given that $\vec{A} \cdot \vec{B} = 0$, can come about in three different ways: $\vec{A} = 0$, $\vec{B} = 0$, or $\vec{A} \perp \vec{B}$.

FIGURE 7–7 Example 7–4. Work done by a force $\vec{F}_P$ acting at an angle θ to the ground is $W = \vec{F}_P \cdot \vec{d}$.

EXAMPLE 7–4 **Using the dot product.** The force shown in Fig. 7–7 has magnitude $F_P = 20\,\text{N}$ and makes an angle of 30° to the ground. Calculate the work done by this force using Eq. 7–4 when the wagon is dragged 100 m along the ground.

APPROACH We choose the x axis horizontal to the right and the y axis vertically upward, and write $\vec{F}_P$ and $\vec{d}$ in terms of unit vectors.
SOLUTION

$$\vec{F}_P = F_x\hat{i} + F_y\hat{j} = \left(F_P\cos 30°\right)\hat{i} + \left(F_P\sin 30°\right)\hat{j} = (17\,\text{N})\hat{i} + (10\,\text{N})\hat{j},$$

whereas $\vec{d} = (100\,\text{m})\hat{i}$. Then, using Eq. 7–4,

$$W = \vec{F}_P \cdot \vec{d} = (17\,\text{N})(100\,\text{m}) + (10\,\text{N})(0) + (0)(0) = 1700\,\text{J}.$$

Note that by choosing the x axis along $\vec{d}$ we simplified the calculation because $\vec{d}$ then has only one component.

7–3 Work Done by a Varying Force

If the force acting on an object is constant, the work done by that force can be calculated using Eq. 7–1. In many cases, however, the force varies in magnitude or direction during a process. For example, as a rocket moves away from Earth, work is done to overcome the force of gravity, which varies as the inverse square of the distance from the Earth's center. Other examples are the force exerted by a spring, which increases with the amount of stretch, or the work done by a varying force exerted to pull a box or cart up an uneven hill.

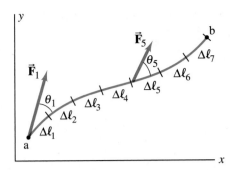

FIGURE 7–8 A particle acted on by a variable force, $\vec{F}$, moves along the path shown from point a to point b.

Figure 7–8 shows the path of an object in the xy plane as it moves from point a to point b. The path has been divided into short intervals each of length $\Delta\ell_1, \Delta\ell_2, \ldots, \Delta\ell_7$. A force $\vec{F}$ acts at each point on the path, and is indicated at two points as $\vec{F}_1$ and $\vec{F}_5$. During each small interval $\Delta\ell$, the force is approximately constant. For the first interval, the force does work ΔW of approximately (see Eq. 7–1)

$$\Delta W \approx F_1 \cos\theta_1 \Delta\ell_1.$$

In the second interval the work done is approximately $F_2 \cos\theta_2 \Delta\ell_2$, and so on. The total work done in moving the particle the total distance $\ell = \Delta\ell_1 + \Delta\ell_2 + \ldots + \Delta\ell_7$ is the sum of all these terms:

$$W \approx \sum_{i=1}^{7} F_i \cos\theta_i \Delta\ell_i. \tag{7–5}$$

We can examine this graphically by plotting $F \cos\theta$ versus the distance ℓ along the path as shown in Fig. 7–9a. The distance ℓ has been subdivided into the same seven intervals (see the vertical dashed lines). The value of $F \cos\theta$ at the center of each interval is indicated by the *horizontal* dashed lines. Each of the shaded rectangles has an area $(F_i \cos\theta)(\Delta\ell_i)$, which is a good estimate of the work done during the interval. The estimate of the work done along the entire path given by Eq. 7–5, equals the sum of the areas of all the rectangles. If we subdivide the distance into a greater number of intervals, so that each $\Delta\ell_i$ is smaller, the estimate of the work done becomes more accurate (the assumption that F is constant over each interval is more accurate). Letting each $\Delta\ell_i$ approach zero (so we approach an infinite number of intervals), we obtain an exact result for the work done:

$$W = \lim_{\Delta\ell_i \to 0} \Sigma F_i \cos\theta_i \Delta\ell_i = \int_a^b F \cos\theta \, d\ell. \tag{7–6}$$

This limit as $\Delta\ell_i \to 0$ is the *integral* of $(F \cos\theta \, d\ell)$ from point a to point b. The symbol for the integral, $\int$, is an elongated S to indicate an infinite sum; and $\Delta\ell$ has been replaced by $d\ell$, meaning an infinitesimal distance. [We also discussed this in the optional Section 2–9.]

In this limit as $\Delta\ell$ approaches zero, the total area of the rectangles (Fig. 7–9a) approaches the area between the $(F \cos\theta)$ curve and the ℓ axis from a to b as shown shaded in Fig. 7–9b. That is, *the work done by a variable force in moving an object between two points is equal to the area under the $(F \cos\theta)$ versus (ℓ) curve between those two points.*

In the limit as $\Delta\ell$ approaches zero, the infinitesimal distance $d\ell$ equals the magnitude of the infinitesimal displacement vector $d\vec{\ell}$. The direction of the vector $d\vec{\ell}$ is along the tangent to the path at that point, so θ is the angle between $\vec{F}$ and $d\vec{\ell}$ at any point. Thus we can rewrite Eq. 7–6, using dot-product notation:

$$W = \int_a^b \vec{F} \cdot d\vec{\ell}. \tag{7–7}$$

This is a *general definition of work*. In this equation, a and b represent two points in space, (x_a, y_a, z_a) and (x_b, y_b, z_b). The integral in Eq. 7–7 is called a *line integral* since it is the integral of $F \cos\theta$ along the line that represents the path of the object. (Equation 7–1 for a constant force is a special case of Eq. 7–7.)

FIGURE 7–9 Work done by a force F is (a) approximately equal to the sum of the areas of the rectangles, (b) exactly equal to the area under the curve of $F \cos\theta$ vs. ℓ.

(a)

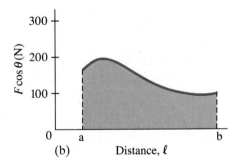

(b)

In rectangular coordinates, any force can be written

$$\vec{F} = F_x\hat{i} + F_y\hat{j} + F_z\hat{k}$$

and the displacement $d\vec{\ell}$ is

$$d\vec{\ell} = dx\hat{i} + dy\hat{j} + dz\hat{k}.$$

Then the work done can be written

$$W = \int_{x_a}^{x_b} F_x\,dx + \int_{y_a}^{y_b} F_y\,dy + \int_{z_a}^{z_b} F_z\,dz.$$

To actually use Eq. 7–6 or 7–7 to calculate the work, there are several options: (1) If $F\cos\theta$ is known as a function of position, a graph like that of Fig. 7–9b can be made and the area determined graphically. (2) Another possibility is to use numerical integration (numerical summing), perhaps with the aid of a computer or calculator. (3) A third possibility is to use the analytical methods of integral calculus, when it is doable. To do so, we must be able to write $\vec{F}$ as a function of position, $F(x, y, z)$, and we must know the path. Let's look at some specific examples.

Work Done by a Spring Force

Let us determine the work needed to stretch or compress a coiled spring, such as that shown in Fig. 7–10. For a person to hold a spring either stretched or compressed an amount x from its normal (relaxed) length requires a force F_P that is directly proportional to x. That is,

$$F_P = kx,$$

where k is a constant, called the *spring constant* (or *spring stiffness constant*), and is a measure of the stiffness of the particular spring. The spring itself exerts a force in the opposite direction (Fig. 7–10b or c):

$$F_S = -kx. \tag{7–8}$$

This force is sometimes called a "restoring force" because the spring exerts its force in the direction opposite the displacement (hence the minus sign), and thus acts to return the spring to its normal length. Equation 7–8 is known as the **spring equation** or **Hooke's law**, and is accurate for springs as long as x is not too great (see Section 12–4) and no permanent deformation occurs.

Let us calculate the work a person does to stretch (or compress) a spring from its normal (unstretched) length, $x_a = 0$, to an extra length, $x_b = x$. We assume the stretching is done slowly, so that the acceleration is essentially zero. The force $\vec{F}_P$ is exerted parallel to the axis of the spring, along the x axis, so $\vec{F}_P$ and $d\vec{\ell}$ are parallel. Hence, since $d\vec{\ell} = dx\hat{i}$ in this case, the work done by the person is[†]

$$W_P = \int_{x_a=0}^{x_b=x} \left[F_P(x)\,\hat{i}\right] \cdot \left[dx\hat{i}\right] = \int_0^x F_P(x)\,dx = \int_0^x kx\,dx = \tfrac{1}{2}kx^2\Big|_0^x = \tfrac{1}{2}kx^2.$$

(As is frequently done, we have used x to represent both the variable of integration, and the particular value of x at the end of the interval $x_a = 0$ to $x_b = x$.) Thus we see that the work needed is proportional to the square of the distance stretched (or compressed), x.

This same result can be obtained by computing the area under the graph of F vs. x (with $\cos\theta = 1$ in this case) as shown in Fig. 7–11. Since the area is a triangle of altitude kx and base x, the work a person does to stretch or compress a spring an amount x is

$$W = \tfrac{1}{2}(x)(kx) = \tfrac{1}{2}kx^2,$$

which is the same result as before. Because $W \propto x^2$, it takes the same amount of work to stretch a spring or compress it the same amount x.

[†]See the Table of Integrals, Appendix B.

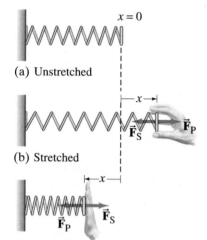

$x = 0$

(a) Unstretched

(b) Stretched

(c) Compressed

FIGURE 7–10 (a) Spring in normal (unstretched) position. (b) Spring is stretched by a person exerting a force $\vec{F}_P$ to the right (positive direction). The spring pulls back with a force $\vec{F}_S$ where $F_S = -kx$. (c) Person compresses the spring ($x < 0$), and the spring pushes back with a force $F_S = -kx$ where $F_S > 0$ because $x < 0$.

FIGURE 7–11 Work done to stretch a spring a distance x equals the triangular area under the curve $F = kx$. The area of a triangle is $\tfrac{1}{2} \times$ base $\times$ altitude, so $W = \tfrac{1}{2}(x)(kx) = \tfrac{1}{2}kx^2$.

EXAMPLE 7–5 **Work done on a spring.** (a) A person pulls on the spring in Fig. 7–10, stretching it 3.0 cm, which requires a maximum force of 75 N. How much work does the person do? (b) If, instead, the person compresses the spring 3.0 cm, how much work does the person do?

APPROACH The force $F = kx$ holds at each point, including x_{max}. Hence F_{max} occurs at $x = x_{max}$.

SOLUTION (a) First we need to calculate the spring constant k:

$$k = \frac{F_{max}}{x_{max}} = \frac{75\,\text{N}}{0.030\,\text{m}} = 2.5 \times 10^3\,\text{N/m}.$$

Then the work done by the person on the spring is

$$W = \tfrac{1}{2}kx_{max}^2 = \tfrac{1}{2}(2.5 \times 10^3\,\text{N/m})(0.030\,\text{m})^2 = 1.1\,\text{J}.$$

(b) The force that the person exerts is still $F_P = kx$, though now both x and F_P are negative (x is positive to the right). The work done is

$$W_P = \int_{x=0}^{x=-0.030\,\text{m}} F_P(x)\,dx = \int_0^{x=-0.030\,\text{m}} kx\,dx = \tfrac{1}{2}kx^2\Big|_0^{-0.030\,\text{m}}$$

$$= \tfrac{1}{2}(2.5 \times 10^3\,\text{N/m})(-0.030\,\text{m})^2 = 1.1\,\text{J},$$

which is the same as for stretching it.

NOTE We cannot use $W = Fd$ (Eq. 7–1) for a spring because the force is not constant.

A More Complex Force Law—Robot Arm

EXAMPLE 7–6 **Force as function of x.** A robot arm that controls the position of a video camera (Fig. 7–12) in an automated surveillance system is manipulated by a motor that exerts a force on the arm. The force is given by

$$F(x) = F_0\left(1 + \frac{1}{6}\frac{x^2}{x_0^2}\right),$$

where $F_0 = 2.0\,\text{N}$, $x_0 = 0.0070\,\text{m}$, and x is the position of the end of the arm. If the arm moves from $x_1 = 0.010\,\text{m}$ to $x_2 = 0.050\,\text{m}$, how much work did the motor do?

APPROACH The force applied by the motor is not a linear function of x. We can determine the integral $\int F(x)\,dx$, or the area under the $F(x)$ curve (shown in Fig. 7–13).

SOLUTION We integrate to find the work done by the motor:

$$W_M = F_0\int_{x_1}^{x_2}\left(1 + \frac{x^2}{6x_0^2}\right)dx = F_0\int_{x_1}^{x_2}dx + \frac{F_0}{6x_0^2}\int_{x_1}^{x_2}x^2\,dx$$

$$= F_0\left(x + \frac{1}{6x_0^2}\frac{x^3}{3}\right)\Big|_{x_1}^{x_2}.$$

We put in the values given and obtain

$$W_M = 2.0\,\text{N}\left[(0.050\,\text{m} - 0.010\,\text{m}) + \frac{(0.050\,\text{m})^3 - (0.010\,\text{m})^3}{(3)(6)(0.0070\,\text{m})^2}\right] = 0.36\,\text{J}.$$

FIGURE 7–12 Robot arm positions a video camera.

FIGURE 7–13 Example 7–6.

7–4 Kinetic Energy and the Work-Energy Principle

Energy is one of the most important concepts in science. Yet we cannot give a simple general definition of energy in only a few words. Nonetheless, each specific type of energy can be defined fairly simply. In this Chapter we define translational kinetic energy; in the next Chapter, we take up potential energy. In later Chapters we will examine other types of energy, such as that related to heat (Chapters 19 and 20). The crucial aspect of all the types of energy is that the sum of all types, the *total energy*, is the same after any process as it was before: that is, energy is a conserved quantity.

For the purposes of this Chapter, we can define energy in the traditional way as "the ability to do work." This simple definition is not very precise, nor is it really valid for all types of energy.[†] It works, however, for mechanical energy which we discuss in this Chapter and the next. We now define and discuss one of the basic types of energy, kinetic energy.

A moving object can do work on another object it strikes. A flying cannonball does work on a brick wall it knocks down; a moving hammer does work on a nail it drives into wood. In either case, a moving object exerts a force on a second object which undergoes a displacement. An object in motion has the ability to do work and thus can be said to have energy. The energy of motion is called **kinetic energy**, from the Greek word *kinetikos*, meaning "motion."

To obtain a quantitative definition for kinetic energy, let us consider a simple rigid object of mass m (treated as a particle) that is moving in a straight line with an initial speed v_1. To accelerate it uniformly to a speed v_2, a constant net force F_{net} is exerted on it parallel to its motion over a displacement d, Fig. 7–14.

FIGURE 7–14 A constant net force F_{net} accelerates a car from speed v_1 to speed v_2 over a displacement d. The net work done is $W_{net} = F_{net} d$.

Then the net work done on the object is $W_{net} = F_{net} d$. We apply Newton's second law, $F_{net} = ma$, and use Eq. 2–12c $(v_2^2 = v_1^2 + 2ad)$, which we rewrite as

$$a = \frac{v_2^2 - v_1^2}{2d},$$

where v_1 is the initial speed and v_2 the final speed. Substituting this into $F_{net} = ma$, we determine the work done:

$$W_{net} = F_{net} d = mad = m\left(\frac{v_2^2 - v_1^2}{2d}\right)d = m\left(\frac{v_2^2 - v_1^2}{2}\right)$$

or

$$W_{net} = \tfrac{1}{2}mv_2^2 - \tfrac{1}{2}mv_1^2. \tag{7–9}$$

Kinetic energy (defined)

We *define* the quantity $\tfrac{1}{2}mv^2$ to be the **translational kinetic energy**, K, of the object:

$$K = \tfrac{1}{2}mv^2. \tag{7–10}$$

(We call this "translational" kinetic energy to distinguish it from rotational kinetic energy, which we discuss in Chapter 10.) Equation 7–9, derived here for one-dimensional motion with a constant force, is valid in general for translational motion of an object in three dimensions and even if the force varies, as we will show at the end of this Section.

[†]Energy associated with heat is often not available to do work, as we will discuss in Chapter 20.

We can rewrite Eq. 7–9 as:

$$W_{\text{net}} = K_2 - K_1$$

or

$$W_{\text{net}} = \Delta K = \tfrac{1}{2}mv_2^2 - \tfrac{1}{2}mv_1^2. \tag{7–11}$$

WORK-ENERGY PRINCIPLE

Equation 7–11 (or Eq. 7–9) is a useful result known as the **work-energy principle**. It can be stated in words:

> **The net work done on an object is equal to the change in the object's kinetic energy.**

WORK-ENERGY PRINCIPLE

Notice that we made use of Newton's second law, $F_{\text{net}} = ma$, where F_{net} is the *net* force—the sum of all forces acting on the object. Thus, the work-energy principle is valid only if W is the *net work* done on the object—that is, the work done by all forces acting on the object.

The work-energy principle is a very useful reformulation of Newton's laws. It tells us that if (positive) net work W is done on an object, the object's kinetic energy increases by an amount W. The principle also holds true for the reverse situation: if the net work W done on an object is negative, the object's kinetic energy decreases by an amount W. That is, a net force exerted on an object opposite to the object's direction of motion decreases its speed and its kinetic energy. An example is a moving hammer (Fig. 7–15) striking a nail. The net force on the hammer ($-\vec{F}$ in Fig. 7–15, where $\vec{F}$ is assumed constant for simplicity) acts toward the left, whereas the displacement $\vec{d}$ of the hammer is toward the right. So the net work done on the hammer, $W_{\text{h}} = (F)(d)(\cos 180°) = -Fd$, is negative and the hammer's kinetic energy decreases (usually to zero).

⚠️ **CAUTION**
Work-energy valid only for net *work*

Figure 7–15 also illustrates how energy can be considered the ability to do work. The hammer, as it slows down, does positive work on the nail: $W_{\text{n}} = (+F)(+d)(\cos 0°) = Fd$ and is positive. The decrease in kinetic energy of the hammer ($= Fd$ by Eq. 7–11) is equal to the work the hammer can do on another object, the nail in this case.

The translational kinetic energy ($= \tfrac{1}{2}mv^2$) is directly proportional to the mass of the object, and it is also proportional to the *square* of the speed. Thus, if the mass is doubled, the kinetic energy is doubled. But if the speed is doubled, the object has four times as much kinetic energy and is therefore capable of doing four times as much work.

FIGURE 7–15 A moving hammer strikes a nail and comes to rest. The hammer exerts a force F on the nail; the nail exerts a force $-F$ on the hammer (Newton's third law). The work done on the nail by the hammer is positive ($W_{\text{n}} = Fd > 0$). The work done on the hammer by the nail is negative ($W_{\text{h}} = -Fd$).

Because of the direct connection between work and kinetic energy, energy is measured in the same units as work: joules in SI units. [The energy unit is ergs in the cgs, and foot-pounds in the British system.] Like work, kinetic energy is a scalar quantity. The kinetic energy of a group of objects is the sum of the kinetic energies of the individual objects.

The work-energy principle can be applied to a particle, and also to an object that can be approximated as a particle, such as an object that is rigid or whose internal motions are insignificant. It is very useful in simple situations, as we will see in the Examples below. The work-energy principle is not as powerful and encompassing as the law of conservation of energy which we treat in the next Chapter, and should not itself be considered a statement of energy conservation.

EXAMPLE 7–7 Kinetic energy and work done on a baseball. A 145-g baseball is thrown so that it acquires a speed of 25 m/s. (*a*) What is its kinetic energy? (*b*) What was the net work done on the ball to make it reach this speed, if it started from rest?

APPROACH We use $K = \tfrac{1}{2}mv^2$, and the work-energy principle, Eq. 7–11.

SOLUTION (*a*) The kinetic energy of the ball after the throw is

$$K = \tfrac{1}{2}mv^2 = \tfrac{1}{2}(0.145\,\text{kg})(25\,\text{m/s})^2 = 45\,\text{J}.$$

(*b*) Since the initial kinetic energy was zero, the net work done is just equal to the final kinetic energy, 45 J.

$v_1 = 20$ m/s $v_2 = 30$ m/s

FIGURE 7–16 Example 7–8.

EXAMPLE 7–8 **ESTIMATE** **Work on a car, to increase its kinetic energy.** How much net work is required to accelerate a 1000-kg car from 20 m/s to 30 m/s (Fig. 7–16)?

APPROACH A car is a complex system. The engine turns the wheels and tires which push against the ground, and the ground pushes back (see Example 4–4). We aren't interested right now in those complications. Instead, we can get a useful result using the work-energy principle, but only if we model the car as a particle or simple rigid object.

SOLUTION The net work needed is equal to the increase in kinetic energy:

$$W = K_2 - K_1 = \tfrac{1}{2}mv_2^2 - \tfrac{1}{2}mv_1^2$$
$$= \tfrac{1}{2}(1000 \text{ kg})(30 \text{ m/s})^2 - \tfrac{1}{2}(1000 \text{ kg})(20 \text{ m/s})^2 = 2.5 \times 10^5 \text{ J}.$$

EXERCISE C (a) Make a guess: will the work needed to accelerate the car in Example 7–8 from rest to 20 m/s be more than, less than, or equal to the work already calculated to accelerate it from 20 m/s to 30 m/s? (b) Make the calculation.

$v_1 = 60$ km/h $v_2 = 0$

$\vec{\mathbf{F}}$

(a)

$\vec{\mathbf{d}}$ ($d = 20$ m)

FIGURE 7–17 Example 7–9.

$v_1 = 120$ km/h $v_2 = 0$

$\vec{\mathbf{F}}$

(b)

$\vec{\mathbf{d}}$ ($d = ?$)

CONCEPTUAL EXAMPLE 7–9 **Work to stop a car.** A car traveling 60 km/h can brake to a stop within a distance d of 20 m (Fig. 7–17a). If the car is going twice as fast, 120 km/h, what is its stopping distance (Fig. 7–17b)? Assume the maximum braking force is approximately independent of speed.

RESPONSE Again we model the car as if it were a particle. Because the net stopping force F is approximately constant, the work needed to stop the car, Fd, is proportional to the distance traveled. We apply the work-energy principle, noting that $\vec{\mathbf{F}}$ and $\vec{\mathbf{d}}$ are in opposite directions and that the final speed of the car is zero:

$$W_{\text{net}} = Fd \cos 180° = -Fd.$$

Then

$$-Fd = \Delta K = \tfrac{1}{2}mv_2^2 - \tfrac{1}{2}mv_1^2$$
$$= 0 - \tfrac{1}{2}mv_1^2.$$

Thus, since the force and mass are constant, we see that the stopping distance, d, increases with the square of the speed:

$$d \propto v^2.$$

PHYSICS APPLIED
Car's stopping distance $\propto$
initial speed squared

If the car's initial speed is doubled, the stopping distance is $(2)^2 = 4$ times as great, or 80 m.

EXERCISE D Can kinetic energy ever be negative?

EXERCISE E (a) If the kinetic energy of an arrow is doubled, by what factor has its speed increased? (b) If its speed is doubled, by what factor does its kinetic energy increase?

EXAMPLE 7–10 **A compressed spring.** A horizontal spring has spring constant $k = 360\,\text{N/m}$. (a) How much work is required to compress it from its uncompressed length $(x = 0)$ to $x = 11.0\,\text{cm}$? (b) If a 1.85-kg block is placed against the spring and the spring is released, what will be the speed of the block when it separates from the spring at $x = 0$? Ignore friction. (c) Repeat part (b) but assume that the block is moving on a table as in Fig. 7–18 and that some kind of constant drag force $F_D = 7.0\,\text{N}$ is acting to slow it down, such as friction (or perhaps your finger).

FIGURE 7–18 Example 7–10.

APPROACH We use our result from Section 7–3 that the net work, W, needed to stretch or compress a spring by a distance x is $W = \frac{1}{2}kx^2$. In (b) and (c) we use the work-energy principle.

SOLUTION (a) The work needed to compress the spring a distance $x = 0.110\,\text{m}$ is

$$W = \tfrac{1}{2}(360\,\text{N/m})(0.110\,\text{m})^2 = 2.18\,\text{J},$$

where we have converted all units to SI.

(b) In returning to its uncompressed length, the spring does 2.18 J of work on the block (same calculation as in part (a), only in reverse). According to the work-energy principle, the block acquires kinetic energy of 2.18 J. Since $K = \frac{1}{2}mv^2$, the block's speed must be

$$v = \sqrt{\frac{2K}{m}}$$

$$= \sqrt{\frac{2(2.18\,\text{J})}{1.85\,\text{kg}}} = 1.54\,\text{m/s}.$$

(c) There are two forces on the block: that exerted by the spring and that exerted by the drag force, $\vec{\mathbf{F}}_D$. Work done by a force such as friction is complicated. For one thing, heat (or, rather, "thermal energy") is produced—try rubbing your hands together. Nonetheless, the product $\vec{\mathbf{F}}_D \cdot \vec{\mathbf{d}}$ for the drag force, even when it is friction, can be used in the work-energy principle to give correct results for a particle-like object. The spring does 2.18 J of work on the block. The work done by the friction or drag force on the block, in the negative x direction, is

$$W_D = -F_D x = -(7.0\,\text{N})(0.110\,\text{m}) = -0.77\,\text{J}.$$

This work is negative because the drag force acts in the direction opposite to the displacement x. The net work done on the block is $W_{\text{net}} = 2.18\,\text{J} - 0.77\,\text{J} = 1.41\,\text{J}$. From the work-energy principle, Eq. 7–11 (with $v_2 = v$ and $v_1 = 0$), we have

$$v = \sqrt{\frac{2W_{\text{net}}}{m}}$$

$$= \sqrt{\frac{2(1.41\,\text{J})}{1.85\,\text{kg}}} = 1.23\,\text{m/s}$$

for the block's speed at the moment it separates from the spring $(x = 0)$.

General Derivation of the Work-Energy Principle

We derived the work-energy principle, Eq. 7–11, for motion in one dimension with a constant force. It is valid even if the force is variable and the motion is in two or three dimensions, as we now show. Suppose the net force $\vec{F}_{net}$ on a particle varies in both magnitude and direction, and the path of the particle is a curve as in Fig. 7–8. The net force may be considered to be a function of ℓ, the distance along the curve. The net work done is (Eq. 7–6):

$$W_{net} = \int \vec{F}_{net} \cdot d\vec{\ell} = \int F_{net} \cos\theta \, d\ell = \int F_{\parallel} \, d\ell,$$

where $F_{\parallel}$ represents the component of the net force parallel to the curve at any point. By Newton's second law,

$$F_{\parallel} = ma_{\parallel} = m\frac{dv}{dt},$$

where $a_{\parallel}$, the component of a parallel to the curve at any point, is equal to the rate of change of speed, dv/dt. We can think of v as a function of ℓ, and using the chain rule for derivatives, we have

$$\frac{dv}{dt} = \frac{dv}{d\ell}\frac{d\ell}{dt} = \frac{dv}{d\ell}v,$$

since $d\ell/dt$ is the speed v. Thus (letting 1 and 2 refer to the initial and final quantities, respectively):

$$W_{net} = \int_1^2 F_{\parallel} \, d\ell = \int_1^2 m\frac{dv}{dt} \, d\ell = \int_1^2 mv\frac{dv}{d\ell} \, d\ell = \int_1^2 mv \, dv,$$

which integrates to

$$W_{net} = \tfrac{1}{2}mv_2^2 - \tfrac{1}{2}mv_1^2 = \Delta K.$$

This is again the work-energy principle, which we have now derived for motion in three dimensions with a variable net force, using the definitions of work and kinetic energy plus Newton's second law.

Notice in this derivation that only the component of $\vec{F}_{net}$ parallel to the motion, $F_{\parallel}$, contributes to the work. Indeed, a force (or component of a force) acting perpendicular to the velocity vector does no work. Such a force changes only the direction of the velocity. It does not affect the magnitude of the velocity. One example of this is uniform circular motion in which an object moving with constant speed in a circle has a ("centripetal") force acting on it toward the center of the circle. This force does no work on the object, because (as we saw in Example 7–3) it is always perpendicular to the object's displacement $d\vec{\ell}$.

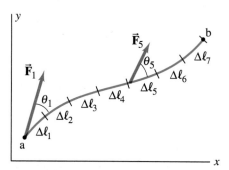

FIGURE 7–8 (repeated)
A particle acted on by a variable force $\vec{F}$, moves along the path shown from point a to point b.

Summary

Work is done on an object by a force when the object moves through a distance, d. The **work** W done by a constant force $\vec{F}$ on an object whose position changes by a displacement $\vec{d}$ is given by

$$W = Fd\cos\theta = \vec{F} \cdot \vec{d}, \qquad (7\text{–}1, 7\text{–}3)$$

where θ is the angle between $\vec{F}$ and $\vec{d}$.

The last expression is called the scalar product of $\vec{F}$ and $\vec{d}$. In general, the **scalar product** of any two vectors $\vec{A}$ and $\vec{B}$ is defined as

$$\vec{A} \cdot \vec{B} = AB\cos\theta \qquad (7\text{–}2)$$

where θ is the angle between $\vec{A}$ and $\vec{B}$. In rectangular coordinates we can also write

$$\vec{A} \cdot \vec{B} = A_x B_x + A_y B_y + A_z B_z. \qquad (7\text{–}4)$$

The work W done by a variable force $\vec{F}$ on an object that moves from point a to point b is

$$W = \int_a^b \vec{F} \cdot d\vec{\ell} = \int_a^b F\cos\theta \, d\ell, \qquad (7\text{–}7)$$

where $d\vec{\ell}$ represents an infinitesimal displacement along the path of the object and θ is the angle between $d\vec{\ell}$ and $\vec{F}$ at each point of the object's path.

The translational **kinetic energy**, K, of an object of mass m moving with speed v is defined to be

$$K = \tfrac{1}{2}mv^2. \qquad (7\text{–}10)$$

The **work-energy principle** states that the net work done on an object by the net resultant force is equal to the change in kinetic energy of the object:

$$W_{net} = \Delta K = \tfrac{1}{2}mv_2^2 - \tfrac{1}{2}mv_1^2. \qquad (7\text{–}11)$$

Questions

1. In what ways is the word "work" as used in everyday language the same as defined in physics? In what ways is it different? Give examples of both.
2. A woman swimming upstream is not moving with respect to the shore. Is she doing any work? If she stops swimming and merely floats, is work done on her?
3. Can a centripetal force ever do work on an object? Explain.
4. Why is it tiring to push hard against a solid wall even though you are doing no work?
5. Does the scalar product of two vectors depend on the choice of coordinate system?
6. Can a dot product ever be negative? If yes, under what conditions?
7. If $\vec{A} \cdot \vec{C} = \vec{B} \cdot \vec{C}$, is it necessarily true that $\vec{A} = \vec{B}$?
8. Does the dot product of two vectors have direction as well as magnitude?
9. Can the normal force on an object ever do work? Explain.
10. You have two springs that are identical except that spring 1 is stiffer than spring 2 $(k_1 > k_2)$. On which spring is more work done: (a) if they are stretched using the same force; (b) if they are stretched the same distance?
11. If the speed of a particle triples, by what factor does its kinetic energy increase?
12. In Example 7–10, it was stated that the block separates from the compressed spring when the spring reached its equilibrium length $(x = 0)$. Explain why separation doesn't take place before (or after) this point.
13. Two bullets are fired at the same time with the same kinetic energy. If one bullet has twice the mass of the other, which has the greater speed and by what factor? Which can do the most work?
14. Does the net work done on a particle depend on the choice of reference frame? How does this affect the work-energy principle?
15. A hand exerts a constant horizontal force on a block that is free to slide on a frictionless surface (Fig. 7–19). The block starts from rest at point A, and by the time it has traveled a distance d to point B it is traveling with speed v_B. When the block has traveled another distance d to point C, will its speed be greater than, less than, or equal to $2v_B$? Explain your reasoning.

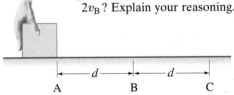

FIGURE 7–19
Question 15.

Problems

7–1 Work, Constant Force

1. (I) How much work is done by the gravitational force when a 280-kg pile driver falls 2.80 m?
2. (I) How high will a 1.85-kg rock go if thrown straight up by someone who does 80.0 J of work on it? Neglect air resistance.
3. (I) A 75.0-kg firefighter climbs a flight of stairs 20.0 m high. How much work is required?
4. (I) A hammerhead with a mass of 2.0 kg is allowed to fall onto a nail from a height of 0.50 m. What is the maximum amount of work it could do on the nail? Why do people not just "let it fall" but add their own force to the hammer as it falls?
5. (II) Estimate the work you do to mow a lawn 10 m by 20 m with a 50-cm wide mower. Assume you push with a force of about 15 N.
6. (II) A lever such as that shown in Fig. 7–20 can be used to lift objects we might not otherwise be able to lift. Show that the ratio of output force, F_O, to input force, F_I, is related to the lengths ℓ_I and ℓ_O from the pivot by $F_O/F_I = \ell_I/\ell_O$. Ignore friction and the mass of the lever, and assume the work output equals work input.

FIGURE 7–20
A lever. Problem 6.

7. (II) What is the minimum work needed to push a 950-kg car 310 m up along a 9.0° incline? Ignore friction.
8. (II) Eight books, each 4.0 cm thick with mass 1.8 kg, lie flat on a table. How much work is required to stack them one on top of another?
9. (II) A box of mass 6.0 kg is accelerated from rest by a force across a floor at a rate of 2.0 m/s^2 for 7.0 s. Find the net work done on the box.
10. (II) (a) What magnitude force is required to give a helicopter of mass M an acceleration of $0.10 g$ upward? (b) What work is done by this force as the helicopter moves a distance h upward?
11. (II) A 380-kg piano slides 3.9 m down a 27° incline and is kept from accelerating by a man who is pushing back on it *parallel to the incline* (Fig. 7–21). Determine: (a) the force exerted by the man, (b) the work done by the man on the piano, (c) the work done by the force of gravity, and (d) the net work done on the piano. Ignore friction.

FIGURE 7–21
Problem 11.

12. (II) A gondola can carry 20 skiers, with a total mass of up to 2250 kg. The gondola ascends at a constant speed from the base of a mountain, at 2150 m, to the summit at 3345 m. (a) How much work does the motor do in moving a full gondola up the mountain? (b) How much work does gravity do on the gondola? (c) If the motor is capable of generating 10% more work than found in (a), what is the acceleration of the gondola?

13. (II) A 17,000-kg jet takes off from an aircraft carrier via a cata-pult (Fig. 7–22a). The gases thrust out from the jet's engines exert a constant force of 130 kN on the jet; the force exerted on the jet by the catapult is plotted in Fig. 7–22b. Determine: (a) the work done on the jet by the gases expelled by its engines during launch of the jet; and (b) the work done on the jet by the catapult during launch of the jet.

(a)

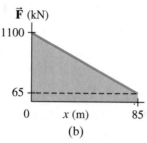
(b)

FIGURE 7–22   Problem 13.

14. (II) A 2200-N crate rests on the floor. How much work is required to move it at constant speed (a) 4.0 m along the floor against a drag force of 230 N, and (b) 4.0 m vertically?

15. (II) A grocery cart with mass of 16 kg is being pushed at constant speed up a flat 12° ramp by a force F_P which acts at an angle of 17° below the horizontal. Find the work done by each of the forces $(m\vec{g}, \vec{F}_N, \vec{F}_P)$ on the cart if the ramp is 15 m long.

7–2 Scalar Product

16. (I) What is the dot product of $\vec{A} = 2.0x^2\hat{i} - 4.0x\hat{j} + 5.0\hat{k}$ and $\vec{B} = 11.0\hat{i} + 2.5x\hat{j}$?

17. (I) For any vector $\vec{V} = V_x\hat{i} + V_y\hat{j} + V_z\hat{k}$ show that
$$V_x = \hat{i} \cdot \vec{V}, \qquad V_y = \hat{j} \cdot \vec{V}, \qquad V_z = \hat{k} \cdot \vec{V}.$$

18. (I) Calculate the angle between the vectors:
$$\vec{A} = 6.8\hat{i} - 3.4\hat{j} - 6.2\hat{k} \quad \text{and} \quad \vec{B} = 8.2\hat{i} + 2.3\hat{j} - 7.0\hat{k}.$$

19. (I) Show that $\vec{A} \cdot (-\vec{B}) = -\vec{A} \cdot \vec{B}$.

20. (I) Vector $\vec{V}_1$ points along the z axis and has magnitude $V_1 = 75$. Vector $\vec{V}_2$ lies in the xz plane, has magnitude $V_2 = 58$, and makes a $-48°$ angle with the x axis (points below x axis). What is the scalar product $\vec{V}_1 \cdot \vec{V}_2$?

21. (II) Given the vector $\vec{A} = 3.0\hat{i} + 1.5\hat{j}$, find a vector $\vec{B}$ that is perpendicular to $\vec{A}$.

22. (II) A constant force $\vec{F} = (2.0\hat{i} + 4.0\hat{j})$ N acts on an object as it moves along a straight-line path. If the object's displace-ment is $\vec{d} = (1.0\hat{i} + 5.0\hat{j})$ m, calculate the work done by $\vec{F}$ using these alternate ways of writing the dot product: (a) $W = Fd\cos\theta$; (b) $W = F_x d_x + F_y d_y$.

23. (II) If $\vec{A} = 9.0\hat{i} - 8.5\hat{j}$, $\vec{B} = -8.0\hat{i} + 7.1\hat{j} + 4.2\hat{k}$, and $\vec{C} = 6.8\hat{i} - 9.2\hat{j}$, determine (a) $\vec{A} \cdot (\vec{B} + \vec{C})$; (b) $(\vec{A} + \vec{C}) \cdot \vec{B}$; (c) $(\vec{B} + \vec{A}) \cdot \vec{C}$.

24. (II) Prove that $\vec{A} \cdot \vec{B} = A_x B_x + A_y B_y + A_z B_z$, starting from Eq. 7–2 and using the distributive property (p. 167, proved in Problem 33).

25. (II) Given vectors $\vec{A} = -4.8\hat{i} + 6.8\hat{j}$ and $\vec{B} = 9.6\hat{i} + 6.7\hat{j}$, determine the vector $\vec{C}$ that lies in the xy plane perpendic-ular to $\vec{B}$ and whose dot product with $\vec{A}$ is 20.0.

26. (II) Show that if two nonparallel vectors have the same magnitude, their sum must be perpendicular to their difference.

27. (II) Let $\vec{V} = 20.0\hat{i} + 22.0\hat{j} - 14.0\hat{k}$. What angles does this vector make with the x, y, and z axes?

28. (II) Use the scalar product to prove the *law of cosines* for a triangle:
$$c^2 = a^2 + b^2 - 2ab\cos\theta,$$
where a, b, and c are the lengths of the sides of a triangle and θ is the angle opposite side c.

29. (II) Vectors $\vec{A}$ and $\vec{B}$ are in the xy plane and their scalar product is 20.0 units. If $\vec{A}$ makes a 27.4° angle with the x axis and has magnitude $A = 12.0$ units, and $\vec{B}$ has magnitude $B = 24.0$ units, what can you say about the direction of $\vec{B}$?

30. (II) $\vec{A}$ and $\vec{B}$ are two vectors in the xy plane that make angles α and β with the x axis respectively. Evaluate the scalar product of $\vec{A}$ and $\vec{B}$ and deduce the following trigonometric identity: $\cos(\alpha - \beta) = \cos\alpha\cos\beta + \sin\alpha\sin\beta$.

31. (II) Suppose $\vec{A} = 1.0\hat{i} + 1.0\hat{j} - 2.0\hat{k}$ and $\vec{B} = -1.0\hat{i} + 1.0\hat{j} + 2.0\hat{k}$, (a) what is the angle between these two vectors? (b) Explain the significance of the sign in part (a).

32. (II) Find a vector of unit length in the xy plane that is perpendicular to $3.0\hat{i} + 4.0\hat{j}$.

33. (III) Show that the scalar product of two vectors is distributive: $\vec{A} \cdot (\vec{B} + \vec{C}) = \vec{A} \cdot \vec{B} + \vec{A} \cdot \vec{C}$. [*Hint*: Use a diagram showing all three vectors in a plane and indicate dot products on the diagram.]

7–3 Work, Varying Force

34. (I) In pedaling a bicycle uphill, a cyclist exerts a downward force of 450 N during each stroke. If the diameter of the circle traced by each pedal is 36 cm, calculate how much work is done in each stroke.

35. (II) A spring has $k = 65$ N/m. Draw a graph like that in Fig. 7–11 and use it to determine the work needed to stretch the spring from $x = 3.0$ cm to $x = 6.5$ cm, where $x = 0$ refers to the spring's unstretched length.

36. (II) If the hill in Example 7–2 (Fig. 7–4) was not an even slope but rather an irregular curve as in Fig. 7–23, show that the same result would be obtained as in Example 7–2: namely, that the work done by gravity depends only on the height of the hill and not on its shape or the path taken.

FIGURE 7–23
Problem 36.

37. (II) The net force exerted on a particle acts in the positive x direction. Its magnitude increases linearly from zero at $x = 0$, to 380 N at $x = 3.0$ m. It remains constant at 380 N from $x = 3.0$ m to $x = 7.0$ m, and then decreases linearly to zero at $x = 12.0$ m. Determine the work done to move the particle from $x = 0$ to $x = 12.0$ m graphically, by determining the area under the F_x versus x graph.

38. (II) If it requires 5.0 J of work to stretch a particular spring by 2.0 cm from its equilibrium length, how much more work will be required to stretch it an additional 4.0 cm?

39. (II) In Fig. 7–9 assume the distance axis is the x axis and that a = 10.0 m and b = 30.0 m. Estimate the work done by this force in moving a 3.50-kg object from a to b.

40. (II) The force on a particle, acting along the x axis, varies as shown in Fig. 7–24. Determine the work done by this force to move the particle along the x axis: (a) from $x = 0.0$ to $x = 10.0$ m; (b) from $x = 0.0$ to $x = 15.0$ m.

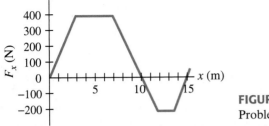

FIGURE 7–24
Problem 40.

41. (II) A child is pulling a wagon down the sidewalk. For 9.0 m the wagon stays on the sidewalk and the child pulls with a horizontal force of 22 N. Then one wheel of the wagon goes off on the grass so the child has to pull with a force of 38 N at an angle of 12° to the side for the next 5.0 m. Finally the wagon gets back on the sidewalk so the child makes the rest of the trip, 13.0 m, with a force of 22 N. How much total work did the child do on the wagon?

42. (II) The resistance of a packing material to a sharp object penetrating it is a force proportional to the fourth power of the penetration depth x; that is, $\vec{F} = kx^4\hat{i}$. Calculate the work done to force a sharp object a distance d into the material.

43. (II) The force needed to hold a particular spring compressed an amount x from its normal length is given by $F = kx + ax^3 + bx^4$. How much work must be done to compress it by an amount X, starting from $x = 0$?

44. (II) At the top of a pole vault, an athlete actually can do work pushing on the pole before releasing it. Suppose the pushing force that the pole exerts back on the athlete is given by $F(x) = (1.5 \times 10^2 \, \text{N/m})x - (1.9 \times 10^2 \, \text{N/m}^2)x^2$ acting over a distance of 0.20 m. How much work is done on the athlete?

45. (II) Consider a force $F_1 = A/\sqrt{x}$ which acts on an object during its journey along the x axis from $x = 0.0$ to $x = 1.0$ m, where $A = 2.0 \, \text{N} \cdot \text{m}^{1/2}$. Show that during this journey, even though F_1 is infinite at $x = 0.0$, the work done on the object by this force is finite.

46. (II) Assume that a force acting on an object is given by $\vec{F} = ax\hat{i} + by\hat{j}$, where the constants $a = 3.0 \, \text{N} \cdot \text{m}^{-1}$ and $b = 4.0 \, \text{N} \cdot \text{m}^{-1}$. Determine the work done on the object by this force as it moves in a straight line from the origin to $\vec{r} = (10.0\hat{i} + 20.0\hat{j})$ m.

47. (II) An object, moving along the circumference of a circle with radius R, is acted upon by a force of constant magnitude F. The force is directed at all times at a 30° angle with respect to the tangent to the circle as shown in Fig. 7–25. Determine the work done by this force when the object moves along the half circle from A to B.

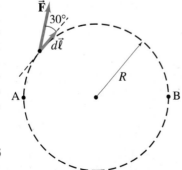

FIGURE 7–25
Problem 47.

48. (III) A 2800-kg space vehicle, initially at rest, falls vertically from a height of 3300 km above the Earth's surface. Determine how much work is done by the force of gravity in bringing the vehicle to the Earth's surface.

49. (III) A 3.0-m-long steel chain is stretched out along the top level of a horizontal scaffold at a construction site, in such a way that 2.0 m of the chain remains on the top level and 1.0 m hangs vertically, Fig. 7–26. At this point, the force on the hanging segment is sufficient to pull the entire chain over the edge. Once the chain is moving, the kinetic friction is so small that it can be neglected. How much work is performed on the chain by the force of gravity as the chain falls from the point where 2.0 m remains on the scaffold to the point where the entire chain has left the scaffold? (Assume that the chain has a linear weight density of 18 N/m.)

FIGURE 7–26
Problem 49.

7–4 Kinetic Energy; Work-Energy Principle

50. (I) At room temperature, an oxygen molecule, with mass of 5.31×10^{-26} kg, typically has a kinetic energy of about 6.21×10^{-21} J. How fast is it moving?

51. (I) (a) If the kinetic energy of a particle is tripled, by what factor has its speed increased? (b) If the speed of a particle is halved, by what factor does its kinetic energy change?

52. (I) How much work is required to stop an electron $(m = 9.11 \times 10^{-31}$ kg) which is moving with a speed of 1.40×10^6 m/s?

53. (I) How much work must be done to stop a 1300-kg car traveling at 95 km/h?

54. (II) Spiderman uses his spider webs to save a runaway train, Fig. 7–27. His web stretches a few city blocks before the 10^4-kg train comes to a stop. Assuming the web acts like a spring, estimate the spring constant.

FIGURE 7–27
Problem 54.

55. (II) A baseball $(m = 145 \, \text{g})$ traveling 32 m/s moves a fielder's glove backward 25 cm when the ball is caught. What was the average force exerted by the ball on the glove?

56. (II) An 85-g arrow is fired from a bow whose string exerts an average force of 105 N on the arrow over a distance of 75 cm. What is the speed of the arrow as it leaves the bow?

57. (II) A mass m is attached to a spring which is held stretched a distance x by a force F (Fig. 7–28), and then released. The spring compresses, pulling the mass. Assuming there is no friction, determine the speed of the mass m when the spring returns: (a) to its normal length ($x = 0$); (b) to half its original extension ($x/2$).

FIGURE 7–28
Problem 57.

58. (II) If the speed of a car is increased by 50%, by what factor will its minimum braking distance be increased, assuming all else is the same? Ignore the driver's reaction time.

59. (II) A 1200-kg car rolling on a horizontal surface has speed $v = 66$ km/h when it strikes a horizontal coiled spring and is brought to rest in a distance of 2.2 m. What is the spring constant of the spring?

60. (II) One car has twice the mass of a second car, but only half as much kinetic energy. When both cars increase their speed by 7.0 m/s, they then have the same kinetic energy. What were the original speeds of the two cars?

61. (II) A 4.5-kg object moving in two dimensions initially has a velocity $\vec{v}_1 = (10.0\hat{i} + 20.0\hat{j})$ m/s. A net force $\vec{F}$ then acts on the object for 2.0 s, after which the object's velocity is $\vec{v}_2 = (15.0\hat{i} + 30.0\hat{j})$ m/s. Determine the work done by $\vec{F}$ on the object.

62. (II) A 265-kg load is lifted 23.0 m vertically with an acceleration $a = 0.150\,g$ by a single cable. Determine (a) the tension in the cable; (b) the net work done on the load; (c) the work done by the cable on the load; (d) the work done by gravity on the load; (e) the final speed of the load assuming it started from rest.

63. (II) (a) How much work is done by the horizontal force $F_P = 150$ N on the 18-kg block of Fig. 7–29 when the force pushes the block 5.0 m up along the 32° frictionless incline? (b) How much work is done by the gravitational force on the block during this displacement? (c) How much work is done by the normal force? (d) What is the speed of the block (assume that it is zero initially) after this displacement? [Hint: Work-energy involves net work done.]

FIGURE 7–29
Problems 63 and 64.

64. (II) Repeat Problem 63 assuming a coefficient of friction $\mu_k = 0.10$.

65. (II) At an accident scene on a level road, investigators measure a car's skid mark to be 98 m long. It was a rainy day and the coefficient of friction was estimated to be 0.38. Use these data to determine the speed of the car when the driver slammed on (and locked) the brakes. (Why does the car's mass not matter?)

66. (II) A 46.0-kg crate, starting from rest, is pulled across a floor with a constant horizontal force of 225 N. For the first 11.0 m the floor is frictionless, and for the next 10.0 m the coefficient of friction is 0.20. What is the final speed of the crate after being pulled these 21.0 m?

67. (II) A train is moving along a track with constant speed v_1 relative to the ground. A person on the train holds a ball of mass m and throws it toward the front of the train with a speed v_2 relative to the train. Calculate the change in kinetic energy of the ball (a) in the Earth frame of reference, and (b) in the train frame of reference. (c) Relative to each frame of reference, how much work was done on the ball? (d) Explain why the results in part (b) are not the same for the two frames—after all, it's the same ball.

68. (III) We usually neglect the mass of a spring if it is small compared to the mass attached to it. But in some applications, the mass of the spring must be taken into account. Consider a spring of unstretched length ℓ and mass M_S uniformly distributed along the length of the spring. A mass m is attached to the end of the spring. One end of the spring is fixed and the mass m is allowed to vibrate horizontally without friction (Fig. 7–30). Each point on the spring moves with a velocity proportional to the distance from that point to the fixed end. For example, if the mass on the end moves with speed v_0, the midpoint of the spring moves with speed $v_0/2$. Show that the kinetic energy of the mass plus spring when the mass is moving with velocity v is

$$K = \tfrac{1}{2}Mv^2$$

where $M = m + \tfrac{1}{3}M_S$ is the "effective mass" of the system. [Hint: Let D be the total length of the stretched spring. Then the velocity of a mass dm of a spring of length dx located at x is $v(x) = v_0(x/D)$. Note also that $dm = dx(M_S/D)$.]

FIGURE 7–30
Problem 68.

69. (III) An elevator cable breaks when a 925-kg elevator is 22.5 m above the top of a huge spring $\left(k = 8.00 \times 10^4 \text{ N/m}\right)$ at the bottom of the shaft. Calculate (a) the work done by gravity on the elevator before it hits the spring; (b) the speed of the elevator just before striking the spring; (c) the amount the spring compresses (note that here work is done by both the spring and gravity).

General Problems

70. (a) A 3.0-g locust reaches a speed of 3.0 m/s during its jump. What is its kinetic energy at this speed? (b) If the locust transforms energy with 35% efficiency, how much energy is required for the jump?

71. In a certain library the first shelf is 12.0 cm off the ground, and the remaining 4 shelves are each spaced 33.0 cm above the previous one. If the average book has a mass of 1.40 kg with a height of 22.0 cm, and an average shelf holds 28 books (standing vertically), how much work is required to fill all the shelves, assuming the books are all laying flat on the floor to start?

72. A 75-kg meteorite buries itself 5.0 m into soft mud. The force between the meteorite and the mud is given by $F(x) = (640 \text{ N/m}^3)x^3$, where x is the depth in the mud. What was the speed of the meteorite when it initially impacted the mud?

73. A 6.10-kg block is pushed 9.25 m up a smooth 37.0° inclined plane by a horizontal force of 75.0 N. If the initial speed of the block is 3.25 m/s up the plane, calculate (a) the initial kinetic energy of the block; (b) the work done by the 75.0-N force; (c) the work done by gravity; (d) the work done by the normal force; (e) the final kinetic energy of the block.

74. The arrangement of atoms in zinc is an example of "hexagonal close-packed" structure. Three of the nearest neighbors are found at the following (x, y, z) coordinates, given in nanometers $(10^{-9}\,\text{m})$: atom 1 is at $(0, 0, 0)$; atom 2 is at $(0.230, 0.133, 0)$; atom 3 is at $(0.077, 0.133, 0.247)$. Find the angle between two vectors: one that connects atom 1 with atom 2 and another that connects atom 1 with atom 3.

75. Two forces, $\vec{F}_1 = (1.50\hat{i} - 0.80\hat{j} + 0.70\hat{k})\,\text{N}$ and $\vec{F}_2 = (-0.70\hat{i} + 1.20\hat{j})\,\text{N}$, are applied on a moving object of mass 0.20 kg. The displacement vector produced by the two forces is $\vec{d} = (8.0\hat{i} + 6.0\hat{j} + 5.0\hat{k})\,\text{m}$. What is the work done by the two forces?

76. The barrels of the 16-in. guns (bore diameter = 16 in. = 41 cm) on the World War II battleship *U.S.S. Massachusetts* were each 15 m long. The shells each had a mass of 1250 kg and were fired with sufficient explosive force to provide them with a muzzle velocity of 750 m/s. Use the work-energy principle to determine the explosive force (assumed to be a constant) that was applied to the shell within the barrel of the gun. Express your answer in both newtons and in pounds.

77. A varying force is given by $F = Ae^{-kx}$, where x is the position; A and k are constants that have units of N and m^{-1}, respectively. What is the work done when x goes from 0.10 m to infinity?

78. The force required to compress an imperfect horizontal spring an amount x is given by $F = 150x + 12x^3$, where x is in meters and F in newtons. If the spring is compressed 2.0 m, what speed will it give to a 3.0-kg ball held against it and then released?

79. A force $\vec{F} = (10.0\hat{i} + 9.0\hat{j} + 12.0\hat{k})\,\text{kN}$ acts on a small object of mass 95 g. If the displacement of the object is $\vec{d} = (5.0\hat{i} + 4.0\hat{j})\,\text{m}$, find the work done by the force. What is the angle between $\vec{F}$ and $\vec{d}$?

80. In the game of paintball, players use guns powered by pressurized gas to propel 33-g gel capsules filled with paint at the opposing team. Game rules dictate that a paintball cannot leave the barrel of a gun with a speed greater than 85 m/s. Model the shot by assuming the pressurized gas applies a constant force F to a 33-g capsule over the length of the 32-cm barrel. Determine F (a) using the work-energy principle, and (b) using the kinematic equations (Eqs. 2–12) and Newton's second law.

81. A softball having a mass of 0.25 kg is pitched horizontally at 110 km/h. By the time it reaches the plate, it may have slowed by 10%. Neglecting gravity, estimate the average force of air resistance during a pitch, if the distance between the plate and the pitcher is about 15 m.

82. An airplane pilot fell 370 m after jumping from an aircraft without his parachute opening. He landed in a snowbank, creating a crater 1.1 m deep, but survived with only minor injuries. Assuming the pilot's mass was 88 kg and his terminal velocity was 45 m/s, estimate: (a) the work done by the snow in bringing him to rest; (b) the average force exerted on him by the snow to stop him; and (c) the work done on him by air resistance as he fell. Model him as a particle.

83. Many cars have "5 mi/h $(8\,\text{km/h})$ bumpers" that are designed to compress and rebound elastically without any physical damage at speeds below 8 km/h. If the material of the bumpers permanently deforms after a compression of 1.5 cm, but remains like an elastic spring up to that point, what must be the effective spring constant of the bumper material, assuming the car has a mass of 1050 kg and is tested by ramming into a solid wall?

84. What should be the spring constant k of a spring designed to bring a 1300-kg car to rest from a speed of 90 km/h so that the occupants undergo a maximum acceleration of $5.0\,g$?

85. Assume a cyclist of weight mg can exert a force on the pedals equal to $0.90\,mg$ on the average. If the pedals rotate in a circle of radius 18 cm, the wheels have a radius of 34 cm, and the front and back sprockets on which the chain runs have 42 and 19 teeth respectively (Fig. 7–31), determine the maximum steepness of hill the cyclist can climb at constant speed. Assume the mass of the bike is 12 kg and that of the rider is 65 kg. Ignore friction. Assume the cyclist's average force is always: (a) downward; (b) tangential to pedal motion.

FIGURE 7–31   Problem 85.

86. A simple pendulum consists of a small object of mass m (the "bob") suspended by a cord of length ℓ (Fig. 7–32) of negligible mass. A force $\vec{F}$ is applied in the horizontal direction (so $\vec{F} = F\hat{i}$), moving the bob very slowly so the acceleration is essentially zero. (Note that the magnitude of $\vec{F}$ will need to vary with the angle θ that the cord makes with the vertical at any moment.) (a) Determine the work done by this force, $\vec{F}$, to move the pendulum from $\theta = 0$ to $\theta = \theta_0$. (b) Determine the work done by the gravitational force on the bob, $\vec{F}_G = m\vec{g}$, and the work done by the force $\vec{F}_T$ that the cord exerts on the bob.

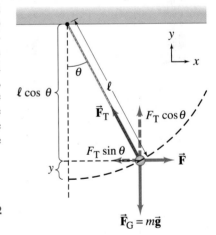

FIGURE 7–32
Problem 86.

87. A car passenger buckles himself in with a seat belt and holds his 18-kg toddler on his lap. Use the work-energy principle to answer the following questions. (a) While traveling 25 m/s, the driver has to make an emergency stop over a distance of 45 m. Assuming constant deceleration, how much force will the arms of the parent need to exert on the child during this deceleration period? Is this force achievable by an average parent? (b) Now assume that the car $(v = 25\,\text{m/s})$ is in an accident and is brought to stop over a distance of 12 m. Assuming constant deceleration, how much force will the parent need to exert on the child? Is this force achievable by an average parent?

88. As an object moves along the x axis from $x = 0.0$ m to $x = 20.0$ m it is acted upon by a force given by $F = (100 - (x - 10)^2)$ N. Determine the work done by the force on the object: (*a*) by first sketching the F vs. x graph and estimating the area under this curve; (*b*) by evaluating the integral $\int_{x=0.0\,\text{m}}^{x=20\,\text{m}} F \, dx$.

89. A cyclist starts from rest and coasts down a 4.0° hill. The mass of the cyclist plus bicycle is 85 kg. After the cyclist has traveled 250 m, (*a*) what was the net work done by gravity on the cyclist? (*b*) How fast is the cyclist going? Ignore air resistance.

90. Stretchable ropes are used to safely arrest the fall of rock climbers. Suppose one end of a rope with unstretched length ℓ is anchored to a cliff and a climber of mass m is attached to the other end. When the climber is a height ℓ above the anchor point, he slips and falls under the influence of gravity for a distance 2ℓ, after which the rope becomes taut and stretches a distance x as it stops the climber (see Fig. 7–33). Assume a stretchy rope behaves as a spring with spring constant k. (*a*) Applying the work-energy principle, show that

$$ x = \frac{mg}{k}\left[1 + \sqrt{1 + \frac{4k\ell}{mg}}\right]. $$

(*b*) Assuming $m = 85$ kg, $\ell = 8.0$ m and $k = 850$ N/m, determine x/ℓ (the fractional stretch of the rope) and kx/mg (the force that the rope exerts on the climber compared to his own weight) at the moment the climber's fall has been stopped.

FIGURE 7–33
Problem 90.

91. A small mass m hangs at rest from a vertical rope of length ℓ that is fixed to the ceiling. A force $\vec{F}$ then pushes on the mass, perpendicular to the taut rope at all times, until the rope is oriented at an angle $\theta = \theta_0$ and the mass has been raised by a vertical distance h (Fig. 7–34). Assume the force's magnitude F is adjusted so that the mass moves at constant speed along its curved trajectory. Show that the work done by $\vec{F}$ during this process equals mgh, which is equivalent to the amount of work it takes to slowly lift a mass m straight up by a height h. [*Hint*: When the angle is increased by $d\theta$ (in radians), the mass moves along an arc length $ds = \ell \, d\theta$.]

FIGURE 7–34
Problem 91.

*Numerical/Computer

*92. (II) The net force along the linear path of a particle of mass 480 g has been measured at 10.0-cm intervals, starting at $x = 0.0$, to be 26.0, 28.5, 28.8, 29.6, 32.8, 40.1, 46.6, 42.2, 48.8, 52.6, 55.8, 60.2, 60.6, 58.2, 53.7, 50.3, 45.6, 45.2, 43.2, 38.9, 35.1, 30.8, 27.2, 21.0, 22.2, and 18.6, all in newtons. Determine the total work done on the particle over this entire range.

*93. (II) When different masses are suspended from a spring, the spring stretches by different amounts as shown in the Table below. Masses are ± 1.0 gram.

Mass (g)	0	50	100	150	200	250	300	350	400
Stretch (cm)	0	5.0	9.8	14.8	19.4	24.5	29.6	34.1	39.2

(*a*) Graph the applied force (in Newtons) versus the stretch (in meters) of the spring, and determine the best-fit straight line. (*b*) Determine the spring constant (N/m) of the spring from the slope of the best-fit line. (*c*) If the spring is stretched by 20.0 cm, estimate the force acting on the spring using the best-fit line.

Answers to Exercises

A: (*c*).

B: (*b*).

C: (*b*) 2.0×10^5 J (i.e., less).

D: No, because the speed v would be the square root of a negative number, which is not real.

E: (*a*) $\sqrt{2}$, (*b*) 4.

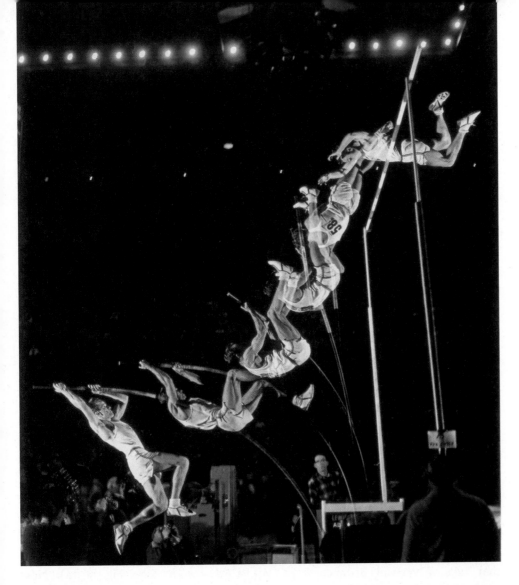

A polevaulter running toward the high bar has kinetic energy. When he plants the pole and puts his weight on it, his kinetic energy gets transformed: first into elastic potential energy of the bent pole and then into gravitational potential energy as his body rises. As he crosses the bar, the pole is straight and has given up all its elastic potential energy to the athlete's gravitational potential energy. Nearly all his kinetic energy has disappeared, also becoming gravitational potential energy of his body at the great height of the bar (world record over 6 m), which is exactly what he wants. In these, and all other energy transformations that continually take place in the world, the total energy is always conserved. Indeed, the conservation of energy is one of the greatest laws of physics, and finds applications in a wide range of other fields.

CHAPTER

8

Conservation of Energy

CHAPTER-OPENING QUESTION—Guess now!

A skier starts at the top of a hill. On which run does her gravitational potential energy change the most: **(a)**, **(b)**, **(c)**, or **(d)**; or are they **(e)** all the same? On which run would her speed at the bottom be the fastest if the runs are icy and we assume no friction? Recognizing that there is always some friction, answer the above two questions again. List your four answers now.

○ Easy
■ Intermediate
♦ Difficult
♦♦ Very difficult

CONTENTS

183

This chapter continues the discussion of the concepts of work and energy begun in Chapter 7 and introduces additional types of energy, in particular potential energy. Now we will see why the concept of energy is so important. The reason, ultimately, is that energy is conserved—the total energy *always* remains constant in any process. That a quantity can be defined which remains constant, as far as our best experiments can tell, is a remarkable statement about nature. The law of conservation of energy is, in fact, one of the great unifying principles of science.

The law of conservation of energy also gives us another tool, another approach, to solving problems. There are many situations for which an analysis based on Newton's laws would be difficult or impossible—the forces may not be known or accessible to measurement. But often these situations can be dealt with using the law of conservation of energy.

In this Chapter we will mainly treat objects as if they were particles or rigid objects that undergo only translational motion, with no internal or rotational motion.

8–1 Conservative and Nonconservative Forces

We will find it important to categorize forces into two types: conservative and nonconservative. By definition, we call any force a **conservative force** if

> the work done by the force on an object moving from one point to another depends only on the initial and final positions of the object, and is independent of the particular path taken.

A conservative force can be a function *only of position*, and cannot depend on other variables like time or velocity.

We can readily show that the force of gravity is a conservative force. The gravitational force on an object of mass m near the Earth's surface is $\vec{\mathbf{F}} = m\vec{\mathbf{g}}$, where $\vec{\mathbf{g}}$ is a constant. The work done by this gravitational force on an object that falls a vertical distance h is $W_G = Fd = mgh$ (see Fig. 8–1a). Now suppose that instead of moving vertically downward or upward, an object follows some arbitrary path in the xy plane, as shown in Fig. 8–1b. The object starts at a vertical height y_1 and reaches a height y_2, where $y_2 - y_1 = h$. To calculate the work done by gravity, W_G, we use Eq. 7–7:

$$W_G = \int_1^2 \vec{\mathbf{F}}_G \cdot d\vec{\boldsymbol{\ell}}$$

$$= \int_1^2 mg \cos\theta \, d\ell.$$

We now let $\phi = 180° - \theta$ be the angle between $d\vec{\boldsymbol{\ell}}$ and its vertical component dy, as shown in Fig. 8–1b. Then, since $\cos\theta = -\cos\phi$ and $dy = d\ell \cos\phi$, we have

$$W_G = -\int_{y_1}^{y_2} mg \, dy$$

$$= -mg(y_2 - y_1). \tag{8-1}$$

Since $(y_2 - y_1)$ is the vertical height h, we see that the work done depends only on the vertical height and does *not* depend on the particular path taken! Hence, by definition, gravity is a conservative force.

Note that in the case shown in Fig. 8–1b, $y_2 > y_1$ and therefore the work done by gravity is negative. If on the other hand $y_2 < y_1$, so that the object is falling, then W_G is positive.

FIGURE 8–1 Object of mass m: (a) falls a height h vertically; (b) is raised along an arbitrary two-dimensional path.

(a)

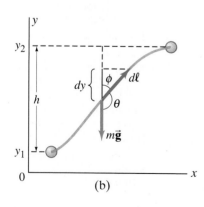

(b)

We can give the definition of a conservative force in another, completely equivalent way:

a force is conservative if the net work done by the force on an object moving around any closed path is zero.

To see why this is equivalent to our earlier definition, consider a small object that moves from point 1 to point 2 via either of two paths labeled A and B in Fig. 8–2a. If we assume a conservative force acts on the object, the work done by this force is the same whether the object takes path A or path B, by our first definition. This work to get from point 1 to point 2 we will call W. Now consider the round trip shown in Fig. 8–2b. The object moves from 1 to 2 via path A and our force does work W. Our object then returns to point 1 via path B. How much work is done during the return? In going from 1 to 2 via path B the work done is W, which by definition equals $\int_1^2 \vec{F} \cdot d\vec{\ell}$. In doing the reverse, going from 2 to 1, the force $\vec{F}$ at each point is the same, but $d\vec{\ell}$ is directed in precisely the opposite direction. Consequently $\vec{F} \cdot d\vec{\ell}$ has the opposite sign at each point so the total work done in making the return trip from 2 to 1 must be $-W$. Hence the total work done in going from 1 to 2 and back to 1 is $W + (-W) = 0$, which proves the equivalence of the two above definitions for a conservative force.

The second definition of a conservative force illuminates an important aspect of such a force: the *work done by a conservative force is recoverable* in the sense that if positive work is done *by* an object (on something else) on one part of a closed path, an equivalent amount of negative work will be done by the object on its return.

As we saw above, the force of gravity is conservative, and it is easy to show that the elastic force $(F = -kx)$ is also conservative.

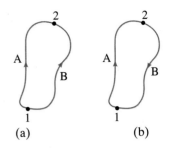

FIGURE 8–2 (a) A tiny object moves between points 1 and 2 via two different paths, A and B. (b) The object makes a round trip, via path A from point 1 to point 2 and via path B back to point 1.

FIGURE 8–3 A crate is pushed at constant speed across a rough floor from position 1 to position 2 via two paths, one straight and one curved. The pushing force $\vec{F}_P$ is always in the direction of motion. (The friction force opposes the motion.) Hence for a constant magnitude pushing force, the work it does is $W = F_P d$, so if d is greater (as for the curved path), then W is greater. The work done does not depend only on points 1 and 2; it also depends on the path taken.

Many forces, such as friction and a push or pull exerted by a person, are **nonconservative forces** since any work they do depends on the path. For example, if you push a crate across a floor from one point to another, the work you do depends on whether the path taken is straight, or is curved. As shown in Fig. 8–3, if a crate is pushed from point 1 to point 2 along the longer semicircular path, you do more work against friction than if you push it along the straight path. This is because the distance is greater and, unlike the gravitational force, the pushing force $\vec{F}_P$ is always in the direction of motion. Thus the work done by the person in Fig. 8–3 does not depend *only* on points 1 and 2; it depends also on the path taken. The force of kinetic friction, also shown in Fig. 8–3, always opposes the motion; it too is a nonconservative force, and we discuss how to treat it later in this Chapter (Section 8–6). Table 8–1 lists a few conservative and nonconservative forces.

TABLE 8–1 Conservative and Nonconservative Forces

Conservative Forces	Nonconservative Forces
Gravitational	Friction
Elastic	Air resistance
Electric	Tension in cord
	Motor or rocket propulsion
	Push or pull by a person

8–2 Potential Energy

In Chapter 7 we discussed the energy associated with a moving object, which is its kinetic energy $K = \frac{1}{2}mv^2$. Now we introduce **potential energy**, which is the energy associated with forces that depend on the position or configuration of objects relative to the surroundings. Various types of potential energy can be defined, and each type is associated with a particular conservative force.

The wound-up spring of a toy is an example of potential energy. The spring acquired its potential energy because work was done *on* it by the person winding the toy. As the spring unwinds, it exerts a force and does work to make the toy move.

Gravitational Potential Energy

FIGURE 8–4 A person exerts an upward force $F_{\text{ext}} = mg$ to lift a brick from y_1 to y_2.

Perhaps the most common example of potential energy is *gravitational potential energy*. A heavy brick held above the ground has potential energy because of its position relative to the Earth. The raised brick has the ability to do work, for if it is released, it will fall to the ground due to the gravitational force, and can do work on a stake, driving it into the ground. Let us seek the form for the gravitational potential energy of an object near the surface of the Earth. For an object of mass m to be lifted vertically, an upward force at least equal to its weight, mg, must be exerted on it, say by a person's hand. To lift it without acceleration a vertical displacement of height h, from position y_1 to y_2 in Fig. 8–4 (upward direction chosen positive), a person must do work equal to the product of the "external" force she exerts, $F_{\text{ext}} = mg$ upward, times the vertical displacement h. That is,

$$W_{\text{ext}} = \vec{\mathbf{F}}_{\text{ext}} \cdot \vec{\mathbf{d}} = mgh \cos 0° = mgh = mg(y_2 - y_1)$$

where both $\vec{\mathbf{F}}_{\text{ext}}$ and $\vec{\mathbf{d}}$ point upward. Gravity is also acting on the object as it moves from y_1 to y_2, and does work on the object equal to

$$W_G = \vec{\mathbf{F}}_G \cdot \vec{\mathbf{d}} = mgh \cos 180° = -mgh = -mg(y_2 - y_1),$$

where $\theta = 180°$ because $\vec{\mathbf{F}}_G$ and $\vec{\mathbf{d}}$ point in opposite directions. Since $\vec{\mathbf{F}}_G$ is downward and $\vec{\mathbf{d}}$ is upward, W_G is negative. If the object follows an arbitrary path, as in Fig. 8–1b, the work done by gravity still depends only on the change in vertical height (Eq. 8–1): $W_G = -mg(y_2 - y_1) = -mgh$.

Next, if we allow the object to start from rest and fall freely under the action of gravity, it acquires a velocity given by $v^2 = 2gh$ (Eq. 2–12c) after falling a height h. It then has kinetic energy $\frac{1}{2}mv^2 = \frac{1}{2}m(2gh) = mgh$, and if it strikes a stake it can do work on the stake equal to mgh.

To summarize, raising an object of mass m to a height h *requires* an amount of work equal to mgh. And once at height h, the object has the *ability* to do an amount of work equal to mgh. Thus we can say that the work done in lifting the object has been stored as gravitational potential energy.

Indeed, we can define the *change in gravitational potential energy U*, when an object moves from a height y_1 to a height y_2, as equal to the work done by a net external force to accomplish this without acceleration:

$$\Delta U = U_2 - U_1 = W_{\text{ext}} = mg(y_2 - y_1).$$

Equivalently, we can define the change in gravitational potential energy as equal to the negative of the work done by gravity itself in the process:

$$\Delta U = U_2 - U_1 = -W_G = mg(y_2 - y_1). \tag{8–2}$$

Equation 8–2 defines the change in gravitational potential energy when an object of mass m moves between two points near the surface of the Earth.[†] The gravitational potential energy, U, at any point a vertical height y above some reference point (the origin of the coordinate system) can be defined as

$$U_{\text{grav}} = mgy. \qquad\qquad \text{[gravity only]} \tag{8–3}$$

Note that the potential energy is associated with the force of gravity between the Earth and the mass m. Hence U_{grav} represents the gravitational potential energy, not simply of the mass m alone, but of the mass–Earth system.

†Section 8–7 deals with the $1/r^2$ dependence of Newton's law of universal gravitation.

Gravitational potential energy depends on the *vertical height* of the object *above some reference level*, $U = mgy$. Sometimes you may wonder from what point to measure y. The gravitational potential energy of a book held high above a table, for example, depends on whether we measure y from the top of the table, from the floor, or from some other reference point. What is physically important in any situation is the *change* in potential energy, ΔU, because that is what is related to the work done, and it is ΔU that can be measured. We can thus choose to measure y from any reference point that is convenient, but we must choose the reference point at the start and be consistent throughout any given calculation. The *change* in potential energy between any two points does not depend on this choice.

Potential energy belongs to a system, and not to a single object alone. Potential energy is associated with a force, and a force on one object is always exerted by some other object. Thus potential energy is a property of the system as a whole. For an object raised to a height y above the Earth's surface, the change in gravitational potential energy is mgy. The system here is the object plus the Earth, and properties of both are involved: object (m) and Earth (g). In general, a *system* is one or more objects that we choose to study. The choice of what makes up a system is always ours, and we often try to choose a simple system. Below, when we deal with the potential energy of an object in contact with a spring, our system will be the object and the spring.

EXERCISE A Return to the Chapter-Opening Question, page 183, and answer it again now. Try to explain why you may have answered differently the first time.

EXAMPLE 8–1 **Potential energy changes for a roller coaster.** A 1000-kg roller-coaster car moves from point 1, Fig. 8–5, to point 2 and then to point 3. (a) What is the gravitational potential energy at points 2 and 3 relative to point 1? That is, take $y = 0$ at point 1. (b) What is the change in potential energy when the car goes from point 2 to point 3? (c) Repeat parts (a) and (b), but take the reference point ($y = 0$) to be at point 3.

APPROACH We are interested in the potential energy of the car–Earth system. We take upward as the positive y direction, and use the definition of gravitational potential energy to calculate the potential energy.

SOLUTION (a) We measure heights from point 1 ($y_1 = 0$), which means initially that the gravitational potential energy is zero. At point 2, where $y_2 = 10\,\text{m}$,

$$U_2 = mgy_2 = (1000\,\text{kg})(9.8\,\text{m/s}^2)(10\,\text{m}) = 9.8 \times 10^4\,\text{J}.$$

At point 3, $y_3 = -15\,\text{m}$, since point 3 is below point 1. Therefore,

$$U_3 = mgy_3 = (1000\,\text{kg})(9.8\,\text{m/s}^2)(-15\,\text{m}) = -1.5 \times 10^5\,\text{J}.$$

(b) In going from point 2 to point 3, the potential energy change ($U_{\text{final}} - U_{\text{initial}}$) is

$$U_3 - U_2 = (-1.5 \times 10^5\,\text{J}) - (9.8 \times 10^4\,\text{J}) = -2.5 \times 10^5\,\text{J}.$$

The gravitational potential energy decreases by $2.5 \times 10^5\,\text{J}$.

(c) Now we set $y_3 = 0$. Then $y_1 = +15\,\text{m}$ at point 1, so the potential energy initially (at point 1) is

$$U_1 = (1000\,\text{kg})(9.8\,\text{m/s}^2)(15\,\text{m}) = 1.5 \times 10^5\,\text{J}.$$

At point 2, $y_2 = 25\,\text{m}$, so the potential energy is

$$U_2 = 2.5 \times 10^5\,\text{J}.$$

At point 3, $y_3 = 0$, so the potential energy is zero. The change in potential energy going from point 2 to point 3 is

$$U_3 - U_2 = 0 - 2.5 \times 10^5\,\text{J} = -2.5 \times 10^5\,\text{J},$$

which is the same as in part (b).

NOTE Work done by gravity depends only on the vertical height, so changes in gravitational potential energy do not depend on the path taken.

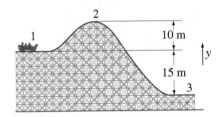

FIGURE 8–5 Example 8–1.

EXERCISE B By how much does the potential energy change when a 1200-kg car climbs to the top of a 300-m-tall hill? (a) $3.6 \times 10^5\,\text{J}$; (b) $3.5 \times 10^6\,\text{J}$; (c) 4 J; (d) 40 J; (e) 39.2 J.

Potential Energy in General

We have defined the change in gravitational potential energy (Eq. 8–2) to be equal to the negative of the work done by gravity when the object moves from height y_1 to y_2, which we now write as

$$\Delta U = -W_G = -\int_1^2 \vec{F}_G \cdot d\vec{\ell}.$$

There are other types of potential energy besides gravitational. In general, we define the *change in potential energy associated with a particular conservative force $\vec{F}$ as the negative of the work done by that force*:

$$\Delta U = U_2 - U_1 = -\int_1^2 \vec{F} \cdot d\vec{\ell} = -W. \tag{8–4}$$

⚠️ **CAUTION**

Potential energy can be defined only for conservative forces

However, we cannot use this definition to define a potential energy for all possible forces. It makes sense only for conservative forces such as gravity, for which the integral depends only on the end points and not on the path taken. It does not apply to nonconservative forces like friction, because the integral in Eq. 8–4 would *not* have a unique value depending on the end points 1 and 2. Thus the concept of potential energy cannot be defined and is meaningless for a nonconservative force.

Elastic Potential Energy

We consider now potential energy associated with elastic materials, which includes a great variety of practical applications.

Consider a simple coil spring as shown in Fig. 8–6, whose mass is so small that we can ignore it. When the spring is compressed and then released, it can do work on a ball (mass m). Thus the spring–ball system has potential energy when compressed (or stretched). Like other elastic materials, a spring is described by Hooke's law (see Section 7–3) as long as the displacement x is not too great. Let us take our coordinate system so the end of the uncompressed spring is at $x = 0$ (Fig. 8–6a) and x is positive to the right. To hold the spring compressed (or stretched) a distance x from its natural (unstretched) length requires the person's hand to exert a force $F_P = kx$ on the spring (Fig. 8–6b), where k is the spring stiffness constant. The spring pushes back with a force (Newton's third law),

$$F_S = -kx,$$

Fig. 8–6c. The negative sign appears because the force $\vec{F}_S$ is in the direction opposite to the displacement x. From Eq. 8–4, the change in potential energy when the spring is compressed or stretched from $x_1 = 0$ (its uncompressed position) to $x_2 = x$ (where x can be $+$ or $-$) is

$$\Delta U = U(x) - U(0) = -\int_1^2 \vec{F}_S \cdot d\vec{\ell} = -\int_0^x (-kx)\, dx = \tfrac{1}{2}kx^2.$$

Here, $U(x)$ means the potential energy at x, and $U(0)$ means U at $x = 0$. It is usually convenient to choose the potential energy at $x = 0$ to be zero: $U(0) = 0$, so the potential energy of a spring compressed or stretched an amount x from equilibrium is

$$U_{el}(x) = \tfrac{1}{2}kx^2. \qquad \text{[elastic spring]} \tag{8–5}$$

FIGURE 8–6 A spring (a) can store energy (elastic potential energy) when compressed (b), which can be used to do work when released (c) and (d).

(a)

(b)

(c)

(d)

Potential Energy Related to Force (1–D)

In the one-dimensional case, where a conservative force can be written as a function of x, say, the potential energy can be written as an indefinite integral

$$U(x) = -\int F(x)\, dx + C, \tag{8–6}$$

where the constant C represents the value of U at $x = 0$; we can sometimes

188 CHAPTER 8 Conservation of Energy

choose $C = 0$. Equation 8–6 tells us how to obtain $U(x)$ when given $F(x)$. If, instead, we are given $U(x)$, we can obtain $F(x)$ by inverting the above equation: that is, we take the derivative of both sides, remembering that integration and differentiation are inverse operations:

$$\frac{d}{dx} \int F(x) \, dx = F(x).$$

Thus

$$F(x) = -\frac{dU(x)}{dx}. \qquad (8\text{–}7)$$

EXAMPLE 8–2 **Determine F from U.** Suppose $U(x) = -ax/(b^2 + x^2)$, where a and b are constants. What is F as a function of x?

APPROACH Since $U(x)$ depends only on x, this is a one-dimensional problem.
SOLUTION Equation 8–7 gives

$$F(x) = -\frac{dU}{dx} = -\frac{d}{dx}\left[-\frac{ax}{b^2 + x^2}\right] = \frac{a}{b^2 + x^2} - \frac{ax}{(b^2 + x^2)^2} 2x = \frac{a(b^2 - x^2)}{(b^2 + x^2)^2}.$$

*Potential Energy in Three Dimensions

In three dimensions, we can write the relation between $\vec{F}(x, y, z)$ and U as:

$$F_x = -\frac{\partial U}{\partial x}, \qquad F_y = -\frac{\partial U}{\partial y}, \qquad F_z = -\frac{\partial U}{\partial z},$$

or

$$\vec{F}(x, y, z) = -\hat{i}\frac{\partial U}{\partial x} - \hat{j}\frac{\partial U}{\partial y} - \hat{k}\frac{\partial U}{\partial z}.$$

Here, $\partial/\partial x$, $\partial/\partial y$ and $\partial/\partial z$ are called partial derivatives; $\partial/\partial x$, for example, means that although U may be a function of x, y, and z, written $U(x, y, z)$, we take the derivative only with respect to x with the other variables held constant.

8–3 Mechanical Energy and Its Conservation

Let us consider a conservative system (meaning only conservative forces do work) in which energy is transformed from kinetic to potential or vice versa. Again, we must consider a system because potential energy does not exist for an isolated object. Our system might be a mass m oscillating on the end of a spring or moving in the Earth's gravitational field.

According to the work-energy principle (Eq. 7–11), the net work W_{net} done on an object is equal to its change in kinetic energy:

$$W_{net} = \Delta K.$$

(If more than one object of our system has work done on it, then W_{net} and ΔK can represent the sum for all of them.) Since we assume a conservative system, we can write the net work done on an object or objects in terms of the change in total potential energy (see Eq. 8–4) between points 1 and 2:

$$\Delta U_{total} = -\int_{1}^{2} \vec{F}_{net} \cdot d\vec{l} = -W_{net}. \qquad (8\text{–}8)$$

We combine the previous two equations, letting U be the total potential energy:

$$\Delta K + \Delta U = 0 \qquad \text{[conservative forces only]} \quad (8\text{–}9a)$$

or

$$(K_2 - K_1) + (U_2 - U_1) = 0. \qquad \text{[conservative forces only]} \quad (8\text{–}9b)$$

We now define a quantity E, called the **total mechanical energy** of our system, as the sum of the kinetic energy plus the potential energy of the system at any moment

$$E = K + U.$$

We can rewrite Eq. 8–9b as

or

$$K_2 + U_2 = K_1 + U_1 \qquad \text{[conservative forces only]} \quad \textbf{(8–10a)}$$

$$E_2 = E_1 = \text{constant.} \qquad \text{[conservative forces only]} \quad \textbf{(8–10b)}$$

Equations 8–10 express a useful and profound principle regarding the total mechanical energy—it is a **conserved quantity**, as long as no nonconservative forces do work; that is, the quantity $E = K + U$ at some initial time 1 is equal to $K + U$ at any later time 2.

To say it another way, consider Eq. 8–9a which tells us $\Delta U = -\Delta K$; that is, if the kinetic energy K increases, then the potential energy U must decrease by an equivalent amount to compensate. Thus the total, $K + U$, remains constant. This is called the **principle of conservation of mechanical energy** for conservative forces:

If only conservative forces are doing work, the total mechanical energy of a system neither increases nor decreases in any process. It stays constant—it is conserved.

We now see the reason for the term "conservative force"—because for such forces, mechanical energy is conserved.

If only one object of a system[†] has significant kinetic energy, then Eqs. 8–10 become

$$E = \tfrac{1}{2}mv^2 + U = \text{constant.} \quad \text{[conservative forces only]} \quad \textbf{(8–11a)}$$

If we let v_1 and U_1 represent the velocity and potential energy at one instant, and v_2 and U_2 represent them at a second instant, then we can rewrite this as

$$\tfrac{1}{2}mv_1^2 + U_1 = \tfrac{1}{2}mv_2^2 + U_2. \qquad \text{[conservative system]} \quad \textbf{(8–11b)}$$

From this equation we can see again that it doesn't make any difference where we choose the potential energy to be zero: adding a constant to U merely adds a constant to both sides of Eq. 8–11b, and these cancel. A constant also doesn't affect the force obtained using Eq. 8–7, $F = -dU/dx$, since the derivative of a constant is zero. Only changes in the potential energy matter.

8–4 Problem Solving Using Conservation of Mechanical Energy

A simple example of the conservation of mechanical energy (neglecting air resistance) is a rock allowed to fall due to Earth's gravity from a height h above the ground, as shown in Fig. 8–7. If the rock starts from rest, all of the initial energy is potential energy. As the rock falls, the potential energy mgy decreases (because y decreases), but the rock's kinetic energy increases to compensate, so that the sum of the two remains constant. At any point along the path, the total mechanical energy is given by

$$E = K + U = \tfrac{1}{2}mv^2 + mgy$$

where y is the rock's height above the ground at a given instant and v is its speed at that point. If we let the subscript 1 represent the rock at one point along its path (for example, the initial point), and the subscript 2 represent it at some other point, then we can write

total mechanical energy at point 1 = total mechanical energy at point 2

or (see also Eq. 8–11b)

$$\tfrac{1}{2}mv_1^2 + mgy_1 = \tfrac{1}{2}mv_2^2 + mgy_2. \qquad \text{[gravity only]} \quad \textbf{(8–12)}$$

Just before the rock hits the ground, where we chose $y = 0$, all of the initial potential energy will have been transformed into kinetic energy.

FIGURE 8–7 The rock's potential energy changes to kinetic energy as it falls. Note bar graphs representing potential energy U and kinetic energy K for the three different positions.

[†]For an object moving under the influence of Earth's gravity, the kinetic energy of the Earth can usually be ignored. For a mass oscillating at the end of a spring, the mass of the spring, and hence its kinetic energy, can often be ignored.

EXAMPLE 8–3 **Falling rock.** If the original height of the rock in Fig. 8–7 is $y_1 = h = 3.0$ m, calculate the rock's speed when it has fallen to 1.0 m above the ground.

APPROACH We apply the principle of conservation of mechanical energy, Eq. 8–12, with only gravity acting on the rock. We choose the ground as our reference level $(y = 0)$.

SOLUTION At the moment of release (point 1) the rock's position is $y_1 = 3.0$ m and it is at rest: $v_1 = 0$. We want to find v_2 when the rock is at position $y_2 = 1.0$ m. Equation 8–12 gives

$$\tfrac{1}{2}mv_1^2 + mgy_1 = \tfrac{1}{2}mv_2^2 + mgy_2.$$

The m's cancel out; setting $v_1 = 0$ and solving for v_2 we find

$$v_2 = \sqrt{2g(y_1 - y_2)} = \sqrt{2(9.8\,\text{m/s}^2)[(3.0\,\text{m}) - (1.0\,\text{m})]} = 6.3\,\text{m/s}.$$

The rock's speed 1.0 m above the ground is 6.3 m/s downward.

NOTE The velocity of the rock is independent of the rock's mass.

EXERCISE C In Example 8–3, what is the rock's speed just before it hits the ground? (a) 6.5 m/s; (b) 7.0 m/s; (c) 7.7 m/s; (d) 8.3 m/s; (e) 9.8 m/s.

Equation 8–12 can be applied to any object moving without friction under the action of gravity. For example, Fig. 8–8 shows a roller-coaster car starting from rest at the top of a hill, and coasting without friction to the bottom and up the hill on the other side. True, there is another force besides gravity acting on the car, the normal force exerted by the tracks. But this "constraint" force acts perpendicular to the direction of motion at each point and so does zero work. We ignore rotational motion of the car's wheels and treat the car as a particle undergoing simple translation. Initially, the car has only potential energy. As it coasts down the hill, it loses potential energy and gains in kinetic energy, but the sum of the two remains constant. At the bottom of the hill it has its maximum kinetic energy, and as it climbs up the other side the kinetic energy changes back to potential energy. When the car comes to rest again at the same height from which it started, all of its energy will be potential energy. Given that the gravitational potential energy is proportional to the vertical height, energy conservation tells us that (in the absence of friction) the car comes to rest at a height equal to its original height. If the two hills are the same height, the car will just barely reach the top of the second hill when it stops. If the second hill is lower than the first, not all of the car's kinetic energy will be transformed to potential energy and the car can continue over the top and down the other side. If the second hill is higher, the car will only reach a height on it equal to its original height on the first hill. This is true (in the absence of friction) no matter how steep the hill is, since potential energy depends only on the vertical height.

FIGURE 8–8 A roller-coaster car moving without friction illustrates the conservation of mechanical energy.

EXAMPLE 8–4 **Roller-coaster car speed using energy conservation.** Assuming the height of the hill in Fig. 8–8 is 40 m, and the roller-coaster car starts from rest at the top, calculate (a) the speed of the roller-coaster car at the bottom of the hill, and (b) at what height it will have half this speed. Take $y = 0$ at the bottom of the hill.

APPROACH We choose point 1 to be where the car starts from rest $(v_1 = 0)$ at the top of the hill $(y_1 = 40$ m). Point 2 is the bottom of the hill, which we choose as our reference level, so $y_2 = 0$. We use conservation of mechanical energy.

SOLUTION (a) We use Eq. 8–12 with $v_1 = 0$ and $y_2 = 0$, which gives

$$mgy_1 = \tfrac{1}{2}mv_2^2$$

or

$$v_2 = \sqrt{2gy_1} = \sqrt{2(9.8\,\text{m/s}^2)(40\,\text{m})} = 28\,\text{m/s}.$$

(b) We again use conservation of energy,

$$\tfrac{1}{2}mv_1^2 + mgy_1 = \tfrac{1}{2}mv_2^2 + mgy_2,$$

but now $v_2 = \tfrac{1}{2}(28\,\text{m/s}) = 14\,\text{m/s}$, $v_1 = 0$, and y_2 is the unknown. Thus

$$y_2 = y_1 - \frac{v_2^2}{2g} = 30\,\text{m}.$$

That is, the car has a speed of 14 m/s when it is 30 *vertical* meters above the lowest point, both when descending the left-hand hill and when ascending the right-hand hill.

The mathematics of the roller-coaster Example 8–4 is almost the same as in Example 8–3. But there is an important difference between them. In Example 8–3 the motion is all vertical and could have been solved using force, acceleration, and the kinematic equations (Eqs. 2–12). But for the roller coaster, where the motion is not vertical, we could not have used Eqs. 2–12 because a is not constant on the curved track; but energy conservation readily gives us the answer.

Paul

Kathleen

h

FIGURE 8–9 Example 8–5.

| **CONCEPTUAL EXAMPLE 8–5** | **Speeds on two water slides.** Two water slides at a pool are shaped differently, but start at the same height h (Fig. 8–9). Two riders, Paul and Kathleen, start from rest at the same time on different slides. (a) Which rider, Paul or Kathleen, is traveling faster at the bottom? (b) Which rider makes it to the bottom first? Ignore friction and assume both slides have the same path length.

RESPONSE (a) Each rider's initial potential energy mgh gets transformed to kinetic energy, so the speed v at the bottom is obtained from $\frac{1}{2}mv^2 = mgh$. The mass cancels and so the speed will be the same, regardless of the mass of the rider. Since they descend the same vertical height, they will finish with the same speed. (b) Note that Kathleen is consistently at a lower elevation than Paul at any instant, until the end. This means she has converted her potential energy to kinetic energy earlier. Consequently, she is traveling faster than Paul for the whole trip, and because the distance is the same, Kathleen gets to the bottom first.

EXERCISE D Two balls are released from the same height above the floor. Ball A falls freely through the air, whereas ball B slides on a curved frictionless track to the floor. How do the speeds of the balls compare when they reach the floor?

You may wonder sometimes whether to approach a problem using work and energy, or instead to use Newton's laws. As a rough guideline, if the force(s) involved are constant, either approach may succeed. If the forces are not constant, and/or the path is not simple, energy is probably the better approach.

There are many interesting examples of the conservation of energy in sports, such as the pole vault illustrated in Fig. 8–10. We often have to make approximations, but the sequence of events in broad outline for the pole vault is as follows. The initial kinetic energy of the running athlete is transformed into elastic potential energy of the bending pole and, as the athlete leaves the ground, into gravitational potential energy. When the vaulter reaches the top and the pole has straightened out again, the energy has all been transformed into gravitational potential energy (if we ignore the vaulter's low horizontal speed over the bar). The pole does not supply any energy, but it acts as a device to *store* energy and thus aid in the transformation of kinetic energy into gravitational potential energy, which is the net result. The energy required to pass over the bar depends on how high the center of mass (CM) of the vaulter must be raised. By bending their bodies, pole vaulters keep their CM so low that it can actually pass slightly beneath the bar (Fig. 8–11), thus enabling them to cross over a higher bar than would otherwise be possible. (Center of mass is covered in Chapter 9.)

PROBLEM SOLVING
Use energy, or Newton's laws?

PHYSICS APPLIED
Sports

FIGURE 8–10 Transformation of energy during a pole vault.

FIGURE 8–11 By bending their bodies, pole vaulters can keep their center of mass so low that it may even pass below the bar. By changing their kinetic energy (of running) into gravitational potential energy ($= mgy$) in this way, vaulters can cross over a higher bar than if the change in potential energy were accomplished without carefully bending the body.

EXAMPLE 8–6 ESTIMATE **Pole vault.** Estimate the kinetic energy and the speed required for a 70-kg pole vaulter to just pass over a bar 5.0 m high. Assume the vaulter's center of mass is initially 0.90 m off the ground and reaches its maximum height at the level of the bar itself.

APPROACH We equate the total energy just before the vaulter places the end of the pole onto the ground (and the pole begins to bend and store potential energy) with the vaulter's total energy when passing over the bar (we ignore the small amount of kinetic energy at this point). We choose the initial position of the vaulter's center of mass to be $y_1 = 0$. The vaulter's body must then be raised to a height $y_2 = 5.0\,\text{m} - 0.9\,\text{m} = 4.1\,\text{m}$.

SOLUTION We use Eq. 8–12,
$$\tfrac{1}{2}mv_1^2 + 0 = 0 + mgy_2$$
so
$$K_1 = \tfrac{1}{2}mv_1^2 = mgy_2 = (70\,\text{kg})(9.8\,\text{m/s}^2)(4.1\,\text{m}) = 2.8 \times 10^3\,\text{J}.$$
The speed is
$$v_1 = \sqrt{\frac{2K_1}{m}} = \sqrt{\frac{2(2800\,\text{J})}{70\,\text{kg}}} = 8.9\,\text{m/s} \approx 9\,\text{m/s}.$$

NOTE This is an approximation because we have ignored such things as the vaulter's speed while crossing over the bar, mechanical energy transformed when the pole is planted in the ground, and work done by the vaulter on the pole. All would increase the needed initial kinetic energy.

As another example of the conservation of mechanical energy, let us consider an object of mass m connected to a horizontal spring (Fig. 8–6) whose own mass can be neglected and whose spring stiffness constant is k. The mass m has speed v at any moment. The potential energy of the system (object plus spring) is $\tfrac{1}{2}kx^2$, where x is the displacement of the spring from its unstretched length. If neither friction nor any other force is acting, conservation of mechanical energy tells us that
$$\tfrac{1}{2}mv_1^2 + \tfrac{1}{2}kx_1^2 = \tfrac{1}{2}mv_2^2 + \tfrac{1}{2}kx_2^2, \qquad \text{[elastic PE only]} \quad \textbf{(8–13)}$$
where the subscripts 1 and 2 refer to the velocity and displacement at two different moments.

EXAMPLE 8–7 **Toy dart gun.** A dart of mass 0.100 kg is pressed against the spring of a toy dart gun as shown in Fig. 8–12a. The spring (with spring stiffness constant $k = 250\,\text{N/m}$ and ignorable mass) is compressed 6.0 cm and released. If the dart detaches from the spring when the spring reaches its natural length $(x = 0)$, what speed does the dart acquire?

APPROACH The dart is initially at rest (point 1), so $K_1 = 0$. We ignore friction and use conservation of mechanical energy; the only potential energy is elastic.

SOLUTION We use Eq. 8–13 with point 1 being at the maximum compression of the spring, so $v_1 = 0$ (dart not yet released) and $x_1 = -0.060\,\text{m}$. Point 2 we choose to be the instant the dart flies off the end of the spring (Fig. 8–12b), so $x_2 = 0$ and we want to find v_2. Thus Eq. 8–13 can be written
$$0 + \tfrac{1}{2}kx_1^2 = \tfrac{1}{2}mv_2^2 + 0.$$
Then
$$v_2^2 = \frac{kx_1^2}{m}$$
and
$$v_2 = \sqrt{\frac{(250\,\text{N/m})(-0.060\,\text{m})^2}{(0.100\,\text{kg})}} = 3.0\,\text{m/s}.$$

NOTE In the horizontal direction, the only force on the dart (neglecting friction) was the force exerted by the spring. Vertically, gravity was counterbalanced by the normal force exerted on the dart by the gun barrel. After it leaves the barrel, the dart will follow a projectile's path under gravity.

FIGURE 8–12 Example 8–7. (a) A dart is pushed against a spring, compressing it 6.0 cm. The dart is then released, and in (b) it leaves the spring at velocity v_2.

(a) $E = \tfrac{1}{2}kx^2$

(b) $E = \tfrac{1}{2}mv^2$

FIGURE 8–13 Example 8–8.

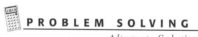

PROBLEM SOLVING

Alternate Solution

EXAMPLE 8–8 **Two kinds of potential energy.** A ball of mass $m = 2.60$ kg, starting from rest, falls a vertical distance $h = 55.0$ cm before striking a vertical coiled spring, which it compresses an amount $Y = 15.0$ cm (Fig. 8–13). Determine the spring stiffness constant of the spring. Assume the spring has negligible mass, and ignore air resistance. Measure all distances from the point where the ball first touches the uncompressed spring ($y = 0$ at this point).

APPROACH The forces acting on the ball are the gravitational pull of the Earth and the elastic force exerted by the spring. Both forces are conservative, so we can use conservation of mechanical energy, including both types of potential energy. We must be careful, however: gravity acts throughout the fall (Fig. 8–13), whereas the elastic force does not act until the ball touches the spring (Fig. 8–13b). We choose y positive upward, and $y = 0$ at the end of the spring in its natural (uncompressed) state.

SOLUTION We divide this solution into two parts. (An alternate solution follows.)
Part 1: Let us first consider the energy changes as the ball falls from a height $y_1 = h = 0.55$ m, Fig. 8–13a, to $y_2 = 0$, just as it touches the spring, Fig. 8–13b. Our system is the ball acted on by gravity plus the spring (which up to this point doesn't do anything). Thus

$$\tfrac{1}{2}mv_1^2 + mgy_1 = \tfrac{1}{2}mv_2^2 + mgy_2$$
$$0 + mgh = \tfrac{1}{2}mv_2^2 + 0.$$

We solve for $v_2 = \sqrt{2gh} = \sqrt{2(9.80 \text{ m/s}^2)(0.550 \text{ m})} = 3.283 \text{ m/s} \approx 3.28 \text{ m/s}$. This is the speed of the ball just as it touches the top of the spring, Fig. 8–13b.
Part 2: As the ball compresses the spring, Figs. 8–13b to c, there are two conservative forces on the ball—gravity and the spring force. So our conservation of energy equation is

$$E_2 \text{ (ball touches spring)} = E_3 \text{ (spring compressed)}$$
$$\tfrac{1}{2}mv_2^2 + mgy_2 + \tfrac{1}{2}ky_2^2 = \tfrac{1}{2}mv_3^2 + mgy_3 + \tfrac{1}{2}ky_3^2.$$

Substituting $y_2 = 0$, $v_2 = 3.283$ m/s, $v_3 = 0$ (the ball comes to rest for an instant), and $y_3 = -Y = -0.150$ m, we have

$$\tfrac{1}{2}mv_2^2 + 0 + 0 = 0 - mgY + \tfrac{1}{2}k(-Y)^2.$$

We know m, v_2, and Y, so we can solve for k:

$$k = \frac{2}{Y^2}\left[\tfrac{1}{2}mv_2^2 + mgY\right] = \frac{m}{Y^2}\left[v_2^2 + 2gY\right]$$
$$= \frac{(2.60 \text{ kg})}{(0.150 \text{ m})^2}\left[(3.283 \text{ m/s})^2 + 2(9.80 \text{ m/s}^2)(0.150 \text{ m})\right] = 1590 \text{ N/m}.$$

Alternate Solution Instead of dividing the solution into two parts, we can do it all at once. After all, we get to choose what two points are used on the left and right of the energy equation. Let us write the energy equation for points 1 and 3 in Fig. 8–13. Point 1 is the initial point just before the ball starts to fall (Fig. 8–13a), so $v_1 = 0$, and $y_1 = h = 0.550$ m. Point 3 is when the spring is fully compressed (Fig. 8–13c), so $v_3 = 0$, $y_3 = -Y = -0.150$ m. The forces on the ball in this process are gravity and (at least part of the time) the spring. So conservation of energy tells us

$$\tfrac{1}{2}mv_1^2 + mgy_1 + \tfrac{1}{2}k(0)^2 = \tfrac{1}{2}mv_3^2 + mgy_3 + \tfrac{1}{2}ky_3^2$$
$$0 + mgh + 0 = 0 - mgY + \tfrac{1}{2}kY^2$$

where we have set $y = 0$ for the spring at point 1 because it is not acting and is not compressed or stretched. We solve for k:

$$k = \frac{2mg(h + y)}{Y^2} = \frac{2(2.60 \text{ kg})(9.80 \text{ m/s}^2)(0.550 \text{ m} + 0.150 \text{ m})}{(0.150 \text{ m})^2} = 1590 \text{ N/m}$$

just as in our first method of solution.

EXAMPLE 8–9 **A swinging pendulum.** The simple pendulum shown in Fig. 8–14 consists of a small bob of mass m suspended by a massless cord of length ℓ. The bob is released (without a push) at $t = 0$, where the cord makes an angle $\theta = \theta_0$ to the vertical. (a) Describe the motion of the bob in terms of kinetic energy and potential energy. Then determine the speed of the bob (b) as a function of position θ as it swings back and forth, and (c) at the lowest point of the swing. (d) Find the tension in the cord, $\vec{F}_T$. Ignore friction and air resistance.

APPROACH We use the law of conservation of mechanical energy (only the conservative force of gravity does work), except in (d) where we use Newton's second law.

SOLUTION (a) At the moment of release, the bob is at rest, so its kinetic energy $K = 0$. As the bob moves down, it loses potential energy and gains kinetic energy. At the lowest point its kinetic energy is a maximum and the potential energy is a minimum. The bob continues its swing until it reaches an equal height and angle (θ_0) on the opposite side, at which point the potential energy is a maximum and $K = 0$. It continues the swinging motion as $U \to K \to U$ and so on, but it can never go higher than $\theta = \pm \theta_0$ (conservation of mechanical energy).

(b) The cord is assumed to be massless, so we need to consider only the bob's kinetic energy, and the gravitational potential energy. The bob has two forces acting on it at any moment: gravity, mg, and the force the cord exerts on it, $\vec{F}_T$. The latter (a constraint force) always acts perpendicular to the motion, so it does no work. We need be concerned only with gravity, for which we can write the potential energy. The mechanical energy of the system is

$$E = \tfrac{1}{2}mv^2 + mgy,$$

where y is the vertical height of the bob at any moment. We take $y = 0$ at the lowest point of the bob's swing. Hence at $t = 0$,

$$y = y_0 = \ell - \ell\cos\theta_0 = \ell(1 - \cos\theta_0)$$

as can be seen from the diagram. At the moment of release

$$E = mgy_0,$$

since $v = v_0 = 0$. At any other point along the swing

$$E = \tfrac{1}{2}mv^2 + mgy = mgy_0.$$

We solve this for v:

$$v = \sqrt{2g(y_0 - y)}.$$

In terms of the angle θ of the cord, we can write

$$(y_0 - y) = (\ell - \ell\cos\theta_0) - (\ell - \ell\cos\theta) = \ell(\cos\theta - \cos\theta_0)$$

so

$$v = \sqrt{2g\ell(\cos\theta - \cos\theta_0)}.$$

(c) At the lowest point, $y = 0$, so

$$v = \sqrt{2gy_0}$$

or

$$v = \sqrt{2g\ell(1 - \cos\theta_0)}.$$

(d) The tension in the cord is the force $\vec{F}_T$ that the cord exerts on the bob. As we've seen, there is no work done by this force, but we can calculate the force simply by using Newton's second law $\Sigma\vec{F} = m\vec{a}$ and by noting that at any point the acceleration of the bob in the inward radial direction is v^2/ℓ, since the bob is constrained to move in an arc of a circle of radius ℓ. In the radial direction, $\vec{F}_T$ acts inward, and a component of gravity equal to $mg\cos\theta$ acts outward. Hence

$$m\frac{v^2}{\ell} = F_T - mg\cos\theta.$$

We solve for F_T and use the result of part (b) for v^2:

$$F_T = m\left(\frac{v^2}{\ell} + g\cos\theta\right) = 2mg(\cos\theta - \cos\theta_0) + mg\cos\theta$$

$$= (3\cos\theta - 2\cos\theta_0)mg.$$

FIGURE 8–14 Example 8–9: a simple pendulum; y is measured positive upward.

8–5 The Law of Conservation of Energy

We now take into account nonconservative forces such as friction, since they are important in real situations. For example, consider again the roller-coaster car in Fig. 8–8, but this time let us include friction. The car will not in this case reach the same height on the second hill as it had on the first hill because of friction.

In this, and in other natural processes, the mechanical energy (sum of the kinetic and potential energies) does not remain constant but decreases. Because frictional forces reduce the mechanical energy (but *not* the total energy), they are called **dissipative forces**. Historically, the presence of dissipative forces hindered the formulation of a comprehensive conservation of energy law until well into the nineteenth century. It was not until then that heat, which is always produced when there is friction (try rubbing your hands together), was interpreted in terms of energy. Quantitative studies in the nineteenth-century (Chapter 19) demonstrated that if heat is considered as a transfer of energy (sometimes called **thermal energy**), then the total energy is conserved in any process. For example, if the roller-coaster car in Fig. 8–8 is subject to frictional forces, then the initial total energy of the car will be equal to the car's kinetic energy plus the potential energy at any subsequent point along its path plus the amount of thermal energy produced in the process. A block sliding freely across a table, for example, comes to rest because of friction. Its initial kinetic energy is all transformed into thermal energy. The block and table are a little warmer as a result of this process: both have absorbed some thermal energy. Another example of the transformation of kinetic energy into thermal energy can be observed by vigorously striking a nail several times with a hammer and then gently touching the nail with your finger.

According to the atomic theory, thermal energy represents kinetic energy of rapidly moving molecules. We shall see in Chapter 18 that a rise in temperature corresponds to an increase in the average kinetic energy of the molecules. Because thermal energy represents the energy of atoms and molecules that make up an object, it is often called **internal energy**. Internal energy, from the atomic point of view, can include not only kinetic energy of molecules but also potential energy (usually electrical in nature) because of the relative positions of atoms within molecules. On a macroscopic level, thermal or internal energy corresponds to nonconservative forces such as friction. But at the atomic level, the energy is partly kinetic, partly potential corresponding to forces that are conservative. For example, the energy stored in food or in a fuel such as gasoline can be regarded as potential energy stored by virtue of the relative positions of the atoms within a molecule due to electric forces between atoms (referred to as chemical bonds). For this energy to be used to do work, it must be released, usually through chemical reactions (Fig. 8–15). This is analogous to a compressed spring which, when released, can do work.

To establish the more general law of conservation of energy, it required nineteenth-century physicists to recognize electrical, chemical, and other forms of energy in addition to heat and to explore if in fact they could fit into a conservation law. For each type of force, conservative or nonconservative, it has always been found possible to define a type of energy that corresponds to the work done by such a force. And it has been found experimentally that the total energy E always remains constant. That is, the change in the total energy, kinetic plus potential plus all other forms of energy, equals zero:

FIGURE 8–15 The burning of fuel (a chemical reaction) releases energy to boil water in this steam engine. The steam produced expands against a piston to do work in turning the wheels.

$$\Delta K + \Delta U + [\text{change in all other forms of energy}] = 0. \qquad \textbf{(8–14)}$$

This is one of the most important principles in physics. It is called the **law of conservation of energy** and can be stated as follows:

LAW OF CONSERVATION OF ENERGY

The total energy is neither increased nor decreased in any process. Energy can be transformed from one form to another, and transferred from one object to another, but the total amount remains constant.

LAW OF CONSERVATION OF ENERGY

For conservative mechanical systems, this law can be derived from Newton's laws (Section 8–3) and thus is equivalent to them. But in its full generality, the validity of the law of conservation of energy rests on experimental observation.

Even though Newton's laws have been found to fail in the submicroscopic world of the atom, the law of conservation of energy has been found to hold there and in every experimental situation so far tested.

8–6 Energy Conservation with Dissipative Forces: Solving Problems

In Section 8–4 we discussed several Examples of the law of conservation of energy for conservative systems. Now let us consider in detail some examples that involve nonconservative forces.

Suppose, for example, that the roller-coaster car rolling on the hills of Fig. 8–8 is subject to frictional forces. In going from some point 1 to a second point 2, the energy dissipated by the friction force $\vec{F}_{fr}$ acting on the car (treating it as a particle) is $\int_1^2 \vec{F}_{fr} \cdot d\vec{\ell}$. If $\vec{F}_{fr}$ is constant in magnitude, the energy dissipated is simply $F_{fr}\ell$, where ℓ is the actual distance along the path traveled by the object from point 1 to point 2. Thus we write our conservation of energy equation, Eq. 8–14, as

$$\Delta K + \Delta U + F_{fr}\ell = 0,$$

or

$$\tfrac{1}{2}m(v_2^2 - v_1^2) + mg(y_2 - y_1) + F_{fr}\ell = 0.$$

We can rewrite this, comparing the initial energy E_1 to the final energy E_2:

$$\tfrac{1}{2}mv_1^2 + mgy_1 = \tfrac{1}{2}mv_2^2 + mgy_2 + F_{fr}\ell. \quad \left[\begin{array}{c}\text{gravity and} \\ \text{friction acting}\end{array}\right] \quad \textbf{(8–15)}$$

That is,

$$E_1 = E_2$$
initial energy $=$ final energy (including thermal energy).

On the left we have the mechanical energy of the system initially. It equals the mechanical energy at any subsequent point along the path plus the amount of thermal (or internal) energy produced in the process.

Other nonconservative forces can be treated similarly. If you are not sure about the sign of the last term $\left(\int \vec{F} \cdot d\vec{\ell}\right)$ on the right, use your intuition: is the mechanical energy increased or decreased in the process.

Work-Energy versus Energy Conservation

The law of conservation of energy is more general and more powerful than the work-energy principle. Indeed, the work-energy principle should *not* be viewed as a statement of conservation of energy. It is nonetheless useful for some mechanical problems; and whether you use it, or use the more powerful conservation of energy, can depend on your *choice of the system* under study. If you choose as your system a particle or rigid object on which external forces do work, then you can use the work-energy principle: the work done by the external forces on your object equals the change in its kinetic energy.

On the other hand, if you choose a system on which no external forces do work, then you need to apply conservation of energy to that system directly.

Consider, for example, a spring connected to a block on a frictionless table (Fig. 8–16). If you choose the block as your system, then the work done on the block by the spring equals the change in kinetic energy of the block: the work-energy principle. (Energy conservation does not apply to this system—the block's energy changes.) If instead you choose the block plus the spring as your system, no external forces do work (since the spring is part of the chosen system). To this system you need to apply conservation of energy: if you compress the spring and then release it, the spring still exerts a force on the block, but the subsequent motion can be discussed in terms of kinetic energy $\left(\tfrac{1}{2}mv^2\right)$ plus potential energy $\left(\tfrac{1}{2}kx^2\right)$, whose total remains constant.

FIGURE 8–16 A spring connected to a block on a frictionless table. If you choose your system to be the block plus spring, then

$$E = \tfrac{1}{2}mv^2 + \tfrac{1}{2}kx^2$$

is conserved.

Problem solving is not a process that can be done by simply following a set of rules. The following Problem Solving Strategy, like all others, is thus *not* a prescription, but is a summary to help you get started solving problems involving energy.

Conservation of Energy

1. **Draw a picture** of the physical situation.

2. Determine **the system** for which you will apply energy conservation: the object or objects and the forces acting.

3. Ask yourself what quantity you are looking for, and **choose initial** (point 1) **and final** (point 2) **positions**.

4. If the object under investigation changes its height during the problem, then **choose a reference frame** with a convenient $y = 0$ level for gravitational potential energy; the lowest point in the problem is often a good choice.

    If springs are involved, choose the unstretched spring position to be x (or y) $= 0$.

5. **Is mechanical energy conserved?** If no friction or other nonconservative forces act, then conservation of mechanical energy holds:

$$K_1 + U_1 = K_2 + U_2.$$

6. **Apply conservation of energy.** If friction (or other nonconservative forces) are present, then an additional term of the form $\int \vec{F} \cdot d\vec{\ell}$ will be needed. For a constant friction force acting over a distance ℓ

$$K_1 + U_1 = K_2 + U_2 + F_{fr}\ell.$$

For other nonconservative forces use your intuition for the sign of $\int \vec{F} \cdot d\vec{\ell}$: is the total mechanical energy increased or decreased in the process?

7. Use the equation(s) you develop to **solve** for the unknown quantity.

FIGURE 8–17 Example 8–10. Because of friction, a roller-coaster car does not reach the original height on the second hill. (Not to scale)

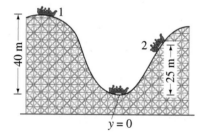

EXAMPLE 8–10 **ESTIMATE** **Friction on the roller-coaster car.** The roller-coaster car in Example 8–4 reaches a vertical height of only 25 m on the second hill before coming to a momentary stop (Fig. 8–17). It traveled a total distance of 400 m. Determine the thermal energy produced and estimate the average friction force (assume it is roughly constant) on the car, whose mass is 1000 kg.

APPROACH We explicitly follow the Problem Solving Strategy above.
SOLUTION

1. **Draw a picture.** See Fig. 8–17.

2. **The system.** The system is the roller-coaster car and the Earth (which exerts the gravitational force). The forces acting on the car are gravity and friction. (The normal force also acts on the car, but does no work, so it does not affect the energy.)

3. **Choose initial and final positions.** We take point 1 to be the instant when the car started coasting (at the top of the first hill), and point 2 to be the instant it stopped 25 m up the second hill.

4. **Choose a reference frame.** We choose the lowest point in the motion to be $y = 0$ for the gravitational potential energy.

5. **Is mechanical energy conserved?** No. Friction is present.

6. **Apply conservation of energy.** There is friction acting on the car, so we use conservation of energy in the form of Eq. 8–15, with $v_1 = 0$, $y_1 = 40$ m, $v_2 = 0$, $y_2 = 25$ m, and $\ell = 400$ m. Thus

$$0 + (1000\,\text{kg})(9.8\,\text{m/s}^2)(40\,\text{m}) = 0 + (1000\,\text{kg})(9.8\,\text{m/s}^2)(25\,\text{m}) + F_{fr}\ell.$$

7. **Solve.** We solve the above equation for $F_{fr}\ell$, the energy dissipated to thermal energy: $F_{fr}\ell = (1000\,\text{kg})(9.8\,\text{m/s}^2)(40\,\text{m} - 25\,\text{m}) = 147{,}000\,\text{J}$. The average force of friction was $F_{fr} = (1.47 \times 10^5\,\text{J})/400\,\text{m} = 370\,\text{N}$. [This result is only a rough average: the friction force at various points depends on the normal force, which varies with slope.]

(a)

(b)

FIGURE 8–18 Example 8–11.

EXAMPLE 8–11 **Friction with a spring.** A block of mass m sliding along a rough horizontal surface is traveling at a speed v_0 when it strikes a massless spring head-on (see Fig. 8–18) and compresses the spring a maximum distance X. If the spring has stiffness constant k, determine the coefficient of kinetic friction between block and surface.

APPROACH At the moment of collision, the block has $K = \frac{1}{2}mv_0^2$ and the spring is presumably uncompressed, so $U = 0$. Initially the mechanical energy of the system is $\frac{1}{2}mv_0^2$. By the time the spring reaches maximum compression, $K = 0$ and $U = \frac{1}{2}kX^2$. In the meantime, the friction force $\left(= \mu_k F_N = \mu_k mg\right)$ has transformed energy $F_{fr}X = \mu_k mgX$ into thermal energy.

SOLUTION Conservation of energy allows us to write

$$\text{energy (initial)} = \text{energy (final)}$$
$$\tfrac{1}{2}mv_0^2 = \tfrac{1}{2}kX^2 + \mu_k mgX.$$

We solve for μ_k and find

$$\mu_k = \frac{v_0^2}{2gX} - \frac{kX}{2mg}.$$

8–7 Gravitational Potential Energy and Escape Velocity

We have been dealing with gravitational potential energy so far in this Chapter assuming the force of gravity is constant, $\vec{F} = m\vec{g}$. This is an accurate assumption for ordinary objects near the surface of the Earth. But to deal with gravity more generally, for points not close to the Earth's surface, we must consider that the gravitational force exerted by the Earth on a particle of mass m decreases inversely as the square of the distance r from the Earth's center. The precise relationship is given by Newton's law of universal gravitation (Sections 6–1 and 6–2):

$$\vec{F} = -G\frac{mM_E}{r^2}\hat{r} \qquad\qquad [r > r_E]$$

where M_E is the mass of the Earth and $\hat{r}$ is a unit vector (at the position of m) directed radially away from the Earth's center. The minus sign indicates that the force on m is directed toward the Earth's center, in the direction opposite to $\hat{r}$. This equation can also be used to describe the gravitational force on a mass m in the vicinity of other heavenly bodies, such as the Moon, a planet, or the Sun, in which case M_E must be replaced by that body's mass.

Suppose an object of mass m moves from one position to another along an arbitrary path (Fig. 8–19) so that its distance from the Earth's center changes from r_1 to r_2. The work done by the gravitational force is

$$W = \int_1^2 \vec{F} \cdot d\vec{\ell} = -GmM_E \int_1^2 \frac{\hat{r} \cdot d\vec{\ell}}{r^2},$$

where $d\vec{\ell}$ represents an infinitesimal displacement. Since $\hat{r} \cdot d\vec{\ell} = dr$ is the component of $d\vec{\ell}$ along $\hat{r}$ (see Fig. 8–19), then

$$W = -GmM_E \int_{r_1}^{r_2} \frac{dr}{r^2} = GmM_E\left(\frac{1}{r_2} - \frac{1}{r_1}\right)$$

or

$$W = \frac{GmM_E}{r_2} - \frac{GmM_E}{r_1}.$$

FIGURE 8–19 Arbitrary path of particle of mass m moving from point 1 to point 2.

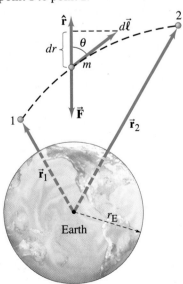

Because the value of the integral depends only on the position of the end points (r_1 and r_2) and not on the path taken, the gravitational force is a conservative force.

We can therefore use the concept of potential energy for the gravitational force. Since the change in potential energy is always defined (Section 8–2) as the negative of the work done by the force, we have

$$\Delta U = U_2 - U_1 = -\frac{GmM_E}{r_2} + \frac{GmM_E}{r_1}. \qquad \textbf{(8–16)}$$

From Eq. 8–16 the potential energy at any distance r from the Earth's center can be written:

$$U(r) = -\frac{GmM_E}{r} + C,$$

where C is a constant. It is usual to choose $C = 0$, so that

$$U(r) = -\frac{GmM_E}{r}. \qquad \left[\begin{array}{c}\text{gravity}\\(r > r_E)\end{array}\right] \quad \textbf{(8–17)}$$

With this choice for C, $U = 0$ at $r = \infty$. As an object approaches the Earth, its potential energy decreases and is always negative (Fig. 8–20).

Equation 8–16 reduces to Eq. 8–2, $\Delta U = mg(y_2 - y_1)$, for objects near the surface of the Earth (see Problem 48).

For a particle of mass m, which experiences only the force of the Earth's gravity, the total energy is conserved because gravity is a conservative force. Therefore we can write

$$\tfrac{1}{2}mv_1^2 - G\frac{mM_E}{r_1} = \tfrac{1}{2}mv_2^2 - G\frac{mM_E}{r_2} = \text{constant.} \quad \left[\begin{array}{c}\text{gravity}\\\text{only}\end{array}\right] \quad \textbf{(8–18)}$$

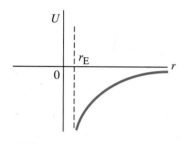

FIGURE 8–20 Gravitational potential energy plotted as a function of r, the distance from Earth's center. Valid only for points $r > r_E$, the radius of the Earth.

EXAMPLE 8–12 **Package dropped from high-speed rocket.** A box of empty film canisters is allowed to fall from a rocket traveling outward from Earth at a speed of 1800 m/s when 1600 km above the Earth's surface. The package eventually falls to the Earth. Estimate its speed just before impact. Ignore air resistance.

APPROACH We use conservation of energy. The package initially has a speed relative to Earth equal to the speed of the rocket from which it falls.

SOLUTION Conservation of energy in this case is expressed by Eq. 8–18:

$$\tfrac{1}{2}mv_1^2 - G\frac{mM_E}{r_1} = \tfrac{1}{2}mv_2^2 - G\frac{mM_E}{r_2}$$

where $v_1 = 1.80 \times 10^3$ m/s, $r_1 = (1.60 \times 10^6 \text{ m}) + (6.38 \times 10^6 \text{ m}) = 7.98 \times 10^6$ m, and $r_2 = 6.38 \times 10^6$ m (the radius of the Earth). We solve for v_2:

$$v_2 = \sqrt{v_1^2 - 2GM_E\left(\frac{1}{r_1} - \frac{1}{r_2}\right)}$$

$$= \sqrt{(1.80 \times 10^3 \text{ m/s})^2 - 2(6.67 \times 10^{-11} \text{ N}\cdot\text{m}^2/\text{kg}^2)(5.98 \times 10^{24} \text{ kg}) \times \left(\frac{1}{7.98 \times 10^6 \text{ m}} - \frac{1}{6.38 \times 10^6 \text{ m}}\right)}$$

$$= 5320 \text{ m/s.}$$

NOTE In reality, the speed will be considerably less than this because of air resistance. Note, incidentally, that the direction of the velocity never entered into this calculation, and this is one of the advantages of the energy method. The rocket could have been heading away from the Earth, or toward it, or at some other angle, and the result would be the same.

Escape Velocity

When an object is projected into the air from the Earth, it will return to Earth unless its speed is very high. But if the speed *is* high enough, it will continue out into space never to return to Earth (barring other forces or collisions). The minimum initial velocity needed to prevent an object from returning to the Earth is called the **escape velocity** from Earth, v_{esc}. To determine v_{esc} from the Earth's surface (ignoring air resistance), we use Eq. 8–18 with $v_1 = v_{esc}$ and $r_1 = r_E = 6.38 \times 10^6$ m, the radius of the Earth. Since we want the minimum speed for escape, we need the object to reach $r_2 = \infty$ with merely zero speed, $v_2 = 0$. Applying Eq. 8–18 we have

$$\tfrac{1}{2}mv_{esc}^2 - G\frac{mM_E}{r_E} = 0 + 0$$

or

$$v_{esc} = \sqrt{2GM_E/r_E} = 1.12 \times 10^4 \text{ m/s} \qquad \textbf{(8–19)}$$

or 11.2 km/s. It is important to note that although a mass can escape from the Earth (or solar system) never to return, the force on it due to the Earth's gravitational field is never actually zero for a finite value of r.

EXAMPLE 8–13 **Escaping the Earth or the Moon.** (*a*) Compare the escape velocities of a rocket from the Earth and from the Moon. (*b*) Compare the energies required to launch the rockets. For the Moon, $M_M = 7.35 \times 10^{22}$ kg and $r_M = 1.74 \times 10^6$ m, and for Earth, $M_E = 5.98 \times 10^{24}$ kg and $r_E = 6.38 \times 10^6$ m.

APPROACH We use Eq. 8–19, replacing M_E and r_E with M_M and r_M for finding v_{esc} from the Moon.

SOLUTION (*a*) Using Eq. 8–19, the ratio of the escape velocities is

$$\frac{v_{esc}(\text{Earth})}{v_{esc}(\text{Moon})} = \sqrt{\frac{M_E\, r_M}{M_M\, r_E}} = 4.7.$$

To escape Earth requires a speed 4.7 times that required to escape the Moon.
(*b*) The fuel that must be burned provides energy proportional to v^2 $(K = \tfrac{1}{2}mv^2)$; so to launch a rocket to escape Earth requires $(4.7)^2 = 22$ times as much energy as to escape from the Moon.

8–8 Power

Power is defined as the *rate at which work is done*. The *average power*, $\overline{P}$, equals the work W done divided by the time t it takes to do it:

$$\overline{P} = \frac{W}{t}. \qquad \textbf{(8–20a)}$$

Since the work done in a process involves the transformation of energy from one type (or object) to another, power can also be defined as the *rate at which energy is transformed*:

$$\overline{P} = \frac{W}{t} = \frac{\text{energy transformed}}{\text{time}}.$$

The *instantaneous power*, P, is

$$P = \frac{dW}{dt}. \qquad \textbf{(8–20b)}$$

The work done in a process is equal to the energy transferred from one object to another. For example, as the potential energy stored in the spring of Fig. 8–6c is transformed to kinetic energy of the ball, the spring is doing work on the ball. Similarly, when you throw a ball or push a grocery cart, *whenever work is done, energy is being transferred from one body to another*. Hence we can also say that power is the *rate at which energy is transformed*:

$$P = \frac{dE}{dt}. \qquad \textbf{(8–20c)}$$

The power of a horse refers to how much work it can do per unit of time.

The power rating of an engine refers to how much chemical or electrical energy can be transformed into mechanical energy per unit of time. In SI units, power is measured in joules per second, and this unit is given a special name, the **watt** (W): $1\,\text{W} = 1\,\text{J/s}$. We are most familiar with the watt for electrical devices: the rate at which an electric lightbulb or heater changes electric energy into light or thermal energy. But the watt is used for other types of energy transformation as well. In the British system, the unit of power is the foot-pound per second ($\text{ft}\cdot\text{lb/s}$). For practical purposes a larger unit is often used, the **horsepower**[†] (hp). One horsepower[†] (hp) is defined as $550\,\text{ft}\cdot\text{lb/s}$, which equals 746 watts. An engine's power is usually specified in hp or in kW $\left(1\,\text{kW} \approx 1\tfrac{1}{3}\text{hp}\right)$.

To see the distinction between energy and power, consider the following example. A person is limited in the work he or she can do, not only by the total energy required, but also by how fast this energy is transformed: that is, by power. For example, a person may be able to walk a long distance or climb many flights of stairs before having to stop because so much energy has been expended. On the other hand, a person who runs very quickly up stairs may feel exhausted after only a flight or two. He or she is limited in this case by power, the rate at which his or her body can transform chemical energy into mechanical energy.

FIGURE 8–21 Example 8–14.

Power needs of a car

EXAMPLE 8–14 **Stair-climbing power.** A 60-kg jogger runs up a long flight of stairs in 4.0 s (Fig. 8–21). The vertical height of the stairs is 4.5 m. (a) Estimate the jogger's power output in watts and horsepower. (b) How much energy did this require?

APPROACH The work done by the jogger is against gravity, and equals $W = mgy$. To find her average output, we divide W by the time it took.

SOLUTION (a) The average power output was

$$\overline{P} = \frac{W}{t} = \frac{mgy}{t} = \frac{(60\,\text{kg})(9.8\,\text{m/s}^2)(4.5\,\text{m})}{4.0\,\text{s}} = 660\,\text{W}.$$

Since there are 746 W in 1 hp, the jogger is doing work at a rate of just under 1 hp. A human cannot do work at this rate for very long.

(b) The energy required is $E = \overline{P}t = (660\,\text{J/s})(4.0\,\text{s}) = 2600\,\text{J}$. This result equals $W = mgy$.

NOTE The person had to transform more energy than this 2600 J. The total energy transformed by a person or an engine always includes some thermal energy (recall how hot you get running up stairs).

Automobiles do work to overcome the force of friction (and air resistance), to climb hills, and to accelerate. A car is limited by the rate it can do work, which is why automobile engines are rated in horsepower. A car needs power most when it is climbing hills and when accelerating. In the next Example, we will calculate how much power is needed in these situations for a car of reasonable size. Even when a car travels on a level road at constant speed, it needs some power just to do work to overcome the retarding forces of internal friction and air resistance. These forces depend on the conditions and speed of the car, but are typically in the range 400 N to 1000 N.

It is often convenient to write the power in terms of the net force $\vec{\mathbf{F}}$ applied to an object and its velocity $\vec{\mathbf{v}}$. Since $P = dW/dt$ and $dW = \vec{\mathbf{F}} \cdot d\vec{\boldsymbol{\ell}}$ (Eq. 7–7), then

$$P = \frac{dW}{dt} = \vec{\mathbf{F}} \cdot \frac{d\vec{\boldsymbol{\ell}}}{dt} = \vec{\mathbf{F}} \cdot \vec{\mathbf{v}}. \tag{8–21}$$

[†]The unit was chosen by James Watt (1736–1819), who needed a way to specify the power of his newly developed steam engines. He found by experiment that a good horse can work all day at an average rate of about $360\,\text{ft}\cdot\text{lb/s}$. So as not to be accused of exaggeration in the sale of his steam engines, he multiplied this by roughly $1\tfrac{1}{2}$ when he defined the hp.

FIGURE 8–22 Example 8–15: Calculation of power needed for a car to climb a hill.

EXAMPLE 8–15 **Power needs of a car.** Calculate the power required of a 1400-kg car under the following circumstances: (*a*) the car climbs a 10° hill (a fairly steep hill) at a steady 80 km/h; and (*b*) the car accelerates along a level road from 90 to 110 km/h in 6.0 s to pass another car. Assume that the average retarding force on the car is $F_R = 700$ N throughout. See Fig. 8–22.

APPROACH First we must be careful not to confuse $\vec{\mathbf{F}}_R$, which is due to air resistance and friction that retards the motion, with the force $\vec{\mathbf{F}}$ needed to accelerate the car, which is the frictional force exerted by the road on the tires—the reaction to the motor-driven tires pushing against the road. We must determine the latter force F before calculating the power.

SOLUTION (*a*) To move at a steady speed up the hill, the car must, by Newton's second law, exert a force F equal to the sum of the retarding force, 700 N, and the component of gravity parallel to the hill, $mg \sin 10°$. Thus

$$F = 700\,\text{N} + mg \sin 10°$$
$$= 700\,\text{N} + (1400\,\text{kg})(9.80\,\text{m/s}^2)(0.174) = 3100\,\text{N}.$$

Since $\bar{v} = 80$ km/h $= 22$ m/s and is parallel to $\vec{\mathbf{F}}$, then (Eq. 8–21) the power is

$$\bar{P} = F\bar{v} = (3100\,\text{N})(22\,\text{m/s}) = 6.80 \times 10^4\,\text{W} = 68.0\,\text{kW} = 91\,\text{hp}.$$

(*b*) The car accelerates from 25.0 m/s to 30.6 m/s (90 to 110 km/h). Thus the car must exert a force that overcomes the 700-N retarding force plus that required to give it the acceleration

$$\bar{a}_x = \frac{(30.6\,\text{m/s} - 25.0\,\text{m/s})}{6.0\,\text{s}} = 0.93\,\text{m/s}^2.$$

We apply Newton's second law with x being the direction of motion:

$$ma_x = \Sigma F_x = F - F_R.$$

We solve for the force required, F:

$$F = ma_x + F_R$$
$$= (1400\,\text{kg})(0.93\,\text{m/s}^2) + 700\,\text{N} = 1300\,\text{N} + 700\,\text{N} = 2000\,\text{N}.$$

Since $P = \vec{\mathbf{F}} \cdot \vec{\mathbf{v}}$, the required power increases with speed and the motor must be able to provide a maximum power output of

$$\bar{P} = (2000\,\text{N})(30.6\,\text{m/s}) = 6.12 \times 10^4\,\text{W} = 61.2\,\text{kW} = 82\,\text{hp}.$$

NOTE Even taking into account the fact that only 60 to 80% of the engine's power output reaches the wheels, it is clear from these calculations that an engine of 75 to 100 kW (100 to 130 hp) is adequate from a practical point of view.

We mentioned in the Example above that only part of the energy output of a car engine reaches the wheels. Not only is some energy wasted in getting from the engine to the wheels, in the engine itself much of the input energy (from the gasoline) does not end up doing useful work. An important characteristic of all engines is their overall efficiency e, defined as the ratio of the useful power output of the engine, P_{out}, to the power input, P_{in}:

$$e = \frac{P_{out}}{P_{in}}.$$

The efficiency is always less than 1.0 because no engine can create energy, and in fact, cannot even transform energy from one form to another without some going to friction, thermal energy, and other nonuseful forms of energy. For example, an automobile engine converts chemical energy released in the burning of gasoline into mechanical energy that moves the pistons and eventually the wheels. But nearly 85% of the input energy is "wasted" as thermal energy that goes into the cooling system or out the exhaust pipe, plus friction in the moving parts. Thus car engines are roughly only about 15% efficient. We discuss efficiency in detail in Chapter 20.

*8–9 Potential Energy Diagrams; Stable and Unstable Equilibrium

FIGURE 8–23 A potential energy diagram.

If only conservative forces do work on an object, we can learn a great deal about its motion simply by examining a potential energy diagram—the graph of $U(x)$ versus x. An example of a potential energy diagram is shown in Fig. 8–23. The rather complex curve represents some complicated potential energy $U(x)$. The total energy $E = K + U$ is constant and can be represented as a horizontal line on this graph. Four different possible values for E are shown, labeled E_0, E_1, E_2, and E_3. What the actual value of E will be for a given system depends on the initial conditions. (For example, the total energy E of a mass oscillating on the end of a spring depends on the amount the spring is initially compressed or stretched.) Kinetic energy $K = \frac{1}{2}mv^2$ cannot be less than zero (v would be imaginary), and because $E = U + K = $ constant, then $U(x)$ must be less than or equal to E for all situations: $U(x) \le E$. Thus the minimum value which the total energy can take for the potential energy shown in Fig. 8–23 is that labeled E_0. For this value of E, the mass can only be at rest at $x = x_0$. The system has potential energy but no kinetic energy at this position.

If the system's total energy E is greater than E_0, say it is E_1 on our plot, the system can have both kinetic and potential energy. Because energy is conserved,

$$K = E - U(x).$$

Since the curve represents $U(x)$ at each x, the kinetic energy at any value of x is represented by the distance between the E line and the curve $U(x)$ at that value of x. In the diagram, the kinetic energy for an object at x_1, when its total energy is E_1, is indicated by the notation K_1.

An object with energy E_1 can oscillate only between the points x_2 and x_3. This is because if $x > x_2$ or $x < x_3$, the potential energy would be greater than E, meaning $K = \frac{1}{2}mv^2 < 0$ and v would be imaginary, and so impossible. At x_2 and x_3 the velocity is zero, since $E = U$ at these points. Hence x_2 and x_3 are called the **turning points** of the motion. If the object is at x_0, say, moving to the right, its kinetic energy (and speed) decreases until it reaches zero at $x = x_2$. The object then reverses direction, proceeding to the left and increasing in speed until it passes x_0 again. It continues to move, decreasing in speed until it reaches $x = x_3$, where again $v = 0$, and the object again reverses direction.

If the object has energy $E = E_2$ in Fig. 8–23, there are four turning points. The object can move in only one of the two potential energy "valleys," depending on where it is initially. It cannot get from one valley to the other because of the barrier between them—for example at a point such as x_4, $U > E_2$, which means v would be imaginary.[†] For energy E_3, there is only one turning point since $U(x) < E_3$ for all $x > x_5$. Thus our object, if moving initially to the left, varies in speed as it passes the potential valleys but eventually stops and turns around at $x = x_5$. It then proceeds to the right indefinitely, never to return.

How do we know the object reverses direction at the turning points? Because of the force exerted on it. The force F is related to the potential energy U by Eq. 8–7, $F = -dU/dx$. The force F is equal to the negative of the slope of the U-versus-x curve at any point x. At $x = x_2$, for example, the slope is positive so the force is negative, which means it acts to the left (toward decreasing values of x).

At $x = x_0$ the slope is zero, so $F = 0$. At such a point the particle is said to be in **equilibrium**. This term means simply that the net force on the object is zero. Hence, its acceleration is zero, and so if it is initially at rest, it remains at rest. If the object at rest at $x = x_0$ were moved slightly to the left or right, a nonzero force would act on it in the direction to move it back toward x_0. An object that returns

[†]Although this is true according to Newtonian physics, modern quantum mechanics predicts that objects can "tunnel" through such a barrier, and such processes have been observed at the atomic and subatomic level.

toward its equilibrium point when displaced slightly is said to be at a point of **stable equilibrium**. Any *minimum* in the potential energy curve represents a point of stable equilibrium.

An object at $x = x_4$ would also be in equilibrium, since $F = -dU/dx = 0$. If the object were displaced a bit to either side of x_4, a force would act to pull the object *away* from the equilibrium point. Points like x_4, where the potential energy curve has a maximum, are points of **unstable equilibrium**. The object will *not* return to equilibrium if displaced slightly, but instead will move farther away.

When an object is in a region over which U is constant, such as near $x = x_6$ in Fig. 8–23, the force is zero over some distance. The object is in equilibrium and if displaced slightly to one side the force is still zero. The object is said to be in **neutral equilibrium** in this region.

Summary

A **conservative force** is one for which the work done by the force in moving an object from one position to another depends only on the two positions and not on the path taken. The work done by a conservative force is recoverable, which is not true for nonconservative forces, such as friction.

Potential energy, U, is energy associated with conservative forces that depend on the position or configuration of objects. Gravitational potential energy is

$$U_{grav} = mgy, \tag{8–3}$$

where the mass m is near the Earth's surface, a height y above some reference point. Elastic potential energy is given by

$$U_{el} = \tfrac{1}{2}kx^2 \tag{8–5}$$

for a spring with stiffness constant k stretched or compressed a displacement x from equilibrium. Other potential energies include chemical, electrical, and nuclear energy.

Potential energy is always associated with a conservative force, and the change in potential energy, ΔU, between two points under the action of a conservative force $\vec{F}$ is defined as the negative of the work done by the force:

$$\Delta U = U_2 - U_1 = -\int_1^2 \vec{F} \cdot d\vec{\ell}. \tag{8–4}$$

Inversely, we can write, for the one-dimensional case,

$$F = -\frac{dU(x)}{dx}. \tag{8–7}$$

Only *changes* in potential energy are physically meaningful, so the position where $U = 0$ can be chosen for convenience.

Potential energy is not a property of an object but is associated with the interaction of two or more objects.

When only conservative forces act, the total **mechanical energy**, E, defined as the sum of kinetic and potential energies, is conserved:

$$E = K + U = \text{constant.} \tag{8–10}$$

If nonconservative forces also act, additional types of energy are involved, such as thermal energy. It has been found experimentally that, when all forms of energy are included, the total energy is conserved. This is the **law of conservation of energy**:

$$\Delta K + \Delta U + \Delta(\text{other energy types}) = 0. \tag{8–14}$$

The gravitational force as described by Newton's law of universal gravitation is a conservative force. The potential energy of an object of mass m due to the gravitational force exerted on it by the Earth is given by

$$U(r) = -\frac{GmM_E}{r}, \tag{8–17}$$

where M_E is the mass of the Earth and r is the distance of the object from the Earth's center ($r \geq$ radius of Earth).

Power is defined as the rate at which work is done or the rate at which energy is transformed from one form to another:

$$P = \frac{dW}{dt} = \frac{dE}{dt}, \tag{8–20}$$

or

$$P = \vec{F} \cdot \vec{v}. \tag{8–21}$$

Questions

1. List some everyday forces that are not conservative, and explain why they aren't.

2. You lift a heavy book from a table to a high shelf. List the forces on the book during this process, and state whether each is conservative or nonconservative.

3. The net force acting on a particle is conservative and increases the kinetic energy by 300 J. What is the change in (a) the potential energy, and (b) the total energy, of the particle?

4. When a "superball" is dropped, can it rebound to a greater height than its original height?

5. A hill has a height h. A child on a sled (total mass m) slides down starting from rest at the top. Does the velocity at the bottom depend on the angle of the hill if (a) it is icy and there is no friction, and (b) there is friction (deep snow)?

6. Why is it tiring to push hard against a solid wall even though no work is done?

7. Analyze the motion of a simple swinging pendulum in terms of energy, (a) ignoring friction, and (b) taking friction into account. Explain why a grandfather clock has to be wound up.

8. Describe precisely what is "wrong" physically in the famous Escher drawing shown in Fig. 8–24.

FIGURE 8–24
Question 8.

9. In Fig. 8–25, water balloons are tossed from the roof of a building, all with the same speed but with different launch angles. Which one has the highest speed when it hits the ground? Ignore air resistance.

FIGURE 8–25
Question 9.

10. A coil spring of mass m rests upright on a table. If you compress the spring by pressing down with your hand and then release it, can the spring leave the table? Explain using the law of conservation of energy.

11. What happens to the gravitational potential energy when water at the top of a waterfall falls to the pool below?

12. Experienced hikers prefer to step over a fallen log in their path rather than stepping on top and jumping down on the other side. Explain.

13. (a) Where does the kinetic energy come from when a car accelerates uniformly starting from rest? (b) How is the increase in kinetic energy related to the friction force the road exerts on the tires?

14. The Earth is closest to the Sun in winter (Northern Hemisphere). When is the gravitational potential energy the greatest?

15. Can the total mechanical energy $E = K + U$ ever be negative? Explain.

16. Suppose that you wish to launch a rocket from the surface of the Earth so that it escapes the Earth's gravitational field. You wish to use minimum fuel in doing this. From what point on the surface of the Earth should you make the launch and in what direction? Do the launch location and direction matter? Explain.

17. Recall from Chapter 4, Example 4–14, that you can use a pulley and ropes to decrease the force needed to raise a heavy load (see Fig. 8–26). But for every meter the load is raised, how much rope must be pulled up? Account for this, using energy concepts.

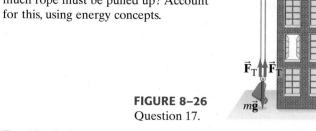

FIGURE 8–26
Question 17.

18. Two identical arrows, one with twice the speed of the other, are fired into a bale of hay. Assuming the hay exerts a constant "frictional" force on the arrows, the faster arrow will penetrate how much farther than the slower arrow? Explain.

19. A bowling ball is hung from the ceiling by a steel wire (Fig. 8–27). The instructor pulls the ball back and stands against the wall with the ball against his nose. To avoid injury the instructor is supposed to release the ball without pushing it. Why?

FIGURE 8–27
Question 19.

20. A pendulum is launched from a point that is a height h above its lowest point in two different ways (Fig. 8–28). During both launches, the pendulum is given an initial speed of 3.0 m/s. On the first launch, the initial velocity of the pendulum is directed upward along the trajectory, and on the second launch it is directed downward along the trajectory. Which launch will cause the highest speed when the pendulum bob passes the lowest point of its swing? Explain.

FIGURE 8–28
Question 20.

21. Describe the energy transformations when a child hops around on a pogo stick.

22. Describe the energy transformations that take place when a skier starts skiing down a hill, but after a time is brought to rest by striking a snowdrift.

23. Suppose you lift a suitcase from the floor to a table. The work you do on the suitcase depends on which of the following: (a) whether you lift it straight up or along a more complicated path, (b) the time the lifting takes, (c) the height of the table, and (d) the weight of the suitcase?

24. Repeat Question 23 for the *power* needed instead of the work.

25. Why is it easier to climb a mountain via a zigzag trail rather than to climb straight up?

26. Figure 8–29 shows a potential energy curve, $U(x)$. (*a*) At which point does the force have greatest magnitude? (*b*) For each labeled point, state whether the force acts to the left or to the right, or is zero. (*c*) Where is there equilibrium and of what type is it?

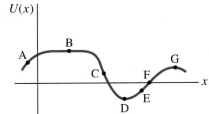

FIGURE 8–29
Question 26.

27. (*a*) Describe in detail the velocity changes of a particle that has energy E_3 in Fig. 8–23 as it moves from x_6 to x_5 and back to x_6. (*b*) Where is its kinetic energy the greatest and the least?

28. Name the type of equilibrium for each position of the balls in Fig. 8–30.

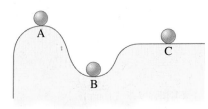

FIGURE 8–30
Question 28.

Problems

8–1 and 8–2 Conservative Forces and Potential Energy

1. (I) A spring has a spring constant k of 82.0 N/m. How much must this spring be compressed to store 35.0 J of potential energy?

2. (I) A 6.0-kg monkey swings from one branch to another 1.3 m higher. What is the change in gravitational potential energy?

3. (II) A spring with $k = 63$ N/m hangs vertically next to a ruler. The end of the spring is next to the 15-cm mark on the ruler. If a 2.5-kg mass is now attached to the end of the spring, where will the end of the spring line up with the ruler marks?

4. (II) A 56.5-kg hiker starts at an elevation of 1270 m and climbs to the top of a 2660-m peak. (*a*) What is the hiker's change in potential energy? (*b*) What is the minimum work required of the hiker? (*c*) Can the actual work done be greater than this? Explain.

5. (II) A 1.60-m tall person lifts a 1.95-kg book off the ground so it is 2.20 m above the ground. What is the potential energy of the book relative to (*a*) the ground, and (*b*) the top of the person's head? (*c*) How is the work done by the person related to the answers in parts (*a*) and (*b*)?

6. (II) A 1200-kg car rolling on a horizontal surface has speed $v = 75$ km/h when it strikes a horizontal coiled spring and is brought to rest in a distance of 2.2 m. What is the spring stiffness constant of the spring?

7. (II) A particular spring obeys the force law $\vec{\mathbf{F}} = \left(-kx + ax^3 + bx^4\right)\hat{\mathbf{i}}$. (*a*) Is this force conservative? Explain why or why not. (*b*) If it is conservative, determine the form of the potential energy function.

8. (II) If $U = 3x^2 + 2xy + 4y^2z$, what is the force, $\vec{\mathbf{F}}$?

9. (II) A particle is constrained to move in one dimension along the x axis and is acted upon by a force given by

$$\vec{\mathbf{F}}(x) = -\frac{k}{x^3}\hat{\mathbf{i}}$$

where k is a constant with units appropriate to the SI system. Find the potential energy function $U(x)$, if U is arbitrarily defined to be zero at $x = 2.0$ m, so that $U(2.0\text{ m}) = 0$.

10. (II) A particle constrained to move in one dimension is subject to a force $F(x)$ that varies with position x as

$$\vec{\mathbf{F}}(x) = A\sin(kx)\hat{\mathbf{i}}$$

where A and k are constants. What is the potential energy function $U(x)$, if we take $U = 0$ at the point $x = 0$?

8–3 and 8–4 Conservation of Mechanical Energy

11. (I) A novice skier, starting from rest, slides down a frictionless 13.0° incline whose vertical height is 125 m. How fast is she going when she reaches the bottom?

12. (I) Jane, looking for Tarzan, is running at top speed (5.0 m/s) and grabs a vine hanging vertically from a tall tree in the jungle. How high can she swing upward? Does the length of the vine affect your answer?

13. (II) In the high jump, the kinetic energy of an athlete is transformed into gravitational potential energy without the aid of a pole. With what minimum speed must the athlete leave the ground in order to lift his center of mass 2.10 m and cross the bar with a speed of 0.70 m/s?

14. (II) A sled is initially given a shove up a frictionless 23.0° incline. It reaches a maximum vertical height 1.12 m higher than where it started. What was its initial speed?

15. (II) A 55-kg bungee jumper leaps from a bridge. She is tied to a bungee cord that is 12 m long when unstretched, and falls a total of 31 m. (*a*) Calculate the spring constant k of the bungee cord assuming Hooke's law applies. (*b*) Calculate the maximum acceleration she experiences.

16. (II) A 72-kg trampoline artist jumps vertically upward from the top of a platform with a speed of 4.5 m/s. (*a*) How fast is he going as he lands on the trampoline, 2.0 m below (Fig. 8–31)? (*b*) If the trampoline behaves like a spring of spring constant 5.8×10^4 N/m, how far does he depress it?

FIGURE 8–31
Problem 16.

17. (II) The total energy E of an object of mass m that moves in one dimension under the influence of only conservative forces can be written as

$$E = \frac{1}{2}mv^2 + U.$$

Use conservation of energy, $dE/dt = 0$, to predict Newton's second law.

18. (II) A 0.40-kg ball is thrown with a speed of 8.5 m/s at an upward angle of 36°. (*a*) What is its speed at its highest point, and (*b*) how high does it go? (Use conservation of energy.)

19. (II) A vertical spring (ignore its mass), whose spring constant is 875 N/m, is attached to a table and is compressed down by 0.160 m. (a) What upward speed can it give to a 0.380-kg ball when released? (b) How high above its original position (spring compressed) will the ball fly?

20. (II) A roller-coaster car shown in Fig. 8–32 is pulled up to point 1 where it is released from rest. Assuming no friction, calculate the speed at points 2, 3, and 4.

FIGURE 8–32
Problems 20 and 34.

21. (II) When a mass m sits at rest on a spring, the spring is compressed by a distance d from its undeformed length (Fig. 8–33a). Suppose instead that the mass is released from rest when it barely touches the undeformed spring (Fig. 8–33b). Find the distance D that the spring is compressed before it is able to stop the mass. Does $D = d$? If not, why not?

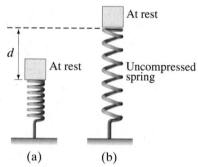

FIGURE 8–33
Problem 21.

(a)   (b)

22. (II) Two masses are connected by a string as shown in Fig. 8–34. Mass $m_A = 4.0$ kg rests on a frictionless inclined plane, while $m_B = 5.0$ kg is initially held at a height of $h = 0.75$ m above the floor. (a) If m_B is allowed to fall, what will be the resulting acceleration of the masses? (b) If the masses were initially at rest, use the kinematic equations (Eqs. 2–12) to find their velocity just before m_B hits the floor. (c) Use conservation of energy to find the velocity of the masses just before m_B hits the floor. You should get the same answer as in part (b).

FIGURE 8–34
Problem 22.

23. (II) A block of mass m is attached to the end of a spring (spring stiffness constant k), Fig. 8–35. The mass is given an initial displacement x_0 from equilibrium, and an initial speed v_0. Ignoring friction and the mass of the spring, use energy methods to find (a) its maximum speed, and (b) its maximum stretch from equilibrium, in terms of the given quantities.

FIGURE 8–35   Problems 23, 37, and 38.

24. (II) A cyclist intends to cycle up a 9.50° hill whose vertical height is 125 m. The pedals turn in a circle of diameter 36.0 cm. Assuming the mass of bicycle plus person is 75.0 kg, (a) calculate how much work must be done against gravity. (b) If each complete revolution of the pedals moves the bike 5.10 m along its path, calculate the average force that must be exerted on the pedals tangent to their circular path. Neglect work done by friction and other losses.

25. (II) A pendulum 2.00 m long is released (from rest) at an angle $\theta_0 = 30.0°$ (Fig. 8–14). Determine the speed of the 70.0-g bob: (a) at the lowest point ($\theta = 0$); (b) at $\theta = 15.0°$, (c) at $\theta = -15.0°$ (i.e., on the opposite side). (d) Determine the tension in the cord at each of these three points. (e) If the bob is given an initial speed $v_0 = 1.20$ m/s when released at $\theta = 30.0°$, recalculate the speeds for parts (a), (b), and (c).

26. (II) What should be the spring constant k of a spring designed to bring a 1200-kg car to rest from a speed of 95 km/h so that the occupants undergo a maximum acceleration of $5.0\,g$?

27. (III) An engineer is designing a spring to be placed at the bottom of an elevator shaft. If the elevator cable breaks when the elevator is at a height h above the top of the spring, calculate the value that the spring constant k should have so that passengers undergo an acceleration of no more than $5.0\,g$ when brought to rest. Let M be the total mass of the elevator and passengers.

28. (III) A skier of mass m starts from rest at the top of a solid sphere of radius r and slides down its frictionless surface. (a) At what angle θ (Fig. 8–36) will the skier leave the sphere? (b) If friction were present, would the skier fly off at a greater or lesser angle?

FIGURE 8–36
Problem 28.

8–5 and 8–6 Law of Conservation of Energy

29. (I) Two railroad cars, each of mass 56,000 kg, are traveling 95 km/h toward each other. They collide head-on and come to rest. How much thermal energy is produced in this collision?

30. (I) A 16.0-kg child descends a slide 2.20 m high and reaches the bottom with a speed of 1.25 m/s. How much thermal energy due to friction was generated in this process?

31. (II) A ski starts from rest and slides down a 28° incline 85 m long. (a) If the coefficient of friction is 0.090, what is the ski's speed at the base of the incline? (b) If the snow is level at the foot of the incline and has the same coefficient of friction, how far will the ski travel along the level? Use energy methods.

32. (II) A 145-g baseball is dropped from a tree 14.0 m above the ground. (a) With what speed would it hit the ground if air resistance could be ignored? (b) If it actually hits the ground with a speed of 8.00 m/s, what is the average force of air resistance exerted on it?

33. (II) A 96-kg crate, starting from rest, is pulled across a floor with a constant horizontal force of 350 N. For the first 15 m the floor is frictionless, and for the next 15 m the coefficient of friction is 0.25. What is the final speed of the crate?

34. (II) Suppose the roller-coaster car in Fig. 8–32 passes point 1 with a speed of 1.70 m/s. If the average force of friction is equal to 0.23 of its weight, with what speed will it reach point 2? The distance traveled is 45.0 m.

35. (II) A skier traveling 9.0 m/s reaches the foot of a steady upward 19° incline and glides 12 m up along this slope before coming to rest. What was the average coefficient of friction?

36. (II) Consider the track shown in Fig. 8–37. The section AB is one quadrant of a circle of radius 2.0 m and is frictionless. B to C is a horizontal span 3.0 m long with a coefficient of kinetic friction $\mu_k = 0.25$. The section CD under the spring is frictionless. A block of mass 1.0 kg is released from rest at A. After sliding on the track, it compresses the spring by 0.20 m. Determine: (a) the velocity of the block at point B; (b) the thermal energy produced as the block slides from B to C; (c) the velocity of the block at point C; (d) the stiffness constant k for the spring.

$m = 1.0$ kg
A
$r = 2.0$ m
B   C   D   **FIGURE 8–37**
Problem 36.

37. (II) A 0.620-kg wood block is firmly attached to a very light horizontal spring $(k = 180 \text{ N/m})$ as shown in Fig. 8–35. This block–spring system, when compressed 5.0 cm and released, stretches out 2.3 cm beyond the equilibrium position before stopping and turning back. What is the coefficient of kinetic friction between the block and the table?

38. (II) A 180-g wood block is firmly attached to a very light horizontal spring, Fig. 8–35. The block can slide along a table where the coefficient of friction is 0.30. A force of 25 N compresses the spring 18 cm. If the spring is released from this position, how far beyond its equilibrium position will it stretch on its first cycle?

39. (II) You drop a ball from a height of 2.0 m, and it bounces back to a height of 1.5 m. (a) What fraction of its initial energy is lost during the bounce? (b) What is the ball's speed just before and just after the bounce? (c) Where did the energy go?

40. (II) A 56-kg skier starts from rest at the top of a 1200-m-long trail which drops a total of 230 m from top to bottom. At the bottom, the skier is moving 11.0 m/s. How much energy was dissipated by friction?

41. (II) How much does your gravitational energy change when you jump as high as you can (say, 1.0 m)?

42. (III) A spring $(k = 75 \text{ N/m})$ has an equilibrium length of 1.00 m. The spring is compressed to a length of 0.50 m and a mass of 2.0 kg is placed at its free end on a frictionless slope which makes an angle of 41° with respect to the horizontal (Fig. 8–38). The spring is then released. (a) If the mass is *not* attached to the spring, how far up the slope will the mass move before coming to rest? (b) If the mass *is* attached to the spring, how far up the slope will the mass move before coming to rest? (c) Now the incline has a coefficient of kinetic friction μ_k. If the block, attached to the spring, is observed to stop just as it reaches the spring's equilibrium position, what is the coefficient of friction μ_k?

FIGURE 8–38
Problem 42.

0.50 m
$\theta = 41°$

43. (III) A 2.0-kg block slides along a horizontal surface with a coefficient of kinetic friction $\mu_k = 0.30$. The block has a speed $v = 1.3$ m/s when it strikes a massless spring head-on (as in Fig. 8–18). (a) If the spring has force constant $k = 120 \text{ N/m}$, how far is the spring compressed? (b) What minimum value of the coefficient of static friction, μ_S, will assure that the spring remains compressed at the maximum compressed position? (c) If μ_S is less than this, what is the speed of the block when it detaches from the decompressing spring? [Hint: Detachment occurs when the spring reaches its natural length $(x = 0)$; explain why.]

44. (III) Early test flights for the space shuttle used a "glider" (mass of 980 kg including pilot). After a horizontal launch at 480 km/h at a height of 3500 m, the glider eventually landed at a speed of 210 km/h. (a) What would its landing speed have been in the absence of air resistance? (b) What was the average force of air resistance exerted on it if it came in at a constant glide angle of 12° to the Earth's surface?

8–7 Gravitational Potential Energy

45. (I) For a satellite of mass m_S in a circular orbit of radius r_S around the Earth, determine (a) its kinetic energy K, (b) its potential energy U ($U = 0$ at infinity), and (c) the ratio K/U.

46. (I) Jill and her friends have built a small rocket that soon after lift-off reaches a speed of 850 m/s. How high above the Earth can it rise? Ignore air friction.

47. (I) The escape velocity from planet A is double that for planet B. The two planets have the same mass. What is the ratio of their radii, r_A/r_B?

48. (II) Show that Eq. 8–16 for gravitational potential energy reduces to Eq. 8–2, $\Delta U = mg(y_2 - y_1)$, for objects near the surface of the Earth.

49. (II) Determine the escape velocity from the Sun for an object (a) at the Sun's surface $(r = 7.0 \times 10^5 \text{ km}, M = 2.0 \times 10^{30} \text{ kg})$, and (b) at the average distance of the Earth $(1.50 \times 10^8 \text{ km})$. Compare to the speed of the Earth in its orbit.

50. (II) Two Earth satellites, A and B, each of mass $m = 950$ kg, are launched into circular orbits around the Earth's center. Satellite A orbits at an altitude of 4200 km, and satellite B orbits at an altitude of 12,600 km. (a) What are the potential energies of the two satellites? (b) What are the kinetic energies of the two satellites? (c) How much work would it require to change the orbit of satellite A to match that of satellite B?

51. (II) Show that the escape velocity for any satellite in a circular orbit is $\sqrt{2}$ times its velocity.

52. (II) (a) Show that the total mechanical energy of a satellite (mass m) orbiting at a distance r from the center of the Earth (mass M_E) is

$$E = -\frac{1}{2}\frac{GmM_E}{r},$$

if $U = 0$ at $r = \infty$. (b) Show that although friction causes the value of E to decrease slowly, kinetic energy must actually increase if the orbit remains a circle.

53. (II) Take into account the Earth's rotational speed (1 rev/day) and determine the necessary speed, with respect to Earth, for a rocket to escape if fired from the Earth at the equator in a direction (a) eastward; (b) westward; (c) vertically upward.

54. (II) (a) Determine a formula for the maximum height h that a rocket will reach if launched vertically from the Earth's surface with speed v_0 ($< v_{esc}$). Express in terms of v_0, r_E, M_E, and G. (b) How high does a rocket go if $v_0 = 8.35$ km/s? Ignore air resistance and the Earth's rotation.

55. (II) (a) Determine the rate at which the escape velocity from the Earth changes with distance from the center of the Earth, dv_{esc}/dr. (b) Use the approximation $\Delta v \approx (dv/dr)\,\Delta r$ to determine the escape velocity for a spacecraft orbiting the Earth at a height of 320 km.

56. (II) A meteorite has a speed of 90.0 m/s when 850 km above the Earth. It is falling vertically (ignore air resistance) and strikes a bed of sand in which it is brought to rest in 3.25 m. (a) What is its speed just before striking the sand? (b) How much work does the sand do to stop the meteorite (mass = 575 kg)? (c) What is the average force exerted by the sand on the meteorite? (d) How much thermal energy is produced?

57. (II) How much work would be required to move a satellite of mass m from a circular orbit of radius $r_1 = 2r_E$ about the Earth to another circular orbit of radius $r_2 = 3r_E$? (r_E is the radius of the Earth.)

58. (II) (a) Suppose we have three masses, m_1, m_2, and m_3, that initially are infinitely far apart from each other. Show that the work needed to bring them to the positions shown in Fig. 8–39 is

$$ W = -G\left(\frac{m_1 m_2}{r_{12}} + \frac{m_1 m_3}{r_{13}} + \frac{m_2 m_3}{r_{23}} \right). $$

(b) Can we say that this formula also gives the potential energy of the system, or the potential energy of one or two of the objects? (c) Is W equal to the binding energy of the system—that is, equal to the energy required to separate the components by an infinite distance? Explain.

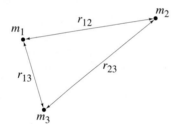

FIGURE 8–39
Problem 58.

59. (II) A NASA satellite has just observed an asteroid that is on a collision course with the Earth. The asteroid has an estimated mass, based on its size, of 5×10^9 kg. It is approaching the Earth on a head-on course with a velocity of 660 m/s relative to the Earth and is now 5.0×10^6 km away. With what speed will it hit the Earth's surface, neglecting friction with the atmosphere?

60. (II) A sphere of radius r_1 has a concentric spherical cavity of radius r_2 (Fig. 8–40). Assume this spherical shell of thickness $r_1 - r_2$ is uniform and has a total mass M. Show that the gravitational potential energy of a mass m at a distance r from the center of the shell ($r > r_1$) is given by

$$ U = -\frac{GmM}{r}. $$

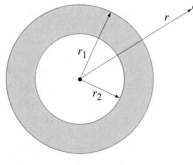

FIGURE 8–40
Problem 60.

61. (III) To escape the solar system, an interstellar spacecraft must overcome the gravitational attraction of both the Earth and Sun. Ignore the effects of other bodies in the solar system. (a) Show that the escape velocity is

$$ v = \sqrt{v_E^2 + (v_S - v_0)^2} = 16.7 \text{ km/s}, $$

where: v_E is the escape velocity from the Earth (Eq. 8–19); $v_S = \sqrt{2GM_S/r_{SE}}$ is the escape velocity from the gravitational field of the Sun at the orbit of the Earth but far from the Earth's influence (r_{SE} is the Sun–Earth distance); and v_0 is the Earth's orbital velocity about the Sun. (b) Show that the energy required is 1.40×10^8 J per kilogram of spacecraft mass. [*Hint*: Write the energy equation for escape from Earth with v' as the velocity, relative to Earth, but far from Earth; then let $v' + v_0$ be the escape velocity from the Sun.]

8–8 Power

62. (I) How long will it take a 1750-W motor to lift a 335-kg piano to a sixth-story window 16.0 m above?

63. (I) If a car generates 18 hp when traveling at a steady 95 km/h, what must be the average force exerted on the car due to friction and air resistance?

64. (I) An 85-kg football player traveling 5.0 m/s is stopped in 1.0 s by a tackler. (a) What is the original kinetic energy of the player? (b) What average power is required to stop him?

65. (II) A driver notices that her 1080-kg car slows down from 95 km/h to 65 km/h in about 7.0 s on the level when it is in neutral. Approximately what power (watts and hp) is needed to keep the car traveling at a constant 80 km/h?

66. (II) How much work can a 3.0-hp motor do in 1.0 h?

67. (II) An outboard motor for a boat is rated at 55 hp. If it can move a particular boat at a steady speed of 35 km/h, what is the total force resisting the motion of the boat?

68. (II) A 1400-kg sports car accelerates from rest to 95 km/h in 7.4 s. What is the average power delivered by the engine?

69. (II) During a workout, football players ran up the stadium stairs in 75 s. The stairs are 78 m long and inclined at an angle of 33°. If a player has a mass of 92 kg, estimate his average power output on the way up. Ignore friction and air resistance.

70. (II) A pump lifts 21.0 kg of water per minute through a height of 3.50 m. What minimum output rating (watts) must the pump motor have?

71. (II) A ski area claims that its lifts can move 47,000 people per hour. If the average lift carries people about 200 m (vertically) higher, estimate the maximum total power needed.

72. (II) A 75-kg skier grips a moving rope that is powered by an engine and is pulled at constant speed to the top of a 23° hill. The skier is pulled a distance $x = 220$ m along the incline and it takes 2.0 min to reach the top of the hill. If the coefficient of kinetic friction between the snow and skis is $\mu_k = 0.10$, what horsepower engine is required if 30 such skiers (max) are on the rope at one time?

73. (III) The position of a 280-g object is given (in meters) by $x = 5.0t^3 - 8.0t^2 - 44t$, where t is in seconds. Determine the net rate of work done on this object (a) at $t = 2.0$ s and (b) at $t = 4.0$ s. (c) What is the average net power input during the interval from $t = 0$ s to $t = 2.0$ s, and in the interval from $t = 2.0$ s to 4.0 s?

74. (III) A bicyclist coasts down a 6.0° hill at a steady speed of 4.0 m/s. Assuming a total mass of 75 kg (bicycle plus rider), what must be the cyclist's power output to climb the same hill at the same speed?

*8–9 Potential Energy Diagrams

*75. (II) Draw a potential energy diagram, U vs. x, and analyze the motion of a mass m resting on a frictionless horizontal table and connected to a horizontal spring with stiffness constant k. The mass is pulled a distance to the right so that the spring is stretched a distance x_0 initially, and then the mass is released from rest.

*76. (II) The spring of Problem 75 has a stiffness constant $k = 160 \text{ N/m}$. The mass $m = 5.0 \text{ kg}$ is released from rest when the spring is stretched $x_0 = 1.0 \text{ m}$ from equilibrium. Determine (a) the total energy of the system; (b) the kinetic energy when $x = \frac{1}{2}x_0$; (c) the maximum kinetic energy; (d) the maximum speed and at what positions it occurs; (e) the maximum acceleration and where it occurs.

*77. (III) The potential energy of the two atoms in a diatomic (two-atom) molecule can be written

$$U(r) = -\frac{a}{r^6} + \frac{b}{r^{12}},$$

where r is the distance between the two atoms and a and b are positive constants. (a) At what values of r is $U(r)$ a minimum? A maximum? (b) At what values of r is $U(r) = 0$? (c) Plot $U(r)$ as a function of r from $r = 0$ to r at a value large enough for all the features in (a) and (b) to show. (d) Describe the motion of one atom with respect to the second atom when $E < 0$, and when $E > 0$. (e) Let F be the force one atom exerts on the other. For what values of r is $F > 0, F < 0, F = 0$? (f) Determine F as a function of r.

*78. (III) The *binding energy* of a two-particle system is defined as the energy required to separate the two particles from their state of lowest energy to $r = \infty$. Determine the binding energy for the molecule discussed in Problem 77.

General Problems

79. What is the average power output of an elevator that lifts 885 kg a vertical height of 32.0 m in 11.0 s?

80. A projectile is fired at an upward angle of 48.0° from the top of a 135-m-high cliff with a speed of 165 m/s. What will be its speed when it strikes the ground below? (Use conservation of energy.)

81. Water flows over a dam at the rate of 580 kg/s and falls vertically 88 m before striking the turbine blades. Calculate (a) the speed of the water just before striking the turbine blades (neglect air resistance), and (b) the rate at which mechanical energy is transferred to the turbine blades, assuming 55% efficiency.

82. A bicyclist of mass 75 kg (including the bicycle) can coast down a 4.0° hill at a steady speed of 12 km/h. Pumping hard, the cyclist can descend the hill at a speed of 32 km/h. Using the same power, at what speed can the cyclist climb the same hill? Assume the force of friction is proportional to the square of the speed v; that is, $F_{fr} = bv^2$, where b is a constant.

83. A 62-kg skier starts from rest at the top of a ski jump, point A in Fig. 8–41, and travels down the ramp. If friction and air resistance can be neglected, (a) determine her speed v_B when she reaches the horizontal end of the ramp at B. (b) Determine the distance s to where she strikes the ground at C.

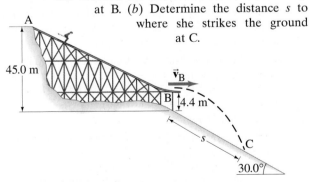

FIGURE 8–41 Problems 83 and 84.

84. Repeat Problem 83, but now assume the ski jump turns upward at point B and gives her a vertical component of velocity (at B) of 3.0 m/s.

85. A ball is attached to a horizontal cord of length ℓ whose other end is fixed, Fig. 8–42. (a) If the ball is released, what will be its speed at the lowest point of its path? (b) A peg is located a distance h directly below the point of attachment of the cord. If $h = 0.80\ell$, what will be the speed of the ball when it reaches the top of its circular path about the peg?

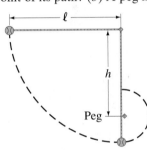

FIGURE 8–42 Problems 85 and 86.

86. Show that h must be greater than 0.60ℓ if the ball in Fig. 8–42 is to make a complete circle about the peg.

87. Show that on a roller coaster with a circular vertical loop (Fig. 8–43), the difference in your apparent weight at the top of the loop and the bottom of the loop is 6 g's—that is, six times your weight. Ignore friction. Show also that as long as your speed is above the minimum needed, this answer doesn't depend on the size of the loop or how fast you go through it.

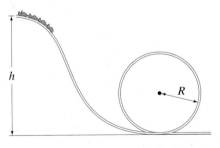

FIGURE 8–43 Problem 87.

88. If you stand on a bathroom scale, the spring inside the scale compresses 0.50 mm, and it tells you your weight is 760 N. Now if you jump on the scale from a height of 1.0 m, what does the scale read at its peak?

89. A 65-kg hiker climbs to the top of a 4200-m-high mountain. The climb is made in 5.0 h starting at an elevation of 2800 m. Calculate (a) the work done by the hiker against gravity, (b) the average power output in watts and in horsepower, and (c) assuming the body is 15% efficient, what rate of energy input was required.

90. The small mass m sliding without friction along the looped track shown in Fig. 8–44 is to remain on the track at all times, even at the very top of the loop of radius r. (a) In terms of the given quantities, determine the minimum release height h. Next, if the actual release height is $2h$, calculate the normal force exerted (b) by the track at the bottom of the loop, (c) by the track at the top of the loop, and (d) by the track after the block exits the loop onto the flat section.

FIGURE 8–44
Problem 90.

91. A 56-kg student runs at 5.0 m/s, grabs a hanging rope, and swings out over a lake (Fig. 8–45). He releases the rope when his velocity is zero. (a) What is the angle θ when he releases the rope? (b) What is the tension in the rope just before he releases it? (c) What is the maximum tension in the rope?

FIGURE 8–45
Problem 91.

92. The nuclear force between two neutrons in a nucleus is described roughly by the Yukawa potential

$$U(r) = -U_0 \frac{r_0}{r} e^{-r/r_0},$$

where r is the distance between the neutrons and U_0 and $r_0 (\approx 10^{-15}\,\text{m})$ are constants. (a) Determine the force $F(r)$. (b) What is the ratio $F(3r_0)/F(r_0)$? (c) Calculate this same ratio for the force between two electrically charged particles where $U(r) = -C/r$, with C constant. Why is the Yukawa force referred to as a "short-range" force?

93. A fire hose for use in urban areas must be able to shoot a stream of water to a maximum height of 33 m. The water leaves the hose at ground level in a circular stream 3.0 cm in diameter. What minimum power is required to create such a stream of water? Every cubic meter of water has a mass of 1.00×10^3 kg.

94. A 16-kg sled starts up a 28° incline with a speed of 2.4 m/s. The coefficient of kinetic friction is $\mu_k = 0.25$. (a) How far up the incline does the sled travel? (b) What condition must you put on the coefficient of static friction if the sled is not to get stuck at the point determined in part (a)? (c) If the sled slides back down, what is its speed when it returns to its starting point?

95. The Lunar Module could make a safe landing if its vertical velocity at impact is 3.0 m/s or less. Suppose that you want to determine the greatest height h at which the pilot could shut off the engine if the velocity of the lander relative to the surface is (a) zero; (b) 2.0 m/s downward; (c) 2.0 m/s upward. Use conservation of energy to determine h in each case. The acceleration due to gravity at the surface of the Moon is $1.62\,\text{m/s}^2$.

96. Proper design of automobile braking systems must account for heat buildup under heavy braking. Calculate the thermal energy dissipated from brakes in a 1500-kg car that descends a 17° hill. The car begins braking when its speed is 95 km/h and slows to a speed of 35 km/h in a distance of 0.30 km measured along the road.

97. Some electric power companies use water to store energy. Water is pumped by reversible turbine pumps from a low reservoir to a high reservoir. To store the energy produced in 1.0 hour by a 180-MW electric power plant, how many cubic meters of water will have to be pumped from the lower to the upper reservoir? Assume the upper reservoir is 380 m above the lower one, and we can neglect the small change in depths of each. Water has a mass of 1.00×10^3 kg for every 1.0 m³.

98. Estimate the energy required from fuel to launch a 1465-kg satellite into orbit 1375 km above the Earth's surface. Consider two cases: (a) the satellite is launched into an equatorial orbit from a point on the Earth's equator, and (b) it is launched from the North Pole into a polar orbit.

99. A satellite is in an elliptic orbit around the Earth (Fig. 8–46). Its speed at the perigee A is 8650 m/s. (a) Use conservation of energy to determine its speed at B. The radius of the Earth is 6380 km. (b) Use conservation of energy to determine the speed at the apogee C.

FIGURE 8–46
Problem 99.

16,460 km
8230 km 8230 km
13,900 km

100. Suppose the gravitational potential energy of an object of mass m at a distance r from the center of the Earth is given by

$$U(r) = -\frac{GMm}{r} e^{-\alpha r}$$

where α is a positive constant and e is the exponential function. (Newton's law of universal gravitation has $\alpha = 0$). (a) What would be the force on the object as a function of r? (b) What would be the object's escape velocity in terms of the Earth's radius R_E?

101. (a) If the human body could convert a candy bar directly into work, how high could a 76-kg man climb a ladder if he were fueled by one bar ($= 1100$ kJ)? (b) If the man then jumped off the ladder, what will be his speed when he reaches the bottom?

102. Electric energy units are often expressed in the form of "kilowatt-hours." (a) Show that one kilowatt-hour (kWh) is equal to 3.6×10^6 J. (b) If a typical family of four uses electric energy at an average rate of 580 W, how many kWh would their electric bill show for one month, and (c) how many joules would this be? (d) At a cost of $0.12 per kWh, what would their monthly bill be in dollars? Does the monthly bill depend on the *rate* at which they use the electric energy?

103. Chris jumps off a bridge with a bungee cord (a heavy stretchable cord) tied around his ankle, Fig. 8–47. He falls for 15 m before the bungee cord begins to stretch. Chris's mass is 75 kg and we assume the cord obeys Hooke's law, $F = -kx$, with $k = 50$ N/m. If we neglect air resistance, estimate how far below the bridge Chris's foot will be before coming to a stop. Ignore the mass of the cord (not realistic, however) and treat Chris as a particle.

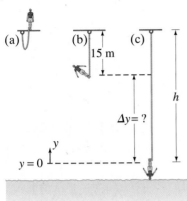

FIGURE 8–47 Problem 103. (a) Bungee jumper about to jump. (b) Bungee cord at its unstretched length. (c) Maximum stretch of cord.

104. In a common test for cardiac function (the "stress test"), the patient walks on an inclined treadmill (Fig. 8–48). Estimate the power required from a 75-kg patient when the treadmill is sloping at an angle of $12°$ and the velocity is 3.3 km/h. (How does this power compare to the power rating of a lightbulb?)

FIGURE 8–48 Problem 104.

105. (a) If a volcano spews a 450-kg rock vertically upward a distance of 320 m, what was its velocity when it left the volcano? (b) If the volcano spews 1000 rocks of this size every minute, estimate its power output.

106. A film of Jesse Owens's famous long jump (Fig. 8–49) in the 1936 Olympics shows that his center of mass rose 1.1 m from launch point to the top of the arc. What minimum speed did he need at launch if he was traveling at 6.5 m/s at the top of the arc?

FIGURE 8–49 Problem 106.

107. An elevator cable breaks when a 920-kg elevator is 24 m above a huge spring $(k = 2.2 \times 10^5$ N/m) at the bottom of the shaft. Calculate (a) the work done by gravity on the elevator before it hits the spring, (b) the speed of the elevator just before striking the spring, and (c) the amount the spring compresses (note that work is done by both the spring and gravity in this part).

108. A particle moves where its potential energy is given by $U(r) = U_0[(2/r^2) - (1/r)]$. (a) Plot $U(r)$ versus r. Where does the curve cross the $U(r) = 0$ axis? At what value of r does the minimum value of $U(r)$ occur? (b) Suppose that the particle has an energy of $E = -0.050U_0$. Sketch in the approximate turning points of the motion of the particle on your diagram. What is the maximum kinetic energy of the particle, and for what value of r does this occur?

109. A particle of mass m moves under the influence of a potential energy

$$U(x) = \frac{a}{x} + bx$$

where a and b are positive constants and the particle is restricted to the region $x > 0$. Find a point of equilibrium for the particle and demonstrate that it is stable.

*Numerical/Computer

***110.** (III) The two atoms in a diatomic molecule exert an attractive force on each other at large distances and a repulsive force at short distances. The magnitude of the force between two atoms in a diatomic molecule can be approximated by the Lennard-Jones force, or $F(r) = F_0[2(\sigma/r)^{13} - (\sigma/r)^7]$, where r is the separation between the two atoms, and σ and F_0 are constant. For an oxygen molecule (which is diatomic) $F_0 = 9.60 \times 10^{-11}$ N and $\sigma = 3.50 \times 10^{-11}$ m. (a) Integrate the equation for $F(r)$ to determine the potential energy $U(r)$ of the oxygen molecule. (b) Find the equilibrium distance r_0 between the two atoms. (c) Graph $F(r)$ and $U(r)$ between $0.9\,r_0$ and $2.5\,r_0$.

Answers to Exercises

A: (e), (e); (e), (c).

B: (b).

C: (c).

D: Equal speeds.

Conservation of linear momentum is another great conservation law of physics. Collisions, such as between billiard or pool balls, illustrate this vector law very nicely: the total vector momentum just before the collision equals the total vector momentum just after the collision. In this photo, the moving cue ball strikes the 11 ball at rest. Both balls move after the collision, at angles, but the sum of their vector momenta equals the initial vector momentum of the incoming cue ball.

We will consider both elastic collisions (where kinetic energy is also conserved) and inelastic collisions. We also examine the concept of center of mass, and how it helps us in the study of complex motion.

$m_2 \vec{\mathbf{v}}'_2$ (after)

$m_1 \vec{\mathbf{v}}_1$ (before)

$m_1 \vec{\mathbf{v}}'_1$ (after)

CHAPTER 9

Linear Momentum

CHAPTER-OPENING QUESTIONS—Guess now!

1. A railroad car loaded with rocks coasts on a level track without friction. A worker on board starts throwing the rocks horizontally backward from the car. Then what happens?

(a) The car slows down.
(b) The car speeds up.
(c) First the car speeds up and then it slows down.
(d) The car's speed remains constant.
(e) None of these.

2. Which answer would you choose if the rocks fall out through a hole in the floor of the car, one at a time?

The law of conservation of energy, which we discussed in the previous Chapter, is one of several great conservation laws in physics. Among the other quantities found to be conserved are linear momentum, angular momentum, and electric charge. We will eventually discuss all of these because the conservation laws are among the most important ideas in science. In this Chapter, we discuss linear momentum, and its conservation. The law of conservation of momentum is essentially a reworking of Newton's laws that gives us tremendous physical insight and problem-solving power.

214

We make use of the laws of conservation of linear momentum and of energy to analyze collisions. Indeed, the law of conservation of momentum is particularly useful when dealing with a system of two or more objects that interact with each other, such as in collisions of ordinary objects or nuclear particles.

Our focus up to now has been mainly on the motion of a single object, often thought of as a "particle" in the sense that we have ignored any rotation or internal motion. In this Chapter we will deal with systems of two or more objects, and toward the end of the Chapter, the concept of center of mass.

9–1 Momentum and Its Relation to Force

The **linear momentum** (or "momentum" for short) of an object is defined as the product of its mass and its velocity. Momentum (plural is *momenta*) is represented by the symbol $\vec{p}$. If we let m represent the mass of an object and $\vec{v}$ represent its velocity, then its momentum $\vec{p}$ is defined as

$$\vec{p} = m\vec{v}. \tag{9–1}$$

Velocity is a vector, so momentum too is a vector. The direction of the momentum is the direction of the velocity, and the magnitude of the momentum is $p = mv$. Because velocity depends on the reference frame, so does momentum; thus the reference frame must be specified. The unit of momentum is that of mass $\times$ velocity, which in SI units is kg·m/s. There is no special name for this unit.

Everyday usage of the term *momentum* is in accord with the definition above. According to Eq. 9–1, a fast-moving car has more momentum than a slow-moving car of the same mass; a heavy truck has more momentum than a small car moving with the same speed. The more momentum an object has, the harder it is to stop it, and the greater effect it will have on another object if it is brought to rest by striking that object. A football player is more likely to be stunned if tackled by a heavy opponent running at top speed than by a lighter or slower-moving tackler. A heavy, fast-moving truck can do more damage than a slow-moving motorcycle.

A force is required to change the momentum of an object, whether it is to increase the momentum, to decrease it, or to change its direction. Newton originally stated his second law in terms of momentum (although he called the product mv the "quantity of motion"). Newton's statement of the **second law of motion**, translated into modern language, is as follows:

The rate of change of momentum of an object is equal to the net force applied to it.

We can write this as an equation,

$$\Sigma\vec{F} = \frac{d\vec{p}}{dt}, \tag{9–2}$$

| NEWTON'S SECOND LAW |

where $\Sigma\vec{F}$ is the net force applied to the object (the vector sum of all forces acting on it). We can readily derive the familiar form of the second law, $\Sigma\vec{F} = m\vec{a}$, from Eq. 9–2 for the case of constant mass. If $\vec{v}$ is the velocity of an object at any moment, then Eq. 9–2 gives

⚠ **CAUTION**
The change in the momentum vector is in the direction of the net force

$$\Sigma\vec{F} = \frac{d\vec{p}}{dt} = \frac{d(m\vec{v})}{dt} = m\frac{d\vec{v}}{dt} = m\vec{a} \qquad \text{[constant mass]}$$

because, by definition, $\vec{a} = d\vec{v}/dt$ and we assume $m = $ constant. Newton's statement, Eq. 9–2, is actually more general than the more familiar one because it includes the situation in which the mass may change. This is important in certain circumstances, such as for rockets which lose mass as they burn fuel (Section 9–10) and in relativity theory (Chapter 36).

EXERCISE A Light carries momentum, so if a light beam strikes a surface, it will exert a force on that surface. If the light is reflected rather than absorbed, the force will be (*a*) the same, (*b*) less, (*c*) greater, (*d*) impossible to tell, (*e*) none of these.

FIGURE 9–1 Example 9–1.

EXAMPLE 9–1 **ESTIMATE** **Force of a tennis serve.** For a top player, a tennis ball may leave the racket on the serve with a speed of 55 m/s (about 120 mi/h), Fig. 9–1. If the ball has a mass of 0.060 kg and is in contact with the racket for about 4 ms (4×10^{-3} s), estimate the average force on the ball. Would this force be large enough to lift a 60-kg person?

APPROACH We write Newton's second law, Eq. 9–2, for the average force as

$$F_{avg} = \frac{\Delta p}{\Delta t} = \frac{mv_2 - mv_1}{\Delta t},$$

where mv_1 and mv_2 are the initial and final momenta. The tennis ball is hit when its initial velocity v_1 is very nearly zero at the top of the throw, so we set $v_1 = 0$, whereas $v_2 = 55$ m/s is in the horizontal direction. We ignore all other forces on the ball, such as gravity, in comparison to the force exerted by the tennis racket.

SOLUTION The force exerted on the ball by the racket is

$$F_{avg} = \frac{\Delta p}{\Delta t} = \frac{mv_2 - mv_1}{\Delta t} = \frac{(0.060 \text{ kg})(55 \text{ m/s}) - 0}{0.004 \text{ s}}$$
$$\approx 800 \text{ N}.$$

This is a large force, larger than the weight of a 60-kg person, which would require a force $mg = (60 \text{ kg})(9.8 \text{ m/s}^2) \approx 600$ N to lift.

NOTE The force of gravity acting on the tennis ball is $mg = (0.060 \text{ kg})(9.8 \text{ m/s}^2) = 0.59$ N, which justifies our ignoring it compared to the enormous force the racket exerts.

NOTE High-speed photography and radar can give us an estimate of the contact time and the velocity of the ball leaving the racket. But a direct measurement of the force is not practical. Our calculation shows a handy technique for determining an unknown force in the real world.

FIGURE 9–2 Example 9–2.

$v = 20$ m/s

x

EXAMPLE 9–2 **Washing a car: momentum change and force.** Water leaves a hose at a rate of 1.5 kg/s with a speed of 20 m/s and is aimed at the side of a car, which stops it, Fig. 9–2. (That is, we ignore any splashing back.) What is the force exerted by the water on the car?

APPROACH The water leaving the hose has mass and velocity, so it has a momentum $p_{initial}$ in the horizontal (x) direction, and we assume gravity doesn't pull the water down significantly. When the water hits the car, the water loses this momentum ($p_{final} = 0$). We use Newton's second law in the momentum form to find the force that the car exerts on the water to stop it. By Newton's third law, the force exerted by the water on the car is equal and opposite. We have a continuing process: 1.5 kg of water leaves the hose in each 1.0-s time interval. So let us write $F = \Delta p/\Delta t$ where $\Delta t = 1.0$ s, and $mv_{initial} = (1.5 \text{ kg})(20 \text{ m/s})$.

SOLUTION The force (assumed constant) that the car must exert to change the momentum of the water is

$$F = \frac{\Delta p}{\Delta t} = \frac{p_{final} - p_{initial}}{\Delta t} = \frac{0 - 30 \text{ kg·m/s}}{1.0 \text{ s}} = -30 \text{ N}.$$

The minus sign indicates that the force exerted by the car on the water is opposite to the water's original velocity. The car exerts a force of 30 N to the left to stop the water, so by Newton's third law, the water exerts a force of 30 N to the right on the car.

NOTE Keep track of signs, although common sense helps too. The water is moving to the right, so common sense tells us the force on the car must be to the right.

EXERCISE B If the water splashes back from the car in Example 9–2, would the force on the car be larger or smaller?

9–2 Conservation of Momentum

The concept of momentum is particularly important because, if no net external force acts on a system, the total momentum of the system is a conserved quantity. Consider, for example, the head-on collision of two billiard balls, as shown in Fig. 9–3. We assume the net external force on this system of two balls is zero—that is, the only significant forces during the collision are the forces that each ball exerts on the other. Although the momentum of each of the two balls changes as a result of the collision, the *sum* of their momenta is found to be the same before as after the collision. If $m_A\vec{v}_A$ is the momentum of ball A and $m_B\vec{v}_B$ the momentum of ball B, both measured just before the collision, then the total momentum of the two balls before the collision is the vector sum $m_A\vec{v}_A + m_B\vec{v}_B$. Immediately after the collision, the balls each have a different velocity and momentum, which we designate by a "prime" on the velocity: $m_A\vec{v}'_A$ and $m_B\vec{v}'_B$. The total momentum after the collision is the vector sum $m_A\vec{v}'_A + m_B\vec{v}'_B$. No matter what the velocities and masses are, experiments show that the total momentum before the collision is the same as afterward, whether the collision is head-on or not, as long as no net external force acts:

momentum before = momentum after

$$m_A\vec{v}_A + m_B\vec{v}_B = m_A\vec{v}'_A + m_B\vec{v}'_B. \qquad [\Sigma\vec{F}_{ext} = 0] \quad \textbf{(9–3)}$$

That is, the total vector momentum of the system of two colliding balls is conserved: it stays constant.

Although the law of conservation of momentum was discovered experimentally, it can be derived from Newton's laws of motion, which we now show.

Let us consider two objects of mass m_A and m_B that have momenta $\vec{p}_A$ and $\vec{p}_B$ before they collide and $\vec{p}'_A$ and $\vec{p}'_B$ after they collide, as in Fig. 9–4. During the collision, suppose that the force exerted by object A on object B at any instant is $\vec{F}$. Then, by Newton's third law, the force exerted by object B on object A is $-\vec{F}$. During the brief collision time, we assume no other (external) forces are acting (or that $\vec{F}$ is much greater than any other external forces acting). Thus we have

$$\vec{F} = \frac{d\vec{p}_B}{dt}$$

and

$$-\vec{F} = \frac{d\vec{p}_A}{dt}.$$

We add these two equations together and find

$$0 = \frac{d(\vec{p}_A + \vec{p}_B)}{dt}$$

which tells us that

$$\vec{p}_A + \vec{p}_B = \text{constant}.$$

The total momentum thus is conserved.

We have put this derivation in the context of a collision. As long as no external forces act, it is valid over any time interval, and conservation of momentum is always valid as long as no external forces act. In the real world, external forces do act: friction on billiard balls, gravity acting on a tennis ball, and so on. So it may seem that conservation of momentum cannot be applied. Or can it? In a collision, the force each object exerts on the other acts only over a very brief time interval, and is very strong relative to the other forces. If we measure the momenta immediately before and after the collision, momentum will be very nearly conserved. We cannot wait for the external forces to produce their effect before measuring $\vec{p}'_A$ and $\vec{p}'_B$.

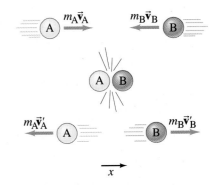

FIGURE 9–3 Momentum is conserved in a collision of two balls, labeled A and B.

FIGURE 9–4 Collision of two objects. Their momenta before collision are $\vec{p}_A$ and $\vec{p}_B$, and after collision are $\vec{p}'_A$ and $\vec{p}'_B$. At any moment during the collision each exerts a force on the other of equal magnitude but opposite direction.

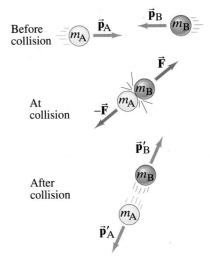

For example, when a racket hits a tennis ball or a bat hits a baseball, both before and after the "collision" the ball moves as a projectile under the action of gravity and air resistance. However, when the bat or racket hits the ball, during this brief time of the collision those external forces are insignificant compared to the collision force the bat or racket exerts on the ball. Momentum is conserved (or very nearly so) as long as we measure $\vec{p}_A$ and $\vec{p}_B$ just before the collision and $\vec{p}'_A$ and $\vec{p}'_B$ immediately after the collision (Eq. 9–3).

Our derivation of the conservation of momentum can be extended to include any number of interacting objects. Let $\vec{P}$ represent the total momentum of a system of n interacting objects which we number from 1 to n:

$$\vec{P} = m_1\vec{v}_1 + m_2\vec{v}_2 + \cdots + m_n\vec{v}_n = \Sigma \vec{p}_i.$$

We differentiate with respect to time:

$$\frac{d\vec{P}}{dt} = \Sigma \frac{d\vec{p}_i}{dt} = \Sigma \vec{F}_i \qquad (9\text{–}4)$$

where $\vec{F}_i$ represents the *net* force on the ith object. The forces can be of two types: (1) *external forces* on objects of the system, exerted by objects outside the system, and (2) *internal forces* that objects within the system exert on other objects in the system. By Newton's third law, the internal forces occur in pairs: if one object exerts a force on a second object, the second exerts an equal and opposite force on the first object. Thus, in the sum over all the forces in Eq. 9–4, all the internal forces cancel each other in pairs. Thus we have

$$\frac{d\vec{P}}{dt} = \Sigma \vec{F}_{\text{ext}}, \qquad (9\text{–}5)$$

where $\Sigma \vec{F}_{\text{ext}}$ is the sum of all external forces acting on our system. If the net external force is zero, then $d\vec{P}/dt = 0$, so $\Delta\vec{P} = 0$ or $\vec{P} = $ constant. Thus we see that

when the net external force on a system of objects is zero, the total momentum of the system remains constant.

This is the **law of conservation of momentum**. It can also be stated as

the total momentum of an isolated system of objects remains constant.

By an **isolated system**, we mean one on which no external forces act—the only forces acting are those between objects of the system.

If a net external force acts on a system, then the law of conservation of momentum will not apply. However, if the "system" can be redefined so as to include the other objects exerting these forces, then the conservation of momentum principle can apply. For example, if we take as our system a falling rock, it does not conserve momentum since an external force, the force of gravity exerted by the Earth, is acting on it and its momentum changes. However, if we include the Earth in the system, the total momentum of rock plus Earth is conserved. (This of course means that the Earth comes up to meet the ball. Since the Earth's mass is so great, its upward velocity is very tiny.)

Although the law of conservation of momentum follows from Newton's second law, as we have seen, it is in fact more general than Newton's laws. In the tiny world of the atom, Newton's laws fail, but the great conservation laws—those of energy, momentum, angular momentum, and electric charge—have been found to hold in every experimental situation tested. It is for this reason that the conservation laws are considered more basic than Newton's laws.

EXAMPLE 9–3 **Railroad cars collide: momentum conserved.** A 10,000-kg railroad car, A, traveling at a speed of 24.0 m/s strikes an identical car, B, at rest. If the cars lock together as a result of the collision, what is their common speed immediately after the collision? See Fig. 9–5.

APPROACH We choose our system to be the two railroad cars. We consider a very brief time interval, from just before the collision until just after, so that external

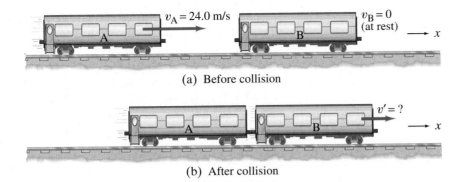

(a) Before collision

(b) After collision

FIGURE 9–5 Example 9–3.

forces such as friction can be ignored. Then we apply conservation of momentum:

$$P_{\text{initial}} = P_{\text{final}}.$$

SOLUTION The initial total momentum is

$$P_{\text{initial}} = m_A v_A + m_B v_B = m_A v_A$$

because car B is at rest initially $(v_B = 0)$. The direction is to the right in the $+x$ direction. After the collision, the two cars become attached, so they will have the same speed, call it v'. Then the total momentum after the collision is

$$P_{\text{final}} = (m_A + m_B)v'.$$

We have assumed there are no external forces, so momentum is conserved:

$$P_{\text{initial}} = P_{\text{final}}$$
$$m_A v_A = (m_A + m_B)v'.$$

Solving for v', we obtain

$$v' = \frac{m_A}{m_A + m_B} v_A = \left(\frac{10,000 \text{ kg}}{10,000 \text{ kg} + 10,000 \text{ kg}} \right)(24.0 \text{ m/s}) = 12.0 \text{ m/s},$$

to the right. Their mutual speed after collision is half the initial speed of car A because their masses are equal.

NOTE We kept symbols until the very end, so we have an equation we can use in other (related) situations.

NOTE We haven't mentioned friction here. Why? Because we are examining speeds just before and just after the very brief time interval of the collision, and during that brief time friction can't do much—it is ignorable (but not for long: the cars will slow down because of friction).

EXERCISE C A 50-kg child runs off a dock at 2.0 m/s (horizontally) and lands in a waiting rowboat of mass 150 kg. At what speed does the rowboat move away from the dock?

EXERCISE D In Example 9–3, what result would you get if (a) $m_B = 3m_A$, (b) m_B is much larger than m_A $(m_B \gg m_A)$, and (c) $m_B \ll m_A$?

The law of conservation of momentum is particularly useful when we are dealing with fairly simple systems such as colliding objects and certain types of "explosions." For example, *rocket propulsion*, which we saw in Chapter 4 can be understood on the basis of action and reaction, can also be explained on the basis of the conservation of momentum. We can consider the rocket and fuel as an isolated system if it is far out in space (no external forces). In the reference frame of the rocket before any fuel is ejected, the total momentum of rocket plus fuel is zero. When the fuel burns, the total momentum remains unchanged: the backward momentum of the expelled gases is just balanced by the forward momentum gained by the rocket itself (see Fig. 9–6). Thus, a rocket can accelerate in empty space. There is no need for the expelled gases to push against the Earth or the air (as is sometimes erroneously thought). Similar examples of (nearly) isolated systems where momentum is conserved are the recoil of a gun when a bullet is fired, and the movement of a rowboat just after a package is thrown from it.

FIGURE 9–6 (a) A rocket, containing fuel, at rest in some reference frame. (b) In the same reference frame, the rocket fires, and gases are expelled at high speed out the rear. The total vector momentum, $\vec{P} = \vec{p}_{\text{gas}} + \vec{p}_{\text{rocket}}$, remains zero.

(a)

$$\vec{P} = 0$$

(b)

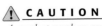

$\vec{p}_{\text{gas}}$ $\vec{p}_{\text{rocket}}$

⊛ **PHYSICS APPLIED**
Rocket propulsion

⚠ **CAUTION**

A rocket pushes on the gases released by the fuel, not on the Earth or other objects

(a) Before shooting (at rest)

(b) After shooting

FIGURE 9–7 Example 9–4.

EXAMPLE 9–4 **Rifle recoil.** Calculate the recoil velocity of a 5.0-kg rifle that shoots a 0.020-kg bullet at a speed of 620 m/s, Fig. 9–7.

APPROACH Our system is the rifle and the bullet, both at rest initially, just before the trigger is pulled. The trigger is pulled, an explosion occurs, and we look at the rifle and bullet just as the bullet leaves the barrel. The bullet moves to the right ($+x$), and the gun recoils to the left. During the very short time interval of the explosion, we can assume the external forces are small compared to the forces exerted by the exploding gunpowder. Thus we can apply conservation of momentum, at least approximately.

SOLUTION Let subscript B represent the bullet and R the rifle; the final velocities are indicated by primes. Then momentum conservation in the x direction gives

$$\text{momentum before} = \text{momentum after}$$
$$m_B v_B + m_R v_R = m_B v'_B + m_R v'_R$$
$$0 + 0 = m_B v'_B + m_R v'_R$$

so

$$v'_R = -\frac{m_B v'_B}{m_R} = -\frac{(0.020\,\text{kg})(620\,\text{m/s})}{(5.0\,\text{kg})} = -2.5\,\text{m/s}.$$

Since the rifle has a much larger mass, its (recoil) velocity is much less than that of the bullet. The minus sign indicates that the velocity (and momentum) of the rifle is in the negative x direction, opposite to that of the bullet.

CONCEPTUAL EXAMPLE 9–5 **Falling on or off a sled.** (a) An empty sled is sliding on frictionless ice when Susan drops vertically from a tree above onto the sled. When she lands, does the sled speed up, slow down, or keep the same speed? (b) Later: Susan falls sideways off the sled. When she drops off, does the sled speed up, slow down, or keep the same speed?

RESPONSE (a) Because Susan falls vertically onto the sled, she has no initial horizontal momentum. Thus the total horizontal momentum afterward equals the momentum of the sled initially. Since the mass of the system (sled + person) has increased, the speed must decrease. (b) At the instant Susan falls off, she is moving with the same horizontal speed as she was while on the sled. At the moment she leaves the sled, she has the same momentum she had an instant before. Because momentum is conserved, the sled keeps the same speed.

EXERCISE E Return to the Chapter-Opening Questions, page 214, and answer them again now. Try to explain why you may have answered differently the first time.

9–3 Collisions and Impulse

Conservation of momentum is a very useful tool for dealing with everyday collision processes, such as a tennis racket or a baseball bat striking a ball, two billiard balls colliding, a hammer hitting a nail. At the subatomic level, scientists learn about the structure of nuclei and their constituents, and about the nature of the forces involved, by careful study of collisions between nuclei and/or elementary particles.

During a collision of two ordinary objects, both objects are deformed, often considerably, because of the large forces involved (Fig. 9–8). When the collision occurs, the force each exerts on the other usually jumps from zero at the moment of contact to a very large value within a very short time, and then abruptly returns to zero again. A graph of the magnitude of the force one object exerts on the other during a collision, as a function of time, is something like the red curve in Fig. 9–9. The time interval Δt is usually very distinct and usually very small.

From Newton's second law, Eq. 9–2, the *net* force on an object is equal to the rate of change of its momentum:

$$\vec{F} = \frac{d\vec{p}}{dt}.$$

(We have written $\vec{F}$ instead of $\Sigma\vec{F}$ for the net force, which we assume is entirely due to the brief but large force that acts during the collision.) This equation applies

FIGURE 9–8 Tennis racket striking a ball. Both the ball and the racket strings are deformed due to the large force each exerts on the other.

FIGURE 9–9 Force as a function of time during a typical collision: F can become very large; Δt is typically milliseconds for macroscopic collisions.

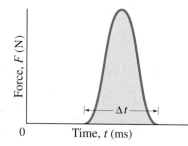

to *each* of the objects in a collision. During the infinitesimal time interval dt, the momentum changes by

$$d\vec{\mathbf{p}} = \vec{\mathbf{F}}\, dt.$$

If we integrate this over the duration of a collision, we have

$$\int_i^f d\vec{\mathbf{p}} = \vec{\mathbf{p}}_f - \vec{\mathbf{p}}_i = \int_{t_i}^{t_f} \vec{\mathbf{F}}\, dt,$$

where $\vec{\mathbf{p}}_i$ and $\vec{\mathbf{p}}_f$ are the initial and final momenta of the object, just before and just after the collision. The integral of the net force over the time interval during which it acts is called the **impulse, $\vec{\mathbf{J}}$**:

$$\vec{\mathbf{J}} = \int_{t_i}^{t_f} \vec{\mathbf{F}}\, dt.$$

Thus the change in momentum of an object, $\Delta\vec{\mathbf{p}} = \vec{\mathbf{p}}_f - \vec{\mathbf{p}}_i$, is equal to the impulse acting on it:

$$\Delta\vec{\mathbf{p}} = \vec{\mathbf{p}}_f - \vec{\mathbf{p}}_i = \int_{t_i}^{t_f} \vec{\mathbf{F}}\, dt = \vec{\mathbf{J}}. \tag{9-6}$$

The units for impulse are the same as for momentum, $kg \cdot m/s$ (or $N \cdot s$) in SI. Since $\vec{\mathbf{J}} = \int \vec{\mathbf{F}}\, dt$, we can state that the impulse $\vec{\mathbf{J}}$ of a force is equal to the area under the F versus t curve, as indicated by the shading in Fig. 9–9.

Equation 9–6 is true only if $\vec{\mathbf{F}}$ is the *net* force on the object. It is valid for *any* net force $\vec{\mathbf{F}}$ where $\vec{\mathbf{p}}_i$ and $\vec{\mathbf{p}}_f$ correspond precisely to the times t_i and t_f. But the impulse concept is really most useful for so-called *impulsive forces*—that is, for a force like that shown in Fig. 9–9, which has a very large magnitude over a very short time interval and is essentially zero outside this time interval. For most collision processes, the impulsive force is much larger than any other force acting, and the others can be neglected. For such an impulsive force, the time interval over which we take the integral in Eq. 9–6 is not critical as long as we start before t_i and end after t_f, since $\vec{\mathbf{F}}$ is essentially zero outside this time interval. (Of course, if the chosen time interval is too large, the effect of the other forces does become significant—such as the flight of a tennis ball which, after the impulsive force administered by the racket, begins to fall under gravity.)

It is sometimes useful to speak of the average force, $\vec{\mathbf{F}}_{avg}$, during a collision, defined as that constant force which, if acting over the same time interval $\Delta t = t_f - t_i$ as the actual force, would produce the same impulse and change in momentum. Thus

$$\vec{\mathbf{F}}_{avg}\,\Delta t = \int_{t_i}^{t_f} \vec{\mathbf{F}}\, dt.$$

Figure 9–10 shows the magnitude of the average force, F_{avg}, for the impulsive force of Fig. 9–9. The rectangular area $F_{avg}\,\Delta t$ equals the area under the impulsive force curve.

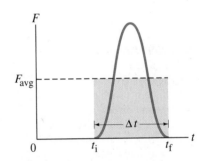

FIGURE 9–10 The average force F_{avg} acting over a very brief time interval Δt gives the same impulse $(F_{avg}\,\Delta t)$ as the actual force.

EXAMPLE 9–6 **ESTIMATE** **Karate blow.** Estimate the impulse and the average force delivered by a karate blow (Fig. 9–11) that breaks a board a few cm thick. Assume the hand moves at roughly $10\,m/s$ when it hits the board.

APPROACH We use the momentum-impulse relation, Eq. 9–6. The hand's speed changes from $10\,m/s$ to zero over a distance of perhaps one cm (roughly how much your hand and the board compress before your hand comes to a stop, or nearly so, and the board begins to give way). The hand's mass should probably include part of the arm, and we take it to be roughly $m \approx 1\,kg$.

FIGURE 9–11 Example 9–6.

SOLUTION The impulse J equals the change in momentum

$$J = \Delta p = (1\,kg)(10\,m/s - 0) = 10\,kg \cdot m/s.$$

We obtain the force from the definition of impulse $F_{avg} = J/\Delta t$; but what is Δt? The hand is brought to rest over the distance of roughly a centimeter: $\Delta x \approx 1\,cm$. The average speed during the impact is $\bar{v} = (10\,m/s + 0)/2 = 5\,m/s$ and equals $\Delta x/\Delta t$. Thus $\Delta t = \Delta x/\bar{v} \approx (10^{-2}\,m)/(5\,m/s) = 2 \times 10^{-3}\,s$ or about $2\,ms$. The force is thus (Eq. 9–6) about

$$F_{avg} = \frac{J}{\Delta t} = \frac{10\,kg \cdot m/s}{2 \times 10^{-3}\,s} \approx 5000\,N = 5\,kN.$$

9–4 Conservation of Energy and Momentum in Collisions

During most collisions, we usually don't know how the collision force varies over time, and so analysis using Newton's second law becomes difficult or impossible. But by making use of the conservation laws for momentum and energy, we can still determine a lot about the motion after a collision, given the motion before the collision. We saw in Section 9–2 that in the collision of two objects such as billiard balls, the total momentum is conserved. If the two objects are very hard and no heat or other form of energy is produced in the collision, then the kinetic energy of the two objects is the same after the collision as before. For the brief moment during which the two objects are in contact, some (or all) of the energy is stored momentarily in the form of elastic potential energy. But if we compare the total kinetic energy just before the collision with the total kinetic energy just after the collision, and they are found to be the same, then we say that the total kinetic energy is conserved. Such a collision is called an **elastic collision**. If we use the subscripts A and B to represent the two objects, we can write the equation for conservation of total kinetic energy as

(a) Approach

(b) Collision

(c) If elastic

(d) If inelastic

FIGURE 9–12 Two equal-mass objects (a) approach each other with equal speeds, (b) collide, and then (c) bounce off with equal speeds in the opposite directions if the collision is elastic, or (d) bounce back much less or not at all if the collision is inelastic.

total kinetic energy before = total kinetic energy after

$$\tfrac{1}{2} m_A v_A^2 + \tfrac{1}{2} m_B v_B^2 = \tfrac{1}{2} m_A v_A'^2 + \tfrac{1}{2} m_B v_B'^2. \qquad \text{[elastic collision]} \quad (9\text{–}7)$$

Primed quantities ($'$) mean after the collision, and unprimed mean before the collision, just as in Eq. 9–3 for conservation of momentum.

At the atomic level the collisions of atoms and molecules are often elastic. But in the "macroscopic" world of ordinary objects, an elastic collision is an ideal that is never quite reached, since at least a little thermal energy (and perhaps sound and other forms of energy) is always produced during a collision. The collision of two hard elastic balls, such as billiard balls, however, is very close to being perfectly elastic, and we often treat it as such.

We do need to remember that even when the kinetic energy is not conserved, the *total* energy is always conserved.

Collisions in which kinetic energy is not conserved are said to be **inelastic collisions**. The kinetic energy that is lost is changed into other forms of energy, often thermal energy, so that the total energy (as always) is conserved. In this case,

$$K_A + K_B = K_A' + K_B' + \text{thermal and other forms of energy.}$$

See Fig. 9–12, and the details in its caption. We discuss inelastic collisions in Section 9–6.

FIGURE 9–13 Two small objects of masses m_A and m_B, (a) before the collision and (b) after the collision.

(a)

(b)

9–5 Elastic Collisions in One Dimension

We now apply the conservation laws for momentum and kinetic energy to an elastic collision between two small objects that collide head-on, so all the motion is along a line. Let us assume that the two objects are moving with velocities v_A and v_B along the x axis before the collision, Fig. 9–13a. After the collision, their velocities are v_A' and v_B', Fig. 9–13b. For any $v > 0$, the object is moving to the right (increasing x), whereas for $v < 0$, the object is moving to the left (toward decreasing values of x).

From conservation of momentum, we have

$$m_A v_A + m_B v_B = m_A v_A' + m_B v_B'.$$

Because the collision is assumed to be elastic, kinetic energy is also conserved:

$$\tfrac{1}{2} m_A v_A^2 + \tfrac{1}{2} m_B v_B^2 = \tfrac{1}{2} m_A v_A'^2 + \tfrac{1}{2} m_B v_B'^2.$$

We have two equations, so we can solve for two unknowns. If we know the masses and velocities before the collision, then we can solve these two equations for the velocities after the collision, v'_A and v'_B. We derive a helpful result by rewriting the momentum equation as

$$m_A(v_A - v'_A) = m_B(v'_B - v_B), \qquad \text{(i)}$$

and we rewrite the kinetic energy equation as

$$m_A(v_A^2 - v'^2_A) = m_B(v'^2_B - v_B^2).$$

Noting that algebraically $(a - b)(a + b) = a^2 - b^2$, we write this last equation as

$$m_A(v_A - v'_A)(v_A + v'_A) = m_B(v'_B - v_B)(v'_B + v_B). \qquad \text{(ii)}$$

We divide Eq. (ii) by Eq. (i), and (assuming $v_A \neq v'_A$ and $v_B \neq v'_B$)[†] obtain

$$v_A + v'_A = v'_B + v_B.$$

We can rewrite this equation as

$$v_A - v_B = v'_B - v'_A$$

or

$$v_A - v_B = -(v'_A - v'_B). \qquad \left[\text{head-on (1-D) elastic collision}\right] \quad \text{(9–8)}$$

Relative speeds (1-D only)

This is an interesting result: it tells us that for any elastic head-on collision, the relative speed of the two objects after the collision has the same magnitude (but opposite direction) as before the collision, no matter what the masses are.

Equation 9–8 was derived from conservation of kinetic energy for elastic collisions, and can be used in place of it. Because the v's are not squared in Eq. 9–8, it is simpler to use in calculations than the conservation of kinetic energy equation (Eq. 9–7) directly.

EXAMPLE 9–7 **Equal masses.** Billiard ball A of mass m moving with speed v_A collides head-on with ball B of equal mass. What are the speeds of the two balls after the collision, assuming it is elastic? Assume (a) both balls are moving initially (v_A and v_B), (b) ball B is initially at rest ($v_B = 0$).

APPROACH There are two unknowns, v'_A and v'_B, so we need two independent equations. We focus on the time interval from just before the collision until just after. No net external force acts on our system of two balls (mg and the normal force cancel), so momentum is conserved. Conservation of kinetic energy applies as well because we are told the collision is elastic.

SOLUTION (a) The masses are equal ($m_A = m_B = m$) so conservation of momentum gives

$$v_A + v_B = v'_A + v'_B.$$

We need a second equation, because there are two unknowns. We could use the conservation of kinetic energy equation, or the simpler Eq. 9–8 derived from it:

$$v_A - v_B = v'_B - v'_A.$$

We add these two equations and obtain

$$v'_B = v_A$$

and then subtract the two equations to obtain

$$v'_A = v_B.$$

That is, the balls exchange velocities as a result of the collision: ball B acquires the velocity that ball A had before the collision, and vice versa.

(b) If ball B is at rest initially, so that $v_B = 0$, we have

$$v'_B = v_A \quad \text{and} \quad v'_A = 0.$$

That is, ball A is brought to rest by the collision, whereas ball B acquires the original velocity of ball A. This result is often observed by billiard and pool players, and is valid only if the two balls have equal masses (and no spin is given to the balls). See Fig. 9–14.

FIGURE 9–14 In this multi-flash photo of a head-on collision between two balls of equal mass, the white cue ball is accelerated from rest by the cue stick and then strikes the red ball, initially at rest. The white ball stops in its tracks and the (equal mass) red ball moves off with the same speed as the white ball had before the collision. See Example 9–7.

[†]Note that Eqs. (i) and (ii), which are the conservation laws for momentum and kinetic energy, are both satisfied by the solution $v'_1 = v_1$ and $v'_2 = v_2$. This is a valid solution, but not very interesting. It corresponds to no collision at all—when the two objects miss each other.

EXAMPLE 9–8 **Unequal masses, target at rest.** A very common practical situation is for a moving object (m_A) to strike a second object (m_B, the "target") at rest ($v_B = 0$). Assume the objects have unequal masses, and that the collision is elastic and occurs along a line (head-on). (a) Derive equations for v'_B and v'_A in terms of the initial velocity v_A of mass m_A and the masses m_A and m_B. (b) Determine the final velocities if the moving object is much more massive than the target ($m_A \gg m_B$). (c) Determine the final velocities if the moving object is much less massive than the target ($m_A \ll m_B$).

APPROACH The momentum equation (with $v_B = 0$) is

$$m_B v'_B = m_A(v_A - v'_A).$$

Kinetic energy is also conserved, and to use it we use Eq. 9–8 and rewrite it as

$$v'_A = v'_B - v_A.$$

SOLUTION (a) We substitute the above v'_A equation into the momentum equation and rearrange to find

$$v'_B = v_A\left(\frac{2m_A}{m_A + m_B}\right).$$

We substitute this value for v'_B back into the equation $v'_A = v'_B - v_A$ to obtain

$$v'_A = v_A\left(\frac{m_A - m_B}{m_A + m_B}\right).$$

To check these two equations we have derived, we let $m_A = m_B$, and we obtain

$$v'_B = v_A \quad \text{and} \quad v'_A = 0.$$

This is the same case treated in Example 9–7, and we get the same result: for objects of equal mass, one of which is initially at rest, the velocity of the one moving initially is completely transferred to the object originally at rest.

(b) We are given $v_B = 0$ and $m_A \gg m_B$. A very heavy moving object strikes a light object at rest, and we have, using the relations for v'_B and v'_A above,

$$v'_B \approx 2v_A$$
$$v'_A \approx v_A.$$

Thus the velocity of the heavy incoming object is practically unchanged, whereas the light object, originally at rest, takes off with twice the velocity of the heavy one. The velocity of a heavy bowling ball, for example, is hardly affected by striking a much lighter bowling pin.

(c) This time we have $v_B = 0$ and $m_A \ll m_B$. A moving light object strikes a very massive object at rest. In this case, using the equations in part (a)

$$v'_B \approx 0$$
$$v'_A \approx -v_A.$$

The massive object remains essentially at rest and the very light incoming object rebounds with essentially its same speed but in the opposite direction. For example, a tennis ball colliding head-on with a stationary bowling ball will hardly affect the bowling ball, but will rebound with nearly the same speed it had initially, just as if it had struck a hard wall.

It can readily be shown (it is given as Problem 40) for any elastic head-on collision that

$$v'_B = v_A\left(\frac{2m_A}{m_A + m_B}\right) + v_B\left(\frac{m_B - m_A}{m_A + m_B}\right)$$

and

$$v'_A = v_A\left(\frac{m_A - m_B}{m_A + m_B}\right) + v_B\left(\frac{2m_B}{m_A + m_B}\right).$$

These general equations, however, should not be memorized. They can always be derived quickly from the conservation laws. For many problems, it is simplest just to start from scratch, as we did in the special cases above and as shown in the next Example.

EXAMPLE 9–9 **A nuclear collision.** A proton (p) of mass 1.01 u (unified atomic mass units) traveling with a speed of 3.60×10^4 m/s has an elastic head-on collision with a helium (He) nucleus $(m_{He} = 4.00\,u)$ initially at rest. What are the velocities of the proton and helium nucleus after the collision? (As mentioned in Chapter 1, $1\,u = 1.66 \times 10^{-27}$ kg, but we won't need this fact.) Assume the collision takes place in nearly empty space.

APPROACH This is an elastic head-on collision. The only external force is Earth's gravity, but it is insignificant compared to the strong force during the collision. So again we use the conservation laws of momentum and of kinetic energy, and apply them to our system of two particles.

SOLUTION Let the proton (p) be particle A and the helium nucleus (He) be particle B. We have $v_B = v_{He} = 0$ and $v_A = v_p = 3.60 \times 10^4$ m/s. We want to find the velocities v'_p and v'_{He} after the collision. From conservation of momentum,

$$m_p v_p + 0 = m_p v'_p + m_{He} v'_{He}.$$

Because the collision is elastic, the kinetic energy of our system of two particles is conserved and we can use Eq. 9–8, which becomes

$$v_p - 0 = v'_{He} - v'_p.$$

Thus

$$v'_p = v'_{He} - v_p,$$

and substituting this into our momentum equation displayed above, we get

$$m_p v_p = m_p v'_{He} - m_p v_p + m_{He} v'_{He}.$$

Solving for v'_{He}, we obtain

$$v'_{He} = \frac{2m_p v_p}{m_p + m_{He}} = \frac{2(1.01\,u)(3.60 \times 10^4\,m/s)}{5.01\,u} = 1.45 \times 10^4\,m/s.$$

The other unknown is v'_p, which we can now obtain from

$$v'_p = v'_{He} - v_p = (1.45 \times 10^4\,m/s) - (3.60 \times 10^4\,m/s) = -2.15 \times 10^4\,m/s.$$

The minus sign for v'_p tells us that the proton reverses direction upon collision, and we see that its speed is less than its initial speed (see Fig. 9–15).

NOTE This result makes sense: the lighter proton would be expected to "bounce back" from the more massive helium nucleus, but not with its full original velocity as from a rigid wall (which corresponds to extremely large, or infinite, mass).

FIGURE 9–15 Example 9–9: (a) before collision, (b) after collision.

9–6 Inelastic Collisions

Collisions in which kinetic energy is not conserved are called **inelastic collisions**. Some of the initial kinetic energy is transformed into other types of energy, such as thermal or potential energy, so the total kinetic energy after the collision is less than the total kinetic energy before the collision. The inverse can also happen when potential energy (such as chemical or nuclear) is released, in which case the total kinetic energy after the interaction can be greater than the initial kinetic energy. Explosions are examples of this type.

Typical macroscopic collisions are inelastic, at least to some extent, and often to a large extent. If two objects stick together as a result of a collision, the collision is said to be **completely inelastic**. Two colliding balls of putty that stick together or two railroad cars that couple together when they collide are examples of completely inelastic collisions. The kinetic energy in some cases is all transformed to other forms of energy in an inelastic collision, but in other cases only part of it is. In Example 9–3, for instance, we saw that when a traveling railroad car collided with a stationary one, the coupled cars traveled off with some kinetic energy. In a completely inelastic collision, the maximum amount of kinetic energy is transformed to other forms consistent with conservation of momentum. Even though kinetic energy is not conserved in inelastic collisions, the total energy is always conserved, and the total vector momentum is also conserved.

EXAMPLE 9–10 **Railroad cars again.** For the completely inelastic collision of two railroad cars that we considered in Example 9–3, calculate how much of the initial kinetic energy is transformed to thermal or other forms of energy.

APPROACH The railroad cars stick together after the collision, so this is a completely inelastic collision. By subtracting the total kinetic energy after the collision from the total initial kinetic energy, we can find how much energy is transformed to other types of energy.

SOLUTION Before the collision, only car A is moving, so the total initial kinetic energy is $\frac{1}{2}m_A v_A^2 = \frac{1}{2}(10{,}000 \text{ kg})(24.0 \text{ m/s})^2 = 2.88 \times 10^6 \text{ J}$. After the collision, both cars are moving with a speed of 12.0 m/s, by conservation of momentum (Example 9–3). So the total kinetic energy afterward is $\frac{1}{2}(20{,}000 \text{ kg})(12.0 \text{ m/s})^2 = 1.44 \times 10^6 \text{ J}$. Hence the energy transformed to other forms is

$$\left(2.88 \times 10^6 \text{ J}\right) - \left(1.44 \times 10^6 \text{ J}\right) = 1.44 \times 10^6 \text{ J},$$

which is half the original kinetic energy.

EXAMPLE 9–11 **Ballistic pendulum.** The *ballistic pendulum* is a device used to measure the speed of a projectile, such as a bullet. The projectile, of mass m, is fired into a large block (of wood or other material) of mass M, which is suspended like a pendulum. (Usually, M is somewhat greater than m.) As a result of the collision, the pendulum and projectile together swing up to a maximum height h, Fig. 9–16. Determine the relationship between the initial horizontal speed of the projectile, v, and the maximum height h.

APPROACH We can analyze the process by dividing it into two parts or two time intervals: (1) the time interval from just before to just after the collision itself, and (2) the subsequent time interval in which the pendulum moves from the vertical hanging position to the maximum height h.

In part (1), Fig. 9–16a, we assume the collision time is very short, so that the projectile comes to rest in the block before the block has moved significantly from its rest position directly below its support. Thus there is effectively no net external force, and we can apply conservation of momentum to this completely inelastic collision. In part (2), Fig. 9–16b, the pendulum begins to move, subject to a net external force (gravity, tending to pull it back to the vertical position); so for part (2), we cannot use conservation of momentum. But we can use conservation of mechanical energy because gravity is a conservative force (Chapter 8). The kinetic energy immediately after the collision is changed entirely to gravitational potential energy when the pendulum reaches its maximum height, h.

SOLUTION In part (1) momentum is conserved:

$$\text{total } P \text{ before} = \text{total } P \text{ after}$$
$$mv = (m + M)v', \tag{i}$$

where v' is the speed of the block and embedded projectile just after the collision, before they have moved significantly.

In part (2), mechanical energy is conserved. We choose $y = 0$ when the pendulum hangs vertically, and then $y = h$ when the pendulum–projectile system reaches its maximum height. Thus we write

$$(K + U) \text{ just after collision} = (K + U) \text{ at pendulum's maximum height}$$

or

$$\tfrac{1}{2}(m + M)v'^2 + 0 = 0 + (m + M)gh. \tag{ii}$$

We solve for v':

$$v' = \sqrt{2gh}.$$

Inserting this result for v' into Eq. (i) above, and solving for v, gives

$$v = \frac{m + M}{m}v' = \frac{m + M}{m}\sqrt{2gh},$$

which is our final result.

(a)

(b)

FIGURE 9–16 Ballistic pendulum. Example 9–11.

NOTE The separation of the process into two parts was crucial. Such an analysis is a powerful problem-solving tool. But how do you decide how to make such a division? Think about the conservation laws. They are your *tools*. Start a problem by asking yourself whether the conservation laws apply in the given situation. Here, we determined that momentum is conserved only during the brief collision, which we called part (1). But in part (1), because the collision is inelastic, the conservation of mechanical energy is not valid. Then in part (2), conservation of mechanical energy is valid, but not conservation of momentum.

Note, however, that if there had been significant motion of the pendulum during the deceleration of the projectile in the block, then there *would* have been an external force (gravity) during the collision, so conservation of momentum would not have been valid in part (1).

PROBLEM SOLVING

Use the conservation laws to analyze a problem

9–7 Collisions in Two or Three Dimensions

Conservation of momentum and energy can also be applied to collisions in two or three dimensions, where the vector nature of momentum is especially important. One common type of non-head-on collision is that in which a moving object (called the "projectile") strikes a second object initially at rest (the "target"). This is the common situation in games such as billiards and pool, and for experiments in atomic and nuclear physics (the projectiles, from radioactive decay or a high-energy accelerator, strike a stationary target nucleus; Fig. 9–17).

Figure 9–18 shows the incoming projectile, m_A, heading along the x axis toward the target object, m_B, which is initially at rest. If these are billiard balls, m_A strikes m_B not quite head-on and they go off at the angles θ'_A and θ'_B, respectively, which are measured relative to m_A's initial direction (the x axis).[†]

Let us apply the law of conservation of momentum to a collision like that of Fig. 9–18. We choose the xy plane to be the plane in which the initial and final momenta lie. Momentum is a vector, and because the total momentum is conserved, its components in the x and y directions also are conserved. The x component of momentum conservation gives

$$p_{Ax} + p_{Bx} = p'_{Ax} + p'_{Bx}$$

or, with $p_{Bx} = m_B v_{Bx} = 0$,

$$m_A v_A = m_A v'_A \cos \theta'_A + m_B v'_B \cos \theta'_B, \qquad \textbf{(9–9a)}$$

where the primes (′) refer to quantities *after* the collision. Because there is no motion in the y direction initially, the y component of the total momentum is zero before the collision. The y component equation of momentum conservation is then

$$p_{Ay} + p_{By} = p'_{Ay} + p'_{By}$$

or

$$0 = m_A v'_A \sin \theta'_A + m_B v'_B \sin \theta'_B. \qquad \textbf{(9–9b)}$$

When we have two independent equations, we can solve for two unknowns, at most.

FIGURE 9–17 A recent color-enhanced version of a cloud-chamber photograph made in the early days (1920s) of nuclear physics. Green lines are paths of helium nuclei (He) coming from the left. One He, highlighted in yellow, strikes a proton of the hydrogen gas in the chamber, and both scatter at an angle; the scattered proton's path is shown in red.

p_x conserved

p_y conserved

FIGURE 9–18 Object A, the projectile, collides with object B, the target. After the collision, they move off with momenta $\vec{p}'_A$ and $\vec{p}'_B$ at angles θ'_A and θ'_B. The objects are shown here as particles, as we would visualize them in atomic or nuclear physics. But they could also be macroscopic pool balls.

[†]The objects may begin to deflect even before they touch if electric, magnetic, or nuclear forces act between them. You might think, for example, of two magnets oriented so that they repel each other: when one moves toward the other, the second moves away before the first one touches it.

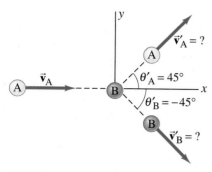

FIGURE 9–19 Example 9–12.

EXAMPLE 9–12 **Billiard ball collision in 2-D.** Billiard ball A moving with speed $v_A = 3.0$ m/s in the $+x$ direction (Fig. 9–19) strikes an equal-mass ball B initially at rest. The two balls are observed to move off at 45° to the x axis, ball A above the x axis and ball B below. That is, $\theta'_A = 45°$ and $\theta'_B = -45°$ in Fig. 9–19. What are the speeds of the two balls after the collision?

APPROACH There is no net external force on our system of two balls, assuming the table is level (the normal force balances gravity). Thus momentum conservation applies, and we apply it to both the x and y components using the xy coordinate system shown in Fig. 9–19. We get two equations, and we have two unknowns, v'_A and v'_B. From symmetry we might guess that the two balls have the same speed. But let us not assume that now. Even though we are not told whether the collision is elastic or inelastic, we can still use conservation of momentum.

SOLUTION We apply conservation of momentum for the x and y components, Eqs. 9–9a and b, and we solve for v'_A and v'_B. We are given $m_A = m_B (= m)$, so

$$\text{(for } x\text{)} \quad mv_A = mv'_A \cos(45°) + mv'_B \cos(-45°)$$

and

$$\text{(for } y\text{)} \quad 0 = mv'_A \sin(45°) + mv'_B \sin(-45°).$$

The m's cancel out in both equations (the masses are equal). The second equation yields [recall that $\sin(-\theta) = -\sin\theta$]:

$$v'_B = -v'_A \frac{\sin(45°)}{\sin(-45°)} = -v'_A \left(\frac{\sin 45°}{-\sin 45°} \right) = v'_A.$$

So they do have equal speeds as we guessed at first. The x component equation gives [recall that $\cos(-\theta) = \cos\theta$]:

$$v_A = v'_A \cos(45°) + v'_B \cos(45°) = 2v'_A \cos(45°),$$

so

$$v'_A = v'_B = \frac{v_A}{2\cos(45°)} = \frac{3.0 \text{ m/s}}{2(0.707)} = 2.1 \text{ m/s}.$$

If we know that a collision is elastic, we can also apply conservation of kinetic energy and obtain a third equation in addition to Eqs. 9–9a and b:

$$K_A + K_B = K'_A + K'_B$$

or, for the collision shown in Fig. 9–18 or 9–19,

$$\tfrac{1}{2}m_A v_A^2 = \tfrac{1}{2}m_A v_A'^2 + \tfrac{1}{2}m_B v_B'^2. \qquad \text{[elastic collision]} \quad \textbf{(9–9c)}$$

If the collision is elastic, we have three independent equations and can solve for three unknowns. If we are given m_A, m_B, v_A (and v_B, if it is not zero), we cannot, for example, predict the final variables, v'_A, v'_B, θ'_A, and θ'_B, because there are four of them. However, if we measure one of these variables, say θ'_A, then the other three variables (v'_A, v'_B, and θ'_B) are uniquely determined, and we can determine them using Eqs. 9–9a, b, and c.

A note of caution: Eq. 9–8 does *not* apply for two-dimensional collisions. It works only when a collision occurs along a line.

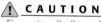

⚠ CAUTION

Equation 9–8 applies only in 1-D

EXAMPLE 9–13 **Proton–proton collision.** A proton traveling with speed 8.2×10^5 m/s collides elastically with a stationary proton in a hydrogen target as in Fig. 9–18. One of the protons is observed to be scattered at a 60° angle. At what angle will the second proton be observed, and what will be the velocities of the two protons after the collision?

APPROACH We saw a two-dimensional collision in Example 9–12, where we needed to use only conservation of momentum. Now we are given less information: we have three unknowns instead of two. Because the collision is elastic, we can use the kinetic energy equation as well as the two momentum equations.

SOLUTION Since $m_A = m_B$, Eqs. 9–9a, b, and c become

$$v_A = v'_A \cos \theta'_A + v'_B \cos \theta'_B \qquad \textbf{(i)}$$
$$0 = v'_A \sin \theta'_A + v'_B \sin \theta'_B \qquad \textbf{(ii)}$$
$$v_A^2 = v'^2_A + v'^2_B, \qquad \textbf{(iii)}$$

where $v_A = 8.2 \times 10^5$ m/s and $\theta'_A = 60°$ are given. In the first and second equations, we move the v'_A terms to the left side and square both sides of the equations:

$$v_A^2 - 2v_A v'_A \cos \theta'_A + v'^2_A \cos^2 \theta'_A = v'^2_B \cos^2 \theta'_B$$
$$v'^2_A \sin^2 \theta'_A = v'^2_B \sin^2 \theta'_B.$$

We add these two equations and use $\sin^2 \theta + \cos^2 \theta = 1$ to get:

$$v_A^2 - 2v_A v'_A \cos \theta'_A + v'^2_A = v'^2_B.$$

Into this equation we substitute $v'^2_B = v_A^2 - v'^2_A$, from equation (iii) above, and get

$$2v'^2_A = 2v_A v'_A \cos \theta'_A$$

or

$$v'_A = v_A \cos \theta'_A = (8.2 \times 10^5 \text{ m/s})(\cos 60°) = 4.1 \times 10^5 \text{ m/s}.$$

To obtain v'_B, we use equation (iii) above (conservation of kinetic energy):

$$v'_B = \sqrt{v_A^2 - v'^2_A} = 7.1 \times 10^5 \text{ m/s}.$$

Finally, from equation (ii), we have

$$\sin \theta'_B = -\frac{v'_A}{v'_B} \sin \theta'_A = -\left(\frac{4.1 \times 10^5 \text{ m/s}}{7.1 \times 10^5 \text{ m/s}}\right)(0.866) = -0.50,$$

so $\theta'_B = -30°$. (The minus sign means particle B moves at an angle below the x axis if particle A is above the axis, as in Fig. 9–19.) An example of such a collision is shown in the bubble chamber photo of Fig. 9–20. Notice that the two trajectories are at right angles to each other after the collision. This can be shown to be true in general for non-head-on elastic collisions of two particles of equal mass, one of which was at rest initially (see Problem 61).

FIGURE 9–20 Photo of a proton–proton collision in a hydrogen bubble chamber (a device that makes visible the paths of elementary particles). The many lines represent incoming protons which can strike the protons of the hydrogen in the chamber.

PROBLEM SOLVING

Momentum Conservation and Collisions

1. Choose your **system**. If the situation is complex, think about how you might break it up into separate parts when one or more conservation laws apply.

2. If a significant **net external force** acts on your chosen system, be sure the time interval Δt is so short that the effect on momentum is negligible. That is, the forces that act between the interacting objects must be the only significant ones if momentum conservation is to be used. [Note: If this is valid for a portion of the problem, you can use momentum conservation only for that portion.]

3. Draw a **diagram** of the initial situation, just before the interaction (collision, explosion) takes place, and represent the momentum of each object with an arrow and a label. Do the same for the final situation, just after the interaction.

4. Choose a **coordinate system** and "+" and "−" directions. (For a head-on collision, you will need only an x axis.) It is often convenient to choose the $+x$ axis in the direction of one object's initial velocity.

5. Apply the **momentum conservation** equation(s):

   total initial momentum = total final momentum.

   You have one equation for each component (x, y, z): only one equation for a head-on collision.

6. If the collision is elastic, you can also write down a **conservation of kinetic energy** equation:

   total initial kinetic energy = total final kinetic energy.

   [Alternately, you could use Eq. 9–8: $v_A - v_B = v'_B - v'_A$, if the collision is one dimensional (head-on).]

7. Solve for the **unknown(s)**.

8. **Check** your work, check the units, and ask yourself whether the results are reasonable.

(a)

(b)

FIGURE 9–21 The motion of the diver is pure translation in (a), but is translation plus rotation in (b). The black dot represents the diver's CM at each moment.

FIGURE 9–23 The center of mass of a two-particle system lies on the line joining the two masses. Here $m_A > m_B$, so the CM is closer to m_A than to m_B.

9–8 Center of Mass (CM)

Momentum is a powerful concept not only for analyzing collisions but also for analyzing the translational motion of real extended objects. Until now, whenever we have dealt with the motion of an extended object (that is, an object that has size), we have assumed that it could be approximated as a point particle or that it undergoes only translational motion. Real extended objects, however, can undergo rotational and other types of motion as well. For example, the diver in Fig. 9–21a undergoes only translational motion (all parts of the object follow the same path), whereas the diver in Fig. 9–21b undergoes both translational and rotational motion. We will refer to motion that is not pure translation as *general motion*.

Observations indicate that even if an object rotates, or several parts of a system of objects move relative to one another, there is one point that moves in the same path that a particle would move if subjected to the same net force. This point is called the **center of mass** (abbreviated CM). The general motion of an extended object (or system of objects) can be considered as *the sum of the translational motion of the CM, plus rotational, vibrational, or other types of motion about the CM*.

As an example, consider the motion of the center of mass of the diver in Fig. 9–21; the CM follows a parabolic path even when the diver rotates, as shown in Fig. 9–21b. This is the same parabolic path that a projected particle follows when acted on only by the force of gravity (projectile motion, Section 3–7). Other points in the rotating diver's body, such as her feet or head, follow more complicated paths.

Figure 9–22 shows a wrench acted on by zero net force, translating and rotating along a horizontal surface. Note that its CM, marked by a red cross, moves in a straight line, as shown by the dashed white line.

FIGURE 9–22 Translation plus rotation: a wrench moving over a horizontal surface. The CM, marked with a red cross, moves in a straight line.

We will show in Section 9–9 that the important properties of the CM follow from Newton's laws if the CM is defined in the following way. We can consider any extended object as being made up of many tiny particles. But first we consider a system made up of only two particles (or small objects), of masses m_A and m_B. We choose a coordinate system so that both particles lie on the x axis at positions x_A and x_B, Fig. 9–23. The center of mass of this system is defined to be at the position x_{CM}, given by

$$x_{CM} = \frac{m_A x_A + m_B x_B}{m_A + m_B} = \frac{m_A x_A + m_B x_B}{M},$$

where $M = m_A + m_B$ is the total mass of the system. The center of mass lies on the line joining m_A and m_B. If the two masses are equal $(m_A = m_B = m)$, then x_{CM} is midway between them, since in this case

$$x_{CM} = \frac{m(x_A + x_B)}{2m} = \frac{(x_A + x_B)}{2}.$$

If one mass is greater than the other, say, $m_A > m_B$, then the CM is closer to the larger mass. If all the mass is concentrated at x_B, so $m_A = 0$, then $x_{CM} = (0 x_A + m_B x_B)/(0 + m_B) = x_B$, as we would expect.

Now let us consider a system consisting of n particles, where n could be very large. This system could be an extended object which we consider as being made up of n tiny particles. If these n particles are all along a straight line (call it the x axis), we define the CM of the system to be located at

$$x_{CM} = \frac{m_1 x_1 + m_2 x_2 + \cdots + m_n x_n}{m_1 + m_2 + \cdots + m_n} = \frac{\sum_{i=1}^{n} m_i x_i}{M}, \qquad \textbf{(9-10)}$$

where $m_1, m_2, \ldots m_n$ are the masses of each particle and $x_1, x_2, \ldots x_n$ are their positions. The symbol $\sum_{i=1}^{n}$ is the summation sign meaning to sum over all the particles, where i takes on integer values from 1 to n. (Often we simply write $\sum m_i x_i$, leaving out the $i = 1$ to n.) The total mass of the system is $M = \sum m_i$.

EXAMPLE 9–14 **CM of three guys on a raft.** Three people of roughly equal masses m on a lightweight (air-filled) banana boat sit along the x axis at positions $x_A = 1.0\,\text{m}$, $x_B = 5.0\,\text{m}$, and $x_C = 6.0\,\text{m}$, measured from the left-hand end as shown in Fig. 9–24. Find the position of the CM. Ignore the boat's mass.

APPROACH We are given the mass and location of the three people, so we use three terms in Eq. 9–10. We approximate each person as a point particle. Equivalently, the location of each person is the position of that person's own CM.

SOLUTION We use Eq. 9–10 with three terms:

$$x_{CM} = \frac{mx_A + mx_B + mx_C}{m + m + m} = \frac{m(x_A + x_B + x_C)}{3m}$$
$$= \frac{(1.0\,\text{m} + 5.0\,\text{m} + 6.0\,\text{m})}{3} = \frac{12.0\,\text{m}}{3} = 4.0\,\text{m}.$$

The CM is 4.0 m from the left-hand end of the boat. This makes sense—it should be closer to the two people in front than the one at the rear.

FIGURE 9–24 Example 9–14.

Note that the coordinates of the CM depend on the reference frame or coordinate system chosen. But the physical location of the CM is independent of that choice.

EXERCISE F Calculate the CM of the three people in Example 9–14 taking the origin at the driver $(x_C = 0)$ on the right. Is the physical location of the CM the same?

If the particles are spread out in two or three dimensions, as for a typical extended object, then we define the coordinates of the CM as

$$x_{CM} = \frac{\sum m_i x_i}{M}, \qquad y_{CM} = \frac{\sum m_i y_i}{M}, \qquad z_{CM} = \frac{\sum m_i z_i}{M}, \qquad \textbf{(9-11)}$$

where x_i, y_i, z_i are the coordinates of the particle of mass m_i and again $M = \sum m_i$ is the total mass.

Although from a practical point of view we usually calculate the components of the CM (Eq. 9–11), it is sometimes convenient (for example, for derivations) to write Eq. 9–11 in vector form. If $\vec{r}_i = x_i\hat{i} + y_i\hat{j} + z_i\hat{k}$ is the position vector of the ith particle, and $\vec{r}_{CM} = x_{CM}\hat{i} + y_{CM}\hat{j} + z_{CM}\hat{k}$ is the position vector of the center of mass, then

$$\vec{r}_{CM} = \frac{\sum m_i \vec{r}_i}{M}. \qquad \textbf{(9-12)}$$

FIGURE 9–25 Example 9–15.

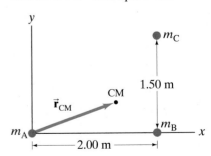

EXAMPLE 9–15 **Three particles in 2-D.** Three particles, each of mass 2.50 kg, are located at the corners of a right triangle whose sides are 2.00 m and 1.50 m long, as shown in Fig. 9–25. Locate the center of mass.

APPROACH We choose our coordinate system as shown (to simplify calculations) with m_A at the origin and m_B on the x axis. Then m_A has coordinates $x_A = y_A = 0$; m_B has coordinates $x_B = 2.0\,\text{m}$, $y_B = 0$; and m_C has coordinates $x_C = 2.0\,\text{m}$, $y_C = 1.5\,\text{m}$.

SOLUTION From Eqs. 9–11,

$$x_{CM} = \frac{(2.50\,\text{kg})(0) + (2.50\,\text{kg})(2.00\,\text{m}) + (2.50\,\text{kg})(2.00\,\text{m})}{3(2.50\,\text{kg})} = 1.33\,\text{m}$$

$$y_{CM} = \frac{(2.50\,\text{kg})(0) + (2.50\,\text{kg})(0) + (2.50\,\text{kg})(1.50\,\text{m})}{7.50\,\text{kg}} = 0.50\,\text{m}.$$

The CM and the position vector $\vec{r}_{CM}$ are shown in Fig. 9–25, inside the "triangle" as we should expect.

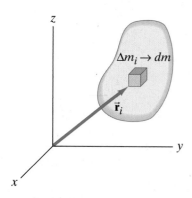

FIGURE 9–26 An extended object, here shown in only two dimensions, can be considered to be made up of many tiny particles (n), each having a mass Δm_i. One such particle is shown located at a point $\vec{\mathbf{r}}_i = x_i\hat{\mathbf{i}} + y_i\hat{\mathbf{j}} + z_i\hat{\mathbf{k}}$. We take the limit of $n \to \infty$ so Δm_i becomes the infinitesimal dm.

FIGURE 9–27 Example 9–16.

EXERCISE G A diver does a high dive involving a flip and a half-pike (legs and arms straight, but body bent in half). What can you say about the diver's center of mass? (*a*) It accelerates with a magnitude of 9.8 m/s^2 (ignoring air friction). (*b*) It moves in a circular path because of the rotation of the diver. (*c*) It must always be roughly located inside the diver's body, somewhere in the geometric center. (*d*) All of the above are true.

It is often convenient to think of an extended object as made up of a continuous distribution of matter. In other words, we consider the object to be made up of n particles, each of mass Δm_i in a tiny volume around a point x_i, y_i, z_i, and we take the limit of n approaching infinity (Fig. 9–26). Then Δm_i becomes the infinitesimal mass dm at points x, y, z. The summations in Eqs. 9–11 and 9–12 become integrals:

$$x_{\text{CM}} = \frac{1}{M} \int x \, dm, \quad y_{\text{CM}} = \frac{1}{M} \int y \, dm, \quad z_{\text{CM}} = \frac{1}{M} \int z \, dm, \quad \textbf{(9–13)}$$

where the sum over all the mass elements is $\int dm = M$, the total mass of the object. In vector notation, this becomes

$$\vec{\mathbf{r}}_{\text{CM}} = \frac{1}{M} \int \vec{\mathbf{r}} \, dm. \qquad \textbf{(9–14)}$$

A concept similar to *center of mass* is **center of gravity** (CG). The CG of an object is that point at which the force of gravity can be considered to act. The force of gravity actually acts on *all* the different parts or particles of an object, but for purposes of determining the translational motion of an object as a whole, we can assume that the entire weight of the object (which is the sum of the weights of all its parts) acts at the CG. There is a conceptual difference between the center of gravity and the center of mass, but for nearly all practical purposes, they are at the same point.[†]

EXAMPLE 9–16 CM of a thin rod. (*a*) Show that the CM of a uniform thin rod of length ℓ and mass M is at its center. (*b*) Determine the CM of the rod assuming its linear mass density λ (its mass per unit length) varies linearly from $\lambda = \lambda_0$ at the left end to double that value, $\lambda = 2\lambda_0$, at the right end.

APPROACH We choose a coordinate system so that the rod lies on the x axis with the left end at $x = 0$, Fig. 9–27. Then $y_{\text{CM}} = 0$ and $z_{\text{CM}} = 0$.

SOLUTION (*a*) The rod is uniform, so its mass per unit length (linear mass density λ) is constant and we write it as $\lambda = M/\ell$. We now imagine the rod as divided into infinitesimal elements of length dx, each of which has mass $dm = \lambda \, dx$. We use Eq. 9–13:

$$x_{\text{CM}} = \frac{1}{M} \int_{x=0}^{\ell} x \, dm = \frac{1}{M} \int_0^{\ell} \lambda x \, dx = \frac{\lambda}{M} \frac{x^2}{2}\Big|_0^{\ell} = \frac{\lambda \ell^2}{2M} = \frac{\ell}{2}$$

where we used $\lambda = M/\ell$. This result, x_{CM} at the center, is what we expected. (*b*) Now we have $\lambda = \lambda_0$ at $x = 0$ and we are told that λ increases linearly to $\lambda = 2\lambda_0$ at $x = \ell$. So we write

$$\lambda = \lambda_0(1 + \alpha x)$$

which satisfies $\lambda = \lambda_0$ at $x = 0$, increases linearly, and gives $\lambda = 2\lambda_0$ at $x = \ell$ if $(1 + \alpha\ell) = 2$. In other words, $\alpha = 1/\ell$. Again we use Eq. 9–13, with $\lambda = \lambda_0(1 + x/\ell)$:

$$x_{\text{CM}} = \frac{1}{M} \int_{x=0}^{\ell} \lambda x \, dx = \frac{1}{M} \lambda_0 \int_0^{\ell} \left(1 + \frac{x}{\ell}\right) x \, dx = \frac{\lambda_0}{M}\left(\frac{x^2}{2} + \frac{x^3}{3\ell}\right)\Big|_0^{\ell} = \frac{5}{6}\frac{\lambda_0}{M}\ell^2.$$

Now let us write M in terms of λ_0 and ℓ. We can write

$$M = \int_{x=0}^{\ell} dm = \int_0^{\ell} \lambda \, dx = \lambda_0 \int_0^{\ell}\left(1 + \frac{x}{\ell}\right) dx = \lambda_0\left(x + \frac{x^2}{2\ell}\right)\Big|_0^{\ell} = \frac{3}{2}\lambda_0\ell.$$

Then

$$x_{\text{CM}} = \frac{5}{6}\frac{\lambda_0}{M}\ell^2 = \frac{5}{9}\ell,$$

which is more than halfway along the rod, as we would expect since there is more mass to the right.

[†]There would be a difference between the CM and CG only in the unusual case of an object so large that the acceleration due to gravity, g, was different at different parts of the object.

For symmetrically shaped objects of uniform composition, such as spheres, cylinders, and rectangular solids, the CM is located at the geometric center of the object. Consider a uniform circular cylinder, such as a solid circular disk. We expect the CM to be at the center of the circle. To show that it is, we first choose a coordinate system whose origin is at the center of the circle with the z axis perpendicular to the disk (Fig. 9–28). When we take the sum $\Sigma m_i x_i$ in Eqs. 9–11, there is as much mass at any $+x_i$ as there is at $-x_i$. So all terms cancel out in pairs and $x_{CM} = 0$. The same is true for y_{CM}. In the vertical (z) direction, the CM must lie halfway between the circular faces: if we choose our origin of coordinates at that point, there is as much mass at any $+z_i$ as at $-z_i$, so $z_{CM} = 0$. For other uniform, symmetrically shaped objects, we can make similar arguments to show that the CM must lie on a line of symmetry. If a symmetric body is *not* uniform, then these arguments do not hold. For example, the CM of a wheel or disk weighted on one side is not at the geometric center but closer to the weighted side.

To locate the center of mass of a group of extended objects, we can use Eqs. 9–11, where the m_i are the masses of these objects and x_i, y_i, and z_i are the coordinates of the CM of each of the objects.

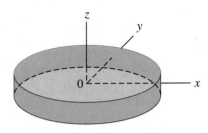

FIGURE 9–28 Cylindrical disk with origin of coordinates at geometric center.

FIGURE 9–29 Example 9–17. This L-shaped object has thickness t (not shown on diagram).

EXAMPLE 9–17 CM of L-shaped flat object. Determine the CM of the uniform thin L-shaped construction brace shown in Fig. 9–29.

APPROACH We can consider the object as two rectangles: rectangle A, which is 2.06 m × 0.20 m, and rectangle B, which is 1.48 m × 0.20 m. We choose the origin at 0 as shown. We assume a uniform thickness t.

SOLUTION The CM of rectangle A is at

$$x_A = 1.03 \text{ m}, \quad y_A = 0.10 \text{ m}.$$

The CM of B is at

$$x_B = 1.96 \text{ m}, \quad y_B = -0.74 \text{ m}.$$

The mass of A, whose thickness is t, is

$$M_A = (2.06 \text{ m})(0.20 \text{ m})(t)(\rho) = (0.412 \text{ m}^2)(\rho t),$$

where ρ is the density (mass per unit volume). The mass of B is

$$M_B = (1.48 \text{ m})(0.20 \text{ m})(\rho t) = (0.296 \text{ m}^2)(\rho t),$$

and the total mass is $M = (0.708 \text{ m}^2)(\rho t)$. Thus

$$x_{CM} = \frac{M_A x_A + M_B x_B}{M} = \frac{(0.412 \text{ m}^2)(1.03 \text{ m}) + (0.296 \text{ m}^2)(1.96 \text{ m})}{(0.708 \text{ m}^2)} = 1.42 \text{ m},$$

where ρt was canceled out in numerator and denominator. Similarly,

$$y_{CM} = \frac{(0.412 \text{ m}^2)(0.10 \text{ m}) + (0.296 \text{ m}^2)(-0.74 \text{ m})}{(0.708 \text{ m}^2)} = -0.25 \text{ m},$$

which puts the CM approximately at the point so labeled in Fig. 9–29. In thickness, $z_{CM} = t/2$, since the object is assumed to be uniform.

Note in this last Example that the CM can actually lie *outside* the object. Another example is a doughnut whose CM is at the center of the hole.

It is often easier to determine the CM or CG of an extended object experimentally rather than analytically. If an object is suspended from any point, it will swing (Fig. 9–30) due to the force of gravity on it, unless it is placed so its CG lies on a vertical line directly below the point from which it is suspended. If the object is two-dimensional, or has a plane of symmetry, it need only be hung from two different pivot points and the respective vertical (plumb) lines drawn. Then the CG will be at the intersection of the two lines, as in Fig. 9–31. If the object doesn't have a plane of symmetry, the CG with respect to the third dimension is found by suspending the object from at least three points whose plumb lines do not lie in the same plane. For symmetrically shaped objects, the CM is located at the geometric center of the object.

FIGURE 9–30 Determining the CM of a flat uniform body.

FIGURE 9–31 Finding the CG.

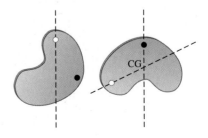

9–9 Center of Mass and Translational Motion

As mentioned in Section 9–8, a major reason for the importance of the concept of center of mass is that the translational motion of the CM for a system of particles (or an extended object) is directly related to the net force acting on the system as a whole. We now show this, by examining the motion of a system of n particles of total mass M, and we assume all the masses remain constant. We begin by rewriting Eq. 9–12 as

$$M\vec{r}_{CM} = \Sigma m_i \vec{r}_i.$$

We differentiate this equation with respect to time:

$$M\frac{d\vec{r}_{CM}}{dt} = \Sigma m_i \frac{d\vec{r}_i}{dt}$$

or

$$M\vec{v}_{CM} = \Sigma m_i \vec{v}_i, \tag{9–15}$$

where $\vec{v}_i = d\vec{r}_i/dt$ is the velocity of the ith particle of mass m_i, and $\vec{v}_{CM}$ is the velocity of the CM. We take the derivative with respect to time again and obtain

$$M\frac{d\vec{v}_{CM}}{dt} = \Sigma m_i \vec{a}_i,$$

where $\vec{a}_i = d\vec{v}_i/dt$ is the acceleration of the ith particle. Now $d\vec{v}_{CM}/dt$ is the acceleration of the CM, $\vec{a}_{CM}$. By Newton's second law, $m_i \vec{a}_i = \vec{F}_i$ where $\vec{F}_i$ is the net force on the ith particle. Therefore

$$M\vec{a}_{CM} = \vec{F}_1 + \vec{F}_2 + \cdots + \vec{F}_n = \Sigma \vec{F}_i. \tag{9–16}$$

That is, the vector sum of all the forces acting on the system is equal to the total mass of the system times the acceleration of its center of mass. Note that our system of n particles could be the n particles that make up one or more extended objects.

The forces $\vec{F}_i$ exerted on the particles of the system can be divided into two types: (1) *external forces* exerted by objects outside the system and (2) *internal forces* that particles within the system exert on one another. By Newton's third law, the internal forces occur in pairs: if one particle exerts a force on a second particle in our system, the second must exert an equal and opposite force on the first. Thus, in the sum over all the forces in Eq. 9–16, these internal forces cancel each other in pairs. We are left, then, with only the external forces on the right side of Eq. 9–16:

$$M\vec{a}_{CM} = \Sigma \vec{F}_{ext}, \qquad \text{[constant } M\text{]} \quad (9\text{–}17)$$

where $\Sigma \vec{F}_{ext}$ is the sum of all the external forces acting on our system, which is the *net force* acting on the system. Thus

> **the sum of all the forces acting on the system is equal to the total mass of the system times the acceleration of its center of mass.**

This is **Newton's second law** for a system of particles. It also applies to an extended object (which can be thought of as a collection of particles), and to a system of objects. Thus we conclude that

> **the center of mass of a system of particles (or objects) with total mass M moves like a single particle of mass M acted upon by the same net external force.**

That is, the system translates as if all its mass were concentrated at the CM and all the external forces acted at that point. We can thus treat the *translational motion* of any object or system of objects as the motion of a particle (see Figs. 9–21 and 9–22).

This result clearly simplifies our analysis of the motion of complex systems and extended objects. Although the motion of various parts of the system may be complicated, we may often be satisfied with knowing the motion of the center of mass. This result also allows us to solve certain types of problems very easily, as illustrated by the following Example.

CONCEPTUAL EXAMPLE 9–18 | **A two-stage rocket.** A rocket is shot into the air as shown in Fig. 9–32. At the moment it reaches its highest point, a horizontal distance d from its starting point, a prearranged explosion separates it into two parts of equal mass. Part I is stopped in midair by the explosion and falls vertically to Earth. Where does part II land? Assume $\vec{\mathbf{g}} = $ constant.

FIGURE 9–32 Example 9–18.

RESPONSE After the rocket is fired, the path of the CM of the system continues to follow the parabolic trajectory of a projectile acted on only by a constant gravitational force. The CM will thus arrive at a point $2d$ from the starting point. Since the masses of I and II are equal, the CM must be midway between them. Therefore, part II lands a distance $3d$ from the starting point.

NOTE If part I had been given a kick up or down, instead of merely falling, the solution would have been somewhat more complicated.

EXERCISE H A woman stands up in a rowboat and walks from one end of the boat to the other. How does the boat move, as seen from the shore?

We can write Eq. 9–17, $M\vec{\mathbf{a}}_{CM} = \Sigma\vec{\mathbf{F}}_{ext}$, in terms of the total momentum $\vec{\mathbf{P}}$ of a system of particles. $\vec{\mathbf{P}}$ is defined, as we saw in Section 9–2 as

$$\vec{\mathbf{P}} = m_1\vec{\mathbf{v}}_1 + m_2\vec{\mathbf{v}}_2 + \cdots + m_n\vec{\mathbf{v}}_n = \Sigma\vec{\mathbf{p}}_i.$$

From Eq. 9–15 $\left(M\vec{\mathbf{v}}_{CM} = \Sigma m_i\vec{\mathbf{v}}_i\right)$ we have

$$\vec{\mathbf{P}} = M\vec{\mathbf{v}}_{CM}. \tag{9–18}$$

Thus, *the total linear momentum of a system of particles is equal to the product of the total mass M and the velocity of the center of mass of the system.* Or, *the linear momentum of an extended object is the product of the object's mass and the velocity of its CM.*

If we differentiate Eq. 9–18 with respect to time, we obtain (assuming the total mass M is constant)

$$\frac{d\vec{\mathbf{P}}}{dt} = M\frac{d\vec{\mathbf{v}}_{CM}}{dt} = M\vec{\mathbf{a}}_{CM}.$$

From Eq. 9–17, we see that

$$\frac{d\vec{\mathbf{P}}}{dt} = \Sigma\vec{\mathbf{F}}_{ext}, \qquad \text{[same as Eq. 9–5]}$$

NEWTON'S SECOND LAW
(for a system)

where $\Sigma\vec{\mathbf{F}}_{ext}$ is the net external force on the system. This is just Eq. 9–5 obtained earlier: **Newton's second law for a system of objects.** It is valid for any definite fixed system of particles or objects. If we know $\Sigma\vec{\mathbf{F}}_{ext}$, we can determine how the total momentum changes.

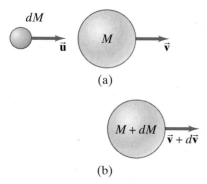

An interesting application is the discovery of nearby stars (see Section 6–5) that seem to "wobble." What could cause such a wobble? It could be that a planet orbits the star, and each exerts a gravitational force on the other. The planets are too small and too far away to have been observed directly by existing telescopes. But the slight wobble in the motion of the star suggests that both the planet and the star (its sun) orbit about their mutual center of mass, and hence the star appears to have a wobble. Irregularities in the star's motion can be obtained to high accuracy, and from the data the size of the planets' orbits can be obtained as well as their masses. See Fig. 6–18 in Chapter 6.

*9–10 Systems of Variable Mass; Rocket Propulsion

We now treat objects or systems whose mass varies. Such systems could be treated as a type of inelastic collision, but it is simpler to use Eq. 9–5, $d\vec{P}/dt = \Sigma\vec{F}_{ext}$, where $\vec{P}$ is the total momentum of the system and $\Sigma\vec{F}_{ext}$ is the net external force exerted on it. Great care must be taken to define the system, and to include all changes in momentum. An important application is to rockets, which propel themselves forward by the ejection of burned gases: the force exerted by the gases on the rocket accelerates the rocket. The mass M of the rocket decreases as it ejects gas, so for the rocket $dM/dt < 0$. Another application is the dropping of material (gravel, packaged goods) onto a conveyor belt. In this situation, the mass M of the loaded conveyor belt increases and $dM/dt > 0$.

To treat the general case of variable mass, let us consider the system shown in Fig. 9–33. At some time t, we have a system of mass M and momentum $M\vec{v}$. We also have a tiny (infinitesimal) mass dM traveling with velocity $\vec{u}$ which is about to enter our system. An infinitesimal time dt later, the mass dM combines with the system. For simplicity we will refer to this as a "collision." So our system has changed in mass from M to $M + dM$ in the time dt. Note that dM can be less than zero, as for a rocket propelled by ejected gases whose mass M thus decreases.

In order to apply Eq. 9–5, $d\vec{P}/dt = \Sigma\vec{F}_{ext}$, we must consider a definite fixed system of particles. That is, in considering the change in momentum, $d\vec{P}$, we must consider the momentum of the same particles initially and finally. We will define our *total system* as including M plus dM. Then initially, at time t, the total momentum is $M\vec{v} + \vec{u}\,dM$ (Fig. 9–33). At time $t + dt$, after dM has combined with M, the velocity of the whole is now $\vec{v} + d\vec{v}$ and the total momentum is $(M + dM)(\vec{v} + d\vec{v})$. So the change in momentum $d\vec{P}$ is

$$d\vec{P} = (M + dM)(\vec{v} + d\vec{v}) - (M\vec{v} + \vec{u}\,dM)$$
$$= M\,d\vec{v} + \vec{v}\,dM + dM\,d\vec{v} - \vec{u}\,dM.$$

The term $dM\,d\vec{v}$ is the product of two differentials and is zero even after we "divide by dt," which we do, and apply Eq. 9–5 to obtain

$$\Sigma\vec{F}_{ext} = \frac{d\vec{P}}{dt} = \frac{M\,d\vec{v} + \vec{v}\,dM - \vec{u}\,dM}{dt}.$$

Thus we get

$$\Sigma\vec{F}_{ext} = M\frac{d\vec{v}}{dt} - (\vec{u} - \vec{v})\frac{dM}{dt}. \qquad \textbf{(9–19a)}$$

Note that the quantity $(\vec{u} - \vec{v})$ is the relative velocity, $\vec{v}_{rel}$, of dM with respect to M.

FIGURE 9–33 (a) At time t, a mass dM is about to be added to our system M. (b) At time $t + dt$, the mass dM has been added to our system.

That is,

$$\vec{v}_{rel} = \vec{u} - \vec{v}$$

is the velocity of the entering mass dM as seen by an observer on M. We can rearrange Eq. 9–19a:

$$M\frac{d\vec{v}}{dt} = \Sigma\vec{F}_{ext} + \vec{v}_{rel}\frac{dM}{dt}. \qquad\qquad \textbf{(9–19b)}$$

We can interpret this equation as follows. $Md\vec{v}/dt$ is the mass times the acceleration of M. The first term on the right, $\Sigma\vec{F}_{ext}$, refers to the external force on the mass M (for a rocket, it would include the force of gravity and air resistance). It does *not* include the force that dM exerts on M as a result of their collision. This is taken care of by the second term on the right, $\vec{v}_{rel}(dM/dt)$, which represents the rate at which momentum is being transferred into (or out of) the mass M because of the mass that is added to (or leaves) it. It can thus be interpreted as the force exerted on the mass M due to the addition (or ejection) of mass. For a rocket this term is called the *thrust*, since it represents the force exerted on the rocket by the expelled gases. For a rocket ejecting burned fuel, $dM/dt < 0$, but so is $\vec{v}_{rel}$ (gases are forced out the back), so the second term in Eq. 9–19b acts to increase $\vec{v}$.

EXAMPLE 9–19 Conveyor belt. You are designing a conveyor system for a gravel yard. A hopper drops gravel at a rate of 75.0 kg/s onto a conveyor belt that moves at a constant speed $v = 2.20$ m/s (Fig. 9–34). (*a*) Determine the additional force (over and above internal friction) needed to keep the conveyor belt moving as gravel falls on it. (*b*) What power output would be needed from the motor that drives the conveyor belt?

APPROACH We assume that the hopper is at rest so $u = 0$, and that the hopper has just begun dropping gravel so $dM/dt = 75.0$ kg/s.

SOLUTION (*a*) The belt needs to move at a constant speed ($dv/dt = 0$), so Eq. 9–19 as written for one dimension, gives:

$$\begin{aligned}
F_{ext} &= M\frac{dv}{dt} - (u - v)\frac{dM}{dt} \\
&= 0 - (0 - v)\frac{dM}{dt} \\
&= v\frac{dM}{dt} = (2.20\,\text{m/s})(75.0\,\text{kg/s}) = 165\,\text{N}.
\end{aligned}$$

(*b*) This force does work at the rate (Eq. 8–21)

$$\begin{aligned}
\frac{dW}{dt} &= \vec{F}_{ext} \cdot \vec{v} = v^2\frac{dM}{dt} \\
&= 363\,\text{W},
\end{aligned}$$

which is the power output required of the motor.

NOTE This work does not all go into kinetic energy of the gravel, since

$$\frac{dK}{dt} = \frac{d}{dt}\left(\frac{1}{2}Mv^2\right) = \frac{1}{2}\frac{dM}{dt}v^2,$$

which is only half the work done by $\vec{F}_{ext}$. The other half of the external work done goes into thermal energy produced by friction between the gravel and the belt (the same friction force that accelerates the gravel).

FIGURE 9–34 Example 9–19. Gravel dropped from hopper onto conveyor belt.

FIGURE 9–35 Example 9–20; $\vec{v}_{rel} = \vec{v}_{gases} - \vec{v}_{rocket}$. M is the mass of the rocket at any instant and is decreasing until burnout.

EXAMPLE 9–20 **Rocket propulsion.** A fully fueled rocket has a mass of 21,000 kg, of which 15,000 kg is fuel. The burned fuel is spewed out the rear at a rate of 190 kg/s with a speed of 2800 m/s relative to the rocket. If the rocket is fired vertically upward (Fig. 9–35) calculate: (*a*) the thrust of the rocket; (*b*) the net force on the rocket at blastoff, and just before burnout (when all the fuel has been used up); (*c*) the rocket's velocity as a function of time, and (*d*) its final velocity at burnout. Ignore air resistance and assume the acceleration due to gravity is constant at $g = 9.80 \text{ m/s}^2$.

APPROACH To begin, the thrust is defined (see discussion after Eq. 9–19b) as the last term in Eq. 9–19b, $v_{rel}(dM/dt)$. The net force [for (*b*)] is the vector sum of the thrust and gravity. The velocity is found from Eq. 9–19b.

SOLUTION (*a*) The thrust is:

$$F_{thrust} = v_{rel}\frac{dM}{dt} = (-2800 \text{ m/s})(-190 \text{ kg/s}) = 5.3 \times 10^5 \text{ N},$$

where we have taken upward as positive so v_{rel} is negative because it is downward, and dM/dt is negative because the rocket's mass is diminishing.

(*b*) $F_{ext} = Mg = (2.1 \times 10^4 \text{ kg})(9.80 \text{ m/s}^2) = 2.1 \times 10^5 \text{ N}$ initially, and at burnout $F_{ext} = (6.0 \times 10^3 \text{ kg})(9.80 \text{ m/s}^2) = 5.9 \times 10^4 \text{ N}$. Hence, the net force on the rocket at blastoff is

$$F_{net} = 5.3 \times 10^5 \text{ N} - 2.1 \times 10^5 \text{ N} = 3.2 \times 10^5 \text{ N}, \quad [\text{blastoff}]$$

and just before burnout it is

$$F_{net} = 5.3 \times 10^5 \text{ N} - 5.9 \times 10^4 \text{ N} = 4.7 \times 10^5 \text{ N}. \quad [\text{burnout}]$$

After burnout, of course, the net force is that of gravity, $-5.9 \times 10^4 \text{ N}$.

(*c*) From Eq. 9–19b we have

$$dv = \frac{F_{ext}}{M}dt + v_{rel}\frac{dM}{M},$$

where $F_{ext} = -Mg$, and M is the mass of the rocket and is a function of time. Since v_{rel} is constant, we can integrate this easily:

$$\int_{v_0}^{v} dv = -\int_0^t g\,dt + v_{rel}\int_{M_0}^{M}\frac{dM}{M}$$

or

$$v(t) = v_0 - gt + v_{rel}\ln\frac{M}{M_0},$$

where $v(t)$ is the rocket's velocity and M its mass at any time t. Note that v_{rel} is negative (-2800 m/s in our case) because it is opposite to the motion, and that $\ln(M/M_0)$ is also negative because $M_0 > M$. Hence, the last term—which represents the thrust—is positive and acts to increase the velocity.

(*d*) The time required to reach burnout is the time needed to use up all the fuel (15,000 kg) at a rate of 190 kg/s; so at burnout,

$$t = \frac{1.50 \times 10^4 \text{ kg}}{190 \text{ kg/s}} = 79 \text{ s}.$$

If we take $v_0 = 0$, then using the result of part (*c*):

$$v = -(9.80 \text{ m/s}^2)(79 \text{ s}) + (-2800 \text{ m/s})\left(\ln\frac{6000 \text{ kg}}{21,000 \text{ kg}}\right) = 2700 \text{ m/s}.$$

Summary

The **linear momentum**, $\vec{p}$, of an object is defined as the product of its mass times its velocity,

$$\vec{p} = m\vec{v}. \qquad (9\text{--}1)$$

In terms of momentum, **Newton's second law** can be written as

$$\Sigma\vec{F} = \frac{d\vec{p}}{dt}. \qquad (9\text{--}2)$$

That is, the rate of change of momentum of an object equals the net force exerted on it.

When the net external force on a system of objects is zero, the total momentum remains constant. This is the **law of conservation of momentum**. Stated another way, the total momentum of an isolated system of objects remains constant.

The law of conservation of momentum is very useful in dealing with the class of events known as **collisions**. In a collision, two (or more) objects interact with each other for a very short time, and the force each exerts on the other during this time interval is very large compared to any other forces acting. The **impulse** of such a force on an object is defined as

$$\vec{J} = \int \vec{F}\, dt$$

and is equal to the change in momentum of the object as long as $\vec{F}$ is the net force on the object:

$$\Delta\vec{p} = \vec{p}_f - \vec{p}_i = \int_{t_i}^{t_f} \vec{F}\, dt = \vec{J}. \qquad (9\text{--}6)$$

Total momentum is conserved in any collision:

$$\vec{p}_A + \vec{p}_B = \vec{p}'_A + \vec{p}'_B.$$

The total energy is also conserved; but this may not be useful unless kinetic energy is conserved, in which case the collision is called an **elastic collision**:

$$\tfrac{1}{2}m_A v_A^2 + \tfrac{1}{2}m_B v_B^2 = \tfrac{1}{2}m_A v_A'^2 + \tfrac{1}{2}m_B v_B'^2. \qquad (9\text{--}7)$$

If kinetic energy is not conserved, the collision is called **inelastic**.

If two colliding objects stick together as the result of a collision, the collision is said to be **completely inelastic**.

For a system of particles, or for an extended object that can be considered as having a continuous distribution of matter, the **center of mass** (CM) is defined as

$$x_{CM} = \frac{\Sigma m_i x_i}{M}, \qquad y_{CM} = \frac{\Sigma m_i y_i}{M}, \qquad z_{CM} = \frac{\Sigma m_i z_i}{M} \qquad (9\text{--}11)$$

or

$$x_{CM} = \frac{1}{M}\int x\, dm, \qquad y_{CM} = \frac{1}{M}\int y\, dm, \qquad z_{CM} = \frac{1}{M}\int z\, dm, \qquad (9\text{--}13)$$

where M is the total mass of the system.

The center of mass of a system is important because this point moves like a single particle of mass M acted on by the same net external force, $\Sigma\vec{F}_{ext}$. In equation form, this is just Newton's second law for a system of particles (or extended objects):

$$M\vec{a}_{CM} = \Sigma\vec{F}_{ext}, \qquad (9\text{--}17)$$

where M is the total mass of the system, $\vec{a}_{CM}$ is the acceleration of the CM of the system, and $\Sigma\vec{F}_{ext}$ is the total (net) external force acting on all parts of the system.

For a system of particles of total linear momentum $\vec{P} = \Sigma m_i \vec{v}_i = M\vec{v}_{CM}$, Newton's second law is

$$\frac{d\vec{P}}{dt} = \Sigma\vec{F}_{ext}. \qquad (9\text{--}5)$$

[*If the mass M of an object is not constant, then

$$M\frac{d\vec{v}}{dt} = \Sigma\vec{F}_{ext} + \vec{v}_{rel}\frac{dM}{dt} \qquad (9\text{--}19b)$$

where $\vec{v}$ is the velocity of the object at any instant and $\vec{v}_{rel}$ is the relative velocity at which mass enters (or leaves) the object.]

Questions

1. We claim that momentum is conserved. Yet most moving objects eventually slow down and stop. Explain.

2. Two blocks of mass m_1 and m_2 rest on a frictionless table and are connected by a spring. The blocks are pulled apart, stretching the spring, and then released. Describe the subsequent motion of the two blocks.

3. A light object and a heavy object have the same kinetic energy. Which has the greater momentum? Explain.

4. When a person jumps from a tree to the ground, what happens to the momentum of the person upon striking the ground?

5. Explain, on the basis of conservation of momentum, how a fish propels itself forward by swishing its tail back and forth.

6. Two children float motionlessly in a space station. The 20-kg girl pushes on the 40-kg boy and he sails away at 1.0 m/s. The girl (a) remains motionless; (b) moves in the same direction at 1.0 m/s; (c) moves in the opposite direction at 1.0 m/s; (d) moves in the opposite direction at 2.0 m/s; (e) none of these.

7. A truck going 15 km/h has a head-on collision with a small car going 30 km/h. Which statement best describes the situation? (a) The truck has the greater change of momentum because it has the greater mass. (b) The car has the greater change of momentum because it has the greater speed. (c) Neither the car nor the truck changes its momentum in the collision because momentum is conserved. (d) They both have the same change in magnitude of momentum because momentum is conserved. (e) None of the above is necessarily true.

8. If a falling ball were to make a perfectly elastic collision with the floor, would it rebound to its original height? Explain.

9. A boy stands on the back of a rowboat and dives into the water. What happens to the rowboat as the boy leaves it? Explain.

10. It is said that in ancient times a rich man with a bag of gold coins was stranded on the surface of a frozen lake. Because the ice was frictionless, he could not push himself to shore and froze to death. What could he have done to save himself had he not been so miserly?

11. The speed of a tennis ball on the return of a serve can be just as fast as the serve, even though the racket isn't swung very fast. How can this be?

12. Is it possible for an object to receive a larger impulse from a small force than from a large force? Explain.

13. How could a force give zero impulse over a nonzero time interval even though the force is not zero for at least a part of that time interval?

14. In a collision between two cars, which would you expect to be more damaging to the occupants: if the cars collide and remain together, or if the two cars collide and rebound backward? Explain.

15. A superball is dropped from a height h onto a hard steel plate (fixed to the Earth), from which it rebounds at very nearly its original speed. (a) Is the momentum of the ball conserved during any part of this process? (b) If we consider the ball and the Earth as our system, during what parts of the process is momentum conserved? (c) Answer part (b) for a piece of putty that falls and sticks to the steel plate.

16. Cars used to be built as rigid as possible to withstand collisions. Today, though, cars are designed to have "crumple zones" that collapse upon impact. What is the advantage of this new design?

17. At a hydroelectric power plant, water is directed at high speed against turbine blades on an axle that turns an electric generator. For maximum power generation, should the turbine blades be designed so that the water is brought to a dead stop, or so that the water rebounds?

18. A squash ball hits a wall at a 45° angle as shown in Fig. 9–36. What is the direction (a) of the change in momentum of the ball, (b) of the force on the wall?

FIGURE 9–36
Question 18.

19. Why can a batter hit a pitched baseball farther than a ball he himself has tossed up in the air?

20. Describe a collision in which all kinetic energy is lost.

21. Inelastic and elastic collisions are similar in that (a) momentum and kinetic energy are conserved in both; (b) momentum is conserved in both; (c) momentum and potential energy are conserved in both; (d) kinetic energy is conserved in both.

22. If a 20-passenger plane is not full, sometimes passengers are told they must sit in certain seats and may not move to empty seats. Why might this be?

23. Why do you tend to lean backward when carrying a heavy load in your arms?

24. Why is the CM of a 1-m length of pipe at its midpoint, whereas this is not true for your arm or leg?

25. Show on a diagram how your CM shifts when you move from a lying position to a sitting position.

26. Describe an analytic way of determining the CM of any thin, triangular-shaped, uniform plate.

27. Place yourself facing the edge of an open door. Position your feet astride the door with your nose and abdomen touching the door's edge. Try to rise on your tiptoes. Why can't this be done?

28. If only an external force can change the momentum of the center of mass of an object, how can the internal force of the engine accelerate a car?

29. A rocket following a parabolic path through the air suddenly explodes into many pieces. What can you say about the motion of this system of pieces?

30. How can a rocket change direction when it is far out in space and essentially in a vacuum?

31. In observations of nuclear β-decay, the electron and recoil nucleus often do not separate along the same line. Use conservation of momentum in two dimensions to explain why this implies the emission of at least one other particle in the disintegration.

32. Bob and Jim decide to play tug-of-war on a frictionless (icy) surface. Jim is considerably stronger than Bob, but Bob weighs 160 lbs while Jim weighs 145 lbs. Who loses by crossing over the midline first?

33. At a carnival game you try to knock over a heavy cylinder by throwing a small ball at it. You have a choice of throwing either a ball that will stick to the cylinder, or a second ball of equal mass and speed that will bounce backward off the cylinder. Which ball is more likely to make the cylinder move?

Problems

9–1 Momentum

1. (I) Calculate the force exerted on a rocket when the propelling gases are being expelled at a rate of 1300 kg/s with a speed of 4.5×10^4 m/s.

2. (I) A constant friction force of 25 N acts on a 65-kg skier for 15 s. What is the skier's change in velocity?

3. (II) The momentum of a particle, in SI units, is given by $\vec{p} = 4.8\, t^2\hat{i} - 8.0\hat{j} - 8.9\, t\hat{k}$. What is the force as a function of time?

4. (II) The force on a particle of mass m is given by $\vec{F} = 26\hat{i} - 12\, t^2\hat{j}$ where F is in N and t in s. What will be the change in the particle's momentum between $t = 1.0$ s and $t = 2.0$ s?

5. (II) A 145-g baseball, moving along the x axis with speed 30.0 m/s, strikes a fence at a 45° angle and rebounds along the y axis with unchanged speed. Give its change in momentum using unit vector notation.

6. (II) A 0.145-kg baseball pitched horizontally at 32.0 m/s strikes a bat and is popped straight up to a height of 36.5 m. If the contact time between bat and ball is 2.5 ms, calculate the average force between the ball and bat during contact.

7. (II) A rocket of total mass 3180 kg is traveling in outer space with a velocity of 115 m/s. To alter its course by 35.0°, its rockets can be fired briefly in a direction perpendicular to its original motion. If the rocket gases are expelled at a speed of 1750 m/s, how much mass must be expelled?

8. (III) Air in a 120-km/h wind strikes head-on the face of a building 45 m wide by 65 m high and is brought to rest. If air has a mass of 1.3 kg per cubic meter, determine the average force of the wind on the building.

9–2 Conservation of Momentum

9. (I) A 7700-kg boxcar traveling 18 m/s strikes a second car. The two stick together and move off with a speed of 5.0 m/s. What is the mass of the second car?

10. (I) A 9150-kg railroad car travels alone on a level frictionless track with a constant speed of 15.0 m/s. A 4350-kg load, initially at rest, is dropped onto the car. What will be the car's new speed?

11. (I) An atomic nucleus at rest decays radioactively into an alpha particle and a smaller nucleus. What will be the speed of this recoiling nucleus if the speed of the alpha particle is 2.8×10^5 m/s? Assume the recoiling nucleus has a mass 57 times greater than that of the alpha particle.

12. (I) A 130-kg tackler moving at 2.5 m/s meets head-on (and tackles) an 82-kg halfback moving at 5.0 m/s. What will be their mutual speed immediately after the collision?

13. (II) A child in a boat throws a 5.70-kg package out horizontally with a speed of 10.0 m/s, Fig. 9–37. Calculate the velocity of the boat immediately after, assuming it was initially at rest. The mass of the child is 24.0 kg and that of the boat is 35.0 kg.

FIGURE 9–37
Problem 13.

$v = 10.0$ m/s

14. (II) An atomic nucleus initially moving at 420 m/s emits an alpha particle in the direction of its velocity, and the remaining nucleus slows to 350 m/s. If the alpha particle has a mass of 4.0 u and the original nucleus has a mass of 222 u, what speed does the alpha particle have when it is emitted?

15. (II) An object at rest is suddenly broken apart into two fragments by an explosion. One fragment acquires twice the kinetic energy of the other. What is the ratio of their masses?

16. (II) A 22-g bullet traveling 210 m/s penetrates a 2.0-kg block of wood and emerges going 150 m/s. If the block is stationary on a frictionless surface when hit, how fast does it move after the bullet emerges?

17. (II) A rocket of mass m traveling with speed v_0 along the x axis suddenly shoots out fuel equal to one-third its mass, perpendicular to the x axis (along the y axis) with speed $2v_0$. Express the final velocity of the rocket in $\hat{\mathbf{i}}, \hat{\mathbf{j}}, \hat{\mathbf{k}}$ notation.

18. (II) The decay of a neutron into a proton, an electron, and a neutrino is an example of a three-particle decay process. Use the vector nature of momentum to show that if the neutron is initially at rest, the velocity vectors of the three must be coplanar (that is, all in the same plane). The result is not true for numbers greater than three.

19. (II) A mass $m_A = 2.0$ kg, moving with velocity $\vec{\mathbf{v}}_A = (4.0\hat{\mathbf{i}} + 5.0\hat{\mathbf{j}} - 2.0\hat{\mathbf{k}})$ m/s, collides with mass $m_B = 3.0$ kg, which is initially at rest. Immediately after the collision, mass m_A is observed traveling at velocity $\vec{\mathbf{v}}_A' = (-2.0\hat{\mathbf{i}} + 3.0\hat{\mathbf{k}})$ m/s. Find the velocity of mass m_B after the collision. Assume no outside force acts on the two masses during the collision.

20. (II) A 925-kg two-stage rocket is traveling at a speed of 6.60×10^3 m/s away from Earth when a predesigned explosion separates the rocket into two sections of equal mass that then move with a speed of 2.80×10^3 m/s relative to each other along the original line of motion. (a) What is the speed and direction of each section (relative to Earth) after the explosion? (b) How much energy was supplied by the explosion? [Hint: What is the change in kinetic energy as a result of the explosion?]

21. (III) A 224-kg projectile, fired with a speed of 116 m/s at a 60.0° angle, breaks into three pieces of equal mass at the highest point of its arc (where its velocity is horizontal). Two of the fragments move with the same speed right after the explosion as the entire projectile had just before the explosion; one of these moves vertically downward and the other horizontally. Determine (a) the velocity of the third fragment immediately after the explosion and (b) the energy released in the explosion.

9–3 Collisions and Impulse

22. (I) A 0.145-kg baseball pitched at 35.0 m/s is hit on a horizontal line drive straight back at the pitcher at 56.0 m/s. If the contact time between bat and ball is 5.00×10^{-3} s, calculate the force (assumed to be constant) between the ball and bat.

23. (II) A golf ball of mass 0.045 kg is hit off the tee at a speed of 45 m/s. The golf club was in contact with the ball for 3.5×10^{-3} s. Find (a) the impulse imparted to the golf ball, and (b) the average force exerted on the ball by the golf club.

24. (II) A 12-kg hammer strikes a nail at a velocity of 8.5 m/s and comes to rest in a time interval of 8.0 ms. (a) What is the impulse given to the nail? (b) What is the average force acting on the nail?

25. (II) A tennis ball of mass $m = 0.060$ kg and speed $v = 25$ m/s strikes a wall at a 45° angle and rebounds with the same speed at 45° (Fig. 9–38). What is the impulse (magnitude and direction) given to the ball?

FIGURE 9–38
Problem 25.

26. (II) A 130-kg astronaut (including space suit) acquires a speed of 2.50 m/s by pushing off with his legs from a 1700-kg space capsule. (a) What is the change in speed of the space capsule? (b) If the push lasts 0.500 s, what is the average force exerted by each on the other? As the reference frame, use the position of the capsule before the push. (c) What is the kinetic energy of each after the push?

27. (II) Rain is falling at the rate of 5.0 cm/h and accumulates in a pan. If the raindrops hit at 8.0 m/s, estimate the force on the bottom of a 1.0 m² pan due to the impacting rain which does not rebound. Water has a mass of 1.00×10^3 kg per m³.

28. (II) Suppose the force acting on a tennis ball (mass 0.060 kg) points in the $+x$ direction and is given by the graph of Fig. 9–39 as a function of time. Use graphical methods to estimate (a) the total impulse given the ball, and (b) the velocity of the ball after being struck, assuming the ball is being served so it is nearly at rest initially.

FIGURE 9–39
Problem 28.

29. (II) With what impulse does a 0.50-kg newspaper have to be thrown to give it a velocity of 3.0 m/s?

30. (II) The force on a bullet is given by the formula $F = [740 - (2.3 \times 10^5 \, \text{s}^{-1})t]$ N over the time interval $t = 0$ to $t = 3.0 \times 10^{-3}$ s. (a) Plot a graph of F versus t for $t = 0$ to $t = 3.0$ ms. (b) Use the graph to estimate the impulse given the bullet. (c) Determine the impulse by integration. (d) If the bullet achieves a speed of 260 m/s as a result of this impulse, given to it in the barrel of a gun, what must the bullet's mass be? (e) What is the recoil speed of the 4.5-kg gun?

31. (II) (a) A molecule of mass m and speed v strikes a wall at right angles and rebounds back with the same speed. If the collision time is Δt, what is the average force on the wall during the collision? (b) If molecules, all of this type, strike the wall at intervals a time t apart (on the average) what is the average force on the wall averaged over a long time?

32. (III) (a) Calculate the impulse experienced when a 65-kg person lands on firm ground after jumping from a height of 3.0 m. (b) Estimate the average force exerted on the person's feet by the ground if the landing is stiff-legged, and again (c) with bent legs. With stiff legs, assume the body moves 1.0 cm during impact, and when the legs are bent, about 50 cm. [*Hint*: The average net force on her which is related to impulse, is the vector sum of gravity and the force exerted by the ground.]

33. (III) A scale is adjusted so that when a large, shallow pan is placed on it, it reads zero. A water faucet at height $h = 2.5$ m above is turned on and water falls into the pan at a rate $R = 0.14$ kg/s. Determine (a) a formula for the scale reading as a function of time t and (b) the reading for $t = 9.0$ s. (c) Repeat (a) and (b), but replace the shallow pan with a tall, narrow cylindrical container of area $A = 20$ cm^2 (the level rises in this case).

9-4 and 9-5 Elastic Collisions

34. (II) A 0.060-kg tennis ball, moving with a speed of 4.50 m/s, has a head-on collision with a 0.090-kg ball initially moving in the same direction at a speed of 3.00 m/s. Assuming a perfectly elastic collision, determine the speed and direction of each ball after the collision.

35. (II) A 0.450-kg hockey puck, moving east with a speed of 4.80 m/s, has a head-on collision with a 0.900-kg puck initially at rest. Assuming a perfectly elastic collision, what will be the speed and direction of each object after the collision?

36. (II) A 0.280-kg croquet ball makes an elastic head-on collision with a second ball initially at rest. The second ball moves off with half the original speed of the first ball. (a) What is the mass of the second ball? (b) What fraction of the original kinetic energy ($\Delta K/K$) gets transferred to the second ball?

37. (II) A ball of mass 0.220 kg that is moving with a speed of 7.5 m/s collides head-on and elastically with another ball initially at rest. Immediately after the collision, the incoming ball bounces backward with a speed of 3.8 m/s. Calculate (a) the velocity of the target ball after the collision, and (b) the mass of the target ball.

38. (II) A ball of mass m makes a head-on elastic collision with a second ball (at rest) and rebounds with a speed equal to 0.350 its original speed. What is the mass of the second ball?

39. (II) Determine the fraction of kinetic energy lost by a neutron ($m_1 = 1.01$ u) when it collides head-on and elastically with a target particle at rest which is (a) $^1_1$H ($m = 1.01$ u); (b) $^2_1$H (heavy hydrogen, $m = 2.01$ u); (c) $^{12}_6$C ($m = 12.00$ u); (d) $^{208}_{82}$Pb (lead, $m = 208$ u).

40. (II) Show that, in general, for any head-on one-dimensional elastic collision, the speeds after collision are

$$v'_B = v_A\left(\frac{2m_A}{m_A + m_B}\right) + v_B\left(\frac{m_B - m_A}{m_A + m_B}\right)$$

and

$$v'_A = v_A\left(\frac{m_A - m_B}{m_A + m_B}\right) + v_B\left(\frac{2m_B}{m_A + m_B}\right),$$

where v_A and v_B are the initial speeds of the two objects of mass m_A and m_B.

41. (III) A 3.0-kg block slides along a frictionless tabletop at 8.0 m/s toward a second block (at rest) of mass 4.5 kg. A coil spring, which obeys Hooke's law and has spring constant $k = 850$ N/m, is attached to the second block in such a way that it will be compressed when struck by the moving block, Fig. 9–40. (a) What will be the maximum compression of the spring? (b) What will be the final velocities of the blocks after the collision? (c) Is the collision elastic? Ignore the mass of the spring.

FIGURE 9–40 Problem 41.

9-6 Inelastic Collisions

42. (I) In a ballistic pendulum experiment, projectile 1 results in a maximum height h of the pendulum equal to 2.6 cm. A second projectile (of the same mass) causes the the pendulum to swing twice as high, $h_2 = 5.2$ cm. The second projectile was how many times faster than the first?

43. (II) (a) Derive a formula for the fraction of kinetic energy lost, $\Delta K/K$, in terms of m and M for the ballistic pendulum collision of Example 9–11. (b) Evaluate for $m = 16.0$ g and $M = 380$ g.

44. (II) A 28-g rifle bullet traveling 210 m/s buries itself in a 3.6-kg pendulum hanging on a 2.8-m-long string, which makes the pendulum swing upward in an arc. Determine the vertical and horizontal components of the pendulum's maximum displacement.

45. (II) An internal explosion breaks an object, initially at rest, into two pieces, one of which has 1.5 times the mass of the other. If 7500 J is released in the explosion, how much kinetic energy does each piece acquire?

46. (II) A 920-kg sports car collides into the rear end of a 2300-kg SUV stopped at a red light. The bumpers lock, the brakes are locked, and the two cars skid forward 2.8 m before stopping. The police officer, estimating the coefficient of kinetic friction between tires and road to be 0.80, calculates the speed of the sports car at impact. What was that speed?

47. (II) You drop a 12-g ball from a height of 1.5 m and it only bounces back to a height of 0.75 m. What was the total impulse on the ball when it hit the floor? (Ignore air resistance).

48. (II) Car A hits car B (initially at rest and of equal mass) from behind while going 35 m/s. Immediately after the collision, car B moves forward at 25 m/s and car A is at rest. What fraction of the initial kinetic energy is lost in the collision?

49. (II) A measure of inelasticity in a head-on collision of two objects is the *coefficient of restitution*, e, defined as

$$e = \frac{v'_A - v'_B}{v_B - v_A},$$

where $v'_A - v'_B$ is the relative velocity of the two objects after the collision and $v_B - v_A$ is their relative velocity before it. (a) Show that $e = 1$ for a perfectly elastic collision, and $e = 0$ for a completely inelastic collision. (b) A simple method for measuring the coefficient of restitution for an object colliding with a very hard surface like steel is to drop the object onto a heavy steel plate, as shown in Fig. 9–41. Determine a formula for e in terms of the original height h and the maximum height h' reached after collision.

FIGURE 9–41 Problem 49. Measurement of coefficient of restitution.

50. (II) A pendulum consists of a mass M hanging at the bottom end of a massless rod of length ℓ, which has a frictionless pivot at its top end. A mass m, moving as shown in Fig. 9–42 with velocity v, impacts M and becomes embedded. What is the smallest value of v sufficient to cause the pendulum (with embedded mass m) to swing clear over the top of its arc?

FIGURE 9–42 Problem 50.

51. (II) A bullet of mass $m = 0.0010\,\text{kg}$ embeds itself in a wooden block with mass $M = 0.999\,\text{kg}$, which then compresses a spring $(k = 120\,\text{N/m})$ by a distance $x = 0.050\,\text{m}$ before coming to rest. The coefficient of kinetic friction between the block and table is $\mu = 0.50$. (a) What is the initial speed of the bullet? (b) What fraction of the bullet's initial kinetic energy is dissipated (in damage to the wooden block, rising temperature, etc.) in the collision between the bullet and the block?

52. (II) A 144-g baseball moving 28.0 m/s strikes a stationary 5.25-kg brick resting on small rollers so it moves without significant friction. After hitting the brick, the baseball bounces straight back, and the brick moves forward at 1.10 m/s. (a) What is the baseball's speed after the collision? (b) Find the total kinetic energy before and after the collision.

53. (II) A 6.0-kg object moving in the $+x$ direction at 5.5 m/s collides head-on with an 8.0-kg object moving in the $-x$ direction at 4.0 m/s. Find the final velocity of each mass if: (a) the objects stick together; (b) the collision is elastic; (c) the 6.0-kg object is at rest after the collision; (d) the 8.0-kg object is at rest after the collision; (e) the 6.0-kg object has a velocity of 4.0 m/s in the $-x$ direction after the collision. Are the results in (c), (d), and (e) "reasonable"? Explain.

9–7 Collisions in Two Dimensions

54. (II) Billiard ball A of mass $m_A = 0.120\,\text{kg}$ moving with speed $v_A = 2.80\,\text{m/s}$ strikes ball B, initially at rest, of mass $m_B = 0.140\,\text{kg}$. As a result of the collision, ball A is deflected off at an angle of $30.0°$ with a speed $v'_A = 2.10\,\text{m/s}$. (a) Taking the x axis to be the original direction of motion of ball A, write down the equations expressing the conservation of momentum for the components in the x and y directions separately. (b) Solve these equations for the speed, v'_B, and angle, θ'_B, of ball B. Do not assume the collision is elastic.

55. (II) A radioactive nucleus at rest decays into a second nucleus, an electron, and a neutrino. The electron and neutrino are emitted at right angles and have momenta of $9.6 \times 10^{-23}\,\text{kg}\cdot\text{m/s}$ and $6.2 \times 10^{-23}\,\text{kg}\cdot\text{m/s}$, respectively. Determine the magnitude and the direction of the momentum of the second (recoiling) nucleus.

56. (II) Two billiard balls of equal mass move at right angles and meet at the origin of an xy coordinate system. Initially ball A is moving upward along the y axis at 2.0 m/s, and ball B is moving to the right along the x axis with speed 3.7 m/s. After the collision (assumed elastic), the second ball is moving along the positive y axis (Fig. 9–43). What is the final direction of ball A, and what are the speeds of the two balls?

FIGURE 9–43 Problem 56. (Ball A after the collision is not shown.)

57. (II) An atomic nucleus of mass m traveling with speed v collides elastically with a target particle of mass $2m$ (initially at rest) and is scattered at $90°$. (a) At what angle does the target particle move after the collision? (b) What are the final speeds of the two particles? (c) What fraction of the initial kinetic energy is transferred to the target particle?

58. (II) A neutron collides elastically with a helium nucleus (at rest initially) whose mass is four times that of the neutron. The helium nucleus is observed to move off at an angle $\theta'_2 = 45°$. Determine the angle of the neutron, θ'_1, and the speeds of the two particles, v'_n and v'_{He}, after the collision. The neutron's initial speed is $6.2 \times 10^5\,\text{m/s}$.

59. (III) A neon atom $(m = 20.0\,\text{u})$ makes a perfectly elastic collision with another atom at rest. After the impact, the neon atom travels away at a $55.6°$ angle from its original direction and the unknown atom travels away at a $-50.0°$ angle. What is the mass (in u) of the unknown atom? [Hint: You could use the law of sines.]

60. (III) For an elastic collision between a projectile particle of mass m_1 and a target particle (at rest) of mass m_2, show that the scattering angle, θ'_1, of the projectile (a) can take any value, 0 to $180°$, for $m_1 < m_2$, but (b) has a maximum angle ϕ given by $\cos^2 \phi = 1 - (m_2/m_1)^2$ for $m_1 > m_2$.

61. (III) Prove that in the elastic collision of two objects of identical mass, with one being a target initially at rest, the angle between their final velocity vectors is always $90°$.

9–8 Center of Mass (CM)

62. (I) The CM of an empty 1250-kg car is 2.50 m behind the front of the car. How far from the front of the car will the CM be when two people sit in the front seat 2.80 m from the front of the car, and three people sit in the back seat 3.90 m from the front? Assume that each person has a mass of 70.0 kg.

63. (I) The distance between a carbon atom ($m = 12\,u$) and an oxygen atom ($m = 16\,u$) in the CO molecule is 1.13×10^{-10} m. How far from the carbon atom is the center of mass of the molecule?

64. (II) Three cubes, of side ℓ_0, $2\ell_0$, and $3\ell_0$, are placed next to one another (in contact) with their centers along a straight line as shown in Fig. 9–44. What is the position, along this line, of the CM of this system? Assume the cubes are made of the same uniform material.

FIGURE 9–44
Problem 64.

65. (II) A square uniform raft, 18 m by 18 m, of mass 6200 kg, is used as a ferryboat. If three cars, each of mass 1350 kg, occupy the NE, SE, and SW corners, determine the CM of the loaded ferryboat relative to the center of the raft.

66. (II) A uniform circular plate of radius $2R$ has a circular hole of radius R cut out of it. The center C' of the smaller circle is a distance $0.80R$ from the center C of the larger circle, Fig. 9–45. What is the position of the center of mass of the plate? [*Hint*: Try subtraction.]

FIGURE 9–45
Problem 66.

67. (II) A uniform thin wire is bent into a semicircle of radius r. Determine the coordinates of its center of mass with respect to an origin of coordinates at the center of the "full" circle.

68. (II) Find the center of mass of the ammonia molecule. The chemical formula is NH_3. The hydrogens are at the corners of an equilateral triangle (with sides 0.16 nm) that forms the base of a pyramid, with nitrogen at the apex (0.037 nm vertically above the plane of the triangle).

69. (III) Determine the CM of a machine part that is a uniform cone of height h and radius R, Fig. 9–46. [*Hint*: Divide the cone into an infinite number of disks of thickness dz, one of which is shown.]

FIGURE 9–46
Problem 69.

70. (III) Determine the CM of a uniform pyramid that has four triangular faces and a square base with equal sides all of length s. [*Hint*: See Problem 69.]

71. (III) Determine the CM of a thin, uniform, semicircular plate.

9–9 CM and Translational Motion

72. (II) Mass $M_A = 35$ kg and mass $M_B = 25$ kg. They have velocities (in m/s) $\vec{v}_A = 12\hat{i} - 16\hat{j}$ and $\vec{v}_B = -20\hat{i} + 14\hat{j}$. Determine the velocity of the center of mass of the system.

73. (II) The masses of the Earth and Moon are 5.98×10^{24} kg and 7.35×10^{22} kg, respectively, and their centers are separated by 3.84×10^{8} m. (*a*) Where is the CM of this system located? (*b*) What can you say about the motion of the Earth–Moon system about the Sun, and of the Earth and Moon separately about the Sun?

74. (II) A mallet consists of a uniform cylindrical head of mass 2.80 kg and a diameter 0.0800 m mounted on a uniform cylindrical handle of mass 0.500 kg and length 0.240 m, as shown in Fig. 9–47. If this mallet is tossed, spinning, into the air, how far above the bottom of the handle is the point that will follow a parabolic trajectory?

FIGURE 9–47
Problem 74.   8.00 cm

75. (II) A 55-kg woman and a 72-kg man stand 10.0 m apart on frictionless ice. (*a*) How far from the woman is their CM? (*b*) If each holds one end of a rope, and the man pulls on the rope so that he moves 2.5 m, how far from the woman will he be now? (*c*) How far will the man have moved when he collides with the woman?

76. (II) Suppose that in Example 9–18 (Fig. 9–32), $m_{II} = 3m_I$. (*a*) Where then would m_{II} land? (*b*) What if $m_I = 3m_{II}$?

77. (II) Two people, one of mass 85 kg and the other of mass 55 kg, sit in a rowboat of mass 78 kg. With the boat initially at rest, the two people, who have been sitting at opposite ends of the boat, 3.0 m apart from each other, now exchange seats. How far and in what direction will the boat move?

78. (III) A 280-kg flatcar 25 m long is moving with a speed of 6.0 m/s along horizontal frictionless rails. A 95-kg worker starts walking from one end of the car to the other in the direction of motion, with speed 2.0 m/s with respect to the car. In the time it takes for him to reach the other end, how far has the flatcar moved?

79. (III) A huge balloon and its gondola, of mass M, are in the air and stationary with respect to the ground. A passenger, of mass m, then climbs out and slides down a rope with speed v, measured with respect to the balloon. With what speed and direction (relative to Earth) does the balloon then move? What happens if the passenger stops?

*9–10 Variable Mass

*80. (II) A 3500-kg rocket is to be accelerated at $3.0\,g$ at take-off from the Earth. If the gases can be ejected at a rate of 27 kg/s, what must be their exhaust speed?

*81. (II) Suppose the conveyor belt of Example 9–19 is retarded by a friction force of 150 N. Determine the required output power (hp) of the motor as a function of time from the moment gravel first starts falling ($t = 0$) until 3.0 s after the gravel begins to be dumped off the end of the 22-m-long conveyor belt.

*82. (II) The jet engine of an airplane takes in 120 kg of air per second, which is burned with 4.2 kg of fuel per second. The burned gases leave the plane at a speed of 550 m/s (relative to the plane). If the plane is traveling 270 m/s (600 mi/h), determine: (*a*) the thrust due to ejected fuel; (*b*) the thrust due to accelerated air passing through the engine; and (*c*) the power (hp) delivered.

*83. (II) A rocket traveling 1850 m/s away from the Earth at an altitude of 6400 km fires its rockets, which eject gas at a speed of 1300 m/s (relative to the rocket). If the mass of the rocket at this moment is 25,000 kg and an acceleration of 1.5 m/s² is desired, at what rate must the gases be ejected?

*84. (III) A sled filled with sand slides without friction down a 32° slope. Sand leaks out a hole in the sled at a rate of 2.0 kg/s. If the sled starts from rest with an initial total mass of 40.0 kg, how long does it take the sled to travel 120 m along the slope?

General Problems

85. A novice pool player is faced with the corner pocket shot shown in Fig. 9–48. Relative dimensions are also shown. Should the player worry that this might be a "scratch shot," in which the cue ball will also fall into a pocket? Give details. Assume equal mass balls and an elastic collision.

FIGURE 9–48
Problem 85.

86. During a Chicago storm, winds can whip horizontally at speeds of 120 km/h. If the air strikes a person at the rate of 45 kg/s per square meter and is brought to rest, calculate the force of the wind on a person. Assume the person is 1.60 m high and 0.50 m wide. Compare to the typical maximum force of friction ($\mu \approx 1.0$) between the person and the ground, if the person has a mass of 75 kg.

87. A ball is dropped from a height of 1.50 m and rebounds to a height of 1.20 m. Approximately how many rebounds will the ball make before losing 90% of its energy?

88. In order to convert a tough split in bowling, it is necessary to strike the pin a glancing blow as shown in Fig. 9–49. Assume that the bowling ball, initially traveling at 13.0 m/s, has five times the mass of a pin and that the pin goes off at 75° from the original direction of the ball. Calculate the speed (a) of the pin and (b) of the ball just after collision, and (c) calculate the angle through which the ball was deflected. Assume the collision is elastic and ignore any spin of the ball.

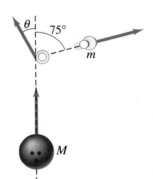

FIGURE 9–49
Problem 88.

89. A gun fires a bullet vertically into a 1.40-kg block of wood at rest on a thin horizontal sheet, Fig. 9–50. If the bullet has a mass of 24.0 g and a speed of 310 m/s, how high will the block rise into the air after the bullet becomes embedded in it?

FIGURE 9–50
Problem 89.

90. A hockey puck of mass $4\,m$ has been rigged to explode, as part of a practical joke. Initially the puck is at rest on a frictionless ice rink. Then it bursts into three pieces. One chunk, of mass m, slides across the ice at velocity $v\hat{\mathbf{i}}$. Another chunk, of mass $2m$, slides across the ice at velocity $2v\hat{\mathbf{j}}$. Determine the velocity of the third chunk.

91. For the completely inelastic collision of two railroad cars that we considered in Example 9–3, calculate how much of the initial kinetic energy is transformed to thermal or other forms of energy.

92. A 4800-kg open railroad car coasts along with a constant speed of 8.60 m/s on a level track. Snow begins to fall vertically and fills the car at a rate of 3.80 kg/min. Ignoring friction with the tracks, what is the speed of the car after 60.0 min? (See Section 9–2.)

*93. Consider the railroad car of Problem 92, which is slowly filling with snow. (a) Determine the speed of the car as a function of time using Eqs. 9–19. (b) What is the speed of the car after 60.0 min? Does this agree with the simpler calculation (Problem 92)?

94. Two blocks of mass m_A and m_B, resting on a frictionless table, are connected by a stretched spring and then released (Fig. 9–51). (a) Is there a net external force on the system? (b) Determine the ratio of their speeds, v_A/v_B. (c) What is the ratio of their kinetic energies? (d) Describe the motion of the CM of this system. (e) How would the presence of friction alter the above results?

FIGURE 9–51   Problem 94.

95. You have been hired as an expert witness in a court case involving an automobile accident. The accident involved car A of mass 1500 kg which crashed into stationary car B of mass 1100 kg. The driver of car A applied his brakes 15 m before he skidded and crashed into car B. After the collision, car A slid 18 m while car B slid 30 m. The coefficient of kinetic friction between the locked wheels and the road was measured to be 0.60. Show that the driver of car A was exceeding the 55-mi/h (90 km/h) speed limit before applying the brakes.

96. A meteor whose mass was about 2.0×10^8 kg struck the Earth $(m_E = 6.0 \times 10^{24}$ kg) with a speed of about 25 km/s and came to rest in the Earth. (a) What was the Earth's recoil speed (relative to Earth at rest before the collision)? (b) What fraction of the meteor's kinetic energy was transformed to kinetic energy of the Earth? (c) By how much did the Earth's kinetic energy change as a result of this collision?

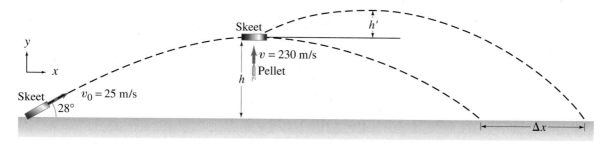

FIGURE 9–54
Problem 103.

97. Two astronauts, one of mass 65 kg and the other 85 kg, are initially at rest in outer space. They then push each other apart. How far apart are they when the lighter astronaut has moved 12 m?

98. A 22-g bullet strikes and becomes embedded in a 1.35-kg block of wood placed on a horizontal surface just in front of the gun. If the coefficient of kinetic friction between the block and the surface is 0.28, and the impact drives the block a distance of 8.5 m before it comes to rest, what was the muzzle speed of the bullet?

99. Two balls, of masses $m_A = 45$ g and $m_B = 65$ g, are suspended as shown in Fig. 9–52. The lighter ball is pulled away to a 66° angle with the vertical and released. (a) What is the velocity of the lighter ball before impact? (b) What is the velocity of each ball after the elastic collision? (c) What will be the maximum height of each ball after the elastic collision?

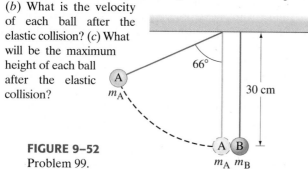

FIGURE 9–52
Problem 99.

100. A block of mass $m = 2.20$ kg slides down a 30.0° incline which is 3.60 m high. At the bottom, it strikes a block of mass $M = 7.00$ kg which is at rest on a horizontal surface, Fig. 9–53. (Assume a smooth transition at the bottom of the incline.) If the collision is elastic, and friction can be ignored, determine (a) the speeds of the two blocks after the collision, and (b) how far back up the incline the smaller mass will go.

FIGURE 9–53 Problems 100 and 101.

101. In Problem 100 (Fig. 9–53), what is the upper limit on mass m if it is to rebound from M, slide up the incline, stop, slide down the incline, and collide with M again?

102. After a completely inelastic collision between two objects of equal mass, each having initial speed, v, the two move off together with speed $v/3$. What was the angle between their initial directions?

103. A 0.25-kg skeet (clay target) is fired at an angle of 28° to the horizon with a speed of 25 m/s (Fig. 9–54). When it reaches the maximum height, h, it is hit from below by a 15-g pellet traveling vertically upward at a speed of 230 m/s. The pellet is embedded in the skeet. (a) How much higher, h', did the skeet go up? (b) How much extra distance, Δx, does the skeet travel because of the collision?

104. A massless spring with spring constant k is placed between a block of mass m and a block of mass $3\,m$. Initially the blocks are at rest on a frictionless surface and they are held together so that the spring between them is compressed by an amount D from its equilibrium length. The blocks are then released and the spring pushes them off in opposite directions. Find the speeds of the two blocks when they detach from the spring.

105. *The gravitational slingshot effect.* Figure 9–55 shows the planet Saturn moving in the negative x direction at its orbital speed (with respect to the Sun) of 9.6 km/s. The mass of Saturn is 5.69×10^{26} kg. A spacecraft with mass 825 kg approaches Saturn. When far from Saturn, it moves in the $+x$ direction at 10.4 km/s. The gravitational attraction of Saturn (a conservative force) acting on the spacecraft causes it to swing around the planet (orbit shown as dashed line) and head off in the opposite direction. Estimate the final speed of the spacecraft after it is far enough away to be considered free of Saturn's gravitational pull.

FIGURE 9–55 Problem 105.

106. Two bumper cars in an amusement park ride collide elastically as one approaches the other directly from the rear (Fig. 9–56). Car A has a mass of 450 kg and car B 490 kg, owing to differences in passenger mass. If car A approaches at 4.50 m/s and car B is moving at 3.70 m/s, calculate (a) their velocities after the collision, and (b) the change in momentum of each.

FIGURE 9–56
Problem 106:
(a) before collision,
(b) after collision.

107. In a physics lab, a cube slides down a frictionless incline as shown in Fig. 9–57 and elastically strikes another cube at the bottom that is only one-half its mass. If the incline is 35 cm high and the table is 95 cm off the floor, where does each cube land? [*Hint:* Both leave the incline moving horizontally.]

FIGURE 9–57
Problem 107.

108. The space shuttle launches an 850-kg satellite by ejecting it from the cargo bay. The ejection mechanism is activated and is in contact with the satellite for 4.0 s to give it a velocity of 0.30 m/s in the z-direction relative to the shuttle. The mass of the shuttle is 92,000 kg. (a) Determine the component of velocity v_f of the shuttle in the minus z-direction resulting from the ejection. (b) Find the average force that the shuttle exerts on the satellite during the ejection.

109. You are the design engineer in charge of the crashworthiness of new automobile models. Cars are tested by smashing them into fixed, massive barriers at 45 km/h. A new model of mass 1500 kg takes 0.15 s from the time of impact until it is brought to rest. (a) Calculate the average force exerted on the car by the barrier. (b) Calculate the average deceleration of the car.

110. Astronomers estimate that a 2.0-km-wide asteroid collides with the Earth once every million years. The collision could pose a threat to life on Earth. (a) Assume a spherical asteroid has a mass of 3200 kg for each cubic meter of volume and moves toward the Earth at 15 km/s. How much destructive energy could be released when it embeds itself in the Earth? (b) For comparison, a nuclear bomb could release about 4.0×10^{16} J. How many such bombs would have to explode simultaneously to release the destructive energy of the asteroid collision with the Earth?

111. An astronaut of mass 210 kg including his suit and jet pack wants to acquire a velocity of 2.0 m/s to move back toward his space shuttle. Assuming the jet pack can eject gas with a velocity of 35 m/s, what mass of gas will need to be ejected?

112. An extrasolar planet can be detected by observing the wobble it produces on the star around which it revolves. Suppose an extrasolar planet of mass m_B revolves around its star of mass m_A. If no external force acts on this simple two-object system, then its CM is stationary. Assume m_A and m_B are in circular orbits with radii r_A and r_B about the system's CM. (a) Show that

$$r_A = \frac{m_B}{m_A} r_B .$$

(b) Now consider a Sun-like star and a single planet with the same characteristics as Jupiter. That is, $m_B = 1.0 \times 10^{-3} m_A$ and the planet has an orbital radius of 8.0×10^{11} m. Determine the radius r_A of the star's orbit about the system's CM. (c) When viewed from Earth, the distant system appears to wobble over a distance of $2 r_A$. If astronomers are able to detect angular displacements θ of about 1 milliarcsec (1 arcsec = $\frac{1}{3600}$ of a degree), from what distance d (in light-years) can the star's wobble be detected ($1 \text{ ly} = 9.46 \times 10^{15}$ m)? (d) The star nearest to our Sun is about 4 ly away. Assuming stars are uniformly distributed throughout our region of the Milky Way Galaxy, about how many stars can this technique be applied to in the search for extrasolar planetary systems?

113. Suppose two asteroids strike head on. Asteroid A ($m_A = 7.5 \times 10^{12}$ kg) has velocity 3.3 km/s before the collision, and asteroid B ($m_B = 1.45 \times 10^{13}$ kg) has velocity 1.4 km/s before the collision in the opposite direction. If the asteroids stick together, what is the velocity (magnitude and direction) of the new asteroid after the collision?

*Numerical/Computer

*114. (III) A particle of mass m_A traveling with speed v_A collides elastically head-on with a stationary particle of smaller mass m_B. (a) Show that the speed of m_B after the collision is

$$v'_B = \frac{2 v_A}{1 + m_B/m_A} .$$

(b) Consider now a third particle of mass m_C at rest between m_A and m_B so that m_A first collides head on with m_C and then m_C collides head on with m_B. Both collisions are elastic. Show that in this case,

$$v'_B = 4 v_A \frac{m_C m_A}{(m_C + m_A)(m_B + m_C)} .$$

(c) From the result of part (b), show that for maximum v'_B, $m_C = \sqrt{m_A m_B}$. (d) Assume $m_B = 2.0$ kg, $m_A = 18.0$ kg and $v_A = 2.0$ m/s. Use a spreadsheet to calculate and graph the values of v'_B from $m_C = 0.0$ kg to $m_C = 50.0$ kg in steps of 1.0 kg. For what value of m_C is the value of v'_B maximum? Does your numerical result agree with your result in part (c)?

Answers to Exercises

A: (c) because the momentum change is greater.

B: Larger (Δp is greater).

C: 0.50 m/s.

D: (a) 6.0 m/s; (b) almost zero; (c) almost 24.0 m/s.

E: (b); (d).

F: $x_{CM} = -2.0$ m; yes.

G: (a).

H: The boat moves in the opposite direction.

You too can experience rapid rotation—if your stomach can take the high angular velocity and centripetal acceleration of some of the faster amusement park rides. If not, try the slower merry-go-round or Ferris wheel. Rotating carnival rides have rotational kinetic energy as well as angular momentum. Angular acceleration is produced by a net torque, and rotating objects have rotational kinetic energy.

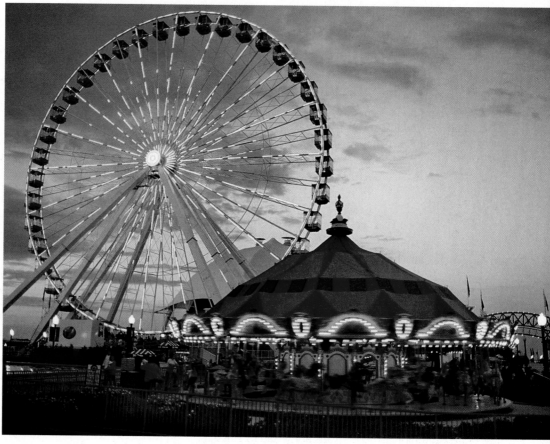

Rotational Motion

CHAPTER-OPENING QUESTION—Guess Now!

A solid ball and a solid cylinder roll down a ramp. They both start from rest at the same time. Which gets to the bottom first?

(a) They get there at the same time.

(b) They get there at almost exactly the same time except for frictional differences.

(c) The ball gets there first.

(d) The cylinder gets there first.

(e) Can't tell without knowing the mass and radius of each.

U ntil now, we have been concerned mainly with translational motion. We discussed the kinematics and dynamics of translational motion (the role of force), and the energy and momentum associated with it. In this Chapter and the next we will deal with rotational motion. We will discuss the kinematics of rotational motion and then its dynamics (involving torque), as well as rotational kinetic energy and angular momentum (the rotational analog of linear momentum). Our understanding of the world around us will be increased significantly—from rotating bicycle wheels and compact discs to amusement park rides, a spinning skater, the rotating Earth, and a centrifuge—and there may be a few surprises.

We will consider mainly the rotation of rigid objects. A **rigid object** is an object with a definite shape that doesn't change, so that the particles composing it stay in fixed positions relative to one another. Any real object is capable of vibrating or deforming when a force is exerted on it. But these effects are often very small, so the concept of an ideal rigid object is very useful as a good approximation.

Our development of rotational motion will parallel our discussion of translational motion: rotational position, angular velocity, angular acceleration, rotational inertia, and the rotational analog of force, "torque."

10–1 Angular Quantities

The motion of a rigid object can be analyzed as the translational motion of its center of mass plus rotational motion *about* its center of mass (Sections 9–8 and 9–9). We have already discussed translational motion in detail, so now we focus our attention on purely rotational motion. By *purely rotational motion* of an object about a fixed axis, we mean that all points in the object move in circles, such as the point P on the rotating wheel of Fig. 10–1, and that the centers of these circles all lie on a line called the **axis of rotation**. In Fig. 10–1 the axis of rotation is perpendicular to the page and passes through point O. We assume the axis is fixed in an inertial reference frame, but we will not always insist that the axis pass through the center of mass.

For a three-dimensional rigid object rotating about a fixed axis, we will use the symbol R to represent the perpendicular distance of a point or particle from the axis of rotation. We do this to distinguish R from r, which will continue to represent the position of a particle with reference to the origin (point) of some coordinate system. This distinction is illustrated in Fig. 10–2. This distinction may seem like a small point, but not being fully aware of it can cause huge errors when working with rotational motion. For a flat, very thin object, like a wheel, with the origin in the plane of the object (at the center of a wheel, for example), R and r will be nearly the same.

Every point in an object rotating about a fixed axis moves in a circle (shown dashed in Fig. 10–1 for point P) whose center is on the axis of rotation and whose radius is R, the distance of that point from the axis of rotation. A straight line drawn from the axis to any point in the object sweeps out the same angle θ in the same time interval.

To indicate the angular position of the object, or how far it has rotated, we specify the angle θ of some particular line in the object (red in Fig. 10–1) with respect to some reference line, such as the x axis in Fig. 10–1. A point in the object, such as P in Fig. 10–1b, moves through an angle θ when it travels the distance ℓ measured along the circumference of its circular path. Angles are commonly stated in degrees, but the mathematics of circular motion is much simpler if we use the *radian* for angular measure. One **radian** (abbreviated rad) is defined as the angle subtended by an arc whose length is equal to the radius. For example, in Fig. 10–1, point P is a distance R from the axis of rotation, and it has moved a distance ℓ along the arc of a circle. The arc length ℓ is said to "subtend" the angle θ. In general, any angle θ is given by

$$\theta = \frac{\ell}{R}, \qquad [\theta \text{ in radians}] \quad \textbf{(10–1a)}$$

where R is the radius of the circle and ℓ is the arc length subtended by the angle θ, which is specified in radians. If $\ell = R$, then $\theta = 1$ rad.

The radian, being the ratio of two lengths, is dimensionless. We thus do not have to mention it in calculations, although it is usually best to include it to remind us the angle is in radians and not degrees. We can rewrite Eq. 10–1a in terms of arc length ℓ:

$$\ell = R\theta. \quad \textbf{(10–1b)}$$

Radians can be related to degrees in the following way. In a complete circle there are 360°, which must correspond to an arc length equal to the circumference of the circle, $\ell = 2\pi R$. Thus $\theta = \ell/R = 2\pi R/R = 2\pi$ rad in a complete circle, so

$$360° = 2\pi \text{ rad}.$$

One radian is therefore $360°/2\pi \approx 360°/6.28 \approx 57.3°$. An object that makes one complete revolution (rev) has rotated through 360°, or 2π radians:

$$1 \text{ rev} = 360° = 2\pi \text{ rad}.$$

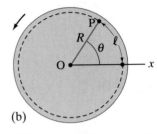

FIGURE 10–1 Looking at a wheel that is rotating counterclockwise about an axis through the wheel's center at O (axis perpendicular to the page). Each point, such as point P, moves in a circular path; ℓ is the distance P travels as the wheel rotates through the angle θ.

FIGURE 10–2 Showing the distinction between $\vec{r}$ (the position vector) and R (the distance from the rotation axis) for a point P on the edge of a cylinder rotating about the z axis.

CAUTION

Use radians in calculating, not degrees

(a)   (b)

FIGURE 10–3 (a) Example 10–1. (b) For small angles, arc length and the chord length (straight line) are nearly equal. For an angle as large as 15°, the error in making this estimate is only 1%. For larger angles the error increases rapidly.

FIGURE 10–4 A wheel rotates from (a) initial position θ_1 to (b) final position θ_2. The angular displacement is $\Delta\theta = \theta_2 - \theta_1$.

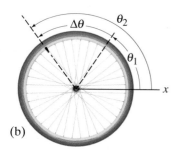

EXAMPLE 10–1 **Birds of prey—in radians.** A particular bird's eye can just distinguish objects that subtend an angle no smaller than about 3×10^{-4} rad. (a) How many degrees is this? (b) How small an object can the bird just distinguish when flying at a height of 100 m (Fig. 10–3a)?

APPROACH For (a) we use the relation $360° = 2\pi$ rad. For (b) we use Eq. 10–1b, $\ell = R\theta$, to find the arc length.

SOLUTION (a) We convert 3×10^{-4} rad to degrees:

$$\left(3 \times 10^{-4}\,\text{rad}\right)\left(\frac{360°}{2\pi\,\text{rad}}\right) = 0.017°,$$

or about 0.02°.

(b) We use Eq. 10–1b, $\ell = R\theta$. For small angles, the arc length ℓ and the chord length are approximately the same (Fig. 10–3b). Since $R = 100$ m and $\theta = 3 \times 10^{-4}$ rad, we find

$$\ell = (100\,\text{m})(3 \times 10^{-4}\,\text{rad}) = 3 \times 10^{-2}\,\text{m} = 3\,\text{cm}.$$

A bird can distinguish a small mouse (about 3 cm long) from a height of 100 m. That is good eyesight.

NOTE Had the angle been given in degrees, we would first have had to convert it to radians to make this calculation. Equation 10–1 is valid *only* if the angle is specified in radians. Degrees (or revolutions) won't work.

To describe rotational motion, we make use of angular quantities, such as angular velocity and angular acceleration. These are defined in analogy to the corresponding quantities in linear motion, and are chosen to describe the rotating object as a whole, so they have the same value for each point in the rotating object. Each point in a rotating object also has translational velocity and acceleration, but they have different values for different points in the object.

When an object, such as the bicycle wheel in Fig. 10–4, rotates from some initial position, specified by θ_1, to some final position, θ_2, its **angular displacement** is

$$\Delta\theta = \theta_2 - \theta_1.$$

The *angular velocity* (denoted by ω, the Greek lowercase letter omega) is defined in analogy with linear (translational) velocity that was discussed in Chapter 2. Instead of linear displacement, we use the angular displacement. Thus the **average angular velocity** of an object rotating about a fixed axis is defined as the time rate of change of angular position:

$$\bar{\omega} = \frac{\Delta\theta}{\Delta t}, \tag{10–2a}$$

where $\Delta\theta$ is the angle through which the object has rotated in the time interval Δt. The **instantaneous angular velocity** is the limit of this ratio as Δt approaches zero:

$$\omega = \lim_{\Delta t \to 0} \frac{\Delta\theta}{\Delta t} = \frac{d\theta}{dt}. \tag{10–2b}$$

Angular velocity has units of radians per second (rad/s). Note that *all points in a rigid object rotate with the same angular velocity*, since every position in the object moves through the same angle in the same time interval.

An object such as the wheel in Fig. 10–4 can rotate about a fixed axis either clockwise or counterclockwise. The direction can be specified with a + or − sign, just as we did in Chapter 2 for linear motion toward the +x or −x direction. The usual convention is to choose the angular displacement $\Delta\theta$ and angular velocity ω as positive when the wheel rotates counterclockwise. If the rotation is clockwise, then θ would decrease, so $\Delta\theta$ and ω would be negative.

Angular acceleration (denoted by α, the Greek lowercase letter alpha), in analogy to linear acceleration, is defined as the change in angular velocity divided by the time required to make this change. The **average angular acceleration** is defined as

$$\bar{\alpha} = \frac{\omega_2 - \omega_1}{\Delta t} = \frac{\Delta \omega}{\Delta t}, \qquad (10\text{–}3a)$$

where ω_1 is the angular velocity initially, and ω_2 is the angular velocity after a time interval Δt. **Instantaneous angular acceleration** is defined as the limit of this ratio as Δt approaches zero:

$$\alpha = \lim_{\Delta t \to 0} \frac{\Delta \omega}{\Delta t} = \frac{d\omega}{dt}. \qquad (10\text{–}3b)$$

Since ω is the same for all points of a rotating object, Eq. 10–3 tells us that α also will be the same for all points. Thus, ω and α are properties of the rotating object as a whole. With ω measured in radians per second and t in seconds, α has units of radians per second squared (rad/s^2).

Each point or particle of a rotating rigid object has, at any instant, a linear velocity v and a linear acceleration a. We can relate the linear quantities at each point, v and a, to the angular quantities, ω and α, of the rotating object. Consider a point P located a distance R from the axis of rotation, as in Fig. 10–5. If the object rotates with angular velocity ω, any point will have a linear velocity whose direction is tangent to its circular path. The magnitude of that point's linear velocity is $v = d\ell/dt$. From Eq. 10–1b, a change in rotation angle $d\theta$ (in radians) is related to the linear distance traveled by $d\ell = R\,d\theta$. Hence

$$v = \frac{d\ell}{dt} = R\frac{d\theta}{dt}$$

or

$$v = R\omega, \qquad (10\text{–}4)$$

where R is a fixed distance from the rotation axis and ω is given in rad/s. Thus, although ω is the same for every point in the rotating object at any instant, the linear velocity v is greater for points farther from the axis (Fig. 10–6). Note that Eq. 10–4 is valid both instantaneously and on the average.

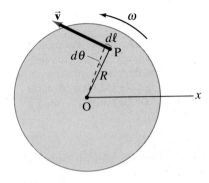

FIGURE 10–5 A point P on a rotating wheel has a linear velocity $\vec{v}$ at any moment.

FIGURE 10–6 A wheel rotating uniformly counterclockwise. Two points on the wheel, at distances R_A and R_B from the center, have the same angular velocity ω because they travel through the same angle θ in the same time interval. But the two points have different linear velocities because they travel different distances in the same time interval. Since $R_B > R_A$, then $v_B > v_A$ (because $v = R\omega$).

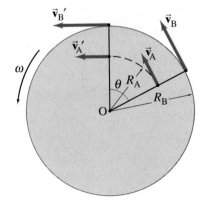

CONCEPTUAL EXAMPLE 10–2 | **Is the lion faster than the horse?** On a rotating carousel or merry-go-round, one child sits on a horse near the outer edge and another child sits on a lion halfway out from the center. (*a*) Which child has the greater linear velocity? (*b*) Which child has the greater angular velocity?

RESPONSE (*a*) The *linear* velocity is the distance traveled divided by the time interval. In one rotation the child on the outer edge travels a longer distance than the child near the center, but the time interval is the same for both. Thus the child at the outer edge, on the horse, has the greater linear velocity. (*b*) The *angular* velocity is the angle of rotation divided by the time interval. In one rotation both children rotate through the same angle $(360° = 2\pi \text{ rad})$. The two children have the same angular velocity.

If the angular velocity of a rotating object changes, the object as a whole—and each point in it—has an angular acceleration. Each point also has a linear acceleration whose direction is tangent to that point's circular path. We use Eq. 10–4 $(v = R\omega)$ to show that the angular acceleration α is related to the tangential linear acceleration a_{tan} of a point in the rotating object by

$$a_{\text{tan}} = \frac{dv}{dt} = R\frac{d\omega}{dt}$$

or

$$a_{\text{tan}} = R\alpha. \qquad (10\text{–}5)$$

In this equation, R is the radius of the circle in which the particle is moving, and the subscript "tan" in a_{tan} stands for "tangential."

FIGURE 10–7 On a rotating wheel whose angular speed is increasing, a point P has both tangential and radial (centripetal) components of linear acceleration. (See also Chapter 5.)

TABLE 10–1
Linear and Rotational Quantities

Linear	Type	Rotational	Relation (θ in radians)
x	displacement	θ	$x = R\theta$
v	velocity	ω	$v = R\omega$
a_{tan}	acceleration	α	$a_{tan} = R\alpha$

FIGURE 10–8 Example 10–3. The total acceleration vector $\vec{a} = \vec{a}_{tan} + \vec{a}_R$, at $t = 8.0$ s.

(a)

(b)

The total linear acceleration of a point at any instant is the vector sum of two components:

$$\vec{a} = \vec{a}_{tan} + \vec{a}_R,$$

where the radial component, $\vec{a}_R$, is the radial or "centripetal" acceleration and its direction is toward the center of the point's circular path; see Fig. 10–7. We saw in Chapter 5 (Eq. 5–1) that a particle moving in a circle of radius R with linear speed v has a radial acceleration $a_R = v^2/R$; we can rewrite this in terms of ω using Eq. 10–4:

$$a_R = \frac{v^2}{R} = \frac{(R\omega)^2}{R} = \omega^2 R. \tag{10–6}$$

Equation 10–6 applies to any particle of a rotating object. Thus the centripetal acceleration is greater the farther you are from the axis of rotation: the children farthest out on a carousel experience the greatest acceleration.

Table 10–1 summarizes the relationships between the angular quantities describing the rotation of an object to the linear quantities for each point of the object.

EXAMPLE 10–3 **Angular and linear velocities and accelerations.** A carousel is initially at rest. At $t = 0$ it is given a constant angular acceleration $\alpha = 0.060 \text{ rad/s}^2$, which increases its angular velocity for 8.0 s. At $t = 8.0$ s, determine the magnitude of the following quantities: (a) the angular velocity of the carousel; (b) the linear velocity of a child (Fig. 10–8a) located 2.5 m from the center, point P in Fig. 10–8b; (c) the tangential (linear) acceleration of that child; (d) the centripetal acceleration of the child; and (e) the total linear acceleration of the child.

APPROACH The angular acceleration α is constant, so we can use $\alpha = \Delta\omega/\Delta t$ to solve for ω after a time $t = 8.0$ s. With this ω and the given α, we determine the other quantities using the relations we just developed, Eqs. 10–4, 10–5, and 10–6.

SOLUTION (a) In Eq. 10–3a, $\bar{\alpha} = (\omega_2 - \omega_1)/\Delta t$, we put $\Delta t = 8.0$ s, $\bar{\alpha} = 0.060 \text{ rad/s}^2$, and $\omega_1 = 0$. Solving for ω_2, we get

$$\omega_2 = \omega_1 + \bar{\alpha}\,\Delta t = 0 + (0.060 \text{ rad/s}^2)(8.0 \text{ s}) = 0.48 \text{ rad/s}.$$

During the 8.0-s interval, the carousel has accelerated from $\omega_1 = 0$ (rest) to $\omega_2 = 0.48 \text{ rad/s}$.

(b) The linear velocity of the child, with $R = 2.5$ m at time $t = 8.0$ s, is found using Eq. 10–4:

$$v = R\omega = (2.5 \text{ m})(0.48 \text{ rad/s}) = 1.2 \text{ m/s}.$$

Note that the "rad" has been dropped here because it is dimensionless (and only a reminder)—it is a ratio of two distances, Eq. 10–1a.

(c) The child's tangential acceleration is given by Eq. 10–5:

$$a_{tan} = R\alpha = (2.5 \text{ m})(0.060 \text{ rad/s}^2) = 0.15 \text{ m/s}^2,$$

and it is the same throughout the 8.0-s acceleration interval.

(d) The child's centripetal acceleration at $t = 8.0$ s is given by Eq. 10–6:

$$a_R = \frac{v^2}{R} = \frac{(1.2 \text{ m/s})^2}{(2.5 \text{ m})} = 0.58 \text{ m/s}^2.$$

(e) The two components of linear acceleration calculated in parts (c) and (d) are perpendicular to each other. Thus the total linear acceleration at $t = 8.0$ s has magnitude

$$a = \sqrt{a_{tan}^2 + a_R^2} = \sqrt{(0.15 \text{ m/s}^2)^2 + (0.58 \text{ m/s}^2)^2} = 0.60 \text{ m/s}^2.$$

Its direction (Fig. 10–8b) is

$$\theta = \tan^{-1}\left(\frac{a_{tan}}{a_R}\right) = \tan^{-1}\left(\frac{0.15 \text{ m/s}^2}{0.58 \text{ m/s}^2}\right) = 0.25 \text{ rad},$$

so $\theta \approx 15°$.

NOTE The linear acceleration at this chosen instant is mostly centripetal, keeping the child moving in a circle with the carousel. The tangential component that speeds up the motion is smaller.

We can relate the angular velocity ω to the frequency of rotation, f. The **frequency** is the number of complete revolutions (rev) per second, as we saw in Chapter 5. One revolution (of a wheel, say) corresponds to an angle of 2π radians, and thus $1\,\text{rev/s} = 2\pi\,\text{rad/s}$. Hence, in general, the frequency f is related to the angular velocity ω by

$$f = \frac{\omega}{2\pi}$$

or

$$\omega = 2\pi f. \tag{10-7}$$

The unit for frequency, revolutions per second (rev/s), is given the special name the hertz (Hz). That is

$$1\,\text{Hz} = 1\,\text{rev/s}.$$

Note that "revolution" is not really a unit, so we can also write $1\,\text{Hz} = 1\,\text{s}^{-1}$.

The time required for one complete revolution is the **period** T, and it is related to the frequency by

$$T = \frac{1}{f}. \tag{10-8}$$

If a particle rotates at a frequency of three revolutions per second, then the period of each revolution is $\frac{1}{3}\,\text{s}$.

EXERCISE A In Example 10–3, we found that the carousel, after 8.0 s, rotates at an angular velocity $\omega = 0.48\,\text{rad/s}$, and continues to do so after $t = 8.0\,\text{s}$ because the acceleration ceased. Determine the frequency and period of the carousel after it has reached a constant angular velocity.

EXAMPLE 10–4 **Hard drive.** The platter of the hard drive of a computer rotates at 7200 rpm (rpm = revolutions per minute = rev/min). (*a*) What is the angular velocity (rad/s) of the platter? (*b*) If the reading head of the drive is located 3.00 cm from the rotation axis, what is the linear speed of the point on the platter just below it? (*c*) If a single bit requires $0.50\,\mu\text{m}$ of length along the direction of motion, how many bits per second can the writing head write when it is 3.00 cm from the axis?

PHYSICS APPLIED
Hard drive and bit speed

APPROACH We use the given frequency f to find the angular velocity ω of the platter and then the linear speed of a point on the platter $(v = R\omega)$. The bit rate is found by dividing the linear speed by the length of one bit $(v = \text{distance/time})$.

SOLUTION (*a*) First we find the frequency in rev/s, given $f = 7200\,\text{rev/min}$:

$$f = \frac{(7200\,\text{rev/min})}{(60\,\text{s/min})} = 120\,\text{rev/s} = 120\,\text{Hz}.$$

Then the angular velocity is

$$\omega = 2\pi f = 754\,\text{rad/s}.$$

(*b*) The linear speed of a point 3.00 cm out from the axis is given by Eq. 10–4:

$$v = R\omega = (3.00 \times 10^{-2}\,\text{m})(754\,\text{rad/s}) = 22.6\,\text{m/s}.$$

(*c*) Each bit requires $0.50 \times 10^{-6}\,\text{m}$, so at a speed of 22.6 m/s, the number of bits passing the head per second is

$$\frac{22.6\,\text{m/s}}{0.50 \times 10^{-6}\,\text{m/bit}} = 45 \times 10^{6}\ \text{bits per second},$$

or 45 megabits/s (Mbps).

EXAMPLE 10–5 **Given ω as function of time.** A disk of radius $R = 3.0\,\text{m}$ rotates at an angular velocity $\omega = (1.6 + 1.2\,t)\,\text{rad/s}$, where t is in seconds. At the instant $t = 2.0\,\text{s}$, determine (a) the angular acceleration, and (b) the speed v and the components of the acceleration a of a point on the edge of the disk.

APPROACH We use $\alpha = d\omega/dt$, $v = R\omega$, $a_{\text{tan}} = R\alpha$, and $a_R = \omega^2 R$, which are Eqs. 10–3b, 10–4, 10–5 and 10–6. We can write ω explicitly showing units of the constants (in case we want to check later): $\omega = [1.6\,\text{s}^{-1} + (1.2\,\text{s}^{-2})t]$ which will give us $\text{s}^{-1}\,(= \text{rad/s})$ for each term.

SOLUTION (a) The angular acceleration is

$$\alpha = \frac{d\omega}{dt} = \frac{d}{dt}(1.6 + 1.2\,t)\,\text{s}^{-1} = 1.2\,\text{rad/s}^2.$$

(b) The speed v of a point $3.0\,\text{m}$ from the center of the rotating disk at $t = 2.0\,\text{s}$ is, using Eq. 10–4,

$$v = R\omega = (3.0\,\text{m})(1.6 + 1.2\,t)\,\text{s}^{-1} = (3.0\,\text{m})(4.0\,\text{s}^{-1}) = 12.0\,\text{m/s}.$$

The components of the linear acceleration of this point at $t = 2.0\,\text{s}$ are

$$a_{\text{tan}} = R\alpha = (3.0\,\text{m})(1.2\,\text{rad/s}^2) = 3.6\,\text{m/s}^2$$

$$a_R = \omega^2 R = [(1.6 + 1.2\,t)\,\text{s}^{-1}]^2(3.0\,\text{m}) = (4.0\,\text{s}^{-1})^2(3.0\,\text{m}) = 48\,\text{m/s}^2.$$

10–2 Vector Nature of Angular Quantities

Both $\vec{\omega}$ and $\vec{\alpha}$ can be treated as vectors, and we define their directions in the following way. Consider the rotating wheel shown in Fig. 10–9a. The linear velocities of different particles of the wheel point in all different directions. The only unique direction in space associated with the rotation is along the axis of rotation, perpendicular to the actual motion. We therefore choose the axis of rotation to be the direction of the angular velocity vector, $\vec{\omega}$. Actually, there is still an ambiguity since $\vec{\omega}$ could point in either direction along the axis of rotation (up or down in Fig. 10–9a). The convention we use, called the **right-hand rule**, is this: when the fingers of the right hand are curled around the rotation axis and point in the direction of the rotation, then the thumb points in the direction of $\vec{\omega}$. This is shown in Fig. 10–9b. Note that $\vec{\omega}$ points in the direction a right-handed screw would move when turned in the direction of rotation. Thus, if the rotation of the wheel in Fig. 10–9a is counterclockwise, the direction of $\vec{\omega}$ is upward as shown in Fig. 10–9b. If the wheel rotates clockwise, then $\vec{\omega}$ points in the opposite direction, downward.[†] Note that no part of the rotating object moves in the direction of $\vec{\omega}$.

If the axis of rotation is fixed in direction, then $\vec{\omega}$ can change only in magnitude. Thus $\vec{\alpha} = d\vec{\omega}/dt$ must also point along the axis of rotation. If the rotation is counterclockwise as in Fig. 10–9a and the magnitude of ω is increasing, then $\vec{\alpha}$ points upward; but if ω is decreasing (the wheel is slowing down), $\vec{\alpha}$ points downward. If the rotation is clockwise, $\vec{\alpha}$ points downward if ω is increasing, and $\vec{\alpha}$ points upward if ω is decreasing.

FIGURE 10–9 (a) Rotating wheel. (b) Right-hand rule for obtaining direction of $\vec{\omega}$.

FIGURE 10–10 (a) Velocity is a true vector. The reflection of $\vec{v}$ points in the same direction. (b) Angular velocity is a pseudovector since it does not follow this rule. As can be seen, the reflection of the wheel rotates in the opposite direction, so the direction of $\vec{\omega}$ is opposite for the reflection.

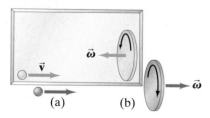

[†]Strictly speaking, $\vec{\omega}$ and $\vec{\alpha}$ are not quite vectors. The problem is that they do not behave like vectors under reflection. Suppose, as we are looking directly into a mirror, a particle moving with velocity $\vec{v}$ to the right passes in front of and parallel to the mirror. In the reflection of the mirror, $\vec{v}$ still points to the right, Fig. 10–10a. Thus a true vector, like velocity, when pointing parallel to the face of the mirror has the same direction in the reflection as in actuality. Now consider a wheel rotating in front of the mirror, so $\vec{\omega}$ points to the right. (We will be looking at the edge of the wheel.) As viewed in the mirror, Fig. 10–10b, the wheel will be rotating in the opposite direction. So $\vec{\omega}$ will point in the opposite direction (to the left) in the mirror. Because $\vec{\omega}$ is different under reflection than a true vector, $\vec{\omega}$ is called a *pseudovector* or *axial vector*. The angular acceleration $\vec{\alpha}$ is also a pseudovector, as are all cross products of true vectors (Section 11–2). The difference between true vectors and pseudovectors is important in elementary particle physics, but will not concern us in this book.

10–3 Constant Angular Acceleration

In Chapter 2, we derived the useful kinematic equations (Eqs. 2–12) that relate acceleration, velocity, distance, and time for the special case of uniform linear acceleration. Those equations were derived from the definitions of linear velocity and acceleration, assuming constant acceleration. The definitions of angular velocity and angular acceleration are the same as those for their linear counterparts, except that θ has replaced the linear displacement x, ω has replaced v, and α has replaced a. Therefore, the angular equations for **constant angular acceleration** will be analogous to Eqs. 2–12 with x replaced by θ, v by ω, and a by α, and they can be derived in exactly the same way. We summarize them here, opposite their linear equivalents (we have chosen $x_0 = 0$ and $\theta_0 = 0$ at the initial time $t = 0$):

Angular	Linear		
$\omega = \omega_0 + \alpha t$	$v = v_0 + at$	[constant α, a]	**(10–9a)**
$\theta = \omega_0 t + \frac{1}{2}\alpha t^2$	$x = v_0 t + \frac{1}{2}at^2$	[constant α, a]	**(10–9b)**
$\omega^2 = \omega_0^2 + 2\alpha\theta$	$v^2 = v_0^2 + 2ax$	[constant α, a]	**(10–9c)**
$\bar{\omega} = \dfrac{\omega + \omega_0}{2}$	$\bar{v} = \dfrac{v + v_0}{2}$	[constant α, a]	**(10–9d)**

Kinematic equations
for constant
angular acceleration
$(x_0 = 0, \ \theta_0 = 0)$

Note that ω_0 represents the angular velocity at $t = 0$, whereas θ and ω represent the angular position and velocity, respectively, at time t. Since the angular acceleration is constant, $\alpha = \bar{\alpha}$.

EXAMPLE 10–6 **Centrifuge acceleration.** A centrifuge rotor is accelerated from rest to 20,000 rpm in 30 s. (*a*) What is its average angular acceleration? (*b*) Through how many revolutions has the centrifuge rotor turned during its acceleration period, assuming constant angular acceleration?

APPROACH To determine $\bar{\alpha} = \Delta\omega/\Delta t$, we need the initial and final angular velocities. For (*b*), we use Eqs. 10–9 (recall that one revolution corresponds to $\theta = 2\pi$ rad).

SOLUTION (*a*) The initial angular velocity is $\omega = 0$. The final angular velocity is

$$\omega = 2\pi f = (2\pi \ \text{rad/rev})\frac{(20{,}000 \ \text{rev/min})}{(60 \ \text{s/min})} = 2100 \ \text{rad/s}.$$

Then, since $\bar{\alpha} = \Delta\omega/\Delta t$ and $\Delta t = 30$ s, we have

$$\bar{\alpha} = \frac{\omega - \omega_0}{\Delta t} = \frac{2100 \ \text{rad/s} - 0}{30 \ \text{s}} = 70 \ \text{rad/s}^2.$$

That is, every second the rotor's angular velocity increases by 70 rad/s, or by $(70/2\pi) = 11$ revolutions per second.

(*b*) To find θ we could use either Eq. 10–9b or 10–9c, or both to check our answer. The former gives

$$\theta = 0 + \tfrac{1}{2}(70 \ \text{rad/s}^2)(30 \ \text{s})^2 = 3.15 \times 10^4 \ \text{rad},$$

where we have kept an extra digit because this is an intermediate result. To find the total number of revolutions, we divide by 2π rad/rev and obtain

$$\frac{3.15 \times 10^4 \ \text{rad}}{2\pi \ \text{rad/rev}} = 5.0 \times 10^3 \ \text{rev}.$$

NOTE Let us calculate θ using Eq. 10–9c:

$$\theta = \frac{\omega^2 - \omega_0^2}{2\alpha} = \frac{(2100 \ \text{rad/s})^2 - 0}{2(70 \ \text{rad/s}^2)} = 3.15 \times 10^4 \ \text{rad}$$

which checks our answer using Eq. 10–9b perfectly.

PHYSICS APPLIED
Centrifuge

10–4 Torque

We have so far discussed rotational kinematics—the description of rotational motion in terms of angular position, angular velocity, and angular acceleration. Now we discuss the dynamics, or causes, of rotational motion. Just as we found analogies between linear and rotational motion for the description of motion, so rotational equivalents for dynamics exist as well.

To make an object start rotating about an axis clearly requires a force. But the direction of this force, and where it is applied, are also important. Take, for example, an ordinary situation such as the overhead view of the door in Fig. 10–11. If you apply a force $\vec{\mathbf{F}}_A$ to the door as shown, you will find that the greater the magnitude, F_A, the more quickly the door opens. But now if you apply the same magnitude force at a point closer to the hinge—say, $\vec{\mathbf{F}}_B$ in Fig. 10–11—the door will not open so quickly. The effect of the force is less: *where* the force acts, as well as its magnitude and direction, affects how quickly the door opens. Indeed, if only this one force acts, the angular acceleration of the door is proportional not only to the magnitude of the force, but is also directly proportional to *the perpendicular distance from the axis of rotation to the line along which the force acts*. This distance is called the **lever arm**, or **moment arm**, of the force, and is labeled R_A and R_B for the two forces in Fig. 10–11. Thus, if R_A in Fig. 10–11 is three times larger than R_B, then the angular acceleration of the door will be three times as great, assuming that the magnitudes of the forces are the same. To say it another way, if $R_A = 3R_B$, then F_B must be three times as large as F_A to give the same angular acceleration. (Figure 10–12 shows two examples of tools whose long lever arms are very effective.)

The angular acceleration, then, is proportional to the product of the *force times the lever arm*. This product is called the *moment of the force* about the axis, or, more commonly, it is called the **torque**, and is represented by τ (Greek lowercase letter tau). Thus, the angular acceleration α of an object is directly proportional to the net applied torque τ:

$$\alpha \propto \tau,$$

and we see that it is torque that gives rise to angular acceleration. This is the rotational analog of Newton's second law for linear motion, $a \propto F$.

We defined the lever arm as the *perpendicular* distance from the axis of rotation to the line of action of the force—that is, the distance which is perpendicular both to the axis of rotation and to an imaginary line drawn along the direction of the force. We do this to take into account the effect of forces acting at an angle. It is clear that a force applied at an angle, such as $\vec{\mathbf{F}}_C$ in Fig. 10–13, will be less effective than the same magnitude force applied perpendicular to the door, such as $\vec{\mathbf{F}}_A$ (Fig. 10–13a). And if you push on the end of the door so that the force is directed at the hinge (the axis of rotation), as indicated by $\vec{\mathbf{F}}_D$, the door will not rotate at all.

The lever arm for a force such as $\vec{\mathbf{F}}_C$ is found by drawing a line along the direction of $\vec{\mathbf{F}}_C$ (this is the "line of action" of $\vec{\mathbf{F}}_C$). Then we draw another line, perpendicular to this line of action, that goes to the axis of rotation and is perpendicular also to it. The length of this second line is the lever arm for $\vec{\mathbf{F}}_C$ and is labeled R_C in Fig. 10–13b. The lever arm for $\vec{\mathbf{F}}_A$ is the full distance from the hinge to the door knob, R_A; thus R_C is much smaller than R_A.

The magnitude of the torque associated with $\vec{\mathbf{F}}_C$ is then $R_C F_C$. This short lever arm R_C and the corresponding smaller torque associated with $\vec{\mathbf{F}}_C$ is consistent with the observation that $\vec{\mathbf{F}}_C$ is less effective in accelerating the door than is $\vec{\mathbf{F}}_A$. When the lever arm is defined in this way, experiment shows that the relation $\alpha \propto \tau$ is valid in general. Notice in Fig. 10–13 that the line of action of the force $\vec{\mathbf{F}}_D$ passes through the hinge, and hence its lever arm is zero. Consequently, zero torque is associated with $\vec{\mathbf{F}}_D$ and it gives rise to no angular acceleration, in accord with everyday experience.

FIGURE 10–11 Top view of a door. Applying the same force with different lever arms, R_A and R_B. If $R_A = 3R_B$, then to create the same effect (angular acceleration), F_B needs to be three times F_A, or $F_A = \frac{1}{3}F_B$.

FIGURE 10–12 (a) A tire iron too can have a long lever arm. (b) A plumber can exert greater torque using a wrench with a long lever arm.

Axis of rotation   Axis of rotation

(a)   (b)

FIGURE 10–13 (a) Forces acting at different angles at the doorknob. (b) The lever arm is defined as the perpendicular distance from the axis of rotation (the hinge) to the line of action of the force.

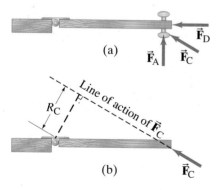

In general, then, we can write the magnitude of the torque about a given axis as

$$\tau = R_\perp F, \qquad\qquad (10\text{--}10a)$$

where $R_\perp$ is the lever arm, and the perpendicular symbol ($\perp$) reminds us that we must use the distance from the axis of rotation that is perpendicular to the line of action of the force (Fig. 10–14a).

An equivalent way of determining the torque associated with a force is to resolve the force into components parallel and perpendicular to the line that connects the axis to the point of application of the force, as shown in Fig. 10–14b. The component $F_\parallel$ exerts no torque since it is directed at the rotation axis (its moment arm is zero). Hence the torque will be equal to $F_\perp$ times the distance R from the axis to the point of application of the force:

$$\tau = R F_\perp. \qquad\qquad (10\text{--}10b)$$

This gives the same result as Eq. 10–10a because $F_\perp = F \sin \theta$ and $R_\perp = R \sin \theta$. So

$$\tau = R F \sin \theta \qquad\qquad (10\text{--}10c)$$

in either case. [Note that θ is the angle between the directions of $\vec{F}$ and R (radial line from the axis to the point where $\vec{F}$ acts)]. We can use any of Eqs. 10–10 to calculate the torque, whichever is easiest.

Because torque is a distance times a force, it is measured in units of m·N in SI units,[†] cm·dyne in the cgs system, and ft·lb in the English system.

When more than one torque acts on an object, the angular acceleration α is found to be proportional to the *net* torque. If all the torques acting on an object tend to rotate it about a fixed axis of rotation in the same direction, the net torque is the sum of the torques. But if, say, one torque acts to rotate an object in one direction, and a second torque acts to rotate the object in the opposite direction (as in Fig. 10–15), the net torque is the difference of the two torques. We normally assign a positive sign to torques that act to rotate the object counterclockwise (just as θ is usually positive counterclockwise), and a negative sign to torques that act to rotate the object clockwise, when the rotation axis is fixed.

EXAMPLE 10–7 **Torque on a compound wheel.** Two thin disk-shaped wheels, of radii $R_A = 30$ cm and $R_B = 50$ cm, are attached to each other on an axle that passes through the center of each, as shown in Fig. 10–15. Calculate the net torque on this compound wheel due to the two forces shown, each of magnitude 50 N.

APPROACH The force $\vec{F}_A$ acts to rotate the system counterclockwise, whereas $\vec{F}_B$ acts to rotate it clockwise. So the two forces act in opposition to each other. We must choose one direction of rotation to be positive—say, counterclockwise. Then $\vec{F}_A$ exerts a positive torque, $\tau_A = R_A F_A$, since the lever arm is R_A. On the other hand, $\vec{F}_B$ produces a negative (clockwise) torque and does not act perpendicular to R_B, so we must use its perpendicular component to calculate the torque it produces: $\tau_B = -R_B F_{B\perp} = -R_B F_B \sin \theta$, where $\theta = 60°$. (Note that θ must be the angle between $\vec{F}_B$ and a radial line from the axis.)

SOLUTION The net torque is

$$\tau = R_A F_A - R_B F_B \sin 60°$$
$$= (0.30\,\text{m})(50\,\text{N}) - (0.50\,\text{m})(50\,\text{N})(0.866) = -6.7\,\text{m·N}.$$

This net torque acts to accelerate the rotation of the wheel in the clockwise direction.

EXERCISE B Two forces ($F_B = 20$ N and $F_A = 30$ N) are applied to a meter stick which can rotate about its left end, Fig. 10–16. Force $\vec{F}_B$ is applied perpendicularly at the midpoint. Which force exerts the greater torque: F_A, F_B, or both the same?

[†]Note that the units for torque are the same as those for energy. We write the unit for torque here as m·N (in SI) to distinguish it from energy (N·m) because the two quantities are very different. An obvious difference is that energy is a scalar, whereas torque has a direction and is a vector (as we will see in Chapter 11). The special name *joule* (1 J = 1 N·m) is used only for energy (and for work), *never* for torque.

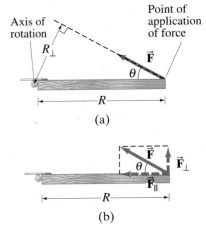

Axis of rotation — Point of application of force

(a)

(b)

FIGURE 10–14 Torque $= R_\perp F = R F_\perp$.

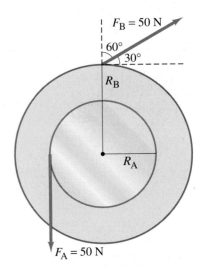

$F_B = 50$ N

$F_A = 50$ N

FIGURE 10–15 Example 10–7. The torque due to $\vec{F}_A$ tends to accelerate the wheel counterclockwise, whereas the torque due to $\vec{F}_B$ tends to accelerate the wheel clockwise.

FIGURE 10–16 Exercise B.

Axis $\quad \vec{F}_B \qquad \vec{F}_A$ 30°

10–5 Rotational Dynamics; Torque and Rotational Inertia

We discussed in Section 10–4 that the angular acceleration α of a rotating object is proportional to the net torque τ applied to it:

$$\alpha \propto \Sigma\tau,$$

where we write $\Sigma\tau$ to remind us that it is the *net* torque (sum of all torques acting on the object) that is proportional to α. This corresponds to Newton's second law for translational motion, $a \propto \Sigma F$, but here torque has taken the place of force, and, correspondingly, the angular acceleration α takes the place of the linear acceleration a. In the linear case, the acceleration is not only proportional to the net force, but it is also inversely proportional to the inertia of the object, which we call its mass, m. Thus we could write $a = \Sigma F/m$. But what plays the role of mass for the rotational case? That is what we now set out to determine. At the same time, we will see that the relation $\alpha \propto \Sigma\tau$ follows directly from Newton's second law, $\Sigma F = ma$.

We first consider a very simple case: a particle of mass m rotating in a circle of radius R at the end of a string or rod whose mass we can ignore compared to m (Fig. 10–17), and we assume that a single force F acts on m tangent to the circle as shown. The torque that gives rise to the angular acceleration is $\tau = RF$. If we use Newton's second law for linear quantities, $\Sigma F = ma$, and Eq. 10–5 relating the angular acceleration to the tangential linear acceleration, $a_{\text{tan}} = R\alpha$, then we have

$$
\begin{aligned}
F &= ma \\
&= mR\alpha,
\end{aligned}
$$

where α is given in rad/s^2. When we multiply both sides of this equation by R, we find that the torque $\tau = RF = R(mR\alpha)$, or

$$\tau = mR^2\alpha. \qquad \text{[single particle]} \quad \textbf{(10–11)}$$

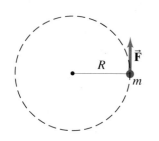

FIGURE 10–17 A mass m rotating in a circle of radius R about a fixed point.

Here at last we have a direct relation between the angular acceleration and the applied torque τ. The quantity mR^2 represents the *rotational inertia* of the particle and is called its *moment of inertia*.

Now let us consider a rotating rigid object, such as a wheel rotating about a fixed axis through its center, such as an axle. We can think of the wheel as consisting of many particles located at various distances from the axis of rotation. We can apply Eq. 10–11 to each particle of the object; that is, we write $\tau_i = m_i R_i^2 \alpha$ for the ith particle of the object. Then we sum over all the particles. The sum of the various torques is just the total torque, $\Sigma\tau$, so we obtain:

$$\Sigma\tau_i = \left(\Sigma m_i R_i^2\right)\alpha \qquad \text{[axis fixed]} \quad \textbf{(10–12)}$$

where we factored out the α since it is the same for all the particles of a rigid object. The resultant torque, $\Sigma\tau$, represents the sum of all internal torques that each particle exerts on another, plus all external torques applied from the outside: $\Sigma\tau = \Sigma\tau_{\text{ext}} + \Sigma\tau_{\text{int}}$. The sum of the internal torques is zero from Newton's third law. Hence $\Sigma\tau$ represents the resultant *external* torque.

The sum $\Sigma m_i R_i^2$ in Eq. 10–12 represents the sum of the masses of each particle in the object multiplied by the square of the distance of that particle from the axis of rotation. If we give each particle a number (1, 2, 3, …), then

$$\Sigma m_i R_i^2 = m_1 R_1^2 + m_2 R_2^2 + m_3 R_3^2 + \cdots.$$

This summation is called the **moment of inertia** (or *rotational inertia*) I of the object:

$$I = \Sigma m_i R_i^2 = m_1 R_1^2 + m_2 R_2^2 + \cdots. \qquad \textbf{(10–13)}$$

Combining Eqs. 10–12 and 10–13, we can write

$$\Sigma\tau = I\alpha. \qquad \left[\begin{array}{c}\text{axis fixed in}\\ \text{inertial reference frame}\end{array}\right] \textbf{(10–14)}$$

This is the rotational equivalent of Newton's second law. It is valid for the rotation of a rigid object about a fixed axis.[†] It can be shown (see Chapter 11) that Eq. 10–14 is valid even when the object is translating with acceleration, as long as I and α are calculated about the center of mass of the object, and the rotation axis through the CM doesn't change direction. (A ball rolling down a ramp is an example.) Then

$$(\Sigma\tau)_{CM} = I_{CM}\alpha_{CM}, \qquad \left[\begin{array}{c}\text{axis fixed in direction,}\\ \text{but may accelerate}\end{array}\right] \textbf{(10–15)}$$

where the subscript CM means "calculated about the center of mass."

We see that the moment of inertia, I, which is a measure of the rotational inertia of an object, plays the same role for rotational motion that mass does for translational motion. As can be seen from Eq. 10–13, the rotational inertia of an object depends not only on its mass, but also on how that mass is distributed with respect to the axis. For example, a large-diameter cylinder will have greater rotational inertia than one of equal mass but smaller diameter (and therefore greater length), Fig. 10–18. The former will be harder to start rotating, and harder to stop. When the mass is concentrated farther from the axis of rotation, the rotational inertia is greater. For rotational motion, the mass of an object *cannot* be considered as concentrated at its center of mass.

NEWTON'S SECOND LAW FOR ROTATION

FIGURE 10–18 A large-diameter cylinder has greater rotational inertia than one of equal mass but smaller diameter.

⚠️ **C A U T I O N**

Mass can not *be considered concentrated at* CM *for rotational motion*

EXAMPLE 10–8 **Two weights on a bar: different axis, different *I*.** Two small "weights," of mass 5.0 kg and 7.0 kg, are mounted 4.0 m apart on a light rod (whose mass can be ignored), as shown in Fig. 10–19. Calculate the moment of inertia of the system (*a*) when rotated about an axis halfway between the weights, Fig. 10–19a, and (*b*) when rotated about an axis 0.50 m to the left of the 5.0-kg mass (Fig. 10–19b).

APPROACH In each case, the moment of inertia of the system is found by summing over the two parts using Eq. 10–13.

SOLUTION (*a*) Both weights are the same distance, 2.0 m, from the axis of rotation. Thus

$$\begin{aligned} I = \Sigma mR^2 &= (5.0\,\text{kg})(2.0\,\text{m})^2 + (7.0\,\text{kg})(2.0\,\text{m})^2 \\ &= 20\,\text{kg}\cdot\text{m}^2 + 28\,\text{kg}\cdot\text{m}^2 = 48\,\text{kg}\cdot\text{m}^2. \end{aligned}$$

(*b*) The 5.0-kg mass is now 0.50 m from the axis, and the 7.0-kg mass is 4.50 m from the axis. Then

$$\begin{aligned} I = \Sigma mR^2 &= (5.0\,\text{kg})(0.50\,\text{m})^2 + (7.0\,\text{kg})(4.5\,\text{m})^2 \\ &= 1.3\,\text{kg}\cdot\text{m}^2 + 142\,\text{kg}\cdot\text{m}^2 = 143\,\text{kg}\cdot\text{m}^2. \end{aligned}$$

NOTE This Example illustrates two important points. First, the moment of inertia of a given system is different for different axes of rotation. Second, we see in part (*b*) that mass close to the axis of rotation contributes little to the total moment of inertia; here, the 5.0-kg object contributed less than 1% to the total.

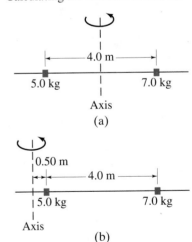

FIGURE 10–19 Example 10–8. Calculating the moment of inertia.

⚠️ **C A U T I O N**

I depends on axis of rotation and on distribution of mass

[†]That is, the axis is fixed relative to the object and is fixed in an inertial reference frame. This includes an axis moving at uniform velocity in an inertial frame, since the axis can be considered fixed in a second inertial frame that moves with respect to the first.

Object	Location of axis		Moment of inertia
(a) **Thin hoop,** radius R_0	Through center	Axis R_0	MR_0^2
(b) **Thin hoop,** radius R_0 width w	Through central diameter	Axis R_0 w	$\frac{1}{2}MR_0^2 + \frac{1}{12}Mw^2$
(c) **Solid cylinder,** radius R_0	Through center	Axis R_0	$\frac{1}{2}MR_0^2$
(d) **Hollow cylinder,** inner radius R_1 outer radius R_2	Through center	Axis R_2 R_1	$\frac{1}{2}M(R_1^2 + R_2^2)$
(e) **Uniform sphere,** radius r_0	Through center	Axis r_0	$\frac{2}{5}Mr_0^2$
(f) **Long uniform rod,** length ℓ	Through center	Axis ℓ	$\frac{1}{12}M\ell^2$
(g) **Long uniform rod,** length ℓ	Through end	Axis ℓ	$\frac{1}{3}M\ell^2$
(h) **Rectangular thin plate,** length ℓ, width w	Through center	Axis w ℓ	$\frac{1}{12}M(\ell^2 + w^2)$

FIGURE 10–20 Moments of inertia for various objects of uniform composition. [We use R for radial distance from an axis, and r for distance from a point (only in e, the sphere), as discussed in Fig. 10–2.]

For most ordinary objects, the mass is distributed continuously, and the calculation of the moment of inertia, ΣmR^2, can be difficult. Expressions can, however, be worked out (using calculus) for the moments of inertia of regularly shaped objects in terms of the dimensions of the objects, as we will discuss in Section 10–7. Figure 10–20 gives these expressions for a number of solids rotated about the axes specified. The only one for which the result is obvious is that for the thin hoop or ring rotated about an axis passing through its center perpendicular to the plane of the hoop (Fig. 10–20a). For this hoop, all the mass is concentrated at the same distance from the axis, R_0. Thus $\Sigma mR^2 = (\Sigma m)R_0^2 = MR_0^2$, where M is the total mass of the hoop.

When calculation is difficult, I can be determined experimentally by measuring the angular acceleration α about a fixed axis due to a known net torque, $\Sigma \tau$, and applying Newton's second law, $I = \Sigma \tau / \alpha$, Eq. 10–14.

10–6 Solving Problems in Rotational Dynamics

When working with torque and angular acceleration (Eq. 10–14), it is important to use a consistent set of units, which in SI is: α in rad/s^2; τ in m·N; and the moment of inertia, I, in kg·m^2.

Rotational Motion

1. As always, draw a clear and complete **diagram**.
2. Choose the object or objects that will be the **system** to be studied.
3. Draw a **free-body diagram** for the object under consideration (or for each object, if more than one), showing all (and only) the forces acting on that object and exactly where they act, so you can determine the torque due to each. Gravity acts at the CG of the object (Section 9–8).
4. Identify the axis of rotation and determine the **torques** about it. Choose positive and negative directions of

rotation (counterclockwise and clockwise), and assign the correct sign to each torque.

5. Apply **Newton's second law for rotation**, $\Sigma\tau = I\alpha$. If the moment of inertia is not given, and it is not the unknown sought, you need to determine it first. Use consistent units, which in SI are: α in rad/s^2; τ in m·N; and I in kg·m^2.
6. Also apply **Newton's second law for translation**, $\Sigma\vec{F} = m\vec{a}$, and **other** laws or principles as needed.
7. **Solve** the resulting equation(s) for the unknown(s).
8. Do a rough **estimate** to determine if your answer is reasonable.

EXAMPLE 10–9 **A heavy pulley.** A 15.0-N force (represented by $\vec{F}_T$) is applied to a cord wrapped around a pulley of mass $M = 4.00$ kg and radius $R_0 = 33.0$ cm, Fig. 10–21. The pulley accelerates uniformly from rest to an angular speed of 30.0 rad/s in 3.00 s. If there is a frictional torque $\tau_{fr} = 1.10$ m·N at the axle, determine the moment of inertia of the pulley. The pulley rotates about its center.

APPROACH We follow the steps of the Problem Solving Strategy above.

SOLUTION

1. **Draw a diagram.** The pulley and the attached cord are shown in Fig. 10–21.
2. **Choose the system:** the pulley.
3. **Draw a free-body diagram.** The cord exerts a force F_T on the pulley as shown in Fig. 10–21. The friction force retards the motion and acts all around the axle in a clockwise direction, as suggested by the arrow $\vec{F}_{fr}$ in Fig. 10–21; we are given its torque, which is all we need. Two other forces should be included in the diagram: the force of gravity mg down, and whatever force holds the axle in place. They do not contribute to the torque (their lever arms are zero) and so we omit them for convenience (or tidiness).
4. **Determine the torques.** The torque exerted by the cord equals $R_0 F_T$ and is counterclockwise, which we choose to be positive. The frictional torque is given as $\tau_{fr} = 1.10$ m·N; it opposes the motion and is negative.
5. **Apply Newton's second law for rotation.** The net torque is
$$\Sigma\tau = R_0 F_T - \tau_{fr} = (0.330\,\text{m})(15.0\,\text{N}) - 1.10\,\text{m·N} = 3.85\,\text{m·N}.$$
The angular acceleration α is found from the given data that it takes 3.0 s to accelerate the pulley from rest to $\omega = 30.0$ rad/s:
$$\alpha = \frac{\Delta\omega}{\Delta t} = \frac{30.0\,\text{rad/s} - 0}{3.00\,\text{s}} = 10.0\,\text{rad/s}^2.$$
We can now solve for I in Newton's second law (see step 7).
6. **Other calculations:** None needed.
7. **Solve for unknowns.** We solve for I in Newton's second law for rotation, $\Sigma\tau = I\alpha$, and insert our values for $\Sigma\tau$ and α:
$$I = \frac{\Sigma\tau}{\alpha} = \frac{3.85\,\text{m·N}}{10.0\,\text{rad/s}^2} = 0.385\,\text{kg·m}^2.$$
8. **Do a rough estimate.** We can do a rough estimate of the moment of inertia by assuming the pulley is a uniform cylinder and using Fig. 10–20c:
$$I \approx \tfrac{1}{2}MR_0^2 = \tfrac{1}{2}(4.00\,\text{kg})(0.330\,\text{m})^2 = 0.218\,\text{kg·m}^2.$$
This is the same order of magnitude as our result, but numerically somewhat less. This makes sense, though, because a pulley is not usually a uniform cylinder but instead has more of its mass concentrated toward the outside edge. Such a pulley would be expected to have a greater moment of inertia than a solid cylinder of equal mass. A thin hoop, Fig. 10–20a, ought to have a greater I than our pulley, and indeed it does: $I = MR_0^2 = 0.436$ kg·m^2.

FIGURE 10–21 Example 10–9.

PROBLEM SOLVING

Usefulness and power of rough estimates

(a)

(b)

FIGURE 10–22 Example 10–10. (a) Pulley and falling bucket of mass m. (b) Free-body diagram for the bucket.

EXAMPLE 10–10 **Pulley and bucket.** Consider again the pulley in Fig. 10–21 and Example 10–9 with the same friction. But this time, instead of a constant 15.0-N force being exerted on the cord, we now have a bucket of weight $w = 15.0$ N (mass $m = w/g = 1.53$ kg) hanging from the cord. See Fig. 10–22a. We assume the cord has negligible mass and does not stretch or slip on the pulley. (a) Calculate the angular acceleration α of the pulley and the linear acceleration a of the bucket. (b) Determine the angular velocity ω of the pulley and the linear velocity v of the bucket at $t = 3.00$ s if the pulley (and bucket) start from rest at $t = 0$.

APPROACH This situation looks a lot like Example 10–9, Fig. 10–21. But there is a big difference: the tension in the cord is now an unknown, and it is no longer equal to the weight of the bucket if the bucket accelerates. Our system has two parts: the bucket, which can undergo translational motion (Fig. 10–22b is its free-body diagram); and the pulley. The pulley does not translate, but it can rotate. We apply the rotational version of Newton's second law to the pulley, $\Sigma \tau = I\alpha$, and the linear version to the bucket, $\Sigma F = ma$.

SOLUTION (a) Let F_T be the tension in the cord. Then a force F_T acts at the edge of the pulley, and we apply Newton's second law, Eq. 10–14, for the rotation of the pulley:

$$I\alpha = \Sigma\tau = R_0 F_T - \tau_{fr}. \qquad \text{[pulley]}$$

Next we look at the (linear) motion of the bucket of mass m. Figure 10–22b, the free-body diagram for the bucket, shows that two forces act on the bucket: the force of gravity mg acts downward, and the tension of the cord F_T pulls upward. Applying Newton's second law, $\Sigma F = ma$, for the bucket, we have (taking downward as positive):

$$mg - F_T = ma. \qquad \text{[bucket]}$$

Note that the tension F_T, which is the force exerted on the edge of the pulley, is *not* equal to the weight of the bucket ($= mg = 15.0$ N). There must be a net force on the bucket if it is accelerating, so $F_T < mg$. Indeed from the last equation above, $F_T = mg - ma$.

To obtain α, we note that the tangential acceleration of a point on the edge of the pulley is the same as the acceleration of the bucket if the cord doesn't stretch or slip. Hence we can use Eq. 10–5, $a_{tan} = a = R_0\alpha$. Substituting $F_T = mg - ma = mg - mR_0\alpha$ into the first equation above (Newton's second law for rotation of the pulley), we obtain

$$I\alpha = \Sigma\tau = R_0 F_T - \tau_{fr} = R_0(mg - mR_0\alpha) - \tau_{fr} = mgR_0 - mR_0^2\alpha - \tau_{fr}.$$

The variable α appears on the left and in the second term on the right, so we bring that term to the left side and solve for α:

$$\alpha = \frac{mgR_0 - \tau_{fr}}{I + mR_0^2}.$$

The numerator $(mgR_0 - \tau_{fr})$ is the net torque, and the denominator $(I + mR_0^2)$ is the total rotational inertia of the system. Then, since $I = 0.385$ kg·m^2, $m = 1.53$ kg, and $\tau_{fr} = 1.10$ m·N (from Example 10–9),

$$\alpha = \frac{(15.0\,\text{N})(0.330\,\text{m}) - 1.10\,\text{m}\cdot\text{N}}{0.385\,\text{kg}\cdot\text{m}^2 + (1.53\,\text{kg})(0.330\,\text{m})^2} = 6.98\,\text{rad/s}^2.$$

The angular acceleration is somewhat less in this case than the 10.0 rad/s^2 of Example 10–9. Why? Because F_T ($= mg - ma$) is less than the 15.0-N weight of the bucket, mg. The linear acceleration of the bucket is

$$a = R_0\alpha = (0.330\,\text{m})(6.98\,\text{rad/s}^2) = 2.30\,\text{m/s}^2.$$

NOTE The tension in the cord F_T is less than mg because the bucket accelerates.

(b) Since the angular acceleration is constant, after 3.00 s

$$\omega = \omega_0 + \alpha t = 0 + (6.98 \text{ rad/s}^2)(3.00 \text{ s}) = 20.9 \text{ rad/s}.$$

The velocity of the bucket is the same as that of a point on the wheel's edge:

$$v = R_0 \omega = (0.330 \text{ m})(20.9 \text{ rad/s}) = 6.91 \text{ m/s}.$$

The same result can also be obtained by using the linear equation $v = v_0 + at = 0 + (2.30 \text{ m/s}^2)(3.00 \text{ s}) = 6.90 \text{ m/s}$. (The difference is due to rounding off.)

EXAMPLE 10–11 **Rotating rod.** A uniform rod of mass M and length ℓ can pivot freely (i.e., we ignore friction) about a hinge or pin attached to the case of a large machine, as in Fig. 10–23. The rod is held horizontally and then released. At the moment of release (when you are no longer exerting a force holding it up), determine (a) the angular acceleration of the rod and (b) the linear acceleration of the tip of the rod. Assume the force of gravity acts at the center of mass of the rod, as shown.

APPROACH (a) The only torque on the rod about the hinge is that due to gravity, which acts with a force $F = Mg$ downward with a lever arm $\ell/2$ at the moment of release (the CM is at the center of a uniform rod). There is also a force on the rod at the hinge, but with the hinge as axis of rotation, the lever arm of this force is zero. The moment of inertia of a uniform rod pivoted about its end is (Fig. 10–20g) $I = \frac{1}{3}M\ell^2$. In part (b) we use $a_{\text{tan}} = R\alpha$.

SOLUTION We use Eq. 10–14, solving for α to obtain the initial angular acceleration of the rod:

$$\alpha = \frac{\tau}{I} = \frac{Mg\dfrac{\ell}{2}}{\frac{1}{3}M\ell^2} = \frac{3}{2}\frac{g}{\ell}.$$

As the rod descends, the force of gravity on it is constant but the torque due to this force is not constant since the lever arm changes. Hence the rod's angular acceleration is not constant.

(b) The linear acceleration of the tip of the rod is found from the relation $a_{\text{tan}} = R\alpha$ (Eq. 10–5) with $R = \ell$:

$$a_{\text{tan}} = \ell\alpha = \frac{3}{2}g.$$

NOTE The tip of the rod falls with an acceleration greater than g! A small object balanced on the tip of the rod would be left behind when the rod is released. In contrast, the CM of the rod, at a distance $\ell/2$ from the pivot, has acceleration $a_{\text{tan}} = (\ell/2)\alpha = \frac{3}{4}g$.

FIGURE 10–23 Example 10–11.

10–7 Determining Moments of Inertia

By Experiment

The moment of inertia of any object about any axis can be determined experimentally, such as by measuring the net torque $\Sigma\tau$ required to give the object an angular acceleration α. Then, from Eq. 10–14, $I = \Sigma\tau/\alpha$. See Example 10–9.

Using Calculus

For simple systems of masses or particles, the moment of inertia can be calculated directly, as in Example 10–8. Many objects can be considered as a continuous distribution of mass. In this case, Eq. 10–13 defining moment of inertia becomes

$$I = \int R^2 \, dm, \tag{10–16}$$

where dm represents the mass of any infinitesimal particle of the object and R is the perpendicular distance of this particle from the axis of rotation. The integral is taken over the whole object. This is easily done only for objects of simple geometric shape.

FIGURE 10–24 Determining the moment of inertia of a hollow cylinder (Example 10–12).

EXAMPLE 10–12 **Cylinder, solid or hollow.** (a) Show that the moment of inertia of a uniform hollow cylinder of inner radius R_1, outer radius R_2, and mass M, is $I = \frac{1}{2}M(R_1^2 + R_2^2)$, as stated in Fig. 10–20d, if the rotation axis is through the center along the axis of symmetry. (b) Obtain the moment of inertia for a solid cylinder.

APPROACH We know that the moment of inertia of a thin ring of radius R is mR^2. So we divide the cylinder into thin concentric cylindrical rings or hoops of thickness dR, one of which is indicated in Fig. 10–24. If the density (mass per unit volume) is ρ, then

$$dm = \rho\, dV,$$

where dV is the volume of the thin ring of radius R, thickness dR, and height h. Since $dV = (2\pi R)(dR)(h)$, we have

$$dm = 2\pi\rho hR\, dR.$$

SOLUTION (a) The moment of inertia is obtained by integrating (summing) over all these rings:

$$I = \int R^2\, dm = \int_{R_1}^{R_2} 2\pi\rho hR^3\, dR = 2\pi\rho h\left[\frac{R_2^4 - R_1^4}{4}\right] = \frac{\pi\rho h}{2}(R_2^4 - R_1^4),$$

where we are given that the cylinder has uniform density, $\rho = $ constant. (If this were not so, we would have to know ρ as a function of R before the integration could be carried out.) The volume V of this hollow cylinder is $V = (\pi R_2^2 - \pi R_1^2)h$, so its mass M is

$$M = \rho V = \rho\pi(R_2^2 - R_1^2)h.$$

Since $(R_2^4 - R_1^4) = (R_2^2 - R_1^2)(R_2^2 + R_1^2)$, we have

$$I = \frac{\pi\rho h}{2}(R_2^2 - R_1^2)(R_2^2 + R_1^2) = \frac{1}{2}M(R_1^2 + R_2^2),$$

as stated in Fig. 10–20d.

(b) For a solid cylinder, $R_1 = 0$ and if we set $R_2 = R_0$, then

$$I = \frac{1}{2}MR_0^2,$$

which is that given in Fig. 10–20c for a solid cylinder of mass M and radius R_0.

The Parallel-Axis Theorem

There are two simple theorems that are helpful in obtaining moments of inertia. The first is called the **parallel-axis theorem**. It relates the moment of inertia I of an object of total mass M about any axis, and its moment of inertia I_{CM} about an axis passing through the center of mass and parallel to the first axis. If the two axes are a distance h apart, then

$$I = I_{CM} + Mh^2. \qquad \text{[parallel axis]} \quad \textbf{(10–17)}$$

Thus, for example, if the moment of inertia about an axis through the CM is known, the moment of inertia about any axis parallel to this axis is easily obtained.

FIGURE 10–25 Example 10–13.

EXAMPLE 10–13 **Parallel axis.** Determine the moment of inertia of a solid cylinder of radius R_0 and mass M about an axis tangent to its edge and parallel to its symmetry axis, Fig. 10–25.

APPROACH We use the parallel-axis theorem with $I_{CM} = \frac{1}{2}MR_0^2$ (Fig. 10–20c).

SOLUTION Since $h = R_0$, Eq. 10–17 gives

$$I = I_{CM} + Mh^2 = \frac{3}{2}MR_0^2.$$

EXERCISE C In Figs. 10–20f and g, the moments of inertia for a thin rod about two different axes are given. Are they related by the parallel-axis theorem? Please show how.

Proof of the Parallel-Axis Theorem

The proof of the parallel-axis theorem is as follows. We choose our coordinate system so the origin is at the CM, and I_{CM} is the moment of inertia about the z axis. Figure 10–26 shows a cross section of an object of arbitrary shape in the xy plane. We let I represent the moment of inertia of the object about an axis parallel to the z axis that passes through the point A in Fig. 10–26 where the point A has coordinates x_A and y_A. Let x_i, y_i, and m_i represent the coordinates and mass of an arbitrary particle of the object. The square of the distance from this point to A is $[(x_i - x_A)^2 + (y_i - y_A)^2]$. So the moment of inertia, I, about the axis through A is

$$I = \Sigma m_i [(x_i - x_A)^2 + (y_i - y_A)^2]$$
$$= \Sigma m_i (x_i^2 + y_i^2) - 2x_A \Sigma m_i x_i - 2y_A \Sigma m_i y_i + (\Sigma m_i)(x_A^2 + y_A^2).$$

The first term on the right is just $I_{CM} = \Sigma m_i (x_i^2 + y_i^2)$ since the CM is at the origin. The second and third terms are zero since, by definition of the CM, $\Sigma m_i x_i = \Sigma m_i y_i = 0$ because $x_{CM} = y_{CM} = 0$. The last term is Mh^2 since $\Sigma m_i = M$ and $(x_A^2 + y_A^2) = h^2$ where h is the distance of A from the CM. Thus we have proved $I = I_{CM} + Mh^2$, which is Eq. 10–17.

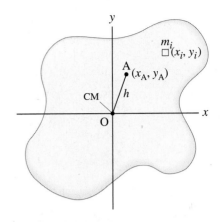

FIGURE 10–26 Derivation of the parallel-axis theorem.

The Perpendicular-Axis Theorem

The parallel-axis theorem can be applied to any object. The second theorem, the **perpendicular-axis theorem**, can be applied only to plane (flat) objects—that is, to two-dimensional objects, or objects of uniform thickness whose thickness can be neglected compared to the other dimensions. This theorem states that the sum of the moments of inertia of a plane object about any two perpendicular axes in the plane of the object, is equal to the moment of inertia about an axis through their point of intersection perpendicular to the plane of the object. That is, if the object is in the xy plane (Fig. 10–27),

$$I_z = I_x + I_y. \qquad \text{[object in } xy \text{ plane]} \quad \textbf{(10–18)}$$

Here I_z, I_x, I_y are moments of inertia about the z, x, and y axes. The proof is simple: since $I_x = \Sigma m_i y_i^2$, $I_y = \Sigma m_i x_i^2$, and $I_z = \Sigma m_i (x_i^2 + y_i^2)$, Eq. 10–18 follows directly.

FIGURE 10–27 The perpendicular-axis theorem.

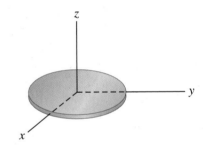

10–8 Rotational Kinetic Energy

The quantity $\frac{1}{2}mv^2$ is the kinetic energy of an object undergoing translational motion. An object rotating about an axis is said to have **rotational kinetic energy**. By analogy with translational kinetic energy, we would expect this to be given by the expression $\frac{1}{2}I\omega^2$ where I is the moment of inertia of the object and ω is its angular velocity. We can indeed show that this is true.

Consider any rigid rotating object as made up of many tiny particles, each of mass m_i. If we let R_i represent the distance of any one particle from the axis of rotation, then its linear velocity is $v_i = R_i \omega$. The total kinetic energy of the whole object will be the sum of the kinetic energies of all its particles:

$$K = \Sigma (\tfrac{1}{2} m_i v_i^2) = \Sigma (\tfrac{1}{2} m_i R_i^2 \omega^2)$$
$$= \tfrac{1}{2} \Sigma (m_i R_i^2) \omega^2.$$

We have factored out the $\frac{1}{2}$ and the ω^2 since they are the same for every particle of a rigid object. Since $\Sigma m_i R_i^2 = I$, the moment of inertia, we see that the kinetic energy, K, of an object rotating about a fixed axis is, as expected,

$$K = \tfrac{1}{2} I \omega^2. \qquad \text{[rotation about a fixed axis]} \quad \textbf{(10–19)}$$

If the axis is not fixed in space, the rotational kinetic energy can take on a more complicated form.

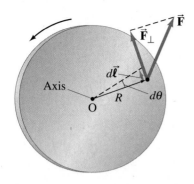

FIGURE 10–28 Calculating the work done by a torque acting on a rigid object rotating about a fixed axis.

The work done on an object rotating about a fixed axis can be written in terms of angular quantities. Suppose a force $\vec{F}$ is exerted at a point whose distance from the axis of rotation is R, as in Fig. 10–28. The work done by this force is

$$W = \int \vec{F} \cdot d\vec{\ell} = \int F_\perp R \, d\theta,$$

where $d\vec{\ell}$ is an infinitesimal distance perpendicular to R with magnitude $d\ell = R \, d\theta$, and $F_\perp$ is the component of $\vec{F}$ perpendicular to R and parallel to $d\vec{\ell}$ (Fig. 10–28). But $F_\perp R$ is the torque about the axis, so

$$W = \int_{\theta_1}^{\theta_2} \tau \, d\theta \tag{10–20}$$

is the work done by a torque τ to rotate an object through the angle $\theta_2 - \theta_1$. The rate of work done, or power P, at any instant is

$$P = \frac{dW}{dt} = \tau \frac{d\theta}{dt} = \tau \omega. \tag{10–21}$$

The work-energy principle holds for rotation of a rigid object about a fixed axis. From Eq. 10–14 we have

$$\tau = I\alpha = I\frac{d\omega}{dt} = I\frac{d\omega}{d\theta}\frac{d\theta}{dt} = I\omega\frac{d\omega}{d\theta},$$

where we used the chain rule and $\omega = d\theta/dt$. Then $\tau \, d\theta = I\omega \, d\omega$ and

$$W = \int_{\theta_1}^{\theta_2} \tau \, d\theta = \int_{\omega_1}^{\omega_2} I\omega \, d\omega = \tfrac{1}{2} I\omega_2^2 - \tfrac{1}{2} I\omega_1^2. \tag{10–22}$$

This is the work-energy principle for a rigid object rotating about a fixed axis. It states that the work done in rotating an object through an angle $\theta_2 - \theta_1$ is equal to the change in rotational kinetic energy of the object.

PHYSICS APPLIED
Energy from a flywheel

EXAMPLE 10–14 **ESTIMATE** **Flywheel.** Flywheels, which are simply large rotating disks, have been suggested as a means of storing energy for solar-powered generating systems. Estimate the kinetic energy that can be stored in an 80,000-kg (80-ton) flywheel with a diameter of 10 m (a three-story building). Assume it could hold together (without flying apart due to internal stresses) at 100 rpm.

APPROACH We use Eq. 10–19, $K = \tfrac{1}{2} I\omega^2$, but only after changing 100 rpm to ω in rad/s.

SOLUTION We are given

$$\omega = 100 \, \text{rpm} = \left(100 \, \frac{\text{rev}}{\text{min}}\right)\left(\frac{1 \, \text{min}}{60 \, \text{sec}}\right)\left(\frac{2\pi \, \text{rad}}{\text{rev}}\right) = 10.5 \, \text{rad/s}.$$

The kinetic energy stored in the disk (for which $I = \tfrac{1}{2} MR_0^2$) is

$$K = \tfrac{1}{2} I\omega^2 = \tfrac{1}{2}\left(\tfrac{1}{2} MR_0^2\right)\omega^2$$
$$= \tfrac{1}{4}(8.0 \times 10^4 \, \text{kg})(5 \, \text{m})^2(10.5 \, \text{rad/s})^2 = 5.5 \times 10^7 \, \text{J}.$$

NOTE In terms of kilowatt-hours $[1 \, \text{kWh} = (1000 \, \text{J/s})(3600 \, \text{s/h})(1 \, \text{h}) = 3.6 \times 10^6 \text{J}]$, this energy is only about 15 kWh, which is not a lot of energy (one 3-kW oven would use it all in 5 h). Thus flywheels seem unlikely for this application.

EXAMPLE 10–15 **Rotating rod.** A rod of mass M is pivoted on a frictionless hinge at one end, as shown in Fig. 10–29. The rod is held at rest horizontally and then released. Determine the angular velocity of the rod when it reaches the vertical position, and the speed of the rod's tip at this moment.

APPROACH We can use the work-energy principle here. The work done is due to gravity, and is equal to the change in gravitational potential energy of the rod.

SOLUTION Since the CM of the rod drops a vertical distance $\ell/2$, the work done by gravity is

$$W = Mg\frac{\ell}{2}.$$

The initial kinetic energy is zero. Hence, from the work-energy principle,

$$\tfrac{1}{2}I\omega^2 = Mg\frac{\ell}{2}.$$

Since $I = \tfrac{1}{3}M\ell^2$ for a rod pivoted about its end (Fig. 10–20g), we can solve for ω:

$$\omega = \sqrt{\frac{3g}{\ell}}.$$

The tip of the rod will have a linear speed (see Eq. 10–4)

$$v = \ell\omega = \sqrt{3g\ell}.$$

NOTE By comparison, an object that falls vertically a height ℓ has a speed $v = \sqrt{2g\ell}$.

FIGURE 10–29 Example 10–15.

EXERCISE D Estimate the energy stored in the rotational motion of a hurricane. Model the hurricane as a uniform cylinder 300 km in diameter and 5 km high, made of air whose mass is 1.3 kg per m³. Estimate the outer edge of the hurricane to move at a speed of 200 km/h.

10–9 Rotational Plus Translational Motion; Rolling

Rolling Without Slipping

The rolling motion of a ball or wheel is familiar in everyday life: a ball rolling across the floor, or the wheels and tires of a car or bicycle rolling along the pavement. Rolling *without slipping* depends on static friction between the rolling object and the ground. The friction is static because the rolling object's point of contact with the ground is at rest at each moment.

Rolling without slipping involves both rotation and translation. There is a simple relation between the linear speed v of the axle and the angular velocity ω of the rotating wheel or sphere: namely, $v = R\omega$, where R is the radius, as we now show. Figure 10–30a shows a wheel rolling to the right without slipping. At the instant shown, point P on the wheel is in contact with the ground and is momentarily at rest. The velocity of the axle at the wheel's center C is $\vec{\mathbf{v}}$. In Fig. 10–30b we have put ourselves in the reference frame of the wheel—that is, we are moving to the right with velocity $\vec{\mathbf{v}}$ relative to the ground. In this reference frame the axle C is at rest, whereas the ground and point P are moving to the left with velocity $-\vec{\mathbf{v}}$ as shown. Here we are seeing pure rotation. We can then use Eq. 10–4 to obtain $v = R\omega$, where R is the radius of the wheel. This is the same v as in Fig. 10–30a, so we see that the linear speed v of the axle relative to the ground is related to the angular velocity ω by

$$v = R\omega, \qquad \text{[rolling without slipping]}$$

This is valid only if there is no slipping.

FIGURE 10–30 (a) A wheel rolling to the right. Its center C moves with velocity $\vec{\mathbf{v}}$. Point P is at rest at this instant. (b) The same wheel as seen from a reference frame in which the axle of the wheel C is at rest—that is, we are moving to the right with velocity $\vec{\mathbf{v}}$ relative to the ground. Point P, which was at rest in (a), here in (b) is moving to the left with velocity $-\vec{\mathbf{v}}$ as shown. (See also Section 3–9 on relative velocity.)

(a)

(b)

(a)

(b)

FIGURE 10–31 (a) A rolling wheel rotates about the instantaneous axis (perpendicular to the page) passing through the point of contact with the ground, P. The arrows represent the instantaneous velocity of each point. (b) Photograph of a rolling wheel. The spokes are more blurred where the speed is greater.

Instantaneous Axis

When a wheel rolls without slipping, the point of contact of the wheel with the ground is instantaneously at rest. It is sometimes useful to think of the motion of the wheel as pure rotation about this "instantaneous axis" passing through that point P (Fig. 10–31a). Points close to the ground have a small linear speed, as they are close to this instantaneous axis, whereas points farther away have a greater linear speed. This can be seen in a photograph of a real rolling wheel (Fig. 10–31b): spokes near the top of the wheel appear more blurry because they are moving faster than those near the bottom of the wheel.

Total Kinetic Energy = K_{CM} + K_{rot}

An object that rotates while its center of mass (CM) undergoes translational motion will have both translational and rotational kinetic energy. Equation 10–19, $K = \frac{1}{2}I\omega^2$, gives the rotational kinetic energy if the rotation axis is fixed. If the object is moving, such as a wheel rolling along the ground, Fig. 10–32, this equation is still valid as long as the rotation axis is fixed in direction. To obtain the total kinetic energy, we note that the rolling wheel undergoes pure rotation about its instantaneous point of contact P, Fig. 10–31. As we saw above relative to the discussion of Fig. 10–30, the speed v of the CM relative to the ground equals the speed of a point on the edge of the wheel relative to its center. Both of these speeds are related to the radius R by $v = \omega R$. Thus the angular velocity ω about point P is the same ω for the wheel about its center, and the total kinetic energy is

$$K_{\text{tot}} = \frac{1}{2}I_P\omega^2,$$

where I_P is the rolling object's moment of inertia about the instantaneous axis at P. We can write K_{tot} in terms of the center of mass using the parallel-axis theorem: $I_P = I_{CM} + MR^2$, where we have substituted $h = R$ in Eq. 10–17. Thus

$$K_{\text{tot}} = \frac{1}{2}I_{CM}\omega^2 + \frac{1}{2}MR^2\omega^2.$$

But $R\omega = v_{CM}$, the speed of the center of mass. So the total kinetic energy of a rolling object is

$$K_{\text{tot}} = \frac{1}{2}I_{CM}\omega^2 + \frac{1}{2}Mv_{CM}^2, \qquad \textbf{(10–23)}$$

where v_{CM} is the linear velocity of the CM, I_{CM} is the moment of inertia about an axis through the CM, ω is the angular velocity about this axis, and M is the total mass of the object.

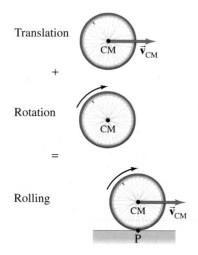

FIGURE 10–32 A wheel rolling without slipping can be considered as translation of the wheel as a whole with velocity $\vec{v}_{CM}$ plus rotation about the CM.

Translation

+

Rotation

=

Rolling

EXAMPLE 10–16 **Sphere rolling down an incline.** What will be the speed of a solid sphere of mass M and radius r_0 when it reaches the bottom of an incline if it starts from rest at a vertical height H and rolls without slipping? See Fig. 10–33. (Assume no slipping occurs because of static friction, which does no work.) Compare your result to that for an object *sliding* down a frictionless incline.

APPROACH We use the law of conservation of energy with gravitational potential energy, now including rotational as well as translational kinetic energy.

SOLUTION The total energy at any point a vertical distance y above the base of the incline is

$$\tfrac{1}{2}Mv^2 + \tfrac{1}{2}I_{CM}\omega^2 + Mgy,$$

where v is the speed of the center of mass, and Mgy is the gravitational potential energy. Applying conservation of energy, we equate the total energy at the top $(y = H, v = 0, \omega = 0)$ to the total energy at the bottom $(y = 0)$:

$$0 + 0 + MgH = \tfrac{1}{2}Mv^2 + \tfrac{1}{2}I_{CM}\omega^2 + 0.$$

The moment of inertia of a solid sphere about an axis through its center of mass is $I_{CM} = \tfrac{2}{5}Mr_0^2$, Fig. 10–20e. Since the sphere rolls without slipping, we have $\omega = v/r_0$ (recall Fig. 10–30). Hence

$$MgH = \tfrac{1}{2}Mv^2 + \tfrac{1}{2}\left(\tfrac{2}{5}Mr_0^2\right)\left(\frac{v^2}{r_0^2}\right).$$

Canceling the M's and r_0's we obtain

$$\left(\tfrac{1}{2} + \tfrac{1}{5}\right)v^2 = gH$$

or

$$v = \sqrt{\tfrac{10}{7}gH}.$$

We can compare this result for the speed of a rolling sphere to that for an object sliding down a plane without rotating and without friction, $\tfrac{1}{2}mv^2 = mgH$ (see our energy equation above, removing the rotational term). For the sliding object, $v = \sqrt{2gH}$, which is greater than for a rolling sphere. An object sliding without friction or rotation transforms its initial potential energy entirely into translational kinetic energy (none into rotational kinetic energy), so the speed of its center of mass is greater.

NOTE Our result for the rolling sphere shows (perhaps surprisingly) that v is independent of both the mass M and the radius r_0 of the sphere.

CONCEPTUAL EXAMPLE 10–17 | **Which is fastest?** Several objects roll without slipping down an incline of vertical height H, all starting from rest at the same moment. The objects are a thin hoop (or a plain wedding band), a spherical marble, a solid cylinder (a D-cell battery), and an empty soup can. In what order do they reach the bottom of the incline? Compare also to a greased box that slides down an incline at the same angle, ignoring sliding friction.

RESPONSE We use conservation of energy with gravitational potential energy plus rotational and translational kinetic energy. The sliding box would be fastest because the potential energy loss (MgH) is transformed completely into translational kinetic energy of the box, whereas for rolling objects the initial potential energy is shared between translational and rotational kinetic energies, and so the speed of the CM is less. For each of the rolling objects we can state that the loss in potential energy equals the increase in kinetic energy:

$$MgH = \tfrac{1}{2}Mv^2 + \tfrac{1}{2}I_{CM}\omega^2.$$

For all our rolling objects, the moment of inertia I_{CM} is a numerical factor times the mass M and the radius R^2 (Fig. 10–20). The mass M is in each term, so the translational speed v_{CM} doesn't depend on M; nor does it depend on the radius R since $\omega = v/R$, so R^2 cancels out for all the rolling objects. Thus the speed v at the bottom depends only on that numerical factor in I_{CM} which expresses how the mass is distributed. The hoop, with all its mass concentrated at radius $R\left(I_{CM} = MR^2\right)$, has the largest moment of inertia; hence it will have the lowest v_{CM} and will arrive at the bottom behind the D-cell $\left(I_{CM} = \tfrac{1}{2}MR^2\right)$, which in turn will be behind the marble $\left(I_{CM} = \tfrac{2}{5}MR^2\right)$. The empty can, which is mainly a hoop plus a small disk, has most of its mass concentrated at R; so it will be a bit faster than the pure hoop but slower than the D-cell. See Fig. 10–34.

NOTE The objects do not have to have the same radius: the speed at the bottom does not depend on the object's mass M or radius R, but only on the shape (and the height of the hill H).

FIGURE 10–33 A sphere rolling down a hill has both translational and rotational kinetic energy. Example 10–16.

⚠ **CAUTION**

Rolling objects go slower than sliding objects because of rotational kinetic energy

FIGURE 10–34 Example 10–17.

Sphere, rolling to the right ⟶

FIGURE 10–35 A sphere rolling to the right on a plane surface. The point in contact with the ground at any moment, point P, is momentarily at rest. Point A to the left of P is moving nearly vertically upward at the instant shown, and point B to the right is moving nearly vertically downward. An instant later, point B will touch the plane and be at rest momentarily. Thus no work is done by the force of static friction.

⚠ **CAUTION**
When is $\Sigma\tau = I\alpha$ *valid?*

FIGURE 10–36 Example 10–18.

If there had been little or no static friction between the rolling objects and the plane in these Examples, the round objects would have slid rather than rolled, or a combination of both. Static friction must be present to make a round object roll. We did not need to take friction into account in the energy equation for the rolling objects because it is *static* friction and does no work—the point of contact of a sphere at each instant does not slide, but moves perpendicular to the plane (first down and then up as shown in Fig. 10–35) as it rolls. Thus, no work is done by the static friction force because the force and the motion (displacement) are perpendicular. The reason the rolling objects in Examples 10–16 and 10–17 move down the slope more slowly than if they were sliding is *not* because friction slows them down. Rather, it is because some of the gravitional potential energy is converted to rotational kinetic energy, leaving less for the translational kinetic energy.

EXERCISE E Return to the Chapter-Opening Question, p. 248, and answer it again now. Try to explain why you may have answered differently the first time.

Using $\Sigma\tau_{CM} = I_{CM}\alpha_{CM}$

We can examine objects rolling down a plane not only from the point of view of kinetic energy, as we did in Examples 10–16 and 10–17, but also in terms of forces and torques. If we calculate torques about an axis fixed in direction (even if the axis is accelerating) which passes through the center of mass of the rolling sphere, then

$$\Sigma\tau_{CM} = I_{CM}\alpha_{CM}$$

is valid, as we discussed in Section 10–5. See Eq. 10–15, whose validity we will show in Chapter 11. Be careful, however: Do not assume $\Sigma\tau = I\alpha$ is always valid. You cannot just calculate τ, I, and α about any axis unless the axis is (1) fixed in an inertial reference frame or (2) fixed in direction but passes through the CM of the object.

EXAMPLE 10–18 Analysis of a sphere on an incline using forces. Analyze the rolling sphere of Example 10–16, Fig. 10–33, in terms of forces and torques. In particular, find the velocity v and the magnitude of the friction force, F_{fr}, Fig. 10–36.

APPROACH We analyze the motion as translation of the CM plus rotation about the CM. F_{fr} is due to static friction and we cannot assume $F_{fr} = \mu_s F_N$, only $F_{fr} \le \mu_s F_N$.

SOLUTION For translation in the x direction we have from $\Sigma F = ma$,

$$Mg\sin\theta - F_{fr} = Ma,$$

and in the y direction

$$F_N - Mg\cos\theta = 0$$

since there is no acceleration perpendicular to the plane. This last equation merely tells us the magnitude of the normal force,

$$F_N = Mg\cos\theta.$$

For the rotational motion about the CM, we use Newton's second law for rotation $\Sigma\tau_{CM} = I_{CM}\alpha_{CM}$ (Eq. 10–15), calculating about an axis passing through the CM but fixed in direction:

$$F_{fr}r_0 = \left(\tfrac{2}{5}Mr_0^2\right)\alpha.$$

The other forces, $\vec{F}_N$ and $M\vec{g}$, point through the axis of rotation (CM), so have lever arms equal to zero and do not appear here. As we saw in Example 10–16 and Fig. 10–30, $\omega = v/r_0$ where v is the speed of the CM. Taking derivatives of $\omega = v/r_0$ with respect to time we have $\alpha = a/r_0$; substituting into the last equation we find

$$F_{fr} = \tfrac{2}{5}Ma.$$

When we substitute this into the top equation, we get

$$Mg\sin\theta - \tfrac{2}{5}Ma = Ma,$$

or

$$a = \tfrac{5}{7}g\sin\theta.$$

We thus see that the acceleration of the CM of a rolling sphere is less than that for an object sliding without friction ($a = g\sin\theta$). The sphere started from rest at the top of the incline (height H). To find the speed v at the bottom we use Eq. 2–12c where the total distance traveled along the plane is $x = H/\sin\theta$ (see Fig. 10–36). Thus

$$v = \sqrt{2ax} = \sqrt{2\left(\frac{5}{7}g\sin\theta\right)\left(\frac{H}{\sin\theta}\right)} = \sqrt{\frac{10}{7}gH}.$$

This is the same result obtained in Example 10–16 although less effort was needed there. To get the magnitude of the force of friction, we use the equations obtained above:

$$F_{fr} = \tfrac{2}{5}Ma = \tfrac{2}{5}M\left(\tfrac{5}{7}g\sin\theta\right) = \tfrac{2}{7}Mg\sin\theta.$$

NOTE If the coefficient of static friction is sufficiently small, or θ sufficiently large so that $F_{fr} > \mu_s F_N$ (that is, if[†] $\tan\theta > \tfrac{7}{2}\mu_s$), the sphere will not simply roll but will slip as it moves down the plane.

*More Advanced Examples

Here we do three more Examples, all of them fun and interesting. When they use $\Sigma\tau = I\alpha$, we must remember that this equation is valid only if τ, α, and I are calculated about an axis that either (1) is fixed in an inertial reference frame, or (2) passes through the CM of the object and remains fixed in direction.

EXAMPLE 10–19 **A falling yo-yo.** String is wrapped around a uniform solid cylinder (something like a yo-yo) of mass M and radius R, and the cylinder starts falling from rest, Fig. 10–37a. As the cylinder falls, find (a) its acceleration and (b) the tension in the string.

APPROACH As always we begin with a free-body diagram, Fig. 10–37b, which shows the weight of the cylinder acting at the CM and the tension of the string $\vec{F}_T$ acting at the edge of the cylinder. We write Newton's second law for the linear motion (down is positive)

$$\begin{aligned} Ma &= \Sigma F \\ &= Mg - F_T. \end{aligned}$$

Since we do not know the tension in the string, we cannot immediately solve for a. So we try Newton's second law for the rotational motion, calculated about the center of mass:

$$\begin{aligned} \Sigma\tau_{CM} &= I_{CM}\alpha_{CM} \\ F_T R &= \tfrac{1}{2}MR^2\alpha. \end{aligned}$$

Because the cylinder "rolls without slipping" down the string, we have the additional relation that $a = \alpha R$ (Eq. 10–5).

SOLUTION The torque equation becomes

$$F_T R = \tfrac{1}{2}MR^2\left(\frac{a}{R}\right) = \tfrac{1}{2}MRa$$

so

$$F_T = \tfrac{1}{2}Ma.$$

Substituting this into the force equation, we obtain

$$\begin{aligned} Ma &= Mg - F_T \\ &= Mg - \tfrac{1}{2}Ma. \end{aligned}$$

Solving for a, we find that $a = \tfrac{2}{3}g$. That is, the linear acceleration is less than what it would be if the cylinder were simply dropped. This makes sense since gravity is not the only vertical force acting; the tension in the string is acting as well. (b) Since $a = \tfrac{2}{3}g$, $F_T = \tfrac{1}{2}Ma = \tfrac{1}{3}Mg$.

(a)

(b)

FIGURE 10–37 Example 10–19.

EXERCISE F Find the acceleration a of a yo-yo whose spindle has radius $\tfrac{1}{2}R$. Assume the moment of inertia is still $\tfrac{1}{2}MR^2$ (ignore the mass of the spindle).

[†] $F_{fr} > \mu_s F_N$ is equivalent to $\tan\theta > \tfrac{7}{2}\mu_s$ because $F_{fr} = \tfrac{2}{7}Mg\sin\theta$ and $\mu_s F_N = \mu_s Mg\cos\theta$.

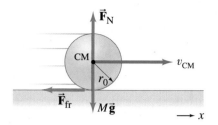

FIGURE 10–38 Example 10–20.

EXAMPLE 10–20 **What if a rolling ball slips?** A bowling ball of mass M and radius r_0 is thrown along a level surface so that initially $(t = 0)$ it slides with a linear speed v_0 but does not rotate. As it slides, it begins to spin, and eventually rolls without slipping. How long does it take to begin rolling without slipping?

APPROACH The free-body diagram is shown in Fig. 10–38, with the ball moving to the right. The friction force does two things: it acts to slow down the translational motion of the CM; and it immediately acts to start the ball rotating clockwise.

SOLUTION Newton's second law for translation gives

$$Ma_x = \Sigma F_x = -F_{fr} = -\mu_k F_N = -\mu_k Mg,$$

where μ_k is the coefficient of kinetic friction because the ball is sliding. Thus $a_x = -\mu_k g$. The velocity of the CM is

$$v_{CM} = v_0 + a_x t = v_0 - \mu_k g t.$$

Next we apply Newton's second law for rotation about the CM, $I_{CM} \alpha_{CM} = \Sigma \tau_{CM}$:

$$\tfrac{2}{5} M r_0^2 \alpha_{CM} = F_{fr} r_0$$
$$= \mu_k Mg r_0.$$

The angular acceleration is thus $\alpha_{CM} = 5\mu_k g / 2r_0$, which is constant. Then the angular velocity of the ball is (Eq. 10–9a)

$$\omega_{CM} = \omega_0 + \alpha_{CM} t = 0 + \frac{5\mu_k g t}{2r_0}.$$

The ball starts rolling immediately after it touches the ground, but it rolls and slips at the same time to begin with. It eventually stops slipping, and then rolls without slipping. The condition for rolling without slipping is that

$$v_{CM} = \omega_{CM} r_0,$$

which is Eq. 10–4, and is *not* valid if there is slipping. This rolling without slipping begins at a time $t = t_1$ given by $v_{CM} = \omega_{CM} r_0$ and we apply the equations for v_{CM} and ω_{CM} above:

$$v_0 - \mu_k g t_1 = \frac{5\mu_k g t_1}{2r_0} r_0$$

so

$$t_1 = \frac{2v_0}{7\mu_k g}.$$

⊛ PHYSICS APPLIED
Braking distribution of a car

FIGURE 10–39 Forces on a braking car (Example 10–21).

EXAMPLE 10–21 **ESTIMATE** **Braking a car.** When the brakes of a car are applied, the front of the car dips down a bit; and the force on the front tires is greater than on the rear tires. To see why, estimate the magnitude of the normal forces, F_{N1} and F_{N2}, on the front and rear tires of the car shown in Fig. 10–39 when the car brakes and decelerates at a rate $a = 0.50 g$. The car has mass $M = 1200\,kg$, the distance between the front and rear axles is 3.0 m, and its CM (where the force of gravity acts) is midway between the axles 75 cm above the ground.

APPROACH Figure 10–39 is the free-body diagram showing all the forces on the car. F_1 and F_2 are the frictional forces that decelerate the car. We let F_1 be the sum of the forces on both front tires, and F_2 likewise for the two rear tires. F_{N1} and F_{N2} are the normal forces the road exerts on the tires and, for our estimate, we assume the static friction force acts the same for all the tires, so that F_1 and F_2 are proportional respectively to F_{N1} and F_{N2}:

$$F_1 = \mu F_{N1} \quad \text{and} \quad F_2 = \mu F_{N2}.$$

SOLUTION The friction forces F_1 and F_2 decelerate the car, so Newton's second law gives

$$F_1 + F_2 = Ma$$
$$= (1200\,\text{kg})(0.50)(9.8\,\text{m/s}^2) = 5900\,\text{N}. \qquad \textbf{(i)}$$

While the car is braking its motion is only translational, so the net torque on the car is zero. If we calculate the torques about the CM as axis, the forces F_1, F_2, and F_{N2} all act to rotate the car clockwise, and only F_{N1} acts to rotate it counterclockwise; so F_{N1} must balance the other three. Hence, F_{N1} must be significantly greater than F_{N2}. Mathematically, we have for the torques calculated about the CM:

$$(1.5\,\text{m})F_{N1} - (1.5\,\text{m})F_{N2} - (0.75\,\text{m})F_1 - (0.75\,\text{m})F_2 = 0.$$

Since F_1 and F_2 are proportional[†] to F_{N1} and F_{N2} $(F_1 = \mu F_{N1},\ F_2 = \mu F_{N2})$, we can write this as

$$(1.5\,\text{m})(F_{N1} - F_{N2}) - (0.75\,\text{m})(\mu)(F_{N1} + F_{N2}) = 0. \qquad \textbf{(ii)}$$

Also, since the car does not accelerate vertically, we have

$$Mg = F_{N1} + F_{N2} = \frac{F_1 + F_2}{\mu}. \qquad \textbf{(iii)}$$

Comparing (iii) to (i), we see that $\mu = a/g = 0.50$. Now we solve (ii) for F_{N1} and use $\mu = 0.50$ to obtain

$$F_{N1} = F_{N2}\left(\frac{2 + \mu}{2 - \mu}\right) = \frac{5}{3}F_{N2}.$$

Thus F_{N1} is $1\frac{2}{3}$ times greater than F_{N2}. Actual magnitudes are determined from (iii) and (i): $F_{N1} + F_{N2} = (5900\,\text{N})/(0.50) = 11{,}800\,\text{N}$ which equals $F_{N2}(1 + \frac{5}{3})$; so $F_{N2} = 4400\,\text{N}$ and $F_{N1} = 7400\,\text{N}$.

NOTE Because the force on the front tires is generally greater than on the rear tires, cars are often designed with larger brake pads on the front wheels than on the rear. Or, to say it another way, if the brake pads are equal, the front ones wear out a lot faster.

*10–10 Why Does a Rolling Sphere Slow Down?

A sphere of mass M and radius r_0 rolling on a horizontal flat surface eventually comes to rest. What force causes it to come to rest? You might think it is friction, but when you examine the problem from a simple straightforward point of view, a paradox seems to arise.

Suppose a sphere is rolling to the right as shown in Fig. 10–40, and is slowing down. By Newton's second law, $\Sigma\vec{F} = M\vec{a}$, there must be a force $\vec{F}$ (presumably frictional) acting to the left as shown, so that the acceleration $\vec{a}$ will also point to the left and v will decrease. Curiously enough, though, if we now look at the torque equation (calculated about the center of mass), $\Sigma\tau_{CM} = I_{CM}\alpha$, we see that the force $\vec{F}$ acts to increase the angular acceleration α, and thus to *increase* the velocity of the sphere. Thus the paradox. The force $\vec{F}$ acts to decelerate the sphere if we look at the translational motion, but speeds it up if we look at the rotational motion.

FIGURE 10–40 Sphere rolling to the right.

[†]Our proportionality constant μ is not equal to μ_s, the static coefficient of friction $(F_{fr} \le \mu_s F_N)$, unless the car is just about to skid.

The resolution of this apparent paradox is that some other force must be acting. The only other forces acting are gravity, $M\vec{g}$, and the normal force $\vec{F}_N$ ($= -M\vec{g}$). These act vertically and hence do not affect the horizontal translational motion. If we assume the sphere and plane are rigid, so the sphere is in contact at only one point, these forces give rise to no torques about the CM either, since they act through the CM.

The only recourse we have to resolve the paradox is to give up our idealization that the objects are rigid. In fact, all objects are deformable to some extent. Our sphere flattens slightly and the level surface also acquires a slight depression where the two are in contact. There is an *area* of contact, not a point. Hence there can be a torque at this area of contact which acts in the opposite direction to the torque associated with $\vec{F}$, and thus acts to slow down the rotation of the sphere. This torque is associated with the normal force $\vec{F}_N$ that the table exerts on the sphere over the whole area of contact. The net effect is that we can consider $\vec{F}_N$ acting vertically a distance ℓ in front of the CM as shown in Fig. 10–41 (where the deformation is greatly exaggerated).

Is it reasonable that the normal force $\vec{F}_N$ should effectively act in *front* of the CM as shown in Fig. 10–41? Yes. The sphere is rolling, and the leading edge strikes the surface with a slight impulse. The table therefore pushes upward a bit more strongly on the front part of the sphere than it would if the sphere were at rest. At the back part of the area of contact, the sphere is starting to move upward and so the table pushes upward on it less strongly than when the sphere is at rest. The table pushing up more strongly on the front part of the area of contact gives rise to the necessary torque and justifies the effective acting point of $\vec{F}_N$ being in front of the CM.

When other forces are present, the tiny torque τ_N due to $\vec{F}_N$ can usually be ignored. For example, when a sphere or cylinder rolls down an incline, the force of gravity has far more influence than τ_N, so the latter can be ignored. For many purposes (but not all), we can assume a hard sphere is in contact with a hard surface at essentially one point.

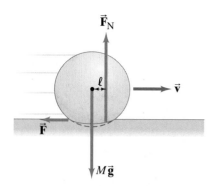

FIGURE 10–41 The normal force, $\vec{F}_N$, exerts a torque that slows down the sphere. The deformation of the sphere and the surface it moves on has been exaggerated for detail.

Summary

When a rigid object rotates about a fixed axis, each point of the object moves in a circular path. Lines drawn perpendicularly from the rotation axis to different points in the object all sweep out the same angle θ in any given time interval.

Angles are conveniently measured in **radians**. One radian is the angle subtended by an arc whose length is equal to the radius, or

$$2\pi \text{ rad} = 360° \quad \text{so} \quad 1 \text{ rad} \approx 57.3°.$$

All parts of a rigid object rotating about a fixed axis have the same **angular velocity** ω and the same **angular acceleration** α at any instant, where

$$\omega = \frac{d\theta}{dt} \tag{10–2b}$$

and

$$\alpha = \frac{d\omega}{dt}. \tag{10–3b}$$

The units of ω and α are rad/s and rad/s².

The linear velocity and acceleration of any point in an object rotating about a fixed axis are related to the angular quantities by

$$v = R\omega \tag{10–4}$$
$$a_{\text{tan}} = R\alpha \tag{10–5}$$
$$a_R = \omega^2 R \tag{10–6}$$

where R is the perpendicular distance of the point from the rotation axis, and a_{tan} and a_R are the tangential and radial

components of the linear acceleration. The frequency f and period T are related to ω (rad/s) by

$$\omega = 2\pi f \tag{10–7}$$
$$T = 1/f. \tag{10–8}$$

Angular velocity and angular acceleration are vectors. For a rigid object rotating about a fixed axis, both $\vec{\omega}$ and $\vec{\alpha}$ point along the rotation axis. The direction of $\vec{\omega}$ is given by the **right-hand rule**.

If a rigid object undergoes uniformly accelerated rotational motion ($\alpha = $ constant), equations analogous to those for linear motion are valid:

$$\omega = \omega_0 + \alpha t; \qquad \theta = \omega_0 t + \tfrac{1}{2}\alpha t^2;$$
$$\omega^2 = \omega_0^2 + 2\alpha\theta; \qquad \bar{\omega} = \frac{\omega + \omega_0}{2}. \tag{10–9}$$

The **torque** due to a force $\vec{F}$ exerted on a rigid object is equal to

$$\tau = R_\perp F = R F_\perp = RF \sin\theta, \tag{10–10}$$

where $R_\perp$, called the **lever arm**, is the perpendicular distance from the axis of rotation to the line along which the force acts, and θ is the angle between $\vec{F}$ and R.

The rotational equivalent of Newton's second law is

$$\Sigma\tau = I\alpha, \tag{10–14}$$

where $I = \Sigma m_i R_i^2$ is the **moment of inertia** of the object about the axis of rotation. This relation is valid for a rigid object rotating about an axis fixed in an inertial reference frame, or when τ, I, and α are calculated about the center of mass of an object even if the CM is moving.

The **rotational kinetic energy** of an object rotating about a fixed axis with angular velocity ω is

$$K = \tfrac{1}{2} I \omega^2. \qquad (10\text{–}19)$$

For an object undergoing both translational and rotational motion, the total kinetic energy is the sum of the translational kinetic energy of the object's CM plus the rotational kinetic energy of the object about its CM:

$$K_{\text{tot}} = \tfrac{1}{2} M v_{\text{CM}}^2 + \tfrac{1}{2} I_{\text{CM}} \omega^2 \qquad (10\text{–}23)$$

as long as the rotation axis is fixed in direction.

The following Table summarizes angular (or rotational) quantities, comparing them to their translational analogs.

Translation	Rotation	Connection
x	θ	$x = R\theta$
v	ω	$v = R\omega$
a	α	$a = R\alpha$
m	I	$I = \Sigma m R^2$
F	τ	$\tau = RF \sin \theta$
$K = \tfrac{1}{2} m v^2$	$\tfrac{1}{2} I \omega^2$	
$W = Fd$	$W = \tau \theta$	
$\Sigma F = ma$	$\Sigma \tau = I\alpha$	

Questions

1. A bicycle odometer (which counts revolutions and is calibrated to report distance traveled) is attached near the wheel hub and is calibrated for 27-inch wheels. What happens if you use it on a bicycle with 24-inch wheels?

2. Suppose a disk rotates at constant angular velocity. Does a point on the rim have radial and/or tangential acceleration? If the disk's angular velocity increases uniformly, does the point have radial and/or tangential acceleration? For which cases would the magnitude of either component of linear acceleration change?

3. Could a nonrigid object be described by a single value of the angular velocity ω? Explain.

4. Can a small force ever exert a greater torque than a larger force? Explain.

5. Why is it more difficult to do a sit-up with your hands behind your head than when your arms are stretched out in front of you? A diagram may help you to answer this.

6. Mammals that depend on being able to run fast have slender lower legs with flesh and muscle concentrated high, close to the body (Fig. 10–42). On the basis of rotational dynamics, explain why this distribution of mass is advantageous.

FIGURE 10–42
Question 6.
A gazelle.

7. If the net force on a system is zero, is the net torque also zero? If the net torque on a system is zero, is the net force zero?

8. Two inclines have the same height but make different angles with the horizontal. The same steel ball is rolled down each incline. On which incline will the speed of the ball at the bottom be greater? Explain.

9. Two spheres look identical and have the same mass. However, one is hollow and the other is solid. Describe an experiment to determine which is which.

10. Two solid spheres simultaneously start rolling (from rest) down an incline. One sphere has twice the radius and twice the mass of the other. Which reaches the bottom of the incline first? Which has the greater speed there? Which has the greater total kinetic energy at the bottom?

11. Why do tightrope walkers (Fig. 10–43) carry a long, narrow beam?

FIGURE 10–43 Question 11.

12. A sphere and a cylinder have the same radius and the same mass. They start from rest at the top of an incline. Which reaches the bottom first? Which has the greater speed at the bottom? Which has the greater total kinetic energy at the bottom? Which has the greater rotational kinetic energy?

13. The moment of inertia of this textbook would be the least about which symmetry axis through its center?

14. The moment of inertia of a rotating solid disk about an axis through its CM is $\tfrac{1}{2} M R^2$ (Fig. 10–20c). Suppose instead that a parallel axis of rotation passes through a point on the edge of the disk. Will the moment of inertia be the same, larger, or smaller?

15. The angular velocity of a wheel rotating on a horizontal axle points west. In what direction is the linear velocity of a point on the top of the wheel? If the angular acceleration points east, describe the tangential linear acceleration of this point at the top of the wheel. Is the angular speed increasing or decreasing?

Problems

10–1 Angular Quantities

1. (I) Express the following angles in radians: (a) 45.0°, (b) 60.0°, (c) 90.0°, (d) 360.0°, and (e) 445°. Give as numerical values and as fractions of π.

2. (I) The Sun subtends an angle of about 0.5° to us on Earth, 150 million km away. Estimate the radius of the Sun.

3. (I) A laser beam is directed at the Moon, 380,000 km from Earth. The beam diverges at an angle θ (Fig. 10–44) of 1.4×10^{-5} rad. What diameter spot will it make on the Moon?

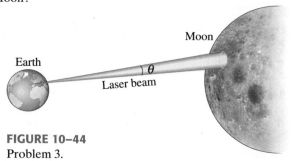

FIGURE 10–44
Problem 3.

4. (I) The blades in a blender rotate at a rate of 6500 rpm. When the motor is turned off during operation, the blades slow to rest in 4.0 s. What is the angular acceleration as the blades slow down?

5. (II) (a) A grinding wheel 0.35 m in diameter rotates at 2500 rpm. Calculate its angular velocity in rad/s. (b) What are the linear speed and acceleration of a point on the edge of the grinding wheel?

6. (II) A bicycle with tires 68 cm in diameter travels 7.2 km. How many revolutions do the wheels make?

7. (II) Calculate the angular velocity of (a) the second hand, (b) the minute hand, and (c) the hour hand, of a clock. State in rad/s. (d) What is the angular acceleration in each case?

8. (II) A rotating merry-go-round makes one complete revolution in 4.0 s (Fig. 10–45). (a) What is the linear speed of a child seated 1.2 m from the center? (b) What is her acceleration (give components)?

FIGURE 10–45
Problem 8.

9. (II) What is the linear speed of a point (a) on the equator, (b) on the Arctic Circle (latitude 66.5° N), and (c) at a latitude of 45.0° N, due to the Earth's rotation?

10. (II) Calculate the angular velocity of the Earth (a) in its orbit around the Sun, and (b) about its axis.

11. (II) How fast (in rpm) must a centrifuge rotate if a particle 7.0 cm from the axis of rotation is to experience an acceleration of 100,000 g's?

12. (II) A 64-cm-diameter wheel accelerates uniformly about its center from 130 rpm to 280 rpm in 4.0 s. Determine (a) its angular acceleration, and (b) the radial and tangential components of the linear acceleration of a point on the edge of the wheel 2.0 s after it has started accelerating.

13. (II) In traveling to the Moon, astronauts aboard the *Apollo* spacecraft put themselves into a slow rotation to distribute the Sun's energy evenly. At the start of their trip, they accelerated from no rotation to 1.0 revolution every minute during a 12-min time interval. The spacecraft can be thought of as a cylinder with a diameter of 8.5 m. Determine (a) the angular acceleration, and (b) the radial and tangential components of the linear acceleration of a point on the skin of the ship 7.0 min after it started this acceleration.

14. (II) A turntable of radius R_1 is turned by a circular rubber roller of radius R_2 in contact with it at their outer edges. What is the ratio of their angular velocities, ω_1/ω_2?

10–2 Vector Nature of $\vec{\omega}$ and $\vec{\alpha}$

15. (II) The axle of a wheel is mounted on supports that rest on a rotating turntable as shown in Fig. 10–46. The wheel has angular velocity $\omega_1 = 44.0$ rad/s about its axle, and the turntable has angular velocity $\omega_2 = 35.0$ rad/s about a vertical axis. (Note arrows showing these motions in the figure.) (a) What are the directions of $\vec{\omega}_1$ and $\vec{\omega}_2$ at the instant shown? (b) What is the resultant angular velocity of the wheel, as seen by an outside observer, at the instant shown? Give the magnitude and direction. (c) What is the magnitude and direction of the angular acceleration of the wheel at the instant shown? Take the z axis vertically upward and the direction of the axle at the moment shown to be the x axis pointing to the right.

FIGURE 10–46
Problem 15.

10–3 Constant Angular Acceleration

16. (I) An automobile engine slows down from 3500 rpm to 1200 rpm in 2.5 s. Calculate (a) its angular acceleration, assumed constant, and (b) the total number of revolutions the engine makes in this time.

17. (I) A centrifuge accelerates uniformly from rest to 15,000 rpm in 220 s. Through how many revolutions did it turn in this time?

18. (I) Pilots can be tested for the stresses of flying high-speed jets in a whirling "human centrifuge," which takes 1.0 min to turn through 20 complete revolutions before reaching its final speed. (a) What was its angular acceleration (assumed constant), and (b) what was its final angular speed in rpm?

19. (II) A cooling fan is turned off when it is running at 850 rev/min. It turns 1350 revolutions before it comes to a stop. (a) What was the fan's angular acceleration, assumed constant? (b) How long did it take the fan to come to a complete stop?

20. (II) Using calculus, derive the angular kinematic equations 10–9a and 10–9b for constant angular acceleration. Start with $\alpha = d\omega/dt$.

21. (II) A small rubber wheel is used to drive a large pottery wheel. The two wheels are mounted so that their circular edges touch. The small wheel has a radius of 2.0 cm and accelerates at the rate of 7.2 rad/s², and it is in contact with the pottery wheel (radius 21.0 cm) without slipping. Calculate (a) the angular acceleration of the pottery wheel, and (b) the time it takes the pottery wheel to reach its required speed of 65 rpm.

22. (II) The angle through which a rotating wheel has turned in time t is given by $\theta = 8.5\,t - 15.0\,t^2 + 1.6\,t^4$, where θ is in radians and t in seconds. Determine an expression (a) for the instantaneous angular velocity ω and (b) for the instantaneous angular acceleration α. (c) Evaluate ω and α at $t = 3.0$ s. (d) What is the average angular velocity, and (e) the average angular acceleration between $t = 2.0$ s and $t = 3.0$ s?

23. (II) The angular acceleration of a wheel, as a function of time, is $\alpha = 5.0\,t^2 - 8.5\,t$, where α is in rad/s² and t in seconds. If the wheel starts from rest ($\theta = 0$, $\omega = 0$, at $t = 0$), determine a formula for (a) the angular velocity ω and (b) the angular position θ, both as a function of time. (c) Evaluate ω and θ at $t = 2.0$ s.

10–4 Torque

24. (I) A 62-kg person riding a bike puts all her weight on each pedal when climbing a hill. The pedals rotate in a circle of radius 17 cm. (a) What is the maximum torque she exerts? (b) How could she exert more torque?

25. (I) Calculate the net torque about the axle of the wheel shown in Fig. 10–47. Assume that a friction torque of 0.40 m·N opposes the motion.

FIGURE 10–47
Problem 25.

26. (II) A person exerts a horizontal force of 32 N on the end of a door 96 cm wide. What is the magnitude of the torque if the force is exerted (a) perpendicular to the door and (b) at a 60.0° angle to the face of the door?

27. (II) Two blocks, each of mass m, are attached to the ends of a massless rod which pivots as shown in Fig. 10–48. Initially the rod is held in the horizontal position and then released. Calculate the magnitude and direction of the net torque on this system when it is first released.

FIGURE 10–48 Problem 27.

28. (II) A wheel of diameter 27.0 cm is constrained to rotate in the xy plane, about the z axis, which passes through its center. A force $\vec{\mathbf{F}} = (-31.0\hat{\mathbf{i}} + 43.4\hat{\mathbf{j}})$ N acts at a point on the edge of the wheel that lies exactly on the x axis at a particular instant. What is the torque about the rotation axis at this instant?

29. (II) The bolts on the cylinder head of an engine require tightening to a torque of 75 m·N. If a wrench is 28 cm long, what force perpendicular to the wrench must the mechanic exert at its end? If the six-sided bolt head is 15 mm across (Fig. 10–49), estimate the force applied near each of the six points by a socket wrench.

FIGURE 10–49 Problem 29.

30. (II) Determine the net torque on the 2.0-m-long uniform beam shown in Fig. 10–50. Calculate about (a) point C, the CM, and (b) point P at one end.

FIGURE 10–50
Problem 30.

10–5 and 10–6 Rotational Dynamics

31. (I) Determine the moment of inertia of a 10.8-kg sphere of radius 0.648 m when the axis of rotation is through its center.

32. (I) Estimate the moment of inertia of a bicycle wheel 67 cm in diameter. The rim and tire have a combined mass of 1.1 kg. The mass of the hub can be ignored (why?).

33. (II) A potter is shaping a bowl on a potter's wheel rotating at constant angular speed (Fig. 10–51). The friction force between her hands and the clay is 1.5 N total. (a) How large is her torque on the wheel, if the diameter of the bowl is 12 cm? (b) How long would it take for the potter's wheel to stop if the only torque acting on it is due to the potter's hand? The initial angular velocity of the wheel is 1.6 rev/s, and the moment of inertia of the wheel and the bowl is 0.11 kg·m².

FIGURE 10–51
Problem 33.

34. (II) An oxygen molecule consists of two oxygen atoms whose total mass is 5.3×10^{-26} kg and whose moment of inertia about an axis perpendicular to the line joining the two atoms, midway between them, is 1.9×10^{-46} kg·m^2. From these data, estimate the effective distance between the atoms.

35. (II) A softball player swings a bat, accelerating it from rest to 2.7 rev/s in a time of 0.20 s. Approximate the bat as a 2.2-kg uniform rod of length 0.95 m, and compute the torque the player applies to one end of it.

36. (II) A grinding wheel is a uniform cylinder with a radius of 8.50 cm and a mass of 0.380 kg. Calculate (a) its moment of inertia about its center, and (b) the applied torque needed to accelerate it from rest to 1750 rpm in 5.00 s if it is known to slow down from 1500 rpm to rest in 55.0 s.

37. (II) A small 650-g ball on the end of a thin, light rod is rotated in a horizontal circle of radius 1.2 m. Calculate (a) the moment of inertia of the ball about the center of the circle, and (b) the torque needed to keep the ball rotating at constant angular velocity if air resistance exerts a force of 0.020 N on the ball. Ignore the rod's moment of inertia and air resistance.

38. (II) The forearm in Fig. 10–52 accelerates a 3.6-kg ball at 7.0 m/s^2 by means of the triceps muscle, as shown. Calculate (a) the torque needed, and (b) the force that must be exerted by the triceps muscle. Ignore the mass of the arm.

FIGURE 10–52
Problems 38 and 39.

39. (II) Assume that a 1.00-kg ball is thrown solely by the action of the forearm, which rotates about the elbow joint under the action of the triceps muscle, Fig. 10–52. The ball is accelerated uniformly from rest to 8.5 m/s in 0.35 s, at which point it is released. Calculate (a) the angular acceleration of the arm, and (b) the force required of the triceps muscle. Assume that the forearm has a mass of 3.7 kg and rotates like a uniform rod about an axis at its end.

40. (II) Calculate the moment of inertia of the array of point objects shown in Fig. 10–53 about (a) the vertical axis, and (b) the horizontal axis. Assume $m = 2.2$ kg, $M = 3.1$ kg, and the objects are wired together by very light, rigid pieces of wire. The array is rectangular and is split through the middle by the horizontal axis. (c) About which axis would it be harder to accelerate this array?

FIGURE 10–53   Problem 40.

41. (II) A merry-go-round accelerates from rest to 0.68 rad/s in 24 s. Assuming the merry-go-round is a uniform disk of radius 7.0 m and mass 31,000 kg, calculate the net torque required to accelerate it.

42. (II) A 0.72-m-diameter solid sphere can be rotated about an axis through its center by a torque of 10.8 m·N which accelerates it uniformly from rest through a total of 180 revolutions in 15.0 s. What is the mass of the sphere?

43. (II) Suppose the force F_T in the cord hanging from the pulley of Example 10–9, Fig. 10–21, is given by the relation $F_T = 3.00t - 0.20t^2$ (newtons) where t is in seconds. If the pulley starts from rest, what is the linear speed of a point on its rim 8.0 s later? Ignore friction.

44. (II) A dad pushes tangentially on a small hand-driven merry-go-round and is able to accelerate it from rest to a frequency of 15 rpm in 10.0 s. Assume the merry-go-round is a uniform disk of radius 2.5 m and has a mass of 760 kg, and two children (each with a mass of 25 kg) sit opposite each other on the edge. Calculate the torque required to produce the acceleration, neglecting frictional torque. What force is required at the edge?

45. (II) Four equal masses M are spaced at equal intervals, ℓ, along a horizontal straight rod whose mass can be ignored. The system is to be rotated about a vertical axis passing through the mass at the left end of the rod and perpendicular to it. (a) What is the moment of inertia of the system about this axis? (b) What minimum force, applied to the farthest mass, will impart an angular acceleration α? (c) What is the direction of this force?

46. (II) Two blocks are connected by a light string passing over a pulley of radius 0.15 m and moment of inertia I. The blocks move (towards the right) with an acceleration of 1.00 m/s^2 along their frictionless inclines (see Fig. 10–54). (a) Draw free-body diagrams for each of the two blocks and the pulley. (b) Determine F_{TA} and F_{TB}, the tensions in the two parts of the string. (c) Find the net torque acting on the pulley, and determine its moment of inertia, I.

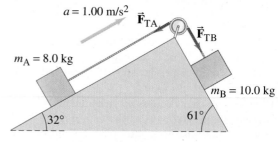

FIGURE 10–54   Problem 46.

47. (II) A helicopter rotor blade can be considered a long thin rod, as shown in Fig. 10–55. (a) If each of the three rotor helicopter blades is 3.75 m long and has a mass of 135 kg, calculate the moment of inertia of the three rotor blades about the axis of rotation. (b) How much torque must the motor apply to bring the blades from rest up to a speed of 5.0 rev/s in 8.0 s?

FIGURE 10–55
Problem 47.

48. (II) A centrifuge rotor rotating at 10,300 rpm is shut off and is eventually brought uniformly to rest by a frictional torque of 1.20 m·N. If the mass of the rotor is 3.80 kg and it can be approximated as a solid cylinder of radius 0.0710 m, through how many revolutions will the rotor turn before coming to rest, and how long will it take?

49. (II) When discussing moments of inertia, especially for unusual or irregularly shaped objects, it is sometimes convenient to work with the **radius of gyration**, k. This radius is defined so that if all the mass of the object were concentrated at this distance from the axis, the moment of inertia would be the same as that of the original object. Thus, the moment of inertia of any object can be written in terms of its mass M and the radius of gyration as $I = Mk^2$. Determine the radius of gyration for each of the objects (hoop, cylinder, sphere, etc.) shown in Fig. 10–20.

50. (II) To get a flat, uniform cylindrical satellite spinning at the correct rate, engineers fire four tangential rockets as shown in Fig. 10–56. If the satellite has a mass of 3600 kg, a radius of 4.0 m, and the rockets each add a mass of 250 kg, what is the required steady force of each rocket if the satellite is to reach 32 rpm in 5.0 min, starting from rest?

FIGURE 10–56
Problem 50.

51. (III) An *Atwood's machine* consists of two masses, m_A and m_B, which are connected by a massless inelastic cord that passes over a pulley, Fig. 10–57. If the pulley has radius R and moment of inertia I about its axle, determine the acceleration of the masses m_A and m_B, and compare to the situation in which the moment of inertia of the pulley is ignored. [*Hint*: The tensions F_{TA} and F_{TB} are not equal. We discussed the Atwood machine in Example 4–13, assuming $I = 0$ for the pulley.]

FIGURE 10–57 Problem 51.
Atwood's machine.

52. (III) A string passing over a pulley has a 3.80-kg mass hanging from one end and a 3.15-kg mass hanging from the other end. The pulley is a uniform solid cylinder of radius 4.0 cm and mass 0.80 kg. (*a*) If the bearings of the pulley were frictionless, what would be the acceleration of the two masses? (*b*) In fact, it is found that if the heavier mass is given a downward speed of 0.20 m/s, it comes to rest in 6.2 s. What is the average frictional torque acting on the pulley?

53. (III) A hammer thrower accelerates the hammer (mass = 7.30 kg) from rest within four full turns (revolutions) and releases it at a speed of 26.5 m/s. Assuming a uniform rate of increase in angular velocity and a horizontal circular path of radius 1.20 m, calculate (*a*) the angular acceleration, (*b*) the (linear) tangential acceleration, (*c*) the centripetal acceleration just before release, (*d*) the net force being exerted on the hammer by the athlete just before release, and (*e*) the angle of this force with respect to the radius of the circular motion. Ignore gravity.

54. (III) A thin rod of length ℓ stands vertically on a table. The rod begins to fall, but its lower end does not slide. (*a*) Determine the angular velocity of the rod as a function of the angle ϕ it makes with the tabletop. (*b*) What is the speed of the tip of the rod just before it strikes the table?

10–7 Moment of Inertia

55. (I) Use the parallel-axis theorem to show that the moment of inertia of a thin rod about an axis perpendicular to the rod at one end is $I = \frac{1}{3}M\ell^2$, given that if the axis passes through the center, $I = \frac{1}{12}M\ell^2$ (Fig. 10–20f and g).

56. (II) Determine the moment of inertia of a 19-kg door that is 2.5 m high and 1.0 m wide and is hinged along one side. Ignore the thickness of the door.

57. (II) Two uniform solid spheres of mass M and radius r_0 are connected by a thin (massless) rod of length r_0 so that the centers are $3r_0$ apart. (*a*) Determine the moment of inertia of this system about an axis perpendicular to the rod at its center. (*b*) What would be the percentage error if the masses of each sphere were assumed to be concentrated at their centers and a very simple calculation made?

58. (II) A ball of mass M and radius r_1 on the end of a thin massless rod is rotated in a horizontal circle of radius R_0 about an axis of rotation AB, as shown in Fig. 10–58. (*a*) Considering the mass of the ball to be concentrated at its center of mass, calculate its moment of inertia about AB. (*b*) Using the parallel-axis theorem and considering the finite radius of the ball, calculate the moment of inertia of the ball about AB. (*c*) Calculate the percentage error introduced by the point mass approximation for $r_1 = 9.0$ cm and $R_0 = 1.0$ m.

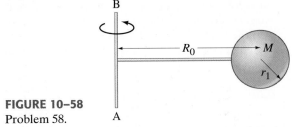

FIGURE 10–58
Problem 58.

59. (II) A thin 7.0-kg wheel of radius 32 cm is weighted to one side by a 1.50-kg weight, small in size, placed 22 cm from the center of the wheel. Calculate (*a*) the position of the center of mass of the weighted wheel and (*b*) the moment of inertia about an axis through its CM, perpendicular to its face.

60. (III) Derive the formula for the moment of inertia of a uniform thin rod of length ℓ about an axis through its center, perpendicular to the rod (see Fig. 10–20f).

61. (III) (*a*) Derive the formula given in Fig. 10–20h for the moment of inertia of a uniform, flat, rectangular plate of dimensions $\ell \times w$ about an axis through its center, perpendicular to the plate. (*b*) What is the moment of inertia about each of the axes through the center that are parallel to the edges of the plate?

10–8 Rotational Kinetic Energy

62. (I) An automobile engine develops a torque of 255 m·N at 3750 rpm. What is the horsepower of the engine?

63. (I) A centrifuge rotor has a moment of inertia of 4.25×10^{-2} kg·m². How much energy is required to bring it from rest to 9750 rpm?

64. (II) A rotating uniform cylindrical platform of mass 220 kg and radius 5.5 m slows down from 3.8 rev/s to rest in 16 s when the driving motor is disconnected. Estimate the power output of the motor (hp) required to maintain a steady speed of 3.8 rev/s.

65. (II) A merry-go-round has a mass of 1640 kg and a radius of 7.50 m. How much net work is required to accelerate it from rest to a rotation rate of 1.00 revolution per 8.00 s? Assume it is a solid cylinder.

66. (II) A uniform thin rod of length ℓ and mass M is suspended freely from one end. It is pulled to the side an angle θ and released. If friction can be ignored, what is its angular velocity, and the speed of its free end, at the lowest point?

67. (II) Two masses, $m_A = 35.0$ kg and $m_B = 38.0$ kg, are connected by a rope that hangs over a pulley (as in Fig. 10–59). The pulley is a uniform cylinder of radius 0.381 m and mass 3.1 kg. Initially m_A is on the ground and m_B rests 2.5 m above the ground. If the system is released, use conservation of energy to determine the speed of m_B just before it strikes the ground. Assume the pulley bearing is frictionless.

FIGURE 10–59
Problem 67.

68. (III) A 4.00-kg mass and a 3.00-kg mass are attached to opposite ends of a thin 42.0-cm-long horizontal rod (Fig. 10–60). The system is rotating at angular speed $\omega = 5.60$ rad/s about a vertical axle at the center of the rod. Determine (a) the kinetic energy K of the system, and (b) the net force on each mass. (c) Repeat parts (a) and (b) assuming that the axle passes through the CM of the system.

4.00 kg

3.00 kg

FIGURE 10–60
Problem 68.

69. (III) A 2.30-m-long pole is balanced vertically on its tip. It starts to fall and its lower end does not slip. What will be the speed of the upper end of the pole just before it hits the ground? [*Hint:* Use conservation of energy.]

10–9 Rotational Plus Translational Motion

70. (I) Calculate the translational speed of a cylinder when it reaches the foot of an incline 7.20 m high. Assume it starts from rest and rolls without slipping.

71. (I) A bowling ball of mass 7.3 kg and radius 9.0 cm rolls without slipping down a lane at 3.7 m/s. Calculate its total kinetic energy.

72. (I) Estimate the kinetic energy of the Earth with respect to the Sun as the sum of two terms, (a) that due to its daily rotation about its axis, and (b) that due to its yearly revolution about the Sun. [Assume the Earth is a uniform sphere with mass $= 6.0 \times 10^{24}$ kg, radius $= 6.4 \times 10^6$ m, and is 1.5×10^8 km from the Sun.]

73. (II) A sphere of radius $r_0 = 24.5$ cm and mass $m = 1.20$ kg starts from rest and rolls without slipping down a 30.0° incline that is 10.0 m long. (a) Calculate its translational and rotational speeds when it reaches the bottom. (b) What is the ratio of translational to rotational kinetic energy at the bottom? Avoid putting in numbers until the end so you can answer: (c) do your answers in (a) and (b) depend on the radius of the sphere or its mass?

74. (II) A narrow but solid spool of thread has radius R and mass M. If you pull up on the thread so that the CM of the spool remains suspended in the air at the same place as it unwinds, (a) what force must you exert on the thread? (b) How much work have you done by the time the spool turns with angular velocity ω?

75. (II) A ball of radius r_0 rolls on the inside of a track of radius R_0 (see Fig. 10–61). If the ball starts from rest at the vertical edge of the track, what will be its speed when it reaches the lowest point of the track, rolling without slipping?

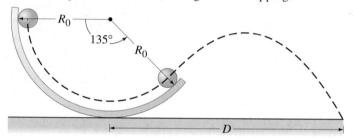

FIGURE 10–61 Problems 75 and 81.

76. (II) A solid rubber ball rests on the floor of a railroad car when the car begins moving with acceleration a. Assuming the ball rolls without slipping, what is its acceleration relative to (a) the car and (b) the ground?

*77. (II) A thin, hollow 0.545-kg section of pipe of radius 10.0 cm starts rolling (from rest) down a 17.5° incline 5.60 m long. (a) If the pipe rolls without slipping, what will be its speed at the base of the incline? (b) What will be its total kinetic energy at the base of the incline? (c) What minimum value must the coefficient of static friction have if the pipe is not to slip?

*78. (II) In Example 10–20, (a) how far has the ball moved down the lane when it starts rolling without slipping? (b) What are its final linear and rotational speeds?

79. (III) The 1100-kg mass of a car includes four tires, each of mass (including wheels) 35 kg and diameter 0.80 m. Assume each tire and wheel combination acts as a solid cylinder. Determine (a) the total kinetic energy of the car when traveling 95 km/h and (b) the fraction of the kinetic energy in the tires and wheels. (c) If the car is initially at rest and is then pulled by a tow truck with a force of 1500 N, what is the acceleration of the car? Ignore frictional losses. (d) What percent error would you make in part (c) if you ignored the rotational inertia of the tires and wheels?

*80. (III) A wheel with rotational inertia $I = \frac{1}{2}MR^2$ about its central axle is set spinning with initial angular speed ω_0 and is then lowered onto the ground so that it touches the ground with no horizontal speed. Initially it slips, but then begins to move forward and eventually rolls without slipping. (a) In what direction does friction act on the slipping wheel? (b) How long does the wheel slip before it begins to roll without slipping? (c) What is the wheel's final translational speed? [*Hint:* Use $\Sigma\vec{F} = m\vec{a}$, $\Sigma\tau_{CM} = I_{CM}\alpha_{CM}$, and recall that only when there is rolling without slipping is $v_{CM} = \omega R$.]

81. (III) A small sphere of radius $r_0 = 1.5$ cm rolls without slipping on the track shown in Fig. 10–61 whose radius is $R_0 = 26.0$ cm. The sphere starts rolling at a height R_0 above the bottom of the track. When it leaves the track after passing through an angle of 135° as shown, (a) what will be its speed, and (b) at what distance D from the base of the track will the sphere hit the ground?

*10–10 Rolling Sphere Slows Down

* 82. (I) A rolling ball slows down because the normal force does not pass exactly through the CM of the ball, but passes in front of the CM. Using Fig. 10–41, show that the torque resulting from the normal force ($\tau_N = \ell F_N$ in Fig. 10–41) is $\frac{7}{5}$ of that due to the frictional force, $\tau_{fr} = r_0 F$ where r_0 is the ball's radius; that is, show that $\tau_N = \frac{7}{5}\tau_{fr}$.

General Problems

83. A large spool of rope rolls on the ground with the end of the rope lying on the top edge of the spool. A person grabs the end of the rope and walks a distance ℓ, holding onto it, Fig. 10–62. The spool rolls behind the person without slipping. What length of rope unwinds from the spool? How far does the spool's center of mass move?

FIGURE 10–62
Problem 83.

84. On a 12.0-cm-diameter audio compact disc (CD), digital bits of information are encoded sequentially along an outward spiraling path. The spiral starts at radius $R_1 = 2.5$ cm and winds its way out to radius $R_2 = 5.8$ cm. To read the digital information, a CD player rotates the CD so that the player's readout laser scans along the spiral's sequence of bits at a constant linear speed of 1.25 m/s. Thus the player must accurately adjust the rotational frequency f of the CD as the laser moves outward. Determine the values for f (in units of rpm) when the laser is located at R_1 and when it is at R_2.

85. (a) A yo-yo is made of two solid cylindrical disks, each of mass 0.050 kg and diameter 0.075 m, joined by a (concentric) thin solid cylindrical hub of mass 0.0050 kg and diameter 0.010 m. Use conservation of energy to calculate the linear speed of the yo-yo just before it reaches the end of its 1.0-m-long string, if it is released from rest. (b) What fraction of its kinetic energy is rotational?

86. A cyclist accelerates from rest at a rate of 1.00 m/s². How fast will a point at the top of the rim of the tire (diameter = 68 cm) be moving after 2.5 s? [*Hint:* At any moment, the lowest point on the tire is in contact with the ground and is at rest—see Fig. 10–63.]

$v = ?$

$a = 1.00$ m/s²

This point on tire at rest momentarily

FIGURE 10–63   Problem 86.

87. Suppose David puts a 0.50-kg rock into a sling of length 1.5 m and begins whirling the rock in a nearly horizontal circle, accelerating it from rest to a rate of 85 rpm after 5.0 s. What is the torque required to achieve this feat, and where does the torque come from?

88. A 1.4-kg grindstone in the shape of a uniform cylinder of radius 0.20 m acquires a rotational rate of 1800 rev/s from rest over a 6.0-s interval at constant angular acceleration. Calculate the torque delivered by the motor.

89. Bicycle gears: (a) How is the angular velocity ω_R of the rear wheel of a bicycle related to the angular velocity ω_F of the front sprocket and pedals? Let N_F and N_R be the number of teeth on the front and rear sprockets, respectively, Fig. 10–64. The teeth are spaced the same on both sprockets and the rear sprocket is firmly attached to the rear wheel. (b) Evaluate the ratio ω_R/ω_F when the front and rear sprockets have 52 and 13 teeth, respectively, and (c) when they have 42 and 28 teeth.

FIGURE 10–64
Problem 89.

90. Figure 10–65 illustrates an H_2O molecule. The O—H bond length is 0.96 nm and the H—O—H bonds make an angle of 104°. Calculate the moment of inertia for the H_2O molecule about an axis passing through the center of the oxygen atom (a) perpendicular to the plane of the molecule, and (b) in the plane of the molecule, bisecting the H—O—H bonds.

FIGURE 10–65
Problem 90.

91. One possibility for a low-pollution automobile is for it to use energy stored in a heavy rotating flywheel. Suppose such a car has a total mass of 1100 kg, uses a uniform cylindrical flywheel of diameter 1.50 m and mass 240 kg, and should be able to travel 350 km without needing a flywheel "spinup." (a) Make reasonable assumptions (average frictional retarding force = 450 N, twenty acceleration periods from rest to 95 km/h, equal uphill and downhill, and that energy can be put back into the flywheel as the car goes downhill), and estimate what total energy needs to be stored in the flywheel. (b) What is the angular velocity of the flywheel when it has a full "energy charge"? (c) About how long would it take a 150-hp motor to give the flywheel a full energy charge before a trip?

92. A hollow cylinder (hoop) is rolling on a horizontal surface at speed $v = 3.3 \, \text{m/s}$ when it reaches a 15° incline. (a) How far up the incline will it go? (b) How long will it be on the incline before it arrives back at the bottom?

93. A wheel of mass M has radius R. It is standing vertically on the floor, and we want to exert a horizontal force F at its axle so that it will climb a step against which it rests (Fig. 10–66). The step has height h, where $h < R$. What minimum force F is needed?

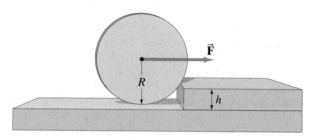

FIGURE 10–66 Problem 93.

94. A marble of mass m and radius r rolls along the looped rough track of Fig. 10–67. What is the minimum value of the vertical height h that the marble must drop if it is to reach the highest point of the loop without leaving the track? (a) Assume $r \ll R$; (b) do not make this assumption. Ignore frictional losses.

FIGURE 10–67 Problem 94.

95. The density (mass per unit length) of a thin rod of length ℓ increases uniformly from λ_0 at one end to $3\lambda_0$ at the other end. Determine the moment of inertia about an axis perpendicular to the rod through its geometric center.

96. If a billiard ball is hit in just the right way by a cue stick, the ball will roll without slipping immediately after losing contact with the stick. Consider a billiard ball (radius r, mass M) at rest on a horizontal pool table. A cue stick exerts a constant horizontal force F on the ball for a time t at a point that is a height h above the table's surface (see Fig. 10–68). Assume that the coefficient of kinetic friction between the ball and table is μ_k. Determine the value for h so that the ball will roll without slipping immediately after losing contact with the stick.

FIGURE 10–68 Problem 96.

97. If the coefficient of static friction between tires and pavement is 0.65, calculate the minimum torque that must be applied to the 66-cm-diameter tire of a 950-kg automobile in order to "lay rubber" (make the wheels spin, slipping as the car accelerates). Assume each wheel supports an equal share of the weight.

98. A cord connected at one end to a block which can slide on an inclined plane has its other end wrapped around a cylinder resting in a depression at the top of the plane as shown in Fig. 10–69. Determine the speed of the block after it has traveled 1.80 m along the plane, starting from rest. Assume (a) there is no friction, (b) the coefficient of friction between all surfaces is $\mu = 0.055$. [*Hint*: In part (b) first determine the normal force on the cylinder, and make any reasonable assumptions needed.]

FIGURE 10–69 Problem 98.

99. The radius of the roll of paper shown in Fig. 10–70 is 7.6 cm and its moment of inertia is $I = 3.3 \times 10^{-3} \, \text{kg} \cdot \text{m}^2$. A force of 2.5 N is exerted on the end of the roll for 1.3 s, but the paper does not tear so it begins to unroll. A constant friction torque of 0.11 m·N is exerted on the roll which gradually brings it to a stop. Assuming that the paper's thickness is negligible, calculate (a) the length of paper that unrolls during the time that the force is applied (1.3 s) and (b) the length of paper that unrolls from the time the force ends to the time when the roll has stopped moving.

FIGURE 10–70 Problem 99.

100. A solid uniform disk of mass 21.0 kg and radius 85.0 cm is at rest flat on a frictionless surface. Figure 10–71 shows a view from above. A string is wrapped around the rim of the disk and a constant force of 35.0 N is applied to the string. The string does not slip on the rim. (a) In what direction does the CM move? When the disk has moved a distance of 5.5 m, determine (b) how fast it is moving, (c) how fast it is spinning (in radians per second), and (d) how much string has unwrapped from around the rim.

FIGURE 10–71 Problem 100, looking down on the disk.

101. When bicycle and motorcycle riders "pop a wheelie," a large acceleration causes the bike's front wheel to leave the ground. Let M be the total mass of the bike-plus-rider system; let x and y be the horizontal and vertical distance of this system's CM from the rear wheel's point of contact with the ground (Fig. 10–72). (a) Determine the horizontal acceleration a required to barely lift the bike's front wheel off of the ground. (b) To minimize the acceleration necessary to pop a wheelie, should x be made as small or as large as possible? How about y? How should a rider position his or her body on the bike in order to achieve these optimal values for x and y? (c) If $x = 35$ cm and $y = 95$ cm, find a.

FIGURE 10–72 Problem 101.

102. A crucial part of a piece of machinery starts as a flat uniform cylindrical disk of radius R_0 and mass M. It then has a circular hole of radius R_1 drilled into it (Fig. 10–73). The hole's center is a distance h from the center of the disk. Find the moment of inertia of this disk (with off-center hole) when rotated about its center, C. [*Hint*: Consider a solid disk and "subtract" the hole; use the parallel-axis theorem.]

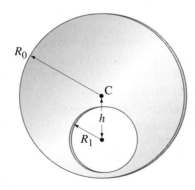

FIGURE 10–73
Problem 102.

103. A thin uniform stick of mass M and length ℓ is positioned vertically, with its tip on a frictionless table. It is released and allowed to fall. Determine the speed of its CM just before it hits the table (Fig. 10–74).

FIGURE 10–74
Problem 103.

$\vec{\mathbf{v}}_{CM}$

*104. (a) For the yo-yo-like cylinder of Example 10–19, we saw that the downward acceleration of its CM was $a = \frac{2}{3}g$. If it starts from rest, what will be the CM velocity after it has fallen a distance h? (b) Now use conservation of energy to determine the cylinder's CM velocity after it has fallen a distance h, starting from rest.

*Numerical/Computer

*105. (II) Determine the torque produced about the support A of the rigid structure, shown in Fig. 10–75, as a function of the leg angle θ if a force $F = 500$ N is applied at the point P perpendicular to the leg end. Graph the values of the torque τ as a function of θ from $\theta = 0°$ to $90°$, in $1°$ increments.

A

30.0 cm $\quad\theta$ 20.0 cm

P

$F = 500$ N

FIGURE 10–75 Problem 105.

*106. (II) Use the expression that was derived in Problem 51 for the acceleration of masses on an Atwood's machine to investigate at what point the moment of inertia of the pulley becomes negligible. Assume $m_A = 0.150$ kg, $m_B = 0.350$ kg, and $R = 0.040$ m. (a) Graph the acceleration as a function of the moment of inertia. (b) Find the acceleration of the masses when the moment of inertia goes to zero. (c) Using your graph to guide you, at what minimum value of I does the calculated acceleration deviate by 2.0% from the acceleration found in part (b)? (d) If the pulley could be thought of as a uniform disk, find the mass of the pulley using the I found in part (c).

Answers to Exercises

A: $f = 0.076$ Hz; $T = 13$ s.

B: $\vec{\mathbf{F}}_A$.

C: Yes; $\frac{1}{12}M\ell^2 + M(\frac{1}{2}\ell)^2 = \frac{1}{3}M\ell^2$.

D: 4×10^{17} J.

E: (c).

F: $a = \frac{1}{3}g$.

This skater is doing a spin. When her arms are spread outward horizontally, she spins less fast than when her arms are held close to the axis of rotation. This is an example of the conservation of angular momentum.

Angular momentum, which we study in this Chapter, is conserved only if no net torque acts on the object or system. Otherwise, the rate of change of angular momentum is proportional to the net applied torque—which, if zero, means the angular momentum is *conserved*. In this Chapter we also examine more complicated aspects of rotational motion.

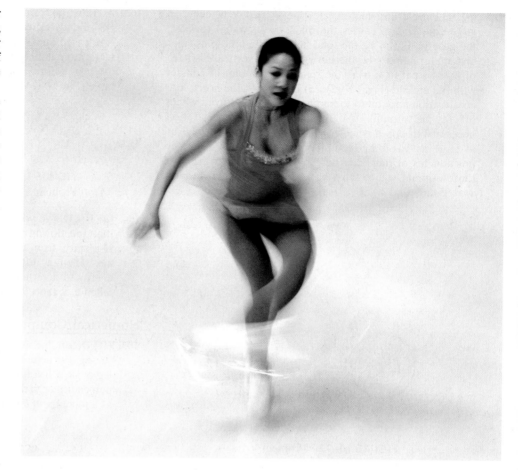

11

Angular Momentum; General Rotation

CONTENTS

CHAPTER-OPENING QUESTION—Guess now!

You are standing on a platform at rest, but that is free to rotate. You hold a spinning bicycle wheel by its axle as shown here. You then flip the wheel over so its axle points down. What happens then?

(a) The platform starts rotating in the direction the bicycle wheel was originally rotating.

(b) The platform starts rotating in the direction opposite to the original rotation of the bicycle wheel.

(c) The platform stays at rest.

(d) The platform turns only while you are flipping the wheel.

(e) None of these is correct.

I n Chapter 10 we dealt with the kinematics and dynamics of the rotation of a rigid object about an axis whose direction is fixed in an inertial reference frame. We analyzed the motion in terms of the rotational equivalent of Newton's laws (torque plays the role that force does for translational motion), as well as rotational kinetic energy.

To keep the axis of a rotating object fixed, the object must usually be constrained by external supports (such as bearings at the end of an axle). The motion of objects that are not constrained to move about a fixed axis is more difficult to describe and analyze. Indeed, the complete analysis of the general rotational motion of an object (or system of objects) is very complicated, and we will only look at some aspects of general rotational motion in this Chapter.

We start this Chapter by introducing the concept of *angular momentum*, which is the rotational analog of linear momentum. We first treat angular momentum and its conservation for an object rotating about a fixed axis. After that, we examine the vector nature of torque and angular momentum. We will derive some general theorems and apply them to some interesting types of motion.

11–1 Angular Momentum—Objects Rotating About a Fixed Axis

In Chapter 10 we saw that if we use the appropriate angular variables, the kinematic and dynamic equations for rotational motion are analogous to those for ordinary linear motion. In like manner, the linear momentum, $p = mv$, has a rotational analog. It is called **angular momentum**, L, and for an object rotating about a fixed axis with angular velocity ω, it is defined as

$$L = I\omega, \tag{11–1}$$

where I is the moment of inertia. The SI units for L are $\text{kg} \cdot \text{m}^2/\text{s}$; there is no special name for this unit.

We saw in Chapter 9 (Section 9–1) that Newton's second law can be written not only as $\Sigma F = ma$, but also more generally in terms of momentum (Eq. 9–2), $\Sigma F = dp/dt$. In a similar way, the rotational equivalent of Newton's second law, which we saw in Eqs. 10–14 and 10–15 can be written as $\Sigma\tau = I\alpha$, can also be written in terms of angular momentum: since the angular acceleration $\alpha = d\omega/dt$ (Eq. 10–3), then $I\alpha = I(d\omega/dt) = d(I\omega)/dt = dL/dt$, so

$$\Sigma\tau = \frac{dL}{dt}. \tag{11–2}$$

NEWTON'S SECOND LAW FOR ROTATION

This derivation assumes that the moment of inertia, I, remains constant. However, Eq. 11–2 is valid even if the moment of inertia changes, and applies also to a system of objects rotating about a fixed axis where $\Sigma\tau$ is the net external torque (discussed in Section 11–4). Equation 11–2 is Newton's second law for rotational motion about a fixed axis, and is also valid for a moving object if its rotation is about an axis passing through its center of mass (as for Eq. 10–15).

Conservation of Angular Momentum

Angular momentum is an important concept in physics because, under certain conditions, it is a conserved quantity. What are the conditions for which it is conserved? From Eq. 11–2 we see immediately that if the net external torque $\Sigma\tau$ on an object (or system of objects) is zero, then

$$\frac{dL}{dt} = 0 \quad \text{and} \quad L = I\omega = \text{constant.} \qquad [\Sigma\tau = 0]$$

This, then, is the **law of conservation of angular momentum** for a rotating object:

The total angular momentum of a rotating object remains constant if the net external torque acting on it is zero.

CONSERVATION OF ANGULAR MOMENTUM

The law of conservation of angular momentum is one of the great conservation laws of physics, along with those for energy and linear momentum.

I large,
ω small

I small,
ω large

(a)　　　　(b)

FIGURE 11–1 A skater doing a spin on ice, illustrating conservation of angular momentum: (a) *I* is large and ω is small; (b) *I* is smaller so ω is larger.

FIGURE 11–2 A diver rotates faster when arms and legs are tucked in than when they are outstretched. Angular momentum is conserved.

FIGURE 11–3 Example 11–1.

When there is zero net torque acting on an object, and the object is rotating about a fixed axis or about an axis through its center of mass whose direction doesn't change, we can write

$$I\omega = I_0\omega_0 = \text{constant}.$$

I_0 and ω_0 are the moment of inertia and angular velocity, respectively, about the axis at some initial time ($t = 0$), and I and ω are their values at some other time. The parts of the object may alter their positions relative to one another, so that I changes. But then ω changes as well and the product $I\omega$ remains constant.

Many interesting phenomena can be understood on the basis of conservation of angular momentum. Consider a skater doing a spin on the tips of her skates, Fig. 11–1. She rotates at a relatively low speed when her arms are outstretched, but when she brings her arms in close to her body, she suddenly spins much faster. From the definition of moment of inertia, $I = \Sigma mR^2$, it is clear that when she pulls her arms in closer to the axis of rotation, R is reduced for the arms so her moment of inertia is reduced. Since the angular momentum $I\omega$ remains constant (we ignore the small torque due to friction), if I decreases, then the angular velocity ω must increase. If the skater reduces her moment of inertia by a factor of 2, she will then rotate with twice the angular velocity.

A similar example is the diver shown in Fig. 11–2. The push as she leaves the board gives her an initial angular momentum about her center of mass. When she curls herself into the tuck position, she rotates quickly one or more times. She then stretches out again, increasing her moment of inertia which reduces the angular velocity to a small value, and then she enters the water. The change in moment of inertia from the straight position to the tuck position can be a factor of as much as $3\frac{1}{2}$.

Note that for angular momentum to be conserved, the net torque must be zero, but the net force does not necessarily have to be zero. The net force on the diver in Fig. 11–2, for example, is not zero (gravity is acting), but the net torque about her CM is zero because the force of gravity acts at her center of mass.

EXAMPLE 11–1 **Object rotating on a string of changing length.** A small mass m attached to the end of a string revolves in a circle on a frictionless tabletop. The other end of the string passes through a hole in the table (Fig. 11–3). Initially, the mass revolves with a speed $v_1 = 2.4\,\text{m/s}$ in a circle of radius $R_1 = 0.80\,\text{m}$. The string is then pulled slowly through the hole so that the radius is reduced to $R_2 = 0.48\,\text{m}$. What is the speed, v_2, of the mass now?

APPROACH There is no net torque on the mass m because the force exerted by the string to keep it moving in a circle is exerted toward the axis; hence the lever arm is zero. We can thus apply conservation of angular momentum.

SOLUTION Conservation of angular momentum gives

$$I_1\omega_1 = I_2\omega_2.$$

Our small mass is essentially a particle whose moment of inertia about the hole is $I = mR^2$ (Eq. 10–11), so we have

$$mR_1^2\omega_1 = mR_2^2\omega_2,$$

or

$$\omega_2 = \omega_1\left(\frac{R_1^2}{R_2^2}\right).$$

Then, since $v = R\omega$, we can write

$$v_2 = R_2\omega_2 = R_2\omega_1\left(\frac{R_1^2}{R_2^2}\right) = R_2\frac{v_1}{R_1}\left(\frac{R_1^2}{R_2^2}\right) = v_1\frac{R_1}{R_2}$$

$$= (2.4\,\text{m/s})\left(\frac{0.80\,\text{m}}{0.48\,\text{m}}\right) = 4.0\,\text{m/s}.$$

The speed increases as the radius decreases.

EXAMPLE 11–2 **Clutch.** A simple clutch consists of two cylindrical plates that can be pressed together to connect two sections of an axle, as needed, in a piece of machinery. The two plates have masses $M_A = 6.0$ kg and $M_B = 9.0$ kg, with equal radii $R_0 = 0.60$ m. They are initially separated (Fig. 11–4). Plate M_A is accelerated from rest to an angular velocity $\omega_1 = 7.2$ rad/s in time $\Delta t = 2.0$ s. Calculate (a) the angular momentum of M_A, and (b) the torque required to have accelerated M_A from rest to ω_1. (c) Next, plate M_B, initially at rest but free to rotate without friction, is placed in firm contact with freely rotating plate M_A, and the two plates both rotate at a constant angular velocity ω_2, which is considerably less than ω_1. Why does this happen, and what is ω_2?

FIGURE 11–4 Example 11–2.

APPROACH We use the definition of angular momentum $L = I\omega$ (Eq. 11–1) plus Newton's second law for rotation, Eq. 11–2.

SOLUTION (a) The angular momentum of M_A will be

$$L_A = I_A \omega_1 = \tfrac{1}{2} M_A R_0^2 \omega_1 = \tfrac{1}{2}(6.0 \text{ kg})(0.60 \text{ m})^2(7.2 \text{ rad/s}) = 7.8 \text{ kg·m}^2/\text{s}.$$

(b) The plate started from rest so the torque, assumed constant, was

$$\tau = \frac{\Delta L}{\Delta t} = \frac{7.8 \text{ kg·m}^2/\text{s} - 0}{2.0 \text{ s}} = 3.9 \text{ m·N}.$$

(c) Initially, M_A is rotating at constant ω_1 (we ignore friction). When plate B comes in contact, why is their joint rotation speed less? You might think in terms of the torque each exerts on the other upon contact. But quantitatively, it's easier to use conservation of angular momentum, since no external torques are assumed to act. Thus

$$\text{angular momentum before} = \text{angular momentum after}$$
$$I_A \omega_1 = (I_A + I_B)\omega_2.$$

Solving for ω_2 we find

$$\omega_2 = \left(\frac{I_A}{I_A + I_B}\right)\omega_1 = \left(\frac{M_A}{M_A + M_B}\right)\omega_1 = \left(\frac{6.0 \text{ kg}}{15.0 \text{ kg}}\right)(7.2 \text{ rad/s}) = 2.9 \text{ rad/s}.$$

EXAMPLE 11–3 **ESTIMATE** **Neutron star.** Astronomers detect stars that are rotating extremely rapidly, known as neutron stars. A neutron star is believed to form from the inner core of a larger star that collapsed, under its own gravitation, to a star of very small radius and very high density. Before collapse, suppose the core of such a star is the size of our Sun $(r \approx 7 \times 10^5 \text{ km})$ with mass 2.0 times as great as the Sun, and is rotating at a frequency of 1.0 revolution every 100 days. If it were to undergo gravitational collapse to a neutron star of radius 10 km, what would its rotation frequency be? Assume the star is a uniform sphere at all times, and loses no mass.

APPROACH We assume the star is isolated (no external forces), so we can use conservation of angular momentum for this process. We use r for the radius of a sphere, as compared to R used for distance from an axis of rotation or cylindrical symmetry: see Fig. 10–2.

SOLUTION From conservation of angular momentum,

$$I_1 \omega_1 = I_2 \omega_2,$$

where the subscripts 1 and 2 refer to initial (normal star) and final (neutron star), respectively. Then, assuming no mass is lost in the process,

$$\omega_2 = \left(\frac{I_1}{I_2}\right)\omega_1 = \left(\frac{\tfrac{2}{5} m_1 r_1^2}{\tfrac{2}{5} m_2 r_2^2}\right)\omega_1 = \frac{r_1^2}{r_2^2}\omega_1.$$

The frequency $f = \omega/2\pi$, so

$$f_2 = \frac{\omega_2}{2\pi} = \frac{r_1^2}{r_2^2} f_1$$

$$= \left(\frac{7 \times 10^5 \text{ km}}{10 \text{ km}}\right)^2 \left(\frac{1.0 \text{ rev}}{100 \text{ d}(24 \text{ h/d})(3600 \text{ s/h})}\right) \approx 6 \times 10^2 \text{ rev/s}.$$

Axis

(a)

$\vec{L}_{person}$

$\vec{L}_{platform}$

(b)

FIGURE 11–5 (a) A person on a circular platform, both initially at rest, begins walking along the edge at speed v. The platform, assumed to be mounted on friction-free bearings, begins rotating in the opposite direction, so that the total angular momentum remains zero, as shown in (b).

Directional Nature of Angular Momentum

Angular momentum is a vector, as we shall discuss later in this Chapter. For now we consider the simple case of an object rotating about a fixed axis, and the direction of $\vec{L}$ is specified by a plus or minus sign, just as we did for one-dimensional linear motion in Chapter 2.

For a symmetrical object rotating about a symmetry axis (such as a cylinder or wheel), the direction of the angular momentum[†] can be taken as the direction of the angular velocity $\vec{\omega}$. That is,

$$\vec{L} = I\vec{\omega}.$$

As a simple example, consider a person standing at rest on a circular platform capable of rotating friction-free about an axis through its center (that is, a simplified merry-go-round). If the person now starts to walk along the edge of the platform, Fig. 11–5a, the platform starts rotating in the opposite direction. Why? One explanation is that the person's foot exerts a force on the platform. Another explanation (and this is the most useful analysis here) is as an example of the conservation of angular momentum. If the person starts walking counterclockwise, the person's angular momentum will be pointed upward along the axis of rotation (remember how we defined the direction of $\vec{\omega}$ using the right-hand rule in Section 10–2). The magnitude of the person's angular momentum will be $L = I\omega = (mR^2)(v/R)$, where v is the person's speed (relative to the Earth, not the platform), R is his distance from the rotation axis, m is his mass, and mR^2 is his moment of inertia if we consider him a particle (mass concentrated at one point). The platform rotates in the opposite direction, so its angular momentum points downward. If the initial total angular momentum was zero (person and platform at rest), it will remain zero after the person starts walking. That is, the upward angular momentum of the person just balances the oppositely directed downward angular momentum of the platform (Fig. 11–5b), so the total vector angular momentum remains zero. Even though the person exerts a force (and torque) on the platform, the platform exerts an equal and opposite torque on the person. So the net torque on the *system* of person plus platform is zero (ignoring friction) and the total angular momentum remains constant.

EXAMPLE 11–4 **Running on a circular platform.** Suppose a 60-kg person stands at the edge of a 6.0-m-diameter circular platform, which is mounted on frictionless bearings and has a moment of inertia of $1800\ kg \cdot m^2$. The platform is at rest initially, but when the person begins running at a speed of 4.2 m/s (with respect to the Earth) around its edge, the platform begins to rotate in the opposite direction as in Fig. 11–5. Calculate the angular velocity of the platform.

APPROACH We use conservation of angular momentum. The total angular momentum is zero initially. Since there is no net torque, $\vec{L}$ is conserved and will remain zero, as in Fig. 11–5. The person's angular momentum is $L_{per} = (mR^2)(v/R)$, and we take this as positive. The angular momentum of the platform is $L_{plat} = -I\omega$.

SOLUTION Conservation of angular momentum gives

$$L = L_{per} + L_{plat}$$
$$0 = mR^2\left(\frac{v}{R}\right) - I\omega.$$

So

$$\omega = \frac{mRv}{I} = \frac{(60\ kg)(3.0\ m)(4.2\ m/s)}{1800\ kg \cdot m^2} = 0.42\ rad/s.$$

NOTE The frequency of rotation is $f = \omega/2\pi = 0.067\ rev/s$ and the period $T = 1/f = 15\ s$ per revolution.

[†]For more complicated situations of objects rotating about a fixed axis, there will be a component of $\vec{L}$ along the direction of $\vec{\omega}$ and its magnitude will be equal to $I\vec{\omega}$, but there could be other components as well. If the total angular momentum is conserved, then the component $I\vec{\omega}$ will also be conserved. So our results here can be applied to any rotation about a fixed axis.

FIGURE 11–6 Example 11–5.

CONCEPTUAL EXAMPLE 11–5 **Spinning bicycle wheel.** Your physics teacher is holding a spinning bicycle wheel while he stands on a stationary frictionless turntable (Fig. 11–6). What will happen if the teacher suddenly flips the bicycle wheel over so that it is spinning in the opposite direction?

RESPONSE We consider the system of turntable, teacher, and bicycle wheel. The total angular momentum initially is $\vec{\mathbf{L}}$ vertically upward. That is also what the system's angular momentum must be afterward, since $\vec{\mathbf{L}}$ is conserved when there is no net torque. Thus, if the wheel's angular momentum after being flipped over is $-\vec{\mathbf{L}}$ downward, then the angular momentum of teacher plus turntable will have to be $+2\vec{\mathbf{L}}$ upward. We can safely predict that the teacher will begin spinning around in the same direction the wheel was spinning originally.

EXERCISE A In Example 11–5, what if he moves the axis only 90° so it is horizontal? (*a*) The same direction and speed as above; (*b*) the same as above, but slower; (*c*) the opposite result.

EXERCISE B Return to the Chapter-Opening Question, page 284, and answer it again now. Try to explain why you may have answered differently the first time.

EXERCISE C Suppose you are standing on the edge of a large freely rotating turntable. If you walk toward the center, (*a*) the turntable slows down; (*b*) the turntable speeds up; (*c*) its rotation speed is unchanged; (*d*) you need to know the walking speed to answer.

11–2 Vector Cross Product; Torque as a Vector

Vector Cross Product

To deal with the vector nature of angular momentum and torque in general, we will need the concept of the *vector cross product* (often called simply the *vector product* or *cross product*). In general, the **vector** or **cross product** of two vectors $\vec{\mathbf{A}}$ and $\vec{\mathbf{B}}$ is defined as another vector $\vec{\mathbf{C}} = \vec{\mathbf{A}} \times \vec{\mathbf{B}}$ *whose magnitude* is

$$C = |\vec{\mathbf{A}} \times \vec{\mathbf{B}}| = AB \sin\theta, \tag{11–3a}$$

where θ is the angle ($< 180°$) *between* $\vec{\mathbf{A}}$ *and* $\vec{\mathbf{B}}$, *and whose direction is perpendicular to both* $\vec{\mathbf{A}}$ *and* $\vec{\mathbf{B}}$ *in the sense of the right-hand rule*, Fig. 11–7. The angle θ is measured between $\vec{\mathbf{A}}$ and $\vec{\mathbf{B}}$ when their tails are together at the same point. According to the right-hand rule, as shown in Fig. 11–7, you orient your right hand so your fingers point along $\vec{\mathbf{A}}$, and when you bend your fingers they point along $\vec{\mathbf{B}}$. When your hand is correctly oriented in this way, your thumb will point along the direction of $\vec{\mathbf{C}} = \vec{\mathbf{A}} \times \vec{\mathbf{B}}$.

The cross product of two vectors, $\vec{\mathbf{A}} = A_x\hat{\mathbf{i}} + A_y\hat{\mathbf{j}} + A_z\hat{\mathbf{k}}$, and $\vec{\mathbf{B}} = B_x\hat{\mathbf{i}} + B_y\hat{\mathbf{j}} + B_z\hat{\mathbf{k}}$, can be written in component form (see Problem 26) as

$$\vec{\mathbf{A}} \times \vec{\mathbf{B}} = \begin{vmatrix} \hat{\mathbf{i}} & \hat{\mathbf{j}} & \hat{\mathbf{k}} \\ A_x & A_y & A_z \\ B_x & B_y & B_z \end{vmatrix} \tag{11–3b}$$

$$= (A_yB_z - A_zB_y)\hat{\mathbf{i}} + (A_zB_x - A_xB_z)\hat{\mathbf{j}} + (A_xB_y - A_yB_x)\hat{\mathbf{k}}. \tag{11–3c}$$

Equation 11–3b is meant to be evaluated using the rules of determinants (and obtain Eq. 11–3c).

FIGURE 11–7 The vector $\vec{\mathbf{C}} = \vec{\mathbf{A}} \times \vec{\mathbf{B}}$ is perpendicular to the plane containing $\vec{\mathbf{A}}$ and $\vec{\mathbf{B}}$; its direction is given by the right-hand rule.

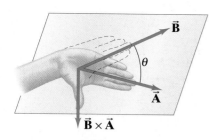

FIGURE 11–8 The vector $\vec{\mathbf{B}} \times \vec{\mathbf{A}}$ equals $-\vec{\mathbf{A}} \times \vec{\mathbf{B}}$; compare to Fig. 11–7.

FIGURE 11–9 Exercise D.

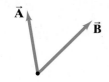

FIGURE 11–10 The torque due to the force $\vec{\mathbf{F}}$ (in the plane of the wheel) starts the wheel rotating counterclockwise so $\vec{\boldsymbol{\omega}}$ and $\vec{\boldsymbol{\alpha}}$ point out of the page.

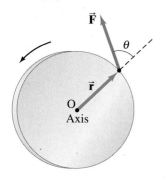

FIGURE 11–11 $\vec{\boldsymbol{\tau}} = \vec{\mathbf{r}} \times \vec{\mathbf{F}}$, where $\vec{\mathbf{r}}$ is the position vector.

Some properties of the cross product are the following:

$$\vec{\mathbf{A}} \times \vec{\mathbf{A}} = 0 \tag{11–4a}$$

$$\vec{\mathbf{A}} \times \vec{\mathbf{B}} = -\vec{\mathbf{B}} \times \vec{\mathbf{A}} \tag{11–4b}$$

$$\vec{\mathbf{A}} \times (\vec{\mathbf{B}} + \vec{\mathbf{C}}) = (\vec{\mathbf{A}} \times \vec{\mathbf{B}}) + (\vec{\mathbf{A}} \times \vec{\mathbf{C}}) \quad \text{[distributive law]} \tag{11–4c}$$

$$\frac{d}{dt}(\vec{\mathbf{A}} \times \vec{\mathbf{B}}) = \frac{d\vec{\mathbf{A}}}{dt} \times \vec{\mathbf{B}} + \vec{\mathbf{A}} \times \frac{d\vec{\mathbf{B}}}{dt}. \tag{11–4d}$$

Equation 11–4a follows from Eqs. 11–3 (since $\theta = 0$). So does Eq. 11–4b, since the magnitude of $\vec{\mathbf{B}} \times \vec{\mathbf{A}}$ is the same as that for $\vec{\mathbf{A}} \times \vec{\mathbf{B}}$, but by the right-hand rule the direction is opposite (see Fig. 11–8). Thus the order of the two vectors is crucial. If you change the order, you change the result. That is, the commutative law does *not* hold for the cross product $(\vec{\mathbf{A}} \times \vec{\mathbf{B}} \neq \vec{\mathbf{B}} \times \vec{\mathbf{A}})$, although it does hold for the dot product of two vectors and for the product of scalars. Note in Eq. 11–4d that the order of quantities in the two products on the right must not be changed (because of Eq. 11–4b).

EXERCISE D For the vectors $\vec{\mathbf{A}}$ and $\vec{\mathbf{B}}$ in the plane of the page as shown in Fig. 11–9, in what direction is (i) $\vec{\mathbf{A}} \cdot \vec{\mathbf{B}}$, (ii) $\vec{\mathbf{A}} \times \vec{\mathbf{B}}$, (iii) $\vec{\mathbf{B}} \times \vec{\mathbf{A}}$? (*a*) Into the page; (*b*) out of the page; (*c*) between $\vec{\mathbf{A}}$ and $\vec{\mathbf{B}}$; (*d*) it is a scalar and has no direction; (*e*) it is zero and has no direction.

The Torque Vector

Torque is an example of a quantity that can be expressed as a cross product. To see this, let us take a simple example: the thin wheel shown in Fig. 11–10 which is free to rotate about an axis through its center at point O. A force $\vec{\mathbf{F}}$ acts at the edge of the wheel, at a point whose position relative to the center O is given by the position vector $\vec{\mathbf{r}}$ as shown. The force $\vec{\mathbf{F}}$ tends to rotate the wheel (assumed initially at rest) counterclockwise, so the angular velocity $\vec{\boldsymbol{\omega}}$ will point out of the page toward the viewer (remember the right-hand rule from Section 10–2). The torque due to $\vec{\mathbf{F}}$ will tend to increase $\vec{\boldsymbol{\omega}}$ so $\vec{\boldsymbol{\alpha}}$ also points outward along the rotation axis. The relation between angular acceleration and torque that we developed in Chapter 10 for an object rotating about a fixed axis is

$$\Sigma \tau = I\alpha,$$

(Eq. 10–14) where I is the moment of inertia. This scalar equation is the rotational equivalent of $\Sigma F = ma$, and we would like to make it a vector equation just as $\Sigma \vec{\mathbf{F}} = m\vec{\mathbf{a}}$ is a vector equation. To do so in the case of Fig. 11–10 we must have the direction of $\vec{\boldsymbol{\tau}}$ point outward along the rotation axis, since $\vec{\boldsymbol{\alpha}}$ $(= d\vec{\boldsymbol{\omega}}/dt)$ has that direction; and the magnitude of the torque must be (see Eqs. 10–10 and Fig. 11–10) $\tau = rF_\perp = rF \sin\theta$. We can achieve this by defining the **torque vector** to be the cross product of $\vec{\mathbf{r}}$ and $\vec{\mathbf{F}}$:

$$\vec{\boldsymbol{\tau}} = \vec{\mathbf{r}} \times \vec{\mathbf{F}}. \tag{11–5}$$

From the definition of the cross product above (Eq. 11–3a) the magnitude of $\vec{\boldsymbol{\tau}}$ will be $rF \sin\theta$ and the direction will be along the axis, as required for this special case.

We will see in Sections 11–3 through 11–5 that if we take Eq. 11–5 as the *general definition of torque*, then the vector relation $\Sigma\vec{\boldsymbol{\tau}} = I\vec{\boldsymbol{\alpha}}$ will hold in general. Thus we state now that Eq. 11–5 is the general definition of torque. It contains both magnitude and direction information. Note that this definition involves the position vector $\vec{\mathbf{r}}$ and thus the torque is being calculated about a point. We can choose that point O as we wish.

For a particle of mass m on which a force $\vec{\mathbf{F}}$ is applied, we define the torque about a point O as

$$\vec{\boldsymbol{\tau}} = \vec{\mathbf{r}} \times \vec{\mathbf{F}}$$

where $\vec{\mathbf{r}}$ is the position vector of the particle relative to O (Fig. 11–11). If we have a system of particles (which could be the particles making up a rigid object) the total torque $\vec{\boldsymbol{\tau}}$ on the system will be the sum of the torques on the individual particles:

$$\vec{\boldsymbol{\tau}} = \Sigma(\vec{\mathbf{r}}_i \times \vec{\mathbf{F}}_i),$$

where $\vec{\mathbf{r}}_i$ is the position vector of the ith particle and $\vec{\mathbf{F}}_i$ is the net force on the ith particle.

EXAMPLE 11–6 **Torque Vector.** Suppose the vector $\vec{r}$ is in the xz plane, as in Fig. 11–11, and is given by $\vec{r} = (1.2\,\text{m})\hat{i} + (1.2\,\text{m})\hat{k}$. Calculate the torque vector $\vec{\tau}$ if $\vec{F} = (150\,\text{N})\hat{i}$.

APPROACH We use the determinant form, Eq. 11–3b.

SOLUTION
$$\vec{\tau} = \vec{r} \times \vec{F} = \begin{vmatrix} \hat{i} & \hat{j} & \hat{k} \\ 1.2\,\text{m} & 0 & 1.2\,\text{m} \\ 150\,\text{N} & 0 & 0 \end{vmatrix} = 0\hat{i} + (180\,\text{m}\cdot\text{N})\hat{j} + 0\hat{k}.$$

So τ has magnitude $180\,\text{m}\cdot\text{N}$ and points along the positive y axis.

EXERCISE E If $\vec{F} = 5.0\,\text{N}\,\hat{i}$ and $\vec{r} = 2.0\,\text{m}\,\hat{j}$, what is $\vec{\tau}$? (a) $10\,\text{mN}$, (b) $-10\,\text{mN}$, (c) $10\,\text{mN}\,\hat{k}$, (d) $-10\,\text{mN}\,\hat{j}$, (e) $-10\,\text{mN}\,\hat{k}$.

11–3 Angular Momentum of a Particle

The most general way of writing Newton's second law for the translational motion of a particle (or system of particles) is in terms of the linear momentum $\vec{p} = m\vec{v}$ as given by Eq. 9–2 (or 9–5):

$$\Sigma\vec{F} = \frac{d\vec{p}}{dt}.$$

The rotational analog of linear momentum is *angular momentum*. Just as the rate of change of $\vec{p}$ is related to the net force $\Sigma\vec{F}$, so we might expect the rate of change of angular momentum to be related to the net torque. Indeed, we saw this was true in Section 11–1 for the special case of a rigid object rotating about a fixed axis. Now we will see it is true in general. We first treat a single particle.

Suppose a particle of mass m has momentum $\vec{p}$ and position vector $\vec{r}$ with respect to the origin O in some chosen inertial reference frame. Then the general definition of the **angular momentum**, $\vec{L}$, of the particle about point O is the vector cross product of $\vec{r}$ and $\vec{p}$:

$$\vec{L} = \vec{r} \times \vec{p}. \qquad \text{[particle]} \quad \textbf{(11–6)}$$

Angular momentum is a vector.[†] Its direction is perpendicular to both $\vec{r}$ and $\vec{p}$ as given by the right-hand rule (Fig. 11–12). Its magnitude is given by

$$L = rp\sin\theta$$

or

$$L = rp_\perp = r_\perp p$$

where θ is the angle between $\vec{r}$ and $\vec{p}$ and $p_\perp (= p\sin\theta)$ and $r_\perp (= r\sin\theta)$ are the components of $\vec{p}$ and $\vec{r}$ perpendicular to $\vec{r}$ and $\vec{p}$, respectively.

Now let us find the relation between angular momentum and torque for a particle. If we take the derivative of $\vec{L}$ with respect to time we have

$$\frac{d\vec{L}}{dt} = \frac{d}{dt}(\vec{r} \times \vec{p}) = \frac{d\vec{r}}{dt} \times \vec{p} + \vec{r} \times \frac{d\vec{p}}{dt}.$$

But

$$\frac{d\vec{r}}{dt} \times \vec{p} = \vec{v} \times m\vec{v} = m(\vec{v} \times \vec{v}) = 0,$$

since $\sin\theta = 0$ for this case. Thus

$$\frac{d\vec{L}}{dt} = \vec{r} \times \frac{d\vec{p}}{dt}.$$

If we let $\Sigma\vec{F}$ represent the resultant force on the particle, then in an inertial reference frame, $\Sigma\vec{F} = d\vec{p}/dt$ and

$$\vec{r} \times \Sigma\vec{F} = \vec{r} \times \frac{d\vec{p}}{dt} = \frac{d\vec{L}}{dt}.$$

But $\vec{r} \times \Sigma\vec{F} = \Sigma\vec{\tau}$ is the net torque on our particle. Hence

$$\Sigma\vec{\tau} = \frac{d\vec{L}}{dt}. \qquad \text{[particle, inertial frame]} \quad \textbf{(11–7)}$$

The time rate of change of angular momentum of a particle is equal to the net torque applied to it. Equation 11–7 is the rotational equivalent of Newton's

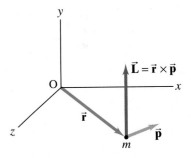

FIGURE 11–12 The angular momentum of a particle of mass m is given by $\vec{L} = \vec{r} \times \vec{p} = \vec{r} \times m\vec{v}$.

[†]Actually a pseudovector; see footnote in Section 10–2.

second law for a particle, written in its most general form. Equation 11–7 is valid only in an inertial frame since only then is it true that $\Sigma \vec{F} = d\vec{p}/dt$, which was used in the proof.

$\vec{p} = m\vec{v}$

FIGURE 11–13 The angular momentum of a particle of mass m rotating in a circle of radius $\vec{r}$ with velocity $\vec{v}$ is $\vec{L} = \vec{r} \times m\vec{v}$ (Example 11–7).

CONCEPTUAL EXAMPLE 11–7 **A particle's angular momentum.** What is the angular momentum of a particle of mass m moving with speed v in a circle of radius r in a counterclockwise direction?

RESPONSE The value of the angular momentum depends on the choice of the point O. Let us calculate $\vec{L}$ with respect to the center of the circle, Fig. 11–13. Then $\vec{r}$ is perpendicular to $\vec{p}$ so $L = |\vec{r} \times \vec{p}| = rmv$. By the right-hand rule, the direction of $\vec{L}$ is perpendicular to the plane of the circle, outward toward the viewer. Since $v = \omega r$ and $I = mr^2$ for a single particle rotating about an axis a distance r away, we can write

$$L = mvr = mr^2\omega = I\omega.$$

11–4 Angular Momentum and Torque for a System of Particles; General Motion

Relation Between Angular Momentum and Torque

Consider a system of n particles which have angular momenta $\vec{L}_1, \vec{L}_2, \ldots, \vec{L}_n$. The system could be anything from a rigid object to a loose assembly of particles whose positions are not fixed relative to each other. The total angular momentum $\vec{L}$ of the system is defined as the vector sum of the angular momenta of all the particles in the system:

$$\vec{L} = \sum_{i=1}^{n} \vec{L}_i. \tag{11–8}$$

The resultant torque acting on the system is the sum of the net torques acting on all the particles:

$$\vec{\tau}_{\text{net}} = \sum \vec{\tau}_i.$$

This sum includes (1) internal torques due to internal forces that particles of the system exert on other particles of the system, and (2) external torques due to forces exerted by objects outside our system. By Newton's third law, the force each particle exerts on another is equal and opposite (and acts along the same line as) the force that the second particle exerts on the first. Hence the sum of all internal torques adds to zero, and

$$\vec{\tau}_{\text{net}} = \sum_i \vec{\tau}_i = \sum \vec{\tau}_{\text{ext}}.$$

Now we take the time derivative of Eq. 11–8 and use Eq. 11–7 for each particle to obtain

$$\frac{d\vec{L}}{dt} = \sum_i \frac{d\vec{L}_i}{dt} = \sum \vec{\tau}_{\text{ext}}$$

or

NEWTON'S SECOND LAW (rotation, system of particles)

$$\frac{d\vec{L}}{dt} = \sum \vec{\tau}_{\text{ext}}. \qquad \text{[inertial reference frame]} \quad \textbf{(11–9a)}$$

This fundamental result states that the time rate of change of the total angular momentum of a system of particles (or a rigid object) equals the resultant external torque on the system. It is the rotational equivalent of Eq. 9–5, $d\vec{P}/dt = \Sigma\vec{F}_{\text{ext}}$ for translational motion. Note that $\vec{L}$ and $\Sigma\vec{\tau}$ must be calculated about the same origin O.

Equation 11–9a is valid when $\vec{L}$ and $\vec{\tau}_{\text{ext}}$ are calculated with reference to a point fixed in an inertial reference frame. (In the derivation, we used Eq. 11–7 which is valid only in this case.) It is also valid when $\vec{\tau}_{\text{ext}}$ and $\vec{L}$ are calculated about a point which is moving uniformly in an inertial reference frame since such a point can be considered the origin of a second inertial reference frame. It is *not* valid in general when $\vec{\tau}_{\text{ext}}$ and $\vec{L}$ are calculated about a point that is *accelerating*, except for one special (and very important) case—when that point is the center of mass (CM) of the system:

$$\frac{d\vec{L}_{\text{CM}}}{dt} = \sum \vec{\tau}_{\text{CM}}. \qquad \text{[even if accelerating]} \quad \textbf{(11–9b)}$$

Equation 11–9b is valid no matter how the CM moves, and $\Sigma\vec{\tau}_{\text{CM}}$ is the net external torque calculated about the center of mass. The derivation is in the optional subsection below.

It is because of the validity of Eq. 11–9b that we are justified in describing the general motion of a system of particles, as we did in Chapter 10, as *translational* motion of the center of mass plus *rotation* about the center of mass. Equations 11–9b plus 9–5 $\left(d\vec{P}_{\text{CM}}/dt = \Sigma\vec{F}_{\text{ext}}\right)$ provide the more general statement of this principle. (See also Section 9–8.)

*Derivation of $d\vec{L}_{\text{CM}}/dt = \Sigma\vec{\tau}_{\text{CM}}$

The proof of Eq. 11–9b is as follows. Let $\vec{r}_i$ be the position vector of the ith particle in an inertial reference frame, and $\vec{r}_{\text{CM}}$ be the position vector of the center of mass of the system in this reference frame. The position of the ith particle with respect to the CM is $\vec{r}_i^*$ where (see Fig. 11–14)

$$\vec{r}_i = \vec{r}_{\text{CM}} + \vec{r}_i^*.$$

If we multiply each term by m_i and take the derivative of this equation, we can write

$$\vec{p}_i = m_i\frac{d\vec{r}_i}{dt} = m_i\frac{d}{dt}\left(\vec{r}_i^* + \vec{r}_{\text{CM}}\right) = m_i\vec{v}_i^* + m_i\vec{v}_{\text{CM}} = \vec{p}_i^* + m_i\vec{v}_{\text{CM}}.$$

The angular momentum with respect to the CM is

$$\vec{L}_{\text{CM}} = \sum_i\left(\vec{r}_i^* \times \vec{p}_i^*\right) = \sum_i\vec{r}_i^* \times \left(\vec{p}_i - m_i\vec{v}_{\text{CM}}\right).$$

Then, taking the time derivative, we have

$$\frac{d\vec{L}_{\text{CM}}}{dt} = \sum_i\left(\frac{d\vec{r}_i^*}{dt} \times \vec{p}_i^*\right) + \sum_i\left(\vec{r}_i^* \times \frac{d\vec{p}_i^*}{dt}\right).$$

The first term on the right is $\vec{v}_i^* \times m\vec{v}_i^*$ and equals zero because $\vec{v}_i^*$ is parallel to itself ($\sin\theta = 0$). Thus

$$\frac{d\vec{L}_{\text{CM}}}{dt} = \sum_i\vec{r}_i^* \times \frac{d}{dt}\left(\vec{p}_i - m_i\vec{v}_{\text{CM}}\right)$$

$$= \sum_i\vec{r}_i^* \times \frac{d\vec{p}_i}{dt} - \left(\sum_i m_i\vec{r}_i^*\right) \times \frac{d\vec{v}_{\text{CM}}}{dt}.$$

The second term on the right is zero since, by Eq. 9–12, $\Sigma m_i\vec{r}_i^* = M\vec{r}_{\text{CM}}^*$, and $\vec{r}_{\text{CM}}^* = 0$ by definition (the position of the CM is at the origin of the CM reference frame). Furthermore, by Newton's second law we have

$$\frac{d\vec{p}_i}{dt} = \vec{F}_i,$$

where $\vec{F}_i$ is the net force on m_i. (Note that $d\vec{p}_i^*/dt \neq \vec{F}_i$ because the CM may be accelerating and Newton's second law does not hold in a noninertial reference frame.) Consequently

$$\frac{d\vec{L}_{\text{CM}}}{dt} = \sum_i\vec{r}_i^* \times \vec{F}_i = \sum_i\left(\vec{\tau}_i\right)_{\text{CM}} = \Sigma\vec{\tau}_{\text{CM}},$$

where $\Sigma\vec{\tau}_{\text{CM}}$ is the resultant external torque on the entire system calculated about the CM. (By Newton's third law, the sum over all the $\vec{\tau}_i$ eliminates the net torque due to internal forces, as we saw on p. 292.) This last equation is Eq. 11–9b, and this concludes its proof.

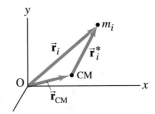

FIGURE 11–14 The position of m_i in the inertial frame is $\vec{r}_i$; with regard to the CM (which could be accelerating) it is $\vec{r}_i^*$, where $\vec{r}_i = \vec{r}_i^* + \vec{r}_{\text{CM}}$ and $\vec{r}_{\text{CM}}$ is the position of the CM in the inertial frame.

Summary

To summarize, the relation

$$\sum \vec{\tau}_{\text{ext}} = \frac{d\vec{L}}{dt}$$

is valid *only* when $\vec{\tau}_{\text{ext}}$ and $\vec{L}$ are calculated with respect to either (1) the origin of an inertial reference frame or (2) the center of mass of a system of particles (or of a rigid object).

11–5 Angular Momentum and Torque for a Rigid Object

Let us now consider the rotation of a rigid object about an axis that has a fixed direction in space, using the general principles just developed.

Let us calculate the component of angular momentum along the rotation axis of the rotating object. We will call this component L_ω since the angular velocity $\vec{\omega}$ points along the rotation axis. For each particle of the object,

$$\vec{L}_i = \vec{r}_i \times \vec{p}_i.$$

Let ϕ be the angle between $\vec{L}_i$ and the rotation axis. (See Fig. 11–15; ϕ is *not* the angle between $\vec{r}_i$ and $\vec{p}_i$, which is 90°). Then the component of $\vec{L}_i$ along the rotation axis is

$$L_{i\omega} = r_i p_i \cos\phi = m_i v_i r_i \cos\phi,$$

where m_i is the mass and v_i the velocity of the ith particle. Now $v_i = R_i \omega$ where ω is the angular velocity of the object and R_i is the perpendicular distance of m_i from the axis of rotation. Furthermore, $R_i = r_i \cos\phi$, as can be seen in Fig. 11–15, so

$$L_{i\omega} = m_i v_i (r_i \cos\phi) = m_i R_i^2 \omega.$$

We sum over all the particles to obtain

$$L_\omega = \sum_i L_{i\omega} = \left(\sum_i m_i R_i^2 \right)\omega.$$

But $\sum m_i R_i^2$ is the moment of inertia I of the object about the axis of rotation. Therefore the component of the total angular momentum along the rotation axis is given by

$$L_\omega = I\omega. \tag{11–10}$$

Note that we would obtain Eq. 11–10 no matter where we choose the point O (for measuring $\vec{r}_i$) as long as it is on the axis of rotation. Equation 11–10 is the same as Eq. 11–1, which we have now proved from the general definition of angular momentum.

If the object rotates about a symmetry axis through the center of mass, then L_ω is the only component of $\vec{L}$, as we now show. For each point on one side of the axis there will be a corresponding point on the opposite side. We can see from Fig. 11–15 that each $\vec{L}_i$ has a component parallel to the axis ($L_{i\omega}$) and a component perpendicular to the axis. The components parallel to the axis add together for each pair of opposite points, but the components perpendicular to the axis for opposite points will have the same magnitude but opposite direction and so will cancel. Hence, for an object rotating about a symmetry axis, the angular momentum vector is parallel to the axis and we can write

$$\vec{L} = I\vec{\omega}, \quad \text{[rotation axis = symmetry axis, through CM]} \tag{11–11}$$

where $\vec{L}$ is measured relative to the center of mass.

The general relation between angular momentum and torque is Eq. 11–9:

$$\sum \vec{\tau} = \frac{d\vec{L}}{dt}$$

where $\sum \vec{\tau}$ and $\vec{L}$ are calculated either about (1) the origin of an inertial reference frame, or (2) the center of mass of the system. This is a vector relation, and must

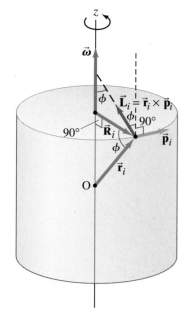

FIGURE 11–15
Calculating $L_\omega = L_z = \Sigma L_{iz}$.
Note that $\vec{L}_i$ is perpendicular to $\vec{r}_i$, and $\vec{R}_i$ is perpendicular to the z axis, so the three angles marked ϕ are equal.

therefore be valid for each component. Hence, for a rigid object, the component along the rotation axis is

$$\Sigma\tau_{axis} = \frac{dL_\omega}{dt} = \frac{d}{dt}(I\omega) = I\frac{d\omega}{dt} = I\alpha,$$

which is valid for a rigid object rotating about an axis fixed relative to the object; also this axis must be either (1) fixed in an inertial system or (2) passing through the CM of the object. This is equivalent to Eqs. 10–14 and 10–15, which we now see are special cases of Eq. 11–9, $\Sigma\vec{\tau} = d\vec{L}/dt$.

EXAMPLE 11–8 **Atwood's machine.** An *Atwood machine* consists of two masses, m_A and m_B, which are connected by an inelastic cord of negligible mass that passes over a pulley, Fig. 11–16. If the pulley has radius R_0 and moment of inertia I about its axle, determine the acceleration of the masses m_A and m_B, and compare to the situation where the moment of inertia of the pulley is ignored.

APPROACH We first determine the angular momentum of the system, and then apply Newton's second law, $\tau = dL/dt$.

SOLUTION The angular momentum is calculated about an axis along the axle through the center O of the pulley. The pulley has angular momentum $I\omega$, where $\omega = v/R_0$ and v is the velocity of m_A and m_B at any instant. The angular momentum of m_A is $R_0 m_A v$ and that of m_B is $R_0 m_B v$. The total angular momentum is

$$L = (m_A + m_B)vR_0 + I\frac{v}{R_0}.$$

The external torque on the system, calculated about the axis O (taking clockwise as positive), is

$$\tau = m_B g R_0 - m_A g R_0.$$

(The force on the pulley exerted by the support on its axle gives rise to no torque because the lever arm is zero.) We apply Eq. 11–9a:

$$\tau = \frac{dL}{dt}$$

$$(m_B - m_A)gR_0 = (m_A + m_B)R_0\frac{dv}{dt} + \frac{I}{R_0}\frac{dv}{dt}.$$

Solving for $a = dv/dt$, we get

$$a = \frac{dv}{dt} = \frac{(m_B - m_A)g}{(m_A + m_B) + I/R_0^2}.$$

If we were to ignore I, $a = (m_B - m_A)g/(m_B + m_A)$ and we see that the effect of the moment of inertia of the pulley is to slow down the system. This is just what we would expect.

FIGURE 11–16 Atwood's machine, Example 11–8. We also discussed this in Example 4–13.

CONCEPTUAL EXAMPLE 11–9 **Bicycle wheel.** Suppose you are holding a bicycle wheel by a handle connected to its axle as in Fig. 11–17a. The wheel is spinning rapidly so its angular momentum $\vec{L}$ points horizontally as shown. Now you suddenly try to tilt the axle upward as shown by the dashed line in Fig. 11–17a (so the CM moves vertically). You expect the wheel to go up (and it would if it weren't rotating), but it unexpectedly swerves to the right! Explain.

RESPONSE To explain this seemingly odd behavior—you may need to do it to believe it—we only need to use the relation $\vec{\tau}_{net} = d\vec{L}/dt$. In the short time Δt, you exert a net torque (about an axis through your wrist) that points along the x axis perpendicular to $\vec{L}$. Thus the change in $\vec{L}$ is

$$\Delta\vec{L} \approx \vec{\tau}_{net}\Delta t;$$

so $\Delta\vec{L}$ must also point (approximately) along the x axis, since $\vec{\tau}_{net}$ does (Fig. 11–17b). Thus the new angular momentum, $\vec{L} + \Delta\vec{L}$, points to the right, looking along the axis of the wheel, as shown in Fig. 11–17b. Since the angular momentum is directed along the axle of the wheel, we see that the axle, which now is along $\vec{L} + \Delta\vec{L}$, must move sideways to the right, which is what we observe.

FIGURE 11–17 When you try to tilt a rotating bicycle wheel vertically upward, it swerves to the side instead.

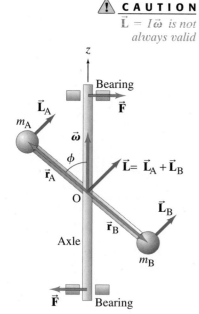

FIGURE 11–18 In this system $\vec{L}$ and $\vec{\omega}$ are not parallel. This is an example of rotational imbalance.

Although Eq. 11–11, $\vec{L} = I\vec{\omega}$, is often very useful, it is not valid in general if the rotation axis is not along a symmetry axis through the center of mass. Nonetheless, it can be shown that every rigid object, no matter what its shape, has three "principal axes" about which Eq. 11–11 is valid (we will not go into the details here). As an example of a case where Eq. 11–11 is not valid, consider the nonsymmetrical object shown in Fig. 11–18. It consists of two equal masses, m_A and m_B, attached to the ends of a rigid (massless) rod which makes an angle ϕ with the axis of rotation. We calculate the angular momentum about the CM at point O. At the moment shown, m_A is coming toward the viewer, and m_B is moving away, so $\vec{L}_A = \vec{r}_A \times \vec{p}_A$ and $\vec{L}_B = \vec{r}_B \times \vec{p}_B$ are as shown. The total angular momentum is $\vec{L} = \vec{L}_A + \vec{L}_B$, which is clearly *not* along $\vec{\omega}$ if $\phi \neq 90°$.

*Rotational Imbalance

Let us go one step further with the system shown in Fig. 11–18, since it is a fine illustration of $\Sigma\vec{\tau} = d\vec{L}/dt$. If the system rotates with constant angular velocity, ω, the magnitude of $\vec{L}$ will not change, but its direction will. As the rod and two masses rotate about the z axis, $\vec{L}$ also rotates about the axis. At the moment shown in Fig. 11–18, $\vec{L}$ is in the plane of the paper. A time dt later, when the rod has rotated through an angle $d\theta = \omega\,dt$, $\vec{L}$ will also have rotated through an angle $d\theta$ (it remains perpendicular to the rod). $\vec{L}$ will then have a component pointing into the page. Thus $d\vec{L}$ points into the page and so must $d\vec{L}/dt$. Because

$$\sum \vec{\tau} = \frac{d\vec{L}}{dt},$$

we see that a net torque, directed into the page at the moment shown, must be applied to the axle on which the rod is mounted. The torque is supplied by bearings (or other constraint) at the ends of the axle. The forces $\vec{F}$ exerted by the bearings on the axle are shown in Fig. 11–18. The direction of each force $\vec{F}$ rotates as the system does, always being in the plane of $\vec{L}$ and $\vec{\omega}$ for this system. If the torque due to these forces were not present, the system would not rotate about the fixed axis as desired.

The axle tends to move in the direction of $\vec{F}$ and thus tends to wobble as it rotates. This has many practical applications, such as the vibrations felt in a car whose wheels are not balanced. Consider an automobile wheel that is symmetrical except for an extra mass m_A on one rim and an equal mass m_B opposite it on the other rim, as shown in Fig. 11–19. Because of the nonsymmetry of m_A and m_B, the wheel bearings would have to exert a force perpendicular to the axle at all times simply to keep the wheel rotating, just as in Fig. 11–18. The bearings would wear excessively and the wobble of the wheel would be felt by occupants of the car. When the wheels are balanced, they rotate smoothly without wobble. This is why "dynamic balancing" of automobile wheels and tires is important. The wheel of Fig. 11–19 would balance *statically* just fine. If equal masses m_C and m_D are added symmetrically, below m_A and above m_B, the wheel will be balanced dynamically as well ($\vec{L}$ will be parallel to $\vec{\omega}$, and $\vec{\tau}_{ext} = 0$).

🏃 **PHYSICS APPLIED**

Automobile wheel balancing

FIGURE 11–19 Unbalanced automobile wheel.

EXAMPLE 11–10 Torque on imbalanced system. Determine the magnitude of the net torque τ_{net} needed to keep the system turning in Fig. 11–18.

APPROACH Figure 11–20 is a view of the angular momentum vector, looking down the rotation axis (z axis) of the object depicted in Fig. 11–18, as it rotates. $L\cos\phi$ is the component of $\vec{L}$ perpendicular to the axle (points to the right in Fig. 11–18). We find dL from Fig. 11–20 and use $\tau_{net} = dL/dt$.

SOLUTION In a time dt, $\vec{L}$ changes by an amount (Fig. 11–20 and Eq. 10–2b)

$$dL = (L \cos \phi) \, d\theta = L \cos \phi \, \omega \, dt,$$

where $\omega = d\theta/dt$. Hence

$$\tau_{\text{net}} = \frac{dL}{dt} = \omega L \cos \phi.$$

Now $L = L_A + L_B = r_A m_A v_A + r_B m_B v_B = r_A m_A(\omega r_A \sin \phi) + r_B m_B(\omega r_B \sin \phi) = (m_A r_A^2 + m_B r_B^2)\omega \sin \phi$. Since $I = (m_A r_A^2 + m_B r_B^2) \sin^2 \phi$ is the moment of inertia about the axis of rotation, then $L = I\omega/\sin \phi$. So

$$\tau_{\text{net}} = \omega L \cos \phi = (m_A r_A^2 + m_B r_B^2)\omega^2 \sin \phi \cos \phi = I\omega^2/\tan \phi.$$

FIGURE 11–20 Angular momentum vector looking down along the rotation axis of the system of Fig. 11–18 as it rotates during a time dt.

The situation of Fig. 11–18 illustrates the usefulness of the vector nature of torque and angular momentum. If we had considered only the components of angular momentum and torque along the rotation axis, we could not have calculated the torque due to the bearings (since the forces $\vec{F}$ act at the axle and hence produce no torque along that axis). By using the concept of vector angular momentum we have a far more powerful technique for understanding and for attacking problems.

11–6 Conservation of Angular Momentum

In Chapter 9 we saw that the most general form of Newton's second law for the translational motion of a particle or system of particles is

$$\sum \vec{F}_{\text{ext}} = \frac{d\vec{P}}{dt},$$

where $\vec{P}$ is the (linear) momentum, defined as $m\vec{v}$ for a particle, or $M\vec{v}_{\text{CM}}$ for a system of particles of total mass M whose CM moves with velocity $\vec{v}_{\text{CM}}$, and $\sum \vec{F}_{\text{ext}}$ is the net external force acting on the particle or system. This relation is valid only in an inertial reference frame.

In this Chapter, we have found a similar relation to describe the general rotation of a system of particles (including rigid objects):

$$\sum \vec{\tau} = \frac{d\vec{L}}{dt},$$

where $\sum \vec{\tau}$ is the net external torque acting on the system, and $\vec{L}$ is the total angular momentum. This relation is valid when $\sum \vec{\tau}$ and $\vec{L}$ are calculated about a point fixed in an inertial reference frame, or about the CM of the system.

For translational motion, if the net force on the system is zero, $d\vec{P}/dt = 0$, so the total linear momentum of the system remains constant. This is the law of conservation of linear momentum. For rotational motion, if the net torque on the system is zero, then

$$\frac{d\vec{L}}{dt} = 0 \quad \text{and} \quad \vec{L} = \text{constant.} \qquad \left[\sum \vec{\tau} = 0\right] \quad \textbf{(11–12)}$$

In words:

The total angular momentum of a system remains constant if the net external torque acting on the system is zero.

CONSERVATION OF ANGULAR MOMENTUM

This is the **law of conservation of angular momentum** in full vector form. It ranks with the laws of conservation of energy and linear momentum (and others to be discussed later) as one of the great laws of physics. In Section 11–1 we saw some Examples of this important law applied to the special case of a rigid object rotating about a fixed axis. Here we have it in general form. We use it now in interesting Examples.

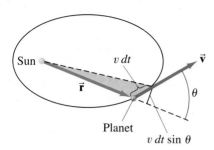

FIGURE 11–21 Kepler's second law of planetary motion (Example 11–11).

EXAMPLE 11–11 **Kepler's second law derived.** Kepler's second law states that each planet moves so that a line from the Sun to the planet sweeps out equal areas in equal times (Section 6–5). Use conservation of angular momentum to show this.

APPROACH We determine the angular momentum of a planet in terms of the area swept out with the help of Fig. 11–21.

SOLUTION The planet moves in an ellipse as shown in Fig. 11–21. In a time dt, the planet moves a distance $v\,dt$ and sweeps out an area dA equal to the area of a triangle of base r and height $v\,dt\sin\theta$ (shown exaggerated in Fig. 11–21). Hence

$$dA = \tfrac{1}{2}(r)(v\,dt\sin\theta)$$

and

$$\frac{dA}{dt} = \tfrac{1}{2}rv\sin\theta.$$

The magnitude of the angular momentum $\vec{\mathbf{L}}$ about the Sun is

$$L = |\vec{\mathbf{r}} \times m\vec{\mathbf{v}}| = mrv\sin\theta,$$

so

$$\frac{dA}{dt} = \frac{1}{2m}L.$$

But $L = \text{constant}$, since the gravitational force $\vec{\mathbf{F}}$ is directed toward the Sun so the torque it produces is zero (we ignore the pull of the other planets). Hence $dA/dt = \text{constant}$, which is what we set out to prove.

FIGURE 11–22 Bullet strikes and becomes embedded in cylinder at its edge (Example 11–12).

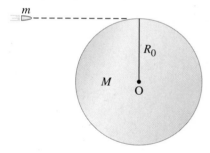

EXAMPLE 11–12 **Bullet strikes cylinder edge.** A bullet of mass m moving with velocity v strikes and becomes embedded at the edge of a cylinder of mass M and radius R_0, as shown in Fig. 11–22. The cylinder, initially at rest, begins to rotate about its symmetry axis, which remains fixed in position. Assuming no frictional torque, what is the angular velocity of the cylinder after this collision? Is kinetic energy conserved?

APPROACH We take as our system the bullet and cylinder, on which there is no net external torque. Thus we can use conservation of angular momentum, and we calculate all angular momenta about the center O of the cylinder.

SOLUTION Initially, because the cylinder is at rest, the total angular momentum about O is solely that of the bullet:

$$L = |\vec{\mathbf{r}} \times \vec{\mathbf{p}}| = R_0 m v,$$

since R_0 is the perpendicular distance of $\vec{\mathbf{p}}$ from O. After the collision, the cylinder $\left(I_{\text{cyl}} = \tfrac{1}{2}MR_0^2\right)$ rotates with the bullet $\left(I_{\text{b}} = mR_0^2\right)$ embedded in it at angular velocity ω:

$$L = I\omega = \left(I_{\text{cyl}} + mR_0^2\right)\omega = \left(\tfrac{1}{2}M + m\right)R_0^2\omega.$$

Hence, because angular momentum is conserved, we find that ω is

$$\omega = \frac{L}{\left(\tfrac{1}{2}M + m\right)R_0^2} = \frac{mvR_0}{\left(\tfrac{1}{2}M + m\right)R_0^2} = \frac{mv}{\left(\tfrac{1}{2}M + m\right)R_0}.$$

Angular momentum is conserved in this collision, but kinetic energy is not:

$$\begin{aligned}
K_{\text{f}} - K_{\text{i}} &= \tfrac{1}{2}I_{\text{cyl}}\omega^2 + \tfrac{1}{2}\left(mR_0^2\right)\omega^2 - \tfrac{1}{2}mv^2 \\
&= \tfrac{1}{2}\left(\tfrac{1}{2}MR_0^2\right)\omega^2 + \tfrac{1}{2}\left(mR_0^2\right)\omega^2 - \tfrac{1}{2}mv^2 \\
&= \tfrac{1}{2}\left(\tfrac{1}{2}M + m\right)\left(\frac{mv}{\tfrac{1}{2}M + m}\right)^2 - \tfrac{1}{2}mv^2 \\
&= -\frac{mM}{2M + 4m}v^2,
\end{aligned}$$

which is less than zero. Hence $K_{\text{f}} < K_{\text{i}}$. This energy is transformed to thermal energy as a result of the inelastic collision.

*11–7 The Spinning Top and Gyroscope

The motion of a rapidly spinning top, or a gyroscope, is an interesting example of rotational motion and of the use of the vector equation

$$\sum \vec{\tau} = \frac{d\vec{L}}{dt}.$$

PHYSICS APPLIED

A spinning top

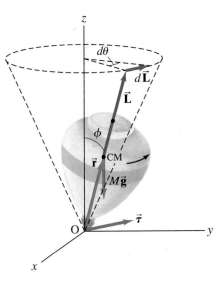

Consider a symmetrical top of mass M spinning rapidly about its symmetry axis, as in Fig. 11–23. The top is balanced on its tip at point O in an inertial reference frame. If the axis of the top makes an angle ϕ to the vertical (z axis), when the top is carefully released its axis will move, sweeping out a cone about the vertical as shown by the dashed lines in Fig. 11–23. This type of motion, in which a torque produces a change in the direction of the rotation axis, is called **precession**. The rate at which the rotation axis moves about the vertical (z) axis is called the angular velocity of precession, Ω (capital Greek omega). Let us now try to understand the reasons for this motion, and calculate Ω.

If the top were not spinning, it would immediately fall to the ground when released due to the pull of gravity. The apparent mystery of a top is that when it is spinning, it does not immediately fall to the ground but instead precesses—it moves slowly sideways. But this is not really so mysterious if we examine it from the point of view of angular momentum and torque, which we calculate about the point O. When the top is spinning with angular velocity ω about its symmetry axis, it has an angular momentum $\vec{L}$ directed along its axis, as shown in Fig. 11–23. (There is also angular momentum due to the precessional motion, so that the total $\vec{L}$ is not exactly along the axis of the top; but if $\Omega \ll \omega$, which is usually the case, we can ignore this.) To change the angular momentum, a torque is required. If no torque were applied to the top, $\vec{L}$ would remain constant in magnitude and direction; the top would neither fall nor precess. But the slightest tip to the side results in a net torque about O, equal to $\vec{\tau}_{net} = \vec{r} \times M\vec{g}$, where $\vec{r}$ is the position vector of the top's center of mass with respect to O, and M is the mass of the top. The direction of $\vec{\tau}_{net}$ is perpendicular to both $\vec{r}$ and $M\vec{g}$ and by the right-hand rule is, as shown in Fig. 11–23, in the horizontal (xy) plane. The change in $\vec{L}$ in a time dt is

$$d\vec{L} = \vec{\tau}_{net}\, dt,$$

which is perpendicular to $\vec{L}$ and horizontal (parallel to $\vec{\tau}_{net}$), as shown in Fig. 11–23. Since $d\vec{L}$ is perpendicular to $\vec{L}$, the magnitude of $\vec{L}$ does not change. Only the direction of $\vec{L}$ changes. Since $\vec{L}$ points along the axis of the top, we see that this axis moves to the right in Fig. 11–23. That is, the upper end of the top's axis moves in a horizontal direction perpendicular to $\vec{L}$. This explains why the top precesses rather than falls. The vector $\vec{L}$ and the top's axis move together in a horizontal circle. As they do so, $\vec{\tau}_{net}$ and $d\vec{L}$ rotate as well so as to be horizontal and perpendicular to $\vec{L}$.

To determine Ω, we see from Fig. 11–23 that the angle $d\theta$ (which is in a horizontal plane) is related to dL by

$$dL = L \sin \phi\, d\theta,$$

since $\vec{L}$ makes an angle ϕ to the z axis. The angular velocity of precession is $\Omega = d\theta/dt$, which becomes (since $d\theta = dL/L \sin \phi$)

$$\Omega = \frac{1}{L \sin \phi} \frac{dL}{dt} = \frac{\tau}{L \sin \phi}. \qquad \text{[spinning top]} \quad \textbf{(11–13a)}$$

But $\tau_{net} = |\vec{r} \times M\vec{g}| = rMg \sin \phi$ [because $\sin(\pi - \phi) = \sin \phi$] so we can also write

$$\Omega = \frac{Mgr}{L}. \qquad \text{[spinning top]} \quad \textbf{(11–13b)}$$

Thus the rate of precession does not depend on the angle ϕ; but it is inversely proportional to the top's angular momentum. The faster the top spins, the greater L is and the slower the top precesses.

FIGURE 11–23 Spinning top.

FIGURE 11–24 A toy gyroscope.

From Eq. 11–1 (or Eq. 11–11) we can write $L = I\omega$, where I and ω are the moment of inertia and angular velocity of the spinning top about its spin axis. Then Eq. 11–13b for the top's precession angular velocity becomes

$$\Omega = \frac{Mgr}{I\omega}.$$ **(11–13c)**

Equations 11–13 apply also to a toy gyroscope, which consists of a rapidly spinning wheel mounted on an axle (Fig. 11–24). One end of the axle rests on a support. The other end of the axle is free and will precess like a top if its "spin" angular velocity ω is large compared to the precession rate ($\omega \gg \Omega$). As ω decreases due to friction and air resistance, the gyroscope will begin to fall, just as does a top.

*11–8 Rotating Frames of Reference; Inertial Forces

Inertial and Noninertial Reference Frames

Up to now, we have examined the motion of objects, including circular and rotational motion, from the outside, as observers fixed on the Earth. Sometimes it is convenient to place ourselves (in theory, if not physically) into a reference frame that is rotating. Let us examine the motion of objects from the point of view, or frame of reference, of persons seated on a rotating platform such as a merry-go-round. It looks to them as if the rest of the world is going around *them*. But let us focus attention on what they observe when they place a tennis ball on the floor of the rotating platform, which we assume is frictionless. If they put the ball down gently, without giving it any push, they will observe that it accelerates from rest and moves outward as shown in Fig. 11–25a. According to Newton's first law, an object initially at rest should stay at rest if no net force acts on it. But, according to the observers on the rotating platform, the ball starts moving even though there is no net force acting on it. To observers on the ground this is all very clear: the ball has an initial velocity when it is released (because the platform is moving), and it simply continues moving in a straight-line path as shown in Fig. 11–25b, in accordance with Newton's first law.

But what shall we do about the frame of reference of the observers on the rotating platform? Since the ball moves without any net force on it, Newton's first law, the law of inertia, does not hold in this rotating frame of reference. For this reason, such a frame is called a **noninertial reference frame**. An **inertial reference frame** (as we discussed in Chapter 4) is one in which the law of inertia—Newton's first law—does hold, and so do Newton's second and third laws. In a noninertial reference frame, such as our rotating platform, Newton's second law also does not hold. For instance in the situation described above, there is no net force on the ball; yet, with respect to the rotating platform, the ball accelerates.

FIGURE 11–25 Path of a ball released on a rotating merry-go-round (a) in the reference frame of the merry-go-round, and (b) in a reference frame fixed on the ground.

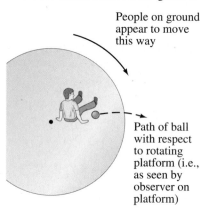

People on ground appear to move this way

Path of ball with respect to rotating platform (i.e., as seen by observer on platform)

(a) Rotating reference frame

Path of ball with respect to ground (i.e., as seen by observers on the ground)

Platform rotating counterclockwise

(b) Inertial reference frame

Fictitious (Inertial) Forces

Because Newton's laws do not hold when observations are made with respect to a rotating frame of reference, calculation of motion can be complicated. However, we can still make use of Newton's laws in such a reference frame if we make use of a trick. The ball on the rotating platform of Fig. 11–25a flies outward when released (even though no force is actually acting on it). So the trick we use is to write down the equation $\Sigma F = ma$ as if a force equal to mv^2/r (or $m\omega^2 r$) were acting radially outward on the object in addition to any other forces that may be acting. This extra force, which might be designated as "centrifugal force" since it *seems* to act outward, is called a **fictitious force** or **pseudoforce**. It is a pseudoforce ("pseudo" means "false") because there is no object that exerts this force. Furthermore, when viewed from an inertial reference frame, the effect doesn't exist at all. We have made up this pseudoforce so that we can make calculations in a noninertial frame using Newton's second law, $\Sigma F = ma$. Thus the observer in the noninertial frame of Fig. 11–25a uses Newton's second law for the ball's outward motion by assuming that a force equal to mv^2/r acts on it. Such pseudoforces are also called **inertial forces** since they arise only because the reference frame is not an inertial one.

The Earth itself is rotating on its axis. Thus, strictly speaking, Newton's laws are not valid on the Earth. However, the effect of the Earth's rotation is usually so small that it can be ignored, although it does influence the movement of large air masses and ocean currents. Because of the Earth's rotation, the material of the Earth is concentrated slightly more at the equator. The Earth is thus not a perfect sphere but is slightly fatter at the equator than at the poles.

*11–9 The Coriolis Effect

In a reference frame that rotates at a constant angular speed ω (relative to an inertial frame), there exists another pseudoforce known as the *Coriolis force*. It appears to act on an object in a rotating reference frame only if the object is moving relative to that rotating reference frame, and it acts to deflect the object sideways. It, too, is an effect of the rotating reference frame being noninertial and hence is referred to as an *inertial force*. It too affects the weather.

To see how the Coriolis force arises, consider two people, A and B, at rest on a platform rotating with angular speed ω, as shown in Fig. 11–26a. They are situated at distances r_A and r_B from the axis of rotation (at O). The woman at A throws a ball with a horizontal velocity $\vec{v}$ (in her reference frame) radially outward toward the man at B on the outer edge of the platform. In Fig. 11–26a, we view the situation from an inertial reference frame. The ball initially has not only the velocity $\vec{v}$ radially outward, but also a tangential velocity $\vec{v}_A$ due to the rotation of the platform. Now Eq. 10–4 tells us that $v_A = r_A \omega$, where r_A is the woman's radial distance from the axis of rotation at O. If the man at B had this same velocity v_A, the ball would reach him perfectly. But his speed is $v_B = r_B \omega$, which is greater than v_A because $r_B > r_A$. Thus, when the ball reaches the outer edge of the platform, it passes a point that the man at B has already gone by because his speed in that direction is greater than the ball's. So the ball passes behind him.

Figure 11–26b shows the situation as seen from the rotating platform as frame of reference. Both A and B are at rest, and the ball is thrown with velocity $\vec{v}$ toward B, but the ball deflects to the right as shown and passes behind B as previously described. This is not a centrifugal-force effect, for the latter acts radially outward. Instead, this effect acts sideways, perpendicular to $\vec{v}$, and is called a **Coriolis acceleration**; it is said to be due to the Coriolis force, which is a fictitious inertial force. Its explanation as seen from an inertial system was given above: it is an effect of being in a rotating system, wherein a point farther from the rotation axis has a higher linear speed. On the other hand, when viewed from the rotating system, we can describe the motion using Newton's second law, $\Sigma \vec{F} = m\vec{a}$, if we add a "pseudoforce" term corresponding to this Coriolis effect.

Let us determine the magnitude of the Coriolis acceleration for the simple case described above. (We assume v is large and distances short, so we can ignore gravity.) We do the calculation from the inertial reference frame (Fig. 11–26a). The ball moves radially outward a distance $r_B - r_A$ at speed v in a short time t given by

$$r_B - r_A = vt.$$

During this time, the ball moves to the side a distance s_A given by

$$s_A = v_A t.$$

The man at B, in this time t, moves a distance

$$s_B = v_B t.$$

The ball therefore passes behind him a distance s (Fig. 11–26a) given by

$$s = s_B - s_A = (v_B - v_A)t.$$

We saw earlier that $v_A = r_A \omega$ and $v_B = r_B \omega$, so

$$s = (r_B - r_A)\omega t.$$

We substitute $r_B - r_A = vt$ (see above) and get

$$s = \omega v t^2. \tag{11–14}$$

This same s equals the sideways displacement as seen from the noninertial rotating system (Fig. 11–26b).

(a) Inertial reference frame

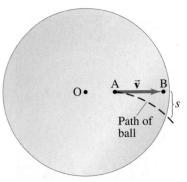

(b) Rotating reference frame

FIGURE 11–26 The origin of the Coriolis effect. Looking down on a rotating platform, (a) as seen from a nonrotating inertial reference frame, and (b) as seen from the rotating platform as frame of reference.

(a)

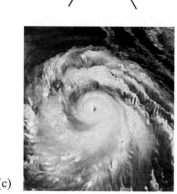

(b)

(c)

We see immediately that Eq. 11–14 corresponds to motion at constant acceleration. For as we saw in Chapter 2 (Eq. 2–12b), $y = \frac{1}{2}at^2$ for a constant acceleration (with zero initial velocity in the y direction). Thus, if we write Eq. 11–14 in the form $s = \frac{1}{2}a_{\text{Cor}}t^2$, we see that the Coriolis acceleration a_{Cor} is

$$a_{\text{Cor}} = 2\omega v. \tag{11–15}$$

This relation is valid for any velocity in the plane of rotation perpendicular to the axis of rotation† (in Fig. 11–26, the axis through point O perpendicular to the page).

Because the Earth rotates, the Coriolis effect has some interesting manifestations on the Earth. It affects the movement of air masses and thus has an influence on weather. In the absence of the Coriolis effect, air would rush directly into a region of low pressure, as shown in Fig. 11–27a. But because of the Coriolis effect, the winds are deflected to the right in the Northern Hemisphere (Fig. 11–27b), since the Earth rotates from west to east. So there tends to be a counterclockwise wind pattern around a low-pressure area. The reverse is true in the Southern Hemisphere. Thus cyclones rotate counterclockwise in the Northern Hemisphere and clockwise in the Southern Hemisphere. The same effect explains the easterly trade winds near the equator: any winds heading south toward the equator will be deflected toward the west (that is, as if coming from the east).

The Coriolis effect also acts on a falling object. An object released from the top of a high tower will not hit the ground directly below the release point, but will be deflected slightly to the east. Viewed from an inertial frame, this happens because the top of the tower revolves with the Earth at a slightly higher speed than the bottom of the tower.

---

†The Coriolis acceleration can be written in general in terms of the vector cross product as $\vec{a}_{\text{Cor}} = -2\vec{\omega} \times \vec{v}$ where $\vec{\omega}$ has direction along the rotation axis; its magnitude is $a_{\text{Cor}} = 2wv_\perp$ where $v_\perp$ is the component of velocity perpendicular to the rotation axis.

Summary

The **angular momentum** $\vec{L}$ of a rigid object rotating about a fixed axis is given by

$$L = I\omega. \tag{11–1}$$

Newton's second law, in terms of angular momentum, is

$$\Sigma\tau = \frac{dL}{dt}. \tag{11–2}$$

If the net torque on an object is zero, $dL/dt = 0$, so $L =$ constant. This is the **law of conservation of angular momentum**.

The **vector product** or **cross product** of two vectors $\vec{A}$ and $\vec{B}$ is another vector $\vec{C} = \vec{A} \times \vec{B}$ whose magnitude is $AB\sin\theta$ and whose direction is perpendicular to both $\vec{A}$ and $\vec{B}$ in the sense of the right-hand rule.

The **torque** $\vec{\tau}$ due to a force $\vec{F}$ is a vector quantity and is always calculated about some point O (the origin of a coordinate system) as follows:

$$\vec{\tau} = \vec{r} \times \vec{F}, \tag{11–5}$$

where $\vec{r}$ is the position vector of the point at which the force $\vec{F}$ acts.

Angular momentum is also a vector. For a particle having momentum $\vec{p} = m\vec{v}$, the angular momentum $\vec{L}$ about some point O is

$$\vec{L} = \vec{r} \times \vec{p}, \tag{11–6}$$

where $\vec{r}$ is the position vector of the particle relative to the point O

at any instant. The net torque $\Sigma\vec{\tau}$ on a particle is related to its angular momentum by

$$\Sigma\vec{\tau} = \frac{d\vec{L}}{dt}. \tag{11–7}$$

For a system of particles, the total angular momentum $\vec{L} = \Sigma\vec{L}_i$. The total angular momentum of the system is related to the total net torque $\Sigma\vec{\tau}$ on the system by

$$\Sigma\vec{\tau} = \frac{d\vec{L}}{dt}. \tag{11–9}$$

This last relation is the vector rotational equivalent of Newton's second law. It is valid when $\vec{L}$ and $\Sigma\vec{\tau}$ are calculated about an origin (1) fixed in an inertial reference system or (2) situated at the CM of the system. For a rigid object rotating about a fixed axis, the component of angular momentum about the rotation axis is given by $L_\omega = I\omega$. If an object rotates about an axis of symmetry, then the vector relation $\vec{L} = I\vec{\omega}$ holds, but this is not true in general.

If the total net torque on a system is zero, then the total vector angular momentum $\vec{L}$ remains constant. This is the important **law of conservation of angular momentum**. It applies to the vector $\vec{L}$ and therefore also to each of its components.

Questions

1. If there were a great migration of people toward the Earth's equator, would the length of the day (a) get longer because of conservation of angular momentum; (b) get shorter because of conservation of angular momentum; (c) get shorter because of conservation of energy; (d) get longer because of conservation of energy; or (e) remain unaffected?

2. Can the diver of Fig. 11–2 do a somersault without having any initial rotation when she leaves the board?

3. Suppose you are sitting on a rotating stool holding a 2-kg mass in each outstretched hand. If you suddenly drop the masses, will your angular velocity increase, decrease, or stay the same? Explain.

4. When a motorcyclist leaves the ground on a jump and leaves the throttle on (so the rear wheel spins), why does the front of the cycle rise up?

5. Suppose you are standing on the edge of a large freely rotating turntable. What happens if you walk toward the center?

6. A shortstop may leap into the air to catch a ball and throw it quickly. As he throws the ball, the upper part of his body rotates. If you look quickly you will notice that his hips and legs rotate in the opposite direction (Fig. 11–28). Explain.

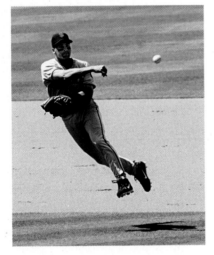

FIGURE 11–28
Question 6. A shortstop in the air, throwing the ball.

7. If all the components of the vectors $\vec{V}_1$ and $\vec{V}_2$ were reversed in direction, how would this alter $\vec{V}_1 \times \vec{V}_2$?

8. Name the four different conditions that could make $\vec{V}_1 \times \vec{V}_2 = 0$.

9. A force $\vec{F} = F\hat{j}$ is applied to an object at a position $\vec{r} = x\hat{i} + y\hat{j} + z\hat{k}$ where the origin is at the CM. Does the torque about the CM depend on x? On y? On z?

10. A particle moves with constant speed along a straight line. How does its angular momentum, calculated about any point not on its path, change in time?

11. If the net force on a system is zero, is the net torque also zero? If the net torque on a system is zero, is the net force zero? Give examples.

12. Explain how a child "pumps" on a swing to make it go higher.

13. Describe the torque needed if the person in Fig. 11–17 is to tilt the axle of the rotating wheel directly upward without it swerving to the side.

14. An astronaut floats freely in a weightless environment. Describe how the astronaut can move her limbs so as to (a) turn her body upside down and (b) turn her body about-face.

15. On the basis of the law of conservation of angular momentum, discuss why a helicopter must have more than one rotor (or propeller). Discuss one or more ways the second propeller can operate in order to keep the helicopter stable.

16. A wheel is rotating freely about a vertical axis with constant angular velocity. Small parts of the wheel come loose and fly off. How does this affect the rotational speed of the wheel? Is angular momentum conserved? Is kinetic energy conserved? Explain.

17. Consider the following vector quantities: displacement, velocity, acceleration, momentum, angular momentum, torque. (a) Which of these are independent of the choice of origin of coordinates? (Consider different points as origin which are at rest with respect to each other.) (b) Which are independent of the velocity of the coordinate system?

18. How does a car make a right turn? Where does the torque come from that is needed to change the angular momentum?

*19. The axis of the Earth precesses with a period of about 25,000 years. This is much like the precession of a top. Explain how the Earth's equatorial bulge gives rise to a torque exerted by the Sun and Moon on the Earth; see Fig. 11–29, which is drawn for the winter solstice (December 21). About what axis would you expect the Earth's rotation axis to precess as a result of the torque due to the Sun? Does the torque exist three months later? Explain.

FIGURE 11–29
Question 19. (Not to scale.)

*20. Why is it that at most locations on the Earth, a plumb bob does not hang precisely in the direction of the Earth's center?

*21. In a rotating frame of reference, Newton's first and second laws remain useful if we assume that a pseudoforce equal to $m\omega^2 r$ is acting. What effect does this assumption have on the validity of Newton's third law?

*22. In the battle of the Falkland Islands in 1914, the shots of British gunners initially fell wide of their marks because their calculations were based on naval battles fought in the Northern Hemisphere. The Falklands are in the Southern Hemisphere. Explain the origin of their problem.

Problems

11–1 Angular Momentum

1. (I) What is the angular momentum of a 0.210-kg ball rotating on the end of a thin string in a circle of radius 1.35 m at an angular speed of 10.4 rad/s?

2. (I) (a) What is the angular momentum of a 2.8-kg uniform cylindrical grinding wheel of radius 18 cm when rotating at 1300 rpm? (b) How much torque is required to stop it in 6.0 s?

3. (II) A person stands, hands at his side, on a platform that is rotating at a rate of 0.90 rev/s. If he raises his arms to a horizontal position, Fig. 11–30, the speed of rotation decreases to 0.70 rev/s. (a) Why? (b) By what factor has his moment of inertia changed?

FIGURE 11–30
Problem 3.

4. (II) A figure skater can increase her spin rotation rate from an initial rate of 1.0 rev every 1.5 s to a final rate of 2.5 rev/s. If her initial moment of inertia was $4.6 \text{ kg} \cdot \text{m}^2$, what is her final moment of inertia? How does she physically accomplish this change?

5. (II) A diver (such as the one shown in Fig. 11–2) can reduce her moment of inertia by a factor of about 3.5 when changing from the straight position to the tuck position. If she makes 2.0 rotations in 1.5 s when in the tuck position, what is her angular speed (rev/s) when in the straight position?

6. (II) A uniform horizontal rod of mass M and length ℓ rotates with angular velocity ω about a vertical axis through its center. Attached to each end of the rod is a small mass m. Determine the angular momentum of the system about the axis.

7. (II) Determine the angular momentum of the Earth (a) about its rotation axis (assume the Earth is a uniform sphere), and (b) in its orbit around the Sun (treat the Earth as a particle orbiting the Sun). The Earth has mass $= 6.0 \times 10^{24} \text{ kg}$ and radius $= 6.4 \times 10^6 \text{ m}$, and is $1.5 \times 10^8 \text{ km}$ from the Sun.

8. (II) (a) What is the angular momentum of a figure skater spinning at 2.8 rev/s with arms in close to her body, assuming her to be a uniform cylinder with a height of 1.5 m, a radius of 15 cm, and a mass of 48 kg? (b) How much torque is required to slow her to a stop in 5.0 s, assuming she does *not* move her arms?

9. (II) A person stands on a platform, initially at rest, that can rotate freely without friction. The moment of inertia of the person plus the platform is I_P. The person holds a spinning bicycle wheel with its axis horizontal. The wheel has moment of inertia I_W and angular velocity ω_W. What will be the angular velocity ω_P of the platform if the person moves the axis of the wheel so that it points (a) vertically upward, (b) at a 60° angle to the vertical, (c) vertically downward? (d) What will ω_P be if the person reaches up and stops the wheel in part (a)?

10. (II) A uniform disk turns at 3.7 rev/s around a frictionless spindle. A nonrotating rod, of the same mass as the disk and length equal to the disk's diameter, is dropped onto the freely spinning disk, Fig. 11–31. They then turn together around the spindle with their centers superposed. What is the angular frequency in rev/s of the combination?

FIGURE 11–31
Problem 10.

11. (II) A person of mass 75 kg stands at the center of a rotating merry-go-round platform of radius 3.0 m and moment of inertia $920 \text{ kg} \cdot \text{m}^2$. The platform rotates without friction with angular velocity 0.95 rad/s. The person walks radially to the edge of the platform. (a) Calculate the angular velocity when the person reaches the edge. (b) Calculate the rotational kinetic energy of the system of platform plus person before and after the person's walk.

12. (II) A potter's wheel is rotating around a vertical axis through its center at a frequency of 1.5 rev/s. The wheel can be considered a uniform disk of mass 5.0 kg and diameter 0.40 m. The potter then throws a 2.6-kg chunk of clay, approximately shaped as a flat disk of radius 8.0 cm, onto the center of the rotating wheel. What is the frequency of the wheel after the clay sticks to it?

13. (II) A 4.2-m-diameter merry-go-round is rotating freely with an angular velocity of 0.80 rad/s. Its total moment of inertia is $1760 \text{ kg} \cdot \text{m}^2$. Four people standing on the ground, each of mass 65 kg, suddenly step onto the edge of the merry-go-round. What is the angular velocity of the merry-go-round now? What if the people were on it initially and then jumped off in a radial direction (relative to the merry-go-round)?

14. (II) A woman of mass m stands at the edge of a solid cylindrical platform of mass M and radius R. At $t = 0$, the platform is rotating with negligible friction at angular velocity ω_0 about a vertical axis through its center, and the woman begins walking with speed v (relative to the platform) toward the center of the platform. (a) Determine the angular velocity of the system as a function of time. (b) What will be the angular velocity when the woman reaches the center?

15. (II) A nonrotating cylindrical disk of moment of inertia I is dropped onto an identical disk rotating at angular speed ω. Assuming no external torques, what is the final common angular speed of the two disks?

16. (II) Suppose our Sun eventually collapses into a white dwarf, losing about half its mass in the process, and winding up with a radius 1.0% of its existing radius. Assuming the lost mass carries away no angular momentum, what would the Sun's new rotation rate be? (Take the Sun's current period to be about 30 days.) What would be its final kinetic energy in terms of its initial kinetic energy of today?

17. (III) Hurricanes can involve winds in excess of 120 km/h at the outer edge. Make a crude estimate of (a) the energy, and (b) the angular momentum, of such a hurricane, approximating it as a rigidly rotating uniform cylinder of air (density 1.3 kg/m^3) of radius 85 km and height 4.5 km.

18. (III) An asteroid of mass $1.0 \times 10^5 \text{ kg}$, traveling at a speed of 35 km/s relative to the Earth, hits the Earth at the equator tangentially, and in the direction of Earth's rotation. Use angular momentum to estimate the percent change in the angular speed of the Earth as a result of the collision.

19. (III) Suppose a 65-kg person stands at the edge of a 6.5-m diameter merry-go-round turntable that is mounted on frictionless bearings and has a moment of inertia of $1850 \text{ kg} \cdot \text{m}^2$. The turntable is at rest initially, but when the person begins running at a speed of 3.8 m/s (with respect to the turntable) around its edge, the turntable begins to rotate in the opposite direction. Calculate the angular velocity of the turntable.

11–2 Vector Cross Product and Torque

20. (I) If vector $\vec{\mathbf{A}}$ points along the negative x axis and vector $\vec{\mathbf{B}}$ along the positive z axis, what is the direction of (a) $\vec{\mathbf{A}} \times \vec{\mathbf{B}}$ and (b) $\vec{\mathbf{B}} \times \vec{\mathbf{A}}$? (c) What is the magnitude of $\vec{\mathbf{A}} \times \vec{\mathbf{B}}$ and $\vec{\mathbf{B}} \times \vec{\mathbf{A}}$?

21. (I) Show that (a) $\hat{\mathbf{i}} \times \hat{\mathbf{i}} = \hat{\mathbf{j}} \times \hat{\mathbf{j}} = \hat{\mathbf{k}} \times \hat{\mathbf{k}} = 0$, (b) $\hat{\mathbf{i}} \times \hat{\mathbf{j}} = \hat{\mathbf{k}}$, $\hat{\mathbf{i}} \times \hat{\mathbf{k}} = -\hat{\mathbf{j}}$, and $\hat{\mathbf{j}} \times \hat{\mathbf{k}} = \hat{\mathbf{i}}$.

22. (I) The directions of vectors $\vec{\mathbf{A}}$ and $\vec{\mathbf{B}}$ are given below for several cases. For each case, state the direction of $\vec{\mathbf{A}} \times \vec{\mathbf{B}}$. (a) $\vec{\mathbf{A}}$ points east, $\vec{\mathbf{B}}$ points south. (b) $\vec{\mathbf{A}}$ points east, $\vec{\mathbf{B}}$ points straight down. (c) $\vec{\mathbf{A}}$ points straight up, $\vec{\mathbf{B}}$ points north. (d) $\vec{\mathbf{A}}$ points straight up, $\vec{\mathbf{B}}$ points straight down.

23. (II) What is the angle θ between two vectors $\vec{\mathbf{A}}$ and $\vec{\mathbf{B}}$, if $|\vec{\mathbf{A}} \times \vec{\mathbf{B}}| = \vec{\mathbf{A}} \cdot \vec{\mathbf{B}}$?

24. (II) A particle is located at $\vec{\mathbf{r}} = (4.0\hat{\mathbf{i}} + 3.5\hat{\mathbf{j}} + 6.0\hat{\mathbf{k}})$ m. A force $\vec{\mathbf{F}} = (9.0\hat{\mathbf{j}} - 4.0\hat{\mathbf{k}})$ N acts on it. What is the torque, calculated about the origin?

25. (II) Consider a particle of a rigid object rotating about a fixed axis. Show that the tangential and radial vector components of the linear acceleration are:

$$\vec{\mathbf{a}}_{tan} = \vec{\boldsymbol{\alpha}} \times \vec{\mathbf{r}} \quad \text{and} \quad \vec{\mathbf{a}}_R = \vec{\boldsymbol{\omega}} \times \vec{\mathbf{v}}.$$

26. (II) (a) Show that the cross product of two vectors, $\vec{\mathbf{A}} = A_x\hat{\mathbf{i}} + A_y\hat{\mathbf{j}} + A_z\hat{\mathbf{k}}$, and $\vec{\mathbf{B}} = B_x\hat{\mathbf{i}} + B_y\hat{\mathbf{j}} + B_z\hat{\mathbf{k}}$ is

$$\vec{\mathbf{A}} \times \vec{\mathbf{B}} = (A_yB_z - A_zB_y)\hat{\mathbf{i}} + (A_zB_x - A_xB_z)\hat{\mathbf{j}}$$

$$+ (A_xB_y - A_yB_x)\hat{\mathbf{k}}.$$

(b) Then show that the cross product can be written

$$\vec{\mathbf{A}} \times \vec{\mathbf{B}} = \begin{vmatrix} \hat{\mathbf{i}} & \hat{\mathbf{j}} & \hat{\mathbf{k}} \\ A_x & A_y & A_z \\ B_x & B_y & B_z \end{vmatrix},$$

where we use the rules for evaluating a determinant. (Note, however, that this is not really a determinant, but a memory aid.)

27. (II) An engineer estimates that under the most adverse expected weather conditions, the total force on the highway sign in Fig. 11–32 will be $\vec{\mathbf{F}} = (\pm 2.4\hat{\mathbf{i}} - 4.1\hat{\mathbf{j}})$ kN, acting at the CM. What torque does this force exert about the base O?

FIGURE 11–32
Problem 27.

28. (II) The origin of a coordinate system is at the center of a wheel which rotates in the xy plane about its axle which is the z axis. A force $F = 215$ N acts in the xy plane, at a $+33.0°$ angle to the x axis, at the point $x = 28.0$ cm, $y = 33.5$ cm. Determine the magnitude and direction of the torque produced by this force about the axis.

29. (II) Use the result of Problem 26 to determine (a) the vector product $\vec{\mathbf{A}} \times \vec{\mathbf{B}}$ and (b) the angle between $\vec{\mathbf{A}}$ and $\vec{\mathbf{B}}$ if $\vec{\mathbf{A}} = 5.4\hat{\mathbf{i}} - 3.5\hat{\mathbf{j}}$ and $\vec{\mathbf{B}} = -8.5\hat{\mathbf{i}} + 5.6\hat{\mathbf{j}} + 2.0\hat{\mathbf{k}}$.

30. (III) Show that the velocity $\vec{\mathbf{v}}$ of any point in an object rotating with angular velocity $\vec{\boldsymbol{\omega}}$ about a fixed axis can be written

$$\vec{\mathbf{v}} = \vec{\boldsymbol{\omega}} \times \vec{\mathbf{r}}$$

where $\vec{\mathbf{r}}$ is the position vector of the point relative to an origin O located on the axis of rotation. Can O be anywhere on the rotation axis? Will $\vec{\mathbf{v}} = \vec{\boldsymbol{\omega}} \times \vec{\mathbf{r}}$ if O is located at a point not on the axis of rotation?

31. (III) Let $\vec{\mathbf{A}}, \vec{\mathbf{B}}$, and $\vec{\mathbf{C}}$ be three vectors, which for generality we assume do not all lie in the same plane. Show that $\vec{\mathbf{A}} \cdot (\vec{\mathbf{B}} \times \vec{\mathbf{C}}) = \vec{\mathbf{B}} \cdot (\vec{\mathbf{C}} \times \vec{\mathbf{A}}) = \vec{\mathbf{C}} \cdot (\vec{\mathbf{A}} \times \vec{\mathbf{B}})$.

11–3 Angular Momentum of a Particle

32. (I) What are the x, y, and z components of the angular momentum of a particle located at $\vec{\mathbf{r}} = x\hat{\mathbf{i}} + y\hat{\mathbf{j}} + z\hat{\mathbf{k}}$ which has momentum $\vec{\mathbf{p}} = p_x\hat{\mathbf{i}} + p_y\hat{\mathbf{j}} + p_z\hat{\mathbf{k}}$?

33. (I) Show that the kinetic energy K of a particle of mass m, moving in a circular path, is $K = L^2/2I$, where L is its angular momentum and I is its moment of inertia about the center of the circle.

34. (I) Calculate the angular momentum of a particle of mass m moving with constant velocity v for two cases (see Fig. 11–33): (a) about origin O, and (b) about O'.

FIGURE 11–33
Problem 34.

35. (II) Two identical particles have equal but opposite momenta, $\vec{\mathbf{p}}$ and $-\vec{\mathbf{p}}$, but they are not traveling along the same line. Show that the total angular momentum of this system does not depend on the choice of origin.

36. (II) Determine the angular momentum of a 75-g particle about the origin of coordinates when the particle is at $x = 4.4$ m, $y = -6.0$ m, and it has velocity $v = (3.2\hat{\mathbf{i}} - 8.0\hat{\mathbf{k}})$ m/s.

37. (II) A particle is at the position $(x, y, z) = (1.0, 2.0, 3.0)$ m. It is traveling with a vector velocity $(-5.0, +2.8, -3.1)$ m/s. Its mass is 3.8 kg. What is its vector angular momentum about the origin?

11–4 and 11–5 Angular Momentum and Torque: General Motion; Rigid Objects

38. (II) An Atwood machine (Fig. 11–16) consists of two masses, $m_A = 7.0$ kg and $m_B = 8.2$ kg, connected by a cord that passes over a pulley free to rotate about a fixed axis. The pulley is a solid cylinder of radius $R_0 = 0.40$ m and mass 0.80 kg. (a) Determine the acceleration a of each mass. (b) What percentage of error in a would be made if the moment of inertia of the pulley were ignored? Ignore friction in the pulley bearings.

39. (II) Four identical particles of mass m are mounted at equal intervals on a thin rod of length ℓ and mass M, with one mass at each end of the rod. If the system is rotated with angular velocity ω about an axis perpendicular to the rod through one of the end masses, determine (a) the kinetic energy and (b) the angular momentum of the system.

40. (II) Two lightweight rods 24 cm in length are mounted perpendicular to an axle and at 180° to each other (Fig. 11–34). At the end of each rod is a 480-g mass. The rods are spaced 42 cm apart along the axle. The axle rotates at 4.5 rad/s. (a) What is the component of the total angular momentum along the axle? (b) What angle does the vector angular momentum make with the axle? [Hint: Remember that the vector angular momentum must be calculated about the same point for both masses, which could be the CM.]

FIGURE 11–34
Problem 40.

41. (II) Figure 11–35 shows two masses connected by a cord passing over a pulley of radius R_0 and moment of inertia I. Mass M_A slides on a frictionless surface, and M_B hangs freely. Determine a formula for (a) the angular momentum of the system about the pulley axis, as a function of the speed v of mass M_A or M_B, and (b) the acceleration of the masses.

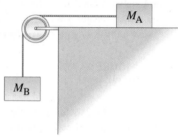

FIGURE 11–35
Problem 41.

42. (III) A thin rod of length ℓ and mass M rotates about a vertical axis through its center with angular velocity ω. The rod makes an angle ϕ with the rotation axis. Determine the magnitude and direction of $\vec{L}$.

43. (III) Show that the total angular momentum $\vec{L} = \Sigma \vec{r}_i \times \vec{p}_i$ of a system of particles about the origin of an inertial reference frame can be written as the sum of the angular momentum about the CM, $\vec{L}^*$ (spin angular momentum), plus the angular momentum of the CM about the origin (orbital angular momentum): $\vec{L} = \vec{L}^* + \vec{r}_{CM} \times M\vec{v}_{CM}$. [Hint: See the derivation of Eq. 11–9b.]

*44. (III) What is the magnitude of the force $\vec{F}$ exerted by each bearing in Fig. 11–18 (Example 11–10)? The bearings are a distance d from point O. Ignore the effects of gravity.

*45. (III) Suppose in Fig. 11–18 that $m_B = 0$; that is, only one mass, m_A, is actually present. If the bearings are each a distance d from O, determine the forces F_A and F_B at the upper and lower bearings respectively. [Hint: Choose an origin—different than O in Fig. 11–18—such that $\vec{L}$ is parallel to $\vec{\omega}$. Ignore effects of gravity.]

*46. (III) Suppose in Fig. 11–18 that $m_A = m_B = 0.60$ kg, $r_A = r_B = 0.30$ m, and the distance between the bearings is 0.23 m. Evaluate the force that each bearing must exert on the axle if $\phi = 34.0°$, $\omega = 11.0$ rad/s?

11–6 Angular Momentum Conservation

47. (II) A thin rod of mass M and length ℓ is suspended vertically from a frictionless pivot at its upper end. A mass m of putty traveling horizontally with a speed v strikes the rod at its CM and sticks there. How high does the bottom of the rod swing?

48. (II) A uniform stick 1.0 m long with a total mass of 270 g is pivoted at its center. A 3.0-g bullet is shot through the stick midway between the pivot and one end (Fig. 11–36). The bullet approaches at 250 m/s and leaves at 140 m/s. With what angular speed is the stick spinning after the collision?

FIGURE 11–36
Problems 48 and 83.

49. (II) Suppose a 5.8×10^{10} kg meteorite struck the Earth at the equator with a speed $v = 2.2 \times 10^4$ m/s, as shown in Fig. 11–37 and remained stuck. By what factor would this affect the rotational frequency of the Earth (1 rev/day)?

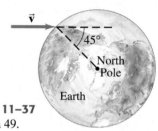

FIGURE 11–37
Problem 49.

50. (III) A 230-kg beam 2.7 m in length slides broadside down the ice with a speed of 18 m/s (Fig. 11–38). A 65-kg man at rest grabs one end as it goes past and hangs on as both he and the beam go spinning down the ice. Assume frictionless motion. (a) How fast does the center of mass of the system move after the collision? (b) With what angular velocity does the system rotate about its CM?

FIGURE 11–38
Problem 50.

51. (III) A thin rod of mass M and length ℓ rests on a frictionless table and is struck at a point $\ell/4$ from its CM by a clay ball of mass m moving at speed v (Fig. 11–39). The ball sticks to the rod. Determine the translational and rotational motion of the rod after the collision.

FIGURE 11–39
Problems 51 and 84.

52. (III) On a level billiards table a cue ball, initially at rest at point O on the table, is struck so that it leaves the cue stick with a center-of-mass speed v_0 and a "reverse" spin of angular speed ω_0 (see Fig. 11–40). A kinetic friction force acts on the ball as it initially skids across the table. (a) Explain why the ball's angular momentum is conserved about point O. (b) Using conservation of angular momentum, find the critical angular speed ω_C such that, if $\omega_0 = \omega_C$, kinetic friction will bring the ball to a complete (as opposed to momentary) stop. (c) If ω_0 is 10% smaller than ω_C, i.e., $\omega_0 = 0.90\,\omega_C$, determine the ball's CM velocity v_{CM} when it starts to roll without slipping. (d) If ω_0 is 10% larger than ω_C, i.e., $\omega_0 = 1.10\,\omega_C$, determine the ball's CM velocity v_{CM} when it starts to roll without slipping. [*Hint*: The ball possesses two types of angular momentum, the first due to the linear speed v_{CM} of its CM relative to point O, the second due to the spin at angular velocity ω about its own CM. The ball's total L about O is the sum of these two angular momenta.]

FIGURE 11–40
Problem 52.

*11–7 Spinning Top

*53. (II) A 220-g top spinning at 15 rev/s makes an angle of 25° to the vertical and precesses at a rate of 1.00 rev per 6.5 s. If its CM is 3.5 cm from its tip along its symmetry axis, what is the moment of inertia of the top?

*54. (II) A toy gyroscope consists of a 170-g disk with a radius of 5.5 cm mounted at the center of a thin axle 21 cm long (Fig. 11–41). The gyroscope spins at 45 rev/s. One end of its axle rests on a stand and the other end precesses horizontally about the stand. (a) How long does it take the gyroscope to precess once around? (b) If all the dimensions of the gyroscope were doubled (radius = 11 cm, axle = 42 cm), how long would it take to precess once?

FIGURE 11–41 A wheel, rotating about a horizontal axle supported at one end, precesses. Problems 54, 55, and 56.

*55. (II) Suppose the solid wheel of Fig. 11–41 has a mass of 300 g and rotates at 85 rad/s; it has radius 6.0 cm and is mounted at the center of a horizontal thin axle 25 cm long. At what rate does the axle precess?

*56. (II) If a mass equal to half the mass of the wheel in Problem 55 is placed at the free end of the axle, what will be the precession rate now? Treat the extra mass as insignificant in size.

*57. (II) A bicycle wheel of diameter 65 cm and mass m rotates on its axle; two 20-cm-long wooden handles, one on each side of the wheel, act as the axle. You tie a rope to a small hook on the end of one of the handles, and then spin the bicycle wheel with a flick of the hand. When you release the spinning wheel, it precesses about the vertical axis defined by the rope, instead of falling to the ground (as it would if it were not spinning). Estimate the rate and direction of precession if the wheel rotates counterclockwise at 2.0 rev/s and its axle remains horizontal.

*11–8 Rotating Reference Frames

*58. (II) If a plant is allowed to grow from seed on a rotating platform, it will grow at an angle, pointing inward. Calculate what this angle will be (put yourself in the rotating frame) in terms of $g, r,$ and ω. Why does it grow inward rather than outward?

*59. (III) Let $\vec{g}'$ be the effective acceleration of gravity at a point on the rotating Earth, equal to the vector sum of the "true" value $\vec{g}$ plus the effect of the rotating reference frame ($m\omega^2 r$ term). See Fig. 11–42. Determine the magnitude and direction of $\vec{g}'$ relative to a radial line from the center of the Earth (a) at the North Pole, (b) at a latitude of 45.0° north, and (c) at the equator. Assume that g (if ω were zero) is a constant 9.80 m/s².

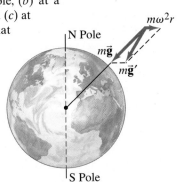

FIGURE 11–42
Problem 59.

*11–9 Coriolis Effect

*60. (II) Suppose the man at B in Fig. 11–26 throws the ball toward the woman at A. (a) In what direction is the ball deflected as seen in the noninertial system? (b) Determine a formula for the amount of deflection and for the (Coriolis) acceleration in this case.

*61. (II) For what directions of velocity would the Coriolis effect on an object moving at the Earth's equator be zero?

*62. (III) We can alter Eqs. 11–14 and 11–15 for use on Earth by considering only the component of $\vec{v}$ perpendicular to the axis of rotation. From Fig. 11–43, we see that this is $v\cos\lambda$ for a vertically falling object, where λ is the latitude of the place on the Earth. If a lead ball is dropped vertically from a 110-m-high tower in Florence, Italy (latitude = 44°), how far from the base of the tower is it deflected by the Coriolis force?

FIGURE 11–43
Problem 62. Object of mass m falling vertically to Earth at a latitude λ.

*63. (III) An ant crawls with constant speed outward along a radial spoke of a wheel rotating at constant angular velocity ω about a vertical axis. Write a vector equation for all the forces (including inertial forces) acting on the ant. Take the x axis along the spoke, y perpendicular to the spoke pointing to the ant's left, and the z axis vertically upward. The wheel rotates counterclockwise as seen from above.

General Problems

64. A thin string is wrapped around a cylindrical hoop of radius R and mass M. One end of the string is fixed, and the hoop is allowed to fall vertically, starting from rest, as the string unwinds. (a) Determine the angular momentum of the hoop about its CM as a function of time. (b) What is the tension in the string as function of time?

65. A particle of mass 1.00 kg is moving with velocity $\vec{v} = (7.0\hat{i} + 6.0\hat{j})$ m/s. (a) Find the angular momentum $\vec{L}$ relative to the origin when the particle is at $\vec{r} = (2.0\hat{j} + 4.0\hat{k})$ m. (b) At position $\vec{r}$ a force of $\vec{F} = 4.0 N\hat{i}$ is applied to the particle. Find the torque relative to the origin.

66. A merry-go-round with a moment of inertia equal to 1260 kg·m^2 and a radius of 2.5 m rotates with negligible friction at 1.70 rad/s. A child initially standing still next to the merry-go-round jumps onto the edge of the platform straight toward the axis of rotation causing the platform to slow to 1.25 rad/s. What is her mass?

67. Why might tall narrow SUVs and buses be prone to "rollover"? Consider a vehicle rounding a curve of radius R on a flat road. When just on the verge of rollover, its tires on the inside of the curve are about to leave the ground, so the friction and normal force on these two tires are zero. The total normal force on the outside tires is F_N and the total friction force is F_{fr}. Assume that the vehicle is not skidding. (a) Analysts define a static stability factor SSF = $w/2h$, where a vehicle's "track width" w is the distance between tires on the same axle, and h is the height of the CM above the ground. Show that the critical rollover speed is

$$v_C = \sqrt{Rg\left(\frac{w}{2h}\right)}.$$

[Hint: Take torques about an axis through the center of mass of the SUV, parallel to its direction of motion.] (b) Determine the ratio of highway curve radii (minimum possible) for a typical passenger car with SSF = 1.40 and an SUV with SSF = 1.05 at a speed of 90 km/h.

68. A spherical asteroid with radius $r = 123$ m and mass $M = 2.25 \times 10^{10}$ kg rotates about an axis at four revolutions per day. A "tug" spaceship attaches itself to the asteroid's south pole (as defined by the axis of rotation) and fires its engine, applying a force F tangentially to the asteroid's surface as shown in Fig. 11–44. If $F = 265$ N, how long will it take the tug to rotate the asteroid's axis of rotation through an angle of 10.0° by this method?

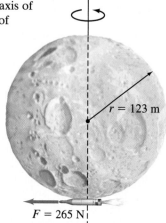

FIGURE 11–44
Problem 68.

$F = 265$ N

69. The time-dependent position of a point object which moves counterclockwise along the circumference of a circle (radius R) in the xy plane with constant speed v is given by

$$\vec{r} = \hat{i}R\cos\omega t + \hat{j}R\sin\omega t$$

where the constant $\omega = v/R$. Determine the velocity $\vec{v}$ and angular velocity $\vec{\omega}$ of this object and then show that these three vectors obey the relation $\vec{v} = \vec{\omega} \times \vec{r}$.

70. The position of a particle with mass m traveling on a helical path (see Fig. 11–45) is given by

$$\vec{r} = R\cos\left(\frac{2\pi z}{d}\right)\hat{i} + R\sin\left(\frac{2\pi z}{d}\right)\hat{j} + z\hat{k}$$

where R and d are the radius and pitch of the helix, respectively, and z has time dependence $z = v_z t$ where v_z is the (constant) component of velocity in the z direction. Determine the time-dependent angular momentum $\vec{L}$ of the particle about the origin.

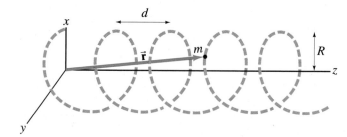

FIGURE 11–45 Problem 70.

71. A boy rolls a tire along a straight level street. The tire has mass 8.0 kg, radius 0.32 m and moment of inertia about its central axis of symmetry of 0.83 kg·m^2. The boy pushes the tire forward away from him at a speed of 2.1 m/s and sees that the tire leans 12° to the right (Fig. 11–46). (a) How will the resultant torque affect the subsequent motion of the tire? (b) Compare the change in angular momentum caused by this torque in 0.20 s to the original magnitude of angular momentum.

FIGURE 11–46
Problem 71.

72. A 70-kg person stands on a tiny rotating platform with arms outstretched. (a) Estimate the moment of inertia of the person using the following approximations: the body (including head and legs) is a 60-kg cylinder, 12 cm in radius and 1.70 m high; and each arm is a 5.0-kg thin rod, 60 cm long, attached to the cylinder. (b) Using the same approximations, estimate the moment of inertia when the arms are at the person's sides. (c) If one rotation takes 1.5 s when the person's arms are outstretched, what is the time for each rotation with arms at the sides? Ignore the moment of inertia of the lightweight platform. (d) Determine the change in kinetic energy when the arms are lifted from the sides to the horizontal position. (e) From your answer to part (d), would you expect it to be harder or easier to lift your arms when rotating or when at rest?

73. Water drives a waterwheel (or turbine) of radius $R = 3.0\,\text{m}$ as shown in Fig. 11–47. The water enters at a speed $v_1 = 7.0\,\text{m/s}$ and exits from the waterwheel at a speed $v_2 = 3.8\,\text{m/s}$. (a) If 85 kg of water passes through per second, what is the rate at which the water delivers angular momentum to the waterwheel? (b) What is the torque the water applies to the waterwheel? (c) If the water causes the waterwheel to make one revolution every 5.5 s, how much power is delivered to the wheel?

FIGURE 11–47
Problem 73.

74. The Moon orbits the Earth such that the same side always faces the Earth. Determine the ratio of the Moon's spin angular momentum (about its own axis) to its orbital angular momentum. (In the latter case, treat the Moon as a particle orbiting the Earth.)

75. A particle of mass m uniformly accelerates as it moves counterclockwise along the circumference of a circle of radius R:
$$\vec{r} = \hat{i}\,R\cos\theta + \hat{j}\,R\sin\theta$$
with $\theta = \omega_0 t + \frac{1}{2}\alpha t^2$, where the constants ω_0 and α are the initial angular velocity and angular acceleration, respectively. Determine the object's tangential acceleration $\vec{a}$ and determine the torque acting on the object using (a) $\vec{\tau} = \vec{r} \times \vec{F}$, (b) $\vec{\tau} = I\vec{\alpha}$.

76. A projectile with mass m is launched from the ground and follows a trajectory given by
$$\vec{r} = (v_{x0}t)\hat{i} + \left(v_{y0}t - \frac{1}{2}gt^2\right)\hat{j}$$
where v_{x0} and v_{y0} are the initial velocities in the x and y direction, respectively, and g is the acceleration due to gravity. The launch position is defined to be the origin. Determine the torque acting on the projectile about the origin using (a) $\vec{\tau} = \vec{r} \times \vec{F}$, (b) $\vec{\tau} = d\vec{L}/dt$.

77. Most of our Solar System's mass is contained in the Sun, and the planets possess almost all of the Solar System's angular momentum. This observation plays a key role in theories attempting to explain the formation of our Solar System. Estimate the fraction of the Solar System's total angular momentum that is possessed by planets using a simplified model which includes only the large outer planets with the most angular momentum. The central Sun (mass $1.99 \times 10^{30}\,\text{kg}$, radius $6.96 \times 10^8\,\text{m}$) spins about its axis once every 25 days and the planets Jupiter, Saturn, Uranus, and Neptune move in nearly circular orbits around the Sun with orbital data given in the Table below. Ignore each planet's spin about its own axis.

Planet	Mean Distance from Sun ($\times 10^6$ km)	Orbital Period (Earth Years)	Mass ($\times 10^{25}$ kg)
Jupiter	778	11.9	190
Saturn	1427	29.5	56.8
Uranus	2870	84.0	8.68
Neptune	4500	165	10.2

78. A bicyclist traveling with speed $v = 9.2\,\text{m/s}$ on a flat road is making a turn with a radius $r = 12\,\text{m}$. The forces acting on the cyclist and cycle are the normal force ($\vec{F}_N$) and friction force ($\vec{F}_{fr}$) exerted by the road on the tires and $m\vec{g}$, the total weight of the cyclist and cycle. Ignore the small mass of the wheels. (a) Explain carefully why the angle θ the bicycle makes with the vertical (Fig. 11–48) must be given by $\tan\theta = F_{fr}/F_N$ if the cyclist is to maintain balance. (b) Calculate θ for the values given. [Hint: Consider the "circular" translational motion of the bicycle and rider.] (c) If the coefficient of static friction between tires and road is $\mu_s = 0.65$, what is the minimum turning radius?

(a)　　　　　　　　(b)

FIGURE 11–48 Problem 78.

79. Competitive ice skaters commonly perform single, double, and triple axle jumps in which they rotate $1\frac{1}{2}, 2\frac{1}{2}$, and $3\frac{1}{2}$ revolutions, respectively, about a vertical axis while airborne. For all these jumps, a typical skater remains airborne for about 0.70 s. Suppose a skater leaves the ground in an "open" position (e.g., arms outstretched) with moment of inertia I_0 and rotational frequency $f_0 = 1.2\,\text{rev/s}$, maintaining this position for 0.10 s. The skater then assumes a "closed" position (arms brought closer) with moment of inertia I, acquiring a rotational frequency f, which is maintained for 0.50 s. Finally, the skater immediately returns to the "open" position for 0.10 s until landing (see Fig. 11–49). (a) Why is angular momentum conserved during the skater's jump? Neglect air resistance. (b) Determine the minimum rotational frequency f during the flight's middle section for the skater to successfully complete a single and a triple axle. (c) Show that, according to this model, a skater must be able to reduce his or her moment of inertia in midflight by a factor of about 2 and 5 in order to complete a single and triple axle, respectively.

FIGURE 11–49 Problem 79.

80. A radio transmission tower has a mass of 80 kg and is 12 m high. The tower is anchored to the ground by a flexible joint at its base, but it is secured by three cables 120° apart (Fig. 11–50). In an analysis of a potential failure, a mechanical engineer needs to determine the behavior of the tower if one of the cables broke. The tower would fall away from the broken cable, rotating about its base. Determine the speed of the top of the tower as a function of the rotation angle θ. Start your analysis with the rotational dynamics equation of motion $d\vec{L}/dt = \vec{\tau}_{\text{net}}$. Approximate the tower as a tall thin rod.

FIGURE 11–50 Problem 80.

81. Suppose a star the size of our Sun, but with mass 8.0 times as great, were rotating at a speed of 1.0 revolution every 9.0 days. If it were to undergo gravitational collapse to a neutron star of radius 12 km, losing $\frac{3}{4}$ of its mass in the process, what would its rotation speed be? Assume the star is a uniform sphere at all times. Assume also that the thrown-off mass carries off either (*a*) no angular momentum, or (*b*) its proportional share $(\frac{3}{4})$ of the initial angular momentum.

82. A baseball bat has a "sweet spot" where a ball can be hit with almost effortless transmission of energy. A careful analysis of baseball dynamics shows that this special spot is located at the point where an applied force would result in pure rotation of the bat about the handle grip. Determine the location of the sweet spot of the bat shown in Fig. 11–51. The linear mass density of the bat is given roughly by $(0.61 + 3.3x^2)$ kg/m, where x is in meters measured from the end of the handle. The entire bat is 0.84 m long. The desired rotation point should be 5.0 cm from the end where the bat is held. [*Hint*: Where is the CM of the bat?]

FIGURE 11–51 Problem 82.

*Numerical/Computer

***83.** (II) A uniform stick 1.00 m long with a total mass of 330 g is pivoted at its center. A 3.0-g bullet is shot through the stick a distance x from the pivot. The bullet approaches at 250 m/s and leaves at 140 m/s (Fig. 11–36). (a) Determine a formula for the angular speed of the spinning stick after the collision as a function of x. (b) Graph the angular speed as a function of x, from $x = 0$ to $x = 0.50$ m.

***84.** (III) Figure 11–39 shows a thin rod of mass M and length ℓ resting on a frictionless table. The rod is struck at a distance x from its CM by a clay ball of mass m moving at speed v. The ball sticks to the rod. (a) Determine a formula for the rotational motion of the system after the collision. (b) Graph the rotational motion of the system as a function of x, from $x = 0$ to $x = \ell/2$, with values of $M = 450$ g, $m = 15$ g, $\ell = 1.20$ m, and $v = 12$ m/s. (c) Does the translational motion depend on x? Explain.

Answers to Exercises

A: (*b*).

B: (*a*).

C: (*b*).

D: (i) (*d*); (ii) (*a*); (iii) (*b*).

E: (*e*).

Our whole built environment, from modern bridges to skyscrapers, has required architects and engineers to determine the forces and stresses within these structures. The object is to keep these structures static—that is, not in motion, especially not falling down.

CONTENTS

C H A P T E R

12

Static Equilibrium; Elasticity and Fracture

CHAPTER-OPENING QUESTION—Guess Now!

The diving board shown here is held by two supports at A and B. Which statement is true about the forces exerted *on* the diving board at A and B?

(a) $\vec{F}_A$ is down, $\vec{F}_B$ is up, and F_B is larger than F_A.
(b) Both forces are up and F_B is larger than F_A.
(c) $\vec{F}_A$ is down, $\vec{F}_B$ is up, and F_A is larger than F_B.
(d) Both forces are down and approximately equal.
(e) $\vec{F}_B$ is down, $\vec{F}_A$ is up, and they are equal.

A   B

W e now study a special case in mechanics—when the net force and the net torque on an object, or system of objects, are both zero. In this case both the linear acceleration and the angular acceleration of the object or system are zero. The object is either at rest, or its center of mass is moving at constant velocity. We will be concerned mainly with the first situation, in which the object or objects are at rest.

We will see how to determine the forces (and torques) that act within a structure. Just how and where these forces act can be very important for buildings, bridges, and other structures, and in the human body.

Statics is concerned with the calculation of the forces acting on and within structures that are in *equilibrium*. Determination of these forces, which occupies us in the first part of this Chapter, then allows a determination of whether the structures can sustain the forces without significant deformation or fracture, since any material can break or buckle if too much force is applied (Fig. 12–1).

FIGURE 12–1 Elevated walkway collapse in a Kansas City hotel in 1981. How a simple physics calculation could have prevented the tragic loss of over 100 lives is considered in Example 12–9.

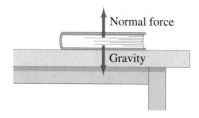

Normal force

Gravity

FIGURE 12–2 The book is in equilibrium; the net force on it is zero.

FIGURE 12–3 Example 12–1.

(a)   200 kg

(b)

12–1 The Conditions for Equilibrium

Objects in daily life have at least one force acting on them (gravity). If they are at rest, then there must be other forces acting on them as well so that the net force is zero. A book at rest on a table, for example, has two forces acting on it, the downward force of gravity and the normal force the table exerts upward on it (Fig. 12–2). Because the book is at rest, Newton's second law tells us the net force on it is zero. Thus the upward force exerted by the table on the book must be equal in magnitude to the force of gravity acting downward on the book. Such an object is said to be in **equilibrium** (Latin for "equal forces" or "balance") under the action of these two forces.

Do not confuse the two forces in Fig. 12–2 with the equal and opposite forces of Newton's third law, which act on different objects. Here, both forces act on the same object; and they add up to zero.

The First Condition for Equilibrium

For an object to be at rest, Newton's second law tells us that the sum of the forces acting on it must add up to zero. Since force is a vector, the components of the net force must each be zero. Hence, a condition for equilibrium is that

$$\Sigma F_x = 0, \qquad \Sigma F_y = 0, \qquad \Sigma F_z = 0. \qquad \textbf{(12–1)}$$

We will mainly be dealing with forces that act in a plane, so we usually need only the x and y components. We must remember that if a particular force component points along the negative x or y axis, it must have a negative sign. Equations 12–1 are called the **first condition for equilibrium**.

EXAMPLE 12–1 **Chandelier cord tension.** Calculate the tensions $\vec{F}_A$ and $\vec{F}_B$ in the two cords that are connected to the vertical cord supporting the 200-kg chandelier in Fig. 12–3. Ignore the mass of the cords.

APPROACH We need a free-body diagram, but for which object? If we choose the chandelier, the cord supporting it must exert a force equal to the chandelier's weight $mg = (200 \text{ kg})(9.8 \text{ m/s}^2) = 1960 \text{ N}$. But the forces $\vec{F}_A$ and $\vec{F}_B$ don't get involved. Instead, let us choose as our object the point where the three cords join (it could be a knot). The free-body diagram is then as shown in Fig. 12–3a. The three forces—$\vec{F}_A$, $\vec{F}_B$, and the tension in the vertical cord equal to the weight of the 200-kg chandelier—act at this point where the three cords join. For this junction point we write $\Sigma F_x = 0$ and $\Sigma F_y = 0$, since the problem is laid out in two dimensions. The directions of $\vec{F}_A$ and $\vec{F}_B$ are known, since tension in a cord can only be along the cord—any other direction would cause the cord to bend, as already pointed out in Chapter 4. Thus, our unknowns are the magnitudes F_A and F_B.

SOLUTION We first resolve $\vec{F}_A$ into its horizontal (x) and vertical (y) components. Although we don't know the value of F_A, we can write (see Fig. 12–3b) $F_{Ax} = -F_A \cos 60°$ and $F_{Ay} = F_A \sin 60°$. $\vec{F}_B$ has only an x component. In the vertical direction, we have the downward force exerted by the vertical cord equal to the weight of the chandelier $= (200 \text{ kg})(g)$, and the vertical component of $\vec{F}_A$ upward:

$$\Sigma F_y = 0$$
$$F_A \sin 60° - (200 \text{ kg})(g) = 0$$

so

$$F_A = \frac{(200 \text{ kg})g}{\sin 60°} = (231 \text{ kg})g = 2260 \text{ N}.$$

In the horizontal direction, with $\Sigma F_x = 0$,

$$\Sigma F_x = F_B - F_A \cos 60° = 0.$$

Thus

$$F_B = F_A \cos 60° = (231 \text{ kg})(g)(0.500) = (115 \text{ kg})g = 1130 \text{ N}.$$

The magnitudes of $\vec{F}_A$ and $\vec{F}_B$ determine the strength of cord or wire that must be used. In this case, the cord must be able to hold more than 230 kg.

EXERCISE A In Example 12–1, F_A has to be greater than the chandelier's weight, mg. Why?

The Second Condition for Equilibrium

Although Eqs. 12–1 are a necessary condition for an object to be in equilibrium, they are not always a sufficient condition. Figure 12–4 shows an object on which the net force is zero. Although the two forces labeled $\vec{F}$ add up to give zero net force on the object, they do give rise to a net torque that will rotate the object. Referring to Eq. 10–14, $\Sigma\tau = I\alpha$, we see that if an object is to remain at rest, the net torque applied to it (calculated about *any* axis) must be zero. Thus we have the **second condition for equilibrium**: that the sum of the torques acting on an object, as calculated about any axis, must be zero:

$$\Sigma\tau = 0. \tag{12-2}$$

This condition will ensure that the angular acceleration, α, about any axis will be zero. If the object is not rotating initially $(\omega = 0)$, it will not start rotating. Equations 12–1 and 12–2 are the only requirements for an object to be in equilibrium.

We will mainly consider cases in which the forces all act in a plane (we call it the xy plane). In such cases the torque is calculated about an axis that is perpendicular to the xy plane. *The choice of this axis is arbitrary.* If the object is at rest, then $\Sigma\tau = 0$ about any axis whatever. Therefore we can choose any axis that makes our calculation easier. Once the axis is chosen, all torques must be calculated about that axis.

FIGURE 12–4 Although the net force on it is zero, the ruler will move (rotate). A pair of equal forces acting in opposite directions but at different points on an object (as shown here) is referred to as a *couple*.

⚠️ **CAUTION**

Axis choice for $\Sigma\tau = 0$ is arbitrary. All torques must be calculated about the same axis.

🚶 **PHYSICS APPLIED**

The lever

FIGURE 12–5 Example 12–2. A lever can "multiply" your force.

CONCEPTUAL EXAMPLE 12–2 | **A lever.** The bar in Fig. 12–5 is being used as a lever to pry up a large rock. The small rock acts as a fulcrum (pivot point). The force F_P required at the long end of the bar can be quite a bit smaller than the rock's weight mg, since it is the *torques* that balance in the rotation about the fulcrum. If, however, the leverage isn't sufficient, and the large rock isn't budged, what are two ways to increase the leverage?

RESPONSE One way is to increase the lever arm of the force F_P by slipping a pipe over the end of the bar and thereby pushing with a longer lever arm. A second way is to move the fulcrum closer to the large rock. This may change the long lever arm R only a little, but it changes the short lever arm r by a substantial fraction and therefore changes the ratio of R/r dramatically. In order to pry the rock, the torque due to F_P must at least balance the torque due to mg; that is, $mgr = F_P R$ and

$$\frac{r}{R} = \frac{F_P}{mg}.$$

With r smaller, the weight mg can be balanced with less force F_P. The ratio of the load force to your applied force ($= mg/F_P$ here) is the **mechanical advantage** of the system, and here equals R/r. A lever is a "simple machine." We discussed another simple machine, the pulley, in Chapter 4, Example 4–14.

EXERCISE B For simplicity, we wrote the equation in Example 12–2 as if the lever were perpendicular to the forces. Would the equation be valid even for a lever at an angle as shown in Fig. 12–5?

12–2 Solving Statics Problems

The subject of statics is important because it allows us to calculate certain forces on (or within) a structure when some of the forces on it are already known. We will mainly consider situations in which all the forces act in a plane, so we can have two force equations (x and y components) and one torque equation, for a total of three equations. Of course, you do not have to use all three equations if they are not needed. When using a torque equation, a torque that tends to rotate the object counterclockwise is usually considered positive, whereas a torque that tends to rotate it clockwise is considered negative. (But the opposite convention would be okay too.)

One of the forces that acts on objects is the force of gravity. As we discussed in Section 9–8, we can consider the force of gravity on an object as acting at its center of gravity (CG) or center of mass (CM), which for practical purposes are the same point. For uniform symmetrically shaped objects, the CG is at the geometric center. For more complicated objects, the CG can be determined as discussed in Section 9–8.

There is no single technique for attacking statics problems, but the following procedure may be helpful.

 PROBLEM SOLVING

$\tau > 0$ *counterclockwise*
$\tau < 0$ *clockwise*

Statics

1. Choose one object at a time for consideration. Make a careful **free-body diagram** by showing all the forces acting on that object, including gravity, and the points at which these forces act. If you aren't sure of the direction of a force, choose a direction; if the actual direction is opposite, your eventual calculation will give a result with a minus sign.

2. Choose a convenient **coordinate system**, and resolve the forces into their components.

3. Using letters to represent unknowns, write down the **equilibrium equations** for the **forces**:

$$\Sigma F_x = 0 \quad \text{and} \quad \Sigma F_y = 0,$$

assuming all the forces act in a plane.

4. For the **torque equation**,

$$\Sigma \tau = 0,$$

choose any axis perpendicular to the xy plane that might make the calculation easier. (For example, you can reduce the number of unknowns in the resulting equation by choosing the axis so that one of the unknown forces acts through that axis; then this force will have zero lever arm and produce zero torque, and so won't appear in the torque equation.) Pay careful attention to determining the lever arm for each force correctly. Give each torque a + or − sign to indicate torque direction. For example, if torques tending to rotate the object counterclockwise are positive, then those tending to rotate it clockwise are negative.

5. **Solve** these equations for the unknowns. Three equations allow a maximum of three unknowns to be solved for. They can be forces, distances, or even angles.

PHYSICS APPLIED

Balancing a seesaw

FIGURE 12–6 (a) Two children on a seesaw, Example 12–3. (b) Free-body diagram of the board.

EXAMPLE 12–3 Balancing a seesaw. A board of mass $M = 2.0\,\text{kg}$ serves as a seesaw for two children, as shown in Fig. 12–6a. Child A has a mass of 30 kg and sits 2.5 m from the pivot point, P (his center of gravity is 2.5 m from the pivot). At what distance x from the pivot must child B, of mass 25 kg, place herself to balance the seesaw? Assume the board is uniform and centered over the pivot.

APPROACH We follow the steps of the Problem Solving Strategy above.

SOLUTION

1. **Free-body diagram.** We choose the board as our object, and assume it is horizontal. Its free-body diagram is shown in Fig. 12–6b. The forces acting on the board are the forces exerted downward on it by each child, $\vec{\mathbf{F}}_A$ and $\vec{\mathbf{F}}_B$, the upward force exerted by the pivot $\vec{\mathbf{F}}_N$, and the force of gravity on the board ($= M\vec{\mathbf{g}}$) which acts at the center of the uniform board.

2. **Coordinate system.** We choose y to be vertical, with positive upward, and x horizontal to the right, with origin at the pivot.

3. **Force equation.** All the forces are in the y (vertical) direction, so

$$\Sigma F_y = 0$$
$$F_N - m_A g - m_B g - Mg = 0,$$

where $F_A = m_A g$ and $F_B = m_B g$ because each child is in equilibrium when the seesaw is balanced.

4. **Torque equation.** Let us calculate the torque about an axis through the board at the pivot point, P. Then the lever arms for F_N and for the weight of the board are zero, and they will contribute zero torque about point P. Thus the torque equation will involve only the forces $\vec{\mathbf{F}}_A$ and $\vec{\mathbf{F}}_B$, which are equal to the weights of the children. The torque exerted by each child will be mg times the appropriate lever arm, which here is the distance of each child from the pivot point. Hence the torque equation is

$$\Sigma \tau = 0$$
$$m_A g(2.5\,\text{m}) - m_B gx + Mg(0\,\text{m}) + F_N(0\,\text{m}) = 0$$

or

$$m_A g(2.5\,\text{m}) - m_B gx = 0,$$

where two terms were dropped because their lever arms were zero.

5. Solve. We solve the torque equation for x and find

$$x = \frac{m_A}{m_B}(2.5\,\text{m}) = \frac{30\,\text{kg}}{25\,\text{kg}}(2.5\,\text{m}) = 3.0\,\text{m}.$$

To balance the seesaw, child B must sit so that her CM is 3.0 m from the pivot point. This makes sense: since she is lighter, she must sit farther from the pivot than the heavier child in order to provide equal torque.

EXERCISE C We did not need to use the force equation to solve Example 12–3 because of our choice of the axis. Use the force equation to find the force exerted by the pivot.

Figure 12–7 shows a uniform beam that extends beyond its support like a diving board. Such a beam is called a **cantilever**. The forces acting on the beam in Fig. 12–7 are those due to the supports, $\vec{F}_A$ and $\vec{F}_B$, and the force of gravity which acts at the CG, 5.0 m to the right of the right-hand support. If you follow the procedure of the last Example and calculate F_A and F_B, assuming they point upward as shown in Fig. 12–7, you will find that F_A comes out negative. If the beam has a mass of 1200 kg and a weight $mg = 12,000\,\text{N}$, then $F_B = 15,000\,\text{N}$ and $F_A = -3000\,\text{N}$ (see Problem 9). Whenever an unknown force comes out negative, it merely means that the force actually points in the opposite direction from what you assumed. Thus in Fig. 12–7, $\vec{F}_A$ actually points downward. With a little reflection it should become clear that the left-hand support must indeed pull downward on the beam (by means of bolts, screws, fasteners and/or glue) if the beam is to be in equilibrium; otherwise the sum of the torques about the CG (or about the point where $\vec{F}_B$ acts) could not be zero.

EXERCISE D Return to the Chapter-Opening Question, p. 311, and answer it again now. Try to explain why you may have answered differently the first time.

PHYSICS APPLIED
Cantilever

PROBLEM SOLVING
If a force comes out negative

FIGURE 12–7 A cantilever. The force vectors shown are hypothetical—one may even have a different direction.

EXAMPLE 12–4 **Force exerted by biceps muscle.** How much force must the biceps muscle exert when a 5.0-kg ball is held in the hand (*a*) with the arm horizontal as in Fig. 12–8a, and (*b*) when the arm is at a 45° angle as in Fig. 12–8b? The biceps muscle is connected to the forearm by a tendon attached 5.0 cm from the elbow joint. Assume that the mass of forearm and hand together is 2.0 kg and their CG is as shown.

APPROACH The free-body diagram for the forearm is shown in Fig. 12–8; the forces are the weights of the arm and ball, the upward force $\vec{F}_M$ exerted by the muscle, and a force $\vec{F}_J$ exerted at the joint by the bone in the upper arm (all assumed to act vertically). We wish to find the magnitude of $\vec{F}_M$, which is done most easily by using the torque equation and by choosing our axis through the joint so that $\vec{F}_J$ contributes zero torque.

SOLUTION (*a*) We calculate torques about the point where $\vec{F}_J$ acts in Fig. 12–8a. The $\Sigma\tau = 0$ equation gives

$$(0.050\,\text{m})F_M - (0.15\,\text{m})(2.0\,\text{kg})g - (0.35\,\text{m})(5.0\,\text{kg})g = 0.$$

We solve for F_M:

$$F_M = \frac{(0.15\,\text{m})(2.0\,\text{kg})g + (0.35\,\text{m})(5.0\,\text{kg})g}{0.050\,\text{m}} = (41\,\text{kg})g = 400\,\text{N}.$$

(*b*) The lever arm, as calculated about the joint, is reduced by the factor cos 45° for all three forces. Our torque equation will look like the one just above, except that each term will have its lever arm reduced by the same factor, which will cancel out. The same result is obtained, $F_M = 400\,\text{N}$.

NOTE The force required of the muscle (400 N) is quite large compared to the weight of the object lifted ($= mg = 49\,\text{N}$). Indeed, the muscles and joints of the body are generally subjected to quite large forces.

PHYSICS APPLIED
Forces in muscles and joints

FIGURE 12–8 Example 12–4.

EXERCISE E How much mass could the person in Example 12–4 hold in the hand with a biceps force of 450 N if the tendon was attached 6.0 cm from the elbow instead of 5.0 cm?

FIGURE 12–9 Example 12–5.

Our next Example involves a beam that is attached to a wall by a hinge and is supported by a cable or cord (Fig. 12–9). It is important to remember that a flexible cable can support a force only along its length. (If there were a component of force perpendicular to the cable, it would bend because it is flexible.) But for a rigid device, such as the hinge in Fig. 12–9, the force can be in any direction and we can know the direction only after solving the problem. The hinge is assumed small and smooth, so it can exert no internal torque (about its center) on the beam.

EXAMPLE 12–5 **Hinged beam and cable.** A uniform beam, 2.20 m long with mass $m = 25.0$ kg, is mounted by a small hinge on a wall as shown in Fig. 12–9. The beam is held in a horizontal position by a cable that makes an angle $\theta = 30.0°$. The beam supports a sign of mass $M = 28.0$ kg suspended from its end. Determine the components of the force $\vec{F}_H$ that the (smooth) hinge exerts on the beam, and the tension F_T in the supporting cable.

APPROACH Figure 12–9 is the free-body diagram for the beam, showing all the forces acting on the beam. It also shows the components of $\vec{F}_T$ and a guess for $\vec{F}_H$. We have three unknowns, F_{Hx}, F_{Hy}, and F_T (we are given θ), so we will need all three equations, $\Sigma F_x = 0$, $\Sigma F_y = 0$, $\Sigma\tau = 0$.

SOLUTION The sum of the forces in the vertical (y) direction is

$$\Sigma F_y = 0$$
$$F_{Hy} + F_{Ty} - mg - Mg = 0. \tag{i}$$

In the horizontal (x) direction, the sum of the forces is

$$\Sigma F_x = 0$$
$$F_{Hx} - F_{Tx} = 0. \tag{ii}$$

For the torque equation, we choose the axis at the point where $\vec{F}_T$ and $M\vec{g}$ act (so our equation then contains only one unknown, F_{Hy}). We choose torques that tend to rotate the beam counterclockwise as positive. The weight mg of the (uniform) beam acts at its center, so we have

$$\Sigma\tau = 0$$
$$-(F_{Hy})(2.20\text{ m}) + mg(1.10\text{ m}) = 0.$$

We solve for F_{Hy}:

$$F_{Hy} = \left(\frac{1.10\text{ m}}{2.20\text{ m}}\right)mg = (0.500)(25.0\text{ kg})(9.80\text{ m/s}^2) = 123\text{ N}. \tag{iii}$$

Next, since the tension $\vec{F}_T$ in the cable acts along the cable ($\theta = 30.0°$), we see from Fig. 12–9 that $\tan\theta = F_{Ty}/F_{Tx}$, or

$$F_{Ty} = F_{Tx}\tan\theta = F_{Tx}(\tan 30.0°) = 0.577\,F_{Tx}. \tag{iv}$$

Equation (i) above gives

$$F_{Ty} = (m + M)g - F_{Hy} = (53.0\text{ kg})(9.80\text{ m/s}^2) - 123\text{ N} = 396\text{ N};$$

Equations (iv) and (ii) give

$$F_{Tx} = F_{Ty}/0.577 = 687\text{ N};$$
$$F_{Hx} = F_{Tx} = 687\text{ N}.$$

The components of $\vec{F}_H$ are $F_{Hy} = 123$ N and $F_{Hx} = 687$ N. The tension in the wire is $F_T = \sqrt{F_{Tx}^2 + F_{Ty}^2} = 793$ N.

Alternate Solution Let us see the effect of choosing a different axis for calculating torques, such as an axis through the hinge. Then the lever arm for F_H is zero, and the torque equation ($\Sigma\tau = 0$) becomes

$$-mg(1.10\text{ m}) - Mg(2.20\text{ m}) + F_{Ty}(2.20\text{ m}) = 0.$$

We solve this for F_{Ty} and find

$$F_{Ty} = \frac{m}{2}g + Mg = (12.5\text{ kg} + 28.0\text{ kg})(9.80\text{ m/s}^2) = 397\text{ N}.$$

We get the same result, within the precision of our significant figures.

NOTE It doesn't matter which axis we choose for $\Sigma\tau = 0$. Using a second axis can serve as a check.

EXAMPLE 12–6 **Ladder.** A 5.0-m-long ladder leans against a smooth wall at a point 4.0 m above a cement floor as shown in Fig. 12–10. The ladder is uniform and has mass $m = 12.0\,\text{kg}$. Assuming the wall is frictionless (but the floor is not), determine the forces exerted on the ladder by the floor and by the wall.

APPROACH Figure 12–10 is the free-body diagram for the ladder, showing all the forces acting on the ladder. The wall, since it is frictionless, can exert a force only perpendicular to the wall, and we label that force $\vec{\mathbf{F}}_W$. The cement floor exerts a force $\vec{\mathbf{F}}_C$ which has both horizontal and vertical force components: F_{Cx} is frictional and F_{Cy} is the normal force. Finally, gravity exerts a force $mg = (12.0\,\text{kg})(9.80\,\text{m/s}^2) = 118\,\text{N}$ on the ladder at its midpoint, since the ladder is uniform.

SOLUTION Again we use the equilibrium conditions, $\Sigma F_x = 0$, $\Sigma F_y = 0$, $\Sigma \tau = 0$. We will need all three since there are three unknowns: F_W, F_{Cx}, and F_{Cy}. The y component of the force equation is

$$\Sigma F_y = F_{Cy} - mg = 0,$$

so immediately we have

$$F_{Cy} = mg = 118\,\text{N}.$$

The x component of the force equation is

$$\Sigma F_x = F_{Cx} - F_W = 0.$$

To determine both F_{Cx} and F_W, we need a torque equation. If we choose to calculate torques about an axis through the point where the ladder touches the cement floor, then $\vec{\mathbf{F}}_C$, which acts at this point, will have a lever arm of zero and so won't enter the equation. The ladder touches the floor a distance $x_0 = \sqrt{(5.0\,\text{m})^2 - (4.0\,\text{m})^2} = 3.0\,\text{m}$ from the wall (right triangle, $c^2 = a^2 + b^2$). The lever arm for mg is half this, or 1.5 m, and the lever arm for F_W is 4.0 m, Fig. 12–10. We get

$$\Sigma \tau = (4.0\,\text{m})F_W - (1.5\,\text{m})mg = 0.$$

Thus

$$F_W = \frac{(1.5\,\text{m})(12.0\,\text{kg})(9.8\,\text{m/s}^2)}{4.0\,\text{m}} = 44\,\text{N}.$$

Then, from the x component of the force equation,

$$F_{Cx} = F_W = 44\,\text{N}.$$

Since the components of $\vec{\mathbf{F}}_C$ are $F_{Cx} = 44\,\text{N}$ and $F_{Cy} = 118\,\text{N}$, then

$$F_C = \sqrt{(44\,\text{N})^2 + (118\,\text{N})^2} = 126\,\text{N} \approx 130\,\text{N}$$

(rounded off to two significant figures), and it acts at an angle to the floor of

$$\theta = \tan^{-1}(118\,\text{N}/44\,\text{N}) = 70°.$$

NOTE The force $\vec{\mathbf{F}}_C$ does *not* have to act along the ladder's direction because the ladder is rigid and not flexible like a cord or cable.

EXERCISE F Why is it reasonable to ignore friction along the wall, but not reasonable to ignore it along the floor?

12–3 Stability and Balance

An object in static equilibrium, if left undisturbed, will undergo no translational or rotational acceleration since the sum of all the forces and the sum of all the torques acting on it are zero. However, if the object is displaced slightly, three outcomes are possible: (1) the object returns to its original position, in which case it is said to be in **stable equilibrium**; (2) the object moves even farther from its original position, and it is said to be in **unstable equilibrium**; or (3) the object remains in its new position, and it is said to be in **neutral equilibrium**.

Consider the following examples. A ball suspended freely from a string is in stable equilibrium, for if it is displaced to one side, it will return to its original position (Fig. 12–11a) due to the net force and torque exerted on it. On the other hand, a pencil standing on its point is in unstable equilibrium. If its center of gravity is directly over its tip (Fig. 12–11b), the net force and net torque on it will be zero. But if it is displaced ever so slightly as shown—say, by a slight vibration or tiny air current—there will be a torque on it, and this torque acts to make the pencil continue to fall in the direction of the original displacement. Finally, an example of an object in neutral equilibrium is a sphere resting on a horizontal tabletop. If it is placed slightly to one side, it will remain in its new position—no net torque acts on it.

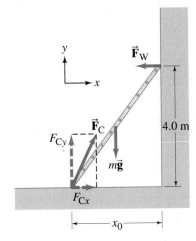

FIGURE 12–10 A ladder leaning against a wall. Example 12–6.

FIGURE 12–11 (a) Stable equilibrium, and (b) unstable equilibrium.

FIGURE 12–12 Equilibrium of a refrigerator resting on a flat floor.

FIGURE 12–13 Humans adjust their posture to achieve stability when carrying loads.

Total CG

In most situations, such as in the design of structures and in working with the human body, we are interested in maintaining stable equilibrium, or *balance*, as we sometimes say. In general, an object whose center of gravity (CG) is below its point of support, such as a ball on a string, will be in stable equilibrium. If the CG is above the base of support, we have a more complicated situation. Consider a standing refrigerator (Fig. 12–12a). If it is tipped slightly, it will return to its original position due to the torque on it as shown in Fig. 12–12b. But if it is tipped too far, Fig. 12–12c, it will fall over. The critical point is reached when the CG shifts from one side of the pivot point to the other. When the CG is on one side, the torque pulls the object back onto its original base of support, Fig. 12–12b. If the object is tipped further, the CG goes past the pivot point and the torque causes the object to topple, Fig. 12–12c. In general, *an object whose center of gravity is above its base of support will be stable if a vertical line projected downward from the CG falls within the base of support.* This is because the normal force upward on the object (which balances out gravity) can be exerted only within the area of contact, so if the force of gravity acts beyond this area, a net torque will act to topple the object.

Stability, then, can be relative. A brick lying on its widest face is more stable than a brick standing on its end, for it will take more of an effort to tip it over. In the extreme case of the pencil in Fig. 12–11b, the base is practically a point and the slightest disturbance will topple it. In general, the larger the base and the lower the CG, the more stable the object.

In this sense, humans are less stable than four-legged mammals, which have a larger base of support because of their four legs, and most also have a lower center of gravity. When walking and performing other kinds of movement, a person continually shifts the body so that its CG is over the feet, although in the normal adult this requires no conscious thought. Even as simple a movement as bending over requires moving the hips backward so that the CG remains over the feet, and you do this repositioning without thinking about it. To see this, position yourself with your heels and back to a wall and try to touch your toes. You won't be able to do it without falling. People carrying heavy loads automatically adjust their posture so that the CG of the total mass is over their feet, Fig. 12–13.

12–4 Elasticity; Stress and Strain

In the first part of this Chapter we studied how to calculate the forces on objects in equilibrium. In this Section we study the effects of these forces: any object changes shape under the action of applied forces. If the forces are great enough, the object will break, or *fracture*, as we will discuss in Section 12–5.

Elasticity and Hooke's Law

If a force is exerted on an object, such as the vertically suspended metal rod shown in Fig. 12–14, the length of the object changes. If the amount of elongation, $\Delta\ell$, is small compared to the length of the object, experiment shows that $\Delta\ell$ is proportional to the force exerted on the object. This proportionality, as we saw in Section 7–3, can be written as an equation:

$$F = k\,\Delta\ell. \tag{12-3}$$

Here F represents the force pulling on the object, $\Delta\ell$ is the change in length, and k is a proportionality constant. Equation 12–3, which is sometimes called **Hooke's law**[†] after Robert Hooke (1635–1703), who first noted it, is found to be valid for almost any solid material from iron to bone—but it is valid only up to a point. For if the force is too great, the object stretches excessively and eventually breaks.

Figure 12–15 shows a typical graph of applied force versus elongation. Up to a point called the **proportional limit**, Eq. 12–3 is a good approximation for many

FIGURE 12–14 Hooke's law: $\Delta\ell \propto$ applied force.

$\Delta\ell$

$m\vec{g}$

m

[†]The term "law" applied to this relation is not really appropriate, since first of all, it is only an approximation, and second, it refers only to a limited set of phenomena. Most physicists prefer to reserve the word "law" for those relations that are deeper and more encompassing and precise, such as Newton's laws of motion or the law of conservation of energy.

common materials, and the curve is a straight line. Beyond this point, the graph deviates from a straight line, and no simple relationship exists between F and $\Delta\ell$. Nonetheless, up to a point farther along the curve called the **elastic limit**, the object will return to its original length if the applied force is removed. The region from the origin to the elastic limit is called the *elastic region*. If the object is stretched beyond the elastic limit, it enters the *plastic region*: it does not return to the original length upon removal of the external force, but remains permanently deformed (such as a bent paper clip). The maximum elongation is reached at the *breaking point*. The maximum force that can be applied without breaking is called the **ultimate strength** of the material (actually, force per unit area, as we discuss in Section 12–5).

Young's Modulus

The amount of elongation of an object, such as the rod shown in Fig. 12–14, depends not only on the force applied to it, but also on the material of which it is made and on its dimensions. That is, the constant k in Eq. 12–3 can be written in terms of these factors.

If we compare rods made of the same material but of different lengths and cross-sectional areas, it is found that for the same applied force, the amount of stretch (again assumed small compared to the total length) is proportional to the original length and inversely proportional to the cross-sectional area. That is, the longer the object, the more it elongates for a given force; and the thicker it is, the less it elongates. These findings can be combined with Eq. 12–3 to yield

$$\Delta\ell = \frac{1}{E}\frac{F}{A}\ell_0, \tag{12–4}$$

where ℓ_0 is the original length of the object, A is the cross-sectional area, and $\Delta\ell$ is the change in length due to the applied force F. E is a constant of proportionality[†] known as the **elastic modulus**, or **Young's modulus**; its value depends only on the material. The value of Young's modulus for various materials is given in Table 12–1 (the shear modulus and bulk modulus in this Table are discussed later in this Section). Because E is a property only of the material and is independent of the object's size or shape, Eq. 12–4 is far more useful for practical calculation than Eq. 12–3.

[†]The fact that E is in the denominator, so $1/E$ is the actual proportionality constant, is merely a convention. When we rewrite Eq. 12–4 to get Eq. 12–5, E is found in the numerator.

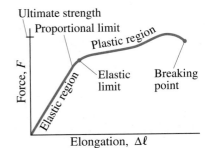

FIGURE 12–15 Applied force vs. elongation for a typical metal under tension.

TABLE 12–1 Elastic Moduli

Material	Young's Modulus, E (N/m²)	Shear Modulus, G (N/m²)	Bulk Modulus, B (N/m²)
Solids			
Iron, cast	100×10^9	40×10^9	90×10^9
Steel	200×10^9	80×10^9	140×10^9
Brass	100×10^9	35×10^9	80×10^9
Aluminum	70×10^9	25×10^9	70×10^9
Concrete	20×10^9		
Brick	14×10^9		
Marble	50×10^9		70×10^9
Granite	45×10^9		45×10^9
Wood (pine) (parallel to grain)	10×10^9		
(perpendicular to grain)	1×10^9		
Nylon	5×10^9		
Bone (limb)	15×10^9	80×10^9	
Liquids			
Water			2.0×10^9
Alcohol (ethyl)			1.0×10^9
Mercury			2.5×10^9
Gases[†]			
Air, H_2, He, CO_2			1.01×10^5

[†]At normal atmospheric pressure; no variation in temperature during process.

EXAMPLE 12-7 **Tension in piano wire.** A 1.60-m-long steel piano wire has a diameter of 0.20 cm. How great is the tension in the wire if it stretches 0.25 cm when tightened?

APPROACH We assume Hooke's law holds, and use it in the form of Eq. 12-4, finding E for steel in Table 12-1.

SOLUTION We solve for F in Eq. 12-4 and note that the area of the wire is $A = \pi r^2 = (3.14)(0.0010 \text{ m})^2 = 3.14 \times 10^{-6} \text{ m}^2$. Then

$$F = E \frac{\Delta\ell}{\ell_0} A = (2.0 \times 10^{11} \text{ N/m}^2)\left(\frac{0.0025 \text{ m}}{1.60 \text{ m}}\right)(3.14 \times 10^{-6} \text{ m}^2) = 980 \text{ N}.$$

NOTE The large tension in all the wires in a piano must be supported by a strong frame.

EXERCISE G Two steel wires have the same length and are under the same tension. But wire A has twice the diameter of wire B. Which of the following is true? (*a*) Wire B stretches twice as much as wire A. (*b*) Wire B stretches four times as much as wire A. (*c*) Wire A stretches twice as much as wire B. (*d*) Wire A stretches four times as much as wire B. (*e*) Both wires stretch the same amount.

Stress and Strain

From Eq. 12-4, we see that the change in length of an object is directly proportional to the product of the object's length ℓ_0 and the force per unit area F/A applied to it. It is general practice to define the force per unit area as the **stress**:

$$\text{stress} = \frac{\text{force}}{\text{area}} = \frac{F}{A},$$

which has SI units of N/m². Also, the **strain** is defined to be the ratio of the change in length to the original length:

$$\text{strain} = \frac{\text{change in length}}{\text{original length}} = \frac{\Delta\ell}{\ell_0},$$

and is dimensionless (no units). Strain is thus the fractional change in length of the object, and is a measure of how much the rod has been deformed. Stress is applied to the material by external agents, whereas strain is the material's response to the stress. Equation 12-4 can be rewritten as

$$\frac{F}{A} = E \frac{\Delta\ell}{\ell_0} \tag{12-5}$$

or

$$E = \frac{F/A}{\Delta\ell/\ell_0} = \frac{\text{stress}}{\text{strain}}.$$

Thus we see that the strain is directly proportional to the stress, in the linear (elastic) region of Fig. 12-15.

FIGURE 12-16 Stress exists *within* the material.

(a)   (b)

Tension, Compression, and Shear Stress

The rod shown in Fig. 12-16a is said to be under *tension* or **tensile stress**. Not only is there a force pulling down on the rod at its lower end, but since the rod is in equilibrium, we know that the support at the top is exerting an equal[†] upward force on the rod at its upper end, Fig. 12-16a. In fact, this tensile stress exists throughout the material. Consider, for example, the lower half of a suspended rod as shown in Fig. 12-16b. This lower half is in equilibrium, so there must be an upward force on it to balance the downward force at its lower end. What exerts this upward force? It must be the upper part of the rod. Thus we see that external forces applied to an object give rise to internal forces, or stress, within the material itself.

[†]Or a greater force if the weight of the rod cannot be ignored compared to F.

Strain or deformation due to tensile stress is but one type of stress to which materials can be subjected. There are two other common types of stress: compressive and shear. **Compressive stress** is the exact opposite of tensile stress. Instead of being stretched, the material is compressed: the forces act inwardly on the object. Columns that support a weight, such as the columns of a Greek temple (Fig. 12–17), are subjected to compressive stress. Equations 12–4 and 12–5 apply equally well to compression and tension, and the values for the modulus E are usually the same.

FIGURE 12–17 This Greek temple, in Agrigento, Sicily, built 2500 years ago, shows the post-and-beam construction. The columns are under compression.

Tension (a) Compression (b) Shear (c)

FIGURE 12–18 The three types of stress for rigid objects.

Figure 12–18 compares tensile and compressive stresses as well as the third type, shear stress. An object under **shear stress** has equal and opposite forces applied *across* its opposite faces. A simple example is a book or brick firmly attached to a tabletop, on which a force is exerted parallel to the top surface. The table exerts an equal and opposite force along the bottom surface. Although the dimensions of the object do not change significantly, the shape of the object does change, Fig. 12–18c. An equation similar to Eq. 12–4 can be applied to calculate shear strain:

$$\Delta\ell = \frac{1}{G}\frac{F}{A}\ell_0, \tag{12–6}$$

but $\Delta\ell$, ℓ_0, and A must be reinterpreted as indicated in Fig. 12–18c. Note that A is the area of the surface *parallel* to the applied force (and not perpendicular as for tension and compression), and $\Delta\ell$ is *perpendicular* to ℓ_0. The constant of proportionality G is called the **shear modulus** and is generally one-half to one-third the value of Young's modulus E (see Table 12–1). Figure 12–19 suggests why $\Delta\ell \propto \ell_0$: the fatter book shifts more for the same shearing force.

FIGURE 12–19 The fatter book (a) shifts more than the thinner book (b) with the same applied shear force.

(a)

Volume Change—Bulk Modulus

If an object is subjected to inward forces from all sides, its volume will decrease. A common situation is an object submerged in a fluid; in this case, the fluid exerts a pressure on the object in all directions, as we shall see in Chapter 13. *Pressure* is defined as force per unit area, and thus is the equivalent of stress. For this situation the change in volume, ΔV, is proportional to the original volume, V_0, and to the change in the pressure, ΔP. We thus obtain a relation of the same form as Eq. 12–4 but with a proportionality constant called the **bulk modulus** B:

(b)

$$\frac{\Delta V}{V_0} = -\frac{1}{B}\Delta P \tag{12–7}$$

or

$$B = -\frac{\Delta P}{\Delta V/V_0}.$$

The minus sign means the volume *decreases* with an increase in pressure.

Values for the bulk modulus are given in Table 12–1. Since liquids and gases do not have a fixed shape, only the bulk modulus (not the Young's or shear moduli) applies to them.

Tension

Compression

Shear

FIGURE 12–20 Fracture as a result of the three types of stress.

12–5 Fracture

If the stress on a solid object is too great, the object fractures, or breaks (Fig. 12–20). Table 12–2 lists the ultimate strengths for tension, compression, and shear for a variety of materials. These values give the maximum force per unit area, or stress, that an object can withstand under each of these three types of stress for various types of material. They are, however, representative values only, and the actual value for a given specimen can differ considerably. It is therefore necessary to maintain a *safety factor* of from 3 to perhaps 10 or more—that is, the actual stresses on a structure should not exceed one-tenth to one-third of the values given in the Table. You may encounter tables of "allowable stresses" in which appropriate safety factors have already been included.

TABLE 12–2 Ultimate Strengths of Materials (force/area)

Material	Tensile Strength (N/m²)	Compressive Strength (N/m²)	Shear Strength (N/m²)
Iron, cast	170×10^6	550×10^6	170×10^6
Steel	500×10^6	500×10^6	250×10^6
Brass	250×10^6	250×10^6	200×10^6
Aluminum	200×10^6	200×10^6	200×10^6
Concrete	2×10^6	20×10^6	2×10^6
Brick		35×10^6	
Marble		80×10^6	
Granite		170×10^6	
Wood (pine) (parallel to grain) (perpendicular to grain)	40×10^6	35×10^6 10×10^6	5×10^6
Nylon	500×10^6		
Bone (limb)	130×10^6	170×10^6	

EXAMPLE 12–8 ESTIMATE Breaking the piano wire. The steel piano wire we discussed in Example 12–7 was 1.60 m long with a diameter of 0.20 cm. Approximately what tension force would break it?

APPROACH We set the tensile stress F/A equal to the tensile strength of steel given in Table 12–2.

SOLUTION The area of the wire is $A = \pi r^2$, where $r = 0.10$ cm $= 1.0 \times 10^{-3}$ m. Table 12–2 tells us

$$\frac{F}{A} = 500 \times 10^6 \, \text{N/m}^2,$$

so the wire would likely break if the force exceeded

$$F = \left(500 \times 10^6 \, \text{N/m}^2\right)(\pi)(1.0 \times 10^{-3} \, \text{m})^2 = 1600 \, \text{N}.$$

As can be seen in Table 12–2, concrete (like stone and brick) is reasonably strong under compression but extremely weak under tension. Thus concrete can be used as vertical columns placed under compression, but is of little value as a beam because it cannot withstand the tensile forces that result from the inevitable sagging of the lower edge of a beam (see Fig. 12–21).

FIGURE 12–21 A beam sags, at least a little (but is exaggerated here), even under its own weight. The beam thus changes shape: the upper edge is compressed, and the lower edge is under tension (elongated). Shearing stress also occurs within the beam.

Compression

Tension

Reinforced concrete, in which iron rods are embedded in the concrete (Fig. 12–22), is much stronger. Stronger still is *prestressed concrete*, which also contains iron rods or a wire mesh, but during the pouring of the concrete, the rods or wire are held under tension. After the concrete dries, the tension on the iron is released, putting the concrete under compression. The amount of compressive stress is carefully predetermined so that when loads are applied to the beam, they reduce the compression on the lower edge, but never put the concrete into tension.

FIGURE 12–22 Steel rods around which concrete is poured for strength.

CONCEPTUAL EXAMPLE 12–9 **A tragic substitution.** Two walkways, one above the other, are suspended from vertical rods attached to the ceiling of a high hotel lobby, Fig. 12–23a. The original design called for single rods 14 m long, but when such long rods proved to be unwieldy to install, it was decided to replace each long rod with two shorter ones as shown schematically in Fig. 12–23b. Determine the net force exerted by the rods on the supporting pin A (assumed to be the same size) for each design. Assume each vertical rod supports a mass m of each bridge.

RESPONSE The single long vertical rod in Fig. 12–23a exerts an upward force equal to mg on pin A to support the mass m of the upper bridge. Why? Because the pin is in equilibrium, and the other force that balances this is the downward force mg exerted on it by the upper bridge (Fig. 12–23c). There is thus a shear stress on the pin because the rod pulls up on one side of the pin, and the bridge pulls down on the other side. The situation when two shorter rods support the bridges (Fig. 12–23b) is shown in Fig. 12–23d, in which only the connections at the upper bridge are shown. The lower rod exerts a force mg downward on the lower of the two pins because it supports the lower bridge. The upper rod exerts a force $2mg$ on the upper pin (labelled A) because the upper rod supports both bridges. Thus we see that when the builders substituted two shorter rods for each single long one, the stress in the supporting pin A was *doubled*. What perhaps seemed like a simple substitution did, in fact, lead to a tragic collapse in 1981 with a loss of life of over 100 people (see Fig. 12–1). Having a feel for physics, and being able to make simple calculations based on physics, can have a great effect, literally, on people's lives.

EXAMPLE 12–10 **Shear on a beam.** A uniform pine beam, 3.6 m long and 9.5 cm × 14 cm in cross section, rests on two supports near its ends, as shown in Fig. 12–24. The beam's mass is 25 kg and two vertical roof supports rest on it, each one-third of the way from the ends. What maximum load force F_L can each of the roof supports exert without shearing the pine beam at its supports? Use a safety factor of 5.0.

APPROACH The symmetry present simplifies our calculation. We first find the shear strength of pine in Table 12–2 and use the safety factor of 5.0 to get F from $F/A \le \frac{1}{5}$ (shear strength). Then we use $\Sigma\tau = 0$ to find F_L.

SOLUTION Each support exerts an upward force F (there is symmetry) that can be at most (see Table 12–2)

$$F = \frac{1}{5}A(5 \times 10^6 \text{ N/m}^2) = \frac{1}{5}(0.095 \text{ m})(0.14 \text{ m})(5 \times 10^6 \text{ N/m}^2) = 13{,}000 \text{ N}.$$

To determine the maximum load force F_L, we calculate the torque about the left end of the beam (counterclockwise positive):

$$\Sigma\tau = -F_L(1.2 \text{ m}) - (25 \text{ kg})(9.8 \text{ m/s}^2)(1.8 \text{ m}) - F_L(2.4 \text{ m}) + F(3.6 \text{ m}) = 0$$

so each of the two roof supports can exert

$$F_L = \frac{(13{,}000 \text{ N})(3.6 \text{ m}) - (250 \text{ N})(1.8 \text{ m})}{(1.2 + 2.4)} = 13{,}000 \text{ N}.$$

The total mass of roof the beam can support is $(2)(13{,}000 \text{ N})/(9.8 \text{ m/s}^2) = 2600 \text{ kg}$.

PHYSICS APPLIED
A tragic collapse

FIGURE 12–23 Example 12–9.

(a)   (b)

(c) Force on pin A exerted by vertical rod

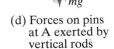

(d) Forces on pins at A exerted by vertical rods

FIGURE 12–24 Example 12–10.

Joints

FIGURE 12–25 A truss bridge.

*12–6 Trusses and Bridges

A beam used to span a wide space, as for a bridge, is subject to strong stresses of all three types as we saw in Fig. 12–21: compression, tension and shear. A basic engineering device to support large spans is the *truss*, an example of which is shown in Fig. 12–25. Wooden truss bridges were first designed by the great architect Andrea Palladio (1518–1580), famous for his design of public buildings and villas. With the introduction of steel in the nineteenth century, much stronger steel trusses came into use, although wood trusses are still used to support the roofs of houses and mountain lodges (Fig. 12–26).

Basically, a **truss** is a framework of rods or struts joined together at their ends by pins or rivets, always arranged as triangles. (Triangles are relatively stable, as compared to a rectangle, which easily becomes a parallelogram under sideways forces and then collapses.) The place where the struts are joined by a pin is called a **joint**.

It is commonly assumed that the struts of a truss are under pure compression or pure tension—that is, the forces act along the length of each strut, Fig. 12–27a. This is an ideal, valid only if a strut has no mass and supports no weight along its length, in which case a strut has only two forces on it, at the ends, as shown in Fig. 12–27a. If the strut is in equilibrium, these two forces must be equal and opposite in direction $(\Sigma \vec{F} = 0)$. But couldn't they be at an angle, as in Fig. 12–27b? No, because then $\Sigma \vec{\tau}$ would not be zero. The two forces *must* act along the strut if the strut is in equilibrium. But in a real case of a strut with mass, there are three forces on the strut, as shown in Fig. 12–27c, and $\vec{F}_1$ and $\vec{F}_2$ do not act along the strut; the vector diagram in Fig. 12–27d shows $\Sigma \vec{F} = \vec{F}_1 + \vec{F}_2 + m\vec{g} = 0$. Can you see why $\vec{F}_1$ and $\vec{F}_2$ both point *above* the strut? (Do $\Sigma \tau$ about each end.)

FIGURE 12–26 A roof truss.

FIGURE 12–27 (a) Each massless strut (or rod) of a truss is assumed to be under tension or compression. (b) The two equal and opposite forces must be along the same line or a net torque would exist. (c) Real struts have mass, so the forces $\vec{F}_1$ and $\vec{F}_2$ at the joints do not act precisely along the strut. (d) Vector diagram of part (c).

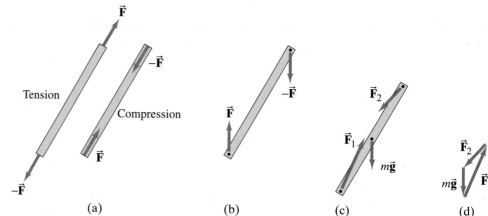

Tension

Compression

(a)   (b)   (c)   (d)

Consider again the simple beam in Example 12–5, Fig. 12–9. The force $\vec{F}_H$ at the pin is *not* along the beam, but acts at an upward angle. If that beam were massless, we see from Eq. (iii) in Example 12–5 with $m = 0$, that $F_{Hy} = 0$, and $\vec{F}_H$ would be along the beam.

The assumption that the forces in each strut of a truss are purely along the strut is still very useful whenever the loads act only at the joints and are much greater than the weight of the struts themselves.

EXAMPLE 12–11 **A truss bridge.** Determine the tension or compression in each of the struts of the truss bridge shown in Fig. 12–28a. The bridge is 64 m long and supports a uniform level concrete roadway whose total mass is 1.40×10^6 kg. Use the **method of joints**, which involves (1) drawing a free-body diagram of the truss as a whole, and (2) drawing a free-body diagram for each of the pins (joints), one by one, and setting $\Sigma \vec{F} = 0$ for each pin. Ignore the mass of the struts. Assume all triangles are equilateral.

APPROACH Any bridge has two trusses, one on each side of the roadway. Consider only one truss, Fig. 12–28a, and it will support half the weight of the roadway. That is, our truss supports a total mass $M = 7.0 \times 10^5$ kg. First we draw a free-body diagram for the entire truss as a single unit, which we assume rests on supports at either end that exert upward forces $\vec{F}_1$ and $\vec{F}_2$, Fig. 12–28b. We assume the mass of the roadway acts entirely at the center, on pin C, as shown. From symmetry we can see that each of the end supports carries half the weight [or do a torque equation about, say, point A: $(F_2)(\ell) - Mg(\ell/2) = 0$], so

$$F_1 = F_2 = \tfrac{1}{2} Mg.$$

SOLUTION We look at pin A and apply $\Sigma \vec{F} = 0$ to it. We label the forces on pin A due to each strut with two subscripts: $\vec{F}_{AB}$ means the force exerted by the strut AB and $\vec{F}_{AC}$ is the force exerted by strut AC. $\vec{F}_{AB}$ and $\vec{F}_{AC}$ act along their respective struts; but not knowing whether each is compressive or tensile, we could draw four different free-body diagrams, as shown in Fig. 12–28c. Only the one on the left could provide $\Sigma \vec{F} = 0$, so we immediately know the directions of $\vec{F}_{AB}$ and $\vec{F}_{AC}$.[†] These forces act on the pin. The force that pin A exerts on strut AB is opposite in direction to $\vec{F}_{AB}$ (Newton's third law), so strut AB is under compression and strut AC is under tension. Now let's calculate the magnitudes of $\vec{F}_{AB}$ and $\vec{F}_{AC}$. At pin A:

$$\Sigma F_x = F_{AC} - F_{AB} \cos 60° = 0$$
$$\Sigma F_y = F_1 - F_{AB} \sin 60° = 0.$$

Thus

$$F_{AB} = \frac{F_1}{\sin 60°} = \frac{\tfrac{1}{2} Mg}{\tfrac{1}{2} \sqrt{3}} = \frac{1}{\sqrt{3}} Mg,$$

which equals $(7.0 \times 10^5 \text{ kg})(9.8 \text{ m/s}^2)/\sqrt{3} = 4.0 \times 10^6$ N; and

$$F_{AC} = F_{AB} \cos 60° = \frac{1}{2\sqrt{3}} Mg.$$

Next we look at pin B, and Fig. 12–28d is the free-body diagram. [Convince yourself that if $\vec{F}_{BD}$ or $\vec{F}_{BC}$ were in the opposite direction, $\Sigma \vec{F}$ could not be zero; note that $\vec{F}_{BA} = -\vec{F}_{AB}$ (and $F_{BA} = F_{AB}$) because now we are at the opposite end of strut AB.] We see that BC is under tension and BD compression. (Recall that the forces on the struts are opposite to the forces shown which are on the pin.) We set $\Sigma \vec{F} = 0$:

$$\Sigma F_x = F_{BA} \cos 60° + F_{BC} \cos 60° - F_{BD} = 0$$
$$\Sigma F_y = F_{BA} \sin 60° - F_{BC} \sin 60° = 0.$$

Then, because $F_{BA} = F_{AB}$, we have

$$F_{BC} = F_{AB} = \frac{1}{\sqrt{3}} Mg,$$

and

$$F_{BD} = F_{AB} \cos 60° + F_{BC} \cos 60° = \frac{1}{\sqrt{3}} Mg \left(\tfrac{1}{2}\right) + \frac{1}{\sqrt{3}} Mg \left(\tfrac{1}{2}\right) = \frac{1}{\sqrt{3}} Mg.$$

The solution is complete. By symmetry, $F_{DE} = F_{AB}$, $F_{CE} = F_{AC}$, and $F_{CD} = F_{BC}$.

NOTE As a check, calculate ΣF_x and ΣF_y for pin C and see if they equal zero. Figure 12–28e shows the free-body diagram.

[†]If we were to choose the direction of a force on a diagram opposite to what it really is, we would get a minus sign.

⊛ **PHYSICS APPLIED**
A truss bridge

▦ **PROBLEM SOLVING**
Method of joints

(a)

(b)

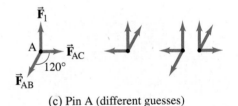

(c) Pin A (different guesses)

(d) Pin B

(e) Pin C

FIGURE 12–28
Example 12–11. (a) A truss bridge. Free-body diagrams:
(b) for the entire truss,
(c) for pin A (different guesses),
(d) for pin B and (e) for pin C.

FIGURE 12–29 (a) Truss with truck of mass m at center of strut AC. (b) Forces on strut AC.

⊕ PHYSICS APPLIED

Suspension bridge

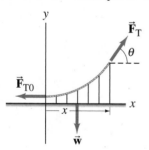

FIGURE 12–30 Suspension bridges (Brooklyn and Manhattan bridges, NY).

FIGURE 12–31 Example 12–12.

Example 12–11 put the roadway load at the center, C. Now consider a heavy load, such as a heavy truck, supported by strut AC at its middle, as shown in Fig. 12–29a. The strut AC sags under this load, telling us there is shear stress in strut AC. Figure 12–29b shows the forces exerted on strut AC: the weight of the truck $m\vec{g}$, and the forces $\vec{F}_A$ and $\vec{F}_C$ that pins A and C exert on the strut. [Note that $\vec{F}_1$ does not appear because it is a force (exerted by external supports) that acts on pin A, not on strut AC.] The forces that pins A and C exert on strut AC will act not only along the strut, but will have vertical components too, perpendicular to the strut to balance the weight of the truck, $m\vec{g}$, creating shear stress. The other struts, not bearing weight, remain under pure tension or compression. Problems 53 and 54 deal with this situation, and an early step in their solution is to calculate the forces $\vec{F}_A$ and $\vec{F}_C$ by using torque equations for the strut.

For very large bridges, truss structures are too heavy. One solution is to build suspension bridges, with the load being carried by relatively light suspension cables under tension, supporting the roadway by means of closely spaced vertical wires, as shown in Fig. 12–30, and in the photo on the first page of this Chapter.

EXAMPLE 12–12 **Suspension bridge.** Determine the shape of the cable between the two towers of a suspension bridge (as in Fig. 12–30), assuming the weight of the roadway is supported uniformly along its length. Ignore the weight of the cable.

APPROACH We take $x = 0$, $y = 0$ at the center of the span, as shown in Fig. 12–31. Let $\vec{F}_{T0}$ be the tension in the cable at $x = 0$; it acts horizontally as shown. Let F_T be the tension in the cable at some other place where the horizontal coordinate is x, as shown. This section of cable supports a portion of the roadway whose weight w is proportional to the distance x, since the roadway is assumed uniform; that is,

$$w = \lambda x$$

where λ is the weight per unit length.

SOLUTION We set $\Sigma \vec{F} = 0$:

$$\Sigma F_x = F_T \cos\theta - F_{T0} = 0$$

$$\Sigma F_y = F_T \sin\theta - w = 0.$$

We divide these two equations,

$$\tan\theta = \frac{w}{F_{T0}} = \frac{\lambda x}{F_{T0}}.$$

The slope of our curve (the cable) at any point is

$$\frac{dy}{dx} = \tan\theta$$

or

$$\frac{dy}{dx} = \frac{\lambda}{F_{T0}} x.$$

We integrate this:

$$\int dy = \frac{\lambda}{F_{T0}} \int x \, dx$$

$$y = Ax^2 + B$$

where we set $A = \lambda/F_{T0}$ and B is a constant of integration. This is just the equation of a parabola.

NOTE Real bridges have cables that do have mass, so the cables hang only approximately as a parabola, although often it is quite close.

*12–7 Arches and Domes

There are various ways that engineers and architects can span a space, such as beams, trusses, and suspension bridges. In this Section we discuss arches and domes.

FIGURE 12–32 Round arches in the Roman Forum. The one in the background is the Arch of Titus.

FIGURE 12–33 An arch is used here to good effect in spanning a chasm on the California coast.

The semicircular **arch** (Figs. 12–32 and 12–33) was introduced by the ancient Romans 2000 years ago. Aside from its aesthetic appeal, it was a tremendous technological innovation. The advantage of the "true" or semicircular arch is that, if well designed, its wedge-shaped stones experience stress which is mainly compressive even when supporting a large load such as the wall and roof of a cathedral. Because the stones are forced to squeeze against one another, they are mainly under compression (see Fig. 12–34). Note, however, that the arch transfers horizontal as well as vertical forces to the supports. A round arch consisting of many well-shaped stones could span a very wide space. However, considerable buttressing on the sides is needed to support the horizontal components of the forces.

PHYSICS APPLIED

Architecture: Beams, arches and domes

FIGURE 12–34 Stones in a round arch (see Fig. 12–32) are mainly under compression.

The pointed arch came into use about A.D. 1100 and became the hallmark of the great Gothic cathedrals. It too was an important technical innovation, and was first used to support heavy loads such as the tower and arch of a cathedral. Apparently the builders realized that, because of the steepness of the pointed arch, the forces due to the weight above could be brought down more nearly vertically, so less horizontal buttressing would be needed. The pointed arch reduced the load on the walls, so there could be more openness and light. The smaller buttressing needed was provided on the outside by graceful flying buttresses (Fig. 12–35).

FIGURE 12–35 Flying buttresses (on the cathedral of Notre Dame, in Paris).

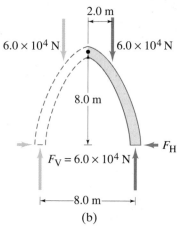

FIGURE 12–36 (a) Forces in a round arch, compared (b) with those in a pointed arch.

To make an accurate analysis of a stone arch is quite difficult in practice. But if we make some simplifying assumptions, we can show why the horizontal component of the force at the base is less for a pointed arch than for a round one. Figure 12–36 shows a round arch and a pointed arch, each with an 8.0-m span. The height of the round arch is thus 4.0 m, whereas that of the pointed arch is larger and has been chosen to be 8.0 m. Each arch supports a weight of 12.0×10^4 N ($= 12{,}000$ kg $\times g$) which, for simplicity, we have divided into two parts (each 6.0×10^4 N) acting on the two halves of each arch as shown. To be in equilibrium, each of the supports must exert an upward force of 6.0×10^4 N. For rotational equilibrium, each support also exerts a horizontal force, F_H, at the base of the arch, and it is this we want to calculate. We focus only on the right half of each arch. We set equal to zero the total torque calculated about the apex of the arch due to the forces exerted on that half arch. For the round arch, the torque equation ($\Sigma \tau = 0$) is (see Fig. 12–36a)

$$(4.0\text{ m})(6.0 \times 10^4\text{ N}) - (2.0\text{ m})(6.0 \times 10^4\text{ N}) - (4.0\text{ m})(F_H) = 0.$$

Thus $F_H = 3.0 \times 10^4$ N for the round arch. For the pointed arch, the torque equation is (see Fig. 12–36b)

$$(4.0\text{ m})(6.0 \times 10^4\text{ N}) - (2.0\text{ m})(6.0 \times 10^4\text{ N}) - (8.0\text{ m})(F_H) = 0.$$

Solving, we find that $F_H = 1.5 \times 10^4$ N—only half as much as for the round arch! From this calculation we can see that the horizontal buttressing force required for a pointed arch is less because the arch is higher, and there is therefore a longer lever arm for this force. Indeed, the steeper the arch, the less the horizontal component of the force needs to be, and hence the more nearly vertical is the force exerted at the base of the arch.

Whereas an arch spans a two-dimensional space, a **dome**—which is basically an arch rotated about a vertical axis—spans a three-dimensional space. The Romans built the first large domes. Their shape was hemispherical and some still stand, such as that of the Pantheon in Rome (Fig. 12–37), built 2000 years ago.

FIGURE 12–37 Interior of the Pantheon in Rome, built almost 2000 years ago. This view, showing the great dome and its central opening for light, was painted about 1740 by Panini. Photographs do not capture its grandeur as well as this painting does.

FIGURE 12–38 The skyline of Florence, showing Brunelleschi's dome on the cathedral.

Fourteen centuries later, a new cathedral was being built in Florence. It was to have a dome 43 m in diameter to rival that of the Pantheon, whose construction has remained a mystery. The new dome was to rest on a "drum" with no external abutments. Filippo Brunelleschi (1377–1446) designed a pointed dome (Fig. 12–38), since a pointed dome, like a pointed arch, exerts a smaller side thrust against its base. A dome, like an arch, is not stable until all the stones are in place. To support smaller domes during construction, wooden frameworks were used. But no trees big enough or strong enough could be found to span the 43-m space required. Brunelleschi decided to try to build the dome in horizontal layers, each bonded to the previous one, holding it in place until the last stone of the circle was placed. Each closed ring was then strong enough to support the next layer. It was an amazing feat. Only in the twentieth century were larger domes built, the largest being that of the Superdome in New Orleans, completed in 1975. Its 200-m diameter dome is made of steel trusses and concrete.

Summary

An object at rest is said to be in **equilibrium**. The subject concerned with the determination of the forces within a structure at rest is called **statics**.

The two necessary conditions for an object to be in equilibrium are that (1) the vector sum of all the forces on it must be zero, and (2) the sum of all the torques (calculated about any arbitrary axis) must also be zero. For a 2-dimensional problem we can write

$$\Sigma F_x = 0, \qquad \Sigma F_y = 0, \qquad \Sigma \tau = 0. \quad \textbf{(12–1, 12–2)}$$

It is important when doing statics problems to apply the equilibrium conditions to only one object at a time.

An object in static equilibrium is said to be in (a) **stable**, (b) **unstable**, or (c) **neutral equilibrium**, depending on whether a slight displacement leads to (a) a return to the original position, (b) further movement away from the original position, or (c) rest in the new position. An object in stable equilibrium is also said to be in **balance**.

Hooke's law applies to many elastic solids, and states that the change in length of an object is proportional to the applied force:

$$F = k \, \Delta \ell. \quad \textbf{(12–3)}$$

If the force is too great, the object will exceed its **elastic limit**, which means it will no longer return to its original shape when the distorting force is removed. If the force is even greater, the **ultimate strength** of the material can be exceeded, and the object will **fracture**. The force per unit area acting on an object is the **stress**, and the resulting fractional change in length is the **strain**. The stress on an object is present within the object and can be of three types: **compression, tension,** or **shear**. The ratio of stress to strain is called the **elastic modulus** of the material. **Young's modulus** applies for compression and tension, and the **shear modulus** for shear. **Bulk modulus** applies to an object whose volume changes as a result of pressure on all sides. All three moduli are constants for a given material when distorted within the elastic region.

Questions

1. Describe several situations in which an object is not in equilibrium, even though the net force on it is zero.

2. A bungee jumper momentarily comes to rest at the bottom of the dive before he springs back upward. At that moment, is the bungee jumper in equilibrium? Explain.

3. You can find the center of gravity of a meter stick by resting it horizontally on your two index fingers, and then slowly drawing your fingers together. First the meter stick will slip on one finger, and then on the other, but eventually the fingers meet at the CG. Why does this work?

4. Your doctor's scale has arms on which weights slide to counter your weight, Fig. 12–39. These weights are much lighter than you are. How does this work?

Weights

FIGURE 12–39
Question 4.

5. A ground retaining wall is shown in Fig. 12–40a. The ground, particularly when wet, can exert a significant force F on the wall. (a) What force produces the torque to keep the wall upright? (b) Explain why the retaining wall in Fig. 12–40b would be much less likely to overturn than that in Fig. 12–40a.

(a)   (b)

FIGURE 12–40   Question 5.

6. Can the sum of the torques on an object be zero while the net force on the object is nonzero? Explain.

7. A ladder, leaning against a wall, makes a 60° angle with the ground. When is it more likely to slip: when a person stands on the ladder near the top or near the bottom? Explain.

8. A uniform meter stick supported at the 25-cm mark is in equilibrium when a 1-kg rock is suspended at the 0-cm end (as shown in Fig. 12–41). Is the mass of the meter stick greater than, equal to, or less than the mass of the rock? Explain your reasoning.

FIGURE 12–41   Question 8.

9. Why do you tend to lean backward when carrying a heavy load in your arms?

10. Figure 12–42 shows a cone. Explain how to lay it on a flat table so that it is in (a) stable equilibrium, (b) unstable equilibrium, (c) neutral equilibrium.

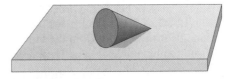

FIGURE 12–42   Question 10.

11. Place yourself facing the edge of an open door. Position your feet astride the door with your nose and abdomen touching the door's edge. Try to rise on your tiptoes. Why can't this be done?

12. Why is it not possible to sit upright in a chair and rise to your feet without first leaning forward?

13. Why is it more difficult to do sit-ups when your knees are bent than when your legs are stretched out?

14. Which of the configurations of brick, (*a*) or (*b*) of Fig. 12–43, is the more likely to be stable? Why?

(a)   (b)

FIGURE 12–43 Question 14. The dots indicate the CG of each brick. The fractions $\frac{1}{4}$ and $\frac{1}{2}$ indicate what portion of each brick is hanging beyond its support.

15. Name the type of equilibrium for each position of the ball in Fig. 12–44.

FIGURE 12–44
Question 15.

16. Is the Young's modulus for a bungee cord smaller or larger than that for an ordinary rope?

17. Examine how a pair of scissors or shears cuts through a piece of cardboard. Is the name "shears" justified? Explain.

18. Materials such as ordinary concrete and stone are very weak under tension or shear. Would it be wise to use such a material for either of the supports of the cantilever shown in Fig. 12–7? If so, which one(s)? Explain.

Problems

12–1 and 12–2 Equilibrium

1. (I) Three forces are applied to a tree sapling, as shown in Fig. 12–45, to stabilize it. If $\vec{F}_A = 385\,\text{N}$ and $\vec{F}_B = 475\,\text{N}$, find $\vec{F}_C$ in magnitude and direction.

FIGURE 12–45
Problem 1.

2. (I) Approximately what magnitude force, F_M, must the extensor muscle in the upper arm exert on the lower arm to hold a 7.3-kg shot put (Fig. 12–46)? Assume the lower arm has a mass of 2.3 kg and its CG is 12.0 cm from the elbow-joint pivot.

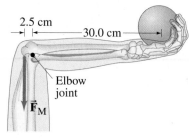

FIGURE 12–46
Problem 2.

3. (I) Calculate the mass *m* needed in order to suspend the leg shown in Fig. 12–47. Assume the leg (with cast) has a mass of 15.0 kg, and its CG is 35.0 cm from the hip joint; the sling is 78.0 cm from the hip joint.

FIGURE 12–47 Problem 3.

4. (I) A tower crane (Fig. 12–48a) must always be carefully balanced so that there is no net torque tending to tip it. A particular crane at a building site is about to lift a 2800-kg air-conditioning unit. The crane's dimensions are shown in Fig. 12–48b. (*a*) Where must the crane's 9500-kg counterweight be placed when the load is lifted from the ground? (Note that the counterweight is usually moved automatically via sensors and motors to precisely compensate for the load.) (*b*) Determine the maximum load that can be lifted with this counterweight when it is placed at its full extent. Ignore the mass of the beam.

(a)

FIGURE 12–48
Problem 4.

(b)

5. (II) Calculate the forces F_A and F_B that the supports exert on the diving board of Fig. 12–49 when a 52-kg person stands at its tip. (*a*) Ignore the weight of the board. (*b*) Take into account the board's mass of 28 kg. Assume the board's CG is at its center.

FIGURE 12–49
Problem 5.

6. (II) Two cords support a chandelier in the manner shown in Fig 12–3 except that the upper cord makes an angle of 45° with the ceiling. If the cords can sustain a force of 1660 N without breaking, what is the maximum chandelier weight that can be supported?

7. (II) The two trees in Fig. 12–50 are 6.6 m apart. A back-packer is trying to lift his pack out of the reach of bears. Calculate the magnitude of the force $\vec{F}$ that he must exert down-ward to hold a 19-kg backpack so that the rope sags at its midpoint by (a) 1.5 m, (b) 0.15 m.

FIGURE 12–50
Problems 7 and 83.

8. (II) A 110-kg horizontal beam is supported at each end. A 320-kg piano rests a quarter of the way from one end. What is the vertical force on each of the supports?

9. (II) Calculate F_A and F_B for the uniform cantilever shown in Fig. 12–7 whose mass is 1200 kg.

10. (II) A 75-kg adult sits at one end of a 9.0-m-long board. His 25-kg child sits on the other end. (a) Where should the pivot be placed so that the board is balanced, ignoring the board's mass? (b) Find the pivot point if the board is uniform and has a mass of 15 kg.

11. (II) Find the tension in the two cords shown in Fig. 12–51. Neglect the mass of the cords, and assume that the angle θ is 33° and the mass m is 190 kg.

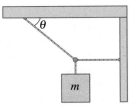

FIGURE 12–51
Problem 11.

12. (II) Find the tension in the two wires supporting the traffic light shown in Fig. 12–52.

FIGURE 12–52
Problem 12.

13. (II) How close to the edge of the 24.0-kg table shown in Fig. 12–53 can a 66.0-kg person sit without tipping it over?

FIGURE 12–53
Problem 13.

14. (II) The force required to pull the cork out of the top of a wine bottle is in the range of 200 to 400 N. A common bottle opener is shown in Fig. 12–54. What range of forces F is required to open a wine bottle with this device?

FIGURE 12–54
Problem 14.

15. (II) Calculate F_A and F_B for the beam shown in Fig. 12–55. The downward forces represent the weights of machinery on the beam. Assume the beam is uniform and has a mass of 280 kg.

FIGURE 12–55
Problem 15.

16. (II) (a) Calculate the magnitude of the force, F_M, required of the "deltoid" muscle to hold up the outstretched arm shown in Fig. 12–56. The total mass of the arm is 3.3 kg. (b) Calculate the magnitude of the force F_J exerted by the shoulder joint on the upper arm and the angle (to the horizontal) at which it acts.

FIGURE 12–56   Problems 16 and 17.

17. (II) Suppose the hand in Problem 16 holds an 8.5-kg mass. What force, F_M, is required of the deltoid muscle, assuming the mass is 52 cm from the shoulder joint?

18. (II) Three children are trying to balance on a seesaw, which includes a fulcrum rock acting as a pivot at the center, and a very light board 3.2 m long (Fig. 12–57). Two play-mates are already on either end. Boy A has a mass of 45 kg, and boy B a mass of 35 kg. Where should girl C, whose mass is 25 kg, place herself so as to balance the seesaw?

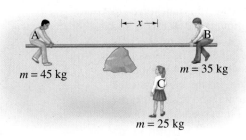

FIGURE 12–57   Problem 18.

19. (II) The Achilles tendon is attached to the rear of the foot as shown in Fig. 12–58. When a person elevates himself just barely off the floor on the "ball of one foot," estimate the tension F_T in the Achilles tendon (pulling upward), and the (downward) force F_B exerted by the lower leg bone on the foot. Assume the person has a mass of 72 kg and D is twice as long as d.

FIGURE 12–58
Problem 19.

20. (II) A shop sign weighing 215 N is supported by a uniform 155-N beam as shown in Fig. 12–59. Find the tension in the guy wire and the horizontal and vertical forces exerted by the hinge on the beam.

FIGURE 12–59
Problem 20.

21. (II) A traffic light hangs from a pole as shown in Fig. 12–60. The uniform aluminum pole AB is 7.20 m long and has a mass of 12.0 kg. The mass of the traffic light is 21.5 kg. Determine (a) the tension in the horizontal massless cable CD, and (b) the vertical and horizontal components of the force exerted by the pivot A on the aluminum pole.

FIGURE 12–60
Problem 21.

22. (II) A uniform steel beam has a mass of 940 kg. On it is resting half of an identical beam, as shown in Fig. 12–61. What is the vertical support force at each end?

FIGURE 12–61
Problem 22.

23. (II) Two wires run from the top of a pole 2.6 m tall that supports a volleyball net. The two wires are anchored to the ground 2.0 m apart, and each is 2.0 m from the pole (Fig. 12–62). The tension in each wire is 115 N. What is the tension in the net, assumed horizontal and attached at the top of the pole?

FIGURE 12–62
Problem 23.

24. (II) A large 62.0-kg board is propped at a 45° angle against the edge of a barn door that is 2.6 m wide. How great a horizontal force must a person behind the door exert (at the edge) in order to open it? Assume that there is negligible friction between the door and the board but that the board is firmly set against the ground.

25. (II) Repeat Problem 24 assuming the coefficient of friction between the board and the door is 0.45.

26. (II) A 0.75-kg sheet hangs from a massless clothesline as shown in Fig. 12–63. The clothesline on either side of the sheet makes an angle of 3.5° with the horizontal. Calculate the tension in the clothesline on either side of the sheet. Why is the tension so much greater than the weight of the sheet?

FIGURE 12–63
Problem 26.

27. (II) A uniform rod AB of length 5.0 m and mass $M = 3.8$ kg is hinged at A and held in equilibrium by a light cord, as shown in Fig. 12–64. A load $W = 22$ N hangs from the rod at a distance x so that the tension in the cord is 85 N. (a) Draw a free-body diagram for the rod. (b) Determine the vertical and horizontal forces on the rod exerted by the hinge. (c) Determine x from the appropriate torque equation.

FIGURE 12–64
Problem 27.

28. (III) A 56.0-kg person stands 2.0 m from the bottom of the stepladder shown in Fig. 12–65. Determine (a) the tension in the horizontal tie rod, which is halfway up the ladder, (b) the normal force the ground exerts on each side of the ladder, and (c) the force (magnitude and direction) that the left side of the ladder exerts on the right side at the hinge on the top. Ignore the mass of the ladder and assume the ground is frictionless. [*Hint:* Consider free-body diagrams for each section of the ladder.]

FIGURE 12–65
Problem 28.

29. (III) A door 2.30 m high and 1.30 m wide has a mass of 13.0 kg. A hinge 0.40 m from the top and another hinge 0.40 m from the bottom each support half the door's weight (Fig. 12–66). Assume that the center of gravity is at the geometrical center of the door, and determine the horizontal and vertical force components exerted by each hinge on the door.

FIGURE 12–66
Problem 29.

30. (III) A cubic crate of side $s = 2.0$ m is top-heavy: its CG is 18 cm above its true center. How steep an incline can the crate rest on without tipping over? What would your answer be if the crate were to slide at constant speed down the plane without tipping over? [*Hint:* The normal force would act at the lowest corner.]

31. (III) A refrigerator is approximately a uniform rectangular solid 1.9 m tall, 1.0 m wide, and 0.75 m deep. If it sits upright on a truck with its 1.0-m dimension in the direction of travel, and if the refrigerator cannot slide on the truck, how rapidly can the truck accelerate without tipping the refrigerator over? [*Hint:* The normal force would act at one corner.]

32. (III) A uniform ladder of mass m and length ℓ leans at an angle θ against a frictionless wall, Fig. 12–67. If the coefficient of static friction between the ladder and the ground is μ_s, determine a formula for the minimum angle at which the ladder will not slip.

FIGURE 12–67
Problem 32.

12–3 Stability and Balance

33. (II) The Leaning Tower of Pisa is 55 m tall and about 7.0 m in diameter. The top is 4.5 m off center. Is the tower in stable equilibrium? If so, how much farther can it lean before it becomes unstable? Assume the tower is of uniform composition.

12–4 Elasticity; Stress and Strain

34. (I) A nylon string on a tennis racket is under a tension of 275 N. If its diameter is 1.00 mm, by how much is it lengthened from its untensioned length of 30.0 cm?

35. (I) A marble column of cross-sectional area 1.4 m² supports a mass of 25,000 kg. (*a*) What is the stress within the column? (*b*) What is the strain?

36. (I) By how much is the column in Problem 35 shortened if it is 8.6 m high?

37. (I) A sign (mass 1700 kg) hangs from the end of a vertical steel girder with a cross-sectional area of 0.012 m². (*a*) What is the stress within the girder? (*b*) What is the strain on the girder? (*c*) If the girder is 9.50 m long, how much is it lengthened? (Ignore the mass of the girder itself.)

38. (II) How much pressure is needed to compress the volume of an iron block by 0.10%? Express your answer in N/m², and compare it to atmospheric pressure $(1.0 \times 10^5 \text{ N/m}^2)$.

39. (II) A 15-cm-long tendon was found to stretch 3.7 mm by a force of 13.4 N. The tendon was approximately round with an average diameter of 8.5 mm. Calculate Young's modulus of this tendon.

40. (II) At depths of 2000 m in the sea, the pressure is about 200 times atmospheric pressure $(1 \text{ atm} = 1.0 \times 10^5 \text{ N/m}^2)$. By what percentage does the interior space of an iron bathysphere's volume change at this depth?

41. (III) A pole projects horizontally from the front wall of a shop. A 6.1-kg sign hangs from the pole at a point 2.2 m from the wall (Fig. 12–68). (*a*) What is the torque due to this sign calculated about the point where the pole meets the wall? (*b*) If the pole is not to fall off, there must be another torque exerted to balance it. What exerts this torque? Use a diagram to show how this torque must act. (*c*) Discuss whether compression, tension, and/or shear play a role in part (*b*).

FIGURE 12–68
Problem 41.

12–5 Fracture

42. (I) The femur bone in the human leg has a minimum effective cross section of about 3.0 cm² $(= 3.0 \times 10^{-4} \text{ m}^2)$. How much compressive force can it withstand before breaking?

43. (II) (*a*) What is the maximum tension possible in a 1.00-mm-diameter nylon tennis racket string? (*b*) If you want tighter strings, what do you do to prevent breakage: use thinner or thicker strings? Why? What causes strings to break when they are hit by the ball?

44. (II) If a compressive force of 3.3×10^4 N is exerted on the end of a 22-cm-long bone of cross-sectional area 3.6 cm², (*a*) will the bone break, and (*b*) if not, by how much does it shorten?

45. (II) (*a*) What is the minimum cross-sectional area required of a vertical steel cable from which is suspended a 270-kg chandelier? Assume a safety factor of 7.0. (*b*) If the cable is 7.5 m long, how much does it elongate?

46. (II) Assume the supports of the uniform cantilever shown in Fig. 12–69 ($m = 2900$ kg) are made of wood. Calculate the minimum cross-sectional area required of each, assuming a safety factor of 9.0.

FIGURE 12–69
Problem 46.

47. (II) An iron bolt is used to connect two iron plates together. The bolt must withstand shear forces up to about 3300 N. Calculate the minimum diameter for the bolt, based on a safety factor of 7.0.

48. (III) A steel cable is to support an elevator whose total (loaded) mass is not to exceed 3100 kg. If the maximum acceleration of the elevator is 1.2 m/s², calculate the diameter of cable required. Assume a safety factor of 8.0.

*12–6 Trusses and Bridges

*49. (II) A heavy load $Mg = 66.0$ kN hangs at point E of the single cantilever truss shown in Fig. 12–70. (*a*) Use a torque equation for the truss as a whole to determine the tension F_T in the support cable, and then determine the force $\vec{F}_A$ on the truss at pin A. (*b*) Determine the force in each member of the truss. Neglect the weight of the trusses, which is small compared to the load.

FIGURE 12–70
Problem 49.

$Mg = 66.0$ kN

*50. (II) Figure 12–71 shows a simple truss that carries a load at the center (C) of 1.35×10^4 N. (*a*) Calculate the force on each strut at the pins, A, B, C, D, and (*b*) determine which struts (ignore their masses) are under tension and which under compression.

FIGURE 12–71
Problem 50.

*51. (II) (*a*) What minimum cross-sectional area must the trusses have in Example 12–11 if they are of steel (and all the same size for looks), using a safety factor of 7.0? (*b*) If at any time the bridge may carry as many as 60 trucks with an average mass of 1.3×10^4 kg, estimate again the area needed for the truss members.

*52. (II) Consider again Example 12–11 but this time assume the roadway is supported uniformly so that $\frac{1}{2}$ its mass M $(= 7.0 \times 10^5 \text{ kg})$ acts at the center and $\frac{1}{4} M$ at each end support (think of the bridge as two spans, AC and CE, so the center pin supports two span ends). Calculate the magnitude of the force in each truss member and compare to Example 12–11.

*53. (III) The truss shown in Fig. 12–72 supports a railway bridge. Determine the compressive or tension force in each strut if a 53-ton $(1 \text{ ton} = 10^3 \text{ kg})$ train locomotive is stopped at the midpoint between the center and one end. Ignore the masses of the rails and truss, and use only $\frac{1}{2}$ the mass of train because there are two trusses (one on each side of the train). Assume all triangles are equilateral. [*Hint*: See Fig. 12–29.]

FIGURE 12–72
Problem 53.

*54. (III) Suppose in Example 12–11, a 23-ton truck $(m = 23 \times 10^3 \text{ kg})$ has its CM located 22 m from the left end of the bridge (point A). Determine the magnitude of the force and type of stress in each strut. [*Hint*: See Fig. 12–29.]

*55. (III) For the "Pratt truss" shown in Fig. 12–73, determine the force on each member and whether it is tensile or compressive. Assume the truss is loaded as shown, and give results in terms of F. The vertical height is a and each of the four lower horizontal spans has length a.

FIGURE 12–73
Problem 55.

12–7 Arches and Domes

*56. (II) How high must a pointed arch be if it is to span a space 8.0 m wide and exert one-third the horizontal force at its base that a round arch would?

General Problems

57. The mobile in Fig. 12–74 is in equilibrium. Object B has mass of 0.748 kg. Determine the masses of objects A, C, and D. (Neglect the weights of the crossbars.)

FIGURE 12–74 Problem 57.

58. A tightly stretched "high wire" is 36 m long. It sags 2.1 m when a 60.0-kg tightrope walker stands at its center. What is the tension in the wire? Is it possible to increase the tension in the wire so that there is no sag?

59. What minimum horizontal force F is needed to pull a wheel of radius R and mass M over a step of height h as shown in Fig. 12–75 $(R > h)$? (*a*) Assume the force is applied at the top edge as shown. (*b*) Assume the force is applied instead at the wheel's center.

FIGURE 12–75
Problem 59.

60. A 28-kg round table is supported by three legs equal distances apart on the edge. What minimum mass, placed on the table's edge, will cause the table to overturn?

61. When a wood shelf of mass 6.6 kg is fastened inside a slot in a vertical support as shown in Fig. 12–76, the support exerts a torque on the shelf. (*a*) Draw a free-body diagram for the shelf, assuming three vertical forces (two exerted by the support slot—explain why). Then calculate (*b*) the magnitudes of the three forces and (*c*) the torque exerted by the support (about the left end of the shelf).

FIGURE 12–76
Problem 61.

32.0 cm

3.0 cm

2.0 cm

62. A 50-story building is being planned. It is to be 180.0 m high with a base 46.0 m by 76.0 m. Its total mass will be about $1.8 \times 10^7 \text{ kg}$, and its weight therefore about $1.8 \times 10^8 \text{ N}$. Suppose a 200-km/h wind exerts a force of 950 N/m² over the 76.0-m-wide face (Fig. 12–77). Calculate the torque about the potential pivot point, the rear edge of the building (where $\vec{F}_E$ acts in Fig. 12–77), and determine whether the building will topple. Assume the total force of the wind acts at the midpoint of the building's face, and that the building is not anchored in bedrock. [*Hint*: $\vec{F}_E$ in Fig. 12–77 represents the force that the Earth would exert on the building in the case where the building would just begin to tip.]

FIGURE 12–77 Forces on a building subjected to wind $(\vec{F}_A)$, gravity $(m\vec{g})$, and the force $\vec{F}_E$ on the building due to the Earth if the building were just about to tip. Problem 62.

63. The center of gravity of a loaded truck depends on how the truck is packed. If it is 4.0 m high and 2.4 m wide, and its CG is 2.2 m above the ground, how steep a slope can the truck be parked on without tipping over (Fig. 12–78)?

FIGURE 12–78
Problem 63.

64. In Fig. 12–79, consider the right-hand (northernmost) section of the Golden Gate Bridge, which has a length $d_1 = 343$ m. Assume the CG of this span is halfway between the tower and anchor. Determine F_{T1} and F_{T2} (which act on the northernmost cable) in terms of mg, the weight of the northernmost span, and calculate the tower height h needed for equilibrium. Assume the roadway is supported only by the suspension cables, and neglect the mass of the cables and vertical wires. [*Hint*: F_{T3} does not act on this section.]

FIGURE 12–79 Problems 64 and 65.

65. Assume that a single-span suspension bridge such as the Golden Gate Bridge has the symmetrical configuration indicated in Fig. 12–79. Assume that the roadway is uniform over the length of the bridge and that each segment of the suspension cable provides the sole support for the roadway directly below it. The ends of the cable are anchored to the ground only, not to the roadway. What must the ratio of d_2 to d_1 be so that the suspension cable exerts no net horizontal force on the towers? Neglect the mass of the cables and the fact that the roadway isn't precisely horizontal.

66. When a mass of 25 kg is hung from the middle of a fixed straight aluminum wire, the wire sags to make an angle of 12° with the horizontal as shown in Fig. 12–80. Determine the radius of the wire.

FIGURE 12–80
Problem 66.

67. The forces acting on a 77,000-kg aircraft flying at constant velocity are shown in Fig. 12–81. The engine thrust, $F_T = 5.0 \times 10^5$ N, acts on a line 1.6 m below the CM. Determine the drag force F_D and the distance above the CM that it acts. Assume $\vec{F}_D$ and $\vec{F}_T$ are horizontal. ($\vec{F}_L$ is the "lift" force on the wing.)

FIGURE 12–81
Problem 67.

68. A uniform flexible steel cable of weight mg is suspended between two points at the same elevation as shown in Fig. 12–82, where $\theta = 56°$. Determine the tension in the cable (*a*) at its lowest point, and (*b*) at the points of attachment. (*c*) What is the direction of the tension force in each case?

FIGURE 12–82
Problem 68.

69. A 20.0-m-long uniform beam weighing 650 N rests on walls A and B, as shown in Fig. 12–83. (*a*) Find the maximum weight of a person who can walk to the extreme end D without tipping the beam. Find the forces that the walls A and B exert on the beam when the person is standing: (*b*) at D; (*c*) at a point 2.0 m to the right of B; (*d*) 2.0 m to the right of A.

FIGURE 12–83
Problem 69.

70. A cube of side ℓ rests on a rough floor. It is subjected to a steady horizontal pull F, exerted a distance h above the floor as shown in Fig. 12–84. As F is increased, the block will either begin to slide, or begin to tip over. Determine the coefficient of static friction μ_s so that (*a*) the block begins to slide rather than tip; (*b*) the block begins to tip. [*Hint*: Where will the normal force on the block act if it tips?]

FIGURE 12–84
Problem 70.

71. A 65.0-kg painter is on a uniform 25-kg scaffold supported from above by ropes (Fig. 12–85). There is a 4.0-kg pail of paint to one side, as shown. Can the painter walk safely to both ends of the scaffold? If not, which end(s) is dangerous, and how close to the end can he approach safely?

FIGURE 12–85
Problem 71.

72. A man doing push-ups pauses in the position shown in Fig. 12–86. His mass $m = 68$ kg. Determine the normal force exerted by the floor (*a*) on each hand; (*b*) on each foot.

FIGURE 12–86
Problem 72.

73. A 23-kg sphere rests between two smooth planes as shown in Fig. 12–87. Determine the magnitude of the force acting on the sphere exerted by each plane.

FIGURE 12–87
Problem 73.

74. A 15.0-kg ball is supported from the ceiling by rope A. Rope B pulls downward and to the side on the ball. If the angle of A to the vertical is 22° and if B makes an angle of 53° to the vertical (Fig. 12–88), find the tensions in ropes A and B.

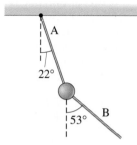

FIGURE 12–88
Problem 74.

75. Parachutists whose chutes have failed to open have been known to survive if they land in deep snow. Assume that a 75-kg parachutist hits the ground with an area of impact of 0.30 m^2 at a velocity of 55 m/s, and that the ultimate strength of body tissue is $5 \times 10^5 \text{ N/m}^2$. Assume that the person is brought to rest in 1.0 m of snow. Show that the person may escape serious injury.

76. A steel wire 2.3 mm in diameter stretches by 0.030% when a mass is suspended from it. How large is the mass?

77. A 2500-kg trailer is attached to a stationary truck at point B, Fig. 12–89. Determine the normal force exerted by the road on the rear tires at A, and the vertical force exerted on the trailer by the support B.

FIGURE 12–89   Problem 77.

78. The roof over a 9.0-m × 10.0-m room in a school has a total mass of 13,600 kg. The roof is to be supported by vertical "2 × 4s" (actually about 4.0 cm × 9.0 cm) equally spaced along the 10.0-m sides. How many supports are required on each side, and how far apart must they be? Consider only compression, and assume a safety factor of 12.

79. A 25-kg object is being lifted by pulling on the ends of a 1.15-mm-diameter nylon cord that goes over two 3.00-m-high poles that are 4.0 m apart, as shown in Fig. 12–90. How high above the floor will the object be when the cord breaks?

FIGURE 12–90   Problem 79.

80. A uniform 6.0-m-long ladder of mass 16.0 kg leans against a smooth wall (so the force exerted by the wall, $\vec{\mathbf{F}}_W$, is perpendicular to the wall). The ladder makes an angle of 20.0° with the vertical wall, and the ground is rough. Determine the coefficient of static friction at the base of the ladder if the ladder is not to slip when a 76.0-kg person stands three-fourths of the way up the ladder.

81. There is a maximum height of a uniform vertical column made of any material that can support itself without buckling, and it is independent of the cross-sectional area (why?). Calculate this height for (a) steel (density $7.8 \times 10^3 \text{ kg/m}^3$), and (b) granite (density $2.7 \times 10^3 \text{ kg/m}^3$).

82. A 95,000-kg train locomotive starts across a 280-m-long bridge at time $t = 0$. The bridge is a uniform beam of mass 23,000 kg and the train travels at a constant 80.0 km/h. What are the magnitudes of the vertical forces, $F_A(t)$ and $F_B(t)$, on the two end supports, written as a function of time during the train's passage?

83. A 23.0-kg backpack is suspended midway between two trees by a light cord as in Fig. 12–50. A bear grabs the backpack and pulls vertically downward with a constant force, so that each section of cord makes an angle of 27° below the horizontal. Initially, without the bear pulling, the angle was 15°; the tension in the cord with the bear pulling is double what it was when he was not. Calculate the force the bear is exerting on the backpack.

84. A uniform beam of mass M and length ℓ is mounted on a hinge at a wall as shown in Fig. 12–91. It is held in a horizontal position by a wire making an angle θ as shown. A mass m is placed on the beam a distance x from the wall, and this distance can be varied. Determine, as a function of x, (a) the tension in the wire and (b) the components of the force exerted by the beam on the hinge.

FIGURE 12–91
Problem 84.

85. Two identical, uniform beams are symmetrically set up against each other (Fig. 12–92) on a floor with which they have a coefficient of friction $\mu_s = 0.50$. What is the minimum angle the beams can make with the floor and still not fall?

FIGURE 12–92
Problem 85.

86. If 35 kg is the maximum mass m that a person can hold in a hand when the arm is positioned with a 105° angle at the elbow as shown in Fig. 12–93, what is the maximum force F_{max} that the biceps muscle exerts on the forearm? Assume the forearm and hand have a total mass of 2.0 kg with a CG that is 15 cm from the elbow, and that the biceps muscle attaches 5.0 cm from the elbow.

FIGURE 12–93
Problem 86.

87. (*a*) Estimate the magnitude of the force $\vec{F}_M$ the muscles exert on the back to support the upper body when a person bends forward. Use the model shown in Fig. 12–94b. (*b*) Estimate the magnitude and direction of the force $\vec{F}_V$ acting on the fifth lumbar vertebra (exerted by the spine below).

$w_H = 0.07w$ (head)

$w_A = 0.12w$ (arms)

$w_T = 0.46w$ (trunk)

$w = $ Total weight of person

(a)　　　　　(b)

FIGURE 12–94 Problem 87.

88. One rod of the square frame shown in Fig. 12–95 contains a turnbuckle which, when turned, can put the rod under tension or compression. If the turnbuckle puts rod AB under a compressive force F, determine the forces produced in the other rods. Ignore the mass of the rods and assume the diagonal rods cross each other freely at the center without friction. [*Hint*: Use the symmetry of the situation.]

FIGURE 12–95
Problem 88.

89. A steel rod of radius $R = 15\,\text{cm}$ and length ℓ_0 stands upright on a firm surface. A 65-kg man climbs atop the rod. (*a*) Determine the percent decrease in the rod's length. (*b*) When a metal is compressed, each atom throughout its bulk moves closer to its neighboring atom by exactly the same fractional amount. If iron atoms in steel are normally $2.0 \times 10^{-10}\,\text{m}$ apart, by what distance did this interatomic spacing have to change in order to produce the normal force required to support the man? [*Note*: Neighboring atoms repel each other, and this repulsion accounts for the observed normal force.]

90. A home mechanic wants to raise the 280-kg engine out of a car. The plan is to stretch a rope vertically from the engine to a branch of a tree 6.0 m above, and back to the bumper (Fig. 12–96). When the mechanic climbs up a stepladder and pulls horizontally on the rope at its midpoint, the engine rises out of the car. (*a*) How much force must the mechanic exert to hold the engine 0.50 m above its normal position? (*b*) What is the system's mechanical advantage?

FIGURE 12–96
Problem 90.

91. A 2.0-m-high box with a 1.0-m-square base is moved across a rough floor as in Fig. 12–97. The uniform box weighs 250 N and has a coefficient of static friction with the floor of 0.60. What minimum force must be exerted on the box to make it slide? What is the maximum height h above the floor that this force can be applied without tipping the box over? Note that as the box tips, the normal force and the friction force will act at the lowest corner.

FIGURE 12–97
Problem 91.

92. You are on a pirate ship and being forced to walk the plank (Fig. 12–98). You are standing at the point marked C. The plank is nailed onto the deck at point A, and rests on the support 0.75 m away from A. The center of mass of the uniform plank is located at point B. Your mass is 65 kg and the mass of the plank is 45 kg. What is the minimum downward force the nails must exert on the plank to hold it in place?

FIGURE 12–98
Problem 92.

93. A uniform sphere of weight mg and radius r_0 is tethered to a wall by a rope of length ℓ. The rope is tied to the wall a distance h above the contact point of the sphere, as shown in Fig. 12–99. The rope makes an angle θ with respect to the wall and is not in line with the ball's center. The coefficient of static friction between the wall and sphere is μ. (*a*) Determine the value of the frictional force on the sphere due to the wall. [*Hint*: A wise choice of axis will make this calculation easy.] (*b*) Suppose the sphere is just on the verge of slipping. Derive an expression for μ in terms of h and θ.

FIGURE 12–99
Problem 93.

*94. Use the method of joints to determine the force in each member of the truss shown in Fig. 12–100. State whether each member is in tension or compression.

FIGURE 12–100
Problem 94.

12,000 N

95. A uniform ladder of mass m and length ℓ leans at an angle θ against a wall, Fig. 12–101. The coefficients of static friction between ladder–ground and ladder–wall are μ_G and μ_W, respectively. The ladder will be on the verge of slipping when both the static friction forces due to the ground and due to the wall take on their maximum values. (a) Show that the ladder will be stable if $\theta \geq \theta_{min}$, where the minimum angle θ_{min} is given by

$$\tan \theta_{min} = \frac{1}{2\mu_G}(1 - \mu_G \mu_W).$$

(b) "Leaning ladder problems" are often analyzed under the seemingly unrealistic assumption that the wall is frictionless (see Example 12–6). You wish to investigate the magnitude of error introduced by modeling the wall as frictionless, if in reality it is frictional. Using the relation found in part (a), calculate the true value of θ_{min} for a frictional wall, taking $\mu_G = \mu_W = 0.40$. Then, determine the approximate value of θ_{min} for the "frictionless wall" model by taking $\mu_G = 0.40$ and $\mu_W = 0$. Finally, determine the percent deviation of the approximate value of θ_{min} from its true value.

FIGURE 12–101
Problem 95.

96. In a mountain-climbing technique called the "Tyrolean traverse," a rope is anchored on both ends (to rocks or strong trees) across a deep chasm, and then a climber traverses the rope while attached by a sling as in Fig. 12–102. This technique generates tremendous forces in the rope and anchors, so a basic understanding of physics is crucial for safety. A typical climbing rope can undergo a tension force of perhaps 29 kN before breaking, and a "safety factor" of 10 is usually recommended. The length of rope used in the Tyrolean traverse must allow for some "sag" to remain in the recommended safety range. Consider a 75-kg climber at the center of a Tyrolean traverse, spanning a 25-m chasm. (a) To be within its recommended safety range, what minimum distance x must the rope sag? (b) If the Tyrolean traverse is set up incorrectly so that the rope sags by only one-fourth the distance found in (a), determine the tension in the rope. Will the rope break?

FIGURE 12–102
Problem 96.

*Numerical/Computer

*97. (III) A metal cylinder has an original diameter of 1.00 cm and a length of 5.00 cm. A tension test was performed on the specimen and the data are listed in the Table. (a) Graph the stress on the specimen vs. the strain. (b) Considering only the elastic region, find the slope of the best-fit straight line and determine the elastic modulus of the metal.

Load (kN)	Elongation (cm)
0	0
1.50	0.0005
4.60	0.0015
8.00	0.0025
11.00	0.0035
11.70	0.0050
11.80	0.0080
12.00	0.0200
16.60	0.0400
20.00	0.1000
21.50	0.2800
19.50	0.4000
18.50	0.4600

*98. (III) Two springs, attached by a rope, are connected as shown in Fig. 12–103. The length AB is 4.0 m and AC = BC. The spring constant of each spring is $k = 20.0$ N/m. A force F acts downward at C on the rope. Graph θ as a function of F from $\theta = 0$ to 75°, assuming the springs are unstretched at $\theta = 0$.

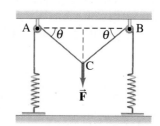

FIGURE 12–103 Problem 98.

Answers to Exercises

A: F_A also has a component to balance the sideways force F_B.

B: Yes: $\cos \theta$ (angle of bar with ground) appears on both sides and cancels out.

C: $F_N = m_A g + m_B g + Mg = 560$ N.

D: (a).

E: 7.0 kg.

F: Static friction at the cement floor $(= F_{Cx})$ is crucial, or else the ladder would slip. At the top, the ladder can move and adjust, so we wouldn't need or expect a strong static friction force there.

G: (b).

$\vec{\mathbf{F}}_{B1}$

B2

$m_2\vec{\mathbf{g}}$

$m_1\vec{\mathbf{g}}$

Underwater divers and sea creatures experience a buoyant force ($\vec{\mathbf{F}}_B$) that closely balances their weight $m\vec{\mathbf{g}}$. The buoyant force is equal to the weight of the volume of fluid displaced (Archimedes' principle) and arises because the pressure increases with depth in the fluid. Sea creatures have a density very close to that of water, so their weight very nearly equals the buoyant force. Humans have a density slightly less than water, so they can float.

When fluids flow, interesting effects occur because the pressure in the fluid is lower where the fluid velocity is higher (Bernoulli's principle).

C H A P T E R

13

Fluids

CHAPTER-OPENING QUESTIONS—Guess now!

1. Which container has the largest pressure at the bottom? Assume each container holds the same volume of water.

(a)　　(b)　　(c)　　(d)　　(e)

The pressures are equal.

2. Two balloons are tied and hang with their nearest edges about 3 cm apart. If you blow between the balloons (not *at* the balloons, but at the opening between them), what will happen?
   (a) Nothing.
   (b) The balloons will move closer together.
   (c) The balloons will move farther apart.

CONTENTS

TABLE 13–1
Densities of Substances[†]

Substance	Density, ρ (kg/m³)
Solids	
Aluminum	2.70×10^3
Iron and steel	$7.8 \ \times 10^3$
Copper	$8.9 \ \times 10^3$
Lead	$11.3 \ \times 10^3$
Gold	$19.3 \ \times 10^3$
Concrete	$2.3 \ \times 10^3$
Granite	$2.7 \ \times 10^3$
Wood (typical)	$0.3 - 0.9 \times 10^3$
Glass, common	$2.4 - 2.8 \times 10^3$
Ice (H_2O)	0.917×10^3
Bone	$1.7 - 2.0 \times 10^3$
Liquids	
Water (4°C)	$1.00 \ \times 10^3$
Blood, plasma	$1.03 \ \times 10^3$
Blood, whole	$1.05 \ \times 10^3$
Sea water	1.025×10^3
Mercury	$13.6 \ \times 10^3$
Alcohol, ethyl	$0.79 \ \times 10^3$
Gasoline	$0.68 \ \times 10^3$
Gases	
Air	1.29
Helium	0.179
Carbon dioxide	1.98
Steam (water, 100°C)	0.598

[†]Densities are given at 0°C and 1 atm pressure unless otherwise specified.

I n previous Chapters we considered objects that were solid and assumed to maintain their shape except for a small amount of elastic deformation. We sometimes treated objects as point particles. Now we are going to shift our attention to materials that are very deformable and can flow. Such "fluids" include liquids and gases. We will examine fluids both at rest (fluid statics) and in motion (fluid dynamics).

13–1 Phases of Matter

The three common **phases**, or **states**, of matter are solid, liquid, and gas. We can distinguish these three phases as follows. A **solid** maintains a fixed shape and a fixed size; even if a large force is applied to a solid, it does not readily change in shape or volume. A **liquid** does not maintain a fixed shape—it takes on the shape of its container—but like a solid it is not readily compressible, and its volume can be changed significantly only by a very large force. A **gas** has neither a fixed shape nor a fixed volume—it will expand to fill its container. For example, when air is pumped into an automobile tire, the air does not all run to the bottom of the tire as a liquid would; it spreads out to fill the whole volume of the tire. Since liquids and gases do not maintain a fixed shape, they both have ability to flow; they are thus often referred to collectively as **fluids**.

The division of matter into three phases is not always simple. How, for example, should butter be classified? Furthermore, a fourth phase of matter can be distinguished, the **plasma** phase, which occurs only at very high temperatures and consists of ionized atoms (electrons separated from the nuclei). Some scientists believe that so-called colloids (suspensions of tiny particles in a liquid) should also be considered a separate phase of matter. **Liquid crystals**, which are used in TV and computer screens, calculators, digital watches, and so on, can be considered a phase of matter intermediate between solids and liquids. However, for our present purposes we will mainly be interested in the three ordinary phases of matter.

13–2 Density and Specific Gravity

It is sometimes said that iron is "heavier" than wood. This cannot really be true since a large log clearly weighs more than an iron nail. What we should say is that iron is more *dense* than wood.

The **density**, ρ, of a substance (ρ is the lowercase Greek letter rho) is defined as its mass per unit volume:

$$\rho = \frac{m}{V},$$ (13–1)

where m is the mass of a sample of the substance and V is its volume. Density is a characteristic property of any pure substance. Objects made of a particular pure substance, such as pure gold, can have any size or mass, but the density will be the same for each.

We will sometimes use the concept of density, Eq. 13–1, to write the mass of an object as

$$m = \rho V,$$

and the weight of an object as

$$mg = \rho V g.$$

The SI unit for density is kg/m³. Sometimes densities are given in g/cm³. Note that since $1 \text{ kg/m}^3 = 1000 \text{ g}/(100 \text{ cm})^3 = 10^3 \text{ g}/10^6 \text{ cm}^3 = 10^{-3} \text{ g/cm}^3$, then a density given in g/cm³ must be multiplied by 1000 to give the result in kg/m³. Thus the density of aluminum is $\rho = 2.70 \text{ g/cm}^3$, which is equal to 2700 kg/m^3. The densities of a variety of substances are given in Table 13–1. The Table specifies temperature and atmospheric pressure because they affect the density of substances (although the effect is slight for liquids and solids). Note that air is roughly 1000 times less dense than water.

EXAMPLE 13–1 **Mass, given volume and density.** What is the mass of a solid iron wrecking ball of radius 18 cm?

APPROACH First we use the standard formula $V = \frac{4}{3}\pi r^3$ (see inside rear cover) to obtain the volume of the sphere. Then Eq. 13–1 and Table 13–1 give us the mass m.

SOLUTION The volume of the sphere is

$$V = \tfrac{4}{3}\pi r^3 = \tfrac{4}{3}(3.14)(0.18\,\text{m})^3 = 0.024\,\text{m}^3.$$

From Table 13–1, the density of iron is $\rho = 7800\,\text{kg/m}^3$, so Eq. 13–1 gives

$$m = \rho V = (7800\,\text{kg/m}^3)(0.024\,\text{m}^3) = 190\,\text{kg}.$$

The **specific gravity** of a substance is defined as the ratio of the density of that substance to the density of water at 4.0°C. Because specific gravity (abbreviated SG) is a ratio, it is a simple number without dimensions or units. The density of water is $1.00\,\text{g/cm}^3 = 1.00 \times 10^3\,\text{kg/m}^3$, so the specific gravity of any substance will be equal numerically to its density specified in g/cm^3, or 10^{-3} times its density specified in kg/m^3. For example (see Table 13–1), the specific gravity of lead is 11.3, and that of alcohol is 0.79.

The concepts of density and specific gravity are especially helpful in the study of fluids because we are not always dealing with a fixed volume or mass.

13–3 Pressure in Fluids

Pressure and force are related, but they are not the same thing. **Pressure** is defined as force per unit area, where the force F is understood to be the magnitude of the force acting perpendicular to the surface area A:

$$\text{pressure} = P = \frac{F}{A}. \tag{13–2}$$

Although force is a vector, pressure is a scalar. Pressure has magnitude only. The SI unit of pressure is N/m^2. This unit has the official name **pascal** (Pa), in honor of Blaise Pascal (see Section 13–5); that is, $1\,\text{Pa} = 1\,\text{N/m}^2$. However, for simplicity, we will often use N/m^2. Other units sometimes used are dynes/cm^2, and lb/in.^2 (abbreviated "psi"). Several other units for pressure are discussed, along with conversions between them, in Section 13–6 (see also the Table inside the front cover).

⚠ **CAUTION**
Pressure is a scalar, not a vector

EXAMPLE 13–2 **Calculating pressure.** The two feet of a 60-kg person cover an area of 500 cm^2. (*a*) Determine the pressure exerted by the two feet on the ground. (*b*) If the person stands on one foot, what will the pressure be under that foot?

APPROACH Assume the person is at rest. Then the ground pushes up on her with a force equal to her weight mg, and she exerts a force mg on the ground where her feet (or foot) contact it. Because $1\,\text{cm}^2 = (10^{-2}\,\text{m})^2 = 10^{-4}\,\text{m}^2$, then $500\,\text{cm}^2 = 0.050\,\text{m}^2$.

SOLUTION (*a*) The pressure on the ground exerted by the two feet is

$$P = \frac{F}{A} = \frac{mg}{A} = \frac{(60\,\text{kg})(9.8\,\text{m/s}^2)}{(0.050\,\text{m}^2)} = 12 \times 10^3\,\text{N/m}^2.$$

(*b*) If the person stands on one foot, the force is still equal to the person's weight, but the area will be half as much, so the pressure will be twice as much: $24 \times 10^3\,\text{N/m}^2$.

Pressure is particularly useful for dealing with fluids. It is an experimental observation that *a fluid exerts pressure in any direction*. This is well known to swimmers and divers who feel the water pressure on all parts of their bodies. At any depth in a fluid at rest, the pressure is the same in all directions at a given depth. To see why, consider a tiny cube of the fluid (Fig. 13–1) which is so small that we can consider it a point and can ignore the force of gravity on it. The pressure on one side of it must equal the pressure on the opposite side. If this weren't true, there would be a net force on the cube and it would start moving. If the fluid is not flowing, then the pressures must be equal.

FIGURE 13–1 Pressure is the same in every direction in a nonmoving fluid at a given depth. If this weren't true, the fluid would be in motion.

FIGURE 13–2 If there were a component of force parallel to the solid surface of the container, the liquid would move in response to it. For a liquid at rest, $F_\parallel = 0$.

FIGURE 13–3 Calculating the pressure at a depth h in a liquid.

FIGURE 13–4 Forces on a flat, slablike volume of fluid for determining the pressure P at a height y in the fluid.

For a fluid at rest, the force due to fluid pressure always acts *perpendicular* to any solid surface it touches. If there were a component of the force parallel to the surface, as shown in Fig. 13–2, then according to Newton's third law the solid surface would exert a force back on the fluid that also would have a component parallel to the surface. Such a component would cause the fluid to flow, in contradiction to our assumption that the fluid is at rest. Thus the force due to the pressure in a fluid at rest is always perpendicular to the surface.

Let us now calculate quantitatively how the pressure in a liquid of uniform density varies with depth. Consider a point at a depth h below the surface of the liquid, as shown in Fig. 13–3 (that is, the surface is a height h above this point). The pressure due to the liquid at this depth h is due to the weight of the column of liquid above it. Thus the force due to the weight of liquid acting on the area A is $F = mg = (\rho V)g = \rho A h g$, where Ah is the volume of the column of liquid, ρ is the density of the liquid (assumed to be constant), and g is the acceleration of gravity. The pressure P due to the weight of liquid is then

$$P = \frac{F}{A} = \frac{\rho A h g}{A}$$

$$P = \rho g h. \qquad\qquad \text{[liquid]} \quad \textbf{(13–3)}$$

Note that the area A doesn't affect the pressure at a given depth. The fluid pressure is directly proportional to the density of the liquid and to the depth within the liquid. In general, the pressure at equal depths within a uniform liquid is the same.

| **EXERCISE A** Return to Chapter-Opening Question 1, page 339, and answer it again now. Try to explain why you may have answered differently the first time.

Equation 13–3 tells us what the pressure is at a depth h in the liquid, due to the liquid itself. But what if there is additional pressure exerted at the surface of the liquid, such as the pressure of the atmosphere or a piston pushing down? And what if the density of the fluid is not constant? Gases are quite compressible and hence their density can vary significantly with depth. Liquids, too, can be compressed, although we can often ignore the variation in density. (One exception is in the depths of the ocean where the great weight of water above significantly compresses the water and increases its density.) To cover these, and other cases, we now treat the general case of determining how the pressure in a fluid varies with depth.

As shown in Fig. 13–4, let us determine the pressure at any height y above some reference point[†] (such as the ocean floor or the bottom of a tank or swimming pool). Within this fluid, at the height y, we consider a tiny, flat, slablike volume of the fluid whose area is A and whose (infinitesimal) thickness is dy, as shown. Let the pressure acting upward on its lower surface (at height y) be P. The pressure acting downward on the top surface of our tiny slab (at height $y + dy$) is designated $P + dP$. The fluid pressure acting on our slab thus exerts a force equal to PA upward on our slab and a force equal to $(P + dP)A$ downward on it. The only other force acting vertically on the slab is the (infinitesimal) force of gravity dF_G, which on our slab of mass dm is

$$dF_G = (dm)g = \rho g \, dV = \rho g A \, dy,$$

where ρ is the density of the fluid at the height y. Since the fluid is assumed to be at rest, our slab is in equilibrium so the net force on it must be zero. Therefore we have

$$PA - (P + dP)A - \rho g A \, dy = 0,$$

which when simplified becomes

$$\frac{dP}{dy} = -\rho g. \qquad\qquad\qquad \textbf{(13–4)}$$

This relation tells us how the pressure within the fluid varies with height above any reference point. The minus sign indicates that the pressure decreases with an increase in height; or that the pressure increases with depth (reduced height).

[†]Now we are measuring y positive upwards, the reverse of what we did to get Eq. 13–3 where we measured the depth (i.e. downward as positive).

If the pressure at a height y_1 in the fluid is P_1, and at height y_2 it is P_2, then we can integrate Eq. 13–4 to obtain

$$\int_{P_1}^{P_2} dP = -\int_{y_1}^{y_2} \rho g \, dy$$

$$P_2 - P_1 = -\int_{y_1}^{y_2} \rho g \, dy, \qquad\qquad (13\text{–}5)$$

where we assume ρ is a function of height y: $\rho = \rho(y)$. This is a general relation, and we apply it now to two special cases: (1) pressure in liquids of uniform density and (2) pressure variations in the Earth's atmosphere.

For liquids in which any variation in density can be ignored, $\rho = $ constant and Eq. 13–5 is readily integrated:

$$P_2 - P_1 = -\rho g(y_2 - y_1). \qquad\qquad (13\text{–}6a)$$

For the everyday situation of a liquid in an open container—such as water in a glass, a swimming pool, a lake, or the ocean—there is a free surface at the top exposed to the atmosphere. It is convenient to measure distances from this top surface. That is, we let h be the *depth* in the liquid where $h = y_2 - y_1$ as shown in Fig. 13–5. If we let y_2 be the position of the top surface, then P_2 represents the atmospheric pressure, P_0, at the top surface. Then, from Eq. 13–6a, the pressure $P (= P_1)$ at a depth h in the fluid is

$$P = P_0 + \rho g h. \qquad \text{[}h \text{ is depth in liquid]} \quad (13\text{–}6b)$$

Note that Eq. 13–6b is simply the liquid pressure (Eq. 13–3) plus the pressure P_0 due to the atmosphere above.

FIGURE 13–5 Pressure at a depth $h = (y_2 - y_1)$ in a liquid of density ρ is $P = P_0 + \rho g h$, where P_0 is the external pressure at the liquid's top surface.

EXAMPLE 13–3 Pressure at a faucet. The surface of the water in a storage tank is 30 m above a water faucet in the kitchen of a house, Fig. 13–6. Calculate the difference in water pressure between the faucet and the surface of the water in the tank.

APPROACH Water is practically incompressible, so ρ is constant even for $h = 30$ m when used in Eq. 13–6b. Only h matters; we can ignore the "route" of the pipe and its bends.

SOLUTION We assume the atmospheric pressure at the surface of the water in the storage tank is the same as at the faucet. So, the water pressure difference between the faucet and the surface of the water in the tank is

$$\Delta P = \rho g h = (1.0 \times 10^3 \text{ kg/m}^3)(9.8 \text{ m/s}^2)(30 \text{ m}) = 2.9 \times 10^5 \text{ N/m}^2.$$

NOTE The height h is sometimes called the **pressure head**. In this Example, the head of water is 30 m at the faucet. The very different diameters of the tank and faucet don't affect the result—only pressure does.

FIGURE 13–6 Example 13–3.

EXAMPLE 13–4 Force on aquarium window. Calculate the force due to water pressure exerted on a 1.0 m × 3.0 m aquarium viewing window whose top edge is 1.0 m below the water surface, Fig. 13–7.

APPROACH At a depth h, the pressure due to the water is given by Eq. 13–6b. Divide the window up into thin horizontal strips of length $\ell = 3.0$ m and thickness dy, as shown in Fig. 13–7. We choose a coordinate system with $y = 0$ at the surface of the water and y is positive downward. (With this choice, the minus sign in Eq. 13–6a becomes plus, or we use Eq. 13–6b with $y = h$.) The force due to water pressure on each strip is $dF = P\,dA = \rho g y \ell \, dy$.

SOLUTION The total force on the window is given by the integral:

$$\int_{y_1 = 1.0 \text{ m}}^{y_2 = 2.0 \text{ m}} \rho g y \ell \, dy = \tfrac{1}{2}\rho g \ell (y_2^2 - y_1^2)$$

$$= \tfrac{1}{2}(1000 \text{ kg/m}^3)(9.8 \text{ m/s}^2)(3.0 \text{ m})[(2.0 \text{ m})^2 - (1.0 \text{ m})^2] = 44,000 \text{ N}.$$

NOTE To check our answer, we can do an estimate: multiply the area of the window (3.0 m^2) times the pressure at the middle of the window $(h = 1.5 \text{ m})$ using Eq. 13–3, $P = \rho g h = (1000 \text{ kg/m}^3)(9.8 \text{ m/s}^2)(1.5 \text{ m}) \approx 1.5 \times 10^4 \text{ N/m}^2$. So $F = PA \approx (1.5 \times 10^4 \text{ N/m}^2)(3.0 \text{ m})(1.0 \text{ m}) \approx 4.5 \times 10^4 \text{ N}$. Good!

FIGURE 13–7 Example 13–4.

EXERCISE B A dam holds back a lake that is 85 m deep at the dam. If the lake is 20 km long, how much thicker should the dam be than if the lake were smaller, only 1.0 km long?

Now let us apply Eq. 13–4 or 13–5 to gases. The density of gases is normally quite small, so the difference in pressure at different heights can usually be ignored if $y_2 - y_1$ is not large (which is why, in Example 13–3, we could ignore the difference in air pressure between the faucet and the top of the storage tank). Indeed, for most ordinary containers of gas, we can assume that the pressure is the same throughout. However, if $y_2 - y_1$ is very large, we cannot make this assumption. An interesting example is the air of Earth's atmosphere, whose pressure at sea level is about $1.013 \times 10^5 \, \text{N/m}^2$ and decreases slowly with altitude.

EXAMPLE 13–5 **Elevation effect on atmospheric pressure.** (*a*) Determine the variation in pressure in the Earth's atmosphere as a function of height y above sea level, assuming g is constant and that the density of the air is proportional to the pressure. (This last assumption is not terribly accurate, in part because temperature and other weather effects are important.) (*b*) At what elevation is the air pressure equal to half the pressure at sea level?

APPROACH We start with Eq. 13–4 and integrate it from the surface of the Earth where $y = 0$ and $P = P_0$, up to height y at pressure P. In (*b*) we choose $P = \frac{1}{2} P_0$.

SOLUTION (*a*) We are assuming that ρ is proportional to P, so we can write

$$\frac{\rho}{\rho_0} = \frac{P}{P_0},$$

where $P_0 = 1.013 \times 10^5 \, \text{N/m}^2$ is atmospheric pressure at sea level and $\rho_0 = 1.29 \, \text{kg/m}^3$ is the density of air at sea level at 0°C (Table 13–1). From the differential change in pressure with height, Eq. 13–4, we have

$$\frac{dP}{dy} = -\rho g = -P\left(\frac{\rho_0}{P_0}\right) g,$$

so

$$\frac{dP}{P} = -\frac{\rho_0}{P_0} g \, dy.$$

We integrate this from $y = 0$ (Earth's surface) and $P = P_0$, to the height y where the pressure is P:

$$\int_{P_0}^{P} \frac{dP}{P} = -\frac{\rho_0}{P_0} g \int_0^y dy$$

$$\ln \frac{P}{P_0} = -\frac{\rho_0}{P_0} gy,$$

since $\ln P - \ln P_0 = \ln(P/P_0)$. Then

$$P = P_0 e^{-(\rho_0 g / P_0) y}.$$

So, based on our assumptions, we find that the air pressure in our atmosphere decreases approximately exponentially with height.

NOTE The atmosphere does not have a distinct top surface, so there is no natural point from which to measure depth in the atmosphere, as we can do for a liquid.

(*b*) The constant $(\rho_0 g / P_0)$ has the value

$$\frac{\rho_0 g}{P_0} = \frac{(1.29 \, \text{kg/m}^3)(9.80 \, \text{m/s}^2)}{(1.013 \times 10^5 \, \text{N/m}^2)} = 1.25 \times 10^{-4} \, \text{m}^{-1}.$$

Then, when we set $P = \frac{1}{2} P_0$ in our expression derived in (*a*), we obtain

$$\tfrac{1}{2} = e^{-(1.25 \times 10^{-4} \, \text{m}^{-1}) y}$$

or, taking natural logarithms of both sides,

$$\ln \tfrac{1}{2} = \left(-1.25 \times 10^{-4} \, \text{m}^{-1}\right) y$$

so (recall $\ln \frac{1}{2} = -\ln 2$, Appendix A–7, Eq. ii)

$$y = (\ln 2.00)/(1.25 \times 10^{-4} \, \text{m}^{-1}) = 5550 \, \text{m}.$$

Thus, at an elevation of about 5500 m (about 18,000 ft), atmospheric pressure drops to half what it is at sea level. It is not surprising that mountain climbers often use oxygen tanks at very high altitudes.

13–4 Atmospheric Pressure and Gauge Pressure

Atmospheric Pressure

The pressure of the air at a given place on Earth varies slightly according to the weather. At sea level, the pressure of the atmosphere on average is $1.013 \times 10^5 \, \text{N/m}^2$ (or $14.7 \, \text{lb/in.}^2$). This value lets us define a commonly used unit of pressure, the **atmosphere** (abbreviated atm):

$$1 \, \text{atm} = 1.013 \times 10^5 \, \text{N/m}^2 = 101.3 \, \text{kPa}.$$

Another unit of pressure sometimes used (in meteorology and on weather maps) is the **bar**, which is defined as

$$1 \, \text{bar} = 1.000 \times 10^5 \, \text{N/m}^2.$$

Thus standard atmospheric pressure is slightly more than 1 bar.

The pressure due to the weight of the atmosphere is exerted on all objects immersed in this great sea of air, including our bodies. How does a human body withstand the enormous pressure on its surface? The answer is that living cells maintain an internal pressure that closely equals the external pressure, just as the pressure inside a balloon closely matches the outside pressure of the atmosphere. An automobile tire, because of its rigidity, can maintain internal pressures much greater than the external pressure.

CONCEPTUAL EXAMPLE 13–6 **Finger holds water in a straw.** You insert a straw of length ℓ into a tall glass of water. You place your finger over the top of the straw, capturing some air above the water but preventing any additional air from getting in or out, and then you lift the straw from the water. You find that the straw retains most of the water (see Fig. 13–8a). Does the air in the space between your finger and the top of the water have a pressure P that is greater than, equal to, or less than the atmospheric pressure P_0 outside the straw?

RESPONSE Consider the forces on the column of water (Fig. 13–8b). Atmospheric pressure outside the straw pushes upward on the water at the bottom of the straw, gravity pulls the water downward, and the air pressure inside the top of the straw pushes downward on the water. Since the water is in equilibrium, the upward force due to atmospheric pressure P_0 must balance the two downward forces. The only way this is possible is for the air pressure inside the straw to be *less than* the atmosphere pressure outside the straw. (When you initially remove the straw from the glass of water, a little water may leave the bottom of the straw, thus increasing the volume of trapped air and reducing its density and pressure.)

FIGURE 13–8 Example 13–6.

Gauge Pressure

It is important to note that tire gauges, and most other pressure gauges, register the pressure above and beyond atmospheric pressure. This is called **gauge pressure**. Thus, to get the **absolute pressure**, P, we must add the atmospheric pressure, P_0, to the gauge pressure, P_G:

$$P = P_0 + P_G.$$

If a tire gauge registers $220 \, \text{kPa}$, the absolute pressure within the tire is $220 \, \text{kPa} + 101 \, \text{kPa} = 321 \, \text{kPa}$, equivalent to about $3.2 \, \text{atm}$ ($2.2 \, \text{atm}$ gauge pressure).

(a)

Master cylinder

Pedal

Brake cylinder

Brake pads

Disk, attached to wheel

(b)

FIGURE 13–9 Applications of Pascal's principle: (a) hydraulic lift; (b) hydraulic brakes in a car.

PHYSICS APPLIED

Hydraulic lift

PHYSICS APPLIED

Hydraulic brakes

13–5 Pascal's Principle

The Earth's atmosphere exerts a pressure on all objects with which it is in contact, including other fluids. External pressure acting on a fluid is transmitted throughout that fluid. For instance, according to Eq. 13–3, the pressure due to the water at a depth of 100 m below the surface of a lake is $P = \rho g h = (1000 \text{ kg/m}^3)(9.8 \text{ m/s}^2)(100 \text{ m}) = 9.8 \times 10^5 \text{ N/m}^2$, or 9.7 atm. However, the total pressure at this point is due to the pressure of water plus the pressure of the air above it. Hence the total pressure (if the lake is near sea level) is 9.7 atm + 1.0 atm = 10.7 atm. This is just one example of a general principle attributed to the French philosopher and scientist Blaise Pascal (1623–1662). **Pascal's principle** states that *if an external pressure is applied to a confined fluid, the pressure at every point within the fluid increases by that amount.*

A number of practical devices make use of Pascal's principle. One example is the hydraulic lift, illustrated in Fig. 13–9a, in which a small input force is used to exert a large output force by making the area of the output piston larger than the area of the input piston. To see how this works, we assume the input and output pistons are at the same height (at least approximately). Then the external input force F_{in}, by Pascal's principle, increases the pressure equally throughout. Therefore, at the same level (see Fig. 13–9a),

$$P_{\text{out}} = P_{\text{in}}$$

where the input quantities are represented by the subscript "in" and the output by "out." Since $P = F/A$, we write the above equality as

$$\frac{F_{\text{out}}}{A_{\text{out}}} = \frac{F_{\text{in}}}{A_{\text{in}}},$$

or

$$\frac{F_{\text{out}}}{F_{\text{in}}} = \frac{A_{\text{out}}}{A_{\text{in}}}.$$

The quantity $F_{\text{out}}/F_{\text{in}}$ is called the **mechanical advantage** of the hydraulic lift, and it is equal to the ratio of the areas. For example, if the area of the output piston is 20 times that of the input cylinder, the force is multiplied by a factor of 20. Thus a force of 200 lb could lift a 4000-lb car.

Figure 13–9b illustrates the brake system of a car. When the driver presses the brake pedal, the pressure in the master cylinder increases. This pressure increase occurs throughout the brake fluid, thus pushing the brake pads against the disk attached to the car's wheel.

13–6 Measurement of Pressure; Gauges and the Barometer

Many devices have been invented to measure pressure, some of which are shown in Fig. 13–10. The simplest is the open-tube *manometer* (Fig 13–10a) which is a U-shaped tube partially filled with a liquid, usually mercury or water. The pressure P being measured is related to the difference in height Δh of the two levels of the liquid by the relation

$$P = P_0 + \rho g \, \Delta h,$$

where P_0 is atmospheric pressure (acting on the top of the liquid in the left-hand tube), and ρ is the density of the liquid. Note that the quantity $\rho g \, \Delta h$ is the gauge pressure—the amount by which P exceeds atmospheric pressure P_0. If the liquid in the left-hand column were lower than that in the right-hand column, P would have to be less than atmospheric pressure (and Δh would be negative).

Instead of calculating the product $\rho g \, \Delta h$, sometimes only the change in height Δh is specified. In fact, pressures are sometimes specified as so many "millimeters of mercury" (mm-Hg) or "mm of water" (mm-H$_2$O). The unit mm-Hg is equivalent to a pressure of 133 N/m^2, since $\rho g \, \Delta h$ for 1 mm = 1.0×10^{-3} m of mercury gives

$$\rho g \, \Delta h = (13.6 \times 10^3 \text{ kg/m}^3)(9.80 \text{ m/s}^2)(1.00 \times 10^{-3} \text{ m}) = 1.33 \times 10^2 \text{ N/m}^2.$$

The unit mm-Hg is also called the **torr** in honor of Evangelista Torricelli (1608–1647), a student of Galileo's who invented the barometer (see next page).

(a) Open-tube manometer

(b) Aneroid gauge (used mainly for air pressure and then called an aneroid barometer)

(c) Tire gauge

FIGURE 13–10 Pressure gauges: (a) open-tube manometer, (b) aneroid gauge, and (c) common tire-pressure gauge.

Conversion factors among the various units of pressure (an incredible nuisance!) are given in Table 13–2. It is important that only $N/m^2 = Pa$, the proper SI unit, be used in calculations involving other quantities specified in SI units.

TABLE 13–2 Conversion Factors Between Different Units of Pressure

In Terms of $1\ Pa = 1\ N/m^2$	1 atm in Different Units
1 atm = $1.013 \times 10^5\ N/m^2$	1 atm = $1.013 \times 10^5\ N/m^2$
= $1.013 \times 10^5\ Pa = 101.3\ kPa$	
1 bar = $1.000 \times 10^5\ N/m^2$	1 atm = 1.013 bar
1 dyne/cm² = $0.1\ N/m^2$	1 atm = 1.013×10^6 dyne/cm²
1 lb/in.² = $6.90 \times 10^3\ N/m^2$	1 atm = 14.7 lb/in.²
1 lb/ft² = $47.9\ N/m^2$	1 atm = 2.12×10^3 lb/ft²
1 cm-Hg = $1.33 \times 10^3\ N/m^2$	1 atm = 76.0 cm-Hg
1 mm-Hg = $133\ N/m^2$	1 atm = 760 mm-Hg
1 torr = $133\ N/m^2$	1 atm = 760 torr
1 mm-H₂O (4°C) = $9.80\ N/m^2$	1 atm = 1.03×10^4 mm-H₂O (4°C)

Another type of pressure gauge is the aneroid gauge (Fig. 13–10b) in which the pointer is linked to the flexible ends of an evacuated thin metal chamber. In an electronic gauge, the pressure may be applied to a thin metal diaphragm whose resulting distortion is translated into an electrical signal by a transducer. A common tire gauge is shown in Fig. 13–10c.

Atmospheric pressure can be measured by a modified kind of mercury manometer with one end closed, called a mercury **barometer** (Fig. 13–11). The glass tube is completely filled with mercury and then inverted into the bowl of mercury. If the tube is long enough, the level of the mercury will drop, leaving a vacuum at the top of the tube, since atmospheric pressure can support a column of mercury only about 76 cm high (exactly 76.0 cm at standard atmospheric pressure). That is, a column of mercury 76 cm high exerts the same pressure as the atmosphere[†]:

$$P = \rho g\,\Delta h$$
$$= \left(13.6 \times 10^3\ kg/m^3\right)\left(9.80\ m/s^2\right)\left(0.760\ m\right) = 1.013 \times 10^5\ N/m^2 = 1.00\ atm.$$

[†]This calculation confirms the entry in Table 13–2, 1 atm = 76.0 cm-Hg.

FIGURE 13–11 A mercury barometer, invented by Torricelli, is shown here when the air pressure is standard atmospheric, 76.0 cm-Hg.

FIGURE 13–12 A water barometer: a full tube of water is inserted into a tub of water, keeping the tube's spigot at the top closed. When the bottom end of the tube is unplugged, some water flows out of the tube into the tub, leaving a vacuum between the water's upper surface and the spigot. Why? Because air pressure can not support a column of water more than 10 m high.

A calculation similar to what we just did will show that atmospheric pressure can maintain a column of water 10.3 m high in a tube whose top is under vacuum (Fig. 13–12). No matter how good a vacuum pump is, water cannot be made to rise more than about 10 m using normal atmospheric pressure. To pump water out of deep mine shafts with a vacuum pump requires multiple stages for depths greater than 10 m. Galileo studied this problem, and his student Torricelli was the first to explain it. The point is that a pump does not really suck water up a tube—it merely reduces the pressure at the top of the tube. Atmospheric air pressure *pushes* the water up the tube if the top end is at low pressure (under a vacuum), just as it is air pressure that pushes (or maintains) the mercury 76 cm high in a barometer. [Force pumps (Section 13–14) that push up from the bottom can exert higher pressure to push water more than 10 m high.]

CONCEPTUAL EXAMPLE 13–7 | **Suction.** A student suggests suction-cup shoes for Space Shuttle astronauts working on the exterior of a spacecraft. Having just studied this Chapter, you gently remind him of the fallacy of this plan. What is it?

RESPONSE Suction cups work by pushing out the air underneath the cup. What holds the suction cup in place is the air pressure outside it. (This can be a substantial force when on Earth. For example, a 10-cm-diameter suction cup has an area of $7.9 \times 10^{-3} \, \text{m}^2$. The force of the atmosphere on it is $(7.9 \times 10^{-3} \, \text{m}^2)(1.0 \times 10^5 \, \text{N/m}^2) \approx 800 \, \text{N}$, about 180 lbs!) But in outer space, there is no air pressure to push the suction cup onto the spacecraft.

We sometimes mistakenly think of suction as something we actively do. For example, we intuitively think that we pull the soda up through a straw. Instead, what we do is lower the pressure at the top of the straw, and the atmosphere *pushes* the soda up the straw.

13–7 Buoyancy and Archimedes' Principle

Objects submerged in a fluid appear to weigh less than they do when outside the fluid. For example, a large rock that you would have difficulty lifting off the ground can often be easily lifted from the bottom of a stream. When the rock breaks through the surface of the water, it suddenly seems to be much heavier. Many objects, such as wood, float on the surface of water. These are two examples of *buoyancy*. In each example, the force of gravity is acting downward. But in addition, an upward *buoyant force* is exerted by the liquid. The buoyant force on fish and underwater divers (as in the Chapter-Opening photo) almost exactly balances the force of gravity downward, and allows them to "hover" in equilibrium.

The buoyant force occurs because the pressure in a fluid increases with depth. Thus the upward pressure on the bottom surface of a submerged object is greater than the downward pressure on its top surface. To see this effect, consider a cylinder of height Δh whose top and bottom ends have an area A and which is completely submerged in a fluid of density ρ_F, as shown in Fig. 13–13. The fluid exerts a pressure $P_1 = \rho_F g h_1$ at the top surface of the cylinder (Eq. 13–3). The force due to this pressure on top of the cylinder is $F_1 = P_1 A = \rho_F g h_1 A$, and it is directed downward. Similarly, the fluid exerts an upward force on the bottom of the cylinder equal to $F_2 = P_2 A = \rho_F g h_2 A$. The net force on the cylinder exerted by the fluid pressure, which is the **buoyant force**, $\vec{F}_B$, acts upward and has the magnitude

$$\begin{aligned} F_B = F_2 - F_1 &= \rho_F g A (h_2 - h_1) \\ &= \rho_F g A \, \Delta h \\ &= \rho_F V g \\ &= m_F g, \end{aligned}$$

FIGURE 13–13 Determination of the buoyant force.

where $V = A \, \Delta h$ is the volume of the cylinder, the product $\rho_F V$ is the mass of the fluid displaced, and $\rho_F V g = m_F g$ is the weight of fluid which takes up a volume equal to the volume of the cylinder. Thus the buoyant force on the cylinder is equal to the weight of fluid displaced by the cylinder.

This result is valid no matter what the shape of the object. Its discovery is credited to Archimedes (287?–212 B.C.), and it is called **Archimedes' principle**: *the buoyant force on an object immersed in a fluid is equal to the weight of the fluid displaced by that object.*

By "fluid displaced," we mean a volume of fluid equal to the submerged volume of the object (or that part of the object that is submerged). If the object is placed in a glass or tub initially filled to the brim with water, the water that flows over the top represents the water displaced by the object.

We can derive Archimedes' principle in general by the following simple but elegant argument. The irregularly shaped object D shown in Fig. 13–14a is acted on by the force of gravity (its weight, $m\vec{g}$, downward) and the buoyant force, $\vec{F}_B$, upward. We wish to determine F_B. To do so, we next consider a body (D′ in Fig. 13–14b), this time made of the fluid itself, with the same shape and size as the original object, and located at the same depth. You might think of this body of fluid as being separated from the rest of the fluid by an imaginary membrane. The buoyant force F_B on this body of fluid will be exactly the same as that on the original object since the surrounding fluid, which exerts F_B, is in exactly the same configuration. This body of fluid D′ is in equilibrium (the fluid as a whole is at rest). Therefore, $F_B = m'g$, where $m'g$ is the weight of the body of fluid. Hence the buoyant force F_B is equal to the weight of the body of fluid whose volume equals the volume of the original submerged object, which is Archimedes' principle.

Archimedes' discovery was made by experiment. What we have done in the last two paragraphs is to show that Archimedes' principle can be derived from Newton's laws.

(a)

(b)

FIGURE 13–14
Archimedes' principle.

CONCEPTUAL EXAMPLE 13–8 **Two pails of water.** Consider two identical pails of water filled to the brim. One pail contains only water, the other has a piece of wood floating in it. Which pail has the greater weight?

RESPONSE Both pails weigh the same. Recall Archimedes' principle: the wood displaces a volume of water with weight equal to the weight of the wood. Some water will overflow the pail, but Archimedes' principle tells us the spilled water has weight equal to the weight of the wood object; so the pails have the same weight.

EXAMPLE 13–9 **Recovering a submerged statue.** A 70-kg ancient statue lies at the bottom of the sea. Its volume is $3.0 \times 10^4 \, cm^3$. How much force is needed to lift it?

APPROACH The force F needed to lift the statue is equal to the statue's weight mg minus the buoyant force F_B. Figure 13–15 is the free-body diagram.

SOLUTION The buoyant force on the statue due to the water is equal to the weight of $3.0 \times 10^4 \, cm^3 = 3.0 \times 10^{-2} \, m^3$ of water (for seawater, $\rho = 1.025 \times 10^3 \, kg/m^3$):

$$F_B = m_{H_2O}g = \rho_{H_2O}Vg$$
$$= (1.025 \times 10^3 \, kg/m^3)(3.0 \times 10^{-2} \, m^3)(9.8 \, m/s^2)$$
$$= 3.0 \times 10^2 \, N.$$

The weight of the statue is $mg = (70 \, kg)(9.8 \, m/s^2) = 6.9 \times 10^2 \, N$. Hence the force F needed to lift it is $690 \, N - 300 \, N = 390 \, N$. It is as if the statue had a mass of only $(390 \, N)/(9.8 \, m/s^2) = 40 \, kg$.

NOTE Here $F = 390 \, N$ is the force needed to lift the statue without acceleration when it is under water. As the statue comes *out* of the water, the force F increases, reaching 690 N when the statue is fully out of the water.

FIGURE 13–15 Example 13–9. The force needed to lift the statue is $\vec{F}$.

Archimedes, is said to have discovered his principle in his bath while thinking how he might determine whether the king's new crown was pure gold or a fake. Gold has a specific gravity of 19.3, somewhat higher than that of most metals, but a determination of specific gravity or density is not readily done directly because, even if the mass is known, the volume of an irregularly shaped object is not easily calculated. However, if the object is weighed in air $(= w)$ and also "weighed" while it is under water $(= w')$, the density can be determined using Archimedes' principle, as the following Example shows. The quantity w' is called the *apparent weight* in water, and is what a scale reads when the object is submerged in water (see Fig. 13–16); w' equals the true weight $(w = mg)$ minus the buoyant force.

EXAMPLE 13–10 Archimedes: Is the crown gold? When a crown of mass 14.7 kg is submerged in water, an accurate scale reads only 13.4 kg. Is the crown made of gold?

APPROACH If the crown is gold, its density and specific gravity must be very high, SG = 19.3 (see Section 13–2 and Table 13–1). We determine the specific gravity using Archimedes' principle and the two free-body diagrams shown in Fig. 13–16.

SOLUTION The apparent weight of the submerged object (the crown) is w' (what the scale reads), and is the force pulling down on the scale hook. By Newton's third law, w' equals the force F'_T that the scale exerts on the crown in Fig. 13–16b. The sum of the forces on the crown is zero, so w' equals the actual weight $w (= mg)$ minus the buoyant force F_B:

$$w' = F'_T = w - F_B$$

so

$$w - w' = F_B.$$

Let V be the volume of the completely submerged object and ρ_O its density (so $\rho_O V$ is its mass), and let ρ_F be the density of the fluid (water). Then $(\rho_F V)g$ is the weight of fluid displaced $(= F_B)$. Now we can write

$$w = mg = \rho_O Vg$$
$$w - w' = F_B = \rho_F Vg.$$

We divide these two equations and obtain

$$\frac{w}{w - w'} = \frac{\rho_O Vg}{\rho_F Vg} = \frac{\rho_O}{\rho_F}.$$

We see that $w/(w - w')$ is equal to the specific gravity of the object if the fluid in which it is submerged is water $(\rho_F = 1.00 \times 10^3 \text{ kg/m}^3)$. Thus

$$\frac{\rho_O}{\rho_{H_2O}} = \frac{w}{w - w'} = \frac{(14.7 \text{ kg})g}{(14.7 \text{ kg} - 13.4 \text{ kg})g} = \frac{14.7 \text{ kg}}{1.3 \text{ kg}} = 11.3.$$

This corresponds to a density of 11,300 kg/m³. The crown is not gold, but seems to be made of lead (see Table 13–1).

FIGURE 13–16 (a) A scale reads the mass of an object in air—in this case the crown of Example 13–10. All objects are at rest, so the tension F_T in the connecting cord equals the weight w of the object: $F_T = mg$. We show the free-body diagram of the crown, and F_T is what causes the scale reading (it is equal to the net downward force on the scale, by Newton's third law). (b) Submerged, the crown has an additional force on it, the buoyant force F_B. The net force is zero, so $F'_T + F_B = mg (= w)$. The scale now reads $m' = 13.4$ kg, where m' is related to the effective weight by $w' = m'g$. Thus $F'_T = w' = w - F_B$.

Archimedes' principle applies equally well to objects that float, such as wood. In general, *an object floats on a fluid if its density* (ρ_O) *is less than that of the fluid* (ρ_F). This is readily seen from Fig. 13–17a, where a submerged log will experience a net upward force and float to the surface if $F_B > mg$; that is, if $\rho_F V g > \rho_O V g$ or $\rho_F > \rho_O$. At equilibrium—that is, when floating—the buoyant force on an object has magnitude equal to the weight of the object. For example, a log whose specific gravity is 0.60 and whose volume is 2.0 m³ has a mass $m = \rho_O V = (0.60 \times 10^3 \text{ kg/m}^3)(2.0 \text{ m}^3) = 1200 \text{ kg}$. If the log is fully submerged, it will displace a mass of water $m_F = \rho_F V = (1000 \text{ kg/m}^3)(2.0 \text{ m}^3) = 2000 \text{ kg}$. Hence the buoyant force on the log will be greater than its weight, and it will float upward to the surface (Fig. 13–17). The log will come to equilibrium when it displaces 1200 kg of water, which means that 1.2 m³ of its volume will be submerged. This 1.2 m³ corresponds to 60% of the volume of the log (1.2/2.0 = 0.60), so 60% of the log is submerged.

In general when an object floats, we have $F_B = mg$, which we can write as (see Fig. 13–18)

$$F_B = mg$$
$$\rho_F V_{displ} g = \rho_O V_O g,$$

where V_O is the full volume of the object and V_{displ} is the volume of fluid it displaces (= volume submerged). Thus

$$\frac{V_{displ}}{V_O} = \frac{\rho_O}{\rho_F}.$$

That is, the fraction of the object submerged is given by the ratio of the object's density to that of the fluid. If the fluid is water, this fraction equals the specific gravity of the object.

(a)

(b)

FIGURE 13–17 (a) The fully submerged log accelerates upward because $F_B > mg$. It comes to equilibrium (b) when $\Sigma F = 0$, so $F_B = mg = (1200 \text{ kg})g$. Thus 1200 kg, or 1.2 m³, of water is displaced.

FIGURE 13–18 An object floating in equilibrium: $F_B = mg$.

EXAMPLE 13–11 **Hydrometer calibration.** A **hydrometer** is a simple instrument used to measure the specific gravity of a liquid by indicating how deeply the instrument sinks in the liquid. A particular hydrometer (Fig. 13–19) consists of a glass tube, weighted at the bottom, which is 25.0 cm long and 2.00 cm² in cross-sectional area, and has a mass of 45.0 g. How far from the end should the 1.000 mark be placed?

APPROACH The hydrometer will float in water if its density ρ is less than $\rho_w = 1.000 \text{ g/cm}^3$, the density of water. The fraction of the hydrometer submerged $(V_{displaced}/V_{total})$ is equal to the density ratio ρ/ρ_w.

SOLUTION The hydrometer has an overall density

$$\rho = \frac{m}{V} = \frac{45.0 \text{ g}}{(2.00 \text{ cm}^2)(25.0 \text{ cm})} = 0.900 \text{ g/cm}^3.$$

Thus, when placed in water, it will come to equilibrium when 0.900 of its volume is submerged. Since it is of uniform cross section, (0.900)(25.0 cm) = 22.5 cm of its length will be submerged. The specific gravity of water is defined to be 1.000, so the mark should be placed 22.5 cm from the weighted end.

FIGURE 13–19 A hydrometer. Example 13–11.

EXERCISE C On the hydrometer of Example 13–11, will the marks above the 1.000 mark represent higher or lower values of density of the liquid in which it is submerged?

Archimedes' principle is also useful in geology. According to the theories of plate tectonics and continental drift, the continents float on a fluid "sea" of slightly deformable rock (mantle rock). Some interesting calculations can be done using very simple models, which we consider in the Problems at the end of the Chapter.

Air is a fluid, and it too exerts a buoyant force. Ordinary objects weigh less in air than they do if weighed in a vacuum. Because the density of air is so small, the effect for ordinary solids is slight. There are objects, however, that *float* in air—helium-filled balloons, for example, because the density of helium is less than the density of air.

PHYSICS APPLIED
Continental drift—plate tectonics

EXERCISE D Which of the following objects, submerged in water, experiences the largest magnitude of the buoyant force? (*a*) A 1-kg helium balloon; (*b*) 1 kg of wood; (*c*) 1 kg of ice; (*d*) 1 kg of iron; (*e*) all the same.

EXERCISE E Which of the following objects, submerged in water, experiences the largest magnitude of the buoyant force? (*a*) A 1-m³ helium balloon; (*b*) 1 m³ of wood; (*c*) 1 m³ of ice; (*d*) 1 m³ of iron; (*e*) all the same.

EXAMPLE 13–12 **Helium balloon.** What volume V of helium is needed if a balloon is to lift a load of 180 kg (including the weight of the empty balloon)?

APPROACH The buoyant force on the helium balloon, F_B, which is equal to the weight of displaced air, must be at least equal to the weight of the helium plus the weight of the balloon and load (Fig. 13–20). Table 13–1 gives the density of helium as 0.179 kg/m³.

SOLUTION The buoyant force must have a minimum value of

$$F_B = (m_{He} + 180 \text{ kg})g.$$

This equation can be written in terms of density using Archimedes' principle:

$$\rho_{air} V g = (\rho_{He} V + 180 \text{ kg})g.$$

Solving now for V, we find

$$V = \frac{180 \text{ kg}}{\rho_{air} - \rho_{He}} = \frac{180 \text{ kg}}{(1.29 \text{ kg/m}^3 - 0.179 \text{ kg/m}^3)} = 160 \text{ m}^3.$$

NOTE This is the minimum volume needed near the Earth's surface, where $\rho_{air} = 1.29$ kg/m³. To reach a high altitude, a greater volume would be needed since the density of air decreases with altitude.

FIGURE 13–20 Example 13–12.

13–8 Fluids in Motion; Flow Rate and the Equation of Continuity

We now turn to the subject of fluids in motion, which is called **fluid dynamics**, or (especially if the fluid is water) **hydrodynamics**.

We can distinguish two main types of fluid flow. If the flow is smooth, such that neighboring layers of the fluid slide by each other smoothly, the flow is said to be **streamline** or **laminar flow**.[†] In streamline flow, each particle of the fluid follows a smooth path, called a **streamline**, and these paths do not cross one another (Fig. 13–21a). Above a certain speed, the flow becomes turbulent. **Turbulent flow** is characterized by erratic, small, whirlpool-like circles called *eddy currents* or *eddies* (Fig. 13–21b). Eddies absorb a great deal of energy, and although a certain amount of internal friction called **viscosity** is present even during streamline flow, it is much greater when the flow is turbulent. A few tiny drops of ink or food coloring dropped into a moving liquid can quickly reveal whether the flow is streamline or turbulent.

[†]The word *laminar* means "in layers."

FIGURE 13–21 (a) Streamline, or laminar, flow; (b) turbulent flow. The photos show airflow around an airfoil or airplane wing (more in Section 13–10).

(a)

(b)

Let us consider the steady laminar flow of a fluid through an enclosed tube or pipe as shown in Fig. 13–22. First we determine how the speed of the fluid changes when the size of the tube changes. The mass **flow rate** is defined as the mass Δm of fluid that passes a given point per unit time Δt:

$$\text{mass flow rate} = \frac{\Delta m}{\Delta t}.$$

FIGURE 13–22 Fluid flow through a pipe of varying diameter.

In Fig. 13–22, the volume of fluid passing point 1 (that is, through area A_1) in a time Δt is $A_1 \Delta \ell_1$, where $\Delta \ell_1$ is the distance the fluid moves in time Δt. Since the velocity[†] of fluid passing point 1 is $v_1 = \Delta \ell_1 / \Delta t$, the mass flow rate through area A_1 is

$$\frac{\Delta m_1}{\Delta t} = \frac{\rho_1 \Delta V_1}{\Delta t} = \frac{\rho_1 A_1 \Delta \ell_1}{\Delta t} = \rho_1 A_1 v_1,$$

where $\Delta V_1 = A_1 \Delta \ell_1$ is the volume of mass Δm_1, and ρ_1 is the fluid density. Similarly, at point 2 (through area A_2), the flow rate is $\rho_2 A_2 v_2$. Since no fluid flows in or out the sides, the flow rates through A_1 and A_2 must be equal. Thus, since

$$\frac{\Delta m_1}{\Delta t} = \frac{\Delta m_2}{\Delta t},$$

then

$$\rho_1 A_1 v_1 = \rho_2 A_2 v_2. \qquad \textbf{(13–7a)}$$

This is called the **equation of continuity**.

If the fluid is incompressible (ρ doesn't change with pressure), which is an excellent approximation for liquids under most circumstances (and sometimes for gases as well), then $\rho_1 = \rho_2$, and the equation of continuity becomes

$$A_1 v_1 = A_2 v_2. \qquad [\rho = \text{constant}] \quad \textbf{(13–7b)}$$

The product Av represents the *volume rate of flow* (volume of fluid passing a given point per second), since $\Delta V / \Delta t = A \Delta \ell / \Delta t = Av$, which in SI units is m^3/s. Equation 13–7b tells us that where the cross-sectional area is large, the velocity is small, and where the area is small, the velocity is large. That this is reasonable can be seen by looking at a river. A river flows slowly through a meadow where it is broad, but speeds up to torrential speed when passing through a narrow gorge.

EXAMPLE 13–13 **ESTIMATE** **Blood flow.** In humans, blood flows from the heart into the aorta, from which it passes into the major arteries. These branch into the small arteries (arterioles), which in turn branch into myriads of tiny capillaries, Fig. 13–23. The blood returns to the heart via the veins. The radius of the aorta is about 1.2 cm, and the blood passing through it has a speed of about 40 cm/s. A typical capillary has a radius of about 4×10^{-4} cm, and blood flows through it at a speed of about 5×10^{-4} m/s. Estimate the number of capillaries that are in the body.

APPROACH We assume the density of blood doesn't vary significantly from the aorta to the capillaries. By the equation of continuity, the volume flow rate in the aorta must equal the volume flow rate through *all* the capillaries. The total area of all the capillaries is given by the area of one capillary multiplied by the total number N of capillaries.

SOLUTION Let A_1 be the area of the aorta and A_2 be the area of *all* the capillaries through which blood flows. Then $A_2 = N\pi r_{cap}^2$, where $r_{cap} \approx 4 \times 10^{-4}$ cm is the estimated average radius of one capillary. From the equation of continuity (Eq. 13–7b), we have

$$v_2 A_2 = v_1 A_1$$
$$v_2 N\pi r_{cap}^2 = v_1 \pi r_{aorta}^2$$

so

$$N = \frac{v_1}{v_2} \frac{r_{aorta}^2}{r_{cap}^2} = \left(\frac{0.40 \text{ m/s}}{5 \times 10^{-4} \text{ m/s}} \right) \left(\frac{1.2 \times 10^{-2} \text{ m}}{4 \times 10^{-6} \text{ m}} \right)^2 \approx 7 \times 10^9,$$

or on the order of 10 billion capillaries.

PHYSICS APPLIED
Blood flow

FIGURE 13–23
Human circulatory system.

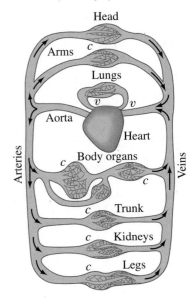

v = valves
c = capillaries

[†]If there were no viscosity, the velocity would be the same across a cross section of the tube. Real fluids have viscosity, and this internal friction causes different layers of the fluid to flow at different speeds. In this case v_1 and v_2 represent the average speeds at each cross section.

FIGURE 13–24   Example 13–14.

EXAMPLE 13–14 **Heating duct to a room.** What area must a heating duct have if air moving 3.0 m/s along it can replenish the air every 15 minutes in a room of volume 300 m³? Assume the air's density remains constant.

APPROACH We apply the equation of continuity at constant density, Eq. 13–7b, to the air that flows through the duct (point 1 in Fig. 13–24) and then into the room (point 2). The volume flow rate in the room equals the volume of the room divided by the 15-min replenishing time.

SOLUTION Consider the room as a large section of the duct, Fig. 13–24, and think of air equal to the volume of the room as passing by point 2 in $t = 15$ min $= 900$ s. Reasoning in the same way we did to obtain Eq. 13–7a (changing Δt to t), we write $v_2 = \ell_2/t$ so $A_2 v_2 = A_2 \ell_2/t = V_2/t$, where V_2 is the volume of the room. Then the equation of continuity becomes $A_1 v_1 = A_2 v_2 = V_2/t$ and

$$A_1 = \frac{V_2}{v_1 t} = \frac{300 \text{ m}^3}{(3.0 \text{ m/s})(900 \text{ s})} = 0.11 \text{ m}^2.$$

If the duct is square, then each side has length $\ell = \sqrt{A} = 0.33$ m, or 33 cm. A rectangular duct 20 cm × 55 cm will also do.

13–9 Bernoulli's Equation

Have you ever wondered why an airplane can fly, or how a sailboat can move against the wind? These are examples of a principle worked out by Daniel Bernoulli (1700–1782) concerning fluids in motion. In essence, **Bernoulli's principle** states that *where the velocity of a fluid is high, the pressure is low, and where the velocity is low, the pressure is high.* For example, if the pressures in the fluid at points 1 and 2 of Fig. 13–22 are measured, it will be found that the pressure is lower at point 2, where the velocity is greater, than it is at point 1, where the velocity is smaller. At first glance, this might seem strange; you might expect that the greater speed at point 2 would imply a higher pressure. But this cannot be the case. For if the pressure in the fluid at point 2 were higher than at point 1, this higher pressure would slow the fluid down, whereas in fact it has sped up in going from point 1 to point 2. Thus the pressure at point 2 must be less than at point 1, to be consistent with the fact that the fluid accelerates.

To help clarify any misconceptions, a faster fluid *would* exert a greater force on an obstacle placed in its path. But that is not what we mean by the pressure in a fluid, and besides we are not considering obstacles that interrupt the flow. We are examining smooth streamline flow. The fluid pressure is exerted on the walls of a tube or pipe, or on the surface of any material the fluid passes over.

Bernoulli developed an equation that expresses this principle quantitatively. To derive Bernoulli's equation, we assume the flow is steady and laminar, the fluid is incompressible, and the viscosity is small enough to be ignored. To be general, we assume the fluid is flowing in a tube of nonuniform cross section that varies in height above some reference level, Fig. 13–25. We will consider the volume of fluid shown in color and calculate the work done to move it from the position shown in Fig. 13–25a to that shown in Fig. 13–25b. In this process, fluid entering area A_1 flows a distance $\Delta \ell_1$ and forces the fluid at area A_2 to move a distance $\Delta \ell_2$. The fluid to the left of area A_1 exerts a pressure P_1 on our section of fluid and does an amount of work

$$W_1 = F_1 \Delta \ell_1 = P_1 A_1 \Delta \ell_1.$$

At area A_2, the work done on our cross section of fluid is

$$W_2 = -P_2 A_2 \Delta \ell_2.$$

The negative sign is present because the force exerted on the fluid is opposite to the motion (thus the fluid shown in color does work on the fluid to the right of point 2). Work is also done on the fluid by the force of gravity. The net effect of the process shown in Fig. 13–25 is to move a mass m of volume $A_1 \Delta \ell_1$ (= $A_2 \Delta \ell_2$, since the

FIGURE 13–25   Fluid flow: for derivation of Bernoulli's equation.

fluid is incompressible) from point 1 to point 2, so the work done by gravity is

$$W_3 = -mg(y_2 - y_1),$$

where y_1 and y_2 are heights of the center of the tube above some (arbitrary) reference level. In the case shown in Fig. 13–25, this term is negative since the motion is uphill against the force of gravity. The net work W done on the fluid is thus

$$W = W_1 + W_2 + W_3$$
$$W = P_1 A_1 \Delta\ell_1 - P_2 A_2 \Delta\ell_2 - mgy_2 + mgy_1.$$

According to the work-energy principle (Section 7–4), the net work done on a system is equal to its change in kinetic energy. Hence

$$\tfrac{1}{2}mv_2^2 - \tfrac{1}{2}mv_1^2 = P_1 A_1 \Delta\ell_1 - P_2 A_2 \Delta\ell_2 - mgy_2 + mgy_1.$$

The mass m has volume $A_1 \Delta\ell_1 = A_2 \Delta\ell_2$ for an incompressible fluid. Thus we can substitute $m = \rho A_1 \Delta\ell_1 = \rho A_2 \Delta\ell_2$, and then divide through by $A_1 \Delta\ell_1 = A_2 \Delta\ell_2$, to obtain

$$\tfrac{1}{2}\rho v_2^2 - \tfrac{1}{2}\rho v_1^2 = P_1 - P_2 - \rho g y_2 + \rho g y_1,$$

which we rearrange to get

$$P_1 + \tfrac{1}{2}\rho v_1^2 + \rho g y_1 = P_2 + \tfrac{1}{2}\rho v_2^2 + \rho g y_2. \qquad \textbf{(13–8)} \qquad \textit{Bernoulli's equation}$$

This is **Bernoulli's equation**. Since points 1 and 2 can be any two points along a tube of flow, Bernoulli's equation can be written as

$$P + \tfrac{1}{2}\rho v^2 + \rho g y = \text{constant}$$

at every point in the fluid, where y is the height of the center of the tube above a fixed reference level. [Note that if there is no flow $(v_1 = v_2 = 0)$, then Eq. 13–8 reduces to the hydrostatic equation, Eq. 13–6a: $P_2 - P_1 = -\rho g(y_2 - y_1)$.]

Bernoulli's equation is an expression of the law of energy conservation, since we derived it from the work-energy principle.

> **EXERCISE F** As water in a level pipe passes from a narrow cross section of pipe to a wider cross section, how does the pressure against the walls change?

EXAMPLE 13–15 **Flow and pressure in a hot-water heating system.** Water circulates throughout a house in a hot-water heating system. If the water is pumped at a speed of 0.50 m/s through a 4.0-cm-diameter pipe in the basement under a pressure of 3.0 atm, what will be the flow speed and pressure in a 2.6-cm-diameter pipe on the second floor 5.0 m above? Assume the pipes do not divide into branches.

APPROACH We use the equation of continuity at constant density to determine the flow speed on the second floor, and then Bernoulli's equation to find the pressure.

SOLUTION We take v_2 in the equation of continuity, Eq. 13–7, as the flow speed on the second floor, and v_1 as the flow speed in the basement. Noting that the areas are proportional to the radii squared $(A = \pi r^2)$, we obtain

$$v_2 = \frac{v_1 A_1}{A_2} = \frac{v_1 \pi r_1^2}{\pi r_2^2} = (0.50 \text{ m/s}) \frac{(0.020 \text{ m})^2}{(0.013 \text{ m})^2} = 1.2 \text{ m/s}.$$

To find the pressure on the second floor, we use Bernoulli's equation (Eq. 13–8):

$$\begin{aligned}
P_2 &= P_1 + \rho g(y_1 - y_2) + \tfrac{1}{2}\rho(v_1^2 - v_2^2) \\
&= (3.0 \times 10^5 \text{ N/m}^2) + (1.0 \times 10^3 \text{ kg/m}^3)(9.8 \text{ m/s}^2)(-5.0 \text{ m}) \\
&\quad + \tfrac{1}{2}(1.0 \times 10^3 \text{ kg/m}^3)\big[(0.50 \text{ m/s})^2 - (1.2 \text{ m/s})^2\big] \\
&= (3.0 \times 10^5 \text{ N/m}^2) - (4.9 \times 10^4 \text{ N/m}^2) - (6.0 \times 10^2 \text{ N/m}^2) \\
&= 2.5 \times 10^5 \text{ N/m}^2 = 2.5 \text{ atm}.
\end{aligned}$$

NOTE The velocity term contributes very little in this case.

FIGURE 13–26 Torricelli's theorem: $v_1 = \sqrt{2g(y_2 - y_1)}$.

FIGURE 13–27 Examples of Bernoulli's principle: (a) atomizer, (b) Ping-Pong ball in jet of air.

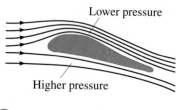

FIGURE 13–28 Lift on an airplane wing. We are in the reference frame of the wing, seeing the air flow by.

⊛ **P H Y S I C S   A P P L I E D**
Airplanes and dynamic lift

13–10 Applications of Bernoulli's Principle: Torricelli, Airplanes, Baseballs, TIA

Bernoulli's equation can be applied to many situations. One example is to calculate the velocity, v_1, of a liquid flowing out of a spigot at the bottom of a reservoir, Fig. 13–26. We choose point 2 in Eq. 13–8 to be the top surface of the liquid. Assuming the diameter of the reservoir is large compared to that of the spigot, v_2 will be almost zero. Points 1 (the spigot) and 2 (top surface) are open to the atmosphere, so the pressure at both points is equal to atmospheric pressure: $P_1 = P_2$. Then Bernoulli's equation becomes

$$\tfrac{1}{2}\rho v_1^2 + \rho g y_1 = \rho g y_2$$

or

$$v_1 = \sqrt{2g(y_2 - y_1)}. \tag{13–9}$$

This result is called **Torricelli's theorem**. Although it is seen to be a special case of Bernoulli's equation, it was discovered a century earlier by Evangelista Torricelli. Equation 13–9 tells us that the liquid leaves the spigot with the same speed that a freely falling object would attain if falling from the same height. This should not be too surprising since Bernoulli's equation relies on the conservation of energy.

Another special case of Bernoulli's equation arises when a fluid is flowing horizontally with no significant change in height; that is, $y_1 = y_2$. Then Eq. 13–8 becomes

$$P_1 + \tfrac{1}{2}\rho v_1^2 = P_2 + \tfrac{1}{2}\rho v_2^2, \tag{13–10}$$

which tells us quantitatively that the speed is high where the pressure is low, and vice versa. It explains many common phenomena, some of which are illustrated in Figs. 13–27 to 13–32. The pressure in the air blown at high speed across the top of the vertical tube of a perfume atomizer (Fig. 13–27a) is less than the normal air pressure acting on the surface of the liquid in the bowl. Thus atmospheric pressure in the bowl pushes the perfume up the tube because of the lower pressure at the top. A Ping-Pong ball can be made to float above a blowing jet of air (some vacuum cleaners can blow air), Fig. 13–27b; if the ball begins to leave the jet of air, the higher pressure in the still air outside the jet pushes the ball back in.

EXERCISE G Return to Chapter-Opening Question 2, page 339, and answer it again now. Try to explain why you may have answered differently the first time. Try it and see.

Airplane Wings and Dynamic Lift

Airplanes experience a "lift" force on their wings, keeping them up in the air, if they are moving at a sufficiently high speed relative to the air and the wing is tilted upward at a small angle (the "attack angle"), as in Fig. 13–28, where streamlines of air are shown rushing by the wing. (We are in the reference frame of the wing, as if sitting on the wing.) The upward tilt, as well as the rounded upper surface of the wing, causes the streamlines to be forced upward and to be crowded together above the wing. The area for air flow between any two streamlines is reduced as the streamlines get closer together, so from the equation of continuity $(A_1 v_1 = A_2 v_2)$, the air speed increases above the wing where the streamlines are squished together. (Recall also how the crowded streamlines in a pipe constriction, Fig. 13–22, indicate the velocity is higher in the constriction.) Because the air speed is greater above the wing than below it, the pressure above the wing is less than the pressure below the wing (Bernoulli's principle). Hence there is a net upward force on the wing called **dynamic lift**. Experiments show that the speed of air above the wing can even be double the speed of the air below it. (Friction between the air and wing exerts a *drag force*, toward the rear, which must be overcome by the plane's engines.)

A flat wing, or one with symmetric cross section, will experience lift as long as the front of the wing is tilted upward (attack angle). The wing shown in Fig. 13–28 can experience lift even if the attack angle is zero, because the rounded upper surface deflects air up, squeezing the streamlines together. Airplanes can fly upside down, experiencing lift, if the attack angle is sufficient to deflect streamlines up and closer together.

Our picture considers streamlines; but if the attack angle is larger than about 15°, turbulence sets in (Fig. 13–21b) leading to greater drag and less lift, causing the wing to "stall" and the plane to drop.

From another point of view, the upward tilt of a wing means the air moving horizontally in front of the wing is deflected downward; the change in momentum of the rebounding air molecules results in an upward force on the wing (Newton's third law).

Sailboats

A sailboat can move *against* the wind, with the aid of the Bernoulli effect, by setting the sails at an angle, as shown in Fig. 13–29. The air travels rapidly over the bulging front surface of the sail, and the relatively still air filling the sail exerts a greater pressure behind the sail, resulting in a net force on the sail, $\vec{\mathbf{F}}_{\text{wind}}$. This force would tend to make the boat move sideways if it weren't for the keel that extends vertically downward beneath the water: the water exerts a force ($\vec{\mathbf{F}}_{\text{water}}$) on the keel nearly perpendicular to the keel. The resultant of these two forces ($\vec{\mathbf{F}}_{\text{R}}$) is almost directly forward as shown.

Baseball Curve

Why a spinning pitched baseball (or tennis ball) curves can also be explained using Bernoulli's principle. It is simplest if we put ourselves in the reference frame of the ball, with the air rushing by, just as we did for the airplane wing. Suppose the ball is rotating counterclockwise as seen from above, Fig. 13–30. A thin layer of air ("boundary layer") is being dragged around by the ball. We are looking down on the ball, and at point A in Fig. 13–30, this boundary layer tends to slow down the oncoming air. At point B, the air rotating with the ball adds its speed to that of the oncoming air, so the air speed is higher at B than at A. The higher speed at B means the pressure is lower at B than at A, resulting in a net force toward B. The ball's path curves toward the left (as seen by the pitcher).

Lack of Blood to the Brain—TIA

In medicine, one of many applications of Bernoulli's principle is to explain a TIA, a transient ischemic attack (meaning a temporary lack of blood supply to the brain). A person suffering a TIA may experience symptoms such as dizziness, double vision, headache, and weakness of the limbs. A TIA can occur as follows. Blood normally flows up to the brain at the back of the head via the two vertebral arteries—one going up each side of the neck—which meet to form the basilar artery just below the brain, as shown in Fig. 13–31. The vertebral arteries issue from the subclavian arteries, as shown, before the latter pass to the arms. When an arm is exercised vigorously, blood flow increases to meet the needs of the arm's muscles. If the subclavian artery on one side of the body is partially blocked, however, as in arteriosclerosis (hardening of the arteries), the blood velocity will have to be higher on that side to supply the needed blood. (Recall the equation of continuity: smaller area means larger velocity for the same flow rate, Eqs. 13–7.) The increased blood velocity past the opening to the vertebral artery results in lower pressure (Bernoulli's principle). Thus blood rising in the vertebral artery on the "good" side at normal pressure can be *diverted down* into the other vertebral artery because of the low pressure on that side, instead of passing upward to the brain. Hence the blood supply to the brain is reduced.

Other Applications

A **venturi tube** is essentially a pipe with a narrow constriction (the throat). The flowing fluid speeds up as it passes through this constriction, so the pressure is lower in the throat. A *venturi meter*, Fig. 13–32, is used to measure the flow speed of gases and liquids, including blood velocity in arteries.

Why does smoke go up a chimney? It's partly because hot air rises (it's less dense and therefore buoyant). But Bernoulli's principle also plays a role. When wind blows across the top of a chimney, the pressure is less there than inside the house. Hence, air and smoke are pushed up the chimney by the higher indoor pressure. Even on an apparently still night there is usually enough ambient air flow at the top of a chimney to assist upward flow of smoke.

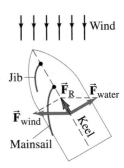

FIGURE 13–29 Sailboat sailing against the wind.

FIGURE 13–30 Looking down on a pitched baseball heading toward home plate. We are in the reference frame of the baseball, with the air flowing by.

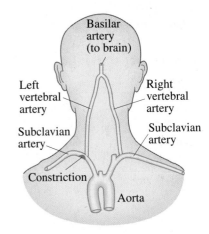

FIGURE 13–31 Rear of the head and shoulders showing arteries leading to the brain and to the arms. High blood velocity past the constriction in the left subclavian artery causes low pressure in the left vertebral artery, in which a reverse (downward) blood flow can then occur, resulting in a TIA, a loss of blood to the brain.

FIGURE 13–32 Venturi meter.

Bernoulli's equation ignores the effects of friction (viscosity) and the compressibility of the fluid. The energy that is transformed to internal (or potential) energy due to compression and to thermal energy by friction can be taken into account by adding terms to Eq. 13–8. These terms are difficult to calculate theoretically and are normally determined empirically. They do not significantly alter the explanations for the phenomena described above.

*13–11 Viscosity

Real fluids have a certain amount of internal friction called **viscosity**, as mentioned in Section 13–8. Viscosity exists in both liquids and gases, and is essentially a frictional force between adjacent layers of fluid as the layers move past one another. In liquids, viscosity is due to the electrical cohesive forces between the molecules. In gases, it arises from collisions between the molecules.

The viscosity of different fluids can be expressed quantitatively by a *coefficient of viscosity*, η (the Greek lowercase letter eta), which is defined in the following way. A thin layer of fluid is placed between two flat plates. One plate is stationary and the other is made to move, Fig. 13–33. The fluid directly in contact with each plate is held to the surface by the adhesive force between the molecules of the liquid and those of the plate. Thus the upper surface of the fluid moves with the same speed v as the upper plate, whereas the fluid in contact with the stationary plate remains stationary. The stationary layer of fluid retards the flow of the layer just above it, which in turn retards the flow of the next layer, and so on. Thus the velocity varies continuously from 0 to v, as shown. The increase in velocity divided by the distance over which this change is made—equal to v/ℓ—is called the *velocity gradient*. To move the upper plate requires a force, which you can verify by moving a flat plate across a puddle of syrup on a table. For a given fluid, it is found that the force required, F, is proportional to the area of fluid in contact with each plate, A, and to the speed, v, and is inversely proportional to the separation, ℓ, of the plates: $F \propto vA/\ell$. For different fluids, the more viscous the fluid, the greater is the required force. Hence the proportionality constant for this equation is defined as the coefficient of viscosity, η:

$$F = \eta A \frac{v}{\ell}. \tag{13–11}$$

Solving for η, we find $\eta = F\ell/vA$. The SI unit for η is $N \cdot s/m^2 = Pa \cdot s$ (pascal·second). In the cgs system, the unit is dyne·s/cm², which is called a *poise* (P). Viscosities are often given in centipoise ($1\ cP = 10^{-2}\ P$). Table 13–3 lists the coefficient of viscosity for various fluids. The temperature is also specified, since it has a strong effect; the viscosity of liquids such as motor oil, for example, decreases rapidly as temperature increases.[†]

FIGURE 13–33
Determination of viscosity.

TABLE 13–3
Coefficients of Viscosity

Fluid (temperature in °C)	Coefficient of Viscosity, η (Pa·s)[†]
Water (0°)	1.8×10^{-3}
(20°)	1.0×10^{-3}
(100°)	0.3×10^{-3}
Whole blood (37°)	$\approx 4 \times 10^{-3}$
Blood plasma (37°)	$\approx 1.5 \times 10^{-3}$
Ethyl alcohol (20°)	1.2×10^{-3}
Engine oil (30°) (SAE 10)	200×10^{-3}
Glycerine (20°)	1500×10^{-3}
Air (20°)	0.018×10^{-3}
Hydrogen (0°)	0.009×10^{-3}
Water vapor (100°)	0.013×10^{-3}

[†] $1\ Pa \cdot s = 10\ P = 1000\ cP$.

*13–12 Flow in Tubes: Poiseuille's Equation, Blood Flow

If a fluid had no viscosity, it could flow through a level tube or pipe without a force being applied. Viscosity acts like a sort of friction (between fluid layers moving at slightly different speeds), so a pressure difference between the ends of a level tube is necessary for the steady flow of any real fluid, be it water or oil in a pipe, or blood in the circulatory system of a human.

The French scientist J. L. Poiseuille (1799–1869), who was interested in the physics of blood circulation (and after whom the poise is named), determined how the variables affect the flow rate of an incompressible fluid undergoing laminar flow in a cylindrical tube. His result, known as *Poiseuille's equation*, is:

$$Q = \frac{\pi R^4 (P_1 - P_2)}{8\eta\ell}, \tag{13–12}$$

where R is the inside radius of the tube, ℓ is the tube length, $P_1 - P_2$ is the pressure

[†]The Society of Automotive Engineers assigns numbers to represent the viscosity of oils: 30 weight (SAE 30) is more viscous than 10 weight. Multigrade oils, such as 20–50, are designed to maintain viscosity as temperature increases; 20–50 means the oil is 20 wt when cool but is like a 50-wt pure oil when it is hot (engine running temperature).

difference between the ends, η is the coefficient of viscosity, and Q is the volume rate of flow (volume of fluid flowing past a given point per unit time which in SI has units of m^3/s). Equation 13–12 applies only to laminar flow.

Poiseuille's equation tells us that the flow rate Q is directly proportional to the "pressure gradient," $(P_1 - P_2)/\ell$, and it is inversely proportional to the viscosity of the fluid. This is just what we might expect. It may be surprising, however, that Q also depends on the *fourth* power of the tube's radius. This means that for the same pressure gradient, if the tube radius is halved, the flow rate is decreased by a factor of 16! Thus the rate of flow, or alternately the pressure required to maintain a given flow rate, is greatly affected by only a small change in tube radius.

An interesting example of this R^4 dependence is *blood flow* in the human body. Poiseuille's equation is valid only for the streamline flow of an incompressible fluid. So it cannot be precisely accurate for blood whose flow is not without turbulence and that contains blood cells (whose diameter is almost equal to that of a capillary). Nonetheless, Poiseuille's equation does give a reasonable first approximation. Because the radius of arteries is reduced as a result of arteriosclerosis (thickening and hardening of artery walls) and by cholesterol buildup, the pressure gradient must be increased to maintain the same flow rate. If the radius is reduced by half, the heart would have to increase the pressure by a factor of about $2^4 = 16$ in order to maintain the same blood-flow rate. The heart must work much harder under these conditions, but usually cannot maintain the original flow rate. Thus, high blood pressure is an indication both that the heart is working harder and that the blood-flow rate is reduced.

PHYSICS APPLIED
Blood flow

*13–13 Surface Tension and Capillarity

The *surface* of a liquid at rest behaves in an interesting way, almost as if it were a stretched membrane under tension. For example, a drop of water on the end of a dripping faucet, or hanging from a thin branch in the early morning dew (Fig. 13–34), forms into a nearly spherical shape as if it were a tiny balloon filled with water. A steel needle can be made to float on the surface of water even though it is denser than the water. The surface of a liquid acts like it is under tension, and this tension, acting along the surface, arises from the attractive forces between the molecules. This effect is called **surface tension**. More specifically, a quantity called the *surface tension*, γ (the Greek letter gamma), is defined as the force F per unit length ℓ that acts perpendicular to any line or cut in a liquid surface, tending to pull the surface closed:

$$\gamma = \frac{F}{\ell}. \tag{13-13}$$

FIGURE 13–34 Spherical water droplets, dew on a blade of grass.

To understand this, consider the U-shaped apparatus shown in Fig. 13–35 which encloses a thin film of liquid. Because of surface tension, a force F is required to pull the movable wire and thus increase the surface area of the liquid. The liquid contained by the wire apparatus is a thin film having both a top and a bottom surface. Hence the total length of the surface being increased is 2ℓ, and the surface tension is $\gamma = F/2\ell$. A delicate apparatus of this type can be used to measure the surface tension of various liquids. The surface tension of water is $0.072 \, N/m$ at $20°C$. Table 13–4 gives the values for several substances. Note that temperature has a considerable effect on the surface tension.

TABLE 13–4 Surface Tension of Some Substances

Substance (temperature in °C)	Surface Tension (N/m)
Mercury (20°)	0.44
Blood, whole (37°)	0.058
Blood, plasma (37°)	0.073
Alcohol, ethyl (20°)	0.023
Water (0°)	0.076
(20°)	0.072
(100°)	0.059
Benzene (20°)	0.029
Soap solution (20°)	≈0.025
Oxygen (−193°)	0.016

(a) Top view

Liquid   Wire
(b) Edge view (magnified)

FIGURE 13–35 U-shaped wire apparatus holding a film of liquid to measure surface tension ($\gamma = F/2\ell$).

FIGURE 13–36 A water strider.

Because of surface tension, some insects (Fig. 13–36) can walk on water, and objects more dense than water, such as a steel needle, can float on the surface. Figure 13–37a shows how the surface tension can support the weight w of an object. Actually, the object sinks slightly into the fluid, so w is the "effective weight" of that object—its true weight less the buoyant force.

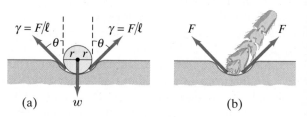

FIGURE 13–37 Surface tension acting on (a) a sphere, and (b) an insect leg. Example 13–16.

EXAMPLE 13–16 ESTIMATE Insect walks on water. The base of an insect's leg is approximately spherical in shape, with a radius of about 2.0×10^{-5} m. The 0.0030-g mass of the insect is supported equally by its six legs. Estimate the angle θ (see Fig. 13–37) for an insect on the surface of water. Assume the water temperature is 20°C.

APPROACH Since the insect is in equilibrium, the upward surface tension force is equal to the pull of gravity downward on each leg. We ignore the buoyant force for this estimate.

SOLUTION For each leg, we assume the surface tension force acts all around a circle of radius r, at an angle θ, as shown in Fig. 13–37a. Only the vertical component, $\gamma \cos \theta$, acts to balance the weight mg. So we set the length ℓ in Eq. 13–13 equal to the circumference of the circle, $\ell \approx 2\pi r$. Then the net upward force due to surface tension is $F_y \approx (\gamma \cos \theta)\ell \approx 2\pi r\gamma \cos \theta$. We set this surface tension force equal to one-sixth the weight of the insect since it has six legs:

$$2\pi r\gamma \cos \theta \approx \tfrac{1}{6}mg$$
$$(6.28)(2.0 \times 10^{-5}\,\text{m})(0.072\,\text{N/m})\cos \theta \approx \tfrac{1}{6}(3.0 \times 10^{-6}\,\text{kg})(9.8\,\text{m/s}^2)$$
$$\cos \theta \approx \frac{0.49}{0.90} = 0.54.$$

So $\theta \approx 57°$. If $\cos \theta$ had come out greater than 1, the surface tension would not be great enough to support the insect's weight.

NOTE Our estimate ignored the buoyant force and ignored any difference between the radius of the insect's "foot" and the radius of the surface depression.

Soaps and detergents lower the surface tension of water. This is desirable for washing and cleaning since the high surface tension of pure water prevents it from penetrating easily between the fibers of material and into tiny crevices. Substances that reduce the surface tension of a liquid are called *surfactants*.

Surface tension plays a role in another interesting phenomenon, capillarity. It is a common observation that water in a glass container rises up slightly where it touches the glass, Fig. 13–38a. The water is said to "wet" the glass. Mercury, on the other hand, is depressed when it touches the glass, Fig. 13–38b; the mercury does not wet the glass. Whether a liquid wets a solid surface is determined by the relative strength of the cohesive forces between the molecules of the liquid compared to the adhesive forces between the molecules of the liquid and those of the container. *Cohesion* refers to the force between molecules of the same type, whereas *adhesion* refers to the force between molecules of different types. Water wets glass because the water molecules are more strongly attracted to the glass molecules than they are to other water molecules. The opposite is true for mercury: the cohesive forces are stronger than the adhesive forces.

In tubes having very small diameters, liquids are observed to rise or fall relative to the level of the surrounding liquid. This phenomenon is called **capillarity**, and such thin tubes are called **capillaries**. Whether the liquid rises or falls (Fig. 13–39) depends on the relative strengths of the adhesive and cohesive forces. Thus water rises in a glass tube, whereas mercury falls. The actual amount of rise (or fall) depends on the surface tension—which is what keeps the liquid surface from breaking apart.

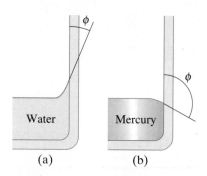

FIGURE 13–38 (a) Water "wets" the surface of glass, whereas (b) mercury does not "wet" the glass.

(a) Water   (b) Mercury

FIGURE 13–39 Capillarity.

(a) Glass tube in water

(b) Glass tube in mercury

*13–14 Pumps, and the Heart

We conclude this Chapter with a brief discussion of pumps, including the heart. Pumps can be classified into categories according to their function. A *vacuum pump* is designed to reduce the pressure (usually of air) in a given vessel. A *force pump*, on the other hand, is a pump that is intended to increase the pressure—for example, to lift a liquid (such as water from a well) or to push a fluid through a pipe. Figure 13–40 illustrates the principle behind a simple reciprocating pump. It could be a vacuum pump, in which case the intake is connected to the vessel to be evacuated. A similar mechanism is used in some force pumps, and in this case the fluid is forced under increased pressure through the outlet.

A centrifugal pump (Fig. 13–41), or any force pump, can be used as a *circulating pump*—that is, to circulate a fluid around a closed path, such as the cooling water or lubricating oil in an automobile.

The heart of a human (and of other animals as well) is essentially a circulating pump. The action of a human heart is shown in Fig. 13–42. There are actually two separate paths for blood flow. The longer path takes blood to the parts of the body, via the arteries, bringing oxygen to body tissues and picking up carbon dioxide, which it carries back to the heart via veins. This blood is then pumped to the lungs (the second path), where the carbon dioxide is released and oxygen is taken up. The oxygen-laden blood is returned to the heart, where it is again pumped to the tissues of the body.

FIGURE 13–40 One kind of pump: the intake valve opens and air (or fluid that is being pumped) fills the empty space when the piston moves to the left. When the piston moves to the right (not shown), the outlet valve opens and fluid is forced out.

FIGURE 13–41 Centrifugal pump: the rotating blades force fluid through the outlet pipe; this kind of pump is used in vacuum cleaners and as a water pump in automobiles.

FIGURE 13–42 (a) In the diastole phase, the heart relaxes between beats. Blood moves into the heart; both atria fill rapidly. (b) When the atria contract, the systole or pumping phase begins. The contraction pushes the blood through the mitral and tricuspid valves into the ventricles. (c) The contraction of the ventricles forces the blood through the semilunar valves into the pulmonary artery, which leads to the lungs, and to the aorta (the body's largest artery), which leads to the arteries serving all the body. (d) When the heart relaxes, the semilunar valves close; blood fills the atria, beginning the cycle again.

Summary

The three common phases of matter are **solid**, **liquid**, and **gas**. Liquids and gases are collectively called **fluids**, meaning they have the ability to flow. The **density** of a material is defined as its mass per unit volume:

$$\rho = \frac{m}{V}. \tag{13–1}$$

Specific gravity is the ratio of the density of the material to the density of water (at 4°C).

Pressure is defined as force per unit area:

$$P = \frac{F}{A}. \tag{13–2}$$

The pressure P at a depth h in a liquid is given by

$$P = \rho g h, \tag{13–3}$$

where ρ is the density of the liquid and g is the acceleration due to gravity. If the density of a fluid is not uniform, the pressure P varies with height y as

$$\frac{dP}{dy} = -\rho g. \tag{13–4}$$

Pascal's principle says that an external pressure applied to a confined fluid is transmitted throughout the fluid.

Pressure is measured using a manometer or other type of gauge. A **barometer** is used to measure atmospheric pressure. Standard **atmospheric pressure** (average at sea level) is $1.013 \times 10^5 \, \text{N/m}^2$. **Gauge pressure** is the total (absolute) pressure less atmospheric pressure.

Archimedes' principle states that an object submerged wholly or partially in a fluid is buoyed up by a force equal to the weight of fluid it displaces $(F_B = m_F g = \rho_F V_{\text{displ}} g)$.

Fluid flow can be characterized either as **streamline** (sometimes called **laminar**), in which the layers of fluid move smoothly and regularly along paths called **streamlines**, or as **turbulent**, in which case the flow is not smooth and regular but is characterized by irregularly shaped whirlpools.

Fluid flow rate is the mass or volume of fluid that passes a given point per unit time. The **equation of continuity** states that for an incompressible fluid flowing in an enclosed tube, the product of the velocity of flow and the cross-sectional area of the tube remains constant:

$$Av = \text{constant.} \qquad \textbf{(13–7b)}$$

Bernoulli's principle tells us that where the velocity of a fluid is high, the pressure in it is low, and where the velocity is low, the pressure is high. For steady laminar flow of an incompressible and nonviscous fluid, **Bernoulli's equation**, which is based on the law of conservation of energy, is

$$P_1 + \tfrac{1}{2}\rho v_1^2 + \rho g y_1 = P_2 + \tfrac{1}{2}\rho v_2^2 + \rho g y_2, \qquad \textbf{(13–8)}$$

for two points along the flow.

[*Viscosity refers to friction within a fluid and is essentially a frictional force between adjacent layers of fluid as they move past one another.]

[*Liquid surfaces hold together as if under tension (**surface tension**), allowing drops to form and objects like needles and insects to stay on the surface.]

Questions

1. If one material has a higher density than another, must the molecules of the first be heavier than those of the second? Explain.

2. Airplane travelers sometimes note that their cosmetics bottles and other containers have leaked during a flight. What might cause this?

3. The three containers in Fig. 13–43 are filled with water to the same height and have the same surface area at the base; hence the water pressure, and the total force on the base of each, is the same. Yet the total weight of water is different for each. Explain this "hydrostatic paradox."

FIGURE 13–43
Question 3.

4. Consider what happens when you push both a pin and the blunt end of a pen against your skin with the same force. Decide what determines whether your skin is cut—the net force applied to it or the pressure.

5. A small amount of water is boiled in a 1-gallon metal can. The can is removed from the heat and the lid put on. As the can cools, it collapses. Explain.

6. When blood pressure is measured, why must the cuff be held at the level of the heart?

7. An ice cube floats in a glass of water filled to the brim. What can you say about the density of ice? As the ice melts, will the water overflow? Explain.

8. Will an ice cube float in a glass of alcohol? Why or why not?

9. A submerged can of Coke® will sink, but a can of Diet Coke® will float. (Try it!) Explain.

10. Why don't ships made of iron sink?

11. Explain how the tube in Fig. 13–44, known as a **siphon**, can transfer liquid from one container to a lower one even though the liquid must flow uphill for part of its journey. (Note that the tube must be filled with liquid to start with.)

FIGURE 13–44
Question 11. A siphon.

12. A barge filled high with sand approaches a low bridge over the river and cannot quite pass under it. Should sand be added to, or removed from, the barge? [*Hint*: Consider Archimedes' principle.]

13. Explain why helium weather balloons, which are used to measure atmospheric conditions at high altitudes, are normally released while filled to only 10–20% of their maximum volume.

14. A row boat floats in a swimming pool, and the level of the water at the edge of the pool is marked. Consider the following situations and explain whether the level of the water will rise, fall, or stay the same. (*a*) The boat is removed from the water. (*b*) The boat in the water holds an iron anchor which is removed from the boat and placed on the shore. (*c*) The iron anchor is removed from the boat and dropped in the pool.

15. Will an empty balloon have precisely the same apparent weight on a scale as a balloon filled with air? Explain.

16. Why do you float higher in salt water than in fresh water?

17. If you dangle two pieces of paper vertically, a few inches apart (Fig. 13–45), and blow between them, how do you think the papers will move? Try it and see. Explain.

FIGURE 13–45
Question 17.

FIGURE 13–46
Question 18. Water coming from a faucet.

18. Why does the stream of water from a faucet become narrower as it falls (Fig. 13–46)?

19. Children are told to avoid standing too close to a rapidly moving train because they might get sucked under it. Is this possible? Explain.

20. A tall Styrofoam cup is filled with water. Two holes are punched in the cup near the bottom, and water begins rushing out. If the cup is dropped so it falls freely, will the water continue to flow from the holes? Explain.

21. Why do airplanes normally take off into the wind?

22. Two ships moving in parallel paths close to one another risk colliding. Why?

23. Why does the canvas top of a convertible bulge out when the car is traveling at high speed? [*Hint*: The windshield deflects air upward, pushing streamlines closer together.]

24. Roofs of houses are sometimes "blown" off (or are they pushed off?) during a tornado or hurricane. Explain using Bernoulli's principle.

Problems

13-2 Density and Specific Gravity

1. (I) The approximate volume of the granite monolith known as El Capitan in Yosemite National Park (Fig. 13–47) is about 10^8 m^3. What is its approximate mass?

FIGURE 13–47 Problem 1.

2. (I) What is the approximate mass of air in a living room $5.6 \text{ m} \times 3.8 \text{ m} \times 2.8 \text{ m}$?

3. (I) If you tried to smuggle gold bricks by filling your backpack, whose dimensions are $56 \text{ cm} \times 28 \text{ cm} \times 22 \text{ cm}$, what would its mass be?

4. (I) State your mass and then estimate your volume. [*Hint:* Because you can swim on or just under the surface of the water in a swimming pool, you have a pretty good idea of your density.]

5. (II) A bottle has a mass of 35.00 g when empty and 98.44 g when filled with water. When filled with another fluid, the mass is 89.22 g. What is the specific gravity of this other fluid?

6. (II) If 5.0 L of antifreeze solution (specific gravity = 0.80) is added to 4.0 L of water to make a 9.0-L mixture, what is the specific gravity of the mixture?

7. (III) The Earth is not a uniform sphere, but has regions of varying density. Consider a simple model of the Earth divided into three regions—inner core, outer core, and mantle. Each region is taken to have a unique constant density (the average density of that region in the real Earth):

Region	Radius (km)	Density (kg/m³)
Inner Core	0–1220	13,000
Outer Core	1220–3480	11,100
Mantle	3480–6371	4,400

(a) Use this model to predict the average density of the entire Earth. (b) The measured radius of the Earth is 6371 km and its mass is $5.98 \times 10^{24} \text{ kg}$. Use these data to determine the actual average density of the Earth and compare it (as a percent difference) with the one you determined in (a).

13-3 to 13-6 Pressure; Pascal's Principle

8. (I) Estimate the pressure needed to raise a column of water to the same height as a 35-m-tall oak tree.

9. (I) Estimate the pressure exerted on a floor by (a) one pointed chair leg (66 kg on all four legs) of area $= 0.020 \text{ cm}^2$, and (b) a 1300-kg elephant standing on one foot (area $= 800 \text{ cm}^2$).

10. (I) What is the difference in blood pressure (mm-Hg) between the top of the head and bottom of the feet of a 1.70-m-tall person standing vertically?

11. (II) How high would the level be in an alcohol barometer at normal atmospheric pressure?

12. (II) In a movie, Tarzan evades his captors by hiding underwater for many minutes while breathing through a long, thin reed. Assuming the maximum pressure difference his lungs can manage and still breathe is -85 mm-Hg, calculate the deepest he could have been.

13. (II) The maximum gauge pressure in a hydraulic lift is 17.0 atm. What is the largest-size vehicle (kg) it can lift if the diameter of the output line is 22.5 cm?

14. (II) The gauge pressure in each of the four tires of an automobile is 240 kPa. If each tire has a "footprint" of 220 cm^2, estimate the mass of the car.

15. (II) (a) Determine the total force and the absolute pressure on the bottom of a swimming pool 28.0 m by 8.5 m whose uniform depth is 1.8 m. (b) What will be the pressure against the *side* of the pool near the bottom?

16. (II) A house at the bottom of a hill is fed by a full tank of water 5.0 m deep and connected to the house by a pipe that is 110 m long at an angle of 58° from the horizontal (Fig. 13–48). (a) Determine the water gauge pressure at the house. (b) How high could the water shoot if it came vertically out of a broken pipe in front of the house?

FIGURE 13–48
Problem 16.

17. (II) Water and then oil (which don't mix) are poured into a U-shaped tube, open at both ends. They come to equilibrium as shown in Fig. 13–49. What is the density of the oil? [*Hint:* Pressures at points a and b are equal. Why?]

FIGURE 13–49
Problem 17.

18. (II) In working out his principle, Pascal showed dramatically how force can be multiplied with fluid pressure. He placed a long, thin tube of radius $r = 0.30 \text{ cm}$ vertically into a wine barrel of radius $R = 21 \text{ cm}$, Fig. 13–50. He found that when the barrel was filled with water and the tube filled to a height of 12 m, the barrel burst. Calculate (a) the mass of water in the tube, and (b) the net force exerted by the water in the barrel on the lid just before rupture.

FIGURE 13–50
Problem 18 (not to scale).

19. (II) What is the normal pressure of the atmosphere at the summit of Mt. Everest, 8850 m above sea level?

20. (II) A hydraulic press for compacting powdered samples has a large cylinder which is 10.0 cm in diameter, and a small cylinder with a diameter of 2.0 cm (Fig. 13–51). A lever is attached to the small cylinder as shown. The sample, which is placed on the large cylinder, has an area of 4.0 cm². What is the pressure on the sample if 350 N is applied to the lever?

FIGURE 13–51 Problem 20.

21. (II) An open-tube mercury manometer is used to measure the pressure in an oxygen tank. When the atmospheric pressure is 1040 mbar, what is the absolute pressure (in Pa) in the tank if the height of the mercury in the open tube is (a) 21.0 cm higher, (b) 5.2 cm lower, than the mercury in the tube connected to the tank?

22. (III) A beaker of liquid accelerates from rest, on a horizontal surface, with acceleration a to the right. (a) Show that the surface of the liquid makes an angle $\theta = \tan^{-1}(a/g)$ with the horizontal. (b) Which edge of the water surface is higher? (c) How does the pressure vary with depth below the surface?

23. (III) Water stands at a height h behind a vertical dam of uniform width b. (a) Use integration to show that the total force of the water on the dam is $F = \frac{1}{2}\rho g h^2 b$. (b) Show that the torque about the base of the dam due to this force can be considered to act with a lever arm equal to $h/3$. (c) For a freestanding concrete dam of uniform thickness t and height h, what minimum thickness is needed to prevent overturning? Do you need to add in atmospheric pressure for this last part? Explain.

24. (III) Estimate the density of the water 5.4 km deep in the sea. (See Table 12–1 and Section 12–4 regarding bulk modulus.) By what fraction does it differ from the density at the surface?

25. (III) A cylindrical bucket of liquid (density ρ) is rotated about its symmetry axis, which is vertical. If the angular velocity is ω, show that the pressure at a distance r from the rotation axis is

$$P = P_0 + \frac{1}{2}\rho\omega^2 r^2,$$

where P_0 is the pressure at $r = 0$.

13–7 Buoyancy and Archimedes' Principle

26. (I) What fraction of a piece of iron will be submerged when it floats in mercury?

27. (I) A geologist finds that a Moon rock whose mass is 9.28 kg has an apparent mass of 6.18 kg when submerged in water. What is the density of the rock?

28. (II) A crane lifts the 16,000-kg steel hull of a sunken ship out of the water. Determine (a) the tension in the crane's cable when the hull is fully submerged in the water, and (b) the tension when the hull is completely out of the water.

29. (II) A spherical balloon has a radius of 7.35 m and is filled with helium. How large a cargo can it lift, assuming that the skin and structure of the balloon have a mass of 930 kg? Neglect the buoyant force on the cargo volume itself.

30. (II) A 74-kg person has an apparent mass of 54 kg (because of buoyancy) when standing in water that comes up to the hips. Estimate the mass of each leg. Assume the body has SG = 1.00.

31. (II) What is the likely identity of a metal (see Table 13–1) if a sample has a mass of 63.5 g when measured in air and an apparent mass of 55.4 g when submerged in water?

32. (II) Calculate the true mass (in vacuum) of a piece of aluminum whose apparent mass is 3.0000 kg when weighed in air.

33. (II) Because gasoline is less dense than water, drums containing gasoline will float in water. Suppose a 230-L steel drum is completely full of gasoline. What total volume of steel can be used in making the drum if the gasoline-filled drum is to float in fresh water?

34. (II) A scuba diver and her gear displace a volume of 65.0 L and have a total mass of 68.0 kg. (a) What is the buoyant force on the diver in seawater? (b) Will the diver sink or float?

35. (II) The specific gravity of ice is 0.917, whereas that of seawater is 1.025. What percent of an iceberg is above the surface of the water?

36. (II) Archimedes' principle can be used not only to determine the specific gravity of a solid using a known liquid (Example 13–10); the reverse can be done as well. (a) As an example, a 3.80-kg aluminum ball has an apparent mass of 2.10 kg when submerged in a particular liquid: calculate the density of the liquid. (b) Derive a formula for determining the density of a liquid using this procedure.

37. (II) (a) Show that the buoyant force F_B on a partially submerged object such as a ship acts at the center of gravity of the fluid before it is displaced. This point is called the **center of buoyancy**. (b) To ensure that a ship is in stable equilibrium, would it be better if its center of buoyancy was above, below, or at the same point as, its center of gravity? Explain. (See Fig. 13–52.)

FIGURE 13–52
Problem 37.

38. (II) A cube of side length 10.0 cm and made of unknown material floats at the surface between water and oil. The oil has a density of 810 kg/m³. If the cube floats so that it is 72% in the water and 28% in the oil, what is the mass of the cube and what is the buoyant force on the cube?

39. (II) How many helium-filled balloons would it take to lift a person? Assume the person has a mass of 75 kg and that each helium-filled balloon is spherical with a diameter of 33 cm.

40. (II) A scuba tank, when fully submerged, displaces 15.7 L of seawater. The tank itself has a mass of 14.0 kg and, when "full," contains 3.00 kg of air. Assuming only a weight and buoyant force act, determine the net force (magnitude and direction) on the fully submerged tank at the beginning of a dive (when it is full of air) and at the end of a dive (when it no longer contains any air).

41. (III) If an object floats in water, its density can be determined by tying a sinker to it so that both the object and the sinker are submerged. Show that the specific gravity is given by $w/(w_1 - w_2)$, where w is the weight of the object alone in air, w_1 is the apparent weight when a sinker is tied to it and the sinker only is submerged, and w_2 is the apparent weight when both the object and the sinker are submerged.

42. (III) A 3.25-kg piece of wood (SG = 0.50) floats on water. What minimum mass of lead, hung from the wood by a string, will cause it to sink?

13–8 to 13–10 Fluid Flow, Bernoulli's Equation

43. (I) A 15-cm-radius air duct is used to replenish the air of a room 8.2 m × 5.0 m × 3.5 m every 12 min. How fast does the air flow in the duct?

44. (I) Using the data of Example 13–13, calculate the average speed of blood flow in the major arteries of the body which have a total cross-sectional area of about 2.0 cm^2.

45. (I) How fast does water flow from a hole at the bottom of a very wide, 5.3-m-deep storage tank filled with water? Ignore viscosity.

46. (II) A fish tank has dimensions 36 cm wide by 1.0 m long by 0.60 m high. If the filter should process all the water in the tank once every 4.0 h, what should the flow speed be in the 3.0-cm-diameter input tube for the filter?

47. (II) What gauge pressure in the water mains is necessary if a firehose is to spray water to a height of 18 m?

48. (II) A $\frac{5}{8}$-in. (inside) diameter garden hose is used to fill a round swimming pool 6.1 m in diameter. How long will it take to fill the pool to a depth of 1.2 m if water flows from the hose at a speed of 0.40 m/s?

49. (II) A 180-km/h wind blowing over the flat roof of a house causes the roof to lift off the house. If the house is 6.2 m × 12.4 m in size, estimate the weight of the roof. Assume the roof is not nailed down.

50. (II) A 6.0-cm-diameter horizontal pipe gradually narrows to 4.5 cm. When water flows through this pipe at a certain rate, the gauge pressure in these two sections is 32.0 kPa and 24.0 kPa, respectively. What is the volume rate of flow?

51. (II) Estimate the air pressure inside a category 5 hurricane, where the wind speed is 300 km/h (Fig. 13–53).

FIGURE 13–53 Problem 51.

52. (II) What is the lift (in newtons) due to Bernoulli's principle on a wing of area 88 m^2 if the air passes over the top and bottom surfaces at speeds of 280 m/s and 150 m/s, respectively?

53. (II) Show that the power needed to drive a fluid through a pipe with uniform cross-section is equal to the volume rate of flow, Q, times the pressure difference, $P_1 - P_2$.

54. (II) Water at a gauge pressure of 3.8 atm at street level flows into an office building at a speed of 0.68 m/s through a pipe 5.0 cm in diameter. The pipe tapers down to 2.8 cm in diameter by the top floor, 18 m above (Fig. 13–54), where the faucet has been left open. Calculate the flow velocity and the gauge pressure in the pipe on the top floor. Assume no branch pipes and ignore viscosity.

FIGURE 13–54
Problem 54.

55. (II) In Fig. 13–55, take into account the speed of the top surface of the tank and show that the speed of fluid leaving the opening at the bottom is

$$v_1 = \sqrt{\frac{2gh}{(1 - A_1^2/A_2^2)}},$$

where $h = y_2 - y_1$, and A_1 and A_2 are the areas of the opening and of the top surface, respectively. Assume $A_1 \ll A_2$ so that the flow remains nearly steady and laminar.

FIGURE 13–55
Problems 55, 56, 58, and 59.

56. (II) Suppose the top surface of the vessel in Fig. 13–55 is subjected to an external gauge pressure P_2. (a) Derive a formula for the speed, v_1, at which the liquid flows from the opening at the bottom into atmospheric pressure, P_0. Assume the velocity of the liquid surface, v_2, is approximately zero. (b) If $P_2 = 0.85$ atm and $y_2 - y_1 = 2.4$ m, determine v_1 for water.

57. (II) You are watering your lawn with a hose when you put your finger over the hose opening to increase the distance the water reaches. If you are pointing the hose at the same angle, and the distance the water reaches increases by a factor of 4, what fraction of the hose opening did you block?

58. (III) Suppose the opening in the tank of Fig. 13–55 is a height h_1 above the base and the liquid surface is a height h_2 above the base. The tank rests on level ground. (a) At what horizontal distance from the base of the tank will the fluid strike the ground? (b) At what other height, h_1', can a hole be placed so that the emerging liquid will have the same "range"? Assume $v_2 \approx 0$.

FIGURE 13–55 (repeated) Problems 55, 56, 58, and 59.

59. (III) (a) In Fig. 13–55, show that Bernoulli's principle predicts that the level of the liquid, $h = y_2 - y_1$, drops at a rate

$$\frac{dh}{dt} = -\sqrt{\frac{2ghA_1^2}{A_2^2 - A_1^2}},$$

where A_1 and A_2 are the areas of the opening and the top surface, respectively, assuming $A_1 \ll A_2$, and viscosity is ignored. (b) Determine h as a function of time by integrating. Let $h = h_0$ at $t = 0$. (c) How long would it take to empty a 10.6-cm-tall cylinder filled with 1.3 L of water if the opening is at the bottom and has a 0.50-cm diameter?

60. (III) (a) Show that the flow speed measured by a venturi meter (see Fig. 13–32) is given by the relation

$$v_1 = A_2 \sqrt{\frac{2(P_1 - P_2)}{\rho(A_1^2 - A_2^2)}}.$$

(b) A venturi meter is measuring the flow of water; it has a main diameter of 3.0 cm tapering down to a throat diameter of 1.0 cm. If the pressure difference is measured to be 18 mm-Hg, what is the speed of the water entering the venturi throat?

61. (III) *Thrust of a rocket.* (a) Use Bernoulli's equation and the equation of continuity to show that the emission speed of the propelling gases of a rocket is

$$v = \sqrt{2(P - P_0)/\rho},$$

where ρ is the density of the gas, P is the pressure of the gas inside the rocket, and P_0 is atmospheric pressure just outside the exit orifice. Assume that the gas density stays approximately constant, and that the area of the exit orifice, A_0, is much smaller than the cross-sectional area, A, of the inside of the rocket (take it to be a large cylinder). Assume also that the gas speed is not so high that significant turbulence or nonsteady flow sets in. (b) Show that the thrust force on the rocket due to the emitted gases is

$$F = 2A_0(P - P_0).$$

62. (III) A fire hose exerts a force on the person holding it. This is because the water accelerates as it goes from the hose through the nozzle. How much force is required to hold a 7.0-cm-diameter hose delivering 450 L/min through a 0.75-cm-diameter nozzle?

*13–11 Viscosity

*63. (II) A viscometer consists of two concentric cylinders, 10.20 cm and 10.60 cm in diameter. A liquid fills the space between them to a depth of 12.0 cm. The outer cylinder is fixed, and a torque of 0.024 m·N keeps the inner cylinder turning at a steady rotational speed of 57 rev/min. What is the viscosity of the liquid?

*64. (III) A long vertical hollow tube with an inner diameter of 1.00 cm is filled with SAE 10 motor oil. A 0.900-cm-diameter, 30.0-cm-long 150-g rod is dropped vertically through the oil in the tube. What is the maximum speed attained by the rod as it falls?

*13–12 Flow in Tubes; Poiseuille's Equation

*65. (I) Engine oil (assume SAE 10, Table 13–3) passes through a fine 1.80-mm-diameter tube that is 8.6 cm long. What pressure difference is needed to maintain a flow rate of 6.2 mL/min?

*66. (I) A gardener feels it is taking too long to water a garden with a $\frac{3}{8}$-in.-diameter hose. By what factor will the time be cut using a $\frac{5}{8}$-in.-diameter hose instead? Assume nothing else is changed.

*67. (II) What diameter must a 15.5-m-long air duct have if the ventilation and heating system is to replenish the air in a room 8.0 m × 14.0 m × 4.0 m every 12.0 min? Assume the pump can exert a gauge pressure of 0.710×10^{-3} atm.

*68. (II) What must be the pressure difference between the two ends of a 1.9-km section of pipe, 29 cm in diameter, if it is to transport oil $(\rho = 950 \text{ kg/m}^3, \eta = 0.20 \text{ Pa·s})$ at a rate of 650 cm³/s?

*69. (II) Poiseuille's equation does not hold if the flow velocity is high enough that turbulence sets in. The onset of turbulence occurs when the **Reynolds number**, Re, exceeds approximately 2000. Re is defined as

$$Re = \frac{2\bar{v}r\rho}{\eta},$$

where $\bar{v}$ is the average speed of the fluid, ρ is its density, η is its viscosity, and r is the radius of the tube in which the fluid is flowing. (a) Determine if blood flow through the aorta is laminar or turbulent when the average speed of blood in the aorta $(r = 0.80 \text{ cm})$ during the resting part of the heart's cycle is about 35 cm/s. (b) During exercise, the blood-flow speed approximately doubles. Calculate the Reynolds number in this case, and determine if the flow is laminar or turbulent.

*70. (II) Assuming a constant pressure gradient, if blood flow is reduced by 85%, by what factor is the radius of a blood vessel decreased?

*71. (III) A patient is to be given a blood transfusion. The blood is to flow through a tube from a raised bottle to a needle inserted in the vein (Fig. 13–56). The inside diameter of the 25-mm-long needle is 0.80 mm, and the required flow rate is 2.0 cm³ of blood per minute. How high h should the bottle be placed above the needle? Obtain ρ and η from the Tables. Assume the blood pressure is 78 torr above atmospheric pressure.

FIGURE 13–56 Problems 71 and 79.

*13–13 Surface Tension and Capillarity

*72. (I) If the force F needed to move the wire in Fig. 13–35 is 3.4×10^{-3} N, calculate the surface tension γ of the enclosed fluid. Assume $\ell = 0.070$ m.

*73. (I) Calculate the force needed to move the wire in Fig. 13–35 if it is immersed in a soapy solution and the wire is 24.5 cm long.

*74. (II) The surface tension of a liquid can be determined by measuring the force F needed to just lift a circular platinum ring of radius r from the surface of the liquid. (a) Find a formula for γ in terms of F and r. (b) At 30°C, if $F = 5.80 \times 10^{-3}$ N and $r = 2.8$ cm, calculate γ for the tested liquid.

*75. (III) Estimate the diameter of a steel needle that can just "float" on water due to surface tension.

*76. (III) Show that inside a soap bubble, there must be a pressure ΔP in excess of that outside equal to $\Delta P = 4\gamma/r$, where r is the radius of the bubble and γ is the surface tension. [*Hint*: Think of the bubble as two hemispheres in contact with each other; and remember that there are two surfaces to the bubble. Note that this result applies to any kind of membrane, where 2γ is the tension per unit length in that membrane.]

*77. (III) A common effect of surface tension is the ability of a liquid to rise up a narrow tube due to what is called capillary action. Show that for a narrow tube of radius r placed in a liquid of density ρ and surface tension γ, the liquid in the tube will reach a height $h = 2\gamma/\rho g r$ above the level of the liquid outside the tube, where g is the gravitational acceleration. Assume that the liquid "wets" the capillary (the liquid surface is vertical at the contact with the inside of the tube).

General Problems

78. A 2.8-N force is applied to the plunger of a hypodermic needle. If the diameter of the plunger is 1.3 cm and that of the needle 0.20 mm, (*a*) with what force does the fluid leave the needle? (*b*) What force on the plunger would be needed to push fluid into a vein where the gauge pressure is 75 mm-Hg? Answer for the instant just before the fluid starts to move.

79. Intravenous infusions are often made under gravity, as shown in Fig. 13–56. Assuming the fluid has a density of $1.00\ \mathrm{g/cm^3}$, at what height h should the bottle be placed so the liquid pressure is (*a*) 55 mm-Hg, and (*b*) 650 mm-H_2O? (*c*) If the blood pressure is 78 mm-Hg above atmospheric pressure, how high should the bottle be placed so that the fluid just barely enters the vein?

80. A beaker of water rests on an electronic balance that reads 998.0 g. A 2.6-cm-diameter solid copper ball attached to a string is submerged in the water, but does not touch the bottom. What are the tension in the string and the new balance reading?

81. Estimate the difference in air pressure between the top and the bottom of the Empire State building in New York City? It is 380 m tall and is located at sea level. Express as a fraction of atmospheric pressure at sea level.

82. A hydraulic lift is used to jack a 920-kg car 42 cm off the floor. The diameter of the output piston is 18 cm, and the input force is 350 N. (*a*) What is the area of the input piston? (*b*) What is the work done in lifting the car 42 cm? (*c*) If the input piston moves 13 cm in each stroke, how high does the car move up for each stroke? (*d*) How many strokes are required to jack the car up 42 cm? (*e*) Show that energy is conserved.

83. When you ascend or descend a great deal when driving in a car, your ears "pop," which means that the pressure behind the eardrum is being equalized to that outside. If this did not happen, what would be the approximate force on an eardrum of area 0.20 $\mathrm{cm^2}$ if a change in altitude of 950 m takes place?

84. Giraffes are a wonder of cardiovascular engineering. Calculate the difference in pressure (in atmospheres) that the blood vessels in a giraffe's head must accommodate as the head is lowered from a full upright position to ground level for a drink. The height of an average giraffe is about 6 m.

85. Suppose a person can reduce the pressure in his lungs to -75 mm-Hg gauge pressure. How high can water then be "sucked" up a straw?

86. Airlines are allowed to maintain a minimum air pressure within the passenger cabin equivalent to that at an altitude of 8000 ft (2400 m) to avoid adverse health effects among passengers due to oxygen deprivation. Estimate this minimum pressure (in atm).

87. A simple model (Fig. 13–57) considers a continent as a block (density $\approx 2800\ \mathrm{kg/m^3}$) floating in the mantle rock around it (density $\approx 3300\ \mathrm{kg/m^3}$). Assuming the continent is 35 km thick (the average thickness of the Earth's continental crust), estimate the height of the continent above the surrounding rock.

FIGURE 13–57
Problem 87.

88. A ship, carrying fresh water to a desert island in the Caribbean, has a horizontal cross-sectional area of 2240 $\mathrm{m^2}$ at the waterline. When unloaded, the ship rises 8.50 m higher in the sea. How many cubic meters of water was delivered?

89. During ascent, and especially during descent, volume changes of trapped air in the middle ear can cause ear discomfort until the middle-ear pressure and exterior pressure are equalized. (*a*) If a rapid descent at a rate of 7.0 m/s or faster commonly causes ear discomfort, what is the maximum rate of decrease in atmospheric pressure (that is, dP/dt) tolerable to most people? (*b*) In a 350-m-tall building, what will be the fastest possible descent time for an elevator traveling from the top to ground floor, assuming the elevator is properly designed to account for human physiology?

90. A raft is made of 12 logs lashed together. Each is 45 cm in diameter and has a length of 6.1 m. How many people can the raft hold before they start getting their feet wet, assuming the average person has a mass of 68 kg? Do *not* neglect the weight of the logs. Assume the specific gravity of wood is 0.60.

91. Estimate the total mass of the Earth's atmosphere, using the known value of atmospheric pressure at sea level.

92. During each heartbeat, approximately 70 $\mathrm{cm^3}$ of blood is pushed from the heart at an average pressure of 105 mm-Hg. Calculate the power output of the heart, in watts, assuming 70 beats per minute.

93. Four lawn sprinkler heads are fed by a 1.9-cm-diameter pipe. The water comes out of the heads at an angle of 35° to the horizontal and covers a radius of 7.0 m. (*a*) What is the velocity of the water coming out of each sprinkler head? (Assume zero air resistance.) (*b*) If the output diameter of each head is 3.0 mm, how many liters of water do the four heads deliver per second? (*c*) How fast is the water flowing inside the 1.9-cm-diameter pipe?

94. A bucket of water is accelerated upward at 1.8 g. What is the buoyant force on a 3.0-kg granite rock (SG = 2.7) submerged in the water? Will the rock float? Why or why not?

95. The stream of water from a faucet decreases in diameter as it falls (Fig. 13–58). Derive an equation for the diameter of the stream as a function of the distance y below the faucet, given that the water has speed v_0 when it leaves the faucet, whose diameter is d.

FIGURE 13–58 Problem 95. Water coming from a faucet.

96. You need to siphon water from a clogged sink. The sink has an area of 0.38 m² and is filled to a height of 4.0 cm. Your siphon tube rises 45 cm above the bottom of the sink and then descends 85 cm to a pail as shown in Fig. 13–59. The siphon tube has a diameter of 2.0 cm. (a) Assuming that the water level in the sink has almost zero velocity, estimate the water velocity when it enters the pail. (b) Estimate how long it will take to empty the sink.

FIGURE 13–59 Problem 96.

97. An airplane has a mass of 1.7×10^6 kg, and the air flows past the lower surface of the wings at 95 m/s. If the wings have a surface area of 1200 m², how fast must the air flow over the upper surface of the wing if the plane is to stay in the air?

98. A drinking fountain shoots water about 14 cm up in the air from a nozzle of diameter 0.60 cm. The pump at the base of the unit (1.1 m below the nozzle) pushes water into a 1.2-cm-diameter supply pipe that goes up to the nozzle. What gauge pressure does the pump have to provide? Ignore the viscosity; your answer will therefore be an underestimate.

99. A hurricane-force wind of 200 km/h blows across the face of a storefront window. Estimate the force on the 2.0 m × 3.0 m window due to the difference in air pressure inside and outside the window. Assume the store is airtight so the inside pressure remains at 1.0 atm. (This is why you should not tightly seal a building in preparation for a hurricane).

100. Blood from an animal is placed in a bottle 1.30 m above a 3.8-cm-long needle, of inside diameter 0.40 mm, from which it flows at a rate of 4.1 cm³/min. What is the viscosity of this blood?

101. Three forces act significantly on a freely floating helium-filled balloon: gravity, air resistance (or drag force), and a buoyant force. Consider a spherical helium-filled balloon of radius r = 15 cm rising upward through 0°C air, and m = 2.8 g is the mass of the (deflated) balloon itself. For all speeds v, except the very slowest ones, the flow of air past a rising balloon is turbulent, and the drag force F_D is given by the relation

$$F_D = \tfrac{1}{2} C_D \rho_{air} \pi r^2 v^2$$

where the constant $C_D = 0.47$ is the "drag coefficient" for a smooth sphere of radius r. If this balloon is released from rest, it will accelerate very quickly (in a few tenths of a second) to its terminal velocity v_T, where the buoyant force is cancelled by the drag force and the balloon's total weight. Assuming the balloon's acceleration takes place over a negligible time and distance, how long does it take the released balloon to rise a distance h = 12 m?

*102. If cholesterol buildup reduces the diameter of an artery by 15%, by what % will the blood flow rate be reduced, assuming the same pressure difference?

103. A two-component model used to determine percent body fat in a human body assumes that a fraction $f (< 1)$ of the body's total mass m is composed of fat with a density of 0.90 g/cm³, and that the remaining mass of the body is composed of fat-free tissue with a density of 1.10 g/cm³. If the specific gravity of the entire body's density is X, show that the percent body fat $(= f \times 100)$ is given by

$$\% \text{ Body fat} = \frac{495}{X} - 450.$$

*Numerical/Computer

*104. (III) Air pressure decreases with altitude. The following data show the air pressure at different altitudes.

Altitude (m)	Pressure (kPa)
0	101.3
1000	89.88
2000	79.50
3000	70.12
4000	61.66
5000	54.05
6000	47.22
7000	41.11
8000	35.65
9000	30.80
10,000	26.50

(a) Determine the best-fit quadratic equation that shows how the air pressure changes with altitude. (b) Determine the best-fit exponential equation that describes the change of air pressure with altitude. (c) Use each fit to find the air pressure at the summit of the mountain K2 at 8611 m, and give the % difference.

Answers to Exercises

A: (d).

B: The same. Pressure depends on depth, not on length.

C: Lower.

D: (a).

E: (e).

F: Increases.

G: (b).

An object attached to a coil spring can exhibit oscillatory motion. Many kinds of oscillatory motion are sinusoidal in time, or nearly so, and are referred to as being simple harmonic motion. Real systems generally have at least some friction, causing the motion to be damped. The automobile spring shown here has a shock absorber (yellow) that purposefully dampens the oscillation to make for a smooth ride. When an external sinusoidal force is exerted on a system able to oscillate, resonance occurs if the driving force is at or near the natural frequency of oscillation.

Oscillations

CHAPTER-OPENING QUESTION—Guess now!

A simple pendulum consists of a mass m (the "bob") hanging on the end of a thin string of length ℓ and negligible mass. The bob is pulled sideways so the string makes a 5.0° angle to the vertical; when released, it oscillates back and forth at a frequency f. If the pendulum was raised to a 10.0° angle instead, its frequency would be

  10.0° ℓ

  m

(a) twice as great.
(b) half as great.
(c) the same, or very close to it.
(d) not quite twice as great.
(e) a bit more than half as great.

CONTENTS

Many objects vibrate or oscillate—an object on the end of a spring, a tuning fork, the balance wheel of an old watch, a pendulum, a plastic ruler held firmly over the edge of a table and gently struck, the strings of a guitar or piano. Spiders detect prey by the vibrations of their webs; cars oscillate up and down when they hit a bump; buildings and bridges vibrate when heavy trucks pass or the wind is fierce. Indeed, because most solids are elastic (see Chapter 12), they vibrate (at least briefly) when given an impulse. Electrical oscillations are necessary in radio and television sets. At the atomic level, atoms vibrate within a molecule, and the atoms of a solid vibrate about their relatively fixed positions. Because it is so common in everyday life and occurs in so many areas of physics, oscillatory motion is of great importance. Mechanical oscillations are fully described on the basis of Newtonian mechanics.

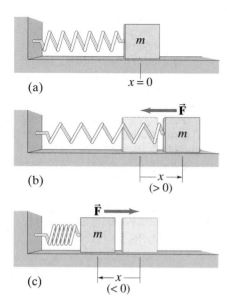

(a) $x = 0$

(b) $\overleftarrow{\vec{F}}$
       m
       x
       (> 0)

(c) $\overrightarrow{\vec{F}}$
       m
       x
       (< 0)

FIGURE 14–1   A mass oscillating at the end of a uniform spring.

FIGURE 14–2   Force on, and velocity of, a mass at different positions of its oscillation cycle on a frictionless surface.

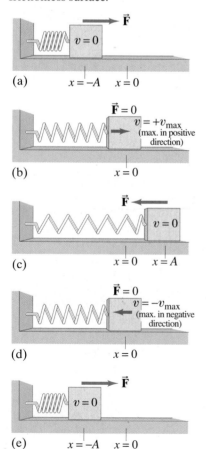

(a) $\vec{F}$
       $v = 0$
       $x = -A$ $x = 0$

(b) $\vec{F} = 0$
       $v = +v_{max}$
       (max. in positive direction)
       $x = 0$

(c) $\vec{F}$
       $v = 0$
       $x = 0$ $x = A$

(d) $\vec{F} = 0$
       $v = -v_{max}$
       (max. in negative direction)
       $x = 0$

(e) $\vec{F}$
       $v = 0$
       $x = -A$ $x = 0$

14–1 Oscillations of a Spring

When an object **vibrates** or **oscillates** back and forth, over the same path, each oscillation taking the same amount of time, the motion is **periodic**. The simplest form of periodic motion is represented by an object oscillating on the end of a uniform coil spring. Because many other types of oscillatory motion closely resemble this system, we will look at it in detail. We assume that the mass of the spring can be ignored, and that the spring is mounted horizontally, as shown in Fig. 14–1a, so that the object of mass m slides without friction on the horizontal surface. Any spring has a natural length at which it exerts no force on the mass m. The position of the mass at this point is called the **equilibrium position**. If the mass is moved either to the left, which compresses the spring, or to the right, which stretches it, the spring exerts a force on the mass that acts in the direction of returning the mass to the equilibrium position; hence it is called a *restoring force*. We consider the common situation where we can assume the restoring force F is directly proportional to the displacement x the spring has been stretched (Fig. 14–1b) or compressed (Fig. 14–1c) from the equilibrium position:

$$F = -kx. \qquad \text{[force exerted by spring]} \quad \textbf{(14–1)}$$

Note that the equilibrium position has been chosen at $x = 0$ and the minus sign in Eq. 14–1 indicates that the restoring force is always in the direction opposite to the displacement x. For example, if we choose the positive direction to the right in Fig. 14–1, x is positive when the spring is stretched (Fig. 14–1b), but the direction of the restoring force is to the left (negative direction). If the spring is compressed, x is negative (to the left) but the force F acts toward the right (Fig. 14–1c).

Equation 14–1 is often referred to as Hooke's law (Sections 7–3, 8–2 and 12–4), and is accurate only if the spring is not compressed to where the coils are close to touching, or stretched beyond the elastic region (see Fig. 12–15). Hooke's law works not only for springs but for other oscillating solids as well; it thus has wide applicability, even though it is valid only over a certain range of F and x values.

The proportionality constant k in Eq. 14–1 is called the *spring constant* for that particular spring, or its *spring stiffness constant*. To stretch the spring a distance x, one has to exert an (external) force on the free end of the spring with a magnitude at least equal to

$$F_{ext} = +kx. \qquad \text{[external force on spring]}$$

The greater the value of k, the greater the force needed to stretch a spring a given distance. That is, the stiffer the spring, the greater the spring constant k.

Note that the force F in Eq. 14–1 is *not* a constant, but varies with position. Therefore the acceleration of the mass m is not constant, so we *cannot* use the equations for constant acceleration developed in Chapter 2.

Let us examine what happens when our uniform spring is initially compressed a distance $x = -A$, as shown in Fig. 14–2a, and then released on the frictionless surface. The spring exerts a force on the mass that pushes it toward the equilibrium position. But because the mass has inertia, it passes the equilibrium position with considerable speed. Indeed, as the mass reaches the equilibrium position, the force on it decreases to zero, but its speed at this point is a maximum, v_{max} (Fig. 14–2b). As the mass moves farther to the right, the force on it acts to slow it down, and it stops for an instant at $x = A$ (Fig. 14–2c). It then begins moving back in the opposite direction, accelerating until it passes the equilibrium point (Fig. 14–2d), and then slows down until it reaches zero speed at the original starting point, $x = -A$ (Fig. 14–2e). It then repeats the motion, moving back and forth symmetrically between $x = A$ and $x = -A$.

EXERCISE A An object is oscillating back and forth. Which of the following statements are true at some time during the course of the motion? (*a*) The object can have zero velocity and, simultaneously, nonzero acceleration. (*b*) The object can have zero velocity and, simultaneously, zero acceleration. (*c*) The object can have zero acceleration and, simultaneously, nonzero velocity. (*d*) The object can have nonzero velocity and nonzero acceleration simultaneously.

EXERCISE B A mass is oscillating on a frictionless surface at the end of a horizontal spring. Where, if anywhere, is the acceleration of the mass zero (see Fig. 14–2)? (*a*) At $x = -A$; (*b*) at $x = 0$; (*c*) at $x = +A$; (*d*) at both $x = -A$ and $x = +A$; (*e*) nowhere.

To discuss oscillatory motion, we need to define a few terms. The distance x of the mass from the equilibrium point at any moment is called the **displacement**. The maximum displacement—the greatest distance from the equilibrium point—is called the **amplitude**, A. One **cycle** refers to the complete to-and-fro motion from some initial point back to that same point—say, from $x = -A$ to $x = A$ and back to $x = -A$. The **period**, T, is defined as the time required to complete one cycle. Finally, the **frequency**, f, is the number of complete cycles per second. Frequency is generally specified in hertz (Hz), where $1\,\text{Hz} = 1$ cycle per second (s^{-1}). It is easy to see, from their definitions, that frequency and period are inversely related, as we saw earlier (Eqs. 5–2 and 10–8):

$$f = \frac{1}{T} \quad \text{and} \quad T = \frac{1}{f}; \qquad\qquad \textbf{(14–2)}$$

for example, if the frequency is 5 cycles per second, then each cycle takes $\frac{1}{5}\,\text{s}$.

The oscillation of a spring hung vertically is essentially the same as that of a horizontal spring. Because of gravity, the length of a vertical spring with a mass m on the end will be longer at equilibrium than when that same spring is horizontal, as shown in Fig. 14–3. The spring is in equilibrium when $\Sigma F = 0 = mg - kx_0$, so the spring stretches an extra amount $x_0 = mg/k$ to be in equilibrium. If x is measured from this new equilibrium position, Eq. 14–1 can be used directly with the same value of k.

EXERCISE C If an oscillating mass has a frequency of 1.25 Hz, it makes 100 oscillations in (*a*) 12.5 s, (*b*) 125 s, (*c*) 80 s, (*d*) 8.0 s.

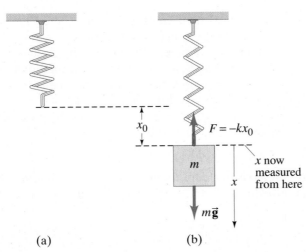

(a) (b)

FIGURE 14–3
(a) Free spring, hung vertically.
(b) Mass m attached to spring in new equilibrium position, which occurs when $\Sigma F = 0 = mg - kx_0$.

EXAMPLE 14–1 **Car springs.** When a family of four with a total mass of 200 kg step into their 1200-kg car, the car's springs compress 3.0 cm. (*a*) What is the spring constant of the car's springs (Fig. 14–4), assuming they act as a single spring? (*b*) How far will the car lower if loaded with 300 kg rather than 200 kg?

APPROACH We use Hooke's law: the weight of the people, mg, causes a 3.0-cm displacement.

SOLUTION (*a*) The added force of $(200\,\text{kg})(9.8\,\text{m/s}^2) = 1960\,\text{N}$ causes the springs to compress $3.0 \times 10^{-2}\,\text{m}$. Therefore (Eq. 14–1), the spring constant is

$$k = \frac{F}{x} = \frac{1960\,\text{N}}{3.0 \times 10^{-2}\,\text{m}} = 6.5 \times 10^4\,\text{N/m}.$$

(*b*) If the car is loaded with 300 kg, Hooke's law gives

$$x = \frac{F}{k} = \frac{(300\,\text{kg})(9.8\,\text{m/s}^2)}{(6.5 \times 10^4\,\text{N/m})} = 4.5 \times 10^{-2}\,\text{m},$$

or 4.5 cm.

NOTE In (*b*), we could have obtained x without solving for k: since x is proportional to F, if 200 kg compresses the spring 3.0 cm, then 1.5 times the force will compress the spring 1.5 times as much, or 4.5 cm.

FIGURE 14–4 Photo of a car's spring. (Also visible is the shock absorber, in blue—see Section 14–7.)

14–2 Simple Harmonic Motion

Any oscillating system for which the net restoring force is directly proportional to the negative of the displacement (as in Eq. 14–1, $F = -kx$) is said to exhibit **simple harmonic motion** (SHM). Such a system is often called a **simple harmonic oscillator** (SHO). We saw in Chapter 12 (Section 12–4) that most solid materials stretch or compress according to Eq. 14–1 as long as the displacement is not too great. Because of this, many natural oscillations are simple harmonic or close to it.

EXERCISE D Which of the following represents a simple harmonic oscillator: (a) $F = -0.5x^2$, (b) $F = -2.3y$, (c) $F = 8.6x$, (d) $F = -4\theta$?

Let us now determine the position x as a function of time for a mass attached to the end of a simple spring with spring constant k. To do so, we make use of Newton's second law, $F = ma$. Since the acceleration $a = d^2x/dt^2$, we have

$$ma = \Sigma F$$
$$m\frac{d^2x}{dt^2} = -kx,$$

where m is the mass[†] which is oscillating. We rearrange this to obtain

$$\frac{d^2x}{dt^2} + \frac{k}{m}x = 0, \qquad \text{[SHM]} \quad (14\text{–}3)$$

which is known as the **equation of motion** for the simple harmonic oscillator. Mathematically it is called a *differential equation*, since it involves derivatives. We want to determine what function of time, $x(t)$, satisfies this equation. We might guess the form of the solution by noting that if a pen were attached to an oscillating mass (Fig. 14–5) and a sheet of paper moved at a steady rate beneath it, the pen would trace the curve shown. The shape of this curve looks a lot like it might be **sinusoidal** (such as cosine or sine) as a function of time, and its height is the amplitude A. Let us then guess that the general solution to Eq. 14–3 can be written in a form such as

$$x = A\cos(\omega t + \phi), \qquad (14\text{–}4)$$

where we include the constant ϕ in the argument to be general.[‡] Let us now put this trial solution into Eq. 14–3 and see if it really works. We need to differentiate the $x = x(t)$ twice:

$$\frac{dx}{dt} = \frac{d}{dt}[A\cos(\omega t + \phi)] = -\omega A\sin(\omega t + \phi)$$
$$\frac{d^2x}{dt^2} = -\omega^2 A\cos(\omega t + \phi).$$

We now put the latter into Eq. 14–3, along with Eq. 14–4 for x:

$$\frac{d^2x}{dt^2} + \frac{k}{m}x = 0$$
$$-\omega^2 A\cos(\omega t + \phi) + \frac{k}{m}A\cos(\omega t + \phi) = 0$$

or

$$\left(\frac{k}{m} - \omega^2\right)A\cos(\omega t + \phi) = 0.$$

Our solution, Eq. 14–4, does indeed satisfy the equation of motion (Eq. 14–3) for

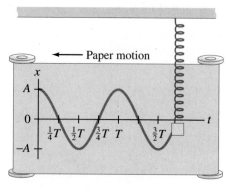

FIGURE 14–5 Sinusoidal nature of SHM as a function of time. In this case, $x = A\cos(2\pi t/T)$.

[†]In the case of a mass, m', on the end of a spring, the spring itself also oscillates and at least a part of its mass must be included. It can be shown—see the Problems—that approximately one-third the mass of the spring, m_s, must be included, so then $m = m' + \frac{1}{3}m_s$ in our equation. Often m_s is small enough to be ignored.

[‡]Another possible way to write the solution is the combination $x = a\cos\omega t + b\sin\omega t$, where a and b are constants. This is equivalent to Eq. 14–4 as can be seen using the trigonometric identity $\cos(A \pm B) = \cos A\cos B \mp \sin A\sin B$.

any time t, but only if $(k/m - \omega^2) = 0$. Hence

$$\omega^2 = \frac{k}{m}. \qquad\qquad (14\text{--}5)$$

Equation 14–4 is the general solution to Eq. 14–3, and it contains two arbitrary constants A and ϕ, which we should expect because the second derivative in Eq. 14–3 implies that two integrations are needed, each yielding a constant. They are "arbitrary" only in a calculus sense, in that they can be anything and still satisfy the differential equation, Eq. 14–3. In real physical situations, however, A and ϕ are determined by the **initial conditions**. Suppose, for example, that the mass is started at its maximum displacement and is released from rest. This is, in fact, what is shown in Fig. 14–5, and for this case $x = A \cos \omega t$. Let us confirm it: we are given $v = 0$ at $t = 0$, where

$$v = \frac{dx}{dt} = \frac{d}{dt}[A\cos(\omega t + \phi)] = -\omega A \sin(\omega t + \phi) = 0. \qquad [\text{at } t = 0]$$

For v to be zero at $t = 0$, then $\sin(\omega t + \phi) = \sin(0 + \phi)$ is zero if $\phi = 0$ (ϕ could also be $\pi, 2\pi$, etc.), and when $\phi = 0$, then

$$x = A\cos \omega t,$$

as we expected. We see immediately that A is the amplitude of the motion, and it is determined initially by how far you pulled the mass m from equilibrium before releasing it.

Consider another interesting case: at $t = 0$, the mass m is at $x = 0$ and is struck, giving it an initial velocity toward increasing values of x. Then at $t = 0$, $x = 0$, so we can write $x = A\cos(\omega t + \phi) = A\cos\phi = 0$, which can happen only if $\phi = \pm\pi/2$ (or $\pm90°$). Whether $\phi = +\pi/2$ or $-\pi/2$ depends on $v = dx/dt = -\omega A\sin(\omega t + \phi) = -\omega A\sin\phi$ at $t = 0$, which we are given as positive ($v > 0$ at $t = 0$); hence $\phi = -\pi/2$ because $\sin(-90°) = -1$. Thus our solution for this case is

$$x = A\cos\left(\omega t - \frac{\pi}{2}\right)$$
$$= A\sin \omega t,$$

where we used $\cos(\theta - \pi/2) = \sin\theta$. The solution in this case is a pure sine wave, Fig. 14–6, where A is still the amplitude.

Many other situations are possible, such as that shown in Fig. 14–7. The constant ϕ is called the **phase angle**, and it tells us how long after (or before) $t = 0$ the peak at $x = A$ is reached. Notice that the value of ϕ does not affect the shape of the $x(t)$ curve, but only affects the displacement at some arbitrary time, $t = 0$. Simple harmonic motion is thus always *sinusoidal*. Indeed, simple harmonic motion is *defined* as motion that is purely sinusoidal.

Since our oscillating mass repeats its motion after a time equal to its period T, it must be at the same position and moving in the same direction at $t = T$ as it was at $t = 0$. Since a sine or cosine function repeats itself after every 2π radians, then from Eq. 14–4, we must have

$$\omega T = 2\pi.$$

Hence

$$\omega = \frac{2\pi}{T} = 2\pi f,$$

where f is the frequency of the motion. Strictly speaking, we call ω the **angular frequency** (units are rad/s) to distinguish it from the frequency f (units are $s^{-1} = $ Hz); sometimes the word "angular" is dropped, so the symbol ω or f needs to be specified. Because $\omega = 2\pi f = 2\pi/T$, we can write Eq. 14–4 as

$$x = A\cos\left(\frac{2\pi t}{T} + \phi\right) \qquad\qquad (14\text{--}6a)$$

or

$$x = A\cos(2\pi f t + \phi). \qquad\qquad (14\text{--}6b)$$

FIGURE 14–6 Special case of SHM where the mass m starts, at $t = 0$, at the equilibrium position $x = 0$ and has initial velocity toward positive values of x ($v > 0$ at $t = 0$).

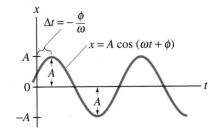

FIGURE 14–7 A plot of $x = A\cos(\omega t + \phi)$ when $\phi < 0$.

Because $\omega = 2\pi f = \sqrt{k/m}$ (Eq. 14–5), then

$$f = \frac{1}{2\pi}\sqrt{\frac{k}{m}}, \tag{14–7a}$$

$$T = 2\pi\sqrt{\frac{m}{k}}. \tag{14–7b}$$

Note that the *frequency and period do not depend on the amplitude*. Changing the amplitude of a simple harmonic oscillator does not affect its frequency. Equation 14–7a tells us that the greater the mass, the lower the frequency; and the stiffer the spring, the higher the frequency. This makes sense since a greater mass means more inertia and therefore a slower response (or acceleration); and larger k means greater force and therefore quicker response. The frequency f (Eq. 14–7a) at which a SHO oscillates naturally is called its **natural frequency** (to distinguish it from a frequency at which it might be forced to oscillate by an outside force, as discussed in Section 14–8).

The simple harmonic oscillator is important in physics because whenever we have a net restoring force proportional to the displacement $(F = -kx)$, which is at least a good approximation for a variety of systems, then the motion is simple harmonic—that is, sinusoidal.

PHYSICS APPLIED
Car springs

EXAMPLE 14–2 **Car springs again.** Determine the period and frequency of the car in Example 14–1a after hitting a bump. Assume the shock absorbers are poor, so the car really oscillates up and down.

APPROACH We put $m = 1400\,\text{kg}$ and $k = 6.5 \times 10^4\,\text{N/m}$ from Example 14–1a into Eqs. 14–7.

SOLUTION From Eq. 14–7b,

$$T = 2\pi\sqrt{\frac{m}{k}} = 2\pi\sqrt{\frac{1400\,\text{kg}}{6.5 \times 10^4\,\text{N/m}}} = 0.92\,\text{s},$$

or slightly less than a second. The frequency $f = 1/T = 1.09\,\text{Hz}$.

EXERCISE E By how much should the mass on the end of a spring be changed to halve the frequency of its oscillations? (*a*) No change; (*b*) doubled; (*c*) quadrupled; (*d*) halved; (*e*) quartered.

EXERCISE F The position of a SHO is given by $x = (0.80\,\text{m})\cos(3.14t - 0.25)$. The frequency is (*a*) 3.14 Hz, (*b*) 1.0 Hz, (*c*) 0.50 Hz, (*d*) 9.88 Hz, (*e*) 19.8 Hz.

Let us continue our analysis of a simple harmonic oscillator. The velocity and acceleration of the oscillating mass can be obtained by differentiation of Eq. 14–4, $x = A\cos(\omega t + \phi)$:

$$v = \frac{dx}{dt} = -\omega A\sin(\omega t + \phi) \tag{14–8a}$$

$$a = \frac{d^2x}{dt^2} = \frac{dv}{dt} = -\omega^2 A\cos(\omega t + \phi). \tag{14–8b}$$

The velocity and acceleration of a SHO also vary sinusoidally. In Fig. 14–8 we plot the displacement, velocity, and acceleration of a SHO as a function of time for the case when $\phi = 0$. As can be seen, the speed reaches its maximum

$$v_{\text{max}} = \omega A = \sqrt{\frac{k}{m}}\,A \tag{14–9a}$$

when the oscillating object is passing through its equilibrium point, $x = 0$. And the speed is zero at points of maximum displacement, $x = \pm A$. This is in accord with our discussion of Fig. 14–2. Similarly, the acceleration has its maximum value

$$a_{\text{max}} = \omega^2 A = \frac{k}{m}\,A \tag{14–9b}$$

which occurs where $x = \pm A$; and a is zero at $x = 0$, as we expect, since $ma = F = -kx$.

FIGURE 14–8 Displacement, x, velocity, dx/dt, and acceleration, d^2x/dt^2, of a simple harmonic oscillator when $\phi = 0$.

(a)

(b)

(c)

For the general case when $\phi \neq 0$, we can relate the constants A and ϕ to the initial values of x, v, and a by setting $t = 0$ in Eqs. 14–4, 14–8, and 14–9:

$$x_0 = x(0) = A \cos \phi$$

$$v_0 = v(0) = -\omega A \sin \phi = -v_{max} \sin \phi$$

$$a_0 = a(0) = -\omega^2 A \cos \phi = -a_{max} \cos \phi.$$

EXAMPLE 14–3 ESTIMATE **A vibrating floor.** A large motor in a factory causes the floor to vibrate at a frequency of 10 Hz. The amplitude of the floor's motion near the motor is about 3.0 mm. Estimate the maximum acceleration of the floor near the motor.

APPROACH Assuming the motion of the floor is roughly SHM we can make an estimate for the maximum acceleration using Eq. 14–9b.

SOLUTION Given $\omega = 2\pi f = (2\pi)(10\,\text{s}^{-1}) = 62.8\,\text{rad/s}$, then Eq. 14–9b gives

$$a_{max} = \omega^2 A = (62.8\,\text{rad/s})^2(0.0030\,\text{m}) = 12\,\text{m/s}^2.$$

NOTE The maximum acceleration is a little over g, so when the floor accelerates down, objects sitting on the floor will actually lose contact momentarily, which will cause noise and serious wear.

PHYSICS APPLIED
Unwanted floor vibrations

EXAMPLE 14–4 **Loudspeaker.** The cone of a loudspeaker (Fig. 14–9) oscillates in SHM at a frequency of 262 Hz ("middle C"). The amplitude at the center of the cone is $A = 1.5 \times 10^{-4}\,\text{m}$, and at $t = 0, x = A$. (*a*) What equation describes the motion of the center of the cone? (*b*) What are the velocity and acceleration as a function of time? (*c*) What is the position of the cone at $t = 1.00\,\text{ms}\,(= 1.00 \times 10^{-3}\,\text{s})$?

APPROACH The motion begins $(t = 0)$ with the cone at its maximum displacement $(x = A$ at $t = 0)$. So we use the cosine function, $x = A \cos \omega t$, with $\phi = 0$.

SOLUTION (*a*) The amplitude $A = 1.5 \times 10^{-4}\,\text{m}$ and

$$\omega = 2\pi f = (6.28\,\text{rad})(262\,\text{s}^{-1}) = 1650\,\text{rad/s}.$$

The motion is described as

$$x = A \cos \omega t = (1.5 \times 10^{-4}\,\text{m}) \cos(1650t),$$

where t is in seconds.

(*b*) The maximum velocity, from Eq. 14–9a, is

$$v_{max} = \omega A = (1650\,\text{rad/s})(1.5 \times 10^{-4}\,\text{m}) = 0.25\,\text{m/s},$$

so

$$v = -(0.25\,\text{m/s}) \sin(1650t).$$

From Eq. 14–9b the maximum acceleration is $a_{max} = \omega^2 A = (1650\,\text{rad/s})^2(1.5 \times 10^{-4}\,\text{m}) = 410\,\text{m/s}^2$, which is more than 40 g's. Then

$$a = -(410\,\text{m/s}^2) \cos(1650t).$$

(*c*) At $t = 1.00 \times 10^{-3}\,\text{s}$,

$$x = A \cos \omega t = (1.5 \times 10^{-4}\,\text{m}) \cos[(1650\,\text{rad/s})(1.00 \times 10^{-3}\,\text{s})]$$

$$= (1.5 \times 10^{-4}\,\text{m}) \cos(1.65\,\text{rad}) = -1.2 \times 10^{-5}\,\text{m}.$$

NOTE Be sure your calculator is set in RAD mode, not DEG mode, for these $\cos \omega t$ calculations.

FIGURE 14–9 Example 14–4. A loudspeaker cone.

⚠ **CAUTION**
Always be sure your calculator is in the correct mode for angles

EXAMPLE 14–5 **Spring calculations.** A spring stretches 0.150 m when a 0.300-kg mass is gently attached to it as in Fig. 14–3b. The spring is then set up horizontally with the 0.300-kg mass resting on a frictionless table as in Fig. 14–2. The mass is pushed so that the spring is compressed 0.100 m from the equilibrium point, and released from rest. Determine: (a) the spring stiffness constant k and angular frequency ω; (b) the amplitude of the horizontal oscillation A; (c) the magnitude of the maximum velocity v_{max}; (d) the magnitude of the maximum acceleration a_{max} of the mass; (e) the period T and frequency f; (f) the displacement x as a function of time; and (g) the velocity at $t = 0.150$ s.

APPROACH When the 0.300-kg mass hangs at rest from the spring as in Fig. 14–3b, we apply Newton's second law for the vertical forces: $\Sigma F = 0 = mg - kx_0$, so $k = mg/x_0$. For the horizontal oscillations, the amplitude is given, and the other quantities can be found from Eqs. 14–4, 14–5, 14–7, and 14–9. We choose x positive to the right.

SOLUTION (a) The spring stretches 0.150 m due to the 0.300-kg load, so

$$k = \frac{F}{x_0} = \frac{mg}{x_0} = \frac{(0.300\,\text{kg})(9.80\,\text{m/s}^2)}{0.150\,\text{m}} = 19.6\,\text{N/m}.$$

From Eq. 14–5,

$$\omega = \sqrt{\frac{k}{m}} = \sqrt{\frac{19.6\,\text{N/m}}{0.300\,\text{kg}}} = 8.08\,\text{s}^{-1}.$$

(b) The spring is now horizontal (on a table). It is compressed 0.100 m from equilibrium and is given no initial speed, so $A = 0.100$ m.

(c) From Eq. 14–9a, the maximum velocity has magnitude

$$v_{max} = \omega A = (8.08\,\text{s}^{-1})(0.100\,\text{m}) = 0.808\,\text{m/s}.$$

(d) Since $F = ma$, the maximum acceleration occurs where the force is greatest—that is, when $x = \pm A = \pm 0.100$ m. Thus its magnitude is

$$a_{max} = \frac{F}{m} = \frac{kA}{m} = \frac{(19.6\,\text{N/m})(0.100\,\text{m})}{0.300\,\text{kg}} = 6.53\,\text{m/s}^2.$$

[This result could also have been obtained directly from Eq. 14–9b, but it is often useful to go back to basics as we did here.]

(e) Equations 14–7b and 14–2 give

$$T = 2\pi\sqrt{\frac{m}{k}} = 2\pi\sqrt{\frac{0.300\,\text{kg}}{19.6\,\text{N/m}}} = 0.777\,\text{s}$$

$$f = \frac{1}{T} = 1.29\,\text{Hz}.$$

(f) The motion begins at a point of maximum compression. If we take x positive to the right in Fig. 14–2, then at $t = 0$, $x = -A = -0.100$ m. So we need a sinusoidal curve that has its maximum negative value at $t = 0$; this is just a negative cosine:

$$x = -A\cos\omega t.$$

To write this in the form of Eq. 14–4 (no minus sign), recall that $\cos\theta = -\cos(\theta - \pi)$. Then, putting in numbers, and recalling $-\cos\theta = \cos(\pi - \theta) = \cos(\theta - \pi)$, we have

$$x = -(0.100\,\text{m})\cos 8.08t$$
$$= (0.100\,\text{m})\cos(8.08t - \pi),$$

where t is in seconds and x is in meters. Note that the phase angle (Eq. 14–4) is $\phi = -\pi$ or $-180°$.

(g) The velocity at any time t is dx/dt (see also part c):

$$v = \frac{dx}{dt} = \omega A\sin\omega t = (0.808\,\text{m/s})\sin 8.08t.$$

At $t = 0.150$ s, $v = (0.808\,\text{m/s})\sin(1.21\,\text{rad}) = 0.756\,\text{m/s}$, and is to the right $(+)$.

EXAMPLE 14–6 **Spring is started with a push.** Suppose the spring of Example 14–5 is compressed 0.100 m from equilibrium $(x_0 = -0.100\text{ m})$ but is given a shove to create a velocity in the $+x$ direction of $v_0 = 0.400\text{ m/s}$. Determine (a) the phase angle ϕ, (b) the amplitude A, and (c) the displacement x as a function of time, $x(t)$.

APPROACH We use Eq. 14–8a, at $t = 0$, to write $v_0 = -\omega A \sin\phi$, and Eq. 14–4 to write $x_0 = A \cos\phi$. Combining these, we can obtain ϕ. We obtain A by using Eq. 14–4 again at $t = 0$. From Example 14–5, $\omega = 8.08\text{ s}^{-1}$.

SOLUTION (a) We combine Eqs. 14–8a and 14–4 at $t = 0$ and solve for the tangent:

$$\tan\phi = \frac{\sin\phi}{\cos\phi} = \frac{(v_0/-\omega A)}{(x_0/A)} = -\frac{v_0}{\omega x_0} = -\frac{0.400\text{ m/s}}{(8.08\text{ s}^{-1})(-0.100\text{ m})} = 0.495.$$

A calculator gives the angle as 26.3°, but we note from this equation that both the sine and cosine are negative, so our angle is in the third quadrant. Hence

$$\phi = 26.3° + 180° = 206.3° = 3.60\text{ rad}.$$

(b) Again using Eq. 14–4 at $t = 0$, as given in the Approach above,

$$A = \frac{x_0}{\cos\phi} = \frac{(-0.100\text{ m})}{\cos(3.60\text{ rad})} = 0.112\text{ m}.$$

(c) $x = A \cos(\omega t + \phi) = (0.112\text{ m})\cos(8.08t + 3.60)$.

14–3 Energy in the Simple Harmonic Oscillator

When forces are not constant, as is the case here with simple harmonic motion, it is often convenient and useful to use the energy approach, as we saw in Chapters 7 and 8.

For a simple harmonic oscillator, such as a mass m oscillating on the end of a massless spring, the restoring force is given by

$$F = -kx.$$

The potential energy function, as we saw in Chapter 8, is given by

$$U = -\int F\,dx = \tfrac{1}{2}kx^2,$$

where we set the constant of integration equal to zero so $U = 0$ at $x = 0$ (the equilibrium position).

The total mechanical energy is the sum of the kinetic and potential energies,

$$E = \tfrac{1}{2}mv^2 + \tfrac{1}{2}kx^2,$$

where v is the velocity of the mass m when it is a distance x from the equilibrium position. SHM can occur only if there is no friction, so the total mechanical energy E remains constant. As the mass oscillates back and forth, the energy continuously changes from potential energy to kinetic energy, and back again (Fig. 14–10). At the extreme points, $x = A$ and $x = -A$, all the energy is stored in the spring as potential energy (and is the same whether the spring is compressed or stretched to the full amplitude). At these extreme points, the mass stops for an instant as it changes direction, so $v = 0$ and:

$$E = \tfrac{1}{2}m(0)^2 + \tfrac{1}{2}kA^2 = \tfrac{1}{2}kA^2. \tag{14–10a}$$

Thus, the *total mechanical energy of a simple harmonic oscillator is proportional to the square of the amplitude.* At the equilibrium point, $x = 0$, all the energy is kinetic:

$$E = \tfrac{1}{2}mv^2 + \tfrac{1}{2}k(0)^2 = \tfrac{1}{2}mv_{\max}^2, \tag{14–10b}$$

where $v_{\max}$ is the maximum velocity during the motion. At intermediate points the energy is part kinetic and part potential, and because energy is conserved

$$E = \tfrac{1}{2}mv^2 + \tfrac{1}{2}kx^2. \tag{14–10c}$$

We can confirm Eqs. 14–10a and b explicitly by inserting Eqs. 14–4 and 14–8a into this last relation:

$$E = \tfrac{1}{2}m\omega^2 A^2 \sin^2(\omega t + \phi) + \tfrac{1}{2}kA^2 \cos^2(\omega t + \phi).$$

Substituting with $\omega^2 = k/m$, or $kA^2 = m\omega^2 A^2 = mv_{\max}^2$, and recalling the important trigonometric identity $\sin^2(\omega t + \phi) + \cos^2(\omega t + \phi) = 1$, we obtain Eqs. 14–10a and b:

$$E = \tfrac{1}{2}kA^2 = \tfrac{1}{2}mv_{\max}^2.$$

[Note that we can check for consistency by solving this equation for $v_{\max}$, and we do get Eq. 14–9a.]

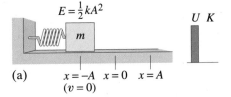

(a) $x = -A$ $x = 0$ $x = A$
$E = \tfrac{1}{2}kA^2$ ($v = 0$)

(b) $x = -A$ $x = 0$ $x = A$
$E = \tfrac{1}{2}mv_{\max}^2$

(c) $x = -A$ $x = 0$ $x = A$
$E = \tfrac{1}{2}kA^2$ ($v = 0$)

(d) $x = -A$ $x = 0$ $x = A$
$E = \tfrac{1}{2}mv^2 + \tfrac{1}{2}kx^2$

FIGURE 14–10 Energy changes from potential energy to kinetic energy and back again as the spring oscillates. Energy bar graphs (on the right) are described in Section 8–4.

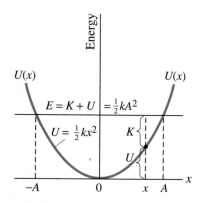

FIGURE 14–11 Graph of potential energy, $U = \frac{1}{2}kx^2$. $K + U = E = $ constant for any point x where $-A \le x \le A$. Values of K and U are indicated for an arbitrary position x.

We can now obtain an equation for the velocity v as a function of x by solving for v in Eq. 14–10c:

$$v = \pm \sqrt{\frac{k}{m}(A^2 - x^2)} \qquad (14\text{–}11a)$$

or, since $v_{max} = A\sqrt{k/m}$,

$$v = \pm v_{max}\sqrt{1 - \frac{x^2}{A^2}}. \qquad (14\text{–}11b)$$

Again we see that v is a maximum at $x = 0$, and is zero at $x = \pm A$.

The potential energy, $U = \frac{1}{2}kx^2$, is plotted in Fig. 14–11 (see also Section 8–9). The upper horizontal line represents a particular value of the total energy $E = \frac{1}{2}kA^2$. The distance between the E line and the U curve represents the kinetic energy, K, and the motion is restricted to x values between $-A$ and $+A$. These results are, of course, consistent with our full solution of the previous Section.

Energy conservation is a convenient way to obtain v, for example, if x is given (or vice versa), without having to deal with time t.

EXAMPLE 14–7 **Energy calculations.** For the simple harmonic oscillation of Example 14–5, determine (a) the total energy, (b) the kinetic and potential energies as a function of time, (c) the velocity when the mass is 0.050 m from equilibrium, (d) the kinetic and potential energies at half amplitude ($x = \pm A/2$).

APPROACH We use conservation of energy for a spring–mass system, Eqs. 14–10 and 14–11.

SOLUTION (a) From Example 14–5, $k = 19.6\,\text{N/m}$ and $A = 0.100\,\text{m}$, so the total energy E from Eq. 14–10a is

$$E = \tfrac{1}{2}kA^2 = \tfrac{1}{2}(19.6\,\text{N/m})(0.100\,\text{m})^2 = 9.80 \times 10^{-2}\,\text{J}.$$

(b) We have, from parts (f) and (g) of Example 14–5, $x = -(0.100\,\text{m})\cos 8.08t$ and $v = (0.808\,\text{m/s})\sin 8.08t$, so

$$U = \tfrac{1}{2}kx^2 = \tfrac{1}{2}(19.6\,\text{N/m})(0.100\,\text{m})^2\cos^2 8.08t = (9.80 \times 10^{-2}\,\text{J})\cos^2 8.08t$$

$$K = \tfrac{1}{2}mv^2 = \tfrac{1}{2}(0.300\,\text{kg})(0.808\,\text{m/s})^2\sin^2 8.08t = (9.80 \times 10^{-2}\,\text{J})\sin^2 8.08t.$$

(c) We use Eq. 14–11b and find

$$v = v_{max}\sqrt{1 - x^2/A^2} = (0.808\,\text{m/s})\sqrt{1 - (\tfrac{1}{2})^2} = 0.70\,\text{m/s}.$$

(d) At $x = A/2 = 0.050\,\text{m}$, we have

$$U = \tfrac{1}{2}kx^2 = \tfrac{1}{2}(19.6\,\text{N/m})(0.050\,\text{m})^2 = 2.5 \times 10^{-2}\,\text{J}$$

$$K = E - U = 7.3 \times 10^{-2}\,\text{J}.$$

CONCEPTUAL EXAMPLE 14–8 **Doubling the amplitude.** Suppose the spring in Fig. 14–10 is stretched twice as far (to $x = 2A$). What happens to (a) the energy of the system, (b) the maximum velocity of the oscillating mass, (c) the maximum acceleration of the mass?

RESPONSE (a) From Eq. 14–10a, the total energy is proportional to the square of the amplitude A, so stretching it twice as far quadruples the energy $(2^2 = 4)$. You may protest, "I did work stretching the spring from $x = 0$ to $x = A$. Don't I do the same work stretching it from A to $2A$?" No. The force you exert is proportional to the displacement x, so for the second displacement, from $x = A$ to $2A$, you do more work than for the first displacement ($x = 0$ to A). (b) From Eq. 14–10b, we can see that when the energy is quadrupled, the maximum velocity must be doubled. $[v_{max} \propto \sqrt{E} \propto A.]$ (c) Since the force is twice as great when we stretch the spring twice as far, the acceleration is also twice as great: $a \propto F \propto x$.

EXERCISE G Suppose the spring in Fig. 14–10 is compressed to $x = -A$, but is given a push to the right so that the initial speed of the mass m is v_0. What effect does this push have on (a) the energy of the system, (b) the maximum velocity, (c) the maximum acceleration?

14–4 Simple Harmonic Motion Related to Uniform Circular Motion

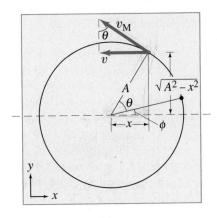

Simple harmonic motion has a simple relationship to a particle rotating in a circle with uniform speed. Consider a mass m rotating in a circle of radius A with speed v_M on top of a table as shown in Fig. 14–12. As viewed from above, the motion is a circle. But a person who looks at the motion from the edge of the table sees an oscillatory motion back and forth, and this corresponds precisely to SHM as we shall now see. What the person sees, and what we are interested in, is the projection of the circular motion onto the x axis, Fig. 14–12. To see that this motion is analogous to SHM, let us calculate the x component of the velocity v_M which is labeled v in Fig. 14–12. The two right triangles indicated in Fig. 14–12 are similar, so

$$\frac{v}{v_M} = \frac{\sqrt{A^2 - x^2}}{A}$$

or

$$v = v_M \sqrt{1 - \frac{x^2}{A^2}}.$$

This is exactly the equation for the speed of a mass oscillating with SHM, Eq. 14–11b, where $v_M = v_{max}$. Furthermore, we can see from Fig. 14–12 that if the angular displacement at $t = 0$ is ϕ, then after a time t the particle will have rotated through an angle $\theta = \omega t$, and so

$$x = A\cos(\theta + \phi) = A\cos(\omega t + \phi).$$

(a)

(b)

But what is ω here? The linear velocity v_M of our particle undergoing rotational motion is related to ω by $v_M = \omega A$, where A is the radius of the circle (see Eq. 10–4, $v = R\omega$). To make one revolution requires a time T, so we also have $v_M = 2\pi A/T$ where $2\pi A$ is the circle's circumference. Hence

$$\omega = \frac{v_M}{A} = \frac{2\pi A/T}{A} = 2\pi/T = 2\pi f$$

FIGURE 14–12 Analysis of simple harmonic motion as a side view (b) of circular motion (a).

where T is the time required for one rotation and f is the frequency. This corresponds precisely to the back-and-forth motion of a simple harmonic oscillator. Thus, the projection on the x axis of a particle rotating in a circle has the same motion as a mass undergoing SHM. Indeed, we can say that the projection of circular motion onto a straight line is simple harmonic motion.

The projection of uniform circular motion onto the y axis is also simple harmonic. Thus uniform circular motion can be thought of as two simple harmonic motions operating at right angles.

FIGURE 14–13 Strobe-light photo of an oscillating pendulum photographed at equal time intervals.

14–5 The Simple Pendulum

A **simple pendulum** consists of a small object (the pendulum bob) suspended from the end of a lightweight cord, Fig. 14–13. We assume that the cord does not stretch and that its mass can be ignored relative to that of the bob. The motion of a simple pendulum moving back and forth with negligible friction resembles simple harmonic motion: the pendulum oscillates along the arc of a circle with equal amplitude on either side of its equilibrium point and as it passes through the equilibrium point (where it would hang vertically) it has its maximum speed. But is it really undergoing SHM? That is, is the restoring force proportional to its displacement? Let us find out.

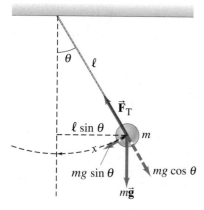

FIGURE 14–14 Simple pendulum.

FIGURE 14–15 The swinging motion of this lamp, hanging by a very long cord from the ceiling of the cathedral at Pisa, is said to have been observed by Galileo and to have inspired him to the conclusion that the period of a pendulum does not depend on amplitude.

Ⓧ PHYSICS APPLIED
Pendulum clock

The displacement of the pendulum along the arc is given by $x = \ell\theta$, where θ is the angle (in radians) that the cord makes with the vertical and ℓ is the length of the cord (Fig. 14–14). If the restoring force is proportional to x or to θ, the motion will be simple harmonic. The restoring force is the net force on the bob, equal to the component of the weight, mg, tangent to the arc:

$$F = -mg\sin\theta,$$

where g is the acceleration of gravity. The minus sign here, as in Eq. 14–1, means that the force is in the direction opposite to the angular displacement θ. Since F is proportional to the sine of θ and not to θ itself, the motion is *not* SHM. However, if θ is small, then $\sin\theta$ is very nearly equal to θ when the latter is specified in radians. This can be seen by looking at the series expansion[†] of $\sin\theta$ (or by looking at the trigonometry Table in Appendix A), or simply by noting in Fig. 14–14 that the arc length $x (= \ell\theta)$ is nearly the same length as the chord $(= \ell\sin\theta)$ indicated by the straight dashed line, *if θ is small*. For angles less than 15°, the difference between θ (in radians) and $\sin\theta$ is less than 1%. Thus, to a very good approximation for small angles,

$$F = -mg\sin\theta \approx -mg\theta.$$

Substituting $x = \ell\theta$ or $\theta = x/\ell$, we have

$$F \approx -\frac{mg}{\ell}x.$$

Thus, for small displacements, the motion is essentially simple harmonic, since this equation fits Hooke's law, $F = -kx$. The effective force constant is $k = mg/\ell$. Thus we can write

$$\theta = \theta_{max}\cos(\omega t + \phi)$$

where θ_{max} is the maximum angular displacement and $\omega = 2\pi f = 2\pi/T$. To obtain ω we use Eq. 14–5, where for k we substitute mg/ℓ: that is,[‡] $\omega = \sqrt{k/m} = \sqrt{(mg/\ell)/m}$, or

$$\omega = \sqrt{\frac{g}{\ell}}. \qquad\qquad [\theta\text{ small}] \quad \textbf{(14–12a)}$$

Then the frequency f is

$$f = \frac{\omega}{2\pi} = \frac{1}{2\pi}\sqrt{\frac{g}{\ell}}, \qquad\qquad [\theta\text{ small}] \quad \textbf{(14–12b)}$$

and the period T is

$$T = \frac{1}{f} = 2\pi\sqrt{\frac{\ell}{g}}. \qquad\qquad [\theta\text{ small}] \quad \textbf{(14–12c)}$$

The mass m of the pendulum bob does not appear in these formulas for T and f. Thus we have the surprising result that the period and frequency of a simple pendulum do not depend on the mass of the pendulum bob. You may have noticed this if you pushed a small child and then a large one on the same swing.

We also see from Eq. 14–12c that the period of a pendulum does not depend on the amplitude (like any SHM, Section 14–2), as long as the amplitude θ is small. Galileo is said to have first noted this fact while watching a swinging lamp in the cathedral at Pisa (Fig. 14–15). This discovery led to the invention of the pendulum clock, the first really precise timepiece, which became the standard for centuries.

Because a pendulum does not undergo *precisely* SHM, the period does depend slightly on the amplitude, the more so for large amplitudes. The accuracy of a pendulum clock would be affected, after many swings, by the decrease in amplitude due to friction. But the mainspring in a pendulum clock (or the falling weight in a grandfather clock) supplies energy to compensate for the friction and to maintain the amplitude constant, so that the timing remains precise.

[†]$\sin\theta = \theta - \dfrac{\theta^3}{3!} + \dfrac{\theta^5}{5!} - \dfrac{\theta^7}{7!} + \cdots.$

[‡]Be careful not to think that $\omega = d\theta/dt$ as in rotational motion. Here θ is the angle of the pendulum at any instant (Fig. 14–14), but we use ω now to represent *not* the rate this angle changes, but rather a constant related to the period, $\omega = 2\pi f = \sqrt{g/\ell}$.

EXERCISE H If a simple pendulum is taken from sea level to the top of a high mountain and started at the same angle of 5°, it would oscillate at the top of the mountain (a) slightly slower, (b) slightly faster, (c) at exactly the same frequency, (d) not at all—it would stop, (e) none of these.

EXERCISE I Return to the Chapter-Opening Question, p. 369, and answer it again now. Try to explain why you may have answered differently the first time.

EXAMPLE 14–9 **Measuring g.** A geologist uses a simple pendulum that has a length of 37.10 cm and a frequency of 0.8190 Hz at a particular location on the Earth. What is the acceleration of gravity at this location?

APPROACH We can use the length ℓ and frequency f of the pendulum in Eq. 14-12b, which contains our unknown, g.

SOLUTION We solve Eq. 14–12b for g and obtain

$$g = (2\pi f)^2 \ell = (6.283 \times 0.8190\,\text{s}^{-1})^2 (0.3710\,\text{m}) = 9.824\,\text{m/s}^2.$$

EXERCISE J (a) Estimate the length of a simple pendulum that makes one swing back and forth per second. (b) What would be the period of a 1.0-m-long pendulum?

*14–6 The Physical Pendulum and the Torsion Pendulum

Physical Pendulum

The term physical pendulum refers to any real extended object which oscillates back and forth, in contrast to the rather idealized simple pendulum where all the mass is assumed concentrated in the tiny pendulum bob. An example of a physical pendulum is a baseball bat suspended from the point O, as shown in Fig. 14–16. The force of gravity acts at the center of gravity (CG) of the object located a distance h from the pivot point O. The physical pendulum is best analyzed using the equations of rotational motion. The torque on a physical pendulum, calculated about point O, is

$$\tau = -mgh \sin\theta.$$

Newton's second law for rotational motion, Eq. 10–14, states that

$$\Sigma\tau = I\alpha = I\frac{d^2\theta}{dt^2},$$

where I is the moment of inertia of the object about the pivot point and $\alpha = d^2\theta/dt^2$ is the angular acceleration. Thus we have

$$I\frac{d^2\theta}{dt^2} = -mgh \sin\theta$$

or

$$\frac{d^2\theta}{dt^2} + \frac{mgh}{I}\sin\theta = 0,$$

where I is calculated about an axis through point O. For small angular amplitude, $\sin\theta \approx \theta$, so we have

$$\frac{d^2\theta}{dt^2} + \left(\frac{mgh}{I}\right)\theta = 0. \qquad \text{[small angular displacement]} \quad \textbf{(14–13)}$$

This is just the equation for SHM, Eq. 14–3, except that θ replaces x and mgh/I replaces k/m. Thus, for small angular displacements, a physical pendulum undergoes SHM, given by

$$\theta = \theta_{max}\cos(\omega t + \phi),$$

where θ_{max} is the maximum angular displacement and $\omega = 2\pi/T$. The period, T, is (see Eq. 14–7b, replacing m/k with I/mgh):

$$T = 2\pi\sqrt{\frac{I}{mgh}}. \qquad \text{[small angular displacement]} \quad \textbf{(14–14)}$$

FIGURE 14–16 A physical pendulum suspended from point O.

FIGURE 14–17 Example 14–10.

FIGURE 14–18 A torsion pendulum. The disc oscillates in SHM between θ_{max} and $-\theta_{max}$.

FIGURE 14–19 Damped harmonic motion. The solid red curve represents a cosine times a decreasing exponential (the dashed curves).

EXAMPLE 14–10 **Moment of inertia measurement.** An easy way to measure the moment of inertia of an object about any axis is to measure the period of oscillation about that axis. (a) Suppose a nonuniform 1.0-kg stick can be balanced at a point 42 cm from one end. If it is pivoted about that end (Fig. 14–17), it oscillates with a period of 1.6 s. What is its moment of inertia about this end? (b) What is its moment of inertia about an axis perpendicular to the stick through its center of mass?

APPROACH We put the given values into Eq. 14–14 and solve for I. For (b) we use the parallel-axis theorem (Section 10–7).

SOLUTION (a) Given $T = 1.6$ s, and $h = 0.42$ m, Eq. 14–14 gives

$$I = mghT^2/4\pi^2 = 0.27 \text{ kg} \cdot \text{m}^2.$$

(b) We use the parallel-axis theorem, Eq. 10–17. The CM is where the stick balanced, 42 cm from the end, so

$$I_{CM} = I - mh^2 = 0.27 \text{ kg} \cdot \text{m}^2 - (1.0 \text{ kg})(0.42 \text{ m})^2 = 0.09 \text{ kg} \cdot \text{m}^2.$$

NOTE Since an object does not oscillate about its CM, we cannot measure I_{CM} directly, but the parallel-axis theorem provides a convenient method to determine I_{CM}.

Torsion Pendulum

Another type of oscillatory motion is a **torsion pendulum**, in which a disc (Fig. 14–18) or a bar (as in Cavendish's apparatus, Fig. 6–3) is suspended from a wire. The twisting (torsion) of the wire serves as the elastic force. The motion here will be SHM since the restoring torque is very closely proportional to the negative of the angular displacement,

$$\tau = -K\theta,$$

where K is a constant that depends on the wire stiffness. Then

$$\omega = \sqrt{K/I}.$$

There is no small angle restriction here, as there is for the physical pendulum (where gravity acts), as long as the wire responds linearly in accordance with Hooke's law.

14–7 Damped Harmonic Motion

The amplitude of any real oscillating spring or swinging pendulum slowly decreases in time until the oscillations stop altogether. Figure 14–19 shows a typical graph of the displacement as a function of time. This is called **damped harmonic motion**. The damping[†] is generally due to the resistance of air and to internal friction within the oscillating system. The energy that is dissipated to thermal energy is reflected in a decreased amplitude of oscillation.

Since natural oscillating systems are damped in general, why do we even talk about (undamped) simple harmonic motion? The answer is that SHM is much easier to deal with mathematically. And if the damping is not large, the oscillations can be thought of as simple harmonic motion on which the damping is superposed, as represented by the dashed curves in Fig. 14–19. Although damping does alter the frequency of vibration, the effect is usually small if the damping is small. Let us look at this in more detail.

The damping force depends on the speed of the oscillating object, and opposes the motion. In some simple cases the damping force can be approximated as being directly proportional to the speed:

$$F_{damping} = -bv,$$

where b is a constant.[‡] For a mass oscillating on the end of a spring, the restoring force of the spring is $F = -kx$; so Newton's second law ($ma = \Sigma F$) becomes

$$ma = -kx - bv.$$

We bring all terms to the left side of the equation and substitute $v = dx/dt$ and

---

[†]To "damp" means to diminish, restrain, or extinguish, as to "dampen one's spirits."

[‡]Such velocity-dependent forces were discussed in Section 5–6.

$a = d^2x/dt^2$ to obtain

$$m\frac{d^2x}{dt^2} + b\frac{dx}{dt} + kx = 0,$$ (14–15)

which is the equation of motion. To solve this equation, we guess at a solution and then check to see if it works. If the damping constant b is small, x as a function of t is as plotted in Fig. 14–19, which looks like a cosine function times a factor (represented by the dashed lines) that decreases in time. A simple function that does this is the exponential, $e^{-\gamma t}$, and the solution that satisfies Eq. 14–15 is

$$x = Ae^{-\gamma t}\cos\omega' t,$$ (14–16)

where A, γ, and ω' are assumed to be constants, and $x = A$ at $t = 0$. We have called the angular frequency ω' (and not ω) because it is not the same as the ω for SHM without damping $\left(\omega = \sqrt{k/m}\right)$.

If we substitute Eq. 14–16 into Eq. 14–15 (we do this in the optional subsection below), we find that Eq. 14–16 is indeed a solution if γ and ω' have the values

$$\gamma = \frac{b}{2m}$$ (14–17)

$$\omega' = \sqrt{\frac{k}{m} - \frac{b^2}{4m^2}}.$$ (14–18)

Thus x as a function of time t for a (lightly) damped harmonic oscillator is

$$x = Ae^{(-b/2m)t}\cos\omega' t.$$ (14–19)

Of course a phase constant, ϕ, can be added to the argument of the cosine in Eq. 14–19. As it stands with $\phi = 0$, it is clear that the constant A in Eq. 14–19 is simply the initial displacement, $x = A$ at $t = 0$. The frequency f is

$$f' = \frac{\omega'}{2\pi} = \frac{1}{2\pi}\sqrt{\frac{k}{m} - \frac{b^2}{4m^2}}.$$ (14–20)

The frequency is lower, and the period longer, than for undamped SHM. (In many practical cases of light damping, however, ω' differs only slightly from $\omega = \sqrt{k/m}$.) This makes sense since we expect damping to slow down the motion. Equation 14–20 reduces to Eq. 14–7a, as it should, when there is no damping ($b = 0$). The constant $\gamma = b/2m$ is a measure of how quickly the oscillations decrease toward zero (Fig. 14–19). The time $t_L = 2m/b$ is the time taken for the oscillations to drop to $1/e$ of the original amplitude; t_L is called the "mean lifetime" of the oscillations. Note that the larger b is, the more quickly the oscillations die away.

The solution, Eq. 14–19, is not valid if b is so large that

$$b^2 > 4mk$$

since then ω' (Eq. 14–18) would become imaginary. In this case the system does not oscillate at all but returns directly to its equilibrium position, as we now discuss.

Three common cases of *heavily damped* systems are shown in Fig. 14–20. Curve C represents the situation when the damping is so large $\left(b^2 \gg 4mk\right)$ that it takes a long time to reach equilibrium; the system is **overdamped**. Curve A represents an **underdamped** situation in which the system makes several swings before coming to rest $\left(b^2 < 4mk\right)$ and corresponds to a more heavily damped version of Eq. 14–19. Curve B represents **critical damping**: $b^2 = 4mk$; in this case equilibrium is reached in the shortest time. These terms all derive from the use of practical damped systems such as door-closing mechanisms and **shock absorbers** in a car (Fig. 14–21), which are usually designed to give critical damping. But as they wear out, underdamping occurs: a door slams and a car bounces up and down several times whenever it hits a bump.

In many systems, the oscillatory motion is what counts, as in clocks and watches, and damping needs to be minimized. In other systems, oscillations are the problem, such as a car's springs, so a proper amount of damping (i.e., critical) is desired. Well-designed damping is needed for all kinds of applications. Large buildings, especially in California, are now built (or retrofitted) with huge dampers to reduce earthquake damage.

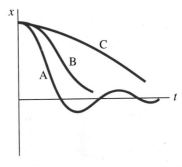

FIGURE 14–20 Underdamped (A), critically damped (B), and overdamped (C) motion.

FIGURE 14–21 Automobile spring and shock absorber provide damping so that a car won't bounce up and down endlessly.

Attached to car frame

Piston

Viscous fluid

Attached to car axle

FIGURE 14–22 Example 14–11.

EXAMPLE 14–11 **Simple pendulum with damping.** A simple pendulum has a length of 1.0 m (Fig. 14–22). It is set swinging with small-amplitude oscillations. After 5.0 minutes, the amplitude is only 50% of what it was initially. (a) What is the value of γ for the motion? (b) By what factor does the frequency, f', differ from f, the undamped frequency?

APPROACH We assume the damping force is proportional to angular speed, $d\theta/dt$. The equation of motion for damped harmonic motion is

$$x = Ae^{-\gamma t}\cos\omega't, \quad \text{where} \quad \gamma = \frac{b}{2m} \quad \text{and} \quad \omega' = \sqrt{\frac{k}{m} - \frac{b^2}{4m^2}},$$

for motion of a mass on the end of a spring. For the simple pendulum without damping, we saw in Section 14–5 that

$$F = -mg\theta$$

for small θ. Since $F = ma$, where a can be written in terms of the angular acceleration $\alpha = d^2\theta/dt^2$ as $a = \ell\alpha = \ell\, d^2\theta/dt^2$, then $F = m\ell\, d^2\theta/dt^2$, and

$$\ell\frac{d^2\theta}{dt^2} + g\theta = 0.$$

Introducing a damping term, $b(d\theta/dt)$, we have

$$\ell\frac{d^2\theta}{dt^2} + b\frac{d\theta}{dt} + g\theta = 0,$$

which is the same as Eq. 14–15 with θ replacing x, and ℓ and g replacing m and k.

SOLUTION (a) We compare Eq. 14–15 with our equation just above and see that our equation $x = Ae^{-\gamma t}\cos\omega't$ becomes an equation for θ with

$$\gamma = \frac{b}{2\ell} \quad \text{and} \quad \omega' = \sqrt{\frac{g}{\ell} - \frac{b^2}{4\ell^2}}.$$

At $t = 0$, we rewrite Eq. 14–16 with θ replacing x as

$$\theta_0 = Ae^{-\gamma\cdot 0}\cos\omega'\cdot 0 = A.$$

Then at $t = 5.0\,\text{min} = 300\,\text{s}$, the amplitude given by Eq. 14–16 has fallen to $0.50\,A$, so

$$0.50A = Ae^{-\gamma(300\,\text{s})}.$$

We solve this for γ and obtain $\gamma = \ln 2.0/(300\,\text{s}) = 2.3 \times 10^{-3}\,\text{s}^{-1}$.

(b) We have $\ell = 1.0\,\text{m}$, so $b = 2\gamma\ell = 2(2.3 \times 10^{-3}\,\text{s}^{-1})(1.0\,\text{m}) = 4.6 \times 10^{-3}\,\text{m/s}$. Thus $(b^2/4\ell^2)$ is very much less than $g/\ell (= 9.8\,\text{s}^{-2})$, and the angular frequency of the motion remains almost the same as that of the undamped motion. Specifically (see Eq. 14–20),

$$f' = \frac{1}{2\pi}\sqrt{\frac{g}{\ell}\left[1 - \frac{\ell}{g}\left(\frac{b^2}{4\ell^2}\right)\right]^{\frac{1}{2}}} \approx \frac{1}{2\pi}\sqrt{\frac{g}{\ell}}\left[1 - \frac{1}{2}\frac{\ell}{g}\left(\frac{b^2}{4\ell^2}\right)\right]$$

where we used the binomial expansion. Then, with $f = (1/2\pi)\sqrt{g/\ell}$ (Eq. 14–12b),

$$\frac{f - f'}{f} \approx \frac{1}{2}\frac{\ell}{g}\left(\frac{b^2}{4\ell^2}\right) = 2.7 \times 10^{-7}.$$

So f' differs from f by less than one part in a million.

*Showing $x = Ae^{-\gamma t}\cos\omega't$ is a Solution

We start with Eq. 14–16, to see if it is a solution to Eq. 14–15. First we take the first and second derivatives

$$\frac{dx}{dt} = -\gamma Ae^{-\gamma t}\cos\omega't - \omega'Ae^{-\gamma t}\sin\omega't$$

$$\frac{d^2x}{dt^2} = \gamma^2Ae^{-\gamma t}\cos\omega't + \gamma A\omega'e^{-\gamma t}\sin\omega't + \omega'\gamma Ae^{-\gamma t}\sin\omega't - \omega'^2Ae^{-\gamma t}\cos\omega't.$$

We next substitute these relations back into Eq. 14–15 and reorganize to

obtain

$$Ae^{-\gamma t}\left[\left(m\gamma^2 - m\omega'^2 - b\gamma + k\right)\cos\omega't + \left(2\omega'\gamma m - b\omega'\right)\sin\omega't\right] = 0. \quad \textbf{(i)}$$

The left side of this equation must equal zero for all times t, but this can only be so for certain values of γ and ω'. To determine γ and ω', we choose two values of t that will make their evaluation easy. At $t = 0$, $\sin\omega't = 0$, so the above relation reduces to $A\left(m\gamma^2 - m\omega'^2 - b\gamma + k\right) = 0$, which means[†] that

$$m\gamma^2 - m\omega'^2 - b\gamma + k = 0. \quad \textbf{(ii)}$$

Then at $t = \pi/2\omega'$, $\cos\omega't = 0$ so Eq. (i) can be valid only if

$$2\gamma m - b = 0. \quad \textbf{(iii)}$$

From Eq. (iii) we have

$$\gamma = \frac{b}{2m}$$

and from Eq. (ii)

$$\omega' = \sqrt{\gamma^2 - \frac{b\gamma}{m} + \frac{k}{m}} = \sqrt{\frac{k}{m} - \frac{b^2}{4m^2}}.$$

Thus we see that Eq. 14–16 is a solution to the equation of motion for the damped harmonic oscillator as long as γ and ω' have these specific values, as already given in Eqs. 14–17 and 14–18.

14–8 Forced Oscillations; Resonance

When an oscillating system is set into motion, it oscillates at its natural frequency (Eqs. 14–7a and 14–12b). However, a system may have an external force applied to it that has its own particular frequency and then we have a **forced oscillation**.

For example, we might pull the mass on the spring of Fig. 14–1 back and forth at a frequency f. The mass then oscillates at the frequency f of the external force, even if this frequency is different from the **natural frequency** of the spring, which we will now denote by f_0 where (see Eqs. 14–5 and 14–7a)

$$\omega_0 = 2\pi f_0 = \sqrt{\frac{k}{m}}.$$

In a forced oscillation the amplitude of oscillation, and hence the energy transferred to the oscillating system, is found to depend on the difference between f and f_0 as well as on the amount of damping, reaching a maximum when the frequency of the external force equals the natural frequency of the system—that is, when $f = f_0$. The amplitude is plotted in Fig. 14–23 as a function of the external frequency f. Curve A represents light damping and curve B heavy damping. The amplitude can become large when the driving frequency f is near the natural frequency, $f \approx f_0$, as long as the damping is not too large. When the damping is small, the increase in amplitude near $f = f_0$ is very large (and often dramatic). This effect is known as **resonance**. The natural frequency f_0 of a system is called its **resonant frequency**.

A simple illustration of resonance is pushing a child on a swing. A swing, like any pendulum, has a natural frequency of oscillation that depends on its length ℓ. If you push on the swing at a random frequency, the swing bounces around and reaches no great amplitude. But if you push with a frequency equal to the natural frequency of the swing, the amplitude increases greatly. At resonance, relatively little effort is required to obtain a large amplitude.

FIGURE 14–23 Resonance for lightly damped (A) and heavily damped (B) systems. (See Fig. 14–26 for a more detailed graph.)

[†]It would also be satisfied by $A = 0$, but this gives the trivial and uninteresting solution $x = 0$ for all t—that is, no oscillation.

FIGURE 14–24 This goblet breaks as it vibrates in resonance to a trumpet call.

FIGURE 14–25 (a) Large-amplitude oscillations of the Tacoma Narrows Bridge, due to gusty winds, led to its collapse (1940). (b) Collapse of a freeway in California, due to the 1989 earthquake.

(a)

(b)

The great tenor Enrico Caruso was said to be able to shatter a crystal goblet by singing a note of just the right frequency at full voice. This is an example of resonance, for the sound waves emitted by the voice act as a forced oscillation on the glass. At resonance, the resulting oscillation of the goblet may be large enough in amplitude that the glass exceeds its elastic limit and breaks (Fig. 14–24).

Since material objects are, in general, elastic, resonance is an important phenomenon in a variety of situations. It is particularly important in structural engineering, although the effects are not always foreseen. For example, it has been reported that a railway bridge collapsed because a nick in one of the wheels of a crossing train set up a resonant oscillation in the bridge. Indeed, marching soldiers break step when crossing a bridge to avoid the possibility that their normal rhythmic march might match a resonant frequency of the bridge. The famous collapse of the Tacoma Narrows Bridge (Fig. 14–25a) in 1940 occurred as a result of strong gusting winds driving the span into large-amplitude oscillatory motion. The Oakland freeway collapse in the 1989 California earthquake (Fig. 14–25b) involved resonant oscillation of a section built on mudfill that readily transmitted that frequency.

We will meet important examples of resonance later. We will also see that vibrating objects often have not one, but many resonant frequencies.

*Equation of Motion and Its Solution

We now look at the equation of motion for a forced oscillation and its solution. Suppose the external force is sinusoidal and can be represented by

$$F_{\text{ext}} = F_0 \cos \omega t,$$

where $\omega = 2\pi f$ is the angular frequency applied externally to the oscillator. Then the equation of motion (with damping) is

$$ma = -kx - bv + F_0 \cos \omega t.$$

This can be written as

$$m\frac{d^2x}{dt^2} + b\frac{dx}{dt} + kx = F_0 \cos \omega t. \tag{14–21}$$

The external force, on the right of the equation, is the only term that does not involve x or one of its derivatives. Problem 68 asks you to show that

$$x = A_0 \sin(\omega t + \phi_0) \tag{14–22}$$

is a solution to Eq. 14–21, by direct substitution, where

$$A_0 = \frac{F_0}{m\sqrt{(\omega^2 - \omega_0^2)^2 + b^2\omega^2/m^2}} \tag{14–23}$$

and

$$\phi_0 = \tan^{-1}\frac{\omega_0^2 - \omega^2}{\omega(b/m)}. \tag{14–24}$$

Actually, the general solution to Eq. 14–21 is Eq. 14–22 plus another term of the form of Eq. 14–19 for the natural damped motion of the oscillator; this second term approaches zero in time, so in many cases we need to be concerned only with Eq. 14–22.

The amplitude of forced harmonic motion, A_0, depends strongly on the difference between the applied and the natural frequency. A plot of A_0 (Eq. 14–23) as a function of the applied frequency, ω, is shown in Fig. 14–26 (a more detailed version of Fig. 14–23) for three specific values of the damping constant b. Curve A $\left(b = \frac{1}{6}m\omega_0\right)$ represents light damping, curve B $\left(b = \frac{1}{2}m\omega_0\right)$ fairly heavy damping, and curve C $\left(b = \sqrt{2}m\omega_0\right)$ overdamped motion. The amplitude can become large when the driving frequency is near the natural frequency, $\omega \approx \omega_0$, as long as the damping is not too large. When the damping is small, the increase in amplitude near $\omega = \omega_0$ is very large and, as we saw, is known as *resonance*.

The natural oscillating frequency $f_0\,(=\omega_0/2\pi)$ of a system is its *resonant frequency*.[†] If $b = 0$, resonance occurs at $\omega = \omega_0$ and the resonant peak (of A_0) becomes infinite; in such a case, energy is being continuously transferred into the system and none is dissipated. For real systems, b is never precisely zero, and the resonant peak is finite. The peak does not occur precisely at $\omega = \omega_0$ (because of the term $b^2\omega^2/m^2$ in the denominator of Eq. 14–23), although it is quite close to ω_0 unless the damping is very large. If the damping is large, there is little or no peak (curve C in Fig. 14–26).

*Q value

The height and narrowness of a resonant peak is often specified by its **quality factor** or **Q value**, defined as

$$Q = \frac{m\omega_0}{b}. \qquad (14\text{–}25)$$

In Fig. 14–26, curve A has $Q = 6$, curve B has $Q = 2$, and curve C has $Q = 1/\sqrt{2}$. The smaller the damping constant b, the larger the Q value becomes, and the higher the resonance peak. The Q value is also a measure of the width of the peak. To see why, let ω_1 and ω_2 be the frequencies where the square of the amplitude A_0 has half its maximum value (we use the square because the power transferred to the system is proportional to A_0^2); then $\Delta\omega = \omega_1 - \omega_2$, which is called the *width* of the resonance peak, is related to Q by

$$\frac{\Delta\omega}{\omega_0} = \frac{1}{Q}. \qquad (14\text{–}26)$$

This relation is accurate only for weak damping. The larger the Q value, the narrower will be the resonance peak relative to its height. Thus a large Q value, representing a system of high quality, has a high, narrow resonance peak.

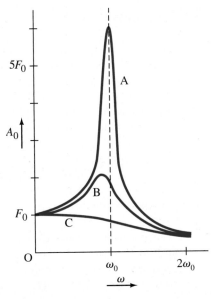

FIGURE 14–26 Amplitude of a forced harmonic oscillator as a function of ω. Curves A, B, and C correspond to light, heavy, and overdamped systems, respectively ($Q = m\omega_0/b = 6, 2, 0.71$).

[†]Sometimes the resonant frequency is defined as the actual value of ω at which the amplitude has its maximum value, and this depends somewhat on the damping constant. Except for very heavy damping, this value is quite close to ω_0.

Summary

An oscillating object undergoes **simple harmonic motion** (SHM) if the restoring force is proportional to the displacement,

$$F = -kx. \qquad (14\text{–}1)$$

The maximum displacement from equilibrium is called the **amplitude**.

The **period**, T, is the time required for one complete cycle (back and forth), and the **frequency**, f, is the number of cycles per second; they are related by

$$f = \frac{1}{T}. \qquad (14\text{–}2)$$

The period of oscillation for a mass m on the end of an ideal massless spring is given by

$$T = 2\pi\sqrt{\frac{m}{k}}. \qquad (14\text{–}7b)$$

SHM is **sinusoidal**, which means that the displacement as a function of time follows a sine or cosine curve. The general solution can be written

$$x = A\cos(\omega t + \phi) \qquad (14\text{–}4)$$

where A is the amplitude, ϕ is the **phase angle**, and

$$\omega = 2\pi f = \sqrt{\frac{k}{m}}. \qquad (14\text{–}5)$$

The values of A and ϕ depend on the **initial conditions** (x and v at $t = 0$).

During SHM, the total energy $E = \frac{1}{2}mv^2 + \frac{1}{2}kx^2$ is continually changing from potential to kinetic and back again.

A **simple pendulum** of length ℓ approximates SHM if its amplitude is small and friction can be ignored. For small amplitudes, its period is given by

$$T = 2\pi\sqrt{\frac{\ell}{g}}, \qquad (14\text{–}12c)$$

where g is the acceleration of gravity.

When friction is present (for all real springs and pendulums), the motion is said to be **damped**. The maximum displacement decreases in time, and the mechanical energy is eventually all transformed to thermal energy. If the friction is very large, so no oscillations occur, the system is said to be **overdamped**. If the friction is small enough that oscillations occur, the system is **underdamped**, and the displacement is given by

$$x = Ae^{-\gamma t}\cos\omega' t, \qquad (14\text{–}16)$$

where γ and ω' are constants. For a **critically damped** system, no oscillations occur and equilibrium is reached in the shortest time.

If an oscillating force is applied to a system capable of vibrating, the amplitude of vibration can be very large if the frequency of the applied force is near the **natural** (or **resonant**) **frequency** of the oscillator; this is called **resonance**.

Questions

1. Give some examples of everyday vibrating objects. Which exhibit SHM, at least approximately?

2. Is the acceleration of a simple harmonic oscillator ever zero? If so, where?

3. Explain why the motion of a piston in an automobile engine is approximately simple harmonic.

4. Real springs have mass. Will the true period and frequency be larger or smaller than given by the equations for a mass oscillating on the end of an idealized massless spring? Explain.

5. How could you double the maximum speed of a simple harmonic oscillator (SHO)?

6. A 5.0-kg trout is attached to the hook of a vertical spring scale, and then is released. Describe the scale reading as a function of time.

7. If a pendulum clock is accurate at sea level, will it gain or lose time when taken to high altitude? Why?

8. A tire swing hanging from a branch reaches nearly to the ground (Fig. 14–27). How could you estimate the height of the branch using only a stopwatch?

FIGURE 14–27 Question 8.

9. For a simple harmonic oscillator, when (if ever) are the displacement and velocity vectors in the same direction? When are the displacement and acceleration vectors in the same direction?

10. A 100-g mass hangs from a long cord forming a pendulum. The mass is pulled a short distance to one side and released from rest. The time to swing over and back is carefully measured to be 2.0 s. If the 100-g mass is replaced by a 200-g mass, which is then pulled over the same distance and released from rest, the time will be (a) 1.0 s, (b) 1.41 s, (c) 2.0 s, (d) 2.82 s, (e) 4.0 s.

11. Two equal masses are attached to separate identical springs next to one another. One mass is pulled so its spring stretches 20 cm and the other is pulled so its spring stretches only 10 cm. The masses are released simultaneously. Which mass reaches the equilibrium point first?

12. Does a car bounce on its springs faster when it is empty or when it is fully loaded?

13. What is the approximate period of your walking step?

14. What happens to the period of a playground swing if you rise up from sitting to a standing position?

*15. A thin uniform rod of mass m is suspended from one end and oscillates with a frequency f. If a small sphere of mass $2m$ is attached to the other end, does the frequency increase or decrease? Explain.

16. A tuning fork of natural frequency 264 Hz sits on a table at the front of a room. At the back of the room, two tuning forks, one of natural frequency 260 Hz and one of 420 Hz are initially silent, but when the tuning fork at the front of the room is set into vibration, the 260-Hz fork spontaneously begins to vibrate but the 420-Hz fork does not. Explain.

17. Why can you make water slosh back and forth in a pan only if you shake the pan at a certain frequency?

18. Give several everyday examples of resonance.

19. Is a rattle in a car ever a resonance phenomenon? Explain.

20. Over the years, buildings have been able to be built out of lighter and lighter materials. How has this affected the natural oscillation frequencies of buildings and the problems of resonance due to passing trucks, airplanes, or by wind and other natural sources of vibration?

Problems

14–1 and 14–2 Simple Harmonic Motion

1. (I) If a particle undergoes SHM with amplitude 0.18 m, what is the total distance it travels in one period?

2. (I) An elastic cord is 65 cm long when a weight of 75 N hangs from it but is 85 cm long when a weight of 180 N hangs from it. What is the "spring" constant k of this elastic cord?

3. (I) The springs of a 1500-kg car compress 5.0 mm when its 68-kg driver gets into the driver's seat. If the car goes over a bump, what will be the frequency of oscillations? Ignore damping.

4. (I) (a) What is the equation describing the motion of a mass on the end of a spring which is stretched 8.8 cm from equilibrium and then released from rest, and whose period is 0.66 s? (b) What will be its displacement after 1.8 s?

5. (II) Estimate the stiffness of the spring in a child's pogo stick if the child has a mass of 35 kg and bounces once every 2.0 seconds.

6. (II) A fisherman's scale stretches 3.6 cm when a 2.4-kg fish hangs from it. (a) What is the spring stiffness constant and (b) what will be the amplitude and frequency of oscillation if the fish is pulled down 2.5 cm more and released so that it oscillates up and down?

7. (II) Tall buildings are designed to sway in the wind. In a 100-km/h wind, for example, the top of the 110-story Sears Tower oscillates horizontally with an amplitude of 15 cm. The building oscillates at its natural frequency, which has a period of 7.0 s. Assuming SHM, find the maximum horizontal velocity and acceleration experienced by a Sears employee as she sits working at her desk located on the top floor. Compare the maximum acceleration (as a percentage) with the acceleration due to gravity.

8. (II) Construct a Table indicating the position x of the mass in Fig. 14–2 at times $t = 0, \frac{1}{4}T, \frac{1}{2}T, \frac{3}{4}T, T,$ and $\frac{5}{4}T$, where T is the period of oscillation. On a graph of x vs. t, plot these six points. Now connect these points with a smooth curve. Based on these simple considerations, does your curve resemble that of a cosine or sine wave?

9. (II) A small fly of mass 0.25 g is caught in a spider's web. The web oscillates predominately with a frequency of 4.0 Hz. (a) What is the value of the effective spring stiffness constant k for the web? (b) At what frequency would you expect the web to oscillate if an insect of mass 0.50 g were trapped?

10. (II) A mass m at the end of a spring oscillates with a frequency of 0.83 Hz. When an additional 680-g mass is added to m, the frequency is 0.60 Hz. What is the value of m?

11. (II) A uniform meter stick of mass M is pivoted on a hinge at one end and held horizontal by a spring with spring constant k attached at the other end (Fig. 14–28). If the stick oscillates up and down slightly, what is its frequency? [Hint: Write a torque equation about the hinge.]

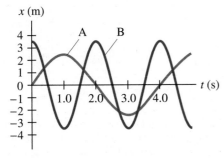

FIGURE 14–28
Problem 11.

12. (II) A balsa wood block of mass 55 g floats on a lake, bobbing up and down at a frequency of 3.0 Hz. (a) What is the value of the effective spring constant of the water? (b) A partially filled water bottle of mass 0.25 kg and almost the same size and shape of the balsa block is tossed into the water. At what frequency would you expect the bottle to bob up and down? Assume SHM.

13. (II) Figure 14–29 shows two examples of SHM, labeled A and B. For each, what is (a) the amplitude, (b) the frequency, and (c) the period? (d) Write the equations for both A and B in the form of a sine or cosine.

FIGURE 14–29   Problem 13.

14. (II) Determine the phase constant ϕ in Eq. 14–4 if, at $t = 0$, the oscillating mass is at (a) $x = -A$, (b) $x = 0$, (c) $x = A$, (d) $x = \frac{1}{2}A$, (e) $x = -\frac{1}{2}A$, (f) $x = A/\sqrt{2}$.

15. (II) A vertical spring with spring stiffness constant 305 N/m oscillates with an amplitude of 28.0 cm when 0.260 kg hangs from it. The mass passes through the equilibrium point ($y = 0$) with positive velocity at $t = 0$. (a) What equation describes this motion as a function of time? (b) At what times will the spring be longest and shortest?

16. (II) The graph of displacement vs. time for a small mass m at the end of a spring is shown in Fig. 14–30. At $t = 0$, $x = 0.43$ cm. (a) If $m = 9.5$ g, find the spring constant, k. (b) Write the equation for displacement x as a function of time.

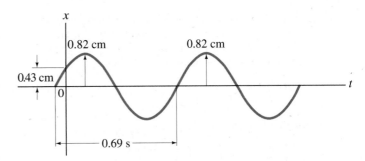

FIGURE 14–30   Problem 16.

17. (II) The position of a SHO as a function of time is given by $x = 3.8 \cos(5\pi t/4 + \pi/6)$ where t is in seconds and x in meters. Find (a) the period and frequency, (b) the position and velocity at $t = 0$, and (c) the velocity and acceleration at $t = 2.0$ s.

18. (II) A tuning fork oscillates at a frequency of 441 Hz and the tip of each prong moves 1.5 mm to either side of center. Calculate (a) the maximum speed and (b) the maximum acceleration of the tip of a prong.

19. (II) An object of unknown mass m is hung from a vertical spring of unknown spring constant k, and the object is observed to be at rest when the spring has extended by 14 cm. The object is then given a slight push and executes SHM. Determine the period T of this oscillation.

20. (II) A 1.25-kg mass stretches a vertical spring 0.215 m. If the spring is stretched an additional 0.130 m and released, how long does it take to reach the (new) equilibrium position again?

21. (II) Consider two objects, A and B, both undergoing SHM, but with different frequencies, as described by the equations $x_A = (2.0\text{ m})\sin(2.0t)$ and $x_B = (5.0\text{ m})\sin(3.0t)$, where t is in seconds. After $t = 0$, find the next three times t at which both objects simultaneously pass through the origin.

22. (II) A 1.60-kg object oscillates from a vertically hanging light spring once every 0.55 s. (a) Write down the equation giving its position y (+ upward) as a function of time t, assuming it started by being compressed 16 cm from the equilibrium position (where $y = 0$), and released. (b) How long will it take to get to the equilibrium position for the first time? (c) What will be its maximum speed? (d) What will be its maximum acceleration, and where will it first be attained?

23. (II) A bungee jumper with mass 65.0 kg jumps from a high bridge. After reaching his lowest point, he oscillates up and down, hitting a low point eight more times in 43.0 s. He finally comes to rest 25.0 m below the level of the bridge. Estimate the spring stiffness constant and the unstretched length of the bungee cord assuming SHM.

24. (II) A block of mass m is supported by two identical parallel vertical springs, each with spring stiffness constant k (Fig. 14–31). What will be the frequency of vertical oscillation?

FIGURE 14–31
Problem 24.

25. (III) A mass m is connected to two springs, with spring constants k_1 and k_2, in two different ways as shown in Fig. 14–32a and b. Show that the period for the configuration shown in part (a) is given by

$$T = 2\pi \sqrt{m\left(\frac{1}{k_1} + \frac{1}{k_2}\right)}$$

and for that in part (b) is given by

$$T = 2\pi \sqrt{\frac{m}{k_1 + k_2}}.$$

Ignore friction.

(a)

FIGURE 14–32
Problem 25.

(b)

26. (III) A mass m is at rest on the end of a spring of spring constant k. At $t = 0$ it is given an impulse J by a hammer. Write the formula for the subsequent motion in terms of $m, k, J,$ and t.

14–3 Energy in SHM

27. (I) A 1.15-kg mass oscillates according to the equation $x = 0.650 \cos 7.40t$ where x is in meters and t in seconds. Determine (a) the amplitude, (b) the frequency, (c) the total energy, and (d) the kinetic energy and potential energy when $x = 0.260$ m.

28. (I) (a) At what displacement of a SHO is the energy half kinetic and half potential? (b) What fraction of the total energy of a SHO is kinetic and what fraction potential when the displacement is one third the amplitude?

29. (II) Draw a graph like Fig. 14–11 for a horizontal spring whose spring constant is 95 N/m and which has a mass of 55 g on the end of it. Assume the spring was started with an initial amplitude of 2.0 cm. Neglect the mass of the spring and any friction with the horizontal surface. Use your graph to estimate (a) the potential energy, (b) the kinetic energy, and (c) the speed of the mass, for $x = 1.5$ cm.

30. (II) A 0.35-kg mass at the end of a spring oscillates 2.5 times per second with an amplitude of 0.15 m. Determine (a) the velocity when it passes the equilibrium point, (b) the velocity when it is 0.10 m from equilibrium, (c) the total energy of the system, and (d) the equation describing the motion of the mass, assuming that at $t = 0$, x was a maximum.

31. (II) It takes a force of 95.0 N to compress the spring of a toy popgun 0.175 m to "load" a 0.160-kg ball. With what speed will the ball leave the gun if fired horizontally?

32. (II) A 0.0125-kg bullet strikes a 0.240-kg block attached to a fixed horizontal spring whose spring constant is 2.25×10^3 N/m and sets it into oscillation with an amplitude of 12.4 cm. What was the initial speed of the bullet if the two objects move together after impact?

33. (II) If one oscillation has 5.0 times the energy of a second one of equal frequency and mass, what is the ratio of their amplitudes?

34. (II) A mass of 240 g oscillates on a horizontal frictionless surface at a frequency of 3.0 Hz and with amplitude of 4.5 cm. (a) What is the effective spring constant for this motion? (b) How much energy is involved in this motion?

35. (II) A mass resting on a horizontal, frictionless surface is attached to one end of a spring; the other end is fixed to a wall. It takes 3.6 J of work to compress the spring by 0.13 m. If the spring is compressed, and the mass is released from rest, it experiences a maximum acceleration of 15 m/s². Find the value of (a) the spring constant and (b) the mass.

36. (II) An object with mass 2.7 kg is executing simple harmonic motion, attached to a spring with spring constant $k = 280$ N/m. When the object is 0.020 m from its equilibrium position, it is moving with a speed of 0.55 m/s. (a) Calculate the amplitude of the motion. (b) Calculate the maximum speed attained by the object.

37. (II) Agent Arlene devised the following method of measuring the muzzle velocity of a rifle (Fig. 14–33). She fires a bullet into a 4.648-kg wooden block resting on a smooth surface, and attached to a spring of spring constant $k = 142.7$ N/m. The bullet, whose mass is 7.870 g, remains embedded in the wooden block. She measures the maximum distance that the block compresses the spring to be 9.460 cm. What is the speed v of the bullet?

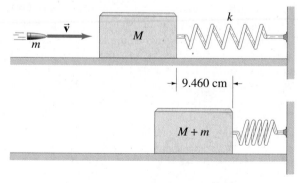

FIGURE 14–33 Problem 37.

38. (II) Obtain the displacement x as a function of time for the simple harmonic oscillator using the conservation of energy, Eqs. 14–10. [*Hint*: Integrate Eq. 14–11a with $v = dx/dt$.]

39. (II) At $t = 0$, a 785-g mass at rest on the end of a horizontal spring ($k = 184$ N/m) is struck by a hammer which gives it an initial speed of 2.26 m/s. Determine (a) the period and frequency of the motion, (b) the amplitude, (c) the maximum acceleration, (d) the position as a function of time, (e) the total energy, and (f) the kinetic energy when $x = 0.40A$ where A is the amplitude.

40. (II) A pinball machine uses a spring launcher that is compressed 6.0 cm to launch a ball up a 15° ramp. Assume that the pinball of mass $m = 25$ g and radius $r = 1.0$ cm is rolling without slipping when it leaves the launcher. If it has a speed of 3.0 m/s, what is the spring constant of the spring that is used in the launcher?

14–5 Simple Pendulum

41. (I) A pendulum has a period of 1.35 s on Earth. What is its period on Mars, where the acceleration of gravity is about 0.37 that on Earth?

42. (I) A pendulum makes 32 oscillations in exactly 50 s. What is its (a) period and (b) frequency?

43. (II) A simple pendulum is 0.30 m long. At $t = 0$ it is released from rest starting at an angle of 13°. Ignoring friction, what will be the angular position of the pendulum at (a) $t = 0.35$ s, (b) $t = 3.45$ s, and (c) $t = 6.00$ s?

44. (II) What is the period of a simple pendulum 53 cm long (a) on the Earth, and (b) when it is in a freely falling elevator?

45. (II) A simple pendulum oscillates with an amplitude of 10.0°. What fraction of the time does it spend between $+5.0°$ and $-5.0°$? Assume SHM.

46. (II) Your grandfather clock's pendulum has a length of 0.9930 m. If the clock loses 26 s per day, how should you adjust the length of the pendulum?

47. (II) Derive a formula for the maximum speed v_{max} of a simple pendulum bob in terms of g, the length ℓ, and the maximum angle of swing θ_{max}.

*14–6 Physical Pendulum and Torsion Pendulum

*48. (II) A pendulum consists of a tiny bob of mass M and a uniform cord of mass m and length ℓ. (a) Determine a formula for the period using the small angle approximation. (b) What would be the fractional error if you use the formula for a simple pendulum, Eq. 14–12c?

*49. (II) The balance wheel of a watch is a thin ring of radius 0.95 cm and oscillates with a frequency of 3.10 Hz. If a torque of 1.1×10^{-5} m·N causes the wheel to rotate 45°, calculate the mass of the balance wheel.

*50. (II) The human leg can be compared to a physical pendulum, with a "natural" swinging period at which walking is easiest. Consider the leg as two rods joined rigidly together at the knee; the axis for the leg is the hip joint. The length of each rod is about the same, 55 cm. The upper rod has a mass of 7.0 kg and the lower rod has a mass of 4.0 kg. (a) Calculate the natural swinging period of the system. (b) Check your answer by standing on a chair and measuring the time for one or more complete back-and-forth swings. The effect of a shorter leg is a shorter swinging period, enabling a faster "natural" stride.

*51. (II) (a) Determine the equation of motion (for θ as a function of time) for a torsion pendulum, Fig. 14–18, and show that the motion is simple harmonic. (b) Show that the period T is $T = 2\pi\sqrt{I/K}$. [The balance wheel of a mechanical watch is an example of a torsion pendulum in which the restoring torque is applied by a coil spring.]

*52. (II) A student wants to use a meter stick as a pendulum. She plans to drill a small hole through the meter stick and suspend it from a smooth pin attached to the wall (Fig. 14–34). Where in the meter stick should she drill the hole to obtain the shortest possible period? How short an oscillation period can she obtain with a meter stick in this way?

FIGURE 14–34
Problem 52.

*53. (II) A meter stick is hung at its center from a thin wire (Fig. 14–35a). It is twisted and oscillates with a period of 5.0 s. The meter stick is sawed off to a length of 70.0 cm. This piece is again balanced at its center and set in oscillation (Fig. 14–35b). With what period does it oscillate?

(a)   (b)

FIGURE 14–35   Problem 53.

*54. (II) An aluminum disk, 12.5 cm in diameter and 375 g in mass, is mounted on a vertical shaft with very low friction (Fig. 14–36). One end of a flat coil spring is attached to the disk, the other end to the base of the apparatus. The disk is set into rotational oscillation and the frequency is 0.331 Hz. What is the torsional spring constant $K(\tau = -K\theta)$?

FIGURE 14–36   Problem 54.

*55. (II) A plywood disk of radius 20.0 cm and mass 2.20 kg has a small hole drilled through it, 2.00 cm from its edge (Fig. 14–37). The disk is hung from the wall by means of a metal pin through the hole, and is used as a pendulum. What is the period of this pendulum for small oscillations?

FIGURE 14–37
Problem 55.

14–7 Damping

56. (II) A 0.835-kg block oscillates on the end of a spring whose spring constant is $k = 41.0 \, \text{N/m}$. The mass moves in a fluid which offers a resistive force $F = -bv$, where $b = 0.662 \, \text{N·s/m}$. (a) What is the period of the motion? (b) What is the fractional decrease in amplitude per cycle? (c) Write the displacement as a function of time if at $t = 0$, $x = 0$, and at $t = 1.00 \, \text{s}$, $x = 0.120 \, \text{m}$.

57. (II) Estimate how the damping constant changes when a car's shock absorbers get old and the car bounces three times after going over a speed bump.

58. (II) A physical pendulum consists of an 85-cm-long, 240-g-mass, uniform wooden rod hung from a nail near one end (Fig. 14–38). The motion is damped because of friction in the pivot; the damping force is approximately proportional to $d\theta/dt$. The rod is set in oscillation by displacing it 15° from its equilibrium position and releasing it. After 8.0 s, the amplitude of the oscillation has been reduced to 5.5°. If the angular displacement can be written as $\theta = Ae^{-\gamma t} \cos \omega' t$, find (a) γ, (b) the approximate period of the motion, and (c) how long it takes for the amplitude to be reduced to $\frac{1}{2}$ of its original value.

FIGURE 14–38
Problem 58.

59. (II) A damped harmonic oscillator loses 6.0% of its mechanical energy per cycle. (a) By what percentage does its frequency differ from the natural frequency $f_0 = (1/2\pi)\sqrt{k/m}$? (b) After how many periods will the amplitude have decreased to $1/e$ of its original value.

60. (II) A vertical spring of spring constant 115 N/m supports a mass of 75 g. The mass oscillates in a tube of liquid. If the mass is initially given an amplitude of 5.0 cm, the mass is observed to have an amplitude of 2.0 cm after 3.5 s. Estimate the damping constant b. Neglect buoyant forces.

61. (III) (a) Show that the total mechanical energy, $E = \frac{1}{2}mv^2 + \frac{1}{2}kx^2$, as a function of time for a lightly damped harmonic oscillator is
$$E = \frac{1}{2}kA^2 e^{-(b/m)t} = E_0 e^{-(b/m)t},$$
where E_0 is the total mechanical energy at $t = 0$. (Assume $\omega' \gg b/2m$.) (b) Show that the fractional energy lost per period is
$$\frac{\Delta E}{E} = \frac{2\pi b}{m\omega_0} = \frac{2\pi}{Q},$$
where $\omega_0 = \sqrt{k/m}$ and $Q = m\omega_0/b$ is called the **quality factor** or **Q value** of the system. A larger Q value means the system can undergo oscillations for a longer time.

62. (III) A glider on an air track is connected by springs to either end of the track (Fig. 14–39). Both springs have the same spring constant, k, and the glider has mass M. (a) Determine the frequency of the oscillation, assuming no damping, if $k = 125 \, \text{N/m}$ and $M = 215 \, \text{g}$. (b) It is observed that after 55 oscillations, the amplitude of the oscillation has dropped to one-half of its initial value. Estimate the value of α, using Eq. 14–16. (c) How long does it take the amplitude to decrease to one-quarter of its initial value?

FIGURE 14–39
Problem 62.

14–8 Forced Oscillations; Resonance

63. (II) (a) For a forced oscillation at resonance $(\omega = \omega_0)$, what is the value of the phase angle ϕ_0 in Eq. 14–22? (b) What, then, is the displacement at a time when the driving force F_{ext} is a maximum, and at a time when $F_{\text{ext}} = 0$? (c) What is the phase difference (in degrees) between the driving force and the displacement in this case?

64. (II) Differentiate Eq. 14–23 to show that the resonant amplitude peaks at
$$\omega = \sqrt{\omega_0^2 - \frac{b^2}{2m^2}}.$$

65. (II) An 1150 kg automobile has springs with $k = 16,000 \, \text{N/m}$. One of the tires is not properly balanced; it has a little extra mass on one side compared to the other, causing the car to shake at certain speeds. If the tire radius is 42 cm, at what speed will the wheel shake most?

*66. (II) Construct an accurate resonance curve, from $\omega = 0$ to $\omega = 2\omega_0$, for $Q = 6.0$.

*67. (II) The amplitude of a driven harmonic oscillator reaches a value of $23.7 F_0/m$ at a resonant frequency of 382 Hz. What is the Q value of this system?

68. (III) By direct substitution, show that Eq. 14–22, with Eqs. 14–23 and 14–24, is a solution of the equation of motion (Eq. 14–21) for the forced oscillator. [*Hint*: To find $\sin \phi_0$ and $\cos \phi_0$ from $\tan \phi_0$, draw a right triangle.]

*69. (III) Consider a simple pendulum (point mass bob) 0.50 m long with a Q of 350. (a) How long does it take for the amplitude (assumed small) to decrease by two-thirds? (b) If the amplitude is 2.0 cm and the bob has mass 0.27 kg, what is the initial energy loss rate of the pendulum in watts? (c) If we are to stimulate resonance with a sinusoidal driving force, how close must the driving frequency be to the natural frequency of the pendulum (give $\Delta f = f - f_0$)?

General Problems

70. A 62-kg person jumps from a window to a fire net 20.0 m below, which stretches the net 1.1 m. Assume that the net behaves like a simple spring. (a) Calculate how much it would stretch if the same person were lying in it. (b) How much would it stretch if the person jumped from 38 m?

71. An energy-absorbing car bumper has a spring constant of 430 kN/m. Find the maximum compression of the bumper if the car, with mass 1300 kg, collides with a wall at a speed of 2.0 m/s (approximately 5 mi/h).

72. The length of a simple pendulum is 0.63 m, the pendulum bob has a mass of 295 g, and it is released at an angle of 15° to the vertical. (a) With what frequency does it oscillate? (b) What is the pendulum bob's speed when it passes through the lowest point of the swing? Assume SHM. (c) What is the total energy stored in this oscillation assuming no losses?

73. A simple pendulum oscillates with frequency f. What is its frequency if the entire pendulum accelerates at $0.50 g$ (a) upward, and (b) downward?

74. A 0.650-kg mass oscillates according to the equation $x = 0.25 \sin(5.50t)$ where x is in meters and t is in seconds. Determine (a) the amplitude, (b) the frequency, (c) the period, (d) the total energy, and (e) the kinetic energy and potential energy when x is 15 cm.

75. (a) A crane has hoisted a 1350-kg car at the junkyard. The crane's steel cable is 20.0 m long and has a diameter of 6.4 mm. If the car starts bouncing at the end of the cable, what is the period of the bouncing? [Hint: Refer to Table 12–1]. (b) What amplitude of bouncing will likely cause the cable to snap? (See Table 12–2, and assume Hooke's law holds all the way up to the breaking point.)

76. An oxygen atom at a particular site within a DNA molecule can be made to execute simple harmonic motion when illuminated by infrared light. The oxygen atom is bound with a spring-like chemical bond to a phosphorus atom, which is rigidly attached to the DNA backbone. The oscillation of the oxygen atom occurs with frequency $f = 3.7 \times 10^{13}$ Hz. If the oxygen atom at this site is chemically replaced with a sulfur atom, the spring constant of the bond is unchanged (sulfur is just below oxygen in the Periodic Table). Predict the frequency for a DNA molecule after the sulfur substitution.

77. A "seconds" pendulum has a period of exactly 2.000 s. That is, each one-way swing takes 1.000 s. What is the length of a seconds pendulum in Austin, Texas, where $g = 9.793 \text{ m/s}^2$? If the pendulum is moved to Paris, where $g = 9.809 \text{ m/s}^2$, by how many millimeters must we lengthen the pendulum? What is the length of a seconds pendulum on the Moon, where $g = 1.62 \text{ m/s}^2$?

78. A 320-kg wooden raft floats on a lake. When a 75-kg man stands on the raft, it sinks 3.5 cm deeper into the water. When he steps off, the raft oscillates for a while. (a) What is the frequency of oscillation? (b) What is the total energy of oscillation (ignoring damping)?

79. At what displacement from equilibrium is the speed of a SHO half the maximum value?

80. A diving board oscillates with simple harmonic motion of frequency 2.5 cycles per second. What is the maximum amplitude with which the end of the board can oscillate in order that a pebble placed there (Fig. 14–40) does not lose contact with the board during the oscillation?

FIGURE 14–40
Problem 80.

81. A rectangular block of wood floats in a calm lake. Show that, if friction is ignored, when the block is pushed gently down into the water and then released, it will then oscillate with SHM. Also, determine an equation for the force constant.

82. A 950-kg car strikes a huge spring at a speed of 25 m/s (Fig. 14–41), compressing the spring 5.0 m. (a) What is the spring stiffness constant of the spring? (b) How long is the car in contact with the spring before it bounces off in the opposite direction?

FIGURE 14–41
Problem 82.

83. A 1.60-kg table is supported on four springs. A 0.80-kg chunk of modeling clay is held above the table and dropped so that it hits the table with a speed of 1.65 m/s (Fig. 14–42). The clay makes an inelastic collision with the table, and the table and clay oscillate up and down. After a long time the table comes to rest 6.0 cm below its original position. (a) What is the effective spring constant of all four springs taken together? (b) With what maximum amplitude does the platform oscillate?

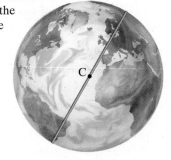

FIGURE 14–42
Problem 83.

84. In some diatomic molecules, the force each atom exerts on the other can be approximated by $F = -C/r^2 + D/r^3$, where r is the atomic separation and C and D are positive constants. (a) Graph F vs. r from $r = 0.8D/C$ to $r = 4D/C$. (b) Show that equilibrium occurs at $r = r_0 = D/C$. (c) Let $\Delta r = r - r_0$ be a small displacement from equilibrium, where $\Delta r \ll r_0$. Show that for such small displacements, the motion is approximately simple harmonic, and (d) determine the force constant. (e) What is the period of such motion? [Hint: Assume one atom is kept at rest.]

85. A mass attached to the end of a spring is stretched a distance x_0 from equilibrium and released. At what distance from equilibrium will it have (a) velocity equal to half its maximum velocity, and (b) acceleration equal to half its maximum acceleration?

86. Carbon dioxide is a linear molecule. The carbon–oxygen bonds in this molecule act very much like springs. Figure 14–43 shows one possible way the oxygen atoms in this molecule can oscillate: the oxygen atoms oscillate symmetrically in and out, while the central carbon atom remains at rest. Hence each oxygen atom acts like a simple harmonic oscillator with a mass equal to the mass of an oxygen atom. It is observed that this oscillation occurs with a frequency of $f = 2.83 \times 10^{13}$ Hz. What is the spring constant of the $C\!-\!O$ bond?

FIGURE 14–43
Problem 86, the
CO_2 molecule.

87. Imagine that a 10-cm-diameter circular hole was drilled all the way through the center of the Earth (Fig. 14–44). At one end of the hole, you drop an apple into the hole. Show that, if you assume that the Earth has a constant density, the subsequent motion of the apple is simple harmonic. How long will the apple take to return? Assume that we can ignore all frictional effects. [Hint: See Appendix D.]

FIGURE 14–44
Problem 87.

88. A thin, straight, uniform rod of length $\ell = 1.00\,\text{m}$ and mass $m = 215\,\text{g}$ hangs from a pivot at one end. (a) What is its period for small-amplitude oscillations? (b) What is the length of a simple pendulum that will have the same period?

89. A mass m is gently placed on the end of a freely hanging spring. The mass then falls 32.0 cm before it stops and begins to rise. What is the frequency of the oscillation?

90. A child of mass m sits on top of a rectangular slab of mass $M = 35\,\text{kg}$, which in turn rests on the frictionless horizontal floor at a pizza shop. The slab is attached to a horizontal spring with spring constant $k = 430\,\text{N/m}$ (the other end is attached to an immovable wall, Fig. 14–45). The coefficient of static friction between the child and the top of the slab is $\mu = 0.40$. The shop owner's intention is that, when displaced from the equilibrium position and released, the slab and child (with no slippage between the two) execute SHM with amplitude $A = 0.50\,\text{m}$. Should there be a weight restriction for this ride? If so, what is it?

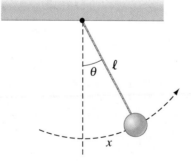

FIGURE 14–45
Problem 90.

91. Estimate the effective spring constant of a trampoline.

92. In Section 14–5, the oscillation of a simple pendulum (Fig. 14–46) is viewed as linear motion along the arc length x and analyzed via $F = ma$. Alternatively, the pendulum's movement can be regarded as rotational motion about its point of support and analyzed using $\tau = I\alpha$. Carry out this alternative analysis and show that

$$\theta(t) = \theta_{\max}\cos\left(\sqrt{\frac{g}{\ell}}\,t + \phi\right),$$

where $\theta(t)$ is the angular displacement of the pendulum from the vertical at time t, as long as its maximum value is less than about 15°.

FIGURE 14–46
Problem 92.

*Numerical/Computer

*93. (II) A mass m on a frictionless surface is attached to a spring with spring constant k as shown in Fig. 14–47. This mass–spring system is then observed to execute simple harmonic motion with a period T. The mass m is changed several times and the associated period T is measured in each case, generating the following data Table:

Mass m (kg)	Period T (s)
0.5	0.445
1.0	0.520
2.0	0.630
3.0	0.723
4.5	0.844

(a) Starting with Eq. 14–7b, show why a graph of T^2 vs. m is expected to yield a straight line. How can k be determined from the straight line's slope? What is the line's y-intercept expected to be? (b) Using the data in the Table, plot T^2 vs. m and show that this graph yields a straight line. Determine the slope and (nonzero) y-intercept. (c) Show that a nonzero y-intercept can be expected in our plot theoretically if, rather than simply using m for the mass in Eq. 14–7b, we use $m + m_0$, where m_0 is a constant. That is, repeat part (a) using $m + m_0$ for the mass in Eq. 14–7b. Then use the result of this analysis to determine k and m_0 from your graph's slope and y-intercept. (d) Offer a physical interpretation for m_0, a mass that appears to be oscillating in addition to the attached mass m.

FIGURE 14–47   Problem 93.

*94. (III) *Damping proportional to* v^2. Suppose the oscillator of Example 14–5 is damped by a force proportional to the square of the velocity, $F_{\text{damping}} = -cv^2$, where $c = 0.275\,\text{kg/m}$ is a constant. Numerically integrate[†] the differential equation from $t = 0$ to $t = 2.00\,\text{s}$ to an accuracy of 2%, and plot your results.

[†]See Section 2–9.

Answers to Exercises

A: (a), (c), (d).

B: (b).

C: (c).

D: (b), (d).

E: (c).

F: (c).

G: All are larger.

H: (a).

I: (c).

J: (a) 25 cm; (b) 2.0 s.

Waves—such as these water waves—spread outward from a source. The source in this case is a small spot of water oscillating up and down briefly where a rock was thrown in (left photo). Other kinds of waves include waves on a cord or string, which also are produced by a vibration. Waves move away from their source, but we also study waves that seem to stand still ("standing waves"). Waves reflect, and they can interfere with each other when they pass through any point at the same time.

Wave Motion

CHAPTER-OPENING QUESTION—Guess now!

You drop a rock into a pond, and water waves spread out in circles.

(a) The waves carry water outward, away from where the rock hit. That moving water carries energy outward.

(b) The waves only make the water move up and down. No energy is carried outward from where the rock hit.

(c). The waves only make the water move up and down, but the waves do carry energy outward, away from where the rock hit.

When you throw a stone into a lake or pool of water, circular waves form and move outward, as shown in the photos above. Waves will also travel along a cord that is stretched out flat on a table if you vibrate one end back and forth as shown in Fig. 15–1. Water waves and waves on a cord are two common examples of **mechanical waves**, which propagate as oscillations of matter. We will discuss other kinds of waves in later Chapters, including electromagnetic waves and light.

CONTENTS

FIGURE 15–1 Wave traveling on a cord. The wave travels to the right along the cord. Particles of the cord oscillate back and forth on the tabletop.

Velocity of cord particle
Velocity of wave

If you have ever watched ocean waves moving toward shore before they break, you may have wondered if the waves were carrying water from far out at sea onto the beach. They don't.† Water waves move with a recognizable velocity. But each particle (or molecule) of the water itself merely oscillates about an equilibrium point. This is clearly demonstrated by observing leaves on a pond as waves move by. The leaves (or a cork) are not carried forward by the waves, but simply oscillate about an equilibrium point because this is the motion of the water itself.

| CONCEPTUAL EXAMPLE 15–1 | **Wave vs. particle velocity.** Is the velocity of a wave moving along a cord the same as the velocity of a particle of the cord? See Fig. 15–1.

RESPONSE No. The two velocities are different, both in magnitude and direction. The wave on the cord of Fig. 15–1 moves to the right along the tabletop, but each piece of the cord only vibrates to and fro. (The cord clearly does not travel in the direction that the wave on it does.)

FIGURE 15–1 (repeated) Wave traveling on a cord. The wave travels to the right along the cord. Particles of the cord oscillate back and forth on the tabletop.

Velocity of cord particle

Velocity of wave

Waves can move over large distances, but the medium (the water or the cord) itself has only a limited movement, oscillating about an equilibrium point as in simple harmonic motion. Thus, although a wave is not matter, the wave pattern can travel in matter. A wave consists of oscillations that move without carrying matter with them.

Waves carry energy from one place to another. Energy is given to a water wave, for example, by a rock thrown into the water, or by wind far out at sea. The energy is transported by waves to the shore. The oscillating hand in Fig. 15–1 transfers energy to the cord, and that energy is transported down the cord and can be transferred to an object at the other end. All forms of traveling waves transport energy.

EXERCISE A Return to the Chapter-Opening Question, page 395, and answer it again now. Try to explain why you may have answered differently the first time.

15–1 Characteristics of Wave Motion

Let us look a little more closely at how a wave is formed and how it comes to "travel." We first look at a single wave bump, or **pulse**. A single pulse can be formed on a cord by a quick up-and-down motion of the hand, Fig. 15–2. The hand pulls up on one end of the cord. Because the end section is attached to adjacent sections, these also feel an upward force and they too begin to move upward. As each succeeding section of cord moves upward, the wave crest moves outward along the cord. Meanwhile, the end section of cord has been returned to its original position by the hand. As each succeeding section of cord reaches its peak position, it too is pulled back down again by tension from the adjacent section of cord. Thus the source of a traveling wave pulse is a disturbance, and cohesive forces between adjacent sections of cord cause the pulse to travel. Waves in other media are created

FIGURE 15–2 Motion of a wave pulse to the right. Arrows indicate velocity of cord particles.

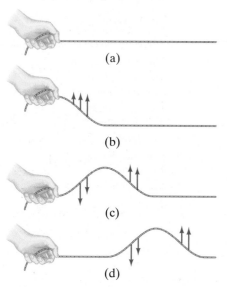

(a)

(b)

(c)

(d)

†Do not be confused by the "breaking" of ocean waves, which occurs when a wave interacts with the ground in shallow water and hence is no longer a simple wave.

and propagate outward in a similar fashion. A dramatic example of a wave pulse is a tsunami or tidal wave that is created by an earthquake in the Earth's crust under the ocean. The bang you hear when a door slams is a sound wave pulse.

A **continuous** or **periodic wave**, such as that shown in Fig. 15–1, has as its source a disturbance that is continuous and oscillating; that is, the source is a *vibration* or *oscillation*. In Fig. 15–1, a hand oscillates one end of the cord. Water waves may be produced by any vibrating object at the surface, such as your hand; or the water itself is made to vibrate when wind blows across it or a rock is thrown into it. A vibrating tuning fork or drum membrane gives rise to sound waves in air. And we will see later that oscillating electric charges give rise to light waves. Indeed, almost any vibrating object sends out waves.

The source of any wave, then, is a vibration. And it is a *vibration* that propagates outward and thus constitutes the wave. If the source vibrates sinusoidally in SHM, then the wave itself—if the medium is perfectly elastic—will have a sinusoidal shape both in space and in time. (1) In space: if you take a picture of the wave in space at a given instant of time, the wave will have the shape of a sine or cosine as a function of position. (2) In time: if you look at the motion of the medium at one place over a long period of time—for example, if you look between two closely spaced posts of a pier or out of a ship's porthole as water waves pass by—the up-and-down motion of that small segment of water will be simple harmonic motion. The water moves up and down sinusoidally in time.

Some of the important quantities used to describe a periodic sinusoidal wave are shown in Fig. 15–3. The high points on a wave are called *crests*; the low points, *troughs*. The **amplitude**, A, is the maximum height of a crest, or depth of a trough, relative to the normal (or equilibrium) level. The total swing from a crest to a trough is twice the amplitude. The distance between two successive crests is called the **wavelength**, λ (the Greek letter lambda). The wavelength is also equal to the distance between *any* two successive identical points on the wave. The **frequency**, f, is the number of crests—or complete cycles—that pass a given point per unit time. The **period**, T, equals $1/f$ and is the time elapsed between two successive crests passing by the same point in space.

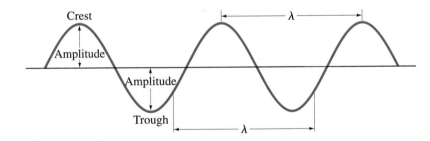

FIGURE 15–3 Characteristics of a single-frequency continuous wave moving through space.

The **wave velocity**, v, is the velocity at which wave crests (or any other part of the waveform) move forward. The wave velocity must be distinguished from the velocity of a particle of the medium itself as we saw in Example 15–1.

A wave crest travels a distance of one wavelength, λ, in a time equal to one period, T. Thus the wave velocity is $v = \lambda/T$. Then, since $1/T = f$,

$$v = \lambda f. \tag{15-1}$$

For example, suppose a wave has a wavelength of 5 m and a frequency of 3 Hz. Since three crests pass a given point per second, and the crests are 5 m apart, the first crest (or any other part of the wave) must travel a distance of 15 m during the 1 s. So the wave velocity is 15 m/s.

EXERCISE B You notice a water wave pass by the end of a pier with about 0.5 s between crests. Therefore (a) the frequency is 0.5 Hz; (b) the velocity is 0.5 m/s; (c) the wavelength is 0.5 m; (d) the period is 0.5 s.

15–2 Types of Waves: Transverse and Longitudinal

When a wave travels down a cord—say, from left to right as in Fig. 15–1—the particles of the cord vibrate up and down in a direction transverse (that is, perpendicular) to the motion of the wave itself. Such a wave is called a **transverse wave** (Fig. 15–4a). There exists another type of wave known as a **longitudinal wave**. In a longitudinal wave, the vibration of the particles of the medium is *along* the direction of the wave's motion. Longitudinal waves are readily formed on a stretched spring or Slinky by alternately compressing and expanding one end. This is shown in Fig. 15–4b, and can be compared to the transverse wave in Fig. 15–4a.

FIGURE 15–4 (a) Transverse wave; (b) longitudinal wave.

A series of compressions and expansions propagate along the spring. The *compressions* are those areas where the coils are momentarily close together. *Expansions* (sometimes called *rarefactions*) are regions where the coils are momentarily far apart. Compressions and expansions correspond to the crests and troughs of a transverse wave.

An important example of a longitudinal wave is a sound wave in air. A vibrating drumhead, for instance, alternately compresses and rarefies the air in contact with it, producing a longitudinal wave that travels outward in the air, as shown in Fig. 15–5.

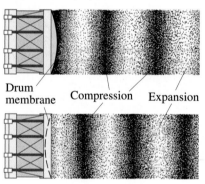

FIGURE 15–5 Production of a sound wave, which is longitudinal, shown at two moments in time about a half period $\left(\frac{1}{2}T\right)$ apart.

As in the case of transverse waves, each section of the medium in which a longitudinal wave passes oscillates over a very small distance, whereas the wave itself can travel large distances. Wavelength, frequency, and wave velocity all have meaning for a longitudinal wave. The wavelength is the distance between successive compressions (or between successive expansions), and frequency is the number of compressions that pass a given point per second. The wave velocity is the velocity with which each compression appears to move; it is equal to the product of wavelength and frequency, $v = \lambda f$ (Eq. 15–1).

A longitudinal wave can be represented graphically by plotting the density of air molecules (or coils of a Slinky) versus position at a given instant, as shown in Fig. 15–6. Such a graphical representation makes it easy to illustrate what is happening. Note that the graph looks much like a transverse wave.

FIGURE 15–6 (a) A longitudinal wave with (b) its graphical representation at a particular instant in time.

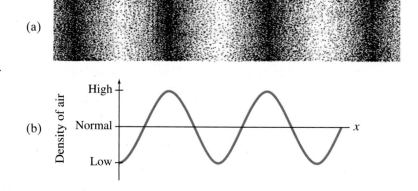

Velocity of Transverse Waves

The velocity of a wave depends on the properties of the medium in which it travels. The velocity of a transverse wave on a stretched string or cord, for example, depends on the tension in the cord, F_T, and on the mass per unit length of the cord, μ (the Greek letter mu, where here $\mu = m/\ell$). For waves of small amplitude, the relationship is

$$v = \sqrt{\frac{F_T}{\mu}}. \qquad \begin{bmatrix} \text{transverse} \\ \text{wave on a cord} \end{bmatrix} \quad \textbf{(15-2)}$$

Before giving a derivation of this formula, it is worth noting that at least qualitatively it makes sense on the basis of Newtonian mechanics. That is, we do expect the tension to be in the numerator and the mass per unit length in the denominator. Why? Because when the tension is greater, we expect the velocity to be greater since each segment of cord is in tighter contact with its neighbor. And, the greater the mass per unit length, the more inertia the cord has and the more slowly the wave would be expected to propagate.

EXERCISE C A wave starts at the left end of a long cord (see Fig. 15–1) when someone shakes the cord back and forth at the rate of 2.0 Hz. The wave is observed to move to the right at 4.0 m/s. If the frequency is increased from 2.0 to 3.0 Hz, the new speed of the wave is (a) 1.0 m/s, (b) 2.0 m/s, (c) 4.0 m/s, (d) 8.0 m/s, (e) 16.0 m/s.

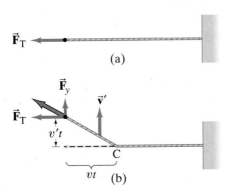

FIGURE 15–7 Diagram of simple wave pulse on a cord for derivation of Eq. 15–2. The vector shown in (b) as the resultant of $\vec{F}_T + \vec{F}_y$ has to be directed along the cord because the cord is flexible. (Diagram is not to scale: we assume $v' \ll v$; the upward angle of the cord is exaggerated for visibility.)

We can make a simple derivation of Eq. 15–2 using a simple model of a cord under a tension F_T as shown in Fig. 15–7a. The cord is pulled upward at a speed v' by the force F_y. As shown in Fig. 15–7b all points of the cord to the left of point C move upward at the speed v', and those to the right are still at rest. The speed of propagation, v, of this wave pulse is the speed of point C, the leading edge of the pulse. Point C moves to the right a distance vt in a time t, whereas the end of the cord moves upward a distance $v't$. By similar triangles we have the approximate relation

$$\frac{F_T}{F_y} = \frac{vt}{v't} = \frac{v}{v'},$$

which is accurate for small displacements ($v't \ll vt$) so that F_T does not change appreciably. As we saw in Chapter 9, the impulse given to an object is equal to its change in momentum. During the time t the total upward impulse is $F_y t = (v'/v)F_T t$. The change in momentum of the cord, Δp, is the mass of cord moving upward times its velocity. Since the upward moving segment of cord has mass equal to the mass per unit length μ times its length vt we have

$$F_y t = \Delta p$$

$$\frac{v'}{v} F_T t = (\mu v t)v'.$$

Solving for v we find $v = \sqrt{F_T/\mu}$ which is Eq. 15–2. Although it was derived for a special case, it is valid for any wave shape since other shapes can be considered to be made up of many tiny such lengths. But it is valid only for small displacements (as was our derivation). Experiment is in accord with this result derived from Newtonian mechanics.

EXAMPLE 15-2 Pulse on a wire. An 80.0-m-long, 2.10-mm-diameter copper wire is stretched between two poles. A bird lands at the center point of the wire, sending a small wave pulse out in both directions. The pulses reflect at the ends and arrive back at the bird's location 0.750 seconds after it landed. Determine the tension in the wire.

APPROACH From Eq. 15–2, the tension is given by $F_T = \mu v^2$. The speed v is distance divided by the time. The mass per unit length μ is calculated from the density of copper and the dimensions of the wire.

SOLUTION Each wave pulse travels 40.0 m to the pole and back again ($= 80.0$ m) in 0.750 s. Thus their speed is $v = (80.0 \text{ m})/(0.750 \text{ s}) = 107$ m/s. We take (Table 13–1) the density of copper as $8.90 \times 10^3 \text{ kg/m}^3$. The volume of copper in the wire is the cross-sectional area (πr^2) times the length ℓ, and the mass of the wire is the volume times the density: $m = \rho(\pi r^2)\ell$ for a wire of radius r and length ℓ. Then $\mu = m/\ell$ is

$$\mu = \rho\pi r^2\ell/\ell = \rho\pi r^2 = (8.90 \times 10^3 \text{ kg/m}^3)\pi(1.05 \times 10^{-3} \text{ m})^2 = 0.0308 \text{ kg/m}.$$

Thus, the tension is $F_T = \mu v^2 = (0.0308 \text{ kg/m})(107 \text{ m/s})^2 = 353$ N.

Velocity of Longitudinal Waves

The velocity of a longitudinal wave has a form similar to that for a transverse wave on a cord (Eq. 15–2); that is,

$$v = \sqrt{\frac{\text{elastic force factor}}{\text{inertia factor}}}.$$

In particular, for a longitudinal wave traveling down a long solid rod,

$$v = \sqrt{\frac{E}{\rho}}, \qquad \begin{bmatrix}\text{longitudinal} \\ \text{wave in a long rod}\end{bmatrix} \quad \textbf{(15–3)}$$

where E is the elastic modulus (Section 12–4) of the material and ρ is its density. For a longitudinal wave traveling in a liquid or gas,

$$v = \sqrt{\frac{B}{\rho}}, \qquad \begin{bmatrix}\text{longitudinal wave} \\ \text{in a fluid}\end{bmatrix} \quad \textbf{(15–4)}$$

where B is the bulk modulus (Section 12–4) and ρ again is the density.

PHYSICS APPLIED
Space perception by animals, using sound waves

EXAMPLE 15-3 Echolocation. Echolocation is a form of sensory perception used by animals such as bats, toothed whales, and dolphins. The animal emits a pulse of sound (a longitudinal wave) which, after reflection from objects, returns and is detected by the animal. Echolocation waves can have frequencies of about 100,000 Hz. (a) Estimate the wavelength of a sea animal's echolocation wave. (b) If an obstacle is 100 m from the animal, how long after the animal emits a wave is its reflection detected?

APPROACH We first compute the speed of longitudinal (sound) waves in sea water, using Eq. 15–4 and Tables 12–1 and 13–1. The wavelength is $\lambda = v/f$.

SOLUTION (a) The speed of longitudinal waves in sea water, which is slightly more dense than pure water, is

$$v = \sqrt{\frac{B}{\rho}} = \sqrt{\frac{2.0 \times 10^9 \text{ N/m}^2}{1.025 \times 10^3 \text{ kg/m}^3}} = 1.4 \times 10^3 \text{ m/s}.$$

Then, using Eq. 15–1, we find

$$\lambda = \frac{v}{f} = \frac{(1.4 \times 10^3 \text{ m/s})}{(1.0 \times 10^5 \text{ Hz})} = 14 \text{ mm}.$$

(b) The time required for the round-trip between the animal and the object is

$$t = \frac{\text{distance}}{\text{speed}} = \frac{2(100 \text{ m})}{1.4 \times 10^3 \text{ m/s}} = 0.14 \text{ s}.$$

NOTE We shall see later that waves can be used to "resolve" (or detect) objects only if the wavelength is comparable to or smaller than the object. Thus, a dolphin can resolve objects on the order of a centimeter or larger in size.

*Deriving Velocity of Wave in a Fluid

We now derive Eq. 15–4. Consider a wave pulse traveling in a fluid in a long tube, so that the wave motion is one dimensional. The tube is fitted with a piston at the end and is filled with a fluid which, at $t = 0$, is of uniform density ρ and at uniform pressure P_0, Fig. 15–8a. At this moment the piston is abruptly made to start moving to the right with speed v', compressing the fluid in front of it. In the (short) time t the piston moves a distance $v't$. The compressed fluid itself also moves with speed v', but the leading edge of the compressed region moves to the right at the characteristic speed v of compression waves in that fluid; we assume the wave speed v is much larger than the piston speed v'. The leading edge of the compression (which at $t = 0$ was at the piston face) thus moves a distance vt in time t as shown in Fig. 15–8b. Let the pressure in the compression be $P_0 + \Delta P$, which is ΔP higher than in the uncompressed fluid. To move the piston to the right requires an external force $(P_0 + \Delta P)S$ acting to the right, where S is the cross-sectional area of the tube. (S for "surface area"; we save A for amplitude.) The *net* force on the compressed region of the fluid is

$$F_{net} = (P_0 + \Delta P)S - P_0 S = S\,\Delta P$$

since the uncompressed fluid exerts a force $P_0 S$ to the left at the leading edge. Hence the impulse given to the compressed fluid, which equals its change in momentum, is

$$F_{net}\,t = \Delta m v'$$
$$S\,\Delta P\,t = (\rho S v t)v',$$

where $(\rho S v t)$ represents the mass of fluid which is given the speed v' (the compressed fluid of area S moves a distance vt, Fig. 15–8, so the volume moved is Svt). Hence we have

$$\Delta P = \rho v v'.$$

From the definition of the bulk modulus, B (Eq. 12–7):

$$B = -\frac{\Delta P}{\Delta V/V_0} = -\frac{\rho v v'}{\Delta V/V_0},$$

where $\Delta V/V_0$ is the fractional change in volume due to compression. The original volume of the compressed fluid is $V_0 = Svt$ (see Fig. 15–8), and it has been compressed by an amount $\Delta V = -Sv't$ (Fig. 15–8b). Thus

$$B = -\frac{\rho v v'}{\Delta V/V_0} = -\rho v v'\left(\frac{Svt}{-Sv't}\right) = \rho v^2,$$

and so

$$v = \sqrt{\frac{B}{\rho}},$$

which is what we set out to show, Eq. 15–4.

The derivation of Eq. 15–3 follows similar lines, but takes into account the expansion of the sides of a rod when the end of the rod is compressed.

Other Waves

Both transverse and longitudinal waves are produced when an **earthquake** occurs. The transverse waves that travel through the body of the Earth are called S waves (S for shear), and the longitudinal waves are called P waves (P for pressure) or *compression* waves. Both longitudinal and transverse waves can travel through a solid since the atoms or molecules can vibrate about their relatively fixed positions in any direction. But only longitudinal waves can propagate through a fluid, because any transverse motion would not experience any restoring force since a fluid is readily deformable. This fact was used by geophysicists to infer that a portion of the Earth's core must be liquid: after an earthquake, longitudinal waves are detected diametrically across the Earth, but not transverse waves.

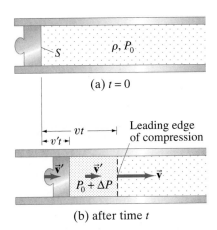

FIGURE 15–8 Determining the speed of a one-dimensional longitudinal wave in a fluid contained in a long narrow tube.

(A) **PHYSICS APPLIED**
Earthquake waves

FIGURE 15–9 A water wave is an example of a *surface wave*, which is a combination of transverse and longitudinal wave motions.

FIGURE 15–10 How a wave breaks. The green arrows represent the local velocity of water molecules.

Besides these two types of waves that can pass through the body of the Earth (or other substance), there can also be *surface waves* that travel along the boundary between two materials. A wave on water is actually a surface wave that moves on the boundary between water and air. The motion of each particle of water at the surface is circular or elliptical (Fig. 15–9), so it is a combination of transverse and longitudinal motions. Below the surface, there is also transverse plus longitudinal wave motion, as shown. At the bottom, the motion is only longitudinal. (When a wave approaches shore, the water drags at the bottom and is slowed down, while the crests move ahead at higher speed (Fig. 15–10) and "spill" over the top.)

Surface waves are also set up on the Earth when an earthquake occurs. The waves that travel along the surface are mainly responsible for the damage caused by earthquakes.

Waves which travel along a line in one dimension, such as transverse waves on a stretched string, or longitudinal waves in a rod or fluid-filled tube, are *linear* or *one-dimensional waves*. Surface waves, such as the water waves pictured at the start of this Chapter, are *two-dimensional waves*. Finally, waves that move out from a source in all directions, such as sound from a loudspeaker or earthquake waves through the Earth, are *three-dimensional waves*.

15–3 Energy Transported by Waves

Waves transport energy from one place to another. As waves travel through a medium, the energy is transferred as vibrational energy from particle to particle of the medium. For a sinusoidal wave of frequency f, the particles move in simple harmonic motion (Chapter 14) as a wave passes, and each particle has energy $E = \frac{1}{2}kA^2$ where A is the maximum displacement (amplitude) of its motion, either transversely or longitudinally (Eq. 14–10a). Using Eq. 14–7a we can write $k = 4\pi^2 mf^2$, where m is the mass of a particle (or small volume) of the medium. Then in terms of the frequency f and amplitude A,

$$E = \tfrac{1}{2}kA^2 = 2\pi^2 mf^2A^2.$$

For three-dimensional waves traveling in an elastic medium, the mass $m = \rho V$, where ρ is the density of the medium and V is the volume of a small slice of the medium. The volume $V = S\ell$ where S is the cross-sectional area through which the wave travels (Fig. 15–11), and we can write ℓ as the distance the wave travels in a time t as $\ell = vt$, where v is the speed of the wave. Thus $m = \rho V = \rho S\ell = \rho Svt$ and

$$E = 2\pi^2 \rho Svtf^2A^2. \tag{15–5}$$

FIGURE 15–11 Calculating the energy carried by a wave moving with velocity v.

From this equation we have the important result that *the energy transported by a wave is proportional to the square of the amplitude, and to the square of the frequency*. The average *rate* of energy transferred is the average power P:

$$\overline{P} = \frac{E}{t} = 2\pi^2 \rho Svf^2A^2. \tag{15–6}$$

Finally, the **intensity**, I, of a wave is defined as the average power transferred across unit area perpendicular to the direction of energy flow:

$$I = \frac{\overline{P}}{S} = 2\pi^2 v\rho f^2A^2. \tag{15–7}$$

If a wave flows out from the source in all directions, it is a three-dimensional wave. Examples are sound traveling in the open air, earthquake waves, and light waves.

If the medium is isotropic (same in all directions), the wave from a point source is a *spherical wave* (Fig. 15–12). As the wave moves outward, the energy it carries is spread over a larger and larger area since the surface area of a sphere of radius r is $4\pi r^2$. Thus the intensity of a wave is

$$I = \frac{\overline{P}}{S} = \frac{\overline{P}}{4\pi r^2}.$$

If the power output $\overline{P}$ is constant, then the intensity decreases as the inverse square of the distance from the source:

$$I \propto \frac{1}{r^2}. \qquad \text{[spherical wave]} \quad \textbf{(15–8a)}$$

If we consider two points at distances r_1 and r_2 from the source, as in Fig. 15–12, then $I_1 = \overline{P}/4\pi r_1^2$ and $I_2 = \overline{P}/4\pi r_2^2$, so

$$\frac{I_2}{I_1} = \frac{\overline{P}/4\pi r_2^2}{\overline{P}/4\pi r_1^2} = \frac{r_1^2}{r_2^2}. \qquad \textbf{(15–8b)}$$

Thus, for example, when the distance doubles $(r_2/r_1 = 2)$, then the intensity is reduced to $\frac{1}{4}$ of its earlier value: $I_2/I_1 = \left(\frac{1}{2}\right)^2 = \frac{1}{4}$.

The amplitude of a wave also decreases with distance. Since the intensity is proportional to the square of the amplitude (Eq. 15–7), $I \propto A^2$, the amplitude A must decrease as $1/r$, so that I can be proportional to $1/r^2$ (Eq. 15–8a). Hence

$$A \propto \frac{1}{r}.$$

To see this directly from Eq. 15–6, consider again two different distances from the source, r_1 and r_2. For constant power output, $S_1 A_1^2 = S_2 A_2^2$ where A_1 and A_2 are the amplitudes of the wave at r_1 and r_2, respectively. Since $S_1 = 4\pi r_1^2$ and $S_2 = 4\pi r_2^2$, we have $\left(A_1^2 r_1^2\right) = \left(A_2^2 r_2^2\right)$, or

$$\frac{A_2}{A_1} = \frac{r_1}{r_2}.$$

When the wave is twice as far from the source, the amplitude is half as large, and so on (ignoring damping due to friction).

EXAMPLE 15–4 **Earthquake intensity.** The intensity of an earthquake P wave traveling through the Earth and detected 100 km from the source is 1.0×10^6 W/m². What is the intensity of that wave if detected 400 km from the source?

APPROACH We assume the wave is spherical, so the intensity decreases as the square of the distance from the source.

SOLUTION At 400 km the distance is 4 times greater than at 100 km, so the intensity will be $\left(\frac{1}{4}\right)^2 = \frac{1}{16}$ of its value at 100 km, or $(1.0 \times 10^6$ W/m²$)/16 = 6.3 \times 10^4$ W/m².

NOTE Using Eq. 15–8b directly gives

$$I_2 = I_1 r_1^2/r_2^2 = (1.0 \times 10^6 \text{ W/m}^2)(100 \text{ km})^2/(400 \text{ km})^2 = 6.3 \times 10^4 \text{ W/m}^2.$$

The situation is different for a one-dimensional wave, such as a transverse wave on a string or a longitudinal wave pulse traveling down a thin uniform metal rod. The area remains constant, so the amplitude A also remains constant (ignoring friction). Thus the amplitude and the intensity do not decrease with distance.

In practice, frictional damping is generally present, and some of the energy is transformed into thermal energy. Thus the amplitude and intensity of a one-dimensional wave decrease with distance from the source. For a three-dimensional wave, the decrease will be greater than that discussed above, although the effect may often be small.

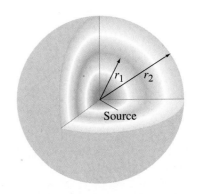

FIGURE 15–12 Wave traveling outward from a point source has spherical shape. Two different crests (or compressions) are shown, of radius r_1 and r_2.

15–4 Mathematical Representation of a Traveling Wave

Let us now consider a one-dimensional wave traveling along the x axis. It could be, for example, a transverse wave on a cord or a longitudinal wave traveling in a rod or in a fluid-filled tube. Let us assume the wave shape is sinusoidal and has a particular wavelength λ and frequency f. At $t = 0$, suppose the wave shape is given by

$$D(x) = A \sin \frac{2\pi}{\lambda} x,$$
(15–9)

FIGURE 15–13 A traveling wave. In time t, the wave moves a distance vt.

as shown by the solid curve in Fig. 15–13: $D(x)$ is the **displacement**[†] of the wave (be it a longitudinal or transverse wave) at position x, and A is the **amplitude** (maximum displacement) of the wave. This relation gives a shape that repeats itself every wavelength, which is needed so that the displacement is the same at $x = 0$, $x = \lambda$, $x = 2\lambda$, and so on (since $\sin 4\pi = \sin 2\pi = \sin 0$).

Now suppose the wave is moving to the right with velocity v. Then, after a time t, each part of the wave (indeed, the whole wave "shape") has moved to the right a distance vt; see the dashed curve in Fig. 15–13. Consider any point on the wave at $t = 0$: say, a crest which is at some position x. After a time t, that crest will have traveled a distance vt so its new position is a distance vt greater than its old position. To describe this same point on the wave shape, the argument of the sine function must be the same, so we replace x in Eq. 15–9 by $(x - vt)$:

1-D wave moving in positive x direction

$$D(x, t) = A \sin\left[\frac{2\pi}{\lambda} (x - vt)\right].$$
(15–10a)

Said another way, if you are riding on a crest, the argument of the sine function, $(2\pi/\lambda)(x - vt)$, remains the same ($= \pi/2, 5\pi/2$, and so on); as t increases, x must increase at the same rate so that $(x - vt)$ remains constant.

Equation 15–10a is the mathematical representation of a sinusoidal wave traveling along the x axis to the right (increasing x). It gives the displacement $D(x, t)$ of the wave at any chosen point x at any time t. The function $D(x, t)$ describes a curve that represents the actual shape of the wave in space at time t. Since $v = \lambda f$ (Eq. 15–1) we can write Eq. 15–10a in other ways that are often convenient:

1-D wave moving in positive x direction

$$D(x, t) = A \sin\left(\frac{2\pi x}{\lambda} - \frac{2\pi t}{T}\right),$$
(15–10b)

where $T = 1/f = \lambda/v$ is the period; and

$$D(x, t) = A \sin(kx - \omega t),$$
(15–10c)

where $\omega = 2\pi f = 2\pi/T$ is the angular frequency and

$$k = \frac{2\pi}{\lambda}$$
(15–11)

⚠ CAUTION

Don't confuse wave number k with spring constant k

is called the **wave number**. (Do not confuse the wave number k with the spring constant k; they are very different quantities.) All three forms, Eqs. 15–10a, b, and c, are equivalent; Eq. 15–10c is the simplest to write and is perhaps the most common. The quantity $(kx - \omega t)$, and its equivalent in the other two equations, is called the **phase** of the wave. The velocity v of the wave is often called the **phase velocity**, since it describes the velocity of the phase (or shape) of the wave and it can be written in terms of ω and k:

$$v = \lambda f = \left(\frac{2\pi}{k}\right)\left(\frac{\omega}{2\pi}\right) = \frac{\omega}{k}.$$
(15–12)

[†]Some books use $y(x)$ in place of $D(x)$. To avoid confusion, we reserve y (and z) for the coordinate positions of waves in two or three dimensions. Our $D(x)$ can stand for pressure (in longitudinal waves), position displacement (transverse mechanical waves), or—as we will see later—electric or magnetic fields (for electromagnetic waves).

For a wave traveling along the x axis to the left (decreasing values of x), we start again with Eq. 15–9 and note that the velocity is now $-v$. A particular point on the wave changes position by $-vt$ in a time t, so x in Eq. 15–9 must be replaced by $(x + vt)$. Thus, for a wave traveling to the left with velocity v,

$$D(x, t) = A \sin\left[\frac{2\pi}{\lambda}(x + vt)\right] \qquad \textbf{(15–13a)}$$

1-D wave moving in negative x direction

$$= A \sin\left(\frac{2\pi x}{\lambda} + \frac{2\pi t}{T}\right) \qquad \textbf{(15–13b)}$$

$$= A \sin(kx + \omega t). \qquad \textbf{(15–13c)}$$

In other words, we simply replace v in Eqs. 15–10 by $-v$.

Let us look at Eq. 15–13c (or, just as well, at Eq. 15–10c). At $t = 0$ we have

$$D(x, 0) = A \sin kx,$$

which is what we started with, a sinusoidal wave shape. If we look at the wave shape in space at a particular later time t_1, then we have

$$D(x, t_1) = A \sin(kx + \omega t_1).$$

That is, if we took a picture of the wave at $t = t_1$, we would see a sine wave with a phase constant ωt_1. Thus, for fixed $t = t_1$, the wave has a sinusoidal shape in space. On the other hand, if we consider a fixed point in space, say $x = 0$, we can see how the wave varies in time:

$$D(0, t) = A \sin \omega t$$

where we used Eq. 15–13c. This is just the equation for simple harmonic motion (Section 14–2). For any other fixed value of x, say $x = x_1$, $D = A \sin(\omega t + kx_1)$ which differs only by a phase constant kx_1. Thus, at any fixed point in space, the displacement undergoes the oscillations of simple harmonic motion in time. Equations 15–10 and 15–13 combine both these aspects to give us the representation for a **traveling sinusoidal wave** (also called a **harmonic wave**).

The argument of the sine in Eqs. 15–10 and 15–13 can in general contain a phase angle ϕ, which for Eq. 15–10c is

$$D(x, t) = A \sin(kx - \omega t + \phi),$$

to adjust for the position of the wave at $t = 0$, $x = 0$, just as in Section 14–2 (see Fig. 14–7). If the displacement is zero at $t = 0$, $x = 0$, as in Fig. 14–6 (or Fig. 15–13), then $\phi = 0$.

Now let us consider a general wave (or wave pulse) of any shape. If frictional losses are small, experiment shows that the wave maintains its shape as it travels. Thus we can make the same arguments as we did right after Eq. 15–9. Suppose our wave has some shape at $t = 0$, given by

$$D(x, 0) = D(x)$$

where $D(x)$ is the displacement of the wave at x and is not necessarily sinusoidal. At some later time, if the wave is traveling to the right along the x axis, the wave will have the same shape but all parts will have moved a distance vt where v is the phase velocity of the wave. Hence we must replace x by $x - vt$ to obtain the amplitude at time t:

$$D(x, t) = D(x - vt). \qquad \textbf{(15–14)}$$

Similarly, if the wave moves to the left, we must replace x by $x + vt$, so

$$D(x, t) = D(x + vt). \qquad \textbf{(15–15)}$$

Thus, any wave traveling along the x axis must have the form of Eq. 15–14 or 15–15.

EXERCISE D A wave is given by $D(x, t) = (5.0\,\text{mm}) \sin(2.0x - 20.0t)$ where x is in meters and t is in seconds. What is the speed of the wave? (a) $10\,\text{m/s}$, (b) $0.10\,\text{m/s}$, (c) $40\,\text{m/s}$, (d) $0.005\,\text{m/s}$, (e) $2.5 \times 10^{-4}\,\text{m/s}$.

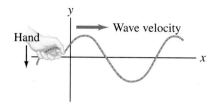

FIGURE 15–14 Example 15–5. The wave at $t = 0$ (the hand is falling). Not to scale.

EXAMPLE 15–5 **A traveling wave.** The left-hand end of a long horizontal stretched cord oscillates transversely in SHM with frequency $f = 250\,\text{Hz}$ and amplitude 2.6 cm. The cord is under a tension of 140 N and has a linear density $\mu = 0.12\,\text{kg/m}$. At $t = 0$, the end of the cord has an upward displacement of 1.6 cm and is falling (Fig. 15–14). Determine (a) the wavelength of waves produced and (b) the equation for the traveling wave.

APPROACH We first find the phase velocity of the transverse wave from Eq. 15–2; then $\lambda = v/f$. In (b), we need to find the phase ϕ using the initial conditions.

SOLUTION (a) The wave velocity is

$$v = \sqrt{\frac{F_T}{\mu}} = \sqrt{\frac{140\,\text{N}}{0.12\,\text{kg/m}}} = 34\,\text{m/s}.$$

Then

$$\lambda = \frac{v}{f} = \frac{34\,\text{m/s}}{250\,\text{Hz}} = 0.14\,\text{m} \quad \text{or} \quad 14\,\text{cm}.$$

(b) Let $x = 0$ at the left-hand end of the cord. The phase of the wave at $t = 0$ is not zero in general as was assumed in Eqs. 15–9, 10, and 13. The general form for a wave traveling to the right is

$$D(x, t) = A \sin(kx - \omega t + \phi),$$

where ϕ is the phase angle. In our case, the amplitude $A = 2.6\,\text{cm}$; and at $t = 0$, $x = 0$, we are given $D = 1.6\,\text{cm}$. Thus

$$1.6 = 2.6 \sin \phi,$$

so $\phi = \sin^{-1}(1.6/2.6) = 38° = 0.66\,\text{rad}$. We also have $\omega = 2\pi f = 1570\,\text{s}^{-1}$ and $k = 2\pi/\lambda = 2\pi/0.14\,\text{m} = 45\,\text{m}^{-1}$. Hence

$$D = (0.026\,\text{m}) \sin\left[(45\,\text{m}^{-1})x - (1570\,\text{s})t + 0.66\right]$$

which we can write more simply as

$$D = 0.026 \sin(45x - 1570t + 0.66),$$

and we specify clearly that D and x are in meters and t in seconds.

*15–5 The Wave Equation

Many types of waves satisfy an important general equation that is the equivalent of Newton's second law of motion for particles. This "equation of motion for a wave" is called the **wave equation**, and we derive it now for waves traveling on a stretched horizontal string.

We assume the amplitude of the wave is small compared to the wavelength so that each point on the string can be assumed to move only vertically and the tension in the string, F_T, does not vary during a vibration. We apply Newton's second law, $\Sigma F = ma$, to the vertical motion of a tiny section of the string as shown in Fig. 15–15. The amplitude of the wave is small, so the angles θ_1 and θ_2 that the string makes with the horizontal are small. The length of this section is then approximately Δx, and its mass is $\mu\,\Delta x$, where μ is the mass per unit length of the string. The net vertical force on this section of string is $F_T \sin \theta_2 - F_T \sin \theta_1$. So Newton's second law applied to the vertical (y) direction gives

$$\Sigma F_y = ma_y$$

$$F_T \sin \theta_2 - F_T \sin \theta_1 = (\mu\,\Delta x)\frac{\partial^2 D}{\partial t^2}. \tag{i}$$

FIGURE 15–15 Deriving the wave equation from Newton's second law: a segment of string under tension F_T.

We have written the acceleration as $a_y = \partial^2 D/\partial t^2$ since the motion is only vertical, and we use the partial derivative notation because the displacement D is a function of both x and t.

Because the angles θ_1 and θ_2 are assumed small, $\sin\theta \approx \tan\theta$ and $\tan\theta$ is equal to the slope s of the string at each point:

$$\sin\theta \;\approx\; \tan\theta \;=\; \frac{\partial D}{\partial x} \;=\; s.$$

Thus our equation (i) at the bottom of the previous page becomes

$$F_T(s_2 - s_1) \;=\; \mu\,\Delta x\,\frac{\partial^2 D}{\partial t^2}$$

or

$$F_T\frac{\Delta s}{\Delta x} \;=\; \mu\,\frac{\partial^2 D}{\partial t^2}, \tag{ii}$$

where $\Delta s = s_2 - s_1$ is the difference in the slope between the two ends of our tiny section. Now we take the limit of $\Delta x \to 0$, so that

$$F_T\lim_{\Delta x\to 0}\frac{\Delta s}{\Delta x} \;=\; F_T\frac{\partial s}{\partial x}$$

$$=\; F_T\frac{\partial}{\partial x}\left(\frac{\partial D}{\partial x}\right) \;=\; F_T\frac{\partial^2 D}{\partial x^2}$$

since the slope $s = \partial D/\partial x$, as we wrote above. Substituting this into the equation labeled (ii) above gives

$$F_T\frac{\partial^2 D}{\partial x^2} \;=\; \mu\,\frac{\partial^2 D}{\partial t^2}$$

or

$$\frac{\partial^2 D}{\partial x^2} \;=\; \frac{\mu}{F_T}\frac{\partial^2 D}{\partial t^2}.$$

We saw earlier in this Chapter (Eq. 15–2) that the velocity of waves on a string is given by $v = \sqrt{F_T/\mu}$, so we can write this last equation as

$$\frac{\partial^2 D}{\partial x^2} \;=\; \frac{1}{v^2}\frac{\partial^2 D}{\partial t^2}. \tag{15–16}$$

This is the **one-dimensional wave equation**, and it can describe not only small amplitude waves on a stretched string, but also small amplitude longitudinal waves (such as sound waves) in gases, liquids, and elastic solids, in which case D can refer to the pressure variations. In this case, the wave equation is a direct consequence of Newton's second law applied to a continuous elastic medium. The wave equation also describes electromagnetic waves for which D refers to the electric or magnetic field, as we shall see in Chapter 31. Equation 15–16 applies to waves traveling in one dimension only. For waves spreading out in three dimensions, the wave equation is the same, with the addition of $\partial^2 D/\partial y^2$ and $\partial^2 D/\partial z^2$ to the left side of Eq. 15–16.

The wave equation is a *linear* equation: the displacement D appears singly in each term. There are no terms that contain D^2, or $D(\partial D/\partial x)$, or the like in which D appears more than once. Thus, if $D_1(x,t)$ and $D_2(x,t)$ are two different solutions of the wave equation, then the linear combination

$$D_3(x,t) \;=\; aD_1(x,t) + bD_2(x,t),$$

where a and b are constants, is also a solution. This is readily seen by direct substitution into the wave equation. This is the essence of the *superposition principle*, which we discuss in the next Section. Basically it says that if two waves pass through the same region of space at the same time, the actual displacement is the sum of the separate displacements. For waves on a string, or for sound waves, this is valid only for small-amplitude waves. If the amplitude is not small enough, the equations for wave propagation may become nonlinear and the principle of superposition would not hold and more complicated effects may occur.

EXAMPLE 15–6 **Wave equation solution.** Verify that the sinusoidal wave of Eq. 15–10c, $D(x, t) = A \sin(kx - \omega t)$, satisfies the wave equation.

APPROACH We substitute Eq. 15–10c into the wave equation, Eq. 15–16.

SOLUTION We take the derivative of Eq. 15–10c twice with respect to t:

$$\frac{\partial D}{\partial t} = -\omega A \cos(kx - \omega t)$$

$$\frac{\partial^2 D}{\partial t^2} = -\omega^2 A \sin(kx - \omega t).$$

With respect to x, the derivatives are

$$\frac{\partial D}{\partial x} = kA \cos(kx - \omega t)$$

$$\frac{\partial^2 D}{\partial x^2} = -k^2 A \sin(kx - \omega t).$$

If we now divide the second derivatives we get

$$\frac{\partial^2 D/\partial t^2}{\partial^2 D/\partial x^2} = \frac{-\omega^2 A \sin(kx - \omega t)}{-k^2 A \sin(kx - \omega t)} = \frac{\omega^2}{k^2}.$$

From Eq. 15–12 we have $\omega^2/k^2 = v^2$, so we see that Eq. 15–10 does satisfy the wave equation (Eq. 15–16).

15–6 The Principle of Superposition

When two or more waves pass through the same region of space at the same time, it is found that for many waves *the actual displacement is the vector* (or *algebraic*) *sum of the separate displacements*. This is called the **principle of superposition**. It is valid for mechanical waves as long as the displacements are not too large and there is a linear relationship between the displacement and the restoring force of the oscillating medium.[†] If the amplitude of a mechanical wave, for example, is so large that it goes beyond the elastic region of the medium, and Hooke's law is no longer operative, the superposition principle is no longer accurate.[‡] For the most part, we will consider systems for which the superposition principle can be assumed to hold.

One result of the superposition principle is that if two waves pass through the same region of space, they continue to move independently of one another. You may have noticed, for example, that the ripples on the surface of water (two-dimensional waves) that form from two rocks striking the water at different places will pass through each other.

Figure 15–16 shows an example of the superposition principle. In this case there are three waves present, on a stretched string, each of different amplitude and frequency. At any time, such as at the instant shown, the actual amplitude at any position x is the algebraic sum of the amplitude of the three waves at that position. The actual wave is not a simple sinusoidal wave and is called a *composite* (or *complex*) *wave*. (Amplitudes are exaggerated in Fig. 15–16.)

It can be shown that any complex wave can be considered as being composed of many simple sinusoidal waves of different amplitudes, wavelengths, and frequencies. This is known as *Fourier's theorem*. A complex periodic wave of period T can be represented as a sum of pure sinusoidal terms whose frequencies are integral multiples of $f = 1/T$. If the wave is not periodic, the sum becomes an integral (called a *Fourier integral*). Although we will not go into the details here, we see the importance of considering sinusoidal waves (and simple harmonic motion): because any other wave shape can be considered a sum of such pure sinusoidal waves.

[†]For electromagnetic waves in vacuum, Chapter 31, the superposition principle always holds.

[‡]Intermodulation distortion in high-fidelity equipment is an example of the superposition principle not holding when two frequencies do not combine linearly in the electronics.

FIGURE 15–16 The superposition principle for one-dimensional waves. Composite wave formed from three sinusoidal waves of different amplitudes and frequencies $(f_0, 2f_0, 3f_0)$ at a certain instant in time. The amplitude of the composite wave at each point in space, at any time, is the algebraic sum of the amplitudes of the component waves. Amplitudes are shown exaggerated; for the superposition principle to hold, they must be small compared to the wavelengths.

CONCEPTUAL EXAMPLE 15–7 | **Making a square wave.** At $t = 0$, three waves are given by $D_1 = A \cos kx$, $D_2 = -\frac{1}{3} A \cos 3kx$, and $D_3 = \frac{1}{5} A \cos 5kx$, where $A = 1.0$ m and $k = 10$ m^{-1}. Plot the sum of the three waves from $x = -0.4$ m to $+0.4$ m. (These three waves are the first three Fourier components of a "square wave.")

RESPONSE The first wave, D_1, has amplitude of 1.0 m and wavelength $\lambda = 2\pi/k = (2\pi/10)$ m $= 0.628$ m. The second wave, D_2, has amplitude of 0.33 m and wavelength $\lambda = 2\pi/3k = (2\pi/30)$ m $= 0.209$ m. The third wave, D_3, has amplitude of 0.20 m and wavelength $\lambda = 2\pi/5k = (2\pi/50)$ m $= 0.126$ m. Each wave is plotted in Fig. 15–17a. The sum of the three waves is shown in Fig. 15–17b. The sum begins to resemble a "square wave," shown in blue in Fig. 15–17b.

When the restoring force is not precisely proportional to the displacement for mechanical waves in some continuous medium, the speed of sinusoidal waves depends on the frequency. The variation of speed with frequency is called **dispersion**. The different sinusoidal waves that compose a complex wave will travel with slightly different speeds in such a case. Consequently, a complex wave will change shape as it travels if the medium is "dispersive." A pure sine wave will not change shape under these conditions, however, except by the influence of friction or dissipative forces. If there is no dispersion (or friction), even a complex linear wave does not change shape.

15–7 Reflection and Transmission

When a wave strikes an obstacle, or comes to the end of the medium in which it is traveling, at least a part of the wave is reflected. You have probably seen water waves reflect off a rock or the side of a swimming pool. And you may have heard a shout reflected from a distant cliff—which we call an "echo."

A wave pulse traveling down a cord is reflected as shown in Fig. 15–18. The reflected pulse returns inverted as in Fig. 15–18a if the end of the cord is fixed; it returns right side up if the end is free as in Fig. 15–18b. When the end is fixed to a support, as in Fig. 15–18a, the pulse reaching that fixed end exerts a force (upward) on the support. The support exerts an equal but opposite force downward on the cord (Newton's third law). This downward force on the cord is what "generates" the inverted reflected pulse.

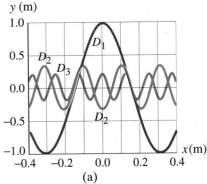

FIGURE 15–17 Example 15–7. Making a square wave.

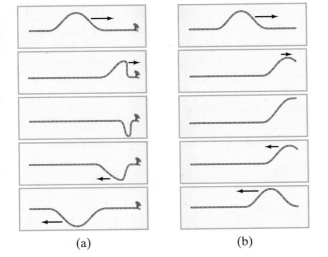

(a)   (b)

FIGURE 15–18 Reflection of a wave pulse on a cord lying on a table. (a) The end of the cord is fixed to a peg. (b) The end of the cord is free to move.

FIGURE 15–19 When a wave pulse traveling to the right along a thin cord (a) reaches a discontinuity where the cord becomes thicker and heavier, then part is reflected and part is transmitted (b).

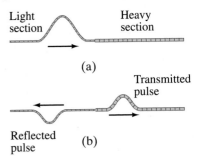

Consider next a pulse that travels down a cord which consists of a light section and a heavy section, as shown in Fig. 15–19. When the wave pulse reaches the boundary between the two sections, part of the pulse is reflected and part is transmitted, as shown. The heavier the second section of the cord, the less the energy that is transmitted. (When the second section is a wall or rigid support, very little is transmitted and most is reflected, as in Fig. 15–18a.) For a periodic wave, the frequency of the transmitted wave does not change across the boundary since the boundary point oscillates at that frequency. Thus if the transmitted wave has a lower speed, its wavelength is also less ($\lambda = v/f$).

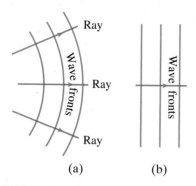

FIGURE 15–20 Rays, signifying the direction of motion, are always perpendicular to the wave fronts (wave crests). (a) Circular or spherical waves near the source. (b) Far from the source, the wave fronts are nearly straight or flat, and are called plane waves.

For a two- or three-dimensional wave, such as a water wave, we are concerned with **wave fronts**, by which we mean all the points along the wave forming the wave crest (what we usually refer to simply as a "wave" at the seashore). A line drawn in the direction of motion, perpendicular to the wave front, is called a **ray**, as shown in Fig. 15–20. Wave fronts far from the source have lost almost all their curvature (Fig. 15–20b) and are nearly straight, as ocean waves often are; they are then called **plane waves**.

For reflection of a two- or three-dimensional plane wave, as shown in Fig. 15–21, the angle that the incoming or *incident wave* makes with the reflecting surface is equal to the angle made by the reflected wave. This is the **law of reflection**:

the angle of reflection equals the angle of incidence.

The "angle of incidence" is defined as the angle (θ_i) the incident ray makes with the perpendicular to the reflecting surface (or the wave front makes with a tangent to the surface), and the "angle of reflection" is the corresponding angle (θ_r) for the reflected wave.

FIGURE 15–21
Law of reflection: $\theta_r = \theta_i$.

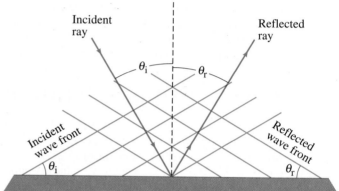

15–8 Interference

Interference refers to what happens when two waves pass through the same region of space at the same time. Consider, for example, the two wave pulses on a cord traveling toward each other as shown in Fig. 15–22. In Fig. 15–22a the two pulses have the same amplitude, but one is a crest and the other a trough; in Fig. 15–22b they are both crests. In both cases, the waves meet and pass right by each other. However, in the region where they overlap, the resultant displacement is the *algebraic sum of their separate displacements* (a crest is considered positive and a trough negative). This is another example of the principle of superposition. In Fig. 15–22a, the two waves have opposite displacements at the instant they pass one another, and they add to zero. The result is called **destructive interference**. In Fig. 15–22b, at the instant the two pulses overlap, they produce a resultant displacement that is greater than the displacement of either separate pulse, and the result is **constructive interference**.

FIGURE 15–22 Two wave pulses pass each other. Where they overlap, interference occurs: (a) destructive, and (b) constructive.

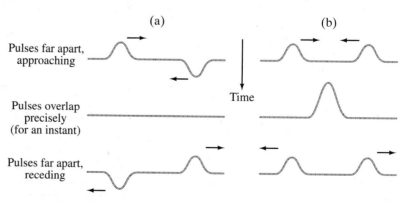

(a)   (b)

Pulses far apart, approaching

Time

Pulses overlap precisely (for an instant)

Pulses far apart, receding

[You may wonder where the energy is at the moment of destructive interference in Fig. 15–22a; the cord may be straight at this instant, but the central parts of it are still moving up or down.]

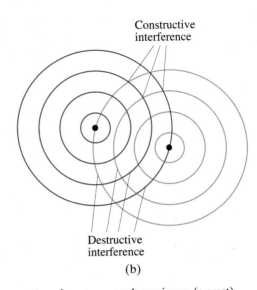

FIGURE 15-23 (a) Interference of water waves. (b) Constructive interference occurs where one wave's maximum (a crest) meets the other's maximum. Destructive interference ("flat water") occurs where one wave's maximum (a crest) meets the other's miminum (a trough).

When two rocks are thrown into a pond simultaneously, the two sets of circular waves interfere with one another as shown in Fig. 15-23a. In some areas of overlap, crests of one wave repeatedly meet crests of the other (and troughs meet troughs), Fig. 15-23b. Constructive interference is occuring at these points, and the water continuously oscillates up and down with greater amplitude than either wave separately. In other areas, destructive interference occurs where the water does not move up and down at all over time. This is where crests of one wave meet troughs of the other, and vice versa. Figure 15-24a shows the displacement of two identical waves graphically as a function of time, as well as their sum, for the case of constructive interference. For constructive interference (Fig. 15-24a), the two waves are **in phase**. At points where destructive interference occurs (Fig. 15-24b) crests of one wave repeatedly meet troughs of the other wave and the two waves are **out of phase** by one-half wavelength or 180°. The crests of one wave occur a half wavelength behind the crests of the other wave. The relative phase of the two water waves in Fig. 15-23 in most areas is intermediate between these two extremes, resulting in *partially* destructive interference, as illustrated in Fig. 15-24c. If the amplitudes of two interfering waves are not equal, fully destructive interference (as in Fig. 15-24b) does not occur.

FIGURE 15-24 Graphs showing two identical waves, and their sum, as a function of time at three locations. In (a) the two waves interfere constructively, in (b) destructively, and in (c) partially destructively.

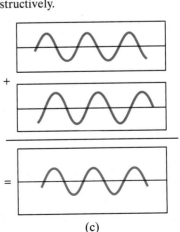

15–9 Standing Waves; Resonance

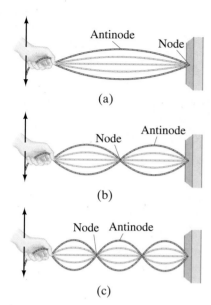

If you shake one end of a cord and the other end is kept fixed, a continuous wave will travel down to the fixed end and be reflected back, inverted, as we saw in Fig. 15–18a. As you continue to vibrate the cord, waves will travel in both directions, and the wave traveling along the cord, away from your hand, will interfere with the reflected wave coming back. Usually there will be quite a jumble. But if you vibrate the cord at just the right frequency, the two traveling waves will interfere in such a way that a large-amplitude **standing wave** will be produced, Fig. 15–25. It is called a "standing wave" because it does not appear to be traveling. The cord simply appears to have segments that oscillate up and down in a fixed pattern. The points of destructive interference, where the cord remains still at all times, are called **nodes**. Points of constructive interference, where the cord oscillates with maximum amplitude, are called **antinodes**. The nodes and antinodes remain in fixed positions for a particular frequency.

Standing waves can occur at more than one frequency. The lowest frequency of vibration that produces a standing wave gives rise to the pattern shown in Fig. 15–25a. The standing waves shown in Figs. 15–25b and 15–25c are produced at precisely twice and three times the lowest frequency, respectively, assuming the tension in the cord is the same. The cord can also vibrate with four loops (four antinodes) at four times the lowest frequency, and so on.

The frequencies at which standing waves are produced are the **natural frequencies** or **resonant frequencies** of the cord, and the different standing wave patterns shown in Fig. 15–25 are different "resonant modes of vibration." A standing wave on a cord is the result of the interference of two waves traveling in opposite directions. A standing wave can also be considered a vibrating object at resonance. Standing waves represent the same phenomenon as the resonance of a vibrating spring or pendulum, which we discussed in Chapter 14. However, a spring or pendulum has only one resonant frequency, whereas the cord has an infinite number of resonant frequencies, each of which is a whole-number multiple of the lowest resonant frequency.

Consider a string stretched between two supports that is plucked like a guitar or violin string, Fig. 15–26a. Waves of a great variety of frequencies will travel in both directions along the string, will be reflected at the ends, and will travel back in the opposite direction. Most of these waves interfere with each other and quickly die out. However, those waves that correspond to the resonant frequencies of the string will persist. The ends of the string, since they are fixed, will be nodes. There may be other nodes as well. Some of the possible resonant modes of vibration (standing waves) are shown in Fig. 15–26b. Generally, the motion will be a combination of these different resonant modes, but only those frequencies that correspond to a resonant frequency will be present.

FIGURE 15–25 Standing waves corresponding to three resonant frequencies.

FIGURE 15–26 (a) A string is plucked. (b) Only standing waves corresponding to resonant frequencies persist for long.

(a)

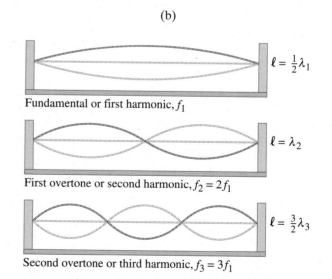

(b)

$\ell = \frac{1}{2}\lambda_1$

Fundamental or first harmonic, f_1

$\ell = \lambda_2$

First overtone or second harmonic, $f_2 = 2f_1$

$\ell = \frac{3}{2}\lambda_3$

Second overtone or third harmonic, $f_3 = 3f_1$

To determine the resonant frequencies, we first note that the wavelengths of the standing waves bear a simple relationship to the length ℓ of the string. The lowest frequency, called the **fundamental frequency**, corresponds to one antinode (or loop). And as can be seen in Fig. 15–26b, the whole length corresponds to one-half wavelength. Thus $\ell = \frac{1}{2}\lambda_1$, where λ_1 stands for the wavelength of the fundamental frequency. The other natural frequencies are called **overtones**; for a vibrating string they are whole-number (integral) multiples of the fundamental, and then are also called **harmonics**, with the fundamental being referred to as the **first harmonic**.[†] The next mode of vibration after the fundamental has two loops and is called the **second harmonic** (or first overtone), Fig. 15–26b. The length of the string ℓ at the second harmonic corresponds to one complete wavelength: $\ell = \lambda_2$. For the third and fourth harmonics, $\ell = \frac{3}{2}\lambda_3$, and $\ell = 2\lambda_4$, respectively, and so on. In general, we can write

$$\ell = \frac{n\lambda_n}{2}, \qquad \text{where } n = 1, 2, 3, \cdots.$$

The integer n labels the number of the harmonic: $n = 1$ for the fundamental, $n = 2$ for the second harmonic, and so on. We solve for λ_n and find

$$\lambda_n = \frac{2\ell}{n}, \qquad n = 1, 2, 3, \cdots. \qquad \left[\begin{array}{l}\text{string fixed} \\ \text{at both ends}\end{array}\right] \quad \textbf{(15–17a)}$$

To find the frequency f of each vibration we use Eq. 15–1, $f = v/\lambda$, and we see that

$$f_n = \frac{v}{\lambda_n} = n\frac{v}{2\ell} = nf_1, \qquad n = 1, 2, 3, \cdots, \qquad \textbf{(15–17b)}$$

where $f_1 = v/\lambda_1 = v/2\ell$ is the fundamental frequency. We see that each resonant frequency is an integer multiple of the fundamental frequency.

Because a standing wave is equivalent to two traveling waves moving in opposite directions, the concept of wave velocity still makes sense and is given by Eq. 15–2 in terms of the tension F_T in the string and its mass per unit length $(\mu = m/\ell)$. That is, $v = \sqrt{F_T/\mu}$ for waves traveling in both directions.

EXAMPLE 15–8 **Piano string.** A piano string is 1.10 m long and has a mass of 9.00 g. (a) How much tension must the string be under if it is to vibrate at a fundamental frequency of 131 Hz? (b) What are the frequencies of the first four harmonics?

APPROACH To determine the tension, we need to find the wave speed using Eq. 15–1 $(v = \lambda f)$, and then use Eq. 15–2, solving it for F_T.

SOLUTION (a) The wavelength of the fundamental is $\lambda = 2\ell = 2.20\,\text{m}$ (Eq. 15–17a with $n = 1$). The speed of the wave on the string is $v = \lambda f = (2.20\,\text{m})(131\,\text{s}^{-1}) = 288\,\text{m/s}$. Then we have (Eq. 15–2)

$$F_T = \mu v^2 = \frac{m}{\ell}v^2 = \left(\frac{9.00 \times 10^{-3}\,\text{kg}}{1.10\,\text{m}}\right)(288\,\text{m/s})^2 = 679\,\text{N}.$$

(b) The frequencies of the second, third, and fourth harmonics are two, three, and four times the fundamental frequency: 262, 393, and 524 Hz, respectively.

NOTE The speed of the wave on the string is *not* the same as the speed of the sound wave that the piano string produces in the air (as we shall see in Chapter 16).

A standing wave does appear to be standing in place (and a traveling wave appears to move). The term "standing" wave is also meaningful from the point of view of energy. Since the string is at rest at the nodes, no energy flows past these points. Hence the energy is not transmitted down the string but "stands" in place in the string.

Standing waves are produced not only on strings, but on any object that is struck, such as a drum membrane or an object made of metal or wood. The resonant frequencies depend on the dimensions of the object, just as for a string they depend on its length. Large objects have lower resonant frequencies than small objects. All musical instruments, from stringed instruments to wind instruments (in which a column of air vibrates as a standing wave) to drums and other percussion instruments, depend on standing waves to produce their particular musical sounds, as we shall see in Chapter 16.

[†]The term "harmonic" comes from music, because such integral multiples of frequencies "harmonize."

Mathematical Representation of a Standing Wave

As we saw, a standing wave can be considered to consist of two traveling waves that move in opposite directions. These can be written (see Eqs. 15–10c and 15–13c)

$$D_1(x, t) = A \sin(kx - \omega t) \quad \text{and} \quad D_2(x, t) = A \sin(kx + \omega t)$$

since, assuming no damping, the amplitudes are equal as are the frequencies and wavelengths. The sum of these two traveling waves produces a standing wave which can be written mathematically as

$$D = D_1 + D_2 = A[\sin(kx - \omega t) + \sin(kx + \omega t)].$$

From the trigonometric identity $\sin \theta_1 + \sin \theta_2 = 2 \sin \frac{1}{2}(\theta_1 + \theta_2) \cos \frac{1}{2}(\theta_1 - \theta_2)$, we can rewrite this as

$$D = 2A \sin kx \cos \omega t. \tag{15–18}$$

If we let $x = 0$ at the left-hand end of the string, then the right-hand end is at $x = \ell$ where ℓ is the length of the string. Since the string is fixed at its two ends (Fig. 15–26), $D(x, t)$ must be zero at $x = 0$ and at $x = \ell$. Equation 15–18 already satisfies the first condition ($D = 0$ at $x = 0$) and satisfies the second condition if $\sin k\ell = 0$ which means

$$k\ell = \pi, 2\pi, 3\pi, \cdots, n\pi, \cdots$$

where $n =$ an integer. Since $k = 2\pi/\lambda$, then $\lambda = 2\ell/n$, which is just Eq. 15–17a.

Equation 15–18, with the condition $\lambda = 2\ell/n$, is the mathematical representation of a standing wave. We see that a particle at any position x vibrates in simple harmonic motion (because of the factor $\cos \omega t$). All particles of the string vibrate with the same frequency $f = \omega/2\pi$, but the amplitude depends on x and equals $2A \sin kx$. (Compare this to a traveling wave for which all particles vibrate with the same amplitude.) The amplitude has a maximum, equal to $2A$, when $kx = \pi/2, 3\pi/2, 5\pi/2$, and so on—that is, at

$$x = \frac{\lambda}{4}, \frac{3\lambda}{4}, \frac{5\lambda}{4}, \cdots.$$

These are, of course, the positions of the antinodes (see Fig. 15–26).

EXAMPLE 15–9 **Wave forms.** Two waves traveling in opposite directions on a string fixed at $x = 0$ are described by the functions

$$D_1 = (0.20 \text{ m}) \sin(2.0x - 4.0t) \quad \text{and} \quad D_2 = (0.20 \text{ m}) \sin(2.0x + 4.0t)$$

(where x is in m, t is in s), and they produce a standing wave pattern. Determine (a) the function for the standing wave, (b) the maximum amplitude at $x = 0.45$ m, (c) where the other end is fixed ($x > 0$), (d) the maximum amplitude, and where it occurs.

APPROACH We use the principle of superposition to add the two waves. The given waves have the form we used to obtain Eq. 15–18, which we thus can use.

SOLUTION (a) The two waves are of the form $D = A \sin(kx \pm \omega t)$, so

$$k = 2.0 \text{ m}^{-1} \quad \text{and} \quad \omega = 4.0 \text{ s}^{-1}.$$

These combine to form a standing wave of the form of Eq. 15–18:

$$D = 2A \sin kx \cos \omega t = (0.40 \text{ m}) \sin(2.0x) \cos(4.0t),$$

where x is in meters and t in seconds.

(b) At $x = 0.45$ m,

$$D = (0.40 \text{ m}) \sin(0.90) \cos(4.0t) = (0.31 \text{ m}) \cos(4.0t).$$

The maximum amplitude at this point is $D = 0.31$ m and occurs when $\cos(4.0t) = 1$.

(c) These waves make a standing wave pattern, so both ends of the string must be nodes. Nodes occur every half wavelength, which for our string is

$$\frac{\lambda}{2} = \frac{1}{2} \frac{2\pi}{k} = \frac{\pi}{2.0} \text{ m} = 1.57 \text{ m}.$$

If the string includes only one loop, its length is $\ell = 1.57$ m. But without more information, it could be twice as long, $\ell = 3.14$ m, or any integral number times 1.57 m, and still provide a standing wave pattern, Fig. 15–27.

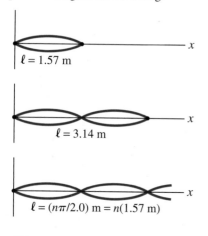

FIGURE 15–27 Example 15–9: possible lengths for the string.

$\ell = 1.57$ m

$\ell = 3.14$ m

$\ell = (n\pi/2.0)$ m $= n(1.57$ m)

(d) The nodes occur at $x = 0, x = 1.57\,\text{m}$, and, if the string is longer than $\ell = 1.57\,\text{m}$, at $x = 3.14\,\text{m}$, $4.71\,\text{m}$, and so on. The maximum amplitude (antinode) is 0.40 m [from part (b) above] and occurs midway between the nodes. For $\ell = 1.57\,\text{m}$, there is only one antinode, at $x = 0.79\,\text{m}$.

*15–10 Refraction†

When any wave strikes a boundary, some of the energy is reflected and some is transmitted or absorbed. When a two- or three-dimensional wave traveling in one medium crosses a boundary into a medium where its speed is different, the transmitted wave may move in a different direction than the incident wave, as shown in Fig. 15–28. This phenomenon is known as **refraction**. One example is a water wave; the velocity decreases in shallow water and the waves refract, as shown in Fig. 15–29 below. [When the wave velocity changes gradually, as in Fig. 15–29, without a sharp boundary, the waves change direction (refract) gradually.]

In Fig. 15–28, the velocity of the wave in medium 2 is less than in medium 1. In this case, the wave front bends so that it travels more nearly parallel to the boundary. That is, the *angle of refraction*, θ_r, is less than the *angle of incidence*, θ_i. To see why this is so, and to help us get a quantitative relation between θ_r and θ_i, let us think of each wave front as a row of soldiers. The soldiers are marching from firm ground (medium 1) into mud (medium 2) and hence are slowed down after the boundary. The soldiers that reach the mud first are slowed down first, and the row bends as shown in Fig. 15–30a. Let us consider the wave front (or row of soldiers) labeled A in Fig. 15–30b. In the same time t that A_1 moves a distance $\ell_1 = v_1 t$, we see that A_2 moves a distance $\ell_2 = v_2 t$. The two right triangles in Fig. 15–30b, shaded yellow and green, have the side labeled a in common. Thus

$$\sin\theta_1 = \frac{\ell_1}{a} = \frac{v_1 t}{a}$$

since a is the hypotenuse, and

$$\sin\theta_2 = \frac{\ell_2}{a} = \frac{v_2 t}{a}.$$

Dividing these two equations, we obtain the **law of refraction**:

$$\frac{\sin\theta_2}{\sin\theta_1} = \frac{v_2}{v_1}. \tag{15–19}$$

Since θ_1 is the angle of incidence (θ_i), and θ_2 is the angle of refraction (θ_r), Eq. 15–19 gives the quantitative relation between the two. If the wave were going in the opposite direction, the geometry would not change; only θ_1 and θ_2 would change roles: θ_1 would be the angle of refraction and θ_2 the angle of incidence. Clearly then, if the wave travels into a medium where it can move faster, it will bend the opposite way, $\theta_r > \theta_i$. We see from Eq. 15–19 that if the velocity increases, the angle increases, and vice versa.

Earthquake waves refract within the Earth as they travel through rock layers of different densities (and therefore the velocity is different) just as water waves do. Light waves refract as well, and when we discuss light, we shall find Eq. 15–19 very useful.

†This Section and the next are covered in more detail in Chapters 32 to 35 on optics.

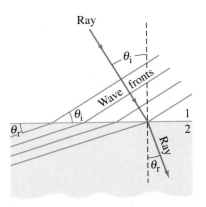

FIGURE 15–28 Refraction of waves passing a boundary.

FIGURE 15–29 Water waves refract gradually as they approach the shore, as their velocity decreases. There is no distinct boundary, as in Fig. 15–28, because the wave velocity changes gradually.

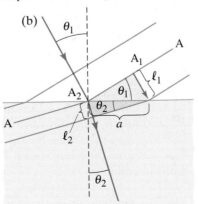

FIGURE 15–30 (a) Soldier analogy to derive (b) law of refraction for waves.

EXAMPLE 15–10 **Refraction of an earthquake wave.** An earthquake P wave passes across a boundary in rock where its velocity increases from 6.5 km/s to 8.0 km/s. If it strikes this boundary at 30°, what is the angle of refraction?

APPROACH We apply the law of refraction, Eq. 15–19, $\sin\theta_2/\sin\theta_1 = v_2/v_1$.

SOLUTION Since $\sin 30° = 0.50$, Eq. 15–19 yields

$$\sin\theta_2 = \frac{(8.0\text{ m/s})}{(6.5\text{ m/s})}(0.50) = 0.62.$$

So $\theta_2 = \sin^{-1}(0.62) = 38°$.

NOTE Be careful with angles of incidence and refraction. As we discussed in Section 15–7 (Fig. 15–21), these angles are between the wave front and the boundary line, or—equivalently—between the ray (direction of wave motion) and the line perpendicular to the boundary. Inspect Fig. 15–30b carefully.

*15–11 Diffraction

Waves spread as they travel. When they encounter an obstacle, they bend around it somewhat and pass into the region behind it, as shown in Fig. 15–31 for water waves. This phenomenon is called **diffraction**.

The amount of diffraction depends on the wavelength of the wave and on the size of the obstacle, as shown in Fig. 15–32. If the wavelength is much larger than the object, as with the grass blades of Fig. 15–32a, the wave bends around them almost as if they are not there. For larger objects, parts (b) and (c), there is more of a "shadow" region behind the obstacle where we might not expect the waves to penetrate—but they do, at least a little. Then notice in part (d), where the obstacle is the same as in part (c) but the wavelength is longer, that there is more diffraction into the shadow region. As a rule of thumb, *only if the wavelength is smaller than the size of the object will there be a significant shadow region*. This rule applies to *reflection* from an obstacle as well. Very little of a wave is reflected unless the wavelength is smaller than the size of the obstacle.

A rough guide to the amount of diffraction is

$$\theta(\text{radians}) \approx \frac{\lambda}{\ell},$$

where θ is roughly the angular spread of waves after they have passed through an opening of width ℓ or around an obstacle of width ℓ.

FIGURE 15–31 Wave diffraction. The waves are coming from the upper left. Note how the waves, as they pass the obstacle, bend around it, into the "shadow region" behind the obstacle.

FIGURE 15–32 Water waves passing objects of various sizes. Note that the longer the wavelength compared to the size of the object, the more diffraction there is into the "shadow region."

(a) Water waves passing blades of grass

(b) Stick in water

(c) Short-wavelength waves passing log

(d) Long-wavelength waves passing log

That waves can bend around obstacles, and thus can carry energy to areas behind obstacles, is very different from energy carried by material particles. A clear example is the following: if you are standing around a corner on one side of a building, you cannot be hit by a baseball thrown from the other side, but you can hear a shout or other sound because the sound waves diffract around the edges of the building.

Summary

Vibrating objects act as sources of **waves** that travel outward from the source. Waves on water and on a string are examples. The wave may be a **pulse** (a single crest) or it may be continuous (many crests and troughs).

The **wavelength** of a continuous wave is the distance between two successive crests (or any two identical points on the wave shape).

The **frequency** is the number of full wavelengths (or crests) that pass a given point per unit time.

The **wave velocity** (how fast a crest moves) is equal to the product of wavelength and frequency,

$$v = \lambda f. \tag{15-1}$$

The **amplitude** of a wave is the maximum height of a crest, or depth of a trough, relative to the normal (or equilibrium) level.

In a **transverse wave**, the oscillations are perpendicular to the direction in which the wave travels. An example is a wave on a string.

In a **longitudinal wave**, the oscillations are along (parallel to) the line of travel; sound is an example.

The velocity of both longitudinal and transverse waves in matter is proportional to the square root of an elastic force factor divided by an inertia factor (or density).

Waves carry energy from place to place without matter being carried. The **intensity** of a wave (energy transported across unit area per unit time) is proportional to the square of the amplitude of the wave.

For a wave traveling outward in three dimensions from a point source, the intensity (ignoring damping) decreases with the square of the distance from the source,

$$I \propto \frac{1}{r^2}. \tag{15-8a}$$

The amplitude decreases linearly with distance from the source.

A one-dimensional transverse wave traveling in a medium to the right along the x axis (x increasing) can be represented by a formula for the displacement of the medium from equilibrium at any point x as a function of time as

$$D(x, t) = A \sin\left[\left(\frac{2\pi}{\lambda}\right)(x - vt)\right] \tag{15-10a}$$

$$= A \sin(kx - \omega t) \tag{15-10c}$$

where

$$k = \frac{2\pi}{\lambda} \tag{15-11}$$

and

$$\omega = 2\pi f.$$

If a wave is traveling toward decreasing values of x,

$$D(x, t) = A \sin(kx + \omega t). \tag{15-13c}$$

[*Waves can be described by the **wave equation**, which in one dimension is $\partial^2 D/\partial x^2 = (1/v^2)\, \partial^2 D/\partial t^2$, Eq. 15–16.]

When two or more waves pass through the same region of space at the same time, the displacement at any given point will be the vector sum of the displacements of the separate waves. This is the **principle of superposition**. It is valid for mechanical waves if the amplitudes are small enough that the restoring force of the medium is proportional to displacement.

Waves reflect off objects in their path. When the **wave front** of a two- or three-dimensional wave strikes an object, the *angle* of *reflection* equals the *angle* of *incidence*, which is the **law of reflection**. When a wave strikes a boundary between two materials in which it can travel, part of the wave is reflected and part is transmitted.

When two waves pass through the same region of space at the same time, they **interfere**. From the superposition principle, the resultant displacement at any point and time is the sum of their separate displacements. This can result in **constructive interference**, **destructive interference**, or something in between depending on the amplitudes and relative phases of the waves.

Waves traveling on a cord of fixed length interfere with waves that have reflected off the end and are traveling back in the opposite direction. At certain frequencies, **standing waves** can be produced in which the waves seem to be standing still rather than traveling. The cord (or other medium) is vibrating as a whole. This is a resonance phenomenon, and the frequencies at which standing waves occur are called **resonant frequencies**. The points of destructive interference (no vibration) are called **nodes**. Points of constructive interference (maximum amplitude of vibration) are called **antinodes**. On a cord of length ℓ fixed at both ends, the wavelengths of standing waves are given by

$$\lambda_n = 2\ell/n \tag{15-17a}$$

where n is an integer.

[*Waves change direction, or **refract**, when traveling from one medium into a second medium where their speed is different. Waves spread, or **diffract**, as they travel and encounter obstacles. A rough guide to the amount of diffraction is $\theta \approx \lambda/\ell$, where λ is the wavelength and ℓ the width of an opening or obstacle. There is a significant "shadow region" only if the wavelength λ is smaller than the size of the obstacle.]

Questions

1. Is the frequency of a simple periodic wave equal to the frequency of its source? Why or why not?

2. Explain the difference between the speed of a transverse wave traveling down a cord and the speed of a tiny piece of the cord.

3. You are finding it a challenge to climb from one boat up onto a higher boat in heavy waves. If the climb varies from 2.5 m to 4.3 m, what is the amplitude of the wave? Assume the centers of the two boats are a half wavelength apart.

4. What kind of waves do you think will travel down a horizontal metal rod if you strike its end (*a*) vertically from above and (*b*) horizontally parallel to its length?

5. Since the density of air decreases with an increase in temperature, but the bulk modulus B is nearly independent of temperature, how would you expect the speed of sound waves in air to vary with temperature?

6. Describe how you could estimate the speed of water waves across the surface of a pond.

7. The speed of sound in most solids is somewhat greater than in air, yet the density of solids is much greater (10^3 to 10^4 times). Explain.

8. Give two reasons why circular water waves decrease in amplitude as they travel away from the source.

9. Two linear waves have the same amplitude and speed, and otherwise are identical, except one has half the wavelength of the other. Which transmits more energy? By what factor?

10. Will any function of $(x - vt)$—see Eq. 15–14—represent a wave motion? Why or why not? If not, give an example.

11. When a sinusoidal wave crosses the boundary between two sections of cord as in Fig. 15–19, the frequency does not change (although the wavelength and velocity do change). Explain why.

12. If a sinusoidal wave on a two-section cord (Fig. 15–19) is inverted upon reflection, does the transmitted wave have a longer or shorter wavelength?

13. Is energy always conserved when two waves interfere? Explain.

14. If a string is vibrating as a standing wave in three segments, are there any places you could touch it with a knife blade without disturbing the motion?

15. When a standing wave exists on a string, the vibrations of incident and reflected waves cancel at the nodes. Does this mean that energy was destroyed? Explain.

16. Can the amplitude of the standing waves in Fig. 15–25 be greater than the amplitude of the vibrations that cause them (up and down motion of the hand)?

17. When a cord is vibrated as in Fig. 15–25 by hand or by a mechanical oscillator, the "nodes" are not quite true nodes (at rest). Explain. [Hint: Consider damping and energy flow from hand or oscillator.]

*18. AM radio signals can usually be heard behind a hill, but FM often cannot. That is, AM signals bend more than FM. Explain. (Radio signals, as we shall see, are carried by electromagnetic waves whose wavelength for AM is typically 200 to 600 m and for FM about 3 m.)

*19. If we knew that energy was being transmitted from one place to another, how might we determine whether the energy was being carried by particles (material objects) or by waves?

Problems

15–1 and 15–2 Characteristics of Waves

1. (I) A fisherman notices that wave crests pass the bow of his anchored boat every 3.0 s. He measures the distance between two crests to be 8.0 m. How fast are the waves traveling?

2. (I) A sound wave in air has a frequency of 262 Hz and travels with a speed of 343 m/s. How far apart are the wave crests (compressions)?

3. (I) Calculate the speed of longitudinal waves in (a) water, (b) granite, and (c) steel.

4. (I) AM radio signals have frequencies between 550 kHz and 1600 kHz (kilohertz) and travel with a speed of 3.0×10^8 m/s. What are the wavelengths of these signals? On FM the frequencies range from 88 MHz to 108 MHz (megahertz) and travel at the same speed. What are their wavelengths?

5. (I) Determine the wavelength of a 5800-Hz sound wave traveling along an iron rod.

6. (II) A cord of mass 0.65 kg is stretched between two supports 8.0 m apart. If the tension in the cord is 140 N, how long will it take a pulse to travel from one support to the other?

7. (II) A 0.40-kg cord is stretched between two supports, 7.8 m apart. When one support is struck by a hammer, a transverse wave travels down the cord and reaches the other support in 0.85 s. What is the tension in the cord?

8. (II) A sailor strikes the side of his ship just below the surface of the sea. He hears the echo of the wave reflected from the ocean floor directly below 2.8 s later. How deep is the ocean at this point?

9. (II) A ski gondola is connected to the top of a hill by a steel cable of length 660 m and diameter 1.5 cm. As the gondola comes to the end of its run, it bumps into the terminal and sends a wave pulse along the cable. It is observed that it took 17 s for the pulse to return. (a) What is the speed of the pulse? (b) What is the tension in the cable?

10. (II) P and S waves from an earthquake travel at different speeds, and this difference helps locate the earthquake "epicenter" (where the disturbance took place). (a) Assuming typical speeds of 8.5 km/s and 5.5 km/s for P and S waves, respectively, how far away did the earthquake occur if a particular seismic station detects the arrival of these two types of waves 1.7 min apart? (b) Is one seismic station sufficient to determine the position of the epicenter? Explain.

11. (II) The wave on a string shown in Fig. 15–33 is moving to the right with a speed of 1.10 m/s. (a) Draw the shape of the string 1.00 s later and indicate which parts of the string are moving up and which down at that instant. (b) Estimate the vertical speed of point A on the string at the instant shown in the Figure.

FIGURE 15–33 Problem 11.

12. (II) A 5.0 kg ball hangs from a steel wire 1.00 mm in diameter and 5.00 m long. What would be the speed of a wave in the steel wire?

13. (II) Two children are sending signals along a cord of total mass 0.50 kg tied between tin cans with a tension of 35 N. It takes the vibrations in the string 0.50 s to go from one child to the other. How far apart are the children?

*14. (II) **Dimensional analysis.** Waves on the surface of the ocean do not depend significantly on the properties of water such as density or surface tension. The primary "return force" for water piled up in the wave crests is due to the gravitational attraction of the Earth. Thus the speed v (m/s) of ocean waves depends on the acceleration due to gravity g. It is reasonable to expect that v might also depend on water depth h and the wave's wavelength λ. Assume the wave speed is given by the functional form $v = C g^\alpha h^\beta \lambda^\gamma$, where α, β, γ, and C are numbers without dimension. (a) In deep water, the water deep below does not affect the motion of waves at the surface. Thus v should be independent of depth h (i.e., $\beta = 0$). Using only dimensional analysis (Section 1–7), determine the formula for the speed of surface waves in deep water. (b) In shallow water, the speed of surface waves is found experimentally to be independent of the wavelength (i.e., $\gamma = 0$). Using only dimensional analysis, determine the formula for the speed of waves in shallow water.

15–3 Energy Transported by Waves

15. (I) Two earthquake waves of the same frequency travel through the same portion of the Earth, but one is carrying 3.0 times the energy. What is the ratio of the amplitudes of the two waves?

16. (II) What is the ratio of (a) the intensities, and (b) the amplitudes, of an earthquake P wave passing through the Earth and detected at two points 15 km and 45 km from the source.

17. (II) Show that if damping is ignored, the amplitude A of circular water waves decreases as the square root of the distance r from the source: $A \propto 1/\sqrt{r}$.

18. (II) The intensity of an earthquake wave passing through the Earth is measured to be $3.0 \times 10^6 \, \text{J/m}^2 \cdot \text{s}$ at a distance of 48 km from the source. (a) What was its intensity when it passed a point only 1.0 km from the source? (b) At what rate did energy pass through an area of $2.0 \, \text{m}^2$ at 1.0 km?

19. (II) A small steel wire of diameter 1.0 mm is connected to an oscillator and is under a tension of 7.5 N. The frequency of the oscillator is 60.0 Hz and it is observed that the amplitude of the wave on the steel wire is 0.50 cm. (a) What is the power output of the oscillator, assuming that the wave is not reflected back? (b) If the power output stays constant but the frequency is doubled, what is the amplitude of the wave?

20. (II) Show that the intensity of a wave is equal to the energy density (energy per unit volume) in the wave times the wave speed.

21. (II) (a) Show that the average rate with which energy is transported along a cord by a mechanical wave of frequency f and amplitude A is

$$\overline{P} = 2\pi^2 \mu v f^2 A^2,$$

where v is the speed of the wave and μ is the mass per unit length of the cord. (b) If the cord is under a tension $F_T = 135 \, \text{N}$ and has mass per unit length 0.10 kg/m, what power is required to transmit 120-Hz transverse waves of amplitude 2.0 cm?

15–4 Mathematical Representation of Traveling Wave

22. (I) A transverse wave on a wire is given by $D(x, t) = 0.015 \sin(25x - 1200t)$ where D and x are in meters and t is in seconds. (a) Write an expression for a wave with the same amplitude, wavelength, and frequency but traveling in the opposite direction. (b) What is the speed of either wave?

23. (I) Suppose at $t = 0$, a wave shape is represented by $D = A \sin(2\pi x/\lambda + \phi)$; that is, it differs from Eq. 15–9 by a constant phase factor ϕ. What then will be the equation for a wave traveling to the left along the x axis as a function of x and t?

24. (II) A transverse traveling wave on a cord is represented by $D = 0.22 \sin(5.6x + 34t)$ where D and x are in meters and t is in seconds. For this wave determine (a) the wavelength, (b) frequency, (c) velocity (magnitude and direction), (d) amplitude, and (e) maximum and minimum speeds of particles of the cord.

25. (II) Consider the point $x = 1.00 \, \text{m}$ on the cord of Example 15–5. Determine (a) the maximum velocity of this point, and (b) its maximum acceleration. (c) What is its velocity and acceleration at $t = 2.50 \, \text{s}$?

26. (II) A transverse wave on a cord is given by $D(x, t) = 0.12 \sin(3.0x - 15.0t)$, where D and x are in m and t is in s. At $t = 0.20 \, \text{s}$, what are the displacement and velocity of the point on the cord where $x = 0.60 \, \text{m}$?

27. (II) A transverse wave pulse travels to the right along a string with a speed $v = 2.0 \, \text{m/s}$. At $t = 0$ the shape of the pulse is given by the function

$$D = 0.45 \cos(2.6x + 1.2),$$

where D and x are in meters. (a) Plot D vs. x at $t = 0$. (b) Determine a formula for the wave pulse at any time t assuming there are no frictional losses. (c) Plot $D(x, t)$ vs. x at $t = 1.0 \, \text{s}$. (d) Repeat parts (b) and (c) assuming the pulse is traveling to the left. Plot all 3 graphs on the same axes for easy comparison.

28. (II) A 524-Hz longitudinal wave in air has a speed of 345 m/s. (a) What is the wavelength? (b) How much time is required for the phase to change by 90° at a given point in space? (c) At a particular instant, what is the phase difference (in degrees) between two points 4.4 cm apart?

29. (II) Write the equation for the wave in Problem 28 traveling to the right, if its amplitude is 0.020 cm, and $D = -0.020 \, \text{cm}$, at $t = 0$ and $x = 0$.

30. (II) A sinusoidal wave traveling on a string in the negative x direction has amplitude 1.00 cm, wavelength 3.00 cm, and frequency 245 Hz. At $t = 0$, the particle of string at $x = 0$ is displaced a distance $D = 0.80 \, \text{cm}$ above the origin and is moving upward. (a) Sketch the shape of the wave at $t = 0$ and (b) determine the function of x and t that describes the wave.

*15–5 The Wave Equation

*31. (II) Determine if the function $D = A \sin kx \cos \omega t$ is a solution of the wave equation.

*32. (II) Show by direct substitution that the following functions satisfy the wave equation: (a) $D(x, t) = A \ln(x + vt)$; (b) $D(x, t) = (x - vt)^4$.

*33. (II) Show that the wave forms of Eqs. 15–13 and 15–15 satisfy the wave equation, Eq. 15–16.

*34. (II) Let two linear waves be represented by $D_1 = f_1(x, t)$ and $D_2 = f_2(x, t)$. If both these waves satisfy the wave equation (Eq. 15–16), show that any combination $D = C_1 D_1 + C_2 D_2$ does as well, where C_1 and C_2 are constants.

*35. (II) Does the function $D(x, t) = e^{-(kx - \omega t)^2}$ satisfy the wave equation? Why or why not?

*36. (II) In deriving Eq. 15–2, $v = \sqrt{F_T/\mu}$, for the speed of a transverse wave on a string, it was assumed that the wave's amplitude A is much less than its wavelength λ. Assuming a sinusoidal wave shape $D = A \sin(kx - \omega t)$, show via the partial derivative $v' = \partial D/\partial t$ that the assumption $A \ll \lambda$ implies that the maximum transverse speed v'_{max} of the string itself is much less than the wave velocity. If $A = \lambda/100$ determine the ratio v'_{max}/v.

15–7 Reflection and Transmission

37. (II) A cord has two sections with linear densities of 0.10 kg/m and 0.20 kg/m, Fig. 15–34. An incident wave, given by $D = (0.050 \, \text{m}) \sin(7.5x - 12.0t)$, where x is in meters and t in seconds, travels along the lighter cord. (a) What is the wavelength on the lighter section of the cord? (b) What is the tension in the cord? (c) What is the wavelength when the wave travels on the heavier section?

$\mu_1 = 0.10 \, \text{kg/m}$ $\mu_2 = 0.20 \, \text{kg/m}$

FIGURE 15–34
Problem 37. $D = (0.050 \, \text{m}) \sin(7.5 \, x - 12.0 \, t)$

38. (II) Consider a sine wave traveling down the stretched two-part cord of Fig. 15–19. Determine a formula (*a*) for the ratio of the speeds of the wave in the two sections, v_H/v_L, and (*b*) for the ratio of the wavelengths in the two sections. (The frequency is the same in both sections. Why?) (*c*) Is the wavelength larger in the heavier cord or the lighter?

39. (II) Seismic reflection prospecting is commonly used to map deeply buried formations containing oil. In this technique, a seismic wave generated on the Earth's surface (for example, by an explosion or falling weight) reflects from the subsurface formation and is detected upon its return to ground level. By placing ground-level detectors at a variety of locations relative to the source, and observing the variation in the source-to-detector travel times, the depth of the subsurface formation can be determined. (*a*) Assume a ground-level detector is placed a distance x away from a seismic-wave source and that a horizontal boundary between overlying rock and a subsurface formation exists at depth D (Fig. 15–35a). Determine an expression for the time t taken by the reflected wave to travel from source to detector, assuming the seismic wave propagates at constant speed v. (*b*) Suppose several detectors are placed along a line at different distances x from the source as in Fig. 15–35b. Then, when a seismic wave is generated, the different travel times t for each detector are measured. Starting with your result from part (*a*), explain how a graph of t^2 vs. x^2 can be used to determine D.

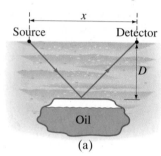

FIGURE 15–35
Problem 39.

40. (III) A cord stretched to a tension F_T consists of two sections (as in Fig. 15–19) whose linear densities are μ_1 and μ_2. Take $x = 0$ to be the point (a knot) where they are joined, with μ_1 referring to that section of cord to the left and μ_2 that to the right. A sinusoidal wave, $D = A \sin[k_1(x - v_1 t)]$, starts at the left end of the cord. When it reaches the knot, part of it is reflected and part is transmitted. Let the equation of the reflected wave be $D_R = A_R \sin[k_1(x + v_1 t)]$ and that for the transmitted wave be $D_T = A_T \sin[k_2(x - v_2 t)]$. Since the frequency must be the same in both sections, we have $\omega_1 = \omega_2$ or $k_1 v_1 = k_2 v_2$. (*a*) Because the cord is continuous, a point an infinitesimal distance to the left of the knot has the same displacement at any moment (due to incident plus reflected waves) as a point just to the right of the knot (due to the transmitted wave). Thus show that $A = A_T + A_R$. (*b*) Assuming that the slope $(\partial D/\partial x)$ of the cord just to the left of the knot is the same as the slope just to the right of the knot, show that the amplitude of the reflected wave is given by

$$A_R = \left(\frac{v_1 - v_2}{v_1 + v_2}\right)A = \left(\frac{k_2 - k_1}{k_2 + k_1}\right)A.$$

(*c*) What is A_T in terms of A?

15–8 Interference

41. (I) The two pulses shown in Fig. 15–36 are moving toward each other. (*a*) Sketch the shape of the string at the moment they directly overlap. (*b*) Sketch the shape of the string a few moments later. (*c*) In Fig. 15–22a, at the moment the pulses pass each other, the string is straight. What has happened to the energy at this moment?

FIGURE 15–36
Problem 41.

42. (II) Suppose two linear waves of equal amplitude and frequency have a phase difference ϕ as they travel in the same medium. They can be represented by

$$D_1 = A \sin(kx - \omega t)$$
$$D_2 = A \sin(kx - \omega t + \phi).$$

(*a*) Use the trigonometric identity $\sin \theta_1 + \sin \theta_2 = 2 \sin\frac{1}{2}(\theta_1 + \theta_2) \cos\frac{1}{2}(\theta_1 - \theta_2)$ to show that the resultant wave is given by

$$D = \left(2A \cos\frac{\phi}{2}\right) \sin\left(kx - \omega t + \frac{\phi}{2}\right).$$

(*b*) What is the amplitude of this resultant wave? Is the wave purely sinusoidal, or not? (*c*) Show that constructive interference occurs if $\phi = 0, 2\pi, 4\pi$, and so on, and destructive interference occurs if $\phi = \pi, 3\pi, 5\pi$, etc. (*d*) Describe the resultant wave, by equation and in words, if $\phi = \pi/2$.

15–9 Standing Waves; Resonance

43. (I) A violin string vibrates at 441 Hz when unfingered. At what frequency will it vibrate if it is fingered one-third of the way down from the end? (That is, only two-thirds of the string vibrates as a standing wave.)

44. (I) If a violin string vibrates at 294 Hz as its fundamental frequency, what are the frequencies of the first four harmonics?

45. (I) In an earthquake, it is noted that a footbridge oscillated up and down in a one-loop (fundamental standing wave) pattern once every 1.5 s. What other possible resonant periods of motion are there for this bridge? What frequencies do they correspond to?

46. (I) A particular string resonates in four loops at a frequency of 280 Hz. Name at least three other frequencies at which it will resonate.

47. (II) A cord of length 1.0 m has two equal-length sections with linear densities of 0.50 kg/m and 1.00 kg/m. The tension in the entire cord is constant. The ends of the cord are oscillated so that a standing wave is set up in the cord with a single node where the two sections meet. What is the ratio of the oscillatory frequencies?

48. (II) The velocity of waves on a string is 96 m/s. If the frequency of standing waves is 445 Hz, how far apart are the two adjacent nodes?

49. (II) If two successive harmonics of a vibrating string are 240 Hz and 320 Hz, what is the frequency of the fundamental?

50. (II) A guitar string is 90.0 cm long and has a mass of 3.16 g. From the bridge to the support post ($= \ell$) is 60.0 cm and the string is under a tension of 520 N. What are the frequencies of the fundamental and first two overtones?

51. (II) Show that the frequency of standing waves on a cord of length ℓ and linear density μ, which is stretched to a tension F_T, is given by

$$f = \frac{n}{2\ell}\sqrt{\frac{F_T}{\mu}}$$

where n is an integer.

52. (II) One end of a horizontal string of linear density 6.6×10^{-4} kg/m is attached to a small-amplitude mechanical 120-Hz oscillator. The string passes over a pulley, a distance $\ell = 1.50$ m away, and weights are hung from this end, Fig. 15–37. What mass m must be hung from this end of the string to produce (a) one loop, (b) two loops, and (c) five loops of a standing wave? Assume the string at the oscillator is a node, which is nearly true.

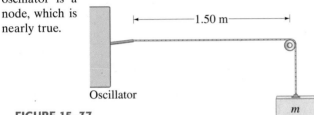

FIGURE 15–37
Problems 52 and 53.

53. (II) In Problem 52, Fig. 15–37, the length of the string may be adjusted by moving the pulley. If the hanging mass m is fixed at 0.070 kg, how many different standing wave patterns may be achieved by varying ℓ between 10 cm and 1.5 m?

54. (II) The displacement of a standing wave on a string is given by $D = 2.4 \sin(0.60x) \cos(42t)$, where x and D are in centimeters and t is in seconds. (a) What is the distance (cm) between nodes? (b) Give the amplitude, frequency, and speed of each of the component waves. (c) Find the speed of a particle of the string at $x = 3.20$ cm when $t = 2.5$ s.

55. (II) The displacement of a transverse wave traveling on a string is represented by $D_1 = 4.2 \sin(0.84x - 47t + 2.1)$, where D_1 and x are in cm and t in s. (a) Find an equation that represents a wave which, when traveling in the opposite direction, will produce a standing wave when added to this one. (b) What is the equation describing the standing wave?

56. (II) When you slosh the water back and forth in a tub at just the right frequency, the water alternately rises and falls at each end, remaining relatively calm at the center. Suppose the frequency to produce such a standing wave in a 45-cm-wide tub is 0.85 Hz. What is the speed of the water wave?

57. (II) A particular violin string plays at a frequency of 294 Hz. If the tension is increased 15%, what will the new frequency be?

58. (II) Two traveling waves are described by the functions

$$D_1 = A\sin(kx - \omega t)$$
$$D_2 = A\sin(kx + \omega t),$$

where $A = 0.15$ m, $k = 3.5$ m^{-1}, and $\omega = 1.8$ s^{-1}. (a) Plot these two waves, from $x = 0$ to a point $x(> 0)$ that includes one full wavelength. Choose $t = 1.0$ s. (b) Plot the sum of the two waves and identify the nodes and antinodes in the plot, and compare to the analytic (mathematical) representation.

59. (II) Plot the two waves given in Problem 58 and their sum, as a function of time from $t = 0$ to $t = T$ (one period). Choose (a) $x = 0$ and (b) $x = \lambda/4$. Interpret your results.

60. (II) A standing wave on a 1.64-m-long horizontal string displays three loops when the string vibrates at 120 Hz. The maximum swing of the string (top to bottom) at the center of each loop is 8.00 cm. (a) What is the function describing the standing wave? (b) What are the functions describing the two equal-amplitude waves traveling in opposite directions that make up the standing wave?

61. (II) On an electric guitar, a "pickup" under each string transforms the string's vibrations directly into an electrical signal. If a pickup is placed 16.25 cm from one of the fixed ends of a 65.00-cm-long string, which of the harmonics from $n = 1$ to $n = 12$ will not be "picked up" by this pickup?

62. (II) A 65-cm guitar string is fixed at both ends. In the frequency range between 1.0 and 2.0 kHz, the string is found to resonate only at frequencies 1.2, 1.5, and 1.8 kHz. What is the speed of traveling waves on this string?

63. (II) Two oppositely directed traveling waves given by $D_1 = (5.0\text{ mm})\cos[(2.0\text{ m}^{-1})x - (3.0\text{ rad/s})t]$ and $D_2 = (5.0\text{ mm})\cos[(2.0\text{ m}^{-1})x + (3.0\text{ rad/s})t]$ form a standing wave. Determine the position of nodes along the x axis.

64. (II) A wire is composed of aluminum with length $\ell_1 = 0.600$ m and mass per unit length $\mu_1 = 2.70$ g/m joined to a steel section with length $\ell_2 = 0.882$ m and mass per unit length $\mu_2 = 7.80$ g/m. This composite wire is fixed at both ends and held at a uniform tension of 135 N. Find the lowest frequency standing wave that can exist on this wire, assuming there is a node at the joint between aluminum and steel. How many nodes (including the two at the ends) does this standing wave possess?

*15–10 Refraction

*65. (I) An earthquake P wave traveling 8.0 km/s strikes a boundary within the Earth between two kinds of material. If it approaches the boundary at an incident angle of 52° and the angle of refraction is 31°, what is the speed in the second medium?

*66. (I) Water waves approach an underwater "shelf" where the velocity changes from 2.8 m/s to 2.5 m/s. If the incident wave crests make a 35° angle with the shelf, what will be the angle of refraction?

*67. (II) A sound wave is traveling in warm air (25°C) when it hits a layer of cold (−15°C) denser air. If the sound wave hits the cold air interface at an angle of 33°, what is the angle of refraction? The speed of sound as a function of temperature can be approximated by $v = (331 + 0.60\,T)$ m/s, where T is in °C.

*68. (II) Any type of wave that reaches a boundary beyond which its speed is increased, there is a maximum incident angle if there is to be a transmitted refracted wave. This maximum incident angle θ_{iM} corresponds to an angle of refraction equal to 90°. If $\theta_i > \theta_{iM}$, all the wave is reflected at the boundary and none is refracted, because this would correspond to $\sin\theta_r > 1$ (where θ_r is the angle of refraction), which is impossible. This phenomenon is referred to as *total internal reflection*. (a) Find a formula for θ_{iM} using the law of refraction, Eq. 15–19. (b) How far from the bank should a trout fisherman stand (Fig. 15–38) so trout won't be frightened by his voice? The speed of sound is about 343 m/s in air and 1440 m/s in water.

FIGURE 15–38
Problem 68b.

*69. (II) A longitudinal earthquake wave strikes a boundary between two types of rock at a 38° angle. As the wave crosses the boundary, the specific gravity of the rock changes from 3.6 to 2.8. Assuming that the elastic modulus is the same for both types of rock, determine the angle of refraction.

*15–11 Diffraction

*70. (II) A satellite dish is about 0.5 m in diameter. According to the user's manual, the dish has to be pointed in the direction of the satellite, but an error of about 2° to either side is allowed without loss of reception. Estimate the wavelength of the electromagnetic waves (speed $= 3 \times 10^8$ m/s) received by the dish.

General Problems

71. A sinusoidal traveling wave has frequency 880 Hz and phase velocity 440 m/s. (a) At a given time, find the distance between any two locations that correspond to a difference in phase of $\pi/6$ rad. (b) At a fixed location, by how much does the phase change during a time interval of 1.0×10^{-4} s?

72. When you walk with a cup of coffee (diameter 8 cm) at just the right pace of about one step per second, the coffee sloshes higher and higher in your cup until eventually it starts to spill over the top, Fig 15–39. Estimate the speed of the waves in the coffee.

FIGURE 15–39 Problem 72.

73. Two solid rods have the same bulk modulus but one is 2.5 times as dense as the other. In which rod will the speed of longitudinal waves be greater, and by what factor?

74. Two waves traveling along a stretched string have the same frequency, but one transports 2.5 times the power of the other. What is the ratio of the amplitudes of the two waves?

75. A bug on the surface of a pond is observed to move up and down a total vertical distance of 0.10 m, lowest to highest point, as a wave passes. (a) What is the amplitude of the wave? (b) If the amplitude increases to 0.15 m, by what factor does the bug's maximum kinetic energy change?

76. A guitar string is supposed to vibrate at 247 Hz, but is measured to actually vibrate at 255 Hz. By what percentage should the tension in the string be changed to get the frequency to the correct value?

77. An earthquake-produced surface wave can be approximated by a sinusoidal transverse wave. Assuming a frequency of 0.60 Hz (typical of earthquakes, which actually include a mixture of frequencies), what amplitude is needed so that objects begin to leave contact with the ground? [Hint: Set the acceleration $a > g$.]

78. A uniform cord of length ℓ and mass m is hung vertically from a support. (a) Show that the speed of transverse waves in this cord is $\sqrt{gh}$, where h is the height above the lower end. (b) How long does it take for a pulse to travel upward from one end to the other?

79. A transverse wave pulse travels to the right along a string with a speed $v = 2.4$ m/s. At $t = 0$ the shape of the pulse is given by the function

$$D = \frac{4.0 \text{ m}^3}{x^2 + 2.0 \text{ m}^2},$$

where D and x are in meters. (a) Plot D vs. x at $t = 0$ from $x = -10$ m to $x = +10$ m. (b) Determine a formula for the wave pulse at any time t assuming there are no frictional losses. (c) Plot $D(x, t)$ vs. x at $t = 1.00$ s. (d) Repeat parts (b) and (c) assuming the pulse is traveling to the left.

80. (a) Show that if the tension in a stretched string is changed by a small amount ΔF_T, the frequency of the fundamental is changed by an amount $\Delta f = \frac{1}{2}(\Delta F_T/F_T)f$. (b) By what percent must the tension in a piano string be increased or decreased to raise the frequency from 436 Hz to 442 Hz? (c) Does the formula in part (a) apply to the overtones as well?

81. Two strings on a musical instrument are tuned to play at 392 Hz (G) and 494 Hz (B). (a) What are the frequencies of the first two overtones for each string? (b) If the two strings have the same length and are under the same tension, what must be the ratio of their masses (m_G/m_A)? (c) If the strings, instead, have the same mass per unit length and are under the same tension, what is the ratio of their lengths (ℓ_G/ℓ_A)? (d) If their masses and lengths are the same, what must be the ratio of the tensions in the two strings?

82. The ripples in a certain groove 10.8 cm from the center of a 33-rpm phonograph record have a wavelength of 1.55 mm. What will be the frequency of the sound emitted?

83. A 10.0-m-long wire of mass 152 g is stretched under a tension of 255 N. A pulse is generated at one end, and 20.0 ms later a second pulse is generated at the opposite end. Where will the two pulses first meet?

84. A wave with a frequency of 220 Hz and a wavelength of 10.0 cm is traveling along a cord. The maximum speed of particles on the cord is the same as the wave speed. What is the amplitude of the wave?

85. A string can have a "free" end if that end is attached to a ring that can slide without friction on a vertical pole (Fig. 15–40). Determine the wavelengths of the resonant vibrations of such a string with one end fixed and the other free.

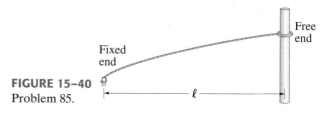

FIGURE 15–40 Problem 85.

86. A highway overpass was observed to resonate as one full loop $(\frac{1}{2}\lambda)$ when a small earthquake shook the ground vertically at 3.0 Hz. The highway department put a support at the center of the overpass, anchoring it to the ground as shown in Fig. 15–41. What resonant frequency would you now expect for the overpass? It is noted that earthquakes rarely do significant shaking above 5 or 6 Hz. Did the modifications do any good? Explain.

Before modification

Added support

After modification

FIGURE 15–41 Problem 86.

87. Figure 15–42 shows the wave shape at two instants of time for a sinusoidal wave traveling to the right. What is the mathematical representation of this wave?

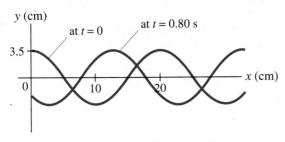

FIGURE 15–42 Problem 87.

88. Estimate the average power of a water wave when it hits the chest of an adult standing in the water at the seashore. Assume that the amplitude of the wave is 0.50 m, the wavelength is 2.5 m, and the period is 4.0 s.

89. A tsunami of wavelength 215 km and velocity 550 km/h travels across the Pacific Ocean. As it approaches Hawaii, people observe an unusual decrease of sea level in the harbors. Approximately how much time do they have to run to safety? (In the absence of knowledge and warning, people have died during tsunamis, some of them attracted to the shore to see stranded fishes and boats.)

90. Two wave pulses are traveling in opposite directions with the same speed of 7.0 cm/s as shown in Fig. 15–43. At $t = 0$, the leading edges of the two pulses are 15 cm apart. Sketch the wave pulses at $t = 1.0, 2.0$ and 3.0 s.

Distance (cm)

FIGURE 15–43 Problem 90.

91. For a spherical wave traveling uniformly away from a point source, show that the displacement can be represented by

$$D = \left(\frac{A}{r}\right)\sin(kr - \omega t),$$

where r is the radial distance from the source and A is a constant.

92. What frequency of sound would have a wavelength the same size as a 1.0-m-wide window? (The speed of sound is 344 m/s at 20°C.) What frequencies would diffract through the window?

*Numerical/Computer

***93.** (II) Consider a wave generated by the periodic vibration of a source and given by the expression $D(x, t) = A\sin^2 k(x - ct)$, where x represents position (in meters), t represents time (in seconds), and c is a positive constant. We choose $A = 5.0$ m and $c = 0.50$ m/s. Use a spreadsheet to make a graph with three curves of $D(x, t)$ from $x = -5.0$ m to $+5.0$ m in steps of 0.050 m at times $t = 0.0, 1.0,$ and 2.0 s. Determine the speed, direction of motion, period, and wavelength of the wave.

***94.** (II) The displacement of a bell-shaped wave pulse is described by a relation that involves the exponential function:

$$D(x, t) = Ae^{-\alpha(x-vt)^2}$$

where the constants $A = 10.0$ m, $\alpha = 2.0\,\text{m}^{-2}$, and $v = 3.0$ m/s. (a) Over the range $-10.0\,\text{m} \le x \le +10.0\,\text{m}$, use a graphing calculator or computer program to plot $D(x, t)$ at each of the three times $t = 0$, $t = 1.0$, and $t = 2.0$ s. Do these three plots demonstrate the wave-pulse shape shifting along the x axis by the expected amount over the span of each 1.0-s interval? (b) Repeat part (a) but assume $D(x, t) = Ae^{-\alpha(x+vt)^2}$.

Answers to Exercises

A: (c).

B: (d).

C: (c).

D: (a).

"If music be the food of physics, play on." [See Shakespeare, *Twelfth Night*, line 1.]

Stringed instruments depend on transverse standing waves on strings to produce their harmonious sounds. The sound of wind instruments originates in longitudinal standing waves of an air column. Percussion instruments create more complicated standing waves.

Besides examining sources of sound, we also study the decibel scale of sound level, sound wave interference and beats, the Doppler effect, shock waves and sonic booms, and ultrasound imaging.

C H A P T E R

16

Sound

CONTENTS

CHAPTER-OPENING QUESTION—Guess now!

A pianist plays the note "middle C." The sound is made by the vibration of the piano string and is propagated outward as a vibration of the air (which can reach your ear). Comparing the vibration on the string to the vibration in the air, which of the following is true?

(a) The vibration on the string and the vibration in the air have the same wavelength.
(b) They have the same frequency.
(c) They have the same speed.
(d) Neither wavelength, frequency, nor speed are the same in the air as on the string.

Sound is associated with our sense of hearing and, therefore, with the physiology of our ears and the psychology of our brain, which interprets the sensations that reach our ears. The term *sound* also refers to the physical sensation that stimulates our ears: namely, longitudinal waves.

We can distinguish three aspects of any sound. First, there must be a *source* for a sound; as with any mechanical wave, the source of a sound wave is a vibrating object. Second, the energy is transferred from the source in the form of longitudinal sound *waves*. And third, the sound is *detected* by an ear or by a microphone. We start this Chapter by looking at some aspects of sound waves themselves.

16–1 Characteristics of Sound

We saw in Chapter 15, Fig. 15–5, how a vibrating drumhead produces a sound wave in air. Indeed, we usually think of sound waves traveling in the air, for normally it is the vibrations of the air that force our eardrums to vibrate. But sound waves can also travel in other materials.

Two stones struck together under water can be heard by a swimmer beneath the surface, for the vibrations are carried to the ear by the water. When you put your ear flat against the ground, you can hear an approaching train or truck. In this case the ground does not actually touch your eardrum, but the longitudinal wave transmitted by the ground is called a sound wave just the same, for its vibrations cause the outer ear and the air within it to vibrate. Sound cannot travel in the absence of matter. For example, a bell ringing inside an evacuated jar cannot be heard, nor does sound travel through the empty reaches of outer space.

The **speed of sound** is different in different materials. In air at 0°C and 1 atm, sound travels at a speed of 331 m/s. We saw in Eq. 15–4 $\left(v = \sqrt{B/\rho}\right)$ that the speed depends on the elastic modulus, B, and the density, ρ, of the material. Thus for helium, whose density is much less than that of air but whose elastic modulus is not greatly different, the speed is about three times as great as in air. In liquids and solids, which are much less compressible and therefore have much greater elastic moduli, the speed is larger still. The speed of sound in various materials is given in Table 16–1. The values depend somewhat on temperature, but this is significant mainly for gases. For example, in air at normal (ambient) temperatures, the speed increases approximately 0.60 m/s for each Celsius degree increase in temperature:

$$v \approx (331 + 0.60\,T)\ \text{m/s}, \qquad \text{[speed of sound in air]}$$

where T is the temperature in °C. Unless stated otherwise, we will assume in this Chapter that $T = 20°C$, so that[†] $v = \left[331 + (0.60)(20)\right]$ m/s = 343 m/s.

TABLE 16–1 Speed of Sound in Various Materials (20°C and 1 atm)

Material	Speed (m/s)
Air	343
Air (0°C)	331
Helium	1005
Hydrogen	1300
Water	1440
Sea water	1560
Iron and steel	≈ 5000
Glass	≈ 4500
Aluminum	≈ 5100
Hardwood	≈ 4000
Concrete	≈ 3000

CONCEPTUAL EXAMPLE 16–1 **Distance from a lightning strike.** A rule of thumb that tells how close lightning has struck is, "one mile for every five seconds before the thunder is heard." Explain why this works, noting that the speed of light is so high (3×10^8 m/s, almost a million times faster than sound) that the time for light to travel to us is negligible compared to the time for the sound.

RESPONSE The speed of sound in air is about 340 m/s, so to travel 1 km = 1000 m takes about 3 seconds. One mile is about 1.6 kilometers, so the time for the thunder to travel a mile is about $(1.6)(3) \approx 5$ seconds.

PHYSICS APPLIED

How far away is the lightning?

| **EXERCISE A** What would be the rule used in Example 16–1 in terms of kilometers?

Two aspects of any sound are immediately evident to a human listener: "loudness" and "pitch." Each refers to a sensation in the consciousness of the listener. But to each of these subjective sensations there corresponds a physically measurable quantity. **Loudness** is related to the intensity (energy per unit time crossing unit area) in the sound wave, and we shall discuss it in Section 16–3.

The **pitch** of a sound refers to whether it is high, like the sound of a piccolo or violin, or low, like the sound of a bass drum or string bass. The physical quantity that determines pitch is the frequency, as was first noted by Galileo. The lower the frequency, the lower the pitch; the higher the frequency, the higher the pitch.[‡] The best human ears can respond to frequencies from about 20 Hz to almost 20,000 Hz. (Recall that 1 Hz is 1 cycle per second.) This frequency range is called the **audible range**. These limits vary somewhat from one individual to another. One general trend is that as people age, they are less able to hear high frequencies, so the high-frequency limit may be 10,000 Hz or less.

[†]We treat the 20°C ("room temperature") as accurate to 2 significant figures.

[‡]Although pitch is determined mainly by frequency, it also depends to a slight extent on loudness. For example, a very loud sound may seem slightly lower in pitch than a quiet sound of the same frequency.

Do not confuse
ultrasonic (high frequency)
with supersonic (high speed)

PHYSICS APPLIED

Autofocusing camera

FIGURE 16–1 Example 16–2. Autofocusing camera emits an ultrasonic pulse. Solid lines represent the wave front of the outgoing wave pulse moving to the right; dashed lines represent the wave front of the pulse reflected off the person's face, returning to the camera. The time information allows the camera mechanism to adjust the lens to focus at the proper distance.

FIGURE 16–2 Longitudinal sound wave traveling to the right, and its graphical representation in terms of pressure.

Sound waves whose frequencies are outside the audible range may reach the ear, but we are not generally aware of them. Frequencies above 20,000 Hz are called **ultrasonic** (do not confuse with *supersonic*, which is used for an object moving with a speed faster than the speed of sound). Many animals can hear ultrasonic frequencies; dogs, for example, can hear sounds as high as 50,000 Hz, and bats can detect frequencies as high as 100,000 Hz. Ultrasonic waves have many useful applications in medicine and other fields, which we discuss later in this Chapter.

EXAMPLE 16–2 **Autofocusing with sound waves.** Older autofocusing cameras determine the distance by emitting a pulse of very high frequency (ultrasonic) sound that travels to the object being photographed, and include a sensor that detects the returning reflected sound, as shown in Fig. 16–1. To get an idea of the time sensitivity of the detector, calculate the travel time of the pulse for an object (*a*) 1.0 m away, and (*b*) 20 m away.

APPROACH If we assume the temperature is about 20°C, then the speed of sound is 343 m/s. Using this speed v and the total distance d back and forth in each case, we can obtain the time ($v = d/t$).

SOLUTION (*a*) The pulse travels 1.0 m to the object and 1.0 m back, for a total of 2.0 m. We solve for t in $v = d/t$:

$$t = \frac{d}{v} = \frac{2.0 \text{ m}}{343 \text{ m/s}} = 0.0058 \text{ s} = 5.8 \text{ ms}.$$

(*b*) The total distance now is $2 \times 20 \text{ m} = 40 \text{ m}$, so

$$t = \frac{40 \text{ m}}{343 \text{ m/s}} = 0.12 \text{ s} = 120 \text{ ms}.$$

NOTE Newer autofocus cameras use infrared light ($v = 3 \times 10^8 \text{ m/s}$) instead of ultrasound, and/or a digital sensor array that detects light intensity differences between adjacent receptors as the lens is automatically moved back and forth, choosing the lens position that provides maximum intensity differences (sharpest focus).

Sound waves whose frequencies are below the audible range (that is, less than 20 Hz) are called **infrasonic**. Sources of infrasonic waves include earthquakes, thunder, volcanoes, and waves produced by vibrating heavy machinery. This last source can be particularly troublesome to workers, for infrasonic waves—even though inaudible—can cause damage to the human body. These low-frequency waves act in a resonant fashion, causing motion and irritation of the body's organs.

16–2 Mathematical Representation of Longitudinal Waves

In Section 15–4, we saw that a one-dimensional sinusoidal wave traveling along the x axis can be represented by the relation (Eq. 15–10c)

$$D = A \sin(kx - \omega t), \tag{16–1}$$

where D is the displacement of the wave at position x and time t, and A is its *amplitude* (maximum value). The wave number k is related to the wavelength λ by $k = 2\pi/\lambda$, and $\omega = 2\pi f$ where f is the frequency. For a transverse wave—such as a wave on a string—the displacement D is perpendicular to the direction of wave propagation along the x axis. But for a longitudinal wave the displacement D is *along the direction of wave propagation*. That is, D is parallel to x and represents the displacement of a tiny volume element of the medium from its equilibrium position.

Longitudinal (sound) waves can also be considered from the point of view of variations in pressure rather than displacement. Indeed, longitudinal waves are often called **pressure waves**. The pressure variation is usually easier to measure than the displacement (see Example 16–7). As can be seen in Fig. 16–2, in a wave "compression" (where molecules are closest together), the pressure is higher than normal, whereas in an expansion (or rarefaction) the pressure is less than normal.

Figure 16–3 shows a graphical representation of a sound wave in air in terms of (a) displacement and (b) pressure. Note that the displacement wave is a quarter wavelength, or 90° ($\pi/2$ rad), out of phase with the pressure wave: where the pressure is a maximum or minimum, the displacement from equilibrium is zero; and where the pressure variation is zero, the displacement is a maximum or minimum.

Pressure Wave Derivation

Let us now derive the mathematical representation of the pressure variation in a traveling longitudinal wave. From the definition of the bulk modulus, B (Eq. 12–7),

$$\Delta P = -B(\Delta V/V),$$

where ΔP represents the pressure difference from the normal pressure P_0 (no wave present) and $\Delta V/V$ is the fractional change in volume of the medium due to the pressure change ΔP. The negative sign reflects the fact that the volume decreases ($\Delta V < 0$) if the pressure is increased. Consider now a layer of fluid through which the longitudinal wave is passing (Fig. 16–4). If this layer has thickness Δx and area S, then its volume is $V = S\,\Delta x$. As a result of pressure variation in the wave, the volume will change by an amount $\Delta V = S\,\Delta D$, where ΔD is the change in thickness of this layer as it compresses or expands. (Remember that D represents the displacement of the medium.) Thus we have

$$\Delta P = -B\frac{S\,\Delta D}{S\,\Delta x}.$$

To be precise, we take the limit of $\Delta x \to 0$, so we obtain

$$\Delta P = -B\frac{\partial D}{\partial x}, \tag{16–2}$$

where we use the partial derivative notation since D is a function of both x and t. If the displacement D is sinusoidal as given by Eq. 16–1, then we have from Eq. 16–2 that

$$\Delta P = -(BAk)\cos(kx - \omega t). \tag{16–3}$$

(Here A is the displacement amplitude, not area which is S.) Thus the pressure varies sinusoidally as well, but is out of phase from the displacement by 90° or a quarter wavelength, as in Fig. 16–3. The quantity BAk is called the **pressure amplitude**, ΔP_M. It represents the maximum and minimum amounts by which the pressure varies from the normal ambient pressure. We can thus write

$$\Delta P = -\Delta P_M \cos(kx - \omega t), \tag{16–4}$$

where, using $v = \sqrt{B/\rho}$ (Eq. 15–4), and $k = \omega/v = 2\pi f/v$ (Eq. 15–12), then

$$\begin{aligned}\Delta P_M &= BAk \\ &= \rho v^2\, Ak \\ &= 2\pi\rho v\, Af. \end{aligned} \tag{16–5}$$

16–3 Intensity of Sound: Decibels

Loudness is a sensation in the consciousness of a human being and is related to a physically measurable quantity, the **intensity** of the wave. Intensity is defined as the energy transported by a wave per unit time across a unit area perpendicular to the energy flow. As we saw in Chapter 15, intensity is proportional to the square of the wave amplitude. Intensity has units of power per unit area, or watts/meter^2 (W/m^2).

The human ear can detect sounds with an intensity as low as 10^{-12} W/m^2 and as high as 1 W/m^2 (and even higher, although above this it is painful). This is an incredibly wide range of intensity, spanning a factor of 10^{12} from lowest to highest. Presumably because of this wide range, what we perceive as loudness is not directly proportional to the intensity. To produce a sound that sounds about twice as loud requires a sound wave that has about 10 times the intensity. This is roughly valid at any sound level for frequencies near the middle of the audible range. For example, a sound wave of intensity 10^{-2} W/m^2 sounds to an average human being like it is about twice as loud as one whose intensity is 10^{-3} W/m^2, and four times as loud as 10^{-4} W/m^2.

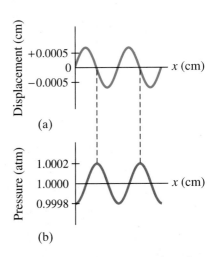

FIGURE 16–3 Representation of a sound wave in space at a given instant in terms of (a) displacement, and (b) pressure.

FIGURE 16–4 Longitudinal wave in a fluid moves to the right. A thin layer of fluid, in a thin cylinder of area S and thickness Δx, changes in volume as a result of pressure variation as the wave passes. At the moment shown, the pressure will increase as the wave moves to the right, so the thickness of our layer will decrease, by an amount ΔD.

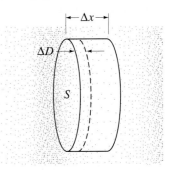

PHYSICS APPLIED
Wide range of human hearing

Sound Level

Because of this relationship between the subjective sensation of loudness and the physically measurable quantity "intensity," sound intensity levels are usually specified on a logarithmic scale. The unit on this scale is a **bel**, after the inventor Alexander Graham Bell, or much more commonly, the **decibel** (dB), which is $\frac{1}{10}$ bel (10 dB = 1 bel). The **sound level**, β, of any sound is defined in terms of its intensity, I, as

$$\beta \text{ (in dB)} = 10 \log \frac{I}{I_0}, \tag{16–6}$$

where I_0 is the intensity of a chosen reference level, and the logarithm is to the base 10. I_0 is usually taken as the minimum intensity audible to a good ear—the "threshold of hearing," which is $I_0 = 1.0 \times 10^{-12} \text{ W/m}^2$. Thus, for example, the sound level of a sound whose intensity $I = 1.0 \times 10^{-10} \text{ W/m}^2$ will be

$$\beta = 10 \log \left(\frac{1.0 \times 10^{-10} \text{ W/m}^2}{1.0 \times 10^{-12} \text{ W/m}^2} \right) = 10 \log 100 = 20 \text{ dB},$$

since log 100 is equal to 2.0. (Appendix A has a brief review of logarithms.) Notice that the sound level at the threshold of hearing is 0 dB. That is, $\beta = 10 \log 10^{-12}/10^{-12} = 10 \log 1 = 0$ since $\log 1 = 0$. Notice too that an increase in intensity by a factor of 10 corresponds to a sound level increase of 10 dB. An increase in intensity by a factor of 100 corresponds to a sound level increase of 20 dB. Thus a 50-dB sound is 100 times more intense than a 30-dB sound, and so on.

Intensities and sound levels for a number of common sounds are listed in Table 16–2.

<table>
<tr><td>⚠ CAUTION</td></tr>
<tr><td>0 dB does not mean zero intensity</td></tr>
</table>

TABLE 16–2
Intensity of Various Sounds

Source of the Sound	Sound Level (dB)	Intensity (W/m²)
Jet plane at 30 m	140	100
Threshold of pain	120	1
Loud rock concert	120	1
Siren at 30 m	100	1×10^{-2}
Truck traffic	90	1×10^{-3}
Busy street traffic	80	1×10^{-4}
Noisy restaurant	70	1×10^{-5}
Talk, at 50 cm	65	3×10^{-6}
Quiet radio	40	1×10^{-8}
Whisper	30	1×10^{-9}
Rustle of leaves	10	1×10^{-11}
Threshold of hearing	0	1×10^{-12}

EXAMPLE 16–3 Sound intensity on the street. At a busy street corner, the sound level is 75 dB. What is the intensity of sound there?

APPROACH We have to solve Eq. 16–6 for intensity I, remembering that $I_0 = 1.0 \times 10^{-12} \text{ W/m}^2$.

SOLUTION From Eq. 16–6

$$\log \frac{I}{I_0} = \frac{\beta}{10},$$

so

$$\frac{I}{I_0} = 10^{\beta/10}.$$

With $\beta = 75$, then

$$I = I_0 10^{\beta/10} = \left(1.0 \times 10^{-12} \text{ W/m}^2\right)\left(10^{7.5}\right) = 3.2 \times 10^{-5} \text{ W/m}^2.$$

NOTE Recall that $x = \log y$ is the same as $y = 10^x$ (Appendix A).

<table>
<tr><td>✸ PHYSICS APPLIED</td></tr>
<tr><td>Loudspeaker response (± 3 dB)</td></tr>
</table>

EXAMPLE 16–4 Loudspeaker response. A high-quality loudspeaker is advertised to reproduce, at full volume, frequencies from 30 Hz to 18,000 Hz with uniform sound level ± 3 dB. That is, over this frequency range, the sound level output does not vary by more than 3 dB for a given input level. By what factor does the intensity change for the maximum change of 3 dB in output sound level?

APPROACH Let us call the average intensity I_1 and the average sound level β_1. Then the maximum intensity, I_2, corresponds to a level $\beta_2 = \beta_1 + 3$ dB. We then use the relation between intensity and sound level, Eq. 16–6.

SOLUTION Equation 16–6 gives

$$\beta_2 - \beta_1 = 10 \log \frac{I_2}{I_0} - 10 \log \frac{I_1}{I_0}$$

$$3\,\text{dB} = 10\left(\log \frac{I_2}{I_0} - \log \frac{I_1}{I_0} \right)$$

$$= 10 \log \frac{I_2}{I_1}$$

because $(\log a - \log b) = \log a/b$ (see Appendix A). This last equation gives

$$\log \frac{I_2}{I_1} = 0.30,$$

or

$$\frac{I_2}{I_1} = 10^{0.30} = 2.0.$$

So $\pm 3\,\text{dB}$ corresponds to a doubling or halving of the intensity.

It is worth noting that a sound-level difference of 3 dB (which corresponds to a doubled intensity, as we just saw) corresponds to only a very small change in the subjective sensation of apparent loudness. Indeed, the average human can distinguish a difference in sound level of only about 1 or 2 dB.

EXERCISE B If an increase of 3 dB means "twice as intense," what does an increase of 6 dB mean?

CONCEPTUAL EXAMPLE 16–5 **Trumpet players.** A trumpeter plays at a sound level of 75 dB. Three equally loud trumpet players join in. What is the new sound level?

RESPONSE The intensity of four trumpets is four times the intensity of one trumpet $(= I_1)$ or $4I_1$. The sound level of the four trumpets would be

$$\beta = 10 \log \frac{4I_1}{I_0} = 10 \log 4 + 10 \log \frac{I_1}{I_0}$$

$$= 6.0\,\text{dB} + 75\,\text{dB} = 81\,\text{dB}.$$

EXERCISE C From Table 16–2, we see that ordinary conversation corresponds to a sound level of about 65 dB. If two people are talking at once, the sound level is (a) 65 dB, (b) 68 dB, (c) 75 dB, (d) 130 dB, (e) 62 dB.

Normally, the loudness or intensity of a sound decreases as you get farther from the source of the sound. In interior rooms, this effect is altered because of reflections from the walls. However, if a source is in the open so that sound can radiate out freely in all directions, the intensity decreases as the inverse square of the distance,

$$I \propto \frac{1}{r^2},$$

as we saw in Section 15–3. Over large distances, the intensity decreases faster than $1/r^2$ because some of the energy is transferred into irregular motion of air molecules. This loss happens more for higher frequencies, so any sound of mixed frequencies will be less "bright" at a distance.

FIGURE 16–5 Example 16–6. Airport worker with sound-intensity-reducing ear covers (headphones).

EXAMPLE 16–6 Airplane roar. The sound level measured 30 m from a jet plane is 140 dB. What is the sound level at 300 m? (Ignore reflections from the ground.)

APPROACH Given the sound level, we can determine the intensity at 30 m using Eq. 16–6. Because intensity decreases as the square of the distance, ignoring reflections, we can find I at 300 m and again apply Eq. 16–6 to obtain the sound level.

SOLUTION The intensity I at 30 m is

$$140 \text{ dB} = 10 \log\left(\frac{I}{10^{-12} \text{ W/m}^2}\right)$$

or

$$14 = \log\left(\frac{I}{10^{-12} \text{ W/m}^2}\right).$$

We raise both sides of this equation to the power 10 (recall $10^{\log x} = x$) and have

$$10^{14} = \frac{I}{10^{-12} \text{ W/m}^2},$$

so $I = (10^{14})(10^{-12} \text{ W/m}^2) = 10^2 \text{ W/m}^2$. At 300 m, 10 times as far, the intensity will be $\left(\frac{1}{10}\right)^2 = 1/100$ as much, or 1 W/m^2. Hence, the sound level is

$$\beta = 10 \log\left(\frac{1 \text{ W/m}^2}{10^{-12} \text{ W/m}^2}\right) = 120 \text{ dB}.$$

Even at 300 m, the sound is at the threshold of pain. This is why workers at airports wear ear covers to protect their ears from damage (Fig. 16–5).

NOTE Here is a simpler approach that avoids Eq. 16–6. Because the intensity decreases as the square of the distance, at 10 times the distance the intensity decreases by $\left(\frac{1}{10}\right)^2 = \frac{1}{100}$. We can use the result that 10 dB corresponds to an intensity change by a factor of 10 (see text just before Example 16–3). Then an intensity change by a factor of 100 corresponds to a sound-level change of $(2)(10 \text{ dB}) = 20 \text{ dB}$. This confirms our result above: $140 \text{ dB} - 20 \text{ dB} = 120 \text{ dB}$.

Intensity Related to Amplitude

The intensity I of a wave is proportional to the square of the wave amplitude, as we saw in Chapter 15. We can therefore relate the amplitude quantitatively to the intensity I or level β, as the following Example shows.

EXAMPLE 16–7 How tiny the displacement is. (*a*) Calculate the displacement of air molecules for a sound having a frequency of 1000 Hz at the threshold of hearing. (*b*) Determine the maximum pressure variation in such a sound wave.

APPROACH In Section 15–3 we found a relation between intensity I and displacement amplitude A of a wave, Eq. 15–7. The amplitude of oscillation of air molecules is what we want to solve for, given the intensity. The pressure is found from Eq. 16–5.

SOLUTION (*a*) At the threshold of hearing, $I = 1.0 \times 10^{-12} \text{ W/m}^2$ (Table 16–2). We solve for the amplitude A in Eq. 15–7:

$$A = \frac{1}{\pi f} \sqrt{\frac{I}{2\rho v}}$$

$$= \frac{1}{(3.14)(1.0 \times 10^3 \text{ s}^{-1})} \sqrt{\frac{1.0 \times 10^{-12} \text{ W/m}^2}{(2)(1.29 \text{ kg/m}^3)(343 \text{ m/s})}}$$

$$= 1.1 \times 10^{-11} \text{ m},$$

where we have taken the density of air to be 1.29 kg/m^3 and the speed of sound in air (assumed 20°C) as 343 m/s.

NOTE We see how incredibly sensitive the human ear is: it can detect displacements of air molecules which are actually less than the diameter of atoms (about 10^{-10} m).

(*b*) Now we are dealing with sound as a pressure wave (Section 16–2). From Eq. 16–5,

$$\Delta P_M = 2\pi\rho v A f$$

$$= 2\pi(1.29 \text{ kg/m}^3)(343 \text{ m/s})(1.1 \times 10^{-11} \text{ m})(1.0 \times 10^3 \text{ s}^{-1}) = 3.1 \times 10^{-5} \text{ Pa}$$

or 3.1×10^{-10} atm. Again we see that the human ear is incredibly sensitive.

By combining Eqs. 15–7 and 16–5, we can write the intensity in terms of the pressure amplitude, ΔP_M:

$$I = 2\pi^2 v\rho f^2 A^2 = 2\pi^2 v\rho f^2 \left(\frac{\Delta P_M}{2\pi\rho v f}\right)^2$$

$$I = \frac{(\Delta P_M)^2}{2v\rho}. \qquad (16-7)$$

The intensity, when given in terms of pressure amplitude, thus does not depend on frequency.

The Ear's Response

The ear is not equally sensitive to all frequencies. To hear the same loudness for sounds of different frequencies requires different intensities. Studies averaged over large numbers of people have produced the curves shown in Fig. 16–6. On this graph, each curve represents sounds that seemed to be equally loud. The number labeling each curve represents the **loudness level** (the units are called *phons*), which is numerically equal to the sound level in dB at 1000 Hz. For example, the curve labeled 40 represents sounds that are heard by an average person to have the same loudness as a 1000-Hz sound with a sound level of 40 dB. From this 40-phon curve, we see that a 100-Hz tone must be at a level of about 62 dB to be perceived as loud as a 1000-Hz tone of only 40 dB.

The lowest curve in Fig. 16–6 (labeled 0) represents the sound level, as a function of frequency, for the *threshold of hearing*, the softest sound that is just audible by a very good ear. Note that the ear is most sensitive to sounds of frequency between 2000 and 4000 Hz, which are common in speech and music. Note too that whereas a 1000-Hz sound is audible at a level of 0 dB, a 100-Hz sound must be nearly 40 dB to be heard. The top curve in Fig. 16–6, labeled 120 phons, represents the *threshold of pain*. Sounds above this level can actually be felt and cause pain.

Figure 16–6 shows that at lower sound levels, our ears are less sensitive to the high and low frequencies relative to middle frequencies. The "loudness" control on some stereo systems is intended to compensate for this low-volume insensitivity. As the volume is turned down, the loudness control boosts the high and low frequencies relative to the middle frequencies so that the sound will have a more "normal-sounding" frequency balance. Many listeners, however, find the sound more pleasing or natural without the loudness control.

FIGURE 16–6 Sensitivity of the human ear as a function of frequency (see text). Note that the frequency scale is "logarithmic" in order to cover a wide range of frequencies.

16–4 Sources of Sound: Vibrating Strings and Air Columns

The source of any sound is a vibrating object. Almost any object can vibrate and hence be a source of sound. We now discuss some simple sources of sound, particularly musical instruments. In musical instruments, the source is set into vibration by striking, plucking, bowing, or blowing. Standing waves are produced and the source vibrates at its natural resonant frequencies. The vibrating source is in contact with the air (or other medium) and pushes on it to produce sound waves that travel outward. The frequencies of the waves are the same as those of the source, but the speed and wavelengths can be different. A drum has a stretched membrane that vibrates. Xylophones and marimbas have metal or wood bars that can be set into vibration. Bells, cymbals, and gongs also make use of a vibrating metal. Many instruments make use of vibrating strings, such as the violin, guitar, and piano, or make use of vibrating columns of air, such as the flute, trumpet, and pipe organ. We have already seen that the pitch of a pure sound is determined by the frequency. Typical frequencies for musical notes on the "equally tempered chromatic scale" are given in Table 16–3 for the octave beginning with middle C. Note that one octave corresponds to a doubling of frequency. For example, middle C has frequency of 262 Hz whereas C′ (C above middle C) has twice that frequency, 524 Hz. [Middle C is the C or "do" note at the middle of a piano keyboard.]

TABLE 16–3 Equally Tempered Chromatic Scale†

Note	Frequency (Hz)
C	262
C# or D♭	277
D	294
D# or E♭	311
E	330
F	349
F# or G♭	370
G	392
G# or A♭	415
A	440
A# or B♭	466
B	494
C′	524

†Only one octave is included.

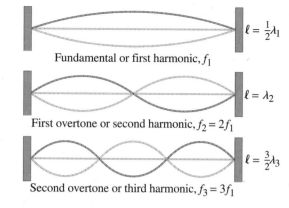

$\ell = \frac{1}{2}\lambda_1$

Fundamental or first harmonic, f_1

$\ell = \lambda_2$

First overtone or second harmonic, $f_2 = 2f_1$

$\ell = \frac{3}{2}\lambda_3$

Second overtone or third harmonic, $f_3 = 3f_1$

FIGURE 16–7 Standing waves on a string—only the lowest three frequencies are shown.

Stringed Instruments

(a)

(b)

FIGURE 16–8 The wavelength of (a) an unfingered string is longer than that of (b) a fingered string. Hence, the frequency of the fingered string is higher. Only one string is shown on this guitar, and only the simplest standing wave, the fundamental, is shown.

We saw in Chapter 15, Fig. 15–26b, how standing waves are established on a string, and we show this again here in Fig. 16–7. Such standing waves are the basis for all stringed instruments. The pitch is normally determined by the lowest resonant frequency, the **fundamental**, which corresponds to nodes occurring only at the ends. The string vibrating up and down as a whole corresponds to a half wavelength as shown at the top of Fig. 16–7; so the wavelength of the fundamental on the string is equal to twice the length of the string. Therefore, the fundamental frequency is $f_1 = v/\lambda = v/2\ell$, where v is the velocity of the wave on the string (*not* in the air). The possible frequencies for standing waves on a stretched string are whole-number multiples of the fundamental frequency:

$$f_n = nf_1 = n\frac{v}{2\ell}, \qquad n = 1, 2, 3, \cdots$$

where $n = 1$ refers to the fundamental and $n = 2, 3, \cdots$ are the overtones. All of the standing waves, $n = 1, 2, 3, \cdots$, are called harmonics,[†] as we saw in Section 15–9.

When a finger is placed on the string of a guitar or violin, the effective length of the string is shortened. So its fundamental frequency, and pitch, is higher since the wavelength of the fundamental is shorter (Fig. 16–8). The strings on a guitar or violin are all the same length. They sound at a different pitch because the strings have different mass per unit length, μ, which affects the velocity on the string, Eq. 15–2,

$$v = \sqrt{F_T/\mu}. \qquad \text{[stretched string]}$$

Thus the velocity on a heavier string is lower and the frequency will be lower for the same wavelength. The tension F_T may also be different. Adjusting the tension is the means for tuning the pitch of each string. In pianos and harps the strings are of different lengths. For the lower notes the strings are not only longer, but heavier as well, and the reason is illustrated in the following Example.

EXAMPLE 16–8 **Piano strings.** The highest key on a piano corresponds to a frequency about 150 times that of the lowest key. If the string for the highest note is 5.0 cm long, how long would the string for the lowest note have to be if it had the same mass per unit length and was under the same tension?

APPROACH Since $v = \sqrt{F_T/\mu}$, the velocity would be the same on each string. So the frequency is inversely proportional to the length ℓ of the string ($f = v/\lambda = v/2\ell$).

SOLUTION We can write, for the fundamental frequencies of each string, the ratio

$$\frac{\ell_L}{\ell_H} = \frac{f_H}{f_L},$$

where the subscripts L and H refer to the lowest and highest notes, respectively. Thus $\ell_L = \ell_H(f_H/f_L) = (5.0 \text{ cm})(150) = 750 \text{ cm}$, or 7.5 m. This would be ridiculously long (≈ 25 ft) for a piano.

NOTE The longer strings of lower frequency are made heavier, of higher mass per unit length, so even on grand pianos the strings are less than 3 m long.

[†]When the resonant frequencies above the fundamental (that is, the overtones) are integral multiples of the fundamental, as here, they are called harmonics. But if the overtones are not integral multiples of the fundamental, as is the case for a vibrating drumhead, for example, they are not harmonics.

EXERCISE D Two strings have the same length and tension, but one is more massive than the other. Which plays the higher note?

EXAMPLE 16–9 **Frequencies and wavelengths in the violin.** A 0.32-m-long violin string is tuned to play A above middle C at 440 Hz. (a) What is the wavelength of the fundamental string vibration, and (b) what are the frequency and wavelength of the sound wave produced? (c) Why is there a difference?

APPROACH The wavelength of the fundamental string vibration equals twice the length of the string (Fig. 16–7). As the string vibrates, it pushes on the air, which is thus forced to oscillate at the same frequency as the string.

SOLUTION (a) From Fig. 16–7 the wavelength of the fundamental is

$$\lambda = 2\ell = 2(0.32\text{ m}) = 0.64\text{ m} = 64\text{ cm}.$$

This is the wavelength of the standing wave on the string.

(b) The sound wave that travels outward in the air (to reach our ears) has the same frequency, 440 Hz. Its wavelength is

$$\lambda = \frac{v}{f} = \frac{343\text{ m/s}}{440\text{ Hz}} = 0.78\text{ m} = 78\text{ cm},$$

where v is the speed of sound in air (assumed at 20°C), Section 16–1.

(c) The wavelength of the sound wave is different from that of the standing wave on the string because the speed of sound in air (343 m/s at 20°C) is different from the speed of the wave on the string ($= f\lambda = 440\text{ Hz} \times 0.64\text{ m} = 280\text{ m/s}$) which depends on the tension in the string and its mass per unit length.

NOTE The frequencies on the string and in the air are the same: the string and air are in contact, and the string "forces" the air to vibrate at the same frequency. But the wavelengths are different because the wave speed on the string is different than that in air.

(a)

(b)

FIGURE 16–9 (a) Piano, showing sounding board to which the strings are attached; (b) sounding box (guitar).

Stringed instruments would not be very loud if they relied on their vibrating strings to produce the sound waves since the strings are too thin to compress and expand much air. Stringed instruments therefore make use of a kind of mechanical amplifier known as a *sounding board* (piano) or *sounding box* (guitar, violin), which acts to amplify the sound by putting a greater surface area in contact with the air (Fig. 16–9). When the strings are set into vibration, the sounding board or box is set into vibration as well. Since it has much greater area in contact with the air, it can produce a more intense sound wave. On an electric guitar, the sounding box is not so important since the vibrations of the strings are amplified electronically.

Wind Instruments

Instruments such as woodwinds, the brasses, and the pipe organ produce sound from the vibrations of standing waves in a column of air within a tube (Fig. 16–10). Standing waves can occur in the air of any cavity, but the frequencies present are complicated for any but very simple shapes such as the uniform, narrow tube of a flute or an organ pipe. In some instruments, a vibrating reed or the vibrating lip of the player helps to set up vibrations of the air column. In others, a stream of air is directed against one edge of the opening or mouthpiece, leading to turbulence which sets up the vibrations. Because of the disturbance, whatever its source, the air within the tube vibrates with a variety of frequencies, but only frequencies that correspond to standing waves will persist.

For a string fixed at both ends, Fig. 16–7, we saw that the standing waves have nodes (no movement) at the two ends, and one or more antinodes (large amplitude of vibration) in between. A node separates successive antinodes. The lowest-frequency standing wave, the *fundamental*, corresponds to a single antinode. The higher-frequency standing waves are called **overtones** or **harmonics**, as we saw in Section 15–9. Specifically, the first harmonic is the fundamental, the second harmonic (= first overtone) has twice the frequency of the fundamental, and so on.

FIGURE 16–10 Wind instruments: flute (left) and clarinet.

TUBE OPEN AT BOTH ENDS

(a) Displacement of air

First harmonic = fundamental
$\ell = \frac{1}{2}\lambda_1$
$f_1 = \frac{v}{2\ell}$

←[motion of air molecules]

Second harmonic
$\ell = \lambda_2$
$f_2 = \frac{v}{\ell} = 2f_1$

Overtones

Third harmonic
$\ell = \frac{3}{2}\lambda_3$
$f_3 = \frac{3v}{2\ell} = 3f_1$

(b) Pressure variation in the air

FIGURE 16–11 Graphs of the three simplest modes of vibration (standing waves) for a uniform tube open at both ends ("open tube"). These simplest modes of vibration are graphed in (a), on the left, in terms of the motion of the air (displacement), and in (b), on the right, in terms of air pressure. Each graph shows the wave format at two times, A and B, a half period apart. The actual motion of molecules for one case, the fundamental, is shown just below the tube at top left.

PHYSICS APPLIED

Wind instruments

The situation is similar for a column of air in a tube of uniform diameter, but we must remember that it is now air itself that is vibrating. We can describe the waves either in terms of the flow of the air—that is, in terms of the *displacement* of air—or in terms of the *pressure* in the air (see Figs. 16–2 and 16–3). In terms of displacement, the air at the closed end of a tube is a displacement node since the air is not free to move there, whereas near the open end of a tube there will be an antinode because the air can move freely in and out. The air within the tube vibrates in the form of longitudinal standing waves. The possible modes of vibration for a tube open at both ends (called an **open tube**) are shown graphically in Fig. 16–11. They are shown for a tube that is open at one end but closed at the other (called a **closed tube**) in Fig. 16–12. [A tube closed at *both* ends, having no connection to the outside air, would be useless as an instrument.] The graphs in part (a) of each Figure (left-hand sides) represent the displacement amplitude of the vibrating air in the tube. Note that these are graphs, and that the air molecules themselves oscillate *horizontally*, parallel to the tube length, as shown by the small arrows in the top diagram of Fig. 16–11a (on the left). The exact position of the antinode near the open end of a tube depends on the diameter of the tube, but if the diameter is small compared to the length, which is the usual case, the antinode occurs very close to the end as shown. We assume this is the case in what follows. (The position of the antinode may also depend slightly on the wavelength and other factors.)

Let us look in detail at the open tube, in Fig. 16–11a, which might be an organ pipe or a flute. An open tube has displacement antinodes at both ends since the air is free to move at open ends. There must be at least one node within an open tube if there is to be a standing wave at all. A single node corresponds to the *fundamental frequency* of the tube. Since the distance between two successive nodes, or between two successive antinodes, is $\frac{1}{2}\lambda$, there is one-half of a wavelength within the length of the tube for the simplest case of the fundamental (top diagram in Fig. 16–11a): $\ell = \frac{1}{2}\lambda$, or $\lambda = 2\ell$. So the fundamental frequency is $f_1 = v/\lambda = v/2\ell$, where v is the velocity of sound in air (the air in the tube). The standing wave with two nodes is the *first overtone* or *second harmonic* and has half the wavelength ($\ell = \lambda$) and twice the frequency of the fundamental. Indeed, in a uniform tube open at both ends, the frequency of each overtone is an integral multiple of the fundamental frequency, as shown in Fig. 16–11a. This is just what is found for a string.

For a closed tube, shown in Fig. 16–12a, which could be an organ pipe, there is always a displacement node at the closed end (because the air is not free to move) and an antinode at the open end (where the air can move freely). Since the distance between a node and the nearest antinode is $\frac{1}{4}\lambda$, we see that the fundamental in a closed tube corresponds to only one-fourth of a wavelength within the length of the tube: $\ell = \lambda/4$, and $\lambda = 4\ell$. The fundamental frequency is thus $f_1 = v/4\ell$, or half that for an open pipe of the same length. There is another

434 CHAPTER 16 Sound

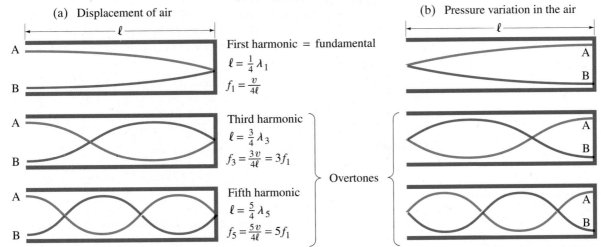

(a) Displacement of air

First harmonic = fundamental
$\ell = \frac{1}{4}\lambda_1$
$f_1 = \frac{v}{4\ell}$

Third harmonic
$\ell = \frac{3}{4}\lambda_3$
$f_3 = \frac{3v}{4\ell} = 3f_1$

Fifth harmonic
$\ell = \frac{5}{4}\lambda_5$
$f_5 = \frac{5v}{4\ell} = 5f_1$

} Overtones

(b) Pressure variation in the air

FIGURE 16–12 Modes of vibration (standing waves) for a tube closed at one end ("closed tube"). See caption for Fig. 16–11.

difference, for as we can see from Fig. 16–12a, only the odd harmonics are present in a closed tube: the overtones have frequencies equal to $3, 5, 7, \ldots$ times the fundamental frequency. There is no way for waves with $2, 4, 6, \ldots$ times the fundamental frequency to have a node at one end and an antinode at the other, and thus they cannot exist as standing waves in a closed tube.

Another way to analyze the vibrations in a uniform tube is to consider a description in terms of the *pressure* in the air, shown in part (b) of Figs. 16–11 and 16–12 (right-hand sides). Where the air in a wave is compressed, the pressure is higher, whereas in a wave expansion (or rarefaction), the pressure is less than normal. The open end of a tube is open to the atmosphere. Hence the pressure variation at an open end must be a *node*: the pressure does not alternate, but remains at the outside atmospheric pressure. If a tube has a closed end, the pressure at that closed end can readily alternate to be above or below atmospheric pressure. Hence there is a pressure *antinode* at a closed end of a tube. There can be pressure nodes and antinodes within the tube. Some of the possible vibrational modes in terms of pressure for an open tube are shown in Fig. 16–11b, and for a closed tube are shown in Fig. 16–12b.

EXAMPLE 16–10 **Organ pipes.** What will be the fundamental frequency and first three overtones for a 26-cm-long organ pipe at 20°C if it is (a) open and (b) closed?

APPROACH All our calculations can be based on Figs. 16–11a and 16–12a.

SOLUTION (a) For the open pipe, Fig. 16–11a, the fundamental frequency is

$$f_1 = \frac{v}{2\ell} = \frac{343 \text{ m/s}}{2(0.26 \text{ m})} = 660 \text{ Hz}.$$

The speed v is the speed of sound in air (the air vibrating in the pipe). The overtones include all harmonics: 1320 Hz, 1980 Hz, 2640 Hz, and so on.

(b) For a closed pipe, Fig. 16–12a, the fundamental frequency is

$$f_1 = \frac{v}{4\ell} = \frac{343 \text{ m/s}}{4(0.26 \text{ m})} = 330 \text{ Hz}.$$

Only odd harmonics are present: the first three overtones are 990 Hz, 1650 Hz, and 2310 Hz.

NOTE The closed pipe plays 330 Hz, which, from Table 16–3, is E above middle C, whereas the open pipe of the same length plays 660 Hz, an octave higher.

Pipe organs use both open and closed pipes, with lengths from a few centimeters to 5 m or more. A flute acts as an open tube, for it is open not only where you blow into it, but also at the opposite end. The different notes on a flute are obtained by shortening the length of the vibrating air column, by uncovering holes along the tube (so a displacement antinode can occur at the hole). The shorter the length of the vibrating air column, the higher the fundamental frequency.

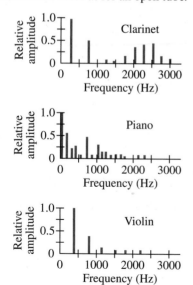

FIGURE 16–13 The amplitudes of the fundamental and first two overtones are added at each point to get the "sum," or composite waveform.

FIGURE 16–14 Sound spectra for different instruments. The spectra change when the instruments play different notes. The clarinet is a bit complicated: it acts like a closed tube at lower frequencies, having only odd harmonics, but at higher frequencies all harmonics occur as for an open tube.

EXAMPLE 16–11 **Flute.** A flute is designed to play middle C (262 Hz) as the fundamental frequency when all the holes are covered. Approximately how long should the distance be from the mouthpiece to the far end of the flute? (This is only approximate since the antinode does not occur precisely at the mouthpiece.) Assume the temperature is 20°C.

APPROACH When all holes are covered, the length of the vibrating air column is the full length. The speed of sound in air at 20°C is 343 m/s. Because a flute is open at both ends, we use Fig. 16–11: the fundamental frequency f_1 is related to the length ℓ of the vibrating air column by $f = v/2\ell$.

SOLUTION Solving for ℓ, we find

$$\ell = \frac{v}{2f} = \frac{343 \text{ m/s}}{2(262 \text{ s}^{-1})} = 0.655 \text{ m} \approx 0.66 \text{ m}.$$

EXERCISE E To see why players of wind instruments "warm up" their instruments (so they will be in tune), determine the fundamental frequency of the flute of Example 16–11 when all holes are covered and the temperature is 10°C instead of 20°C.

EXERCISE F Return to the Chapter-Opening Question, page 424, and answer it again now. Try to explain why you may have answered differently the first time.

*16–5 Quality of Sound, and Noise; Superposition

Whenever we hear a sound, particularly a musical sound, we are aware of its loudness, its pitch, and also of a third aspect called its *timbre* or "quality." For example, when a piano and then a flute play a note of the same loudness and pitch (say, middle C), there is a clear difference in the overall sound. We would never mistake a piano for a flute. This is what is meant by the timbre or *quality* of a sound. For musical instruments, the term *tone color* is also used.

Just as loudness and pitch can be related to physically measurable quantities, so too can quality. The quality of a sound depends on the presence of overtones—their number and their relative amplitudes. Generally, when a note is played on a musical instrument, the fundamental as well as overtones are present simultaneously. Figure 16–13 illustrates how the *principle of superposition* (Section 15–6) applies to three wave forms, in this case the fundamental and first two overtones (with particular amplitudes): they add together at each point to give a composite *waveform*. Normally, more than two overtones are present. [Any complex wave can be analyzed into a superposition of sinusoidal waves of appropriate amplitudes, wavelengths, and frequencies—see Section 15–6. Such an analysis is called a *Fourier analysis*.]

The relative amplitudes of the overtones for a given note are different for different musical instruments, which is what gives each instrument its characteristic quality or timbre. A bar graph showing the relative amplitudes of the harmonics for a given note produced by an instrument is called a *sound spectrum*. Several typical examples for different musical instruments are shown in Fig. 16–14. The fundamental usually has the greatest amplitude, and its frequency is what is heard as the pitch.

The manner in which an instrument is played strongly influences the sound quality. Plucking a violin string, for example, makes a very different sound than pulling a bow across it. The sound spectrum at the very start (or end) of a note (as when a hammer strikes a piano string) can be very different from the subsequent sustained tone. This too affects the subjective tone quality of an instrument.

An ordinary sound, like that made by striking two stones together, is a noise that has a certain quality, but a clear pitch is not discernible. Such a noise is a mixture of many frequencies which bear little relation to one another. A sound spectrum made of that noise would not show discrete lines like those of Fig. 16–14. Instead it would show a continuous, or nearly continuous, spectrum of frequencies. Such a sound we call "noise" in comparison with the more harmonious sounds which contain frequencies that are simple multiples of the fundamental.

16–6 Interference of Sound Waves; Beats

Interference in Space

We saw in Section 15–8 that when two waves simultaneously pass through the same region of space, they interfere with one another. Interference also occurs with sound waves.

Consider two large loudspeakers, A and B, a distance d apart on the stage of an auditorium as shown in Fig. 16–15. Let us assume the two speakers are emitting sound waves of the same single frequency and that they are in phase: that is, when one speaker is forming a compression, so is the other. (We ignore reflections from walls, floor, etc.) The curved lines in the diagram represent the crests of sound waves from each speaker at one instant in time. We must remember that for a sound wave, a crest is a compression in the air whereas a trough—which falls between two crests—is a rarefaction. A human ear or detector at a point such as C, which is the same distance from each speaker, will experience a loud sound because the interference will be constructive—two crests reach it at one moment, two troughs reach it a moment later. On the other hand, at a point such as D in the diagram, little if any sound will be heard because destructive interference occurs—compressions of one wave meet rarefactions of the other and vice versa (see Fig. 15–24 and the related discussion on water waves in Section 15–8).

An analysis of this situation is perhaps clearer if we graphically represent the waveforms as in Fig. 16–16. In Fig. 16–16a, it can be seen that at point C, constructive interference occurs since both waves simultaneously have crests or simultaneously have troughs when they arrive at C. In Fig. 16–16b we see that, to reach point D, the wave from speaker B must travel a greater distance than the wave from A. Thus the wave from B lags behind that from A. In this diagram, point E is chosen so that the distance ED is equal to AD. Thus we see that if the distance BE is equal to precisely one-half the wavelength of the sound, the two waves will be exactly out of phase when they reach D, and destructive interference occurs. This then is the criterion for determining at what points destructive interference occurs: destructive interference occurs at any point whose distance from one speaker is one-half wavelength greater than its distance from the other speaker. Notice that if this extra distance (BE in Fig. 16–16b) is equal to a whole wavelength (or 2, 3, ... wavelengths), then the two waves will be in phase and *constructive interference* occurs. If the distance BE equals $\frac{1}{2}, 1\frac{1}{2}, 2\frac{1}{2}, \ldots$ wavelengths, *destructive interference* occurs.

It is important to realize that a person at point D in Fig. 16–15 or 16–16 hears nothing at all (or nearly so), yet sound is coming from both speakers. Indeed, if one of the speakers is turned off, the sound from the other speaker will be clearly heard.

If a loudspeaker emits a whole range of frequencies, only specific wavelengths will destructively interfere completely at a given point.

EXAMPLE 16–12 **Loudspeakers' interference.** Two loudspeakers are 1.00 m apart. A person stands 4.00 m from one speaker. How far must this person be from the second speaker to detect destructive interference when the speakers emit an 1150-Hz sound? Assume the temperature is 20°C.

APPROACH To sense destructive interference, the person must be one-half wavelength closer to or farther from one speaker than from the other—that is, at a distance $= 4.00$ m $\pm \lambda/2$. We can determine λ since we know f and v.

SOLUTION The speed of sound at 20°C is 343 m/s, so the wavelength of this sound is (Eq. 15–1)

$$\lambda = \frac{v}{f} = \frac{343 \text{ m/s}}{1150 \text{ Hz}} = 0.30 \text{ m}.$$

For destructive interference to occur, the person must be one-half wavelength farther from one loudspeaker than from the other, or 0.15 m. Thus the person must be 3.85 m or 4.15 m from the second speaker.

NOTE If the speakers are less than 0.15 m apart, there will be no point that is 0.15 m farther from one speaker than the other, and there will be no point where destructive interference could occur.

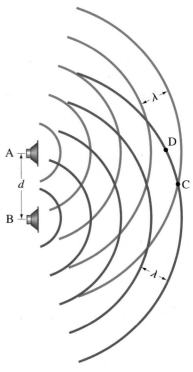

FIGURE 16–15 Sound waves from two loudspeakers interfere.

FIGURE 16–16 Sound waves of a single frequency from loudspeakers A and B (see Fig. 16–15) constructively interfere at C and destructively interfere at D. [Shown here are graphical representations, not the actual longitudinal sound waves.]

(a)

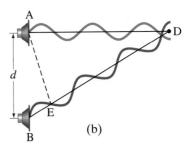

(b)

FIGURE 16–17 Beats occur as a result of the superposition of two sound waves of slightly different frequency.

In the figure, labels read: $f_A = 50$ Hz, $f_B = 60$ Hz, t = 0, t = 0.05 s, t = 0.10 s, t = 0.15 s, sum, beat period (0.10 s).

Beats—Interference in Time

We have been discussing interference of sound waves that takes place in space. An interesting and important example of interference that occurs in time is the phenomenon known as **beats**: If two sources of sound—say, two tuning forks—are close in frequency but not exactly the same, sound waves from the two sources interfere with each other. The sound level at a given position alternately rises and falls in time, because the two waves are sometimes in phase and sometimes out of phase due to their different wavelengths. The regularly spaced intensity changes are called beats.

To see how beats arise, consider two equal-amplitude sound waves of frequency $f_A = 50$ Hz and $f_B = 60$ Hz, respectively. In 1.00 s, the first source makes 50 vibrations whereas the second makes 60. We now examine the waves at one point in space equidistant from the two sources. The waveforms for each wave as a function of time, at a fixed position, are shown on the top graph of Fig. 16–17; the red line represents the 50-Hz wave, and the blue line represents the 60-Hz wave. The lower graph in Fig. 16–17 shows the sum of the two waves as a function of time. At time $t = 0$ the two waves are shown to be in phase and interfere constructively. Because the two waves vibrate at different rates, at time $t = 0.05$ s they are completely out of phase and interfere destructively. At $t = 0.10$ s, they are again in phase and the resultant amplitude again is large. Thus the resultant amplitude is large every 0.10 s and drops drastically in between. This rising and falling of the intensity is what is heard as beats.[†] In this case the beats are 0.10 s apart. That is, the **beat frequency** is ten per second, or 10 Hz. This result, that the beat frequency equals the difference in frequency of the two waves, is valid in general, as we now show.

Let the two waves, of frequencies f_1 and f_2, be represented at a fixed point in space by

$$D_1 = A \sin 2\pi f_1 t$$

and

$$D_2 = A \sin 2\pi f_2 t.$$

The resultant displacement, by the principle of superposition, is

$$D = D_1 + D_2 = A(\sin 2\pi f_1 t + \sin 2\pi f_2 t).$$

Using the trigonometric identity $\sin \theta_1 + \sin \theta_2 = 2 \sin\frac{1}{2}(\theta_1 + \theta_2) \cos\frac{1}{2}(\theta_1 - \theta_2)$, we have

$$D = \left[2A \cos 2\pi\left(\frac{f_1 - f_2}{2}\right)t \right] \sin 2\pi\left(\frac{f_1 + f_2}{2}\right)t. \qquad \textbf{(16–8)}$$

We can interpret Eq. 16–8 as follows. The superposition of the two waves results in a wave that vibrates at the average frequency of the two components, $(f_1 + f_2)/2$. This vibration has an amplitude given by the expression in brackets, and this amplitude varies in time, from zero to a maximum of $2A$ (the sum of the separate amplitudes), with a frequency of $(f_1 - f_2)/2$. A beat occurs whenever $\cos 2\pi[(f_1 - f_2)/2]t$ equals $+1$ or -1 (see Fig. 16–17); that is, two beats occur per cycle, so the beat frequency is twice $(f_1 - f_2)/2$ which is just $f_1 - f_2$, the difference in frequency of the component waves.

[†]Beats will be heard even if the amplitudes are not equal, as long as the difference in amplitude is not great.

The phenomenon of beats can occur with any kind of wave and is a very sensitive method for comparing frequencies. For example, to tune a piano, a piano tuner listens for beats produced between his standard tuning fork and that of a particular string on the piano, and knows it is in tune when the beats disappear. The members of an orchestra tune up by listening for beats between their instruments and that of a standard tone (usually A above middle C at 440 Hz) produced by a piano or an oboe. A beat frequency is perceived as an intensity modulation (a wavering between loud and soft) for beat frequencies below 20 Hz or so, and as a separate low tone for higher beat frequencies (audible if the tones are strong enough).

PHYSICS APPLIED
Tuning a piano

EXAMPLE 16–13 **Beats.** A tuning fork produces a steady 400-Hz tone. When this tuning fork is struck and held near a vibrating guitar string, twenty beats are counted in five seconds. What are the possible frequencies produced by the guitar string?

APPROACH For beats to occur, the string must vibrate at a frequency different from 400 Hz by whatever the beat frequency is.

SOLUTION The beat frequency is

$$f_{beat} = 20 \text{ vibrations}/5 \text{ s} = 4 \text{ Hz}.$$

This is the difference of the frequencies of the two waves. Because one wave is known to be 400 Hz, the other must be either 404 Hz or 396 Hz.

16–7 Doppler Effect

You may have noticed that you hear the pitch of the siren on a speeding fire truck drop abruptly as it passes you. Or you may have noticed the change in pitch of a blaring horn on a fast-moving car as it passes by you. The pitch of the engine noise of a racecar changes as the car passes an observer. When a source of sound is moving toward an observer, the pitch the observer hears is higher than when the source is at rest; and when the source is traveling away from the observer, the pitch is lower. This phenomenon is known as the **Doppler effect**[†] and occurs for all types of waves. Let us now see why it occurs, and calculate the difference between the perceived and source frequencies when there is relative motion between source and observer.

Consider the siren of a fire truck at rest, which is emitting sound of a particular frequency in all directions as shown in Fig. 16–18a. The sound waves are moving at the speed of sound in air, v_{snd}, which is independent of the velocity of the source or observer. If our source, the fire truck, is moving, the siren emits sound at the same frequency as it does at rest. But the sound wavefronts it emits forward, in front of it, are closer together than when the fire truck is at rest, as shown in Fig. 16–18b. This is because the fire truck, as it moves, is "chasing" the previously emitted wavefronts, and emits each crest closer to the previous one. Thus an observer on the sidewalk in front of the truck will detect more wave crests passing per second, so the frequency heard is higher. The wavefronts emitted behind the truck, on the other hand, are farther apart than when the truck is at rest because the truck is speeding away from them. Hence, fewer wave crests per second pass by an observer behind the moving truck (Fig. 16–18b) and the perceived pitch is lower.

(a) At rest

(b) Fire truck moving

FIGURE 16–18 (a) Both observers on the sidewalk hear the same frequency from a fire truck at rest. (b) Doppler effect: observer toward whom the fire truck moves hears a higher-frequency sound, and observer behind the fire truck hears a lower-frequency sound.

[†] After J. C. Doppler (1803–1853).

(a) Source fixed

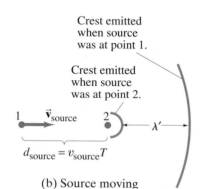

Crest emitted when source was at point 1.

Crest emitted when source was at point 2.

(b) Source moving

FIGURE 16–19 Determination of the frequency shift in the Doppler effect (see text). The red dot is the source.

We can calculate the frequency shift perceived by making use of Fig. 16–19, and we assume the air (or other medium) is at rest in our reference frame. (The stationary observer is off to the right.) In Fig. 16–19a, the source of the sound is shown as a red dot, and is at rest. Two successive wave crests are shown, the second of which has just been emitted and so is still near the source. The distance between these crests is λ, the wavelength. If the frequency of the source is f, then the time between emissions of wave crests is

$$T = \frac{1}{f} = \frac{\lambda}{v_{snd}}.$$

In Fig. 16–19b, the source is moving with a velocity v_{source} toward the observer. In a time T (as just defined), the first wave crest has moved a distance $d = v_{snd} T = \lambda$, where v_{snd} is the velocity of the sound wave in air (which is the same whether the source is moving or not). In this same time, the source has moved a distance $d_{source} = v_{source} T$. Then the distance between successive wave crests, which is the wavelength λ' the observer will perceive, is

$$\begin{aligned}
\lambda' &= d - d_{source} \\
&= \lambda - v_{source} T \\
&= \lambda - v_{source} \frac{\lambda}{v_{snd}} \\
&= \lambda \left(1 - \frac{v_{source}}{v_{snd}} \right).
\end{aligned}$$

We subtract λ from both sides of this equation and find that the shift in wavelength, $\Delta\lambda$, is

$$\Delta\lambda = \lambda' - \lambda = -\lambda \frac{v_{source}}{v_{snd}}.$$

So the shift in wavelength is directly proportional to the source speed v_{source}. The frequency f' that will be perceived by our stationary observer on the ground is given by

$$f' = \frac{v_{snd}}{\lambda'} = \frac{v_{snd}}{\lambda \left(1 - \dfrac{v_{source}}{v_{snd}} \right)}.$$

Since $v_{snd}/\lambda = f$, then

$$f' = \frac{f}{\left(1 - \dfrac{v_{source}}{v_{snd}} \right)}. \qquad \begin{bmatrix} \text{source moving toward} \\ \text{stationary observer} \end{bmatrix} \quad \textbf{(16–9a)}$$

Because the denominator is less than 1, the observed frequency f' is greater than the source frequency f. That is, $f' > f$. For example, if a source emits a sound of frequency 400 Hz when at rest, then when the source moves toward a fixed observer with a speed of 30 m/s, the observer hears a frequency (at 20°C) of

$$f' = \frac{400\ \text{Hz}}{1 - \dfrac{30\ \text{m/s}}{343\ \text{m/s}}} = 438\ \text{Hz}.$$

Now consider a source moving *away* from the stationary observer at a speed v_{source}. Using the same arguments as above, the wavelength λ' perceived by our observer will have the minus sign on d_{source} (second equation on this page) changed to plus:

$$\begin{aligned}
\lambda' &= d + d_{source} \\
&= \lambda \left(1 + \frac{v_{source}}{v_{snd}} \right).
\end{aligned}$$

The difference between the observed and emitted wavelengths will be $\Delta\lambda = \lambda' - \lambda = +\lambda(v_{source}/v_{snd})$. The observed frequency of the wave, $f' = v_{snd}/\lambda'$, will be

$$f' = \frac{f}{\left(1 + \dfrac{v_{source}}{v_{snd}} \right)}. \qquad \begin{bmatrix} \text{source moving away from} \\ \text{stationary observer} \end{bmatrix} \quad \textbf{(16–9b)}$$

If a source emitting at 400 Hz is moving away from a fixed observer at 30 m/s, the observer hears a frequency $f' = (400\ \text{Hz})/[1 + (30\ \text{m/s})/(343\ \text{m/s})] = 368\ \text{Hz}$.

FIGURE 16–20 Observer moving with speed v_{obs} toward a stationary source detects wave crests passing at speed $v' = v_{snd} + v_{obs}$ where v_{snd} is the speed of the sound waves in air.

The Doppler effect also occurs when the source is at rest and the observer is in motion. If the observer is traveling *toward* the source, the pitch heard is higher than that of the emitted source frequency. If the observer is traveling *away* from the source, the pitch heard is lower. Quantitatively the change in frequency is different than for the case of a moving source. With a fixed source and a moving observer, the distance between wave crests, the wavelength λ, is not changed. But the velocity of the crests with respect to the observer *is* changed. If the observer is moving toward the source, Fig. 16–20, the speed v' of the waves relative to the observer is a simple addition of velocities: $v' = v_{snd} + v_{obs}$, where v_{snd} is the velocity of sound in air (we assume the air is still) and v_{obs} is the velocity of the observer. Hence, the frequency heard is

$$f' = \frac{v'}{\lambda} = \frac{v_{snd} + v_{obs}}{\lambda}.$$

Because $\lambda = v_{snd}/f$, then

$$f' = \frac{(v_{snd} + v_{obs})f}{v_{snd}},$$

or

$$f' = \left(1 + \frac{v_{obs}}{v_{snd}}\right)f. \qquad \begin{bmatrix} \text{observer moving toward} \\ \text{stationary source} \end{bmatrix} \quad \textbf{(16–10a)}$$

If the observer is moving away from the source, the relative velocity is $v' = v_{snd} - v_{obs}$, so

$$f' = \left(1 - \frac{v_{obs}}{v_{snd}}\right)f. \qquad \begin{bmatrix} \text{observer moving away} \\ \text{from stationary source} \end{bmatrix} \quad \textbf{(16–10b)}$$

EXAMPLE 16–14 **A moving siren.** The siren of a police car at rest emits at a predominant frequency of 1600 Hz. What frequency will you hear if you are at rest and the police car moves at 25.0 m/s (*a*) toward you, and (*b*) away from you?

APPROACH The observer is fixed, and the source moves, so we use Eqs. 16–9. The frequency you (the observer) hear is the emitted frequency f divided by the factor $(1 \pm v_{source}/v_{snd})$ where v_{source} is the speed of the police car. Use the minus sign when the car moves toward you (giving a higher frequency); use the plus sign when the car moves away from you (lower frequency).

SOLUTION (*a*) The car is moving toward you, so (Eq. 16–9a)

$$f' = \frac{f}{\left(1 - \dfrac{v_{source}}{v_{snd}}\right)} = \frac{1600 \text{ Hz}}{\left(1 - \dfrac{25.0 \text{ m/s}}{343 \text{ m/s}}\right)} = 1726 \text{ Hz} \approx 1730 \text{ Hz}.$$

(*b*) The car is moving away from you, so (Eq. 16–9b)

$$f' = \frac{f}{\left(1 + \dfrac{v_{source}}{v_{snd}}\right)} = \frac{1600 \text{ Hz}}{\left(1 + \dfrac{25.0 \text{ m/s}}{343 \text{ m/s}}\right)} = 1491 \text{ Hz} \approx 1490 \text{ Hz}.$$

EXERCISE G Suppose the police car of Example 16–14 is at rest and emits at 1600 Hz. What frequency would you hear if you were moving at 25.0 m/s (*a*) toward it, and (*b*) away from it?

When a sound wave is reflected from a moving obstacle, the frequency of the reflected wave will, because of the Doppler effect, be different from that of the incident wave. This is illustrated in the following Example.

EXAMPLE 16–15 **Two Doppler shifts.** A 5000-Hz sound wave is emitted by a stationary source. This sound wave reflects from an object moving 3.50 m/s toward the source (Fig. 16–21). What is the frequency of the wave reflected by the moving object as detected by a detector at rest near the source?

APPROACH There are actually two Doppler shifts in this situation. First, the moving object acts like an observer moving toward the source with speed $v_{obs} = 3.50$ m/s (Fig. 16–21a) and so "detects" a sound wave of frequency (Eq. 16–10a) $f' = f[1 + (v_{obs}/v_{snd})]$. Second, reflection of the wave from the moving object is equivalent to the object reemitting the wave, acting effectively as a moving source with speed $v_{source} = 3.50$ m/s (Fig. 16–21b). The final frequency detected, f'', is given by $f'' = f'/[1 - v_{source}/v_{snd}]$, Eq. 16–9a.

SOLUTION The frequency f' that is "detected" by the moving object is (Eq. 16–10a):

$$f' = \left(1 + \frac{v_{obs}}{v_{snd}}\right)f = \left(1 + \frac{3.50 \text{ m/s}}{343 \text{ m/s}}\right)(5000 \text{ Hz}) = 5051 \text{ Hz}.$$

The moving object now "emits" (reflects) a sound of frequency (Eq. 16–9a)

$$f'' = \frac{f'}{\left(1 - \dfrac{v_{source}}{v_{snd}}\right)} = \frac{5051 \text{ Hz}}{\left(1 - \dfrac{3.50 \text{ m/s}}{343 \text{ m/s}}\right)} = 5103 \text{ Hz}.$$

Thus the frequency shifts by 103 Hz.

NOTE Bats use this technique to be aware of their surroundings. This is also the principle behind Doppler radar as speed-measuring devices for vehicles and other objects.

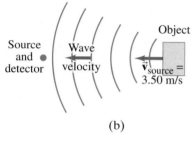

FIGURE 16–21 Example 16–15.

The incident wave and the reflected wave in Example 16–15, when mixed together (say, electronically), interfere with one another and beats are produced. The beat frequency is equal to the difference in the two frequencies, 103 Hz. This Doppler technique is used in a variety of medical applications, usually with ultrasonic waves in the megahertz frequency range. For example, ultrasonic waves reflected from red blood cells can be used to determine the velocity of blood flow. Similarly, the technique can be used to detect the movement of the chest of a young fetus and to monitor its heartbeat.

PHYSICS APPLIED

Doppler blood-flow meter and other medical uses

For convenience, we can write Eqs. 16–9 and 16–10 as a single equation that covers all cases of both source and observer in motion:

$$f' = f\left(\frac{v_{snd} \pm v_{obs}}{v_{snd} \mp v_{source}}\right). \qquad \begin{bmatrix} \text{source and} \\ \text{observer moving} \end{bmatrix} \quad \textbf{(16–11)}$$

PROBLEM SOLVING

Getting the signs right

To get the signs right, recall from your own experience that the frequency is higher when observer and source approach each other, and lower when they move apart. Thus the upper signs in numerator and denominator apply if source and/or observer move toward each other; the lower signs apply if they are moving apart.

EXERCISE H How fast would a source have to approach an observer for the observed frequency to be one octave above (twice) the produced frequency? (a) $\frac{1}{2}v_{snd}$, (b) v_{snd}, (c) $2v_{snd}$, (d) $4v_{snd}$.

Doppler Effect for Light

The Doppler effect occurs for other types of waves as well. Light and other types of electromagnetic waves (such as radar) exhibit the Doppler effect: although the formulas for the frequency shift are not identical to Eqs. 16–9 and 16–10, as discussed in Chapter 44, the effect is similar. One important application is for weather forecasting using radar. The time delay between the emission of radar pulses and their reception after being reflected off raindrops gives the position of precipitation. Measuring the Doppler shift in frequency (as in Example 16–15) tells how fast the storm is moving and in which direction.

Another important application is to astronomy, where the velocities of distant galaxies can be determined from the Doppler shift. Light from distant galaxies is shifted toward lower frequencies, indicating that the galaxies are moving away from us. This is called the **redshift** since red has the lowest frequency of visible light. The greater the frequency shift, the greater the velocity of recession. It is found that the farther the galaxies are from us, the faster they move away. This observation is the basis for the idea that the universe is expanding, and is one basis for the idea that the universe began as a great explosion, affectionately called the "Big Bang" (Chapter 44).

PHYSICS APPLIED

Doppler effect for EM waves and weather forecasting

PHYSICS APPLIED

Redshift in cosmology

*16–8 Shock Waves and the Sonic Boom

An object such as an airplane traveling faster than the speed of sound is said to have a **supersonic speed**. Such a speed is often given as a **Mach**[†] number, which is defined as the ratio of the speed of the object to the speed of sound in the surrounding medium. For example, a plane traveling 600 m/s high in the atmosphere, where the speed of sound is only 300 m/s, has a speed of Mach 2.

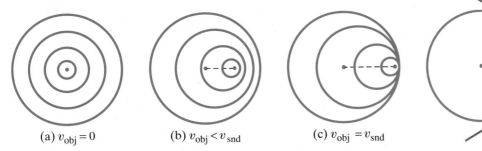

(a) $v_{obj} = 0$ (b) $v_{obj} < v_{snd}$ (c) $v_{obj} = v_{snd}$ (d) $v_{obj} > v_{snd}$

FIGURE 16–22 Sound waves emitted by an object (a) at rest or (b, c, and d) moving. (b) If the object's velocity is less than the velocity of sound, the Doppler effect occurs; (d) if its velocity is greater than the velocity of sound, a shock wave is produced.

When a source of sound moves at subsonic speeds (less than the speed of sound), the pitch of the sound is altered as we have seen (the Doppler effect); see also Fig. 16–22a and b. But if a source of sound moves faster than the speed of sound, a more dramatic effect known as a **shock wave** occurs. In this case, the source is actually "outrunning" the waves it produces. As shown in Fig. 16–22c, when the source is traveling *at* the speed of sound, the wave fronts it emits in the forward direction "pile up" directly in front of it. When the object moves faster, at a supersonic speed, the wave fronts pile up on one another along the sides, as shown in Fig. 16–22d. The different wave crests overlap one another and form a single very large crest which is the shock wave. Behind this very large crest there is usually a very large trough. A shock wave is essentially the result of constructive interference of a large number of wave fronts. A shock wave in air is analogous to the bow wave of a boat traveling faster than the speed of the water waves it produces, Fig. 16–23.

FIGURE 16–23 Bow waves produced by a boat.

[†]After the Austrian physicist Ernst Mach (1838–1916).

(a)　　　　　　　　　　　　　　　　　　　　　　　(b)

FIGURE 16–24 (a) The (double) sonic boom has already been heard by person A on the left. The front shock wave is just being heard by person B in the center. And it will shortly be heard by person C on the right. (b) Special photo of supersonic aircraft showing shock waves produced in the air. (Several closely spaced shock waves are produced by different parts of the aircraft.)

When an airplane travels at supersonic speeds, the noise it makes and its disturbance of the air form into a shock wave containing a tremendous amount of sound energy. When the shock wave passes a listener, it is heard as a loud *sonic boom*. A sonic boom lasts only a fraction of a second, but the energy it contains is often sufficient to break windows and cause other damage. Actually, a sonic boom is made up of two or more booms since major shock waves can form at the front and the rear of the aircraft, as well as at the wings, etc. (Fig. 16–24). Bow waves of a boat are also multiple, as can be seen in Fig. 16–23.

When an aircraft approaches the speed of sound, it encounters a barrier of sound waves in front of it (see Fig. 16–22c). To exceed the speed of sound, the aircraft needs extra thrust to pass through this "sound barrier." This is called "breaking the sound barrier." Once a supersonic speed is attained, this barrier no longer impedes the motion. It is sometimes erroneously thought that a sonic boom is produced only at the moment an aircraft is breaking through the sound barrier. Actually, a shock wave follows the aircraft at all times it is traveling at supersonic speeds. A series of observers on the ground will each hear a loud "boom" as the shock wave passes, Fig. 16–24. The shock wave consists of a cone whose apex is at the aircraft. The angle of this cone, θ (see Fig. 16–22d), is given by

$$\sin \theta = \frac{v_{\text{snd}}}{v_{\text{obj}}}, \qquad\qquad\textbf{(16–12)}$$

where v_{obj} is the velocity of the object (the aircraft) and v_{snd} is the velocity of sound in the medium. (The proof is left as Problem 75.)

*16–9 Applications: Sonar, Ultrasound, and Medical Imaging

*Sonar

The reflection of sound is used in many applications to determine distance. The **sonar**[†] or pulse-echo technique is used to locate underwater objects. A transmitter sends out a sound pulse through the water, and a detector receives its reflection, or echo, a short time later. This time interval is carefully measured, and from it the distance to the reflecting object can be determined since the speed of sound in water is known. The depth of the sea and the location of reefs, sunken ships, submarines, or schools of fish can be determined in this way. The interior structure of the Earth is studied in a similar way by detecting reflections of waves traveling through the Earth whose source was a deliberate explosion (called "soundings"). An analysis of waves reflected from various structures and boundaries within the Earth reveals characteristic patterns that are also useful in the exploration for oil and minerals.

[†]Sonar stands for "*so*und *na*vigation *r*anging."

Sonar generally makes use of **ultrasonic** frequencies: that is, waves whose frequencies are above 20 kHz, beyond the range of human detection. For sonar, the frequencies are typically in the range 20 kHz to 100 kHz. One reason for using ultrasound waves, other than the fact that they are inaudible, is that for shorter wavelengths there is less diffraction (Section 15–11) so the beam spreads less and smaller objects can be detected.

*Ultrasound Medical Imaging

The diagnostic use of ultrasound in medicine, in the form of images (sometimes called *sonograms*), is an important and interesting application of physical principles. A **pulse-echo technique** is used, much like sonar, except that the frequencies used are in the range of 1 to 10 MHz $(1\,\text{MHz} = 10^6\,\text{Hz})$. A high-frequency sound pulse is directed into the body, and its reflections from boundaries or interfaces between organs and other structures and lesions in the body are then detected. Tumors and other abnormal growths, or pockets of fluid, can be distinguished; the action of heart valves and the development of a fetus can be examined; and information about various organs of the body, such as the brain, heart, liver, and kidneys, can be obtained. Although ultrasound does not replace X-rays, for certain kinds of diagnosis it is more helpful. Some kinds of tissue or fluid are not detected in X-ray photographs, but ultrasound waves are reflected from their boundaries. "Real-time" ultrasound images are like a movie of a section of the interior of the body.

The pulse-echo technique for medical imaging works as follows. A brief pulse of ultrasound is emitted by a transducer that transforms an electrical pulse into a sound-wave pulse. Part of the pulse is reflected as echoes at each interface in the body, and most of the pulse (usually) continues on, Fig. 16–25a. The detection of reflected pulses by the same transducer can then be displayed on the screen of a display terminal or monitor. The time elapsed from when the pulse is emitted to when each reflection (echo) is received is proportional to the distance to the reflecting surface. For example, if the distance from transducer to the vertebra is 25 cm, the pulse travels a round-trip distance of $2 \times 25\,\text{cm} = 0.50\,\text{m}$. The speed of sound in human tissue is about 1540 m/s (close to that of sea water), so the time taken is

$$t = \frac{d}{v} = \frac{(0.50\,\text{m})}{(1540\,\text{m/s})} = 320\,\mu s.$$

The *strength* of a reflected pulse depends mainly on the difference in density of the two materials on either side of the interface and can be displayed as a pulse or as a dot (Figs. 16–25b and c). Each echo dot (Fig. 16–25c) can be represented as a point whose position is given by the time delay and whose brightness depends on the strength of the echo.

PHYSICS APPLIED
Ultrasound medical imaging

FIGURE 16–25 (a) Ultrasound pulse passes through the abdomen, reflecting from surfaces in its path. (b) Reflected pulses plotted as a function of time when received by transducer. The vertical dashed lines point out which reflected pulse goes with which surface. (c) Dot display for the same echoes: brightness of each dot is related to signal strength.

FIGURE 16–26 (a) Ten traces are made across the abdomen by moving the transducer, or by using an array of transducers. (b) The echoes are plotted as dots to produce the image. More closely spaced traces would give a more detailed image.

(a)

Body organs

(b)

FIGURE 16–27 Ultrasound image of a human fetus within the uterus.

A two-dimensional image can then be formed out of these dots from a series of scans. The transducer is moved, or an array of transducers is used, each of which sends out a pulse at each position and receives echoes as shown in Fig. 16–26a. Each trace can be plotted, spaced appropriately one below the other, to form an image on a monitor screen as shown in Fig. 16–26b. Only 10 lines are shown in Fig. 16–26, so the image is crude. More lines give a more precise image.[†] An ultrasound image is shown in Fig. 16–27.

[†] *Radar* used for aircraft involves a similar pulse-echo technique except that it uses electromagnetic (EM) waves, which, like light, travel with a speed of 3×10^8 m/s.

Summary

Sound travels as a longitudinal wave in air and other materials. In air, the speed of sound increases with temperature; at 20°C, it is about 343 m/s.

The **pitch** of a sound is determined by the frequency; the higher the frequency, the higher the pitch.

The **audible range** of frequencies for humans is roughly 20 Hz to 20,000 Hz (1 Hz = 1 cycle per second).

The **loudness** or **intensity** of a sound is related to the amplitude squared of the wave. Because the human ear can detect sound intensities from 10^{-12} W/m^2 to over 1 W/m^2, sound levels are specified on a logarithmic scale. The **sound level** β, specified in decibels, is defined in terms of intensity I as

$$\beta(\text{in dB}) = 10 \log\left(\frac{I}{I_0}\right), \quad \textbf{(16–6)}$$

where the reference intensity I_0 is usually taken to be 10^{-12} W/m^2.

Musical instruments are simple sources of sound in which *standing waves* are produced.

The strings of a stringed instrument may vibrate as a whole with nodes only at the ends; the frequency at which this standing wave occurs is called the **fundamental**. The fundamental frequency corresponds to a wavelength equal to twice the length of the string, $\lambda_1 = 2\ell$. The string can also vibrate at higher frequencies, called **overtones** or **harmonics**, in which there are one or more additional nodes. The frequency of each harmonic is a whole-number multiple of the fundamental.

In wind instruments, standing waves are set up in the column of air within the tube.

The vibrating air in an **open tube** (open at both ends) has displacement antinodes at both ends. The fundamental frequency corresponds to a wavelength equal to twice the tube length: $\lambda_1 = 2\ell$. The harmonics have frequencies that are 1, 2, 3, 4, ... times the fundamental frequency, just as for strings.

For a **closed tube** (closed at one end), the fundamental corresponds to a wavelength four times the length of the tube: $\lambda_1 = 4\ell$. Only the odd harmonics are present, equal to 1, 3, 5, 7, ... times the fundamental frequency.

Sound waves from different sources can interfere with each other. If two sounds are at slightly different frequencies, **beats** can be heard at a frequency equal to the difference in frequency of the two sources.

The **Doppler effect** refers to the change in pitch of a sound due to the motion either of the source or of the listener. If source and listener are approaching each other, the perceived pitch is higher; if they are moving apart, the perceived pitch is lower.

[*Shock waves and a sonic boom occur when an object moves at a supersonic speed—faster than the speed of sound. Ultrasonic-frequency (higher than 20 kHz) sound waves are used in many applications, including sonar and medical imaging.]

Questions

1. What is the evidence that sound travels as a wave?

2. What is the evidence that sound is a form of energy?

3. Children sometimes play with a homemade "telephone" by attaching a string to the bottoms of two paper cups. When the string is stretched and a child speaks into one cup, the sound can be heard at the other cup (Fig. 16–28). Explain clearly how the sound wave travels from one cup to the other.

FIGURE 16–28 Question 3.

4. When a sound wave passes from air into water, do you expect the frequency or wavelength to change?

5. What evidence can you give that the speed of sound in air does not depend significantly on frequency?

6. The voice of a person who has inhaled helium sounds very high-pitched. Why?

7. What is the main reason the speed of sound in hydrogen is greater than the speed of sound in air?

8. Two tuning forks oscillate with the same amplitude, but one has twice the frequency. Which (if either) produces the more intense sound?

9. How will the air temperature in a room affect the pitch of organ pipes?

10. Explain how a tube might be used as a filter to reduce the amplitude of sounds in various frequency ranges. (An example is a car muffler.)

11. Why are the frets on a guitar (Fig. 16–29) spaced closer together as you move up the fingerboard toward the bridge?

FIGURE 16–29
Question 11.

12. A noisy truck approaches you from behind a building. Initially you hear it but cannot see it. When it emerges and you do see it, its sound is suddenly "brighter"—you hear more of the high-frequency noise. Explain. [*Hint*: See Section 15–11 on diffraction.]

13. Standing waves can be said to be due to "interference in space," whereas beats can be said to be due to "interference in time." Explain.

14. In Fig. 16–15, if the frequency of the speakers is lowered, would the points D and C (where destructive and constructive interference occur) move farther apart or closer together?

15. Traditional methods of protecting the hearing of people who work in areas with very high noise levels have consisted mainly of efforts to block or reduce noise levels. With a relatively new technology, headphones are worn that do not block the ambient noise. Instead, a device is used which detects the noise, inverts it electronically, then feeds it to the headphones *in addition to* the ambient noise. How could adding *more* noise reduce the sound levels reaching the ears?

16. Consider the two waves shown in Fig. 16–30. Each wave can be thought of as a superposition of two sound waves with slightly different frequencies, as in Fig. 16–17. In which of the waves, (a) or (b), are the two component frequencies farther apart? Explain.

(a)

(b)

FIGURE 16–30 Question 16.

17. Is there a Doppler shift if the source and observer move in the same direction, with the same velocity? Explain.

18. If a wind is blowing, will this alter the frequency of the sound heard by a person at rest with respect to the source? Is the wavelength or velocity changed?

19. Figure 16–31 shows various positions of a child on a swing moving toward a person on the ground who is blowing a whistle. At which position, A through E, will the child hear the highest frequency for the sound of the whistle? Explain your reasoning.

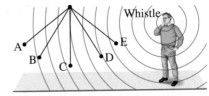

FIGURE 16–31
Question 19.

20. Approximately how many octaves are there in the human audible range?

21. At a race track, you can estimate the speed of cars just by listening to the difference in pitch of the engine noise between approaching and receding cars. Suppose the sound of a certain car drops by a full octave (frequency halved) as it goes by on the straightaway. How fast is it going?

Problems

[Unless stated otherwise, assume $T = 20°C$ and $v_{sound} = 343$ m/s in air.]

16-1 Characteristics of Sound

1. (I) A hiker determines the length of a lake by listening for the echo of her shout reflected by a cliff at the far end of the lake. She hears the echo 2.0 s after shouting. Estimate the length of the lake.

2. (I) A sailor strikes the side of his ship just below the waterline. He hears the echo of the sound reflected from the ocean floor directly below 2.5 s later. How deep is the ocean at this point? Assume the speed of sound in sea water is 1560 m/s (Table 16–1) and does not vary significantly with depth.

3. (I) (a) Calculate the wavelengths in air at 20°C for sounds in the maximum range of human hearing, 20 Hz to 20,000 Hz. (b) What is the wavelength of a 15-MHz ultrasonic wave?

4. (I) On a warm summer day (27°C), it takes 4.70 s for an echo to return from a cliff across a lake. On a winter day, it takes 5.20 s. What is the temperature on the winter day?

5. (II) A *motion sensor* can accurately measure the distance d to an object repeatedly via the sonar technique used in Example 16–2. A short ultrasonic pulse is emitted and reflects from any objects it encounters, creating echo pulses upon their arrival back at the sensor. The sensor measures the time interval t between the emission of the original pulse and the arrival of the first echo. (a) The smallest time interval t that can be measured with high precision is 1.0 ms. What is the smallest distance (at 20°C) that can be measured with the motion sensor? (b) If the motion sensor makes 15 distance measurements every second (that is, it emits 15 sound pulses per second at evenly spaced time intervals), the measurement of t must be completed within the time interval between the emissions of successive pulses. What is the largest distance (at 20°C) that can be measured with the motion sensor? (c) Assume that during a lab period the room's temperature increases from 20°C to 23°C. What percent error will this introduce into the motion sensor's distance measurements?

6. (II) An ocean fishing boat is drifting just above a school of tuna on a foggy day. Without warning, an engine backfire occurs on another boat 1.35 km away (Fig. 16–32). How much time elapses before the backfire is heard (a) by the fish, and (b) by the fishermen?

FIGURE 16–32
Problem 6.

7. (II) A stone is dropped from the top of a cliff. The splash it makes when striking the water below is heard 3.0 s later. How high is the cliff?

8. (II) A person, with his ear to the ground, sees a huge stone strike the concrete pavement. A moment later two sounds are heard from the impact: one travels in the air and the other in the concrete, and they are 0.75 s apart. How far away did the impact occur? See Table 16–1.

9. (II) Calculate the percent error made over one mile of distance by the "5-second rule" for estimating the distance from a lightning strike if the temperature is (a) 30°C, and (b) 10°C.

16-2 Mathematical Representation of Waves

10. (I) The pressure amplitude of a sound wave in air ($\rho = 1.29$ kg/m³) at 0°C is 3.0×10^{-3} Pa. What is the displacement amplitude if the frequency is (a) 150 Hz and (b) 15 kHz?

11. (I) What must be the pressure amplitude in a sound wave in air (0°C) if the air molecules undergo a maximum displacement equal to the diameter of an oxygen molecule, about 3.0×10^{-10} m? Assume a sound-wave frequency of (a) 55 Hz and (b) 5.5 kHz.

12. (II) Write an expression that describes the pressure variation as a function of x and t for the waves described in Problem 11.

13. (II) The pressure variation in a sound wave is given by
$$\Delta P = 0.0035 \sin(0.38\pi x - 1350\pi t),$$
where ΔP is in pascals, x in meters, and t in seconds. Determine (a) the wavelength, (b) the frequency, (c) the speed, and (d) the displacement amplitude of the wave. Assume the density of the medium to be $\rho = 2.3 \times 10^3$ kg/m³.

16-3 Intensity of Sound; Decibels

14. (I) What is the intensity of a sound at the pain level of 120 dB? Compare it to that of a whisper at 20 dB.

15. (I) What is the sound level of a sound whose intensity is 2.0×10^{-6} W/m²?

16. (I) What are the lowest and highest frequencies that an ear can detect when the sound level is 40 dB? (See Fig. 16–6.)

17. (II) Your auditory system can accommodate a huge range of sound levels. What is the ratio of highest to lowest intensity at (a) 100 Hz, (b) 5000 Hz? (See Fig. 16–6.)

18. (II) You are trying to decide between two new stereo amplifiers. One is rated at 100 W per channel and the other is rated at 150 W per channel. In terms of dB, how much louder will the more powerful amplifier be when both are producing sound at their maximum levels?

19. (II) At a painfully loud concert, a 120-dB sound wave travels away from a loudspeaker at 343 m/s. How much sound wave energy is contained in each 1.0-cm³ volume of air in the region near this loudspeaker?

20. (II) If two firecrackers produce a sound level of 95 dB when fired simultaneously at a certain place, what will be the sound level if only one is exploded?

21. (II) A person standing a certain distance from an airplane with four equally noisy jet engines is experiencing a sound level of 130 dB. What sound level would this person experience if the captain shut down all but one engine?

22. (II) A cassette player is said to have a signal-to-noise ratio of 62 dB, whereas for a CD player it is 98 dB. What is the ratio of intensities of the signal and the background noise for each device?

23. (II) (a) Estimate the power output of sound from a person speaking in normal conversation. Use Table 16–2. Assume the sound spreads roughly uniformly over a sphere centered on the mouth. (b) How many people would it take to produce a total sound output of 75 W of ordinary conversation? [Hint: Add intensities, not dBs.]

24. (II) A 50-dB sound wave strikes an eardrum whose area is 5.0×10^{-5} m². (a) How much energy is received by the eardrum per second? (b) At this rate, how long would it take your eardrum to receive a total energy of 1.0 J?

25. (II) Expensive amplifier A is rated at 250 W, while the more modest amplifier B is rated at 45 W. (a) Estimate the sound level in decibels you would expect at a point 3.5 m from a loudspeaker connected in turn to each amp. (b) Will the expensive amp sound twice as loud as the cheaper one?

26. (II) At a rock concert, a dB meter registered 130 dB when placed 2.2 m in front of a loudspeaker on the stage. (a) What was the power output of the speaker, assuming uniform spherical spreading of the sound and neglecting absorption in the air? (b) How far away would the sound level be a somewhat reasonable 85 dB?

27. (II) A fireworks shell explodes 100 m above the ground, creating a colorful display of sparks. How much greater is the sound level of the explosion for a person standing at a point directly below the explosion than for a person a horizontal distance of 200 m away (Fig. 16–33)?

FIGURE 16–33
Problem 27.

28. (II) If the amplitude of a sound wave is made 2.5 times greater, (a) by what factor will the intensity increase? (b) By how many dB will the sound level increase?

29. (II) Two sound waves have equal displacement amplitudes, but one has 2.6 times the frequency of the other. (a) Which has the greater pressure amplitude and by what factor is it greater? (b) What is the ratio of their intensities?

30. (II) What would be the sound level (in dB) of a sound wave in air that corresponds to a displacement amplitude of vibrating air molecules of 0.13 mm at 380 Hz?

31. (II) (a) Calculate the maximum displacement of air molecules when a 330-Hz sound wave passes whose intensity is at the threshold of pain (120 dB). (b) What is the pressure amplitude in this wave?

32. (II) A jet plane emits 5.0×10^5 J of sound energy per second. (a) What is the sound level 25 m away? Air absorbs sound at a rate of about 7.0 dB/km; calculate what the sound level will be (b) 1.00 km and (c) 7.50 km away from this jet plane, taking into account air absorption.

16–4 Sources of Sound: Strings and Air Columns

33. (I) What would you estimate for the length of a bass clarinet, assuming that it is modeled as a closed tube and that the lowest note that it can play is a D♭ whose frequency is 69.3 Hz?

34. (I) The A string on a violin has a fundamental frequency of 440 Hz. The length of the vibrating portion is 32 cm, and it has a mass of 0.35 g. Under what tension must the string be placed?

35. (I) An organ pipe is 124 cm long. Determine the fundamental and first three audible overtones if the pipe is (a) closed at one end, and (b) open at both ends.

36. (I) (a) What resonant frequency would you expect from blowing across the top of an empty soda bottle that is 21 cm deep, if you assumed it was a closed tube? (b) How would that change if it was one-third full of soda?

37. (I) If you were to build a pipe organ with open-tube pipes spanning the range of human hearing (20 Hz to 20 kHz), what would be the range of the lengths of pipes required?

38. (II) Estimate the frequency of the "sound of the ocean" when you put your ear very near a 20-cm-diameter seashell (Fig. 16–34).

FIGURE 16–34
Problem 38.

39. (II) An unfingered guitar string is 0.73 m long and is tuned to play E above middle C (330 Hz). (a) How far from the end of this string must a fret (and your finger) be placed to play A above middle C (440 Hz)? (b) What is the wavelength on the string of this 440-Hz wave? (c) What are the frequency and wavelength of the sound wave produced in air at 25°C by this fingered string?

40. (II) (a) Determine the length of an open organ pipe that emits middle C (262 Hz) when the temperature is 15°C. (b) What are the wavelength and frequency of the fundamental standing wave in the tube? (c) What are λ and f in the traveling sound wave produced in the outside air?

41. (II) An organ is in tune at 22.0°C. By what percent will the frequency be off at 5.0°C?

42. (II) How far from the mouthpiece of the flute in Example 16–11 should the hole be that must be uncovered to play F above middle C at 349 Hz?

43. (II) A bugle is simply a tube of fixed length that behaves as if it is open at both ends. A bugler, by adjusting his lips correctly and blowing with proper air pressure, can cause a harmonic (usually other than the fundamental) of the air column within the tube to sound loudly. Standard military tunes like *Taps* and *Reveille* require only four musical notes: G4 (392 Hz), C5 (523 Hz), E5 (659 Hz), and G5 (784 Hz). (a) For a certain length ℓ, a bugle will have a sequence of four consecutive harmonics whose frequencies very nearly equal those associated with the notes G4, C5, E5, and G5. Determine this ℓ. (b) Which harmonic is each of the (approximate) notes G4, C5, E5, and G5 for the bugle?

44. (II) A particular organ pipe can resonate at 264 Hz, 440 Hz, and 616 Hz, but not at any other frequencies in between. (a) Show why this is an open or a closed pipe. (b) What is the fundamental frequency of this pipe?

45. (II) When a player's finger presses a guitar string down onto a fret, the length of the vibrating portion of the string is shortened, thereby increasing the string's fundamental frequency (see Fig. 16–35). The string's tension and mass per unit length remain unchanged. If the unfingered length of the string is ℓ = 65.0 cm, determine the positions x of the first six frets, if each fret raises the pitch of the fundamental by one musical note in comparison to the neighboring fret. On the equally tempered chromatic scale, the ratio of frequencies of neighboring notes is $2^{1/12}$.

FIGURE 16–35
Problem 45.

ℓ = 65.0 cm

46. (II) A uniform narrow tube 1.80 m long is open at both ends. It resonates at two successive harmonics of frequencies 275 Hz and 330 Hz. What is (a) the fundamental frequency, and (b) the speed of sound in the gas in the tube?

47. (II) A pipe in air at 23.0°C is to be designed to produce two successive harmonics at 240 Hz and 280 Hz. How long must the pipe be, and is it open or closed?

48. (II) How many overtones are present within the audible range for a 2.48-m-long organ pipe at 20°C (a) if it is open, and (b) if it is closed?

49. (II) Determine the fundamental and first overtone frequencies for an 8.0-m-long hallway with all doors closed. Model the hallway as a tube closed at both ends.

50. (II) In a *quartz oscillator*, used as a stable clock in electronic devices, a transverse (shear) standing sound wave is excited across the thickness d of a quartz disk and its frequency f is detected electronically. The parallel faces of the disk are unsupported and so behave as "free ends" when the sound wave reflects from them (see Fig. 16–36). If the oscillator is designed to operate with the first harmonic, determine the required disk thickness if $f = 12.0$ MHz. The density and shear modulus of quartz are $\rho = 2650$ kg/m^3 and $G = 2.95 \times 10^{10}$ N/m^2.

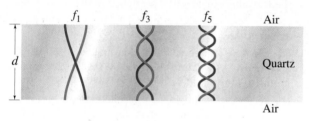

FIGURE 16–36 Problem 50.

51. (III) The human ear canal is approximately 2.5 cm long. It is open to the outside and is closed at the other end by the eardrum. Estimate the frequencies (in the audible range) of the standing waves in the ear canal. What is the relationship of your answer to the information in the graph of Fig. 16–6?

*16–5 Quality of Sound, Superposition

*52. (II) Approximately what are the intensities of the first two overtones of a violin compared to the fundamental? How many decibels softer than the fundamental are the first and second overtones? (See Fig. 16–14.)

16–6 Interference; Beats

53. (I) A piano tuner hears one beat every 2.0 s when trying to adjust two strings, one of which is sounding 370 Hz. How far off in frequency is the other string?

54. (I) What is the beat frequency if middle C (262 Hz) and C$^{\#}$ (277 Hz) are played together? What if each is played two octaves lower (each frequency reduced by a factor of 4)?

55. (II) A guitar string produces 4 beats/s when sounded with a 350-Hz tuning fork and 9 beats/s when sounded with a 355-Hz tuning fork. What is the vibrational frequency of the string? Explain your reasoning.

56. (II) The two sources of sound in Fig. 16–15 face each other and emit sounds of equal amplitude and equal frequency (294 Hz) but 180° out of phase. For what minimum separation of the two speakers will there be some point at which (a) complete constructive interference occurs and (b) complete destructive interference occurs. (Assume $T = 20$°C.)

57. (II) How many beats will be heard if two identical flutes, each 0.66 m long, try to play middle C (262 Hz), but one is at 5.0°C and the other at 28°C?

58. (II) Two loudspeakers are placed 3.00 m apart, as shown in Fig. 16–37. They emit 494-Hz sounds, in phase. A microphone is placed 3.20 m distant from a point midway between the two speakers, where an intensity maximum is recorded. (a) How far must the microphone be moved to the right to find the first intensity minimum? (b) Suppose the speakers are reconnected so that the 494-Hz sounds they emit are exactly out of phase. At what positions are the intensity maximum and minimum now?

FIGURE 16–37
Problem 58.

59. (II) Two piano strings are supposed to be vibrating at 220 Hz, but a piano tuner hears three beats every 2.0 s when they are played together. (a) If one is vibrating at 220.0 Hz, what must be the frequency of the other (is there only one answer)? (b) By how much (in percent) must the tension be increased or decreased to bring them in tune?

60. (II) A source emits sound of wavelengths 2.64 m and 2.72 m in air. (a) How many beats per second will be heard? (Assume $T = 20$°C.) (b) How far apart in space are the regions of maximum intensity?

16–7 Doppler Effect

61. (I) The predominant frequency of a certain fire truck's siren is 1350 Hz when at rest. What frequency do you detect if you move with a speed of 30.0 m/s (a) toward the fire truck, and (b) away from it?

62. (I) A bat at rest sends out ultrasonic sound waves at 50.0 kHz and receives them returned from an object moving directly away from it at 30.0 m/s. What is the received sound frequency?

63. (II) (a) Compare the shift in frequency if a 2300-Hz source is moving toward you at 18 m/s, versus you moving toward it at 18 m/s. Are the two frequencies exactly the same? Are they close? (b) Repeat the calculation for 160 m/s and then again (c) for 320 m/s. What can you conclude about the asymmetry of the Doppler formulas? (d) Show that at low speeds (relative to the speed of sound), the two formulas—source approaching and detector approaching—yield the same result.

64. (II) Two automobiles are equipped with the same single-frequency horn. When one is at rest and the other is moving toward the first at 15 m/s, the driver at rest hears a beat frequency of 4.5 Hz. What is the frequency the horns emit? Assume $T = 20$°C.

65. (II) A police car sounding a siren with a frequency of 1280 Hz is traveling at 120.0 km/h. (a) What frequencies does an observer standing next to the road hear as the car approaches and as it recedes? (b) What frequencies are heard in a car traveling at 90.0 km/h in the opposite direction before and after passing the police car? (c) The police car passes a car traveling in the same direction at 80.0 km/h. What two frequencies are heard in this car?

66. (II) A bat flies toward a wall at a speed of 7.0 m/s. As it flies, the bat emits an ultrasonic sound wave with frequency 30.0 kHz. What frequency does the bat hear in the reflected wave?

67. (II) In one of the original Doppler experiments, a tuba was played on a moving flat train car at a frequency of 75 Hz, and a second identical tuba played the same tone while at rest in the railway station. What beat frequency was heard in the station if the train car approached the station at a speed of 12.0 m/s?

68. (II) If a speaker mounted on an automobile broadcasts a song, with what speed (km/h) does the automobile have to move toward a stationary listener so that the listener hears the song with each musical note shifted up by one note in comparison to the song heard by the automobile's driver? On the equally tempered chromatic scale, the ratio of frequencies of neighboring notes is $2^{1/12}$.

69. (II) A wave on the surface of the ocean with wavelength 44 m is traveling east at a speed of 18 m/s relative to the ocean floor. If, on this stretch of ocean surface, a powerboat is moving at 15 m/s (relative to the ocean floor), how often does the boat encounter a wave crest, if the boat is traveling (a) west, and (b) east?

70. (III) A factory whistle emits sound of frequency 720 Hz. When the wind velocity is 15.0 m/s from the north, what frequency will observers hear who are located, at rest, (a) due north, (b) due south, (c) due east, and (d) due west, of the whistle? What frequency is heard by a cyclist heading (e) north or (f) west, toward the whistle at 12.0 m/s? Assume $T = 20°C$.

71. (III) The Doppler effect using ultrasonic waves of frequency 2.25×10^6 Hz is used to monitor the heartbeat of a fetus. A (maximum) beat frequency of 260 Hz is observed. Assuming that the speed of sound in tissue is 1.54×10^3 m/s, calculate the maximum velocity of the surface of the beating heart.

*16–8 Shock Waves; Sonic Boom

*72. (II) An airplane travels at Mach 2.0 where the speed of sound is 310 m/s. (a) What is the angle the shock wave makes with the direction of the airplane's motion? (b) If the plane is flying at a height of 6500 m, how long after it is directly overhead will a person on the ground hear the shock wave?

*73. (II) A space probe enters the thin atmosphere of a planet where the speed of sound is only about 45 m/s. (a) What is the probe's Mach number if its initial speed is 15,000 km/h? (b) What is the angle of the shock wave relative to the direction of motion?

*74. (II) A meteorite traveling 8800 m/s strikes the ocean. Determine the shock wave angle it produces (a) in the air just before entering the ocean, and (b) in the water just after entering. Assume $T = 20°C$.

*75. (II) Show that the angle θ a sonic boom makes with the path of a supersonic object is given by Eq. 16–12.

*76. (II) You look directly overhead and see a plane exactly 1.25 km above the ground flying faster than the speed of sound. By the time you hear the sonic boom, the plane has traveled a horizontal distance of 2.0 km. See Fig. 16–38. Determine (a) the angle of the shock cone, θ, and (b) the speed of the plane (the Mach number). Assume the speed of sound is 330 m/s.

FIGURE 16–38
Problem 76.

*77. (II) A supersonic jet traveling at Mach 2.2 at an altitude of 9500 m passes directly over an observer on the ground. Where will the plane be relative to the observer when the latter hears the sonic boom? (See Fig. 16–39.)

FIGURE 16–39 Problem 77.

General Problems

78. A fish finder uses a sonar device that sends 20,000-Hz sound pulses downward from the bottom of the boat, and then detects echoes. If the maximum depth for which it is designed to work is 75 m, what is the minimum time between pulses (in fresh water)?

79. A science museum has a display called a sewer pipe symphony. It consists of many plastic pipes of various lengths, which are open on both ends. (a) If the pipes have lengths of 3.0 m, 2.5 m, 2.0 m, 1.5 m and 1.0 m, what frequencies will be heard by a visitor's ear placed near the ends of the pipes? (b) Why does this display work better on a noisy day than on a quiet day?

80. A single mosquito 5.0 m from a person makes a sound close to the threshold of human hearing (0 dB). What will be the sound level of 100 such mosquitoes?

81. What is the resultant sound level when an 82-dB sound and an 89-dB sound are heard simultaneously?

82. The sound level 9.00 m from a loudspeaker, placed in the open, is 115 dB. What is the acoustic power output (W) of the speaker, assuming it radiates equally in all directions?

83. A stereo amplifier is rated at 175 W output at 1000 Hz. The power output drops by 12 dB at 15 kHz. What is the power output in watts at 15 kHz?

84. Workers around jet aircraft typically wear protective devices over their ears. Assume that the sound level of a jet airplane engine, at a distance of 30 m, is 130 dB, and that the average human ear has an effective radius of 2.0 cm. What would be the power intercepted by an unprotected ear at a distance of 30 m from a jet airplane engine?

85. In audio and communications systems, the *gain*, β, in decibels is defined as

$$\beta = 10 \log \left(\frac{P_{out}}{P_{in}} \right),$$

where P_{in} is the power input to the system and P_{out} is the power output. A particular stereo amplifier puts out 125 W of power for an input of 1.0 mW. What is its gain in dB?

86. For large concerts, loudspeakers are sometimes used to amplify a singer's sound. The human brain interprets sounds that arrive within 50 ms of the original sound as if they came from the same source. Thus if the sound from a loudspeaker reaches a listener first, it would sound as if the loudspeaker is the source of the sound. Conversely, if the singer is heard first and the loudspeaker adds to the sound within 50 ms, the sound would seem to come from the singer, who would now seem to be singing louder. The second situation is desired. Because the signal to the loudspeaker travels at the speed of light $(3 \times 10^8 \text{ m/s})$, which is much faster than the speed of sound, a delay is added to the signal sent to the loudspeaker. How much delay must be added if the loudspeaker is 3.0 m behind the singer and we want its sound to arrive 30 ms after the singer's?

87. Manufacturers typically offer a particular guitar string in a choice of diameters so that players can tune their instruments with a preferred string tension. For example, a nylon high-E string is available in a low- and high-tension model with diameter 0.699 mm and 0.724 mm, respectively. Assuming the density ρ of nylon is the same for each model, compare (as a ratio) the tension in a tuned high- and low-tension string.

88. The high-E string on a guitar is fixed at both ends with length $\ell = 65.0$ cm and fundamental frequency $f_1 = 329.6$ Hz. On an acoustic guitar, this string typically has a diameter of 0.33 mm and is commonly made of brass (7760 kg/m^3), while on an electric guitar it has a diameter of 0.25 mm and is made of nickel-coated steel (7990 kg/m^3). Compare (as a ratio) the high-E string tension on an acoustic versus an electric guitar.

89. The A string of a violin is 32 cm long between fixed points with a fundamental frequency of 440 Hz and a mass per unit length of 7.2×10^{-4} kg/m. (a) What are the wave speed and tension in the string? (b) What is the length of the tube of a simple wind instrument (say, an organ pipe) closed at one end whose fundamental is also 440 Hz if the speed of sound is 343 m/s in air? (c) What is the frequency of the first over-tone of each instrument?

90. A tuning fork is set into vibration above a vertical open tube filled with water (Fig. 16–40). The water level is allowed to drop slowly. As it does so, the air in the tube above the water level is heard to resonate with the tuning fork when the distance from the tube opening to the water level is 0.125 m and again at 0.395 m. What is the frequency of the tuning fork?

FIGURE 16–40
Problem 90.

0.125 m

0.395 m

91. Two identical tubes, each closed at one end, have a fundamental frequency of 349 Hz at 25.0°C. The air temperature is increased to 30.0°C in one tube. If the two pipes are sounded together now, what beat frequency results?

92. Each string on a violin is tuned to a frequency $1\frac{1}{2}$ times that of its neighbor. The four equal-length strings are to be placed under the same tension; what must be the mass per unit length of each string relative to that of the lowest string?

93. The diameter D of a tube does affect the node at the open end of a tube. The end correction can be roughly approximated as adding $D/3$ to the effective length of the tube. For a closed tube of length 0.60 m and diameter 3.0 cm, what are the first four harmonics, taking the end correction into consideration?

94. A person hears a pure tone in the 500 to 1000-Hz range coming from two sources. The sound is loudest at points equidistant from the two sources. To determine exactly what the frequency is, the person moves about and finds that the sound level is minimal at a point 0.36 m farther from one source than the other. What is the frequency of the sound?

95. The frequency of a steam train whistle as it approaches you is 552 Hz. After it passes you, its frequency is measured as 486 Hz. How fast was the train moving (assume constant velocity)?

96. Two trains emit 516-Hz whistles. One train is stationary. The conductor on the stationary train hears a 3.5-Hz beat frequency when the other train approaches. What is the speed of the moving train?

97. Two loudspeakers are at opposite ends of a railroad car as it moves past a stationary observer at 10.0 m/s, as shown in Fig. 16–41. If the speakers have identical sound frequencies of 348 Hz, what is the beat frequency heard by the observer when (a) he listens from the position A, in front of the car, (b) he is between the speakers, at B, and (c) he hears the speakers after they have passed him, at C?

$v = 10.0$ m/s

C   B   A

FIGURE 16–41 Problem 97.

98. Two open organ pipes, sounding together, produce a beat frequency of 8.0 Hz. The shorter one is 2.40 m long. How long is the other?

99. A bat flies toward a moth at speed 7.5 m/s while the moth is flying toward the bat at speed 5.0 m/s. The bat emits a sound wave of 51.35 kHz. What is the frequency of the wave detected by the bat after that wave reflects off the moth?

100. If the velocity of blood flow in the aorta is normally about 0.32 m/s, what beat frequency would you expect if 3.80-MHz ultrasound waves were directed along the flow and reflected from the red blood cells? Assume that the waves travel with a speed of 1.54×10^3 m/s.

101. A bat emits a series of high-frequency sound pulses as it approaches a moth. The pulses are approximately 70.0 ms apart, and each is about 3.0 ms long. How far away can the moth be detected by the bat so that the echo from one pulse returns before the next pulse is emitted?

102. (a) Use the binomial expansion to show that Eqs. 16–9a and 16–10a become essentially the same for small relative velocity between source and observer. (b) What percent error would result if Eq. 16–10a were used instead of Eq. 16–9a for a relative velocity of 18.0 m/s?

103. Two loudspeakers face each other at opposite ends of a long corridor. They are connected to the same source which produces a pure tone of 282 Hz. A person walks from one speaker toward the other at a speed of 1.4 m/s. What "beat" frequency does the person hear?

104. A *Doppler flow meter* is used to measure the speed of blood flow. Transmitting and receiving elements are placed on the skin, as shown in Fig. 16–42. Typical sound-wave frequencies of about 5.0 MHz are used, which have a reasonable chance of being reflected from red blood cells. By measuring the frequency of the reflected waves, which are Doppler-shifted because the red blood cells are moving, the speed of the blood flow can be deduced. "Normal" blood flow speed is about 0.1 m/s. Suppose that an artery is partly constricted, so that the speed of the blood flow is increased, and the flow meter measures a Doppler shift of 780 Hz. What is the speed of blood flow in the constricted region? The effective angle between the sound waves (both transmitted and reflected) and the direction of blood flow is 45°. Assume the velocity of sound in tissue is 1540 m/s.

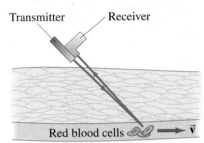

FIGURE 16–42
Problem 104.

105. The wake of a speedboat is 15° in a lake where the speed of the water wave is 2.2 km/h. What is the speed of the boat?

106. A source of sound waves (wavelength λ) is a distance ℓ from a detector. Sound reaches the detector directly, and also by reflecting off an obstacle, as shown in Fig. 16–43. The obstacle is equidistant from source and detector. When the obstacle is a distance d to the right of the line of sight between source and detector, as shown, the two waves arrive in phase. How much farther to the right must the obstacle be moved if the two waves are to be out of phase by $\frac{1}{2}$ wavelength, so destructive interference occurs? (Assume $\lambda \ll \ell, d$.)

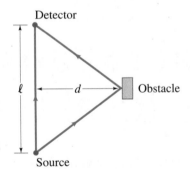

FIGURE 16–43
Problem 106.

107. A dramatic demonstration, called "singing rods," involves a long, slender aluminum rod held in the hand near the rod's midpoint. The rod is stroked with the other hand. With a little practice, the rod can be made to "sing," or emit a clear, loud, ringing sound. For a 75-cm-long rod, (*a*) what is the fundamental frequency of the sound? (*b*) What is its wavelength in the rod, and (*c*) what is the wavelength of the sound in air at 20°C?

108. Assuming that the maximum displacement of the air molecules in a sound wave is about the same as that of the speaker cone that produces the sound (Fig. 16–44), estimate by how much a loudspeaker cone moves for a fairly loud (105 dB) sound of (*a*) 8.0 kHz, and (*b*) 35 Hz.

FIGURE 16–44
Problem 108.

*Numerical/Computer

*109. (III) The manner in which a string is plucked determines the mixture of harmonic amplitudes in the resulting wave. Consider a string exactly $\frac{1}{2}$-m long that is fixed at both its ends located at $x = 0.0$ and $x = \frac{1}{2}$ m. The first five harmonics of this string have wavelengths $\lambda_1 = 1.0$ m, $\lambda_2 = \frac{1}{2}$ m, $\lambda_3 = \frac{1}{3}$ m, $\lambda_4 = \frac{1}{4}$ m, and $\lambda_5 = \frac{1}{5}$ m. According to Fourier's theorem, any shape of this string can be formed by a sum of its harmonics, with each harmonic having its own unique amplitude A. We limit the sum to the first five harmonics in the expression

$$D(x) = A_1 \sin\left(\frac{2\pi}{\lambda_1} x\right) + A_2 \sin\left(\frac{2\pi}{\lambda_2} x\right)$$
$$+ A_3 \sin\left(\frac{2\pi}{\lambda_3} x\right) + A_4 \sin\left(\frac{2\pi}{\lambda_4} x\right) + A_5 \sin\left(\frac{2\pi}{\lambda_5} x\right),$$

and D is the displacement of the string at a time $t = 0$. Imagine plucking this string at its midpoint (Fig. 16–45a) or at a point two-thirds from the left end (Fig. 16–45b). Using a graphing calculator or computer program, show that the above expression can fairly accurately represent the shape in: (*a*) Fig. 16–45a, if
$A_1 = 1.00$,
$A_2 = 0.00$,
$A_3 = -0.11$,
$A_4 = 0.00$, and
$A_5 = 0.040$; and in
(*b*) Fig. 16–45b, if
$A_1 = 0.87$,
$A_2 = -0.22$,
$A_3 = 0.00$,
$A_4 = 0.054$, and
$A_5 = -0.035$.

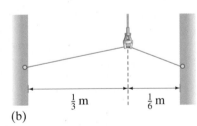

FIGURE 16–45
Problem 109.   (b)

Answers to Exercises

A: 1 km for every 3 s before the thunder is heard.

B: 4 times as intense.

C: (*b*).

D: The less-massive one.

E: 257 Hz.

F: (*b*).

G: (*a*) 1717 Hz, (*b*) 1483 Hz.

H: (*a*).

Heating the air inside a "hot-air" balloon raises the air's temperature, causing it to expand, and forces air out the opening at the bottom. The reduced amount of air inside means its density is lower than the outside air, so there is a net buoyant force upward on the balloon. In this Chapter we study temperature and its effects on matter: thermal expansion and the gas laws.

C H A P T E R

17

Temperature, Thermal Expansion, and the Ideal Gas Law

CONTENTS

CHAPTER-OPENING QUESTION—Guess now!

A hot-air balloon, open at one end (see photos above), rises when the air inside is heated by a flame. For the following properties, is the air inside the balloon higher, lower, or the same as for the air outside the balloon?

(i) Temperature,
(ii) Pressure,
(iii) Density.

In the next four Chapters, Chapters 17 through 20, we study temperature, heat and thermodynamics, and the kinetic theory of gases.

We will often consider a particular **system**, by which we mean a particular object or set of objects; everything else in the universe is called the "environment." We can describe the **state** (or condition) of a particular system—such as a gas in a container—from either a microscopic or macroscopic point of view. A **microscopic** description would involve details of the motion of all the atoms or molecules making up the system, which could be very complicated. A **macroscopic** description is given in terms of quantities that are detectable directly by our senses and instruments, such as volume, mass, pressure, and temperature.

The description of processes in terms of macroscopic quantities is the field of **thermodynamics**. Quantities that can be used to describe the state of a system are called **state variables**. To describe the state of a pure gas in a container, for example, requires only three state variables, which are typically the volume, the pressure, and the temperature. More complex systems require more than three state variables to describe them.

The emphasis in this Chapter is on the concept of temperature. We begin, however, with a brief discussion of the theory that matter is made up of atoms and that these atoms are in continual random motion. This theory is called *kinetic theory* ("kinetic," you may recall, is Greek for "moving"), and we discuss it in more detail in Chapter 18.

17–1 Atomic Theory of Matter

The idea that matter is made up of atoms dates back to the ancient Greeks. According to the Greek philosopher Democritus, if a pure substance—say, a piece of iron—were cut into smaller and smaller bits, eventually a smallest piece of that substance would be obtained which could not be divided further. This smallest piece was called an **atom**, which in Greek means "indivisible."[†]

Today the atomic theory is universally accepted. The experimental evidence in its favor, however, came mainly in the eighteenth, nineteenth, and twentieth centuries, and much of it was obtained from the analysis of chemical reactions.

We will often speak of the relative masses of individual atoms and molecules—what we call the **atomic mass** or **molecular mass**, respectively.[‡] These are based on arbitrarily assigning the abundant carbon atom, ^{12}C, the atomic mass of exactly 12.0000 **unified atomic mass units** (u). In terms of kilograms,

$$1\,u = 1.6605 \times 10^{-27}\,kg.$$

The atomic mass of hydrogen is then 1.0078 u, and the values for other atoms are as listed in the Periodic Table inside the back cover of this book, and also in Appendix F. The molecular mass of a compound is the sum of atomic masses of the atoms making up the molecules of that compound.[§]

An important piece of evidence for the atomic theory is called **Brownian motion**, named after the biologist Robert Brown, who is credited with its discovery in 1827. While he was observing tiny pollen grains suspended in water under his microscope, Brown noticed that the tiny grains moved about in tortuous paths (Fig. 17–1), even though the water appeared to be perfectly still. The atomic theory easily explains Brownian motion if the further reasonable assumption is made that the atoms of any substance are continually in motion. Then Brown's tiny pollen grains are jostled about by the vigorous barrage of rapidly moving molecules of water.

In 1905, Albert Einstein examined Brownian motion from a theoretical point of view and was able to calculate from the experimental data the approximate size and mass of atoms and molecules. His calculations showed that the diameter of a typical atom is about 10^{-10} m.

FIGURE 17–1 Path of a tiny particle (pollen grain, for example) suspended in water. The straight lines connect observed positions of the particle at equal time intervals.

[†]Today we do not consider the atom as indivisible, but rather as consisting of a nucleus (containing protons and neutrons) and electrons.

[‡]The terms *atomic weight* and *molecular weight* are sometimes used for these quantities, but properly speaking we are comparing masses.

[§]An *element* is a substance, such as gold, iron, or copper, that cannot be broken down into simpler substances by chemical means. *Compounds* are substances made up of elements, and can be broken down into them; examples are carbon dioxide and water. The smallest piece of an element is an atom; the smallest piece of a compound is a molecule. Molecules are made up of atoms; a molecule of water, for example, is made up of two atoms of hydrogen and one of oxygen; its chemical formula is H_2O.

(a)

(b)

(c)

FIGURE 17–2
Atomic arrangements in
(a) a crystalline solid, (b) a liquid,
and (c) a gas.

At the start of Chapter 13, we distinguished the three common phases (or states) of matter—solid, liquid, gas—based on **macroscopic**, or "large-scale," properties. Now let us see how these three phases of matter differ, from the atomic or **microscopic** point of view. Clearly, atoms and molecules must exert attractive forces on each other. For how else could a brick or a block of aluminum hold together in one piece? The attractive forces between molecules are of an electrical nature (more on this in later Chapters). When molecules come too close together, the force between them must become repulsive (electric repulsion between their outer electrons), for how else could matter take up space? Thus molecules maintain a minimum distance from each other. In a solid material, the attractive forces are strong enough that the atoms or molecules move only slightly (oscillate) about relatively fixed positions, often in an array known as a crystal lattice, as shown in Fig. 17–2a. In a liquid, the atoms or molecules are moving more rapidly, or the forces between them are weaker, so that they are sufficiently free to pass around one another, as in Fig. 17–2b. In a gas, the forces are so weak, or the speeds so high, that the molecules do not even stay close together. They move rapidly every which way, Fig. 17–2c, filling any container and occasionally colliding with one another. On average, the speeds are sufficiently high in a gas that when two molecules collide, the force of attraction is not strong enough to keep them close together and they fly off in new directions.

EXAMPLE 17–1 **ESTIMATE** **Distance between atoms.** The density of copper is $8.9 \times 10^3 \text{ kg/m}^3$, and each copper atom has a mass of 63 u. Estimate the average distance between the centers of neighboring copper atoms.

APPROACH We consider a cube of copper 1 m on a side. From the given density ρ we can calculate the mass m of a cube of volume $V = 1 \text{ m}^3$ ($m = \rho V$). We divide this by the mass of one atom (63 u) to obtain the number of atoms in 1 m^3. We assume the atoms are in a uniform array, and we let N be the number of atoms in a 1-m length; then $(N)(N)(N) = N^3$ equals the total number of atoms in 1 m^3.

SOLUTION The mass of 1 copper atom is $63 \text{ u} = 63 \times 1.66 \times 10^{-27} \text{ kg} = 1.05 \times 10^{-25} \text{ kg}$. This means that in a cube of copper 1 m on a side (volume $= 1 \text{ m}^3$), there are

$$\frac{8.9 \times 10^3 \text{ kg/m}^3}{1.05 \times 10^{-25} \text{ kg/atom}} = 8.5 \times 10^{28} \text{ atoms/m}^3.$$

The volume of a cube of side ℓ is $V = \ell^3$, so on one edge of the 1-m-long cube there are $(8.5 \times 10^{28})^{\frac{1}{3}}$ atoms $= 4.4 \times 10^9$ atoms. Hence the distance between neighboring atoms is

$$\frac{1 \text{ m}}{4.4 \times 10^9 \text{ atoms}} = 2.3 \times 10^{-10} \text{ m}.$$

NOTE Watch out for units. Even though "atoms" is not a unit, it is helpful to include it to make sure you calculate correctly.

FIGURE 17–3 Expansion joint on a bridge.

17–2 Temperature and Thermometers

In everyday life, **temperature** is a measure of how hot or cold something is. A hot oven is said to have a high temperature, whereas the ice of a frozen lake is said to have a low temperature.

Many properties of matter change with temperature. For example, most materials expand when their temperature is increased.[†] An iron beam is longer when hot than when cold. Concrete roads and sidewalks expand and contract slightly according to temperature, which is why compressible spacers or expansion joints (Fig. 17–3) are placed at regular intervals. The electrical resistance of matter changes with temperature (Chapter 25). So too does the color radiated by objects, at least at high temperatures: you may have noticed that the heating element of an electric stove glows with a red color when hot.

[†]Most materials expand when their temperature is raised, but not all. Water, for example, in the range 0°C to 4°C contracts with an increase in temperature (see Section 17–4).

At higher temperatures, solids such as iron glow orange or even white. The white light from an ordinary incandescent lightbulb comes from an extremely hot tungsten wire. The surface temperatures of the Sun and other stars can be measured by the predominant color (more precisely, wavelengths) of light they emit.

Instruments designed to measure temperature are called **thermometers**. There are many kinds of thermometers, but their operation always depends on some property of matter that changes with temperature. Many common thermometers rely on the expansion of a material with an increase in temperature. The first idea for a thermometer, by Galileo, made use of the expansion of a gas. Common thermometers today consist of a hollow glass tube filled with mercury or with alcohol colored with a red dye, as were the earliest usable thermometers (Fig. 17–4).

Inside a common liquid-in-glass thermometer, the liquid expands more than the glass when the temperature is increased, so the liquid level rises in the tube (Fig. 17–5a). Although metals also expand with temperature, the change in length of a metal rod, say, is generally too small to measure accurately for ordinary changes in temperature. However, a useful thermometer can be made by bonding together two dissimilar metals whose rates of expansion are different (Fig. 17–5b). When the temperature is increased, the different amounts of expansion cause the bimetallic strip to bend. Often the bimetallic strip is in the form of a coil, one end of which is fixed while the other is attached to a pointer, Fig. 17–6. This kind of thermometer is used as ordinary air thermometers, oven thermometers, automatic off switches in electric coffeepots, and in room thermostats for determining when the heater or air conditioner should go on or off. Very precise thermometers make use of electrical properties (Chapter 25), such as resistance thermometers, thermocouples, and thermistors, often with a digital readout.

Temperature Scales

In order to measure temperature quantitatively, some sort of numerical scale must be defined. The most common scale today is the **Celsius** scale, sometimes called the **centigrade** scale. In the United States, the **Fahrenheit** scale is also common. The most important scale in scientific work is the absolute, or Kelvin, scale, and it will be discussed later in this Chapter.

One way to define a temperature scale is to assign arbitrary values to two readily reproducible temperatures. For both the Celsius and Fahrenheit scales these two fixed points are chosen to be the freezing point and the boiling point[†] of water, both taken at standard atmospheric pressure. On the Celsius scale, the freezing point of water is chosen to be 0°C ("zero degrees Celsius") and the boiling point 100°C. On the Fahrenheit scale, the freezing point is defined as 32°F and the boiling point 212°F. A practical thermometer is calibrated by placing it in carefully prepared environments at each of the two temperatures and marking the position of the liquid or pointer. For a Celsius scale, the distance between the two marks is divided into one hundred equal intervals representing each degree between 0°C and 100°C (hence the name "centigrade scale" meaning "hundred steps"). For a Fahrenheit scale, the two points are labeled 32°F and 212°F and the distance between them is divided into 180 equal intervals. For temperatures below the freezing point of water and above the boiling point of water, the scales may be extended using the same equally spaced intervals. However, thermometers can be used only over a limited temperature range because of their own limitations—for example, the liquid mercury in a mercury-in-glass thermometer solidifies at some point, below which the thermometer will be useless. It is also rendered useless above temperatures where the fluid, such as alcohol, vaporizes. For very low or very high temperatures, specialized thermometers are required, some of which we will mention later.

[†]The freezing point of a substance is defined as that temperature at which the solid and liquid phases coexist in equilibrium—that is, without any net liquid changing into the solid or vice versa. Experimentally, this is found to occur at only one definite temperature, for a given pressure. Similarly, the boiling point is defined as that temperature at which the liquid and gas coexist in equilibrium. Since these points vary with pressure, the pressure must be specified (usually it is 1 atm).

FIGURE 17–4 Thermometers built by the Accademia del Cimento (1657–1667) in Florence, Italy, are among the earliest known. These sensitive and exquisite instruments contained alcohol, sometimes colored, like many thermometers today.

FIGURE 17–5 (a) Mercury- or alcohol-in-glass thermometer; (b) bimetallic strip.

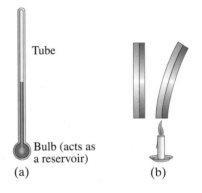

Tube

Bulb (acts as a reservoir)

(a)   (b)

FIGURE 17–6 Photograph of a thermometer using a coiled bimetallic strip.

FIGURE 17–7 Celsius and Fahrenheit scales compared.

⚠ **CAUTION**

Convert temperature by remembering $0°C = 32°F$ *and a change of* $5\,C° = 9\,F°$

FIGURE 17–8 Constant-volume gas thermometer.

Every temperature on the Celsius scale corresponds to a particular temperature on the Fahrenheit scale, Fig. 17–7. It is easy to convert from one to the other if you remember that 0°C corresponds to 32°F and that a range of 100° on the Celsius scale corresponds to a range of 180° on the Fahrenheit scale. Thus, one Fahrenheit degree (1 F°) corresponds to $100/180 = \frac{5}{9}$ of a Celsius degree (1 C°). That is, $1\,F° = \frac{5}{9}C°$. (Notice that when we refer to a specific temperature, we say "degrees Celsius," as in 20°C; but when we refer to a *change* in temperature or a temperature interval, we say "Celsius degrees," as in "2 C°.") The conversion between the two temperature scales can be written

$$T(°C) = \tfrac{5}{9}\big[T(°F) - 32\big]$$

or

$$T(°F) = \tfrac{9}{5}T(°C) + 32.$$

Rather than memorizing these relations (it would be easy to confuse them), it is usually easier simply to remember that $0°C = 32°F$ and that a change of $5\,C° =$ a change of $9\,F°$.

EXAMPLE 17–2 **Taking your temperature.** Normal body temperature is 98.6°F. What is this on the Celsius scale?

APPROACH We recall that $0°C = 32°F$ and $5\,C° = 9\,F°$.

SOLUTION First we relate the given temperature to the freezing point of water (0°C). That is, 98.6°F is $98.6 - 32.0 = 66.6\,F°$ above the freezing point of water. Since each F° is equal to $\frac{5}{9}C°$, this corresponds to $66.6 \times \frac{5}{9} = 37.0$ Celsius degrees above the freezing point. The freezing point is 0°C, so the temperature is 37.0°C.

EXERCISE A Determine the temperature at which both scales give the same numerical reading $(T_C = T_F)$.

Different materials do not expand in quite the same way over a wide temperature range. Consequently, if we calibrate different kinds of thermometers exactly as described above, they will not usually agree precisely. Because of how we calibrate them, they will agree at 0°C and at 100°C. But because of different expansion properties, they may not agree precisely at intermediate temperatures (remember we arbitrarily divided the thermometer scale into 100 equal divisions between 0°C and 100°C). Thus a carefully calibrated mercury-in-glass thermometer might register 52.0°C, whereas a carefully calibrated thermometer of another type might read 52.6°C. Discrepancies below 0°C and above 100°C can also be significant.

Because of such discrepancies, some standard kind of thermometer must be chosen so that all temperatures can be precisely defined. The chosen standard for this purpose is the **constant-volume gas thermometer**. As shown in the simplified diagram of Fig. 17–8, this thermometer consists of a bulb filled with a dilute gas connected by a thin tube to a mercury manometer (Section 13–6). The volume of the gas is kept constant by raising or lowering the right-hand tube of the manometer so that the mercury in the left-hand tube coincides with the reference mark. An increase in temperature causes a proportional increase in pressure in the bulb. Thus the tube must be lifted higher to keep the gas volume constant. The height of the mercury in the right-hand column is then a measure of the temperature. This thermometer gives the same results for all gases in the limit of reducing the gas pressure in the bulb toward zero. The resulting scale serves as a basis for the standard temperature scale (Section 17–10).

17–3 Thermal Equilibrium and the Zeroth Law of Thermodynamics

We are all familiar with the fact that if two objects at different temperatures are placed in thermal contact (meaning thermal energy can transfer from one to the other), the two objects will eventually reach the same temperature. They are then said to be in **thermal equilibrium**. For example, you leave a fever thermometer in your mouth until it comes into thermal equilibrium with that environment, and then you read it. Two objects are defined to be in thermal equilibrium if, when placed in thermal contact, no net energy flows from one to the other, and their temperatures don't change. Experiments indicate that

> **if two systems are in thermal equilibrium with a third system, then they are in thermal equilibrium with each other.**

This postulate is called the **zeroth law of thermodynamics**. It has this unusual name because it was not until after the great first and second laws of thermodynamics (Chapters 19 and 20) were worked out that scientists realized that this apparently obvious postulate needed to be stated first.

Temperature is a property of a system that determines whether the system will be in thermal equilibrium with other systems. When two systems are in thermal equilibrium, their temperatures are, by definition, equal, and no net thermal energy will be exchanged between them. This is consistent with our everyday notion of temperature, since when a hot object and a cold one are put into contact, they eventually come to the same temperature. Thus the importance of the zeroth law is that it allows a useful definition of temperature.

17–4 Thermal Expansion

Most substances expand when heated and contract when cooled. However, the amount of expansion or contraction varies, depending on the material.

Linear Expansion

Experiments indicate that the change in length $\Delta \ell$ of almost all solids is, to a good approximation, directly proportional to the change in temperature ΔT, as long as ΔT is not too large. The change in length is also proportional to the original length of the object, ℓ_0. That is, for the same temperature increase, a 4-m-long iron rod will increase in length twice as much as a 2-m-long iron rod. We can write this proportionality as an equation:

$$\Delta \ell = \alpha \ell_0 \Delta T, \tag{17-1a}$$

where α, the proportionality constant, is called the *coefficient of linear expansion* for the particular material and has units of $(C°)^{-1}$. We write $\ell = \ell_0 + \Delta \ell$, Fig. 17–9, and rewrite this equation as $\ell = \ell_0 + \Delta \ell = \ell_0 + \alpha \ell_0 \Delta T$, or

$$\ell = \ell_0(1 + \alpha \Delta T), \tag{17-1b}$$

where ℓ_0 is the length initially, at temperature T_0, and ℓ is the length after heating or cooling to a temperature T. If the temperature change $\Delta T = T - T_0$ is negative, then $\Delta \ell = \ell - \ell_0$ is also negative; the length shortens as the temperature decreases.

FIGURE 17–9 A thin rod of length ℓ_0 at temperature T_0 is heated to a new uniform temperature T and acquires length ℓ, where $\ell = \ell_0 + \Delta \ell$.

TABLE 17–1 Coefficients of Expansion, near 20°C

Material	Coefficient of Linear Expansion, α (C°)$^{-1}$	Coefficient of Volume Expansion, β (C°)$^{-1}$
Solids		
Aluminum	25×10^{-6}	75×10^{-6}
Brass	19×10^{-6}	56×10^{-6}
Copper	17×10^{-6}	50×10^{-6}
Gold	14×10^{-6}	42×10^{-6}
Iron or steel	12×10^{-6}	35×10^{-6}
Lead	29×10^{-6}	87×10^{-6}
Glass (Pyrex®)	3×10^{-6}	9×10^{-6}
Glass (ordinary)	9×10^{-6}	27×10^{-6}
Quartz	0.4×10^{-6}	1×10^{-6}
Concrete and brick	$\approx 12 \times 10^{-6}$	$\approx 36 \times 10^{-6}$
Marble	$1.4–3.5 \times 10^{-6}$	$4–10 \times 10^{-6}$
Liquids		
Gasoline		950×10^{-6}
Mercury		180×10^{-6}
Ethyl alcohol		1100×10^{-6}
Glycerin		500×10^{-6}
Water		210×10^{-6}
Gases		
Air (and most other gases at atmospheric pressure)		3400×10^{-6}

The values of α for various materials at 20°C are listed in Table 17–1. Actually, α does vary slightly with temperature (which is why thermometers made of different materials do not agree precisely). However, if the temperature range is not too great, the variation can usually be ignored.

PHYSICS APPLIED

Expansion in structures

EXAMPLE 17–3 **Bridge expansion.** The steel bed of a suspension bridge is 200 m long at 20°C. If the extremes of temperature to which it might be exposed are −30°C to +40°C, how much will it contract and expand?

APPROACH We assume the bridge bed will expand and contract linearly with temperature, as given by Eq. 17–1a.

SOLUTION From Table 17–1, we find that $\alpha = 12 \times 10^{-6}$(C°)$^{-1}$ for steel. The increase in length when it is at 40°C will be

$$\Delta\ell = \alpha\ell_0\Delta T = (12 \times 10^{-6}/\text{C}°)(200\,\text{m})(40°\text{C} - 20°\text{C}) = 4.8 \times 10^{-2}\,\text{m},$$

or 4.8 cm. When the temperature decreases to −30°C, $\Delta T = -50$ C°. Then

$$\Delta\ell = (12 \times 10^{-6}/\text{C}°)(200\,\text{m})(-50\,\text{C}°) = -12.0 \times 10^{-2}\,\text{m},$$

or a decrease in length of 12 cm. The total range the expansion joints must accommodate is 12 cm + 4.8 cm ≈ 17 cm (Fig. 17–3).

CONCEPTUAL EXAMPLE 17–4 **Do holes expand or contract?** If you heat a thin, circular ring (Fig. 17–10a) in the oven, does the ring's hole get larger or smaller?

RESPONSE You might guess that the metal expands into the hole, making the hole smaller. But it is not so. Imagine the ring is solid, like a coin (Fig. 17–10b). Draw a circle on it with a pen as shown. When the metal expands, the material inside the circle will expand along with the rest of the metal; so the circle expands. Cutting the metal where the circle is makes clear to us that the hole in Fig. 17–10a increases in diameter.

FIGURE 17–10 Example 17–4.

(a)   (b)

EXAMPLE 17-5 **Ring on a rod.** An iron ring is to fit snugly on a cylindrical iron rod. At 20°C, the diameter of the rod is 6.445 cm and the inside diameter of the ring is 6.420 cm. To slip over the rod, the ring must be slightly larger than the rod diameter by about 0.008 cm. To what temperature must the ring be brought if its hole is to be large enough so it will slip over the rod?

APPROACH The hole in the ring must be increased from a diameter of 6.420 cm to 6.445 cm + 0.008 cm = 6.453 cm. The ring must be heated since the hole diameter will increase linearly with temperature (Example 17–4).

SOLUTION We solve for ΔT in Eq. 17–1a and find

$$\Delta T = \frac{\Delta \ell}{\alpha \ell_0} = \frac{6.453 \text{ cm} - 6.420 \text{ cm}}{(12 \times 10^{-6}/\text{C}°)(6.420 \text{ cm})} = 430 \text{ C}°.$$

So it must be raised at least to $T = (20°C + 430 \text{ C}°) = 450°C$.

NOTE In doing Problems, do not forget the last step, adding in the initial temperature (20°C here).

CONCEPTUAL EXAMPLE 17-6 | **Opening a tight jar lid.** When the lid of a glass jar is tight, holding the lid under hot water for a short time will often make it easier to open (Fig. 17–11). Why?

RESPONSE The lid may be struck by the hot water more directly than the glass and so expand sooner. But even if not, metals generally expand more than glass for the same temperature change (α is greater—see Table 17–1).

NOTE If you put a hard-boiled egg in cold water immediately after cooking it, it is easier to peel: the different thermal expansions of the shell and egg cause the egg to separate from the shell.

(A) **PHYSICS APPLIED**
Opening a tight lid

(A) **PHYSICS APPLIED**
Peeling a hard-boiled egg

FIGURE 17-11 Example 17–6.

Volume Expansion

The change in *volume* of a material which undergoes a temperature change is given by a relation similar to Eq. 17–1a, namely,

$$\Delta V = \beta V_0 \Delta T, \qquad (17-2)$$

where ΔT is the change in temperature, V_0 is the original volume, ΔV is the change in volume, and β is the *coefficient of volume expansion*. The units of β are $(\text{C}°)^{-1}$.

Values of β for various materials are given in Table 17–1. Notice that for solids, β is normally equal to approximately 3α. To see why, consider a rectangular solid of length ℓ_0, width W_0, and height H_0. When its temperature is changed by ΔT, its volume changes from $V_0 = \ell_0 W_0 H_0$ to

$$V = \ell_0 (1 + \alpha \Delta T) W_0 (1 + \alpha \Delta T) H_0 (1 + \alpha \Delta T),$$

using Eq. 17–1b and assuming α is the same in all directions. Thus,

$$\Delta V = V - V_0 = V_0 (1 + \alpha \Delta T)^3 - V_0 = V_0 [3\alpha \Delta T + 3(\alpha \Delta T)^2 + (\alpha \Delta T)^3].$$

If the amount of expansion is much smaller than the original size of the object, then $\alpha \Delta T \ll 1$ and we can ignore all but the first term and obtain

$$\Delta V \approx (3\alpha) V_0 \Delta T.$$

This is Eq. 17–2 with $\beta \approx 3\alpha$. For solids that are not isotropic (having the same properties in all directions), however, the relation $\beta \approx 3\alpha$ is not valid. Note also that linear expansion has no meaning for liquids and gases since they do not have fixed shapes.

EXERCISE B A long thin bar of aluminum at 0°C is 1.0 m long and has a volume of $1.0000 \times 10^{-3} \text{ m}^3$. When heated to 100°C, the length of the bar becomes 1.0025 m. What is the approximate volume of the bar at 100°C? (a) $1.0000 \times 10^{-3} \text{ m}^3$; (b) $1.0025 \times 10^{-3} \text{ m}^3$; (c) $1.0050 \times 10^{-3} \text{ m}^3$; (d) $1.0075 \times 10^{-3} \text{ m}^3$; (e) $2.5625 \times 10^{-3} \text{ m}^3$.

Equations 17–1 and 17–2 are accurate only if $\Delta \ell$ (or ΔV) is small compared to ℓ_0 (or V_0). This is of particular concern for liquids and even more so for gases because of the large values of β. Furthermore, β itself varies substantially with temperature for gases. Therefore, a more convenient way of dealing with gases is needed, and will be discussed starting in Section 17–6.

EXAMPLE 17–7 **Gas tank in the Sun.** The 70-liter (L) steel gas tank of a car is filled to the top with gasoline at 20°C. The car sits in the Sun and the tank reaches a temperature of 40°C (104°F). How much gasoline do you expect to overflow from the tank?

APPROACH Both the gasoline and the tank expand as the temperature increases, and we assume they do so linearly as described by Eq. 17–2. The volume of overflowing gasoline equals the volume increase of the gasoline minus the increase in volume of the tank.

SOLUTION The gasoline expands by

$$\Delta V = \beta V_0 \Delta T = (950 \times 10^{-6}/C°)(70\,L)(40°C - 20°C) = 1.3\,L.$$

The tank also expands. We can think of it as a steel shell that undergoes volume expansion $(\beta \approx 3\alpha = 36 \times 10^{-6}/C°)$. If the tank were solid, the surface layer (the shell) would expand just the same. Thus the tank increases in volume by

$$\Delta V = (36 \times 10^{-6}/C°)(70\,L)(40°C - 20°C) = 0.050\,L,$$

so the tank expansion has little effect. More than a liter of gas could spill out.

NOTE Want to save a few pennies? You pay for gas by volume, so fill your gas tank when it is cool and the gas is denser—more molecules for the same price. But don't fill the tank quite all the way.

Anomalous Behavior of Water Below 4°C

Most substances expand more or less uniformly with an increase in temperature, as long as no phase change occurs. Water, however, does not follow the usual pattern. If water at 0°C is heated, it actually *decreases* in volume until it reaches 4°C. Above 4°C water behaves normally and expands in volume as the temperature is increased, Fig. 17–12. Water thus has its greatest density at 4°C. This anomalous behavior of water is of great importance for the survival of aquatic life during cold winters. When the water in a lake or river is above 4°C and begins to cool by contact with cold air, the water at the surface sinks because of its greater density. It is replaced by warmer water from below. This mixing continues until the temperature reaches 4°C. As the surface water cools further, it remains on the surface because it is less dense than the 4°C water below. Water then freezes first at the surface, and the ice remains on the surface since ice (specific gravity = 0.917) is less dense than water. The water at the bottom remains liquid unless it is so cold that the whole body of water freezes. If water were like most substances, becoming more dense as it cools, the water at the bottom of a lake would be frozen first. Lakes would freeze solid more easily since circulation would bring the warmer water to the surface to be efficiently cooled. The complete freezing of a lake would cause severe damage to its plant and animal life. Because of the unusual behavior of water below 4°C, it is rare for any large body of water to freeze completely, and this is helped by the layer of ice on the surface which acts as an insulator to reduce the flow of heat out of the water into the cold air above. Without this peculiar but wonderful property of water, life on this planet as we know it might not have been possible.

Not only does water expand as it cools from 4°C to 0°C, it expands even more as it freezes to ice. This is why ice cubes float in water and pipes break when water inside them freezes.

FIGURE 17–12 Behavior of water as a function of temperature near 4°C. (a) Volume of 1.00000 g of water, as a function of temperature. (b) Density vs. temperature. [Note the break in each axis.]

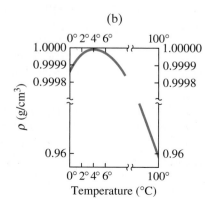

*17–5 Thermal Stresses

In many situations, such as in buildings and roads, the ends of a beam or slab of material are rigidly fixed, which greatly limits expansion or contraction. If the temperature should change, large compressive or tensile stresses, called *thermal stresses*, will occur. The magnitude of such stresses can be calculated using the concept of elastic modulus developed in Chapter 12. To calculate the internal stress, we can think of this process as occurring in two steps: (1) the beam tries to expand (or contract) by an amount $\Delta\ell$ given by Eq. 17–1; (2) the solid in contact with the beam exerts a force to compress (or expand) it, keeping it at its original length. The force F required is given by Eq. 12–4:

$$\Delta\ell = \frac{1}{E}\frac{F}{A}\ell_0,$$

where E is Young's modulus for the material. To calculate the internal stress, F/A, we then set $\Delta\ell$ in Eq. 17–1a equal to $\Delta\ell$ in the equation above and find

$$\alpha\ell_0\Delta T = \frac{1}{E}\frac{F}{A}\ell_0.$$

Hence, the stress

$$\frac{F}{A} = \alpha E\,\Delta T.$$

EXAMPLE 17–8 **Stress in concrete on a hot day.** A highway is to be made of blocks of concrete 10 m long placed end to end with no space between them to allow for expansion. If the blocks were placed at a temperature of 10°C, what compressive stress would occur if the temperature reached 40°C? The contact area between each block is 0.20 m². Will fracture occur?

(Ⓧ) **PHYSICS APPLIED**
Highway buckling

APPROACH We use the expression for the stress F/A we just derived, and find the value of E from Table 12–1. To see if fracture occurs, we compare this stress to the ultimate strength of concrete in Table 12–2.

SOLUTION

$$\frac{F}{A} = \alpha E\,\Delta T = (12\times10^{-6}/\text{C}°)(20\times10^{9}\,\text{N/m}^2)(30\,\text{C}°) = 7.2\times10^{6}\,\text{N/m}^2.$$

This stress is not far from the ultimate strength of concrete under compression (Table 12–2) and exceeds it for tension and shear. If the concrete is not perfectly aligned, part of the force will act in shear, and fracture is likely. This is why soft spacers or expansion joints (Fig. 17–3) are used in concrete sidewalks, highways, and bridges.

EXERCISE C How much space would you allow between the 10-m-long concrete blocks if you expected a temperature range of 0°F to 110°F?

17–6 The Gas Laws and Absolute Temperature

Equation 17–2 is not very useful for describing the expansion of a gas, partly because the expansion can be so great, and partly because gases generally expand to fill whatever container they are in. Indeed, Eq. 17–2 is meaningful only if the pressure is kept constant. The volume of a gas depends very much on the pressure as well as on the temperature. It is therefore valuable to determine a relation between the volume, the pressure, the temperature, and the mass of a gas. Such a relation is called an **equation of state**. (By the word *state*, we mean the physical condition of the system.)

If the state of a system is changed, we will always wait until the pressure and temperature have reached the same values throughout. We thus consider only **equilibrium states** of a system—when the variables that describe it (such as temperature and pressure) are the same throughout the system and are not changing in time. We also note that the results of this Section are accurate only for gases that are not too dense (the pressure is not too high, on the order of an atmosphere or less) and not close to the liquefaction (boiling) point.

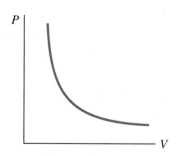

FIGURE 17–13 Pressure vs. volume of a fixed amount of gas at a constant temperature, showing the inverse relationship as given by Boyle's law: as the pressure decreases, the volume increases.

FIGURE 17–14 Volume of a fixed amount of gas as a function of (a) Celsius temperature, and (b) Kelvin temperature, when the pressure is kept constant.

For a given quantity of gas it is found experimentally that, to a good approximation, *the volume of a gas is inversely proportional to the absolute pressure applied to it when the temperature is kept constant.* That is,

$$V \propto \frac{1}{P}, \qquad \text{[constant } T\text{]}$$

where P is the absolute pressure (*not* "gauge pressure"—see Section 13–4). For example, if the pressure on a gas is doubled, the volume is reduced to half its original volume. This relation is known as **Boyle's law**, after Robert Boyle (1627–1691), who first stated it on the basis of his own experiments. A graph of P vs. V for a fixed temperature is shown in Fig. 17–13. Boyle's law can also be written

$$PV = \text{constant.} \qquad \text{[constant } T\text{]}$$

That is, at constant temperature, if either the pressure or volume of a fixed amount of gas is allowed to vary, the other variable also changes so that the product PV remains constant.

Temperature also affects the volume of a gas, but a quantitative relationship between V and T was not found until more than a century after Boyle's work. The Frenchman Jacques Charles (1746–1823) found that when the pressure is not too high and is kept constant, the volume of a gas increases with temperature at a nearly linear rate, as shown in Fig. 17–14a. However, all gases liquefy at low temperatures (for example, oxygen liquefies at −183°C), so the graph cannot be extended below the liquefaction point. Nonetheless, the graph is essentially a straight line and if projected to lower temperatures, as shown by the dashed line, it crosses the axis at about −273°C.

Such a graph can be drawn for any gas, and the straight line always projects back to −273°C at zero volume. This seems to imply that if a gas could be cooled to −273°C, it would have zero volume, and at lower temperatures a negative volume, which makes no sense. It could be argued that −273°C is the lowest temperature possible; indeed, many other more recent experiments indicate that this is so. This temperature is called the **absolute zero** of temperature. Its value has been determined to be −273.15°C.

Absolute zero forms the basis of a temperature scale known as the **absolute scale** or **Kelvin scale**, and it is used extensively in scientific work. On this scale the temperature is specified as degrees Kelvin or, preferably, simply as *kelvins* (K) without the degree sign. The intervals are the same as for the Celsius scale, but the zero on this scale (0 K) is chosen as absolute zero. Thus the freezing point of water (0°C) is 273.15 K, and the boiling point of water is 373.15 K. Indeed, any temperature on the Celsius scale can be changed to kelvins by adding 273.15 to it:

$$T(\text{K}) = T(°\text{C}) + 273.15.$$

Now let us look at Fig. 17–14b, where the graph of the volume of a gas versus absolute temperature is a straight line that passes through the origin. Thus, to a good approximation, *the volume of a given amount of gas is directly proportional to the absolute temperature when the pressure is kept constant.* This is known as **Charles's law**, and is written

$$V \propto T. \qquad \text{[constant } P\text{]}$$

A third gas law, known as **Gay-Lussac's law**, after Joseph Gay-Lussac (1778–1850), states that *at constant volume, the absolute pressure of a gas is directly proportional to the absolute temperature:*

$$P \propto T. \qquad \text{[constant } V\text{]}$$

The laws of Boyle, Charles, and Gay-Lussac are not really laws in the sense that we use this term today (precise, deep, wide-ranging validity). They are really only approximations that are accurate for real gases only as long as the pressure and density of the gas are not too high, and the gas is not too close to liquefaction (condensation). The term *law* applied to these three relationships has become traditional, however, so we have stuck with that usage.

CONCEPTUAL EXAMPLE 17–9 | **Why you should not throw a closed glass jar into a campfire.** What can happen if you did throw an empty glass jar, with the lid on tight, into a fire, and why?

RESPONSE The inside of the jar is not empty. It is filled with air. As the fire heats the air inside, its temperature rises. The volume of the glass jar changes only slightly due to the heating. According to Gay-Lussac's law the pressure P of the air inside the jar can increase dramatically, enough to cause the jar to explode, throwing glass pieces outward.

17–7 The Ideal Gas Law

The gas laws of Boyle, Charles, and Gay-Lussac were obtained by means of a very useful scientific technique: namely, holding one or more variables constant to see clearly the effects on one variable due to changing one other variable. These laws can now be combined into a single more general relation between the absolute pressure, volume, and absolute temperature of a fixed quantity of gas:

$$PV \propto T.$$

This relation indicates how any of the quantities P, V, or T will vary when the other two quantities change. This relation reduces to Boyle's, Charles's, or Gay-Lussac's law when either T, P, or V, respectively, is held constant.

Finally, we must incorporate the effect of the amount of gas present. Anyone who has blown up a balloon knows that the more air forced into the balloon, the bigger it gets (Fig. 17–15). Indeed, careful experiments show that at constant temperature and pressure, the volume V of an enclosed gas increases in direct proportion to the mass m of gas present. Hence we write

$$PV \propto mT.$$

This proportion can be made into an equation by inserting a constant of proportionality. Experiment shows that this constant has a different value for different gases. However, the constant of proportionality turns out to be the same for all gases if, instead of the mass m, we use the number of *moles*.

One **mole** (abbreviated mol) is defined as the amount of substance that contains as many atoms or molecules as there are in precisely 12 grams of carbon 12 (whose atomic mass is exactly 12 u). A simpler but equivalent definition is this: 1 mol is that quantity of substance whose mass in grams is numerically equal to the molecular mass (Section 17–1) of the substance. For example, the molecular mass of hydrogen gas (H_2) is 2.0 u (since each molecule contains two atoms of hydrogen and each atom has an atomic mass of 1.0 u). Thus 1 mol of H_2 has a mass of 2.0 g. Similarly, 1 mol of neon gas has a mass of 20 g, and 1 mol of CO_2 has a mass of $[12 + (2 \times 16)] = 44$ g since oxygen has atomic mass of 16 (see Periodic Table inside the rear cover). The mole is the official unit of amount of substance in the SI system. In general, the number of moles, n, in a given sample of a pure substance is equal to the mass of the sample in grams divided by the molecular mass specified as grams per mole:

$$n \text{ (mole)} = \frac{\text{mass (grams)}}{\text{molecular mass (g/mol)}}.$$

For example, the number of moles in 132 g of CO_2 (molecular mass 44 u) is

$$n = \frac{132 \text{ g}}{44 \text{ g/mol}} = 3.0 \text{ mole}.$$

FIGURE 17–15 Blowing up a balloon means putting more air (more air molecules) into the balloon, which increases its volume. The pressure is nearly constant (atmospheric) except for the small effect of the balloon's elasticity.

We can now write the proportion discussed above ($PV \propto mT$) as an equation:

$$PV = nRT,$$

(17–3)

where n represents the number of moles and R is the constant of proportionality. R is called the **universal gas constant** because its value is found experimentally to be the same for all gases. The value of R, in several sets of units (only the first is the proper SI unit), is

$$R = 8.314 \, \text{J}/(\text{mol} \cdot \text{K}) \qquad \text{[SI units]}$$
$$= 0.0821 \, (\text{L} \cdot \text{atm})/(\text{mol} \cdot \text{K})$$
$$= 1.99 \, \text{calories}/(\text{mol} \cdot \text{K}).^{\dagger}$$

Equation 17–3 is called the **ideal gas law**, or the **equation of state for an ideal gas**. We use the term "ideal" because real gases do not follow Eq. 17–3 precisely, particularly at high pressure (and density) or when the gas is near the liquefaction point (= boiling point). However, at pressures less than an atmosphere or so, and when T is not close to the liquefaction point of the gas, Eq. 17–3 is quite accurate and useful for real gases.

Always remember, when using the ideal gas law, that temperatures must be given in kelvins (K) and that the pressure P must always be *absolute* pressure, not gauge pressure (Section 13–4).

⚠ **CAUTION**

Always give T in kelvins and P as absolute (not gauge) pressure

EXERCISE D Return to the Chapter-Opening Question, page 454, and answer it again now. Try to explain why you may have answered differently the first time.

EXERCISE E An ideal gas is contained in a steel sphere at 27.0°C and 1.00 atm absolute pressure. If no gas is allowed to escape and the temperature is raised to 127°C, what will be the new pressure? (*a*) 1.33 atm; (*b*) 0.75 atm; (*c*) 4.7 atm; (*d*) 0.21 atm; (*e*) 1.00 atm.

17–8 Problem Solving with the Ideal Gas Law

The ideal gas law is an extremely useful tool, and we now consider some Examples. We will often refer to "standard conditions" or **standard temperature and pressure (STP)**, which means:

STP
$$T = 273 \, \text{K} \, (0°\text{C}) \qquad \text{and} \qquad P = 1.00 \, \text{atm} = 1.013 \times 10^5 \, \text{N/m}^2 = 101.3 \, \text{kPa}.$$

EXAMPLE 17–10 **Volume of one mole at STP.** Determine the volume of 1.00 mol of any gas, assuming it behaves like an ideal gas, at STP.

APPROACH We use the ideal gas law, solving for V.

SOLUTION We solve for V in Eq. 17–3:

$$V = \frac{nRT}{P} = \frac{(1.00 \, \text{mol})(8.314 \, \text{J/mol} \cdot \text{K})(273 \, \text{K})}{(1.013 \times 10^5 \, \text{N/m}^2)} = 22.4 \times 10^{-3} \, \text{m}^3.$$

Since 1 liter (L) is $1000 \, \text{cm}^3 = 1.00 \times 10^{-3} \, \text{m}^3$, 1.00 mol of any (ideal) gas has volume $V = 22.4 \, \text{L}$ at STP.

PROBLEM SOLVING

1 mol of gas at STP has $V = 22.4 \, \text{L}$

The value of 22.4 L for the volume of 1 mol of an ideal gas at STP is worth remembering, for it sometimes makes calculation simpler.

EXERCISE F What is the volume of 1.00 mol of ideal gas at 546 K (= 2 × 273 K) and 2.0 atm absolute pressure? (*a*) 11.2 L, (*b*) 22.4 L, (*c*) 44.8 L, (*d*) 67.2 L, (*e*) 89.6 L.

---

†Calories will be defined in Section 19–1; sometimes it is useful to use R as given in terms of calories.

EXAMPLE 17–11 **Helium balloon.** A helium party balloon, assumed to be a perfect sphere, has a radius of 18.0 cm. At room temperature (20°C), its internal pressure is 1.05 atm. Find the number of moles of helium in the balloon and the mass of helium needed to inflate the balloon to these values.

APPROACH We can use the ideal gas law to find n, since we are given P and T, and can find V from the given radius.

SOLUTION We get the volume V from the formula for a sphere:

$$V = \tfrac{4}{3}\pi r^3$$
$$= \tfrac{4}{3}\pi (0.180 \text{ m})^3 = 0.0244 \text{ m}^3.$$

The pressure is given as $1.05 \text{ atm} = 1.064 \times 10^5 \text{ N/m}^2$. The temperature must be expressed in kelvins, so we change 20°C to $(20 + 273)\text{K} = 293 \text{ K}$. Finally, we use the value $R = 8.314 \text{ J/(mol·K)}$ because we are using SI units. Thus

$$n = \frac{PV}{RT} = \frac{(1.064 \times 10^5 \text{ N/m}^2)(0.0244 \text{ m}^3)}{(8.314 \text{ J/mol·K})(293 \text{ K})} = 1.066 \text{ mol}.$$

The mass of helium (atomic mass = 4.00 g/mol as given in the Periodic Table or Appendix F) can be obtained from

$$\text{mass} = n \times \text{molecular mass} = (1.066 \text{ mol})(4.00 \text{ g/mol}) = 4.26 \text{ g}$$

or 4.26×10^{-3} kg.

EXAMPLE 17–12 **ESTIMATE** **Mass of air in a room.** Estimate the mass of air in a room whose dimensions are 5 m × 3 m × 2.5 m high, at STP.

APPROACH First we determine the number of moles n using the given volume. Then we can multiply by the mass of one mole to get the total mass.

SOLUTION Example 17–10 told us that 1 mol of a gas at 0°C has a volume of 22.4 L. The room's volume is 5 m × 3 m × 2.5 m, so

$$n = \frac{(5 \text{ m})(3 \text{ m})(2.5 \text{ m})}{22.4 \times 10^{-3} \text{ m}^3} \approx 1700 \text{ mol}.$$

Air is a mixture of about 20% oxygen (O_2) and 80% nitrogen (N_2). The molecular masses are $2 \times 16 \text{ u} = 32 \text{ u}$ and $2 \times 14 \text{ u} = 28 \text{ u}$, respectively, for an average of about 29 u. Thus, 1 mol of air has a mass of about $29 \text{ g} = 0.029 \text{ kg}$, so our room has a mass of air

$$m \approx (1700 \text{ mol})(0.029 \text{ kg/mol}) \approx 50 \text{ kg}.$$

NOTE That is roughly 100 lb of air!

PHYSICS APPLIED
Mass (and weight)
of the air in a room

EXERCISE G At 20°C, would there be (a) more, (b) less, or (c) the same air mass in a room than at 0°C?

Frequently, volume is specified in liters and pressure in atmospheres. Rather than convert these to SI units, we can instead use the value of R given in Section 17–7 as $0.0821 \text{ L·atm/mol·K}$.

In many situations it is not necessary to use the value of R at all. For example, many problems involve a change in the pressure, temperature, and volume of a fixed amount of gas. In this case, $PV/T = nR = \text{constant}$, since n and R remain constant. If we now let P_1, V_1, and T_1 represent the appropriate variables initially, and P_2, V_2, T_2 represent the variables after the change is made, then we can write

PROBLEM SOLVING
Using the ideal gas law as a ratio

$$\frac{P_1 V_1}{T_1} = \frac{P_2 V_2}{T_2}.$$

If we know any five of the quantities in this equation, we can solve for the sixth. Or, if one of the three variables is constant ($V_1 = V_2$, or $P_1 = P_2$, or $T_1 = T_2$) then we can use this equation to solve for one unknown when given the other three quantities.

FIGURE 17–16 Example 17–13.

EXAMPLE 17–13 **Check tires cold.** An automobile tire is filled (Fig. 17–16) to a gauge pressure of 200 kPa at 10°C. After a drive of 100 km, the temperature within the tire rises to 40°C. What is the pressure within the tire now?

APPROACH We do not know the number of moles of gas, or the volume of the tire, but we assume they are constant. We use the ratio form of the ideal gas law.

SOLUTION Since $V_1 = V_2$, then

$$\frac{P_1}{T_1} = \frac{P_2}{T_2}.$$

This is, incidentally, a statement of Gay-Lussac's law. Since the pressure given is the gauge pressure (Section 13–4), we must add atmospheric pressure ($= 101\,\text{kPa}$) to get the absolute pressure $P_1 = (200\,\text{kPa} + 101\,\text{kPa}) = 301\,\text{kPa}$. We convert temperatures to kelvins by adding 273 and solve for P_2:

$$P_2 = P_1\left(\frac{T_2}{T_1}\right) = (3.01 \times 10^5\,\text{Pa})\left(\frac{313\,\text{K}}{283\,\text{K}}\right) = 333\,\text{kPa}.$$

Subtracting atmospheric pressure, we find the resulting gauge pressure to be 232 kPa, which is a 16% increase. This Example shows why car manuals suggest checking tire pressure when the tires are cold.

17–9 Ideal Gas Law in Terms of Molecules: Avogadro's Number

The fact that the gas constant, R, has the same value for all gases is a remarkable reflection of simplicity in nature. It was first recognized, although in a slightly different form, by the Italian scientist Amedeo Avogadro (1776–1856). Avogadro stated that *equal volumes of gas at the same pressure and temperature contain equal numbers of molecules*. This is sometimes called **Avogadro's hypothesis**. That this is consistent with R being the same for all gases can be seen as follows. From Eq. 17–3, $PV = nRT$, we see that for the same number of moles, n, and the same pressure and temperature, the volume will be the same for all gases as long as R is the same. Second, the number of molecules in 1 mole is the same for all gases.[†] Thus Avogadro's hypothesis is equivalent to R being the same for all gases.

The number of molecules in one mole of any pure substance is known as **Avogadro's number**, N_A. Although Avogadro conceived the notion, he was not able to actually determine the value of N_A. Indeed, precise measurements were not done until the twentieth century.

A number of methods have been devised to measure N_A, and the accepted value today is

Avogadro's number

$$N_A = 6.02 \times 10^{23}. \qquad \text{[molecules/mole]}$$

Since the total number of molecules, N, in a gas is equal to the number per mole times the number of moles $(N = nN_A)$, the ideal gas law, Eq. 17–3, can be written in terms of the number of molecules present:

$$PV = nRT = \frac{N}{N_A}RT,$$

or

IDEAL GAS LAW
(in terms of molecules)

$$PV = NkT, \qquad \textbf{(17–4)}$$

where $k = R/N_A$ is called the **Boltzmann constant** and has the value

$$k = \frac{R}{N_A} = \frac{8.314\,\text{J/mol}\cdot\text{K}}{6.02 \times 10^{23}/\text{mol}} = 1.38 \times 10^{-23}\,\text{J/K}.$$

[†]For example, the molecular mass of H_2 gas is 2.0 atomic mass units (u), whereas that of O_2 gas is 32.0 u. Thus 1 mol of H_2 has a mass of 0.0020 kg and 1 mol of O_2 gas, 0.0320 kg. The number of molecules in a mole is equal to the total mass M of a mole divided by the mass m of one molecule; since this ratio (M/m) is the same for all gases by definition of the mole, a mole of any gas must contain the same number of molecules.

EXAMPLE 17–14 **Hydrogen atom mass.** Use Avogadro's number to determine the mass of a hydrogen atom.

APPROACH The mass of one atom equals the mass of 1 mol divided by the number of atoms in 1 mol, N_A.

SOLUTION One mole of hydrogen atoms (atomic mass = 1.008 u, Section 17–1 or Appendix F) has a mass of 1.008×10^{-3} kg and contains 6.02×10^{23} atoms. Thus one atom has a mass

$$m = \frac{1.008 \times 10^{-3} \text{ kg}}{6.02 \times 10^{23}}$$

$$= 1.67 \times 10^{-27} \text{ kg}.$$

EXAMPLE 17–15 **ESTIMATE** **How many molecules in one breath?** Estimate how many molecules you breathe in with a 1.0-L breath of air.

PHYSICS APPLIED
Molecules in a breath

APPROACH We determine what fraction of a mole 1.0 L is by using the result of Example 17–10 that 1 mole has a volume of 22.4 L at STP, and then multiply that by N_A to get the number of molecules in this number of moles.

SOLUTION One mole corresponds to 22.4 L at STP, so 1.0 L of air is $(1.0 \text{ L})/(22.4 \text{ L/mol}) = 0.045$ mol. Then 1.0 L of air contains

$$(0.045 \text{ mol})(6.02 \times 10^{23} \text{ molecules/mol}) \approx 3 \times 10^{22} \text{ molecules}.$$

*17–10 Ideal Gas Temperature Scale— a Standard

It is important to have a very precisely defined temperature scale so that measurements of temperature made at different laboratories around the world can be accurately compared. We now discuss such a scale that has been accepted by the general scientific community.

The standard thermometer for this scale is the constant-volume gas thermometer discussed in Section 17–2. The scale itself is called the **ideal gas temperature scale**, since it is based on the property of an ideal gas that the pressure is directly proportional to the absolute temperature (Gay-Lussac's law). A real gas, which would need to be used in any real constant-volume gas thermometer, approaches this ideal at low density. In other words, the temperature at any point in space is *defined* as being proportional to the pressure in the (nearly) ideal gas used in the thermometer. To set up a scale we need two fixed points. One fixed point will be $P = 0$ at $T = 0$ K. The second fixed point is chosen to be the **triple point** of water, which is that point where water in the solid, liquid, and gas states can coexist in equilibrium. This occurs only at a unique temperature and pressure,[†] and can be reproduced at different laboratories with great precision. The pressure at the triple point of water is 4.58 torr and the temperature is 0.01°C. This temperature corresponds to 273.16 K, since absolute zero is about −273.15°C. In fact, the triple point is now *defined* to be exactly 273.16 K.

[†]Liquid water and steam can coexist (the boiling point) at a range of temperatures depending on the pressure. Water boils at a lower temperature when the pressure is less, such as high in the mountains. The triple point represents a more precisely reproducible fixed point than does either the freezing point or boiling point of water at, say, 1 atm. See Section 18–3 for further discussion.

FIGURE 17–17 Temperature readings of a constant-volume gas thermometer for the boiling point of water at 1.00 atm are plotted, for different gases, as a function of the gas pressure in the thermometer at the triple point (P_{tp}). Note that as the amount of gas in the thermometer is reduced, so that $P_{tp} \rightarrow 0$, all gases give the same reading, 373.15 K. For pressure less than 0.10 atm (76 torr), the variation shown is less than 0.07 K.

The absolute or Kelvin temperature T at any point is then defined, using a constant-volume gas thermometer for an ideal gas, as

$$T = (273.16 \text{ K})\left(\frac{P}{P_{tp}}\right). \qquad \text{[ideal gas; constant volume]} \quad \textbf{(17–5a)}$$

In this relation, P_{tp} is the pressure of the gas in the thermometer at the triple point temperature of water, and P is the pressure in the thermometer when it is at the point where T is being determined. Note that if we let $P = P_{tp}$ in this relation, then $T = 273.16 \text{ K}$, as it must.

The definition of temperature, Eq. 17–5a, with a constant-volume gas thermometer filled with a real gas is only approximate because we find that we get different results for the temperature depending on the type of gas that is used in the thermometer. Temperatures determined in this way also vary depending on the amount of gas in the bulb of the thermometer: for example, the boiling point of water at 1.00 atm is found from Eq. 17–5a to be 373.87 K when the gas is O_2 and $P_{tp} = 1000$ torr. If the amount of O_2 in the bulb is reduced so that at the triple point $P_{tp} = 500$ torr, the boiling point of water from Eq. 17–5a is then found to be 373.51 K. If H_2 gas is used instead, the corresponding values are 373.07 K and 373.11 K (see Fig. 17–17). But now suppose we use a particular real gas and make a series of measurements in which the amount of gas in the thermometer bulb is reduced to smaller and smaller amounts, so that P_{tp} becomes smaller and smaller. It is found experimentally that an extrapolation of such data to $P_{tp} = 0$ always gives the *same value* for the temperature of a given system (such as $T = 373.15 \text{ K}$ for the boiling point of water at 1.00 atm) as shown in Fig. 17–17. Thus the temperature T at any point in space, determined using a constant-volume gas thermometer containing a real gas, is defined using this limiting process:

$$T = (273.16 \text{ K}) \lim_{P_{tp} \rightarrow 0}\left(\frac{P}{P_{tp}}\right). \qquad \text{[constant volume]} \quad \textbf{(17–5b)}$$

This defines the **ideal gas temperature scale**. One of the great advantages of this scale is that the value for T does not depend on the kind of gas used. But the scale does depend on the properties of gases in general. Helium has the lowest condensation point of all gases; at very low pressures it liquefies at about 1 K, so temperatures below this cannot be defined on this scale.

Summary

The atomic theory of matter postulates that all matter is made up of tiny entities called **atoms**, which are typically 10^{-10} m in diameter.

Atomic and **molecular masses** are specified on a scale where ordinary carbon (^{12}C) is arbitrarily given the value 12.0000 u (atomic mass units).

The distinction between solids, liquids, and gases can be attributed to the strength of the attractive forces between the atoms or molecules and to their average speed.

Temperature is a measure of how hot or cold something is. **Thermometers** are used to measure temperature on the **Celsius** (°C), **Fahrenheit** (°F), and **Kelvin** (K) scales. Two standard points on each scale are the freezing point of water (0°C, 32°F, 273.15 K) and the boiling point of water (100°C, 212°F, 373.15 K). A one-kelvin change in temperature equals a change of one Celsius degree or $\frac{9}{5}$ Fahrenheit degrees. Kelvins are related to °C by $T(\text{K}) = T(°\text{C}) + 273.15$.

The change in length, $\Delta \ell$, of a solid, when its temperature changes by an amount ΔT, is directly proportional to the temperature change and to its original length ℓ_0. That is,

$$\Delta \ell = \alpha \ell_0 \Delta T, \qquad \textbf{(17–1a)}$$

where α is the *coefficient of linear expansion*.

The change in volume of most solids, liquids, and gases is proportional to the temperature change and to the original volume V_0:

$$\Delta V = \beta V_0 \Delta T. \qquad \textbf{(17–2)}$$

The *coefficient of volume expansion*, β, is approximately equal to 3α for uniform solids.

Water is unusual because, unlike most materials whose volume increases with temperature, its volume actually decreases as the temperature increases in the range from 0°C to 4°C.

The **ideal gas law**, or **equation of state for an ideal gas**, relates the pressure P, volume V, and temperature T (in kelvins) of n moles of gas by the equation

$$PV = nRT, \qquad \textbf{(17–3)}$$

where $R = 8.314 \text{ J/mol·K}$ for all gases. Real gases obey the ideal gas law quite accurately if they are not at too high a pressure or near their liquefaction point.

One **mole** is that amount of a substance whose mass in grams is numerically equal to the atomic or molecular mass of that substance.

Avogadro's number, $N_A = 6.02 \times 10^{23}$, is the number of atoms or molecules in 1 mol of any pure substance.

The ideal gas law can be written in terms of the number of molecules N in the gas as

$$PV = NkT, \qquad \textbf{(17–4)}$$

where $k = R/N_A = 1.38 \times 10^{-23} \text{ J/K}$ is Boltzmann's constant.

Questions

1. Which has more atoms: 1 kg of iron or 1 kg of aluminum? See the Periodic Table or Appendix F.

2. Name several properties of materials that could be exploited to make a thermometer.

3. Which is larger, $1C°$ or $1F°$?

4. If system A is in equilibrium with system B, but B is not in equilibrium with system C, what can you say about the temperatures of A, B, and C?

5. Suppose system C is not in equilibrium with system A nor in equilibrium with system B. Does this imply that A and B are not in equilibrium? What can you infer regarding the temperatures of A, B, and C?

6. In the relation $\Delta\ell = \alpha\ell_0\Delta T$, should ℓ_0 be the initial length, the final length, or does it matter?

7. A flat bimetallic strip consists of a strip of aluminum riveted to a strip of iron. When heated, the strip will bend. Which metal will be on the outside of the curve? Why?

8. Long steam pipes that are fixed at the ends often have a section in the shape of a ∪. Why?

9. A flat uniform cylinder of lead floats in mercury at 0°C. Will the lead float higher or lower if the temperature is raised?

10. Figure 17–18 shows a diagram of a simple **thermostat** used to control a furnace (or other heating or cooling system). The bimetallic strip consists of two strips of different metals bonded together. The electric switch (attached to the bimetallic strip) is a glass vessel containing liquid mercury that conducts electricity when it can flow to touch both contact wires. Explain how this device controls the furnace and how it can be set at different temperatures.

Temperature setting lever
Bimetallic strip
Liquid mercury switch
Wires to heater
Liquid mercury

FIGURE 17–18 A thermostat (Question 10).

11. Explain why it is advisable to add water to an overheated automobile engine only slowly, and only with the engine running.

12. The units for the coefficients of expansion α are $(C°)^{-1}$, and there is no mention of a length unit such as meters. Would the expansion coefficient change if we used feet or millimeters instead of meters?

13. When a cold mercury-in-glass thermometer is first placed in a hot tub of water, the mercury initially descends a bit and then rises. Explain.

14. The principal virtue of Pyrex glass is that its coefficient of linear expansion is much smaller than that for ordinary glass (Table 17–1). Explain why this gives rise to the higher resistance to heat of Pyrex.

15. Will a grandfather clock, accurate at 20°C, run fast or slow on a hot day (30°C)? The clock uses a pendulum supported on a long thin brass rod.

16. Freezing a can of soda will cause its bottom and top to bulge so badly the can will not stand up. What has happened?

17. Why might you expect an alcohol-in-glass thermometer to be more precise than a mercury-in-glass thermometer?

18. Will the buoyant force on an aluminum sphere submerged in water increase, decrease, or remain the same, if the temperature is increased from 20°C to 40°C?

19. If an atom is measured to have a mass of 6.7×10^{-27} kg, what atom do you think it is?

20. From a practical point of view, does it really matter what gas is used in a constant-volume gas thermometer? If so, explain. [Hint: See Fig. 17–17.]

21. A ship loaded in sea water at 4°C later sailed up a river into fresh water where it sank in a storm. Explain why a ship might be more likely to sink in fresh water than on the open sea. [Hint: Consider the buoyant force due to water.]

Problems

17–1 Atomic Theory

1. (I) How does the number of atoms in a 21.5-g gold ring compare to the number in a silver ring of the same mass?

2. (I) How many atoms are there in a 3.4-g copper penny?

17–2 Temperature and Thermometers

3. (I) (a) "Room temperature" is often taken to be 68°F. What is this on the Celsius scale? (b) The temperature of the filament in a lightbulb is about 1900°C. What is this on the Fahrenheit scale?

4. (I) Among the highest and lowest natural air temperatures recorded are 136°F in the Libyan desert and −129°F in Antarctica. What are these temperatures on the Celsius scale?

5. (I) A thermometer tells you that you have a fever of 39.4°C. What is this in Fahrenheit?

6. (II) In an alcohol-in-glass thermometer, the alcohol column has length 11.82 cm at 0.0°C and length 21.85 cm at 100.0°C. What is the temperature if the column has length (a) 18.70 cm, and (b) 14.60 cm?

17–4 Thermal Expansion

7. (I) The Eiffel Tower (Fig. 17–19) is built of wrought iron approximately 300 m tall. Estimate how much its height changes between January (average temperature of 2°C) and July (average temperature of 25°C). Ignore the angles of the iron beams and treat the tower as a vertical beam.

FIGURE 17–19 Problem 7. The Eiffel Tower in Paris.

8. (I) A concrete highway is built of slabs 12 m long (20°C). How wide should the expansion cracks between the slabs be (at 15°C) to prevent buckling if the range of temperature is −30°C to +50°C?

9. (I) Super Invar™, an alloy of iron and nickel, is a strong material with a very low coefficient of thermal expansion $(0.20 \times 10^{-6}/C°)$. A 1.6-m-long tabletop made of this alloy is used for sensitive laser measurements where extremely high tolerances are required. How much will this alloy table expand along its length if the temperature increases 5.0 C°? Compare to tabletops made of steel.

10. (II) To what temperature would you have to heat a brass rod for it to be 1.0% longer than it is at 25°C?

11. (II) The density of water at 4°C is $1.00 \times 10^3 \text{ kg/m}^3$. What is water's density at 94°C? Assume a constant coefficient of volume expansion.

12. (II) At a given latitude, ocean water in the so-called "mixed layer" (from the surface to a depth of about 50 m) is at approximately the same temperature due to the mixing action of waves. Assume that because of global warming, the temperature of the mixed layer is everywhere increased by 0.5°C, while the temperature of the deeper portions of the ocean remains unchanged. Estimate the resulting rise in sea level. The ocean covers about 70% of the Earth's surface.

13. (II) To make a secure fit, rivets that are larger than the rivet hole are often used and the rivet is cooled (usually in dry ice) before it is placed in the hole. A steel rivet 1.872 cm in diameter is to be placed in a hole 1.870 cm in diameter in a metal at 20°C. To what temperature must the rivet be cooled if it is to fit in the hole?

14. (II) A uniform rectangular plate of length ℓ and width w has a coefficient of linear expansion α. Show that, if we neglect very small quantities, the change in area of the plate due to a temperature change ΔT is $\Delta A = 2\alpha\ell w\,\Delta T$. See Fig. 17–20.

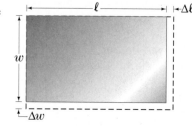

FIGURE 17–20
Problem 14.
A rectangular plate is heated.

15. (II) An aluminum sphere is 8.75 cm in diameter. What will be its change in volume if it is heated from 30°C to 180°C?

16. (II) A typical car has 17 L of liquid coolant circulating at a temperature of 93°C through the engine's cooling system. Assume that, in this normal condition, the coolant completely fills the 3.5-L volume of the aluminum radiator and the 13.5-L internal cavities within the steel engine. When a car overheats, the radiator, engine, and coolant expand and a small reservoir connected to the radiator catches any resultant coolant overflow. Estimate how much coolant overflows to the reservoir if the system is heated from 93°C to 105°C. Model the radiator and engine as hollow shells of aluminum and steel, respectively. The coefficient of volume expansion for coolant is $\beta = 410 \times 10^{-6}/\text{C°}$.

17. (II) It is observed that 55.50 mL of water at 20°C completely fills a container to the brim. When the container and the water are heated to 60°C, 0.35 g of water is lost. (a) What is the coefficient of volume expansion of the container? (b) What is the most likely material of the container? Density of water at 60°C is 0.98324 g/mL.

18. (II) (a) A brass plug is to be placed in a ring made of iron. At 15°C, the diameter of the plug is 8.753 cm and that of the inside of the ring is 8.743 cm. They must both be brought to what common temperature in order to fit? (b) What if the plug were iron and the ring brass?

19. (II) If a fluid is contained in a long narrow vessel so it can expand in essentially one direction only, show that the effective coefficient of linear expansion α is approximately equal to the coefficient of volume expansion β.

20. (II) (a) Show that the change in the density ρ of a substance, when the temperature changes by ΔT, is given by $\Delta\rho = -\beta\rho\,\Delta T$. (b) What is the fractional change in density of a lead sphere whose temperature decreases from 25°C to −55°C?

21. (II) Wine bottles are never completely filled: a small volume of air is left in the glass bottle's cylindrically shaped neck (inner diameter $d = 18.5$ mm) to allow for wine's fairly large coefficient of thermal expansion. The distance H between the surface of the liquid contents and the bottom of the cork is called the "headspace height" (Fig. 17–21), and is typically $H = 1.5$ cm for a 750-mL bottle filled at 20°C. Due to its alcoholic content, wine's coefficient of volume expansion is about double that of water; in comparison, the thermal expansion of glass can be neglected. Estimate H if the bottle is kept (a) at 10°C, (b) at 30°C.

FIGURE 17–21
Problem 21.

22. (III) (a) Determine a formula for the change in surface area of a uniform solid sphere of radius r if its coefficient of linear expansion is α (assumed constant) and its temperature is changed by ΔT. (b) What is the increase in area of a solid iron sphere of radius 60.0 cm if its temperature is raised from 15°C to 275°C?

23. (III) The pendulum in a grandfather clock is made of brass and keeps perfect time at 17°C. How much time is gained or lost in a year if the clock is kept at 28°C? (Assume the frequency dependence on length for a simple pendulum applies.)

24. (III) A 28.4-kg solid aluminum cylindrical wheel of radius 0.41 m is rotating about its axle in frictionless bearings with angular velocity $\omega = 32.8$ rad/s. If its temperature is then raised from 20.0°C to 95.0°C, what is the fractional change in ω?

*17–5 Thermal Stresses

*25. (I) An aluminum bar has the desired length when at 18°C. How much stress is required to keep it at this length if the temperature increases to 35°C?

*26. (II) (a) A horizontal steel I-beam of cross-sectional area 0.041 m² is rigidly connected to two vertical steel girders. If the beam was installed when the temperature was 25°C, what stress is developed in the beam when the temperature drops to −25°C? (b) Is the ultimate strength of the steel exceeded? (c) What stress is developed if the beam is concrete and has a cross-sectional area of 0.13 m²? Will it fracture?

*27. (III) A barrel of diameter 134.122 cm at 20°C is to be enclosed by an iron band. The circular band has an inside diameter of 134.110 cm at 20°C. It is 9.4 cm wide and 0.65 cm thick. (a) To what temperature must the band be heated so that it will fit over the barrel? (b) What will be the tension in the band when it cools to 20°C?

17–6 Gas Laws; Absolute Temperature

28. (I) What are the following temperatures on the Kelvin scale: (a) 66°C, (b) 92°F, (c) −55°C, (d) 5500°C?

29. (I) Absolute zero is what temperature on the Fahrenheit scale?

30. (II) Typical temperatures in the interior of the Earth and Sun are about 4000°C and 15×10^6 °C, respectively. (a) What are these temperatures in kelvins? (b) What percent error is made in each case if a person forgets to change °C to K?

17–7 and 17–8 Ideal Gas Law

31. (I) If 3.80 m^3 of a gas initially at STP is placed under a pressure of 3.20 atm, the temperature of the gas rises to 38.0°C. What is the volume?

32. (I) In an internal combustion engine, air at atmospheric pressure and a temperature of about 20°C is compressed in the cylinder by a piston to $\frac{1}{8}$ of its original volume (compression ratio = 8.0). Estimate the temperature of the compressed air, assuming the pressure reaches 40 atm.

33. (II) Calculate the density of nitrogen at STP using the ideal gas law.

34. (II) If 14.00 mol of helium gas is at 10.0°C and a gauge pressure of 0.350 atm, calculate (a) the volume of the helium gas under these conditions, and (b) the temperature if the gas is compressed to precisely half the volume at a gauge pressure of 1.00 atm.

35. (II) A stoppered test tube traps 25.0 cm^3 of air at a pressure of 1.00 atm and temperature of 18°C. The cylindrically shaped stopper at the test tube's mouth has a diameter of 1.50 cm and will "pop off" the test tube if a net upward force of 10.0 N is applied to it. To what temperature would one have to heat the trapped air in order to "pop off" the stopper? Assume the air surrounding the test tube is always at a pressure of 1.00 atm.

36. (II) A storage tank contains 21.6 kg of nitrogen (N_2) at an absolute pressure of 3.85 atm. What will the pressure be if the nitrogen is replaced by an equal mass of CO_2 at the same temperature?

37. (II) A storage tank at STP contains 28.5 kg of nitrogen (N_2). (a) What is the volume of the tank? (b) What is the pressure if an additional 25.0 kg of nitrogen is added without changing the temperature?

38. (II) A scuba tank is filled with air to a pressure of 204 atm when the air temperature is 29°C. A diver then jumps into the ocean and, after a short time treading water on the ocean surface, checks the tank's pressure and finds that it is only 194 atm. Assuming the diver has inhaled a negligible amount of air from the tank, what is the temperature of the ocean water?

39. (II) What is the pressure inside a 38.0-L container holding 105.0 kg of argon gas at 20.0°C?

40. (II) A tank contains 30.0 kg of O_2 gas at a gauge pressure of 8.20 atm. If the oxygen is replaced by helium at the same temperature, how many kilograms of the latter will be needed to produce a gauge pressure of 7.00 atm?

41. (II) A sealed metal container contains a gas at 20.0°C and 1.00 atm. To what temperature must the gas be heated for the pressure to double to 2.00 atm? (Ignore expansion of the container.)

42. (II) A tire is filled with air at 15°C to a gauge pressure of 250 kPa. If the tire reaches a temperature of 38°C, what fraction of the original air must be removed if the original pressure of 250 kPa is to be maintained?

43. (II) If 61.5 L of oxygen at 18.0°C and an absolute pressure of 2.45 atm are compressed to 48.8 L and at the same time the temperature is raised to 56.0°C, what will the new pressure be?

44. (II) A helium-filled balloon escapes a child's hand at sea level and 20.0°C. When it reaches an altitude of 3600 m, where the temperature is 5.0°C and the pressure only 0.68 atm, how will its volume compare to that at sea level?

45. (II) A sealed metal container can withstand a pressure difference of 0.50 atm. The container initially is filled with an ideal gas at 18°C and 1.0 atm. To what temperature can you cool the container before it collapses? (Ignore any changes in the container's volume due to thermal expansion.)

46. (II) You buy an "airtight" bag of potato chips packaged at sea level, and take the chips on an airplane flight. When you take the potato chips out of your luggage, you notice it has noticeably "puffed up." Airplane cabins are typically pressurized at 0.75 atm, and assuming the temperature inside an airplane is about the same as inside a potato chip processing plant, by what percentage has the bag "puffed up" in comparison to when it was packaged?

47. (II) A typical scuba tank, when fully charged, contains 12 L of air at 204 atm. Assume an "empty" tank contains air at 34 atm and is connected to an air compressor at sea level. The air compressor intakes air from the atmosphere, compresses it to high pressure, and then inputs this high-pressure air into the scuba tank. If the (average) flow rate of air from the atmosphere into the intake port of the air compressor is 290 L/min, how long will it take to fully charge the scuba tank? Assume the tank remains at the same temperature as the surrounding air during the filling process.

48. (III) A sealed container containing 4.0 mol of gas is squeezed, changing its volume from 0.020 m^3 to 0.018 m^3. During this process, the temperature decreases by 9.0 K while the pressure increases by 450 Pa. What was the original pressure and temperature of the gas in the container?

49. (III) Compare the value for the density of water vapor at exactly 100°C and 1 atm (Table 13–1) with the value predicted from the ideal gas law. Why would you expect a difference?

50. (III) An air bubble at the bottom of a lake 37.0 m deep has a volume of 1.00 cm^3. If the temperature at the bottom is 5.5°C and at the top 18.5°C, what is the volume of the bubble just before it reaches the surface?

17–9 Ideal Gas Law in Terms of Molecules; Avogadro's Number

51. (I) Calculate the number of molecules/m^3 in an ideal gas at STP.

52. (I) How many moles of water are there in 1.000 L at STP? How many molecules?

53. (II) What is the pressure in a region of outer space where there is 1 molecule/cm^3 and the temperature is 3 K?

54. (II) Estimate the number of (a) moles and (b) molecules of water in all the Earth's oceans. Assume water covers 75% of the Earth to an average depth of 3 km.

55. (II) The lowest pressure attainable using the best available vacuum techniques is about $10^{-12} \, \text{N/m}^2$. At such a pressure, how many molecules are there per cm^3 at $0°\text{C}$?

56. (II) Is a gas mostly empty space? Check by assuming that the spatial extent of common gas molecules is about $\ell_0 = 0.3 \, \text{nm}$ so one gas molecule occupies an approximate volume equal to ℓ_0^3. Assume STP.

57. (III) Estimate how many molecules of air are in each 2.0-L breath you inhale that were also in the last breath Galileo took. [*Hint*: Assume the atmosphere is about $10 \, \text{km}$ high and of constant density.]

*17–10 Ideal Gas Temperature Scale

*58. (I) In a constant-volume gas thermometer, what is the limiting ratio of the pressure at the boiling point of water at $1 \, \text{atm}$ to that at the triple point? (Keep five significant figures.)

*59. (I) At the boiling point of sulfur ($444.6°\text{C}$) the pressure in a constant-volume gas thermometer is 187 torr. Estimate (a) the pressure at the triple point of water, (b) the temperature when the pressure in the thermometer is 118 torr.

*60. (II) Use Fig. 17–17 to determine the inaccuracy of a constant-volume gas thermometer using oxygen if it reads a pressure $P = 268$ torr at the boiling point of water at $1 \, \text{atm}$. Express answer (a) in kelvins and (b) as a percentage.

*61. (III) A constant-volume gas thermometer is being used to determine the temperature of the melting point of a substance. The pressure in the thermometer at this temperature is 218 torr; at the triple point of water, the pressure is 286 torr. Some gas is now released from the thermometer bulb so that the pressure at the triple point of water becomes 163 torr. At the temperature of the melting substance, the pressure is 128 torr. Estimate, as accurately as possible, the melting-point temperature of the substance.

General Problems

62. A Pyrex measuring cup was calibrated at normal room temperature. How much error will be made in a recipe calling for $350 \, \text{mL}$ of cool water, if the water and the cup are hot, at $95°\text{C}$, instead of at room temperature? Neglect the glass expansion.

63. A precise steel tape measure has been calibrated at $15°\text{C}$. At $36°\text{C}$, (a) will it read high or low, and (b) what will be the percentage error?

64. A cubic box of volume $6.15 \times 10^{-2} \, \text{m}^3$ is filled with air at atmospheric pressure at $15°\text{C}$. The box is closed and heated to $185°\text{C}$. What is the net force on each side of the box?

65. The gauge pressure in a helium gas cylinder is initially $32 \, \text{atm}$. After many balloons have been blown up, the gauge pressure has decreased to $5 \, \text{atm}$. What fraction of the original gas remains in the cylinder?

66. If a rod of original length ℓ_1 has its temperature changed from T_1 to T_2, determine a formula for its new length ℓ_2 in terms of T_1, T_2, and α. Assume (a) $\alpha = \text{constant}$, (b) $\alpha = \alpha(T)$ is some function of temperature, and (c) $\alpha = \alpha_0 + bT$ where α_0 and b are constants.

67. If a scuba diver fills his lungs to full capacity of $5.5 \, \text{L}$ when $8.0 \, \text{m}$ below the surface, to what volume would his lungs expand if he quickly rose to the surface? Is this advisable?

68. (a) Use the ideal gas law to show that, for an ideal gas at constant pressure, the coefficient of volume expansion is equal to $\beta = 1/T$, where T is the kelvin temperature. Compare to Table 17–1 for gases at $T = 293 \, \text{K}$. (b) Show that the bulk modulus (Section 12–4) for an ideal gas held at constant temperature is $B = P$, where P is the pressure.

69. A house has a volume of $870 \, \text{m}^3$. (a) What is the total mass of air inside the house at $15°\text{C}$? (b) If the temperature drops to $-15°\text{C}$, what mass of air enters or leaves the house?

70. Assume that in an alternate universe, the laws of physics are very different from ours and that "ideal" gases behave as follows: (i) At constant temperature, pressure is inversely proportional to the square of the volume. (ii) At constant pressure, the volume varies directly with the $\frac{2}{3}$ power of the temperature. (iii) At $273.15 \, \text{K}$ and $1.00 \, \text{atm}$ pressure, 1.00 mole of an ideal gas is found to occupy $22.4 \, \text{L}$. Obtain the form of the ideal gas law in this alternate universe, including the value of the gas constant R.

71. An iron cube floats in a bowl of liquid mercury at $0°\text{C}$. (a) If the temperature is raised to $25°\text{C}$, will the cube float higher or lower in the mercury? (b) By what percent will the fraction of volume submerged change?

72. (a) The tube of a mercury thermometer has an inside diameter of $0.140 \, \text{mm}$. The bulb has a volume of $0.275 \, \text{cm}^3$. How far will the thread of mercury move when the temperature changes from $10.5°\text{C}$ to $33.0°\text{C}$? Take into account expansion of the Pyrex glass. (b) Determine a formula for the change in length of the mercury column in terms of relevant variables. Ignore tube volume compared to bulb volume.

73. From the known value of atmospheric pressure at the surface of the Earth, estimate the total number of air molecules in the Earth's atmosphere.

74. Estimate the percent difference in the density of iron at STP, and when it is a solid deep in the Earth where the temperature is $2000°\text{C}$ and under $5000 \, \text{atm}$ of pressure. Assume the bulk modulus ($90 \times 10^9 \, \text{N/m}^2$) and the coefficient of volume expansion do not vary with temperature and are the same as at STP.

75. What is the average distance between nitrogen molecules at STP?

76. A helium balloon, assumed to be a perfect sphere, has a radius of $22.0 \, \text{cm}$. At room temperature ($20°\text{C}$), its internal pressure is $1.06 \, \text{atm}$. Determine the number of moles of helium in the balloon, and the mass of helium needed to inflate the balloon to these values.

77. A standard cylinder of oxygen used in a hospital has gauge pressure $= 2000 \, \text{psi}$ ($13,800 \, \text{kPa}$) and volume $= 14 \, \text{L}$ ($0.014 \, \text{m}^3$) at $T = 295 \, \text{K}$. How long will the cylinder last if the flow rate, measured at atmospheric pressure, is constant at $2.4 \, \text{L/min}$?

78. A brass lid screws tightly onto a glass jar at $15°\text{C}$. To help open the jar, it can be placed into a bath of hot water. After this treatment, the temperatures of the lid and the jar are both $75°\text{C}$. The inside diameter of the lid is $8.0 \, \text{cm}$. Find the size of the gap (difference in radius) that develops by this procedure.

79. The density of gasoline at $0°\text{C}$ is $0.68 \times 10^3 \, \text{kg/m}^3$. (a) What is the density on a hot day, when the temperature is $35°\text{C}$? (b) What is the percent change in density?

80. A helium balloon has volume V_0 and temperature T_0 at sea level where the pressure is P_0 and the air density is ρ_0. The balloon is allowed to float up in the air to altitude y where the temperature is T_1. (a) Show that the volume occupied by the balloon is then $V = V_0(T_1/T_0)e^{+cy}$ where $c = \rho_0 g/P_0 = 1.25 \times 10^{-4}\,\text{m}^{-1}$. (b) Show that the buoyant force does not depend on altitude y. Assume that the skin of the balloon maintains the helium pressure at a constant factor of 1.05 times greater than the outside pressure. [Hint: Assume that the pressure change with altitude is $P = P_0 e^{-cy}$, as in Example 13–5, Chapter 13.]

81. The first length standard, adopted in the eighteenth century, was a platinum bar with two very fine marks separated by what was defined to be exactly one meter. If this standard bar was to be accurate to within $\pm 1.0\,\mu\text{m}$, how carefully would the trustees have needed to control the temperature? The coefficient of linear expansion is $9 \times 10^{-6}/\text{C}°$.

82. A scuba tank when fully charged has a pressure of 180 atm at 20°C. The volume of the tank is 11.3 L. (a) What would the volume of the air be at 1.00 atm and at the same temperature? (b) Before entering the water, a person consumes 2.0 L of air in each breath, and breathes 12 times a minute. At this rate, how long would the tank last? (c) At a depth of 20.0 m in sea water at a temperature of 10°C, how long would the same tank last assuming the breathing rate does not change?

83. A temperature controller, designed to work in a steam environment, involves a bimetallic strip constructed of brass and steel, connected at their ends by rivets. Each of the metals is 2.0 mm thick. At 20°C, the strip is 10.0 cm long and straight. Find the radius of curvature r of the assembly at 100°C. See Fig. 17–22.

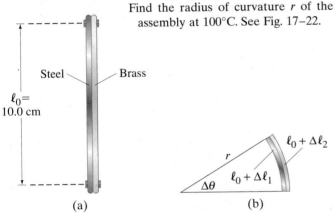

FIGURE 17–22   Problem 83.

84. A copper wire sags 50.0 cm between two utility poles 30.0 m apart when the temperature is −15°C. Estimate the amount of sag when the temperature is +35°C. [Hint: An estimate can be made by assuming the shape of the wire is approximately an arc of a circle; hard equations can sometimes be solved by guessing values.]

85. Snorkelers breathe through short tubular "snorkels" while swimming under water very near the surface. One end of the snorkel attaches to the snorkeler's mouth while the other end protrudes above the water's surface. Unfortunately, snorkels cannot support breathing to any great depth: it is said that a typical snorkeler below a water depth of only about 30 cm cannot draw a breath through a snorkel. Based on this claim, what is the approximate fractional change in a typical person's lung volume when drawing a breath? Assume, in equilibrium, the air pressure in a snorkeler's lungs matches that of the surrounding water pressure.

*Numerical/Computer

*86. (II) A thermocouple consists of a junction of two different types of materials that produces a voltage depending on its temperature. A thermocouple's voltages were recorded when at different temperatures as follows:

Temperature (°C)	50	100	200	300
Voltage (mV)	1.41	2.96	5.90	8.92

Use a spreadsheet to fit these data to a cubic equation and determine the temperature when the thermocouple produces 3.21 mV. Get a second value of the temperature by fitting the data to a quadratic equation.

*87. (III) You have a vial of an unknown liquid which might be octane (gasoline), water, glycerin, or ethyl alcohol. You are trying to determine its identity by studying how its volume changes with temperature changes. You fill a Pyrex graduated cylinder to 100.00 mL with the liquid when the liquid and the cylinder are at 0.000°C. You raise the temperature in five-degree increments, allowing the graduated cylinder and liquid to come to equilibrium at each temperature. You read the volumes listed below off the graduated cylinder at each temperature. Take into account the expansion of the Pyrex glass cylinder. Graph the data, possibly using a spreadsheet program, and determine the slope of the line to find the effective (combined) coefficient of volume expansion β. Then determine β for the liquid and which liquid is in the vial.

Temperature (°C)	Volume Reading (apparent mL)
0.000	100.00
5.000	100.24
10.000	100.50
15.000	100.72
20.000	100.96
25.000	101.26
30.000	101.48
35.000	101.71
40.000	101.97
45.000	102.20
50.000	102.46

Answers to Exercises

A: −40°.

B: (d).

C: 8 mm.

D: (i) Higher, (ii) same, (iii) lower.

E: (a).

F: (b).

G: (b) Less.

In this winter scene in Yellowstone Park, we recognize the three states of matter for water: as a liquid, as a solid (snow and ice), and as a gas (steam). In this Chapter we examine the microscopic theory of matter as atoms or molecules that are always in motion, which we call kinetic theory. We will see that the temperature of a gas is directly related to the average kinetic energy of its molecules. We will consider ideal gases, but we will also look at real gases and how they change phase, including evaporation, vapor pressure, and humidity.

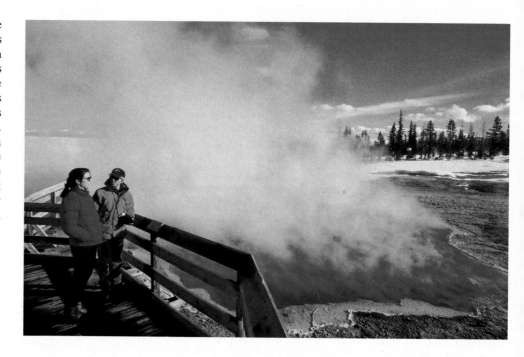

C H A P T E R

18

Kinetic Theory of Gases

CHAPTER-OPENING QUESTION—Guess now!

The typical speed of an air molecule at room temperature (20°C) is

  (a) nearly at rest (<10 km/h).
  (b) on the order of 10 km/h.
  (c) on the order of 100 km/h.
  (d) on the order of 1000 km/h.
  (e) nearly the speed of light.

CONTENTS

T he analysis of matter in terms of atoms in continuous random motion is called **kinetic theory**. We now investigate the properties of a gas from the point of view of kinetic theory, which is based on the laws of classical mechanics. But to apply Newton's laws to each one of the vast number of molecules in a gas ($>10^{25}/m^3$ at STP) is far beyond the capability of any present computer. Instead we take a statistical approach and determine averages of certain quantities, and these averages correspond to macroscopic variables. We will, of course, demand that our microscopic description correspond to the macroscopic properties of gases; otherwise our theory would be of little value. Most importantly, we will arrive at an important relation between the average kinetic energy of molecules in a gas and the absolute temperature.

18–1 The Ideal Gas Law and the Molecular Interpretation of Temperature

We make the following assumptions about the molecules in a gas. These assumptions reflect a simple view of a gas, but nonetheless the results they predict correspond well to the essential features of real gases that are at low pressures and far from the liquefaction point. Under these conditions real gases follow the ideal gas law quite closely, and indeed the gas we now describe is referred to as an **ideal gas**.

The assumptions, which represent the basic postulates of the kinetic theory for an ideal gas, are

1. There are a large number of molecules, N, each of mass m, moving in random directions with a variety of speeds. This assumption is in accord with our observation that a gas fills its container and, in the case of air on Earth, is kept from escaping only by the force of gravity.

2. The molecules are, on average, far apart from one another. That is, their average separation is much greater than the diameter of each molecule.

3. The molecules are assumed to obey the laws of classical mechanics, and are assumed to interact with one another only when they collide. Although molecules exert weak attractive forces on each other between collisions, the potential energy associated with these forces is small compared to the kinetic energy, and we ignore it for now.

4. Collisions with another molecule or the wall of the vessel are assumed to be perfectly elastic, like the collisions of perfectly elastic billiard balls (Chapter 9). We assume the collisions are of very short duration compared to the time between collisions. Then we can ignore the potential energy associated with collisions in comparison to the kinetic energy between collisions.

We can see immediately how this kinetic view of a gas can explain Boyle's law (Section 17–6). The pressure exerted on a wall of a container of gas is due to the constant bombardment of molecules. If the volume is reduced by (say) half, the molecules are closer together and twice as many will be striking a given area of the wall per second. Hence we expect the pressure to be twice as great, in agreement with Boyle's law.

Now let us calculate quantitatively the pressure a gas exerts on its container as based on kinetic theory. We imagine that the molecules are inside a rectangular container (at rest) whose ends have area A and whose length is ℓ, as shown in Fig. 18–1a. The pressure exerted by the gas on the walls of its container is, according to our model, due to the collisions of the molecules with the walls. Let us focus our attention on the wall, of area A, at the left end of the container and examine what happens when one molecule strikes this wall, as shown in Fig. 18–1b. This molecule exerts a force on the wall, and according to Newton's third law the wall exerts an equal and opposite force back on the molecule. The magnitude of this force on the molecule, according to Newton's second law, is equal to the molecule's rate of change of momentum, $F = dp/dt$ (Eq. 9–2). Assuming the collision is elastic, only the x component of the molecule's momentum changes, and it changes from $-mv_x$ (it is moving in the negative x direction) to $+mv_x$. Thus the change in the molecule's momentum, $\Delta(mv)$, which is the final momentum minus the initial momentum, is

$$\Delta(mv) = mv_x - (-mv_x) = 2mv_x$$

for one collision. This molecule will make many collisions with the wall, each separated by a time Δt, which is the time it takes the molecule to travel across the container and back again, a distance (x component) equal to 2ℓ. Thus $2\ell = v_x \Delta t$, or

$$\Delta t = \frac{2\ell}{v_x}.$$

The time Δt between collisions is very small, so the number of collisions per second is very large. Thus the average force—averaged over many collisions—will be equal to the momentum change during one collision divided by the time between collisions (Newton's second law):

$$F = \frac{\Delta(mv)}{\Delta t} = \frac{2mv_x}{2\ell/v_x} = \frac{mv_x^2}{\ell}. \qquad \text{[due to one molecule]}$$

During its passage back and forth across the container, the molecule may collide with the tops and sides of the container, but this does not alter its x component of momentum and thus does not alter our result. It may also collide with other molecules, which may change its v_x. However, any loss (or gain) of momentum is acquired by other molecules, and because we will eventually sum over all the molecules, this effect will be included. So our result above is not altered.

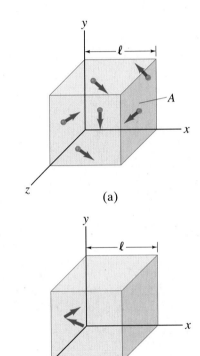

(a)

(b)

FIGURE 18–1 (a) Molecules of a gas moving about in a rectangular container. (b) Arrows indicate the momentum of one molecule as it rebounds from the end wall.

The actual force due to one molecule is intermittent, but because a huge number of molecules are striking the wall per second, the force is, on average, nearly constant. To calculate the force due to *all* the molecules in the container, we have to add the contributions of each. Thus the net force on the wall is

$$F = \frac{m}{\ell}\left(v_{x1}^2 + v_{x2}^2 + \cdots + v_{xN}^2\right),$$

where v_{x1} means v_x for molecule number 1 (we arbitrarily assign each molecule a number) and the sum extends over the total number of molecules N in the container. The average value of the square of the x component of velocity is

$$\overline{v_x^2} = \frac{v_{x1}^2 + v_{x2}^2 + \cdots + v_{xN}^2}{N},$$ (18–1)

where the overbar ($^-$) means "average." Thus we can write the force as

$$F = \frac{m}{\ell} N \overline{v_x^2}.$$

We know that the square of any vector is equal to the sum of the squares of its components (theorem of Pythagoras). Thus $v^2 = v_x^2 + v_y^2 + v_z^2$ for any velocity v. Taking averages, we obtain

$$\overline{v^2} = \overline{v_x^2} + \overline{v_y^2} + \overline{v_z^2}.$$

Since the velocities of the molecules in our gas are assumed to be random, there is no preference to one direction or another. Hence

$$\overline{v_x^2} = \overline{v_y^2} = \overline{v_z^2}.$$

Combining this relation with the one just above, we get

$$\overline{v^2} = 3\overline{v_x^2}.$$

We substitute this into the equation for net force F:

$$F = \frac{m}{\ell} N \frac{\overline{v^2}}{3}.$$

The pressure on the wall is then

$$P = \frac{F}{A} = \frac{1}{3}\frac{Nm\overline{v^2}}{A\ell}$$

or

$$P = \frac{1}{3}\frac{Nm\overline{v^2}}{V},$$ (18–2)

where $V = \ell A$ is the volume of the container. This is the result we were seeking, the pressure exerted by a gas on its container expressed in terms of molecular properties.

Equation 18–2 can be rewritten in a clearer form by multiplying both sides by V and rearranging the right-hand side:

$$PV = \tfrac{2}{3} N\left(\tfrac{1}{2}m\overline{v^2}\right).$$ (18–3)

The quantity $\tfrac{1}{2}m\overline{v^2}$ is the average kinetic energy $\overline{K}$ of the molecules in the gas. If we compare Eq. 18–3 with Eq. 17–4, the ideal gas law $PV = NkT$, we see that the two agree if

$$\tfrac{2}{3}\left(\tfrac{1}{2}m\overline{v^2}\right) = kT,$$

TEMPERATURE RELATED TO AVERAGE KINETIC ENERGY OF MOLECULES

or

$$\overline{K} = \tfrac{1}{2}m\overline{v^2} = \tfrac{3}{2}kT.$$ [ideal gas] (18–4)

This equation tells us that

the average translational kinetic energy of molecules in random motion in an ideal gas is directly proportional to the absolute temperature of the gas.

The higher the temperature, according to kinetic theory, the faster the molecules are moving on the average. This relation is one of the triumphs of the kinetic theory.

EXAMPLE 18–1 **Molecular kinetic energy.** What is the average translational kinetic energy of molecules in an ideal gas at 37°C?

APPROACH We use the absolute temperature in Eq. 18–4.

SOLUTION We change 37°C to 310 K and insert into Eq. 18–4:

$$\bar{K} = \tfrac{3}{2}kT = \tfrac{3}{2}(1.38 \times 10^{-23}\,\text{J/K})(310\,\text{K}) = 6.42 \times 10^{-21}\,\text{J}.$$

NOTE A mole of molecules would have a total translational kinetic energy equal to $(6.42 \times 10^{-21}\,\text{J})(6.02 \times 10^{23}) = 3860\,\text{J}$, which equals the kinetic energy of a 1-kg stone traveling almost 90 m/s.

EXERCISE A In a mixture of the gases oxygen and helium, which statement is valid: (*a*) the helium molecules will be moving faster than the oxygen molecules, on average; (*b*) both kinds of molecules will be moving at the same speed; (*c*) the oxygen molecules will, on average, be moving more rapidly than the helium molecules; (*d*) the kinetic energy of the helium will exceed that of the oxygen; (*e*) none of the above.

Equation 18–4 holds not only for gases, but also applies reasonably accurately to liquids and solids. Thus the result of Example 18–1 would apply to molecules within living cells at body temperature (37°C).

We can use Eq. 18–4 to calculate how fast molecules are moving on the average. Notice that the average in Eqs. 18–1 through 18–4 is over the *square* of the speed. The square root of $\overline{v^2}$ is called the **root-mean-square** speed, v_{rms} (since we are taking the square *root* of the *mean* of the *square* of the speed):

$$v_{\text{rms}} = \sqrt{\overline{v^2}} = \sqrt{\frac{3kT}{m}}. \qquad\qquad (18\text{–}5)$$

EXAMPLE 18–2 **Speeds of air molecules.** What is the rms speed of air molecules (O_2 and N_2) at room temperature (20°C)?

APPROACH To obtain v_{rms}, we need the masses of O_2 and N_2 molecules and then apply Eq. 18–5 to oxygen and nitrogen separately, since they have different masses.

SOLUTION The masses of one molecule of O_2 (molecular mass = 32 u) and N_2 (molecular mass = 28 u) are (where 1 u = 1.66×10^{-27} kg)

$$m(O_2) = (32)(1.66 \times 10^{-27}\,\text{kg}) = 5.3 \times 10^{-26}\,\text{kg},$$
$$m(N_2) = (28)(1.66 \times 10^{-27}\,\text{kg}) = 4.6 \times 10^{-26}\,\text{kg}.$$

Thus, for oxygen

$$v_{\text{rms}} = \sqrt{\frac{3kT}{m}} = \sqrt{\frac{(3)(1.38 \times 10^{-23}\,\text{J/K})(293\,\text{K})}{(5.3 \times 10^{-26}\,\text{kg})}} = 480\,\text{m/s},$$

and for nitrogen the result is $v_{\text{rms}} = 510$ m/s. These speeds are more than 1700 km/h or 1000 mi/h, and are greater than the speed of sound ≈ 340 m/s at 20°C (Chapter 16).

NOTE The speed v_{rms} is a magnitude only. The *velocity* of molecules averages to zero: the velocity has direction, and as many molecules move to the right as to the left, as many up as down, as many inward as outward.

EXERCISE B Now you can return to the Chapter-Opening Question, page 476, and answer it correctly. Try to explain why you may have answered differently the first time.

EXERCISE C If you double the volume of a gas while keeping the pressure and number of moles constant, the average (rms) speed of the molecules (*a*) doubles, (*b*) quadruples, (*c*) increases by $\sqrt{2}$, (*d*) is half, (*e*) is $\frac{1}{4}$.

EXERCISE D By what factor must the absolute temperature change to double v_{rms}? (*a*) $\sqrt{2}$; (*b*) 2; (*c*) $2\sqrt{2}$; (*d*) 4; (*e*) 16.

CONCEPTUAL EXAMPLE 18–3 | **Less gas in the tank.** A tank of helium is used to fill balloons. As each balloon is filled, the number of helium atoms remaining in the tank decreases. How does this affect the rms speed of molecules remaining in the tank?

RESPONSE The rms speed is given by Eq. 18–5: $v_{rms} = \sqrt{3kT/m}$. So only the temperature matters, not pressure P or number of moles n. If the tank remains at a constant (ambient) temperature, then the rms speed remains constant even though the pressure of helium in the tank decreases.

In a collection of molecules, the **average speed**, $\bar{v}$, is the average of the magnitudes of the speeds themselves; $\bar{v}$ is generally not equal to v_{rms}. To see the difference between the average speed and the rms speed, consider the following Example.

EXAMPLE 18–4 | **Average speed and rms speed.** Eight particles have the following speeds, given in m/s: 1.0, 6.0, 4.0, 2.0, 6.0, 3.0, 2.0, 5.0. Calculate (*a*) the average speed and (*b*) the rms speed.

APPROACH In (*a*) we sum the speeds and divide by $N = 8$. In (*b*) we square each speed, sum the squares, divide by $N = 8$, and take the square root.

SOLUTION (*a*) The average speed is

$$\bar{v} = \frac{1.0 + 6.0 + 4.0 + 2.0 + 6.0 + 3.0 + 2.0 + 5.0}{8} = 3.6\,\text{m/s}.$$

(*b*) The rms speed is (Eq. 18–1):

$$v_{rms} = \sqrt{\frac{(1.0)^2 + (6.0)^2 + (4.0)^2 + (2.0)^2 + (6.0)^2 + (3.0)^2 + (2.0)^2 + (5.0)^2}{8}}\,\text{m/s}$$

$$= 4.0\,\text{m/s}.$$

We see in this Example that $\bar{v}$ and v_{rms} are not necessarily equal. In fact, for an ideal gas they differ by about 8%. We will see in the next Section how to calculate $\bar{v}$ for an ideal gas. We already have the tool to calculate v_{rms} (Eq. 18–5).

*Kinetic Energy Near Absolute Zero

Equation 18–4, $\bar{K} = \frac{3}{2}kT$, implies that as the temperature approaches absolute zero, the kinetic energy of molecules approaches zero. Modern quantum theory, however, tells us this is not quite so. Instead, as absolute zero is approached, the kinetic energy approaches a very small nonzero minimum value. Even though all real gases become liquid or solid near 0 K, molecular motion does not cease, even at absolute zero.

18–2 Distribution of Molecular Speeds

The Maxwell Distribution

The molecules in a gas are assumed to be in random motion, which means that many molecules have speeds less than the average speed and others have speeds greater than the average. In 1859, James Clerk Maxwell (1831–1879) worked out a formula for the most probable distribution of speeds in a gas containing N molecules. We will not give a derivation here but merely quote his result:

$$f(v) = 4\pi N \left(\frac{m}{2\pi kT}\right)^{\frac{3}{2}} v^2 e^{-\frac{1}{2}\frac{mv^2}{kT}} \tag{18–6}$$

where $f(v)$ is called the **Maxwell distribution of speeds**, and is plotted in Fig. 18–2. The quantity $f(v)\,dv$ represents the number of molecules that have speed between v and $v + dv$. Notice that $f(v)$ does not give the number of molecules with speed v; $f(v)$ must be multiplied by dv to give the number of molecules (the number of molecules depends on the "width" or "range" of velocities included, dv). In the formula for $f(v)$, m is the mass of a single molecule, T is the absolute temperature, and k is the Boltzmann constant. Since N is the total number of molecules in the gas,

FIGURE 18–2 Distribution of speeds of molecules in an ideal gas. Note that $\bar{v}$ and v_{rms} are not at the peak of the curve. This is because the curve is skewed to the right: it is not symmetrical. The speed at the peak of the curve is the "most probable speed," v_p.

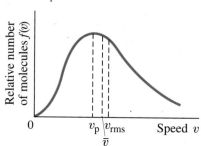

when we sum over all the molecules in the gas we must get N; thus we must have

$$\int_0^\infty f(v)\, dv = N.$$

(Problem 22 is an exercise to show that this is true.)

Experiments to determine the distribution of speeds in real gases, starting in the 1920s, confirmed with considerable accuracy the Maxwell distribution (for gases at not too high a pressure) and the direct proportion between average kinetic energy and absolute temperature, Eq. 18–4.

The Maxwell distribution for a given gas depends only on the absolute temperature. Figure 18–3 shows the distributions for two different temperatures. Just as v_{rms} increases with temperature, so the whole distribution curve shifts to the right at higher temperatures.

Figure 18–3 illustrates how kinetic theory can be used to explain why many chemical reactions, including those in biological cells, take place more rapidly as the temperature increases. Most chemical reactions take place in a liquid solution, and the molecules in a liquid have a distribution of speeds close to the Maxwell distribution. Two molecules may chemically react only if their kinetic energy is great enough so that when they collide, they partially penetrate into each other. The minimum energy required is called the *activation energy*, E_A, and it has a specific value for each chemical reaction. The molecular speed corresponding to a kinetic energy of E_A for a particular reaction is indicated in Fig. 18–3. The relative number of molecules with energy greater than this value is given by the area under the curve to the right of $v(E_A)$, shown in Fig. 18–3 by the two different shadings. We see that the number of molecules that have kinetic energies in excess of E_A increases greatly for only a small increase in temperature. The rate at which a chemical reaction occurs is proportional to the number of molecules with energy greater than E_A, and thus we see why reaction rates increase rapidly with increased temperature.

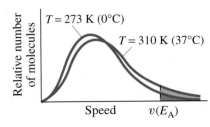

FIGURE 18–3 Distribution of molecular speeds for two different temperatures.

PHYSICS APPLIED
How chemical reactions depend on temperature

*Calculations Using the Maxwell Distribution

Let us see how the Maxwell distribution can be used to obtain some interesting results.

EXAMPLE 18–5 **Determining $\bar{v}$ and v_p.** Determine formulas for (a) the average speed, $\bar{v}$, and (b) the most probable speed, v_p, of molecules in an ideal gas at temperature T.

APPROACH (a) The average value of any quantity is found by multiplying each possible value of the quantity (here, speed) by the number of molecules that have that value, and then summing all these numbers and dividing by N (the total number). For (b), we want to find where the curve of Fig. 18–2 has zero slope; so we set $df/dv = 0$.

SOLUTION (a) We are given a continuous distribution of speeds (Eq. 18–6), so the sum over the speeds becomes an integral over the product of v and the number $f(v)\, dv$ that have speed v:

$$\bar{v} = \frac{\int_0^\infty v\, f(v)\, dv}{N} = 4\pi \left(\frac{m}{2\pi kT} \right)^{\frac{3}{2}} \int_0^\infty v^3 e^{-\frac{1}{2}\frac{mv^2}{kT}}\, dv.$$

We can integrate by parts or look up the definite integral in a Table, and obtain

$$\bar{v} = 4\pi \left(\frac{m}{2\pi kT} \right)^{\frac{3}{2}} \left(\frac{2k^2 T^2}{m^2} \right) = \sqrt{\frac{8}{\pi} \frac{kT}{m}} \approx 1.60 \sqrt{\frac{kT}{m}}.$$

(b) The *most probable speed* is that speed which occurs more than any others, and thus is that speed where $f(v)$ has its maximum value. At the maximum of the curve, the slope is zero: $df(v)/dv = 0$. Taking the derivative of Eq. 18–6 gives

$$\frac{df(v)}{dv} = 4\pi N \left(\frac{m}{2\pi kT} \right)^{\frac{3}{2}} \left(2v e^{-\frac{mv^2}{2kT}} - \frac{2mv^3}{2kT} e^{-\frac{mv^2}{2kT}} \right) = 0.$$

Solving for v, we find

$$v_p = \sqrt{\frac{2kT}{m}} \approx 1.41 \sqrt{\frac{kT}{m}}.$$

(Another solution is $v = 0$, but this corresponds to a minimum, not a maximum.)

In summary,

Most probable speed, v_p

$$v_p = \sqrt{2\frac{kT}{m}} \approx 1.41\sqrt{\frac{kT}{m}} \qquad \textbf{(18–7a)}$$

Average speed, $\bar{v}$

$$\bar{v} = \sqrt{\frac{8}{\pi}\frac{kT}{m}} \approx 1.60\sqrt{\frac{kT}{m}} \qquad \textbf{(18–7b)}$$

and from Eq. 18–5

rms speed, v_{rms}

$$v_{rms} = \sqrt{3\frac{kT}{m}} \approx 1.73\sqrt{\frac{kT}{m}}.$$

These are all indicated in Fig. 18–2. From Eq. 18–6 and Fig. 18–2, it is clear that the speeds of molecules in a gas vary from zero up to many times the average speed, but as can be seen from the graph, most molecules have speeds that are not far from the average. Less than 1% of the molecules exceed four times v_{rms}.

18–3 Real Gases and Changes of Phase

The ideal gas law

$$PV = NkT$$

is an accurate description of the behavior of a real gas as long as the pressure is not too high and as long as the temperature is far from the liquefaction point. But what happens when these two criteria are not satisfied? First we discuss real gas behavior, and then we examine how kinetic theory can help us understand this behavior.

Let us look at a graph of pressure plotted against volume for a given amount of gas. On such a "PV diagram," Fig. 18–4, each point represents an equilibrium state of the given substance. The various curves (labeled A, B, C, and D) show how the pressure varies, as the volume is changed at constant temperature, for several different values of the temperature. The dashed curve A' represents the behavior of a gas as predicted by the ideal gas law; that is, PV = constant. The solid curve A represents the behavior of a real gas at the same temperature. Notice that at high pressure, the volume of a real gas is less than that predicted by the ideal gas law. The curves B and C in Fig. 18–4 represent the gas at successively lower temperatures, and we see that the behavior deviates even more from the curves predicted by the ideal gas law (for example, B'), and the deviation is greater the closer the gas is to liquefying.

To explain this, we note that at higher pressure we expect the molecules to be closer together. And, particularly at lower temperatures, the potential energy associated with the attractive forces between the molecules (which we ignored before) is no longer negligible compared to the now reduced kinetic energy of the molecules. These attractive forces tend to pull the molecules closer together so that at a given pressure, the volume is less than expected from the ideal gas law, as in Fig. 18–4. At still lower temperatures, these forces cause liquefaction, and the molecules become very close together. Section 18–5 discusses in more detail the effect of these attractive molecular forces, as well as the effect of the volume which the molecules themselves occupy.

Curve D represents the situation when liquefaction occurs. At low pressure on curve D (on the right in Fig. 18–4), the substance is a gas and occupies a large volume. As the pressure is increased, the volume decreases until point b is reached. Beyond b, the volume decreases with no change in pressure; the substance is gradually changing from the gas to the liquid phase. At point a, all of the substance has changed to

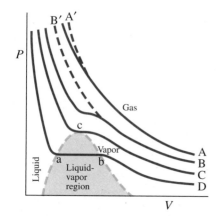

FIGURE 18–4 PV diagram for a real substance. Curves A, B, C, and D represent the same substance at different temperatures ($T_A > T_B > T_C > T_D$).

liquid. Further increase in pressure reduces the volume only slightly—liquids are nearly incompressible—so on the left the curve is very steep as shown. The colored area under the dashed line represents the region where the gas and liquid phases exist together in equilibrium.

Curve C in Fig. 18–4 represents the behavior of the substance at its **critical temperature**; the point c (the one point where curve C is horizontal) is called the **critical point**. At temperatures less than the critical temperature (and this is the definition of the term), a gas will change to the liquid phase if sufficient pressure is applied. Above the critical temperature, no amount of pressure can cause a gas to change phase and become a liquid. The critical temperatures for various gases are given in Table 18–1. Scientists tried for many years to liquefy oxygen without success. Only after the discovery of the behavior of substances associated with the critical point was it realized that oxygen can be liquefied only if first cooled below its critical temperature of −118°C.

Often a distinction is made between the terms "gas" and "vapor": a substance below its critical temperature in the gaseous state is called a **vapor**; above the critical temperature, it is called a **gas**.

The behavior of a substance can be diagrammed not only on a PV diagram but also on a PT diagram. A PT diagram, often called a **phase diagram**, is particularly convenient for comparing the different phases of a substance. Figure 18–5 is the phase diagram for water. The curve labeled ℓ-v represents those points where the liquid and vapor phases are in equilibrium—it is thus a graph of the boiling point versus pressure. Note that the curve correctly shows that at a pressure of 1 atm the boiling point is 100°C and that the boiling point is lowered for a decreased pressure. The curve s-ℓ represents points where solid and liquid exist in equilibrium and thus is a graph of the freezing point versus pressure. At 1 atm, the freezing point of water is 0°C, as shown. Notice also in Fig. 18–5 that at a pressure of 1 atm, the substance is in the liquid phase if the temperature is between 0°C and 100°C, but is in the solid or vapor phase if the temperature is below 0°C or above 100°C. The curve labeled s-v is the *sublimation point* versus pressure curve. **Sublimation** refers to the process whereby at low pressures a solid changes directly into the vapor phase without passing through the liquid phase. For water, sublimation occurs if the pressure of the water vapor is less than 0.0060 atm. Carbon dioxide, which in the solid phase is called dry ice, sublimates even at atmospheric pressure (Fig. 18–6).

The intersection of the three curves (in Fig. 18–5) is the **triple point**. For water this occurs at $T = 273.16\,\text{K}$ and $P = 6.03 \times 10^{-3}\,\text{atm}$. It is only at the triple point that the three phases can exist together in equilibrium. Because the triple point corresponds to a unique value of temperature and pressure, it is precisely reproducible and is often used as a point of reference. For example, the standard of temperature is usually specified as exactly 273.16 K at the triple point of water, rather than 273.15 K at the freezing point of water at 1 atm.

Notice that the solid liquid (s-ℓ) curve for water slopes upward to the left. This is true only of substances that *expand* upon freezing: at a higher pressure, a lower temperature is needed to cause the liquid to freeze. More commonly, substances contract upon freezing and the s-ℓ curve slopes upward to the right, as shown for carbon dioxide (CO_2) in Fig. 18–6.

The phase transitions we have been discussing are the common ones. Some substances, however, can exist in several forms in the solid phase. A transition from one phase to another occurs at a particular temperature and pressure, just like ordinary phase changes. For example, ice has been observed in at least eight forms at very high pressure. Ordinary helium has two distinct liquid phases, called helium I and II. They exist only at temperatures within a few degrees of absolute zero. Helium II exhibits very unusual properties referred to as **superfluidity**. It has essentially zero viscosity and exhibits strange properties such as climbing up the sides of an open container. Also interesting are **liquid crystals** (used for TV and computer monitors, Section 35–11) which can be considered to be in a phase between liquid and solid.

TABLE 18–1 Critical Temperatures and Pressures

Substance	Critical Temperature °C	Critical Temperature K	Critical Pressure (atm)
Water	374	647	218
CO_2	31	304	72.8
Oxygen	−118	155	50
Nitrogen	−147	126	33.5
Hydrogen	−239.9	33.3	12.8
Helium	−267.9	5.3	2.3

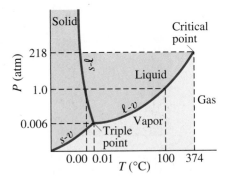

FIGURE 18–5 Phase diagram for water (note that the scales are not linear).

FIGURE 18–6 Phase diagram for carbon dioxide.

PHYSICS APPLIED
Liquid crystals

18–4 Vapor Pressure and Humidity

Evaporation

If a glass of water is left out overnight, the water level will have dropped by morning. We say the water has evaporated, meaning that some of the water has changed to the vapor or gas phase.

This process of **evaporation** can be explained on the basis of kinetic theory. The molecules in a liquid move past one another with a variety of speeds that follow, approximately, the Maxwell distribution. There are strong attractive forces between these molecules, which is what keeps them close together in the liquid phase. A molecule near the surface of the liquid may, because of its speed, leave the liquid momentarily. But just as a rock thrown into the air returns to the Earth, so the attractive forces of the other molecules can pull the vagabond molecule back to the liquid surface—that is, if its velocity is not too large. A molecule with a high enough velocity, however, will escape the liquid entirely (like an object leaving Earth with a high enough speed, Section 8–7), and become part of the gas phase. Only those molecules that have kinetic energy above a particular value can escape to the gas phase. We have already seen that kinetic theory predicts that the relative number of molecules with kinetic energy above a particular value (such as E_A in Fig. 18–3) increases with temperature. This is in accord with the well-known observation that the evaporation rate is greater at higher temperatures.

Because it is the fastest molecules that escape from the surface, the average speed of those remaining is less. When the average speed is less, the absolute temperature is less. Thus kinetic theory predicts that *evaporation is a cooling process.* You have no doubt noticed this effect when you stepped out of a warm shower and felt cold as the water on your body began to evaporate; and after working up a sweat on a hot day, even a slight breeze makes you feel cool through evaporation.

FIGURE 18–7 Vapor appears above a liquid in a closed container.

Vapor Pressure

Air normally contains water vapor (water in the gas phase), and it comes mainly from evaporation. To look at this process in a little more detail, consider a closed container that is partially filled with water (or another liquid) and from which the air has been removed (Fig. 18–7). The fastest moving molecules quickly evaporate into the empty space above the liquid's surface. As they move about, some of these molecules strike the liquid surface and again become part of the liquid phase: this is called **condensation**. The number of molecules in the vapor increases until a point is reached when the number of molecules returning to the liquid equals the number leaving in the same time interval. Equilibrium then exists, and the space above the liquid surface is said to be *saturated.* The pressure of the vapor when it is saturated is called the **saturated vapor pressure** (or sometimes simply the vapor pressure).

The saturated vapor pressure does not depend on the volume of the container. If the volume above the liquid were reduced suddenly, the density of molecules in the vapor phase would be increased temporarily. More molecules would then be striking the liquid surface per second. There would be a net flow of molecules back to the liquid phase until equilibrium was again reached, and this would occur at the same value of the saturated vapor pressure, as long as the temperature had not changed.

The saturated vapor pressure of any substance depends on the temperature. At higher temperatures, more molecules have sufficient kinetic energy to break from the liquid surface into the vapor phase. Hence equilibrium will be reached at a higher pressure. The saturated vapor pressure of water at various temperatures is given in Table 18–2. Notice that even solids—for example, ice—have a measurable saturated vapor pressure.

In everyday situations, evaporation from a liquid takes place into the air above it rather than into a vacuum. This does not materially alter the discussion above relating to Fig. 18–7. Equilibrium will still be reached when there are sufficient molecules in the gas phase that the number reentering the liquid equals the number leaving. The concentration of particular molecules (such as water) in the gas phase

TABLE 18–2 Saturated Vapor Pressure of Water

Temperature (°C)	Saturated Vapor Pressure	
	torr (= mm-Hg)	Pa (= N/m²)
−50	0.030	4.0
−10	1.95	2.60×10^2
0	4.58	6.11×10^2
5	6.54	8.72×10^2
10	9.21	1.23×10^3
15	12.8	1.71×10^3
20	17.5	2.33×10^3
25	23.8	3.17×10^3
30	31.8	4.24×10^3
40	55.3	7.37×10^3
50	92.5	1.23×10^4
60	149	1.99×10^4
70†	234	3.12×10^4
80	355	4.73×10^4
90	526	7.01×10^4
100‡	760	1.01×10^5
120	1489	1.99×10^5
150	3570	4.76×10^5

†Boiling point on summit of Mt. Everest.
‡Boiling point at sea level.

is not affected by the presence of air, although collisions with air molecules may lengthen the time needed to reach equilibrium. Thus equilibrium occurs at the same value of the saturated vapor pressure as if air were not there.

If the container is large or is not closed, all the liquid may evaporate before saturation is reached. And if the container is not sealed—as, for example, a room in your house—it is not likely that the air will become saturated with water vapor (unless it is raining outside).

Boiling

The saturated vapor pressure of a liquid increases with temperature. When the temperature is raised to the point where the saturated vapor pressure at that temperature equals the external pressure, **boiling** occurs (Fig. 18–8). As the boiling point is approached, tiny bubbles tend to form in the liquid, which indicate a change from the liquid to the gas phase. However, if the vapor pressure inside the bubbles is less than the external pressure, the bubbles immediately are crushed. As the temperature is increased, the saturated vapor pressure inside a bubble eventually becomes equal to or exceeds the external pressure. The bubble will then not collapse but can rise to the surface. Boiling has then begun. *A liquid boils when its saturated vapor pressure equals the external pressure.* This occurs for water at a pressure of 1 atm (760 torr) at 100°C, as can be seen from Table 18–2.

The boiling point of a liquid clearly depends on the external pressure. At high elevations, the boiling point of water is somewhat less than at sea level since the air pressure is less up there. For example, on the summit of Mt. Everest (8850 m) the air pressure is about one-third of what it is at sea level, and from Table 18–2 we can see that water will boil at about 70°C. Cooking food by boiling takes longer at high elevations, since the temperature is less. Pressure cookers, however, reduce cooking time, because they build up a pressure as high as 2 atm, allowing higher boiling temperatures to be attained.

FIGURE 18–8 Boiling: bubbles of water vapor float upward from the bottom (where the temperature is highest).

Partial Pressure and Humidity

When we refer to the weather as being dry or humid, we are referring to the water vapor content of the air. In a gas such as air, which is a mixture of several types of gases, the total pressure is the sum of the *partial pressures* of each gas present.[†] By **partial pressure**, we mean the pressure each gas would exert if it alone were present. The partial pressure of water in the air can be as low as zero and can vary up to a maximum equal to the saturated vapor pressure of water at the given temperature. Thus, at 20°C, the partial pressure of water cannot exceed 17.5 torr (see Table 18–2). The **relative humidity** is defined as the ratio of the partial pressure of water vapor to the saturated vapor pressure at a given temperature. It is usually expressed as a percentage:

$$\text{Relative humidity} = \frac{\text{partial pressure of } H_2O}{\text{saturated vapor pressure of } H_2O} \times 100\%.$$

Thus, when the humidity is close to 100%, the air holds nearly all the water vapor it can.

EXAMPLE 18–6 **Relative humidity.** On a particular hot day, the temperature is 30°C and the partial pressure of water vapor in the air is 21.0 torr. What is the relative humidity?

APPROACH From Table 18–2, we see that the saturated vapor pressure of water at 30°C is 31.8 torr.

SOLUTION The relative humidity is thus

$$\frac{21.0 \text{ torr}}{31.8 \text{ torr}} \times 100\% = 66\%.$$

[†]For example, 78% (by volume) of air molecules are nitrogen and 21% oxygen, with much smaller amounts of water vapor, argon, and other gases. At an air pressure of 1 atm, oxygen exerts a partial pressure of 0.21 atm and nitrogen 0.78 atm.

FIGURE 18–9 Fog or mist settling around a castle where the temperature has dropped below the dew point.

Humans are sensitive to humidity. A relative humidity of 40–50% is generally optimum for both health and comfort. High humidity, particularly on a hot day, reduces the evaporation of moisture from the skin, which is one of the body's vital mechanisms for regulating body temperature. Very low humidity, on the other hand, can dry the skin and mucous membranes.

Air is saturated with water vapor when the partial pressure of water in the air is equal to the saturated vapor pressure at that temperature. If the partial pressure of water exceeds the saturated vapor pressure, the air is said to be **supersaturated**. This situation can occur when a temperature decrease occurs. For example, suppose the temperature is 30°C and the partial pressure of water is 21 torr, which represents a humidity of 66% as we saw in Example 18–6. Suppose now that the temperature falls to, say, 20°C, as might happen at nightfall. From Table 18–2 we see that the saturated vapor pressure of water at 20°C is 17.5 torr. Hence the relative humidity would be greater than 100%, and the supersaturated air cannot hold this much water. The excess water may condense and appear as dew, or as fog or rain (Fig. 18–9).

When air containing a given amount of water is cooled, a temperature is reached where the partial pressure of water equals the saturated vapor pressure. This is called the **dew point**. Measurement of the dew point is the most accurate means of determining the relative humidity. One method uses a polished metal surface in contact with air, which is gradually cooled down. The temperature at which moisture begins to appear on the surface is the dew point, and the partial pressure of water can then be obtained from saturated vapor pressure Tables. If, for example, on a given day the temperature is 20°C and the dew point is 5°C, then the partial pressure of water (Table 18–2) in the 20°C air is 6.54 torr, whereas its saturated vapor pressure is 17.5 torr; hence the relative humidity is $6.54/17.5 = 37\%$.

EXERCISE E As the air warms up in the afternoon, how would the relative humidity change if there were no further evaporation? It would (*a*) increase, (*b*) decrease, (*c*) stay the same.

CONCEPTUAL EXAMPLE 18–7 **Dryness in winter.** Why does the air inside heated buildings seem very dry on a cold winter day?

RESPONSE Suppose the relative humidity outside on a −10°C day is 50%. Table 18–2 tells us the partial pressure of water in the air is about 1.0 torr. If this air is brought indoors and heated to +20°C, the relative humidity is $(1.0\,\text{torr})/(17.5\,\text{torr}) = 5.7\%$. Even if the outside air were saturated at a partial pressure of 1.95 torr, the inside relative humidity would be at a low 11%.

*18–5 Van der Waals Equation of State

In Section 18–3, we discussed how real gases deviate from ideal gas behavior, particularly at high densities or when near condensing to a liquid. We would like to understand these deviations using a microscopic (molecular) point of view. J. D. van der Waals (1837–1923) analyzed this problem and in 1873 arrived at an equation of state which fits real gases more accurately than the ideal gas law. His analysis is based on kinetic theory but takes into account: (1) the finite size of molecules (we previously neglected the actual volume of the molecules themselves, compared to the total volume of the container, and this assumption becomes poorer as the density increases and molecules become closer together); (2) the range of the forces between molecules may be greater than the size of the molecules (we previously assumed that intermolecular forces act only during collisions, when the molecules are "in contact"). Let us now look at this analysis and derive the van der Waals equation of state.

Assume the molecules in a gas are spherical with radius r. If we assume these molecules behave like hard spheres, then two molecules collide and bounce off one another if the distance between their centers (Fig. 18–10) gets as small as $2r$. Thus the actual volume in which the molecules can move about is somewhat less than the volume V of the container holding the gas. The amount of "unavailable volume" depends on the number of molecules and on their size. Let b represent the "unavailable volume per mole" of gas.

FIGURE 18–10 Molecules, of radius r, colliding.

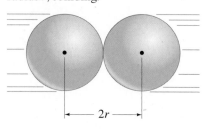

Then in the ideal gas law we replace V by $(V - nb)$, where n is the number of moles, and we obtain

$$P(V - nb) = nRT.$$

If we divide through by n, we have

$$P\left(\frac{V}{n} - b\right) = RT. \tag{18–8}$$

This relation (sometimes called the **Clausius equation of state**) predicts that for a given temperature T and volume V, the pressure P will be greater than for an ideal gas. This makes sense since the reduced "available" volume means the number of collisions with the walls is increased.

Next we consider the effects of attractive forces between molecules, which are responsible for holding molecules in the liquid and solid states at lower temperatures. These forces are electrical in nature and although they act even when molecules are not touching, we assume their range is small—that is, they act mainly between nearest neighbors. Molecules at the edge of the gas, headed toward a wall of the container, are slowed down by a net force pulling them back into the gas. Thus these molecules will exert less force and less pressure on the wall than if there were no attractive forces. The reduced pressure will be proportional to the density of molecules in the layer of gas at the surface, and also to the density in the next layer, which exerts the inward force.[†] Therefore we expect the pressure to be reduced by a factor proportional to the density squared $(n/V)^2$, here written as moles per volume. If the pressure P is given by Eq. 18–8, then we should reduce this by an amount $a(n/V)^2$ where a is a proportionality constant. Thus we have

$$P = \frac{RT}{(V/n) - b} - \frac{a}{(V/n)^2}$$

or

$$\left(P + \frac{a}{(V/n)^2}\right)\left(\frac{V}{n} - b\right) = RT, \tag{18–9}$$

which is the **van der Waals equation of state**.

The constants a and b in the van der Waals equation are different for different gases and are determined by fitting to experimental data for each gas. For CO_2 gas, the best fit is obtained for $a = 0.36 \text{ N} \cdot \text{m}^4/\text{mol}^2$ and $b = 4.3 \times 10^{-5} \text{ m}^3/\text{mol}$. Figure 18–11 shows a typical PV diagram for Eq. 18–9 (a "van der Waals gas") for four different temperatures, with detailed caption, and it should be compared to Fig. 18–4 for real gases.

Neither the van der Waals equation of state nor the many other equations of state that have been proposed are accurate for all gases under all conditions. Yet Eq. 18–9 is a very useful relation. And because it is quite accurate for many situations, its derivation gives us further insight into the nature of gases at the microscopic level. Note that at low densities, $a/(V/n)^2 \ll P$ and $b \ll V/n$, so that the van der Waals equation reduces to the equation of state for an ideal gas, $PV = nRT$.

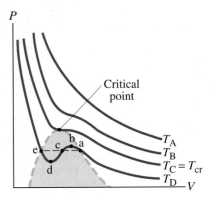

FIGURE 18–11 PV diagram for a van der Waals gas, shown for four different temperatures. For T_A, T_B, and T_C (T_C is chosen equal to the critical temperature), the curves fit experimental data very well for most gases. The curve labeled T_D, a temperature below the critical point, passes through the liquid–vapor region. The maximum (point b) and minimum (point d) would seem to be artifacts, since we usually see constant pressure, as indicated by the horizontal dashed line (and Fig. 18–4). However, for very pure supersaturated vapors or supercooled liquids, the sections ab and ed, respectively, have been observed. (The section bd would be unstable and has not been observed.)

*18–6 Mean Free Path

If gas molecules were truly point particles, they would have zero cross-section and never collide with one another. If you opened a perfume bottle, you would be able to smell it almost instantaneously across the room, since molecules travel hundreds of meters per second. In reality, it takes time before you detect an odor and, according to kinetic theory, this must be due to collisions between molecules of nonzero size.

If we were to follow the path of a particular molecule, we would expect to see it follow a zigzag path as shown in Fig. 18–12. Between each collision the molecule would move in a straight-line path. (Not quite true if we take account of the small intermolecular forces that act between collisions.) An important parameter for a given situation is the **mean free path**, which is defined as the average distance a molecule travels between collisions. We would expect that the greater the gas density, and the larger the molecules, the shorter the mean free path would be. We now determine the nature of this relationship for an ideal gas.

FIGURE 18–12 Zigzag path of a molecule colliding with other molecules.

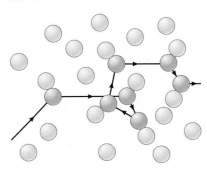

[†]This is similar to the gravitational force in which the force on mass m_1 due to mass m_2 is proportional to the product of their masses (Newton's law of universal gravitation, Chapter 6).

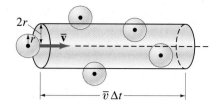

FIGURE 18–13 Molecule at left moves to the right with speed $\bar{v}$. It collides with any molecule whose center is within the cylinder of radius $2r$.

Suppose our gas is made up of molecules which are hard spheres of radius r. A collision will occur whenever the centers of two molecules come within a distance $2r$ of one another. Let us follow a molecule as it traces a straight-line path. In Fig. 18–13, the dashed line represents the path of our particle if it makes no collisions. Also shown is a cylinder of radius $2r$. If the center of another molecule lies within this cylinder, a collision will occur. (Of course, when a collision occurs the particle's path would change direction, as would our imagined cylinder, but our result won't be altered by unbending a zigzag cylinder into a straight one for purposes of calculation.) Assume our molecule is an average one, moving at the mean speed $\bar{v}$ in the gas. For the moment, let us assume that the other molecules are not moving, and that the concentration of molecules (number per unit volume) is N/V. Then the number of molecules whose centers lie within the cylinder of Fig. 18–13 is N/V times the volume of this cylinder, and this also represents the number of collisions that will occur. In a time Δt, our molecule travels a distance $\bar{v}\,\Delta t$, so the length of the cylinder is $\bar{v}\,\Delta t$ and its volume is $\pi(2r)^2\,\bar{v}\,\Delta t$. Hence the number of collisions that occur in a time Δt is $(N/V)\pi(2r)^2\,\bar{v}\,\Delta t$. We define the **mean free path**, ℓ_M, as the average distance between collisions. This distance is equal to the distance traveled ($\bar{v}\,\Delta t$) in a time Δt divided by the number of collisions made in time Δt:

$$\ell_M = \frac{\bar{v}\,\Delta t}{(N/V)\pi(2r)^2\,\bar{v}\,\Delta t} = \frac{1}{4\pi r^2 (N/V)}. \qquad \textbf{(18–10a)}$$

Thus we see that ℓ_M is inversely proportional to the cross-sectional area ($= \pi r^2$) of the molecules and to their concentration (number/volume), N/V. However, Eq. 18–10a is not fully correct since we assumed the other molecules are all at rest. In fact, they are moving, and the number of collisions in a time Δt must depend on the *relative* speed of the colliding molecules, rather than on $\bar{v}$. Hence the number of collisions per second is $(N/V)\pi(2r)^2 v_{rel}\,\Delta t$ (rather than $(N/V)\pi(2r)^2\,\bar{v}\,\Delta t$), where v_{rel} is the average relative speed of colliding molecules. A careful calculation shows that for a Maxwellian distribution of speeds $v_{rel} = \sqrt{2}\bar{v}$. Hence the mean free path is

Mean free path

$$\ell_M = \frac{1}{4\pi\sqrt{2}r^2(N/V)}. \qquad \textbf{(18–10b)}$$

EXAMPLE 18–8 ESTIMATE Mean free path of air molecules at STP. Estimate the mean free path of air molecules at STP, standard temperature and pressure (0°C, 1 atm). The diameter of O_2 and N_2 molecules is about 3×10^{-10} m.

APPROACH We saw in Example 17–10 that 1 mol of an ideal gas occupies a volume of 22.4×10^{-3} m^3 at STP. We can thus determine N/V and apply Eq. 18–10b.

SOLUTION

$$\frac{N}{V} = \frac{6.02 \times 10^{23} \text{ molecules}}{22.4 \times 10^{-3} \text{ m}^3} = 2.69 \times 10^{25} \text{ molecules/m}^3.$$

Then

$$\ell_M = \frac{1}{4\pi\sqrt{2}\,(1.5 \times 10^{-10}\,\text{m})^2 (2.7 \times 10^{25}\,\text{m}^{-3})} \approx 9 \times 10^{-8}\,\text{m}.$$

NOTE This is about 300 times the diameter of an air molecule.

At very low densities, such as in an evacuated vessel, the concept of mean free path loses meaning since collisions with the container walls may occur more frequently than collisions with other molecules. For example, in a cubical box that is (say) 20 cm on a side containing air at 10^{-7} torr ($\approx 10^{-10}$ atm), the mean free path would be about 900 m, which means many more collisions are made with the walls than with other molecules. (Note, nonetheless, that the box contains over 10^{12} molecules.) If the concept of mean free path included also collision with the walls, it would be closer to 0.2 m than to the 900 m calculated from Eq. 18–10b.

*18–7 Diffusion

If you carefully place a few drops of food coloring in a container of water as in Fig. 18–14, you will find that the color spreads throughout the water. The process may take some time (assuming you do not shake the glass), but eventually the color will become uniform. This mixing, known as **diffusion**, takes place because of the random movement of the molecules. Diffusion occurs in gases too. Common examples include perfume or smoke (or the odor of something cooking on the stove) diffusing in air, although convection (moving air currents) often plays a greater role in spreading odors than does diffusion. Diffusion depends on concentration, by which we mean the number of molecules or moles per unit volume. In general, *the diffusing substance moves from a region where its concentration is high to one where its concentration is low.*

(a)

(b)

(c)

FIGURE 18–14 A few drops of food coloring (a) dropped into water, (b) spreads slowly throughout the water, eventually (c) becoming uniform.

Diffusion can be readily understood on the basis of kinetic theory and the random motion of molecules. Consider a tube of cross-sectional area A containing molecules in a higher concentration on the left than on the right, Fig. 18–15. We assume the molecules are in random motion. Yet there will be a net flow of molecules to the right. To see why this is true, let us consider the small section of tube of length Δx as shown. Molecules from both regions 1 and 2 cross into this central section as a result of their random motion. The more molecules there are in a region, the more will strike a given area or cross a boundary. Since there is a greater concentration of molecules in region 1 than in region 2, more molecules cross into the central section from region 1 than from region 2. There is, then, a net flow of molecules from left to right, from high concentration toward low concentration. The net flow becomes zero only when the concentrations become equal.

You might expect that the greater the difference in concentration, the greater the flow rate. Indeed, the rate of diffusion, J (number of molecules or moles or kg per second), is directly proportional to the difference in concentration per unit distance, $(C_1 - C_2)/\Delta x$ (which is called the **concentration gradient**), and to the cross-sectional area A (see Fig. 18–15):

$$J = DA \frac{C_1 - C_2}{\Delta x},$$

or, in terms of derivatives,

$$J = DA \frac{dC}{dx}. \tag{18–11}$$

D is a constant of proportionality called the **diffusion constant**. Equation 18–11 is known as the **diffusion equation**, or **Fick's law**. If the concentrations are given in mol/m^3, then J is the number of moles passing a given point per second. If the concentrations are given in kg/m^3, then J is the mass movement per second (kg/s). The length Δx is given in meters. The values of D for a variety of substances are given in Table 18–3.

FIGURE 18–15 Diffusion occurs from a region of high concentration to one of lower concentration (only one type of molecule is shown).

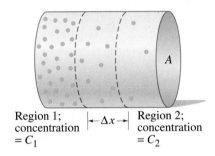

Region 1; concentration $= C_1$ |←Δx→| Region 2; concentration $= C_2$

TABLE 18–3 Diffusion Constants, D (20°C, 1 atm)

Diffusing Molecules	Medium	D (m²/s)
H_2	Air	6.3×10^{-5}
O_2	Air	1.8×10^{-5}
O_2	Water	100×10^{-11}
Blood hemoglobin	Water	6.9×10^{-11}
Glycine (an amino acid)	Water	95×10^{-11}
DNA (mass 6×10^6 u)	Water	0.13×10^{-11}

EXAMPLE 18–9 ESTIMATE **Diffusion of ammonia in air.** To get an idea of the time required for diffusion, estimate how long it might take for ammonia (NH_3) to be detected 10 cm from a bottle after it is opened, assuming only diffusion is occurring.

APPROACH This will be an order-of-magnitude calculation. The rate of diffusion J can be set equal to the number of molecules N diffusing across area A in a time t: $J = N/t$. Then the time $t = N/J$, where J is given by Eq. 18–11. We will have to make some assumptions and rough approximations about concentrations to use Eq. 18–11.

SOLUTION Using Eq. 18–11, we find

$$t = \frac{N}{J} = \frac{N}{DA} \frac{\Delta x}{\Delta C}.$$

The average concentration (midway between bottle and nose) can be approximated by $\overline{C} \approx N/V$, where V is the volume over which the molecules move and is roughly of the order of $V \approx A \, \Delta x$, where Δx is $10 \, \text{cm} = 0.10 \, \text{m}$. We substitute $N = \overline{C}V = \overline{C}A \, \Delta x$ into the above equation:

$$t \approx \frac{(\overline{C}A \, \Delta x) \, \Delta x}{DA \, \Delta C} = \frac{\overline{C}}{\Delta C} \frac{(\Delta x)^2}{D}.$$

The concentration of ammonia is high near the bottle (C) and low near the detecting nose (≈ 0), so $\overline{C} \approx C/2 \approx \Delta C/2$, or $(\overline{C}/\Delta C) \approx \frac{1}{2}$. Since NH_3 molecules have a size somewhere between H_2 and O_2, from Table 18–3 we can estimate $D \approx 4 \times 10^{-5} \, \text{m}^2/\text{s}$. Then

$$t \approx \frac{1}{2} \frac{(0.10 \, \text{m})^2}{(4 \times 10^{-5} \, \text{m}^2/\text{s})} \approx 100 \, \text{s},$$

or about a minute or two.

NOTE This result seems rather long from experience, suggesting that air currents (convection) are more important than diffusion for transmitting odors.

CONCEPTUAL EXAMPLE 18–10 **Colored rings on a paper towel.** A child colors a small spot on a wet paper towel with a brown marker. Later, she discovers that instead of a brown spot, there are concentric colored rings around the marked spot. What happened?

RESPONSE The ink in a brown marker is composed of several different inks that mix to make brown. These inks each diffuse at different rates through the wet paper towel. After a period of time the inks have diffused far enough that the differences in distances traveled is sufficient to separate the different colors. Chemists and biochemists use a similar technique, called *chromatography*, to separate substances based on their diffusion rates through a medium.

Summary

According to the **kinetic theory** of gases, which is based on the idea that a gas is made up of molecules that are moving rapidly and at random, the average kinetic energy of the molecules is proportional to the Kelvin temperature T:

$$\overline{K} = \tfrac{1}{2}m\overline{v^2} = \tfrac{3}{2}kT \qquad (18\text{–}4)$$

where k is Boltzmann's constant.

At any moment, there exists a wide distribution of molecular speeds within a gas. The **Maxwell distribution of speeds** is derived from simple kinetic theory assumptions, and is in good accord with experiment for gases at not too high a pressure.

The behavior of real gases at high pressure, and/or when near their liquefaction point, deviates from the ideal gas law due to the finite size of molecules and to the attractive forces between molecules.

Below the **critical temperature**, a gas can change to a liquid if sufficient pressure is applied; but if the temperature is higher than the critical temperature, no amount of pressure will cause a liquid surface to form.

The **triple point** of a substance is that unique temperature and pressure at which all three phases—solid, liquid, and gas—can coexist in equilibrium. Because of its precise reproducibility, the triple point of water is often taken as a standard reference point.

Evaporation of a liquid is the result of the fastest moving molecules escaping from the surface. Because the average molecular velocity is less after the fastest molecules escape, the temperature decreases when evaporation takes place.

Saturated vapor pressure refers to the pressure of the vapor above a liquid when the two phases are in equilibrium. The vapor pressure of a substance (such as water) depends strongly on temperature and is equal to atmospheric pressure at the boiling point.

Relative humidity of air at a given place is the ratio of the partial pressure of water vapor in the air to the saturated vapor pressure at that temperature; it is usually expressed as a percentage.

[*The **van der Waals equation of state** takes into account the finite volume of molecules, and the attractive forces between molecules, to better approximate the behavior of real gases.]

[*The **mean free path** is the average distance a molecule moves between collisions with other molecules.]

[*Diffusion** is the process whereby molecules of a substance move (on average) from one area to another because of a difference in that substance's concentration.]

Questions

1. Why doesn't the size of different molecules enter into the ideal gas law?

2. When a gas is rapidly compressed (say, by pushing down a piston) its temperature increases. When a gas expands against a piston, it cools. Explain these changes in temperature using the kinetic theory, in particular noting what happens to the momentum of molecules when they strike the moving piston.

3. In Section 18–1 we assumed the gas molecules made perfectly elastic collisions with the walls of the container. This assumption is not necessary as long as the walls are at the same temperature as the gas. Why?

4. Explain in words how Charles's law follows from kinetic theory and the relation between average kinetic energy and the absolute temperature.

5. Explain in words how Gay-Lussac's law follows from kinetic theory.

6. As you go higher in the Earth's atmosphere, the ratio of N_2 molecules to O_2 molecules increases. Why?

7. Can you determine the temperature of a vacuum?

8. Is temperature a macroscopic or microscopic variable?

9. Explain why the peak of the curve for 310 K in Fig. 18–3 is not as high as for 273 K. (Assume the total number of molecules is the same for both.)

10. Escape velocity for the Earth refers to the minimum speed an object must have to leave the Earth and never return. (a) The escape velocity for the Moon is about one-fifth what it is for the Earth due to the Moon's smaller mass; explain why the Moon has practically no atmosphere. (b) If hydrogen was once in the Earth's atmosphere, why would it have probably escaped?

11. If a container of gas is at rest, the average velocity of molecules must be zero. Yet the average speed is not zero. Explain.

12. If the pressure in a gas is doubled while its volume is held constant, by what factor do (a) v_{rms} and (b) $\bar{v}$ change?

13. What everyday observation would tell you that not all molecules in a material have the same speed?

14. We saw that the saturated vapor pressure of a liquid (say, water) does not depend on the external pressure. Yet the temperature of boiling does depend on the external pressure. Is there a contradiction? Explain.

15. Alcohol evaporates more quickly than water at room temperature. What can you infer about the molecular properties of one relative to the other?

16. Explain why a hot humid day is far more uncomfortable than a hot dry day at the same temperature.

17. Is it possible to boil water at room temperature (20°C) without heating it? Explain.

18. What exactly does it mean when we say that oxygen boils at −183°C?

19. A length of thin wire is placed over a block of ice (or an ice cube) at 0°C and weights are hung from the ends of the wire. It is found that the wire cuts its way through the ice cube, but leaves a solid block of ice behind it. This process is called *regelation*. Explain how this happens by inferring how the freezing point of water depends on pressure.

20. Consider two days when the air temperature is the same but the humidity is different. Which is more dense, the dry air or the humid air at the same T? Explain.

21. (a) Why does food cook faster in a pressure cooker? (b) Why does pasta or rice need to boil longer at high altitudes? (c) Is it harder to boil water at high altitudes?

22. How do a gas and a vapor differ?

23. (a) At suitable temperatures and pressures, can ice be melted by applying pressure? (b) At suitable temperatures and pressures, can carbon dioxide be melted by applying pressure?

24. Why does dry ice not last long at room temperature?

25. Under what conditions can liquid CO_2 exist? Be specific. Can it exist as a liquid at normal room temperature?

26. Why does exhaled air appear as a little white cloud in the winter (Fig. 18–16)?

FIGURE 18–16
Question 26.

*27. Discuss why sound waves can travel in a gas only if their wavelength is somewhat larger than the mean free path.

*28. Name several ways to reduce the mean free path in a gas.

Problems

18–1 Molecular Interpretation of Temperature

1. (I) (a) What is the average translational kinetic energy of an oxygen molecule at STP? (b) What is the total translational kinetic energy of 1.0 mol of O_2 molecules at 25°C?

2. (I) Calculate the rms speed of helium atoms near the surface of the Sun at a temperature of about 6000 K.

3. (I) By what factor will the rms speed of gas molecules increase if the temperature is increased from 0°C to 180°C?

4. (I) A gas is at 20°C. To what temperature must it be raised to triple the rms speed of its molecules?

5. (I) What speed would a 1.0-g paper clip have if it had the same kinetic energy as a molecule at 15°C?

6. (I) A 1.0-mol sample of hydrogen gas has a temperature of 27°C. (a) What is the total kinetic energy of all the gas molecules in the sample? (b) How fast would a 65-kg person have to run to have the same kinetic energy?

7. (I) Twelve molecules have the following speeds, given in arbitrary units: 6.0, 2.0, 4.0, 6.0, 0.0, 4.0, 1.0, 8.0, 5.0, 3.0, 7.0, and 8.0. Calculate (a) the mean speed, and (b) the rms speed.

8. (II) The rms speed of molecules in a gas at 20.0°C is to be increased by 2.0%. To what temperature must it be raised?

9. (II) If the pressure in a gas is tripled while its volume is held constant, by what factor does $v_{\rm rms}$ change?

10. (II) Show that the rms speed of molecules in a gas is given by $v_{\rm rms} = \sqrt{3P/\rho}$, where P is the pressure in the gas, and ρ is the gas density.

11. (II) Show that for a mixture of two gases at the same temperature, the ratio of their rms speeds is equal to the inverse ratio of the square roots of their molecular masses.

12. (II) What is the rms speed of nitrogen molecules contained in an 8.5-m^3 volume at 3.1 atm if the total amount of nitrogen is 1800 mol?

13. (II) (a) For an ideal gas at temperature T show that
$$\frac{dv_{\rm rms}}{dT} = \frac{1}{2}\frac{v_{\rm rms}}{T},$$
and using the approximation $\Delta v_{\rm rms} \approx \dfrac{dv_{\rm rms}}{dT}\Delta T$, show that
$$\frac{\Delta v_{\rm rms}}{v_{\rm rms}} \approx \frac{1}{2}\frac{\Delta T}{T}.$$
(b) If the average air temperature changes from −5°C in winter to 25°C in summer, estimate the percent change in the rms speed of air molecules between these seasons.

14. (II) What is the average distance between oxygen molecules at STP?

15. (II) Two isotopes of uranium, ^{235}U and ^{238}U (the superscripts refer to their atomic masses), can be separated by a gas diffusion process by combining them with fluorine to make the gaseous compound UF_6. Calculate the ratio of the rms speeds of these molecules for the two isotopes, at constant T. Use Appendix F for masses.

16. (II) Can pockets of vacuum persist in an ideal gas? Assume that a room is filled with air at 20°C and that somehow a small spherical region of radius 1 cm within the room becomes devoid of air molecules. Estimate how long it will take for air to refill this region of vacuum. Assume the atomic mass of air is 29 u.

17. (II) Calculate (a) the rms speed of a nitrogen molecule at 0°C and (b) determine how many times per second it would move back and forth across a 5.0-m-long room on the average, assuming it made very few collisions with other molecules.

18. (III) Estimate how many air molecules rebound from a wall in a typical room per second, assuming an ideal gas of N molecules contained in a cubic room with sides of length ℓ at temperature T and pressure P. (a) Show that the frequency f with which gas molecules strike a wall is
$$f = \frac{\bar{v}_x}{2}\frac{P}{kT}\ell^2$$
where $\bar{v}_x$ is the average x component of the molecule's velocity. (b) Show that the equation can then be written as
$$f \approx \frac{P\ell^2}{\sqrt{4mkT}}$$
where m is the mass of a gas molecule. (c) Assume a cubic air-filled room is at sea level, has a temperature 20°C, and has sides of length $\ell = 3\,\text{m}$. Determine f.

18–2 Distribution of Molecular Speeds

19. (I) If you double the mass of the molecules in a gas, is it possible to change the temperature to keep the velocity distribution from changing? If so, what do you need to do to the temperature?

20. (I) A group of 25 particles have the following speeds: two have speed 10 m/s, seven have 15 m/s, four have 20 m/s, three have 25 m/s, six have 30 m/s, one has 35 m/s, and two have 40 m/s. Determine (a) the average speed, (b) the rms speed, and (c) the most probable speed.

21. (II) A gas consisting of 15,200 molecules, each of mass 2.00×10^{-26} kg, has the following distribution of speeds, which crudely mimics the Maxwell distribution:

Number of Molecules	Speed (m/s)
1600	220
4100	440
4700	660
3100	880
1300	1100
400	1320

(a) Determine $v_{\rm rms}$ for this distribution of speeds. (b) Given your value for $v_{\rm rms}$, what (effective) temperature would you assign to this gas? (c) Determine the mean speed $\bar{v}$ of this distribution and use this value to assign an (effective) temperature to the gas. Is the temperature you find here consistent with the one you determined in part (b)?

22. (III) Starting from the Maxwell distribution of speeds, Eq. 18–6, show (a) $\int_0^\infty f(v)\,dv = N$, and (b)
$$\int_0^\infty v^2 f(v)\,dv/N = 3kT/m.$$

18–3 Real Gases

23. (I) CO_2 exists in what phase when the pressure is 30 atm and the temperature is 30°C (Fig. 18–6)?

24. (I) (a) At atmospheric pressure, in what phases can CO_2 exist? (b) For what range of pressures and temperatures can CO_2 be a liquid? Refer to Fig. 18–6.

25. (I) Water is in which phase when the pressure is 0.01 atm and the temperature is (a) 90°C, (b) −20°C?

26. (II) You have a sample of water and are able to control temperature and pressure arbitrarily. (a) Using Fig. 18–5, describe the phase changes you would see if you started at a temperature of 85°C, a pressure of 180 atm, and decreased the pressure down to 0.004 atm while keeping the temperature fixed. (b) Repeat part (a) with the temperature at 0.0°C. Assume that you held the system at the starting conditions long enough for the system to stabilize before making further changes.

18–4 Vapor Pressure and Humidity

27. (I) What is the partial pressure of water vapor at 30°C if the humidity is 85%?

28. (I) What is the partial pressure of water on a day when the temperature is 25°C and the relative humidity is 55%?

29. (I) What is the air pressure at a place where water boils at 80°C?

30. (II) What is the dew point if the humidity is 75% on a day when the temperature is 25°C?

31. (II) If the air pressure at a particular place in the mountains is 0.75 atm, estimate the temperature at which water boils.

32. (II) What is the mass of water in a closed room 5.0 m × 6.0 m × 2.4 m when the temperature is 24.0°C and the relative humidity is 65%?

33. (II) What is the approximate pressure inside a pressure cooker if the water is boiling at a temperature of 120°C? Assume no air escaped during the heating process, which started at 12°C.

34. (II) If the humidity in a room of volume 440 m³ at 25°C is 65%, what mass of water can still evaporate from an open pan?

35. (II) A **pressure cooker** is a sealed pot designed to cook food with the steam produced by boiling water somewhat above 100°C. The pressure cooker in Fig. 18–17 uses a weight of mass m to allow steam to escape at a certain pressure through a small hole (diameter d) in the cooker's lid. If $d = 3.0$ mm, what should m be in order to cook food at 120°C? Assume that atmospheric pressure outside the cooker is 1.01×10^5 Pa.

Weight (mass m)

←Diameter d

Steam

Water

FIGURE 18–17
Problem 35.

36. (II) When using a mercury barometer (Section 13–6), the vapor pressure of mercury is usually assumed to be zero. At room temperature mercury's vapor pressure is about 0.0015 mm-Hg. At sea level, the height h of mercury in a barometer is about 760 mm. (a) If the vapor pressure of mercury is neglected, is the true atmospheric pressure greater or less than the value read from the barometer? (b) What is the percent error? (c) What is the percent error if you use a water barometer and ignore water's saturated vapor pressure at STP?

37. (II) If the humidity is 45% at 30.0°C, what is the dew point? Use linear interpolation to find the temperature of the dew point to the nearest degree.

38. (III) Air that is at its dew point of 5°C is drawn into a building where it is heated to 20°C. What will be the relative humidity at this temperature? Assume constant pressure of 1.0 atm. Take into account the expansion of the air.

39. (III) What is the mathematical relation between water's boiling temperature and atmospheric pressure? (a) Using the data from Table 18–2, in the temperature range from 50°C to 150°C, plot $\ln P$ versus $(1/T)$, where P is water's saturated vapor pressure (Pa) and T is temperature on the Kelvin scale. Show that a straight-line plot results and determine the slope and y-intercept of the line. (b) Show that your result implies

$$P = Be^{-A/T}$$

where A and B are constants. Use the slope and y-intercept from your plot to show that $A \approx 5000$ K and $B \approx 7 \times 10^{10}$ Pa.

*18–5 Van der Waals Equation of State

*40. (II) In the van der Waals equation of state, the constant b represents the amount of "unavailable volume" occupied by the molecules themselves. Thus V is replaced by $(V - nb)$, where n is the number of moles. For oxygen, b is about 3.2×10^{-5} m³/mol. Estimate the diameter of an oxygen molecule.

*41. (II) For oxygen gas, the van der Waals equation of state achieves its best fit for $a = 0.14$ N·m⁴/mol² and $b = 3.2 \times 10^{-5}$ m³/mol. Determine the pressure in 1.0 mol of the gas at 0°C if its volume is 0.70 L, calculated using (a) the van der Waals equation, (b) the ideal gas law.

*42. (III) A 0.5-mol sample of O_2 gas is in a large cylinder with a movable piston on one end so it can be compressed. The initial volume is large enough that there is not a significant difference between the pressure given by the ideal gas law and that given by the van der Waals equation. As the gas is slowly compressed at constant temperature (use 300 K), at what volume does the van der Waals equation give a pressure that is 5% different than the ideal gas law pressure? Let $a = 0.14$ N·m⁴/mol² and $b = 3.2 \times 10^{-5}$ m³/mol.

*43. (III) (a) From the van der Waals equation of state, show that the critical temperature and pressure are given by

$$T_{cr} = \frac{8a}{27bR}, \qquad P_{cr} = \frac{a}{27b^2}.$$

[Hint: Use the fact that the P versus V curve has an inflection point at the critical point so that the first and second derivatives are zero.] (b) Determine a and b for CO_2 from the measured values of $T_{cr} = 304$ K and $P_{cr} = 72.8$ atm.

*44. (III) How well does the ideal gas law describe the pressurized air in a scuba tank? (a) To fill a typical scuba tank, an air compressor intakes about 2300 L of air at 1.0 atm and compresses this gas into the tank's 12-L internal volume. If the filling process occurs at 20°C, show that a tank holds about 96 mol of air. (b) Assume the tank has 96 mol of air at 20°C. Use the ideal gas law to predict the air's pressure within the tank. (c) Use the van der Waals equation of state to predict the air's pressure within the tank. For air, the van der Waals constants are $a = 0.1373$ N·m⁴/mol² and $b = 3.72 \times 10^{-5}$ m³/mol. (d) Taking the van der Waals pressure as the true air pressure, show that the ideal gas law predicts a pressure that is in error by only about 3%.

*18–6 Mean Free Path

*45. (II) At about what pressure would the mean free path of air molecules be (a) 0.10 m and (b) equal to the diameter of air molecules, $\approx 3 \times 10^{-10}$ m? Assume $T = 20°C$.

*46. (II) Below a certain threshold pressure, the air molecules (0.3-nm diameter) within a research vacuum chamber are in the "collision-free regime," meaning that a particular air molecule is as likely to cross the container and collide first with the opposite wall, as it is to collide with another air molecule. Estimate the threshold pressure for a vacuum chamber of side 1.0 m at 20°C.

*47. (II) A very small amount of hydrogen gas is released into the air. If the air is at 1.0 atm and 15°C, estimate the mean free path for a H_2 molecule. What assumptions did you make?

*48. (II) (a) The mean free path of CO_2 molecules at STP is measured to be about 5.6×10^{-8} m. Estimate the diameter of a CO_2 molecule. (b) Do the same for He gas for which $\ell_M \approx 25 \times 10^{-8}$ at STP.

*49. (II) (a) Show that the number of collisions a molecule makes per second, called the *collision frequency*, f, is given by $f = \bar{v}/\ell_M$, and thus $f = 4\sqrt{2}\,\pi r^2\,\bar{v}N/V$. (b) What is the collision frequency for N_2 molecules in air at $T = 20°C$ and $P = 1.0 \times 10^{-2}$ atm?

*50. (II) We saw in Example 18–8 that the mean free path of air molecules at STP, ℓ_M, is about 9×10^{-8} m. Estimate the collision frequency f, the number of collisions per unit time.

*51. (II) A cubic box 1.80 m on a side is evacuated so the pressure of air inside is 10^{-6} torr. Estimate how many collisions molecules make with each other for each collision with a wall (0°C).

*52. (III) Estimate the maximum allowable pressure in a 32-cm-long cathode ray tube if 98% of all electrons must hit the screen without first striking an air molecule.

*18–7 Diffusion

*53. (I) Approximately how long would it take for the ammonia of Example 18–9 to be detected 1.0 m from the bottle after it is opened? What does this suggest about the relative importance of diffusion and convection for carrying odors?

*54. (II) Estimate the time needed for a glycine molecule (see Table 18–3) to diffuse a distance of 15 μm in water at 20°C if its concentration varies over that distance from 1.00 mol/m^3 to 0.50 mol/m^3? Compare this "speed" to its rms (thermal) speed. The molecular mass of glycine is about 75 u.

*55. (II) Oxygen diffuses from the surface of insects to the interior through tiny tubes called tracheae. An average trachea is about 2 mm long and has cross-sectional area of 2×10^{-9} m^2. Assuming the concentration of oxygen inside is half what it is outside in the atmosphere, (a) show that the concentration of oxygen in the air (assume 21% is oxygen) at 20°C is about 8.7 mol/m^3, then (b) calculate the diffusion rate J, and (c) estimate the average time for a molecule to diffuse in. Assume the diffusion constant is 1×10^{-5} m^2/s.

General Problems

56. A sample of ideal gas must contain at least $N = 10^6$ molecules in order for the Maxwell distribution to be a valid description of the gas, and to assign it a meaningful temperature. For an ideal gas at STP, what is the smallest length scale ℓ (volume $V = \ell^3$) for which a valid temperature can be assigned?

57. In outer space the density of matter is about one atom per cm^3, mainly hydrogen atoms, and the temperature is about 2.7 K. Calculate the rms speed of these hydrogen atoms, and the pressure (in atmospheres).

58. Calculate approximately the total translational kinetic energy of all the molecules in an *E. coli* bacterium of mass 2.0×10^{-15} kg at 37°C. Assume 70% of the cell, by weight, is water, and the other molecules have an average molecular mass on the order of 10^5 u.

59. (a) Estimate the rms speed of an amino acid, whose molecular mass is 89 u, in a living cell at 37°C. (b) What would be the rms speed of a protein of molecular mass 85,000 u at 37°C?

60. The escape speed from the Earth is 1.12×10^4 m/s, so that a gas molecule travelling away from Earth near the outer boundary of the Earth's atmosphere would, at this speed, be able to escape from the Earth's gravitational field and be lost to the atmosphere. At what temperature is the average speed of (a) oxygen molecules, and (b) helium atoms equal to 1.12×10^4 m/s? (c) Can you explain why our atmosphere contains oxygen but not helium?

61. The second postulate of kinetic theory is that the molecules are, on the average, far apart from one another. That is, their average separation is much greater than the diameter of each molecule. Is this assumption reasonable? To check, calculate the average distance between molecules of a gas at STP, and compare it to the diameter of a typical gas molecule, about 0.3 nm. If the molecules were the diameter of ping-pong balls, say 4 cm, how far away would the next ping-pong ball be on average?

62. A sample of liquid cesium is heated in an oven to 400°C and the resulting vapor is used to produce an atomic beam. The volume of the oven is 55 cm^3, the vapor pressure of Cs at 400°C is 17 mm-Hg, and the diameter of cesium atoms in the vapor is 0.33 nm. (a) Calculate the mean speed of cesium atoms in the vapor. (b) Determine the number of collisions a single Cs atom undergoes with other cesium atoms per second. (c) Determine the total number of collisions per second between all of the cesium atoms in the vapor. Note that a collision involves two Cs atoms and assume the ideal gas law holds.

63. Consider a container of oxygen gas at a temperature of 20°C that is 1.00 m tall. Compare the gravitational potential energy of a molecule at the top of the container (assuming the potential energy is zero at the bottom) with the average kinetic energy of the molecules. Is it reasonable to neglect the potential energy?

64. In humid climates, people constantly *dehumidify* their cellars to prevent rot and mildew. If the cellar in a house (kept at 20°C) has 115 m² of floor space and a ceiling height of 2.8 m, what is the mass of water that must be removed from it in order to drop the humidity from 95% to a more reasonable 40%?

65. Assuming a typical nitrogen or oxygen molecule is about 0.3 nm in diameter, what percent of the room you are sitting in is taken up by the volume of the molecules themselves?

66. A scuba tank has a volume of 3100 cm³. For very deep dives, the tank is filled with 50% (by volume) pure oxygen and 50% pure helium. (*a*) How many molecules are there of each type in the tank if it is filled at 20°C to a gauge pressure of 12 atm? (*b*) What is the ratio of the average kinetic energies of the two types of molecule? (*c*) What is the ratio of the rms speeds of the two types of molecule?

67. A space vehicle returning from the Moon enters the atmosphere at a speed of about 42,000 km/h. Molecules (assume nitrogen) striking the nose of the vehicle with this speed correspond to what temperature? (Because of this high temperature, the nose of a space vehicle must be made of special materials; indeed, part of it does vaporize, and this is seen as a bright blaze upon reentry.)

68. At room temperature, it takes approximately 2.45×10^3 J to evaporate 1.00 g of water. Estimate the average speed of evaporating molecules. What multiple of v_{rms} (at 20°C) for water molecules is this? (Assume Eq. 18–4 holds.)

69. Calculate the total water vapor pressure in the air on the following two days: (*a*) a hot summer day, with the temperature 30°C and the relative humidity at 65%; (*b*) a cold winter day, with the temperature 5°C and the relative humidity at 75%.

* 70. At 300 K, an 8.50-mol sample of carbon dioxide occupies a volume of 0.220 m³. Calculate the gas pressure, first by assuming the ideal gas law, and then by using the van der Waals equation of state. (The values for a and b are given in Section 18–5.) In this range of pressure and volume, the van der Waals equation is very accurate. What percent error did you make in assuming ideal-gas-law behavior?

* 71. The density of atoms, mostly hydrogen, in interstellar space is about one per cubic centimeter. Estimate the mean free path of the hydrogen atoms, assuming an atomic diameter of 10^{-10} m.

* 72. Using the ideal gas law, find an expression for the mean free path ℓ_M that involves pressure and temperature instead of (N/V). Use this expression to find the mean free path for nitrogen molecules at a pressure of 7.5 atm and 300 K.

73. A sauna has 8.5 m³ of air volume, and the temperature is 90°C. The air is perfectly dry. How much water (in kg) should be evaporated if we want to increase the relative humidity from 0% to 10%? (See Table 18–2.)

74. A 0.50-kg trash-can lid is suspended against gravity by tennis balls thrown vertically upward at it. How many tennis balls per second must rebound from the lid elastically, assuming they have a mass of 0.060 kg and are thrown at 12 m/s?

* 75. Sound waves in a gas can only propagate if the gas molecules collide with each other on the time scale of the sound wave's period. Thus the highest possible frequency f_{max} for a sound wave in a gas is approximately equal to the inverse of the average collision time between molecules. Assume a gas, composed of molecules with mass m and radius r, is at pressure P and temperature T. (*a*) Show that

$$f_{max} \approx 16Pr^2 \sqrt{\frac{\pi}{mkT}}.$$

(*b*) Determine f_{max} for 20°C air at sea level. How many times larger is f_{max} than the highest frequency in the human audio range (20 kHz)?

* Numerical/Computer

* 76. (II) Use a spreadsheet to calculate and graph the fraction of molecules in each 50-m/s speed interval from 100 m/s to 5000 m/s if $T = 300$ K.

* 77. (II) Use numerical integration [Section 2–9] to estimate (within 2%) the fraction of molecules in air at 1.00 atm and 20°C that have a speed greater than 1.5 times the most probable speed.

* 78. (II) For oxygen gas the van der Waals constants are $a = 0.14$ N·m⁴/mol² and $b = 3.2 \times 10^{-5}$ m³/mol. Using these values, graph six curves of pressure vs. volume between $V = 2 \times 10^{-5}$ m³ to 2.0×10^{-4} m³, for 1 mol of oxygen gas at temperatures of 80 K, 100 K, 120 K, 130 K, 150 K, and 170 K. From the graphs determine approximately the critical temperature for oxygen.

Answers to Exercises

A: (*a*).

B: (*d*).

C: (*c*).

D: (*d*).

E: (*b*).

When it is cold, warm clothes act as insulators to reduce heat loss from the body to the environment by conduction and convection. Heat radiation from a campfire can warm you and your clothes. The fire can also transfer energy directly by heat convection and conduction to what you are cooking. Heat, like work, represents a transfer of energy. Heat is defined as a transfer of energy due to a difference of temperature. Work is a transfer of energy by mechanical means, not due to a temperature difference. The first law of thermodynamics links the two in a general statement of energy conservation: the heat Q added to a system minus the net work W done by the system equals the change in internal energy ΔE_{int} of the system: $\Delta E_{int} = Q - W$. Internal energy E_{int} is the sum total of all the energy of the molecules of the system.

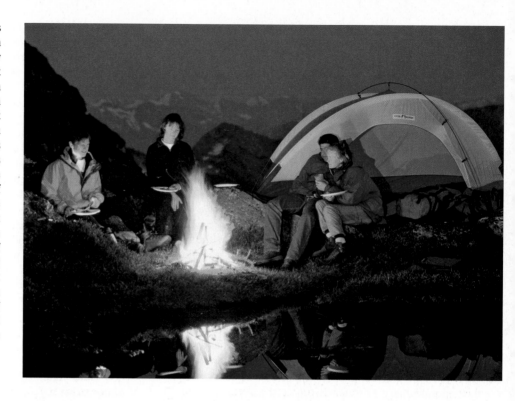

19

Heat and the First Law of Thermodynamics

CONTENTS

CHAPTER-OPENING QUESTION—Guess now!

A 5-kg cube of warm iron (60°C) is put in thermal contact with a 10-kg cube of cold iron (15°C). Which statement is valid:

(a) Heat flows spontaneously from the warm cube to the cold cube until both cubes have the same heat content.

(b) Heat flows spontaneously from the warm cube to the cold cube until both cubes have the same temperature.

(c) Heat can flow spontaneously from the warm cube to the cold cube, but can also flow spontaneously from the cold cube to the warm cube.

(d) Heat can never flow from a cold object or area to a hot object or area.

(e) Heat flows from the larger cube to the smaller one because the larger one has more internal energy.

When a pot of cold water is placed on a hot burner of a stove, the temperature of the water increases. We say that heat "flows" from the hot burner to the cold water. When two objects at different temperatures are put in contact, heat spontaneously flows from the hotter one to the colder one. The spontaneous flow of heat is in the direction tending to equalize the temperature. If the two objects are kept in contact long enough for their temperatures to become equal, the objects are said to be in thermal equilibrium, and there is no further heat flow between them. For example, when a fever thermometer is first placed in your mouth, heat flows from your mouth to the thermometer. When the thermometer reaches the same temperature as the inside of your mouth, the thermometer and your mouth are then in equilibrium, and no more heat flows.

Heat and temperature are often confused. They are very different concepts and we will make the clear distinction between them. We start this Chapter by defining and using the concept of heat. We also begin our discussion of thermodynamics, which is the name we give to the study of processes in which energy is transferred as heat and as work.

19–1 Heat as Energy Transfer

We use the term "heat" in everyday life as if we knew what we meant. But the term is often used inconsistently, so it is important for us to define heat clearly, and to clarify the phenomena and concepts related to heat.

We commonly speak of the flow of heat—heat flows from a stove burner to a pot of soup, from the Sun to the Earth, from a person's mouth into a fever thermometer. Heat flows spontaneously from an object at higher temperature to one at lower temperature. Indeed, an eighteenth-century model of heat pictured heat flow as movement of a fluid substance called *caloric*. However, the caloric fluid could never be detected. In the nineteenth century, it was found that the various phenomena associated with heat could be described consistently using a new model that views heat as akin to work, as we will discuss in a moment. First we note that a common unit for heat, still in use today, is named after caloric. It is called the **calorie** (cal) and is defined as *the amount of heat necessary to raise the temperature of 1 gram of water by 1 Celsius degree*. [To be precise, the particular temperature range from 14.5°C to 15.5°C is specified because the heat required is very slightly different at different temperatures. The difference is less than 1% over the range 0 to 100°C, and we will ignore it for most purposes.] More often used than the calorie is the **kilocalorie** (kcal), which is 1000 calories. Thus *1 kcal is the heat needed to raise the temperature of 1 kg of water by 1 C°*. Often a kilocalorie is called a **Calorie** (with a capital C), and this Calorie (or the kJ) is used to specify the energy value of food. In the British system of units, heat is measured in British thermal units (Btu). One Btu is defined as the heat needed to raise the temperature of 1 lb of water by 1 F°. It can be shown (Problem 4) that $1 \text{ Btu} = 0.252 \text{ kcal} = 1056 \text{ J}$.

The idea that heat is related to energy transfer was pursued by a number of scientists in the 1800s, particularly by an English brewer, James Prescott Joule (1818–1889). One of Joule's experiments is shown (simplified) in Fig. 19–1. The falling weight causes the paddle wheel to turn. The friction between the water and the paddle wheel causes the temperature of the water to rise slightly (barely measurable, in fact, by Joule). In this and many other experiments (some involving electrical energy), Joule determined that a given amount of work done was always equivalent to a particular amount of heat input. Quantitatively, 4.186 joules (J) of work was found to be equivalent to 1 calorie (cal) of heat. This is known as the **mechanical equivalent of heat**:

$$4.186 \text{ J} = 1 \text{ cal};$$
$$4.186 \text{ kJ} = 1 \text{ kcal}.$$

As a result of these and other experiments, scientists came to interpret heat not as a substance, and not exactly as a form of energy. Rather, heat refers to a *transfer of energy*: when heat flows from a hot object to a cooler one, it is energy that is being transferred from the hot to the cold object. Thus, **heat** is *energy transferred from one object to another because of a difference in temperature*. In SI units, the unit for heat, as for any form of energy, is the joule. Nonetheless, calories and kcal are still sometimes used. Today the calorie is *defined* in terms of the joule (via the mechanical equivalent of heat, above), rather than in terms of the properties of water, as given previously. The latter is still handy to remember: 1 cal raises 1 g of water by 1 C°, or 1 kcal raises 1 kg of water by 1 C°.

Joule's result was crucial because it extended the work-energy principle to include processes involving heat. It also led to the establishment of the law of conservation of energy, which we shall discuss more fully later in this Chapter.

⚠ **CAUTION**
Heat is not a fluid

Weight

FIGURE 19–1 Joule's experiment on the mechanical equivalent of heat.

⚠ **CAUTION**
Heat is energy transferred because of a ΔT

EXAMPLE 19–1 **ESTIMATE** **Working off the extra calories.** Suppose you throw caution to the wind and eat too much ice cream and cake on the order of 500 Calories. To compensate, you want to do an equivalent amount of work climbing stairs or a mountain. How much total height must you climb?

APPROACH The work W you need to do in climbing stairs equals the change in gravitational potential energy: $W = \Delta PE = mgh$, where h is the vertical height climbed. For this estimate, approximate your mass as $m \approx 60\,\text{kg}$.

SOLUTION 500 food Calories are 500 kcal, which in joules is

$$(500\,\text{kcal})(4.186 \times 10^3\,\text{J/kcal}) = 2.1 \times 10^6\,\text{J}.$$

The work done to climb a vertical height h is $W = mgh$. We solve for h:

$$h = \frac{W}{mg} = \frac{2.1 \times 10^6\,\text{J}}{(60\,\text{kg})(9.80\,\text{m/s}^2)} = 3600\,\text{m}.$$

This is a huge elevation change (over 11,000 ft).

NOTE The human body does not transform food energy with 100% efficiency—it is more like 20% efficient. As we shall discuss in the next Chapter, some energy is always "wasted," so you would actually have to climb only about $(0.2)(3600\,\text{m}) \approx 700\,\text{m}$, which is more reasonable (about 2300 ft of elevation gain).

19–2 Internal Energy

The sum total of all the energy of all the molecules in an object is called its **internal energy**. (Sometimes **thermal energy** is used to mean the same thing.) We introduce the concept of internal energy now since it will help clarify ideas about heat.

Distinguishing Temperature, Heat, and Internal Energy

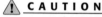

CAUTION

Distinguish heat from internal energy and from temperature

Using the kinetic theory, we can make a clear distinction between temperature, heat, and internal energy. Temperature (in kelvins) is a measure of the *average* kinetic energy of individual molecules. Internal energy refers to the *total* energy of all the molecules within the object. (Thus two equal-mass hot ingots of iron may have the same temperature, but two of them have twice as much internal energy as one does.) Heat, finally, refers to a *transfer* of energy from one object to another because of a difference in temperature.

CAUTION

Direction of heat flow depends on temperature (not on amount of internal energy)

Notice that the direction of heat flow between two objects depends on their temperatures, not on how much internal energy each has. Thus, if 50 g of water at 30°C is placed in contact (or mixed) with 200 g of water at 25°C, heat flows *from* the water at 30°C *to* the water at 25°C even though the internal energy of the 25°C water is much greater because there is so much more of it.

EXERCISE A Return to the Chapter-Opening Question, page 496, and answer it again now. Try to explain why you may have answered differently the first time.

Internal Energy of an Ideal Gas

Let us calculate the internal energy of n moles of an ideal monatomic (one atom per molecule) gas. The internal energy, E_{int}, is the sum of the translational kinetic energies of all the atoms.[†] This sum is equal to the average kinetic energy per molecule times the total number of molecules, N:

$$E_{\text{int}} = N\left(\tfrac{1}{2}m\overline{v^2}\right).$$

Using Eq. 18–4, $\overline{K} = \tfrac{1}{2}m\overline{v^2} = \tfrac{3}{2}kT$, we can write this as

$$E_{\text{int}} = \tfrac{3}{2}NkT \tag{19–1a}$$

[†]The symbol U is used in some books for internal energy. The use of E_{int} avoids confusion with U which stands for potential energy (Chapter 8).

or (recall Section 17–9)

$$E_{int} = \tfrac{3}{2}nRT, \qquad \text{[ideal monatomic gas]} \quad \text{(19–1b)}$$

where n is the number of moles. Thus, the internal energy of an ideal gas depends only on temperature and the number of moles of gas.

If the gas molecules contain more than one atom, then the rotational and vibrational energy of the molecules (Fig. 19–2) must also be taken into account. The internal energy will be greater at a given temperature than for a monatomic gas, but it will still be a function only of temperature for an ideal gas.

The internal energy of real gases also depends mainly on temperature, but where real gases deviate from ideal gas behavior, their internal energy depends also somewhat on pressure and volume (due to atomic potential energy).

The internal energy of liquids and solids is quite complicated, for it includes electrical potential energy associated with the forces (or "chemical" bonds) between atoms and molecules.

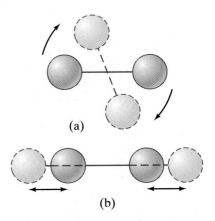

FIGURE 19–2 Besides translational kinetic energy, molecules can have (a) rotational kinetic energy, and (b) vibrational energy (both kinetic and potential).

19–3 Specific Heat

If heat flows into an object, the object's temperature rises (assuming no phase change). But how much does the temperature rise? That depends. As early as the eighteenth century, experimenters had recognized that the amount of heat Q required to change the temperature of a given material is proportional to the mass m of the material present and to the temperature change ΔT. This remarkable simplicity in nature can be expressed in the equation

$$Q = mc\,\Delta T, \qquad\qquad\qquad \text{(19–2)}$$

where c is a quantity characteristic of the material called its **specific heat**. Because $c = Q/m\,\Delta T$, specific heat is specified in units[†] of J/kg·C° (the proper SI unit) or kcal/kg·C°. For water at 15°C and a constant pressure of 1 atm, $c = 4.19 \times 10^3$ J/kg·C° or 1.00 kcal/kg·C°, since, by definition of the cal and the joule, it takes 1 kcal of heat to raise the temperature of 1 kg of water by 1 C°. Table 19–1 gives the values of specific heat for other solids and liquids at 20°C. The values of c for solids and liquids depend to some extent on temperature (as well as slightly on pressure), but for temperature changes that are not too great, c can often be considered constant.[‡] Gases are more complicated and are treated in Section 19–8.

EXAMPLE 19–2 How heat transferred depends on specific heat. (a) How much heat input is needed to raise the temperature of an empty 20-kg vat made of iron from 10°C to 90°C? (b) What if the vat is filled with 20 kg of water?

APPROACH We apply Eq. 19–2 to the different materials involved.

SOLUTION (a) Our system is the iron vat alone. From Table 19–1, the specific heat of iron is 450 J/kg·C°. The change in temperature is (90°C − 10°C) = 80 C°. Thus,

$$Q = mc\,\Delta T = (20\,\text{kg})(450\,\text{J/kg·C°})(80\,\text{C°}) = 7.2 \times 10^5\,\text{J} = 720\,\text{kJ}.$$

(b) Our system is the vat plus the water. The water alone would require

$$Q = mc\,\Delta T = (20\,\text{kg})(4186\,\text{J/kg·C°})(80\,\text{C°}) = 6.7 \times 10^6\,\text{J} = 6700\,\text{kJ},$$

or almost 10 times what an equal mass of iron requires. The total, for the vat plus the water, is 720 kJ + 6700 kJ = 7400 kJ.

NOTE In (b), the iron vat and the water underwent the same temperature change, $\Delta T = 80$ C°, but their specific heats are different.

[†]Note that J/kg·C° means $\dfrac{\text{J}}{\text{kg·C°}}$ and *not* (J/kg)·C° = J·C°/kg (otherwise we would have written it that way).

[‡]To take into account the dependence of c on T, we can write Eq. 19–2 in differential form: $dQ = mc(T)\,dT$, where $c(T)$ means c is a function of temperature T. Then the heat Q required to change the temperature from T_1 to T_2 is

$$Q = \int_{T_1}^{T_2} mc(T)\,dT.$$

TABLE 19–1 Specific Heats
(at 1 atm constant pressure and 20°C unless otherwise stated)

Substance	Specific Heat, c	
	kcal/kg·C° (= cal/g·C°)	J/kg·C°
Aluminum	0.22	900
Alcohol (ethyl)	0.58	2400
Copper	0.093	390
Glass	0.20	840
Iron or steel	0.11	450
Lead	0.031	130
Marble	0.21	860
Mercury	0.033	140
Silver	0.056	230
Wood	0.4	1700
Water		
Ice (−5°C)	0.50	2100
Liquid (15°C)	1.00	4186
Steam (110°C)	0.48	2010
Human body (average)	0.83	3470
Protein	0.4	1700

If the iron vat in part (*a*) of Example 19–2 had been *cooled* from 90°C to 10°C, 720 kJ of heat would have flowed *out* of the iron. In other words, Eq. 19–2 is valid for heat flow either in or out, with a corresponding increase or decrease in temperature. We saw in part (*b*) that water requires almost 10 times as much heat as an equal mass of iron to make the same temperature change. Water has one of the highest specific heats of all substances, which makes it an ideal substance for hot-water space-heating systems and other uses that require a minimal drop in temperature for a given amount of heat transfer. It is the water content, too, that causes the apples rather than the crust in hot apple pie to burn our tongues, through heat transfer.

19–4 Calorimetry—Solving Problems

In discussing heat and thermodynamics, we shall often refer to particular systems. As already mentioned in earlier Chapters, a **system** is any object or set of objects that we wish to consider. Everything else in the universe we will refer to as its "environment" or the "surroundings." There are several categories of systems. A **closed system** is one for which no mass enters or leaves (but energy may be exchanged with the environment). In an **open system**, mass may enter or leave (as may energy). Many (idealized) systems we study in physics are closed systems. But many systems, including plants and animals, are open systems since they exchange materials (food, oxygen, waste products) with the environment. A closed system is said to be **isolated** if no energy in any form passes across its boundaries; otherwise it is not isolated.

When different parts of an isolated system are at different temperatures, heat will flow (energy is transferred) from the part at higher temperature to the part at lower temperature—that is, within the system. If the system is truely isolated, no energy is transferred into or out of it. So the *conservation of energy* again plays an important role for us: the heat lost by one part of the system is equal to the heat gained by the other part:

$$\text{heat lost } = \text{ heat gained}$$

or

$$\text{energy out of one part } = \text{ energy into another part.}$$

These simple relations are very useful, but depend on the (often very good) approximation that the whole system is isolated (no other energy transfers occur). Let us take an Example.

EXAMPLE 19–3 **The cup cools the tea.** If 200 cm³ of tea at 95°C is poured into a 150-g glass cup initially at 25°C (Fig. 19–3), what will be the common final temperature T of the tea and cup when equilibrium is reached, assuming no heat flows to the surroundings?

APPROACH We apply conservation of energy to our system of tea plus cup, which we are assuming is isolated: all of the heat that leaves the tea flows into the cup. We use the specific heat equation, Eq. 19–2, to determine how the heat flow is related to the temperature changes.

SOLUTION Because tea is mainly water, its specific heat is 4186 J/kg·C° (Table 19–1), and its mass m is its density times its volume $(V = 200\,\text{cm}^3 = 200 \times 10^{-6}\,\text{m}^3)$: $m = \rho V = (1.0 \times 10^3\,\text{kg/m}^3)(200 \times 10^{-6}\,\text{m}^3) = 0.20\,\text{kg}$. We use Eq. 19–2, apply conservation of energy, and let T be the as yet unknown final temperature:

$$\text{heat lost by tea } = \text{ heat gained by cup}$$
$$m_{\text{tea}} c_{\text{tea}}(95°C - T) = m_{\text{cup}} c_{\text{cup}}(T - 25°C).$$

Putting in numbers and using Table 19–1 ($c_{\text{cup}} = 840\,\text{J/kg·C}°$ for glass), we solve for T, and find

$$(0.20\,\text{kg})(4186\,\text{J/kg·C}°)(95°C - T) = (0.15\,\text{kg})(840\,\text{J/kg·C}°)(T - 25°C)$$
$$79{,}500\,\text{J} - (837\,\text{J/C}°)T = (126\,\text{J/C}°)T - 3150\,\text{J}$$
$$T = 86°C.$$

The tea drops in temperature by 9 C° by coming into equilibrium with the cup.

NOTE The cup increases in temperature by $86°C - 25°C = 61\,\text{C}°$. Its much greater change in temperature (compared with that of the tea water) is due to its much smaller specific heat compared to that of water.

FIGURE 19–3 Example 19–3.

95°C

25°C

$T = ?$

(a)   (b)

NOTE In this calculation, the ΔT (of Eq. 19–2, $Q = mc\,\Delta T$) is a positive quantity on both sides of our conservation of energy equation. On the left is "heat lost" and ΔT is the initial minus the final temperature ($95°C - T$), whereas on the right is "heat gained" and ΔT is the final minus the initial temperature. But consider the following alternate approach.

Alternate Solution We can set up this Example (and others) by a different approach. We can write that the total heat transferred into or out of the isolated system is zero:

$$\Sigma Q = 0.$$

Then each term is written as $Q = mc(T_f - T_i)$, and $\Delta T = T_f - T_i$ is always the final minus the initial temperature, and each ΔT can be positive or negative. In the present Example:

$$\Sigma Q = m_{cup}c_{cup}(T - 25°C) + m_{tea}c_{tea}(T - 95°C) = 0.$$

The second term is negative because T will be less than 95°C. Solving the algebra gives the same result.

FIGURE 19–4 Simple water calorimeter.

The exchange of energy, as exemplified in Example 19–3, is the basis for a technique known as **calorimetry**, which is the quantitative measurement of heat exchange. To make such measurements, a **calorimeter** is used; a simple water calorimeter is shown in Fig. 19–4. It is very important that the calorimeter be well insulated so that almost no heat is exchanged with the surroundings. One important use of the calorimeter is in the determination of specific heats of substances. In the technique known as the "method of mixtures," a sample of a substance is heated to a high temperature, which is accurately measured, and then quickly placed in the cool water of the calorimeter. The heat lost by the sample will be gained by the water and the calorimeter cup. By measuring the final temperature of the mixture, the specific heat can be calculated, as illustrated in the following Example.

EXAMPLE 19–4 **Unknown specific heat determined by calorimetry.**
An engineer wishes to determine the specific heat of a new metal alloy. A 0.150-kg sample of the alloy is heated to 540°C. It is then quickly placed in 0.400 kg of water at 10.0°C, which is contained in a 0.200-kg aluminum calorimeter cup. (We do not need to know the mass of the insulating jacket since we assume the air space between it and the cup insulates it well, so that its temperature does not change significantly.) The final temperature of the system is 30.5°C. Calculate the specific heat of the alloy.

APPROACH We apply conservation of energy to our system, which we take to be the alloy sample, the water, and the calorimeter cup. We assume this system is isolated, so the energy lost by the hot alloy equals the energy gained by the water and calorimeter cup.

SOLUTION The heat lost equals the heat gained:

$$\left(\begin{array}{c}\text{heat lost}\\\text{by alloy}\end{array}\right) = \left(\begin{array}{c}\text{heat gained}\\\text{by water}\end{array}\right) + \left(\begin{array}{c}\text{heat gained by}\\\text{calorimeter cup}\end{array}\right)$$

$$m_a c_a \Delta T_a = m_w c_w \Delta T_w + m_{cal} c_{cal} \Delta T_{cal}$$

where the subscripts a, w, and cal refer to the alloy, water, and calorimeter, respectively, and each $\Delta T > 0$. When we put in values and use Table 19–1, this equation becomes

$$(0.150\,kg)(c_a)(540°C - 30.5°C) = (0.400\,kg)(4186\,J/kg \cdot C°)(30.5°C - 10.0°C)$$
$$+ (0.200\,kg)(900\,J/kg \cdot C°)(30.5°C - 10.0°C)$$
$$(76.4\,kg \cdot C°)\,c_a = (34{,}300 + 3690)\,J$$
$$c_a = 497\,J/kg \cdot C°.$$

In making this calculation, we have ignored any heat transferred to the thermometer and the stirrer (which is used to quicken the heat transfer process and thus reduce heat loss to the outside). It can be taken into account by adding additional terms to the right side of the above equation and will result in a slight correction to the value of c_a.

In all Examples and Problems of this sort, be sure to include *all* objects that gain or lose heat (within reason). On the "heat loss" side here, it is only the hot metal alloy. On the "heat gain" side, it is both the water and the aluminum calorimeter cup. For simplicity, we have ignored very small masses, such as the thermometer and the stirrer, which will affect the energy balance only very slightly.

FIGURE 19–5 Temperature as a function of the heat added to bring 1.0 kg of ice at −40°C to steam above 100°C.

19–5 Latent Heat

When a material changes phase from solid to liquid, or from liquid to gas (see also Section 18–3), a certain amount of energy is involved in this **change of phase**. For example, let us trace what happens when a 1.0-kg block of ice at −40°C is heated at a slow steady rate until all the ice has changed to water, then the (liquid) water is heated to 100°C and changed to steam, and heated further above 100°C, all at 1 atm pressure. As shown in the graph of Fig. 19–5, as the ice is heated starting at −40°C, its temperature rises at a rate of about 2 C°/kcal of heat added (for ice, $c \approx 0.50 \, \text{kcal/kg} \cdot \text{C}°$). However, when 0°C is reached, the temperature stops increasing even though heat is still being added. The ice gradually changes to water in the liquid state, with no change in temperature. After about 40 kcal has been added at 0°C, half the ice remains and half has changed to water. After about 80 kcal, or 330 kJ, has been added, all the ice has changed to water, still at 0°C. Continued addition of heat causes the water's temperature to again increase, now at a rate of 1 C°/kcal. When 100°C is reached, the temperature again remains constant as the heat added changes the liquid water to vapor (steam). About 540 kcal (2260 kJ) is required to change the 1.0 kg of water completely to steam, after which the graph rises again, indicating that the temperature of the steam rises as heat is added.

The heat required to change 1.0 kg of a substance from the solid to the liquid state is called the **heat of fusion**; it is denoted by L_F. The heat of fusion of water is 79.7 kcal/kg or, in proper SI units, 333 kJ/kg $(= 3.33 \times 10^5 \, \text{J/kg})$. The heat required to change a substance from the liquid to the vapor phase is called the **heat of vaporization**, L_V. For water it is 539 kcal/kg or 2260 kJ/kg. Other substances follow graphs similar to Fig. 19–5, although the melting-point and boiling-point temperatures are different, as are the specific heats and heats of fusion and vaporization. Values for the heats of fusion and vaporization, which are also called the **latent heats**, are given in Table 19–2 for a number of substances.

The heats of vaporization and fusion also refer to the amount of heat *released* by a substance when it changes from a gas to a liquid, or from a liquid to a solid. Thus, steam releases 2260 kJ/kg when it changes to water, and water releases 333 kJ/kg when it becomes ice.

The heat involved in a change of phase depends not only on the latent heat but also on the total mass of the substance. That is,

$$Q = mL, \tag{19–3}$$

where L is the latent heat of the particular process and substance, m is the mass of the substance, and Q is the heat added or released during the phase change. For example, when 5.00 kg of water freezes at 0°C, $(5.00 \, \text{kg})(3.33 \times 10^5 \, \text{J/kg}) = 1.67 \times 10^6 \, \text{J}$ of energy is released.

TABLE 19–2 Latent Heats (at 1 atm)

Substance	Melting Point (°C)	Heat of Fusion kcal/kg[†]	Heat of Fusion kJ/kg	Boiling Point (°C)	Heat of Vaporization kcal/kg[†]	Heat of Vaporization kJ/kg
Oxygen	−218.8	3.3	14	−183	51	210
Nitrogen	−210.0	6.1	26	−195.8	48	200
Ethyl alcohol	−114	25	104	78	204	850
Ammonia	−77.8	8.0	33	−33.4	33	137
Water	0	79.7	333	100	539	2260
Lead	327	5.9	25	1750	208	870
Silver	961	21	88	2193	558	2300
Iron	1808	69.1	289	3023	1520	6340
Tungsten	3410	44	184	5900	1150	4800

[†]Numerical values in kcal/kg are the same in cal/g.

EXERCISE B A pot of water is boiling on a gas stove, and then you turn up the heat. What happens? (a) The temperature of the water starts increasing. (b) There is a tiny decrease in the rate of water loss by evaporation. (c) The rate of water loss by boiling increases. (d) There is an appreciable increase in both the rate of boiling and the temperature of the water. (e) None of these.

Calorimetry sometimes involves a change of state, as the following Examples show. Indeed, latent heats are often measured using calorimetry.

EXAMPLE 19–5 **Will all the ice melt?** A 0.50-kg chunk of ice at −10°C is placed in 3.0 kg of "iced" tea at 20°C. At what temperature and in what phase will the final mixture be? The tea can be considered as water. Ignore any heat flow to the surroundings, including the container.

APPROACH Before we can write down an equation applying conservation of energy, we must first check to see if the final state will be all ice, a mixture of ice and water at 0°C, or all water. To bring the 3.0 kg of water at 20°C down to 0°C would require an energy release of (Eq. 19–2)

$$m_w c_w (20°C - 0°C) = (3.0 \text{ kg})(4186 \text{ J/kg} \cdot C°)(20 \text{ C}°) = 250 \text{ kJ.}$$

On the other hand, to raise the ice from −10°C to 0°C would require

$$m_{ice} c_{ice}[0°C - (-10°C)] = (0.50 \text{ kg})(2100 \text{ J/kg} \cdot C°)(10 \text{ C}°) = 10.5 \text{ kJ,}$$

and to change the ice to water at 0°C would require (Eq. 19–3)

$$m_{ice} L_F = (0.50 \text{ kg})(333 \text{ kJ/kg}) = 167 \text{ kJ,}$$

for a total of 10.5 kJ + 167 kJ = 177 kJ. This is not enough energy to bring the 3.0 kg of water at 20°C down to 0°C, so we see that the mixture must end up all water, somewhere between 0°C and 20°C.

SOLUTION To determine the final temperature T, we apply conservation of energy and write: heat gain = heat loss,

$$\begin{pmatrix} \text{heat to raise} \\ \text{0.50 kg of ice} \\ \text{from } -10°C \\ \text{to } 0°C \end{pmatrix} + \begin{pmatrix} \text{heat to change} \\ \text{0.50 kg} \\ \text{of ice} \\ \text{to water} \end{pmatrix} + \begin{pmatrix} \text{heat to raise} \\ \text{0.50 kg of water} \\ \text{from } 0°C \\ \text{to } T \end{pmatrix} = \begin{pmatrix} \text{heat lost by} \\ \text{3.0 kg of} \\ \text{water cooling} \\ \text{from } 20°C \text{ to } T \end{pmatrix}.$$

Using some of the results from above, we obtain

$$10.5 \text{ kJ} + 167 \text{ kJ} + (0.50 \text{ kg})(4186 \text{ J/kg} \cdot C°)(T - 0°C)$$
$$= (3.0 \text{ kg})(4186 \text{ J/kg} \cdot C°)(20°C - T).$$

Solving for T we obtain

$$T = 5.0°C.$$

EXERCISE C How much more ice at −10°C would be needed in Example 19–5 to bring the tea down to 0°C, while just melting all the ice?

PROBLEM SOLVING
First determine (or estimate) the final state

PROBLEM SOLVING
Then determine the final temperature

Calorimetry

1. Be sure you have sufficient information to apply energy conservation. Ask yourself: **is the system isolated** (or very nearly so, enough to get a good estimate)? Do we know or can we calculate all significant sources of energy transfer?

2. Apply **conservation of energy**:

   heat gained = heat lost.

   For each substance in the system, a heat (energy) term will appear on either the left or right side of this equation. [Alternatively, use $\Sigma Q = 0$.]

3. If **no phase changes** occur, each term in the energy conservation equation (above) will have the form

   $$Q(\text{gain}) = mc(T_f - T_i)$$

   or

   $$Q(\text{lost}) = mc(T_i - T_f)$$

   where T_i and T_f are the initial and final temperatures

of the substance, and m and c are its mass and specific heat, respectively.

4. If **phase changes** do or might occur, there may be terms in the energy conservation equation of the form $Q = mL$, where L is the latent heat. But *before* applying energy conservation, determine (or estimate) in which phase the final state will be, as we did in Example 19–5 by calculating the different contributing values for heat Q.

5. Be sure each term appears on the correct side of the **energy equation** (heat gained or heat lost) and that each ΔT is positive.

6. Note that when the system reaches thermal **equilibrium**, the final **temperature** of each substance will have the *same* value. There is only one T_f.

7. **Solve** your energy equation for the unknown.

EXAMPLE 19–6 **Determining a latent heat.** The specific heat of liquid mercury is $140\,\text{J/kg}\cdot\text{C}°$. When 1.0 kg of solid mercury at its melting point of $-39°\text{C}$ is placed in a 0.50-kg aluminum calorimeter filled with 1.2 kg of water at $20.0°\text{C}$, the mercury melts and the final temperature of the combination is found to be $16.5°\text{C}$. What is the heat of fusion of mercury in J/kg?

APPROACH We follow the Problem Solving Strategy above.

SOLUTION

1. **Is the system isolated?** The mercury is placed in a calorimeter, which we assume is well insulated. Our isolated system is the calorimeter, the water, and the mercury.

2. **Conservation of energy.** The heat gained by the mercury = the heat lost by the water and calorimeter.

3. and **4. Phase changes.** There is a phase change (of mercury), plus we use specific heat equations. The heat gained by the mercury (Hg) includes a term representing the melting of the Hg,

$$Q(\text{melt solid Hg}) = m_{Hg}L_{Hg},$$

plus a term representing the heating of the liquid Hg from $-39°\text{C}$ to $+16.5°\text{C}$:

$$Q(\text{heat liquid Hg}) = m_{Hg}c_{Hg}[16.5°\text{C} - (-39°\text{C})]$$
$$= (1.0\,\text{kg})(140\,\text{J/kg}\cdot\text{C}°)(55.5\,\text{C}°) = 7770\,\text{J}.$$

All of this heat gained by the mercury is obtained from the water and calorimeter, which cool down:

$$Q_{cal} + Q_w = m_{cal}c_{cal}(20.0°\text{C} - 16.5°\text{C}) + m_w c_w(20.0°\text{C} - 16.5°\text{C})$$
$$= (0.50\,\text{kg})(900\,\text{J/kg}\cdot\text{C}°)(3.5\,\text{C}°) + (1.2\,\text{kg})(4186\,\text{J/kg}\cdot\text{C}°)(3.5\,\text{C}°)$$
$$= 19{,}200\,\text{J}.$$

5. **Energy equation.** The conservation of energy tells us the heat lost by the water and calorimeter cup must equal the heat gained by the mercury:

$$Q_{cal} + Q_w = Q(\text{melt solid Hg}) + Q(\text{heat liquid Hg})$$

or

$$19{,}200\,\text{J} = m_{Hg}L_{Hg} + 7770\,\text{J}.$$

6. **Equilibrium temperature.** It is given as $16.5°\text{C}$, and we already used it.

7. Solve. The only unknown in our energy equation (point 5) is L_{Hg}, the latent heat of fusion (or melting) of mercury. We solve for it, putting in $m_{Hg} = 1.0 \, kg$:

$$L_{Hg} = \frac{19{,}200 \, J - 7770 \, J}{1.0 \, kg} = 11{,}400 \, J/kg \approx 11 \, kJ/kg,$$

where we rounded off to 2 significant figures.

Evaporation

The latent heat to change a liquid to a gas is needed not only at the boiling point. Water can change from the liquid to the gas phase even at room temperature. This process is called **evaporation** (see also Section 18–4). The value of the heat of vaporization of water increases slightly with a decrease in temperature: at 20°C, for example, it is 2450 kJ/kg (585 kcal/kg) compared to 2260 kJ/kg (= 539 kcal/kg) at 100°C. When water evaporates, the remaining liquid cools, because the energy required (the latent heat of vaporization) comes from the water itself; so its internal energy, and therefore its temperature, must drop.[†]

Evaporation of water from the skin is one of the most important methods the body uses to control its temperature. When the temperature of the blood rises slightly above normal, the hypothalamus region of the brain detects this temperature increase and sends a signal to the sweat glands to increase their production. The energy (latent heat) required to vaporize this water comes from the body, and hence the body cools.

PHYSICS APPLIED
Body temperature

Kinetic Theory of Latent Heats

We can make use of kinetic theory to see why energy is needed to melt or vaporize a substance. At the melting point, the latent heat of fusion does not act to increase the average kinetic energy (and the temperature) of the molecules in the solid, but instead is used to overcome the potential energy associated with the forces between the molecules. That is, work must be done against these attractive forces to break the molecules loose from their relatively fixed positions in the solid so they can freely roll over one another in the liquid phase. Similarly, energy is required for molecules held close together in the liquid phase to escape into the gaseous phase. This process is a more violent reorganization of the molecules than is melting (the average distance between the molecules is greatly increased), and hence the heat of vaporization is generally much greater than the heat of fusion for a given substance.

19–6 The First Law of Thermodynamics

Up to now in this Chapter we have discussed internal energy and heat. But work too is often involved in thermodynamic processes.

In Chapter 8 we saw that work is done when energy is transferred from one object to another by mechanical means. In Section 19–1 we saw that heat is a transfer of energy from one object to a second one at a lower temperature. Thus, heat is much like work. To distinguish them, *heat* is defined as a *transfer of energy due to a difference in temperature*, whereas work is a transfer of energy that is not due to a temperature difference.

In Section 19–2, we defined the internal energy of a system as the sum total of all the energy of the molecules within the system. We would expect that the internal energy of a system would be increased if work was done on the system, or if heat were added to it. Similarly the internal energy would be decreased if heat flowed out of the system or if work were done by the system on something in the surroundings.

[†]According to kinetic theory, evaporation is a cooling process because it is the fastest-moving molecules that escape from the surface. Hence the average speed of the remaining molecules is less, so by Eq. 18–4 the temperature is less.

Thus it is reasonable to extend conservation of energy and propose an important law: the change in internal energy of a closed system, ΔE_{int}, will be equal to the energy added to the system by heating minus the work done by the system on the surroundings. In equation form we write

$$\Delta E_{int} = Q - W \tag{19–4}$$

where Q is the net heat *added* to the system and W is the net work done *by* the system.[†] We must be careful and consistent in following the sign conventions for Q and W. Because W in Eq. 19–4 is the work done *by* the system, then if work is done *on* the system, W will be negative and E_{int} will increase. Similarly, Q is positive for heat added to the system, so if heat leaves the system, Q is negative.

Equation 19–4 is known as the **first law of thermodynamics**. It is one of the great laws of physics, and its validity rests on experiments (such as Joule's) to which no exceptions have been seen. Since Q and W represent energy transferred into or out of the system, the internal energy changes accordingly. Thus, the first law of thermodynamics is a great and broad statement of the *law of conservation of energy*.

It is worth noting that the conservation of energy law was not formulated until the nineteenth century, for it depended on the interpretation of heat as a transfer of energy.

Equation 19–4 applies to a closed system. It also applies to an open system (Section 19–4) if we take into account the change in internal energy due to the increase or decrease in the amount of matter. For an isolated system (p. 500), no work is done and no heat enters or leaves the system, so $W = Q = 0$, and hence $\Delta E_{int} = 0$.

A given system at any moment is in a particular state and can be said to have a certain amount of internal energy, E_{int}. But a system does not "have" a certain amount of heat or work. Rather, when work is done on a system (such as compressing a gas), or when heat is added or removed from a system, the state of the system *changes*. Thus, work and heat are involved in *thermodynamic processes* that can change the system from one state to another; they are not characteristic of the state itself. Quantities which describe the state of a system, such as internal energy E_{int}, pressure P, volume V, temperature T, and mass m or number of moles n, are called **state variables**. Q and W are *not* state variables.

Because E_{int} is a *state variable*, which depends only on the state of the system and not on how the system arrived in that state, we can write

$$\Delta E_{int} = E_{int,2} - E_{int,1} = Q - W$$

where $E_{int,1}$ and $E_{int,2}$ represent the internal energy of the system in states 1 and 2, and Q and W are the heat added to the system and work done by the system in going from state 1 to state 2.

It is sometimes useful to write the first law of thermodynamics in differential form:

$$dE_{int} = dQ - dW.$$

Here, dE_{int} represents an infinitesimal change in internal energy when an infinitesimal amount of heat dQ is added to the system, and the system does an infinitesimal amount of work dW.[‡]

[†]This convention relates historically to steam engines: the interest was in the heat *input* and the work *output*, both regarded as positive. In other books you may see the first law of thermodynamics written as $\Delta E_{int} = Q + W$, in which case W refers to the work done *on* the system.

[‡]The differential form of the first law is often written

$$dE_{int} = đQ - đW,$$

where the bars on the differential sign ($đ$) are used to remind us that W and Q are not functions of the state variables (such as P, V, T, n). Internal energy, E_{int}, is a function of the state variables, and so dE_{int} represents the differential (called an *exact differential*) of some function E_{int}. The differentials $đW$ and $đQ$ are not exact differentials (they are not the differential of some mathematical function); they thus only represent infinitesimal amounts. This issue won't really be of concern in this book.

EXAMPLE 19–7 **Using the first law.** 2500 J of heat is added to a system, and 1800 J of work is done on the system. What is the change in internal energy of the system?

APPROACH We apply the first law of thermodynamics, Eq. 19–4, to our system.

SOLUTION The heat added to the system is $Q = 2500$ J. The work W done *by* the system is -1800 J. Why the minus sign? Because 1800 J done *on* the system (as given) equals -1800 J done *by* the system, and it is the latter we need for the sign conventions we used in Eq. 19–4. Hence

$$\Delta E_{int} = 2500\,\text{J} - (-1800\,\text{J}) = 2500\,\text{J} + 1800\,\text{J} = 4300\,\text{J}.$$

You may have intuitively thought that the 2500 J and the 1800 J would need to be added together, since both refer to energy added to the system. You would have been right.

EXERCISE D What would be the internal energy change in Example 19–7 if 2500 J of heat is added to the system and 1800 J of work is done *by* the system (i.e., as output)?

*The First Law of Thermodynamics Extended

To write the first law of thermodynamics in its complete form, consider a system that has kinetic energy K (there is motion) as well as potential energy U. Then the first law of thermodynamics would have to include these terms and would be written as

$$\Delta K + \Delta U + \Delta E_{int} = Q - W. \tag{19–5}$$

EXAMPLE 19–8 **Kinetic energy transformed to thermal energy.** A 3.0-g bullet traveling at a speed of 400 m/s enters a tree and exits the other side with a speed of 200 m/s. Where did the bullet's lost kinetic energy go, and what was the energy transferred?

APPROACH Take the bullet and tree as our system. No potential energy is involved. No work is done on (or by) the system by outside forces, nor is any heat added because no energy was transferred to or from the system due to a temperature difference. Thus the kinetic energy gets transformed into internal energy of the bullet and tree.

SOLUTION From the first law of thermodynamics as given in Eq. 19–5, we are given $Q = W = \Delta U = 0$, so we have

$$\Delta K + \Delta E_{int} = 0$$

or, using subscripts i and f for initial and final velocities

$$\Delta E_{int} = -\Delta K = -(K_f - K_i) = \tfrac{1}{2}m(v_i^2 - v_f^2)$$
$$= \tfrac{1}{2}(3.0 \times 10^{-3}\,\text{kg})[(400\,\text{m/s})^2 - (200\,\text{m/s})^2] = 180\,\text{J}.$$

NOTE The internal energy of the bullet and tree both increase, as both experience a rise in temperature. If we had chosen the bullet alone as our system, work would be done on it and heat transfer would occur.

19–7 The First Law of Thermodynamics Applied; Calculating the Work

Let us analyze some simple processes in the light of the first law of thermodynamics.

Isothermal Processes ($\Delta T = 0$)

First we consider an idealized process that is carried out at constant temperature. Such a process is called an **isothermal** process (from the Greek meaning "same temperature"). If the system is an ideal gas, then $PV = nRT$ (Eq. 17–3), so for a fixed amount of gas kept at constant temperature, $PV = \text{constant}$. Thus the process follows a curve like AB on the PV diagram shown in Fig. 19–6, which is a curve for $PV = \text{constant}$. Each point on the curve, such as point A, represents the state of the system at a given moment—that is, its pressure P and volume V. At a lower temperature, another isothermal process would be represented by a curve like A'B' in Fig. 19–6 (the product $PV = nRT = \text{constant}$ is less when T is less). The curves shown in Fig. 19–6 are referred to as *isotherms.*

FIGURE 19–6 *PV* diagram for an ideal gas undergoing isothermal processes at two different temperatures.

FIGURE 19–7 An ideal gas in a cylinder fitted with a movable piston.

Let us assume that the gas is enclosed in a container fitted with a movable piston, Fig. 19–7, and that the gas is in contact with a **heat reservoir** (a body whose mass is so large that, ideally, its temperature does not change significantly when heat is exchanged with our system). We also assume that the process of compression (volume decreases) or expansion (volume increases) is done **quasistatically** ("almost statically"), by which we mean extremely slowly, so that all of the gas moves between a series of equilibrium states each of which are at the same constant temperature. If an amount of heat Q is added to the system and temperature is to remain constant, the gas will expand and do an amount of work W on the environment (it exerts a force on the piston and moves it through a distance). The temperature and mass are kept constant so, from Eq. 19–1, the internal energy does not change: $\Delta E_{int} = \frac{3}{2}nR\,\Delta T = 0$. Hence, by the first law of thermodynamics, Eq. 19–4, $\Delta E_{int} = Q - W = 0$, so $W = Q$: the work done by the gas in an isothermal process equals the heat added to the gas.

Adiabatic Processes ($Q = 0$)

An **adiabatic** process is one in which no heat is allowed to flow into or out of the system: $Q = 0$. This situation can occur if the system is extremely well insulated, or the process happens so quickly that heat—which flows slowly—has no time to flow in or out. The very rapid expansion of gases in an internal combustion engine is one example of a process that is very nearly adiabatic. A slow adiabatic expansion of an ideal gas follows a curve like that labeled AC in Fig. 19–8. Since $Q = 0$, we have from Eq. 19–4 that $\Delta E_{int} = -W$. That is, the internal energy decreases if the gas expands; hence the temperature decreases as well (because $\Delta E_{int} = \frac{3}{2}nR\,\Delta T$). This is evident in Fig. 19–8 where the product $PV (= nRT)$ is less at point C than at point B (curve AB is for an isothermal process, for which $\Delta E_{int} = 0$ and $\Delta T = 0$). In the reverse operation, an adiabatic compression (going from C to A, for example), work is done *on* the gas, and hence the internal energy increases and the temperature rises. In a diesel engine, the fuel–air mixture is rapidly compressed adiabatically by a factor of 15 or more; the temperature rise is so great that the mixture ignites spontaneously.

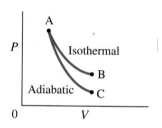

FIGURE 19–8 *PV* diagram for adiabatic (AC) and isothermal (AB) processes on an ideal gas.

Isobaric and Isovolumetric Processes

Isothermal and adiabatic processes are just two possible processes that can occur. Two other simple thermodynamic processes are illustrated on the *PV* diagrams of Fig. 19–9: (a) an **isobaric** process is one in which the pressure is kept constant, so the process is represented by a horizontal straight line on the *PV* diagram, Fig. 19–9a; (b) an **isovolumetric** (or *isochoric*) process is one in which the volume does not change (Fig. 19–9b). In these, and in all other processes, the first law of thermodynamics holds.

FIGURE 19–9 (a) Isobaric ("same pressure") process. (b) Isovolumetric ("same volume") process.

Work Done in Volume Changes

We often want to calculate the work done in a process. Suppose we have a gas confined to a cylindrical container fitted with a movable piston (Fig. 19–10). We must always be careful to define exactly what our system is. In this case we choose our system to be the gas; so the container's walls and the piston are parts of the environment. Now let us calculate the work done by the gas when it expands quasistatically, so that P and T are defined for the system at all instants.[†] The gas expands against the piston, whose area is A. The gas exerts a force $F = PA$ on the piston, where P is the pressure in the gas. The work done by the gas to move the piston an infinitesimal displacement $d\vec{\ell}$ is

$$dW = \vec{F} \cdot d\vec{\ell} = PA\,d\ell = P\,dV \tag{19–6}$$

FIGURE 19–10 The work done by a gas when its volume increases by $dV = A\,d\ell$ is $dW = P\,dV$.

since the infinitesimal increase in volume is $dV = A\,d\ell$. If the gas was *compressed* so that $d\vec{\ell}$ pointed into the gas, the volume would decrease and $dV < 0$. The work done by the gas in this case would then be negative, which is equivalent to saying that positive work was done *on* the gas, not by it. For a finite change in volume

---

[†]If the gas expands or is compressed quickly, there would be turbulence and different parts would be at different pressure (and temperature).

from V_A to V_B, the work W done by the gas will be

$$W = \int dW = \int_{V_A}^{V_B} P\,dV. \qquad (19\text{-}7)$$

Equations 19–6 and 19–7 are valid for the work done in any volume change—by a gas, a liquid, or a solid—as long as it is done quasistatically.

In order to integrate Eq. 19–7, we need to know how the pressure varies during the process, and this depends on the type of process. Let us first consider a quasistatic isothermal expansion of an ideal gas. This process is represented by the curve between points A and B on the PV diagram of Fig. 19–11. The work done by the gas in this process, according to Eq. 19–7, is just the area between the PV curve and the V axis, and is shown shaded in Fig. 19–11. We can do the integral in Eq. 19–7 for an ideal gas by using the ideal gas law, $P = nRT/V$. The work done at constant T is

$$W = \int_{V_A}^{V_B} P\,dV = nRT \int_{V_A}^{V_B} \frac{dV}{V} = nRT \ln\frac{V_B}{V_A}. \quad \begin{bmatrix}\text{isothermal process;}\\ \text{ideal gas}\end{bmatrix} \quad (19\text{-}8)$$

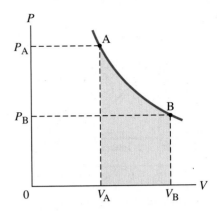

FIGURE 19–11 Work done by an ideal gas in an isothermal process equals the area under the PV curve. Shaded area equals the work done by the gas when it expands from V_A to V_B.

FIGURE 19–12 Process ADB consists of an isovolumetric (AD) and an isobaric (DB) process.

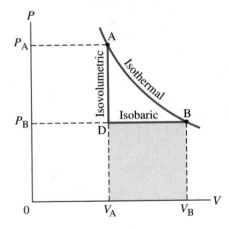

Let us next consider a different way of taking an ideal gas between the same states A and B. This time, let us lower the pressure in the gas from P_A to P_B, as indicated by the line AD in Fig. 19–12. (In this *isovolumetric* process, heat must be allowed to flow out of the gas so its temperature drops.) Then let the gas expand from V_A to V_B at constant pressure $(= P_B)$, which is indicated by the line DB in Fig. 19–12. (In this *isobaric* process, heat is added to the gas to raise its temperature.) No work is done in the isovolumetric process AD, since $dV = 0$:

$$W = 0. \qquad \text{[isovolumetric process]}$$

In the isobaric process DB the pressure remains constant, so

$$W = \int_{V_A}^{V_B} P\,dV = P_B(V_B - V_A) = P\,\Delta V. \qquad \text{[isobaric process]} \quad (19\text{-}9a)$$

The work done is again represented on the PV diagram by the area between the curve (ADB) and the V axis, as indicated by the shading in Fig. 19–12. Using the ideal gas law, we can also write

$$W = P_B(V_B - V_A) = nRT_B\left(1 - \frac{V_A}{V_B}\right). \qquad \begin{bmatrix}\text{isobaric process;}\\ \text{ideal gas}\end{bmatrix} \quad (19\text{-}9b)$$

As can be seen from the shaded areas in Figs. 19–11 and 19–12, or by putting in numbers in Eqs. 19–8 and 19–9 (try it for $V_B = 2V_A$), the work done in these two processes is different. This is a general result. *The work done in taking a system from one state to another depends not only on the initial and final states but also on the type of process (or "path").*

This result reemphasizes the fact that work cannot be considered a property of a system. The same is true of heat. The heat input required to change the gas from state A to state B depends on the process; for the isothermal process of Fig. 19–11, the heat input turns out to be greater than for the process ADB of Fig. 19–12. In general, *the amount of heat added or removed in taking a system from one state to another depends not only on the initial and final states but also on the path or process.*

CONCEPTUAL EXAMPLE 19–9 **Work in isothermal and adiabatic processes.** In Fig. 19–8 we saw the PV diagrams for a gas expanding in two ways, isothermally and adiabatically. The initial volume V_A was the same in each case, and the final volumes were the same $(V_B = V_C)$. In which process was more work done by the gas?

RESPONSE Our system is the gas. More work was done by the gas in the isothermal process, which we can see in two simple ways by looking at Fig. 19–8. First, the "average" pressure was higher during the isothermal process AB, so $W = \overline{P}\,\Delta V$ was greater (ΔV is the same for both processes). Second, we can look at the area under each curve: the area under curve AB, which represents the work done, was greater (since curve AB is higher) than that under AC.

EXERCISE E Is the work done by the gas in process ADB of Fig. 19–12 greater than, less than, or equal to the work done in the isothermal process AB?

FIGURE 19–13 Example 19–10.

EXAMPLE 19–10 **First law in isobaric and isovolumetric processes.** An ideal gas is slowly compressed at a constant pressure of 2.0 atm from 10.0 L to 2.0 L. This process is represented in Fig. 19–13 as the path B to D. (In this process, some heat flows out of the gas and the temperature drops.) Heat is then added to the gas, holding the volume constant, and the pressure and temperature are allowed to rise (line DA) until the temperature reaches its original value $(T_A = T_B)$. Calculate (a) the total work done by the gas in the process BDA, and (b) the total heat flow into the gas.

APPROACH (a) Work is done only in the compression process BD. In process DA, the volume is constant so $\Delta V = 0$ and no work is done. (b) We use the first law of thermodynamics, Eq. 19–4.

SOLUTION (a) During the compression BD, the pressure is 2.0 atm $= 2(1.01 \times 10^5 \, \text{N/m}^2)$ and the work done is (since $1 \, \text{L} = 10^3 \, \text{cm}^3 = 10^{-3} \, \text{m}^3$)

$$W = P \Delta V = (2.02 \times 10^5 \, \text{N/m}^2)(2.0 \times 10^{-3} \, \text{m}^3 - 10.0 \times 10^{-3} \, \text{m}^3)$$
$$= -1.6 \times 10^3 \, \text{J}.$$

The total work done by the gas is $-1.6 \times 10^3 \, \text{J}$, where the minus sign means that $+1.6 \times 10^3 \, \text{J}$ of work is done on the gas.

(b) Because the temperature at the beginning and at the end of process BDA is the same, there is no change in internal energy: $\Delta E_{\text{int}} = 0$. From the first law of thermodynamics we have

$$0 = \Delta E_{\text{int}} = Q - W$$

so $Q = W = -1.6 \times 10^3 \, \text{J}$. Because Q is negative, 1600 J of heat flows out of the gas for the whole process, BDA.

EXERCISE F In Example 19–10, if the heat lost from the gas in the process BD is $8.4 \times 10^3 \, \text{J}$, what is the change in internal energy of the gas during process BD?

EXAMPLE 19–11 **Work done in an engine.** In an engine, 0.25 mol of an ideal monatomic gas in the cylinder expands rapidly and adiabatically against the piston. In the process, the temperature of the gas drops from 1150 K to 400 K. How much work does the gas do?

APPROACH We take the gas as our system (the piston is part of the surroundings). The pressure is not constant, and its varying value is not given. Instead, we can use the first law of thermodynamics because we can determine ΔE_{int} given $Q = 0$ (the process is adiabatic).

SOLUTION We determine ΔE_{int} from Eq. 19–1 for the internal energy of an ideal monatomic gas, using subscripts f and i for final and initial states:

$$\Delta E_{\text{int}} = E_{\text{int, f}} - E_{\text{int, i}} = \tfrac{3}{2} n R (T_f - T_i)$$
$$= \tfrac{3}{2} (0.25 \, \text{mol})(8.314 \, \text{J/mol·K})(400 \, \text{K} - 1150 \, \text{K})$$
$$= -2300 \, \text{J}.$$

Then, from the first law of thermodynamics, Eq. 19–4,

$$W = Q - \Delta E_{\text{int}} = 0 - (-2300 \, \text{J}) = 2300 \, \text{J}.$$

Table 19–3 gives a brief summary of the processes we have discussed.

Free Expansion

One type of adiabatic process is a so-called **free expansion** in which a gas is allowed to expand in volume adiabatically without doing any work. The apparatus to

TABLE 19–3 Simple Thermodynamic Processes and the First Law

Process	What is constant:	The first law predicts:
Isothermal	$T = $ constant	$\Delta T = 0$ makes $\Delta E_{\text{int}} = 0$, so $Q = W$
Isobaric	$P = $ constant	$Q = \Delta E_{\text{int}} + W = \Delta E_{\text{int}} + P \Delta V$
Isovolumetric	$V = $ constant	$\Delta V = 0$ makes $W = 0$, so $Q = \Delta E_{\text{int}}$
Adiabatic	$Q = 0$	$\Delta E_{\text{int}} = -W$

accomplish a free expansion is shown in Fig. 19–14. It consists of two well-insulated compartments (to ensure no heat flow in or out) connected by a valve or stopcock. One compartment is filled with gas, the other is empty. When the valve is opened, the gas expands to fill both containers. No heat flows in or out $(Q = 0)$, and no work is done because the gas does not move any other object. Thus $Q = W = 0$ and by the first law of thermodynamics, $\Delta E_{int} = 0$. *The internal energy of a gas does not change in a free expansion.* For an ideal gas, $\Delta T = 0$ also, since E_{int} depends only on T (Section 19–2). Experimentally, the free expansion has been used to determine if the internal energy of *real gases* depends only on T. The experiments are very difficult to do accurately, but it has been found that the temperature of a real gas drops very slightly in a free expansion. Thus the internal energy of real gases does depend, a little, on pressure or volume as well as on temperature.

A free expansion can not be plotted on a PV diagram, because the process is rapid, not quasistatic. The intermediate states are not equilibrium states, and hence the pressure (and even the volume at some instants) is not clearly defined.

FIGURE 19–14 Free expansion.

19–8 Molar Specific Heats for Gases, and the Equipartition of Energy

In Section 19–3 we discussed the concept of specific heat and applied it to solids and liquids. Much more than for solids and liquids, the values of the specific heat for gases depends on how the process is carried out. Two important processes are those in which either the volume or the pressure is kept constant. Although for solids and liquids it matters little, Table 19–4 shows that the specific heats of gases at constant volume (c_V) and at constant pressure (c_P) are quite different.

Molar Specific Heats for Gases

The difference in specific heats for gases is nicely explained in terms of the first law of thermodynamics and kinetic theory. Our discussion is simplified if we use **molar specific heats**, C_V and C_P, which are defined as the heat required to raise 1 mol of the gas by 1 C° at constant volume and at constant pressure, respectively. That is, in analogy to Eq. 19–2, the heat Q needed to raise the temperature of n moles of gas by ΔT is

$$Q = nC_V\Delta T \qquad \text{[volume constant]} \quad \textbf{(19–10a)}$$
$$Q = nC_P\Delta T. \qquad \text{[pressure constant]} \quad \textbf{(19–10b)}$$

It is clear from the definition of molar specific heat (or by comparing Eqs. 19–2 and 19–10) that

$$C_V = Mc_V$$
$$C_P = Mc_P,$$

where M is the molecular mass of the gas ($M = m/n$ in grams/mol). The values for molar specific heats are included in Table 19–4, and we see that the values are nearly the same for different gases that have the same number of atoms per molecule.

TABLE 19–4 Specific Heats of Gases at 15°C

Gas	Specific heats (kcal/kg · K)		Molar specific heats (cal/mol · K)		$C_P - C_V$ (cal/mol · K)	$\gamma = \dfrac{C_P}{C_V}$
	c_V	c_P	C_V	C_P		
Monatomic						
He	0.75	1.15	2.98	4.97	1.99	1.67
Ne	0.148	0.246	2.98	4.97	1.99	1.67
Diatomic						
N_2	0.177	0.248	4.96	6.95	1.99	1.40
O_2	0.155	0.218	5.03	7.03	2.00	1.40
Triatomic						
CO_2	0.153	0.199	6.80	8.83	2.03	1.30
H_2O (100°C)	0.350	0.482	6.20	8.20	2.00	1.32

Now we use kinetic theory and imagine that an ideal gas is slowly heated via two different processes—first at constant volume, and then at constant pressure. In both of these processes, we let the temperature increase by the same amount, ΔT. In the process done at constant volume, no work is done since $\Delta V = 0$. Thus, according to the first law of thermodynamics, the heat added (which we denote by Q_V) all goes into increasing the internal energy of the gas:

$$Q_V = \Delta E_{int}.$$

In the process carried out at constant pressure, work is done, and hence the heat added, Q_P, must not only increase the internal energy but also is used to do the work $W = P\,\Delta V$. Thus, more heat must be added in this process at constant pressure than in the first process at constant volume. For the process at constant pressure, we have from the first law of thermodynamics

$$Q_P = \Delta E_{int} + P\,\Delta V.$$

Since ΔE_{int} is the same in the two processes (ΔT was chosen to be the same), we can combine the two above equations:

$$Q_P - Q_V = P\,\Delta V.$$

From the ideal gas law, $V = nRT/P$, so for a process at constant pressure we have $\Delta V = nR\,\Delta T/P$. Putting this into the above equation and using Eqs. 19–10, we find

$$nC_P\Delta T - nC_V\Delta T = P\left(\frac{nR\,\Delta T}{P}\right)$$

or, after cancellations,

$$C_P - C_V = R. \tag{19–11}$$

Since the gas constant $R = 8.314\ \text{J/mol}\cdot\text{K} = 1.99\ \text{cal/mol}\cdot\text{K}$, our prediction is that C_P will be larger than C_V by about $1.99\ \text{cal/mol}\cdot\text{K}$. Indeed, this is very close to what is obtained experimentally, as shown in the next to last column of Table 19–4.

Now we calculate the molar specific heat of a monatomic gas using kinetic theory. In a process carried out at constant volume, no work is done; so the first law of thermodynamics tells us that if heat Q is added to the gas, the internal energy of the gas changes by

$$\Delta E_{int} = Q.$$

For an ideal monatomic gas, the internal energy E_{int} is the total kinetic energy of all the molecules,

$$E_{int} = N\left(\tfrac{1}{2}m\overline{v^2}\right) = \tfrac{3}{2}nRT$$

as we saw in Section 19–2. Then, using Eq. 19–10a, we can write $\Delta E_{int} = Q$ in the form

$$\Delta E_{int} = \tfrac{3}{2}nR\,\Delta T = nC_V\,\Delta T \tag{19–12}$$

or

$$C_V = \tfrac{3}{2}R. \tag{19–13}$$

Since $R = 8.314\ \text{J/mol}\cdot\text{K} = 1.99\ \text{cal/mol}\cdot\text{K}$, kinetic theory predicts that $C_V = 2.98\ \text{cal/mol}\cdot\text{K}$ for an ideal monatomic gas. This is very close to the experimental values for monatomic gases such as helium and neon (Table 19–4). From Eq. 19–11, C_P is predicted to be about $4.97\ \text{cal/mol}\cdot\text{K}$, also in agreement with experiment.

Equipartition of Energy

The measured molar specific heats for more complex gases (Table 19–4), such as diatomic (two atoms) and triatomic (three atoms) gases, increase with the increased number of atoms per molecule. We can explain this by assuming that the internal energy includes not only translational kinetic energy but other forms of energy as well. Take, for example, a diatomic gas. As shown in Fig. 19–15 the two atoms can rotate about two different axes (but rotation about a third axis passing through the two atoms would give rise to very little energy since the moment of inertia is so small). The molecules can have rotational as well as translational kinetic energy. It is useful to introduce the idea of **degrees of freedom**, by which we mean the number of independent ways molecules can possess energy. For example, a monatomic gas is said to have three degrees of freedom, since an atom can have velocity along the x axis, the y axis, and the z axis. These are considered to be three independent motions because a change in any one of the components would not affect any of the others. A diatomic molecule has the same three degrees of freedom associated with translational kinetic

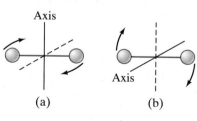

FIGURE 19–15 A diatomic molecule can rotate about two different axes.

Axis

Axis

(a)   (b)

energy plus two more degrees of freedom associated with rotational kinetic energy, for a total of five degrees of freedom. A quick look at Table 19–4 indicates that the C_V for diatomic gases is about $\frac{5}{3}$ times as great as for a monatomic gas—that is, in the same ratio as their degrees of freedom. This result led nineteenth-century physicists to an important idea, the **principle of equipartition of energy**. This principle states that energy is shared equally among the active degrees of freedom, and in particular each active degree of freedom of a molecule has on the average an energy equal to $\frac{1}{2}kT$. Thus, the average energy for a molecule of a monatomic gas would be $\frac{3}{2}kT$ (which we already knew) and of a diatomic gas $\frac{5}{2}kT$. Hence the internal energy of a diatomic gas would be $E_{int} = N(\frac{5}{2}kT) = \frac{5}{2}nRT$, where n is the number of moles. Using the same argument we did for monatomic gases, we see that for diatomic gases the molar specific heat at constant volume would be $\frac{5}{2}R = 4.97$ cal/mol·K, in accordance with measured values. More complex molecules have even more degrees of freedom and thus greater molar specific heats.

The situation was complicated, however, by measurements that showed that for diatomic gases at very low temperatures, C_V has a value of only $\frac{3}{2}R$, as if it had only three degrees of freedom. And at very high temperatures, C_V was about $\frac{7}{2}R$, as if there were seven degrees of freedom. The explanation is that at low temperatures, nearly all molecules have only translational kinetic energy. That is, no energy goes into rotational energy, so only three degrees of freedom are "active." At very high temperatures, on the other hand, all five degrees of freedom are active plus two additional ones. We can interpret the two new degrees of freedom as being associated with the two atoms vibrating as if they were connected by a spring, as shown in Fig. 19–16. One degree of freedom comes from the kinetic energy of the vibrational motion, and the second comes from the potential energy of vibrational motion ($\frac{1}{2}kx^2$). At room temperature, these two degrees of freedom are apparently not active. See Fig. 19–17.

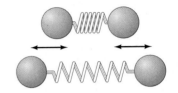

FIGURE 19–16 A diatomic molecule can vibrate, as if the two atoms were connected by a spring. Of course they are not connected by a spring; rather they exert forces on each other that are electrical in nature, but of a form that resembles a spring force.

FIGURE 19–17 Molar specific heat C_V as a function of temperature for hydrogen molecules (H_2). As the temperature is increased, some of the translational kinetic energy can be transferred in collisions into rotational kinetic energy and, at still higher temperature, into vibrational kinetic energy. [Note: H_2 dissociates into two atoms at about 3200 K, so the last part of the curve is shown dashed.]

Just why fewer degrees of freedom are "active" at lower temperatures was eventually explained by Einstein using the quantum theory. [According to quantum theory, energy does not take on continuous values but is quantized—it can have only certain values, and there is a certain minimum energy. The minimum rotational and vibrational energies are higher than for simple translational kinetic energy, so at lower temperatures and lower translational kinetic energy, there is not enough energy to excite the rotational or vibrational kinetic energy.] Calculations based on kinetic theory and the principle of equipartition of energy (as modified by the quantum theory) give numerical results in accord with experiment.

*Solids

The principle of equipartition of energy can be applied to solids as well. The molar specific heat of any solid, at high temperature, is close to $3R$ (6.0 cal/mol·K), Fig. 19–18. This is called the *Dulong and Petit* value after the scientists who first measured it in 1819. (Note that Table 19–1 gives the specific heats per kilogram, not per mole.) At high temperatures, each atom apparently has six degrees of freedom, although some are not active at low temperatures. Each atom in a crystalline solid can vibrate about its equilibrium position as if it were connected by springs to each of its neighbors. Thus it can have three degrees of freedom for kinetic energy and three more associated with potential energy of vibration in each of the x, y, and z directions, which is in accord with the measured values.

FIGURE 19–18 Molar specific heats of solids as a function of temperature.

19–9 Adiabatic Expansion of a Gas

The PV curve for the quasistatic (slow) adiabatic expansion ($Q = 0$) of an ideal gas was shown in Fig. 19–8 (curve AC). It is somewhat steeper than for an isothermal process ($\Delta T = 0$), which indicates that for the same change in volume the change in pressure will be greater. Hence the temperature of the gas must drop during an adiabatic expansion. Conversely, the temperature rises during an adiabatic compression.

We can derive the relation between the pressure P and the volume V of an ideal gas that is allowed to slowly expand adiabatically. We begin with the first law of thermodynamics, written in differential form:

$$dE_{int} = dQ - dW = -dW = -P\,dV,$$

since $dQ = 0$ for an adiabatic process. Equation 19–12 gives us a relation between ΔE_{int} and C_V, which is valid for any ideal gas process since E_{int} is a function only of T for an ideal gas. We write this in differential form:

$$dE_{int} = nC_V\,dT.$$

When we combine these last two equations, we obtain

$$nC_V\,dT + P\,dV = 0.$$

We next take the differential of the ideal gas law, $PV = nRT$, allowing P, V, and T to vary:

$$P\,dV + V\,dP = nR\,dT.$$

We solve for dT in this relation and substitute it into the previous relation and get

$$nC_V\left(\frac{P\,dV + V\,dP}{nR}\right) + P\,dV = 0$$

or, multiplying through by R and rearranging,

$$(C_V + R)P\,dV + C_V V\,dP = 0.$$

We note from Eq. 19–11 that $C_V + R = C_P$, so we have

$$C_P P\,dV + C_V V\,dP = 0,$$

or

$$\frac{C_P}{C_V}P\,dV + V\,dP = 0.$$

We define

$$\gamma = \frac{C_P}{C_V} \tag{19–14}$$

so that our last equation becomes

$$\frac{dP}{P} + \gamma\frac{dV}{V} = 0.$$

This is integrated to become

$$\ln P + \gamma \ln V = \text{constant}.$$

This simplifies (using the rules for addition and multiplication of logarithms) to

$$PV^\gamma = \text{constant}. \qquad \begin{bmatrix}\text{quasistatic adiabatic}\\\text{process; ideal gas}\end{bmatrix} \tag{19–15}$$

This is the relation between P and V for a quasistatic adiabatic expansion or contraction. We will find it very useful when we discuss heat engines in the next Chapter. Table 19–4 (p. 511) gives values of γ for some real gases. Figure 19–8 compares an adiabatic expansion (Eq. 19–15) in curve AC to an isothermal expansion ($PV = \text{constant}$) in curve AB. It is important to remember that the ideal gas law, $PV = nRT$, continues to hold even for an adiabatic expansion ($PV^\gamma = \text{constant}$); clearly PV is not constant, meaning T is not constant.

EXAMPLE 19–12 **Compressing an ideal gas.** An ideal monatomic gas is compressed starting at point A in the PV diagram of Fig. 19–19, where $P_A = 100\,\text{kPa}$, $V_A = 1.00\,\text{m}^3$, and $T_A = 300\,\text{K}$. The gas is first compressed adiabatically to state B ($P_B = 200\,\text{kPa}$). The gas is then further compressed from point B to point C ($V_C = 0.50\,\text{m}^3$) in an isothermal process. (a) Determine V_B. (b) Calculate the work done *on* the gas for the whole process.

APPROACH Volume V_B is obtained using Eq. 19–15. The work done *by* a gas is given by Eq. 19–7, $W = \int P \, dV$. The work *on* the gas is the negative of this: $W_{on} = -\int P \, dV$.

SOLUTION In the adiabatic process, Eq. 19–15 tells us $PV^\gamma = $ constant. Therefore, $PV^\gamma = P_A V_A^\gamma = P_B V_B^\gamma$ where for a monatomic gas $\gamma = C_P/C_V = (5/2)/(3/2) = \frac{5}{3}$.

(*a*) Eq. 19–15 gives $V_B = V_A (P_A/P_B)^{\frac{1}{\gamma}} = (1.00 \text{ m}^3)(100 \text{ kPa}/200 \text{ kPa})^{\frac{3}{5}} = 0.66 \text{ m}^3$.

(*b*) The pressure P at any instant during the adiabatic process is given by $P = P_A V_A^\gamma V^{-\gamma}$. The work done on the gas in going from V_A to V_B is

$$W_{AB} = -\int_A^B P \, dV = -P_A V_A^\gamma \int_{V_A}^{V_B} V^{-\gamma} \, dV = -P_A V_A^\gamma \left(\frac{1}{-\gamma + 1} \right)\left(V_B^{1-\gamma} - V_A^{1-\gamma} \right).$$

Since $\gamma = \frac{5}{3}$, then $-\gamma + 1 = 1 - \gamma = -\frac{2}{3}$, so

$$W_{AB} = -\left(P_A V_A^{\frac{5}{3}} \right)\left(-\frac{3}{2} \right)\left(V_A^{-\frac{2}{3}} \right)\left[\left(\frac{V_B}{V_A} \right)^{-\frac{2}{3}} - 1 \right] = +\frac{3}{2} P_A V_A \left[\left(\frac{V_B}{V_A} \right)^{-\frac{2}{3}} - 1 \right]$$

$$= +\frac{3}{2}(100 \text{ kPa})(1.00 \text{ m}^3)\left[(0.66)^{-\frac{2}{3}} - 1 \right] = +48 \text{ kJ}.$$

For the isothermal process from B to C, the work is done at constant temperature, so the pressure at any instant during the process is $P = nRT_B/V$ and

$$W_{BC} = -\int_B^C P \, dV = -nRT_B \int_{V_B}^{V_C} \frac{dV}{V} = -nRT_B \ln \frac{V_C}{V_B} = -P_B V_B \ln \frac{V_C}{V_B} = +37 \text{ kJ}.$$

The total work done on the gas is $48 \text{ kJ} + 37 \text{ kJ} = 85 \text{ kJ}$.

FIGURE 19–19 Example 19–12.

19–10 Heat Transfer: Conduction, Convection, Radiation

Heat is transferred from one place or body to another in three different ways: by *conduction*, *convection*, and *radiation*. We now discuss each of these in turn; but in practical situations, any two or all three may be operating at the same time. We start with conduction.

Conduction

When a metal poker is put in a hot fire, or a silver spoon is placed in a hot bowl of soup, the exposed end of the poker or spoon soon becomes hot as well, even though it is not directly in contact with the source of heat. We say that heat has been conducted from the hot end to the cold end.

Heat **conduction** in many materials can be visualized as being carried out via molecular collisions. As one end of an object is heated, the molecules there move faster and faster. As they collide with slower-moving neighbors, they transfer some of their kinetic energy to these other molecules, which in turn transfer energy by collision with molecules still farther along the object. In metals, collisions of free electrons are mainly responsible for conduction.

Heat conduction from one point to another takes place only if there is a difference in temperature between the two points. Indeed, it is found experimentally that the rate of heat flow through a substance is proportional to the difference in temperature between its ends. The rate of heat flow also depends on the size and shape of the object. To investigate this quantitatively, let us consider the heat flow through a uniform cylinder, as illustrated in Fig. 19–20. It is found experimentally that the heat flow ΔQ over a time interval Δt is given by the relation

$$\frac{\Delta Q}{\Delta t} = kA \frac{T_1 - T_2}{\ell} \tag{19–16a}$$

where A is the cross-sectional area of the object, ℓ is the distance between the two ends, which are at temperatures T_1 and T_2, and k is a proportionality constant called the **thermal conductivity** which is characteristic of the material. From Eq. 19–16a, we see that the rate of heat flow (units of J/s) is directly proportional to the cross-sectional area and to the temperature gradient $(T_1 - T_2)/\ell$.

FIGURE 19–20 Heat conduction between areas at temperatures T_1 and T_2. If T_1 is greater than T_2, the heat flows to the right; the rate is given by Eq. 19–16a.

TABLE 19–5
Thermal Conductivities

| Substance | Thermal conductivity, k | |
	kcal / (s·m·C°)	J / (s·m·C°)
Silver	10×10^{-2}	420
Copper	9.2×10^{-2}	380
Aluminum	5.0×10^{-2}	200
Steel	1.1×10^{-2}	40
Ice	5×10^{-4}	2
Glass	2.0×10^{-4}	0.84
Brick	2.0×10^{-4}	0.84
Concrete	2.0×10^{-4}	0.84
Water	1.4×10^{-4}	0.56
Human tissue	0.5×10^{-4}	0.2
Wood	0.3×10^{-4}	0.1
Fiberglass	0.12×10^{-4}	0.048
Cork	0.1×10^{-4}	0.042
Wool	0.1×10^{-4}	0.040
Goose down	0.06×10^{-4}	0.025
Polyurethane	0.06×10^{-4}	0.024
Air	0.055×10^{-4}	0.023

FIGURE 19–21 Example 19–13.

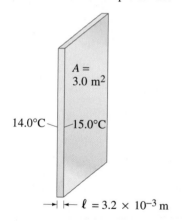

$A = 3.0\ m^2$

$14.0°C$ — $15.0°C$

$\ell = 3.2 \times 10^{-3}\ m$

PHYSICS APPLIED
Thermal windows

In some cases (such as when k or A cannot be considered constant) we need to consider the limit of an infinitesimally thin slab of thickness dx. Then Eq. 19–16a becomes

$$\frac{dQ}{dt} = -kA\frac{dT}{dx},$$ (19–16b)

where dT/dx is the temperature gradient† and the negative sign is included since the heat flow is in the direction opposite to the temperature gradient.

The thermal conductivities, k, for a variety of substances are given in Table 19–5. Substances for which k is large conduct heat rapidly and are said to be good thermal **conductors**. Most metals fall in this category, although there is a wide range even among them, as you may observe by holding the ends of a silver spoon and a stainless-steel spoon immersed in the same hot cup of soup. Substances for which k is small, such as wool, fiberglass, polyurethane, and goose down, are poor conductors of heat and are therefore good thermal **insulators**.

The relative magnitudes of k can explain simple phenomena such as why a tile floor is much colder on the feet than a rug-covered floor at the same temperature. Tile is a better conductor of heat than the rug. Heat that flows from your foot to the rug is not conducted away rapidly, so the rug's surface quickly warms up to the temperature of your foot and feels good. But the tile conducts the heat away rapidly and thus can take more heat from your foot quickly, so your foot's surface temperature drops.

EXAMPLE 19–13 **Heat loss through windows.** A major source of heat loss from a house is through the windows. Calculate the rate of heat flow through a glass window $2.0\ m \times 1.5\ m$ in area and $3.2\ mm$ thick, if the temperatures at the inner and outer surfaces are $15.0°C$ and $14.0°C$, respectively (Fig. 19–21).

APPROACH Heat flows by conduction through the glass from the higher inside temperature to the lower outside temperature. We use the heat conduction equation, Eq. 19–16a.

SOLUTION Here $A = (2.0\ m)(1.5\ m) = 3.0\ m^2$ and $\ell = 3.2 \times 10^{-3}\ m$. Using Table 19–5 to get k, we have

$$\frac{\Delta Q}{\Delta t} = kA\frac{T_1 - T_2}{\ell} = \frac{(0.84\ J/s \cdot m \cdot C°)(3.0\ m^2)(15.0°C - 14.0°C)}{(3.2 \times 10^{-3}\ m)}$$
$$= 790\ J/s.$$

NOTE This rate of heat flow is equivalent to $(790\ J/s)/(4.19 \times 10^3\ J/kcal) = 0.19\ kcal/s$, or $(0.19\ kcal/s) \times (3600\ s/h) = 680\ kcal/h$.

You might notice in Example 19–13 that 15°C is not very warm for the living room of a house. The room itself may indeed be much warmer, and the outside might be colder than 14°C. But the temperatures of 15°C and 14°C were specified as those at the window surfaces, and there is usually a considerable drop in temperature of the air in the vicinity of the window both on the inside and the outside. That is, the layer of air on either side of the window acts as an insulator, and normally the major part of the temperature drop between the inside and outside of the house takes place across the air layer. If there is a heavy wind, the air outside a window will constantly be replaced with cold air; the temperature gradient across the glass will be greater and there will be a much greater rate of heat loss. Increasing the width of the air layer, such as using two panes of glass separated by an air gap, will reduce the heat loss more than simply increasing the glass thickness, since the thermal conductivity of air is much less than that for glass.

The insulating properties of clothing come from the insulating properties of air. Without clothes, our bodies in still air would heat the air in contact with the skin and would soon become reasonably comfortable because air is a very good insulator.

---

†Equations 19–16 are quite similar to the relations describing diffusion (Section 18–7) and the flow of fluids through a pipe (Section 13–12). In those cases, the flow of matter was found to be proportional to the concentration gradient dC/dx, or to the pressure gradient $(P_1 - P_2)/\ell$. This close similarity is one reason we speak of the "flow" of heat. Yet we must keep in mind that no substance is flowing in the case of heat—it is energy that is being transferred.

But since air moves—there are breezes and drafts, and people move about—the warm air would be replaced by cold air, thus increasing the temperature difference and the heat loss from the body. Clothes keep us warm by trapping air so it cannot move readily. It is not the cloth that insulates us, but the air that the cloth traps. Goose down is a very good insulator because even a small amount of it fluffs up and traps a great amount of air.

[For practical purposes the thermal properties of building materials, particularly when considered as insulation, are usually specified by *R*-values (or "thermal resistance"), defined for a given thickness ℓ of material as:

$$R = \ell/k.$$

The *R*-value of a given piece of material combines the thickness ℓ and the thermal conductivity *k* in one number. In the United States, *R*-values are given in British units as $\text{ft}^2 \cdot \text{h} \cdot \text{F}°/\text{Btu}$ (for example, *R*-19 means $R = 19 \, \text{ft}^2 \cdot \text{h} \cdot \text{F}°/\text{Btu}$). Table 19–6 gives *R*-values for some common building materials. *R*-values increase directly with material thickness: for example, 2 inches of fiberglass is *R*-6, whereas 4 inches is *R*-12.]

PHYSICS APPLIED
R-values of thermal insulation

TABLE 19–6 *R*-values

Material	Thickness	*R*-value ($\text{ft}^2 \cdot \text{h} \cdot \text{F}° / \text{Btu}$)
Glass	$\frac{1}{8}$ inch	1
Brick	$3\frac{1}{2}$ inches	0.6–1
Plywood	$\frac{1}{2}$ inch	0.6
Fiberglass insulation	4 inches	12

Convection

Although liquids and gases are generally not very good conductors of heat, they can transfer heat quite rapidly by convection. **Convection** is the process whereby heat flows by the mass movement of molecules from one place to another. Whereas conduction involves molecules (and/or electrons) moving only over small distances and colliding, convection involves the movement of large numbers of molecules over large distances.

A forced-air furnace, in which air is heated and then blown by a fan into a room, is an example of *forced convection*. *Natural convection* occurs as well, and one familiar example is that hot air rises. For instance, the air above a radiator (or other type of heater) expands as it is heated (Chapter 17), and hence its density decreases. Because its density is less than that of the surrounding cooler air, it rises, just as a log submerged in water floats upward because its density is less than that of water. Warm or cold ocean currents, such as the balmy Gulf Stream, represent natural convection on a global scale. Wind is another example of convection, and weather in general is strongly influenced by convective air currents.

When a pot of water is heated (Fig. 19–22), convection currents are set up as the heated water at the bottom of the pot rises because of its reduced density. That heated water is replaced by cooler water from above. This principle is used in many heating systems, such as the hot-water radiator system shown in Fig. 19–23. Water is heated in the furnace, and as its temperature increases, it expands and rises as shown. This causes the water to circulate in the heating system. Hot water then enters the radiators, heat is transferred by conduction to the air, and the cooled water returns to the furnace. Thus, the water circulates because of convection; pumps are sometimes used to improve circulation. The air throughout the room also becomes heated as a result of convection. The air heated by the radiators rises and is replaced by cooler air, resulting in convective air currents, as shown by the green arrows in Fig. 19–23.

Other types of furnaces also depend on convection. Hot-air furnaces with registers (openings) near the floor often do not have fans but depend on natural convection, which can be appreciable. In other systems, a fan is used. In either case, it is important that cold air can return to the furnace so that convective currents circulate throughout the room if the room is to be uniformly heated. Convection is not always favorable. Much of the heat from a fireplace, for example, goes up the chimney and not out into the room.

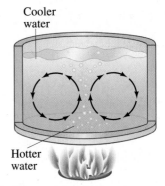

Cooler water

Hotter water

FIGURE 19–22 Convection currents in a pot of water being heated on a stove.

PHYSICS APPLIED
Convective home heating

FIGURE 19–23 Convection plays a role in heating a house. The circular arrows show the convective air currents in the rooms.

Radiator

Hot water

Furnace

Cool

Radiation

Convection and conduction require the presence of matter as a medium to carry the heat from the hotter to the colder region. But a third type of heat transfer occurs without any medium at all. All life on Earth depends on the transfer of energy from the Sun, and this energy is transferred to the Earth over empty (or nearly empty) space. This form of energy transfer is heat—since the Sun's surface temperature is much higher (6000 K) than Earth's—and is referred to as **radiation**. The warmth we receive from a fire is mainly radiant energy.

As we shall see in later Chapters, radiation consists essentially of electromagnetic waves. Suffice it to say for now that radiation from the Sun consists of visible light plus many other wavelengths that the eye is not sensitive to, including infrared (IR) radiation.

The rate at which an object radiates energy has been found to be proportional to the fourth power of the Kelvin temperature, T. That is, an object at 2000 K, as compared to one at 1000 K, radiates energy at a rate $2^4 = 16$ times as much. The rate of radiation is also proportional to the area A of the emitting object, so the rate at which energy leaves the object, $\Delta Q/\Delta t$, is

$$\frac{\Delta Q}{\Delta t} = \epsilon \sigma A T^4. \tag{19–17}$$

This is called the **Stefan-Boltzmann equation**, and σ is a universal constant called the **Stefan-Boltzmann constant** which has the value

$$\sigma = 5.67 \times 10^{-8}\,\text{W/m}^2\cdot\text{K}^4.$$

The factor ϵ (Greek letter epsilon), called the **emissivity**, is a number between 0 and 1 that is characteristic of the surface of the radiating material. Very black surfaces, such as charcoal, have emissivity close to 1, whereas shiny metal surfaces have ϵ close to zero and thus emit correspondingly less radiation. The value depends somewhat on the temperature of the material.

Not only do shiny surfaces emit less radiation, but they absorb little of the radiation that falls upon them (most is reflected). Black and very dark objects are good emitters ($\epsilon \approx 1$), and they also absorb nearly all the radiation that falls on them—which is why light-colored clothing is usually preferable to dark clothing on a hot day. Thus, **a good absorber is also a good emitter.**

Any object not only emits energy by radiation but also absorbs energy radiated by other objects. If an object of emissivity ϵ and area A is at a temperature T_1, it radiates energy at a rate $\epsilon \sigma A T_1^4$. If the object is surrounded by an environment at temperature T_2, the rate at which the surroundings radiate energy is proportional to T_2^4, and the rate that energy is absorbed by the object is proportional to T_2^4. The *net* rate of radiant heat flow from the object is given by

$$\frac{\Delta Q}{\Delta t} = \epsilon \sigma A\left(T_1^4 - T_2^4\right), \tag{19–18}$$

where A is the surface area of the object, T_1 its temperature and ϵ its emissivity (at temperature T_1), and T_2 is the temperature of the surroundings. This equation is consistent with the experimental fact that equilibrium between the object and its surroundings is reached when they come to the same temperature. That is, $\Delta Q/\Delta t$ must equal zero when $T_1 = T_2$, so ϵ must be the same for emission and absorption. This confirms the idea that a good emitter is a good absorber. Because both the object and its surroundings radiate energy, there is a net transfer of energy from one to the other unless everything is at the same temperature.

EXAMPLE 19–14 **ESTIMATE** **Cooling by radiation.** An athlete is sitting unclothed in a locker room whose dark walls are at a temperature of 15°C. Estimate his rate of heat loss by radiation, assuming a skin temperature of 34°C and $\epsilon = 0.70$. Take the surface area of the body not in contact with the chair to be 1.5 m^2.

APPROACH We use Eq. 19–18, with Kelvin temperatures.

SOLUTION We have

$$\frac{\Delta Q}{\Delta t} = \epsilon \sigma A\left(T_1^4 - T_2^4\right)$$

$$= (0.70)\left(5.67 \times 10^{-8}\,\text{W/m}^2\cdot\text{K}^4\right)\left(1.5\,\text{m}^2\right)\left[(307\,\text{K})^4 - (288\,\text{K})^4\right] = 120\,\text{W}.$$

NOTE The "output" of this resting person is a bit more than what a 100-W lightbulb uses.

A resting person naturally produces heat internally at a rate of about 100 W, which is less than the heat loss by radiation as calculated in Example 19–14. Hence, the person's temperature would drop, causing considerable discomfort. The body responds to excessive heat loss by increasing its metabolic rate, and shivering is one method by which the body increases its metabolism. Naturally, clothes help a lot. Example 19–14 illustrates that a person may be uncomfortable even if the temperature of the air is, say, 25°C, which is quite a warm room. If the walls or floor are cold, radiation to them occurs no matter how warm the air is. Indeed, it is estimated that radiation accounts for about 50% of the heat loss from a sedentary person in a normal room. Rooms are most comfortable when the walls and floor are warm and the air is not so warm. Floors and walls can be heated by means of hot-water conduits or electric heating elements. Such first-rate heating systems are becoming more common today, and it is interesting to note that 2000 years ago the Romans, even in houses in the remote province of Great Britain, made use of hot-water and steam conduits in the floor to heat their houses.

Heating of an object by radiation from the Sun cannot be calculated using Eq. 19–18 since this equation assumes a uniform temperature, T_2, of the environment surrounding the object, whereas the Sun is essentially a point source. Hence the Sun must be treated as a separate source of energy. Heating by the Sun is calculated using the fact that about 1350 J of energy strikes the atmosphere of the Earth from the Sun per second per square meter of area at right angles to the Sun's rays. This number, 1350 W/m², is called the **solar constant**. The atmosphere may absorb as much as 70% of this energy before it reaches the ground, depending on the cloud cover. On a clear day, about 1000 W/m² reaches the Earth's surface. An object of emissivity ϵ with area A facing the Sun absorbs energy from the Sun at a rate, in watts, of about

$$\frac{\Delta Q}{\Delta t} = \left(1000 \text{ W/m}^2\right)\epsilon A \cos\theta, \qquad (19\text{--}19)$$

where θ is the angle between the Sun's rays and a line perpendicular to the area A (Fig. 19–24). That is, $A\cos\theta$ is the "effective" area, at right angles to the Sun's rays.

The explanation for the **seasons** and the polar ice caps (see Fig. 19–25) depends on this $\cos\theta$ factor in Eq. 19–19. The seasons are *not* a result of how close the Earth is to the Sun—in fact, in the Northern Hemisphere, summer occurs when the Earth is farthest from the Sun. It is the angle (i.e., $\cos\theta$) that really matters. Furthermore, the reason the Sun heats the Earth more at midday than at sunrise or sunset is also related to this $\cos\theta$ factor.

An interesting application of thermal radiation to diagnostic medicine is **thermography**. A special instrument, the thermograph, scans the body, measuring the intensity of radiation from many points and forming a picture that resembles an X-ray (Fig. 19–26). Areas where metabolic activity is high, such as in tumors, can often be detected on a thermogram as a result of their higher temperature and consequent increased radiation.

FIGURE 19–24 Radiant energy striking a body at an angle θ.

FIGURE 19–25 (a) Earth's seasons arise from the $23\frac{1}{2}°$ angle Earth's axis makes with its orbit around the Sun. (b) June sunlight makes an angle of about 23° with the equator. Thus θ in the southern United States (A) is near 0° (direct summer sunlight), whereas in the Southern Hemisphere (B), θ is 50° or 60°, and less heat can be absorbed—hence it is winter. Near the poles (C), there is never strong direct sunlight; $\cos\theta$ varies from about $\frac{1}{2}$ in summer to 0 in winter; so with little heating, ice can form.

(a)   (b)

FIGURE 19–26 Thermograms of a healthy person's arms and hands (a) before and (b) after smoking a cigarette, showing a temperature decrease due to impaired blood circulation associated with smoking. The thermograms have been color-coded according to temperature; the scale on the right goes from blue (cold) to white (hot).

EXAMPLE 19–15 **ESTIMATE** **Star radius.** The giant star Betelgeuse emits radiant energy at a rate 10^4 times greater than our Sun, whereas its surface temperature is only half (2900 K) that of our Sun. Estimate the radius of Betelgeuse, assuming $\epsilon = 1$ for both. The Sun's radius is $r_S = 7 \times 10^8$ m.

APPROACH We assume both Betelgeuse and the Sun are spherical, with surface area $4\pi r^2$.

SOLUTION We solve Eq. 19–17 for A:

$$4\pi r^2 = A = \frac{(\Delta Q/\Delta t)}{\epsilon \sigma T^4}.$$

Then

$$\frac{r_B^2}{r_S^2} = \frac{(\Delta Q/\Delta t)_B}{(\Delta Q/\Delta t)_S} \cdot \frac{T_S^4}{T_B^4} = (10^4)(2^4) = 16 \times 10^4.$$

Hence $r_B = \sqrt{16 \times 10^4}\, r_S = (400)(7 \times 10^8 \text{ m}) \approx 3 \times 10^{11}$ m.

NOTE If Betelgeuse were our Sun, it would envelop us (Earth is 1.5×10^{11} m from the Sun).

EXERCISE G Fanning yourself on a hot day cools you by (a) increasing the radiation rate of the skin; (b) increasing conductivity; (c) decreasing the mean free path of air; (d) increasing the evaporation of perspiration; (e) none of these.

Summary

Internal energy, E_{int}, refers to the total energy of all the molecules in an object. For an ideal monatomic gas,

$$E_{int} = \tfrac{3}{2}NkT = \tfrac{3}{2}nRT \tag{19–1}$$

where N is the number of molecules or n is the number of moles.

Heat refers to the transfer of energy from one object to another because of a difference of temperature. Heat is thus measured in energy units, such as joules.

Heat and internal energy are also sometimes specified in calories or kilocalories (kcal), where

$$1 \text{ kcal} = 4.186 \text{ kJ}$$

is the amount of heat needed to raise the temperature of 1 kg of water by 1 C°.

The **specific heat**, c, of a substance is defined as the energy (or heat) required to change the temperature of unit mass of substance by 1 degree; as an equation,

$$Q = mc\,\Delta T, \tag{19–2}$$

where Q is the heat absorbed or given off, ΔT is the temperature increase or decrease, and m is the mass of the substance.

When heat flows between parts of an isolated system, conservation of energy tells us that the heat gained by one part of the system is equal to the heat lost by the other part of the system. This is the basis of **calorimetry**, which is the quantitative measurement of heat exchange.

Exchange of energy occurs, without a change in temperature, whenever a substance changes phase. The **heat of fusion** is the heat required to melt 1 kg of a solid into the liquid phase; it is also equal to the heat given off when the substance changes from liquid to solid. The **heat of vaporization** is the energy required to change 1 kg of a substance from the liquid to the vapor phase; it is also the energy given off when the substance changes from vapor to liquid.

The **first law of thermodynamics** states that the change in internal energy ΔE_{int} of a system is equal to the heat *added to* the system, Q, minus the work, W, done *by* the system:

$$\Delta E_{int} = Q - W. \tag{19–4}$$

This important law is a broad restatement of the conservation of energy and is found to hold for all processes.

Two simple thermodynamic processes are **isothermal**, which is a process carried out at constant temperature, and **adiabatic**, a process in which no heat is exchanged. Two more are **isobaric** (a process carried out at constant pressure) and **isovolumetric** (a process at constant volume).

The work done by (or on) a gas to change its volume by dV is $dW = P\,dV$, where P is the pressure.

Work and heat are not functions of the state of a system (as are P, V, T, n, and E_{int}) but depend on the type of process that takes a system from one state to another.

The **molar specific heat** of an ideal gas at constant volume, C_V, and at constant pressure, C_P, are related by

$$C_P - C_V = R, \tag{19–11}$$

where R is the gas constant. For a monatomic ideal gas, $C_V = \tfrac{3}{2}R$.

For ideal gases made up of diatomic or more complex molecules, C_V is equal to $\tfrac{1}{2}R$ times the number of **degrees of freedom** of the molecule. Unless the temperature is very high, some of the degrees of freedom may not be active and so do not contribute. According to the **principle of equipartition of energy**, energy is shared equally among the active degrees of freedom in an amount $\tfrac{1}{2}kT$ per molecule on average.

When an ideal gas expands (or contracts) adiabatically ($Q = 0$), the relation $PV^\gamma = \text{constant}$ holds, where

$$\gamma = \frac{C_P}{C_V}. \tag{19–14}$$

Heat is transferred from one place (or object) to another in three different ways: conduction, convection, and radiation.

In **conduction**, energy is transferred by collisions between molecules or electrons with higher kinetic energy to slower-moving neighbors.

Convection is the transfer of energy by the mass movement of molecules over considerable distances.

Radiation, which does not require the presence of matter, is energy transfer by electromagnetic waves, such as from the Sun.

All objects radiate energy in an amount that is proportional to the fourth power of their Kelvin temperature (T^4) and to their surface area. The energy radiated (or absorbed) also depends on the nature of the surface, which is characterized by the emissivity, ϵ (dark surfaces absorb and radiate more than do bright shiny ones).

Radiation from the Sun arrives at the surface of the Earth on a clear day at a rate of about 1000 W/m^2.

Questions

1. What happens to the work done on a jar of orange juice when it is vigorously shaken?

2. When a hot object warms a cooler object, does temperature flow between them? Are the temperature changes of the two objects equal? Explain.

3. (a) If two objects of different temperature are placed in contact, will heat naturally flow from the object with higher internal energy to the object with lower internal energy? (b) Is it possible for heat to flow even if the internal energies of the two objects are the same? Explain.

4. In warm regions where tropical plants grow but the temperature may drop below freezing a few times in the winter, the destruction of sensitive plants due to freezing can be reduced by watering them in the evening. Explain.

5. The specific heat of water is quite large. Explain why this fact makes water particularly good for heating systems (that is, hot-water radiators).

6. Why does water in a canteen stay cooler if the cloth jacket surrounding the canteen is kept moist?

7. Explain why burns caused by steam at 100°C on the skin are often more severe than burns caused by water at 100°C.

8. Explain why water cools (its temperature drops) when it evaporates, using the concepts of latent heat and internal energy.

9. Will potatoes cook faster if the water is boiling more vigorously?

10. Very high in the Earth's atmosphere the temperature can be 700°C. Yet an animal there would freeze to death rather than roast. Explain.

11. What happens to the internal energy of water vapor in the air that condenses on the outside of a cold glass of water? Is work done or heat exchanged? Explain.

12. Use the conservation of energy to explain why the temperature of a well-insulated gas increases when it is compressed—say, by pushing down on a piston—whereas the temperature decreases when the gas expands.

13. In an isothermal process, 3700 J of work is done by an ideal gas. Is this enough information to tell how much heat has been added to the system? If so, how much?

14. Explorers on failed Arctic expeditions have survived by covering themselves with snow. Why would they do that?

15. Why is wet sand at the beach cooler to walk on than dry sand?

16. When hot-air furnaces are used to heat a house, why is it important that there be a vent for air to return to the furnace? What happens if this vent is blocked by a bookcase?

17. Is it possible for the temperature of a system to remain constant even though heat flows into or out of it? If so, give examples.

18. Discuss how the first law of thermodynamics can apply to metabolism in humans. In particular, note that a person does work W, but very little heat Q is added to the body (rather, it tends to flow out). Why then doesn't the internal energy drop drastically in time?

19. Explain in words why C_P is greater than C_V.

20. Explain why the temperature of a gas increases when it is adiabatically compressed.

21. An ideal monatomic gas is allowed to expand slowly to twice its volume (1) isothermally; (2) adiabatically; (3) isobarically. Plot each on a PV diagram. In which process is ΔE_{int} the greatest, and in which is ΔE_{int} the least? In which is W the greatest and the least? In which is Q the greatest and the least?

22. Ceiling fans are sometimes reversible, so that they drive the air down in one season and pull it up in another season. Which way should you set the fan for summer? For winter?

23. Goose down sleeping bags and parkas are often specified as so many inches or centimeters of *loft*, the actual thickness of the garment when it is fluffed up. Explain.

24. Microprocessor chips nowadays have a "heat sink" glued on top that looks like a series of fins. Why is it shaped like that?

25. Sea breezes are often encountered on sunny days at the shore of a large body of water. Explain, assuming the temperature of the land rises more rapidly than that of the nearby water.

26. The Earth cools off at night much more quickly when the weather is clear than when cloudy. Why?

27. Explain why air-temperature readings are always taken with the thermometer in the shade.

28. A premature baby in an incubator can be dangerously cooled even when the air temperature in the incubator is warm. Explain.

29. The floor of a house on a foundation under which the air can flow is often cooler than a floor that rests directly on the ground (such as a concrete slab foundation). Explain.

30. Why is the liner of a **thermos bottle** silvered (Fig. 19–27), and why does it have a vacuum between its two walls?

FIGURE 19–27
Question 30.

31. A 22°C day is warm, while a swimming pool at 22°C feels cool. Why?

32. In the Northern Hemisphere the amount of heat required to heat a room where the windows face north is much higher than that required where the windows face south. Explain.

33. Heat loss occurs through windows by the following processes: (1) ventilation around edges; (2) through the frame, particularly if it is metal; (3) through the glass panes; and (4) radiation. (a) For the first three, what is (are) the mechanism(s): conduction, convection, or radiation? (b) Heavy curtains reduce which of these heat losses? Explain in detail.

34. Early in the day, after the Sun has reached the slope of a mountain, there tends to be a gentle upward movement of air. Later, after a slope goes into shadow, there is a gentle downdraft. Explain.

35. A piece of wood lying in the Sun absorbs more heat than a piece of shiny metal. Yet the metal feels hotter than the wood when you pick it up. Explain.

36. An "emergency blanket" is a thin shiny (metal-coated) plastic foil. Explain how it can help to keep an immobile person warm.

37. Explain why cities situated by the ocean tend to have less extreme temperatures than inland cities at the same latitude.

Problems

19–1 Heat as Energy Transfer

1. (I) To what temperature will 8700 J of heat raise 3.0 kg of water that is initially at 10.0°C?

2. (II) When a diver jumps into the ocean, water leaks into the gap region between the diver's skin and her wetsuit, forming a water layer about 0.5 mm thick. Assuming the total surface area of the wetsuit covering the diver is about 1.0 m², and that ocean water enters the suit at 10°C and is warmed by the diver to skin temperature of 35°C, estimate how much energy (in units of candy bars = 300 kcal) is required by this heating process.

3. (II) An average active person consumes about 2500 Cal a day. (a) What is this in joules? (b) What is this in kilowatt-hours? (c) If your power company charges about 10 ¢ per kilowatt-hour, how much would your energy cost per day if you bought it from the power company? Could you feed yourself on this much money per day?

4. (II) A British thermal unit (Btu) is a unit of heat in the British system of units. One Btu is defined as the heat needed to raise 1 lb of water by 1 F°. Show that

$$1 \text{ Btu} = 0.252 \text{ kcal} = 1056 \text{ J}.$$

5. (II) How many joules and kilocalories are generated when the brakes are used to bring a 1200-kg car to rest from a speed of 95 km/h?

6. (II) A small immersion heater is rated at 350 W. Estimate how long it will take to heat a cup of soup (assume this is 250 mL of water) from 15°C to 75°C.

19–3 and 19–4 Specific Heat; Calorimetry

7. (I) An automobile cooling system holds 18 L of water. How much heat does it absorb if its temperature rises from 15°C to 95°C?

8. (I) What is the specific heat of a metal substance if 135 kJ of heat is needed to raise 5.1 kg of the metal from 18.0°C to 37.2°C?

9. (II) (a) How much energy is required to bring a 1.0-L pot of water at 20°C to 100°C? (b) For how long could this amount of energy run a 100-W lightbulb?

10. (II) Samples of copper, aluminum, and water experience the same temperature rise when they absorb the same amount of heat. What is the ratio of their masses?

11. (II) How long does it take a 750-W coffeepot to bring to a boil 0.75 L of water initially at 8.0°C? Assume that the part of the pot which is heated with the water is made of 280 g of aluminum, and that no water boils away.

12. (II) A hot iron horseshoe (mass = 0.40 kg), just forged (Fig. 19–28), is dropped into 1.05 L of water in a 0.30-kg iron pot initially at 20.0°C. If the final equilibrium temperature is 25.0°C, estimate the initial temperature of the hot horseshoe.

FIGURE 19–28
Problem 12.

13. (II) A 31.5-g glass thermometer reads 23.6°C before it is placed in 135 mL of water. When the water and thermometer come to equilibrium, the thermometer reads 39.2°C. What was the original temperature of the water? [Hint: Ignore the mass of fluid inside the glass thermometer.]

14. (II) Estimate the Calorie content of 65 g of candy from the following measurements. A 15-g sample of the candy is placed in a small aluminum container of mass 0.325 kg filled with oxygen. This container is placed in 2.00 kg of water in an aluminum calorimeter cup of mass 0.624 kg at an initial temperature of 15.0°C. The oxygen-candy mixture in the small container is ignited, and the final temperature of the whole system is 53.5°C.

15. (II) When a 290-g piece of iron at 180°C is placed in a 95-g aluminum calorimeter cup containing 250 g of glycerin at 10°C, the final temperature is observed to be 38°C. Estimate the specific heat of glycerin.

16. (II) The heat capacity, C, of an object is defined as the amount of heat needed to raise its temperature by 1 C°. Thus, to raise the temperature by ΔT requires heat Q given by

$$Q = C \Delta T.$$

(a) Write the heat capacity C in terms of the specific heat, c, of the material. (b) What is the heat capacity of 1.0 kg of water? (c) Of 35 kg of water?

17. (II) The 1.20-kg head of a hammer has a speed of 7.5 m/s just before it strikes a nail (Fig. 19–29) and is brought to rest. Estimate the temperature rise of a 14-g iron nail generated by 10 such hammer blows done in quick succession. Assume the nail absorbs all the energy.

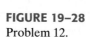

FIGURE 19–29
Problem 17.

19–5 Latent Heat

18. (I) How much heat is needed to melt 26.50 kg of silver that is initially at 25°C?

19. (I) During exercise, a person may give off 180 kcal of heat in 25 min by evaporation of water from the skin. How much water has been lost?

20. (II) A 35-g ice cube at its melting point is dropped into an insulated container of liquid nitrogen. How much nitrogen evaporates if it is at its boiling point of 77 K and has a latent heat of vaporization of 200 kJ/kg? Assume for simplicity that the specific heat of ice is a constant and is equal to its value near its melting point.

21. (II) High-altitude mountain climbers do not eat snow, but always melt it first with a stove. To see why, calculate the energy absorbed from your body if you (a) eat 1.0 kg of −10°C snow which your body warms to body temperature of 37°C. (b) You melt 1.0 kg of −10°C snow using a stove and drink the resulting 1.0 kg of water at 2°C, which your body has to warm to 37°C.

22. (II) An iron boiler of mass 180 kg contains 730 kg of water at 18°C. A heater supplies energy at the rate of 52,000 kJ/h. How long does it take for the water (a) to reach the boiling point, and (b) to all have changed to steam?

23. (II) In a hot day's race, a bicyclist consumes 8.0 L of water over the span of 3.5 hours. Making the approximation that all of the cyclist's energy goes into evaporating this water as sweat, how much energy in kcal did the rider use during the ride? (Since the efficiency of the rider is only about 20%, most of the energy consumed does go to heat, so our approximation is not far off.)

24. (II) The specific heat of mercury is 138 J/kg·C°. Determine the latent heat of fusion of mercury using the following calorimeter data: 1.00 kg of solid Hg at its melting point of −39.0°C is placed in a 0.620-kg aluminum calorimeter with 0.400 kg of water at 12.80°C; the resulting equilibrium temperature is 5.06°C.

25. (II) At a crime scene, the forensic investigator notes that the 7.2-g lead bullet that was stopped in a doorframe apparently melted completely on impact. Assuming the bullet was shot at room temperature (20°C), what does the investigator calculate as the minimum muzzle velocity of the gun?

26. (II) A 58-kg ice-skater moving at 7.5 m/s glides to a stop. Assuming the ice is at 0°C and that 50% of the heat generated by friction is absorbed by the ice, how much ice melts?

19–6 and 19–7 First Law of Thermodynamics

27. (I) Sketch a PV diagram of the following process: 2.0 L of ideal gas at atmospheric pressure are cooled at constant pressure to a volume of 1.0 L, and then expanded isothermally back to 2.0 L, whereupon the pressure is increased at constant volume until the original pressure is reached.

28. (I) A gas is enclosed in a cylinder fitted with a light frictionless piston and maintained at atmospheric pressure. When 1250 kcal of heat is added to the gas, the volume is observed to increase slowly from 12.0 m³ to 18.2 m³. Calculate (a) the work done by the gas and (b) the change in internal energy of the gas.

29. (II) The pressure in an ideal gas is cut in half slowly, while being kept in a container with rigid walls. In the process, 365 kJ of heat left the gas. (a) How much work was done during this process? (b) What was the change in internal energy of the gas during this process?

30. (II) A 1.0-L volume of air initially at 3.5 atm of (absolute) pressure is allowed to expand isothermally until the pressure is 1.0 atm. It is then compressed at constant pressure to its initial volume, and lastly is brought back to its original pressure by heating at constant volume. Draw the process on a PV diagram, including numbers and labels for the axes.

31. (II) Consider the following two-step process. Heat is allowed to flow out of an ideal gas at constant volume so that its pressure drops from 2.2 atm to 1.4 atm. Then the gas expands at constant pressure, from a volume of 5.9 L to 9.3 L, where the temperature reaches its original value. See Fig. 19–30. Calculate (a) the total work done by the gas in the process, (b) the change in internal energy of the gas in the process, and (c) the total heat flow into or out of the gas.

FIGURE 19–30
Problem 31.

32. (II) The PV diagram in Fig. 19–31 shows two possible states of a system containing 1.55 moles of a monatomic ideal gas. ($P_1 = P_2 = 455$ N/m², $V_1 = 2.00$ m³, $V_2 = 8.00$ m³.) (a) Draw the process which depicts an isobaric expansion from state 1 to state 2, and label this process A. (b) Find the work done by the gas and the change in internal energy of the gas in process A. (c) Draw the two-step process which depicts an isothermal expansion from state 1 to the volume V_2, followed by an isovolumetric increase in temperature to state 2, and label this process B. (d) Find the change in internal energy of the gas for the two-step process B.

FIGURE 19–31
Problem 32.

33. (II) Suppose 2.60 mol of an ideal gas of volume $V_1 = 3.50$ m³ at $T_1 = 290$ K is allowed to expand isothermally to $V_2 = 7.00$ m³ at $T_2 = 290$ K. Determine (a) the work done by the gas, (b) the heat added to the gas, and (c) the change in internal energy of the gas.

34. (II) In an engine, an almost ideal gas is compressed adiabatically to half its volume. In doing so, 2850 J of work is done on the gas. (a) How much heat flows into or out of the gas? (b) What is the change in internal energy of the gas? (c) Does its temperature rise or fall?

35. (II) One and one-half moles of an ideal monatomic gas expand adiabatically, performing 7500 J of work in the process. What is the change in temperature of the gas during this expansion?

36. (II) Determine (a) the work done and (b) the change in internal energy of 1.00 kg of water when it is all boiled to steam at 100°C. Assume a constant pressure of 1.00 atm.

37. (II) How much work is done by a pump to slowly compress, isothermally, 3.50 L of nitrogen at 0°C and 1.00 atm to 1.80 L at 0°C?

38. (II) When a gas is taken from a to c along the curved path in Fig. 19–32, the work done by the gas is $W = -35$ J and the heat added to the gas is $Q = -63$ J. Along path abc, the work done is $W = -54$ J. (a) What is Q for path abc? (b) If $P_c = \frac{1}{2}P_b$, what is W for path cda? (c) What is Q for path cda? (d) What is $E_{int, a} - E_{int, c}$? (e) If $E_{int, d} - E_{int, c} = 12$ J, what is Q for path da?

FIGURE 19–32
Problems 38, 39, and 40.

39. (III) In the process of taking a gas from state a to state c along the curved path shown in Fig. 19–32, 85 J of heat leaves the system and 55 J of work is done on the system. (a) Determine the change in internal energy, $E_{int, a} - E_{int, c}$. (b) When the gas is taken along the path cda, the work done by the gas is $W = 38$ J. How much heat Q is added to the gas in the process cda? (c) If $P_a = 2.2P_d$, how much work is done by the gas in the process abc? (d) What is Q for path abc? (e) If $E_{int, a} - E_{int, b} = 15$ J, what is Q for the process bc? Here is a summary of what is given:

$$Q_{a \to c} = -85 \text{ J}$$
$$W_{a \to c} = -55 \text{ J}$$
$$W_{cda} = 38 \text{ J}$$
$$E_{int, a} - E_{int, b} = 15 \text{ J}$$
$$P_a = 2.2P_d.$$

40. (III) Suppose a gas is taken clockwise around the rectangular cycle shown in Fig. 19–32, starting at b, then to a, to d, to c, and returning to b. Using the values given in Problem 39, (a) describe each leg of the process, and then calculate (b) the net work done during the cycle, (c) the total internal energy change during the cycle, and (d) the net heat flow during the cycle. (e) What percentage of the intake heat was turned into usable work: i.e., how efficient is this "rectangular" cycle (give as a percentage)?

*41. (III) Determine the work done by 1.00 mol of a van der Waals gas (Section 18–5) when it expands from volume V_1 to V_2 isothermally.

19–8 Molecular Specific Heat for Gases; Equipartition of Energy

42. (I) What is the internal energy of 4.50 mol of an ideal diatomic gas at 645 K, assuming all degrees of freedom are active?

43. (I) If a heater supplies 1.8×10^6 J/h to a room 3.5 m $\times$ 4.6 m $\times$ 3.0 m containing air at 20°C and 1.0 atm, by how much will the temperature rise in one hour, assuming no losses of heat or air mass to the outside? Assume air is an ideal diatomic gas with molecular mass 29.

44. (I) Show that if the molecules of a gas have n degrees of freedom, then theory predicts $C_V = \frac{1}{2}nR$ and $C_P = \frac{1}{2}(n + 2)R$.

45. (II) A certain monatomic gas has specific heat $c_V = 0.0356$ kcal/kg·C°, which changes little over a wide temperature range. What is the atomic mass of this gas? What gas is it?

46. (II) Show that the work done by n moles of an ideal gas when it expands adiabatically is $W = nC_V(T_1 - T_2)$, where T_1 and T_2 are the initial and final temperatures, and C_V is the molar specific heat at constant volume.

47. (II) An audience of 1800 fills a concert hall of volume 22,000 m³. If there were no ventilation, by how much would the temperature of the air rise over a period of 2.0 h due to the metabolism of the people (70 W/person)?

48. (II) The specific heat at constant volume of a particular gas is 0.182 kcal/kg·K at room temperature, and its molecular mass is 34. (a) What is its specific heat at constant pressure? (b) What do you think is the molecular structure of this gas?

49. (II) A 2.00 mole sample of N_2 gas at 0°C is heated to 150°C at constant pressure (1.00 atm). Determine (a) the change in internal energy, (b) the work the gas does, and (c) the heat added to it.

50. (III) A 1.00-mol sample of an ideal diatomic gas at a pressure of 1.00 atm and temperature of 420 K undergoes a process in which its pressure increases linearly with temperature. The final temperature and pressure are 720 K and 1.60 atm. Determine (a) the change in internal energy, (b) the work done by the gas, and (c) the heat added to the gas. (Assume five active degrees of freedom.)

19–9 Adiabatic Expansion of a Gas

51. (I) A 1.00-mol sample of an ideal diatomic gas, originally at 1.00 atm and 20°C, expands adiabatically to 1.75 times its initial volume. What are the final pressure and temperature for the gas? (Assume no molecular vibration.)

52. (II) Show, using Eqs. 19–6 and 19–15, that the work done by a gas that slowly expands adiabatically from pressure P_1 and volume V_1, to P_2 and V_2, is given by $W = (P_1V_1 - P_2V_2)/(\gamma - 1)$.

53. (II) A 3.65-mol sample of an ideal diatomic gas expands adiabatically from a volume of 0.1210 m³ to 0.750 m³. Initially the pressure was 1.00 atm. Determine: (a) the initial and final temperatures; (b) the change in internal energy; (c) the heat lost by the gas; (d) the work done on the gas. (Assume no molecular vibration.)

54. (II) An ideal monatomic gas, consisting of 2.8 mol of volume 0.086 m³, expands adiabatically. The initial and final temperatures are 25°C and −68°C. What is the final volume of the gas?

55. (III) A 1.00-mol sample of an ideal monatomic gas, originally at a pressure of 1.00 atm, undergoes a three-step process: (1) it is expanded adiabatically from $T_1 = 588$ K to $T_2 = 389$ K; (2) it is compressed at constant pressure until its temperature reaches T_3; (3) it then returns to its original pressure and temperature by a constant-volume process. (a) Plot these processes on a PV diagram. (b) Determine T_3. (c) Calculate the change in internal energy, the work done by the gas, and the heat added to the gas for each process, and (d) for the complete cycle.

56. (III) Consider a **parcel of air** moving to a different altitude y in the Earth's atmosphere (Fig. 19–33). As the parcel changes altitude it acquires the pressure P of the surrounding air. From Eq. 13–4 we have

$$\frac{dP}{dy} = -\rho g$$

where ρ is the parcel's altitude-dependent mass density.

FIGURE 19–33
Problem 56.

During this motion, the parcel's volume will change and, because air is a poor heat conductor, we assume this expansion or contraction will take place adiabatically. (*a*) Starting with Eq. 19–15, $PV^\gamma = $ constant, show that for an ideal gas undergoing an adiabatic process, $P^{1-\gamma}T^\gamma = $ constant. Then show that the parcel's pressure and temperature are related by

$$(1 - \gamma)\frac{dP}{dy} + \gamma\frac{P}{T}\frac{dT}{dy} = 0$$

and thus

$$(1 - \gamma)(-\rho g) + \gamma\frac{P}{T}\frac{dT}{dy} = 0.$$

(*b*) Use the ideal gas law with the result from part (*a*) to show that the change in the parcel's temperature with change in altitude is given by

$$\frac{dT}{dy} = \frac{1 - \gamma}{\gamma}\frac{mg}{k}$$

where m is the average mass of an air molecule and k is the Boltzmann constant. (*c*) Given that air is a diatomic gas with an average molecular mass of 29, show that $dT/dy = -9.8$ C°/km. This value is called the **adiabatic lapse rate** for dry air. (*d*) In California, the prevailing westerly winds descend from one of the highest elevations (the 4000-m Sierra Nevada mountains) to one of the lowest elevations (Death Valley, -100 m) in the continental United States. If a dry wind has a temperature of -5°C at the top of the Sierra Nevada, what is the wind's temperature after it has descended to Death Valley?

19–10 Conduction, Convection, Radiation

57. (I) (*a*) How much power is radiated by a tungsten sphere (emissivity $\epsilon = 0.35$) of radius 16 cm at a temperature of 25°C? (*b*) If the sphere is enclosed in a room whose walls are kept at -5°C, what is the *net* flow rate of energy out of the sphere?

58. (I) One end of a 45-cm-long copper rod with a diameter of 2.0 cm is kept at 460°C, and the other is immersed in water at 22°C. Calculate the heat conduction rate along the rod.

59. (II) How long does it take the Sun to melt a block of ice at 0°C with a flat horizontal area 1.0 m² and thickness 1.0 cm? Assume that the Sun's rays make an angle of 35° with the vertical and that the emissivity of ice is 0.050.

60. (II) *Heat conduction to skin.* Suppose 150 W of heat flows by conduction from the blood capillaries beneath the skin to the body's surface area of 1.5 m². If the temperature difference is 0.50 C°, estimate the average distance of capillaries below the skin surface.

61. (II) A ceramic teapot ($\epsilon = 0.70$) and a shiny one ($\epsilon = 0.10$) each hold 0.55 L of tea at 95°C. (*a*) Estimate the rate of heat loss from each, and (*b*) estimate the temperature drop after 30 min for each. Consider only radiation, and assume the surroundings are at 20°C.

62. (II) A copper rod and an aluminum rod of the same length and cross-sectional area are attached end to end (Fig. 19–34). The copper end is placed in a furnace maintained at a constant temperature of 225°C. The aluminum end is placed in an ice bath held at constant temperature of 0.0°C. Calculate the temperature at the point where the two rods are joined.

Cu	Al
225°C	$T = ?$	0.0°C

FIGURE 19–34 Problem 62.

63. (II) (*a*) Using the solar constant, estimate the rate at which the whole Earth receives energy from the Sun. (*b*) Assume the Earth radiates an equal amount back into space (that is, the Earth is in equilibrium). Then, assuming the Earth is a perfect emitter ($\epsilon = 1.0$), estimate its average surface temperature. [*Hint*: Use area $A = 4\pi r_E^2$, and state why.]

64. (II) A 100-W lightbulb generates 95 W of heat, which is dissipated through a glass bulb that has a radius of 3.0 cm and is 0.50 mm thick. What is the difference in temperature between the inner and outer surfaces of the glass?

65. (III) A house thermostat is normally set to 22°C, but at night it is turned down to 12°C for 9.0 h. Estimate how much more heat would be needed (state as a percentage of daily usage) if the thermostat were not turned down at night. Assume that the outside temperature averages 0°C for the 9.0 h at night and 8°C for the remainder of the day, and that the heat loss from the house is proportional to the difference in temperature inside and out. To obtain an estimate from the data, you will have to make other simplifying assumptions; state what these are.

66. (III) Approximately how long should it take 9.5 kg of ice at 0°C to melt when it is placed in a carefully sealed Styrofoam ice chest of dimensions 25 cm × 35 cm × 55 cm whose walls are 1.5 cm thick? Assume that the conductivity of Styrofoam is double that of air and that the outside temperature is 34°C.

67. (III) A cylindrical pipe has inner radius R_1 and outer radius R_2. The interior of the pipe carries hot water at temperature T_1. The temperature outside is T_2 ($< T_1$). (a) Show that the rate of heat loss for a length L of pipe is

$$\frac{dQ}{dt} = \frac{2\pi k(T_1 - T_2)L}{\ln(R_2/R_1)},$$

where k is the thermal conductivity of the pipe. (b) Suppose the pipe is steel with $R_1 = 3.3$ cm, $R_2 = 4.0$ cm, and $T_2 = 18°C$. If the pipe holds still water at $T_1 = 71°C$, what will be the initial rate of change of its temperature? (c) Suppose water at 71°C enters the pipe and moves at a speed of 8.0 cm/s. What will be its temperature drop per centimeter of travel?

68. (III) Suppose the insulating qualities of the wall of a house come mainly from a 4.0-in. layer of brick and an R-19 layer of insulation, as shown in Fig. 19–35. What is the total rate of heat loss through such a wall, if its total area is 195 ft^2 and the temperature difference across it is 12 F°?

FIGURE 19–35 Problem 68. Two layers insulating a wall.

General Problems

69. A soft-drink can contains about 0.20 kg of liquid at 5°C. Drinking this liquid can actually consume some of the fat in the body, since energy is needed to warm the liquid to body temperature (37°C). How many food Calories should the drink have so that it is in perfect balance with the heat needed to warm the liquid (essentially water)?

70. (a) Find the total power radiated into space by the Sun, assuming it to be a perfect emitter at $T = 5500$ K. The Sun's radius is 7.0×10^8 m. (b) From this, determine the power per unit area arriving at the Earth, 1.5×10^{11} m away.

71. To get an idea of how much thermal energy is contained in the world's oceans, estimate the heat liberated when a cube of ocean water, 1 km on each side, is cooled by 1 K. (Approximate the ocean water as pure water for this estimate.)

72. A mountain climber wears a goose-down jacket 3.5 cm thick with total surface area 0.95 m^2. The temperature at the surface of the clothing is $-18°C$ and at the skin is 34°C. Determine the rate of heat flow by conduction through the jacket (a) assuming it is dry and the thermal conductivity k is that of goose down, and (b) assuming the jacket is wet, so k is that of water and the jacket has matted to 0.50 cm thickness.

73. During light activity, a 70-kg person may generate 200 kcal/h. Assuming that 20% of this goes into useful work and the other 80% is converted to heat, estimate the temperature rise of the body after 30 min if none of this heat is transferred to the environment.

74. Estimate the rate at which heat can be conducted from the interior of the body to the surface. Assume that the thickness of tissue is 4.0 cm, that the skin is at 34°C and the interior at 37°C, and that the surface area is 1.5 m^2. Compare this to the measured value of about 230 W that must be dissipated by a person working lightly. This clearly shows the necessity of convective cooling by the blood.

75. A marathon runner has an average metabolism rate of about 950 kcal/h during a race. If the runner has a mass of 55 kg, estimate how much water she would lose to evaporation from the skin for a race that lasts 2.2 h.

76. A house has well-insulated walls 19.5 cm thick (assume conductivity of air) and area 410 m^2, a roof of wood 5.5 cm thick and area 280 m^2, and uncovered windows 0.65 cm thick and total area 33 m^2. (a) Assuming that heat is lost only by conduction, calculate the rate at which heat must be supplied to this house to maintain its inside temperature at 23°C if the outside temperature is $-15°C$. (b) If the house is initially at 12°C, estimate how much heat must be supplied to raise the temperature to 23°C within 30 min. Assume that only the air needs to be heated and that its volume is 750 m^3. (c) If natural gas costs $0.080 per kilogram and its heat of combustion is 5.4×10^7 J/kg, how much is the monthly cost to maintain the house as in part (a) for 24 h each day, assuming 90% of the heat produced is used to heat the house? Take the specific heat of air to be 0.24 kcal/kg·C°.

77. In a typical game of squash (Fig. 19–36), two people hit a soft rubber ball at a wall until they are about to drop due to dehydration and exhaustion. Assume that the ball hits the wall at a velocity of 22 m/s and bounces back with a velocity of 12 m/s, and that the kinetic energy lost in the process heats the ball. What will be the temperature increase of the ball after one bounce? (The specific heat of rubber is about 1200 J/kg·C°.)

FIGURE 19–36 Problem 77.

78. A bicycle pump is a cylinder 22 cm long and 3.0 cm in diameter. The pump contains air at 20.0°C and 1.0 atm. If the outlet at the base of the pump is blocked and the handle is pushed in very quickly, compressing the air to half its original volume, how hot does the air in the pump become?

79. A microwave oven is used to heat 250 g of water. On its maximum setting, the oven can raise the temperature of the liquid water from 20°C to 100°C in 1 min 45 s ($= 105$ s). (a) At what rate does the oven input energy to the liquid water? (b) If the power input from the oven to the water remains constant, determine how many grams of water will boil away if the oven is operated for 2 min (rather than just 1 min 45 s).

80. The temperature within the Earth's crust increases about 1.0 C° for each 30 m of depth. The thermal conductivity of the crust is 0.80 W/C°·m. (a) Determine the heat transferred from the interior to the surface for the entire Earth in 1.0 h. (b) Compare this heat to the amount of energy incident on the Earth in 1.0 h due to radiation from the Sun.

81. An ice sheet forms on a lake. The air above the sheet is at -18°C, whereas the water is at 0°C. Assume that the heat of fusion of the water freezing on the lower surface is conducted through the sheet to the air above. How much time will it take to form a sheet of ice 15 cm thick?

82. An iron meteorite melts when it enters the Earth's atmosphere. If its initial temperature was -105°C outside of Earth's atmosphere, calculate the minimum velocity the meteorite must have had before it entered Earth's atmosphere.

83. A scuba diver releases a 3.60-cm-diameter (spherical) bubble of air from a depth of 14.0 m. Assume the temperature is constant at 298 K, and that the air behaves as an ideal gas. (a) How large is the bubble when it reaches the surface? (b) Sketch a PV diagram for the process. (c) Apply the first law of thermodynamics to the bubble, and find the work done by the air in rising to the surface, the change in its internal energy, and the heat added or removed from the air in the bubble as it rises. Take the density of water to be 1000 kg/m^3.

84. A reciprocating compressor is a device that compresses air by a back-and-forth straight-line motion, like a piston in a cylinder. Consider a reciprocating compressor running at 150 rpm. During a compression stroke, 1.00 mol of air is compressed. The initial temperature of the air is 390 K, the engine of the compressor is supplying 7.5 kW of power to compress the air, and heat is being removed at the rate of 1.5 kW. Calculate the temperature change per compression stroke.

85. The temperature of the glass surface of a 75-W lightbulb is 75°C when the room temperature is 18°C. Estimate the temperature of a 150-W lightbulb with a glass bulb the same size. Consider only radiation, and assume that 90% of the energy is emitted as heat.

86. Suppose 3.0 mol of neon (an ideal monatomic gas) at STP are compressed slowly and isothermally to 0.22 the original volume. The gas is then allowed to expand quickly and adiabatically back to its original volume. Find the highest and lowest temperatures and pressures attained by the gas, and show on a PV diagram where these values occur.

87. At very low temperatures, the molar specific heat of many substances varies as the cube of the absolute temperature:

$$C = k \frac{T^3}{T_0^3},$$

which is sometimes called Debye's law. For rock salt, $T_0 = 281$ K and $k = 1940$ J/mol·K. Determine the heat needed to raise 2.75 mol of salt from 22.0 K to 48.0 K.

88. A diesel engine accomplishes ignition without a spark plug by an adiabatic compression of air to a temperature above the ignition temperature of the diesel fuel, which is injected into the cylinder at the peak of the compression. Suppose air is taken into the cylinder at 280 K and volume V_1 and is compressed adiabatically to 560°C (≈ 1000°F) and volume V_2. Assuming that the air behaves as an ideal gas whose ratio of C_P to C_V is 1.4, calculate the compression ratio V_1/V_2 of the engine.

89. When 6.30×10^5 J of heat is added to a gas enclosed in a cylinder fitted with a light frictionless piston maintained at atmospheric pressure, the volume is observed to increase from 2.2 m^3 to 4.1 m^3. Calculate (a) the work done by the gas, and (b) the change in internal energy of the gas. (c) Graph this process on a PV diagram.

90. In a cold environment, a person can lose heat by conduction and radiation at a rate of about 200 W. Estimate how long it would take for the body temperature to drop from 36.6°C to 35.6°C if metabolism were nearly to stop. Assume a mass of 70 kg. (See Table 19–1.)

*Numerical/Computer

*91. (II) Suppose 1.0 mol of steam at 100°C of volume 0.50 m^3 is expanded isothermally to volume 1.00 m^3. Assume steam obeys the van der Waals equation $(P + n^2 a/V^2)(V/n - b) = RT$, Eq. 18–9, with $a = 0.55$ N·m^4/mol^2 and $b = 3.0 \times 10^{-5}$ m^3/mol. Using the expression $dW = P \, dV$, determine numerically the total work done W. Your result should agree within 2% of the result obtained by integrating the expression for dW.

Answers to Exercises

A: (b).

B: (c).

C: 0.21 kg.

D: 700 J.

E: Less.

F: -6.8×10^3 J.

G: (d).

There are many uses for a heat engine, such as old steam trains and modern coal-burning power plants. Steam engines produce steam which does work: on turbines to generate electricity, and on a piston that moves linkage to turn locomotive wheels. The efficiency of any engine—no matter how carefully engineered—is limited by nature as described in the second law of thermodynamics. This great law is best stated in terms of a quantity called entropy, which is unlike any other. Entropy is *not* conserved, but instead is constrained always to increase in any real process. Entropy is a measure of disorder. The second law of thermodynamics tells us that as time moves forward, the disorder in the universe increases.

We discuss many practical matters including heat engines, heat pumps, and refrigeration.

CHAPTER 20

Second Law of Thermodynamics

CHAPTER-OPENING QUESTION—Guess now!

Fossil-fuel electric generating plants produce "thermal pollution." Part of the heat produced by the burning fuel is not converted to electric energy. The reason for this waste is

(a) The efficiency is higher if some heat is allowed to escape.

(b) Engineering technology has not yet reached the point where 100% waste heat recovery is possible.

(c) Some waste heat *must* be produced: this is a fundamental property of nature when converting heat to useful work.

(d) The plants rely on fossil fuels, not nuclear fuel.

(e) None of the above.

I n this final Chapter on heat and thermodynamics, we discuss the famous second law of thermodynamics, and the quantity "entropy" that arose from this fundamental law and is its quintessential expression. We also discuss heat engines—the engines that transform heat into work in power plants, trains, and motor vehicles—because they first showed us that a new law was needed. Finally, we briefly discuss the third law of thermodynamics.

20–1 The Second Law of Thermodynamics—Introduction

The first law of thermodynamics states that energy is conserved. There are, however, many processes we can imagine that conserve energy but are not observed to occur in nature. For example, when a hot object is placed in contact with a cold object, heat flows from the hotter one to the colder one, never spontaneously the reverse. If heat were to leave the colder object and pass to the hotter one, energy could still be conserved. Yet it does not happen spontaneously.[†] As a second example, consider what happens when you drop a rock and it hits the ground. The initial potential energy of the rock changes to kinetic energy as the rock falls. When the rock hits the ground, this energy in turn is transformed into internal energy of the rock and the ground in the vicinity of the impact; the molecules move faster and the temperature rises slightly. But have you seen the reverse happen—a rock at rest on the ground suddenly rise up in the air because the thermal energy of molecules is transformed into kinetic energy of the rock as a whole? Energy could be conserved in this process, yet we never see it happen.

There are many other examples of processes that occur in nature but whose reverse does not. Here are two more. (1) If you put a layer of salt in a jar and cover it with a layer of similar-sized grains of pepper, when you shake it you get a thorough mixture. But no matter how long you shake it, the mixture does not separate into two layers again. (2) Coffee cups and glasses break spontaneously if you drop them. But they do not go back together spontaneously (Fig. 20–1).

(a) Initial state. (b) Later: cup reassembles and rises up. (c) Later still: cup lands on table.

FIGURE 20–1 Have you ever observed this process, a broken cup spontaneously reassembling and rising up onto a table? This process could conserve energy and other laws of mechanics.

The first law of thermodynamics (conservation of energy) would not be violated if any of these processes occurred in reverse. To explain this lack of reversibility, scientists in the latter half of the nineteenth century formulated a new principle known as the second law of thermodynamics.

The **second law of thermodynamics** is a statement about which processes occur in nature and which do not. It can be stated in a variety of ways, all of which are equivalent. One statement, due to R. J. E. Clausius (1822–1888), is that

> **heat can flow spontaneously from a hot object to a cold object; heat will not flow spontaneously from a cold object to a hot object.**

SECOND LAW OF THERMODYNAMICS (Clausius statement)

Since this statement applies to one particular process, it is not obvious how it applies to other processes. A more general statement is needed that will include other possible processes in a more obvious way.

The development of a general statement of the second law of thermodynamics was based partly on the study of heat engines. A **heat engine** is any device that changes thermal energy into mechanical work, such as a steam engine or automobile engine. We now examine heat engines, both from a practical point of view and to show their importance in developing the second law of thermodynamics.

[†]By spontaneously, we mean by itself without input of work of some sort. (A refrigerator does move heat from a cold environment to a warmer one, but only because its motor does work—Section 20–4.)

FIGURE 20–2 Schematic diagram
of energy transfers for a heat engine.

⚠ **CAUTION**

New sign convention:
$Q_H > 0, Q_L > 0, W > 0$

PHYSICS APPLIED

Engines

20–2 Heat Engines

It is easy to produce thermal energy by doing work—for example, by simply rubbing your hands together briskly, or indeed by any frictional process. But to get work from thermal energy is more difficult, and a practical device to do this was invented only about 1700 with the development of the steam engine.

The basic idea behind any heat engine is that mechanical energy can be obtained from thermal energy only when heat is allowed to flow from a high temperature to a low temperature. In the process, some of the heat can then be transformed to mechanical work, as diagrammed schematically in Fig. 20–2. We will be interested only in engines that run in a repeating *cycle* (that is, the system returns repeatedly to its starting point) and thus can run continuously. In each cycle the change in internal energy of the system is $\Delta E_{int} = 0$ because it returns to the starting state. Thus a heat input Q_H at a high temperature T_H is partly transformed into work W and partly exhausted as heat Q_L at a lower temperature T_L (Fig. 20–2). By conservation of energy, $Q_H = W + Q_L$. The high and low temperatures, T_H and T_L, are called the **operating temperatures** of the engine. Note carefully that we are now using a new sign convention: we take Q_H, Q_L, and W as always positive. The direction of each energy transfer is shown by the arrow on the applicable diagram, such as Fig. 20–2.

Steam Engine and Internal Combustion Engine

The operation of a steam engine is illustrated in Fig. 20–3. Steam engines are of two main types, each making use of steam heated by combustion of coal, oil or gas, or by nuclear energy. In a reciprocating engine, Fig. 20–3a, the heated steam passes through the intake valve and expands against a piston, forcing it to move. As the piston returns to its original position, it forces the gases out the exhaust valve. A steam turbine, Fig. 20–3b, is very similar except that the reciprocating piston is replaced by a rotating turbine that resembles a paddlewheel with many sets of blades. Most of our electricity today is generated using steam turbines.[†] The material that is heated and cooled, steam in this case, is called the **working substance**.

[†]Even nuclear power plants utilize steam turbines; the nuclear fuel—uranium—merely serves as fuel to heat the steam.

FIGURE 20–3 Steam engines.

(a) Reciprocating type

(b) Turbine (boiler and
condenser not shown)

Intake valve (open) · Exhaust valve (closed) · Gas–air mixture from carburetor · Connecting rod · Crankshaft

Both valves closed · Cylinder · Piston

Both valves closed · Spark plug (firing)

Both valves closed · Rings

Exhaust valve (open) · To exhaust pipe

(a) Intake (b) Compression (c) Ignition (d) Expansion (power stroke) (e) Exhaust

In an internal combustion engine (used in most automobiles), the high temperature is achieved by burning the gasoline–air mixture in the cylinder itself (ignited by the spark plug), as described in Fig. 20–4.

Why a ΔT Is Needed to Drive a Heat Engine

To see why a *temperature difference* is required to run an engine, let us examine the steam engine. In the reciprocating engine, for example, suppose there were no condenser or pump (Fig. 20–3a), and that the steam was at the same temperature throughout the system. This would mean that the pressure of the gas being exhausted would be the same as that on intake. Thus, although work would be done by the gas *on* the piston when it expanded, an equal amount of work would have to be done *by* the piston to force the steam out the exhaust; hence, no net work would be done. In a real engine, the exhausted gas is cooled to a lower temperature and condensed so that the exhaust pressure is less than the intake pressure. Thus, although the piston must do work on the gas to expel it on the exhaust stroke, it is less than the work done by the gas on the piston during the intake. So a net amount of work can be obtained—but only if there is a difference of temperature. Similarly, in the gas turbine if the gas isn't cooled, the pressure on each side of the blades would be the same. By cooling the gas on the exhaust side, the pressure on the back side of the blade is less and hence the turbine turns.

Efficiency and the Second Law

The **efficiency**, e, of any heat engine can be defined as the ratio of the work it does, W, to the heat input at the high temperature, Q_H (Fig. 20–2):

$$e = \frac{W}{Q_H}.$$

This is a sensible definition since W is the output (what you get from the engine), whereas Q_H is what you put in and pay for in burned fuel. Since energy is conserved, the heat input Q_H must equal the work done plus the heat that flows out at the low temperature (Q_L):

$$Q_H = W + Q_L.$$

Thus $W = Q_H - Q_L$, and the efficiency of an engine is

$$e = \frac{W}{Q_H} \qquad\qquad (20\text{--}1a)$$

$$= \frac{Q_H - Q_L}{Q_H} = 1 - \frac{Q_L}{Q_H}. \qquad\qquad (20\text{--}1b)$$

To give the efficiency as a percent, we multiply Eqs. 20–1 by 100. Note that e could be 1.0 (or 100%) only if Q_L were zero—that is, only if no heat were exhausted to the environment.

FIGURE 20–4 Four-stroke-cycle internal combustion engine: (a) the gasoline–air mixture flows into the cylinder as the piston moves down; (b) the piston moves upward and compresses the gas; (c) the brief instant when firing of the spark plug ignites the highly compressed gasoline–air mixture, raising it to a high temperature; (d) the gases, now at high temperature and pressure, expand against the piston in this, the power stroke; (e) the burned gases are pushed out to the exhaust pipe; when the piston reaches the top, the exhaust valve closes and the intake valve opens, and the whole cycle repeats. (a), (b), (d), and (e) are the four strokes of the cycle.

EXAMPLE 20–1 **Car efficiency.** An automobile engine has an efficiency of 20% and produces an average of 23,000 J of mechanical work per second during operation. (a) How much heat input is required, and (b) how much heat is discharged as waste heat from this engine, per second?

APPROACH We want to find the heat input Q_H as well as the heat output Q_L, given $W = 23,000$ J each second and an efficiency $e = 0.20$. We can use the definition of efficiency, Eq. 20–1 in its various forms, to find first Q_H and then Q_L.

SOLUTION (a) From Eq. 20–1a, $e = W/Q_H$, we solve for Q_H:

$$Q_H = \frac{W}{e} = \frac{23,000 \text{ J}}{0.20}$$
$$= 1.15 \times 10^5 \text{ J} = 115 \text{ kJ}.$$

The engine requires $115 \text{ kJ/s} = 115 \text{ kW}$ of heat input.

(b) Now we use Eq. 20–1b $(e = 1 - Q_L/Q_H)$ and solve for Q_L:

$$Q_L = (1 - e)Q_H = (0.80)115 \text{ kJ} = 92 \text{ kJ}.$$

The engine discharges heat to the environment at a rate of $92 \text{ kJ/s} = 92 \text{ kW}$.

NOTE Of the 115 kJ that enters the engine per second, only 23 kJ does useful work whereas 92 kJ is wasted as heat output.

NOTE The problem was stated in terms of energy per unit time. We could just as well have stated it in terms of power, since $1 \text{ J/s} = 1 \text{ W}$.

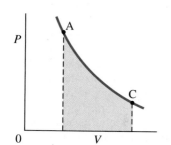

FIGURE 20–5 Adiabatic process, Exercise A.

EXERCISE A An adiabatic process is defined as one in which no heat flows in or out of the system. If an ideal gas expands as shown in Fig. 20–5 (see also Fig. 19–8), the work W done in this expansion equals the area under the graph, shown shaded. The efficiency of this process would be $e = W/Q$, much greater than 100% ($= \infty$ since $Q = 0$). Is this a violation of the second law?

It is clear from Eq. 20–1b, $e = 1 - Q_L/Q_H$, that the efficiency of an engine will be greater if Q_L can be made small. However, from experience with a wide variety of systems, it has not been found possible to reduce Q_L to zero. If Q_L could be reduced to zero we would have a 100% efficient engine, as diagrammed in Fig. 20–6.

FIGURE 20–6 Schematic diagram of a hypothetical perfect heat engine in which all the heat input would be used to do work.

That such a perfect engine (running continuously in a cycle) is not possible is another way of expressing the second law of thermodynamics:

No device is possible whose sole effect is to transform a given amount of heat completely into work.

This is known as the **Kelvin-Planck statement of the second law of thermodynamics.** Said another way, *there can be no perfect (100% efficient) heat engine* such as that diagrammed in Fig. 20–6.

If the second law were not true, so that a perfect engine could be built, some rather remarkable things could happen. For example, if the engine of a ship did not need a low-temperature reservoir to exhaust heat into, the ship could sail across the ocean using the vast resources of the internal energy of the ocean water. Indeed, we would have no fuel problems at all!

20–3 Reversible and Irreversible Processes; the Carnot Engine

In the early nineteenth century, the French scientist N. L. Sadi Carnot (1796–1832) studied in detail the process of transforming heat into mechanical energy. His aim had been to determine how to increase the efficiency of heat engines, but his studies soon led him to investigate the foundations of thermodynamics itself. In 1824, Carnot invented (on paper) an idealized type of engine which we now call the *Carnot engine*. No Carnot engine actually exists, but as a theoretical idea it played an important role in the establishment and understanding of the second law of thermodynamics.

Reversible and Irreversible Processes

The Carnot engine involves *reversible processes*, so before we discuss it we must discuss what is meant by reversible and irreversible processes. A **reversible process** is one that is carried out infinitely slowly, so that the process can be considered as a series of equilibrium states, and the whole process could be done in reverse with no change in magnitude of the work done or heat exchanged. For example, a gas contained in a cylinder fitted with a tight, movable, but frictionless piston could be compressed isothermally in a reversible way if done infinitely slowly. Not all very slow (quasistatic) processes are reversible, however. If there is friction present, for example (as between the movable piston and cylinder just mentioned), the work done in one direction (going from some state A to state B) will not be the negative of the work done in the reverse direction (state B to state A). Such a process would not be considered reversible. A perfectly reversible process is not possible in reality because it would require an infinite time; reversible processes can be approached arbitrarily closely, however, and they are very important theoretically.

All real processes are **irreversible**: they are not done infinitely slowly. There could be turbulence in the gas, friction would be present, and so on. Any process could not be done precisely in reverse since the heat lost to friction would not reverse itself, the turbulence would be different, and so on. For any given volume there would not be a well-defined pressure P and temperature T since the system would not always be in an equilibrium state. Thus a real, irreversible, process cannot be plotted on a PV diagram, except insofar as it may approach an ideal reversible process. But a reversible process (since it is a quasistatic series of equilibrium states) always can be plotted on a PV diagram; and a reversible process, when done in reverse, retraces the same path on a PV diagram. Although all real processes are irreversible, reversible processes are conceptually important, just as the concept of an ideal gas is.

Carnot's Engine

Now let us look at Carnot's idealized engine. The **Carnot engine** makes use of a **reversible cycle**, by which we mean a series of reversible processes that take a given substance (the *working substance*) from an initial equilibrium state through many other equilibrium states and returns it again to the same initial state. In particular, the Carnot engine utilizes the **Carnot cycle**, which is illustrated in Fig. 20–7, with the working substance assumed to be an ideal gas. Let us take point a as the initial state. The gas is first expanded isothermally and reversibly, path ab, at temperature T_H, as heat Q_H is added to it. Next the gas is expanded adiabatically and reversibly, path bc; no heat is exchanged and the temperature of the gas is reduced to T_L. The third step is a reversible isothermal compression, path cd, during which heat Q_L flows out of the working substance. Finally, the gas is compressed adiabatically, path da, back to its original state. Thus a Carnot cycle consists of two isothermal and two adiabatic processes.

The net work done in one cycle by a Carnot engine (or any other type of engine using a reversible cycle) is equal to the area enclosed by the curve representing the cycle on the PV diagram, the curve abcd in Fig. 20–7. (See Section 19–7.)

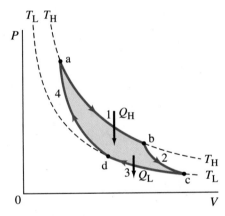

FIGURE 20–7 The Carnot cycle. Heat engines work in a cycle, and the cycle for the Carnot engine begins at point a on this PV diagram. (1) The gas is first expanded isothermally, with the addition of heat Q_H, along the path ab at temperature T_H. (2) Next the gas expands adiabatically from b to c—no heat is exchanged, but the temperature drops to T_L. (3) The gas is then compressed at constant temperature T_L, path cd, and heat Q_L flows out. (4) Finally, the gas is compressed adiabatically, path da, back to its original state. No Carnot engine actually exists, but as a theoretical idea it played an important role in the development of thermodynamics.

Carnot Efficiency and the Second Law of Thermodynamics

The efficiency of a Carnot engine, like any heat engine, is given by Eq. 20–1b:

$$e = 1 - \frac{Q_L}{Q_H}.$$

For a Carnot engine using an ideal gas, however, we can show that the efficiency depends only on the temperatures of the heat reservoirs, T_H and T_L. In the first isothermal process ab in Fig. 20–7, the work done by the gas is (see Eq. 19–8)

$$W_{ab} = nRT_H \ln \frac{V_b}{V_a},$$

where n is the number of moles of the ideal gas used as working substance. Because the internal energy of an ideal gas does not change when the temperature remains constant, the first law of thermodynamics tells us that the heat added to the gas equals the work done by the gas:

$$Q_H = nRT_H \ln \frac{V_b}{V_a}.$$

Similarly, the heat lost by the gas in the isothermal process cd is

$$Q_L = nRT_L \ln \frac{V_c}{V_d}.$$

The paths bc and da are adiabatic, so we have from Eq. 19–15:

$$P_b V_b^\gamma = P_c V_c^\gamma \quad \text{and} \quad P_d V_d^\gamma = P_a V_a^\gamma,$$

where $\gamma = C_P/C_V$ is the ratio of molar specific heats (Eq. 19–14). Also, from the ideal gas law,

$$\frac{P_b V_b}{T_H} = \frac{P_c V_c}{T_L} \quad \text{and} \quad \frac{P_d V_d}{T_L} = \frac{P_a V_a}{T_H}.$$

When we divide these last equations, term by term, into the corresponding set of equations on the line above, we obtain

$$T_H V_b^{\gamma-1} = T_L V_c^{\gamma-1} \quad \text{and} \quad T_L V_d^{\gamma-1} = T_H V_a^{\gamma-1}.$$

Next we divide the equation on the left by the one on the right and obtain

$$\left(\frac{V_b}{V_a}\right)^{\gamma-1} = \left(\frac{V_c}{V_d}\right)^{\gamma-1}$$

or

$$\frac{V_b}{V_a} = \frac{V_c}{V_d}.$$

Inserting this result in our equations for Q_H and Q_L above, we obtain

$$\frac{Q_L}{Q_H} = \frac{T_L}{T_H}. \qquad \text{[Carnot cycle]} \quad \textbf{(20–2)}$$

Hence the efficiency of a reversible Carnot engine can now be written

$$e_{\text{ideal}} = 1 - \frac{Q_L}{Q_H}$$

or

$$e_{\text{ideal}} = 1 - \frac{T_L}{T_H}. \qquad \left[\begin{array}{l}\text{Carnot efficiency;}\\\text{Kelvin temperatures}\end{array}\right] \quad \textbf{(20–3)}$$

The temperatures T_L and T_H are the absolute or Kelvin temperatures as measured on the ideal gas temperature scale. Thus the efficiency of a Carnot engine depends only on the temperatures T_L and T_H.

We could imagine other possible reversible cycles that could be used for an ideal reversible engine. According to a theorem stated by Carnot:

All reversible engines operating between the same two constant temperatures T_H and T_L have the same efficiency. Any irreversible engine operating between the same two fixed temperatures will have an efficiency less than this.

This is known as **Carnot's theorem.**[†] It tells us that Eq. 20–3, $e = 1 - (T_L/T_H)$, applies to any ideal reversible engine with fixed input and exhaust temperatures, T_H and T_L, and that this equation represents a maximum possible efficiency for a real (i.e., irreversible) engine.

In practice, the efficiency of real engines is always less than the Carnot efficiency. Well-designed engines reach perhaps 60% to 80% of Carnot efficiency.

EXAMPLE 20–2 A phony claim? An engine manufacturer makes the following claims: An engine's heat input per second is 9.0 kJ at 435 K. The heat output per second is 4.0 kJ at 285 K. Do you believe these claims?

APPROACH The engine's efficiency can be calculated from the definition, Eqs. 20–1. It must be less than the maximum possible, Eq. 20–3.

SOLUTION The claimed efficiency of the engine is (Eq. 20–1b)

$$e = 1 - \frac{Q_L}{Q_H} = 1 - \frac{4.0\,\text{kJ}}{9.0\,\text{kJ}} = 0.56,$$

or 56%. However, the maximum possible efficiency is given by the Carnot efficiency, Eq. 20–3:

$$e_{\text{ideal}} = 1 - \frac{T_L}{T_H} = 1 - \frac{285\,\text{K}}{435\,\text{K}} = 0.34,$$

or 34%. The manufacturer's claims violate the second law of thermodynamics and cannot be believed.

EXERCISE B A motor is running with an intake temperature $T_H = 400\,\text{K}$ and an exhaust temperature $T_L = 300\,\text{K}$. Which of the following is *not* a possible efficiency for the engine? (*a*) 0.10; (*b*) 0.16; (*c*) 0.24; (*d*) 0.30.

It is clear from Eq. 20–3 that a 100% efficient engine is not possible. Only if the exhaust temperature, T_L, were at absolute zero would 100% efficiency be obtainable. But reaching absolute zero is a practical (as well as theoretical) impossibility.[‡] Thus we can state, as we already did in Section 20–2, that **no device is possible whose sole effect is to transform a given amount of heat completely into work**. As we saw in Section 20–2, this is known as the *Kelvin-Planck statement of the second law of thermodynamics*. It tells us that there can be no perfect (100% efficient) heat engine such as the one diagrammed in Fig. 20–6.

EXERCISE C Return to the Chapter-Opening Question, page 528, and answer it again now. Try to explain why you may have answered differently the first time.

*The Otto Cycle

The operation of an automobile internal combustion engine (Fig. 20–4) can be approximated by a reversible cycle known as the *Otto cycle*, whose PV diagram is shown in Fig. 20–8. Unlike the Carnot cycle, the input and exhaust temperatures of the Otto cycle are *not* constant. Paths ab and cd are adiabatic, and paths bc and da are at constant volume. The gas (gasoline-air mixture) enters the cylinder at point a and is compressed adiabatically (compression stroke) to point b. At b ignition occurs (spark plug) and the burning of the gas adds heat Q_H to the system at constant volume (approximately in a real engine). The temperature and pressure rise, and then in the power stroke, cd, the gas expands adiabatically. In the exhaust stroke, da, heat Q_L is ejected to the environment (in a real engine, the gas leaves the engine and is replaced by a new mixture of air and fuel).

FIGURE 20–8 The Otto cycle.

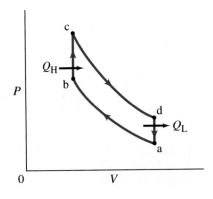

[†] Carnot's theorem can be shown to follow directly from either the Clausius or Kelvin-Planck statements of the second law of thermodynamics.

[‡] This result is known as the *third law of thermodynamics*, as discussed in Section 20–10.

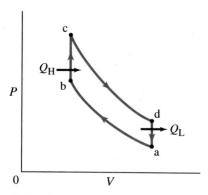

FIGURE 20–8 (repeated for Example 20–3) The Otto cycle.

FIGURE 20–9 Schematic diagram of energy transfers for a refrigerator or air conditioner.

EXAMPLE 20–3 **The Otto cycle.** (a) Show that for an ideal gas as working substance, the efficiency of an Otto cycle engine is

$$e = 1 - \left(\frac{V_a}{V_b}\right)^{1-\gamma}$$

where γ is the ratio of specific heats ($\gamma = C_P/C_V$, Eq. 19–14) and V_a/V_b is the *compression ratio*. (b) Calculate the efficiency for a compression ratio $V_a/V_b = 8.0$ assuming a diatomic gas like O_2 and N_2.

APPROACH We use the original definition of efficiency and the results from Chapter 19 for constant volume and adiabatic processes (Sections 19–8 and 19–9).

SOLUTION The heat exchanges take place at constant volume in the ideal Otto cycle, so from Eq. 19–10a:

$$Q_H = nC_V(T_c - T_b) \quad \text{and} \quad Q_L = nC_V(T_d - T_a).$$

Then from Eq. 20–1b,

$$e = 1 - \frac{Q_L}{Q_H} = 1 - \left[\frac{T_d - T_a}{T_c - T_b}\right].$$

To get this in terms of the compression ratio, V_a/V_b, we use the result from Section 19–9, Eq. 19–15, $PV^\gamma = $ constant during the adiabatic processes ab and cd. Thus

$$P_a V_a^\gamma = P_b V_b^\gamma \quad \text{and} \quad P_c V_c^\gamma = P_d V_d^\gamma.$$

We use the ideal gas law, $P = nRT/V$, and substitute P into these two equations

$$T_a V_a^{\gamma-1} = T_b V_b^{\gamma-1} \quad \text{and} \quad T_c V_c^{\gamma-1} = T_d V_d^{\gamma-1}.$$

Then the efficiency (see above) is

$$e = 1 - \left[\frac{T_d - T_a}{T_c - T_b}\right] = 1 - \left[\frac{T_c(V_c/V_d)^{\gamma-1} - T_b(V_b/V_a)^{\gamma-1}}{T_c - T_b}\right].$$

But processes bc and da are at constant volume, so $V_c = V_b$ and $V_d = V_a$. Hence $V_c/V_d = V_b/V_a$ and

$$e = 1 - \left[\frac{(V_b/V_a)^{\gamma-1}(T_c - T_b)}{T_c - T_b}\right] = 1 - \left(\frac{V_b}{V_a}\right)^{\gamma-1} = 1 - \left(\frac{V_a}{V_b}\right)^{1-\gamma}.$$

(b) For diatomic molecules (Section 19–8), $\gamma = C_P/C_V = 1.4$ so

$$e = 1 - (8.0)^{1-\gamma} = 1 - (8.0)^{-0.4} = 0.56.$$

Real engines do not reach this high efficiency because they do not follow perfectly the Otto cycle, plus there is friction, turbulence, heat loss and incomplete combustion of the gases.

20–4 Refrigerators, Air Conditioners, and Heat Pumps

The operating principle of refrigerators, air conditioners, and heat pumps is just the reverse of a heat engine. Each operates to transfer heat out of a cool environment into a warm environment. As diagrammed in Fig. 20–9, by doing work W, heat is taken from a low-temperature region, T_L (such as inside a refrigerator), and a greater amount of heat is exhausted at a high temperature, T_H (the room). You can often feel this heated air blowing out beneath a refrigerator. The work W is usually done by an electric motor which compresses a fluid, as illustrated in Fig. 20–10.

(a)

FIGURE 20–10 (a) Typical refrigerator system. The electric compressor motor forces a gas at high pressure through a heat exchanger (condenser) on the rear outside wall of the refrigerator where Q_H is given off and the gas cools to become liquid. The liquid passes from a high-pressure region, via a valve, to low-pressure tubes on the inside walls of the refrigerator; the liquid evaporates at this lower pressure and thus absorbs heat (Q_L) from the inside of the refrigerator. The fluid returns to the compressor where the cycle begins again. (b) Schematic diagram, like Fig. 20–9.

A perfect **refrigerator**—one in which no work is required to take heat from the low-temperature region to the high-temperature region—is not possible. This is the **Clausius statement of the second law of thermodynamics**, already mentioned in Section 20–1, which can be stated formally as

No device is possible whose sole effect is to transfer heat from one system at a temperature T_L into a second system at a higher temperature T_H.

To make heat flow from a low-temperature object (or system) to one at a higher temperature, work must be done. Thus, *there can be no perfect refrigerator.*

The **coefficient of performance** (COP) of a refrigerator is defined as the heat Q_L removed from the low-temperature area (inside a refrigerator) divided by the work W done to remove the heat (Fig. 20–9 or 20–10b):

$$\text{COP} = \frac{Q_L}{W}. \qquad \begin{bmatrix} \text{refrigerator and} \\ \text{air conditioner} \end{bmatrix} \quad \textbf{(20–4a)}$$

This makes sense because the more heat Q_L that can be removed from the inside of the refrigerator for a given amount of work, the better (more efficient) the refrigerator is. Energy is conserved, so from the first law of thermodynamics we can write (see Fig. 20–9 or 20–10b) $Q_L + W = Q_H$, or $W = Q_H - Q_L$. Then Eq. 20–4a becomes

$$\text{COP} = \frac{Q_L}{W} = \frac{Q_L}{Q_H - Q_L}. \qquad \begin{bmatrix} \text{refrigerator} \\ \text{air conditioner} \end{bmatrix} \quad \textbf{(20–4b)}$$

For an ideal refrigerator (not a perfect one, which is impossible), the best we could do would be

$$\text{COP}_{\text{ideal}} = \frac{T_L}{T_H - T_L}, \qquad \begin{bmatrix} \text{refrigerator} \\ \text{air conditioner} \end{bmatrix} \quad \textbf{(20–4c)}$$

analogous to an ideal (Carnot) engine (Eqs. 20–2 and 20–3).

An **air conditioner** works very much like a refrigerator, although the actual construction details are different: an air conditioner takes heat Q_L from inside a room or building at a low temperature, and deposits heat Q_H outside to the environment at a higher temperature. Equations 20–4 also describe the coefficient of performance for an air conditioner.

SECOND LAW OF THERMODYNAMICS (Clausius statement)

 PHYSICS APPLIED
Refrigerator

PHYSICS APPLIED
Air conditioner

EXAMPLE 20–4 **Making ice.** A freezer has a COP of 3.8 and uses 200 W of power. How long would it take this otherwise empty freezer to freeze an ice-cube tray that contains 600 g of water at 0°C?

APPROACH In Eq. 20–4b, Q_L is the heat that must be transferred out of the water so it will become ice. To determine Q_L, we use the latent heat of fusion L of water and Eq. 19–3, $Q = mL$.

SOLUTION From Table 19–2, $L = 333 \text{ kJ/kg}$. Hence $Q = mL = (0.600 \text{ kg})(3.33 \times 10^5 \text{ J/kg}) = 2.0 \times 10^5 \text{ J}$ is the total energy that needs to be removed from the water. The freezer does work at the rate of $200 \text{ W} = 200 \text{ J/s} = W/t$, which is the work W it can do in t seconds. We solve for t: $t = W/(200 \text{ J/s})$. For W, we use Eq. 20–4a: $W = Q_L/\text{COP}$. Thus

$$t = \frac{W}{200 \text{ J/s}} = \frac{Q_L/\text{COP}}{200 \text{ J/s}}$$

$$= \frac{(2.0 \times 10^5 \text{ J})/(3.8)}{200 \text{ J/s}} = 260 \text{ s},$$

or about $4\frac{1}{2}$ min.

FIGURE 20–11 A heat pump uses an electric motor to "pump" heat from the cold outside to the warm inside of a house.

Heat naturally flows from high temperature to low temperature. Refrigerators and air conditioners do work to accomplish the opposite: to make heat flow from cold to hot. We might say they "pump" heat from cold areas to hotter areas, against the natural tendency of heat to flow from hot to cold, just as water can be pumped uphill, against the natural tendency to flow downhill. The term **heat pump** is usually reserved for a device that can heat a house in winter by using an electric motor that does work W to take heat Q_L from the outside at low temperature and delivers heat Q_H to the warmer inside of the house; see Fig. 20–11. As in a refrigerator, there is an indoor and an outdoor heat exchanger (coils of the refrigerator) and an electric compressor motor. The operating principle is like that for a refrigerator or air conditioner; but the objective of a heat pump is to heat (deliver Q_H), rather than to cool (remove Q_L). Thus, the coefficient of performance of a heat pump is defined differently than for an air conditioner because it is the heat Q_H delivered to the inside of the house that is important now:

$$\text{COP} = \frac{Q_H}{W}. \qquad \text{[heat pump]} \quad \textbf{(20–5)}$$

The COP is necessarily greater than 1. Most heat pumps can be "turned around" and used as air conditioners in the summer.

EXAMPLE 20–5 **Heat pump.** A heat pump has a coefficient of performance of 3.0 and is rated to do work at 1500 W. (*a*) How much heat can it add to a room per second? (*b*) If the heat pump were turned around to act as an air conditioner in the summer, what would you expect its coefficient of performance to be, assuming all else stays the same?

APPROACH We use the definitions of coefficient of performance, which are different for the two devices in (*a*) and (*b*).

SOLUTION (*a*) We use Eq. 20–5 for the heat pump, and, since our device does 1500 J of work per second, it can pour heat into the room at a rate of

$$Q_H = \text{COP} \times W = 3.0 \times 1500 \text{ J} = 4500 \text{ J}$$

per second, or at a rate of 4500 W.

(b) If our device is turned around in summer, it can take heat Q_L from inside the house, doing 1500 J of work per second to then dump $Q_H = 4500$ J per second to the hot outside. Energy is conserved, so $Q_L + W = Q_H$ (see Fig. 20–11, but reverse the inside and outside of the house). Then

$$Q_L = Q_H - W = 4500 \text{ J} - 1500 \text{ J} = 3000 \text{ J}.$$

The coefficient of performance as an air conditioner would thus be (Eq. 20–4a)

$$\text{COP} = \frac{Q_L}{W} = \frac{3000 \text{ J}}{1500 \text{ J}} = 2.0.$$

20–5 Entropy

Thus far we have stated the second law of thermodynamics for specific situations. What we really need is a general statement of the second law of thermodynamics that will cover all situations, including those discussed earlier in this Chapter that are not observed in nature even though they would not violate the first law of thermodynamics. It was not until the latter half of the nineteenth century that the second law of thermodynamics was finally stated in a general way—namely, in terms of a quantity called **entropy**, introduced by Clausius in the 1860s. In Section 20–7 we will see that entropy can be interpreted as a measure of the order or disorder of a system.

When we deal with entropy—as with potential energy—it is the *change* in entropy during a process that is important, not the absolute amount. According to Clausius, the change in entropy S of a system, when an amount of heat Q is *added* to it by a reversible process at constant temperature, is given by

$$\Delta S = \frac{Q}{T}, \tag{20–6}$$

where T is the kelvin temperature.

If the temperature is not constant, we define entropy S by the relation

$$dS = \frac{dQ}{T}. \tag{20–7}$$

Then the change in entropy of a system taken reversibly between two states a and b is given by[†]

$$\Delta S = S_b - S_a = \int_a^b dS = \int_a^b \frac{dQ}{T}. \quad \text{[reversible process]} \tag{20–8}$$

A careful analysis (see next page) shows that the change in entropy when a system moves by a reversible process from any state a to another state b does not depend on the process. That is, $\Delta S = S_b - S_a$ depends only on the states a and b of the system. Thus entropy (unlike heat) is a *state variable*. Any system in a given state has a temperature, a volume, a pressure, and also has a particular value of entropy.

It is easy to see why entropy is a state variable for a Carnot cycle. In Eq. 20–2 we saw that $Q_L/Q_H = T_L/T_H$, which we rewrite as

$$\frac{Q_L}{T_L} = \frac{Q_H}{T_H}.$$

In the PV diagram for a Carnot cycle, Fig. 20–7, the entropy change $\Delta S = Q/T$ in going from state a to state c along path abc $(= Q_H/T_H + 0)$ is thus the same as going along the path adc. That is, the change in entropy is path independent—it depends only on the initial and final states of the system.

[†]Equation 20–8 says nothing about the absolute value of S; it only gives the change in S. This is much like potential energy (Chapter 8). However, one form of the so-called *third law of thermodynamics* (see also Section 20–10) states that as $T \to 0$, $S \to 0$.

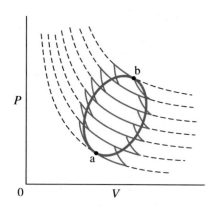

FIGURE 20–12 Any reversible cycle can be approximated as a series of Carnot cycles. (The dashed lines represent isotherms.)

*Showing Entropy Is a State Variable

In our study of the Carnot cycle we found (Eq. 20–2) that $Q_L/Q_H = T_L/T_H$. We rewrite this as

$$\frac{Q_H}{T_H} = \frac{Q_L}{T_L}.$$

In this relation, both Q_H and Q_L are positive. But now let us recall our original convention as used in the first law (Section 19–6), that Q is positive when it represents a heat flow into the system (as Q_H) and negative for a heat flow out of the system (as $-Q_L$). Then this relation becomes

$$\frac{Q_H}{T_H} + \frac{Q_L}{T_L} = 0. \qquad \text{[Carnot cycle]} \quad \textbf{(20–9)}$$

Now consider *any* reversible cycle, as represented by the smooth (oval-shaped) curve in Fig. 20–12. Any reversible cycle can be approximated as a series of Carnot cycles. Figure 20–12 shows only six—the isotherms (dashed lines) are connected by adiabatic paths for each—and the approximation becomes better and better if we increase the number of Carnot cycles. Equation 20–9 is valid for each of these cycles, so we can write

$$\Sigma \frac{Q}{T} = 0 \qquad \text{[Carnot cycles]} \quad \textbf{(20–10)}$$

for the sum of all these cycles. But note that the heat output Q_L of one cycle crosses the boundary below it and is approximately equal to the negative of the heat input, Q_H, of the cycle below it (actual equality occurs in the limit of an infinite number of infinitely thin Carnot cycles). Hence the heat flows on the inner paths of all these Carnot cycles cancel out, so the net heat transferred, and the work done, is the same for the series of Carnot cycles as for the original cycle. Hence, in the limit of infinitely many Carnot cycles, Eq. 20–10 applies to any reversible cycle. In this case Eq. 20–10 becomes

$$\oint \frac{dQ}{T} = 0, \qquad \text{[reversible cycle]} \quad \textbf{(20–11)}$$

where dQ represents an infinitesimal heat flow.[†] The symbol $\oint$ means that the integral is taken around a closed path; the integral can be started at any point on the path such as at a or b in Fig. 20–12, and proceed in either direction. If we divide the cycle of Fig. 20–12 into two parts as indicated in Fig. 20–13, then

$$\int_{a}^{b}{}_{I} \frac{dQ}{T} + \int_{b}^{a}{}_{II} \frac{dQ}{T} = 0.$$

The first term is the integral from point a to point b along path I in Fig. 20–13, and the second term is the integral from b back to a along path II. If path II is taken in reverse, dQ at each point becomes $-dQ$, since the path is reversible. Therefore

$$\int_{a}^{b}{}_{I} \frac{dQ}{T} = \int_{a}^{b}{}_{II} \frac{dQ}{T}. \qquad \text{[reversible paths]} \quad \textbf{(20–12)}$$

FIGURE 20–13 The integral, $\oint dS$, of the entropy for a reversible cycle is zero. Hence the difference in entropy between states a and b, $S_b - S_a = \int_a^b dS$, is the same for path I as for path II.

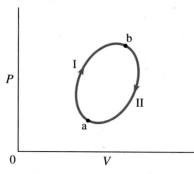

The integral of dQ/T between any two equilibrium states, a and b, does not depend on the path of the process. By defining entropy as $dS = dQ/T$ (Eq. 20–7), we see from Eq. 20–12 that the change in entropy between any two states along a reversible path is *independent of the path between two points a and b*. Thus *entropy is a state variable*—its value depends only on the state of the system, and not on the process or the past history of how it got there.[‡] This is in clear distinction to Q and W which are *not* state variables; their values do depend on the processes undertaken.

[†]dQ is often written $\bar{d}Q$: see footnote at the end of Section 19–6.
[‡]Real processes are irreversible. Because entropy is a state variable, the change in entropy ΔS for an irreversible process can be determined by calculating ΔS for a reversible process between the same two states.

20–6 Entropy and the Second Law of Thermodynamics

We have defined a new quantity, S, the entropy, which can be used to describe the state of the system, along with P, T, V, E_{int}, and n. But what does this rather abstract quantity have to do with the second law of thermodynamics? To answer this, let us take some examples in which we calculate the entropy changes during particular processes. But note first that Eq. 20–8 can be applied only to reversible processes. How then do we calculate $\Delta S = S_b - S_a$ for a real process that is irreversible? What we can do is this: we figure out some other *reversible* process that takes the system between the same two states, and calculate ΔS for this reversible process. This will equal ΔS for the irreversible process since ΔS depends only on the initial and final states of the system.

If the temperature varies during a process, a summation of the heat flow over the changing temperature can often be calculated using calculus or a computer. However, if the temperature change is not too great, a reasonable approximation can be made using the average value of the temperature, as indicated in the next Example.

EXAMPLE 20–6 **ESTIMATE** **Entropy change when mixing water.** A sample of 50.0 kg of water at 20.00°C is mixed with 50.0 kg of water at 24.00°C. Estimate the change in entropy.

APPROACH The final temperature of the mixture will be 22.00°C, since we started with equal amounts of water. We use the specific heat of water and the methods of calorimetry (Sections 19–3 and 19–4) to determine the heat transferred. Then we use the average temperature of each sample of water to estimate the entropy change $(\Delta Q/T)$.

SOLUTION A quantity of heat,

$$Q = mc\,\Delta T = (50.0\,\text{kg})(4186\,\text{J/kg}\cdot\text{C}°)(2.00\,\text{C}°) = 4.186 \times 10^5\,\text{J},$$

flows out of the hot water as it cools down from 24°C to 22°C, and this heat flows into the cold water as it warms from 20°C to 22°C. The total change in entropy, ΔS, will be the sum of the changes in entropy of the hot water, ΔS_H, and that of the cold water, ΔS_C:

$$\Delta S = \Delta S_H + \Delta S_C.$$

We estimate entropy changes by writing $\Delta S = Q/\overline{T}$, where $\overline{T}$ is an "average" temperature for each process, which ought to give a reasonable estimate since the temperature change is small. For the hot water we use an average temperature of 23°C (296 K), and for the cold water an average temperature of 21°C (294 K). Thus

$$\Delta S_H \approx -\frac{4.186 \times 10^5\,\text{J}}{296\,\text{K}} = -1414\,\text{J/K}$$

which is negative because this heat flows out, whereas heat is added to the cold water:

$$\Delta S_C \approx \frac{4.186 \times 10^5\,\text{J}}{294\,\text{K}} = 1424\,\text{J/K}.$$

The entropy of the hot water (S_H) decreases since heat flows out of the hot water. But the entropy of the cold water (S_C) increases by a greater amount. The total change in entropy is

$$\Delta S = \Delta S_H + \Delta S_C \approx -1414\,\text{J/K} + 1424\,\text{J/K} \approx 10\,\text{J/K}.$$

We see that although the entropy of one part of the system decreased, the entropy of the other part increased by a greater amount so that the net change in entropy of the whole system is positive.

We can now show in general that for an isolated system of two objects, the flow of heat from the higher-temperature (T_H) object to the lower-temperature (T_L) object always results in an increase in the total entropy. The two objects eventually come to some intermediate temperature, T_M. The heat lost by the hotter object $(Q_H = -Q$, where Q is positive) is equal to the heat gained by the colder one $(Q_L = Q)$, so the total change in entropy is

$$\Delta S = \Delta S_H + \Delta S_L = -\frac{Q}{T_{HM}} + \frac{Q}{T_{LM}},$$

where T_{HM} is some intermediate temperature between T_H and T_M for the hot object as it cools from T_H to T_M, and T_{LM} is the counterpart for the cold object. Since the temperature of the hot object is, at all times during the process, greater than that of the cold object, then $T_{HM} > T_{LM}$. Hence

$$\Delta S = Q \left(\frac{1}{T_{LM}} - \frac{1}{T_{HM}} \right) > 0.$$

One object decreases in entropy, while the other gains in entropy, but the *total* change is positive.

EXAMPLE 20–7 **Entropy changes in a free expansion.** Consider the *adiabatic free expansion* of n moles of an ideal gas from volume V_1 to volume V_2, where $V_2 > V_1$ as was discussed in Section 19–7, Fig. 19–14. Calculate the change in entropy (*a*) of the gas and (*b*) of the surrounding environment. (*c*) Evaluate ΔS for 1.00 mole, with $V_2 = 2.00\,V_1$.

APPROACH We saw in Section 19–7 that the gas is initially in a closed container of volume V_1, and, with the opening of a valve, it expands adiabatically into a previously empty container. The total volume of the two containers is V_2. The whole apparatus is thermally insulated from the surroundings, so no heat flows into the gas, $Q = 0$. The gas does no work, $W = 0$, so there is no change in internal energy, $\Delta E_{int} = 0$, and the temperature of the initial and final states is the same, $T_2 = T_1 = T$. The process takes place very quickly, and so is irreversible. Thus we cannot apply Eq. 20–8 to this process. Instead we must think of a reversible process that will take the gas from volume V_1 to V_2 at the same temperature, and use Eq. 20–8 on this reversible process to get ΔS. A reversible isothermal process will do the trick; in such a process, the internal energy does not change, so from the first law,

$$dQ = dW = P\,dV.$$

SOLUTION (*a*) For the gas,

$$\Delta S_{gas} = \int \frac{dQ}{T} = \frac{1}{T} \int_{V_1}^{V_2} P\,dV.$$

The ideal gas law tells us $P = nRT/V$, so

$$\Delta S_{gas} = \frac{nRT}{T} \int_{V_1}^{V_2} \frac{dV}{V} = nR \ln \frac{V_2}{V_1}.$$

Since $V_2 > V_1$, $\Delta S_{gas} > 0$.

(*b*) Since no heat is transferred to the surrounding environment, there is no change of the state of the environment due to this process. Hence $\Delta S_{env} = 0$. Note that the total change in entropy, $\Delta S_{gas} + \Delta S_{env}$, is greater than zero.

(*c*) Since $n = 1.00$ and $V_2 = 2.00\,V_1$, then $\Delta S_{gas} = R \ln 2.00 = 5.76\,\text{J/K}$.

EXAMPLE 20–8 **Heat transfer.** A red-hot 2.00-kg piece of iron at temperature $T_1 = 880$ K is thrown into a huge lake whose temperature is $T_2 = 280$ K. Assume the lake is so large that its temperature rise is insignificant. Determine the change in entropy (a) of the iron and (b) of the surrounding environment (the lake).

APPROACH The process is irreversible, but the same entropy change will occur for a reversible process, and we use the concept of specific heat, Eq. 19–2.

SOLUTION (a) We assume the specific heat of the iron is constant at $c = 450$ J/kg·K. Then $dQ = mc \, dT$ and in a quasistatic reversible process

$$\Delta S_{\text{iron}} = \int \frac{dQ}{T} = mc \int_{T_1}^{T_2} \frac{dT}{T} = mc \ln \frac{T_2}{T_1} = -mc \ln \frac{T_1}{T_2}.$$

Putting in numbers, we find

$$\Delta S_{\text{iron}} = -(2.00 \text{ kg})(450 \text{ J/kg·K}) \ln \frac{880 \text{ K}}{280 \text{ K}} = -1030 \text{ J/K}.$$

(b) The initial and final temperatures of the lake are the same, $T = 280$ K. The lake receives from the iron an amount of heat

$$Q = mc(T_2 - T_1) = (2.00 \text{ kg})(450 \text{ J/kg·K})(880 \text{ K} - 280 \text{ K}) = 540 \text{ kJ}.$$

Strictly speaking, this is an irreversible process (the lake heats up locally before equilibrium is reached), but is equivalent to a reversible isothermal transfer of heat $Q = 540$ kJ at $T = 280$ K. Hence

$$\Delta S_{\text{env}} = \frac{540 \text{ kJ}}{280 \text{ K}} = 1930 \text{ J/K}.$$

Thus, although the entropy of the iron actually decreases, the *total* change in entropy of iron plus environment is positive: 1930 J/K − 1030 J/K = 900 J/K.

EXERCISE D A 1.00-kg piece of ice at 0°C melts very slowly to water at 0°C. Assume the ice is in contact with a heat reservoir whose temperature is only infinitesimally greater than 0°C. Determine the entropy change of (a) the ice cube and (b) the heat reservoir.

In each of these Examples, the entropy of our system plus that of the environment (or surroundings) either stayed constant or increased. For any *reversible* process, such as that in Exercise D, the total entropy change is zero. This can be seen in general as follows: any reversible process can be considered as a series of quasistatic isothermal transfers of heat ΔQ between a system and the environment, which differ in temperature only by an infinitesimal amount. Hence the change in entropy of either the system or environment is $\Delta Q/T$ and that of the other is $-\Delta Q/T$, so the total is

$$\Delta S = \Delta S_{\text{syst}} + \Delta S_{\text{env}} = 0. \qquad \text{[any reversible process]}$$

In Examples 20–6, 20–7, and 20–8, we found that the total entropy of system plus environment increases. Indeed, it has been found that for all real (irreversible) processes, the total entropy increases. No exceptions have been found. We can thus make the *general statement of the second law of thermodynamics* as follows:

> **The entropy of an isolated system never decreases. It either stays constant (reversible processes) or increases (irreversible processes).**

Since all real processes are irreversible, we can equally well state the second law as:

> **The total entropy of any system plus that of its environment increases as a result of any natural process:**

$$\Delta S = \Delta S_{\text{syst}} + \Delta S_{\text{env}} > 0. \qquad (20\text{–}13)$$

SECOND LAW OF THERMODYNAMICS (general statement)

Although the entropy of one part of the universe may decrease in any process (see the Examples above), the entropy of some other part of the universe always increases by a greater amount, so the total entropy always increases.

Now that we finally have a quantitative general statement of the second law of thermodynamics, we can see that it is an unusual law. It differs considerably from other laws of physics, which are typically equalities (such as $F = ma$) or conservation laws (such as for energy and momentum). The second law of thermodynamics introduces a new quantity, the entropy, S, but does not tell us it is conserved. Quite the opposite. Entropy is not conserved in natural processes. Entropy always increases in time.

"Time's Arrow"

The second law of thermodynamics summarizes which processes are observed in nature, and which are not. Or, said another way, it tells us about the *direction* processes go. For the reverse of any of the processes in the last few Examples, the entropy would decrease; and we never observe them. For example, we never observe heat flowing spontaneously from a cold object to a hot object, the reverse of Example 20–8. Nor do we ever observe a gas spontaneously compressing itself into a smaller volume, the reverse of Example 20–7 (gases always expand to fill their containers). Nor do we see thermal energy transform into kinetic energy of a rock so the rock rises spontaneously from the ground. Any of these processes would be consistent with the first law of thermodynamics (conservation of energy). But they are not consistent with the second law of thermodynamics, and this is why we need the second law. If you were to see a movie run backward, you would probably realize it immediately because you would see odd occurrences—such as rocks rising spontaneously from the ground, or air rushing in from the atmosphere to fill an empty balloon (the reverse of free expansion). When watching a movie or video, we are tipped off to a faked reversal of time by observing whether entropy is increasing or decreasing. Hence entropy has been called **time's arrow,** for it can tell us in which direction time is going.

20–7 Order to Disorder

The concept of entropy, as we have discussed it so far, may seem rather abstract. But we can relate it to the more ordinary concepts of *order* and *disorder*. In fact, the entropy of a system can be considered a *measure of the disorder of the system.* Then the second law of thermodynamics can be stated simply as:

Natural processes tend to move toward a state of greater disorder.

Exactly what we mean by disorder may not always be clear, so we now consider a few examples. Some of these will show us how this very general statement of the second law applies beyond what we usually consider as thermodynamics.

Let us look at the simple processes mentioned in Section 20–1. First, a jar containing separate layers of salt and pepper is more orderly than when the salt and pepper are all mixed up. Shaking a jar containing separate layers results in a mixture, and no amount of shaking brings the layers back again. The natural process is from a state of relative order (layers) to one of relative disorder (a mixture), not the reverse. That is, disorder increases. Next, a solid coffee cup is a more "orderly" and useful object than the pieces of a broken cup. Cups break when they fall, but they do not spontaneously mend themselves (as faked in Fig. 20–1). Again, the normal course of events is an increase of disorder.

Let us consider some processes for which we have actually calculated the entropy change, and see that an increase in entropy results in an increase in disorder (or vice versa). When ice melts to water at 0°C, the entropy of the water increases (Exercise D). Intuitively, we can think of solid water, ice, as being more ordered than the less orderly fluid state which can flow all over the place. This change from order to disorder can be seen more clearly from the molecular point of view: the orderly arrangement of water molecules in an ice crystal has changed to the disorderly and somewhat random motion of the molecules in the fluid state.

When a hot object is put in contact with a cold object, heat flows from the high temperature to the low until the two objects reach the same intermediate temperature. At the beginning of the process we can distinguish two classes of molecules: those with a high average kinetic energy (the hot object), and those with a low average kinetic energy (the cooler object). After the process in which heat flows, all the molecules are in one class with the same average kinetic energy. We no longer have the more orderly arrangement of molecules in two classes. Order has gone to disorder. Furthermore, the separate hot and cold objects could serve as the hot- and cold-temperature regions of a heat engine, and thus could be used to obtain useful work. But once the two objects are put in contact and reach the same temperature, no work can be obtained. Disorder has increased, since a system that has the ability to perform work must surely be considered to have a higher order than a system no longer able to do work.

When a stone falls to the ground, its macroscopic kinetic energy is transformed to thermal energy. Thermal energy is associated with the disorderly random motion of molecules, but the molecules in the falling stone all have the same velocity downward in addition to their own random velocities. Thus, the more orderly kinetic energy of the stone as a whole is changed to disordered thermal energy when the stone strikes the ground. Disorder increases in this process, as it does in all processes that occur in nature.

*Biological Evolution

An interesting example of the increase in entropy relates to biological evolution and to growth of organisms. Clearly, a human being is a highly ordered organism. The theory of evolution describes the process from the early macromolecules and simple forms of life to *Homo sapiens*, which is a process of increasing order. So, too, the development of an individual from a single cell to a grown person is a process of increasing order. Do these processes violate the second law of thermodynamics? No, they do not. In the processes of evolution and growth, and even during the mature life of an individual, waste products are eliminated. These small molecules that remain as a result of metabolism are simple molecules without much order. Thus they represent relatively higher disorder or entropy. Indeed, the total entropy of the molecules cast aside by organisms during the processes of evolution and growth is greater than the decrease in entropy associated with the order of the growing individual or evolving species.

PHYSICS APPLIED
Biological Evolution and Development

20–8 Unavailability of Energy; Heat Death

In the process of heat conduction from a hot object to a cold one, we have seen that entropy increases and that order goes to disorder. The separate hot and cold objects could serve as the high- and low-temperature regions for a heat engine and thus could be used to obtain useful work. But after the two objects are put in contact with each other and reach the same uniform temperature, no work can be obtained from them. With regard to being able to do useful work, order has gone to disorder in this process.

The same can be said about a falling rock that comes to rest upon striking the ground. Before hitting the ground, all the kinetic energy of the rock could have been used to do useful work. But once the rock's mechanical kinetic energy becomes thermal energy, doing useful work is no longer possible.

Both these examples illustrate another important aspect of the second law of thermodynamics:

in any natural process, some energy becomes unavailable to do useful work.

In any process, no energy is ever lost (it is always conserved). Rather, energy becomes less useful—it can do less useful work. As time goes on, **energy is degraded**, in a sense; it goes from more orderly forms (such as mechanical) eventually to the least orderly form, internal, or thermal, energy. Entropy is a factor here because the amount of energy that becomes unavailable to do work is proportional to the change in entropy during any process.[†]

[†]It can be shown that the amount of energy that becomes unavailable to do useful work is equal to $T_L \Delta S$, where T_L is the lowest available temperature and ΔS is the total increase in entropy during the process.

A natural outcome of the degradation of energy is the prediction that as time goes on, the universe should approach a state of maximum disorder. Matter would become a uniform mixture, and heat will have flowed from high-temperature regions to low-temperature regions until the whole universe is at one temperature. No work could then be done. All the energy of the universe would have degraded to thermal energy. This prediction, called the **heat death** of the universe, has been much discussed, but would lie very far in the future. It is a complicated subject, and some scientists question whether thermodynamic modeling of the universe is possible or appropriate.

*20–9 Statistical Interpretation of Entropy and the Second Law

The ideas of entropy and disorder are made clearer with the use of a statistical or probabilistic analysis of the molecular state of a system. This statistical approach, which was first applied toward the end of the nineteenth century by Ludwig Boltzmann (1844–1906), makes a clear distinction between the "macrostate" and the "microstate" of a system. The **microstate** of a system would be specified in giving the position and velocity of every particle (or molecule). The **macrostate** of a system is specified by giving the macroscopic properties of the system—the temperature, pressure, number of moles, and so on. In reality, we can know only the macrostate of a system. We could not possibily know the velocity and position of every one of the huge number of molecules in a system at a given moment. Nonetheless, we can hypothesize a great many different microstates that can correspond to the *same* macrostate.

Let us take a very simple example. Suppose you repeatedly shake four coins in your hand and drop them on a table. Specifying the number of heads and the number of tails that appear on a given throw is the macrostate of this system. Specifying each coin as being a head or a tail is the microstate of the system. In the following Table we see how many microstates correspond to each macrostate:

Macrostate	Possible Microstates (H = heads, T = tails)	Number of Microstates
4 heads	H H H H	1
3 heads, 1 tail	H H H T, H H T H, H T H H, T H H H	4
2 heads, 2 tails	H H T T, H T H T, T H H T, H T T H, T H T H, T T H H	6
1 head, 3 tails	T T T H, T T H T, T H T T, H T T T	4
4 tails	T T T T	1

A basic assumption behind the statistical approach is that *each microstate is equally probable*. Thus the number of microstates that give the same macrostate corresponds to the relative probability of that macrostate occurring. The macrostate of two heads and two tails is the most probable one in our case of tossing four coins; out of the total of 16 possible microstates, six correspond to two heads and two tails, so the probability of throwing two heads and two tails is 6 out of 16, or 38%. The probability of throwing one head and three tails is 4 out of 16, or 25%. The probability of four heads is only 1 in 16, or 6%. If you threw the coins 16 times, you might not find that two heads and two tails appear exactly 6 times, or four tails exactly once. These are only probabilities or averages. But if you made 1600 throws, very nearly 38% of them would be two heads and two tails. The greater the number of tries, the closer the percentages are to the calculated probabilities.

EXERCISE E In the Table above, what is the probability that there will be at least two heads? (a) $\frac{1}{2}$; (b) $\frac{1}{16}$; (c) $\frac{1}{8}$; (d) $\frac{3}{8}$; (e) $\frac{11}{16}$.

If we toss more coins—say, 100 all at the same time—the relative probability of throwing all heads (or all tails) is greatly reduced. There is only one microstate corresponding to all heads. For 99 heads and 1 tail, there are 100 microstates since each of the coins could be the one tail. The relative probabilities for other macrostates are given in Table 20–1. About 1.3×10^{30} microstates are possible.[†] Thus the relative probability of finding all heads is 1 in 10^{30}, an incredibly unlikely event! The probability of obtaining 50 heads and 50 tails (see Table 20–1) is $(1.0 \times 10^{29})/1.3 \times 10^{30} = 0.08$ or 8%. The probability of obtaining anything between 45 and 55 heads is over 70%.

TABLE 20–1 Probabilities of Various Macrostates for 100 Coin Tosses

Macrostate		Number of Microstates	Probability
Heads	Tails		
100	0	1	7.9×10^{-31}
99	1	1.0×10^{2}	7.9×10^{-29}
90	10	1.7×10^{13}	1.4×10^{-17}
80	20	5.4×10^{20}	4.2×10^{-10}
60	40	1.4×10^{28}	0.01
55	45	6.1×10^{28}	0.05
50	50	1.0×10^{29}	0.08
45	55	6.1×10^{28}	0.05
40	60	1.4×10^{28}	0.01
20	80	5.4×10^{20}	4.2×10^{-10}
10	90	1.7×10^{13}	1.4×10^{-17}
1	99	1.0×10^{2}	7.9×10^{-29}
0	100	1	7.9×10^{-31}

Thus we see that as the number of coins increases, the probability of obtaining the most orderly arrangement (all heads or all tails) becomes extremely unlikely. The least orderly arrangement (half heads, half tails) is the most probable, and the probability of being within, say, 5% of the most probable arrangement greatly increases as the number of coins increases. These same ideas can be applied to the molecules of a system. For example, the most probable state of a gas (say, the air in a room) is one in which the molecules take up the whole space and move about randomly; this corresponds to the Maxwellian distribution, Fig. 20–14a (and see Section 18–2). On the other hand, the very orderly arrangement of all the molecules located in one corner of the room and all moving with the same velocity (Fig. 20–14b) is extremely unlikely.

From these examples, it is clear that probability is directly related to disorder and hence to entropy. That is, the most probable state is the one with greatest entropy or greatest disorder and randomness. Boltzmann showed, consistent with Clausius's definition $(dS = dQ/T)$, that the entropy of a system in a given (macro) state can be written

$$S = k \ln \mathcal{W}, \tag{20–14}$$

where k is Boltzmann's constant $(k = R/N_A = 1.38 \times 10^{-23} \text{ J/K})$ and $\mathcal{W}$ is the number of microstates corresponding to the given macrostate. That is, $\mathcal{W}$ is proportional to the probability of occurrence of that state. $\mathcal{W}$ is called the **thermodynamic probability**, or, sometimes, the disorder parameter.

[†]Each coin has two possibilities, heads or tails. Then the possible number of microstates is $2 \times 2 \times 2 \times \cdots = 2^{100} = 1.27 \times 10^{30}$ (using a calculator or logarithms).

FIGURE 20–14 (a) Most probable distribution of molecular speeds in a gas (Maxwellian, or random); (b) orderly, but highly unlikely, distribution of speeds in which all molecules have nearly the same speed.

(a)

(b)

EXAMPLE 20–9 **Free expansion—statistical determination of entropy.**
Use Eq. 20–14 to determine the change in entropy for the adiabatic *free expansion* of a gas, a calculation we did macroscopically in Example 20–7. Assume $\mathcal{W}$, the number of microstates for each macrostate, is the number of possible positions.

APPROACH We assume the number of moles is $n = 1$, and then the number of molecules is $N = nN_A = N_A$. We let the volume double, just as in Example 20–7. Because the volume doubles, the number of possible positions for each molecule doubles.

SOLUTION When the volume doubles, each molecule has two times as many positions (microstates) available. For two molecules, the number of total microstates increases by $2 \times 2 = 2^2$. For N_A molecules, the total number of microstates increases by a factor of $2 \times 2 \times 2 \times \cdots = 2^{N_A}$. That is

$$\frac{\mathcal{W}_2}{\mathcal{W}_1} = 2^{N_A}.$$

The change in entropy is, from Eq. 20–14,

$$\Delta S = S_2 - S_1 = k(\ln \mathcal{W}_2 - \ln \mathcal{W}_1) = k \ln \frac{\mathcal{W}_2}{\mathcal{W}_1} = k \ln 2^{N_A} = kN_A \ln 2 = R \ln 2$$

which is the same result we obtained in Example 20–7.

In terms of probability, the second law of thermodynamics—which tells us that entropy increases in any process—reduces to the statement that those processes occur which are most probable. The second law thus becomes a trivial statement. However there is an additional element now. The second law in terms of probability does not *forbid* a decrease in entropy. Rather, it says the probability is extremely low. It is not impossible that salt and pepper could separate spontaneously into layers, or that a broken tea cup could mend itself. It is even possible that a lake could freeze over on a hot summer day (that is, heat flow out of the cold lake into the warmer surroundings). But the probability for such events occurring is extremely small. In our coin examples, we saw that increasing the number of coins from 4 to 100 drastically reduced the probability of large deviations from the average, or most probable, arrangement. In ordinary systems we are not dealing with 100 molecules, but with incredibly large numbers of molecules: in 1 mol alone there are 6×10^{23} molecules. Hence the probability of deviation far from the average is incredibly tiny. For example, it has been calculated that the probability that a stone resting on the ground should transform 1 cal of thermal energy into mechanical energy and rise up into the air is much less likely than the probability that a group of monkeys typing randomly would by chance produce the complete works of Shakespeare.

*20–10 Thermodynamic Temperature; Third Law of Thermodynamics

In Section 20–3 we saw for a Carnot cycle that the ratio of the heat absorbed Q_H from the high-temperature reservoir and the heat exhausted Q_L to the low-temperature reservoir is directly related to the ratio of the temperatures of the two reservoirs (Eq. 20–2):

$$\frac{Q_L}{Q_H} = \frac{T_L}{T_H}.$$

This result is valid for any reversible engine and does not depend on the working substance. It can thus serve as the basis for the **Kelvin** or **thermodynamic temperature scale**.

We use this relation and the ideal gas temperature scale (Section 17–10) to complete the definition of the thermodynamic scale: we assign the value

$T_{tp} = 273.16 \, \text{K}$ to the triple point of water so that

$$T = (273.16 \, \text{K})\left(\frac{Q}{Q_{tp}}\right),$$

where Q and Q_{tp} are the magnitudes of the heats exchanged by a Carnot engine with reservoirs at temperatures T and T_{tp}. Then the thermodynamic scale is identical to the ideal gas scale over the latter's range of validity.

Very low temperatures are difficult to obtain experimentally. The closer the temperature is to absolute zero, the more difficult it is to reduce the temperature further, and it is generally accepted that *it is not possible to reach absolute zero in any finite number of processes*. This last statement is one way to state[†] the **third law of thermodynamics**. Since the maximum efficiency that any heat engine can have is the Carnot efficiency

$$e = 1 - \frac{T_L}{T_H},$$

and since T_L can never be zero, we see that a 100% efficient heat engine is not possible.

FIGURE 20–15 (a) An array of mirrors focuses sunlight on a boiler to produce steam at a solar energy installation. (b) A fossil-fuel steam plant (this one uses forest waste products, biomass). (c) Large cooling towers at an electric generating plant.

(a)

(b)

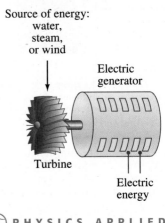
(c)

*20–11 Thermal Pollution, Global Warming, and Energy Resources

FIGURE 20–16 Mechanical energy is transformed to electric energy with a turbine and generator.

Much of the energy we utilize in everyday life—from motor vehicles to most of the electricity produced by power plants—makes use of a heat engine. Electricity produced by falling water at dams, by windmills, or by solar cells (Fig. 20–15a) does not involve a heat engine. But over 90% of the electric energy produced in the U.S. is generated at fossil-fuel steam plants (coal, oil, or gas—see Fig. 20–15b), and they make use of a heat engine (essentially steam engines). In electric power plants, the steam drives the turbines and generators (Fig. 20–16) whose output is electric energy. The various means to turn the turbine are discussed briefly in Table 20–2 (next page), along with some of the advantages and disadvantages of each. Even nuclear power plants use nuclear fuel to run a steam engine.

The heat output Q_L from every heat engine, from power plants to cars, is referred to as **thermal pollution** because this heat (Q_L) must be absorbed by the environment—such as by water from rivers or lakes, or by the air using large cooling towers (Fig. 20–15c). When water is the coolant, this heat raises the temperature of the water, altering the natural ecology of aquatic life (largely because warmer water holds less oxygen). In the case of air cooling towers, the output heat Q_L raises the temperature of the atmosphere, which affects the weather.

PHYSICS APPLIED
Heat engines and thermal pollution

[†] See also the statement in the footnote on p. 539.

TABLE 20–2 Electric Energy Resources

Form of Electric Energy Production	% of Production (approx.)		Advantages	Disadvantages
	U.S.	World		
Fossil-fuel steam plants: burn coal, oil, or natural gas to boil water, producing high-pressure steam that turns a turbine of a generator (Figs. 20–3b, 20–16); uses heat engine.	71	66	We know how to build them; for now relatively inexpensive.	Air pollution; thermal pollution; limited efficiency; land devastation from extraction of raw materials (mining); global warming; accidents such as oil spills at sea; limited fuel supply (estimates range from a couple of decades to a few centuries).
Nuclear energy:				
Fission: nuclei of uranium or plutonium atoms split ("fission") with release of energy (Chapter 42) that heats steam; uses heat engine.	20	16	Normally almost no air pollution; less contribution to global warming; relatively inexpensive.	Thermal pollution; accidents can release damaging radioactivity; difficult disposal of radioactive by-products; possible diversion of nuclear material by terrorists; limited fuel supply.
Fusion: energy released when isotopes of hydrogen (or other small nuclei) combine or "fuse" (Chapter 42).	0	0	Relatively "clean"; vast fuel supply (hydrogen in water molecules in oceans); less contribution to global warming.	Not yet workable.
Hydroelectric: falling water turns turbines at the base of a dam.	7	16	No heat engine needed; no air, water, or thermal pollution; relatively inexpensive; high efficiency; dams can control flooding.	Reservoirs behind dams inundate scenic or inhabited land; dams block upstream migration of salmon and other fish for reproduction; few locations remain for new dams; drought.
Geothermal: natural steam from inside the Earth comes to the surface (hot springs, geysers, steam vents); or cold water passed down into contact with hot, dry rock is heated to steam.	<1	<1	No heat engine needed; little air pollution; good efficiency; relatively inexpensive and "clean."	Few appropriate sites; small production; mineral content of spent hot water can pollute.
Wind power: 3-kW to 5-MW windmills (vanes up to 50 m wide) turn a generator.	<1	<1	No heat engine; no air, water, or thermal pollution; relatively inexpensive.	Large array of big windmills might affect weather and be eyesores; hazardous to migratory birds; winds not always strong.
Solar energy:	<1	<1		
Active solar heating: rooftop solar panels absorb the Sun's rays, which heat water in tubes for space heating and hot water supply.			No heat engine needed; no air or thermal pollution; unlimited fuel supply.	Space limitations; may require back-up; relatively expensive; less effective when cloudy.
Passive solar heating: architectural devices—windows along southern exposure, sunshade over windows to keep Sun's rays out in summer.			No heat engine needed; no air or thermal pollution; relatively inexpensive.	Almost none, but other methods needed too.
Solar cells (photovoltaic cells): convert sunlight directly into electricity without use of heat engine.			No heat engine; thermal, air, and water pollution very low; good efficiency (>30% and improving).	Expensive; chemical pollution at manufacture; large land area needed as Sun's energy not concentrated.

Air pollution—by which we mean the chemicals released in the burning of fossil fuels in cars, power plants, and industrial furnaces—gives rise to smog and other problems. Another much talked about issue is the buildup of CO_2 in the Earth's atmosphere due to the burning of fossil fuels. CO_2 absorbs some of the infrared radiation that the Earth naturally emits (Section 19–10) and thus can contribute to **global warming**. Limiting the burning of fossil fuels can help these problems.

Thermal pollution, however, is unavoidable. Engineers can try to design and build engines that are more efficient, but they cannot surpass the Carnot efficiency and must live with T_L being at best the ambient temperature of water or air. The second law of thermodynamics tells us the limit imposed by nature. What we can do, in the light of the second law of thermodynamics, is use less energy and conserve our fuel resources.

PROBLEM SOLVING

Thermodynamics

1. Define the **system** you are dealing with; distinguish the system under study from its surroundings.

2. Be careful of **signs** associated with **work** and **heat**. In the first law, work done *by* the system is positive; work done *on* the system is negative. Heat *added* to the system is positive, but heat *removed* from it is negative. With heat engines, we usually consider the heat intake, the heat exhausted, and the work done as positive.

3. Watch the **units** used for work and heat; work is most often expressed in joules, and heat can be in calories, kilocalories, or joules. Be consistent: choose only one unit for use throughout a given problem.

4. **Temperatures** must generally be expressed in kelvins; temperature *differences* may be expressed in C° or K.

5. **Efficiency** (or coefficient of performance) is a ratio of two energy transfers: useful output divided by required input. Efficiency (but *not* coefficient of performance) is always less than 1 in value, and hence is often stated as a percentage.

6. The **entropy** of a system increases when heat is added to the system, and decreases when heat is removed. If heat is transferred from system A to system B, the change in entropy of A is negative and the change in entropy of B is positive.

Summary

A **heat engine** is a device for changing thermal energy, by means of heat flow, into useful work.

The **efficiency** of a heat engine is defined as the ratio of the work W done by the engine to the heat input Q_H. Because of conservation of energy, the work output equals $Q_H - Q_L$, where Q_L is the heat exhausted to the environment; hence the efficiency

$$e = \frac{W}{Q_H} = 1 - \frac{Q_L}{Q_H}. \qquad (20\text{-}1)$$

Carnot's (idealized) engine consists of two isothermal and two adiabatic processes in a reversible cycle. For a **Carnot engine**, or any reversible engine operating between two temperatures, T_H and T_L (in kelvins), the efficiency is

$$e_{\text{ideal}} = 1 - \frac{T_L}{T_H}. \qquad (20\text{-}3)$$

Irreversible (real) engines always have an efficiency less than this.

The operation of **refrigerators** and **air conditioners** is the reverse of that of a heat engine: work is done to extract heat from a cool region and exhaust it to a region at a higher temperature. The coefficient of performance (COP) for either is

$$\text{COP} = \frac{Q_L}{W}, \qquad \begin{bmatrix} \text{refrigerator or} \\ \text{air conditioner} \end{bmatrix} \quad (20\text{-}4a)$$

where W is the work needed to remove heat Q_L from the area with the low temperature.

A **heat pump** does work W to bring heat Q_L from the cold outside and deliver heat Q_H to warm the interior. The coefficient of performance of a heat pump is

$$\text{COP} = \frac{Q_H}{W}. \qquad \text{[heat pump]} \quad (20\text{-}5)$$

The **second law of thermodynamics** can be stated in several equivalent ways:

(a) heat flows spontaneously from a hot object to a cold one, but not the reverse;

(b) there can be no 100% efficient heat engine—that is, one that can change a given amount of heat completely into work;

(c) natural processes tend to move toward a state of greater disorder or greater **entropy**.

Statement (c) is the most general statement of the second law of thermodynamics, and can be restated as: the total entropy, S, of any system plus that of its environment increases as a result of any natural process:

$$\Delta S > 0. \qquad (20\text{-}13)$$

Entropy, which is a state variable, is a quantitative measure of the disorder of a system. The change in entropy of a system during a reversible process is given by $\Delta S = \int dQ/T$.

The second law of thermodynamics tells us in which direction processes tend to proceed; hence entropy is called "time's arrow."

As time goes on, energy is degraded to less useful forms—that is, it is less available to do useful work.

[*All heat engines give rise to **thermal pollution** because they exhaust heat to the environment.]

Questions

1. Can mechanical energy ever be transformed completely into heat or internal energy? Can the reverse happen? In each case, if your answer is no, explain why not; if yes, give one or two examples.

2. Can you warm a kitchen in winter by leaving the oven door open? Can you cool the kitchen on a hot summer day by leaving the refrigerator door open? Explain.

3. Would a definition of heat engine efficiency as $e = W/Q_L$ be useful? Explain.

4. What plays the role of high-temperature and low-temperature areas in (a) an internal combustion engine, and (b) a steam engine? Are they, strictly speaking, heat reservoirs?

5. Which will give the greater improvement in the efficiency of a Carnot engine, a $10\,C°$ increase in the high-temperature reservoir, or a $10\,C°$ decrease in the low-temperature reservoir? Explain.

6. The oceans contain a tremendous amount of thermal (internal) energy. Why, in general, is it not possible to put this energy to useful work?

7. Discuss the factors that keep real engines from reaching Carnot efficiency.

8. The expansion valve in a refrigeration system, Fig. 20–10, is crucial for cooling the fluid. Explain how the cooling occurs.

9. Describe a process in nature that is nearly reversible.

10. (a) Describe how heat could be added to a system reversibly. (b) Could you use a stove burner to add heat to a system reversibly? Explain.

11. Suppose a gas expands to twice its original volume (a) adiabatically, (b) isothermally. Which process would result in a greater change in entropy? Explain.

12. Give three examples, other than those mentioned in this Chapter, of naturally occurring processes in which order goes to disorder. Discuss the observability of the reverse process.

13. Which do you think has the greater entropy, 1 kg of solid iron or 1 kg of liquid iron? Why?

14. (a) What happens if you remove the lid of a bottle containing chlorine gas? (b) Does the reverse process ever happen? Why or why not? (c) Can you think of two other examples of irreversibility?

15. You are asked to test a machine that the inventor calls an "in-room air conditioner": a big box, standing in the middle of the room, with a cable that plugs into a power outlet. When the machine is switched on, you feel a stream of cold air coming out of it. How do you know that this machine cannot cool the room?

16. Think up several processes (other than those already mentioned) that would obey the first law of thermodynamics, but, if they actually occurred, would violate the second law.

17. Suppose a lot of papers are strewn all over the floor; then you stack them neatly. Does this violate the second law of thermodynamics? Explain.

18. The first law of thermodynamics is sometimes whimsically stated as, "You can't get something for nothing," and the second law as, "You can't even break even." Explain how these statements could be equivalent to the formal statements.

19. Powdered milk is very slowly (quasistatically) added to water while being stirred. Is this a reversible process? Explain.

20. Two identical systems are taken from state a to state b by two different *irreversible* processes. Will the change in entropy for the system be the same for each process? For the environment? Answer carefully and completely.

21. It can be said that the *total change in entropy during a process is a measure of the irreversibility of the process.* Discuss why this is valid, starting with the fact that $\Delta S_{total} = \Delta S_{system} + \Delta S_{environment} = 0$ for a reversible process.

22. Use arguments, other than the principle of entropy increase, to show that for an adiabatic process, $\Delta S = 0$ if it is done reversibly and $\Delta S > 0$ if done irreversibly.

Problems

20–2 Heat Engines

1. (I) A heat engine exhausts 7800 J of heat while performing 2600 J of useful work. What is the efficiency of this engine?

2. (I) A certain power plant puts out 580 MW of electric power. Estimate the heat discharged per second, assuming that the plant has an efficiency of 35%.

3. (II) A typical compact car experiences a total drag force at 55 mi/h of about 350 N. If this car gets 35 miles per gallon of gasoline at this speed, and a liter of gasoline (1 gal = 3.8 L) releases about 3.2×10^7 J when burned, what is the car's efficiency?

4. (II) A four-cylinder gasoline engine has an efficiency of 0.22 and delivers 180 J of work per cycle per cylinder. The engine fires at 25 cycles per second. (a) Determine the work done per second. (b) What is the total heat input per second from the gasoline? (c) If the energy content of gasoline is 130 MJ per gallon, how long does one gallon last?

5. (II) The burning of gasoline in a car releases about 3.0×10^4 kcal/gal. If a car averages 38 km/gal when driving 95 km/h, which requires 25 hp, what is the efficiency of the engine under those conditions?

6. (II) Figure 20–17 is a PV diagram for a reversible heat engine in which 1.0 mol of argon, a nearly ideal monatomic gas, is initially at STP (point a). Points b and c are on an isotherm at $T = 423$ K. Process ab is at constant volume, process ac at constant pressure. (a) Is the path of the cycle carried out clockwise or counterclockwise? (b) What is the efficiency of this engine?

FIGURE 20–17
Problem 6.

7. (III) The operation of a *diesel engine* can be idealized by the cycle shown in Fig. 20–18. Air is drawn into the cylinder during the intake stroke (not part of the idealized cycle). The air is compressed adiabatically, path ab. At point b diesel fuel is injected into the cylinder which immediately burns since the temperature is very high. Combustion is slow, and during the first part of the power stroke, the gas expands at (nearly) constant pressure, path bc. After burning, the rest of the power stroke is adiabatic, path cd. Path da corresponds to the exhaust stroke. (*a*) Show that, for a quasistatic reversible engine undergoing this cycle using an ideal gas, the ideal efficiency is

$$e = 1 - \frac{(V_a/V_c)^{-\gamma} - (V_a/V_b)^{-\gamma}}{\gamma[(V_a/V_c)^{-1} - (V_a/V_b)^{-1}]},$$

where V_a/V_b is the "compression ratio", V_a/V_c is the "expansion ratio", and γ is defined by Eq. 19–14. (*b*) If $V_a/V_b = 16$ and $V_a/V_c = 4.5$, calculate the efficiency assuming the gas is diatomic (like N_2 and O_2) and ideal.

FIGURE 20–18
Problem 7.

20–3 Carnot Engine

8. (I) What is the maximum efficiency of a heat engine whose operating temperatures are 550°C and 365°C?

9. (I) It is not necessary that a heat engine's hot environment be hotter than ambient temperature. Liquid nitrogen (77 K) is about as cheap as bottled water. What would be the efficiency of an engine that made use of heat transferred from air at room temperature (293 K) to the liquid nitrogen "fuel" (Fig. 20–19)?

FIGURE 20–19
Problem 9.

10. (II) A heat engine exhausts its heat at 340°C and has a Carnot efficiency of 38%. What exhaust temperature would enable it to achieve a Carnot efficiency of 45%?

11. (II) (*a*) Show that the work done by a Carnot engine is equal to the area enclosed by the Carnot cycle on a *PV* diagram, Fig. 20–7. (See Section 19–7.) (*b*) Generalize this to any reversible cycle.

12. (II) A Carnot engine's operating temperatures are 210°C and 45°C. The engine's power output is 950 W. Calculate the rate of heat output.

13. (II) A nuclear power plant operates at 65% of its maximum theoretical (Carnot) efficiency between temperatures of 660°C and 330°C. If the plant produces electric energy at the rate of 1.2 GW, how much exhaust heat is discharged per hour?

14. (II) A Carnot engine performs work at the rate of 520 kW with an input of 950 kcal of heat per second. If the temperature of the heat source is 560°C, at what temperature is the waste heat exhausted?

15. (II) Assume that a 65 kg hiker needs 4.0×10^3 kcal of energy to supply a day's worth of metabolism. Estimate the maximum height the person can climb in one day, using only this amount of energy. As a rough prediction, treat the person as an isolated heat engine, operating between the internal temperature of 37°C (98.6°F) and the ambient air temperature of 20°C.

16. (II) A particular car does work at the rate of about 7.0 kJ/s when traveling at a steady 20.0 m/s along a level road. This is the work done against friction. The car can travel 17 km on 1 L of gasoline at this speed (about 40 mi/gal). What is the minimum value for T_H if T_L is 25°C? The energy available from 1 L of gas is 3.2×10^7 J.

17. (II) A heat engine utilizes a heat source at 580°C and has a Carnot efficiency of 32%. To increase the efficiency to 38%, what must be the temperature of the heat source?

18. (II) The working substance of a certain Carnot engine is 1.0 mol of an ideal monatomic gas. During the isothermal expansion portion of this engine's cycle, the volume of the gas doubles, while during the adiabatic expansion the volume increases by a factor of 5.7. The work output of the engine is 920 J in each cycle. Compute the temperatures of the two reservoirs between which this engine operates.

19. (III) A Carnot cycle, shown in Fig. 20–7, has the following conditions: $V_a = 7.5$ L, $V_b = 15.0$ L, $T_H = 470$°C, and $T_L = 260$°C. The gas used in the cycle is 0.50 mol of a diatomic gas, $\gamma = 1.4$. Calculate (*a*) the pressures at a and b; (*b*) the volumes at c and d. (*c*) What is the work done along process ab? (*d*) What is the heat lost along process cd? (*e*) Calculate the net work done for the whole cycle. (*f*) What is the efficiency of the cycle, using the definition $e = W/Q_H$? Show that this is the same as given by Eq. 20–3.

20. (III) One mole of monatomic gas undergoes a Carnot cycle with $T_H = 350$°C and $T_L = 210$°C. The initial pressure is 8.8 atm. During the isothermal expansion, the volume doubles. (*a*) Find the values of the pressure and volume at the points a, b, c, and d (see Fig. 20–7). (*b*) Determine Q, W, and ΔE_{int} for each segment of the cycle. (*c*) Calculate the efficiency of the cycle using Eqs. 20–1 and 20–3.

* 21. (III) In an engine that approximates the Otto cycle (Fig. 20–8), gasoline vapor must be ignited at the end of the cylinder's adiabatic compression by the spark from a spark plug. The ignition temperature of 87-octane gasoline vapor is about 430°C and, assuming that the working gas is diatomic and enters the cylinder at 25°C, determine the maximum compression ratio of the engine.

20–4 Refrigerators, Air Conditioners, Heat Pumps

22. (I) If an ideal refrigerator keeps its contents at 3.0°C when the house temperature is 22°C, what is its coefficient of performance?

23. (I) The low temperature of a freezer cooling coil is −15°C and the discharge temperature is 33°C. What is the maximum theoretical coefficient of performance?

24. (II) An ideal (Carnot) engine has an efficiency of 38%. If it were possible to run it backward as a heat pump, what would be its coefficient of performance?

25. (II) An ideal heat pump is used to maintain the inside temperature of a house at $T_{in} = 22°C$ when the outside temperature is T_{out}. Assume that when it is operating, the heat pump does work at a rate of 1500 W. Also assume that the house loses heat via conduction through its walls and other surfaces at a rate given by $(650 \text{ W/C°})(T_{in} - T_{out})$. (*a*) For what outside temperature would the heat pump have to operate at all times in order to maintain the house at an inside temperature of 22°C? (*b*) If the outside temperature is 8°C, what percentage of the time does the heat pump have to operate in order to maintain the house at an inside temperature of 22°C?

26. (II) A restaurant refrigerator has a coefficient of performance of 5.0. If the temperature in the kitchen outside the refrigerator is 32°C, what is the lowest temperature that could be obtained inside the refrigerator if it were ideal?

27. (II) A heat pump is used to keep a house warm at 22°C. How much work is required of the pump to deliver 3100 J of heat into the house if the outdoor temperature is (*a*) 0°C, (*b*) −15°C? Assume ideal (Carnot) behavior.

28. (II) (*a*) Given that the coefficient of performance of a refrigerator is defined (Eq. 20–4a) as

$$\text{COP} = \frac{Q_L}{W},$$

show that for an ideal (Carnot) refrigerator,

$$\text{COP}_{ideal} = \frac{T_L}{T_H - T_L}.$$

(*b*) Write the COP in terms of the efficiency *e* of the reversible heat engine obtained by running the refrigerator backward. (*c*) What is the coefficient of performance for an ideal refrigerator that maintains a freezer compartment at −18°C when the condenser's temperature is 24°C?

29. (II) A "Carnot" refrigerator (reverse of a Carnot engine) absorbs heat from the freezer compartment at a temperature of −17°C and exhausts it into the room at 25°C. (*a*) How much work must be done by the refrigerator to change 0.40 kg of water at 25°C into ice at −17°C? (*b*) If the compressor output is 180 W, what minimum time is needed to take 0.40 kg of 25°C water and freeze it at 0°C?

30. (II) A central heat pump operating as an air conditioner draws 33,000 Btu per hour from a building and operates between the temperatures of 24°C and 38°C. (*a*) If its coefficient of performance is 0.20 that of a Carnot air conditioner, what is the effective coefficient of performance? (*b*) What is the power (kW) required of the compressor motor? (*c*) What is the power in terms of hp?

31. (II) What volume of water at 0°C can a freezer make into ice cubes in 1.0 h, if the coefficient of performance of the cooling unit is 7.0 and the power input is 1.2 kilowatt?

20–5 and 20–6 Entropy

32. (I) What is the change in entropy of 250 g of steam at 100°C when it is condensed to water at 100°C?

33. (I) A 7.5-kg box having an initial speed of 4.0 m/s slides along a rough table and comes to rest. Estimate the total change in entropy of the universe. Assume all objects are at room temperature (293 K).

34. (I) What is the change in entropy of 1.00 m³ of water at 0°C when it is frozen to ice at 0°C?

35. (II) If 1.00 m³ of water at 0°C is frozen and cooled to −10°C by being in contact with a great deal of ice at −10°C, estimate the total change in entropy of the process.

36. (II) If 0.45 kg of water at 100°C is changed by a reversible process to steam at 100°C, determine the change in entropy of (*a*) the water, (*b*) the surroundings, and (*c*) the universe as a whole. (*d*) How would your answers differ if the process were irreversible?

37. (II) An aluminum rod conducts 9.50 cal/s from a heat source maintained at 225°C to a large body of water at 22°C. Calculate the rate at which entropy increases in this process.

38. (II) A 2.8-kg piece of aluminum at 43.0°C is placed in 1.0 kg of water in a Styrofoam container at room temperature (20°C). Estimate the net change in entropy of the system.

39. (II) An ideal gas expands isothermally ($T = 410 \text{ K}$) from a volume of 2.50 L and a pressure of 7.5 atm to a pressure of 1.0 atm. What is the entropy change for this process?

40. (II) When 2.0 kg of water at 12.0°C is mixed with 3.0 kg of water at 38.0°C in a well-insulated container, what is the change in entropy of the system? (*a*) Make an estimate; (*b*) use the integral $\Delta S = \int dQ/T$.

41. (II) (*a*) An ice cube of mass *m* at 0°C is placed in a large 20°C room. Heat flows (from the room to the ice cube) such that the ice cube melts and the liquid water warms to 20°C. The room is so large that its temperature remains nearly 20°C at all times. Calculate the change in entropy for the (water + room) system due to this process. Will this process occur naturally? (*b*) A mass *m* of liquid water at 20°C is placed in a large 20°C room. Heat flows (from the water to the room) such that the liquid water cools to 0°C and then freezes into a 0°C ice cube. The room is so large that its temperature remains 20°C at all times. Calculate the change in entropy for the (water + room) system due to this process. Will this process occur naturally?

42. (II) The temperature of 2.0 mol of an ideal diatomic gas goes from 25°C to 55°C at a constant volume. What is the change in entropy? Use $\Delta S = \int dQ/T$.

43. (II) Calculate the change in entropy of 1.00 kg of water when it is heated from 0°C to 75°C. (*a*) Make an estimate; (*b*) use the integral $\Delta S = \int dQ/T$. (*c*) Does the entropy of the surroundings change? If so, by how much?

44. (II) An ideal gas of *n* moles undergoes the reversible process ab shown in the PV diagram of Fig. 20–20. The temperature *T* of the gas is the same at points a and b. Determine the change in entropy of the gas due to this process.

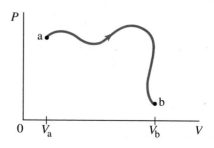

FIGURE 20–20
Problem 44.

45. (II) Two samples of an ideal gas are initially at the same temperature and pressure. They are each compressed reversibly from a volume V to volume $V/2$, one isothermally, the other adiabatically. (a) In which sample is the final pressure greater? (b) Determine the change in entropy of the gas for each process by integration. (c) What is the entropy change of the environment for each process?

46. (II) A 150-g insulated aluminum cup at 15°C is filled with 215 g of water at 100°C. Determine (a) the final temperature of the mixture, and (b) the total change in entropy as a result of the mixing process (use $\Delta S = \int dQ/T$).

47. (II) (a) Why would you expect the total entropy change in a Carnot cycle to be zero? (b) Do a calculation to show that it is zero.

48. (II) 1.00 mole of nitrogen (N_2) gas and 1.00 mole of argon (Ar) gas are in separate, equal-sized, insulated containers at the same temperature. The containers are then connected and the gases (assumed ideal) allowed to mix. What is the change in entropy (a) of the system and (b) of the environment? (c) Repeat part (a) but assume one container is twice as large as the other.

49. (II) Thermodynamic processes are sometimes represented on TS (temperature–entropy) diagrams, rather than PV diagrams. Determine the slope of a constant-volume process on a TS diagram when a system with n moles of an ideal gas with constant-volume molar specific heat C_V is at temperature T.

50. (III) The specific heat per mole of potassium at low temperatures is given by $C_V = aT + bT^3$, where $a = 2.08 \text{ mJ/mol·K}^2$ and $b = 2.57 \text{ mJ/mol·K}^4$. Determine (by integration) the entropy change of 0.15 mol of potassium when its temperature is lowered from 3.0 K to 1.0 K.

51. (III) Consider an ideal gas of n moles with molar specific heats C_V and C_P. (a) Starting with the first law, show that when the temperature and volume of this gas are changed by a reversible process, its change in entropy is given by

$$dS = nC_V \frac{dT}{T} + nR \frac{dV}{V}.$$

(b) Show that the expression in part (a) can be written as

$$dS = nC_V \frac{dP}{P} + nC_P \frac{dV}{V}.$$

(c) Using the expression from part (b), show that if $dS = 0$ for the reversible process (that is, the process is adiabatic), then $PV^\gamma = \text{constant}$, where $\gamma = C_P/C_V$.

20-8 Unavailability of Energy

52. (III) A general theorem states that the amount of energy that becomes unavailable to do useful work in any process is equal to $T_L \Delta S$, where T_L is the lowest temperature available and ΔS is the total change in entropy during the process. Show that this is valid in the specific cases of (a) a falling rock that comes to rest when it hits the ground; (b) the free adiabatic expansion of an ideal gas; and (c) the conduction of heat, Q, from a high-temperature (T_H) reservoir to a low-temperature (T_L) reservoir. [Hint: In part (c) compare to a Carnot engine.]

53. (III) Determine the work available in a 3.5-kg block of copper at 490 K if the surroundings are at 290 K. Use results of Problem 52.

*20-9 Statistical Interpretation of Entropy

*54. (I) Use Eq. 20–14 to determine the entropy of each of the five macrostates listed in the Table on page 546.

*55. (II) Suppose that you repeatedly shake six coins in your hand and drop them on the floor. Construct a table showing the number of microstates that correspond to each macrostate. What is the probability of obtaining (a) three heads and three tails and (b) six heads?

*56. (II) Calculate the relative probabilities, when you throw two dice, of obtaining (a) a 7, (b) an 11, (c) a 4.

*57. (II) (a) Suppose you have four coins, all with tails up. You now rearrange them so two heads and two tails are up. What was the change in entropy of the coins? (b) Suppose your system is the 100 coins of Table 20–1; what is the change in entropy of the coins if they are mixed randomly initially, 50 heads and 50 tails, and you arrange them so all 100 are heads? (c) Compare these entropy changes to ordinary thermodynamic entropy changes, such as Examples 20–6, 20–7, and 20–8.

*58. (III) Consider an isolated gas-like system consisting of a box that contains $N = 10$ distinguishable atoms, each moving at the same speed v. The number of unique ways that these atoms can be arranged so that N_L atoms are within the left-hand half of the box and N_R atoms are within the right-hand half of the box is given by $N!/N_L!N_R!$, where, for example, the factorial $4! = 4 \cdot 3 \cdot 2 \cdot 1$ (the only exception is that $0! = 1$). Define each unique arrangement of atoms within the box to be a microstate of this system. Now imagine the following two possible macrostates: state A where all of the atoms are within the left-hand half of the box and none are within the right-hand half; and state B where the distribution is uniform (that is, there is the same number in each half). See Fig. 20–21. (a) Assume the system is initially in state A and, at a later time, is found to be in state B. Determine the system's change in entropy. Can this process occur naturally? (b) Assume the system is initially in state B and, at a later time, is found to be in state A. Determine the system's change in entropy. Can this process occur naturally?

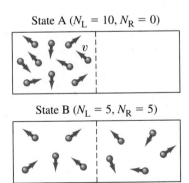

State A ($N_L = 10$, $N_R = 0$)

State B ($N_L = 5$, $N_R = 5$)

FIGURE 20–21 Problem 58.

*20-11 Energy Resources

*59. (II) Energy may be stored for use during peak demand by pumping water to a high reservoir when demand is low and then releasing it to drive turbines when needed. Suppose water is pumped to a lake 135 m above the turbines at a rate of $1.35 \times 10^5 \text{ kg/s}$ for 10.0 h at night. (a) How much energy (kWh) is needed to do this each night? (b) If all this energy is released during a 14-h day, at 75% efficiency, what is the average power output?

*60. (II) Solar cells (Fig. 20–22) can produce about 40 W of electricity per square meter of surface area if directly facing the Sun. How large an area is required to supply the needs of a house that requires 22 kWh/day? Would this fit on the roof of an average house? (Assume the Sun shines about 9 h/day.)

FIGURE 20–22 Problem 60.

*61. (II) Water is stored in an artificial lake created by a dam (Fig. 20–23). The water depth is 38 m at the dam, and a steady flow rate of 32 m^3/s is maintained through hydroelectric turbines installed near the base of the dam. How much electrical power can be produced?

FIGURE 20–23 Problem 61.

General Problems

62. It has been suggested that a heat engine could be developed that made use of the temperature difference between water at the surface of the ocean and water several hundred meters deep. In the tropics, the temperatures may be 27°C and 4°C, respectively. (a) What is the maximum efficiency such an engine could have? (b) Why might such an engine be feasible in spite of the low efficiency? (c) Can you imagine any adverse environmental effects that might occur?

63. A heat engine takes a diatomic gas around the cycle shown in Fig. 20–24. (a) Using the ideal gas law, determine how many moles of gas are in this engine. (b) Determine the temperature at point c. (c) Calculate the heat input into the gas during the constant volume process from points b to c. (d) Calculate the work done by the gas during the isothermal process from points a to b. (e) Calculate the work done by the gas during the adiabatic process from points c to a. (f) Determine the engine's efficiency. (g) What is the maximum efficiency possible for an engine working between T_a and T_c?

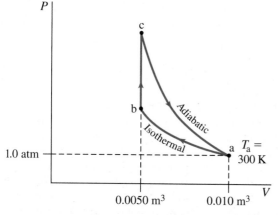

FIGURE 20–24 Problem 63.

64. A 126.5-g insulated aluminum cup at 18.00°C is filled with 132.5 g of water at 46.25°C. After a few minutes, equilibrium is reached. Determine (a) the final temperature, and (b) the total change in entropy.

65. (a) At a steam power plant, steam engines work in pairs, the heat output of the first one being the approximate heat input of the second. The operating temperatures of the first are 710°C and 430°C, and of the second 415°C and 270°C. If the heat of combustion of coal is 2.8×10^7 J/kg, at what rate must coal be burned if the plant is to put out 950 MW of power? Assume the efficiency of the engines is 65% of the ideal (Carnot) efficiency. (b) Water is used to cool the power plant. If the water temperature is allowed to increase by no more than 5.5 C°, estimate how much water must pass through the plant per hour.

66. (II) Refrigeration units can be rated in "tons." A 1-ton air conditioning system can remove sufficient energy to freeze 1 British ton (2000 pounds = 909 kg) of 0°C water into 0°C ice in one 24-h day. If, on a 35°C day, the interior of a house is maintained at 22°C by the continuous operation of a 5-ton air conditioning system, how much does this cooling cost the homeowner per hour? Assume the work done by the refrigeration unit is powered by electricity that costs $0.10 per kWh and that the unit's coefficient of performance is 15% that of an ideal refrigerator. 1 kWh = 3.60×10^6 J.

67. A 35% efficient power plant puts out 920 MW of electrical power. Cooling towers are used to take away the exhaust heat. (a) If the air temperature (15°C) is allowed to rise 7.0 C°, estimate what volume of air (km^3) is heated per day. Will the local climate be heated significantly? (b) If the heated air were to form a layer 150 m thick, estimate how large an area it would cover for 24 h of operation. Assume the air has density 1.2 kg/m^3 and that its specific heat is about 1.0 kJ/kg·C° at constant pressure.

68. (a) What is the coefficient of performance of an ideal heat pump that extracts heat from 11°C air outside and deposits heat inside your house at 24°C? (b) If this heat pump operates on 1400 W of electrical power, what is the maximum heat it can deliver into your house each hour?

69. The operation of a certain heat engine takes an ideal monatomic gas through a cycle shown as the rectangle on the PV diagram of Fig. 20–25. (a) Determine the efficiency of this engine. Let Q_H and Q_L be the total heat input and total heat exhausted during one cycle of this engine. (b) Compare (as a ratio) the efficiency of this engine to that of a Carnot engine operating between T_H and T_L, where T_H and T_L are the highest and lowest temperatures achieved.

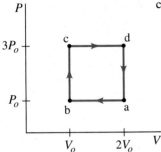

FIGURE 20–25
Problem 69.

70. A car engine whose output power is 155 hp operates at about 15% efficiency. Assume the engine's water temperature of 95°C is its cold-temperature (exhaust) reservoir and 495°C is its thermal "intake" temperature (the temperature of the exploding gas–air mixture). (a) What is the ratio of its efficiency relative to its maximum possible (Carnot) efficiency? (b) Estimate how much power (in watts) goes into moving the car, and how much heat, in joules and in kcal, is exhausted to the air in 1.0 h.

71. Suppose a power plant delivers energy at 850 MW using steam turbines. The steam goes into the turbines superheated at 625 K and deposits its unused heat in river water at 285 K. Assume that the turbine operates as an ideal Carnot engine. (a) If the river's flow rate is $34 \, \mathrm{m^3/s}$, estimate the average temperature increase of the river water immediately downstream from the power plant. (b) What is the entropy increase per kilogram of the downstream river water in J/kg·K?

72. 0.75 mole of an ideal monatomic gas at STP first undergoes an isothermal expansion so that the volume at b is 2.5 times the volume at a (Fig. 20–26). Next, heat is extracted at a constant volume so that the pressure drops. The gas is then compressed adiabatically back to the original state. (a) Calculate the pressures at b and c. (b) Determine the temperature at c. (c) Determine the work done, heat input or extracted, and the change in entropy for each process. (d) What is the efficiency of this cycle?

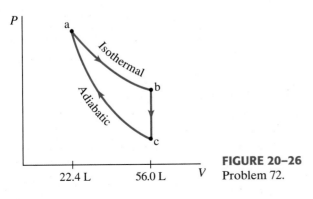

FIGURE 20–26
Problem 72.

73. Two 1100-kg cars are traveling 75 km/h in opposite directions when they collide and are brought to rest. Estimate the change in entropy of the universe as a result of this collision. Assume $T = 15°C$.

74. Metabolizing 1.0 kg of fat results in about 3.7×10^7 J of internal energy in the body. (a) In one day, how much fat does the body burn to maintain the body temperature of a person staying in bed and metabolizing at an average rate of 95 W? (b) How long would it take to burn 1.0-kg of fat this way assuming there is no food intake?

75. A cooling unit for a new freezer has an inner surface area of $6.0 \, \mathrm{m^2}$, and is bounded by walls 12 cm thick with a thermal conductivity of 0.050 W/m·K. The inside must be kept at $-10°C$ in a room that is at 20°C. The motor for the cooling unit must run no more than 15% of the time. What is the minimum power requirement of the cooling motor?

76. An ideal air conditioner keeps the temperature inside a room at 21°C when the outside temperature is 32°C. If 3.3 kW of power enters a room through the windows in the form of direct radiation from the Sun, how much electrical power would be saved if the windows were shaded so only 500 W came through them?

77. The *Stirling cycle*, shown in Fig. 20–27, is useful to describe external combustion engines as well as solar-power systems. Find the efficiency of the cycle in terms of the parameters shown, assuming a monatomic gas as the working substance. The processes ab and cd are isothermal whereas bc and da are at constant volume. How does it compare to the Carnot efficiency?

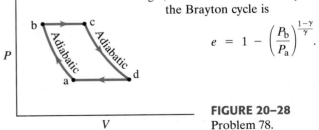

FIGURE 20–27
Problem 77.

78. A gas turbine operates under the *Brayton cycle*, which is depicted in the PV diagram of Fig. 20–28. In process ab the air–fuel mixture undergoes an adiabatic compression. This is followed, in process bc, with an isobaric (constant pressure) heating, by combustion. Process cd is an adiabatic expansion with expulsion of the products to the atmosphere. The return step, da, takes place at constant pressure. If the working gas behaves like an ideal gas, show that the efficiency of the Brayton cycle is

$$e = 1 - \left(\frac{P_b}{P_a} \right)^{\frac{1-\gamma}{\gamma}}.$$

FIGURE 20–28
Problem 78.

79. Thermodynamic processes can be represented not only on PV and PT diagrams; another useful one is a TS (temperature–entropy) diagram. (a) Draw a TS diagram for a Carnot cycle. (b) What does the area within the curve represent?

80. An aluminum can, with negligible heat capacity, is filled with 450 g of water at 0°C and then is brought into thermal contact with a similar can filled with 450 g of water at 50°C. Find the change in entropy of the system if no heat is allowed to exchange with the surroundings. Use $\Delta S = \int dQ/T$.

81. A dehumidifier is essentially a "refrigerator with an open door." The humid air is pulled in by a fan and guided to a cold coil, whose temperature is less than the dew point, and some of the air's water condenses. After this water is extracted, the air is warmed back to its original temperature and sent into the room. In a well-designed dehumidifier, the heat is exchanged between the incoming and outgoing air. Thus the heat that is removed by the refrigerator coil mostly comes from the condensation of water vapor to liquid. Estimate how much water is removed in 1.0 h by an ideal dehumidifier, if the temperature of the room is 25°C, the water condenses at 8°C, and the dehumidifier does work at the rate of 650 W of electrical power.

* **82.** A bowl contains a large number of red, orange, and green jelly beans. You are to make a line of three jelly beans. (*a*) Construct a table showing the number of microstates that correspond to each macrostate. Then determine the probability of (*b*) all 3 beans red, and (*c*) 2 greens, 1 orange.

* Numerical/Computer

* **83.** (II) At low temperature the specific heat of diamond varies with absolute temperature T according to the Debye equation $C_V = 1.88 \times 10^3 \left(T/T_D\right)^3$ J·mol^{-1}·K^{-1} where the Debye temperature for diamond is $T_D = 2230$ K. Use a spreadsheet and numerical integration to determine the entropy change of 1.00 mol of diamond when it is heated at constant volume from 4 K to 40 K. Your result should agree within 2% of the result obtained by integrating the expression for dS. [*Hint*: $dS = nC_V \, dT/T$, where n is the number of moles.]

Answers to Exercises

A: No. Efficiency makes no sense for a single process. It is defined (Eqs. 20–1) only for cyclic processes that return to the initial state.

B: (*d*).

C: (*c*).

D: 1220 J/K; −1220 J/K. (Note that the *total* entropy change, $\Delta S_{\text{ice}} + \Delta S_{\text{res}}$, is zero.)

E: (*e*).

Mathematical Formulas

A–1 Quadratic Formula

If $\qquad ax^2 + bx + c = 0$

then $\qquad x = \dfrac{-b \pm \sqrt{b^2 - 4ac}}{2a}$

A–2 Binomial Expansion

$$(1 \pm x)^n = 1 \pm nx + \frac{n(n-1)}{2!}x^2 \pm \frac{n(n-1)(n-2)}{3!}x^3 + \cdots$$

$$(x + y)^n = x^n\left(1 + \frac{y}{x}\right)^n = x^n\left(1 + n\frac{y}{x} + \frac{n(n-1)}{2!}\frac{y^2}{x^2} + \cdots\right)$$

A–3 Other Expansions

$$e^x = 1 + x + \frac{x^2}{2!} + \frac{x^3}{3!} + \cdots$$

$$\ln(1 + x) = x - \frac{x^2}{2} + \frac{x^3}{3} - \frac{x^4}{4} + \cdots$$

$$\sin\theta = \theta - \frac{\theta^3}{3!} + \frac{\theta^5}{5!} - \cdots$$

$$\cos\theta = 1 - \frac{\theta^2}{2!} + \frac{\theta^4}{4!} - \cdots$$

$$\tan\theta = \theta + \frac{\theta^3}{3} + \frac{2}{15}\theta^5 + \cdots \qquad |\theta| < \frac{\pi}{2}$$

In general: $\quad f(x) = f(0) + \left(\dfrac{df}{dx}\right)_0 x + \left(\dfrac{d^2f}{dx^2}\right)_0 \dfrac{x^2}{2!} + \cdots$

A–4 Exponents

$$(a^n)(a^m) = a^{n+m} \qquad\qquad \frac{1}{a^n} = a^{-n}$$

$$(a^n)(b^n) = (ab)^n \qquad\qquad a^n a^{-n} = a^0 = 1$$

$$(a^n)^m = a^{nm} \qquad\qquad a^{\frac{1}{2}} = \sqrt{a}$$

A–5 Areas and Volumes

Object	Surface area	Volume
Circle, radius r	πr^2	—
Sphere, radius r	$4\pi r^2$	$\frac{4}{3}\pi r^3$
Right circular cylinder, radius r, height h	$2\pi r^2 + 2\pi rh$	$\pi r^2 h$
Right circular cone, radius r, height h	$\pi r^2 + \pi r\sqrt{r^2 + h^2}$	$\frac{1}{3}\pi r^2 h$

A–6 Plane Geometry

1. *Equal angles:*

FIGURE A–1 If line a_1 is parallel to line a_2, then $\theta_1 = \theta_2$.

2. *Equal angles:*

FIGURE A–2 If $a_1 \perp a_2$ and $b_1 \perp b_2$, then $\theta_1 = \theta_2$.

3. The sum of the angles in any plane triangle is 180°.

4. *Pythagorean theorem:*

FIGURE A–3

In any right triangle (one angle = 90°) of sides a, b, and c:

$$a^2 + b^2 = c^2$$

where c is the length of the hypotenuse (opposite the 90° angle).

5. *Similar triangles:* Two triangles are said to be similar if all three of their angles are equal (in Fig. A–4, $\theta_1 = \phi_1$, $\theta_2 = \phi_2$, and $\theta_3 = \phi_3$). Similar triangles can have different sizes and different orientations.

(*a*) Two triangles are similar if any two of their angles are equal. (This follows because the third angles must also be equal since the sum of the angles of a triangle is 180°.)

(*b*) The ratios of corresponding sides of two similar triangles are equal (Fig. A–4):

$$\frac{a_1}{b_1} = \frac{a_2}{b_2} = \frac{a_3}{b_3}.$$

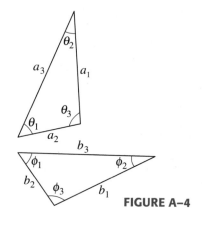

FIGURE A–4

6. *Congruent triangles:* Two triangles are congruent if one can be placed precisely on top of the other. That is, they are similar triangles and they have the same size. Two triangles are congruent if any of the following holds:

(*a*) The three corresponding sides are equal.

(*b*) Two sides and the enclosed angle are equal ("side-angle-side").

(*c*) Two angles and the enclosed side are equal ("angle-side-angle").

A–7 Logarithms

Logarithms are defined in the following way:

$$\text{if} \quad y = A^x, \quad \text{then} \quad x = \log_A y.$$

That is, the logarithm of a number y to the base A is that number which, as the exponent of A, gives back the number y. For **common logarithms**, the base is 10, so

$$\text{if} \quad y = 10^x, \quad \text{then} \quad x = \log y.$$

The subscript 10 on $\log_{10}$ is usually omitted when dealing with common logs. Another important base is the exponential base $e = 2.718\dots$, a natural number. Such logarithms are called **natural logarithms** and are written ln. Thus,

$$\text{if} \quad y = e^x, \quad \text{then} \quad x = \ln y.$$

For any number y, the two types of logarithm are related by

$$\ln y = 2.3026 \log y.$$

Some simple rules for logarithms are as follows:

$$\log(ab) = \log a + \log b, \tag{i}$$

which is true because if $a = 10^n$ and $b = 10^m$, then $ab = 10^{n+m}$. From the definition of logarithm, $\log a = n$, $\log b = m$, and $\log(ab) = n + m$; hence, $\log(ab) = n + m = \log a + \log b$. In a similar way, we can show that

$$\log\left(\frac{a}{b}\right) = \log a - \log b \tag{ii}$$

and

$$\log a^n = n \log a. \tag{iii}$$

These three rules apply to any kind of logarithm.

If you do not have a calculator that calculates logs, you can easily use a **log table**, such as the small one shown here (Table A–1): the number N whose log we want is given to two digits. The first digit is in the vertical column to the left, the second digit is in the horizontal row across the top. For example, Table A–1 tells us that $\log 1.0 = 0.000$, $\log 1.1 = 0.041$, and $\log 4.1 = 0.613$. Table A–1 does not include the decimal point. The Table gives logs for numbers between 1.0 and 9.9. For larger or smaller numbers, we use rule (i) above, $\log(ab) = \log a + \log b$. For example, $\log(380) = \log(3.8 \times 10^2) = \log(3.8) + \log(10^2)$. From the Table, $\log 3.8 = 0.580$; and from rule (iii) above $\log(10^2) = 2 \log(10) = 2$, since $\log(10) = 1$. [This follows from the definition of the logarithm: if $10 = 10^1$, then $1 = \log(10)$.] Thus,

$$\begin{aligned}\log(380) &= \log(3.8) + \log(10^2) \\ &= 0.580 + 2 \\ &= 2.580.\end{aligned}$$

Similarly,

$$\begin{aligned}\log(0.081) &= \log(8.1) + \log(10^{-2}) \\ &= 0.908 - 2 = -1.092.\end{aligned}$$

The reverse process of finding the number N whose log is, say, 2.670, is called "taking the **antilogarithm**." To do so, we separate our number 2.670 into two parts, making the separation at the decimal point:

$$\begin{aligned}\log N &= 2.670 = 2 + 0.670 \\ &= \log 10^2 + 0.670.\end{aligned}$$

We now look at Table A–1 to see what number has its log equal to 0.670; none does, so we must **interpolate**: we see that $\log 4.6 = 0.663$ and $\log 4.7 = 0.672$. So the number we want is between 4.6 and 4.7, and closer to the latter by $\frac{7}{9}$. Approximately we can say that $\log 4.68 = 0.670$. Thus

$$\begin{aligned}\log N &= 2 + 0.670 \\ &= \log(10^2) + \log(4.68) = \log(4.68 \times 10^2),\end{aligned}$$

so $N = 4.68 \times 10^2 = 468$.

If the given logarithm is negative, say, -2.180, we proceed as follows:

$$\begin{aligned}\log N &= -2.180 = -3 + 0.820 \\ &= \log 10^{-3} + \log 6.6 = \log 6.6 \times 10^{-3},\end{aligned}$$

so $N = 6.6 \times 10^{-3}$. Notice that we added to our given logarithm the next largest integer (3 in this case) so that we have an integer, plus a decimal number between 0 and 1.0 whose antilogarithm can be looked up in the Table.

TABLE A–1 Short Table of Common Logarithms

N	0.0	0.1	0.2	0.3	0.4	0.5	0.6	0.7	0.8	0.9
1	000	041	079	114	146	176	204	230	255	279
2	301	322	342	362	380	398	415	431	447	462
3	477	491	505	519	531	544	556	568	580	591
4	602	613	623	633	643	653	663	672	681	690
5	699	708	716	724	732	740	748	756	763	771
6	778	785	792	799	806	813	820	826	833	839
7	845	851	857	863	869	875	881	886	892	898
8	903	908	914	919	924	929	935	940	944	949
9	954	959	964	968	973	978	982	987	991	996

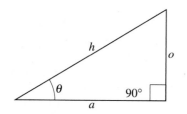

FIGURE A–5

A–8 Vectors

Vector addition is covered in Sections 3–2 to 3–5.
Vector multiplication is covered in Sections 3–3, 7–2, and 11–2.

A–9 Trigonometric Functions and Identities

The trigonometric functions are defined as follows (see Fig. A–5, o = side opposite, a = side adjacent, h = hypotenuse. Values are given in Table A–2):

$$\sin \theta = \frac{o}{h} \qquad \csc \theta = \frac{1}{\sin \theta} = \frac{h}{o}$$

$$\cos \theta = \frac{a}{h} \qquad \sec \theta = \frac{1}{\cos \theta} = \frac{h}{a}$$

$$\tan \theta = \frac{o}{a} = \frac{\sin \theta}{\cos \theta} \qquad \cot \theta = \frac{1}{\tan \theta} = \frac{a}{o}$$

and recall that

$$a^2 + o^2 = h^2 \qquad \text{[Pythagorean theorem]}.$$

Figure A–6 shows the signs (+ or −) that cosine, sine, and tangent take on for angles θ in the four quadrants (0° to 360°). Note that angles are measured counterclockwise from the x axis as shown; negative angles are measured from *below* the x axis, clockwise: for example, $-30° = +330°$, and so on.

The following are some useful identities among the trigonometric functions:

$$\sin^2 \theta + \cos^2 \theta = 1$$

$$\sec^2 \theta - \tan^2 \theta = 1, \quad \csc^2 \theta - \cot^2 \theta = 1$$

$$\sin 2\theta = 2 \sin \theta \cos \theta$$

$$\cos 2\theta = \cos^2 \theta - \sin^2 \theta = 2 \cos^2 \theta - 1 = 1 - 2 \sin^2 \theta$$

$$\tan 2\theta = \frac{2 \tan \theta}{1 - \tan^2 \theta}$$

$$\sin(A \pm B) = \sin A \cos B \pm \cos A \sin B$$

$$\cos(A \pm B) = \cos A \cos B \mp \sin A \sin B$$

$$\tan(A \pm B) = \frac{\tan A \pm \tan B}{1 \mp \tan A \tan B}$$

$$\sin(180° - \theta) = \sin \theta$$

$$\cos(180° - \theta) = -\cos \theta$$

$$\sin(90° - \theta) = \cos \theta$$

$$\cos(90° - \theta) = \sin \theta$$

$$\sin(-\theta) = -\sin \theta$$

$$\cos(-\theta) = \cos \theta$$

$$\tan(-\theta) = -\tan \theta$$

$$\sin\tfrac{1}{2}\theta = \sqrt{\frac{1 - \cos \theta}{2}}, \quad \cos\tfrac{1}{2}\theta = \sqrt{\frac{1 + \cos \theta}{2}}, \quad \tan\tfrac{1}{2}\theta = \sqrt{\frac{1 - \cos \theta}{1 + \cos \theta}}$$

$$\sin A \pm \sin B = 2 \sin\left(\frac{A \pm B}{2}\right) \cos\left(\frac{A \mp B}{2}\right).$$

For any triangle (see Fig. A–7):

$$\frac{\sin \alpha}{a} = \frac{\sin \beta}{b} = \frac{\sin \gamma}{c} \qquad \text{[Law of sines]}$$

$$c^2 = a^2 + b^2 - 2ab \cos \gamma. \qquad \text{[Law of cosines]}$$

Values of sine, cosine, tangent are given in Table A–2.

FIGURE A–6

First Quadrant
(0° to 90°)
$x > 0$
$y > 0$

$\sin \theta = y/r > 0$
$\cos \theta = x/r > 0$
$\tan \theta = y/x > 0$

Second Quadrant
(90° to 180°)
$x < 0$
$y > 0$

$\sin \theta > 0$
$\cos \theta < 0$
$\tan \theta < 0$

Third Quadrant
(180° to 270°)
$x < 0$
$y < 0$

$\sin \theta < 0$
$\cos \theta < 0$
$\tan \theta > 0$

Fourth Quadrant
(270° to 360°)
$x > 0$
$y < 0$

$\sin \theta < 0$
$\cos \theta > 0$
$\tan \theta < 0$

FIGURE A–7

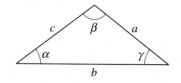

TABLE A–2 Trigonometric Table: Numerical Values of Sin, Cos, Tan

Angle in Degrees	Angle in Radians	Sine	Cosine	Tangent	Angle in Degrees	Angle in Radians	Sine	Cosine	Tangent
0°	0.000	0.000	1.000	0.000					
1°	0.017	0.017	1.000	0.017	46°	0.803	0.719	0.695	1.036
2°	0.035	0.035	0.999	0.035	47°	0.820	0.731	0.682	1.072
3°	0.052	0.052	0.999	0.052	48°	0.838	0.743	0.669	1.111
4°	0.070	0.070	0.998	0.070	49°	0.855	0.755	0.656	1.150
5°	0.087	0.087	0.996	0.087	50°	0.873	0.766	0.643	1.192
6°	0.105	0.105	0.995	0.105	51°	0.890	0.777	0.629	1.235
7°	0.122	0.122	0.993	0.123	52°	0.908	0.788	0.616	1.280
8°	0.140	0.139	0.990	0.141	53°	0.925	0.799	0.602	1.327
9°	0.157	0.156	0.988	0.158	54°	0.942	0.809	0.588	1.376
10°	0.175	0.174	0.985	0.176	55°	0.960	0.819	0.574	1.428
11°	0.192	0.191	0.982	0.194	56°	0.977	0.829	0.559	1.483
12°	0.209	0.208	0.978	0.213	57°	0.995	0.839	0.545	1.540
13°	0.227	0.225	0.974	0.231	58°	1.012	0.848	0.530	1.600
14°	0.244	0.242	0.970	0.249	59°	1.030	0.857	0.515	1.664
15°	0.262	0.259	0.966	0.268	60°	1.047	0.866	0.500	1.732
16°	0.279	0.276	0.961	0.287	61°	1.065	0.875	0.485	1.804
17°	0.297	0.292	0.956	0.306	62°	1.082	0.883	0.469	1.881
18°	0.314	0.309	0.951	0.325	63°	1.100	0.891	0.454	1.963
19°	0.332	0.326	0.946	0.344	64°	1.117	0.899	0.438	2.050
20°	0.349	0.342	0.940	0.364	65°	1.134	0.906	0.423	2.145
21°	0.367	0.358	0.934	0.384	66°	1.152	0.914	0.407	2.246
22°	0.384	0.375	0.927	0.404	67°	1.169	0.921	0.391	2.356
23°	0.401	0.391	0.921	0.424	68°	1.187	0.927	0.375	2.475
24°	0.419	0.407	0.914	0.445	69°	1.204	0.934	0.358	2.605
25°	0.436	0.423	0.906	0.466	70°	1.222	0.940	0.342	2.747
26°	0.454	0.438	0.899	0.488	71°	1.239	0.946	0.326	2.904
27°	0.471	0.454	0.891	0.510	72°	1.257	0.951	0.309	3.078
28°	0.489	0.469	0.883	0.532	73°	1.274	0.956	0.292	3.271
29°	0.506	0.485	0.875	0.554	74°	1.292	0.961	0.276	3.487
30°	0.524	0.500	0.866	0.577	75°	1.309	0.966	0.259	3.732
31°	0.541	0.515	0.857	0.601	76°	1.326	0.970	0.242	4.011
32°	0.559	0.530	0.848	0.625	77°	1.344	0.974	0.225	4.331
33°	0.576	0.545	0.839	0.649	78°	1.361	0.978	0.208	4.705
34°	0.593	0.559	0.829	0.675	79°	1.379	0.982	0.191	5.145
35°	0.611	0.574	0.819	0.700	80°	1.396	0.985	0.174	5.671
36°	0.628	0.588	0.809	0.727	81°	1.414	0.988	0.156	6.314
37°	0.646	0.602	0.799	0.754	82°	1.431	0.990	0.139	7.115
38°	0.663	0.616	0.788	0.781	83°	1.449	0.993	0.122	8.144
39°	0.681	0.629	0.777	0.810	84°	1.466	0.995	0.105	9.514
40°	0.698	0.643	0.766	0.839	85°	1.484	0.996	0.087	11.43
41°	0.716	0.656	0.755	0.869	86°	1.501	0.998	0.070	14.301
42°	0.733	0.669	0.743	0.900	87°	1.518	0.999	0.052	19.081
43°	0.750	0.682	0.731	0.933	88°	1.536	0.999	0.035	28.636
44°	0.768	0.695	0.719	0.966	89°	1.553	1.000	0.017	57.290
45°	0.785	0.707	0.707	1.000	90°	1.571	1.000	0.000	∞

B Derivatives and Integrals

B–1 Derivatives: General Rules

(See also Section 2–3.)

$$\frac{dx}{dx} = 1$$

$$\frac{d}{dx}[af(x)] = a\frac{df}{dx} \qquad (a = \text{constant})$$

$$\frac{d}{dx}[f(x) + g(x)] = \frac{df}{dx} + \frac{dg}{dx}$$

$$\frac{d}{dx}[f(x)g(x)] = \frac{df}{dx}g + f\frac{dg}{dx}$$

$$\frac{d}{dx}[f(y)] = \frac{df}{dy}\frac{dy}{dx} \qquad [\text{chain rule}]$$

$$\frac{dx}{dy} = \frac{1}{\left(\dfrac{dy}{dx}\right)} \qquad \text{if } \frac{dy}{dx} \neq 0.$$

B–2 Derivatives: Particular Functions

$$\frac{da}{dx} = 0 \qquad (a = \text{constant})$$

$$\frac{d}{dx}x^n = nx^{n-1}$$

$$\frac{d}{dx}\sin ax = a\cos ax$$

$$\frac{d}{dx}\cos ax = -a\sin ax$$

$$\frac{d}{dx}\tan ax = a\sec^2 ax$$

$$\frac{d}{dx}\ln ax = \frac{1}{x}$$

$$\frac{d}{dx}e^{ax} = ae^{ax}$$

B–3 Indefinite Integrals: General Rules

(See also Section 7–3.)

$$\int dx = x$$

$$\int af(x)\,dx = a\int f(x)\,dx \qquad (a = \text{constant})$$

$$\int [f(x) + g(x)]\,dx = \int f(x)\,dx + \int g(x)\,dx$$

$$\int u\,dv = uv - \int v\,du \qquad (\text{integration by parts})$$

B–4 Indefinite Integrals: Particular Functions

(An arbitrary constant can be added to the right side of each equation.)

$$\int a\,dx = ax \qquad (a = \text{constant})$$

$$\int x^m\,dx = \frac{1}{m+1}x^{m+1} \qquad (m \neq -1)$$

$$\int \sin ax\,dx = -\frac{1}{a}\cos ax$$

$$\int \cos ax\,dx = \frac{1}{a}\sin ax$$

$$\int \tan ax\,dx = \frac{1}{a}\ln|\sec ax|$$

$$\int \frac{1}{x}\,dx = \ln x$$

$$\int e^{ax}\,dx = \frac{1}{a}e^{ax}$$

$$\int \frac{dx}{x^2+a^2} = \frac{1}{a}\tan^{-1}\frac{x}{a}$$

$$\int \frac{dx}{x^2-a^2} = \frac{1}{2a}\ln\left(\frac{x-a}{x+a}\right) \qquad (x^2 > a^2)$$

$$= -\frac{1}{2a}\ln\left(\frac{a+x}{a-x}\right) \qquad (x^2 < a^2)$$

$$\int \frac{dx}{\sqrt{x^2\pm a^2}} = \ln\left(x + \sqrt{x^2\pm a^2}\right)$$

$$\int \frac{dx}{\sqrt{a^2-x^2}} = \sin^{-1}\left(\frac{x}{a}\right) = -\cos^{-1}\left(\frac{x}{a}\right) \qquad \text{if } x^2 \leq a^2$$

$$\int \frac{dx}{(x^2\pm a^2)^{\frac{3}{2}}} = \frac{\pm x}{a^2\sqrt{x^2\pm a^2}}$$

$$\int \frac{x\,dx}{(x^2\pm a^2)^{\frac{3}{2}}} = \frac{-1}{\sqrt{x^2\pm a^2}}$$

$$\int \sin^2 ax\,dx = \frac{x}{2} - \frac{\sin 2ax}{4a}$$

$$\int xe^{-ax}\,dx = -\frac{e^{-ax}}{a^2}(ax+1)$$

$$\int x^2e^{-ax}\,dx = -\frac{e^{-ax}}{a^3}(a^2x^2 + 2ax + 2)$$

B–5 A Few Definite Integrals

$$\int_0^\infty x^n e^{-ax}\,dx = \frac{n!}{a^{n+1}} \qquad \int_0^\infty x^2 e^{-ax^2}\,dx = \sqrt{\frac{\pi}{16a^3}}$$

$$\int_0^\infty e^{-ax^2}\,dx = \sqrt{\frac{\pi}{4a}} \qquad \int_0^\infty x^3 e^{-ax^2}\,dx = \frac{1}{2a^2}$$

$$\int_0^\infty xe^{-ax^2}\,dx = \frac{1}{2a} \qquad \int_0^\infty x^{2n} e^{-ax^2}\,dx = \frac{1\cdot 3\cdot 5\cdots(2n-1)}{2^{n+1}a^n}\sqrt{\frac{\pi}{a}}$$

More on Dimensional Analysis

An important use of dimensional analysis (Section 1–7) is to obtain the *form* of an equation: how one quantity depends on others. To take a concrete example, let us try to find an expression for the period T of a simple pendulum. First, we try to figure out what T could depend on, and make a list of these variables. It might depend on its length ℓ, on the mass m of the bob, on the angle of swing θ, and on the acceleration due to gravity, g. It might also depend on air resistance (we would use the viscosity of air), the gravitational pull of the Moon, and so on; but everyday experience suggests that the Earth's gravity is the major force involved, so we ignore the other possible forces. So let us assume that T is a function of ℓ, m, θ, and g, and that each of these factors is present to some power:

$$T = C\ell^w m^x \theta^y g^z.$$

C is a dimensionless constant, and w, x, y, and z are exponents we want to solve for. We now write down the dimensional equation (Section 1–7) for this relationship:

$$[T] = [L]^w[M]^x[L/T^2]^z.$$

Because θ has no dimensions (a radian is a length divided by a length—see Eq. 10–1a), it does not appear. We simplify and obtain

$$[T] = [L]^{w+z}[M]^x[T]^{-2z}$$

To have dimensional consistency, we must have

$$1 = -2z$$
$$0 = w + z$$
$$0 = x.$$

We solve these equations and find that $z = -\frac{1}{2}$, $w = \frac{1}{2}$, and $x = 0$. Thus our desired equation must be

$$T = C\sqrt{\ell/g}\, f(\theta), \tag{C–1}$$

where $f(\theta)$ is some function of θ that we cannot determine using this technique. Nor can we determine in this way the dimensionless constant C. (To obtain C and f, we would have to do an analysis such as that in Chapter 14 using Newton's laws, which reveals that $C = 2\pi$ and $f \approx 1$ for small θ). But look what we *have* found, using only dimensional consistency. We obtained the form of the expression that relates the period of a simple pendulum to the major variables of the situation, ℓ and g (see Eq. 14–12c), and saw that it does not depend on the mass m.

How did we do it? And how useful is this technique? Basically, we had to use our intuition as to which variables were important and which were not. This is not always easy, and often requires a lot of insight. As to usefulness, the final result in our example could have been obtained from Newton's laws, as in Chapter 14. But in many physical situations, such a derivation from other laws cannot be done. In those situations, dimensional analysis can be a powerful tool.

In the end, any expression derived by the use of dimensional analysis (or by any other means, for that matter) must be checked against experiment. For example, in our derivation of Eq. C–1, we can compare the periods of two pendulums of different lengths, ℓ_1 and ℓ_2, whose amplitudes (θ) are the same. For, using Eq. C–1, we would have

$$\frac{T_1}{T_2} = \frac{C\sqrt{\ell_1/g}\, f(\theta)}{C\sqrt{\ell_2/g}\, f(\theta)} = \sqrt{\frac{\ell_1}{\ell_2}}.$$

Because C and $f(\theta)$ are the same for both pendula, they cancel out, so we can experimentally determine if the ratio of the periods varies as the ratio of the square roots of the lengths. This comparison to experiment checks our derivation, at least in part; C and $f(\theta)$ could be determined by further experiments.

D Gravitational Force due to a Spherical Mass Distribution

In Chapter 6 we stated that the gravitational force exerted by or on a uniform sphere acts as if all the mass of the sphere were concentrated at its center, if the other object (exerting or feeling the force) is outside the sphere. In other words, the gravitational force that a uniform sphere exerts on a particle outside it is

$$F = G\frac{mM}{r^2}, \qquad [m \text{ outside sphere of mass } M]$$

where m is the mass of the particle, M the mass of the sphere, and r the distance of m from the center of the sphere. Now we will derive this result. We will use the concepts of infinitesimally small quantities and integration.

First we consider a very thin, uniform spherical shell (like a thin-walled basketball) of mass M whose thickness t is small compared to its radius R (Fig. D–1). The force on a particle of mass m at a distance r from the center of the shell can be calculated as the vector sum of the forces due to all the particles of the shell. We imagine the shell divided up into thin (infinitesimal) circular strips so that all points on a strip are equidistant from our particle m. One of these circular strips, labeled AB, is shown in Fig. D–1. It is $R\,d\theta$ wide, t thick, and has a radius $R\sin\theta$. The force on our particle m due to a tiny piece of the strip at point A is represented by the vector $\vec{F}_A$ shown. The force due to a tiny piece of the strip at point B, which is diametrically opposite A, is the force $\vec{F}_B$. We take the two pieces at A and B to be of equal mass, so $F_A = F_B$. The horizontal components of $\vec{F}_A$ and $\vec{F}_B$ are each equal to

$$F_A \cos\phi$$

and point toward the center of the shell. The vertical components of $\vec{F}_A$ and $\vec{F}_B$ are of equal magnitude and point in opposite directions, and so cancel. Since for every point on the strip there is a corresponding point diametrically opposite (as with A and B), we see that the net force due to the entire strip points toward the center of the shell. Its magnitude will be

$$dF = G\frac{m\,dM}{\ell^2}\cos\phi,$$

where dM is the mass of the entire circular strip and ℓ is the distance from all points on the strip to m, as shown. We write dM in terms of the density ρ; by density we mean the mass per unit volume (Section 13–2). Hence, $dM = \rho\,dV$, where dV is the volume of the strip and equals $(2\pi R\sin\theta)(t)(R\,d\theta)$. Then the force dF due to the circular strip shown is

$$dF = G\frac{m\rho 2\pi R^2 t \sin\theta\, d\theta}{\ell^2}\cos\phi. \qquad \textbf{(D–1)}$$

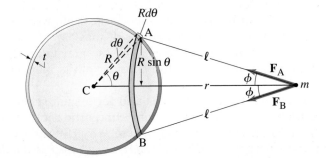

FIGURE D–1 Calculating the gravitational force on a particle of mass m due to a uniform spherical shell of radius R and mass M.

A-9

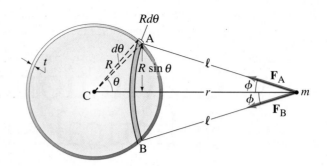

FIGURE D–1 (repeated)
Calculating the gravitational force on a particle of mass m due to a uniform spherical shell of radius R and mass M.

To get the total force F that the entire shell exerts on the particle m, we must integrate over all the circular strips: that is, we integrate

$$dF = G\frac{m\rho 2\pi R^2 t \sin\theta\, d\theta}{\ell^2}\cos\phi \qquad (\textbf{D–1})$$

from $\theta = 0°$ to $\theta = 180°$. But our expression for dF contains ℓ and ϕ, which are functions of θ. From Fig. D–1 we can see that

$$\ell\cos\phi = r - R\cos\theta.$$

Furthermore, we can write the law of cosines for triangle CmA:

$$\cos\theta = \frac{r^2 + R^2 - \ell^2}{2rR}. \qquad (\textbf{D–2})$$

With these two expressions we can reduce our three variables (ℓ, θ, ϕ) to only one, which we take to be ℓ. We do two things with Eq. D–2: (1) We put it into the equation for $\ell\cos\phi$ above:

$$\cos\phi = \frac{1}{\ell}(r - R\cos\theta) = \frac{r^2 + \ell^2 - R^2}{2r\ell}.$$

and (2) we take the differential of both sides of Eq. D–2 (because $\sin\theta\, d\theta$ appears in the expression for dF, Eq. D–1), considering r and R to be constants when summing over the strips:

$$-\sin\theta\, d\theta = -\frac{2\ell\, d\ell}{2rR} \qquad \text{or} \qquad \sin\theta\, d\theta = \frac{\ell\, d\ell}{rR}.$$

We insert these into Eq. D–1 for dF and find

$$dF = Gm\rho\pi t\frac{R}{r^2}\left(1 + \frac{r^2 - R^2}{\ell^2}\right)d\ell.$$

Now we integrate to get the net force on our thin shell of radius R. To integrate over all the strips ($\theta = 0°$ to $180°$), we must go from $\ell = r - R$ to $\ell = r + R$ (see Fig. D–1). Thus,

$$F = Gm\rho\pi t\frac{R}{r^2}\left[\ell - \frac{r^2 - R^2}{\ell}\right]_{\ell = r - R}^{\ell = r + R}$$

$$= Gm\rho\pi t\frac{R}{r^2}(4R).$$

The volume V of the spherical shell is its area $(4\pi R^2)$ times the thickness t. Hence the mass $M = \rho V = \rho 4\pi R^2 t$, and finally

$$F = G\frac{mM}{r^2}. \qquad \left[\begin{array}{c}\text{particle of mass } m \text{ outside a}\\ \text{thin uniform spherical shell of mass } M\end{array}\right]$$

This result gives us the force a thin shell exerts on a particle of mass m a distance r from the center of the shell, and *outside* the shell. We see that the force is the same as that between m and a particle of mass M at the center of the shell. In other words, for purposes of calculating the gravitational force exerted on or by a uniform spherical shell, we can consider all its mass concentrated at its center.

What we have derived for a shell holds also for a solid sphere, since a solid sphere can be considered as made up of many concentric shells, from $R = 0$ to $R = R_0$, where R_0 is the radius of the solid sphere. Why? Because if each shell has mass dM, we write for each shell, $dF = Gm\, dM/r^2$, where r is the distance from the center C to mass m and is the same for all shells. Then the total force equals the sum or integral over dM, which gives the total mass M. Thus the result

$$F = G\frac{mM}{r^2} \qquad \begin{bmatrix} \text{particle of mass } m \text{ outside} \\ \text{solid sphere of mass } M \end{bmatrix} \quad \textbf{(D–3)}$$

is valid for a solid sphere of mass M even if the density varies with distance from the center. (It is not valid if the density varies within each shell—that is, depends not only on R.) Thus the gravitational force exerted on or by spherical objects, including nearly spherical objects like the Earth, Sun, and Moon, can be considered to act as if the objects were point particles.

This result, Eq. D–3, is true only if the mass m is outside the sphere. Let us next consider a point mass m that is located inside the spherical shell of Fig. D–1. Here, r would be less than R, and the integration over ℓ would be from $\ell = R - r$ to $\ell = R + r$, so

$$\left[\ell - \frac{r^2 - R^2}{\ell} \right]_{R-r}^{R+r} = 0.$$

Thus the force on any mass inside the shell would be zero. This result has particular importance for the electrostatic force, which is also an inverse square law. For the gravitational situation, we see that at points within a solid sphere, say 1000 km below the Earth's surface, only the mass up to that radius contributes to the net force. The outer shells beyond the point in question contribute zero net gravitational effect.

The results we have obtained here can also be reached using the gravitational analog of Gauss's law for electrostatics (Chapter 22).

Differential Form of Maxwell's Equations

Maxwell's equations can be written in another form that is often more convenient than Eqs. 31–5. This material is usually covered in more advanced courses, and is included here simply for completeness.

We quote here two theorems, without proof, that are derived in vector analysis textbooks. The first is called **Gauss's theorem** or the **divergence theorem**. It relates the integral over a surface of any vector function $\vec{\mathbf{F}}$ to a volume integral over the volume enclosed by the surface:

$$\oint_{\text{Area } A} \vec{\mathbf{F}} \cdot d\vec{\mathbf{A}} = \int_{\text{Volume } V} \vec{\mathbf{\nabla}} \cdot \vec{\mathbf{F}} \, dV.$$

The operator $\vec{\mathbf{\nabla}}$ is the **del operator**, defined in Cartesian coordinates as

$$\vec{\mathbf{\nabla}} = \hat{\mathbf{i}} \frac{\partial}{\partial x} + \hat{\mathbf{j}} \frac{\partial}{\partial y} + \hat{\mathbf{k}} \frac{\partial}{\partial z}.$$

The quantity

$$\vec{\mathbf{\nabla}} \cdot \vec{\mathbf{F}} = \frac{\partial F_x}{\partial x} + \frac{\partial F_y}{\partial y} + \frac{\partial F_z}{\partial z}$$

is called the **divergence** of $\vec{\mathbf{F}}$. The second theorem is **Stokes's theorem**, and relates a line integral around a closed path to a surface integral over any surface enclosed by that path:

$$\oint_{\text{Line}} \vec{\mathbf{F}} \cdot d\vec{\boldsymbol{\ell}} = \int_{\text{Area } A} \vec{\mathbf{\nabla}} \times \vec{\mathbf{F}} \cdot d\vec{\mathbf{A}}.$$

The quantity $\vec{\mathbf{\nabla}} \times \vec{\mathbf{F}}$ is called the **curl** of $\vec{\mathbf{F}}$. (See Section 11–2 on the vector product.)

We now use these two theorems to obtain the differential form of Maxwell's equations in free space. We apply Gauss's theorem to Eq. 31–5a (Gauss's law):

$$\oint_A \vec{\mathbf{E}} \cdot d\vec{\mathbf{A}} = \int \vec{\mathbf{\nabla}} \cdot \vec{\mathbf{E}} \, dV = \frac{Q}{\epsilon_0}.$$

Now the charge Q can be written as a volume integral over the charge density ρ: $Q = \int \rho \, dV.$ Then

$$\int \vec{\mathbf{\nabla}} \cdot \vec{\mathbf{E}} \, dV = \frac{1}{\epsilon_0} \int \rho \, dV.$$

Both sides contain volume integrals over the same volume, and for this to be true over *any* volume, whatever its size or shape, the integrands must be equal:

$$\vec{\mathbf{\nabla}} \cdot \vec{\mathbf{E}} = \frac{\rho}{\epsilon_0}. \tag{E–1}$$

This is the differential form of Gauss's law. The second of Maxwell's equations, $\oint \vec{\mathbf{B}} \cdot d\vec{\mathbf{A}} = 0,$ is treated in the same way, and we obtain

$$\vec{\mathbf{\nabla}} \cdot \vec{\mathbf{B}} = 0. \tag{E–2}$$

Next, we apply Stokes's theorem to the third of Maxwell's equations,

$$\oint \vec{\mathbf{E}} \cdot d\vec{\boldsymbol{\ell}} \;=\; \int \vec{\boldsymbol{\nabla}} \times \vec{\mathbf{E}} \cdot d\vec{\mathbf{A}} \;=\; -\frac{d\Phi_B}{dt}.$$

Since the magnetic flux $\Phi_B = \int \vec{\mathbf{B}} \cdot d\vec{\mathbf{A}}$, we have

$$\int \vec{\boldsymbol{\nabla}} \times \vec{\mathbf{E}} \cdot d\vec{\mathbf{A}} \;=\; -\frac{\partial}{\partial t} \int \vec{\mathbf{B}} \cdot d\vec{\mathbf{A}}$$

where we use the partial derivative, $\partial\vec{\mathbf{B}}/\partial t$, since B may also depend on position. These are surface integrals over the same area, and to be true over any area, even a very small one, we must have

$$\vec{\boldsymbol{\nabla}} \times \vec{\mathbf{E}} \;=\; -\frac{\partial\vec{\mathbf{B}}}{\partial t}. \qquad\qquad \textbf{(E–3)}$$

This is the third of Maxwell's equations in differential form. Finally, to the last of Maxwell's equations,

$$\oint \vec{\mathbf{B}} \cdot d\vec{\boldsymbol{\ell}} \;=\; \mu_0 I + \mu_0 \epsilon_0 \frac{d\Phi_E}{dt},$$

we apply Stokes's theorem and write $\Phi_E = \int \vec{\mathbf{E}} \cdot d\vec{\mathbf{A}}$:

$$\int \vec{\boldsymbol{\nabla}} \times \vec{\mathbf{B}} \cdot d\vec{\mathbf{A}} \;=\; \mu_0 I + \mu_0 \epsilon_0 \frac{\partial}{\partial t} \int \vec{\mathbf{E}} \cdot d\vec{\mathbf{A}}.$$

The conduction current I can be written in terms of the current density $\vec{\mathbf{j}}$, using Eq. 25–12:

$$I \;=\; \int \vec{\mathbf{j}} \cdot d\vec{\mathbf{A}}.$$

Then Maxwell's fourth equation becomes:

$$\int \vec{\boldsymbol{\nabla}} \times \vec{\mathbf{B}} \cdot d\vec{\mathbf{A}} \;=\; \mu_0 \int \vec{\mathbf{j}} \cdot d\vec{\mathbf{A}} + \mu_0 \epsilon_0 \frac{\partial}{\partial t} \int \vec{\mathbf{E}} \cdot d\vec{\mathbf{A}}.$$

For this to be true over any area A, whatever its size or shape, the integrands on each side of the equation must be equal:

$$\vec{\boldsymbol{\nabla}} \times \vec{\mathbf{B}} \;=\; \mu_0 \vec{\mathbf{j}} + \mu_0 \epsilon_0 \frac{\partial\vec{\mathbf{E}}}{\partial t}. \qquad\qquad \textbf{(E–4)}$$

Equations E–1, 2, 3, and 4 are Maxwell's equations in differential form for free space. They are summarized in Table E–1.

TABLE E–1 Maxwell's Equations in Free Space[†]	
Integral form	**Differential form**
$\oint \vec{\mathbf{E}} \cdot d\vec{\mathbf{A}} = \dfrac{Q}{\epsilon_0}$	$\vec{\boldsymbol{\nabla}} \cdot \vec{\mathbf{E}} = \dfrac{\rho}{\epsilon_0}$
$\oint \vec{\mathbf{B}} \cdot d\vec{\mathbf{A}} = 0$	$\vec{\boldsymbol{\nabla}} \cdot \vec{\mathbf{B}} = 0$
$\oint \vec{\mathbf{E}} \cdot d\vec{\boldsymbol{\ell}} = -\dfrac{d\Phi_B}{dt}$	$\vec{\boldsymbol{\nabla}} \times \vec{\mathbf{E}} = -\dfrac{\partial\vec{\mathbf{B}}}{\partial t}$
$\oint \vec{\mathbf{B}} \cdot d\vec{\boldsymbol{\ell}} = \mu_0 I + \mu_0 \epsilon_0 \dfrac{d\Phi_E}{dt}$	$\vec{\boldsymbol{\nabla}} \times \vec{\mathbf{B}} = \mu_0 \vec{\mathbf{j}} + \mu_0 \epsilon_0 \dfrac{\partial\vec{\mathbf{E}}}{\partial t}$

[†]$\vec{\boldsymbol{\nabla}}$ stands for the *del operator* $\vec{\boldsymbol{\nabla}} = \hat{\mathbf{i}}\dfrac{\partial}{\partial x} + \hat{\mathbf{j}}\dfrac{\partial}{\partial y} + \hat{\mathbf{k}}\dfrac{\partial}{\partial z}$ in Cartesian coordinates.

Selected Isotopes

(1) Atomic Number Z	(2) Element	(3) Symbol	(4) Mass Number A	(5) Atomic Mass†	(6) % Abundance (or Radioactive Decay‡ Mode)	(7) Half-life (if radioactive)
0	(Neutron)	n	1	1.008665	β^-	10.23 min
1	Hydrogen	H	1	1.007825	99.9885%	
	Deuterium	d or D	2	2.014102	0.0115%	
	Tritium	t or T	3	3.016049	β^-	12.33 yr
2	Helium	He	3	3.016029	0.000137%	
			4	4.002603	99.999863%	
3	Lithium	Li	6	6.015123	7.59%	
			7	7.016005	92.41%	
4	Beryllium	Be	7	7.016930	EC, γ	53.22 days
			9	9.012182	100%	
5	Boron	B	10	10.012937	19.9%	
			11	11.009305	80.1%	
6	Carbon	C	11	11.011434	β^+, EC	20.39 min
			12	12.000000	98.93%	
			13	13.003355	1.07%	
			14	14.003242	β^-	5730 yr
7	Nitrogen	N	13	13.005739	β^+, EC	9.965 min
			14	14.003074	99.632%	
			15	15.000109	0.368%	
8	Oxygen	O	15	15.003066	β^+, EC	122.24 s
			16	15.994915	99.757%	
			18	17.999161	0.205%	
9	Fluorine	F	19	18.998403	100%	
10	Neon	Ne	20	19.992440	90.48%	
			22	21.991385	9.25%	
11	Sodium	Na	22	21.994436	β^+, EC, γ	2.6027 yr
			23	22.989769	100%	
			24	23.990963	β^-, γ	14.959 h
12	Magnesium	Mg	24	23.985042	78.99%	
13	Aluminum	Al	27	26.981539	100%	
14	Silicon	Si	28	27.976927	92.2297%	
			31	30.975363	β^-, γ	157.3 min
15	Phosphorus	P	31	30.973762	100%	
			32	31.973907	β^-	14.262 days

† The masses given in column (5) are those for the neutral atom, including the Z electrons.

‡ Chapter 41; EC = electron capture.

(1) Atomic Number Z	(2) Element	(3) Symbol	(4) Mass Number A	(5) Atomic Mass	(6) % Abundance (or Radioactive Decay Mode)	(7) Half-life (if radioactive)
16	Sulfur	S	32	31.972071	94.9%	
			35	34.969032	β^-	87.51 days
17	Chlorine	Cl	35	34.968853	75.78%	
			37	36.965903	24.22%	
18	Argon	Ar	40	39.962383	99.600%	
19	Potassium	K	39	38.963707	93.258%	
			40	39.963998	0.0117%	
					β^-, EC, γ, β^+	1.248×10^9 yr
20	Calcium	Ca	40	39.962591	96.94%	
21	Scandium	Sc	45	44.955912	100%	
22	Titanium	Ti	48	47.947946	73.72%	
23	Vanadium	V	51	50.943960	99.750%	
24	Chromium	Cr	52	51.940508	83.789%	
25	Manganese	Mn	55	54.938045	100%	
26	Iron	Fe	56	55.934938	91.75%	
27	Cobalt	Co	59	58.933195	100%	
			60	59.933817	β^-, γ	5.2711 yr
28	Nickel	Ni	58	57.935343	68.077%	
			60	59.930786	26.223%	
29	Copper	Cu	63	62.929598	69.17%	
			65	64.927790	30.83%	
30	Zinc	Zn	64	63.929142	48.6%	
			66	65.926033	27.9%	
31	Gallium	Ga	69	68.925574	60.108%	
32	Germanium	Ge	72	71.922076	27.5%	
			74	73.921178	36.3%	
33	Arsenic	As	75	74.921596	100%	
34	Selenium	Se	80	79.916521	49.6%	
35	Bromine	Br	79	78.918337	50.69%	
36	Krypton	Kr	84	83.911507	57.00%	
37	Rubidium	Rb	85	84.911790	72.17%	
38	Strontium	Sr	86	85.909260	9.86%	
			88	87.905612	82.58%	
			90	89.907738	β^-	28.79 yr
39	Yttrium	Y	89	88.905848	100%	
40	Zirconium	Zr	90	89.904704	51.4%	
41	Niobium	Nb	93	92.906378	100%	
42	Molybdenum	Mo	98	97.905408	24.1%	
43	Technetium	Tc	98	97.907216	β^-, γ	4.2×10^6 yr
44	Ruthenium	Ru	102	101.904349	31.55%	
45	Rhodium	Rh	103	102.905504	100%	
46	Palladium	Pd	106	105.903486	27.33%	
47	Silver	Ag	107	106.905097	51.839%	
			109	108.904752	48.161%	
48	Cadmium	Cd	114	113.903359	28.7%	
49	Indium	In	115	114.903878	95.71%; β^-	4.41×10^{14} yr
50	Tin	Sn	120	119.902195	32.58%	
51	Antimony	Sb	121	120.903816	57.21%	

(1) Atomic Number Z	(2) Element	(3) Symbol	(4) Mass Number A	(5) Atomic Mass	(6) % Abundance (or Radioactive Decay Mode)	(7) Half-life (if radioactive)
52	Tellurium	Te	130	129.906224	34.1%; $\beta^-\beta^-$	$>7.9 \times 10^{22}$ yr
53	Iodine	I	127	126.904473	100%	
			131	130.906125	β^-, γ	8.0207 days
54	Xenon	Xe	132	131.904154	26.89%	
			136	135.907219	8.87%; $\beta^-\beta^-$	$>3.6 \times 10^{20}$ yr
55	Cesium	Cs	133	132.905452	100%	
56	Barium	Ba	137	136.905827	11.232%	
			138	137.905247	71.70%	
57	Lanthanum	La	139	138.906353	99.910%	
58	Cerium	Ce	140	139.905439	88.45%	
59	Praseodymium	Pr	141	140.907653	100%	
60	Neodymium	Nd	142	141.907723	27.2%	
61	Promethium	Pm	145	144.912749	EC, α	17.7 yr
62	Samarium	Sm	152	151.919732	26.75%	
63	Europium	Eu	153	152.921230	52.19%	
64	Gadolinium	Gd	158	157.924104	24.84%	
65	Terbium	Tb	159	158.925347	100%	
66	Dysprosium	Dy	164	163.929175	28.2%	
67	Holmium	Ho	165	164.930322	100%	
68	Erbium	Er	166	165.930293	33.6%	
69	Thulium	Tm	169	168.934213	100%	
70	Ytterbium	Yb	174	173.938862	31.8%	
71	Lutetium	Lu	175	174.940772	97.41%	
72	Hafnium	Hf	180	179.946550	35.08%	
73	Tantalum	Ta	181	180.947996	99.988%	
74	Tungsten (wolfram)	W	184	183.950931	30.64%; α	$>3 \times 10^{17}$ yr
75	Rhenium	Re	187	186.955753	62.60%; β^-	4.35×10^{10} yr
76	Osmium	Os	191	190.960930	β^-, γ	15.4 days
			192	191.961481	40.78%	
77	Iridium	Ir	191	190.960594	37.3%	
			193	192.962926	62.7%	
78	Platinum	Pt	195	194.964791	33.832%	
79	Gold	Au	197	196.966569	100%	
80	Mercury	Hg	199	198.968280	16.87%	
			202	201.970643	29.9%	
81	Thallium	Tl	205	204.974428	70.476%	
82	Lead	Pb	206	205.974465	24.1%	
			207	206.975897	22.1%	
			208	207.976652	52.4%	
			210	209.984188	β^-, γ, α	22.20 yr
			211	210.988737	β^-, γ	36.1 min
			212	211.991898	β^-, γ	10.64 h
			214	213.999805	β^-, γ	26.8 min
83	Bismuth	Bi	209	208.980399	100%	
			211	210.987269	α, γ, β^-	2.14 min
84	Polonium	Po	210	209.982874	α, γ, EC	138.376 days
			214	213.995201	α, γ	164.3 μs
85	Astatine	At	218	218.008694	α, β^-	1.5 s

(1) Atomic Number Z	(2) Element	(3) Symbol	(4) Mass Number A	(5) Atomic Mass	(6) % Abundance (or Radioactive Decay Mode)	(7) Half-life (if radioactive)
86	Radon	Rn	222	222.017578	α, γ	3.8235 days
87	Francium	Fr	223	223.019736	β^-, γ, α	22.00 min
88	Radium	Ra	226	226.025410	α, γ	1600 yr
89	Actinium	Ac	227	227.027752	β^-, γ, α	21.772 yr
90	Thorium	Th	228	228.028741	α, γ	1.9116 yr
			232	232.038055	100%; α, γ	1.405×10^{10} yr
91	Protactinium	Pa	231	231.035884	α, γ	3.276×10^4 yr
92	Uranium	U	232	232.037156	α, γ	68.9 yr
			233	233.039635	α, γ	1.592×10^5 yr
			235	235.043930	0.720%; α, γ	7.04×10^8 yr
			236	236.045568	α, γ	2.342×10^7 yr
			238	238.050788	99.274%; α, γ	4.468×10^9 yr
			239	239.054293	β^-, γ	23.45 min
93	Neptunium	Np	237	237.048173	α, γ	2.144×10^6 yr
			239	239.052939	β^-, γ	2.356 days
94	Plutonium	Pu	239	239.052163	α, γ	24,110 yr
			244	244.064204	α	8.00×10^7 yr
95	Americium	Am	243	243.061381	α, γ	7370 yr
96	Curium	Cm	247	247.070354	α, γ	1.56×10^7 yr
97	Berkelium	Bk	247	247.070307	α, γ	1380 yr
98	Californium	Cf	251	251.079587	α, γ	898 yr
99	Einsteinium	Es	252	252.082980	α, EC, γ	471.7 days
100	Fermium	Fm	257	257.095105	α, γ	100.5 days
101	Mendelevium	Md	258	258.098431	α, γ	51.5 days
102	Nobelium	No	259	259.10103	α, EC	58 min
103	Lawrencium	Lr	262	262.10963	α, EC, fission	≈ 4 h
104	Rutherfordium	Rf	263	263.11255	fission	10 min
105	Dubnium	Db	262	262.11408	α, fission, EC	35 s
106	Seaborgium	Sg	266	266.12210	α, fission	≈ 21 s
107	Bohrium	Bh	264	264.12460	α	≈ 0.44 s
108	Hassium	Hs	269	269.13406	α	≈ 10 s
109	Meitnerium	Mt	268	268.13870	α	21 ms
110	Darmstadtium	Ds	271	271.14606	α	≈ 70 ms
111	Roentgenium	Rg	272	272.15360	α	3.8 ms
112		Uub	277	277.16394	α	≈ 0.7 ms

Preliminary evidence (unconfirmed) has been reported for elements 113, 114, 115, 116 and 118.

Answers to Odd-Numbered Problems

CHAPTER 1

1. (a) 1.4×10^{10} y;
   (b) 4.4×10^{17} s.
3. (a) 1.156×10^0;
   (b) 2.18×10^1;
   (c) 6.8×10^{-3};
   (d) 3.2865×10^2;
   (e) 2.19×10^{-1};
   (f) 4.44×10^2.
5. 4.6%.
7. 1.00×10^5 s.
9. 0.24 rad.
11. (a) 0.2866 m;
    (b) 0.000085 V;
    (c) 0.00076 kg;
    (d) 0.0000000000600 s;
    (e) 0.0000000000000225 m;
    (f) 2,500,000,000 V.
13. $5'10'' = 1.8$ m, 165 lbs $= 75.2$ kg.
15. (a) $\dfrac{0.111 \text{ yd}^2}{1 \text{ ft}^2}$;
    (b) $\dfrac{10.8 \text{ ft}^2}{1 \text{ m}^2}$.
17. (a) 3.9×10^{-9} in.;
    (b) 1.0×10^8 atoms.
19. (a) $\dfrac{0.621 \text{ mi/h}}{1 \text{ km/h}}$;
    (b) $\dfrac{3.28 \text{ ft/s}}{1 \text{ m/s}}$;
    (c) $\dfrac{0.278 \text{ m/s}}{1 \text{ km/h}}$.
21. (a) 9.46×10^{15} m;
    (b) 6.31×10^4 AU;
    (c) 7.20 AU/h.
23. (a) 3.80×10^{13} m^2;
    (b) 13.4.
25. 6×10^5 books.
27. 5×10^4 L.
29. (a) 1800.
31. 5×10^4 m.
33. 6.5×10^6 m.
35. $[M/L^3]$.
37. (a) Cannot;
    (b) can;
    (c) can.
39. (1×10^{-5})%, 8 significant figures.

41. (a) 3.16×10^7 s;
    (b) 3.16×10^{16} ns;
    (c) 3.17×10^{-8} y.
43. 2×10^{-4} m.
45. 1×10^{11} gal/y.
47. 9 cm/y.
49. 2×10^9 kg/y.
51. 75 min.
53. 4×10^5 metric tons, 1×10^8 gal.
55. 1×10^3 days
57. 210 yd, 190 m.
59. (a) 0.10 nm;
    (b) 1.0×10^5 fm;
    (c) 1.0×10^{10} Å;
    (d) 9.5×10^{25} Å.
61. (a) 3%, 3%;
    (b) 0.7%, 0.2%.
63. 8×10^{-2} m^3.
65. L/m, L/y, L.
67. (a) 13.4;
    (b) 49.3.
69. 4×10^{51} kg.

CHAPTER 2

1. 61 m.
3. 0.65 cm/s, no.
5. 300 m/s, 1 km every 3 sec.
7. (a) 9.26 m/s;
   (b) 3.1 m/s.
9. (a) 0.3 m/s;
   (b) 1.2 m/s;
   (c) 0.30 m/s;
   (d) 1.4 m/s;
   (e) -0.95 m/s.
11. 2.0×10^1 s.
13. (a) 5.4×10^3 m;
    (b) 72 min.
15. (a) 61 km/h;
    (b) 0.
17. (a) 16 m/s;
    (b) $+5$ m/s.
19. 6.73 m/s.
21. 5 s.
23. (a) 48 s;
    (b) 90 s to 108 s;
    (c) 0 to 42 s, 65 s to 83 s, 90 s to 108 s;
    (d) 65 s to 83 s.

25. (a) 21.2 m/s;
    (b) 2.00 m/s^2.
27. 17.0 m/s^2.
29. (a) m/s, m/s^2;
    (b) $2B$ m/s^2;
    (c) $(A + 10B)$ m/s, $2B$ m/s^2;
    (d) $A - 3Bt^{-4}$.
31. 1.5 m/s^2, 99 m.
33. 240 m/s^2.
35. 4.41 m/s^2, 2.61 s.
37. 45.0 m.
39. (a) 560 m;
    (b) 47 s;
    (c) 23 m, 21 m.
41. (a) 96 m;
    (b) 76 m.
43. 27 m/s.
45. 117 km/h.
47. 0.49 m/s^2.
49. 1.6 s.
51. (a) 20 m;
    (b) 4 s.
53. 1.16 s.
55. 5.18 s.
57. (a) 25 m/s;
    (b) 33 m;
    (c) 1.2 s;
    (d) 5.2 s.
59. (a) 14 m/s;
    (b) fifth floor.
61. 1.3 m.
63. 18.8 m/s, 18.1 m.
65. 52 m.
67. 106 m.
69. (a) $\dfrac{g}{k}\left(1 - e^{-kt}\right)$;
    (b) $\dfrac{g}{k}$.
71. 6.
73. 1.3 m.
75. (b) 10 m;
    (c) 40 m.
77. 5.2×10^{-2} m/s^2.
79. 4.6 m/s to 5.4 m/s, 5.8 m/s to 6.7 m/s, smaller range of velocities.
81. (a) 5.39 s;
    (b) 40.3 m/s;
    (c) 90.9 m.

83. (*a*) 8.7 min;
  (*b*) 7.3 min.
85. 2.3.
87. Stop.
89. 1.5 poles.
91. 0.44 m/min, 2.9 burgers/min.
93. (*a*) Where the slopes are the same;
  (*b*) bicycle A;
  (*c*) when the two graphs cross; first crossing, B passing A; second crossing, A passing B;
  (*d*) B until the slopes are equal, A after that;
  (*e*) same.
95. (*c*)

97. (*b*) 6.8 m.

CHAPTER 3

1. 286 km, 11° south of west.

3. 10.1, −39.4°.
5. (*a*)

(*b*) −22.8, 9.85;
  (*c*) 24.8, 23.4° above the −x axis.
7. (*a*) 625 km/h, 553 km/h;
  (*b*) 1560 km, 1380 km.
9. (*a*) 4.2 at 315°;
  (*b*) $1.0\hat{\mathbf{i}} - 5.0\hat{\mathbf{j}}$ or 5.1 at 280°.
11. (*a*) $-53.7\hat{\mathbf{i}} + 1.31\hat{\mathbf{j}}$ or 53.7 at 1.4° above −x axis;
  (*b*) $53.7\hat{\mathbf{i}} - 1.31\hat{\mathbf{j}}$ or 53.7 at 1.4° below +x axis, they are opposite.
13. (*a*) $-92.5\hat{\mathbf{i}} - 19.4\hat{\mathbf{j}}$ or 94.5 at 11.8° below −x axis;
  (*b*) $122\hat{\mathbf{i}} - 86.6\hat{\mathbf{j}}$ or 150 at 35.3° below +x axis.
15. $(-2450\text{ m})\hat{\mathbf{i}} + (3870\text{ m})\hat{\mathbf{j}} + (2450\text{ m})\hat{\mathbf{k}}$, 5190 m.
17. $(9.60\hat{\mathbf{i}} - 2.00t\hat{\mathbf{k}})$ m/s, $(-2.00\hat{\mathbf{k}})$ m/s^2.
19. Parabola.
21. (*a*) 4.0*t* m/s, 3.0*t* m/s;
  (*b*) 5.0*t* m/s;
  (*c*) $(2.0t^2\hat{\mathbf{i}} + 1.5t^2\hat{\mathbf{j}})$ m;
  (*d*) $v_x = 8.0$ m/s, $v_y = 6.0$ m/s, $v = 10.0$ m/s, $\vec{\mathbf{r}} = (8.0\hat{\mathbf{i}} + 6.0\hat{\mathbf{j}})$ m.
23. (*a*) $(3.16\hat{\mathbf{i}} + 2.78\hat{\mathbf{j}})$ cm/s;
  (*b*) 4.21 cm/s at 41.3°.
25. (*a*) $(6.0t\hat{\mathbf{i}} - 18.0t^2\hat{\mathbf{j}})$ m/s, $(6.0\hat{\mathbf{i}} - 36.0t\hat{\mathbf{j}})$ m/s^2;
  (*b*) $(19\hat{\mathbf{i}} - 94\hat{\mathbf{j}})$ m, $(15\hat{\mathbf{i}} - 110\hat{\mathbf{j}})$ m/s.
27. 414 m at −65.0°.
29. 44 m, 6.9 m.
31. 18°, 72°.

33. 2.26 s.
35. 22.3 m.
37. 39 m.
41. (*a*) 12 s;
  (*b*) 62 m.
43. 5.5 s.

45. (*a*) $(2.3\hat{\mathbf{i}} + 2.5\hat{\mathbf{j}})$ m/s;
  (*b*) 5.3 m;
  (*c*) $(2.3\hat{\mathbf{i}} - 10.2\hat{\mathbf{j}})$ m/s.
47. No, 0.76 m too low; 4.5 m to 34.7 m.
51. $\tan^{-1} gt/v_0$.
53. (*a*) 50.0 m;
  (*b*) 6.39 s;
  (*c*) 221 m;
  (*d*) 38.3 m/s at 25.7°.
55. $\frac{1}{2}\tan^{-1}\left(-\frac{1}{\tan\phi}\right) = \frac{\phi}{2} + \frac{\pi}{4}$.
57. $(10.5$ m/s$)\hat{\mathbf{i}}$, $(6.5$ m/s$)\hat{\mathbf{i}}$.
59. 1.41 m/s.
61. 23 s, 23 m.
63. (*a*) 11.2 m/s, 27° above the horizontal;
  (*b*) 11.2 m/s, 27° below the horizontal.
65. 6.3°, west of south.
67. (*a*) 46 m;
  (*b*) 92 s.
69. (*a*) 1.13 m/s;
  (*b*) 3.20 m/s.
71. 43.6° north of east.
73. $(66$ m$)\hat{\mathbf{i}} - (35$ m$)\hat{\mathbf{j}} - (12$ m$)\hat{\mathbf{k}}$, 76 m, 28° south of east, 9° below the horizontal.
75. 131 km/h, 43.1° north of east.
77. 7.0 m/s.
79. 1.8 m/s^2.
81. 1.9 m/s, 2.7 s.
83. (*a*) $\dfrac{Dv}{(v^2 - u^2)}$;
  (*b*) $\dfrac{D}{\sqrt{v^2 - u^2}}$.
85. 54°.
87. $[(1.5$ m$)\hat{\mathbf{i}} - (2.0t$ m$)\hat{\mathbf{i}}] + [(-3.1$ m$)\hat{\mathbf{j}} + (1.75t^2$ m$)\hat{\mathbf{j}}$, $(3.5$ m/s$^2)\hat{\mathbf{j}}$, parabolic.
89. Row at an angle of 24.9° upstream and run 104 m along the bank in a total time of 862 seconds.
91. 69.9° north of east.
93. (*a*) 13 m;
  (*b*) 31° below the horizontal.
95. 5.1 s.
97. (*a*) 13 m/s, 12 m/s;
  (*b*) 33 m.
99. (*a*) $x = (3.03t - 0.0265)$ m, 3.03 m/s;
  (*b*) $y = (0.158 - 0.855t + 6.09t^2)$ m, 12.2 m/s^2.

CHAPTER 4

1. 77 N.
3. (a) 6.7×10^2 N;
   (b) 1.2×10^2 N;
   (c) 2.5×10^2 N;
   (d) 0.
5. 1.3×10^6 N, 39%, 1.3×10^6 N.
7. 2.1×10^2 N.
9. $m > 1.5$ kg.
11. 89.8 N.
13. 1.8 m/s², up.
15. Descend with $a \geq 2.2$ m/s².
17. -2800 m/s², 280 g's, 1.9×10^5 N.
19. (a) 7.5 s, 13 s, 7.5 s;
    (b) 12%, 0%, -12%;
    (c) 55%.
21. (a) 3.1 m/s²;
    (b) 25 m/s;
    (c) 78 s.
23. 3.3×10^3 N.
25. (a) 150 N;
    (b) 14.5 m/s.
27. (a) 47.0 N;
    (b) 17.0 N;
    (c) 0.
29. (a)   (b)

31. (a) 1.5 m;
    (b) 11.5 kN, no.
33. (a) 31 N, 63 N;
    (b) 35 N, 71 N.
35. 6.3×10^3 N, 8.4×10^3 N.
37. (a) 19.0 N at 237.5°, 1.03 m/s² at 237.5°;
    (b) 14.0 N at 51.0°, 0.758 m/s² at 51.0°.
39. $\dfrac{5}{2}\dfrac{F_0}{m}t_0^2$.
41. 4.0×10^2 m.
43. 12°.
45. (a) 9.9 N;
    (b) 260 N.
47. (a) $m_E g - F_T = m_E a$;
       $F_T - m_C g = m_C a$;
    (b) 0.68 m/s², 10,500 N.
49. (a) 2.8 m;
    (b) 2.5 s.

51. (a)

(b) $g\,\dfrac{m_B}{m_A + m_B}$, $g\,\dfrac{m_A\,m_B}{m_A + m_B}$.

53. $g\,\dfrac{m_B + \dfrac{\ell_B}{\ell_A + \ell_B}\,m_C}{m_A + m_B + m_C}$.

55. $(m + M)g\tan\theta$.
57. 1.52 m/s², 18.3 N, 19.8 N.
59. $\dfrac{(m_A + m_B + m_C)m_B}{\sqrt{(m_A^2 - m_B^2)}}\,g$.

61. (a) $\left(\dfrac{2y}{\ell} - 1\right)g$;
    (b) $\sqrt{2gy_0\left(1 - \dfrac{y_0}{\ell}\right)}$;
    (c) $\dfrac{2}{3}\sqrt{g\ell}$.

63. 6.3 N.
65. 2.0 s, no change.
67. (a) $g\,\dfrac{(m_A\sin\theta - m_B)}{(m_A + m_B)}$;
    (b) $m_A\sin\theta > m_B$
       (m_A down the plane),
       $m_A\sin\theta < m_B$
       (m_A up the plane).
69. (a) $\dfrac{m_B\sin\theta_B - m_A\sin\theta_A}{m_A + m_B}\,g$;
    (b) 6.8 kg, 26 N;
    (c) 0.74.
71. 9.9°.
73. (a) $41\ \dfrac{N}{m/s}$;
    (b) 1.4×10^2 N.
75. (a) $Mg/2$;
    (b) $Mg/2$, $Mg/2$, $3Mg/2$, Mg.
77. 8.7×10^2 N,
    72° above the horizontal.
79. (a) 0.6 m/s²;
    (b) 1.5×10^5 N.
81. 1.76×10^4 N.
83. 3.8×10^2 N, 7.6×10^2 N.
85. 3.4 m/s.
87. (a) 23 N;
    (b) 3.8 N.

89. (a) $g\sin\theta$, $\sqrt{\dfrac{2\ell}{g\sin\theta}}$,
    $\sqrt{2\ell g\sin\theta}$, $mg\cos\theta$;
    (b)

The graphs are all consistent with the results of the limiting cases.

CHAPTER 5

1. 65 N, 0.
3. 0.20.
5. 8.8 m/s².
7. 1.0×10^2 N, 0.48.
9. 0.51.
11. 4.2 m.
13. 1.2×10^3 N.
15. (a) 0.67;
    (b) 6.8 m/s;
    (c) 16 m/s.
17. (a) 1.7 m/s²;
    (b) 4.3×10^2 N;
    (c) 1.7 m/s², 2.2×10^2 N.
19. (a) 0.80 m;
    (b) 1.3 s.
21. (a) A will pull B along;
    (b) B will eventually catch up to A;

(c) $\mu_A < \mu_B : a =$

$$g\left[\frac{(m_A + m_B)\sin\theta - (\mu_A m_A + \mu_B m_B)\cos\theta}{(m_A + m_B)}\right],$$

$$F_T = g\frac{m_A m_B}{(m_A + m_B)}(\mu_B - \mu_A)\cos\theta,$$

$$\mu_A > \mu_B : a_A = g(\sin\theta - \mu_A\cos\theta),$$

$$a_B = g(\sin\theta - \mu_B\cos\theta), F_T = 0.$$

23. (a) 5.0 kg;
   (b) 6.7 kg.

25. (a) $\dfrac{v_0^2}{2dg\cos\theta} - \tan\theta;$

   (b) $\mu_s \geq \tan\theta.$

27. (a) 0.22 s;
   (b) 0.16 m.

29. 0.51.

31. (a) 82 N;
   (b) 4.5 m/s².

33. $(M + m)g\dfrac{(\sin\theta + \mu\cos\theta)}{(\cos\theta - \mu\sin\theta)}.$

35. (a) 1.41 m/s²;
   (b) 31.7 N.

37. $\sqrt{rg}.$

39. 30 m.

41. 31 m/s.

43. 0.9 g's.

45. 9.0 rev/min.

47. (a) 1.9×10^3 m;
   (b) 5.4×10^3 N;
   (c) 3.8×10^3 N.

49. 3.0×10^2 N.

51. 0.164.

53. (a) 7960 N;
   (b) 588 N;
   (c) 29.4 m/s.

55. 6.2 m/s.

57. (b) $\vec{v} = (-6.0\,\text{m/s})\sin(3.0\,\text{rad/s}\,t)\,\hat{i}$
    $+ (6.0\,\text{m/s})\cos(3.0\,\text{rad/s}\,t)\,\hat{j},$
    $\vec{a} = (-18\,\text{m/s}^2)\cos(3.0\,\text{rad/s}\,t)\,\hat{i}$
    $+ (-18\,\text{m/s}^2)\sin(3.0\,\text{rad/s}\,t)\,\hat{j};$
   (c) $v = 6.0$ m/s, $a = 18$ m/s².

59. $17\,\text{m/s} \leq v \leq 32\,\text{m/s}.$

61. (a) $a_t = (\pi/2)$ m/s², $a_c = 0$;
   (b) $a_t = (\pi/2)$ m/s²,
    $a_c = (\pi^2/8)$ m/s²;
   (c) $a_t = (\pi/2)$ m/s²,
    $a_c = (\pi^2/2)$ m/s².

63. (a) 1.64 m/s;
   (b) 3.45 m/s.

65. $m/b.$

67. (a) $\dfrac{mg}{b} + \left(v_0 - \dfrac{mg}{b}\right)e^{-\frac{b}{m}t};$

   (b) $-\dfrac{mg}{b} + \left(v_0 + \dfrac{mg}{b}\right)e^{-\frac{b}{m}t}.$

69. (a) 14 kg/m;
   (b) 570 N.

71. $\dfrac{mg}{b}\left[t + \dfrac{m}{b}\left(e^{-\frac{b}{m}t} - 1\right)\right], ge^{-\frac{b}{m}t}.$

75. 10 m.

77. 0.46.

79. 102 N, 0.725.

81. Yes, 14 m/s.

83. 28.3 m/s, 0.410 rev/s.

85. 3500 N, 1900 N.

87. 35°.

89. 132 m.

91. (a) 55 s;
   (b) centripetal component of the
    normal force.

93. (a) $\theta = \cos^{-1}\dfrac{g}{4\pi^2 rf^2};$

   (b) 73.6°;
   (c) no.

95. 82°.

97. (a) 16 m/s;
   (b) 13 m/s.

99. (a) 0.88 m/s²;
   (b) 0.98 m/s².

101. (a) 42.2 m/s;
   (b) 35.6 m, 52.6 m.

103. (a)

(b)

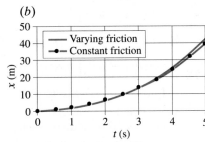

(c) speed: −12%, position: −6.6%.

CHAPTER 6

1. 1610 N.

3. 1.9 m/s².

5. $\frac{2}{9}$.

7. 0.91 g's.

9. 1.4×10^{-8} N at 45°.

11. $Gm^2\left\{\left[\dfrac{2}{x_0^2} + \dfrac{3x_0}{(x_0^2 + y_0^2)^{3/2}}\right]\hat{i}\right.$

    $\left. + \left[\dfrac{4}{y_0^2} + \dfrac{3y_0}{(x_0^2 + y_0^2)^{3/2}}\right]\hat{j}\right\}.$

13. $2^{1/3}$ times larger.

15. 3.46×10^8 m from the center of the
   Earth.

19. (b) g decreases as r increases;
   (c) 9.42 m/s² approximate,
    9.43 m/s² exact.

21. 9.78 m/s², 0.099° south of radially
   inward.

23. 7.52×10^3 m/s.

25. 1.7 m/s² upward.

27. 7.20×10^3 s.

29. (a) 520 N;
   (b) 520 N;
   (c) 690 N;
   (d) 350 N;
   (e) 0.

31. (a) 59 N, toward the Moon;
   (b) 110 N, away from the Moon.

33. (a) They are executing centripetal
    motion;
   (b) 9.6×10^{29} kg.

35. $\sqrt{\dfrac{GM}{\ell}}.$

37. 5070 s, or 84.5 min.

39. 160 y.

41. 2×10^8 y.

43. Europa: 671×10^3 km, Ganymede:
   1070×10^3 km, Callisto:
   1880×10^3 km.

45. (a) 180 AU;
   (b) 360 AU;
   (c) 360/1.

47. (a) $\log T = \frac{3}{2}\log r + \frac{1}{2}\log\left(\dfrac{4\pi^2}{Gm_J}\right),$

    slope $= \frac{3}{2}$,

    y-intercept $= \frac{1}{2}\log\left(\dfrac{4\pi^2}{Gm_J}\right);$

   (b)

   slope $= 1.50$ as predicted,
    $m_J = 1.97 \times 10^{27}$ kg.

49. (a) 5.95×10^{-3} m/s²;
   (b) no, only by about 0.06%.

51. 2.64×10^6 m.

53. (a) 4.38×10^7 m/s²;
   (b) 2.8×10^9 N;
   (c) 9.4×10^3 m/s.

55. $T_{\text{inner}} = 2.0 \times 10^4$ s,
   $T_{\text{outer}} = 7.1 \times 10^4$ s.

57. 5.4×10^{12} m, it is still in the solar system, nearest to Pluto's orbit.

59. 2.3 g's.

61. 7.4×10^{36} kg, 3.7×10^6 M_{Sun}.

65. 1.21×10^6 m.

67. $V_{\text{deposit}} = 5 \times 10^7$ m³,
$r_{\text{deposit}} = 200$ m;
$m_{\text{deposit}} = 4 \times 10^{10}$ kg.

69. 8.99 days.

71. $0.44r$.

73. (a) 53 N;
(b) 3.1×10^{26} kg.

77. 1×10^{-10} m³/kg·s².

79. (a)

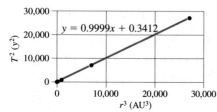

$y = 0.9999x + 0.3412$

(b) 39.44 AU.

CHAPTER 7

1. 7.7×10^3 J.

3. 1.47×10^4 J.

5. 6000 J.

7. 4.5×10^5 J.

9. 590 J.

11. (a) 1700 N;
(b) -6600 J;
(c) 6600 J;
(d) 0.

13. (a) 1.1×10^7 J;
(b) 5.0×10^7 J.

15. -490 J, 0, 490 J.

21. $1.5\hat{\mathbf{i}} - 3.0\hat{\mathbf{j}}$.

23. (a) 7.1;
(b) -250;
(c) 2.0×10^1.

25. $-1.4\hat{\mathbf{i}} + 2.0\hat{\mathbf{j}}$.

27. $52.5°$, $48.0°$, $115°$.

29. $113.4°$ or $301.4°$.

31. (a) $130°$;
(b) negative sign says that the angle is obtuse.

35. 0.11 J.

37. 3.0×10^3 J.

39. 2800 J.

41. 670 J.

43. $\frac{1}{2}kX^2 + \frac{1}{4}aX^4 + \frac{1}{5}bX^5$.

45. 4.0 J.

47. $\frac{\sqrt{3}\pi RF}{2}$.

49. 72 J.

51. (a) $\sqrt{3}$;
(b) $\frac{1}{4}$.

53. -4.5×10^5 J.

55. 3.0×10^2 N.

57. (a) $\sqrt{\dfrac{Fx}{m}}$;
(b) $\sqrt{\dfrac{3Fx}{4m}}$.

59. 8.3×10^4 N/m.

61. 1400 J.

63. (a) 640 J;
(b) -470 J;
(c) 0;
(d) 4.3 m/s.

65. 27 m/s.

67. (a) $\frac{1}{2}mv_2^2\left(1 + 2\dfrac{v_1}{v_2}\right)$;
(b) $\frac{1}{2}mv_2^2$;
(c) $\frac{1}{2}mv_2^2\left(1 + 2\dfrac{v_1}{v_2}\right)$ relative to Earth, $\frac{1}{2}mv_2^2$ relative to train;
(d) the ball moves different distances during the throwing process in the two frames of reference.

69. (a) 2.04×10^5 J;
(b) 21.0 m/s;
(c) 2.37 m.

71. 1710 J.

73. (a) 32.2 J;
(b) 554 J;
(c) -333 J;
(d) 0;
(e) 253 J.

75. 12.3 J.

77. $\dfrac{A}{k}e^{-0.10k}$.

79. 86 kJ, $42°$.

81. 1.5 N.

83. 2×10^7 N/m.

85. $6.7°$, $10°$.

87. (a) 130 N, yes (≈ 29 lbs);
(b) 470 N, perhaps not (≈ 110 lbs).

89. (a) 1.5×10^4 J;
(b) 18 m/s.

93. (a) $F = 10.0x$;
(b) 10.0 N/m;
(c) 2.00 N.

CHAPTER 8

1. 0.924 m.

3. 54 cm.

5. (a) 42.0 J;
(b) 11 J;
(c) same as part (a), unrelated to part (b).

7. (a) Yes, the expression for the work depends only on the endpoints;
(b) $U(x) = \frac{1}{2}kx^2 - \frac{1}{4}ax^4 - \frac{1}{5}bx^5 + C$.

9. $U(x) = -\dfrac{k}{2x^2} + \dfrac{k}{8\,m^2}$.

11. 49 m/s.

13. 6.5 m/s.

15. (a) 93 N/m;
(b) 22 m/s².

19. (a) 7.47 m/s;
(b) 3.01 m.

21. No, $D = 2d$.

23. (a) $\sqrt{v_0^2 + \dfrac{k}{m}x_0^2}$;
(b) $\sqrt{x_0^2 + \dfrac{m}{k}v_0^2}$.

25. (a) 2.29 m/s;
(b) 1.98 m/s;
(c) 1.98 m/s;
(d) 0.870 N, 0.800 N, 0.800 N;
(e) 2.59 m/s, 2.31 m/s, 2.31 m/s.

27. $k = \dfrac{12Mg}{h}$.

29. 3.9×10^7 J.

31. (a) 25 m/s;
(b) 370 m.

33. 12 m/s.

35. 0.020.

37. 0.40.

39. (a) 25%;
(b) 6.3 m/s, 5.4 m/s;
(c) primarily into heat energy.

41. For a mass of 75 kg, the energy change is 740 J.

43. (a) 0.13 m;
(b) 0.77;
(c) 0.5 m/s.

45. (a) $\dfrac{GMm_s}{2r_s}$;
(b) $-\dfrac{GMm_s}{r_s}$;
(c) $-\dfrac{1}{2}$.

47. $\dfrac{1}{4}$.

49. (a) 6.2×10^5 m/s;
(b) 4.2×10^4 m/s,
$v_{\text{esc at Earth orbit}} = \sqrt{2}\,v_{\text{Earth orbit}}$.

53. (a) 1.07×10^4 m/s;
(b) 1.16×10^4 m/s;
(c) 1.12×10^4 m/s.

55. (a) $-\sqrt{\dfrac{GM_E}{2r^3}}$;
(b) 1.09×10^4 m/s.

57. $\dfrac{GMm}{12r_E}$.

59. 1.12×10^4 m/s.

63. 510 N.

65. 2.9×10^4 W or 38 hp.

67. 4.2×10^3 N, opposing the velocity.

69. 510 W.

71. 2×10^6 W.

73. (a) -2.0×10^2 W;
(b) 3800 W;
(c) -120 W;
(d) 1200 W.

75. The mass oscillates between $+x_0$ and $-x_0$, with a maximum speed at $x = 0$.

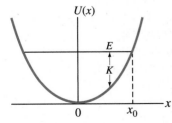

77. (a) $r_{U\text{min}} = \left(\dfrac{2b}{a}\right)^{\frac{1}{6}}$, $r_{U\text{max}} = 0$;

(b) $r_{U=0} = \left(\dfrac{b}{a}\right)^{\frac{1}{6}}$;

(c)

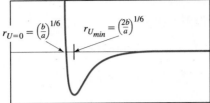

(d) $E < 0$: bound oscillatory motion between two turning points, $E > 0$: unbounded;

(e) $r_{F>0} < \left(\dfrac{2b}{a}\right)^{\frac{1}{6}}$,
$r_{F<0} > \left(\dfrac{2b}{a}\right)^{\frac{1}{6}}$,
$r_{F=0} = \left(\dfrac{2b}{a}\right)^{\frac{1}{6}}$;

(f) $F(r) = \dfrac{12b}{r^{13}} - \dfrac{6a}{r^7}$.

79. 2.52×10^4 W.

81. (a) 42 m/s;
(b) 2.6×10^5 W.

83. (a) 28.2 m/s;
(b) 116 m.

85. (a) $\sqrt{2g\ell}$;
(b) $\sqrt{1.2g\ell}$.

89. (a) 8.9×10^5 J;
(b) 5.0×10^1 W, 6.6×10^{-2} hp;
(c) 330 W, 0.44 hp.

91. (a) 29°;
(b) 480 N;
(c) 690 N.

93. 5800 W or 7.8 hp.

95. (a) 2.8 m;
(b) 1.5 m;
(c) 1.5 m.

97. 1.7×10^5 m^3.

99. (a) 5220 m/s;
(b) 3190 m/s.

101. (a) 1500 m;
(b) 170 m/s.

103. 60 m.

105. (a) 79 m/s;
(b) 2.4×10^7 W.

107. (a) 2.2×10^5 J;
(b) 22 m/s;
(c) -1.4 m.

109. $x = \sqrt{\dfrac{a}{b}}$.

CHAPTER 9

1. 5.9×10^7 N.

3. $(9.6t\hat{\mathbf{i}} - 8.9\hat{\mathbf{k}})$ N.

5. 4.35 kg·m/s $(\hat{\mathbf{j}} - \hat{\mathbf{i}})$.

7. 1.40×10^2 kg.

9. 2.0×10^4 kg.

11. 4.9×10^3 m/s.

13. -0.966 m/s.

15. 1:2.

17. $\frac{3}{2}v_0\hat{\mathbf{i}} - v_0\hat{\mathbf{j}}$.

19. $(4.0\hat{\mathbf{i}} + 3.3\hat{\mathbf{j}} - 3.3\hat{\mathbf{k}})$ m/s.

21. (a) $(116\hat{\mathbf{i}} + 58.0\hat{\mathbf{j}})$ m/s;
(b) 5.02×10^5 J.

23. (a) 2.0 kg·m/s, forward;
(b) 5.8×10^2 N, forward.

25. 2.1 kg·m/s, to the left.

27. 0.11 N.

29. 1.5 kg·m/s.

31. (a) $\dfrac{2mv}{\Delta t}$;
(b) $\dfrac{2mv}{t}$.

33. (a) 0.98 N $+ (1.4$ N/s$)t$;
(b) 13.3 N;
(c) $\left[(0.62 \text{ N/m}^2) \times \sqrt{2.5 \text{ m} - (0.070 \text{ m/s})t}\,\right]$
$+ (1.4 \text{ N/s})t$, 13.2 N.

35. 1.60 m/s (west), 3.20 m/s (east).

37. (a) 3.7 m/s;
(b) 0.67 kg.

39. (a) 1.00;
(b) 0.890;
(c) 0.286;
(d) 0.0192.

41. (a) 0.37 m;
(b) -1.6 m/s, 6.4 m/s;
(c) yes.

43. (a) $\dfrac{-M}{m + M}$;
(b) -0.96.

45. 3.0×10^3 J, 4.5×10^3 J.

47. 0.11 kg·m/s, upward.

49. (b) $e = \sqrt{\dfrac{h'}{h}}$.

51. (a) 890 m/s;
(b) 0.999 of initial kinetic energy lost.

53. (a) 7.1×10^{-2} m/s;
    (b) -5.4 m/s, 4.1 m/s;
    (c) $0, 0.13$ m/s, reasonable;
    (d) 0.17 m/s, 0, not reasonable;
    (e) in this case, -4.0 m/s, 3.1 m/s, reasonable.

55. 1.14×10^{-22} kg·m/s, $147°$ from the electron's momentum, $123°$ from the neutrino's momentum.

57. (a) $30°$;
    (b) $v'_A = v'_B = \dfrac{v}{\sqrt{3}}$;
    (c) $\frac{2}{3}$.

59. 39.9 u.

63. 6.5×10^{-11} m.

65. $(1.2\,\text{m})\hat{\mathbf{i}} - (1.2\,\text{m})\hat{\mathbf{j}}$.

67. $0\hat{\mathbf{i}} + \dfrac{2r}{\pi}\hat{\mathbf{j}}$.

69. $0\hat{\mathbf{i}} + 0\hat{\mathbf{j}} + \frac{3}{4}h\hat{\mathbf{k}}$.

71. $0\hat{\mathbf{i}} + \dfrac{4R}{3\pi}\hat{\mathbf{j}}$.

73. (a) 4.66×10^6 m from the center of the Earth.

75. (a) 5.7 m;
    (b) 4.2 m;
    (c) 4.3 m.

77. 0.41 m toward the initial position of the 85-kg person.

79. $v\dfrac{m}{m + M}$, upward, balloon also stops.

81. 0.93 hp.

83. -76 m/s.

85. Good possibility of a "scratch" shot.

87. 11 bounces.

89. 1.4 m.

91. 50%.

93. (a) $v = \dfrac{M_0 v_0}{M_0 + \dfrac{dM}{dt}t}$;
    (b) 8.2 m/s, yes.

95. 112 km/h or 70 mi/h.

97. 21 m.

99. (a) 1.9 m/s;
    (b) -0.3 m/s, 1.5 m/s;
    (c) 0.6 cm, 12 cm.

101. $m < \frac{1}{3}M$ or $m < 2.33$ kg.

103. (a) 8.3×10^{-13} m/s;
    (b) 3.3×10^{-17};
    (c) 2.1 J.

105. 29.6 km/s.

107. 0.38 m, 1.5 m.

109. (a) 1.3×10^5 N;
    (b) -83 m/s^2.

111. 12 kg.

113. 0.2 km/s, in the original direction of m_A.

CHAPTER 10

1. (a) $\dfrac{\pi}{4}$ rad, 0.785 rad;
   (b) $\dfrac{\pi}{3}$ rad, 1.05 rad;
   (c) $\dfrac{\pi}{2}$ rad, 1.57 rad;
   (d) 2π rad, 6.283 rad;
   (e) $\dfrac{89\pi}{36}$ rad, 7.77 rad.

3. 5.3×10^3 m.

5. (a) 260 rad/s;
   (b) 46 m/s, 1.2×10^4 m/s^2.

7. (a) 1.05×10^{-1} rad/s;
   (b) 1.75×10^{-3} rad/s;
   (c) 1.45×10^{-1} rad/s;
   (d) 0.

9. (a) 464 m/s;
   (b) 185 m/s;
   (c) 328 m/s.

11. $36,000$ rev/min.

13. (a) 1.5×10^{-4} rad/s^2;
    (b) 1.6×10^{-2} m/s^2, 6.2×10^{-4} m/s^2.

15. (a) $-\hat{\mathbf{i}}, \hat{\mathbf{k}}$;
    (b) 56.2 rad/s, $38.5°$ from $-x$ axis towards $+z$ axis;
    (c) 1540 rad/s^2, $-\hat{\mathbf{j}}$.

17. $28,000$ rev.

19. (a) -0.47 rad/s^2;
    (b) 190 s.

21. (a) 0.69 rad/s^2;
    (b) 9.9 s.

23. (a) $\omega = \frac{1}{3}5.0t^3 - \frac{1}{2}8.5t^2$;
    (b) $\theta = \frac{1}{12}5.0t^4 - \frac{1}{6}8.5t^3$;
    (c) $\omega(2.0\,\text{s}) = -4$ rad/s, $\theta(2.0\,\text{s}) = -5$ rad.

25. 1.4 m·N, clockwise.

27. $mg(\ell_2 - \ell_1)$, clockwise.

29. 270 N, 1700 N.

31. 1.81 kg·m^2.

33. (a) 9.0×10^{-2} m·N;
    (b) 12 s.

35. 56 m·N.

37. (a) 0.94 kg·m^2;
    (b) 2.4×10^{-2} m·N.

39. (a) 78 rad/s^2;
    (b) 670 N.

41. 2.2×10^4 m·N.

43. 17.5 m/s.

45. (a) $14M\ell^2$;
    (b) $\frac{14}{3}M\ell\alpha$;
    (c) perpendicular to the rod and the axis.

47. (a) 1.90×10^3 kg·m^2;
    (b) 7.5×10^3 m·N.

49. (a) R_0;
    (b) $\sqrt{\frac{1}{2}R_0^2 + \frac{1}{12}w^2}$;
    (c) $\sqrt{\frac{1}{2}}R_0$;
    (d) $\sqrt{\frac{1}{2}(R_1^2 + R_2^2)}$;
    (e) $\sqrt{\frac{2}{5}}r_0$;
    (f) $\sqrt{\frac{1}{12}}\ell$;
    (g) $\sqrt{\frac{1}{3}}\ell$;
    (h) $\sqrt{\frac{1}{12}(\ell^2 + w^2)}$.

51. $a = \dfrac{(m_B - m_A)}{(m_A + m_B + I/r^2)}g$ vs. $a_{I=0} = \dfrac{(m_B - m_A)}{(m_A + m_B)}g$.

53. (a) 9.70 rad/s^2;
    (b) 11.6 m/s^2;
    (c) 585 m/s^2;
    (d) 4.27×10^3 N;
    (e) $1.14°$.

57. (a) $5.3Mr_0^2$;
    (b) -15%.

59. (a) 3.9 cm from center along line connecting the small weight and the center;
    (b) 0.42 kg·m^2.

61. (b) $\frac{1}{12}M\ell^2, \frac{1}{12}Mw^2$.

63. $22,200$ J.

65. $14,200$ J.

67. 1.4 m/s.

69. 8.22 m/s.

71. 7.0×10^1 J.

73. (a) 8.37 m/s, 41.8 rad/s;
    (b) $\frac{5}{2}$;
    (c) the translational speed and the energy relationship are independent of both mass and radius, but the rotational speed depends on the radius.

75. $\sqrt{\frac{10}{7}g(R_0 - r_0)}$.

77. (a) 4.06 m/s;
    (b) 8.99 J;
    (c) 0.158.

79. (a) 4.1×10^5 J;
    (b) 18%;
    (c) 1.3 m/s^2;
    (d) 6%.

81. (a) 1.6 m/s;
    (b) 0.48 m.

83. $\dfrac{\ell}{2}, \dfrac{\ell}{2}$.

85. (a) 0.84 m/s;
(b) 96%.

87. 2.0 m·N, from the arm swinging the sling.

89. (a) $\dfrac{\omega_R}{\omega_F} = \dfrac{N_F}{N_R}$;
(b) 4.0;
(c) 1.5.

91. (a) 1.7×10^8 J;
(b) 2.2×10^3 rad/s;
(c) 25 min.

93. $\dfrac{Mg\sqrt{2Rh - h^2}}{R - h}$.

95. $\dfrac{\lambda_0 \ell^3}{8}$.

97. 5.0×10^2 m·N.

99. (a) 1.6 m;
(b) 1.1 m.

101. (a) $\dfrac{x}{y} g$;
(b) x should be as small as possible, y should be as large as possible, and the rider should move upward and toward the rear of the bicycle;
(c) 3.6 m/s².

103. $\sqrt{\dfrac{3g\ell}{4}}$.

105.

CHAPTER 11

1. 3.98 kg·m²/s.

3. (a) L is conserved: If I increases, ω must decrease;
(b) increased by a factor of 1.3.

5. 0.38 rev/s.

7. (a) 7.1×10^{33} kg·m²/s;
(b) 2.7×10^{40} kg·m²/s.

9. (a) $-\dfrac{I_W}{I_P}\omega_W$;
(b) $-\dfrac{I_W}{2I_P}\omega_W$;
(c) $\omega_W \dfrac{I_W}{I_P}$;
(d) 0.

11. (a) 0.55 rad/s;
(b) 420 J, 240 J.

13. 0.48 rad/s, 0.80 rad/s.

15. $\frac{1}{2}\omega$.

17. (a) 3.7×10^{16} J;
(b) 1.9×10^{20} kg·m²/s.

19. -0.32 rad/s.

23. 45°.

27. $(25\hat{\mathbf{i}} \pm 14\hat{\mathbf{j}} \mp 19\hat{\mathbf{k}})$ m·N.

29. (a) $-7.0\hat{\mathbf{i}} - 11\hat{\mathbf{j}} + 0.5\hat{\mathbf{k}}$;
(b) 170°.

37. $(-55\hat{\mathbf{i}} - 45\hat{\mathbf{j}} + 49\hat{\mathbf{k}})$ kg·m²/s.

39. (a) $(\frac{1}{6}M + \frac{7}{9}m)\ell^2\omega^2$;
(b) $(\frac{1}{3}M + \frac{14}{9}m)\ell^2\omega$.

41. (a) $\left[(M_A + M_B)R_0 + \dfrac{I}{R_0}\right]v$;
(b) $\dfrac{M_B g}{M_A + M_B + \dfrac{I}{R_0^2}}$.

45. $F_A = \dfrac{(d + r_A\cos\phi)m_A r_A \omega^2 \sin\phi}{2d}$,
$F_B = \dfrac{(d - r_A\cos\phi)m_A r_A \omega^2 \sin\phi}{2d}$.

47. $\dfrac{m^2 v^2}{g(m + M)(m + \frac{4}{3}M)}$.

49. $\Delta\omega/\omega_0 = -8.4 \times 10^{-13}$.

51. $v_{CM} = \dfrac{m}{M + m}v$,
ω (about CM) $= \left(\dfrac{12m}{4M + 7m}\right)\dfrac{v}{\ell}$.

53. 8.3×10^{-4} kg·m².

55. 8.0 rad/s.

57. 14 rev/min, CCW when viewed from above.

59. (a) 9.80 m/s², along a radial line;
(b) 9.78 m/s², 0.0988° south from a radial line;
(c) 9.77 m/s², along a radial line.

61. Due north or due south.

63. $(mr\omega^2 - F_{fr})\hat{\mathbf{i}}$
$+ (F_{spoke} - 2m\omega v)\hat{\mathbf{j}}$
$+ (F_N - mg)\hat{\mathbf{k}}$.

65. (a) $(-24\hat{\mathbf{i}} + 28\hat{\mathbf{j}} - 14\hat{\mathbf{k}})$ kg·m²/s;
(b) $(16\hat{\mathbf{j}} - 8.0\hat{\mathbf{k}})$ m·N.

67. (b) 0.750.

69. $v[-\sin(\omega t)\hat{\mathbf{i}} + \cos(\omega t)\hat{\mathbf{j}}]$,
$\vec{\omega} = \left(\dfrac{v}{R}\right)\hat{\mathbf{k}}$.

71. (a) The wheel will turn to the right;
(b) $\Delta L/L_0 = 0.19$.

73. (a) 820 kg·m²/s²;
(b) 820 m·N;
(c) 930 W.

75. $\vec{\mathbf{a}}_{tan} = -R\alpha\sin\theta\hat{\mathbf{i}} + R\alpha\cos\theta\hat{\mathbf{j}}$;
(a) $mR^2\alpha\hat{\mathbf{k}}$;
(b) $mR^2\alpha\hat{\mathbf{k}}$.

77. 0.965.

79. (a) There is zero net torque exerted about any axis through the skater's center of mass;
(b) $f_{single\ axel} = 2.5$ rad/s, $f_{triple\ axel} = 6.5$ rad/s.

81. (a) 17,000 rev/s;
(b) 4300 rev/s.

83. (a) $\omega = \left(12\,\dfrac{\text{rad/s}}{\text{m}}\right)x$;
(b)

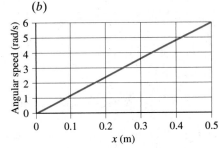

CHAPTER 12

1. 528 N, $(1.20 \times 10^2)°$ clockwise from $\vec{\mathbf{F}}_A$.

3. 6.73 kg.

5. (a) $F_A = 1.5 \times 10^3$ N down, $F_B = 2.0 \times 10^3$ N up;
(b) $F_A = 1.8 \times 10^3$ N down, $F_B = 2.6 \times 10^3$ N up.

7. (a) 230 N;
(b) 2100 N.

9. -2.9×10^3 N, 1.5×10^4 N.

11. 3400 N, 2900 N.

13. 0.28 m.

15. 6300 N, 6100 N.

17. 1600 N.

19. 1400 N, 2100 N.

21. (a) 410 N;
(b) 410 N, 328 N.

23. 120 N.

25. 550 N.

27. (a)

(b) $F_{AH} = 51$ N, $F_{AV} = -9$ N;
(c) 2.4 m.

29. $F_{\text{top}} = 55.2\,\text{N}$ right, $63.7\,\text{N}$ up,
$F_{\text{bottom}} = 55.2\,\text{N}$ left, $63.7\,\text{N}$ up.

31. $5.2\,\text{m/s}^2$.

33. $2.5\,\text{m}$ at the top.

35. (a) $1.8 \times 10^5\,\text{N/m}^2$;
(b) 3.5×10^{-6}.

37. (a) $1.4 \times 10^6\,\text{N/m}^2$;
(b) 6.9×10^{-6};
(c) $6.6 \times 10^{-5}\,\text{m}$.

39. $9.6 \times 10^6\,\text{N/m}^2$.

41. (a) $1.3 \times 10^2\,\text{m}\cdot\text{N}$, clockwise;
(b) the wall;
(c) all three are present.

43. (a) $393\,\text{N}$;
(b) thicker.

45. (a) $3.7 \times 10^{-5}\,\text{m}^2$;
(b) $2.7 \times 10^{-3}\,\text{m}$.

47. $1.3\,\text{cm}$.

49. (a) $150\,\text{kN}$, $170\,\text{kN}$ $23°$ above AC;
(b) tension: F_{DE}, F_{BD}, F_{BC};
compression:
F_{CE}, F_{CD}, F_{AC}, F_{AB}.

51. (a) $5.5 \times 10^{-2}\,\text{m}^2$;
(b) $8.6 \times 10^{-2}\,\text{m}^2$.

53. $F_{\text{AB}} = F_{\text{BD}} = F_{\text{DE}} = 7.5 \times 10^4\,\text{N}$,
compression;
$F_{\text{BC}} = F_{\text{CD}} = 7.5 \times 10^4\,\text{N}$, tension;
$F_{\text{CE}} = F_{\text{AC}} = 3.7 \times 10^4\,\text{N}$, tension.

55. $F_{\text{AB}} = F_{\text{JG}} = \dfrac{3\sqrt{2}}{2} F$, compression;
$F_{\text{AC}} = F_{\text{JH}} = F_{\text{CE}} = F_{\text{HE}} = \frac{3}{2}F$,
tension; $F_{\text{BC}} = F_{\text{GH}} = F$, tension;
$F_{\text{BE}} = F_{\text{GE}} = \dfrac{\sqrt{2}}{2} F$, tension;
$F_{\text{BD}} = F_{\text{GD}} = 2F$, compression;
$F_{\text{DE}} = 0$.

57. $0.249\,\text{kg}$, $0.194\,\text{kg}$, $0.0554\,\text{kg}$.

59. (a) $Mg\sqrt{\dfrac{h}{2R - h}}$;
(b) $Mg\,\dfrac{\sqrt{h(2R - h)}}{R - h}$.

61. (a)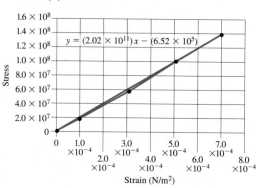
(b) $mg = 65\,\text{N}$, $F_{\text{right}} = 550\,\text{N}$,
$F_{\text{left}} = 490\,\text{N}$;
(c) $11\,\text{m}\cdot\text{N}$.

63. $29°$.

65. 3.8.

67. $5.0 \times 10^5\,\text{N}$, $3.2\,\text{m}$.

69. (a) $650\,\text{N}$;
(b) $F_{\text{A}} = 0$, $F_{\text{B}} = 1300\,\text{N}$;
(c) $F_{\text{A}} = 160\,\text{N}$, $F_{\text{B}} = 1140\,\text{N}$;
(d) $F_{\text{A}} = 810\,\text{N}$, $F_{\text{B}} = 940\,\text{N}$.

71. He can walk only $0.95\,\text{m}$ to the right
of the right support, and $0.83\,\text{m}$ to
the left of the left support.

73. $F_{\text{left}} = 120\,\text{N}$, $F_{\text{right}} = 210\,\text{N}$.

75. $F/A =$
$3.8 \times 10^5\,\text{N/m}^2 <$ tissue strength.

77. $F_{\text{A}} = 1.7 \times 10^4\,\text{N}$,
$F_{\text{B}} = 7.7 \times 10^3\,\text{N}$.

79. $2.5\,\text{m}$.

81. (a) $6500\,\text{m}$;
(b) $6400\,\text{m}$.

83. $650\,\text{N}$.

85. $45°$.

87. (a) $2.4w$;
(b) $2.6w$, $31°$ above the horizontal.

89. (a) $(4.5 \times 10^{-6})\%$;
(b) $9.0 \times 10^{-18}\,\text{m}$.

91. $0.83\,\text{m}$.

93. (a) $mg\left(1 - \dfrac{r_0}{h}\cot\theta\right)$;
(b) $\dfrac{h}{r_0} - \cot\theta$.

95. (b) $46°$, $51°$, 11%.

97. (a)

(b)

CHAPTER 13

1. $3 \times 10^{11}\,\text{kg}$.

3. $6.7 \times 10^2\,\text{kg}$.

5. 0.8547.

7. (a) $5510\,\text{kg/m}^3$;
(b) $5520\,\text{kg/m}^3$, 0.3%.

9. (a) $8.1 \times 10^7\,\text{N/m}^2$;
(b) $2 \times 10^5\,\text{N/m}^2$.

11. $13\,\text{m}$.

13. $6990\,\text{kg}$.

15. (a) $2.8 \times 10^7\,\text{N}$, $1.2 \times 10^5\,\text{N/m}^2$;
(b) $1.2 \times 10^5\,\text{N/m}^2$.

17. $683\,\text{kg/m}^3$.

19. $3.36 \times 10^4\,\text{N/m}^2$.

21. (a) $1.32 \times 10^5\,\text{Pa}$;
(b) $9.7 \times 10^4\,\text{Pa}$.

23. (c) $0.38h$, no.

27. $2990\,\text{kg/m}^3$.

29. $920\,\text{kg}$.

31. Iron or steel.

33. $1.1 \times 10^{-2}\,\text{m}^3$.

35. 10.5%.

37. Above.

39. 3600 balloons.

43. $2.8\,\text{m/s}$.

45. $1.0 \times 10^1\,\text{m/s}$.

47. $1.8 \times 10^5\,\text{N/m}^2$.

49. $1.2 \times 10^5\,\text{N}$.

51. $9.7 \times 10^4\,\text{Pa}$.

57. (b) $\frac{1}{2}$.

59. (b) $h = \left[\sqrt{h_0} - t\sqrt{\dfrac{gA_1^2}{2(A_2^2 - A_1^2)}}\,\right]^2$
(c) $92\,\text{s}$.

63. $7.9 \times 10^{-2}\,\text{Pa}\cdot\text{s}$.

65. $6.9 \times 10^3\,\text{Pa}$.

67. $0.10\,\text{m}$.

69. (a) Laminar;
(b) turbulent.

71. $1.0\,\text{m}$.

73. $0.012\,\text{N}$.

75. $1.5\,\text{mm}$.

79. (a) $0.75\,\text{m}$;
(b) $0.65\,\text{m}$;
(c) $1.1\,\text{m}$.

81. $0.047\,\text{atm}$.

83. $0.24\,\text{N}$.

85. $1.0\,\text{m}$.

87. $5.3\,\text{km}$.

89. (a) $-88\,\text{Pa/s}$;
(b) $5.0 \times 10^1\,\text{s}$.

91. $5 \times 10^{18}\,\text{kg}$.

93. (a) 8.5 m/s;

    (b) 0.24 L/s;

    (c) 0.85 m/s.

95. $d\left(\dfrac{v_0^2}{v_0^2 + 2gy}\right)^{\frac{1}{4}}$.

97. 170 m/s.

99. 1.2×10^4 N.

101. 4.9 s.

CHAPTER 14

1. 0.72 m.

3. 1.5 Hz.

5. 350 N/m.

7. 0.13 m/s, 0.12 m/s², 1.2%.

9. (a) 0.16 N/m;

    (b) 2.8 Hz.

11. $\dfrac{\sqrt{3k/M}}{2\pi}$.

13. (a) 2.5 m, 3.5 m;

    (b) 0.25 Hz, 0.50 Hz;

    (c) 4.0 s, 2.0 s;

    (d) $x_A = (2.5 \text{ m}) \sin(\frac{1}{2}\pi t)$,

      $x_B = (3.5 \text{ m}) \cos(\pi t)$.

15. (a) $y(t) =$

    $(0.280 \text{ m}) \sin[(34.3 \text{ rad/s})t]$;

    (b) $t_{\text{longest}} =$

      4.59×10^{-2} s $+ n(0.183 \text{ s})$,

      $n = 0, 1, 2, \ldots$;

      $t_{\text{shortest}} =$

      1.38×10^{-1} s $+ n(0.183 \text{ s})$,

      $n = 0, 1, 2, \ldots$.

17. (a) 1.6 s, $\frac{5}{8}$ Hz;

    (b) 3.3 m, -7.5 m/s;

    (c) -13 m/s, 29 m/s².

19. 0.75 s.

21. 3.1 s, 6.3 s, 9.4 s.

23. 88 N/m, 17.8 m.

27. (a) 0.650 m;

    (b) 1.18 Hz;

    (c) 13.3 J;

    (d) 11.2 J, 2.1 J.

29.

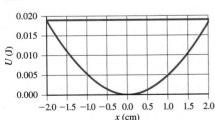

    (a) 0.011 J;

    (b) 0.0083 J;

    (c) 0.55 m/s.

31. 10.2 m/s.

33. $A_{\text{high energy}} = \sqrt{5} A_{\text{low energy}}$.

35. (a) 430 N/m;

    (b) 3.7 kg.

37. 309.8 m/s.

39. (a) 0.410 s, 2.44 Hz;

    (b) 0.148 m;

    (c) 34.6 m/s²;

    (d) $x = (0.148 \text{ m}) \sin(4.87\pi t)$;

    (e) 2.00 J;

    (f) 1.68 J.

41. 2.2 s.

43. (a) $-5.4°$;

    (b) 8.4°;

    (c) $-13°$.

45. $\frac{1}{3}$.

47. $\sqrt{2g\ell(1 - \cos\theta)}$.

49. 0.41 g.

51. (a) $\theta = \theta_0 \cos(\omega t + \phi)$, $\omega = \sqrt{\dfrac{K}{I}}$.

53. 2.9 s.

55. 1.08 s.

57. Decreased by a factor of 6.

59. (a) $(-1.21 \times 10^{-3})\%$;

    (b) 32.3 periods.

63. (a) 0°;

    (b) 0, $\pm A$;

    (c) $\frac{1}{2}\pi$ or 90°.

65. 1.6 m/s.

67. 1.37×10^8.

69. (a) 170 s;

    (b) 1.3×10^{-5} W;

    (c) 1.0×10^{-3} Hz on either side.

71. 0.11 m.

73. (a) $1.22 f$;

    (b) $0.71 f$.

75. (a) 0.41 s;

    (b) 9 mm.

77. 0.9922 m, 1.6 mm, 0.164 m.

79. $x = \pm \dfrac{\sqrt{3}A}{2} \approx \pm 0.866A$.

81. $\rho_{\text{water}} g(\text{area}_{\text{bottom side}})$.

83. (a) 130 N/m;

    (b) 0.096 m.

85. (a) $x = \pm \dfrac{\sqrt{3}x_0}{2} \approx \pm 0.866x_0$;

    (b) $x = \pm \frac{1}{2}x_0$.

87. 84.5 min.

89. 1.25 Hz.

91. ~ 3000 N/m.

93. (a) $k = \dfrac{4\pi^2}{\text{slope}}$, y-intercept $= 0$;

    (b) slope $= 0.13$ s²/kg,

      y-intercept $= 0.14$ s²

    (c) $k = \dfrac{4\pi^2}{\text{slope}} = 310$ N/m,

      y-intercept $= \dfrac{4\pi^2 m_0}{k}$,

      $m_0 = 1.1$ kg;

    (d) portion of spring's mass that is effectively oscillating.

CHAPTER 15

1. 2.7 m/s.

3. (a) 1400 m/s;

    (b) 4100 m/s;

    (c) 5100 m/s.

5. 0.62 m.

7. 4.3 N.

9. (a) 78 m/s;

    (b) 8300 N.

11. (a)

    (b) -5 cm/s.

13. 18 m.

15. $A_{\text{more energy}} / A_{\text{less energy}} = \sqrt{3}$.

19. (a) 0.38 W;

    (b) 0.25 cm.

21. (b) 420 W.

23. $D = A \sin\left[2\pi\left(\dfrac{x}{\lambda} + \dfrac{t}{T}\right) + \phi\right]$.

25. (a) 41 m/s;

    (b) 6.4×10^4 m/s²;

    (c) 35 m/s, 3.2×10^4 m/s².

27. (b) $D =$
(0.45 m) cos[2.6(x − 2.0t) + 1.2];
(d) $D =$
(0.45 m) cos[2.6(x + 2.0t) + 1.2].

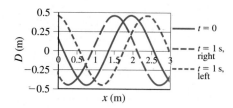

29. $D = (0.020\,\text{cm}) \times$
$\sin[(9.54\,\text{m}^{-1})x - (3290\,\text{rad/s})t + \frac{3}{2}\pi]$

31. Yes, it is a solution.

35. Yes, it is a solution.

37. (a) 0.84 m;
(b) 0.26 N;
(c) 0.59 m.

39. (a) $T = \dfrac{2}{v}\sqrt{D^2 + \left(\dfrac{x}{2}\right)^2}$;

(b) slope $= \dfrac{1}{v^2}$,

y-intercept $= \dfrac{4}{v^2}D^2$.

41. (a)

(b)

(c) all kinetic energy.

43. 662 Hz.

45. $T_n = \dfrac{(1.5\,\text{s})}{n}$, $n = 1, 2, 3, \dots$,

$f_n = n(0.67\,\text{Hz})$, $n = 1, 2, 3, \dots$.

47. $f_{0.50}/f_{1.00} = \sqrt{2}$.

49. 80 Hz.

53. 11.

55. (a) $D_2 = 4.2\sin(0.84x + 47t + 2.1)$;
(b) $8.4\sin(0.84x + 2.1)\cos(47t)$.

57. 315 Hz.

59. (a)

(b)

61. $n = 4$, $n = 8$, and $n = 12$.

63. $x = \left[\pm\left(n + \frac{1}{2}\right)\frac{1}{4}\pi\right]$ m,
$n = 1, 2, 3, \dots$.

65. 5.2 km/s.

67. $(3.0 \times 10^1)°$.

69. 44°.

71. (a) 0.042 m;
(b) 0.55 radians.

73. The speed is greater in the less
dense rod, by a factor of
$\sqrt{2.5} = 1.6$.

75. (a) 0.05 m;
(b) 2.25.

77. 0.69 m.

79. (a) $t = 0$ s;

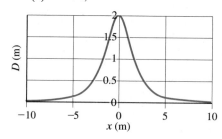

(b) $D = \dfrac{4.0\,\text{m}^3}{(x - 2.4t)^2 + 2.0\,\text{m}^2}$

(c) $t = 1.0$ s, moving right;

(d) $D = \dfrac{4.0\,\text{m}^3}{(x + 2.4t)^2 + 2.0\,\text{m}^2}$,

$t = 1.0$ s, moving left.

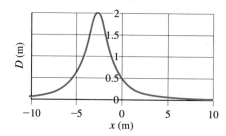

81. (a) G: 784 Hz, 1180 Hz, B: 988 Hz,
1480 Hz;
(b) 1.59;
(c) 1.26;
(d) 0.630.

83. 6.3 m from the end where the first
pulse originated.

85. $\lambda = \dfrac{4\ell}{2n - 1}$, $n = 1, 2, 3, \dots$.

87. $D(x, t) =$
(3.5 cm) cos(0.10πx − 1.5πt), with
x in cm and t in s.

89. 12 min.

93. speed = 0.50 m/s; direction
of motion = +x, period = 2π s,
wavelength = π m.

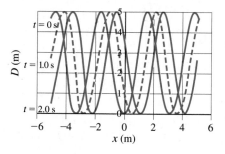

CHAPTER 16

1. (a) 1.7 cm to 17 m;
(b) 2.3×10^{-5} m.

3. (a) 17 cm to 17 m;
(b) 3.4×10^{-5} m.

5. (a) 0.17 m;
(b) 11 m;
(c) 0.5%.

7. 41 m.

9. (a) 8%;
(b) 4%.

11. (a) 4.4×10^{-5} Pa;
(b) 4.4×10^{-3} Pa.

13. (a) 5.3 m;
(b) 675 Hz;
(c) 3600 m/s;
(d) 1.0×10^{-13} m.

15. 63 dB.

17. (a) 10^9;
(b) 10^{12}.

19. 2.9×10^{-9} J.

21. 124 dB.

23. (a) 9.4×10^{-6} W;
(b) 8.0×10^6 people.

25. (a) 122 dB, 115 dB;
(b) no.

27. 7 dB.

29. (a) The higher frequency wave, 2.6;
(b) 6.8.

31. (a) 3.2×10^{-5} m;
(b) 3.0×10^{1} Pa.

33. 1.24 m.

35. (a) 69.2 Hz, 207 Hz, 346 Hz, 484 Hz;
(b) 138 Hz, 277 Hz, 415 Hz, 553 Hz.

37. 8.6 mm to 8.6 m.

39. (a) 0.18 m;
(b) 1.1 m;
(c) 440 Hz, 0.79 m.

41. -3.0%.

43. (a) 1.31 m;
(b) 3, 4, 5, 6.

45. 3.65 cm, 7.09 cm, 10.3 cm, 13.4 cm, 16.3 cm, 19.0 cm.

47. 4.3 m, open.

49. 21 Hz, 43 Hz.

51. 3430 Hz, 10,300 Hz, 17,200 Hz, relatively sensitive frequencies.

53. ± 0.50 Hz.

55. 346 Hz.

57. 10 beats/s.

59. (a) 221.5 Hz or 218.5 Hz;
(b) 1.4% increase or decrease.

61. (a) 1470 Hz;
(b) 1230 Hz.

63. (a) 2430 Hz, 2420 Hz, difference of 10 Hz;
(b) 4310 Hz, 3370 Hz, difference of 940 Hz;
(c) 34,300 Hz, 4450 Hz, difference of 29,900 Hz;
(d) $f'_{\text{source moving}} \approx f'_{\text{observer moving}}$
$= f\left(1 + \dfrac{v_{\text{object}}}{v_{\text{sound}}}\right)$.

65. (a) 1420 Hz, 1170 Hz;
(b) 1520 Hz, 1080 Hz;
(c) 1330 Hz, 1240 Hz.

67. 3 Hz.

69. (a) Every 1.3 s;
(b) every 15 s.

71. 8.9 cm/s.

73. (a) 93;
(b) 0.62°.

77. 19 km.

79. (a) 57 Hz, 69 Hz, 86 Hz, 110 Hz, 170 Hz.

81. 90 dB.

83. 11 W.

85. 51 dB.

87. 51.

89. (a) 280 m/s, 57 N;
(b) 0.19 m;
(c) 880 Hz, 1320 Hz.

91. 3 Hz.

93. 141 Hz, 422 Hz, 703 Hz, 984 Hz.

95. 22 m/s.

97. (a) No beats;
(b) 20 Hz;
(c) no beats.

99. 55.2 kHz.

101. 11.5 m.

103. 2.3 Hz.

105. 17 km/h.

107. (a) 3400 Hz;
(b) 1.50 m;
(c) 0.10 m.

109. (a)

(b)

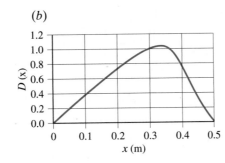

CHAPTER 17

1. $N_{\text{Au}} = 0.548 N_{\text{Ag}}$.

3. (a) 20°C;
(b) 3500°F.

5. 102.9°F.

7. 0.08 m.

9. 1.6×10^{-6} m for Super Invar™, 9.6×10^{-5} m for steel, steel is $60\times$ as much.

11. 981 kg/m³.

13. -69°C.

15. 3.9 cm³.

17. (a) $5.0 \times 10^{-5}/\text{C}°$;
(b) copper.

21. (a) 2.7 cm;
(b) 0.3 cm.

23. 55 min.

25. 3.0×10^{7} N/m².

27. (a) 27°C;
(b) 5500 N.

29. -459.67°F.

31. 1.35 m³.

33. 1.25 kg/m³.

35. 181°C.

37. (a) 22.8 m³;
(b) 1.88 atm.

39. 1660 atm.

41. 313°C.

43. 3.49 atm.

45. -130°C.

47. 7.0 min.

49. Actual = 0.588 m³, ideal = 0.598 m³, nonideal behavior.

51. 2.69×10^{25} molecules/m³.

53. 4×10^{-17} Pa.

55. 300 molecules/cm³.

57. 19 molecules/breath.

59. (a) 71.2 torr;
(b) 180°C.

61. 223 K.

63. (a) Low;
(b) 0.025%.

65. 20%.

67. 9.9 L, not advisable.

69. (a) 1100 kg;
(b) 100 kg.

71. (a) Lower;
(b) 0.36%.

73. 1.1×10^{44} molecules.

75. 3.34 nm.

77. 13 h.

79. (a) 0.66×10^{3} kg/m³;
(b) -3%.

81. ± 0.11 C°.

83. 3.6 m.

85. 3% decrease.

87.

Slope of the line: 4.92×10^{-2} ml/°C,
relative β: 492×10^{-6}/°C,
β for the liquid: 501×10^{-6}/°C,
which liquid: glycerin.

CHAPTER 18

1. (a) 5.65×10^{-21} J;
   (b) 3.7×10^3 J.

3. 1.29.

5. 3.5×10^{-9} m/s.

7. (a) 4.5;
   (b) 5.2.

9. $\sqrt{3}$.

13. 5.6%.

15. 1.004.

17. (a) 461 m/s;
   (b) 26 round trips/s.

19. Double the temperature.

21. (a) 642 m/s;
   (b) 199 K;
   (c) 595 m/s, 201 K, yes.

23. Vapor.

25. (a) Vapor;
   (b) solid.

27. 3600 Pa.

29. 355 torr or 4.73×10^4 Pa or
   0.466 atm.

31. 92°C.

33. 1.99×10^5 Pa or 1.97 atm.

35. 70 g.

37. 16.6°C.

39. (a) Slope $= -5.00 \times 10^3$ K,
   intercept $= 24.9$.

41. (a) 3.1×10^6 Pa;
   (b) 3.2×10^6 Pa.

43. (b) $a = 0.365$ N·m⁴/mol²,
   $b = 4.28 \times 10^{-5}$ m³/mol.

45. (a) 1.0×10^{-2} Pa;
   (b) 3×10^7 Pa.

47. 2.1×10^{-7} m, stationary targets,
   effective radius of $r_{H_2} + r_{air}$.

49. (b) 4.7×10^7 s⁻¹.

51. $\frac{1}{40}$.

53. 3.5 h, convection is much more
   important than diffusion.

55. (b) 4×10^{-11} mol/s;
   (c) 0.6 s.

57. 260 m/s, 3.7×10^{-22} atm.

59. (a) 290 m/s;
   (b) 9.5 m/s.

61. 80 cm.

63. Kinetic energy $= 6.07 \times 10^{-21}$ J,
   potential energy $= 5.21 \times 10^{-25}$ J,
   yes, potential energy can be
   neglected.

65. 0.01%.

67. 1.5×10^5 K.

69. (a) 2800 Pa;
   (b) 650 Pa.

71. 2×10^{13} m.

73. 0.36 kg.

75. (b) 4.6×10^9 Hz,
   2×10^5 times larger.

77. 0.21.

CHAPTER 19

1. 10.7°C.

3. (a) 1.0×10^7 J;
   (b) 2.9 kWh;
   (c) $0.29 per day, no.

5. 4.2×10^5 J, 1.0×10^2 kcal.

7. 6.0×10^6 J.

9. (a) 3.3×10^5 J;
   (b) 56 min.

11. 6.9 min.

13. 40.0°C.

15. 2.3×10^3 J/kg·C°.

17. 54 C°.

19. 0.31 kg.

21. (a) 5.1×10^5 J;
   (b) 1.5×10^5 J.

23. 4700 kcal.

25. 360 m/s.

27.

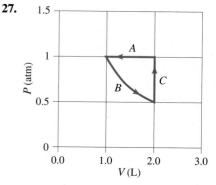

29. (a) 0;
   (b) −365 kJ.

31. (a) 480 J;
   (b) 0;
   (c) 480 J into gas.

33. (a) 4350 J;
   (b) 4350 J;
   (c) 0.

35. -4.0×10^2 K.

37. 236 J.

39. (a) 3.0×10^1 J;
   (b) 68 J;
   (c) −84 J;
   (d) −114 J;
   (e) −15 J.

41. $RT \ln \dfrac{(V_2 - b)}{(V_1 - b)} + a\left(\dfrac{1}{V_2} - \dfrac{1}{V_1}\right)$.

43. 22 C°.

45. 83.7 g/mol, krypton.

47. 48 C°.

49. (a) 6230 J;
   (b) 2490 J;
   (c) 8720 J.

51. 0.457 atm, −39°C.

53. (a) 404 K, 195 K;
   (b) -1.59×10^4 J;
   (c) 0;
   (d) -1.59×10^4 J.

55. (a)

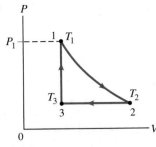

(b) 209 K;

(c) $Q_{1\to2} = 0$,
$\Delta E_{1\to2} = -2480\,\text{J}$,
$W_{1\to2} = 2480\,\text{J}$;
$Q_{2\to3} = -3740\,\text{J}$,
$\Delta E_{2\to3} = -2240\,\text{J}$,
$W_{2\to3} = -1490\,\text{J}$;
$Q_{3\to1} = 4720\,\text{J}$,
$\Delta E_{3\to1} = 4720\,\text{J}$,
$W_{3\to1} = 0$;

(d) $Q_{\text{cycle}} = 990\,\text{J}$,
$\Delta E_{\text{cycle}} = 0$,
$W_{\text{cycle}} = 990\,\text{J}$.

57. (a) $5.0 \times 10^1\,\text{W}$;

(b) 17 W.

59. 21 h.

61. (a) Ceramic: 14 W, shiny: 2.0 W;

(b) ceramic: 11 C°, shiny: 1.6 C°.

63. (a) $1.73 \times 10^{17}\,\text{W}$;

(b) 278 K or 5°C.

65. 22%.

67. (b) 4.8 C°/s;

(c) 0.60 C°/cm.

69. 6.4 Cal.

71. $4 \times 10^{15}\,\text{J}$.

73. 1 C°.

75. 3.6 kg.

77. 0.14 C°.

79. (a) 640 W;

(b) 4.2 g.

81. 1.1 days.

83. (a) $r = 2.390\,\text{cm}$;

(b)

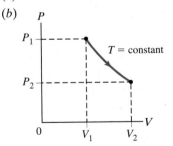

(c) $Q = 4.99\,\text{J}$, $\Delta E = 0$, $W = 4.99\,\text{J}$.

85. 110°C.

87. 305 J.

89. (a) $1.9 \times 10^5\,\text{J}$;

(b) $-1.3 \times 10^5\,\text{J}$;

(c)

91. 2200 J.

CHAPTER 20

1. 0.25.

3. 0.16.

5. 0.21.

7. 0.55.

9. 0.74.

13. $1.4 \times 10^{13}\,\text{J/h}$.

15. 1400 m.

17. 660°C.

19. (a) $4.1 \times 10^5\,\text{Pa}$, $2.1 \times 10^5\,\text{Pa}$;

(b) 34 L, 17 L;

(c) 2100 J;

(d) −1500 J;

(e) 600 J;

(f) 0.3.

21. 8.54.

23. 5.4.

25. (a) −4°C;

(b) 29%.

27. (a) 230 J;

(b) 390 J.

29. (a) $3.1 \times 10^4\,\text{J}$;

(b) 2.7 min.

31. 91 L.

33. 0.20 J/K.

35. $5 \times 10^4\,\text{J/K}$.

37. $5.50 \times 10^{-2}\,\dfrac{\text{J/K}}{\text{s}}$.

39. 9.3 J/K.

41. (a) $9.3m\,\text{J/K}$, yes;

(b) $-9.3m\,\text{J/K}$, no.

43. (a) 1010 J/K;

(b) 1020 J/K;

(c) $-9.0 \times 10^2\,\text{J/K}$.

45. (a) Adiabatic;

(b) $\Delta S_{\text{adiabatic}} = 0$,
$\Delta S_{\text{isothermal}} = -nR\ln 2$;

(c) $\Delta S_{\text{environment adiabatic}} = 0$,
$\Delta S_{\text{environment isothermal}} = nR\ln 2$.

47. (a) All processes are reversible.

49. $\dfrac{T}{nC_V}$.

53. $2.1 \times 10^5\,\text{J}$.

55. (a) $\frac{5}{16}$;

(b) $\frac{1}{64}$.

57. (a) $2.47 \times 10^{-23}\,\text{J/K}$;

(b) $-9.2 \times 10^{-22}\,\text{J/K}$;

(c) these are many orders of magnitude smaller, due to the relatively small number of microstates for the coins.

59. (a) $1.79 \times 10^6\,\text{kWh}$;

(b) $9.6 \times 10^4\,\text{kW}$.

61. 12 MW.

63. (a) 0.41 mol;

(b) 396 K;

(c) 810 J;

(d) −700 J;

(e) 810 J;

(f) 0.13;

(g) 0.24.

65. (a) 110 kg/s;

(b) $9.3 \times 10^7\,\text{gal/h}$.

67. (a) $18\,\text{km}^3/\text{days}$;

(b) $120\,\text{km}^2$.

69. (a) 0.19;

(b) 0.23.

71. (a) 5.0 C°;

(b) 72.8 J/kg·K.

73. 1700 J/K.

75. 57 W or 0.076 hp.

77. $e_{\text{Sterling}} =$

$$\left(\frac{T_H - T_L}{T_H}\right)\left[\frac{\ln\left(\dfrac{V_b}{V_a}\right)}{\ln\left(\dfrac{V_b}{V_a}\right) + \dfrac{3}{2}\left(\dfrac{T_H - T_L}{T_H}\right)}\right],$$

$e_{\text{Sterling}} < e_{\text{Carnot}}$.

79. (a)

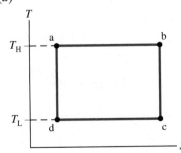

(b) W_{net}.

81. 16 kg.

83. $3.61 \times 10^{-2}\,\text{J/K}$.

Index

Note: The abbreviation *defn* means the page cited gives the definition of the term; *fn* means the reference is in a footnote; *pr* means it is found in a Problem or Question; *ff* means "also the following pages."

Absolute pressure, 345
Absolute temperature scale, 457, 464
Absolute zero, 464, 480
Accelerating reference frames, 85, 88, 155–56
Acceleration, 24–43, 60–62
  angular, 251–54
  average, 24–26
  centripetal, 120 and *ff*
  constant, 28–31, 62
  Coriolis, 301–02
  in g's, 37
  due to gravity, 34–35, 143
  instantaneous, 27–28, 60–61
  of the Moon, 140
  motion at constant, 28–39, 62–71
  radial, 120 *ff*, 128
  related to force, 86–88
  tangential, 128–29, 251–52
  uniform, 28–29, 62–71
  variable, 39–43
Accelerometer, 100
Accuracy, 3–5
  precision vs., 5–6
Action at a distance, 154
Action-reaction (Newton's third law), 89–91
Activation energy, 481
Active solar heating, 550
Addition of vectors, 52–58
Addition of velocities, 71–74
Adhesion (*defn*), 360
Adiabatic expansion of a gas, 514–15
Adiabatic lapse rate, 525 *pr*
Adiabatic processes, 508
Air bags, 31
Air columns, vibrations of, 431 *ff*
Air conditioners, 537–38
Airplane wing, 356–57
Air parcel, 525 *pr*
Air pollution, 551
Air resistance, 34–35, 129

Amplitude, 371, 397, 404
  intensity related to, 430
  pressure, 427
  of vibration, 371
  of wave, 397, 402, 404, 426, 430
Aneroid barometer, 347
Aneroid gauge, 347
Angle, 7 *fn*, 249
  of incidence, 410, 415
  phase, 373, 405
  radian measure of, 249
  of reflection, 410
  of refraction, 415
  solid, 7 *fn*
Angstrom (unit), 17 *pr*
Angular acceleration, 251–55
Angular displacement, 250, 381
Angular frequency, 373
Angular momentum, 285–89, 291–300
  conservation, law of, 285–87, 297–98
  directional nature of, 288–89, 291
  of a particle, 291–92
  relation between torque and, 292–97
  vector, 288, 291
Angular position, 249
Angular quantities, 249–54
  vector nature, 254
Angular velocity, 250–54
  of precession, 299
Antilogarithm, A3
Antinodes, 412, 433, 434, 435
Apparent weight, 148–49, 350
Apparent weightlessness, 148–49
Approximations, 9–12
Arches, 327–28
Archimedes, 349–50
Archimedes' principle, 348–52
  geology and, 351
Areas, 9, A1, inside rear cover
  under a curve or graph, 169–71
Aristotle, 2, 84
Arteriosclerosis, 359
Asteroids, 159 *pr*, 162 *pr*, 210 *pr*, 247 *pr*, 308 *pr*
Atmosphere (unit), 345
Atmospheric pressure, 344–48
  decrease with altitude, 344
Atomic mass, 455

Atomic mass unit, 7, 455
Atomic theory of matter, 455–56
Atomic weight, 455 *fn*
Atoms, 455 *ff*, 476 *ff* (*see also* Kinetic theory)
  distance between, 456
Attack angle, 356
Atwood's machine, 99, 279 *pr*, 294
Audible range, 425
Autofocusing camera, 426
Average acceleration, 24–26
Average speed, 20, 480–82
Average velocity, 20–22
Average velocity vector, 60
Avogadro, Amedeo, 468
Avogadro's hypothesis, 468
Avogadro's number, 468–69
Axial vector, 254 *fn*
Axis, instantaneous, 268
Axis of rotation (*defn*), 249

Back, forces in, 337 *pr*
Balance, 317–18
Balance a car wheel, 296
Ballistic pendulum, 226
Balloons, hot air, 454
Banking of curves, 126–27
Bar (unit), 345
Barometer, 347
Base quantities, 7
Base units (*defn*), 7
Baseball, 82 *pr*, 163, 303 *pr*, 310 *pr*, 357
Baseball curve, Bernoulli's principle and, 357
Basketball, 105 *pr*
Beams, 322, 323
Beat frequency, 438–39
Beats, 438–39
Bel (unit), 428
Bell, Alexander Graham, 428
Bernoulli, Daniel, 354
Bernoulli's equation, 354–58
Bernoulli's principle, 354–57
Bicycle, 181 *pr*, 281 *pr*, 283 *pr*, 289, 295, 309 *pr*
Big Bang, 13
Billiard balls, 214, 223, 228, 245 *pr*, 282 *pr*, 307 *pr*
Bimetallic-strip thermometer, 457

Photo Credits

Useful Geometry Formulas—Areas, Volumes

Circumference of circle $\quad C = \pi d = 2\pi r$

Area of circle $\qquad\qquad A = \pi r^2 = \dfrac{\pi d^2}{4}$

Area of rectangle $\qquad\quad A = \ell w$

Area of parallelogram $\quad A = bh$

Area of triangle $\qquad\quad A = \frac{1}{2}hb$

Right triangle
(Pythagoras) $\quad c^2 = a^2 + b^2$

Sphere: surface area $\quad A = 4\pi r^2$
volume $\qquad\qquad V = \frac{4}{3}\pi r^3$

Rectangular solid:
volume $\qquad\qquad V = \ell wh$

Cylinder (right):
surface area $\qquad A = 2\pi r\ell + 2\pi r^2$
volume $\qquad\qquad V = \pi r^2 \ell$

Right circular cone:
surface area $\qquad A = \pi r^2 + \pi r \sqrt{r^2 + h^2}$
volume $\qquad\qquad V = \frac{1}{3}\pi r^2 h$

Quadratic Formula

Equation with unknown x, in the form

$$ax^2 + bx + c = 0,$$

has solutions

$$x = \frac{-b \pm \sqrt{b^2 - 4ac}}{2a}.$$

Exponents

$(a^n)(a^m) = a^{n+m}$ [Example: $(a^3)(a^2) = a^5$]
$(a^n)(b^n) = (ab)^n$ [Example: $(a^3)(b^3) = (ab)^3$]
$(a^n)^m = a^{nm}$ $\left[\begin{array}{l}\text{Example: } (a^3)^2 = a^6 \\ \text{Example: } (a^{\frac{1}{4}})^4 = a\end{array}\right]$

$a^{-1} = \dfrac{1}{a} \qquad a^{-n} = \dfrac{1}{a^n} \qquad a^0 = 1$

$a^{\frac{1}{2}} = \sqrt{a} \qquad\quad a^{\frac{1}{4}} = \sqrt{\sqrt{a}}$

$(a^n)(a^{-m}) = \dfrac{a^n}{a^m} = a^{n-m}$ [Ex.: $(a^5)(a^{-2}) = a^3$]

$\dfrac{a^n}{b^n} = \left(\dfrac{a}{b}\right)^n$

Logarithms [Appendix A–7; Table A–1]

If $y = 10^x$, then $x = \log_{10} y = \log y$.
If $y = e^x$, then $x = \log_e y = \ln y$.

$\log(ab) = \log a + \log b$

$\log\left(\dfrac{a}{b}\right) = \log a - \log b$

$\log a^n = n \log a$

Some Derivations and Integrals†

$\dfrac{d}{dx}x^n = nx^{n-1} \qquad\qquad \displaystyle\int \sin ax\, dx = -\dfrac{1}{a}\cos ax$

$\dfrac{d}{dx}\sin ax = a\cos ax \qquad \displaystyle\int \cos ax\, dx = \dfrac{1}{a}\sin ax$

$\dfrac{d}{dx}\cos ax = -a\sin ax \qquad \displaystyle\int \dfrac{1}{x}dx = \ln x$

$\displaystyle\int x^m\, dx = \dfrac{1}{m+1}x^{m+1} \qquad \displaystyle\int e^{ax}\, dx = \dfrac{1}{a}e^{ax}$

†See Appendix B for more.

Binomial Expansion

$(1 \pm x)^n = 1 \pm nx + \dfrac{n(n-1)}{2\cdot 1}x^2 \pm \dfrac{n(n-1)(n-2)}{3\cdot 2\cdot 1}x^3 + \cdots$ [for $x^2 < 1$]

$\approx 1 \pm nx$ [for $x \ll 1$]

Trigonometric Formulas [Appendix A–9]

$\sin\theta = \dfrac{\text{opp}}{\text{hyp}}$

$\cos\theta = \dfrac{\text{adj}}{\text{hyp}}$

$\tan\theta = \dfrac{\text{opp}}{\text{adj}}$

$\text{adj}^2 + \text{opp}^2 = \text{hyp}^2$ (Pythagorean theorem)

$\tan\theta = \dfrac{\sin\theta}{\cos\theta}$

$\sin^2\theta + \cos^2\theta = 1$

$\sin 2\theta = 2\sin\theta\cos\theta$

$\cos 2\theta = (\cos^2\theta - \sin^2\theta) = (1 - 2\sin^2\theta) = (2\cos^2\theta - 1)$

$\sin(180° - \theta) = \sin\theta \qquad\qquad \cos(180° - \theta) = -\cos\theta$

$\sin(90° - \theta) = \cos\theta$

$\cos(90° - \theta) = \sin\theta$

$\sin\frac{1}{2}\theta = \sqrt{(1 - \cos\theta)/2} \qquad \cos\frac{1}{2}\theta = \sqrt{(1 + \cos\theta)/2}$

$\sin\theta \approx \theta$ [for small $\theta \lesssim 0.2$ rad]

$\cos\theta \approx 1 - \dfrac{\theta^2}{2}$ [for small $\theta \lesssim 0.2$ rad]

$\sin(A \pm B) = \sin A \cos B \pm \cos A \sin B$

$\cos(A \pm B) = \cos A \cos B \mp \sin A \sin B$

For any triangle:

$c^2 = a^2 + b^2 - 2ab\cos\gamma$ (law of cosines)

$\dfrac{\sin\alpha}{a} = \dfrac{\sin\beta}{b} = \dfrac{\sin\gamma}{c}$ (law of sines)

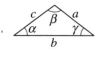